Proceedings of the
8th International Conference

# CYTOCHROME P450
## Biochemistry, Biophysics and Molecular Biology

8th International Conference
Lisbon, Portugal
24-28 October 1993

# CYTOCHROME P450
# Biochemistry, Biophysics and Molecular Biology

Edited by :
Maria Celeste LECHNER

Gulbenkian Institute of Science

**British Library Cataloguing in Publication Data**
A catalogue record for this book is available from the British Library.

ISBN 2-7420-0050-X

**Éditions John Libbey Eurotext**
6, rue Blanche, 92120 Montrouge, France. Tél. : (1) 47.35.85.52.
**John Libbey and Company Ltd**
13, Smiths Yard, Summerley Street, London SW18 4HR, England.
Tél. : (01) 947.27.77.
**John Libbey CIC**
Via L. Spallanzani, 11, 00161 Rome, Italy. Tél. : (06) 862.289.

©John Libbey Eurotext, 1994, Paris

Il est interdit de reproduire intégralement ou partiellement le présent ouvrage — loi du 11 mars 1957 — sans autorisation de l'éditeur ou du Centre Français du Copyright, 6 bis, rue Gabriel-Laumain, 75010 Paris

# Preface

The present volume is derived from the "**8th International Conference on Cytochrome P450. Biochemistry, Biophysics and Molecular Biology**" held at the Calouste Gulbenkian Foundation in Lisbon, the 24-28 October 1993.

During recent years the field of Cytochrome P450 research has diversified into numerous new areas of increasingly important scientific investigation. The demonstration of the wide superfamily of CYP genes and their ubiquity in all living organisms together with the recognition of the diversity of biological roles played by these enzymes has converted Cytochrome P450 from a restricted topic – as a pharmacologicaly important atypical cytochrome, showing peculiar structural and functional properties – into a vast and central scientific domain.

The explosion of knowledge observed in the past few years and the multifaceted characters of Cytochrome P450 science justify more than ever the organisation of thematic meetings of this series so successfully started 17 years ago.
Conversely to the situation in 1976 we are now faced with a vast fan of topics of which importance and degree of interpenetration do not adapt to a simplified option in the scope of the most traditional areas of Cytochrome P450 science. Actually, Cytochrome P450 research, besides providing invaluable models for physico-chemical studies of hemoproteins, has been launching new challenges on the molecular analysis of the evolution of gene superfamilies, mechanisms of positive and negative regulation of gene expression, development, differentiation and signal transduction pathways.
These topics constitute some of the most critical questions to a modern understanding of the general life processes. Moreover P450 investigation has been recently leading to most fundamental breakthroughs in carcinogenesis, mutagenesis, endocrinology, toxicology and immunotoxicology, simultaneously creating new opportunities for applications in bioproduction, entomology, botany and ecology.

The program that was established for the 8th International Conference on Cytochrome P450 reflects the dynamics and main vectors through which this increasingly important area of the Biological Sciences research is developing. We believe that it demonstrated that no rigid boundaries can be established to separate what used to be classified as the areas of Physico-Chemistry, or of Enzymology, and those more specifically Physiological or Genetical. We should remember that only the advent of Recombinant DNA Technology has been providing the necessary tools for a satisfactory resolution of the complex puzzle of multiplicity of the mono-oxygenases, and that it is now possible – through the heterologous expression of cloned individual P450s – to identify the molecular basis of the catalytic properties and tridimensional structure of each genetically defined entity.
From an evolutionary standpoint there are evidences that P450 genes existed on our planet for more than 3 billion years – an indication of the fundamental roles played by these proteins in all living organisms.

The emerging concept of the importance of Cytochrome P450 enzymes for the maintenance of the steady-state levels of critical life molecules places P450 at a very central position of biological investigation.

This rapid evolution of the basic concepts has been shading off the boundaries between the fundamental areas and the applications of P450 science, in biotechnology as well as in health and environmental sciences in such a way that it is becoming more and more vital to keep a permanent exchange of information and experiences between the different specialists.

The programme of this Conference aimed to correspond to this need by proposing a thematic meeting ahead of the traditionally defined disciplines, based on the more fundamental scientific questions that are raised at present – in a convergence of the different scientific disciplines involved – therefore bringing together specialists from distinct related areas.

I would like to address my very special thanks to the members of the International Scientific Committee who so generously have collaborated to the establishment of a program that has fulfilled these ambitious objectives in such a way that it has indeed attracted a high number of distinguished specialists to participate and debate ideas. We thank all the Chairmen for their effort and enthusiasm in holding the different sessions and stimulating discussions, as well as all participants for their fundamental contribution to the success of the Conference.

It is our absolute duty to thank all our collaborators and members of the Local Organizing Committee who have contributed their best personal effort to give shape to this important international event.

We believe that the Lisbon meeting was a demonstration that the fast-moving scientific field of Cytochrome P450, and its importance and complexity, require that scientists in many diverse areas of expertise meet together to share their views and new experimental findings. We hope that this Conference made a positive contribution to stimulate Cytochrome P450 specialists from convergent areas who will be responsible for the rate at which this field is rapidly developing and moving forward.

*Maria Celeste Lechner*
Oeiras, December 1993

*We would like to address a special thank to the so numerous Japanese colleagues who visited Portugal precisely when the arrival of the Portuguese navigators in Japan, **450 years ago**, is commemorated. It is for us, Portuguese, an important historical reference to have been the first Europeans to establish contact with the remote Japanese civilization, originating in the appearance of new artistic tendencies, as well as important scientific exchanges in which, at that time, we had the role of transferring agents of technology and scientific knowledge from Europe to Japan. I hope that, **450 years later,** this **P450 Congress** also serves to re-intensify the scientific cooperation between our two countries, that Portuguese would presently have so much to gain, given the well known pioneer role of Japanese scientists in this specific area.*

# INTERNATIONAL SCIENTIFIC COMMITTEE AND PROGRAM COORDINATORS

| | |
|---|---|
| Maria Celeste LECHNER | Portugal *(Chairperson)* |
| Alexander ARCHAKOV | Russia |
| Minor J. COONX | USA |
| Ronald ESTABROOK | USA |
| Franck GONZALEZ | USA |
| Jan-Ake GUSTAFSSON | Sweden |
| Osmo HANNINEN | Finland |
| Ryuichi LATO | Japan |
| Anthony LU | USA |
| Daniel NEBERT | USA |
| Franz OESCH | Germany |
| Tsuneo OMURA | Japan |
| Klaus RUCKPAUL | Germany |
| John B. SCHENKMAN | USA |
| Stephen B. SLIGAR | USA |
| Sergei USANOV | Belarus |
| Michael WATERMAN | USA |

## Local Organizing Committee

Maria Celeste LECHNER
Maria Margarida TELHADA
Vera RIBEIRO
Maria João GAMA
Teresa Maria PEREIRA
Aline BETTENCOURT
Maria Manuela CRODEIRO
Maria José MARINHO*

Gulbenkian Institute of Science
Laboratory of Biochemistry
* Advanced Courses Department

## Acknowledgments

We wish to thank all sponsoring Organizations for their support.

This Conference was organized under the auspices of the IUBMB (International Union of Biochemistry and Molecular Biology), and with national support from the JNICT (Junta Nacional de Investigação Científica e Tecnológica).

We express our deep gratitude for the financial support and cooperation provided by:

Commission of the European Communities-DGXII.
Fundação Calouste Gulbenkian, Lisboa.
Fundação Luso-Americana para o Desenvolvimento, Lisboa.
Fundação Oriente, Lisboa.
International Science Foundation, N.Y.
Sociedade Portuguesa de Bioquímica.
Camara Municipal de Oeiras.
Camara Municipal de Lisboa.
Instituto do Vinho do Porto.
Lisbon Convention Bureau, GICOL.

Ciba- Geigy AG.
Eisai Co.
F. Hoffman-La Roche AG.
Merck and Co.
Nihon Schering Co.
Rhône-Poulenc S.A.
Roussel Uclaf.
Taiho Pharmaceuticals Co.

Takeda Chemicals Industry.
Satis, Amersham Int., Portugal.
Boehringer Manheim, Portugal.
Década, S.A., Bioanal. Div., Portugal
Isaza, Applied Biosystems, Portugal.
Polaroid, Portugal.
Quimigranel, Portugal.
Trieme, Portugal

# Foreword

This book is based on the presentations given during the **8th International Conference on Cytochrome P450. Biochemistry, Biophysics and Molecular Biology.**

The program included 7 Plenary Lectures and 40 Lectures presented in a total of 8 Symposia, by invited speakers selected by the International Scientific Committee for their high competence in some of the most representative topics of P450 science. Two hundred and three communications were presented within 6 thematic Colloquia and two Poster sessions held during the whole period of the Conference.
The Opening Lecture was given by Dr Anthony Lu, in commemoration of the 25th anniversary of his publication on the first solubilisation, resolution and reconstitution of the microsomal hepatic system of the Cytochrome P450; an important land mark for the identification of the diversity of P450 proteins.

In this volume the Lectures, Plenary and Symposia were combined with Colloquia communications in 6 Chapters, which do not necessarily follow the schedule of the Conference.
The order of presentation which was adopted aims to promote the new concepts and aspects of P450 science more intensively investigated since the previous meeting of the series, held in Moscow in 1991.
Hence, the first 3 Chapters concern topics on the evolution of Cytochromes P450, the diversity of regulatory mechanisms involved in biogenesis and their importance in environment, cancer risk assessment and immunotoxicity. Chapter 4 deals with Biotechnology and novel P450 functions, and includes the Opening Plenary Lecture on basic P450 research and applications in drug development. Chapters 5 and 6 present the most recent progress in Cytochrome P450 molecular structure and electron flow mechanisms. Finally Chapter 7 consists in a compilation of selected papers contributed by the participants in the areas of the Physico-Chemistry and the Molecular Biology of Cytochrome P450.

The Volume ends with the closing plenary lecture given by Professor Ronald Estabrook on the Cytochrome P450, Agenda for Action Adaptation and Application as a summary of the state of the art and a laying out for future developments in the area.

# Contents

| | |
|---|---|
| V | Preface |
| VII | Committee |
| VIII | Acknowledgements |
| IX | Foreword |

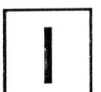

## I  EVOLUTION AND PHYLOGENETIC DIVERSITY OF P450 GENES

**3**    Evolutionary argument for a connection between drug metabolism and signal transduction.
D.W. NEBERT, R. FEYEREISEN

**15**    Hierarchical clustering and proposed classification of the P450 superfamily using consensus sequences.
A.I. ARCHAKOV, K.N. DEGTYARENKO

**23**    Function and diversity of plant cytochrome P450.
F. DURST, I. BENVENISTE, J.-P. SALAÜN, D. WERCK

**31**    Insect cytochrome P450: functions, regulation and diversity.
R. FEYEREISEN, J.F. ANDERSEN, F.A. CARIÑO, M.B. COHEN, J.F. KOENER, S. REPECKO, J.A. SCOTT, M.J. SNYDER

**37**    The role of conserved 5'-flanking (Barbie Box) DNA sequences and barbiturate-responsive DNA-binding proteins in the mechanism of induction by barbiturates of P450 cytochromes in *Bacillus megaterium* and other prokaryotic and eukaryotic organisms.
A.J. FULCO, J.-S. HE, Q. LIANG

**43**    Nitric oxide synthase: mechanisms and relationship to cytochrome P450.
K.A. WHITE, R.A. PUFAHL, N.M. OLKEN, J.M. HEVEL, M.K. RICHARDS, M.A. MARLETTA

# II CONTROL MECHANISMS IN THE BIOGENESIS OF CYTOCHROME P450'S

51 Growth hormone regulation of liver sexual differentiation.
J.Å. GUSTAFSSON, P. TOLLET, A. MODE

59 Mechanisms of liver-specific expression of the *CYP2D5* gene.
Y.-H. LEE, F.J. GONZALEZ

67 Expression and function of Ad4BP, a steroidogenic cell-specific transcription factor.
K.-I. MOROHASHI, S.-I. HONDA, M. NOMURA, K. TAKAYAMA, M. HARA, T. OMURA

75 DNA-binding regulatory factors and inducible expression of the P4501A1 gene.
K. SOGAWA, N. MATSUSHITA, M. EMA, Y. FUJII-KURIYAMA

81 Induction of brain P450 by solvents and pharmaceuticals.
M. WARNER, M. STRÖMSTEDT, A. WYSS, S. YOSHIDA, J.Å. GUSTAFSSON

89 Hormonal influence on developmental changes in hepatic P450 in rat liver.
Y. YAMAZOE, N. MURAYAMA, M. SHIMADA, K. NAGATA, H. ISHIKAWA, H. HASHIMOTO, K. NAKAYAMA, K. SOGAWA, Y. FUJII-KURIYAMA, R. KATO

97 Mechanisms of enhancement and inhibition of cytochrome P450 catalytic activities.
F.P. GUENGERICH, B.-R. KIM, E.M.J. GILLAM, T. SHIMADA

103 Role of individual human liver P450s and other enzymes in anti-cancer drug metabolism: drug activation and drug resistance mechanisms.
T.K.H. CHANG, G. CHEN, D.J. WAXMAN

109 Comparative analysis of cytochrome P450 3A expression in cultures of rat, rabbit and human hepatocytes.
P.S. GUZELIAN

115 Receptor-dependent coregulation of CYP4A genes and peroxisome proliferation.
G.G. GIBSON

121 Phenobarbital induction of P450 cytochromes in rat liver: genetic and endocrine control of a common regulatory mechanism.
M. LARSEN, F. IKEGWUONU, P. BRAKE, C. JEFCOATE

125 Structural and functional characteristics of CYP3A genes related to the development-dependent phenotypic expression and induction by Dexamethasone.
M.C. LECHNER, M.M. TELHADA, V. RIBEIRO, T.M. PEREIRA, M.J. GAMA

129 Expression of xenobiotic-metabolizing cytochrome P450 forms in human adult and fetal liver.
J. HAKKOLA, M. PASANEN, R. PURKUNEN, S. SAARIKOSKI, O. PELKONEN, J. MÄENPÄÄ, A. RANE, H. RAUNIO

133 Transcriptional regulation of CYP2B1/B2 gene in rat liver.
G. PADMANABAN, C. NIRODI, L. PRABHU, N. RAM, S. SULTANA

137 Identification of DNA sequences involved in basal and phenobarbital induced transcription of a CYP2B2 gene.
A. SHERVINGTON, I.R. PHILLIPS, E.A. SHEPHARD

141 Tissue-specific expression of cytochrome P450arom gene by alternative use of multiple exons 1 and promoters, and a switching of tissue-specific exons 1.
N. HARADA, S.-I. HONDA, Y. TAKAGI

# CYTOCHROME P450 IN CANCER AND IN IMMUNOTOXYCOLOGY

147 Importance of individual enzymes in the control of ultimate carcinogens.
F. OESCH, B. OESCH-BARTLOMOWICZ, H. GLATT, K.-L. PLATT, M. ARAND

155 Genetic polymorphism of P450 and human cancer.
K. KAWAJIRI, J. WATANABE, S.-I. HAYASHI

163 Polymorphisms of cytochrome P450 CYP2D6 as risk factor in carcinogenesis.
U.A. MEYER

173 CYP1A1-dependent benzo[a]pyrene metabolism activates genes regulated by the human immunodeficiency virus 1 long terminal repeat DNA sequences.
A. PUGA, Y. YAO

183 Cytochrome P450 2B4 antigenic determinants revealed by pepscan.
E.F. KOLESANOVA, S.A. KOZIN, A.I. ARCHAKOV

189    Specificity of covalent binding and autoantibodies in drug-induced immunoallergic hepatotoxicity.
S. LECOEUR, D. CHALLINE, J.C. GAUTIER, F. BALLET, P. BEAUNE

193    Metabolic activation of the thiophene ring as a primary event in immunoallergic hepatitis to tienilic acid.
P.M. DANSETTE, M.P. LOPEZ-GARCIA, C. AMAR, C. PONS, P. VALADON, E. BONIERBALE, P.H. BEAUNE, J.C. HOMBERG, D. MANSUY

197    Anti-cytochrome P450 antibodies in patients with aromatic anticonvulsant-induced hypersensitivity reactions.
J.S. LEEDER, V.A. COOK, R.J. RILEY

201    Presence of cytochrome P450 on human hepatocyte plasma membrane. Recognition by several autoantibodies.
J. LOEPER, V. DESCATOIRE, M. MAURICE, P. BEAUNE, G. FELDMANN, F.P. GUENGERICH, D. PESSAYRE

205    Activation of promutagens by rat CYP P450 expressing V79 cells.
A.S. RODRIGUES, I. DUARTE SILVA, M.J. MONTEIRO, H. CARIA, A. LAIRES, T. CHAVECA, J. RUEFF

209    P450 and azole tolerance in fungi.
S.L. KELLY, D.C. LAMB, B.C. BALDWIN, D.E. KELLY

213    Cytochrome P4502E1-dependent metabolism as a biomarker for genetic susceptibility to chemically-induced cancers: a pilot study.
S.L. CAMPLEMAN, S.J. TAMAKI, B.-A. HOENER, M.T. SMITH

217    Comparison of the principal drug-metabolizing enzyme systems in human tumors and corresponding peritumoral tissues: importance for sensitivity and/or resistance to anticancer drugs.
G.G. CHABOT, L. MASSAAD, C. TOUSSAINT, N. ALBIN, O. PARISE, V. RIBRAG, F. JANOT, A. GOUYETTE

 **BIOTECHNOLOGY AND NOVEL P450 FUNCTIONS. FROM BASIC P450 RESEARCH TO APPLICATIONS IN DRUG DESIGN**

225    Molecular engineering steroid hydroxylase activity of mammalian P450s.
M. NEGISHI, M. IWASAKI, T. SUEYOSHI, T.A. DARDEN, L.G. PEDERSEN

233    Co-expression of mammalian, plant or yeast P450s and P450 reductases in *Saccharomyces cerevisiae* as cloning and bioconversion tools.
D. POMPON, G. TRUAN, A. BELLAMINE, M. KAZMAIER, P. URBAN

241    Expression in *E. coli* of enzymatically active fusion proteins containing cytochrome P450 and cytochrome P450 reductase.
C.W. FISHER, M.S. SHET, P.L. HOLMANS, R.W. ESTABROOK

249    The role of P450s in insect-plant interactions.
H. PRAPAIPONG, C.-F. HUNG, M.R. BERENBAUM, M.A. SCHULER

257    From basic cytochrome P450 research to applications in drug development.
A.Y.H. LU

## BIOSYNTHESIS AND MEMBRANE TOPOLOGY OF MICROSOMAL P450. FUNCTIONAL IMPLICATIONS OF P450 STRUCTURE

265    Biosynthesis and membrane topology of microsomal cytochrome P450.
M. SAKAGUCHI, K. MIHARA, T. OMURA

271    Structure and function in P450s: are bacterial P450s good models for eukaryotic P450s?
J.A. PETERSON, J. DEISENHOFER, C. HASEMANN, K.G. RAVICHANDRAN, S.S. BODDUPALLI, S. GRAHAM-LORENCE

279    Structural and functional mapping of P450 proteins inferred from multiple sequence alignment.
O. GOTOH

285    Role of the $NH_2$-terminal region of microsomal P450s in catalysis and membrane binding.
S.J. PERNECKY, V.S. FUJITA, L.L. BESTERVELT, M.J. COON

293    Structure-function analysis of eukaryotic P450s expressed in *E. coli*.
I.A. PIKULEVA, C.M. JENKINS, M.R. WATERMAN

299    Proton and electron transfer mechanism in dioxygen activation by cytochrome P450cam.
H. SHIMADA, R. MAKINO, M. UNNO, T. HORIUCHI, Y. ISHIMURA

307    High level expression of integral membrane proteins induces proliferation of the endoplasmic reticulum.
R. MENZEL, E. KÄRGEL, C. WOLFF, F. VOGEL, W.-H. SCHUNK

311    Membrane topology and complex formation of cytochrome P450SCC in proteoliposomes studied by delayed fluorescence depolarization.
D. SCHWARZ, A. CHERNOGOLOV, V. KRÜGER, S. USANOV, A. STIER

315     Core glycosylation of cytochrome P450(arom): evidence for localization of N-terminus of microsomal cytochrome P450 in the lumen.
O. SHIMOZAWA, M. SAKAGUCHI, H. OGAWA, N. HARADA, K. MIHARA, T. OMURA

319     Systematic analysis of N-terminal transmembrane segments of microsomal P450s.
K.N. DEGTYARENKO, A.I. ARCHAKOV

323     Membrane topology of P450c21.
L.-C. HSU, B.-C. CHUNG

327     Study of the membrane topology and functional importance of the central region of cytochrome P450scc (CYP11A1) molecule by selective chemical modification of $Cys_{264}$.
A. CHERNOGOLOV, S. USANOV, R. KRAFT, D. SCHWARZ

## VI. DETERMINANTS OF COMPONENT INTERACTIONS IN P450 SYSTEMS. ELECTRON TRANSFER AND OXYGEN ACTIVATION

333     The importance of one-electron transfers in the mechanism of cytochrome P450.
R.E. WHITE

341     Studies of the interactions of microsomal cytochrome P450 reductase with cytochromes P450.
H.W. STROBEL, S. SHEN

349     On the nature of the NADPH-cytochrome P450 reductase-cytochrome P450 electron transfer complex.
A.I. VOZNESENSKY, J.B. SCHENKMAN

357     Electron pathways in adrenal mitochondrial cytochrome P450 systems: relative rates of leakage and hydroxylation.
R. RAPOPORT, M. RAIKHINSTEIN, D. SKLAN, I. HANUKOGLU

365     Reaction mechanism of P450 dependent steroidogenesis: regulation of catalytic activity of bovine adrenal P450 (11β).
S. TAKEMORI, S. KOMINAMI, S.I. IKUSHIRO, T. YAMAZAKI, D. HARADA

373     Electron transfer associated dioxygen activation in P450 systems.
S.G. SLIGAR, J. AIKENS, N. GERBER, M. MCLEAN, K. SUSLICK, D. BENSON

| | |
|---|---|
| 379 | Conformational control of interprotein electron transfer.<br>F. MILLETT, B. DURHAM, A. WILLIE, S. HAHM, L. GEREN, R. LIU, P. STAYTON, M. McLEAN, M. MILLER, S. SLIGAR, G. PEILAK, J. KRAUT |
| 387 | Studies on electron transfer pathways in cytochrome P450 systems.<br>R. BERNHARDT, V. BECKERT, H. UHLMANN, S.G. SLIGAR |
| 395 | Electron transfer and protein-protein interactions in soluble reconstituted liver monooxygenase systems.<br>G.I. BACHMANOVA, I.P. KANAEVA, I.F. SEVRUKOVA, O.V. NIKITYUK, N.V. STEPANOVA, T.V. KNUSHKO, Y.M. KOEN, A.I. ARCHAKOV |
| 403 | Thermodynamic and kinetic studies of the process of camphor entry into cytochrome P450cam.<br>G. HUI BON HOA, E. DEPREZ |
| 409 | Cytochrome P450: topology and catalysis.<br>P.R. ORTIZ DE MONTELLANO, N. SHIRANE, Z. SUI, J. FRUETEL, J.A. PETERSON, J.J. DE VOSS |
| 417 | Conclusive evidence of a quarternary cluster model for cholesterol side chain cleavage reaction catalyzed by cytochrome P-450scc.<br>T. HARA, M. TAKESHIMA |
| 421 | Structural features of the bovine mitochondrial electron transfer system CYP11A1.<br>R.-J. KUBAN, A. MARG, K. RUCKPAUL |
| 425 | An integrated, phase-theoretic approach to cytochrome P450c17 (CYP17) catalysis: probabilities of hydroxyprogesterone stabilization vs decomposition depend on reductase "arrival" frequencies in stochastic relation to intermediate "departure" frequencies at the active site.<br>W.N. KÜHN-VELTEN |
| 429 | Kinetic studies on a genetically engineered fused enzyme between rat cytochrome P4501A1 and yeast NADPH-P450 reductase.<br>T. SAKAKI, S. KOMINAMI, S. TAKEMORI, H. OHKAWA, M. AKIYOSHI-SHIBATA, Y. YABUSAKI |
| 433 | Overexpression of yeast cytochrome $b_5$ gene can supress ketoconazole hypersensitivity of a NADPH-P450 reductase deficient strain.<br>G. TRUAN, J.-C. EPINAT, C. ROUGEULLE, C. CULLIN, D. POMPON |
| 437 | Structure-function interrelationships in mitochondrial and microsomal cytochrome P-450-dependent monooxygenases.<br>S.A. USANOV, M. IWASAKI, M. NEGISHI |

441     Direct evidence of the electron releasing character of the cysteinate proximal ligand of cytochrome P450 from X-ray absorption near edge-spectroscopy.
J.H. DAWSON, H.I. LIU, M. SONO, S. KADKHODAYAN, B. HEDMAN, K.O. HODGSON

445     Electron transfer in cytochrome P450 containing semi-artificial enzymes.
V.Y. UVAROV, V.V. SHUMYANTSEVA, E.A. BYKHOVSKAYA, L.N. KOLYADA, A.I. ARCHAKOV

451     Segment directed mutagenesis of human cytochrome P450 1A2 by PCR and effect on the catalytic activity and substrate selectivity
R. LAINÉ, P. URBAN, D. POMPON

455     Specific inhibition of the last steps of the aldosterone biosynthesis by 18-vinylprogesterone in bovine adrenocortical cells.
G. DEFAYE, A. PIFFETEAU, C. DELORME, A. MARQUET

459     Mechanism based inactivation of nitric oxide synthase, a P450 like enzyme, by xenobiotics.
Y. OSAWA, J.C. DAVILA, C.A. MEYER, M. NAKATSUKA

463     Molecular recognition in cytochrome P450: control of uncoupling reactions via site-directed mutagenesis.
P.J. LOIDA, S.G. SLIGAR

467     Metabolism of two dopaminergic ergot derivatives in genetically engineered V79-cells expressing CYP450-enzymes.
H. GIESCHEN, M. HILDEBRAND, B. SALOMON

471     Mutants affecting substrate recognition and substrate binding in P450BM-P.
S. GRAHAM-LORENCE, D. SANDERS, J.A. PETERSON

## CYTOCHROME P450.
## BIOCHEMISTRY, BIOPHYSICS AND MOLECULAR BIOLOGY

477     The use of synthetic peptides for the investigation of components interaction in monooxygenase rabbit liver system.
A.M. KRITSKY, I.P. KANAEVA, D.R. DAVYDOV, N.V. STEPANOVA, G.I. BACHMANOVA

481     Multiple forms of NADPH-cytochrome P450 reductase in higher plants.
A. LESOT, C. BÈGUE-KIRN, M.P. HASENFRATZ, F. DURST, I. BENVENISTE

| | |
|---|---|
| 485 | Strain- and tissue-specific expression of rat CYP2B.<br>A.Y. GRISHANOVA, L. GULYAEVA, E.V. PETCHKOVSKI, V.V. LYAKHOVICH |
| 489 | The studies of cytochrome P450scc in liposomes by means of limited trypsinolysis.<br>A.V. KRIVOSHEEV, O.N. MURASHKO, S.A. USANOV |
| 493 | Computer modeling of cytochrome P450cam and $b_5$ interaction.<br>A.S. IVANOV, V.S. SKVORTSOV, Y.A. LYULKIN, A.B. RUMYANTSEV |
| 497 | Wild mice as bioindicators for pollution monitoring.<br>A. BHATIA, K. MAZZUCCO |
| 503 | Expression of functional human microsomal epoxide hydrolase in *Escherichia coli*<br>A. GAEDIGK, H. NAKAMURA, D.M. GRANT |
| 507 | Regulation of CYP2B gene expression in the intestinal mucosa.<br>A.P. ELIA, E.A. SHEPHARD, I.R. PHILLIPS |
| 511 | Clofibrate decreases CYP2A5 in male DBA/2N mice.<br>A. KOJO, S. LAITINEN, P. PELLINEN, R. JUVONEN, H. RAUNIO, O. PELKONEN, M. PASANEN |
| 515 | Exchange of N- or C- terminal segments between human P4501A1 and 1A2 leads respectively to high stability and high activity chimeras.<br>A. BELLAMINE, D. POMPON |
| 519 | Assessment of CYP2A6 and CYP3A4 activities *in vivo* in different diseases in man.<br>A. RAUTIO, E. SALMELA, P. ARVELA, O. PELKONEN, E.A. SOTANIEMI |
| 523 | Putative active site model for CYP2C9 (tolbutamide hydroxylase).<br>B.C. JONES, G. HAWKSWORTH, V. HORNE, A. NEWLANDS, M. TUTE, D.A. SMITH |
| 527 | Expression of a bovine P450c17 cDNA in the yeast *Saccharomyces cerevisiae*.<br>B. DUMAS, G. CAUET, E. DEGRYSE, R. SPAGNOLI, T. ACHSTETTER |
| 531 | Conformational dynamics of cytochrome $P450_{cam}$ as monitored by photoacoustic calorimetry.<br>C. DI PRIMO, G. HUI BON HOA, E. DEPREZ, P. DOUZOU, S.G. SLIGAR |
| 535 | The molecular basis of aromatase deficiency in an adult female due to mutations in the CYP19 gene.<br>C.R. FISHER, Y. ITO, F.A. CONTE, M.M. GRUMBACH, E.R. SIMPSON |

539   The loss of cytochromes P450 (CYPs) in rat liver cell culture is triggered during hepatocyte isolation and again during the first 4 hours of culture.
C.R.W. PADGHAM, X.-J. WANG, A.J. PAINE

543   Conformational states and substates of cytochrome $P450_{cam}$ – insight in protein dynamics and folding.
C. JUNG, W. PFEIL, K. KÖPKE, H. SCHULZE, O. RISTAU

547   Hepatic cytochrome P450 profile in BB rats with spontaneous insulin-dependent diabetes mellitus.
C.R. BARNETT, P.R. FLATT, A.J. BONE, C. IONNIDES

551   2,3,7,8-tetrachlorodibenzo-p-dioxin versus 3-methylcholanthrene. Ah receptor binding, transformation and induction of CYP1A1.
D.S. RIDDICK, P.A. HARPER, A.B. OKEY

555   Reconstitution of cytochrome P450scc by octylglucoside dialysis/adsorption into (large) unilamellar phospholipid vesicle. Structural and functional characterization.
D. SCHWARZ, A. CHERNOGOLOV, K. GAST, W. RICHTER

559   Pressure-induced transitions in cytochrome P450 IIB4 (LM2): evidence of the conformational inhomogeneity in the oligomers in solution and in proteoliposomes.
D.R. DAVYDOV, G. HUI BON HOA

563   The effect of phenobarbitone and β-naphthoflavone on selected biochemical liver parameters following subchronic administration to female beagle dogs.
E. MOLITOR, H. THOMAS, J. WEYMANN, F. WAECHTER

567   1,3-butadiene exposure of rats: hemoglobin adducts of 1,2-epoxybutene and cytochrome P450 related changes in styrene metabolism.
E. ELOVAARA, S. OSTERMAN-GOLKAR, T. NAKAJIMA, H.V. GELBOIN, V. RIIHIMÄKI, M. SORSA

571   Tritium planigraphy study on cytochrome P450 2B4 structure.
E.V. APLETALINA, V. YU. UVAROV, O.M. IPATOVA

575   Carcinogenic activation of cooked food heterocyclic amines by cytochrome P450 1A2. A theoretical study.
F. SANZ, J.J. LOZANO, E. LÓPEZ-DE-BRIÑAS, R. DE-LA-TORRE, F. MANAUT

579   Three-dimensional structure of active-site of cytochrome P450 1A2.
F. MANAUT, E. LÓPEZ-DE-BRIÑAS, J.J. LOZANO, F. SANZ

| | |
|---|---|
| 583 | Expression of a bovine $P450_{11\beta}$ cDNA in the yeast *Saccharomyces cerevisiae*.<br>G. CAUET, B. DUMAS, E. DEGRYSE, R. SPAGNOLI, T. ACHSTETTER |
| 587 | Induction by phenobarbital of CYP2B gene expression in primary rat hepatocytes and in rat hepatoma cell line.<br>G. CIARAMELLA, M. EDWARDS, E.A. SHEPHARD, I.R. PHILLIPS |
| 591 | Molecular cloning of cDNA for a renal 27-hydroxylating cytochrome P450.<br>H. POSTLIND |
| 595 | Low temperature behaviour of the high-spin/low-spin equilibrium of cytochrome $P450_{cam}$: the role of the cooling rate and temperature induced changes of the protonic activity.<br>H. SCHULZE, O. RISTAU, C. JUNG |
| 599 | Induction of cytochrome P450IA1 mRNA in aryl hydrocarbon hydroxylase-deficient mutants of mouse hepatoma line, Hepa-1.<br>H. KIKUCHI, M. USUDA, I. SAGAMI, S. IKAWA, M. WATANABE |
| 603 | Antigenic determinants of cytochrome P450 superfamily and their classification.<br>I.R. DEDINSKY, S.A. KOZIN, A.I. ARCHAKOV |
| 611 | Vitamin D3 hydroxylases in human skin.<br>I. SCHUSTER, H. EGGER, G. HERZIG, G. VORISEK |
| 615 | Two cytochrome P450 aldosterone synthase mRNAs are present in the hamster adrenal.<br>J.-G. LeHOUX, H. BERNARD, L. DUCHARME, A. LEFEBVRE, J. LeHOUX, S. VÉRONNEAU |
| 619 | Purification of the insect P450 enzyme, ecdysone 20-hydroxylase, from the migratory locust, *Locusta migratoria*, by a new affinity method.<br>J. WINTER, H. KAYSER |
| 623 | Effect of carrageenan-induced granuloma on the apoprotein content and activities of cytochrome P450 isozymes in liver and nasal mucosa in rats.<br>J. MUNTANE, V. LONGO, M.T. MITJAVILA, P.G. GERVASI, M. INGELMAN-SUNDBERG |
| 627 | Interaction of ethanol and type I binding substrate n-heptane with liver cytochrome P450 from male rats.<br>J.S. RAKOTO, M.M. SZUTOWSKI |

| | |
|---|---|
| 631 | Coumarin and methoxsalen metabolism by CYP2A6 and CYP2A-5 isoforms in man and mouse.<br>J. MÄENPÄÄ, R. JUVONEN, H. RAUNIO, A. RAUTIO, O. PELKONEN |
| 635 | Accelerated caffeine metabolism after omeprazole treatment: measurements in breath, plasma and urine.<br>K.L. ROST, I. ROOTS |
| 639 | The effect of dexamethasone and phenobarbital coadministration on the activity of rat liver P450 system.<br>K. MONOSTORY, L. VERECZKEY |
| 643 | Polymerase chain reaction mediated detection of cytochrome P450 gene in the yeast *Candida apicola*.<br>K. LOTTERMOSER, O. ASPERGER, W.-H. SCHUNCK |
| 647 | Chicken cytochrome $P450_{17\alpha}$: cloning of active type E2 and mutated type A2 defective in 17,20-lyase activity.<br>K. NISHIMORI, H. IKUSHIMA, S. MIZUNO. |
| 651 | Expression of aromatase cytochrome P450 in choliocarcinoma cells.<br>K. YAMADA, S. KATOH, N. HARADA, Y. TAKAGI |
| 655 | Immortalization of hepatocytes *via* a transgenic approach.<br>K. KRAMER, G. CIARAMELLA, M. EDWARDS, E.A. SHEPHARD, I.R. PHILLIPS |
| 659 | Analyses of promoter regions of rat CYP11B genes involved in mineralo- and glucocorticoid syntheses.<br>K. MUKAI, H. SHIMADA, Y. ISHIMURA |
| 663 | Regulatory proteins involved in the tissue-specific expression and induction of a CYP2B2 gene.<br>L.M. FERNANDEZ, L.A. FORREST, A. SHERVINGTON, E.A. SHEPHARD, I.R. PHILLIPS |
| 667 | Breast cancer aromatase: identification and expression of cytochrome P450 aromatase mRNA in T47D breast cancer cells.<br>L. TAN, S.I. SADEKOVA, T.Y.K. CHOW |
| 671 | Spectral map analysis of cytochrome P450 isoenzymes.<br>L.M.H. KOYMANS, H. MOEREELS, P.J. LEWI, P.A.J. JANSSEN |
| 677 | Oxidative dealkylation of N,N-dialkylamides.<br>L. CONSTANTINO, J. ILEY |
| 681 | Influence of isoamyl alcohol and ethanol on the oesophageal and hepatic cytochrome P450 metabolism of nitrosamines and their carcinogenicity.<br>L.F. RIBEIRO PINTO, P.F. SWANN |

| | |
|---|---|
| 685 | Regulation of the rat liver and lung CYP1A expression by different inducers.<br>L.F. GULYAEVA, A.Y. GRISHANOVA, E.V. PETCHKOVSKI, V.V. LYAKHOVICH |
| 689 | Expression study of CYP genes in drosophila strains resistant or sensitive to insecticides.<br>M. AMICHOT, A. BRUN, A. CUANY, C. HELVIG, J.P. SALAUN, F. DURST, J.B. BERGE |
| 693 | Metabolism of selected sex steroids by rat CYP1A1, 1A2 and 2B1 expressed in V79 cell lines.<br>M. HILDEBRAND, H. GIESCHEN, B. SALOMON |
| 697 | Thiophene derivatives as new mechanism-based inhibitors of cytochromes P450: inactivation of yeast expressed human liver cytochrome P450 2C9 by tienilic acid.<br>M.P. LOPEZ-GARCIA, P.M. DANSETTE, D. MANSUY |
| 701 | Cholesterol-binding region of cytochrome P450scc (P-450XIA1): identification and amino acid sequence of the cholesterol binding domain.<br>M. TSUJITA, Y. ICHIKAWA |
| 705 | Inhibition of cytochrome P450 enzyme activity using specific inhibitory antibodies.<br>M. HATCHER, R. BROWN, M.J. O'SULLIVAN, S. SWINBURNE |
| 709 | Mammalian cell lines stably expressing bovine adrenal 11β-hydroxylase cDNA.<br>M. HUSEMANN, T. PETRI |
| 713 | Expression of cDNAs encoding human and marmoset CYP2As.<br>M. NANJI, P. CLAIR, I.R. PHILLIPS, E.A. SHEPHARD |
| 717 | Localization and distribution of CYP2B1 in the rat lung by immunocytochemistry and by *in situ* hybridization.<br>M.H. FIGUEIREDO, V. RIBEIRO, D. HENRIQUE, M.C. LECHNER, V. BAIROS |
| 721 | Site directed mutagenesis on bovine adrenodoxin to target amino acids involved in electron transfer and binding to adrenodoxin reductase and cytochrome P450$_{scc}$.<br>M.-F. PALIN, J. SYGUSCH, J.-G. LEHOUX |
| 725 | Cytochrome P450 obtusifoliol 14α-methyl demethylase from *Zea mays*: enzymology and inhibition.<br>M. TATON, F. SALMON, A. RAHIER |

| | |
|---|---|
| 729 | Mechanism of 9-hydroxy ellipticine induced inhibition of hepatic aryl hydrocarbon hydroxylase.<br>M.J. OLNES, P.B. DePETRILLO, R.N. KURL |
| 733 | Induction of rat liver cytochrome P450 3A1 by metyrapone.<br>M.C. WRIGHT, A.J. PAINE |
| 737 | Long-term preservation and induction of cytochromes P450 in co-cultured rat hepatocytes.<br>M. AKRAWI, V. ROGIERS, A. VERCRUYSSE, I.R. PHILLIPS, E.A. SHEPHARD |
| 741 | Development and prevalidation of an *Escherichia coli* tester strain for genotoxins.<br>M. KRANENDONK, M. RUAS, A. LAIRES, J. RUEFF |
| 745 | Coumarin 7-hydroxylation (CYP2A6) polymorphism in a Turkish population.<br>M. ISCAN, H. ROSTAMI, M. ISCAN, T. GÜRAY, A. RAUTIO, O. PELKONEN |
| 749 | Exon-7 point mutation (m2; 4889A→G) in human CYP1A1 gene as susceptibility factor for lung cancer.<br>N. DRAKOULIS, I. CASCORBI, J. BROCKMÖLLER, C.R. GROSS, I. ROOTS |
| 753 | A role for Asp251 in the activation of oxygen by cytochrome $P450_{cam}$.<br>N.C. GERBER, S.G. SLIGAR |
| 757 | Activation and regulation of aromatic hydrocarbon-inducible *Cyp*1a1 and *Cyp*1a2 gene expression in mouse hepatocytes in primary culture.<br>N. NEMOTO, J. SAKURAI |
| 761 | Species differences in the *in vitro* metabolism of probe substrates for CYP2 enzymes.<br>O.M. CORCORAN, R. HYLAND, B.C. JONES, D.A. SMITH |
| 765 | Resonance raman spectra of NADPH cytochrome P450 reductase and its complex with cytochrome P450 2B4.<br>O. ALEXANDROVA, Y. IVANOV, V. UVAROV |
| 769 | Effects of acute and chronic acetone administration on cytochrome P450 2E1 mRNA in murine liver using *in situ* hybridization.<br>P.G. FORKERT, A.C. JACKSON, S. CHEN, S.P. TAM |
| 773 | Downregulation of the Ah receptor following 2, 3, 7, 8-tetrachlorodibenzo-*p*-dioxin (TCDD) treatment in Hepa-1 cells in culture.<br>P.A. HARPER, J.V. GIANNONE, M.R. PROBST, A.B. OKEY |
| 777 | Lidocaine metabolism as a liver function test.<br>P. ARVELA, A. RAUTIO, E.A. SOTANIEMI |

| | |
|---|---|
| 781 | Formation of cytochrome P450 metabolite complexes: isozyme specificity and use for isozyme quantitation.<br>P.H. ROOS, G. GÜNTHER, W.G. HANSTEIN |
| 785 | Risk assessment and bioavailability for mammals of soil-bound polycyclic aromatic hydrocarbons.<br>P.H. ROOS, W.G. HANSTEIN, D. STROTKAMP, M. VAN AFFERDEN, D. TAPPE |
| 789 | Expression of components of the cytochrome P450 mediated mono-oxygenase system using recombinant baculoviruses.<br>P. CLAIR, E.A. SHEPHARD, I.R. PHILLIPS |
| 793 | Advantages of the human over yeast NADPH-P450 reductase on supporting human P450 activities in yeast.<br>P. URBAN, A. PERRET, G. TRUAN, D. POMPON |
| 797 | Inducing effect of porphyrinogenic agents on mouse liver Cyp2a-5 expression.<br>P. SALONPÄÄ, K. KRAUSE, O. PELKONEN, H. RAUNIO |
| 801 | Function of residue-365 in mouse P4502A5 (P450coh): site directed mutagenesis, spectral and enzymatic properties.<br>R.O. JUVONEN, M. IWASAKI, M. NEGISHI |
| 805 | Looking for the gene encoding the fungal cytochrome P450 involved in 11β-hydroxylation of steroids: application of PCR techniques.<br>R. KOMEL, D. ROZMAN, M. VITAS, K. DROBIC, S.L. KELLY |
| 809 | Interaction of verapamil and cimetidine in rat liver microsomes.<br>R. WACKE, U. FISCHER, J. AHREND, B. DREWELOW |
| 813 | Stopped flow kinetics under high pressure: an approach to explore the transition state of haemoproteins in elementary reactions.<br>R. LANGE, I. HEIBER-LANGER, N. BEC, C. BALNY |
| 817 | A three-dimensional model of P450arom using P450BM-P as a template.<br>S. GRAHAM-LORENCE, J.A. PETERSON, E.R. SIMPSON |
| 821 | *Ortho*- and *para*-hydroxylation of pentachlorophenol in aquatic plants: possible involvement of cytochrome P450.<br>S. ROY, O. HÄNNINEN |
| 825 | Cytochrome P450 1A1 expression during the differentiation process in the Caco-2 cell line.<br>S. CHARRASSE, I. FABRE, M. DAUJAT, P. MAUREL, C. LARROQUE |

| | |
|---|---|
| 829 | Substrate-regulated, cAMP-dependent phosphorylation of cytochrome P450 3A1.<br>S. MKRTCHIAN, E. ELIASSON, J.R. HALPERT, M. INGELMAN-SUNDBERG |
| 833 | Growth hormone regiospecificically represses the expression of CYP2B1/2 and CYP3A in liver.<br>T. OINONEN, K.O. LINDROS |
| 837 | Developmentally programmed changes in CYP3A1 induction by dexamethasone in male rat liver.<br>T. PEREIRA, M.C. LECHNER |
| 841 | Different mechanisms of regioselection of fatty acid hydroxylation by two laurate ($\omega$-1) – hydroxylating P450s.<br>T. FUKUDA, Y. IMAI, M. KOMORI, M. NAKAMURA, E. KUSUNOSE, K. SATOUCHI, M. KUSUNOSE |
| 845 | Use of the paraxathine/caffeine ratio in plasma and in saliva for CYP1A2 phenotyping.<br>U. FUHR, K.L. ROST, R. ENGELHARDT, A. HORST STAIB |
| 849 | Human CYP1A2 activity: relationship between caffeine clearance *in vivo* and its 3-demethylation *in vitro*.<br>U. FUHR, R. ENGELHARDT, M. SACHS, K.L. ROST, P. SCHYMANSKY, C. HUSCHKA, D. LIERMANN, H. WISSING, A. HORST STAIB |
| 853 | Prediction and experimental confirmation of the cytochromes b5 and P450 peptide mapping.<br>V. ZGODA, D. PROSCHLYAKOV, O. IPATOVA, V. PROSOROVSKY, G. BACHMANOVA |
| 857 | Modulation of CYP3A1 expression by heme in transfected COS-1 cells.<br>V. RIBEIRO, M.C. LECHNER |
| 861 | Limited proteolysis of cytochrome P450c21 from bovine adrenal cortex microsomes.<br>V.M. GUZOV, I.N. ZELKO, S.A. BYLINSKAYA, S.A. USANOV |
| 865 | Interaction of FITC with P450s' N-terminus.<br>V.V. SHUMYANTSEVA, G.P. KUZNETSOVA, V.Y. UVAROV, A.I. ARCHAKOV |
| 869 | Structural requirements for ferredoxin associated electron transfer evaluated by analysis of Tyr-82 and His-56 bovine adrenodoxin mutants.<br>V. BECKERT, R. DETTMER, R. BERNHARDT |

873     Nonchemical induction of rat liver CYP1A1.
V.V. LYAKHOVICH, A.Y. GRISHANOVA, O.A. GROMOVA, L.F. GULYAEVA

877     Inhibition of cyclosporine metabolism in mouse and human microsomes after cytochrome P450 inactivation by Cannabidiol.
W. JAEGER, L.M. BORNHEIM, M.A. CORREIA, L.Z. BENET

881     Ethinyl estradiol-mediated induction of a cyclosporin A metabolizing 2C6 isozyme in rats.
W. JAEGER, M.A. CORREIA, L.M. BORNHEIM, L.Z. BENET

885     Heterologous expression and structure-function analysis of CYP2A10 and CYP2A11, which differ in only eight amino acids but have strikingly different activities toward testosterone and coumarin.
X. DING, H.M. PENG, M.J. COON

889     Dynamic topology and electron transfer interactions of microsomal cytochrome P450 in liver and adrenal cortex: protein rotation study.
Y. OHTA, M. YAMADA, S. KAWATO

893     Diverse functions of aromatase cytochrome P450: catecholestrogen synthesis, cocaine N-demethylation, and other selective drug metabolisms.
Y. OSAWA, T. HIGASHIYAMA, C. YARBOROUGH

897     Computer modeling of cytochromes P450cam interaction with camphor and its derivatives by Monte Carlo method.
Y.A. LYULKIN, A.S. IVANOV, V.S. SKVORTSOV, A.B. RUMYANTSEV

899     Inhibition of cytochrome P450 enzymes by cytokines in human hepatocyte cultures.
Z.A. RAZZAK, L. CORCOS, A. GUILLOUZO

## VIII   CYTOCHROME P450'S : AGENDA FOR ACTION, ADAPTATION AND APPLICATION

905     The P450s: agenda for action, adaptation, and application.
R.W. ESTABROOK

# I | EVOLUTION AND PHYLOGENETIC DIVERSITY OF P450 GENES

# Evolutionary argument for a connection between drug metabolism and signal transduction

Daniel W. Nebert[1], René Feyereisen[2]

[1]Department of Environmental Health, University of Cincinnati Medical Center, Cincinnati, Ohio 45267-0056, USA. [2]Department of Entomology, The University of Arizona, Tucson, Arizona, 85721, USA

## Summary

It appears likely that present-day so-called 'drug-metabolizing enzymes,' and the receptors controlling them, play a key role in *regulating the steady-state levels of ligands* important in the transcription of genes involved in cell division, differentiation, apoptosis, homeostasis, and neuroendocrine functions. In this Chapter we present a reasonable scenario for appearance of the earliest P450 having a *reductase* function during times of high temperatures and anaerobic conditions. The ancestral P450 then became important in the *detoxification of atmospheric oxygen*. Later, P450 enzyme-substrate complexes became another way to prevent peroxide and superoxide formation through *peroxidase*, *terminal oxidase*, and *monooxygenase* functions. Early P450 substrates were probably sterols and other lipids, followed by other important endogenous compounds required for gene regulation, homeostasis, growth and intercellular communication. The *synthesis* and *degradation* of plant metabolites by drug-metabolizing enzymes--both in plants, and in plant-eating animals--were undoubtedly precursors to the present-day metabolic *activation* and *detoxification* of innumerable environmental pollutants, carcinogens and drugs.

## INTRODUCTION

During the recent explosion in molecular biology, the knowledge gained about genes encoding drug-metabolizing enzymes has given us much greater insight than we had 10 years ago, about how these genes evolved. First, we propose a scenario of how ancestral P450 might have arisen under anaerobic conditions on this planet, and then became important to protect the organism against diatomic oxygen and especially its extremely reactive intermediates. Second, we describe the evolution of P450 enzyme-substrate complexes and the efficient coupling (*e.g.* cytochrome $b_5$ and iron-sulfur containing proteins) of P450-mediated reactions important to terminal oxidase and monooxygenase functions. Third, the multitude of innumerable drug-metabolizing enzymes other than P450 enzymes is reviewed. Fourth, we summarize the likelihood that present-day drug-metabolizing enzymes and their receptors play critical roles in maintaining the steady-state levels of endogenous ligands that effect ligand-modulated transcription of genes which participate in signal transduction pathways, as well as the levels of metabolites involved in second-messenger pathways. Lastly, we expand upon the role of the arachidonic acid cascade as an example of these second-messenger pathways.

## RESULTS AND DISCUSSION

### When, and for what reason, might cytochrome P450 have first appeared?

It is well known that the earliest prokaryotes were thermophilic, anaerobic and autotrophic. Photosynthetic bacteria were probably present by 3,500 million years ago (MYA). Photosystem I, found in present-day anaerobic photosynthetic (chlorophyll-containing) bacteria, does not produce oxygen and depends upon $H_2S$, $H_2$ or organic molecules to donate electrons (**Fig. 1**). In contrast, photosystem II, an invention of the early Archaean cyanobacteria about 3,600 MYA, uses water as the electron donor and is the source of today's oxygen. Initially, sulfides, $Fe^{++}$, and organic material (including biogenic methane) were oxidized. Our planet literally rusted. The adaptation to aerobic life is clearly a prokaryotic event, and may have occurred as early as the Archaean Period 3,500 to 3,000 MYA (Towe, 1990; Schopf, 1993). A great increase in the $pO_2$ probably occurred in the early Proterozoic Period (about 2,200 MYA), but eukaryotes as symbiotic organisms may not have proliferated until 500 MY later.

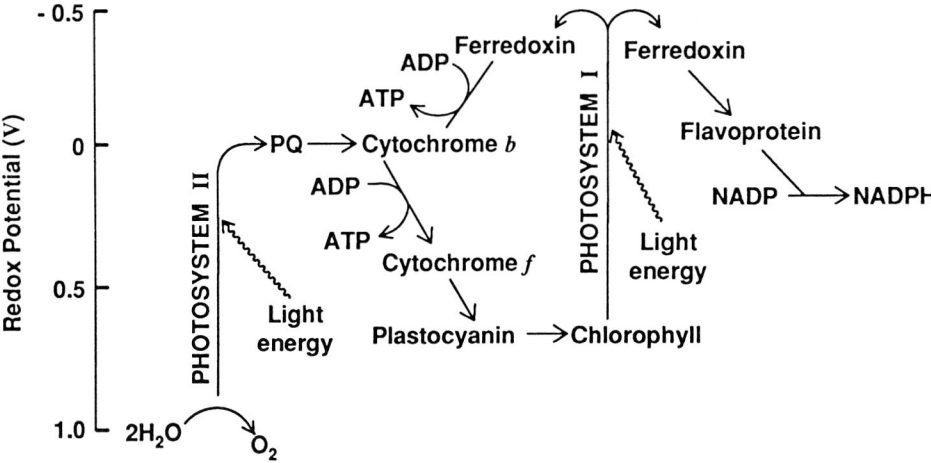

**Fig. 1. Photosystems I and II.** PS I (at *right*) receives electrons from the cytochrome *b/f* complex via plastocyanin and activates them to an energy level at which they can flow through ferredoxin, eventually to reduce $NADP^+$. PS II (at *left*) has a sufficiently low redox potential (volts) when electrons are donated to plastoquinone (PQ), thereby leading to replacement electrons being picked up from the surrounding water; electrons must be excited a second time to reach the energy level from which they can be donated to ferredoxin. The protein that oxidizes water is highly conserved across billions of years and is 50% identical between cyanobacteria and green plants. [*Modified and reproduced, with permission from* Loomis, 1988].

When the classical stoichiometry of P450 monooxygenases is discussed today, emphasis is usually placed upon the insertion of one atom of oxygen into a substrate. One possibility, however, is that the earliest P450 might not have metabolized atmospheric oxygen at all. Perhaps the ancestral P450 might have been an electron carrier. Experimental evidence has shown, for example, that only a limited number of mutations is needed to convert an electron carrier such as cytochrome *c* to an oxygen-binding protein (Wallace & Clark-Lewis, 1992) or cytochrome $b_5$ to a peroxidase (Sligar *et al.*, 1987). If we examine some of the catalytic functions of present-day P450 systems, perhaps this would shed some light on its ancestral functions.

Reductive metabolism by P450 enzymes

In addition to monooxygenase, oxidase and peroxidase functions, present-day P450 reactions include reductase functions. The reductive metabolism by P450 enzymes may represent the earliest ancestral function. The present-day P450-mediated anaerobic metabolism of azodyes and arylamines, for example, is very efficient, with turnover rates that are manyfold higher than those of P450 enzymes as monooxygenases. Similarly, Nakahara *et al.* (1993) suggested that fungal CYP55A1, a soluble nitric oxide reductase having a very high turnover rate, might represent one of the oldest functions of P450, well before the rise in atmospheric oxygen. Anaerobic respiration such as denitrification is usually seen in prokaryotes, and the Archaean atmosphere might have contained significant amounts of nitric oxide. The presence of CYP55A1 in a fungus might be an anomaly that could be explained by a lateral gene transfer event from a prokaryote (Kizawa *et al.*, 1991). It will be of interest to see if other denitrifying P450 enzymes can be found in bacteria.

The enzyme ribulose 1,5-*bis*phosphate carboxylase (Rubisco), which fixes $CO_2$ in photosynthesis, is also the starting point of photorespiration, where Rubisco acts as an oxygenase. Molecular oxygen and $CO_2$ compete for the enzyme, which is not particularly good at discriminating between the two gases. Thus, photosynthetic organisms were probably selected to find ways of removing molecular oxygen, the product of photosystem II (**Fig. 1**), not only because singlet oxygen formation or 1- or 2-electron reductions of $O_2$ cause toxicity, but also because oxygen competes with $CO_2$ at the Rubisco active-site.

Other P450 reactions that do not need atmospheric oxygen or a reducing moiety

Other modern P450 reactions, which might represent early P450 functions, include the reactions with hypervalent oxygen substrates that require no molecular oxygen, and no reductant (*e.g.* allene oxide synthase, thromboxane synthase, prostacyclin synthase). These enzymes, which have very high turnover numbers, are the isomerases of endoperoxides and hydroperoxides. Perhaps the early P450 enzymes were well suited to modify lipoidal products of chemical or enzymatic peroxidation. Significantly, endoperoxides and allylic hydroperoxides are easily formed from conjugated dienes and olefins, respectively, by reaction with singlet oxygen that results from the reaction of light with oxygen in the ground (triplet) state. The early environment of photosystem II (**Fig. 1**) had all the necessary ingredients to produce such peroxides: light, molecular oxygen, photosensitizers and membrane lipids. It might also be significant that present-day P450 proteins have maintained their peroxidative function. The presence of oxygen-consuming P450 enzymes in **Rhizobium** is not understood, but in view of the oxygen sensitivity of nitrogen fixation, hemoproteins that bind oxygen appear to act as "scavengers" and may have an adaptive function. For example, it is possible that a P450 of an oxygen-evolving photoautotrophic *and* nitrogen-fixing cyanobacterium (CYP110 from *Anabaena sp.*) may be involved in the ω-hydroxylation of the glycolipid layer which protects nitrogenase from oxygen in heterocysts (Lammers *et al.*, 1990). Interestingly, in terms of evolutionary relatedness, *CYP110* is quite close to *CYP102* from **Bacillus megaterium** (Nelson *et al.*, 1993), which exhibits ω-hydroxylase activity.

Detoxification of atmospheric oxygen

The evolution of life on this planet represents the progression of various pathways by which organic molecules can capture photons from the sun and harness this energy into productive metabolic pathways without generating serious levels of toxic intermediates. Atmospheric oxygen is capable of generating life-threatening intermediates. We suggest that early P450 evolved as a defense mechanism in order to detoxify atmospheric oxygen by a 4-electron reduction, thereby bypassing the 1-electron steps that produce toxic intermediates (**Fig. 2**).

$$O_2 \xrightarrow{e^-} \cdot O_2^- \xrightarrow{e^-} H_2O_2 \xrightarrow{e^-} \cdot OH \xrightarrow{e^-} H_2O$$

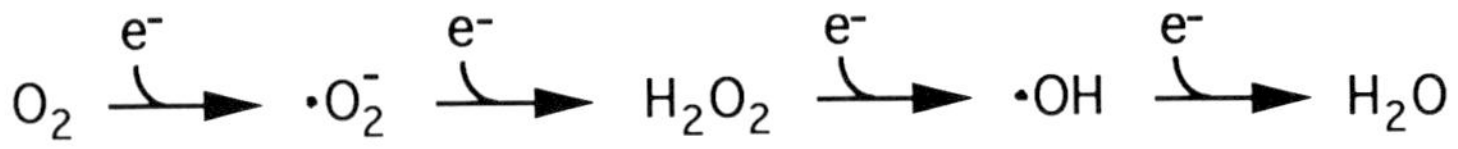

**Fig. 2. Intermediates from normal metabolism involving atmospheric oxygen.** The formation of $\cdot O_2^-$, $H_2O_2$, and $\cdot OH$ occurs by successive 1-electron reductions.

As mentioned above, emphasis on modern-day P450 reactions is usually given to the insertion of one atom of oxygen into a substrate. Clearly, this capacity to modify an endless list of substrates is what makes P450 enzymes so fascinating to pharmacologists and toxicologists. On the other hand, the mixed-function oxidation part of the stoichiometry (*i.e.* reduction of the other atom of oxygen to water) is common to *all* monooxygenation reactions. From an evolutionary point of view, we believe that this mixed-function oxidase reaction has not often been considered. P450 enzymes can operate at very low $pO_2$, and will autoxidize at less than 1 ppm $O_2$. Therefore, because it is widely believed that molecular oxygen was not a major constituent of the Earth's primitive atmosphere, the reduction of oxygen to water by the ancient P450 enzymes should take on new significance.

Adaptation of life to molecular oxygen is an obvious success story (Babcock & Wikstrom, 1992), which may obscure the fact that, amongst the 220 or so oxidases (Keevil & Mason, 1978), quite a few may initially have competed with the ancestral cytochrome oxidase for this new gas. It is likely that oxidases were modifications of existing proteins, rather than novel macromolecules (Mason, 1968; Wilkinson, 1980).

Wickramasinghe & Villee (1975) proposed that P450 might have evolved before the advent of significant levels of atmospheric oxygen. Diatomic oxygen, if allowed to degenerate in 1-electron steps, goes through three violently reactive and toxic intermediates (**Fig. 2**), before reaching water. Oxidases (*e.g.* cytochrome oxidase) can add 4 electrons quite efficiently during energy production, but some of these reactive intermediates become inevitable byproducts. P450 probably appeared as part of the multifactorial *defense system*--which also included peroxidases, catalase and superoxide dismutases--designed to combat the toxicity of oxygen intermediates. P450 hydroxylase activity would have been an unlikely early function, but P450 might have been useful in mopping up traces of oxygen. The low redox potential of P450 enzymes would have allowed them to function in reductive metabolism under an anaerobic environment, and P450 enzymes are likely to have evolved along separate lines in different prokaryotes before the formation of eukaryotic cells (Wickramasinghe & Villee, 1975). Our current understanding of the structure and functions of P450 enzymes supports many of these hypotheses.

Subsequent development of monooxygenase functions--perhaps because of the requirement for detoxification of atmospheric oxygen

In studying the diverse structures and functions of P450, we must realize that bacterial P450 enzymes studied today are not "ancestral." In fact, any and all extant P450 enzymes have been subject to natural selection. For instance, camphor hydroxylation by P450cam is a highly specialized function.

Tightly coupled reactions (*e.g.* camphor hydroxylation by CYP101) can be significantly "uncoupled" by a single amino acid (Thr-252) substitution (Imai *et al.*, 1989); this observation suggests that the "primordial" P450 could not possibly have been tightly coupled, and that 1-, 2- and 4-electron oxidase activities should be regarded as metabolic functions that have evolved along the way, during the development of modern P450 monooxygenases. Significantly, the Thr-252 residue is almost universally conserved in all extant P450 sequences, and its role in the catalytic site to stabilize reduced oxygen intermediates is well documented. Thr-252, therefore, may have been an early feature for the arrival of the oxidase function of P450.

In the presence of superoxide dismutase and/or catalase, the ancestral P450 enzyme might have participated in oxygen detoxification. It is likely that the early attempts at P450 monooxygenation of substrates were accompanied by substantial uncoupling, *i.e.* leakiness with the generation of superoxide and/or hydrogen peroxide. This might explain the evolution of cytochrome $b_5$ involvement. The role of cytochrome $b_5$ in electron transport (Staudt *et al.*, 1974) has been interpreted as a protective mechanism against the generation of these potentially harmful byproducts, and cytochrome $b_5$ function would lead to the "safe" generation of water. Interestingly, present-day P450 enzymes such as CYP2E1 are still capable of functioning as a "terminal oxidase" and converting oxygen to water. Although this process is wasteful (costing two NADPH), it does result in the elimination of molecular oxygen without generating superoxide and/or hydrogen peroxide.

## Subsequent to oxygen detoxification, substrate binding developed as an adaptation

It is well documented that the binding of substrate to P450 causes a change in the heme iron spin state that is associated with an increase in redox potential, thereby favoring reduction of the complex. Not only substrate, but also electron donors--cytochrome $b_5$ (Tamburini & Gibson, 1983) and adrenodoxin (Light & Orme-Johnson, 1981)--cause a change in the P450 heme iron spin state and shift in the redox potential. It is noteworthy that the oxyferro complex of P450 and substrate is less prone to autoxidation than the substrate-depleted complex (Tukey & Kamin, 1982). Light & Orme-Johnson (1981) noted that the more stable, reduced enzyme-substrate complex is a "simple and efficacious way" to prevent futile cycles which may produce cytotoxic peroxides or superoxide.

Binding of substrates was therefore a selective force in the evolution of P450, and the inefficient metabolism of oxygen by P450, instead of becoming extinct, was used to synthesize new molecules. This newly found function was preserved and improved upon, despite its very low catalytic efficiency.

## What were the early biological substrates for P450?

The solubility of oxygen is low at higher temperatures, oxygen is produced by membrane-bound enzymes, and the electron carriers (ferredoxins, flavoproteins and cytochromes) were probably associated with, or bound to, membranes. Therefore, it is very likely that the early ligands of P450 might have been *lipids*--and *sterols* in particular. Sterols have been suggested as some of the earliest substrates for P450 monooxygenases (Nelson & Strobel, 1987). Indeed, the C14 demethylation of lanosterol (in animals and fungi) or obtusifoliol (in plants) is an ancient function, because 14-desmethyl sterols are found in cyanobacteria and in ***Methylococcus capsulatus.*** Removal of the C14 methyl group from the cyclization product of squalene "streamlines" the α-face of the sterol (Bloch, 1983), and enhances the rigidity of membranes by favoring van der Waals interactions with the acyl chains of phospholipids. In eubacteria, bacteriohopanoids (such as bacteriohopanetetrol) may play this role in building membrane sterols. There would have been a strong selective pressure in eukaryotic cells--having no cell wall like bacteria but having an internal cytoskeleton--to optimize the 14-demethylase and, therefore, the membrane function of sterols. This optimization, *i.e.* the natural selection of mutants with increased affinity or catalytic efficiency, probably led to a "specific" P450 enzyme to fill this role. However, during evolution of such a specialized function, the early 14-demethylase probably was not a perfect monooxygenase, but must have continued to operate as a 1- or 2-electron oxidase as well--in other words, as an "uncoupled" monooxygenase. Perhaps this might explain why, in *Saccharomyces cerevisiae*, the lanosterol 14α-demethylase (***CYP51***) and $Mn^{++}$ superoxide dismutase (***Mn-SOD***) genes are still adjacent to one another (Turi *et al.*, 1991).

Role of P450 in the synthesis and detoxification of plant metabolites and environmental chemicals

The divergence of animals and plants occurred about 1,200 MYA, and the role of P450 in animal-plant interactions has been extensively reviewed (Gonzalez & Nebert, 1990). As animals began to ingest plants, the defensive response of plants was to make their tissues more toxic or less palatable to the animals--by way of gene duplication events and genetic drift to develop unique catalytic functions by the new gene products. In turn, animal P450 genes countered by similar mechanisms, so as to be able to continue ingesting plants. The plant, and especially the animal, P450 enzymes and other drug-metabolizing enzymes have continued to diversify until the present--in both synthesizing and degrading noxious chemicals derived from innumerable sources of environmental pollution, carcinogens and drugs. An additional area of importance concerns the mutual needs of animals and plants for some aspects of growth and reproduction (*e.g.* larval development, pollination, dispersal of seeds), in which the synthesis and degradation of plant metabolites by P450 and other drug-metabolizing enzymes play a major role in controlling the rate of development of the animal (***reviewed in*** Nebert, 1994).

Definition of drug-metabolizing enzymes (DMEs)

Enzymes that metabolize drugs, carcinogens, and other environmental pollutants have collectively and classically been called "drug-metabolizing enzymes" (DMEs). Most of the general classes of DMEs are listed in **Table 1**. The total number of substrates for the P450 enzymes and other DMEs includes virtually everything in the Merck Index. It is noteworthy that every DME has one or several endogenous substrates. There is not a single DME which metabolizes drugs or environmental chemicals only.

**Table 1.** List of different categories of "drug-metabolizing" enzymes

---
P450s, flavin-containing monooxygenases (FMOs), hydroxylases,
  oxidases, monoamine oxidases (MAOs), dioxygenases, reductases
UDP glucuronosyl-, GSH-, sulfo- transferases
Transaminases, acetyltransferases, methyltransferases
Quinone reductases, aldoketoreductases, carboxylesterases
NAD- & NADP-dependent alcohol, and steroid dehydrogenases
Glycosylases, glucuronidases, various hydrolases and esterases

---

What would have been substrates for the earliest DMEs? Very early endogenous substrates undoubtedly included simple alcohols, aldehydes, lipids, purines, pyrimidines and amines. Therefore, the earliest DMEs would have been alcohol and aldehyde dehydrogenases and various purine-, pyrimidine- and amine-metabolizing enzymes. It seems highly probable that DME metabolism of many of these endogenous substrates has evolved into multiple important roles for gene regulation and intercellular communication. For example, the importance of the interrelationships among, and the critical roles of, catecholamines and cyclic AMP, ATP derivatives, GMP- and GTP-binding proteins, and the GTPase superfamily in both prokaryotes and eukaryotes have been well characterized (*reviewed in* Nebert, 1994).

In the regulation of gene expression by P450-mediated metabolic products as ligands for receptors and transcription factors, what was the *initial* adaptive advantage? Perhaps the lanosterol 14-demethylase--because it was already available for the function of stabilizing membrane moieties--began to participate with other DMEs in the production of cholesterol and other steroids involved in gene regulation.

DMEs evolved to control steady-state levels of endogenous ligands
--------

Organisms capable of oxygenating substrates would clearly display an adaptive advantage for synthesizing and degrading intermediates required for gene regulation and intercellular communication. Non-peptide ligands which bind to endogenous receptors appear to be involved in ligand-modulated transcription of genes that participate in various signal transduction pathways effecting growth, morphogenesis, apoptosis, homeostasis, neuroendocrine functions, and cell type-specific proliferation including tumor promotion (**Fig. 3**). An early role for DMEs in prokaryotes probably included energy substrate utilization: splitting of the molecule, starting with various inaccessible carbon and other food sources, thereby rendering them accessible to further metabolism. Another relatively early role for these DMEs in prokaryotes and early eukaryotes was likely related to their metabolic ability to control the steady-state levels of the ligands that modulate cell division, growth, morphogenesis, and mating. It seems very reasonable that these roles have diversified into a very large number of additional complex signal transduction pathways that exist today in all eukaryotes (Nebert, 1990; 1991; 1994), and that the amount of diversification became particularly striking in the past 500 MY.

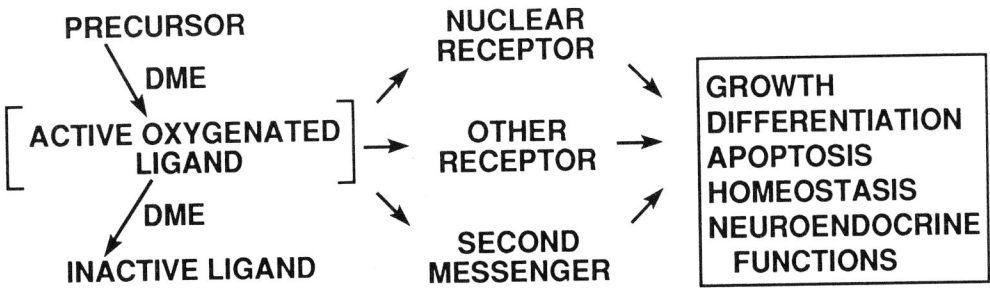

Fig. 3. **Illustrated summary of the probable relationship between drug-metabolizing enzymes (DMEs) and signal transduction pathways effecting critical functions of the cell.** Examples of a "nuclear receptor" would be the glucocorticoid receptor and the vitamin $D_3$ receptor (O'Malley, 1990). Examples of "other receptors" would include the recently discovered Per/ARNT/Sim (PAS) superfamily of transcription factors (Swanson & Bradfield, 1993) and the odorant receptor superfamily that appears specific to the olfactory epithelium (Buck & Axel, 1991). An example of "second messenger" would be the arachidonic acid cascade, described briefly below. [*Modified and reproduced, with permission from* Nebert, 1991].

The hypothesis summarized in **Fig. 3** is based on a number of experimental observations. (**a**) Endogenous non-peptide ligands for receptors involved in growth regulation and homeostasis are small organic molecules ($M_r = 250 \pm 200$). (**b**) Synthesis and degradation of these molecules always involve DMEs. (**c**) Binding of the natural ligand to the receptor always appears to be associated with cell- and/or developmental-specific increases (or decreases) in particular subsets of DMEs (Nebert & Gonzalez, 1987). (**d**) Foreign chemicals that induce these enzymes also appear to bind to endogenous receptors, acting as either agonists or antagonists of receptor function.

This hypothesis would also suggest that, when an "inducer of drug metabolism" is administered in large quantities to a cell or the intact animal, the cell or organism "senses" an abnormally elevated concentration of what is perceived to be an endogenous effector molecule; the cellular response thus would include the turning on of DME genes in order to degrade the excessive amounts of this chemical signal (Nebert, 1991; 1994). Therefore, a more appropriate name for "drug-metabolizing enzymes" might be "effector ligand-metabolizing enzymes."

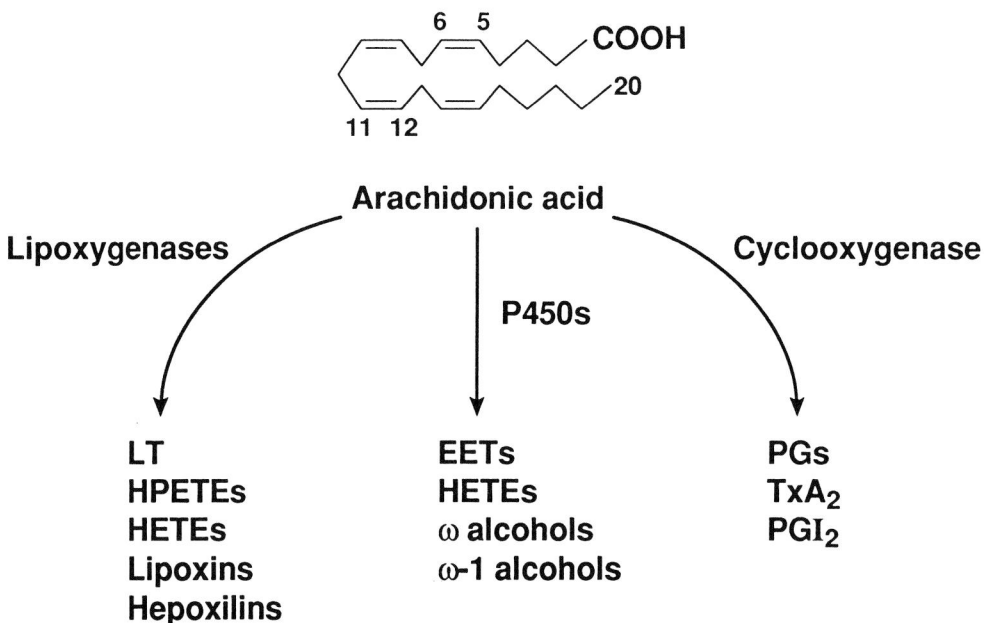

Fig. 4. Formation of arachidonic acid metabolites by lipoxygenases (*left*), cyclooxygenase (*right*), and cytochromes P450 (*center*). At least six P450 enzymes have been shown to be capable of forming EETs (*reviewed in* Capdevila *et al.*, 1992). LT, leukotriene. HPETEs, hydroperoxyeicosatetraenoic acids. HETEs, hydroxyeicosatetraenoic acids. EETs, epoxyeicosatrienoic acids. PGs, prostaglandins. $TxA_2$, thromboxane $A_2$. $PGI_2$, prostaglandin $I_2$.

## "Late" DME genes need not carry out the same functions as "early" DME genes

It should be emphasized that genes encoding DMEs that existed 1,000 or 3,500 MYA will, in all likelihood, differ in function from those that have only existed since 10 or 35 MYA. New genes continue to evolve--due to gene duplication by unequal crossing-over, gene inactivations or terminal mutational events, replication slippage, insertions, and gene conversions, perhaps responding to selective dietary pressures. It seems reasonable to presume that a "new" DME gene that has just appeared during the past several dozen million years of evolution will not have the same probability of carrying out a critical life function as a DME gene that existed 1,000 or 3,500 MYA. This would explain the recently described, numerous human polymorphisms in many DME genes, for example, which may or may not be important in terms of cell division or cell death, differentiation, homeostasis, or neuroendocrine functions (*reviewed in* Nebert, 1994).

Link between P450-mediated endoperoxide formation and arachidonic acid metabolites

A particular case in point about the ancestral P450-mediated endoperoxidative function and present-day mediation of second- messenger pathways concerns the mammalian arachidonic acid cascade (**Fig. 4**). All enzymes in the arachidonic acid cascade may be considered as DMEs, including at least six different P450 enzymes (Nebert, 1991; 1994). Although generally regarded as manifesting pronounced effects in the kidney and lung (Capdevila *et al.*, 1992), arachidonic acid metabolites undoubtedly exist in all mammalian cells and extracellular spaces. The metabolism of arachidonic acid by P450 enzymes leads to formation of EETs, HETEs, and ω and ω-1 alcohols. As summarized in **Table 2**, these metabolites are known to possess a broad spectrum of physiologic effects on growth, differentiation and homeostasis. The involvement of P450 enzymes and other DMEs with the arachidonic acid cascade might have set into motion, especially during the last 600 MY, the evolution of increasingly complex ligands for activating genes effecting growth, differentiation, apoptosis, homeostasis and neuroendocrine functions.

**Table 2.** Physiologic and subcellular effects attributed to P450-mediated metabolites of arachidonic acid*

---

*Effects of EETs*:
  Bronchodilation, renal vasoconstriction, intestinal vasodilation, inhibition of cyclooxygenase, mitogenesis, inhibition of platelet aggregation, modulation of ion transport, enhanced peptide hormone secretion, mobilization of intracellular $Ca^{++}$

*Effects of HETEs*:
  Inhibition of Na,K ATPase, vasodilation, chemotaxis of neutrophils

*Effects of ω- and ω-1 alcohols*:
  Stimulation of Na,K ATPase, vasoconstriction, bronchoconstriction

---

*EETs, epoxyeicosatrienoic acids. HETEs, hydroxyeicosatetraenoic acids.

Closing remarks

We have reviewed evidence for the likelihood that the earliest P450 probably arose on this planet when $pO_2$ levels were virtually nil. Reductive metabolism under anaerobic conditions might have been the function of the earliest P450. Next, an important ancestral P450 function that evolved was the detoxification of diatomic oxygen. With the development of improved substrate binding and efficiently coupled oxidase and monooxygenase functions, thereby avoiding significant leakage of highly reactive $O_2$ intermediates, P450 then took on endogenous substrates important in cell division and intercellular communication. Animal P450 enzymes and other DMEs evolved in order to survive while eating plants in particular ecological niches, and plant DMEs evolved to defend themselves against animals. The diversification of these functions of synthesis and degradation undoubtedly exploded in the past 500 MY to include a vast array of environmental pollutant and drug substrates.

ACKNOWLEDGMENTS

We thank our colleagues for valuable discussions and critical reading of this manuscript. Supported by NIH Grants R01 AG09235, R01 ES06321, and P30 ES06096 (D.W.N.), and R01 GM39014 (R.F.).

REFERENCES

Babcock, G.T. & Wikstrom, M. (1992): Oxygen activation and the conservation of energy in cell respiration. *Nature* **356**, 301-309.

Bloch, K.E. (1983): Sterol structure and membrane function. *CRC Crit. Rev. Biochem.* **14**, 47-92.

Buck, L. & Axel, R. (1991): A novel multigene family may encode odorant receptors: A molecular basis for odor recognition. *Cell* **65**, 175-187.

Capdevila, J.H., Falck, J.R. & Estabrook, R.W. (1992): Cytochrome P450 and the arachidonate cascade. *FASEB J.* **6**, 731-736.

Gonzalez, F.J. & Nebert, D.W. (1990): Evolution of the P450 gene superfamily: Animal-plant "warfare," molecular drive, and human genetic differences in drug oxidation. *Trends Genet.* **6**, 182-186.

Imai, M., Shimada, H., Watanabe, Y., Matsushima-Hibaya, Y., Makino, R., Koga, H., Horiushi, T. & Ishimura, Y. (1989): Uncoupling of the cytochrome P-450cam monooxygenase reaction by a single mutation, threonine-252 to alanine or valine: a possible role of the hydroxy amino acid in oxygen activation. *Proc. Natl. Acad. Sci. U.S.A.* **86**, 7823-2827.

Keevil, T. & Mason, H.S. (1978): Molecular oxygen in biological oxidations -- an overview. *Meth. Enzymol.* **52**, 3-40.

Kizawa, H., Tomura, D., Oda, M., Fukamisu, A., Hoshino, T., Gotoh, O., Yasui, T. & Shoun, H. (1991): Nucleotide sequence of the unique nitrate/nitrite-inducible cytochrome P450 cDNA from *Fusarium oxysporum*. *J. Biol. Chem.* **266**, 10632-10637.

Lammers, P.J., McLaughlin, S., Papin, S., Trujillo-Provenco, C. & Ryncarz, A.J. (1990): Developmental rearrangement of cyanobacterial *nif* genes. *J. Bacteriol.* **172**, 6981-6990.

Light, D.R. & Orme-Johnson, N.R. (1981): Beef adrenal cortical cytochrome P-450 which catalyzes the conversion of cholesterol to pregnenolone. *J. Biol. Chem.* **256**, 343-350.

Loomis, W.F. (1988): *Four Billion Years: An Essay on the Evolution of Genes and Organisms*, Sinauer Associates Inc., Massachusetts: Sunderland Publishers.

Mason, H.S. (1968): Homology among oxidases. In *Homologous Enzymes and Biochemical Evolution*, eds. N.V. Thoai & J. Roche, pp. 69-91, New York: Gordon & Breach.

Nakahara, K., Tanimoto, T., Hatano, K.I., Usuda, K. & Shoun, H. (1993): Cytochrome P450 55A1 (P450dNIR) acts as nitric oxide reductase employing NADH as the direct electron donor. *J. Biol. Chem.* **268**, 8350-8355.

Nebert, D.W. (1990): Growth signal pathways. *Nature* **347**, 709-710.

Nebert, D.W. (1991): Proposed role of drug-metabolizing enzymes: Regulation of steady-state levels of the ligands that effect growth, homeostasis, differentiation, and neuroendocrine functions. *Mol. Endocrinol.* **5**, 1203-1214.

Nebert, D.W. (1994): Drug-metabolizing enzymes in ligand-modulated transcription. *Biochem. Pharmacol*, in press.

Nebert, D.W. & Gonzalez, F.J. (1987): P450 genes: Structure, evolution and regulation. *Annu. Rev. Biochem.* **56**, 945-993.

Nelson, D.R., Kamataki, T., Waxman, D.J., Guengerich, F.P., Estabrook, R.W., Feyereisen, R., Gonzalez, F.J., Coon, M.J., Gunsalus, I.C., Gotoh, O., Okuda, K. & Nebert, D.W. (1993): The P450 superfamily: Update on new sequences, gene mapping, accession numbers, early trivial names, and nomenclature. *DNA Cell Biology* **12**, 1-51.

Nelson, D.R. & Strobel, H.W. (1987): Evolution of cytochrome P-450 proteins. *Mol. Biol. Evol.* **4**, 572-593.

O'Malley, B. (1990): The steroid receptor superfamily: More excitement predicted for the future. *Mol. Endocrinol.* **4**, 363-369.

Schopf, J.W. (1993): Microfossils of the early Archaean apex chart: new evidence of the antiquity of life. *Science* **260**, 640-646.

Sligar, S.G., Egeberg, K.D., Sage, J.T., Morikis, D. & Champion, P.M. (1987): Alteration of heme axial ligands by site-directed mutagenesis: a cytochrome becomes a catalytic demethylase. *J. Am. Chem. Soc.* **109**, 7896-7897.

Staudt, H., Lichtenberger, F. & Ulrich, V. (1974): The role of NADH in uncoupled microsomal monooxygenations. *Eur. J. Biochem.* **46**, 99-106.

Swanson, H.I. & Bradfield, C.A. (1993): The Ah receptor: Genetics, structure and function. *Pharmacogenetics* **3**, 213-230.

Tamburini, P.P. & Gibson, G.G. (1983): Thermodynamic studies of the protein-protein interactions between cytochrome P450 and cytochrome $b_5$. *J. Biol. Chem.* **258**, 13444-13452.

Towe, K.M. (1990): Aerobic respiration in the Archaean. *Nature* **348**, 54-56.

Tukey, R.C. & Kamin, H. (1982): The oxyferro complex of adrenal cytochrome P-450scc. *J. Biol. Chem.* **257**, 9309-9314.

Turi, T.G., Kalb, V.F. & Loper, J.C. (1991): Cytochrome P450 lanosterol 14α-demethylase and manganese superoxide dismutase are adjacent genes in *Saccharomyces cerevisiae*. *Yeast* **7**, 627-630.

Wallace, C.J.A. & Clark-Lewis, I. (1992): Functional role of heme ligation in cytochrome *c*. *J. Biol. Chem.* **267**, 3852-3861.

Wickramasinghe, R.H. & Villee, C.A. (1975): Early role during chemical evolution for cytochrome P-450 in oxygen detoxification. *Nature* **256**, 509-511.

Wilkinson, C.F. (1980): The metabolism of xenobiotics: a study in biochemical evolution. In *The Scientific Basis of Toxicity Assessment*, ed. H. Witschi, pp. 251-268, Amsterdam: Elsevier.

# Hierarchical clustering and proposed classification of the P450 superfamily using consensus sequences

Alexander I. Archakov, Kirill N. Degtyarenko

*Institute of Biomedical Chemistry, Pogodinskaya 10, 119832 Moscow, Russia*

The method for clustering and classification of P450s has been developed basing on consecutive multiple alignments of consensus sequences. On the basis of this multiple alignment the novel evolutionary tree of P450 superfamily has been constructed. "Strong" and "weak" hierarchies in the clustering of P450 sequences were revealed. 181 complete amino acid sequences of P450 were classified using the proposed algorithm. A code was assigned to each P450 amino acid sequence reflecting its relative distance from the corresponding consensus sequences at three hierarchical levels. Thus, only the principle of sequence similarity is chosen for classification without involving such criteria as enzymatic properties, phylogenetic assignment or chronology.

## INTRODUCTION

Extant classification of P450s (Nelson et al., 1993) is based on the concept of divergent evolution of this superfamily. This classification plays a significant role in understanding of P450 superfamily evolution. Undoubtedly, ousting the trivial names of P450 enzymes, it promoted the ordering knowledge in the P450 field. However, some principles of this classification could be found arbitrary. Originally, the names of families and subfamilies were chosen on the basis of some P450 proteins trivial names. Thus, the naming of families has been carried out either chronologically (the later the sequence was published the higher number was assigned) or in a manner emphasizing their enzymatic specificity. For P450 enzymes involved in steroidogenesis and steroid metabolism, the family names correspond to the numbering of reacting carbon atoms in steroid ring (families CYP7, 11, 17, 19, 21, 24, 27). Finally, family names reflect certain taxonomy although molecular evolution of P450s, in contrast to that of many protein families, might not follow phylogeny (Degtyarenko & Archakov, 1993). The families CYP1 through CYP49 are reserved for animals, fungal families are CYP51 to CYP69, plants occupy families CYP71 to CYP99 whereas bacterial families start at CYP101 (Nebert & Nelson, 1991). Newly discovered proteins (or genes) were subsequently placed in the corresponding families and subfamilies following the sequence similarity. However, the names for new subfamilies and individual proteins were assigned, as a rule, on a chronological basis. Therefore, the proposed nomenclature cover only the composition of P450 families and subfamilies but does not take into account structural resemblance between and within these groups.

Multiple sequence alignments of the P450 proteins have been constructed using various methods (Zvelebil et al., 1991; Gotoh, 1992; Lewis & Moereels, 1992). The comparison of dendrograms for representative members of P450 superfamily obtained by different authors (Nebert & Nelson, 1991; Zvelebil et al., 1991; Gotoh, 1992) shows that their branching topologies are similar. Almost all prokaryotic P450s and fungal CYP55 form one major class, "B-class", and all eukaryotic P450s and bacterial CYP102 belong to the other major class, "E-class" (we have adopted these terms from Gotoh, 1992).

According to the neutral theory of molecular evolution (Kimura, 1991), only few amino acid residues in the key regions of a protein molecule are highly conserved because they are functionally important. Thus, the rate of protein evolution is limited by conserved regions of molecule, i.e. it can be determined as the rate of consensus sequence evolution. That is why a "two-step" multiple alignment procedure can be performed after preliminary clustering. The result of such a two-step technique combines the equal representation of each P450 subfamily in the final multiple alignment with the preferential comparison of conserved regions and the reduction of random matches between subfamilies. The choice of a scoring matrix or a gap penalty does not dramatically affect the alignment within a single subfamily.

## MATERIALS AND METHODS

*Sequences.* Amino acid sequences of 181 P450s have been obtained from the Cytochrome P450 DataBase, release 1.1 (Archakov et al. 1992). In this study, each *CYP* gene product has been represented by a single amino acid sequence (allelic variants were not considered). In case of CYP102, only N-terminal part (1-472) which corresponds to the haem domain of P450BM-3 has been used.

*Programs.* The MULTALIN program (Corpet, 1988) has been used for performing consecutive multiple sequence alignments and for consensus sequence construction. This program was also used to calculate pairwise similarity scores between query sequences and appropriate consensus sequences. Each set of amino acid sequences was previously randomized for reduction of the arrangement effect on the order of clustering. With MULTALIN, each alignment cycle was repeated until the clustering of the sequences remained the same. The RANDOMIC program for randomization of the sequences set was kindly provided by Dr. T.A.Kulikova (Institute of Medical Technics, Moscow). All computations were performed on a 33 MHz IBM AT 386 compatible microcomputer with 640 Kb of memory.

*A "Two-Step" Multiple Alignment.* At the first step, the sequences were aligned within each subfamily (family) by the methods described above. For each of these multiple alignments the corresponding consensus sequences were obtained. At the second step, the obtained consensus sequences and those sequences which are sole representatives of their subfamilies (families) were aligned together. The scoring marix of Dayhoff et al. (1978) was supplemented with the elements related to the classes of amino acids applied for consensus sequence building, such as aliphatic (Ile, Leu, Met, Val), aromatic (Phe, Trp, Tyr), polar (Ser, Thr), tiny (Ala, Gly), positively charged (Arg, His, Lys), and negatively charged amino acids and their amides (Asp, Asn, Glu, Gln). Each row (column) relevant to the given class was calculated as average of the corresponding amino acid rows (columns) of Dayhoff et al. matrix.

## RESULTS AND DISCUSSION

*Hierarchical clustering of P450 superfamily.* For the construction of evolutionary tree of P450s, a two-step multiple alignment procedure has been used. The use of consensus sequences both speeds up the multiple alignment procedure and facilitates the perception of results. Thus, instead of 181 full-length P450 sequences only 24 consensus sequences and 31 unique sequences were aligned. The obtained dendrogram

(Fig. 1 ) resembles other published trees of the P450 superfamily (Nebert & Nelson, 1991; Gotoh, 1992) in general. Surprisingly, in this case two yeast families, CYP51 and CYP52, were clustered. It is confusing that the consensus of three *Drosophila* sequences classified previously as the CYP4E subfamily (Nelson et al., 1993) does not form a cluster with the other representatives of CYP4 family but is most similar to CYP7. In contrast to Gotoh (1992), cyanobacterial CYP110 is not included into E-class but appears to be the most distant sequence in the present P450 superfamily. Two plant P450s, CYP71 and CYP73, are related to the {CYP1, CYP2, CYP17, CYP21} cluster whereas CYP72 is close to the CYP4 family cluster. There are some other differences in branching topologies of the trees, but these details need not to be described here.

Hierarchies in P450 superfamily could be divided into "weak" and "strong" (McMorris & Powers, 1991). Indeed, the composition of stable clusters described above is largely insensitive to any changes in the sequence arrangement as well as to the multiple alignment method and replacement of one substitution matrix by another. This is an evident case of strong hierarchy. On the contrary, small differences in lengths of ancient branches can lead to different hierarchies on the same sequence set. The relative positions of five major branches in the E-class together with those of most families in the B-class are extremely variable depending on algorithm used for multiple alignment and tree building. In other words, these clusters form weak hierarchy. Another case of weak hierarchy can be observed *within* any stable P450 group where the order of sequence clustering also varies significantly.

*Structural classification of P450 superfamily.* The three-position coding principle is proposed to classify P450s on the basis of their sequence similarity. The first two-digit position is assigned with family, the next one - with subfamily, and the last - with the unique protein (gene). The use of consecutive multiple alignment technique allows us to arrange all the individual P450s, subfamilies, and families according to their relative distance from their subfamily, family, and superfamily consensus sequence, respectively.

At the first hierarchical level, the sequences were aligned within P450 subfamilies and families, containing sole subfamily ({CYP1A}; {CYP2A, CYP2B, CYP2C, CYP2D, CYP2E, CYP2F, CYP2G, CYP2H}; {CYP3A}; {CYP4A, CYP4B, CYP4E}; {CYP6A}; {CYP7}; {CYP11A, CYP11B}; {CYP17}; {CYP19}; {CYP21A}; {CYP27}; {CYP51}; {CYP52A, CYP52C}). The corresponding (sub)family consensus sequences were built up for each of these multiple alignments. At the second hierarchical level, the P450 subfamily consensus sequences and those sequences which being the sole representatives of their subfamilies were aligned together ({CYP2A÷, CYP2B÷, CYP2C÷, CYP2D÷, CYP2E÷, CYP2F÷, CYP2G÷, CYP2H÷, CYP2J1}; {CYP4A÷, CYP4B÷, CYP4E÷, CYP4C1, CYP4D1, CYP4F1}; {CYP6A÷, CYP6B1}; {CYP11A÷, CYP11B÷}; {CYP52A÷, CYP52C÷, CYP52B1, CYP2D1}; {CYP105A1, CYP105B1, CYP105C1, CYP105D1}; {CYP107A1, CYP107B1}). Thus, consensus sequences for

---

**Fig. 1.** Dendrogram for the P450 superfamily obtained with a multiple alignment of 24 consensus sequences and 31 unique sequences by MULTALIN program. Consensus sequences are indicated by ÷. Systematic names of P450s correspond to last nomenclature (Nelson et al. 1993).

Abbreviations: chi = chicken; hum = human; rab = rabbit; rat = rat; tro = trout; bdi = *Blaberus discoidalis* (cockroach); dme = *Drosophila melanogaster* (fruit fly); ppx = *Papilio polyxenes* (butterfly); lst = *Lymnaea stagnalis* (pond snail); cro = *Catharanthus roseus* (Madagascar periwinkle); htu = *Helianthus tuberosus* (Jerusalem artichoke); pam = *Persea americana* (avocado); ani = *Aspergillus niger*; cma = *Candida maltosa*; ctr = *Candida tropicalis*; fox = *Fusarium oxysporum*; sce = *Saccharomyces cerevisiae*; ana = *Anabaena* spp.; atu = *Agrobacterium tumefaciens*; bja = *Bradyrhizobium japonicum*; bme = *Bacillus megaterium*; bsu = *Bacillus subtilis*; ppu = *Pseudomonas putida*; pse = *Pseudomonas* spp.; ser = *Saccharopolyspora erythraea*; sgl = *Streptomyces griseolus*; sgr = *Streptomyces griseus*; str = *Streptomyces* spp.

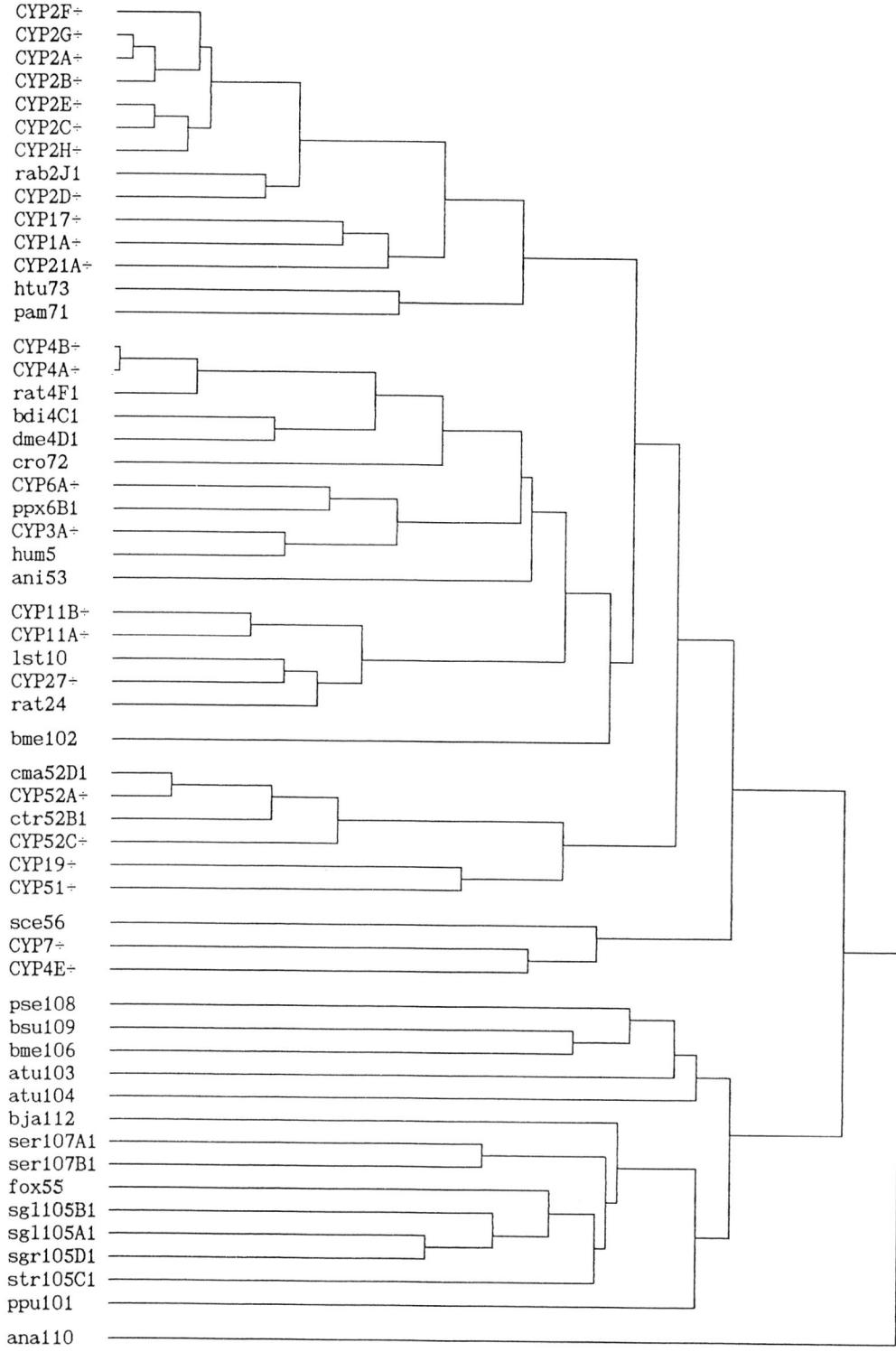

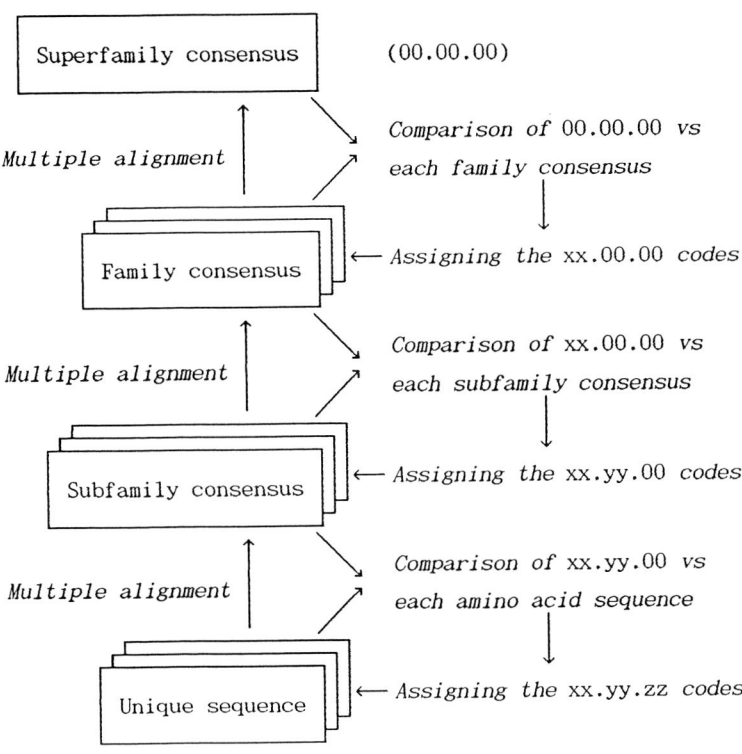

Fig. 2. The scheme of an algorithm for classification of P450 superfamily (see text).

each family consisting of more than one subfamily have been constructed.

At last, 15 P450 family consensus sequences (CYP1÷, CYP2÷, CYP3÷, CYP4÷, CYP6÷, CYP7÷, CYP11÷, CYP17÷, CYP19÷, CYP21÷, CYP27÷, CYP51÷, CYP52÷, CYP105÷, CYP107÷) and 18 sole representatives of their families (CYP5, CYP10, CYP24, CYP53, CYP55, CYP56, CYP71, CYP72, CYP73, CYP101, CYP102, CYP103, CYP104, CYP106, CYP108, CYP109, CYP110, CYP112) were aligned together. As a result, consensus sequence for the whole superfamily has been built up.

Fig. 2 represents the principle of an algorithm used for the construction of structural classification of P450s. It is based on the pairwise alignment scores between a given sequence and consensus sequence for the corresponding cluster. In our coding scheme, the first, second and third two-digit positions are intended for family, subfamily, and individual sequence codes, respectively. The consensus sequence for entire P450 superfamily is designated as 00.00.00. The family consensus sequence that gets maximal score with 00.00.00 is designated as 01.00.00. In the case of family comprised of a single subfamily, code 01 is implied to be assigned to the second position; similarly, for (sub)family represented by sole sequence, code 01 should be ascribed to the third position.

The results obtained are summarized in Tab. 1. It was found that the nearest family to 00.00.00 is CYP52 (01.00.00), the most remote family is CYP110 (33.00.00). Since CYP110 family is presented by unique sequence, this sequence should be encoded as

Table 1. PROPOSED CODES FOR P450s.

| SysName | Spp. | Code | SysName | Spp. | Code | SysName | Spp. | Code |
|---|---|---|---|---|---|---|---|---|
| CYP1A1 | dog | 11.01.04 | CYP2D9 | mus | 06.08.05 | CYP17 | bov | 12.01.01 |
| CYP1A1 | gpi | 11.01.05 | CYP2D10 | mus | 06.08.03 | CYP17 | chi | 12.01.06 |
| CYP1A1 | ham | 11.01.03 | CYP2D11 | mus | 06.08.06 | CYP17 | hum | 12.01.02 |
| CYP1A1 | hum | 11.01.07 | CYP2D14 | bov | 06.08.10 | CYP17 | mus | 12.01.04 |
| CYP1A1 | mon | 11.01.06 | CYP2E1 | hum | 06.06.02 | CYP17 | pig | 12.01.05 |
| CYP1A1 | mus | 11.01.02 | CYP2E1 | mus | 06.06.04 | CYP17 | rat | 12.01.03 |
| CYP1A1 | rab | 11.01.08 | CYP2E1 | rab | 06.06.03 | CYP17 | tro | 12.01.07 |
| CYP1A1 | rat | 11.01.01 | CYP2E1 | rat | 06.06.01 | CYP19 | chi | 09.01.04 |
| CYP1A1 | tro | 11.01.14 | CYP2E2 | rab | 06.06.05 | CYP19 | hum | 09.01.03 |
| CYP1A2 | ham | 11.01.13 | CYP2F1 | hum | 06.05.01 | CYP19 | mus | 09.01.01 |
| CYP1A2 | hum | 11.01.11 | CYP2F2 | mus | 06.05.02 | CYP19 | rat | 09.01.02 |
| CYP1A2 | mus | 11.01.10 | CYP2G1 | rab | 06.03.02 | CYP19 | tro | 09.01.05 |
| CYP1A2 | rab | 11.01.12 | CYP2G1 | rat | 06.03.01 | CYP21A1 | bov | 18.01.02 |
| CYP1A2 | rat | 11.01.09 | CYP2H1 | chi | 06.07.01 | CYP21A2 | hum | 18.01.04 |
| CYP2A1 | rat | 06.04.07 | CYP2H2 | chi | 06.07.02 | CYP21A1 | mus | 18.01.05 |
| CYP2A2 | rat | 06.04.08 | CYP2J1 | rab | 06.09.01 | CYP21A1 | pig | 18.01.03 |
| CYP2A3 | rat | 06.04.04 | CYP3A1 | rat | 16.01.04 | CYP21A1 | she | 18.01.01 |
| CYP2A4 | mus | 06.04.02 | CYP3A2 | rat | 16.01.06 | CYP24 | rat | 13.01.01 |
| CYP2A5 | mus | 06.04.01 | CYP3A3 | hum | 16.01.02 | CYP27 | hum | 04.01.02 |
| CYP2A6 | hum | 06.04.03 | CYP3A4 | hum | 16.01.01 | CYP27 | rab | 04.01.01 |
| CYP2A7 | hum | 06.04.05 | CYP3A5 | hum | 16.01.05 | CYP27 | rat | 04.01.03 |
| CYP2A8 | ham | 06.04.06 | CYP3A6 | rab | 16.01.08 | CYP51 | cal | 07.01.01 |
| CYP2B1 | rat | 06.01.02 | CYP3A7 | hum | 16.01.09 | CYP51 | ctr | 07.01.02 |
| CYP2B2 | rat | 06.01.03 | CYP3A11 | mus | 16.01.03 | CYP51 | sce | 07.01.03 |
| CYP2B3 | rat | 06.01.11 | CYP3A13 | mus | 16.01.07 | CYP52A1 | ctr | 01.01.01 |
| CYP2B4 | rab | 06.01.04 | CYP4A1 | rat | 02.02.06 | CYP52A2 | ctr | 01.01.03 |
| CYP2B5 | rab | 06.01.05 | CYP4A2 | rat | 02.02.08 | CYP52A3 | cma | 01.01.07 |
| CYP2B6 | hum | 06.01.07 | CYP4A3 | rat | 02.02.07 | CYP52A4 | cma | 01.01.04 |
| CYP2B7 | hum | 06.01.08 | CYP4A4 | rab | 02.02.05 | CYP52A5 | cma | 01.01.02 |
| CYP2B9 | mus | 06.01.10 | CYP4A5 | rab | 02.02.01 | CYP52A6 | ctr | 01.01.08 |
| CYP2B10 | mus | 06.01.01 | CYP4A6 | rab | 02.02.02 | CYP52A7 | ctr | 01.01.10 |
| CYP2B11 | dog | 06.01.06 | CYP4A7 | rab | 02.02.04 | CYP52A8 | ctr | 01.01.09 |
| CYP2B12 | rat | 06.01.09 | CYP4A8 | rat | 02.02.09 | CYP52A10 | cma | 01.01.06 |
| CYP2C1 | rab | 06.02.08 | CYPA11 | hum | 02.02.03 | CYP52A11 | cma | 01.01.05 |
| CYP2C2 | rab | 06.02.15 | CYP4B1 | hum | 02.03.02 | CYP52B1 | ctr | 01.04.01 |
| CYP2C3 | rab | 06.02.16 | CYP4B1 | rab | 02.03.03 | CYP52C1 | ctr | 01.02.02 |
| CYP2C4 | rab | 06.02.09 | CYP4B1 | rat | 02.03.01 | CYP52C2 | cma | 01.02.01 |
| CYP2C5 | rab | 06.02.11 | CYP4C1 | bdi | 02.04.01 | CYP52D1 | cma | 01.03.01 |
| CYP2C6 | rat | 06.02.06 | CYP4D1 | dme | 02.05.01 | CYP53 | ani | 15.01.01 |
| CYP2C7 | rat | 06.02.13 | CYP4E1h | dme | 02.06.01 | CYP55 | fox | 32.01.01 |
| CYP2C8 | hum | 06.02.14 | CYP4E1l | dme | 02.06.02 | CYP56 | sce | 20.01.01 |
| CYP2C9 | hum | 06.02.03 | CYP4E1d | dme | 02.06.03 | CYP71 | pam | 22.01.01 |
| CYP2C10 | hum | 06.02.01 | CYP4F1 | rat | 02.01.01 | CYP72 | cro | 10.01.01 |
| CYP2C11 | rat | 06.02.02 | CYP5 | hum | 05.01.01 | CYP73 | htu | 17.01.01 |
| CYP2C12 | rat | 06.02.18 | CYP6A1 | mdo | 14.01.01 | CYP101 | ppu | 26.01.01 |
| CYP2C13 | rat | 06.02.17 | CYP6A2 | dme | 14.01.02 | CYP102 | bme | 21.01.01 |
| CYP2C14 | rab | 06.02.10 | CYP6B1 | ppx | 14.02.01 | CYP103 | atu | 24.01.01 |
| CYP2C16 | rab | 06.02.07 | CYP7 | hum | 19.01.01 | CYP104 | atu | 30.01.01 |
| CYP2C22 | rat | 06.02.19 | CYP7 | rat | 19.01.02 | CYP105A1 | sgl | 25.02.01 |
| CYP2C23 | rat | 06.02.20 | CYP10 | lst | 03.01.01 | CYP105B1 | sgl | 25.03.01 |
| CYP2C26 | ham | 06.02.04 | CYP11A1 | bov | 08.01.04 | CYP105C1 | str | 25.04.01 |
| CYP2C27 | ham | 06.02.05 | CYP11A1 | hum | 08.01.03 | CYP105D1 | stg | 25.01.01 |
| CYP2C28 | ham | 06.02.12 | CYP11A1 | pig | 08.01.01 | CYP106 | bme | 27.01.01 |
| CYP2D1 | rat | 06.08.01 | CYP11A1 | rat | 08.01.02 | CYP108 | pse | 23.01.01 |
| CYP2D2 | rat | 06.08.08 | CYP11B1 | bov | 08.02.07 | CYP107A1 | ser | 28.01.01 |
| CYP2D3 | rat | 06.08.04 | CYP11B1 | hum | 08.02.06 | CYP107B1 | ser | 28.02.01 |
| CYP2D4 | rat | 06.08.07 | CYP11B1 | mus | 08.02.04 | CYP109 | bsu | 29.01.01 |
| CYP2D5 | rat | 06.08.02 | CYP11B1 | rat | 08.02.03 | CYP110 | ana | 33.01.01 |
| CYP2D6 | hum | 06.08.09 | CYP11B2 | hum | 08.02.05 | CYP112 | bja | 31.01.01 |
| CYP2D7 | hum | 06.08.11 | CYP11B2 | mus | 08.02.02 | | | |
| CYP2D8P | hum | 06.08.12 | CYP11B3 | rat | 08.02.01 | | | |

33.01.01. Analogously, a family consensus sequence (xx.00.00) within the relevant family is compared with each subfamily consensus sequence or unique subfamily sequence. For example, the most similar sequence to 01.00.00 is CYP52A subfamily consensus sequence (01.01.00), the second is CYP52C subfamily consensus sequence (01.02.00), CYP52B1 is the most distant sequence (01.04.01). Thus, the next two positions are filled. Finally, the individual protein sequences are compared with a subfamily consensus sequence (xx.yy.00). For example, the nearest sequence to 01.01.00 is CYP52A1 (01.01.01), whereas CYP52A7 represents the most distant sequence (01.01.10).

Such an algorithm of classification is based on the unified concept: all the (consensus) sequences belonging to any group are arranged in order of decreasing similarity to the consensus sequence with refer to the corresponding group. Thus, "01" in any position designates maximal similarity of a given (consensus) sequence to the appropriate consensus sequence of higher order. It does not imply that any succession of sequences could be simply directed along the "distance" axis. The distances from sequences $P$ and $Q$ to the corresponding consensus sequence say nothing about the distance between $P$ and $Q$. For instance, CYP52, CYP4, and CYP10 families in our classification occupy the first, second, and third top positions, respectively, that does not mean high similarity between these families. The structure of any cluster of related sequences may be conceived rather as a set of points in multidimensional space.

In conclusion, when characterizing the proposed structural classification, one more important peculiarity should be emphasized. Both P450 family consensus sequences and unique sequences can be found throughout the wide range of similarity scores with P450 superfamily consensus sequence (Tab. 1). However, there is a substantial disparity between consensus and unique sequences in respect of "stability" of the proposed classification. If the family presently comprising only one sequence is augmented with new sequences in future, this could result in drastic change of family arrangement. On the contrary, addition of the new P450 sequences to the families, which have been already well represented, should not affect family position significantly. The same considerations would be valid for the subfamily level but not for an individual sequence level. In the last case, the third code position appears to be varying most frequently. It stands to reason that the newly discovered P450 families/subfamilies would shuffle family/subfamily codes as well. Hence, our system would change along with the growth of the sequence information on P450 superfamily. In order to avoid possible confusion, it is essential that the proposed code names should be accompanied by systematic names according to Nelson et al. (1993) nomenclature.


## REFERENCES

Archakov, A.I., Bachmanova, G.I., Sandler, M.K., Tutochkin, I.Yu., and Lisitsa, A.V. (1992): Cytochrome P-450 Database and its scientific application. In *Proceedings of 7th International Conference on Biochemistry and Biophysics of Cytochrome P-450*, eds. A.I. Archakov and G.I. Bachmanova, pp. 673-679. Moscow: INCO-TNC Publishers.

Corpet, F. (1988): Multiple sequence alignment with hierarchical clustering. *Nucleic Acids Res.* 16, 10881-10890.

Degtyarenko, K.N., and Archakov, A.I. (1993): Molecular evolution of P450 superfamily and P450-containing monooxygenase systems. *FEBS Lett.* (in press).

Dayhoff, M.O., Schwartz, R.M., and Orcutt, B.C. (1978) A model of evolutionary change in proteins. In *Atlas of Protein Sequence and Structure*, ed. M.O. Dayhoff, Vol. 5, Suppl. 3, pp. 345-352. National Biomedical Research Foundation, Washington, DC.

Gotoh, O. (1992): Alignment of protein sequences as a tool for evolutionary and structural investigation of cytochrome P-450. In *Proceedings of 7th International Conference on Biochemistry and Biophysics of Cytochrome P-450*, eds. A.I. Archakov and G.I. Bachmanova, pp. 686-691. Moscow: INCO-TNC Publishers.

Feng, D.-F., and Doolittle, R. (1987): Progressive sequence alignment as a prerequisite to correct phylogenetic trees. *J. Mol. Evol.* **25**, 351-360.

Kimura, M. (1991): The neutral theory of molecular evolution: a review of recent evidence. *Jpn. J. Genet.* **66**, 367-386.

Lewis, D.F.V., and Moereels, H. (1992): The sequence homologies of cytochromes P-450 and active-site geometries. *J. Computer-Aided Mol. Design* **6**, 235-252.

McMorris, F.R., and Powers, R.C. (1991): Consensus weak hierarchies. *Bull. Math. Biol.* **53**, 679-684.

Nebert, D.W., and Nelson, D.R. (1991): P450 gene nomenclature based on evolution. *Methods Enzymol.* **206**, 3-11.

Nelson, D.R., Kamataki, T., Waxman, D.J., Guengerich, F.P., Estabrook, R.W., Feyereisen, R., Gonzalez, F.J., Coon, M.J., Gunsalus, I.C., Gotoh, O., Okuda, K., and Nebert, D.W. (1993): The P450 superfamily: update on new sequences, gene mapping, accession numbers, early trivial names of enzymes, and nomenclature. *DNA Cell Biol.* **12**, 1-51.

Zvelebil, M.J.J.M., Wolf, C.R., and Sternberg, M.G.E. (1991): A predicted three-dimensional structure of human cytochrome P450: implications for substrate specificity. *Protein Engineering* **4**, 271-282.

# Function and diversity of plant cytochrome P450

Francis Durst, Irène Benveniste, Jean-Pierre Salaün, Danièle Werck

*Institut de Biologie Moléculaire des Plantes, Département d'Enzymologie Moléculaire et Cellulaire, CNRS, 28, rue Goethe, 67083 Strasbourg Cedex, France*

The two first reports of cytochrome P450 in membrane fractions from higher plants appeared in the late 60's, and interestingly they were concerned with the oxidation of a physiological substrate (kaurene) on one hand (Murphy & West, 1969), and a pesticide (monuron) on the other hand (Frear *et al.*, 1969). Nowadays, there is conclusive evidence for P450 involvement in the oxidation of over 30 physiological substrates and over 20 xenobiotics (Durst & Benveniste, 1993), and there is indirect evidence, mainly from *in vivo* induction and inhibition studies, for several dozen more substrates. P450 is probably involved in a significant number of the oxidation reactions required for the biosynthesis of secondary plant metabolites. Since it is generally believed that the more than 20000 secondary metabolites that are presently identified amount to only 5 to 10% of the total (Swain, 1977), the total number of P450-catalyzed reactions may be enormous. It is likely that it is this wide spectrum of 'physiological' P450's rather than the existence of some specialized broad substrate range isoforms that accounts for the role of these oxygenases in the biotransformation of pesticides and environmental chemicals. In the following lines, we will give a short account of the diversity of these plant P450's: diversity of reactions, of the functions and a first study of their relationships. The diversity of the plant NADPH-P450 reductases is documented in the paper by Lesot *et al*, in this volume.

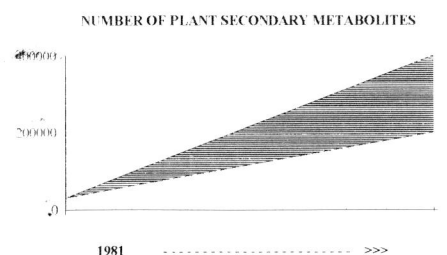

Figure 1: 200000 to 400000 secondary metabolites may exist in higher plants

**Diversity of reactions:** All plant P450 monooxygenases described so far are microsomal. Most sub cellular fractionation studies have shown P450 linked to the ER although association with plasma membranes and provacuolar fractions have also been reported (Durst, 1991). There is no consistent evidence for mitochondrial or chloroplastic P450 in plants. The distribution, induction and function of these enzymes has been covered by recent reviews (Durst, 1991, Durst & Benveniste, 1993). Besides hydroxylation and epoxidation, the plant P450's catalyze an impressive number of isomerization, ring migration, cyclization, carbon-carbon coupling and oxidative desaturation. Apparently these activities do not rely on some special characteristics of the plant hemoprotein but rather on the chemical reactivity of the substrates and of the catalytic intermediates generated in the active site. A notable exception is the allene oxide synthase (AOS), an $O_2$- and NADPH-independent hemoprotein purified from flaxseed by Song & Brash (1991). This enzyme catalyzes the dehydration of 13-hydroperoxylinoleate (Fig. 1) to form an allene oxide which may undergo cyclization to jasmonic acid, a plant stress hormone, or be hydrolyzed to the corresponding α- and γ-ketols (1, 2 and 3 in fig.1).

The cDNA deduced peptide sequence of AOS (CYP74) which has been determined recently (Brash, unpublished), shows severe deviation from the canonical P450 structure in helices I and K and in the heme-binding domain. It is particularly interesting that this affects the 'oxygen groove' centred around the highly conserved threonine (T252 in P450cam) which is missing in AOS.

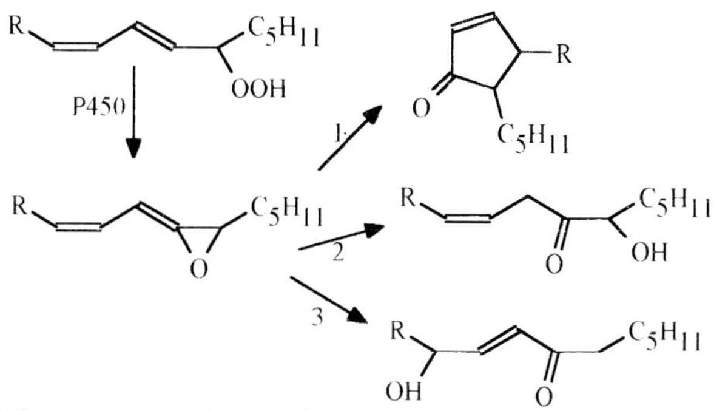

Figure 1: Reaction catalyzed by the flaxseed allene oxide synthase (AOS)

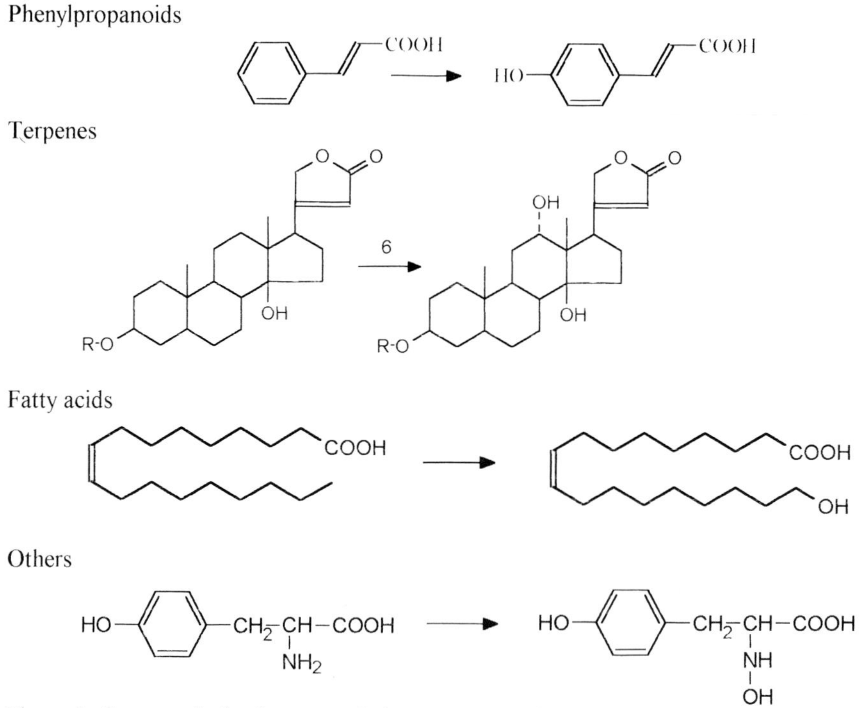

Figure 2: Some typical substrates of plant P450 : *tr*-cinnamic acid, digitoxin, oleic acid and tyrosine

Most physiological activities described so far fall into 3 main pathways (Fig. 2.) of the so-called secondary plant metabolism: phenylpropanoids (lignins, pigments, UV-protectants, defence molecules...), terpenes (sterols, hormones, defence molecules, aroma...), fatty acids (cutin and suberin precursors, defence molecules...). Furthermore, there is growing evidence that P450 is involved in an extreme diversity of species-specific metabolisms like arginine hydroxylation in cyanogenic species (Halkier & Möller, 1991), para-ortho carbon-carbon coupling in opioid synthesizing species (Gerardy & Zenk, 1993), etc... Beside these physiological reactions, P450 plays an important role in the biotransformation of xenobiotics: pesticides, pollutants. For several herbicides it has been established that P450 is the prime factor for selectivity, and it is also implicated in the appearance of resistant weed biotypes.

**Induction of P450 in plants:** Several studies, mostly from this laboratory, have well documented the selective induction of several P450 isoforms by physiological, physical and chemical agents (Table 1).

| Class | inducer |
|---|---|
| Physical Factors: | light (phytochrome), UV, wounding, oxygenation |
| Challenge by pathogens: | fungi, viruses, elicitors (fungal membrane extracts, etc) |
| Chemicals: | manganese, cadmium, mercury, |
| | clofibrate (free acid, ester, and analogs) |
| | phenobarbital, aminopyrine, ethanol |
| | biphenyl, arochlor, ß-naphtoflavone, isosaffrole |
| | herbicides (2,4-D, monuron, chlortoluron, mecoprop...) |
| | safeners (NA, cyometrinil, oxabetrinil, CGA 133205...) |
| | azoles |

Table 1: Summary of factors that induce P450 in plants.

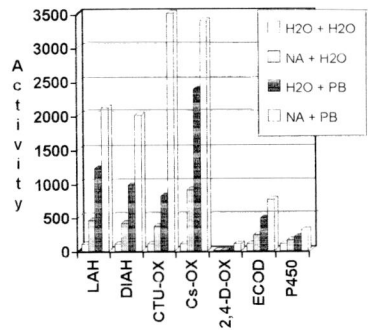

Figure 3: Induction of laurate hydroxylase (LAH), diclofop hydroxylase (DIAH), chlortoluron oxidase (CTU-OX), chlorsulfuron oxidase (Cs-OX), 2,4-dichlorophenoxyacetate oxidase (2,4-D-OX), ECOD and P450 in wheat seedlings treated with various combinations of naphtalene anhydride (NA) and phenobarbital (PB).

An interesting case is that of the so-called safeners, chemicals which are used to protect certain crops against pesticides. Several safeners are in field use. They were developed by screening, but recent work by several groups, notably the laboratory of S. Frear at Fargo, suggest that the molecular basis for safener action is the induction of P450 and GST isoforms. The following example is taken from the work of Zimmerlin & Durst (1992) in our laboratory (Fig. 3). Part at least of the tolerance of wheat towards herbicides is due to its capability to catalyze the aryl-hydroxylation of various chemical structures.

We have achieved very high induction by combining, in a sequential treatment, the effects of phenobarbital (PB) and naphtalic anhydride (NA), a well documented safener. In the experiment of Fig.

3, we had coated wheat seeds with NA, and, after 48h germination, treated the seedlings with PB. This combination produced a 20 to 35 fold increase of the oxidation of diclofop, a diphenylether, chlorsulfuron, a sulfonylurea and chlortoluron, a phenylurea. Furthermore, the 2,4-D oxidase which was not detectable in control seedlings became measurable.

The finding that the same or similar chemicals induce P450 in animals and plants raises the question whether this is coincidental or may rely on common mechanisms. The studies on induction of plant fatty acid hydroxylases by Salaün and coworkers (Benveniste *et al*, 1982, Salaün *et al*, 1981, Zimmerlin *et al*, 1992) in our laboratory suggest that at least PB and clofibrate may function in a similar manner in plant and animal tissues. Their results show that PB and clofibrate induce selectively the medium and long chain fatty acid $\omega$- and the ($\omega$-1)-hydroxylases, respectively.

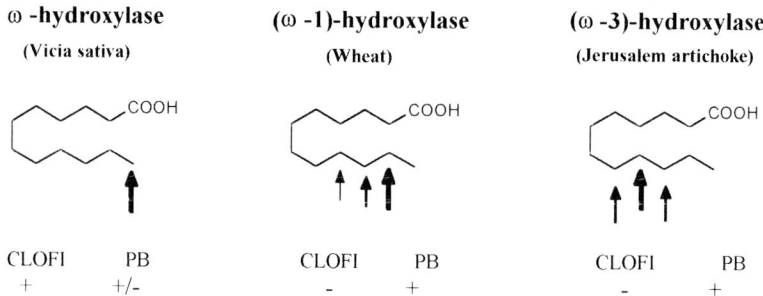

Figure 4: Regioselectivity and induction of the 3 laurate hydroxylase systems from higher plants

In some plant species, principally leguminosae, fatty acids are hydroxylated exclusively at the methyl terminus ($\omega$ position) by microsomal P450 (Benveniste et al., 1982). As in mammals, clofibrate and other peroxisome proliferators induce selectively these plant $\omega$-hydroxylases (Salaün *et al*, 1987, ) and produce peroxisome proliferation. Remarkably, microsomes from most other plant species catalyze in-chain-oxidation. In microsomes from wheat seedlings, lauric acid is hydroxylated at carbons 11, 10 and 9 by an ($\omega$-1)-LAH (Zimmerlin *et al*, 1992). As shown in Fig. 3, this laurate ($\omega$-1)-hydroxylase is induced 11 fold by PB. In Jerusalem artichoke, 10-, 9- and 8-hydroxylauric acids are formed by IC-LAH, which is also strongly stimulated by PB (Salaün et al., 1981, 1993). It is noteworthy that none of the microsomal fractions analysed to date was able to hydroxylate both terminal and sub terminal carbons. This contrasts markedly with the animals systems where both reactions occur in the same microsomal fractions.

**Diversity of the plant P450's:** since almost two decades it is clear that the different P450-dependent reactions described in plants involve several isozymes. However, due to the great difficulty in preparing pure isoforms that retain significant activity, the question of the substrate specificity of a plant P450 isoform had been barely addressed. Recently Werck-Reichhart and coworkers in this laboratory purified (Gabriac *et al*, 1991) and cloned (Teutsch *et al*, 1993) the cinnamic acid 4-hydroxylase, a major P450 in plants (Fig. 5).

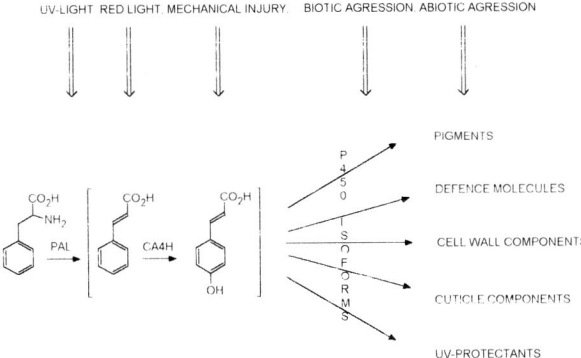

Figure 5: Involvement of CA4H and other P450's in the phenylpropanoid pathway.

This whole pathway is highly inducible by a range of biotic and abiotic factors, and CA4H is particularly susceptible to several like light, wounding, infection and exposure to various chemicals. In a collaborative effort with D. Pompon (CNRS, Gif) and M. Kazmaier (Orsan, Paris), this enzyme was expressed with very high efficiency (turn-over=400) in yeast cells. This provides an ideal system for detecting the metabolism of low affinity substrates. We have thus studied the catalytic capability of CA4H towards all the physiological substrates of plant P450 identified so far, plus a wide range of xenobiotics (Werck-Reichhart et al., unpublished). Interestingly, none of the physiological compounds tested, even phenylpropanoids that display the cinnamate motif, was metabolized. In contrast, 7-ethoxy and 7-methoxycoumarin, the herbicide chlortoluron and PCMA were substrate of CA4H (Tab. 2).

| REACTION ASSAYED | SUBSTRATE | CA4H Yeast |
| --- | --- | --- |
| CINNAMATE 4-HYDROXYLASE | (cinnamic acid) | ++++ |
| 7-ETHOXYCOUMARIN O-DEETHYLASE | (7-ethoxycoumarin) | ++ |
| 7-METHOXYCOUMARIN O-DEMETHYLASE | CH$_3$O— | + |
| p-CHLORO-N-METHYL-ANILINE N-DEMETHYLASE | (p-chloro-N-methyl-aniline) | ++ |
| CHLOROTOLURON HYDROXYLASE | (chlortoluron) | + |

Table 2: P450 reactions in microsomes from yeast transformed with CA4H from Jerusalem artichoke

These results show that the enzyme, while discriminating efficiently against endogenous compounds that may be present in the same subcellular compartment, is well able to accommodate and catabolize xenobiotics. It is interesting that PCMA is also metabolized by CYP71 from avocado (Bozak et al., 1992), thus showing that a same chemical may be substrate of different isoforms.

```
                 1                                                                      80
   CA4H.Con    MDLL..L1EK tLlalF.Aai .AiviSKLRG K.fKLPPGPi PvPIFGnWLQ VGDDLNHRNL td.AKrFG.i fLLRMGQRNL
Pheprop.Con    MdLL...Llek tlialfaaii iatliSklrG krfkLPPGPi pvPifGnwLq vgddlnHrnL td.AK.fG.i fLLrmGqrnl
                81                                                                     160
   CA4H.Con    VVVSSP.LaK EVLHTQgVEF GSRTRNVVFD IFTGkGQDMV FTVYGEHWRK MRRIMTVPFF TNKVVQQyRy GWEaEAa.VV
Pheprop.Con    vVvSsPdlaK evLhTqgveF gsRtrNvvfd iftgkgQDMV FtvYGehWrk mRrimtvpff tnKvvqqyr. gwe.Eaa.vv
               161                                                                     240
   CA4H.Con    dDVkkNpdaa ..GiViR.RL QLMMYNnM.R IMFDRRFESE .dPLflkLKA LNGERSRLAQ SFEYNYGDFI PIL.RPFL.g
Pheprop.Con    .dvkknp.aa .egiVir.rL qlmMyNnm.r imfdrRfese .dplf.klKa lngErsrlAq sFeyNyGDFI PiL.rpfL.g
               241                                                                     320
   CA4H.Con    YLKiCkeVKd rR.qLFKdYF VDERKkigSt Ks.tnnegLK CAIDHIL.Ae kKGE...INe DNVLYIVENI NVAAIETTLW
Pheprop.Con    ylKickevkd kr.aLfk.yF vDErK.iast k..tgn.glk caiDhileae .kgE...ine dNvlyiveNi nvAaieTtlw
               321                                                                     400
   CA4H.Con    SIEWGIAELV NHpeIQ.K.R .ElD.vLGpG .QvTEPDl.k LPYLQAVvKE TLRLRMAIPL LVPHMNLHDa KLgG.DIPAE
Pheprop.Con    sIEWgiAElv nhpeIqkK.r .EmD.vlGpg .qvtEpDi.. LPYLqAvvKE TlRlrmaiPL lvPhmnlhda kl.GydIPae
               401                                                                     480
   CA4H.Con    SKILVNAWWL aNNPahWKKP EEFRPERF.E E.EshVEANG NDFRY1PFGV GRRSCPGIIL ALPILGITiG RlVQNfELLP
Pheprop.Con    skilVNaWwl annPa.WkkP eEFrPERFle e.es.veanG NDFryIPFGv GRRsCpGiil alpilgit.G rlVqnfellp
               481              523
   CA4H.Con    PPGQSkiDTs EKGGQFSLHI LkHStivaKP Rsf.......
Pheprop.Con    Ppgqskidts ekgqqFsLhi lkhstivakp rsf.......
```

Figure 6: the top row (CYP73.Con) shows the consensus among the CYP73 from 4 different plants: Jerusalem artichoke, A. thaliana, mung bean and alfalfa), the bottom line (Pheprop.Con) shows the consensus between the CYP73 and CYP75.

At present, there is very little information on the relationship between plant P450's engaged in a same pathway, the reason being that although cDNA sequences are becoming available, very few (3 until now) have been linked to a physiological reaction. In fig. 6, we show the consensus between CA4H from Jerusalem artichoke and 3 other CYP73 sequences from alfalfa, mung bean and *A. thaliana*. The positional aminoacid identity over 80% is very high considering the evolutionary distance between compositae, fabaceae and brassicaceae. This is witness to the central role of CA4H which catalyzes the first oxidative step in the phenylpropanoid pathway. The bottom row in fig. 6 shows the consensus between the cinnamate hydroxylases and 2 flavonoid 3'5'-hydroxylases (CYP75) which oxidize flavonoids that contain the cinnamate motif. Only few residues, however clustered at some positions, are conserved between the 2 families. It will be interesting, at the light of the recent cristal structures of P450BM-3 and P450scc, to determine which may be implicated in substrate recognition.

The following dendrogram (Fig. 7) shows the pairwise relationships among 26 plant P450. Only the 4

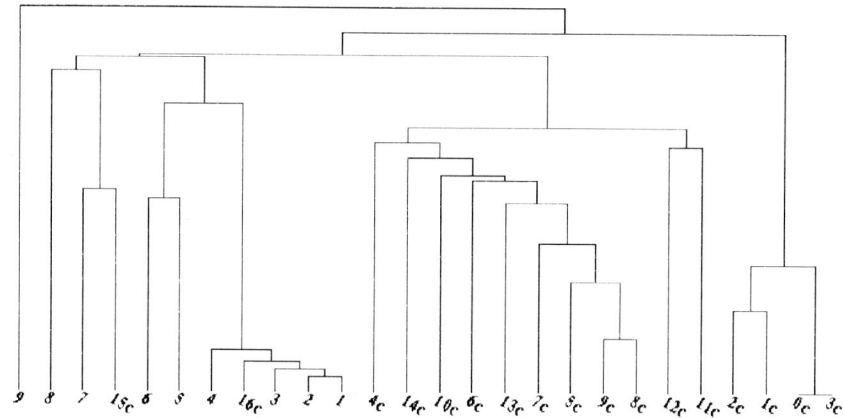

Figure 7: Dendrogram of 0c: CYP72, 1c to 16c: 16 non-identified P450 from C. roseus, 1,2,3,4: Cyp73 from 4 species, 5: flavonoid 3'5'-hydroxylase, 6: CYP71, 7,8,9: non-identified plant P450's

CYP73 (cinnamate hydroxylase) and the flavonoid 3'5'-hydroxylase code for a defined physiological function. 16c is clearly a novel member of the CYP73 family. It is noteworthy that not a single of these P450 is orthologous to an animal or microbial P450. Sequences 1c to 16c have been obtained from *Catharantus roseus* by PCR using a heme-binding region deduced primer (Meijer et al., 1993). Since sequences 1c to 16c are very partial (from heme-binding Cys to carboxy terminus), the dendrogram is based on a very partial alignment and should be taken as such.

During the past few month, the use of molecular biology techniques, in particular degenerate PCR, has onset a mounting flow of clones and sequences encoding plant P450 of unknown functions. We are faced now with the difficult task of assigning a role to these genes, for it seems evident that the 40 and some activities identified so far are only a minute fraction of the P450 enzymes at work in plants.

**References**

Benveniste, I., Salaun, J-P., Simon, A., Reichhart, D., Durst, F. (1982): Cytochrome P-450 dependent w-hydroxylation of lauric acid by microsomes from pea seedlings. *Plant. Physiol.* 70, 122-126.

Bozak, K.,R., OKeefe, .DP., Christoffersen, R.E. (1992): Expression of a Ripening-Related Avocado (Persea americana) Cytochrome P450 in Yeast. *Plant. Physiol.* 100, 1976-1981.

Durst, F. (1991): Biochemistry and physiology of plant cytochrome P450. In *Frontiers in Biotransformation*, eds K. Ruckpaul & H. Rein, Berlin: Akademie Verlag, vol. 4, pp. 193-232

Durst, F., and Benveniste, I. (1993): Cytochrome P450 in plants. In *Handbook of experimental pharmacology*, eds J.B. Schenkman & H. Greim, vol. 105, pp. 294-310. Berlin-Heidelberg: Springer Verlag.

Frear, D.S., Swanson, H.R., Tanaka, F.S. (1969): N-demethylation of substituted 3-(phenyl)1-methyl-ureas : isolation and characterization of a microsomal mixed function oxidase from cotton. *Phytochemistry* 8, 2157-2169

Gabriac, B., Werck-Reichhart, D., Teutsch, H., Durst, F. (1991): Purification and immunocharacterization of a plant cytochrome P450: The cinnamic acid 4-hydroxylase. *Arch. Biochem. Biophys.* 288, 302-309.

Gerardy, R., Zenk, M., (1993): Formation of Salutaridine from (R)-Reticuline by a Membrane-Bound Cytochrome-P-450 Enzyme from Papaver somniferum. *Phytochemistry.* 32, 79-86.

Halkier, B.A., Möller, B.L., (1991): Involvement of cytochrome P450 in the biosynthesis of dhurrin in Sorghum bicolor (L.) Moensch. *Plant. Physiol.* 96, 10-17.

Meijer, A.H., Souer, E., Verpoorte, R., Hoge, J.H.C. (1993): Isolation of Cytochrome-P-450 cDNA Clones from the Higher Plant Catharantus roseus by a PCR Strategy. *Plant. Mol. Biol.* 22, 379-383.

Murphy, P.J., West, C.A. (1969): The role of mixed function oxidases in kaurene metabolism. *Arch. Biochem. Biophys.*, 133, 395-407

Salaün, J-P., Benveniste, I., Reichhart, D., Durst, F. (1981): Induction and specificity of a cytochrome P-450 dependent laurate in-chain hydroxylase from higher plant microsomes. *Eur. J. Biochem.* 119, 651-655.

Salaün, J-P., Simon, A., Durst, F. (1987): Specific induction of lauric acid w-hydroxylase by clofibrate, diethylhexyl-phtalate and 2,4-dichlorophenoxyacetic acid in higher plants. *Lipids*. 21, 776-779.

Salaün, J-P., Weissbart, D., Helvig, C., Durst, F., Mioskowski, C. (1993): Regioselective hydroxylation and epoxidation of lauric acid and unsaturated analogues by cytochrome P450 in Jerusalem artichoke microsomes. *Plant. Physiol. Biochem.* 31, 285-293.

Song, W,C., Brash, A.R., (1991): Purification of an Allene Oxide Synthase and Identification of the Enzyme As a Cytochrome-P-450. *Science*. 253, 781-784.

Swain, T. (1977): Secondary metabolites as protective agents. *Ann Rev Plant Physiol*. 28, 479-501

Teutsch, G., Hasenfratz, M-P., Lesot, A., Stoltz, C., Garnier, J-M., Jeltsch, J-M., Durst, F., Werck-Reichhart, D. (1993): Isolation and sequence of a cDNA encoding the Jerusalem artichoke cinnamate 4-hydroxylase, a major plant cytochrome P450 involved in the general phenylpropanoid pathway. *Proc. Natl. Acad. Sci. USA*. 90, 4102-4106.

Zimmerlin, A., Durst, F. (1992): Aryl Hydroxylation of the Herbicide Diclofop by a Wheat Cytochrome-P-450 Monooxygenase - Substrate Specificity and Physiological Activity. *Plant. Physiol.* 100, 874-881.

Zimmerlin, A., Salaün, J-P., Durst, F., Mioskowski, C. (1992): Cytochrome P-450 dependent hydroxylation of lauric acid at the subterminal position and oxidation of unsaturated analogs in wheat microsomes. *Plant. Physiol.* 100, 863-873.

# Insect cytochrome P450: functions, regulation and diversity

René Feyereisen[1], John F. Andersen[1], Flerida A. Cariño[2], Michael B. Cohen[3], Josette F. Koener[1], Stacy Repecko[1], Julie A. Scott[1], Mark J. Snyder[1]

[1]Department of Entomology and Center for Insect Science, University of Arizona, Tucson, Arizona 85721, USA. [2]Department of Chemistry, Ateneo de Manila University, Quezon City, Philippines. [3]International Rice Research Institute, Los Baños, Philippines

## INTRODUCTION

In general terms, insect P450 systems appear to be similar to vertebrate P450 with regard to their function and catalytic versatility. Both mitochondrial and microsomal P450 systems have been described in insects. Endogenous functions of insect P450 enzymes include the metabolism of molting hormones (the ecdysteroids, see Winter and Kayser, this volume), juvenile hormones (sesquiterpenoids that control metamorphosis and reproduction) and pheromones. Insect microsomal P450 enzymes are best known for their role in insecticide detoxification (and resistance) and in the metabolism of plant chemicals (Reviews in Hodgson, 1985; Brattsten and Ahmad, 1986). However, insect P450 enzymes also activate certain types of insecticides, for instance the conversion of phosphorothioates (P=S) to phosphates (P=O) as in the case of the parathion to paraoxon reaction. This can result in an increased potency for inhibition of acetylcholinesterase by 3 or 4 orders of magnitude. P450 levels are a function of the insect species, sex, genetic makeup, developmental stage (including hormonal influences), and immediate past environmental (including dietary) exposure. For each insect tissue, these factors determine which P450 genes are active and at what level. It has clearly been established that P450 enzymes can metabolize host plant chemicals and that *induction* of P450 enzymes by host plant chemicals can lead to altered (usually decreased) toxicity of both plant toxins and man-made insecticides (Review in Brattsten and Ahmad, 1986). For instance, dietary nicotine induces several P450 activities in the midgut of the tobacco hornworm, *Manduca sexta*, including P450s that metabolize the inducer to cotinine-*N*-oxide. The inducible metabolism of nicotine allows this insect to thrive on tobacco plants (Snyder et al., 1993a,b). The study of insect P450 has been hampered by many difficulties (Hodgson, 1985), and it is only recently that insect P450s have been cloned and sequenced. Thus, our knowledge about the structural aspects of insect P450 is lagging far behind the vertebrate P450 field. Studies on insecticide resistance led to the cloning and sequencing of *CYP6A1* and *CYP6A2* from the house fly, *Musca domestica* and the fruit fly *Drosophila melanogaster* (Feyereisen et al., 1989; Waters et al., 1992) whereas a study of host-plant adaptation of the black swallowtail butterfly (*Papilio polyxenes*) led to the cloning of *CYP6B1* (Cohen et al., 1992, and Schuler et al., this volume). In contrast, all other P450 sequences now available were obtained seredipitously. *CYP4E1* from *Drosophila melanogaster* had been sequenced partially by Snyder and Davidson (1983), but was "discovered" independently in 1991 by D. Nero (Cornell University) and T. Holton (University of Melbourne) by an exhaustive search of GenBank. *CYP4C1* from the cockroach, *Blaberus discoidalis*, was obtained in a differential hybridization screen of fat body cDNAs to detect transcripts induced by the peptide hypertrehalosemic hormone (Bradfield et al., 1991). *CYP4D1* from *Drosophila melanogaster* has been obtained during a chromosomal walk through the 2D region of the X chromosome (Gandhi et al., 1992). It is clear that the mapping and sequencing of the *Drosophila* genome will provide more P450 sequences, and perhaps this organism will be the first higher eukaryote for which the complete complement of P450 genes will be known. We provide here a short review of our current efforts in the study of insect P450.

## CLONING, SEQUENCE AND REGULATION OF THE *CYP6A1* GENE

The *CYP6A1* cDNA was obtained by antibody screening of a cDNA library from phenobarbital-treated house flies of the insecticide-resistant strain "Rutgers" (Feyereisen et al., 1989). The closest relative to family 6 is family 3 (Nelson et al., 1993). A genomic clone for *CYP6A1* was obtained and sequenced (Cohen, Koener and Feyereisen, unpublished results). The *CYP6A1* gene is located on chromosome V of the house fly, and its transcription start site was mapped by primer extension analysis. The structure of the gene shows putative TATA and CAAT boxes upstream of the transcription start site. Also present upstream of the transcription start site is a sequence that appears homologous to the barbie box of *CYP102* (Shaw and Fulco, 1993). Although *CYP6A1* is rapidly and highly inducible by phenobarbital in the house fly (Cariño et al., 1992), the significance of the putative barbie box has not yet been studied experimentally. The coding sequence of *CYP6A1* is interrupted at the Glu364 codon by a single intron of 60 bp (Cohen, Koener and Feyereisen, unpublished results). No intron has been observed in vertebrate P450 genes at a homologous position.

## CONSTITUTIVE OVEREXPRESSION OF THE *CYP6A1* GENE IN INSECTICIDE-RESISTANT STRAINS OF THE HOUSE FLY

The *CYP6A1* gene was shown to be constitutively overexpressed in seven insecticide-resistant strains, including the multi-resistant Rutgers strain (Cariño et al., 1992). In larvae and adults of this strain, mRNA levels for *CYP6A1* were at least ten times higher than in a reference susceptible strain. This high constitutive expression is not caused by an amplification of the *CYP6A1* gene. Crosses between the Rutgers strain and a multiply-marked strain revealed that the overexpression of the *CYP6A1* gene is controlled by an incompletely dominant locus on chromosome II, both in larvae and in adults (Cariño et al., 1993). This shows that overexpression is the result of a mutation affecting a *trans*-acting factor that regulates *CYP6A1* expression on chromosome V.

Metabolic resistance to insecticides (organophosphorus compounds, carbamates etc.) has repeatedly been mapped to chromosome II in the house fly by genetic methods, and our results support the hypothesis (Plapp, 1984) that a major resistance gene on chromosome II is a regulatory gene.

## RELATED *CYP6* GENES IN THE HOUSE FLY

A cluster of five *CYP6* genes has been discovered on a 30 kb span of the house fly chromosome V (Cohen and Feyereisen, unpublished results). These genes are all interrupted by a short (56-133 bp) intron located at the same position as the intron in *CYP6A1*, and a fragment of a gene comprising only the coding sequence downstream of the common intron is also found in this cluster. The relative orientation of the genes suggests that an original gene was duplicated twice and that the three genes were then duplicated and inverted to yield a cluster of 6 genes of which one may have been split during the inversion event. Most remarkably, the genes in this tight cluster are highly divergent (56-83% identity amongst the 5 intact genes). This suggests that some evolutionary pressure has kept the genes clustered despite their great divergence. Transcripts for each of these genes can be found either by northern analysis or by reverse transcription followed by PCR.

## FUNCTIONAL EXPRESSION OF *CYP6A1* IN *E. COLI*

The catalytic activity of CYP6A1 was analyzed in a reconstituted system of house fly P450 reductase and CYP6A1. The cDNAs were expressed in *E. coli* (Andersen et al., 1993) as shown on Fig. 1. The house fly P450 reductase had been cloned previously by antibody screening of a cDNA library and sequenced

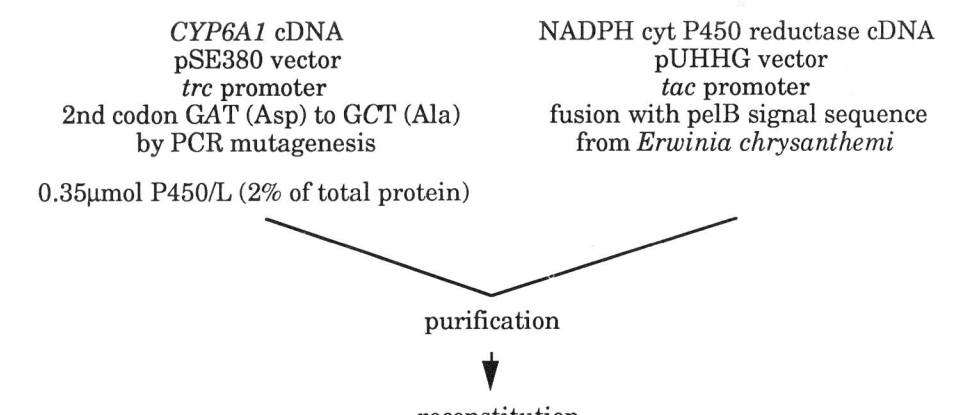

Fig.1  Expression of CYP6A1 and P450 reductase in *E. coli*.

(Koener et al., 1993). It is 55% identical to the rat protein. The expressed CYP6A1 protein shows a typical CO-reduced difference spectrum with a peak at 448 nm and produces type I binding spectra with several ligands. However, in the reconstituted system we could only observe a high turnover with the cyclodiene insecticides heptachlor and aldrin, which were epoxidized to heptachlor epoxide and dieldrin respectively. Immunoinhibition studies show that CYP6A1 is the major cyclodiene epoxidase in house fly microsomes.

## DIVERSITY OF INSECT P450 GENES

The remarkable conservation of sequence of CYP4 proteins in a region that corresponds to the I helix of P450cam suggested that the polymerase chain reaction (PCR) could be used to identify new P450 genes in insects. The strategy depicted in Fig. 2 was used to detect expressed CYP4 genes in several insect

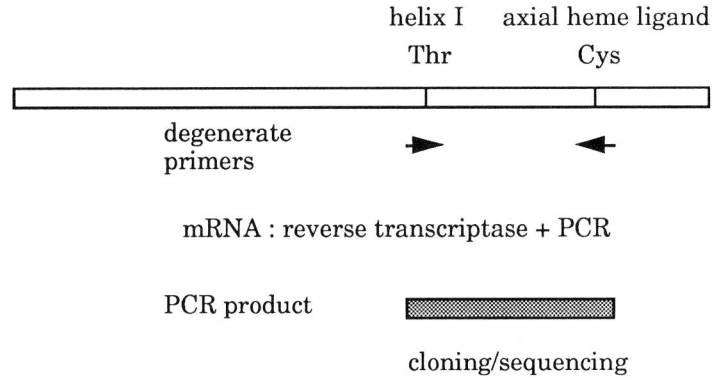

Fig.2. Strategy for identification of new *CYP4* genes from insects

species. Sequencing of cloned PCR products revealed a great diversity of new insect P450s (more than 10 CYP4 sequences in each species) in the mosquito *Anopheles albimanus*, in the house fly *Musca domestica*, in the cockroach *Diploptera punctata* and in the tobacco hornworm *Manduca sexta*. In the latter two species, mRNA from specific tissues (fat body and midgut) was used, and this allowed us to identify quickly those P450 genes that are selectively or more abundantly expressed in a particular tissue. The relative number of clones obtained for each P450 is a crude indication of the relative level of expression of each gene. For instance, a sequence represented in six independent clones is likely to represent a more abundant mRNA than a sequence represented by a single clone. This project is still in progress, and the strategy used for CYP4 genes may also be adapted for CYP6 genes. When more sequence information becomes available, it may become possible to screen rapidly for the major P450 expressed in any species or tissue. Although the catalytic activity of these new CYP4-like proteins needs to be studied by heterologous expression of their full-length cDNAs, it is tempting to suggest that they represent xenobiotic-metabolizing P450s in insects. This suggestion is strengthened by the observation (Amichot et al., this volume) that *Drosophila* CYP4 genes are overexpressed in insecticide resistant strains.

We have currently only a glimpse of the diversity of P450 genes in insects. As plants and microorganisms evolved new P450s to synthesize more diverse and more complex allelochemicals, it will not be surprising to find an equivalent diversity of insect P450s that have evolved to metabolize these chemicals. *CYP6B1* is a first example of such a P450 (Cohen et al., 1992).
The evolutionary relationship of steroid-metabolizing P450s in vertebrates and invertebrates will also be a challenging area of future research. The *CYP10* gene of the mollusc *Lymnaea stagnalis* is expressed in the dorsal bodies, an endocrine organ that produces the (unidentified) gonadotropic hormone (Teunissen et al., 1992). This P450 is most similar to P450scc and may therefore represent a steroid-metabolizing P450. Perhaps beyond the P450 superfamily, but certainly a heme-thiolate P450 protein, the insect nitric oxide synthase (Ribeiro and Nussenzveig, 1993) may also contain in its sequence some important traces of the evolution of these proteins.

ACKNOWLEDGEMENTS : The work reviewed here was supported by grants from the NIH (GM39014) and the USDA (89-37263-4960).

# REFERENCES

Andersen, J. F., Utermohlen, J.G. and Feyereisen, R. (1993): Expression of house fly CYP6A1 and NADPH cytochrome P450 reductase in *E. coli* and reconstitution of an insecticide metabolizing P450 system. *Biochemistry* (in press).

Bradfield, J. Y., Lee, Y.H., and Keeley, L. L. (1991): Cytochrome P450 family 4 in a cockroach: molecular cloning and regulation by hypertrehalosemic hormone. *Proc. Natl. Acad. Sci. USA* 88, 4558-4562.

Brattsten, L. B and Ahmad, S. (1986): *Molecular aspects of insect-plant interactions.* Plenum Press, New York.

Cariño, F., Koener, J. F., Plapp, F. W. Jr. and Feyereisen, R. (1992): Expression of the cytochrome P450 gene *CYP6A1* in the house fly, *Musca domestica. ACS Symp. Ser.*505, 31-40.

Cariño, F., Koener, J. F., Plapp, F. W. Jr. and Feyereisen, R. (1993): Constitutive overexpression of the cytochrome P450 gene *CYP6A1* in a house fly strain with metabolic resistance to insecticides. *Insect Biochem. Mol. Biol.* (in press).

Cohen, M. B., Schuler, M. A., and Berenbaum, M. R. (1992): A host-inducible cytochrome P450 from a host-specific caterpillar: molecular cloning and evolution *Proc. Natl. Acad. Sci. USA* 89,10920-10924.

Feyereisen, R., Koener, J. F., Farnsworth, D. E. and Nebert, D. W. (1989): Isolation and sequence of cDNA encoding a cytochrome P-450 from an insecticide-resistant strain of the house fly, *Musca domestica. Proc. Natl. Acad. Sci.* 86, 1465-1469.

Gandhi, R., Varak, E. and Goldberg, M. L. (1992): Molecular analysis of a cytochrome P450 gene of family 4 on the *Drosophila* X chromosome. DNA and Cell Biology, 11, 397-404.

Hodgson, E. (1985): Microsomal mono-oxygenases. In *Comprehensive Insect Physiology, Biochemistry, and Physiology,* (Ed. by G.A. Kerkut and L.I. Gilbert) Vol. 11.pp. 225-321, Pergamon Press, Oxford.

Koener, J. F., Cariño, F. A., and Feyereisen, R. (1993): The cDNA and deduced protein sequence of house fly NADPH-cytochrome P450 reductase. *Insect Biochem. Mol. Biol.* 23, 439-447.

Plapp, F. W. Jr. (1984): The genetic basis of insecticide resistance in the house fly: evidence that a single locus plays a major role in metabolic resistance to insecticides. *Pest. Biochem. Physiol.* 22, 194-201.

Ribeiro, J.M.C., and Nussenzveig, R.H. (1993): Nitric-oxide synthase activity from a hematophagous insect salivary gland. *FEBS Letters* 330, 165-168.

Shaw, G.C. and Fulco, A. J. (1993): Inhibition by barbiturates of the binding of Bm3R1 repressor to its operator site on the barbiturate-inducible cytochrome P450BM-3 gene of *Bacillus megaterium. J. Biol. Chem.* 268, 2997-3004.

Snyder, M. and Davidson, N. (1983): Two gene families clustered in a small region of the *Drosophila* genome. *J. Mol. Biol.* 166,101-118.

Snyder, M. J., Hsu, E. L., and Feyereisen, R. (1993a): Induction of cytochrome P450 activities by nicotine in the tobacco hornworm, *Manduca sexta. J. Chem. Ecol.* (in press).

Snyder, M. J., Walding, J. K., and Feyereisen, R. (1993b): Metabolic fate of the allelochemical nicotine in the tobacco hornworm, *Manduca sexta. Insect Biochem. Mol. Biol.* (in press).

Teunissen, Y., Geraerts, W.P.M., van Heerikhuizen, H., Planta, R. J. and Joosse, J. (1992): Molecular cloning of a cDNA encoding a member of a novel cytochrome P450 family in the mollusc, *Lymnaea stagnalis. J. Biochem.* 112,249-252.

Waters, L. C., Zelhof, A. C., Shaw, B. J. and Ch'ang, L. Y. (1992): Possible involvement of the long terminal repeat of transposable element 17.6 in regulating expression of an insecticide resistance-associated P450 gene in *Drosophila. Proc. Natl. Acad. Sci USA* 89, 4855-4859.

# The role of conserved 5'-flanking (Barbie Box) DNA sequences and barbiturate-responsive DNA-binding proteins in the mechanism of induction by barbiturates of P450 cytochromes in *Bacillus megaterium* and other prokaryotic and eukaryotic organisms

Armand J. Fulco[1,2], Jian-Sen He[2], Qianwa Liang[2]

[1]Department of Biological Chemistry and [2]Laboratory of Structural Biology and Molecular Medicine, School of Medicine, University of California, Los Angeles, California, USA

In our laboratory, two barbiturate-inducible P450 cytochromes from *Bacillus megaterium* were discovered and characterized and then cloned and expressed in *E. coli*. The genes encoding them, including the apparently complete regulatory regions, have been sequenced and studied with respect to the barbiturate-mediated induction mechanisms. (For a review of this work see Fulco, 1991). $P450_{BM-1}$ (Mol. Wt. 47.5 kD) which shows sequence similarity to $P450_{CAM}$ and to several other bacterial P450s, can be moderately induced by barbiturates (~20 fold) in *B. megaterium*. $P450_{BM-3}$ (Mol. Wt. 117.7 kD), a catalytically self-sufficient fatty acid monooxygenase that shows significant structural homology to the 2-component microsomal P450-reductase systems of eukaryotes but not to known bacterial systems, can be induced several hundred fold by barbiturates. The barbiturate-mediated inductions of cytochromes $P450_{BM-1}$ and $P450_{BM-3}$ involve coordinately regulated and mechanistically related derepressions (He and Fulco, 1991; Shaw and Fulco, 1992, 1993; Fulco et al, 1993). In the absence of barbiturates, transcription is inhibited by the binding of a repressor protein, Bm3R1 to operator sites and ancillary binding sites ("Barbie Boxes") located in the 5'-flanking regions of the P450 structural genes; in the presence of inducer barbiturates, Bm3R1 no longer binds and transcription is turned on. Bm3R1, which contains a helix-turn-helix DNA binding motif, can be displaced from the operator and ancillary sequences by the concerted effect of two separate mechanisms. One involves the direct interaction of the barbiturate with Bm3R1 in such a manner that the Bm3R1 can no longer bind to its operator (Shaw and Fulco, 1993). In the second mechanism, the Bm3R1 is displaced by barbiturate-responsive proteins that apparently bind to the Barbie box, a 15-17 bp DNA sequence located at or near a Bm3R1 binding site as well as to other sites in the regulatory region and probably to Bm3R1 itself (i.e. protein-protein interaction). The binding proteins that displace Bm3R1 at the Barbie box site may be induced or, as we have demonstrated in rat and *B. megaterium*, activated in some way by barbiturates (He and Fulco, 1991). The induction of $P450_{BM-1}$ and $P450_{BM-3}$ in *B. megaterium* appears to be mechanistically related to the analogous induction of cytochromes P450b and e of the rat (He and Fulco, 1991) and, based on sequence homologies, probably to other barbiturate-inducible proteins from both eukaryotes and other prokaryotes. The potential importance of these sequences is highlighted by our observation (Shaw and Fulco, 1993) that essentially all of the genes that encode barbiturate-inducible proteins and whose 5'-flanking regions have been sequenced contain sequences that are highly homologous to the 15 bp Barbie box sequence of the $P450_{BM-3}$ gene and to a derived consensus Barbie box sequence that strongly binds the barbiturate-regulated proteins of rat and *B. megaterium*. Recent unpublished findings from our laboratory involving the site-specific mutagenesis of the Barbie box sequences of the genes encoding $P450_{BM-1}$ and $P450_{BM-3}$ demonstrate that these sequences are critically involved in the regulation of barbiturate-mediated expression of the *B. megaterium* P450s (Liang and Fulco, 1993). Results from other laboratories suggest that the Barbie box sequences of barbiturate-induced P450s of the rat (Upadhya et al, 1992) and the bacterium,

*Streptomyces griseolus* (Patel and Omer, 1992), may also be involved in the induction process since portions of these sequences are apparently protected from DNase digestion by the binding of barbiturate-responsive proteins, presumably analogous to repressor and positive regulator proteins of *B. megaterium*. Table I (below) shows the various Barbie box sequences that have been recognized to date in the 5'-flanking regions of genes encoding barbiturate-inducible P450s and ancillary enzymes.

**TABLE I**
**The Occurrence of Barbie Box sequences in the 5' Flanking Regions of Barbiturate-Inducible Genes**

| GENE | 5' POS | 15 BASE SEQUENCE | ID |
|---|---|---|---|
| **Consensus sequence** | - - - - | A T C A A A A G C T G G A G G | 15 |
| $P450_{BM-3}$ (*B. megaterium*) | -227 | A T C A A A A G C T G G t G G | 14 |
| $P450_{BM-1}$ (*B. megaterium*) | -302 | A T a A A A A G C T G G t G c | 12 |
| P450b [CYP2B1] (*rat*) | -104 | A g C t A A A G C a G G A G G | 12 |
| P450e [CYP2B2] (*rat*) | -100 | A g C c A A A G C a G G A G G | 12 |
| P450 CYP2C1 (*rabbit*) | -228 | t T C A A A A G a g G G c t t | 9 |
| Epoxide hydrolase (*rat*) | -84 | g T C t A A A G t c c a g G G | 8 |
| Glutathione-S-trans. (*mouse*) | -166 | A g g g A A A G g T G G t G G | 10 |
| Aldehyde dehydrogen. (*rat*) | -168 | A T t t A A A G g c a a a G G | 9 |
| P450 CYP6A1 (*house fly*) | -188 | A a a A A A A G C T G a A t G | 11 |
| Aldehyde dehydrog. (*mouse*) | -574 | A T C A A A A G C T t G g G a | 13 |
|  | -58 | A T a A A A A G g a G c A a G | 10 |
| P450 CYP3A2 (*rat*) | -1166 | A T a g A A A G C a t t c t G | 8 |
|  | -1007 | c a t t A A A G C c t G t G G | 8 |
|  | -48 | t c C c A A A G C T G t g t G | 9 |
| SU1 [128bp] (*S. griseolus*) | -23 | c T t t A A A G g T G a g a a | 7 |

## BARBITURATE-MEDIATED INTERACTION OF PROTEINS WITH BARBIE BOX SEQUENCES

The evidence implicating Bm3R1 as the repressor regulating the expression of $P450_{BM-3}$ is well established (Shaw and Fulco, 1992; 1993). We found that an open reading frame just upstream from the *B. megaterium* cytochrome $P450_{BM-3}$ structural gene encodes a protein, designated Bm3R1, which contains a helix-turn-helix DNA-binding motif and negatively controls the expression of the $P450_{BM-3}$ gene at the transcriptional level as well as its own expression; it does so by binding to a 20-bp perfect palindromic operator site located between the promoter sequences and the *bm3R1* structural gene. A *B. megaterium* mutant that constitutively produced cytochrome $P450_{BM-3}$ was isolated and complementation by a DNA fragment containing the wild-type *bm3R1* gene showed that the mutation in this locus was *trans*-dominant. We identified a single base change that resulted in a glycine to glutamate substitution in the β-turn region of the DNA-binding motif and totally eliminated the binding of this mutant Bm3R1 to the 20 bp palindromic operator. As determined by gel retardation and footprinting experiments, in the presence of inducing concentrations of barbiturates *in vitro*, Bm3R1 binding to the operator site was strongly inhibited or eliminated. Thus the barbiturate-mediated induction of cytochrome $P450_{BM-3}$ appeared, in part, to involve the release of inhibition by Bm3R1 by interfering with its binding to the palindromic putative operator sequence and perhaps to other sites on the regulatory region of the gene encoding cytochrome $P450_{BM-3}$. An unexpected bonus obtained from a detailed study of the G->E Bm3R1 mutant (Fulco et al, 1993) was the finding that the expression of cytochrome $P450_{BM-1}$ was also constitutive. This indicated that the expression of the gene encoding this protein was under the control of the Bm3R1 repressor. Indeed, preliminary evidence from our lab now indicates that, in the $P450_{BM-1}$ gene regulatory region, Bm3R1 binds to the Barbie box region and represses transcription; in other words, the barbiturate-responsive negative regulator protein for $P450_{BM-1}$ expression originally reported by He and Fulco (1991) is actually Bm3R1 and it is this protein that dissociates from the Barbie box in the presence of inducer barbiturates. This does not mean, however, that a simple interaction between barbiturates and Bm3R1 is the sole basis for barbiturate-mediated induction of the *B. megaterium* P450s; indeed, it is not. There appears to be a concerted effect involving not only dissociation of the Bm3R1 repressor from operator sites by direct interaction with barbiturates but also inhibition of the binding of Bm3R1 to regulatory regions by its interaction with two smaller proteins that are produced or activated in the presence of barbiturates, possibly by the release of repression by Bm3R1. One or both of these proteins, designated Bm1P1 and Bm1P2, may also bind to the Barbie box regions and function in this manner to displace Bm3R1. It seems plausible that barbiturate inducers might resemble and mimic the mode of action of an as yet unidentified endogenous inducer(s) in *B. megaterium* that functions in the same manner. The general characteristics of the barbiturate-responsive binding proteins involved in the regulation of the *B. megaterium* P450s are presented in Table II (below).

## *TABLE II*
## Barbiturate-responsive proteins involved in the regulation of expression of the *P450BM-1* and *P450BM-3* genes of *B. megaterium*

| PROTEIN | MOL. WT. | RESIDUES | P.I. | NET CHARGE | PROTEIN TYPE | FUNCTION |
|---|---|---|---|---|---|---|
| Bm3R1 | 21,884 | 192 | 5.13 | -7 | HTH motif; binds to $P450_{BM-3}$ operator & to both Barbie boxes. | Repressor regulating expression of itself, $P450_{BM-3}$ and $P450_{BM-1}$. |
| Bm3R1 (39G>E) Altered Bm3R1 protein in a *B. megaterium* mutant constitutive for $P450_{BM-3}$. | 21,924 | 192 | 5.00 | -8 | The turn of the HTH motif is lost with Gly to Glu change at residue 39. Does not bind to Barbie boxes or operator. | No longer a repressor for its own expression or the expression of $P450_{BM-3}$ and $P450_{BM-1}$. |
| Bm1P1 | 11,506 | 98 | 6.11 | -1 | HTH motif; barbiturate-responsive binds to 5' flank. regions of $P450_{BM-1}$; may bind to Barbie box and/or interact w. Bm3R1 | In concert with Bm1P1 increases $P450_{BM-1}$ and $P450_{BM-3}$ expression; inhibits Bm3R1 binding. |
| Bm1P2 | 10,297 | 88 (includes 11 lys and 10 ser) | 9.55 | +6 | DNA binding; binds to 5' flank. regions of $P450_{BM-1}$ but probably not to Barbie boxes; may interact with Bm3R1 | In concert with Bm1P2 increases $P450_{BM-1}$ and $P450_{BM-3}$ expression; inhibits Bm3R1 binding. |
| Other proteins | ? | ? | ? | ? | Bind to 5' flanking regions of $P450_{BM-1}$ & $P450_{BM-3}$. | Regulation of $P450_{BM-1}$ and $P450_{BM-3}$ genes. |

Although we ourselves have not characterized analogous proteins from eukaryotic sources other than the Barbie box binding protein detected by He and Fulco (1991) in nuclear extracts obtained from rats treated with phenobarbital, evidence is now accumulating from other laboratories that, in the rat at least, a protein analogous to Bm3R1 is also involved in the barbiturate-mediated regulation of the P450b (CYP 2B) gene. In addition to work already mentioned (Upadhya et al, 1992) that seems to suggest the existence of a repressor-type protein that is responsive to barbiturates, Dr. Ronald Lubet of the National Cancer Institute (personal communication) and his coworkers have detected a protein in rat liver nuclear extracts from animals untreated by barbiturates that binds to the 30 mer containing the P450b Barbie box sequence (see Table I) used by Upadhaya et al (1992) in their footprinting assays. This protein disappears however and is replaced by a smaller Barbie box binding protein in extracts from phenobarbital-treated animals. This suggests that a repressor-positive regulator system responsive to barbiturates operates in the rat that is similar to that observed in *B. megaterium*.

Based on our most recent findings, the regulation of the $P450_{BM-1}$ and $P450_{BM-3}$ genes by Bm3R1 repressor protein can be visualized as shown in Figures 1 and 2 (below).

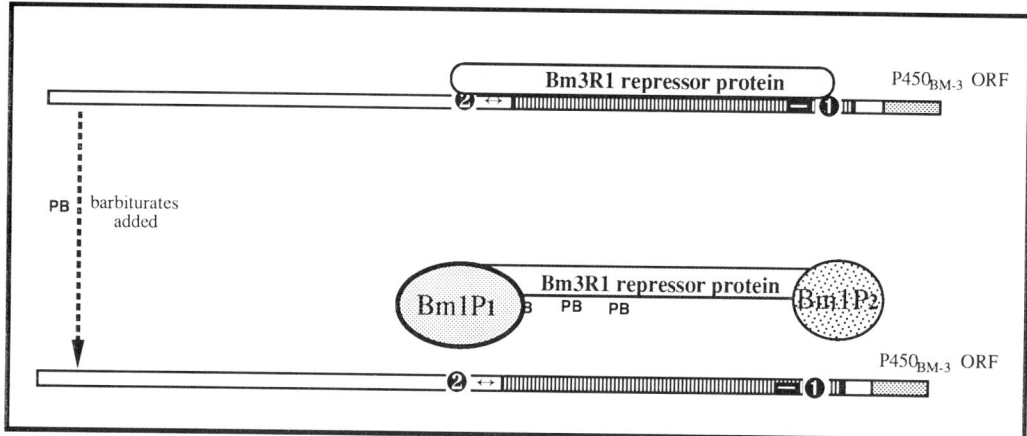

Fig. 1. *Barbiturate-mediated regulation of the expression of cytochrome $P450_{BM-3}$*. In the scheme above, ❶ is a weak promoter inhibited by barbiturates in the growth medium; ❷ is the promoter driving the expression of both the *Bm3R1* and *$P450_{BM-3}$* genes and is strongly activated by barbiturate inducers; ↔ is the 20 bp palindromic bicistronic operator sequence that tightly binds Bm3R1 in the absence but not in the presence of strongly-inducing barbiturates; the 15 bp Barbie box sequence that binds to Bm3R1 in the absence but not in the presence of inducer barbiturates is indicated by the black bordered rectangle ▬ just to the left of ❶, the weak promoter. Bm1P1 and Bm1P2 are proteins encoded in a large segment of DNA 5' to the $P450_{BM-1}$ structural gene that are activated and/or induced in the presence of barbiturates. The vertically hatched area ▨▨▨▨ is the open reading frame encoding Bm3R1, a region of the DNA that, as shown, also contains the weak promoter and the Barbie box sequence.

---

Some of the evidence for the scheme shown for $P450_{BM-3}$ induction is as follows: *(1)* Bm3R1 repressor binds specifically to the 20 bp palindromic operator sequence and to the Barbie box region but a G39E mutant of Bm3R1 does not. The strain containing this mutation is constitutive for the expression of cytochrome $P450_{BM-3}$ and $P450_{BM-1}$. Furthermore, a 26 bp region including the entire operator sequence is protected from DNase I digestion by the binding of Bm3R1 but the mutant protein affords no protection. *(2)* Barbiturates that induce P450BM-3 *in vivo* inhibit the binding of Bm3R1 with the operator *in vitro* and also eliminate protection of the operator against DNase. Tests in cell cultures indicated that the relative *in vivo* inducer potencies of the tested barbiturates correlate with their effectiveness, *in vitro*, in inhibiting the binding of Bm3R1 to its operator. *(3)* In gel retardation assays, binding of Bm3R1 to the Barbie sequences is inhibited by the presence of Bm1P1 and Bm1P2.

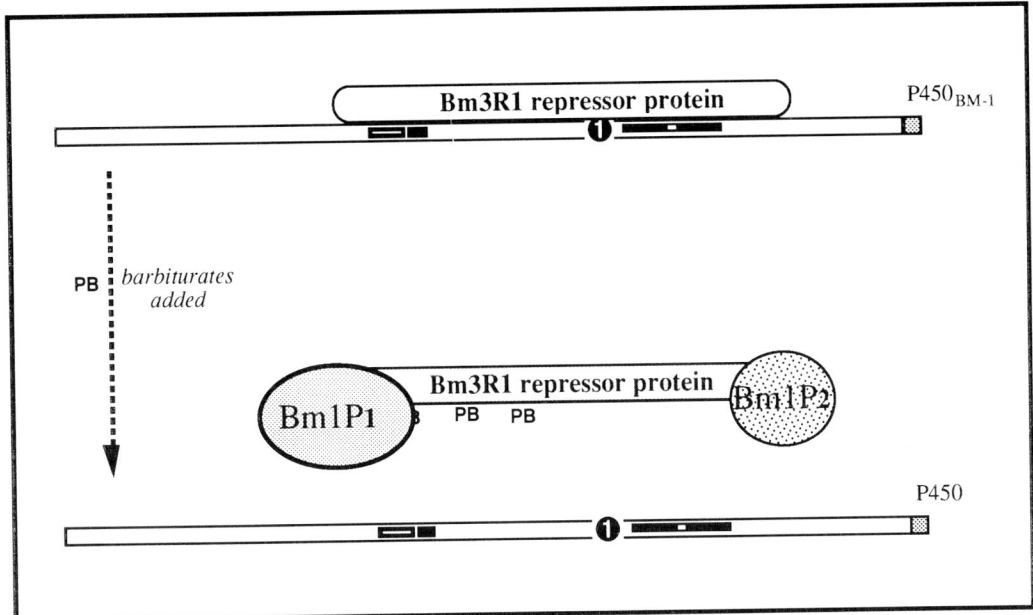

Fig. 2. *Barbiturate-mediated regulation of the expression of cytochrome* $P450_{BM-1}$. In the scheme above, ❶ is the promoter activated by barbiturates; ▭ is the 15 bp Barbie box sequence that binds Bm3R1 in the absence of barbiturates; ▬ is a 6-8 bp sequence 3' to and contiguous with the Barbie box that very strongly binds to Bm3R1 in the absence of barbiturates; when this site is eliminated from the gene, expression of $P450_{BM-1}$ is constitutive. ▬▭▬ is a 24 bp inverted repeat separated by a 5 bp (stem and loop structure) that is the binding site for several proteins.

---

Some of the evidence for the scheme shown in Fig. 2 for the regulation of the $P450_{BM-1}$ gene is based primarily on gel-retardation assays and footprinting studies and is similar to that discussed (above) for the barbiturate-mediated expression of $P450_{BM-3}$. The results of such studies are summarized in Table II. In addition, it should be noted that Bm1P1 and Bm1P2 are encoded in the 5'-flanking region of the $P450_{BM-1}$ gene. Indeed, one of these proteins is encoded on a strand of DNA complementary to that encoding $P450_{BM-1}$ and has a promoter region (-10 and -35 sequences) that overlaps the analogous region in the opposite strand encoding $P450_{BM-1}$. Thus, if the Bm3R1 repressor inhibits transcription of the $P450_{BM-1}$ gene by overlapping or binding in this region it probably inhibits transcription of *Bm1P1* as well. Finally, since the presence of both Bm1P1 and Bm1P2 are necessary for the type of interaction with Bm3R1 necessary to inhibit the binding of this repressor to the 20-bp operator sequence of the $P450_{BM-3}$ gene and to the Barbie boxes of both genes, it would not be surprising to discover that the expression of the two positive regulators was also coordinately regulated. Our current research goal is to determine, in detail, how the various elements of these complex systems interact to account for barbiturate-mediated induction.

## A GENERAL MECHANISM FOR BARBITURATE-MEDIATED ENZYME INDUCTION

The apparently universal presence of Barbie boxes in the regulatory regions of barbiturate-inducible genes of prokaryotes and eukaryotes coupled with the recognition that analogous barbiturate-responsive proteins may be interacting with these unique DNA sequences to regulate gene expression perhaps justifies our proposal of a general hypothesis, as follows:

<u>In the absence of barbiturates</u>, a repressor protein (Bm3R1 or an analog) binds strongly to one or more operator sites and a Barbie box sequence in the 5'-flanking region of the structural gene encoding the

P450 (or other) enzyme and *inhibits transcription*. Binding to the Barbie box alone probably does not inhibit transcription.

In the presence of inducer barbiturates, the repressor no longer binds to the operator(s) or Barbie box. In part, this involves interaction of the barbiturate with the repressor in such a manner that the binding affinity of the repressor for its operator is significantly reduced. In theory, barbiturates could also interact directly with the operator DNA to inhibit repressor binding. At the same time, the repressor interacts (forms complexes) with one or more barbiturate-responsive proteins, a process that inhibits its binding to the Barbie box. The "positive regulator" protein (s) which may be induced or simply activated in some way by barbiturates (directly or via barbiturate-mediated changes in the repressor), may also directly bind to the Barbie box. Under these conditions, *the rate of transcription of the P450 mRNA is greatly accelerated* and the encoded P450 is thus induced by barbiturates.

## REFERENCES

Fulco, A. J., He, J.-S. and Liang, Q. (1993) unpublished experiments

Fulco, A. J. (1991) P450$_{BM-3}$ and Other Inducible Bacterial P450 Cytochromes: Biochemistry and Regulation. *Annu. Rev. Pharm. Toxicol.* 31, 177-203.

Fulco, A. J., He, J.-S. and Liang, Q. (1993) unpublished experiments.

He, J.-S. and Fulco, A.J. (1991) A barbiturate-regulated protein binding to a common sequence in the cytochrome P450 genes of rodents and bacteria. *J. Biol. Chem.* 266, 7864-7869.

Liang, Q. and Fulco, A. J .(1993) unpublished experiments.

Patel, N. V. and Omer, C. A. (1992) Phenobarbital and sulfonylurea-inducible operons encoding herbicide metabolizing cytochromes P-450 in *Streptomyces griseolus Gene*, 112: 67-76.

Shaw, G.-C. and Fulco, A. J. (1992) Barbiturate-mediated regulation of expression of the cytochrome P450$_{BM-3}$ gene of *Bacillus megaterium* by Bm3R1 protein. *J. Biol. Chem.* 267, 5515-5526.

Shaw, G.-C. and Fulco, A. J. (1993) Inhibition by barbiturates of the binding of Bm3R1 repressor to its operator site on the barbiturate-inducible cytochrome P450$_{BM-3}$ gene of Bacillus megaterium. *J. Biol. Chem.* 268, 2997-3004.

Upadhya, P. ,Venkateswara Rao, M., Venkateswar, V., Rangarajan, P. N. and Padmanaban, G. (1992) Identification and functional characterization of a *cis*-acting positive DNA element regulating CYP 2B1/B2 gene transcription in rat liver. Nucleic Acids Research 20: 557-562.

*Cytochrome P450. 8th International Conference.* Ed. M.C. Lechner. John Libbey Eurotext, Paris © 1994, pp. 43-48.

# Nitric oxide synthase: mechanisms and relationship to cytochrome P450

Kimberly A. White[1], Robert A. Pufahl[2], Norman M. Olken[2], Joan M. Hevel[2], Michael K. Richards[1], Michael A. Marletta[1,2]

*[1]Department of Biological Chemistry, University of Michigan Medical School and [2]Interdepartmental Program in Medicinal Chemistry, College of Pharmacy, The University of Michigan, Ann Arbor, MI 48109-1065 USA*

The discovery of nitric oxide (•NO) as a mammalian metabolic intermediate has spawned a swiftly moving new field of research. The intense interest and rapid progress have developed primarily because •NO controls and influences a number of critical physiological processes. The best characterized examples of biological reactions controlled by •NO include vasodilation and regulation of normal vascular tone, inhibition of platelet aggregation, neuronal transmission and cytostasis [for reviews, see (Garthwaite, 1991; Ignarro, 1990; Marletta, 1993; Marletta **et al.**, 1990; Moncada **et al.**, 1991; Nathan, 1992)]. In addition to these activities it appears that abnormally high levels of •NO are also involved in the hypotension associated with endotoxic shock (Kilbourn **et al.**, 1990), inflammatory response-induced tissue injury (Mulligan **et al.**, 1991), mutagenesis (Nguyen **et al.**, 1992; Wink **et al.**, 1991), and the formation of carcinogenic N-nitrosamines (Miwa **et al.**, 1987). In tissues where •NO controls normal vascular tone, low levels can also be deleterious such as in the case of pulmonary hypertension in infants and adults. •NO administered via ventilator gas in low concentrations has very recently been found to be useful in the treatment of this type of hypertension.

## NOS: GENERAL CHARACTERIZATION

The formation of •NO and citrulline from L-arginine in mammalian cells is catalyzed by the enzyme •NO synthase (NOS; EC 1.14.13.39) (Figure 1). Co-substrates for the reaction include NADPH and molecular oxygen. As discussed below, NOS also contains a number of enzyme-bound redox active cofactors. The family of NOS isoforms fall into two general categories: (*i*) a constitutive form strictly regulated by $Ca^{2+}$ and calmodulin, and (*ii*) a cytokine-inducible form that shows no apparent post-transcriptional regulation (Marletta, 1993). However, a recent report has showed that calmodulin co-purifies with the murine macrophage inducible NOS (Cho **et al.**, 1992) and further studies have shown that partial dissociation of this bound calmodulin leads to a loss in enzyme activity that is restored upon calmodulin re-binding (Stevens-Truss & Marletta, unpublished results).

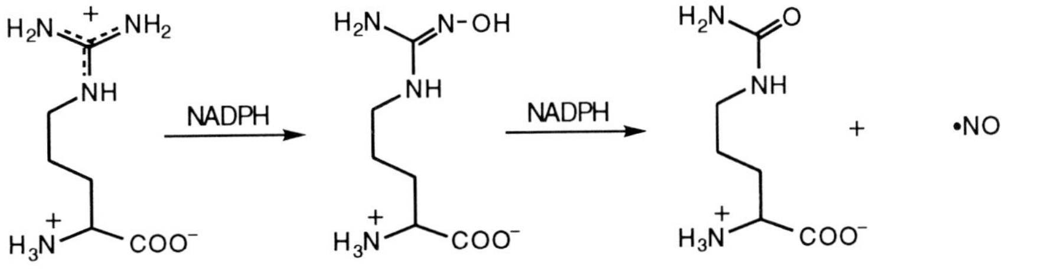


L-arginine — $N^G$-hydroxy-L-arginine — citrulline — nitric oxide

Fig. 1. Reaction catalyzed by NOS. The reaction is known to require NADPH and has been shown to involve $N^G$-hydroxy-L-arginine as an intermediate.

Enzyme-bound cofactor identification resulted from findings directed toward purification and efforts to isolate the cDNAs encoding the respective isoforms. Early purification studies showed a dependence on a low molecular weight substance that was later determined to be (6R)-tetrahydro-L-biopterin ($H_4B$) (Tayeh & Marletta, 1989). The function of this cofactor in the reaction is still not firmly established, but recent results towards understanding its role will be described below. The isolation of the cDNA from the rat cerebellar constitutive NOS showed that the C-terminus showed significant homology to NADPH cytochrome P-450 reductase (Bredt **et al.**, 1991). The highly conserved NADPH, FAD and FMN binding sites were clearly visible. The same level of homology was subsequently found in the cDNAs for the constitutive isoform from bovine endothelial cells (Lamas **et al.**, 1992) and the inducible isoform from murine macrophages (Lowenstein **et al.**, 1992; Lyons **et al.**, 1992; Xie **et al.**, 1992). Several other isoforms that have been cloned and sequenced show the same type of homology. As the cDNA sequences predicted, one equivalent each of FAD and FMN were found associated with the inducible NOS (Hevel **et al.**, 1991) which has since been demonstrated for the other isoforms.

As illustrated in Fig. 1, $N^G$-hydroxy-L-arginine (NHA) has been shown to be an intermediate in the reaction (Pufahl **et al.**, 1992; Stuehr **et al.**, 1991). Speculation on the potential involvement for this, at the time unknown, compound was made a number of years before (Iyengar **et al.**, 1987). This finding coupled with the reductase homology noted above, suggested that at least part of the reaction mechanism might involve a cytochrome P-450 type hydroxylation step. This led to the finding that NOS contained a cytochrome P-450 type chromophore. First shown for the murine macrophage inducible isoform (White & Marletta, 1992) and subsequently for the cerebellar isoforms (McMillan **et al.**, 1992; Stuehr & Ikeda-Saito, 1992), NOS treated with dithionite under CO forms a chromophore with a $\lambda_{max}$ of 445-447 nm. The heme was identified as an iron-protoporphyrin IX by the pyridine hemochromogen assay (McMillan **et al.**, 1992; White and Marletta, 1992). The absolute spectrum of the enzyme shows a Soret maximum ~ 400 nm, indicative a mixed, but predominately high spin state heme. The CO-treated enzyme is catalytically inhibited either when L-arginine is the substrate (White and Marletta, 1992) or when NHA is the substrate (Pufahl & Marletta, 1993) suggesting the iron-protoporphyrin IX heme is involved in both currently defined steps of the reaction. Overall, the evidence supports a view a electron transfer as illustrated in Fig. 2.

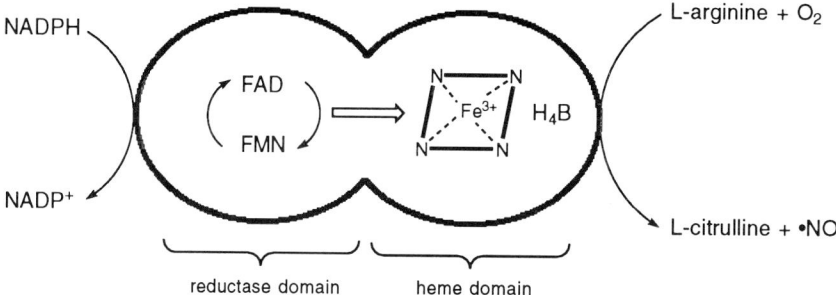

Fig. 2. General view of electron transfer in NOS. Reducing equivalents are assumed to be transferred from NADPH to the heme via the flavins. Based on CO inhibition experiments, the oxidative chemistry is shown taking place at the heme domain. Also shown is $H_4B$, although the binding site for this cofactor has yet to be established.

As mentioned above, the function of the reduced pterin is still not completely clear. The role of $H_4B$ has been most thoroughly studied with the murine inducible NOS. The picture that is emerging suggests an unusual role for this reduced pterin whose typical function is to act as a co-substrate in providing two electrons in the aromatic amino acid hydroxylases. $H_4B$ forms a 1:1 complex with NOS (Hevel & Marletta, 1992). The reduced pterin will dissociate from the enzyme especially if the purification is carried out without $H_4B$ in the buffers (Hevel and Marletta, 1992). The amount of pterin that dissociates must be empirically determined, but NOS isolated in this manner is typically about 20-50% saturated with pterin (Hevel and Marletta, 1992). To date several interesting effects of $H_4B$ have been observed. Macrophage NOS isolated in the absence of $H_4B$ has a significantly altered Soret peak of the heme spectrum. The $\lambda_{max}$ of this so-called pterin-deficient NOS is ~ 418 nm, indicative of a low-spin state heme (Hevel & Marletta, 1993). As mentioned above the pterin-saturated NOS $\lambda_{max}$ is ~ 400nm.

NOS assayed in the presence of H$_4$B is more tightly coupled as well (Hevel et al., 1992). These results suggested that H$_4$B had an effect on the structure of the enzyme and led us to examine the influence of H$_4$B on the monomer (130 kDa) to dimer (260 kDa) ratio, since all NOS isoforms appear to be active as dimers. Indeed, using native gels to analyze the monomer : dimer ratio, we found that a significant portion of pterin-deficient NOS exists in the inactive monomeric state and that the ratio is shifted to dimer with the addition of H$_4$B (Hevel and Marletta, 1993). These observations as well as several others regarding the function reduced pterin are under continuing investigation.

## NOS: MECHANISM

Any reaction proposed for NOS must include some step that involves single electron chemistry in order to account for the formation of •NO. The CO inhibition results described above are consistent with the heme playing a key role in catalysis. This has led to the mechanism illustrated in Figure 3.

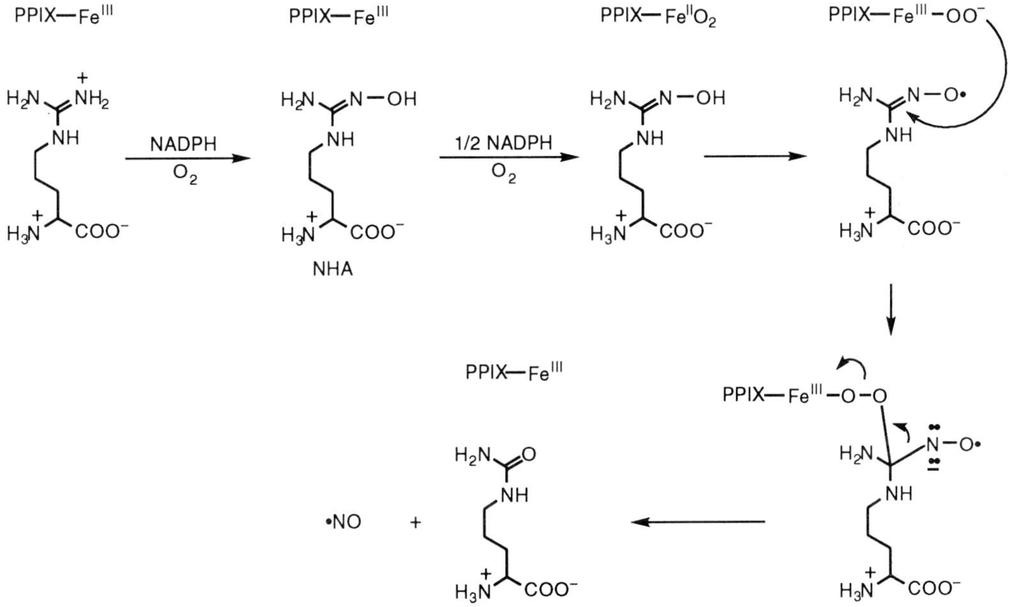

Fig. 3. Current speculation on the NOS reaction mechanism. The first step in the reaction involves the experimentally determined formation of NHA. The chemistry shown beyond that, while based on precedent, has not been proven.

Involvement of the heme in the hydroxylation step is a typical P-450 type reaction. Using the heme in the further oxidation of NHA is shown as (i) reduction of the ferric heme to ferrous via the flavin reductase domain, (ii) single electron oxidation of NHA by the ferrous oxy complex, (iii) attack the generated ferric peroxy complex at the guanido carbon of •NHA, and (iv) collapse of this intermediate to form products. The key step is the single electron oxidation of NHA by the ferrous oxy complex. It was shown this way and not in the reverse since we have shown that NHA does not directly reduce NOS (Pufahl and Marletta, 1993). Ferrous peroxide chemistry has been shown to occur in P-450 catalyzed reactions (Vaz et al., 1991). Studies are underway to furhter study the NOS reaction mechanism.

## NOS: INHIBITION

A number of inhibitors of NOS have been reported and many of them have been simple arginine derivatives with alkyl substitutions on the one of the terminal guanido nitrogens. N$^G$-Methyl-L-arginine

(NMA), first reported by Hibbs and colleagues to inhibit $NO_2^-$ / $NO_3^-$ synthesis in activated macrophages (Hibbs **et al.**, 1987), was since shown to be a competitive type inhibitor of NOS (Olken **et al.**, 1991). However further studies showed that when NMA was pre-incubated with NOS, the enzyme was irreversibly inactivated (Feldman **et al.**, 1993; Olken **et al.**, 1991). This irreversible inactivation requires enzymatic turnover and the products of NMA turnover are those that would have been predicted from typical P-450 catalyzed chemistry (Olken & Marletta, 1993). The NOS catalyzed reaction on NMA includes N-demethylation and N-hydroxylation of the methyl bearing carbon. This is shown below in Figure 4.

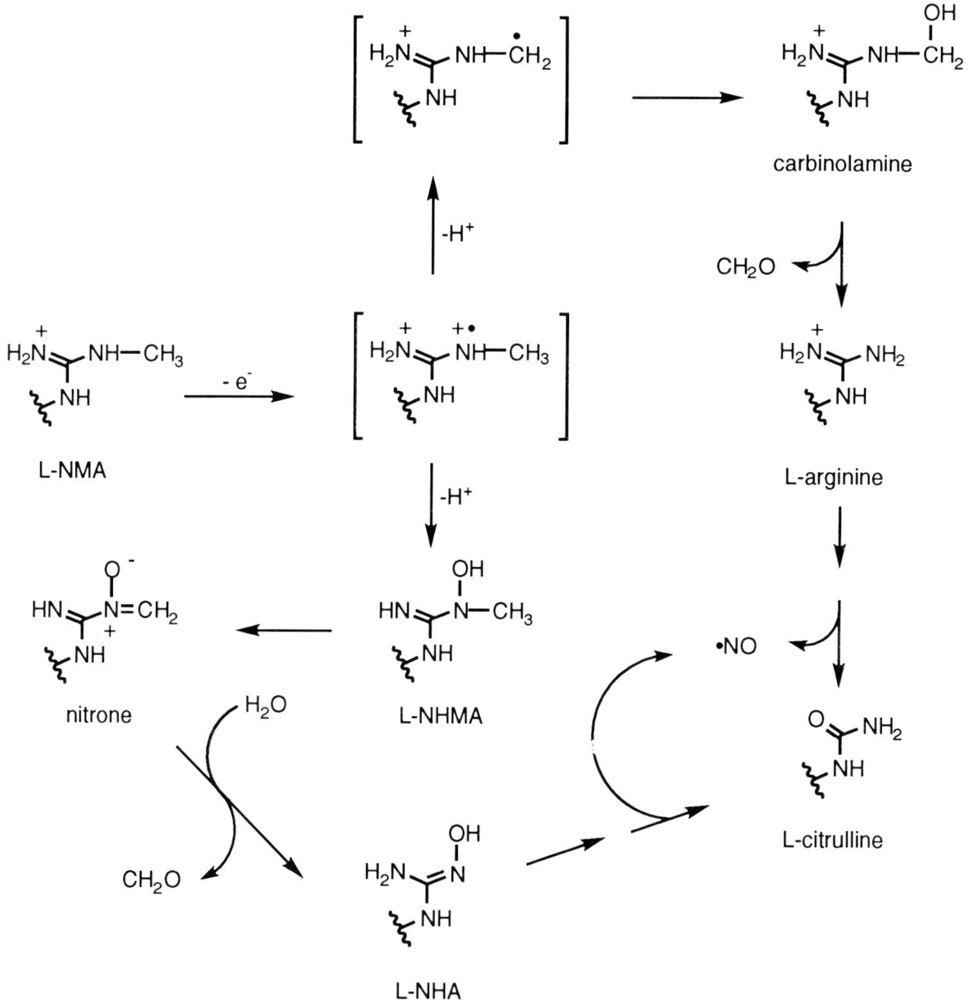

Fig. 4. NOS catalyzed oxidation of NMA. NHMA, NHA, arginine, citrulline and $CH_2O$ have all been identified as products.

NHMA has been synthesized and found to be a more potent irreversible inhibitor of NOS (Pufahl **et al.**, 1992). The formation of NHA from NHMA has lead to the speculation of the nitrone intermediate shown. Similarly, the carbanolamine has not been observed but would be expected, if formed, to generate $CH_2O$ and arginine. All of the reactions observed are consistent with those expected for P-450 catalyzed reactions. Interestingly, no reaction at the unsubstituted nitrogen was observed.

CONCLUSIONS

The characterization of NOS has shown some striking parallels with cytochrome P-450. Although it appears to lie outside the P450 super gene family from sequence considerations, the optical properties, reactions catalyzed and co-substrates place it in a P-450 related category. The exact relationship to the well-established P-450 family of proteins will become more clear with studies that are underway.

ACKNOWLEDGMENTS

Support from the NIH (CA50414 and CA26731), and the Hitchings Award of the Burroughs-Wellcome Fund is gratefully acknowledged.

REFERENCES

Bredt, D. S., Hwang, P. M., Glatt, C. E., Lowenstein, C., Reed, R. R., & Snyder, S. H. (1991): Cloned and expressed nitric oxide synthase structurally resembles cytochrome P-450 reductase. *Nature (London)* 351, 714-718.

Cho, H. J., Xie, Q.-W., Calaycay, J., Mumford, R. A., Swiderek, K. M., Lee, T. D., & Nathan, C. (1992): Calmodulin is a subunit of nitric oxide synthase from macrophages. *J. Exp. Med.* 176, 599-604.

Feldman, P. L., Griffith, O. W., Hong, H., & Stuehr, D. J. (1993): Irreversible inactivation of macrophage and brain nitric oxide synthase by L-$N^G$-methylarginine requires NADPH-dependent hydroxylation. *J. Med. Chem.* 36, 491-496.

Garthwaite, J. (1991): Glutamate, nitric oxide and cell-cell signalling in the nervous system. *Trends Neurosci.* 14, 60-67.

Hevel, J. M., & Marletta, M. A. (1992): Macrophage nitric oxide synthase: Relationship between enzyme-bound tetrahydrobiopterin and synthase activity. *Biochem.* 31, 7160-7165.

Hevel, J. M., & Marletta, M. A. (1993): Influence of tetrahydrobiopterin on the structure of macrophage nitric oxide synthase. , submitted for publication.

Hevel, J. M., White, K. A., & Marletta, M. A. (1991): Purification of the inducible murine macrophage nitric oxide synthase. Identification as a flavoprotein. *J. Biol. Chem.* 266, 22789-22791.

Hevel, J. M., White, K. A., & Marletta, M. A. (1992). Purification of the inducible murine macrophage nitric oxide synthase: identification as a flavoprotein and detection of enzyme-bound tetrahydrobiopterin. London: Portland Press.

Hibbs, J. B. J., Vavrin, Z., & Taintor, R. R. (1987): L-Arginine is required for expression of the activated macrophage effector mechanism causing selective metabolic inhibition in target cells. *J. Immunol.* 138, 550-565.

Ignarro, L. J. (1990): Biosynthesis and metabolism of endothelium-derived relaxing nitric oxide. *Ann. Rev. of Pharmacol. Toxicol.* 30, 535-560.

Iyengar, R., Stuehr, D. J., & Marletta, M. A. (1987): Macrophage synthesis of nitrite, nitrate, and N-nitrosamines: precursors and role of the respiratory burst. *Proc. Natl. Acad. Sci. USA* 84, 6369-6373.

Kilbourn, R. G., Gross, S. S., Jubran, A., Adams, J., Griffith, O. W., Levi, R., & Lodato, R. F. (1990): $N^G$-methyl-L-arginine inhibits tumor necrosis factor-induced hypotension: Implications for the involvement of nitric oxide. *Proc. Natl. Acad. Sci. USA* 87, 3629-3632.

Lamas, S., Marsden, P. A., Li, G. K., Tempst, P., & Michel, T. (1992): Endothelial nitric oxide synthase: Molecular cloning and characterization of a distinct constitutive enzyme isoform. *Proc. Natl. Acad. Sci. USA* 89, 6348-6352.

Lowenstein, C. J., Glatt, C. S., Bredt, D. S., & Snyder, S. H. (1992): Cloned and expressed macrophage nitric oxide synthase contrasts with the brain enzyme. *Proc. Natl. Acad. Sci. USA* 89, 6711-6715.

Lyons, C. R., Orloff, G. J., & Cunningham, J. M. (1992): Molecular cloning and functional expression of an inducible nitric oxide synthase from a murine macrophage cell line. *J. Biol. Chem.* 267(9), 6370-6374.

Marletta, M. A. (1993): Nitric oxide synthase structure and mechanism. *J. Biol. Chem.* 268, 12231-12234.

Marletta, M. A., Tayeh, M. A., & Hevel, J. M. (1990): Unraveling the biological significance of nitric oxide. *BioFactors* 2, 219-225.

McMillan, K., Bredt, D. S., Hirsch, D. J., Snyder, S. H., Clark, J. E., & Masters, B. S. S. (1992): Cloned, expressed rat cerebellar nitric oxide synthase contains stoichiometric amounts of heme, which binds carbon monoxide. *Proc. Natl. Acad. Sci. USA* 89, 11141-11145.

Miwa, M., Stuehr, D. J., Marletta, M. A., Wishnok, J. S., & Tannenbaum, S. R. (1987): Nitrosation of amines by stimulated macrophages. *Carcinogenesis* 8, 955-958.

Moncada, S., Palmer, R. M. J., & Higgs, E. A. (1991): Nitric oxide: physiology, pathophysiology, and pharmacology. *Pharmacol. Rev.* 43, 109-142.

Mulligan, M. S., Hevel, J. M., Marletta, M. A., & Ward, P. A. (1991): Tissue injury caused by deposition of immune complexes is L-arginine dependent. *Proc. Natl. Acad. Sci. USA* 88, 6338-6342.

Nathan, C. (1992): Nitric oxide as a secretory product of mammalian cells. *FASEB J.* 6, 3051-3064.

Nguyen, T., Brunson, D., Crespi, C. L., Penman, B. W., Wishnok, J. S., & Tannenbaum, S. R. (1992): DNA damage and mutation in human cells exposed to nitric oxide *in vitro*. *Proc. Natl. Acad. Sci. USA* 89, 3030-3034.

Olken, N. M., & Marletta, M. A. (1993): $N^G$-Methyl-L-arginine functions as an alternate substrate and mechanism-based inhibitor of nitric oxide synthase. *Biochem* 32, 9677-9685.

Olken, N. M., Rusche, K. M., Richards, M. K., & Marletta, M. A. (1991): Inactivation of macrophage nitric oxide synthase activity by $N^G$-methyl-L-arginine. *Biochem. Biophys. Res. Commun.* 177(in press), 828-833.

Pufahl, R. A., & Marletta, M. A. (1993): Oxidation of $N^G$-Hydroxy-L-arginine by nitric oxide synthase: evidence for the involvement of the heme in catalysis. *Biochem. Biophys. Res. Commun.* 193, 963-970.

Pufahl, R. A., Nanjappan, P. G., Woodard, R. W., & Marletta, M. A. (1992): Mechanistic probes of N-hydroxylation of L-arginine by the inducible nitric oxide synthase from murine macrophages. *Biochem* 31, 6822-6828.

Stuehr, D. J., & Ikeda-Saito, M. (1992): Spectral Characterization of brain and macrophage nitric oxide synthases. *J. Biol. Chem.* 267, 20547-20550.

Stuehr, D. J., Kwon, N. S., Nathan, C. F., Griffith, O. W., Feldman, P. L., & Wiseman, J. (1991): $N^\omega$-hydroxy-L-arginine is an intermediate in the biosynthesis of nitric oxide from L-arginine. *J. Biol. Chem.* 266(10), 6259-6263.

Tayeh, M. A., & Marletta, M. A. (1989): Macrophage oxidation of L-arginine to nitric oxide, nitrite, and nitrate. Tetrahydrobiopterin is required as a cofactor. *J. Biol. Chem.* 264, 19654-19658.

Vaz, A. D. N., Roberts, E. S., & Coon, M. J. (1991): Olefin Formation in the oxidative deformylation of aldehydes by cytochrome P-450. Mechanistic implications for catalysis by oxygen-derived peroxide. *J. Am. Chem. Soc.* 113, 5886-5887.

White, K. A., & Marletta, M. A. (1992): Nitric oxide synthase is a cytochrome P-450 type hemoprotein. *Biochem.* 31, 6627-6631.

Wink, D. A., Kasprzak, K. S., Maragos, C. M., Elespuru, R. K., Misra, M., Dunams, T. M., Cebula, T. A., Koch, W. H., Andrews, A. W., Allen, J. S., & Keefer, L. K. (1991): DNA deaminating ability and genotoxicity of nitric oxide and its progenitors. *Science* 254, 1001-1003.

Xie, Q.-W., Cho, H. J., Calaycay, J., Mumford, R. A., Swiderek, K. M., Lee, T. D., Ding, A., Troso, T., & Nathan, C. (1992): Cloning and characterization of inducible nitric oxide synthase from mouse macrophages. *Science* 256, 225-228.

# II | CONTROL MECHANISMS IN THE BIOGENESIS OF CYTOCHROME P450'S

# Growth hormone regulation of liver sexual differentiation

Jan-Åke Gustafsson, Petra Tollet, Agneta Mode

*Department of Medical Nutrition, Novum, Karolinska Institute, Huddinge University Hospital, S-141 86 Huddinge, Sweden*

It has been known for a long time that rodents display large sexual differences with regard to metabolism of steroids and xenobiotics (for references, see Gustafsson et al., 1983; Zaphiropoulos et al., 1989). This difference in metabolic rate has pharmacological consequences. For instance, male rats sleep for a considerably shorter time when given phenobarbital than their female counterparts and this is due to a more rapid elimination of the drug through a cytochrome P450 catalyzed metabolic inactivation. The sexually differentiated metabolism of corticosterone in the rat was first shown when steroid metabolites in excreta from rats were identified with gas chromatography-mass spectrometry. The most conspicuous difference was the exclusive excretion in the bile, urine and feces of female rats of 15ß-hydroxylated sulphurylated corticosterone derivatives (Gustafsson et al., 1968; Eriksson et al., 1971). Male rats, on the other hand, excreted significant quantities of polyhydroxylated corticosterone derivatives (Cronholm et al., 1972). The sexually dimorphic liver metabolism is imprinted at birth by testicular androgens (Einarsson et al., 1973; Gustafsson and Stenberg, 1974a, b). For instance, adult male rats which have been neonatally orchidectomized, develop a completely feminine pattern of metabolism but when orchidectomy is postponed until seven days of age a significant masculinization is seen in the adult animal. This neonatal imprinting by testicular androgens can be mimicked by one single administration of androgen to neonatally orchidectomized male rats.

The sex differences observed *in vivo* could also be found *in vitro* in liver microsomal preparations (Gustafsson et al., 1968; Berg and Gustafsson, 1973; Morgan et al., 1985). The enzymes responsible for these differences in steroid hydroxylation have since been purified and their cDNAs and genes have been cloned (MacGeoch et al., 1987; Zaphiropoulos et al., 1988; Ström et al., 1988; Zaphiropoulos et al., 1990 a,b; Westin et al., 1990; Eguchi et al., 1991).

In an effort to understand the endocrine regulatory mechanisms which gave rise to the quite astounding sexual dimorphism of rat liver, a large number of endocrine ablations and substitutions were made and growth hormone was identified as the responsible hormone. It could be shown that an intact pituitary was a prerequisite for effects of sex steroids on liver metabolism. Normally, estrogens feminized the male liver (i.e. induced 15ß-hydroxylation of steroid sulphates) and, conversely, androgens may masculinize female liver metabolism. None of these effects were seen in hypophysectomized rats (Gustafsson and Stenberg, 1976). Thus the role of the hypothalamo-pituitary axis in controling sexually differentiated liver metabolism was confirmed.

Normally, male rats secrete growth hormone in a burst-like fashion with peaks every four-six hours and deep troughs in between whereas female rats display a more continuous secretory pattern with no pronounced peaks so that the plasma levels never drop to zero. When these secretion patterns were mimicked in hyphophysectomized rats, it was found that the male specific 2C11 was controlled by the male secretory pattern of growth hormone and female specific 2C12 by the female-type growth hormone secretory pattern ( Mode et al., 1981). One and the same hormone has completely different effects depending upon in which way it is presented to its target tissue.

It seems that all sexually dimorphic metabolic phenomena in the rat liver are controlled by the growth hormone pulsatile pattern. These include regulation of $\alpha_2$-urinary globulin (Husman et al., 1985), steroid 5α-reductase and LDL receptors (Rudling et al., 1992).

The molecular mechanisms involved in growth hormone effects in the liver is the subject of this presentation.

RESULTS AND DISCUSSION

Investigations into the mechanism of growth hormone action are done with primary hepatocytes cultured on matrigel. As target genes we have used the male specific 2C11 and the female specific 2C12 but we also use two other P450 isozymes, the male specific 2C13 and the female dominant 2C7.

After two days in culture, the hepatocytes respond promptly to added growth hormone by induction of 2C12 mRNA, which reaches plateau levels after about twelve hours (Guzelian et al., 1988). As a comparison we have also measured growth hormone induction of IGF-1 mRNA. Only growth hormone, not prolactin, induced 2C12 and IGF-1 mRNA. Insulin was shown to augment the effect of growth hormone on 2C12 mRNA, and it could be demonstrated that insulin exerted this effect by increasing growth hormone receptor levels. The effect of insulin was not transcriptional but translational or posttranslational.

When the regulation of IGF-1 is compared with P4502C12, it is evident that growth hormone acts on these two genes by completely different pathways. Active protein biosynthesis is required for growth hormone control of 2C12 mRNA whereas this is not the case for growth hormone regulation of IGF-1. This could be clearly shown with the use of the protein synthesis inhibitor, cycloheximide (Tollet et al., 1990,1991). Interestingly, IGF-1 which mediates several of the actions of growth hormone, had no effect on 2C12 mRNA accumulation.

In order to elucidate the mechanism involved in transduction of the growth hormone signal in hepatocytes, the role of several of the known intracellular signal transduction pathways has been investigated. Phospholipase $A_2$ is involved since mepacrine, an inhibitor of phospholipase $A_2$, can completely abolish the growth hormone induction of P450 2C12 mRNA. The role of the phospholipase $A_2$ seems to be to provide arachidonic acid which is then metabolized to the ultimate mediator of growth hormone action on P450 2C12 induction. Surprisingly, with the use of specific inhibitors of the various arachidonic acid metabolic pathways, it was found that this activation is mediated by a cytochrome P450 and not a lipoxygenase or a cyclooxygenase.

Arachidonic acid itself cannot substitute for growth hormone action so it is evident that other factors are needed in concert with the putative arachidonic acid pathway to result in an effect on P450 2C12 transcription. Other pathways which seem to be involved are phospholipase C, protein kinase C and tyrosine kinase. When growth hormone is administered to hepatocytes, there is an early transient intracellular accumulation of diacylglycerol and a release of phosphocholine. There is no measurable change in phosphoinositols. This suggests that phosphatidylcholine is a substrate of a growth hormone-activated phospholipase C. Protein kinase C is also involved in the signal transduction pathway since inhibitors of this kinase (high concentrations of phorbol esters and staurosporine), inhibit the action of growth hormone on P450 2C12 induction. However, diacylglycerol does not mimic the effects of growth hormone on P450 2C12 when it is added to hepatocytes. Tyrosine kinase inhibitors, tyrphostin, herbimycin A and methyl-2,3-dihydroxycinnamate, can also block the effects of growth hormone on P450 2C12 induction. Recent data from other laboratories show that protein kinase Jak 2 as well as MAP-kinase are involved in growth hormone action.

Numerous attempts have been made to define growth hormone responsive elements in genomic sequences of 2C11, 2C12, 2C13 and 2C7. We have looked for sex dependent DNAse I footprints, DNAse I hypersensitive regions as well as growth hormone responsive elements as defined by transient transfection assays using constructs consisting of various parts of the 5'flanks and first introns of 2C11, 2C12, 2C13 and 2C7 genes coupled to reporter genes such as chloramphenicol acetyltransferase and luciferase. Whereas it is clear that growth hormone influences the endogenous 2C11, 2C12, 2C13 and 2C7 genes at the transcriptional level (Legraverend et al., 1992) and that many liver specific transcription factors have been shown to play important roles in liver specific expression of these P450 genes and many sites for various transcription factors have been defined on the P450 genes, we have not obtained any convincing data for the existence of growth hormone responsive elements so far. It is not inconceivable that regions very far upstream may be involved (we have hitherto covered regions up to 6-8 kbs upstream) or that the chromatin structure might be important.

CONCLUSION

Sexual differentiation of rat liver is regulated through the hypothalamo-pituitary-liver axis and it is through this axis that sex steroids exert their effects on liver metabolism. A single hormone, namely, growth hormone is responsible for the sex differences in liver metabolism and these differences can be reproduced in hypophysectomized rats by administration of the hormone in the male or female secretion pattern. The mechanism through which growth hormone produces these effects is unknown but there is evidence for the involvement of several inteacellular pathways including protein kinase C, phospholipase A2 , a cytochrome P450 and a tyrosine kinase. It is currently somewhat unclear why such a sophisticated loop is operating to control sexual dimorphism in rat liver metabolic events. Many of the steroid metabolites produced during the P450 catalyzed processes in rat liver have long been thought to represent biologically inactive catabolic products but it is not inconceivable that at least some of these compounds could have a function in special tissues and that the physiological role of hepatic P450 dependent steroid hormone metabolism may have been underestimated.

Acknowledgements

These studies have been supported by a grant from the Swedish Medical Research Council (No. 13x2819).

References

Berg A and Gustafsson J-Å. (1973) Regulation of hydroxylation of $5\alpha$-androstane-$3\alpha,17\beta$-diol in liver microsomes from male and female rats. *J. Biol. Chem.* 248: 6559-6567.

Cronholm T, Eriksson H and Gustafsson J-Å. (1972) Excretion of steroid hormone metabolites in bile of male rats. *Steroids* 19: 455-470.

Eguchi H, Westin S, Ström A, Gustafsson J-Å, and Zaphiropoulos PG. (1991) Gene structure and expression of the rat cytochrome P450IIC13, a polymorphic, male-specific cytochrome in the P450IIC subfamily. *Biochemistry* 30: 10844-10849.

Einarsson K, Gustafsson J-Å and Stenberg Å. (1973) Neonatal imprinting of liver microsomal hydroxylation and reduction of steroids. *J. Biol. Chem.* 248: 4987-4997.

Eriksson H, Gustafsson J-Å and Sjövall J. (1971) Studies on the structure, biosynthesis and bacterial metabolism of 15-hydroxylated steroids in the female rat. *Eur. J. Biochem.* 19: 433-441.

Gustafsson J-Å and Sjövall J. (1968) Steroids in germfree and conventional rats. VI. Identification of $15\alpha$- and 21-hydroxylated $C_{21}$ steroids in faeces from germfree rats. *Eur J. Biochem.* 6: 236-247.

Gustafsson J-Å and Stenberg Å. (1974) Irreversible androgenic programming at birth of microsomal and soluble liver enzymes active on 4-androstene-3,17-dione and 5α-androstane-3α,17ß-diol. *J. Biol. Chem.* 249: 711-718.

Gustafsson J-Å and Stenberg Å. (1974) Neonatal programming of androgen responsiveness of liver of adult rats. *J. Biol. Chem.* 249: 719-723.

Gustafsson J-Å and Stenberg Å. (1976) On the obligatory role of the hypophysis in sexual differentiation of hepatic metabolism in rats. *Proc. Natl. Acad. Sci. USA* 73: 1462-1465.

Gustafsson J-Å, Lisboa BP and Sjövall J. (1968) Studies on the metabolism of $C_{19}$ steroids in rat liver. III. Isolation and biosynthesis of 5α-androstanetriols in rat liver microsomes. *Eur. J. Biochem.* 6: 317-324.

Gustafsson J-Å, Mode A, Norstedt G and Skett P. (1983) Sex steroid induced changes in hepatic enzymes. *Ann. Rev. Physiol.* 45: 51-60.

Guzelian PS, Li D, Schuetz EG, Thomas P, Levin W, Mode M and Gustafsson J-Å. (1988) Sex change in cytochrome P-450 phenotype by growth hormone treatment of adult rat hepatocytes maintained in a culture system on matrigel. *Proc. Natl. Acad. Sci. USA* 85: 9783-9787.

Husman B, Norstedt G, Mode A and Gustafsson J-Å. (1985) The mode of growth hormone administration is of major importance for the excretion of the major male rat urinary proteins. *Mol. Cell. Endocrinol.* 40: 205-210.

Legraverend C, Mode A, Westin S, Ström A, Eguchi H, Zaphiropoulos PG and Gustafsson J-Å. (1992) Transcriptional regulation of rat P-450 2C gene subfamily members by the sexually dimorphic pattern of growth hormone secretion. *Mol. Endocrinology* 259-266.

MacGeoch C, Morgan ET, Cordell B and Gustafsson J-Å. (1987) Growth hormone regulates expression of rat liver cytochrome $P-450_{15\beta}$ at a pretranslational level. *Biochem. Biophys. Res. Comm.* 143: 782-788.

Mode A, Norstedt G, Simic B, Eneroth P and Gustafsson J-Å. (1981) Continuous infusion of growth hormone feminizes hepatic steroid metabolism in the rat. *Endocrinology* 108: 2103-2108.

Morgan ET, MacGeoch C and Gustafsson J-Å. (1985) Sexual differentiation of cytochrome P-450 in rat liver. Evidence for a constitutive isozyme as the male-specific 16α–hydroxylase. *Mol. Pharmacol.* 27: 471-479.

Rudling M, Norstedt G, Olivecrona H, Reihnér E, Gustafsson J-Å and Angelin B. (1992) Importance of growth hormone for the induction of hepatic low density lipoprotein receptors. *Proc.Natl.Acad.Sci.* 89: 6983-6987.

Ström A, Mode A, Zaphiropoulos P, Nilsson A-G, Morgan E and Gustafsson J-Å. (1988) Cloning and pretranslational hormonal regulation of testosterone 16a-hydroxylase (P-$450_{16\alpha}$) in male rat liver. *Acta Endocrinologica* 118: 314-320.

Tollet P, Enberg B, and Mode A. (1990) Growth hormone (GH) regulation of cytochrome P-450IIC12, insulin-like growth factor-I (IGF-I), and GH receptor messenger RNA expression in primary rat hepatocytes: A hormonal interplay with insulin, IGF-I, and thyroid hormone. *Mol. Endocrinol.* 4: 1934-1942.

Tollet P, Legraverend C, Gustafsson J-Å, and Mode A. (1991) A role for protein kinases in the growth hormone regulation of cytochrome P4502C12 and insulin-like growth factor-I messenger RNA expression in primary adult rat hepatocytes. *Mol. Endocrinol.* 5: 1351-1358.

Westin S, Ström S, Gustafsson J-Å, and Zaphiropoulos PG. (1990) Growth hormone regulation of the cytochrome P-450IIC subfamily in the rat: Inductive, repressive, and transcriptional effects on P-450f (IIC7) and P-$450_{PB1}$ (IIC6) gene expression. *Molecular Pharmacology* 38: 192-197.

Zaphiropoulos PG, Mode A, Norstedt G and Gustafsson J-Å. (1989) Regulation of sexual differentiation in drug and steroid metabolism. *Trends in Pharmacological Sciences* 10: 149-153.

Zaphiropoulos PG, Mode M, Ström A, Möller C, Fernandez C and Gustafsson J-Å. (1988) cDNA cloning, sequence, and regulation of a major female-specific and growth hormone-inducible rat liver cytochrome P-450 active in 15ß-hydroxylation of steroid sulfates. *Proc. Natl. Acad. Sci. USA* 85: 4214-4217.

Zaphiropoulos PG, Ström A, Robertson JA and Gustafsson J-Å. (1990) Structural and regulatory analysis of the male-specific rat liver cytochrome P-450g: Repression by continuous growth hormone administration. *Mol. Endocrinol.* 4: 53-58.

Zaphiropoulos PG, Westin S, Ström A, Mode A and Gustafsson J-Å. (1990) Structural and regulatory analysis of a cytochrome P-450 gene (CYP2C12) expressed predominantly in female rat liver. *DNA and Cell Biology* 9: 49-56.

# Mechanisms of liver-specific expression of the *CYP2D5* gene

Ying-Hue Lee, Frank J. Gonzalez

*Laboratory of Molecular Carcinogenesis, National Cancer Institute, National Institutes of Health, Bethesda, Maryland, USA 20892*

Introduction:

Eukaryote genes are transcribed through the action of protein *trans*-acting "factors" that bind to regulatory *cis*-acting elements usually located upstream of the transcription start site. Two general types of regulation have been described, inducer-dependent and constitutive. Inducers can activate gene expression through two mechanisms. They can bind to a cell surface receptor that, through a cascade of events, result in activation of transcription factors which in turn stimulate gene transcription. Inducers can also stimulate transcription through direct binding to cytoplasmic receptors. Examples include the steroid receptor superfamily, the peroxisome proliferator activated receptor and the *Ah* receptor. The inducer ligand binds to the receptor displacing the heat shock protein 90 and causing transformation of the receptor into a form that can enter the nucleus and bind to regulatory regions of target genes. The *cis*-acting regions are sometimes inducible enhancers that can function in an orientation-independent and somewhat distance-independent manner.

Constitutively expressed genes are regulated by transcription factors that are inducer-independent although these genes can have other *cis* elements that interact with inducer-dependent factors. It is generally believed that tissue-enriched transcription factors govern tissue-specific expression of genes. Little is known about what regulates transcription of the genes encoding tissue-enriched transcription factors and sparse information is available on how the receptors and transcription factors interact with the RNA polymerase II complex to enhance transcription.

The cytochromes P450 are regulated by both inducer-dependent and inducer-independent mechanisms (Gonzalez, 1992; Gonzalez *et al.*, 1993). The most actively studied inducible P450 gene is *CYP1A1* which is regulated by the Ah receptor (Hankinson, 1993; Swanson & Bradfield, 1993).

Most xenobiotic-metabolizing P450s are expressed in liver, although low level expression is found in extrahepatic tissues such as lung and gastrointestinal tract. Rodents have a number of hepatic P450s that are under developmental and sex-dependent control (Table I). The sex-dependent expression is due to differences in circulating growth hormone patterns between males and females (Gonzalez, 1992), a phenomenon that is not found in humans or rabbits.

Table I. Modes of Regulation of Rat Hepatic Cytochrome P450 Genes

| P450 | Expression Pattern |
| --- | --- |
| CYP2A1<br>CYP3A2 | activated after birth in both sexes<br>then repressed in females at puberty |
| CYP2E1<br>CYP2D1<br>CYP2D2<br>CYP2D5 | activated after birth in both sexes |
| CYP2C6<br>CYP2C7<br>CYP2D3 | activated in both sexes at puberty |
| CYP2A2<br>CYP2C11<br>CYP2C13<br>CYP2C22<br>CYP4A2 | activated in males at puberty |
| CYP2C12 | activated in females at puberty |

Liver-specific expression of genes has been the most actively studied in the area of tissue-specific gene regulation owing to the abundance and cellular homogeneity of this tissue. Genes encoding serum albumin, transferrin, α1-antitrypsin, fibrinogen, amylase, glucose 6-phosphate dehydrogenase and other liver enzymes and proteins have been intensively studied. A number of transcription factors have been identified that are enriched in liver tissue (Table II).

Several families of factors have been identified that are distinguished based on their conserved DNA-binding domains. The domains homeo and forkhead were first described in *Drosophila*. The zinc finger domain of HNF-4 is similar to that found in the steroid receptor superfamily. The bZIP domain was described in the C/EBP proteins isolated from rat liver. DBP is a member of the C/EBP family but lacks a leucine zipper domain.

The factors listed in Table II are enriched in liver. However, they can also be expressed to some degree in extrahepatic tissues. For example, some of the C/EBP factors are found in adipose tissue and lung. It should also be noted, that other members of these families exist that may not be expressed in liver. For example, a number of HNF-3-like forkhead homologues have been isolated that are expressed in brain, kidney, lung and intestine (Clevidence et al., 1993).

Table II. Hepatocyte-Enriched Transcription Factors

| Family (DNA-binding Motif) | Transcription Factor |
|---|---|
| POU-Homeodomain | HNF-1α (LF-B1, APF) |
|  | HNF-1β (vHNF-1) |
| Forkhead | HNF-3α |
|  | HNF-3β |
|  | HNF-3γ |
| Zinc Finger | HNF-4 |
| bZIP (basic region, leucine zipper) | C/EBPα |
|  | C/EBPβ |
|  | C/EBPγ |
|  | DBP |

Taken from Lai and Darnell (1991) and Williams et al., (1991). Abbreviations are HNF Hepatocyte Nuclear Factor; C/EBP CCAAT Enhancer Binding Protein; CRB C/EBP Related Protein; DBP Albumin D site Binding Protein.

It is not unreasonable to assume that the liver-enriched transcription factors that have been characterized using other liver-specific genes are also involved in expression of the P450s. Such has been demonstrated by recent studies described below.

Cytochromes P450 have complex and diverse regulatory control circuits. Certain P450 forms are expressed at very low levels and are induced by foreign compounds such as polycyclic aromatic hydrocarbons, phenobarbital, and clofibrate. Other P450s are constitutively expressed. These enzymes are enriched in liver tissues and their expression is activated during development. The rat has been used as an experimental model to investigate the mechanisms of tissue-specific, developmentally-programmed transcription of P450s in the CYP2 family. A variety of experimental approaches were used including *in vitro* transcription, heterologous promoter transcription assays, *trans*-activation transcription assays, and *in vitro* DNA binding assays including gel mobility shift, DNase I footprinting and methylation interference to determine the mechanism of regulation of the *CYP2C6*, *CYP2E1*, and *CYP2D5* genes. These studies have shown that the *CYP2C6* promoter is controlled by the liver-enriched transcription factor DBP but is capable of binding to C/EBPα, C/EBPβ and DBP (Yano et al., 1992). Even though the C/EBP proteins are capable of binding the regulatory region they can not activate the *CYP2C6* promoter. *CYP2E1* is under control of the factor HNF-1α (Ueno & Gonzalez, 1990) and *CYP2D5* is regulated in part by Sp1 and C/EBPβ (Lee et al., 1994). These data indicate that the developmentally-activated expression of P450 genes in rat liver is due in part to the presence and activity of liver-enriched transcription factors.

Regulation of the *CYP2D5* gene:

The *CYP2D5* gene becomes transcriptionally active soon after birth and is maximally active when rats reach puberty. Co-transfection of HepG2 cells of a 2D5-CAT construct with various transcription factor expression vectors revealed that C/EBPβ plays a role in *CYP2D5* transcription (Figure 1). DBP, HNF-1α and C/EBPα were ineffective at *trans*-activation.

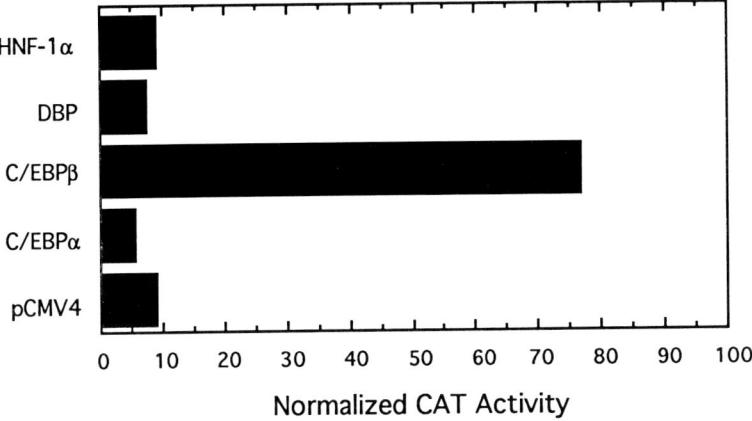

Fig. 1. Co-transfection assay with the *CYP2D5* promoter (-301/+73)-CAT reporter plasmid DNA and various expression vectors containing coding sequences for different liver-enriched transcription factors. The pSVOA.2D5 (-301/+73)-CAT construct (10 μg) was co-transfected with 10 μg of different expression vectors as indicated and 2 μg of pSV232A.1Δ5'(luciferase control vector), and the CAT/luciferase values were determined.

Deletion analysis revealed that a segment of DNA between -120 and -98 was required for both constitutive and C/EBPβ-augmented activity (Figure 2). DNase I footprinting confirmed that a protein(s) in HepG2 and adult rat liver nuclear extracts specifically-bound to the -108 to -90 region.

Recombinant C/EBPβ was unable to elicit a footprint even though it was able to activate transcription of the *CYP2D5* promoter. In addition, a double-stranded oligonucleotide corresponding to the region between -120 and -78 that is involved in transcription activation by C/EBPβ was unable to bind the recombinant factor. C/EBPβ was, however, able to bingd a consensus C/EBP binding site. Suprisingly, in the presence of recombinant Sp1, C/EBPβ was able to interact with the the specific oligonucleotide and Sp1 to form a ternary complex (Figure 3). These data suggest that C/EBPβ requires Sp1 or a factor exhibiting similar binding characteristics to Sp1, for liver-specific expression of the *CYP2D5* gene. Such an interaction has previously not been described for a liver-enriched transcription factor. A model depicting the role of Sp1 and C/EBPβ in regulation of the *CYP2D5* gene is shown in Figure 4.

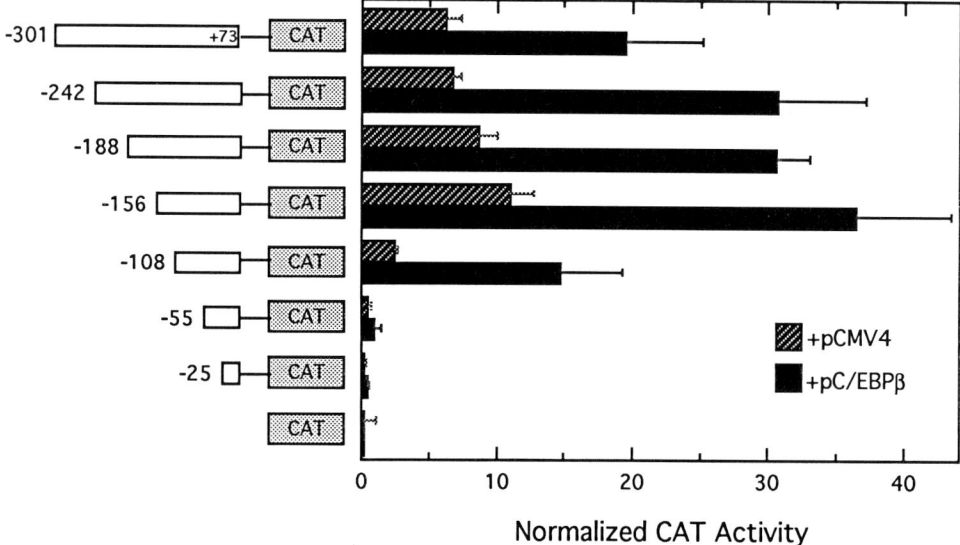

Fig. 2. Activity of various deletions in the *CYP2D5* promoter and the co-transfection assay with C/EBPβ in HepG2 cells. Various deletions of the pSVOA.2D5-CAT constructs (10 μg) were cotransfected with 10 μg of pCMV-4 or pMEX-C/EBPβ and 2 μg of pSV232AL-AΔ5' (luciferase-expressing transfection control plasmid), and the CAT/luciferase values were determined. The bar values represent the average of three independent transfections, and their standard deviations are indicate by the lines.

Fig. 3. Gel mobility shift analysis of the ternary complex between the

## presence of C/EBPβ

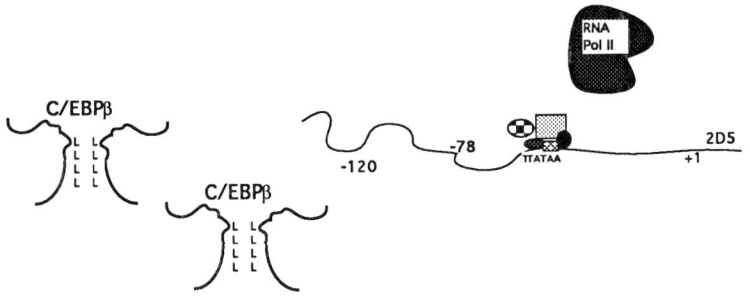

## presence of Sp1

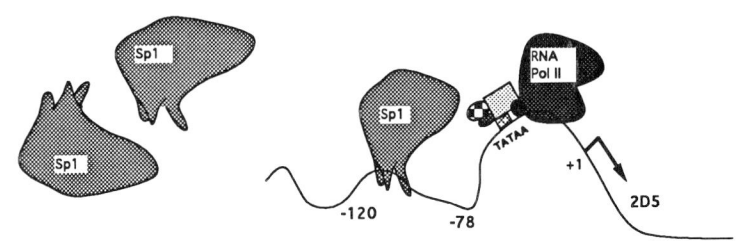

## presence of both Sp1 and C/EBPβ

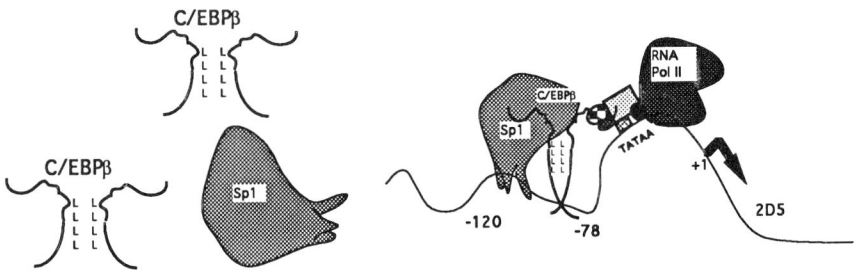

Fig. 4. Model for gene regulation by Sp1 and C/EBPβ.

References:

Clevidence, D.E., Overdier, D.G., Tao, W., Qian, X., Pani, L., Lai, E., and Costa, R.H. (1993): Identification of nine tissue-specific transcription factors of the hepatocyte nuclear factor 3 forkhead DNA-binding domain family. *Proc. Natl. Acad. Sci. USA* 90: 3948-3952.

Gonzalez, F.J. (1992): Control of constitutively-expressed developmentally-activated rat hepatic cytochrome P450 genes. *Keio J. Med.* 41: 68-75.

Gonzalez, F.J., Liu, S.Y. and Yano, M. (1993): Regulation of cytochrome P450 genes: molecular mechanisms .*Pharmacogenetics* 3, 51-57, 1993.

Hankinson, O. (1993): Research on the aryl hydrocarbon (Dioxin) receptor is primed to take off. *Arch. Biochem. Biophys.* 300: 1-5.

Lai, E. and Darnell, J.E. (1991): Transcriptional control in hepatocytes: a window on development. *Trends Biol. Sci.* 16: 427-430.

Lee, Y-H., Yano, M., Liu, S.-Y., Matsunaga, E., Johnson, P.F. and Gonzalez, F.J. (1994): A novel *cis*-acting element controlling the rat CYP2D5 gene requiring cooperativity between C/EBPβ and an Sp1 factor. *Mol. Cell Biol.* (In press).

Swanson, H.I. and Bradfield, C.A. (1993): The AH-receptor:genetics, structure and function. *Pharmacogenetics* (In press)

Ueno, T. and Gonzalez, F.J. (1990): Transcriptional control of the rat hepatic CYP2E1 gene. *Mol. Cell Biol.* 10: 4495-4505.

Williams, S.C., Cantwell, C.A., Johnson, P.F. (1991): A family of C/EBP-related proteins capable of forming covalently linked leucine zipper dimers in vitro. *Genes & Dev.* 5: 1553-1567.

Yano, M., Falvey, E. and Gonzalez, F.J. (1992): Role of the liver-enriched transcription factor DBP in expression of the cytochrome P450 *CYP2C6* gene. *Mol. Cell Biol.* 12: 2847-2854.

# Expression and function of Ad4BP, a steroidogenic cell-specific transcription factor

Ken-Ichirou Morohashi, Shin-Ichiro Honda, Masatoshi Nomura, Koichi Takayama, Mizuki Hara, Tsuneo Omura

*Department of Molecular Biology, Graduate School of Medical Science, Kyushu University, Higashi-ku, Fukuoka 812, Japan*

Steroid hormones are synthesized from cholesterol through the multi-step reactions catalyzed by six forms of steroidogenic P-450s (Simpson & Waterman, 1988). We have investigated the following two important problems in the regulation of these P-450 (CYP) genes. (i) The mechanism for tissue-specific expression of the steroidogenic CYP genes. The CYP genes are expressed only in the steroidogenic tissues, such as adrenal cortex, ovary, and testis, but not in non-steroidogenic tissues. (ii) The mechanism for cAMP-dependent transcription of the genes. Several sequences have been reported to confer cAMP-dependent transcriptional activity to the steroidogenic CYP genes. As will be described below, the steroidogenic tissue-specific and cAMP-dependent transcription of these CYP genes can be explained mainly by the function of a tissue-specific transcription factor, Ad4BP.

(I) *Ad4 site in the upstream regions of all the steroidogenic P-450 genes.* The gene structures of all the steroidogenic CYP genes have been reported. Comparison of the nucleotide sequences of the upstream regions clearly indicated that several sequences are conserved among several animal species. As shown in Fig. 1 (upper panel), we identified six conserved sequences, Ad1 (CRE), Ad2, Ad3, Ad4, Ad5, and Ad6, in the upstream regions to -400 bp of the CYP11B genes with an exception of the rat CYP11B1 gene (Honda *et al.*, 1990; Morohashi *et al.*, 1990; Nomura *et al*, 1993). In the case of the CYP11A genes, conserved sequences, Ad4, -58/-50 element (Takayama *et al,*), and a region around -110 bp containing SF-3 (Rice *et al.*, 1990) and ASP/SP-1 sites (Momoi *et al.*, 1993), were also found in the upstream (Fig. 1, lower panel). Further analyses with all the other steroidogenic CYP genes revealed that Ad4 sites are present in all the steroidogenic CYP gene promoters and are conserved remarkably among animal species (Kitajima *et al.* unpublished). In our recent observation, Ad4 sites are also present in the distal promoters of the human CYP11A (-1.8 kb to -1.5 kb) and the bovine CYP11B (-1.5 kb to -1.1 kb) genes, which show their transcriptional activities in a steroidogenic cells-specific manner (Takayama *et al.*, unpublished). These distal promoters have a similar structure in which a b-ZIP binding site (CRE in the CYP11A and C/EBP in the CYP11B) are located between two Ad4 sites (Fig. 2). These

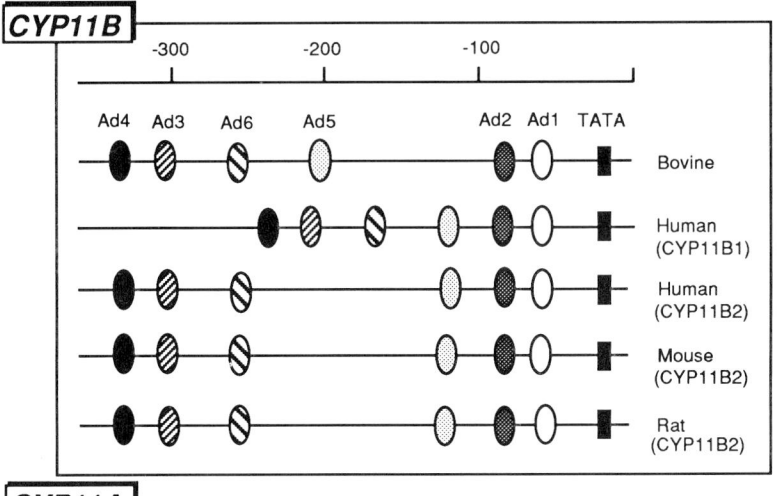

Fig. 1, Schematic presentation of the CYP11B and CYP11A gene promoters. (upper) The conserved nucleotide sequences, Ad1(CRE), Ad2, Ad3, Ad4, Ad5, and Ad6, were identified by an aliginment of the nucleotide sequences of the bovine (Kirita et al., 1990), human (Kawamoto et al., 1992), mouse (Mouw et al., 1989), and rat CYP11B genes. Ad4 site is also reffered as SF-1 (Rice et al. 1990). (lower) The nucleotide sequences of the human (Morohashi et al., 1987), mouse (Rice et al., 1990), rat (Oonk et al., 1990), and bovine (Ahlgren et al., 1990) were aligined to search conserved sequences. Cis-elements, -58/-50 and Ad4 in the human gene, SF-1, -2, and -3 in the mouse gene, and SP1 and ASP in the bovine gene were reported by Takayama et al., (submitted), by Rice et al. (1990), and Momoi et al. (1992), respectively. The bovine SP1 region was overlapped with the mouse SF-3.

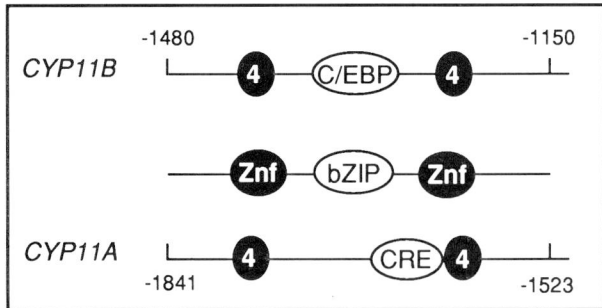

Fig. 2, Schematic presentation of the distal promoters of the human CYP11A and bovine CYP11B genes. The distal promoter of the CYP11A gene have two Ad4 sites and CRE, while that of the CYP11B also have two Ad4 sites and C/EBP site. These two distal promoters have similar structure containing b-ZIP binding sites between Ad4 sites.

observations strongly suggest the central role of Ad4 site in the transcriptional regulation of the steroidogenic CYP genes.

(II)  *Structure and Steroidogenic Cell-specific Expression of Ad4BP and ELP*
A cDNA clone coding for bovine Ad4BP was isolated from an adrenal cortex cDNA library (Honda et al., 1993). The nucleotide sequence showed that Ad4BP belongs to a steroid/thyroid hormone receptor super genefamily, which has a Zn-finger structure as the DNA binding domain (Evans, 1988). Comparison of the primary structures of the members of the family indicated that Ad4BP shows a striking homology with ELP (Tsukiyama et al., 1992) (Fig. 3). ELP was cloned from a cDNA library of mouse embryonal carcinoma (EC) cells as a mammalian counterpart of *FTZ-F1* (Lavorgna et al., 1991), which governs the segmentation of *Drosophila* embryo by regulating a *Fushitarazu* gene as a transcription factor in the early developmental stage. These two proteins, Ad4BP and ELP, have an almost identical long span of peptide at the central portion. Apparent differences can be seen in the N- and C-terminal portions, indicating that ELP lacks a phosphorylation site of protein kinase A and a heptad repeat conserved in the Zn-finger proteins. Determination of the genomic structure of the rat (Nomura et al. unpublished) and mouse (Ikeda et al., 1993) Ad4BP revealed that the two proteins are produced from a single gene by an alternative usage of the promoters and an alternative splicing.
As described above, Ad4BP is involved in the transcriptional regulation of all the steroidogenic CYP genes as the essential transcription factor, which strongly suggests the simultaneous expression of the P-450s and Ad4BP in the steroidogenic cells. The mRNAs prepared from several tissues were investigated by RT-PCR to see whether the tissues express Ad4BP. As expected, Ad4BP transcript was detected in all the steroidogenic tissues such as adrenal cortex, ovary, testis, brain, placenta, and adipocyte (Honda et al., 1993). Immunochemical and immunohistochemical studies using the antiserum raised to Ad4BP were performed with adrenal cortex, ovary and testis to confirm the steroidogenic cell-specific expressionof Ad4BP. As shown in Fig. 4, the steroidogenic cells in the three zones of adrenal cortex were positive for immunodetection (Morohashi et al., unpublished). Futher investigations with ovary and testis confirmed that granulosa and theca cells in follicle and corpus luteum, and Leydig cells also expressed Ad4BP.
As described above, Ad4BP and ELP are transcribed from a single gene. Although the ELP trascript was detected in mouse EC cells, the transcript has not been detected in animal tissues. Therefore, RT-PCR was performed with ELP-specific primer. Interestingly three steroidogenic tissues of adult rats examined so far expressed ELP as well as Ad4BP. A larger amount of Ad4BP than ELP was detected even in EC cells.

(III)  *Transcriptional activity of Ad4BP and ELP.*
To elucidate the function of Ad4BP in the steroidogenic cell-specific and cAMP-dependent transcription of the CYP genes, we carried out trans-activation experiments using a CAT reporter gene under the control of the CYP11B gene promoter, and an Ad4BP expression vector. These plasmids were cotrans-

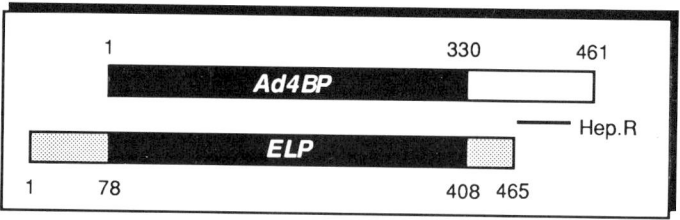

Fig. 3, Comparison of the primary structure between Ad4BP and ELP. Regions from 1st to 330th amino acids of Ad4BP and 78th to 408th amino acids of ELP are transcribed from the same exons (black boxes). Regions from 331st to 461st amino acids of Ad4BP indicated by a open box, and 1st to 77th and 409th to 465th amino acids of ELP indicated by shadowed boxes are resulted from an alternative usage of the promoters and alternative splicing. Heptad repeat is indicated by a bar.

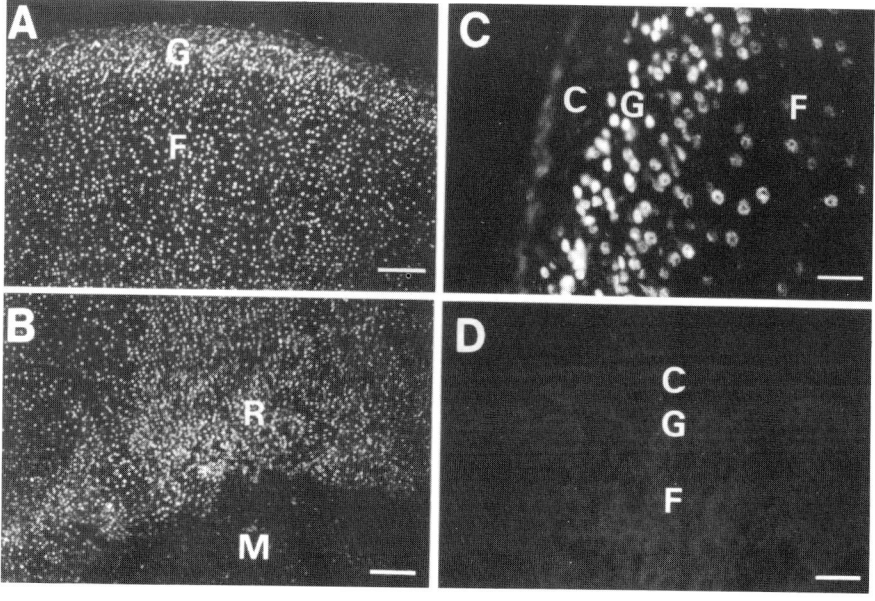

Fig. 4, Localization of Ad4BP in the rat adrenal. Frozen section of an adrenal gland of rat were stained with the antiserum to Ad4BP (panels A, B, and C). Pre-immune serum was used as the control (panel D). The outer cortical region in panel A is shown as higher magnification in C and D. G, F, R, M, and C denote the zona glomerulosa, zona fasciculata, zona reticularis, the medulla, and capsule, respectively. Bars=100 μm in panel A and B, and 25 μm in panel C and D.

fected into non-steroidogenic CV-1 cells in the presence or absence of the expression vector of protein kinase A (PKA) (Mellon et al. 1989) (Fig. 5). The CAT activity driven by the CYP11B promoter is low in the absence of Ad4BP expression vector. The activity was induced gradually, showing a good correlation with the increasing amounts of the expression vector of Ad4BP. PKA was necessary for the function of Ad4BP. Almost identical results were also obtained when the CYP11A gene promoter was used.

CRE (Montminy et al., 1986) is a representative *cis*-element regulating cAMP-dependent transcription as reported in many genes. Since CRE is in the upstream regions of both the CYP genes, the function of the element was examined with various OVEC reporter constructs (Morohashi et al., 1993). Interestingly, Ad4 site was found to have little transcriptional activity by itself. Copresence of Ad1(CRE) is essential for the Ad4 site to express the cAMP-dependent transcriptional activity. Recently, Watanabe et al. (1993) drew an identical conclusion concerning the cooperative function of Ad4 and CRE using the human CYP21 gene. These observations clearly indicated that Ad4BP is essential for the expression of the steroidogenic CYP genes; Ad4BP endowed steroidogenic cell-like ability to the non-steriodgenic CV-1 cells, and CRE was necessary for the cAMP (PKA)-dependent transcriptional activity of Ad4 site. As described above, comparison of the primary structures of Ad4BP and ELP revealed that they share an identical DNA binding domain inaccordance with the identical DNA sequences recognized by these two factors. Since ELP was also expressed in the steroidogenic cells, the transcriptional activity was investigated in CV-1 cells with the CYP11B gene promoter. Ad4BP activated the transcription in a PKA-dependent manner, whereas ELP failed to activate the transcription both in the presence or absence of PKA expression vector (Fig. 6). Several transcription factors were reported to have negative regulators encoded by a same gene. We, therefore, investigated the possibility whether ELP inhibits the transcription driven by Ad4BP. As shown in Fig. 6, the transcription activity of Ad4BP was inhibited by the copresence of ELP. Copresence of the transcriptional activator and repressor in the same cells provides an interesting mechanism of transcriptional regulation, in which the transcription is modulated by the competitive binding of the two factors to the same site.

We described here that Ad4 site and the binding protein, Ad4BP, are essential for the steroidogenic cell-specific and cAMP-dependent transcription of all the steroidogenic CYP genes. The critical role of Ad4BP as a steroidogenic cell-specific transcription factor was strongly indicated by the simultaneous location of Ad4BP and steroidogenic P-450s. However, the regulatory mechanisms of the steroidogenic CYP genes can not be fully understood only by the function of Ad4BP. Actually, ELP is present in the same steroidogenic cells as a negative regulator, and CREB/ATF proteins binding to Ad1(CRE) and PKA are necessary for the function of Ad4BP. Different regulatory mechanisms by other transcription factors such as NGFI-B (Wilson et al., 1993) or ASP (Kagawa et al., 1992) are proposed in the regulation of the mouse and human CYP21 genes. Consideration of the participation of these transcription factors

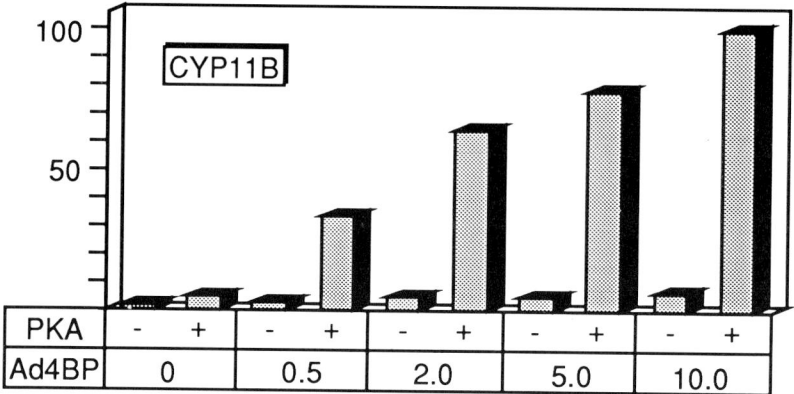

Fig. 5, Transcriptional activation of CYP11B gene by Ad4BP. Five µg of p7β1.5H-CAT (carring 1.5 kb upstream region of the bovine CYP11B gene) were transfected into non-steroidogenic CV-1 cells with incrreasing amount of an Ad4BP expression vector in the presence or absence of an expression vector of protein kinase A (PKA). The amounts of the acetylated products were indicated.

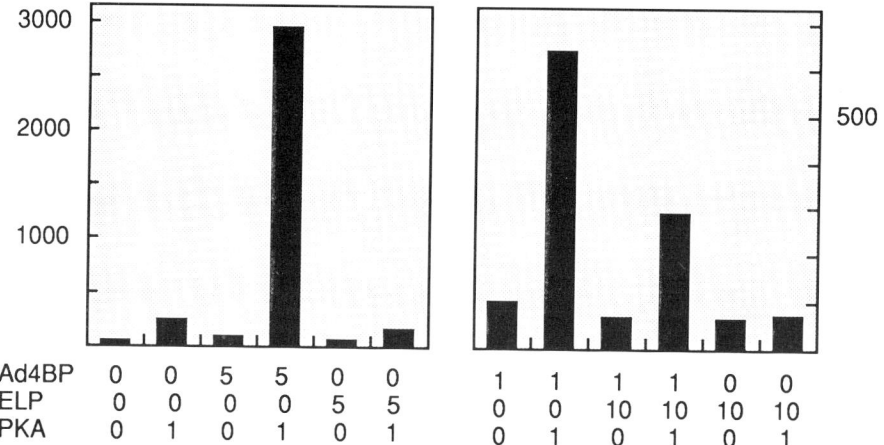

Fig. 6, Comparison of the transcriptional activities between Ad4BP and ELP. (Left panel) The indicated amounts of the expression vectors of Ad4BP and ELP were transcfected with or without PKA expression vector into CV-1 cells. The CAT gene controled by the bovine CYP11B gene promoter was used to assess the expressed transcriptional activities of Ad4BP and ELP.
(Right panel) The suppressive activity of ELP was examined. The indicated amounts of the expression vectors were cotransfected. The CAT activity in the presence of Ad4BP but in the absence of ELP and PKA was chose as one hundred units.

as well as Ad4BP is essential for deeper understanding of the regulatory mechanisms of the steroidogenic CYP genes.

Acknowledgment
The experiments described in this article using OVEC were performed as a cooperative study with Drs. U. M. Zanger (University of Basel) and M. R. Waterman (Vanderbilt University). Immunohistochemical study was done by Drs. H. Iida and Y. Shibata (Kyushu University). ELP expression vector was supplied by Dr. O. Niwa (Hiroshima University).

References
Ahlgren, R., Simpson, E. R., Waterman, M. R., & Lund, J. (1990): Characterization of the promoter/regulatory region of the bovine CYP11A (P-450scc) Gene. *J. Biol. Chem.* 265, 3313-3319.
Evans R. M. (1988): The steroid and thyroid hormone receptor superfamily. *Science* 240, 889-895.
Honda, S., Morohashi, K., & Omura, T. (1990): Novel cAMP Regulatory Elements in the promoter region of bovine P-450(11b) gene. *J. Biochem.* 108, 1042-1049.
Honda, S., Morohashi. K., Nomura, M., Takeya, M., Kitajima, M., & Omura, T. (1993): Ad4BP regulating steroidogenic P-450 genes is a member of steroid hormone receptor superfamily. *J. Biol. Chem.* 268, 7494-7502.
Ikeda, Y., Lala, D. S., Luo, X., Kim, E., Moisan, M., & Parker, K. L. (1993): Characterization of the mouse *FTZ-F1* gene, which encodes a key regulator of steroid hydroxylase gene expression. *Mol. Endocrinol.* 7, 852-860.
Kagawa, N. & Waterman, M. R. (1992): Purification and characterization of a transcription factor which appears to regulate cAMP responsiveness of the human CYP21B gene. *J. Biol. Chem.* 267, 25213-25219.
Kawamoto, T., Mitsuuchi, Y., Toda, K., Yokoyama, Y., Miyahara, K., Miura, S., Ohnishi, T., Ichikawa, Y., Nakao, K., Imura, H., Ulick, S., & Shizuta Y. (1992): Role of steroid 11β-hydroxylase and steroid 18-hydroxylase in the biosynthesis of glucocorticoids and mineralocorticoids in humans. *Proc. Natl. Acad. Sci. USA* 89, 1458-1462.
Lavorgna, G., Ueda, H., Clos, J., & Wu, C. (1991): FTZ-F1, a steroid hormone receptor-like protein implicated in the activation of *fushi tarazu* . *Science* 252, 848-851
Mellon, P. L., Clegg, C. H., Correll, L. A. & McKnight G. S. (1989): Regulation of transcription by cyclic AMP-dependent protein kinase. *Proc. Natl. Acad. Sci. USA* 86, 4887-4891.
Momoi, K., Waterman, M. R., Simpson, E. R., & Zanger, U. M. (1992): 3', 5'-Cyclic adenosine monophosphate-dependent transcription of the CYP11A (cholesterol side chain cleavage cytochrome P-450) gene involves a DNA response element containing a putative binding site for transcription factor Sp1. *Mol. Endocrinol.* 6, 1682-1690.
Montminy, M. R., Sevarino, K. A., Wagner, J. A., Mandel, G., & Goodman, R. H. (1986): Identification of a cyclic-AMP responsive element within the rat somatostatin gene. *Proc. Natl. Acad. Sci. USA* 83, 6682-6686.

Morohashi, K., Sogawa, K., Omura, T., & Fujii-Kuriyama, Y. (1987): Gene structure of human cytochrome P-450(SCC), cholesterol desmolase. *J. Biochem.* 101, 879-887.

Morohashi, K. & Omura, T. (1990): Tissue-specific transcription of P-450(11β gene *in vitro*. *J. Biochem.* 108, 1050-1056.

Morohashi, K., Honda, S., Inomata, Y., Handa, H., & Omura, T. (1992): A common *trans*-acting factor, Ad4-binding protein, to the promoters of steroidogenic P-450s. *J. Biol. Chem.* 267, 17913-17919.

Morohashi, K., Zanger, U. M., Honda, S., Hara, M., Waterman, M. R., & Omura, T. (1993): Activation of CYP11A and CYP11B gene promoters by the steroidogenic cell-specific transcription factor, Ad4BP. *Mol. Endocrinol.* in press.

Morohashi, K., Iida, H., Nomura, M., Honda, S., Tsukiyama, T., Niwa, O., Hara, T., Shibata, Y., & Omura, T. Functional difference between Ad4BP and ELP, and their distribution in the steroidogenic cells. *submitted*

Mouw, A. R., Rice, D. A., Meade, J. C., Chua, S. C., White, P. C., Schimmer B. P., & Parker, K. L. (1989): Structural and functional analysis of the promoter region of the gene encoding mouse steroid 11βhydroxylase gene. *J. Biol. Chem.* 264, 1305-1309.

Nomura, M., Morohashi, K., Kirita, S., Nonaka, Y., Okamoto, M., Nawata, H., & Omura, T. (1993): Three forms of rat CYP11B genes. *J. Biochem.* 113, 144-152.

Oonk, R. B., Parker, K. L., Gibson, J. L., & Richards, J. S. (1990): Rat cholesterol side-chain cleavage cytochrome P-450 (P-450scc) gene. *J. Biol. Chem.* 265, 22392-22401.

Rice, D. A., Kirkman M. S., Aitken, L. D., Mouw, A. R., Schimmer, B. P., & Parker, K. L. (1990): Analysis of the promoter region of the gene encoding mouse cholesterol side-chain cleavage enzyme. *J. Biol. Chem.* 265, 1173-11720.

Simpson, E. R. and Waterman, M. R. (1988): Regulation of the synthesis of steroidogenic enzymes in adrenal cortical cells by ACTH. *Annu Rev Physiol.* 50, 427-440.

Takayama, K., Morohashi, K., Honda, S., Hara, N., & Omura, T. Characterization of the transcription regulatory regions identified in the human CYP11A and bovine CYP11B gene promoter. *submitted*

Tsukiyama, T., Ueda, H., Hirose, S., & Niwa, O. (1992): Embryonal long terminal repeat-binding protein is a murine homolog of FTZ-F1, a member of the steroid receptor superfamily. *Mol. Cell Biol.* 12, 1286-1291

Watanabe, N., Kitazume, M., Fujisawa, J., Yoshida, M., & Fujii-Kuriyama, Y. (1993): A novel cAMP-regulatory region including a sequence like the cAMP-responsive element, far upstream of the human CYP21A2 gene. *Eur. J. Biochem.* 214, 521-531.

Wilson, T. E., Mouw, A. R., Weaver, C. A., Milbrandt, J., & Parker, K. L. (1993): The orphan nuclear receptor NGFI-B regulates expression of the gene encoding steroid 21-hydroxylase. *Mol. Cell. Biol.* 13, 861-868.

# DNA-binding regulatory factors and inducible expression of the P4501A1 gene

Kazuhiro Sogawa, Natsuki Matsushita, Masatugu Ema, Yoshiaki Fujii-Kuriyama

*Department of Chemistry, Faculty of Science, Tohoku University, Sendai, Japan*

Ah receptor and Arnt are helix-loop-helix proteins and are involved in the inducible expression of the P4501A1 gene. Association of the Ah receptor and Arnt was examined in vitro using $^{35}$S-labeled Arnt and C4 cell cytosol fraction. Cotransfection of the expression plasmids of Ah receptor and Arnt showed more activated transcription from a reporter gene consisting of the P4501A1 gene upstream sequence and the CAT structural gene than transfection with either of the two plasmids. The transcription enhancing activity of the Ah receptor and Arnt was investigated. The transcription activation domain of the two proteins was localized in the C-terminal region of the molecules.

The P4501A1 gene contains at least two kinds of cis-acting DNA elements. One is BTE, a GC box sequence, localized immediately upstream of the TATA box, and is necessary for high level of the gene expression (Yanagida et al., 1990). As the binding factor, BTEB and BTEB2 as well as Sp1 were reported (Imataka et al., 1992; Sogawa et al., 1993). The other is XRE. XRE functions as inducible enhancer and is presumed to be a binding site of the Ah receptor. The position and orientation of XREs in the human and rat P4501A1 genes are conserved well. (Fig.1) (Kubota et al., 1991). Recently Arnt (Ah receptor nuclear translocator) is found to be a binding component to XRE (Reyes et al., 1992). Structural analysis of the factors revealed that both proteins contain helix-loop-helix domain at the N-terminal region, suggesting strongly that the two proteins may form a heterodimer (Hoffman et al., 1991; Ema et al., 1992; Burbach et al., 1992).
We examined this possibility in vitro using synthesized $^{35}$S-labeled Arnt and Hepa-1 C4 cell cytosol fraction. Fig. 2A shows that radiolabeled Arnt was precipitated with Ah receptor by treatment of the anti-Ah receptor antibody, when the cytosolic fraction from nontreated C4 cells was incubated with in vitro synthesized Arnt in the presence of 3-MC. This reaction mixture also gave the

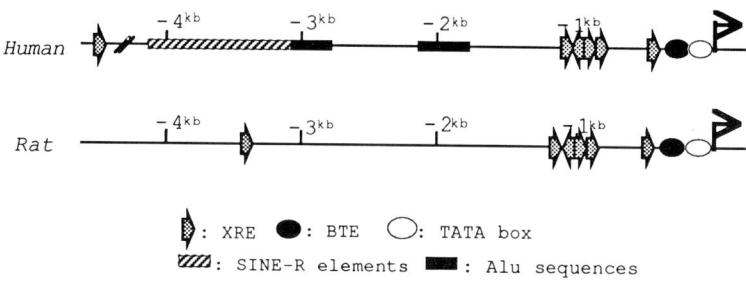

Fig. 1. Schematic representation of various elements in the 5'-flanking region of the human and rat P4501A1 gene. Positions of the XRE, BTE, Alu sequences, and SINE-R elements are indicated on the line. The most upstream XRE of the human P4501A1 gene is putative. Right-angled arrows show the transcription initiation site. Sizes are in kb.

retarded band with the XRE sequence in the gel mobility shift assay (data not shown, Matsushita et al., 1993). Since C4 mutant cells were reported to have the functionally normal Ah receptor and defective Arnt (Hankinson, 1983; Legraverend et al., 1982), these results clearly indicate that in vitro synthesized Arnt was able to be substituted for the cellular defective Arnt in the C4 cells to form the XRE-binding complex in association with the cellular Ah receptor. In contrast, the complex formation of the Ah receptor and Arnt was not observed in the incubation in the absence of the inducer, because Arnt was not immunoprecipitated with anti-Ah receptor antibody (Fig. 2B).

We constructed exression plasmids of the Ah receptor and Arnt proteins under the control of the cytomegalovirus promoter and enhancer and transfected them either separately or in combination into CV-1 cells with a reporter plasmid pMC6.3k consisting of the P4501A1 upstream sequence and CAT structural gene (Fujisawa et al., 1986). The cell transfected with these expression plasmids and the reporter plasmid were cultured for 40 h in the presence or absence of the inducer and then the expressed CAT activity was determined as shown in Fig. 3. The CAT activity expressed in the cells transfected with both the effector plasmids was synergistically increased as compared with those expressed in the cells transfected with either alone. Although the synergistic effect was very distinct and the inducibility was reproducible, the inducibility of the CAT expression seemed lower (2.2-fold) than that observed

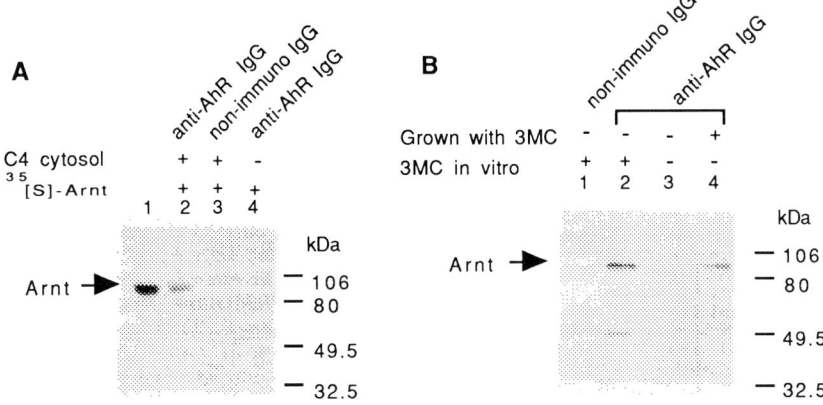

Fig. 2. Coimmunoprecipitation of in vitro synthesized Arnt by anti-Ah receptor antibody. A, $^{35}$S-labeled Arnt (lane 1) synthesized in vitro by reticulocyte lysates was incubated with (lane 2 and 3) or without (lane 4) cytosolic extracts prepared from nontreated C4 cells in the presence of 3-MC (1μM). Immunoprecipitates by anti-Ah receptor IgG (lane 2 and 4) or nonimmune IgG (lane 3) were analyzed by 7.5 % SDS-PAGE. B, $^{35}$S-labeled Arnt was incubated with cytosolic extracts of nontreated C4 cells in the presence (lane 1 and 2) or absence (lane 3) of 3-MC as above, and incubated with cytosolic extracts of 3-MC-treated C4 cells (lane 4). Immunoprecipitates by anti-Ah receptor IgG (lanes 2-4) or nonimmune IgG (lane 1) were analyzed by 7.5 % SDS-PAGE.

with Hepa-1 cells. The reason for this low inducibility was not known and remains to be seen. Anyhow, this result strongly suggests that the two proteins function cooperatively as a transcription activator in a complex formation.

In order to investigate the transcription enhancing activity of Arnt, we have constructed a chimeric plasmids which encodes GAL4 DNA binding domain (1-147) (Sadowski and Ptashne, 1989) and Arnt, and transfected the plasmid into CV-1 or COS cells with a reporter plasmid containing five GAL4 DNA binding sites in its promoter. The chimeric protein enhanced the CAT activity more than 100-fold over the control, suggesting that Arnt has a ranscription enhancing activity. Deletion analysis of the chimeric plasmid revealed that the transcription activation domain of Arnt was localized in the sequence of 35 amino acids in the C-terminal region. Fig. 4 shows amino acid sequence of the C-terminal region. Seven out of 35 amino acids are acidic amino acids, and

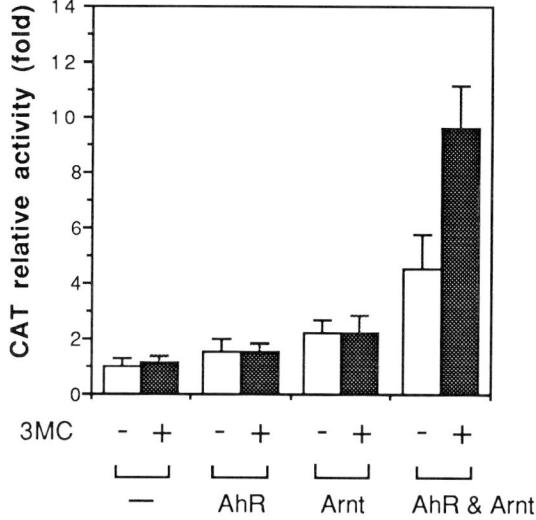

Fig. 3. Effects of co-transfection of Ah receptor and Arnt expression plasmids. The cells were incubated in the presence (+) or absence (-) of 1μM 3-MC for 40 h and then the expressed CAT activity was assayed. Results are average of three independent experiments. The data are expressed in ratios of the expressed CAT activities to that of pCMSV expressed in the absence of 3-MC. Bars on the top of histgrams indicate standard errors.

755QPEVFQEMLSMLGDQSNS

YNNEEFPDLTMFPPFSE789

Fig 4. Amino acid sequence of the C-terminal region of the Arnt protein. Acidic amino acids are outlined.

no basic amino acids are present. This feature of the sequence suggests that the transcription activation domain of the Arnt protein may be classified into acidic activation domain (Hahn, 1993). We also constructed a GAL4-Ah receptor chimeric plasmid for analysis of the transcription enhancing activity. The Ah receptor possessed transcription activation domains in the C-terminal half of the molecule.

This work was supported in part by a Grant-in Aid for Scientific Research on Priority Areas from the Ministry of Education, Science and Culture of Japan.

Burbach, K. M., Poland, A., and Bradfield, C. A. (1992): Cloning of the Ah-receptor cDNA reveals a distinctive ligand-activated transcription factor. Proc. Natl. Acad. Sci. USA. 89, 8185-8189.

Ema,, M., Sogawa, K., Watanabe, N., Chujoh, Y., Matsushita, N., Gotoh, O., Funae, Y., and Fujii-Kuriyama, Y. (1992): cDNA cloning and structure of mouse putative Ah receptor. Biochem. Biophys. Res. Commun. 184, 246-253.

Fujisawa-Sehara, A., Sogawa, K., Nishi, C., and Fujii-Kuriyama, Y. (1986): Regulatory DNA elements localized remotely upstream from the drug-metabolizing cytochrome P-450c gene. Nucleic Acids Res. 14, 1465-1477.

Hahn, S. (1993): Structure (?) and function of acidic transcription activators. Cell 72, 481-483.

Hankinson, O. (1983): Dominant and recessive aryl hydrocarbon hydroxylase-deficient mutants of mouse hepatoma line, Hepa-1, and assignment of recessive mutants to three complementation groups. Somatic Cell Genet. 9, 497-514.

Hoffman, E. C., Reyes, H., Chu, F.-F., Sander, F., Conley, L. H., Brooks, B. A., and Hankinson, O. (1991): Cloning of a factor required for activity of the Ah (Dioxin) receptor. Science, 252, 954-958.

Imataka, H., Kazuhiro, K., Yasumoto, K., Kikuchi, Y., Sasano, K., Kobayashi, A., Hayami, M., and Fujii-Kuriyama, Y. (1992): Two regulatory proteins that bind to the basic transcription element (BTE), a GC box sequence in the promoter region of the rat P-4501A1 gene. EMBO J. 11, 3663-3671.

Kubota, M., Sogawa, K., Kaizu, Y., Sawaya, T., Watanabe, J., Kawajiri, K., Gotoh, O., and Fujii-Kuriyama, Y. (1991): Xenobiotic responsive element in the 5'-upstream region of the human P-450c gene. J. Biochem. 110, 232-236.

Legraverend, C., Hannah, R. R., Eisen, H. J., Owens, I. S., Nebert, D. W., and Hankinson, O. (1982): Regulatory gene product of the Ah locus. Characterization of receptor mutants among mouse hepatoma clones. J. Biol. Chem. 257, 6402-6407.

Matsushita, N., Sogawa, K., Ema, M., Yoshida, A., and Fujii-Kuriyama, Y. (1993): A factor binding to the xenobiotic responsive element (XRE) of P-4501A1 gene consists of at least two helix-loop-helix proteins, Ah receptor and Arnt. J. Biol. Chem. 268, 21002-21006.

Reyes, H., Reisz-Porszasz, S., and Hankinson, O. (1992): Identification of the Ah receptor nuclear translocator protein (Arnt) protein as a component of the DNA binding form of the Ah receptor. Science 256, 1193-1195.

Sadowski, I. and Ptashne, M. (1989): A vector for expressing GAL4(1-147) fusions in mammalian cells. Nucleic Acids Res. 17, 7539.

Sogawa, K., Imataka, H., Yamasaki, Y., Kusume, H., Abe, H., and Fujii-Kuriyama, Y. (1993): cDNA cloning and transcriptional properties of a novel GC box-binding protein, BTEB2. Nucleic Acids Res. 21, 1527-1532.

Yanagida, A., Sogawa, K., Yasumoto, K., and Fujii-Kuriyama, Y. (1990): A novel cis-acting DNA element required for a high level of inducible expression of the rat P-450c gene. Mol. Cell. Biol. 10, 1470-1475.

# Induction of brain P450 by solvents and pharmaceuticals

Margaret Warner, Maria Strömstedt, Adrian Wyss, Shigetaka Yoshida, Jan-Åke Gustafsson

*Department of Medical Nutrition, Karolinska Institute, Novum, F60 141 86 Huddinge, Sweden*

The discovery of 1-methyl-4-phenyl-1,2,3,6-tetrahydropyridine (MPTP) as a Parkinsonian neurotoxin (Langston et al., 1983) has clearly established that xenobiotics can be activated by enzymes within the central nervous system (CNS) to neurotoxins and has further stimulated interest in the role of brain P450 in degenerative diseases of the CNS. The presence of P450 in any tissue is associated with a risk of toxicity because during the catalytic cycle reactive intermediates can be formed from oxygen or from substrates and these can be cytotoxic and/ or carcinogenic. P450 is present in the brain but there is some uncertainty about the quantity and the presence of hepatic forms of P450 in the brains of untreated animals. The available data are summarized in Table 1

Some epidemiological studies indicate a correlation between poor metabolizers of debrisoquine (DB) and early onset of Parkinson's disease (Armstrong et al.,1992; Comella et al.,1987). This would suggest a role of P450 2D in detoxification of some as yet unidentified neurotoxin. However, this correlation is not a general finding (Kallio et al.,1991). Both MPTP and tetrahydroisoquinoline (TIQ), another substance which induces Parkinsonism when administered to monkeys (Ohta,1989; Yoshida et al., 1990), are metabolised by 2D. In rats, the metabolism of MPTP by 2D P450 reduces its CNS toxicity (Jimenez-Jimenez et al., 1991). Although a link between 2D phenotype and Parkinson's disease is not certain, it remains possible that some endogenous and/or exogenous compound capable of causing degenerative diseases in the CNS is activated or rendered inactive through metabolism by cytochrome P450.

Table 1 CATALYTIC ACTIVITIES IN BRAIN MICROSOMES OF UNTREATED RATS

| Catalytic activity n mol/mg microsomal protein /min | Reference # | Brain | Liver | Brain P450 pmol/mg |
|---|---|---|---|---|
| aminopyrine N-demethylase | Marietta et al. (1979) | 0.026 | 2.93 | 35 |
| | Anandatheerhavarada et al. (1990) | 173 | 107 | |
| meperidine N-demethylase | Marietta et al. (1979) | 0.0023 | 6.8 | |
| hexobarbital hydroxylase | Marietta et al. (1979) | 0.0056 | 21.3 | |
| benzo(a)pyrene hydroxylase | Guengerich and Mason (1979) | 0.004 | 0.22 | 17 |
| | Marietta et al. (1979) | 0.0035 | 0.039 | |
| | Anandatheerhavarada et al. (1990) | 0.01 (1A) | 1.78 | |
| naphthalene hydroxylase | Mesnil et al. (1988) | 0.246 | x 200 | |
| ethoxyresorufin | Walther et al. (1987) | 0.0036 | | 17 |
| AHH | Das et al. (1982) | 0.0016 | | |
| morphine N-demethylase | Anandatheerhavarada et al. (1990) | 128 (2B) | 74 | |
| ethoxycoumarin O-deethylase | Näslund et al. (1988) | 0.001 (2B) | | |
| | Guengerich et al. (1979) | 0.0017 | 1.2 | |
| | Qato and Maines (1985) | 0.0088 | | 3-5 |
| | Rouet et al. (1981) | 0.17 | 3.1 | |
| | Ravindranath et al. (1990) | 1.6 | 6.5 | 88 |
| | Srivastava and Seth (1983) | 0.0016 | 0.353 | |
| aniline hydroxylase | Anandatheerhavarada et al. (1993) | 1.4 (2E1) | 36 | |
| p-nitrophenol hydroxylase | Anandatheerhavarada et al. (1993) | 0.07 (2E1) | 0.78 | |
| N-nitroso-dimethylamine N-demethylase | Anandatheerhavarada et al. (1993) | 0.8 (2E1) | 0.64 | |
| pentoxyresorufin O-dealkylase | Anandatheerhavarada et al. (1993) | 0.012 (2B) | | |
| codein N-demethylase | Kodaira & Spector (1989) | 0.00023 | 0.170 | |
| aromatase | Roselli et al. (1984) * | 0.0006 | | |
| 5α-androstane-3β,17β-diol hydroxylase. | Warner et al (1989) | 0.20 | - | |
| DHEA 7α- hydroxylase | Akwa et al. (1992) | 0.322 | | |
| bufuralol hydroxylase | Fonne-Pfister (1987) | 0.00084 | 1.29 | |
| estrogen 2/4-hydroxylase | Brown et al. (1985) | 0.0002 | 2.95 | |
| Testosterone hydroxylations 6α, 15β, 15α, 6β, 16β, 1β, 2β | Jayyosi et al. (1992) | 0.00016-0.002323 | | |

*Activity was expressed per mg tissue in the paper and has been converted using the value of 10 mg microsomal protein / g tissue

It is possible that forms of P450 other than 2D can activate or inactivate neurotoxins and that it is the pattern of P450 isozymes and not only the DB phenotype which is critical. Identification of the constitutive forms of P450 in the brain as well as inducibility by diet, pharmaceuticals, and xenobiotics of forms not normally present in the brain is, therefore, of importance. In this study we show that several hepatic forms of P450, which are not normally detectable in the brain, can be induced in the brain of rats upon exposure of animals to solvents and pharmaceuticals and when liver function is compromised. The induction of these forms of P450 in the brain, particularly the effects of ethanol, may influence xenobiotic-induced CNS toxicity and may be a factor in age-related degenerative diseases of the CNS.

## Results and Discussion

QUANTITATION AND IDENTIFICATION OF THE FORMS OF P450 IN THE CONTROL RAT BRAIN

P450 was partially purified from solubilized brain membranes and spectrally quantitated according to Omura and Sato (1964). The yield of P450 was 30-50 pmol/g wet weight of tissue. We have used antibodies against P450 1A1, 2A, 2B1, 2C11, 2C12, 2E2, and 3A on Western blots and shown that none of these antibodies recognise bands of the correct molecular weight in brain P450 preparations even when 100 pmol is loaded per lane. All of the antibodies recognise bands of the correct size when 10 pmol of liver P450 is loaded in each lane. We conclude from these results that these hepatic forms of P450, if they are present in the brain, account for a minor part of brain P450.

Despite this, when the same antibodies are used in immunohistochemical studies, we find extensive staining of the brain. The staining pattern is similar to that we have reported previously (Warner et al., 1988) and that observed in other laboratories (Kapitulnik et al.,1987; Hansson et al., 1990; Cammer et al.,1991).

There are two possible explanations for this apparent discrepancy: 1) The P450 isolated in our studies is not representative of the overall P450 in the brain and there is selective loss of certain forms during purification. 2) The widespread immunohistochemical staining is not a reflection of P450 but recognition of epitopes on unrelated proteins. In view of the numerous publications on immunohistochemical localization of P450 in the brain and the reports from some laboratories of the abundance of P450 1A, 2B, and 2E1 in the rat brain (Table 1), we have tried to resolve this issue.

If the problem of lack of detection of the hepatic forms of P450 in the brain is due to selective loss of specific forms of P450 during purification, then these forms should be detectable on Western blots with brain microsomes. Indeed, in some laboratories P450 1A, 2B, and 2E are detectable on Western blots with brain microsomes. We routinely find that the yield of microsomal protein from the brain is approximately 10mg/ g wet weight of tissue. When 200mg of brain microsomal protein is loaded per lane, and the blots are probed with the above mentioned antibodies, no signals are visible. This is not surprising if the content of brain P450 is 3-5 pmol/mg microsomal protein since it means that 0.6-1 pmol P450 were loaded per lane and in this case only the most abundant forms of P450 would be detectable. However, if brain microsomes contain 100 pmol P450 /mg microsomal protein as found in some labs, there would be 20 pmol/ lane and the most abundant forms would be detected.

The reason for the high level of P450 in the rat brains in some laboratories is not clear. It could be a difference in rat strain, diet or some other environmental factors. We can only conclude that the Wistar and Sprague Dawley rats used in our laboratory, do not have high levels of hepatic forms of P450 in their brains.

INDUCTION OF HEPATIC FORMS OF P450 IN THE BRAIN

Ethanol and isopropanol when administered IP (1ml/kg) cause a 3-5 fold increase in the level of P450 in the brain. On Western blots, with 20 pmol of this P450, several hepatic forms can be detected. After ethanol treatment these are 2C, 2D, 2E1 and 4A and after isopropanol they are 1A, 2C, 2E1, 2D, 3A and 4A. The N-terminals of the proteins induced after ethanol treatment have been sequenced and the identities of the P450s confirmed. These studies indicate that the hepatic forms of P450 can survive the purification procedure. Interestingly, when immunohistochemistry is done with the same antibodies, there is no difference in the staining pattern in the brain between the controls and the ethanol-treated animals. These results indicate that the immunohistochemical staining does not reflect the presence of P450 in the brain but is more likely the reaction of the antibodies with epitopes on unrelated proteins.

The level of P450 in the brain is also increased by long-term treatment with the neuroleptic drug, chlorpromazine, and upon removal of 60% of the liver. In this case of chlorpromazine treatment, P450 3A was induced and 16h after removal of 60% of the liver P450 1A, 2C, 2D, and 4A were induced.

The question remains, therefore, about the identity of the major constitutive forms of P450 in the rat brain. We have identified three of these forms. One is P450 4A2, another is 2D 1/5 and the third is 5$\alpha$-androstane-3$\beta$,17$\beta$-diol hydroxylase.

P450 4A2 can be detected on Western blots with P450 partially purified from brain membranes. By comparison with the liver 4A signals, it can be concluded that 4A represents less than 5% of brain P450. The N-terminal of P450 2D1or 2D5 was sequenced from proteins isolated from the brains of control rats. Because of the similarity between these two proteins it was not possible to determine which of the enzymes is present. The third constitutive form of brain P450 (3$\beta$-diol hydroxylase) can account for up to 50% of the P450 in some brain areas (Warner et al., 1989). A novel function for this P450 in the brain has been found. It is the regulation of the levels of the potent anaesthetic steroid $3\alpha,5\alpha$-tetrahydro progesterone ($3\alpha,5\alpha$-THP). We have demonstrated that $3\beta,5\alpha$-tetrahydro progesterone ($3\beta,5\alpha$-THP) is a very good substrate for this constitutive form of brain P450. In analogy with the elimination of androgens from target tissues the $3\alpha$-steroid represents a pool of active steroid while the $3\beta$-steroids are rapidly hydroxylated and represent the major route of elimination. When the hydroxylation of $3\beta,5\alpha$-THP is inhibited, the duration of anaesthesia of an administered dose of $3\alpha,5\alpha$-THP is prolonged. We conclude from these results that a major form of P450 in the brain has an important role in regulating the level of steroid which influences mood, sleep/wakefulness and other functions related to the $GABA_A$ receptor.

Both 4A P450 and $3\beta$-diol hydroxylase are present in the brains of newborn rats ( 2-7days old). Interestingly no P450 2B was detected in brain P450 in induced or uninduced rats even though this has been found to be a major form of brain P450 in other laboratories (Table 1).

Conclusions
Many forms of P450, not normally expressed in the brain, can be induced upon intake of solvents, pharmaceuticals and when liver function is compromised. P450 in the brain may, therefore, be important in the in situ activation and/ or inactivation of xenobiotics and must be considered as a potentially important factor in xenobiotic induction of degenerative diseases of the CNS.

Acknowledgements
This work was supported by grants from the Work Health Environment Fund and from the Swedish Medical Research Council (Nos. 13x-06807 and B92-03P-08561-04A).


References
Akwa, Y., Morfin, R.F., Robel, P. & Baulieu, E-E (1992): Neurosteroid metabolism. $7\alpha$-hydroxylation of dehydroepiandrosterone and pregnenolone by rat brain microsomes. *Biochem. J.* 288, 959-964.
Anandatheerthavarada, H. K., Shankar,S.K., & Ravindranath, V. (1990): Rat brain P-450 : catalytic, immunochemical properties and inducibility of multiple forms. *Brain Res.* 536, 339-343.
Anandatheerthavarada, H.K., Shankar,S.K., Bhamre, S., Boyd, M.R., Song, B-J. & Ravindranath, V. (1993): Induction of brain cytochrome P450 2E1 by chronic ethanol treatment. *Brain Research* 601, 279-285.
Anandatheerthavarada, H.K., Williams, J.F. & Wecker, L. (1993): The chronic

administration of nicotine induces cytochrome P450 in rat brain. *J. Neurochem.* 60 1941-1944.

Armstrong, M., Daly, A.K., Cholerton, S., Bateman,D.N., Idle, J.R. (1992): Mutant debrisoquine hydroxylation genes in Parkinson's disease. *Lancet* 339, 1017-1018.

Baron, J.A. (1986): Cigarette smoking and Parkinson's disease. *Neurology* 36, 1490-1496.

Brown, C. G., White, N. & Jefcoate (1985): Oestrogen-2/4-hydroxylase activity in the rat brain during lactation and throughout the oestrus cycle. *J. Endocrinol.* 107, 191-196.

Cammer, W., Downing, M., Clarke, & W., Schenkman, J. B. (1991): Immunocytochemical staining of the RLM6 form of cytochrome P450 in oligodendrocytes and myelin of rat brain. *J. Histochem. Cytochem.* 39, 1089-1094.

Comella,C., Tanner, C.M., Goetz, C.G., Gans, S., Rapp, D., D., Fischer, J.(1987): Debrisoquine metabolism in Parkinson's disease. *Neurology* 37, 261-262.

Das, M., Seth, P.K., Dixit, R. & Mukhtar, H. (1982): Aryl hydrocarbon hydroxylase of rat brain mitochondria: properties of and effect of inhibitors and inducers on enzyme activity. *Arch. Biochem.Biophys.* 217, 205-215.

Fonne-Pfister, R., Bargetzi, M.J., & Meyer, U. (1987): MPTP, the neurotoxin inducing Parkinson's disease, is a potent competitive inhibitor of human and rat cytochrome P450 isozymes (P450 buf1, P450 db1) catalyzing debrisoquine 4-hydroxylation. *Biochem. Biophys. Res. Commun.* 148, 1144-1150.

Guengerich, F.P. & Mason, P.S. (1979): Immunological comparison of hepatic and extrahepatic cytochromes P450. *Mol. Pharmacol.* 15, 154-164.

Hansson, T., Tindberg, N., Ingelman-Sundberg, M. & Köhler, C. (1990): Regional distribution of ethanol-inducible cytochrome P450 2E1 in the rat central nervous system. *Neuroscience* 34, 451-463.

Jayyosi, Z.A., Cooper, K.O.A. & Thomas, P.E. (1992): Brain cytochrome P450 and testosterone metabolism by rat brain subcellular fractions: presence of cytochrome P450 3A immunoreactive protein in rat brain mitochondria. *Archives of Biochemistry and Biophysics* 298, 265-270.

Jimenez-Jimenez, F.J., Tabernero, C., Mena, M.A., Garcia de Yebenes, J., Garcia de Yebenes, M.J., Casarejos, M.J., Pardo, B., Garcia-Agundez, J.A., Benitez, J., Martinez, A. (1991): Acute effects of 1-methyl-4-phenyl-1,2,3,6-tetrahydropyridine in a model of rat designated a poor metabolizer of debrisoquine. *J Neurochem* 57, 81-87.

Kallio, J., Marttila, R.J., Rinne, U.K., Sonninen,V., Syvalahti, E. (1991): Debrisoquine oxidation and Parkinson's disease. *Acta Neurol Scand* 83, 194-197.

Kapitulnik, J., Gelboin, H.V., Guengerich, F.P. and Jacobowitz, D. M. (1987). Immunohistochemical localization of cytochrome P450 in rat brain. Neuroscience 20, 829-833.

Kodaira, H. & Spector, S. (1988): Transformation of thebane to oripavane, codeine, and morphine by rat liver, kidney and brain microsomes. *Proc. Natl. Acad. Sci. USA.* 85, 1267-1271.

Köhler, C., Eriksson, L.G., Hansson, T., Warner, M. and Gustafsson, J-Å (1988): Immunohistochemical localization of cytochrome P450 in rat brain. *Neuroscience Letters* 20, 829-832.

Langston, J.W., Ballard, P., Tetrud, J.W., and Irwin, I. (1983): Chronic Parkinsonism in humans due to a product of meperidine-analog synthesis. *Science* 219, 979-980

Marietta, M.P., Vesell, E.S., Hartman, R.D., Weisz, J. & Dvorchik, B.H. (1979): Characterization of aminopyrine N-demethylation in rat brain: Comparison with hepatic aminopyrine N-demethylation. *J. Pharm. Exp. Ther.* 208, 271-279.

Mensil, M., Testa, B., & Jenner, P. (1988): In vitro inhibition by stiripentol of rat brain cytochrome P450- mediated naphthalene hydroxylation. *Xenobiotica* 18, 1097-1106.

Näslund, B., Glauman, H., Warner, M., Gustafsson, J-Å., and Hansson, T. (1987): Cytochrome P450 b and c in the rat brain and pituitary gland. *Mol. Pharmacol.* 33, 31-37.

Ohta, S. (1989): Tetrahydroisoquinolines in connection with Parkinson's disease. *Rinsho Shinkeigaku* 12, 1504-1506.

Omura, T. and Sato, R. (1964): The carbon monoxide-binding pigment of liver microsomes I. Evidence for its hemoprotein nature. *J. Biol. Chem.* 239, 2370-2378.

Qato, M.K.& Maines, M.D. (1985): Regulation of heme and drug metabolism activities in the brain by manganese. *Biochem. Biophys. Res. Commun.* 128, 18-24.

Ravindranath, V., Anandatheerhavarada, H.K., & Shankar,S. K. (1990): NADPH cytochrome P450 reductase in rat, mouse and human brain. *Biochem. Pharmacol.* 39, 1013-1018.

Roselli, C.E., Ellinwood, E.E. & Resko, J.A. (1984): Regulation of brain aromatase activity in rats. *Endocrinology* 114, 192-200.

Rouet, P., Alexandrov, K., Markovits, P., Frayssinet, C, & Dansette, P.M. (1981): Metabolism of benzo(a) pyrene by brain microsomes of fetal and adult rats and mice. Induction by 5,6-benzoflavone, comparison with liver and lung microsomal activities. *Carcinogenesis* 2, 919-926.

Srivastava, S.P. & Seth, P.K. (1983): 7-Ethoxycoumarin O-deethylase activity in rat brain microsomes. *Biochem. Pharmacol.* 32, 3657-3660.

Walther, B., Gheresi-Egea, J. M., Minn, A., and Siest, G. (1986): Subcellular distribution of cytochrome P450 in the brain. *Brain Res.* 375, 338-344.

Warner, M., Köhler, C., Hansson, T. and Gustafsson, J-Å. (1988): Regional distribution of cytochrome P450 in the rat brain:Spectral quantitation and contribution of P450 b,e and c,d. *J. Neurochem.* 50, 1057-1065.

Warner, M., Strömstedt, M., Möller, L. and Gustafsson, J-Å (1989 ): Distribution and regulation of 5α-androstane-3β,17β-diol hydroxylase in the rat central nervous system. *Endocrinology* 124, 2699-2706.

Yoshida, M., Niwa, T., Nagatsu, T. (1990): Parkinsonism in monkeys produced by chronic administration of an endogenous substance of the brain, tetrahydroisoquinoline: the behavioral and biochemical changes. *Neurosci Lett ers.* 119, 109-113.

# Hormonal influence on developmental changes in hepatic P450 in rat liver

Yasushi Yamazoe[1], Norie Murayama[1], Miki Shimada[1], Kiyoshi Nagata[1], Hiroshi Ishikawa[2], Hisashi Hashimoto[2], Kazuo Nakayama[3], Kazuhiro Sogawa[3], Yoshiaki Fujii-Kuriyama[3], Ryuichi Kato[1]

[1]Department of Pharmacology, School of Medicine, Keio University, Tokyo. [2]Department of Anatomy, Jikei University, School of Medicine, Tokyo. [3]Department of Chemistry, Faculty of Sciences, Tohoku University Sendai, Japan

Hepatic levels of cytochrome P450 form vary dramatically during the maturation of rats. The ontogenic profiles differ among individual forms, but may be divided into two types, neonatal P450 and pubertal P450, from the onset in the liver. The former group contains several P450s belonging to gene (sub)families of CYP1A, CYP2A, CYP2B, CYP2E, and CYP3A, while many members of CYP2C are included in the latter. Consistent with these ontogenic differences, both groups of hepatic P450s show opposite susceptibility toward pituitary growth hormone (GH) in hypophysectomized rats and also in spontaneous dwarf rats, which contain no detectable level of GH in the pituitary. These results, together with reported data on an age-related expression of the GH receptor in the liver, indicate that switching of hepatic P450 forms during puberty is mainly caused by the age-dependent changes in a signal transduction system mediated by the GH-GH receptor. In addition, GH and thyroid hormone were shown to interact with the CYP2B1 gene at the transcriptional step to suppress the activation using a CYP2B1-CAT fused gene.

## METHODS

Anti-P450 antibodies used were the same as described in our previous papers (Kamataki, T., et al., 1983a;1983b; Yamazoe, Y., et al., 1986;1987;1988;1989;1990). Liver microsomes were prepared from both sexes of the Sprague-Dawley strain of rats and from spontaneous dwarf rats. Specific P450 amounts were determined by Western blots using a tetramethylbenzidine-peroxidase or alkaline-phosphate staining method. Ovine growth hormone and human growth hormone were used to determine the effect on hepatic P450s (Kato, R., et al., 1986). For expression of CYP2B1-CAT(chloramphenicol acetyltransferase)-fused gene, constructs containing different lengths of 5'-flanking regions of the CYP2B1 gene were transfected in isolated hepatocytes of male rats.

RESULTS AND DISCUSSION

Developmental differences of hepatic P450

Total P450 amounts differ depending on the age and sex of the animals in livers: Total P450 contents determined by the CO-difference spectra are low in neonate, increase rapidly in puberty, and attain the mature levels at around 9 weeks of age. More than twenty forms of P450 have been isolated from rats. Developmental changes of each individual P450 form are not pleiotropic but rather differ in their profiles. Typical developmental profiles of P450 are shown in Fig. 1. Each P450 level is normalized into the maximum as 100% in this profile. Several P450 forms such as CYP2A1, CYP2B1 and CYP3A2 were detectable within 24 hours after birth in the livers of male (top) and female rats (bottom) (Yamazoe, Y., et al., 1987;

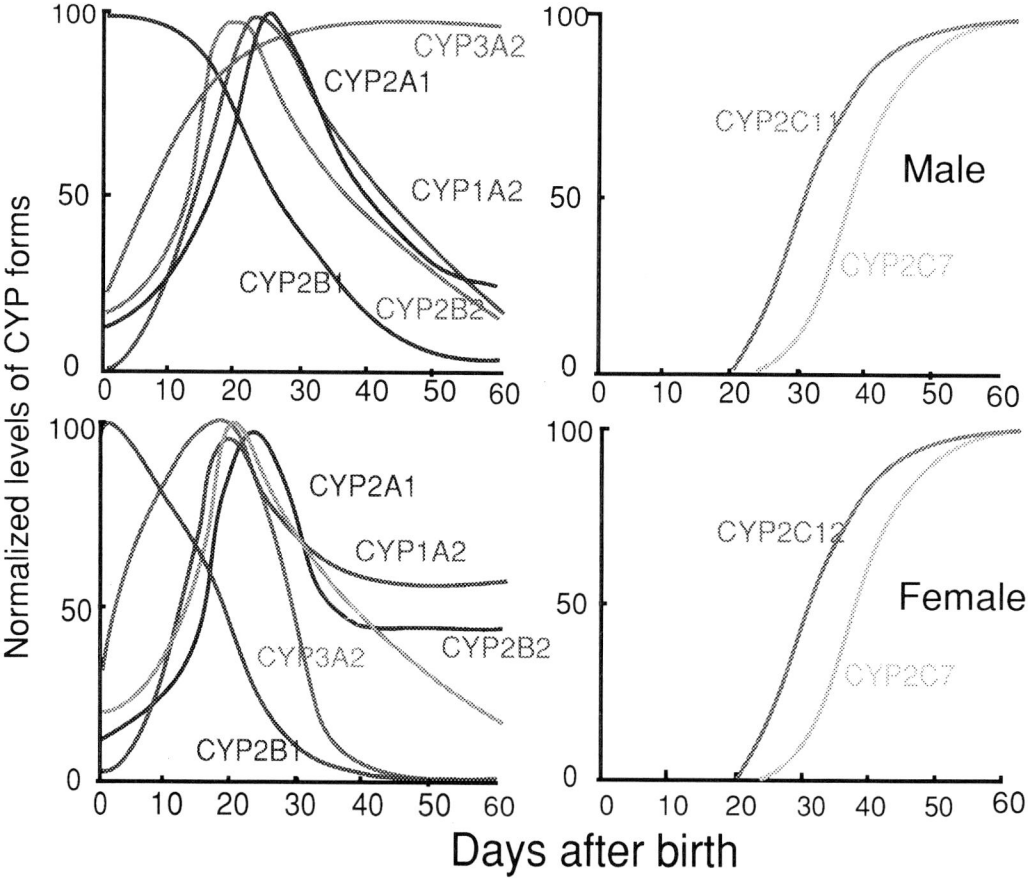

Fig. 1 Age-dependent changes of cytochrome P450 in livers of male and female rats

1988;1990). These P450s were increased rapidly and reached their maxima within 3 weeks of age. In pubertal periods, hepatic levels of CYP1A2, CYP2A1, CYP2B1, CYP2B2 and CYP3A2 were diminished or almost completely abolished in both male and female rats, except that CYP3A2 maintained high levels in the male animals (Nagata, K., et al., 1990a). A major CYP2C subfamily of P450s such as CYP2C7, CYP2C11 and CYP2C12 was, in contrast, undetected in neonate, but increased abruptly during 4 to 9 weeks of age (Kamataki, T., et al., 1983b; Sasamura, H., et al., 1990; Shimada, M., et al., 1987). These data indicate that hepatic P450s can possibly be divided into two classes, neonatal P450 and pubertal P450, from the period of the appearance in the liver. Thus, CYP1A2, CYP2A1, CYP2B1, CYP2B2, CYP2E1, CYP3A1 and CYP3A2 are classified in neonatal P450s, while several CYP2C P450s such as male-specific CYP2C11 and female-specific CYP2C12 in rat livers constitute pubertal P450.

Association between ontogenicity and growth hormone susceptibility

Studies on hormonal influences showed that pituitary GH affected hepatic levels of over ten different forms of P450s in hypophysectomized rat models (Kato, R., et al., 1986; MacGeoch, C., et al., 1985; Morgan, E.T., et al., 1985; Shimada, M., et al., 1989a;1989b; Yamazoe, Y., et al., 1986; 1989;1990). Microsomal contents of the pubertal P450s, CYP2C11 and CYP2C12, were decreased by hypophysectomy of male and female rats, respectively. Intermittent injection (mimicking the male secretory pattern) and continuous infusion of GH (mimicking the female secretory pattern) restored, respectively, CYP2C11 and CYP2C12 contents in hypophysectomized rats. Pituitary influence was not restricted on pubertal P450s. As we reported previously (Yamazoe, Y., et al., 1986;1987), neonatal P450s such as CYP2B1 and CYP3A2 were suppressed by the administration of GH to hypophysectomized rats. These results, along with experiments with other different P450s, indicate the association of the age-dependent appearance in the liver and GH susceptibility of rat P450s. However, hypophysectomy is a drastic procedure which disrupts many endocrine functions besides GH. A report using Lewis strain-derived GH-deficient rats also cast doubt on the role of GH on the constitutive expression of rat P450s in livers (Bullock, P., et al., 1991; Legraverend, C., et al., 1992). Distinct spontaneous dwarf rats were found from a colony of the Sprague-Dawley strain in Japan (Nogami, H., et al., 1989; Takeuchi, T., et al., 1990). These rats have roughly one-third of the body weight of normal SD rats, but are reproductively competent. Homozygous dwarf male and female rats made available by mating of dwarf parents had no detectable level of GH protein in the pituitary, although other pituitary hormones such as prolactin and ACTH are present. The cause of the GH defect has been characterized as a point mutation of the GH gene, which occurred at a junction of the third intron and fourth exon (Takeuchi, T., et al., 1990). These results indicate the specific defect of GH among endocrines in this spontaneous dwarf rat. To assess the role of GH on hepatic expression of P450s in rat, we studied and compared levels of several P450 forms between 2 months of age of normal and dwarf rats. As shown in Fig. 2, a major pubertal/male-specific P450, CYP2C11, was detected in dwarf male and female rats. The expressed levels were one-third to one-sixth of those in normal male rats. In addition, a clear strain dif-

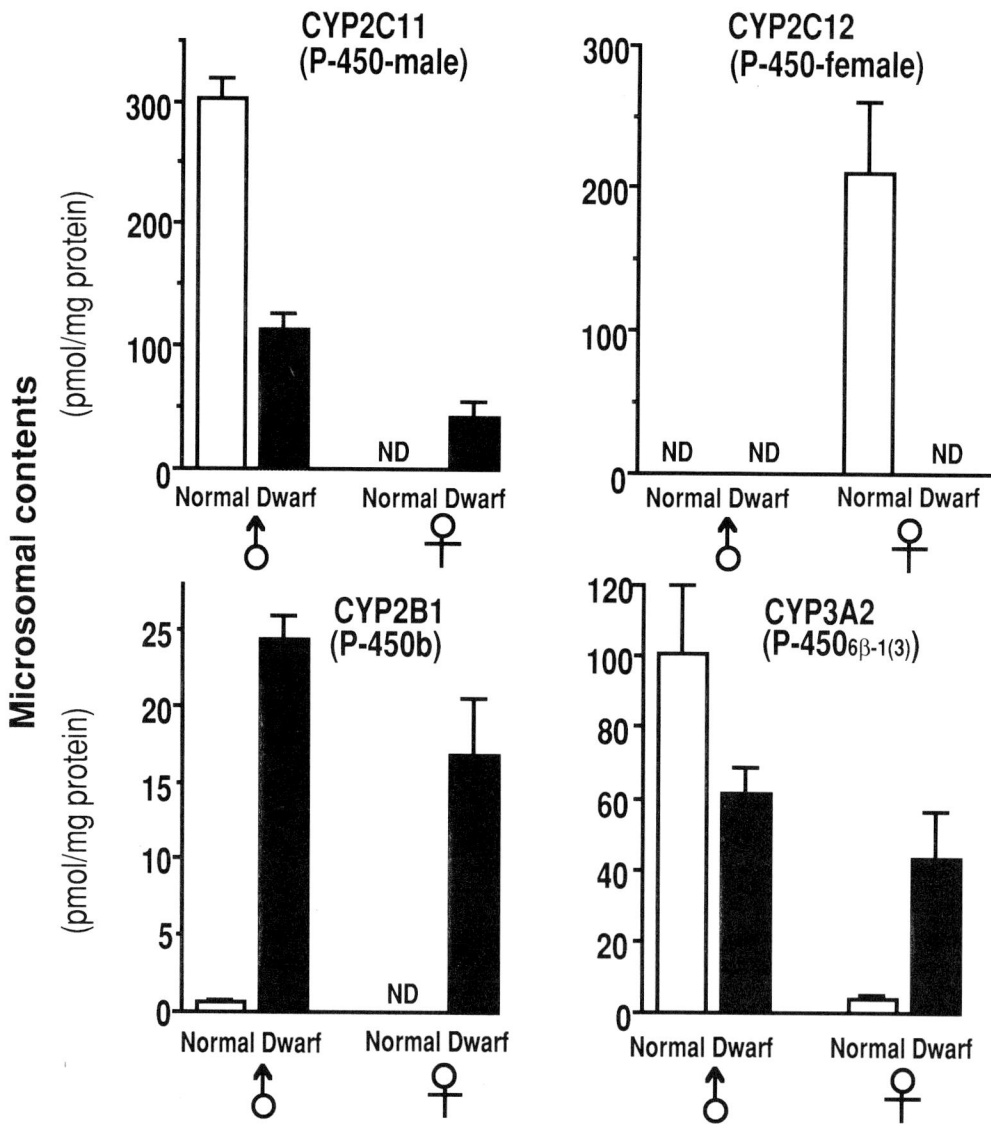

Fig. 2 Specific CYP contents in livers of normal and spontaneous dwarf rats

ference was observed in a female-specific form, CYP2C12. This pubertal form was a major constitutive form in the normal SD strain of female rats, but was not detectable in the liver of dwarf females as well as dwarf male rats. The mode of expression and amounts of CYP2C11 and CYP2C12 expressed in dwarf rats are quite similar to those observed in hypophysectomized SD rats.

Hepatic levels of two major neonatal P450 forms are also shown in this figure. A major phenobarbital-inducible form, CYP2B1, was constitutively expressed, but in trace amounts, in normal mature male rats and was undetectable in mature female rats (Yamazoe, Y., et al., 1987). Amounts of this neonatal form were markedly high in dwarf rats of both sexes. Another neonatal P450, CYP3A2, was expressed in both sexes of neonatal SD rats, but sex-specifically in the mature male rats. Hepatic levels of CYP3A2 in dwarf rats were high and not different between both sexes. These results were roughly consistent with the data obtained from hypophysectomized rat models, except that hepatic levels of CYP2B1 and CYP3A2 were slightly lower in dwarf rats than in hypophysectomized SD rats. To assess clearly the role of GH on hepatic P450s, dwarf rats were treated with GH in dual modes of administration. Treatment of dwarf male rats with S.C. injection of GH (once a day for 7 days) caused the elevation of pubertal CYP2C11 to a level comparable with that in normal SD male rats. Continuous infusion of GH, which mimicked the female secretory pattern, evoked CYP2C12 in both sexes of dwarf rats, while it abolished a neonatal form, CYP3A2. These results clearly confirm the results obtained from hypophysectomized rat models, and also indicate the positive and negative regulatory role of GH on pubertal and neonatal P450s, respectively.

In rats, plasma levels of GH vary, depending on the age and the secretory pattern, and become sex-differentiated during puberty (Edén, S., 1979; Jansson, J., et al., 1985). Higher levels of GH were reported in neonate than in mature rats. This is rather in contrast to the GH susceptibility of neonatal P450s. However, GH effects on hepatic drug-metabolizing activities (Wilson, J.T., 1970) and insulin-like growth factor (Mathews, L.S., et al., 1986;1989) are known to have delays on their onsets. The GH signal is not transferred to liver enzymes until puberty, because of the absence or low amounts of the GH receptor in the liver (Maes, M., et al., 1983; Tiong, T.S. and Herington, A.C., 1992): The level of the somatogenic receptor in the liver is reported to start to increase three weeks after birth and to attain the mature level at two months of age in rats. These results may explain why neonatal P450s, which appear in the liver soon after birth, are not suppressed by GH. Neonatal P450 levels are mostly decreased during puberty, which could also be possibly interpreted as the start of GH-signal surge in the liver during this period. These data clearly indicate the suppressive role of GH on neonatal P450s, and suggest that plasma GH level/GH receptor development is the main limiting factor on the developmental changes of neonatal P450s.

Although showing different dependencies on GH-secretion modes, several pubertal P450s including CYP2C7, CYP2C11, CYP2C12, and CYP2C22 (Nagata, K., et al., 1990b) are expressed to their maximal levels in the presence of GH. These pubertal P450s appear in the liver at around 4 weeks of age and continue to increase until 9 weeks of age. These data indicate that GH is a major stimulative factor for the maintenance of several pubertal forms. What is the factor limiting the pubertal onset of pubertal P450s? The start of GH-signal surge could be a factor for the expression of a female-specific CYP2C12, since no CYP2C12 protein and mRNA are detected in the absence of GH in rats. CYP2C11 protein and mRNA are detectable in the GH-deficient animals, hypophysectomized rats and dwarf rats, although both levels are increased by the intermittent injection of GH. Interestingly, a preliminary experiment with pubertal female dwarf rats suggest the similarity of the ontogenic profiles of CYP2C11 between normal male SD and dwarf female rats. A GH-independent CYP2C6 is reported to be regulated developmentally by a nuclear binding factor DBP in rats (Yano, M., et al., 1992). These results suggest the existence of an unknown and non-GH-dependent factor which governs the ontogenicity of CYP2C11 in rats.

## Mechanism of hormonal suppression on neonatal P450s

Several studies showed that GH and thyroid hormone regulate neonatal and pubertal P450s, mainly through changing the levels of their specific mRNAs in rat livers (Rutgers, M., et al., 1989; Sasamura, H., et al., 1990; Shimada, M., et al., 1989). Investigation on P450 genes suggests that an interaction of the 5'-flanking regions with several nuclear components is necessary for the expression in cells. To define the mechanism of the interaction of these hormones with hepatic P450s, we have expressed a CAT-fusion gene of CYP2B1 in primary cultured hepatocytes. Introduction of the gene corresponding to a 7 Kb length of the 5'-flanking region resulted in expression of CAT activity in the presence of phenobarbital. The expression was further suppressed by the addition of GH or triiodothyronine, indicating that these hormones suppress CYP2B1 expression by interacting at the stage of transcriptional activation.

## REFERENCES

Bullock, P., Gemzik, B., Johnson, D., Thomas, P., and Parkinson, A. (1991): Evidence from dwarf rats that growth hormone may not regulate the sexual differentiation of liver cytochrome P450 enzymes and steroid 5 alpha-reductase. Proc. Natl. Acad. Sci. USA 88, 5227-31.

Edén, S. (1979): Age- and sex-related differences in episodic growth hormone secretion in the rat. Endocrinology 105, 555-560.

Jansson, J., Edén, S., and Isaksson, O. (1985): Sexual dimorphism in the control of growth hormone secretion. Endocrine Rev. 6, 128-150.

Kamataki, T., Maeda, K., Yamazoe, Y., Matsuda, N., Ishii, K., and Kato, R. (1983a): A high-spin form of cytochrome P-450 highly purified from polychlorinated biphenyl-treated rats. Mol. Pharmacol. 24, 146-155.

Kamataki, T., Maeda, K., Yamazoe, Y., Nagai, T., and Kato, R. (1983b): Sex-difference of cytochrome P-450 in the rat: Purification, characterization, and quantitation of constitutive forms of cytochrome P-450 from liver microsomes of male and female rats. Arch. Biochem. Biophys. 225, 758-770.

Kato, R., Yamazoe, Y., Shimada, M., Murayama, N., and Kamataki, T. (1986): Effect of growth hormone and ectopic transplantation of pituitary gland on sex-specific forms of cytochrome P-450 and testosterone and drug oxidations in rat liver. J. Biochem. 100, 895-902.

Legraverend, C., Mode, A., Wells, T., Robinson, I., and Gustafsson, J.Å. (1992): Hepatic steroid hydroxylating enzymes are controlled by the sexually dimorphic pattern of growth hormone secretion in normal and dwarf rats. FASEB J. 6, 711-718.

MacGeoch, C., Morgan, E.T., and Gustafsson, J.-Å. (1985): Hypothalamo-pituitary regulation of cytochrome P-450 15β apoprotein levels in rat liver. Endocrinology 117, 2085-2092.

Maes, M., de Hertogh, R., Watrin-Granger, P., and Ketelslegers, J.M. (1983): Ontogeny of liver somatotropic and lactogenic binding sites in male and female rats. Endocrinology 113, 1325-1332.

Mathews, L.S., Enberg, B., and Norstedt, G. (1989): Regulation of rat growth hormone receptor gene expression. J. Biol. Chem. 264, 9905-9910.

Mathews, L.S., Norstedt, G., and Palmiter, R.D. (1986): Regulation of insulin-like growth factor I gene expression by growth hormone. Proc. Natl. Acad. Sci.USA 83, 9343-9347.

Morgan, E.T., MacGeoch, C., and Gustafsson, J.-Å (1985): Sexual differentiation of cytochrome P-450 in rat liver Evidence for a constitutive isozyme as the male-specific 16α-hydroxylase. Mol. Pharmacol. 27, 471-479.

Nagata, K., Gonzalez, F.J., Yamazoe, Y., and Kato, R. (1990a): Purification and characterization of four catalytically active testosterone 6β-hydroxylase P-450s from rat liver microsomes: Comparison of a novel form with three structurally and functionally related forms. J. Biochem. 107, 718-725.

Nagata, K., Sasamura, H., Miyata, M., Shimada, M., Yamazoe, Y., and Kato, R. (1990b): cDNA and deduced amino acid sequences of a male dominant P-450Md mRNA in rats. Nucleic Acids Res. 18, 4934.

Nogami, H., Takeuchi, T., Suzuki, K., Okuma, S., and Ishikawa, H. (1989): Studies on prolactin and growth hormone gene expression in the pituitary gland of spontaneous dwarf rats. Endocrinology 125, 964-970.

Rutgers, M., Heusdens, F.A., and Visser, T.J. (1989): Metabolism of triiodothyroacetic acid ($TA_3$) in rat liver. I. Deiodination of $TA_3$ and $TA_3$ sulfate by microsomes. Endocrinology 125, 424-432.

Sasamura, H., Nagata, K., Yamazoe, Y., Shimada, M., Saruta, T., and Kato, R. (1990): Effect of growth hormone on rat hepatic cytochrome P-450f mRNA: a new mode of regulation. Mol. Cell. Endocrinol. 68, 53-60.

Shimada, M., Murayama, N., Yamauchi, K., Yamazoe, Y., and Kato, R. (1989): Suppression in the expression of a male-specific cytochrome P-450, P-450-male: Difference in the effect of chemical inducers on P450-male mRNA and protein in rat livers. Arch. Biochem. Biophys. 270, 578-587.

Shimada, M., Murayama, N., Yamazoe, Y., Kamataki, T., and Kato, R. (1987): Further studies on the persistence of neonatal androgen imprinting on sex-specific cytochrome P-450, testosterone and drug oxidations. Japan. J. Pharmacol. 45, 467-478.

Shimada, M., Nagata, K., Murayama, N., Yamazoe, Y., and Kato, R. (1989): Role of growth hormone in modulating the constitutive and phenobarbital-induced levels of two P-4506β (Testosterone 6β-hydroxylase) mRNAs in rat livers. J. Biochem. 106, 1030-1034.

Takeuchi, T., Suzuki, H., Sakurai, S., Nogami, H., Okuma, S., and Ishikawa, H. (1990): Molecular mechanism of growth hormone (GH) deficiency in the spontaneous dwarf rat: Detection of abnormal splicing of GH messenger ribonucleic acid by the polymerase chain reaction. Endocrinology 126, 31-38.

Tiong, T.S., and Herington, A.C. (1992): Ontogeny of messenger RNA for the rat growth hormone receptor and serum binding protein. Mol. Cell. Endocrinol. 83, 133-141.

Wilson, J.T. (1970): Alteration of normal development of drug metabolism by injection of growth hormone. Nature 225, 861-863.

Yamazoe, Y., Ling, X., Murayama, N., Gong, D., Nagata, K., and Kato, R. (1990): Modulation of hepatic level of microsomal testosterone 7α-hydroxylase, P-450a (P450IIA), by thyroid hormone and growth hormone in rat liver. J. Biochem. 108, 599-603.

Yamazoe, Y., Murayama, N., Shimada, M., Imaoka, S., Funae, Y., and Kato, R. (1989): Suppression of hepatic levels of an ethanol-inducible P-450DM/j by growth hormone: Relationship between the increased level of P-450DM/j and depletion of growth hormone in diabetes. Mol. Pharmacol. 36, 716-722.

Yamazoe, Y., Murayama, N., Shimada, M., Yamauchi, K., Nagata, K., Imaoka, S., Funae, Y., and Kato, R. (1988): A sex-specific form of cytochrome P-450 catalyzing propoxycoumarin O-depropylation and its identity with testosterone 6β-hydroxylase in untreated rat livers: Reconstitution of the activity with microsomal lipids. J. Biochem. 104, 785-790.

Yamazoe, Y., Shimada, M., Kamataki, T., and Kato, R. (1986): Effect of hypophysectomy and growth hormone treatment on sex-specific forms of cytochrome P-450 in relation to drug and steroid metabolisms in rat liver microsomes. Japan. J. Pharmacol. 42, 371-382.

Yamazoe, Y., Shimada, M., Murayama, N., and Kato, R. (1987): Suppression of levels of phenobarbital-inducible rat liver cytochrome P-450 by pituitary hormone. J. Biol. Chem. 262, 7423-7428.

Yamazoe, Y., Shimada, M., Murayama, N., Kawano, S., and Kato, R. (1986): The regulation by growth hormone of microsomal testosterone 6β-hydroxylase in male rat livers. J. Biochem. 100, 1095-1097.

Yamazoe, Y., Shimada, M., Yamauchi, K., Murayama, N., and Kato, R. (1988): Alteration of hepatic drug metabolizing activities and contents of cytochrome P-450 isozymes by neonatal monosodium glutamate treatment. Biochem. Pharmacol. 37, 1687-1691.

Yano, M., Falvey, E., and Gonzalez, F.J. (1992): Role of the liver-enriched transcription factor DBP in expression of the cytochrome P450 CYP2C6 gene. Mol. Cell. Biol. 12, 2847-2854.

This work was supported in part by Grants-in Aid from the Ministry of Education, Science and Culture, and by grants from the Japan Health Science Foundation and the Cell Science Research Foundation.

# Mechanisms of enhancement and inhibition of cytochrome P450 catalytic activities

F.P. Guengerich, Bok-R. Kim, Elizabeth M.J. Gillam, Tsutomu Shimada

*Department of Biochemistry and Center in Molecular Toxicology, Vanderbilt University School of Medicine, Nashville, Tennessee 37232-0146, USA*

Considerable attention has been given to the regulation of cytochrome P450 (P450) activity by enhancement of levels of mRNA and proteins. Another mechanism for increased activity is through direct stimulation of P450 enzymes. This phenomenon was first observed with substrate solvents and certain natural products (Wiebel *et al.*, 1971; Cinti, 1978). Conney and his associates showed that certain flavonoids could dramatically enhance the mutagenicity of carcinogens (Buening *et al.*, 1981) and that this phenomenon could be demonstrated to occur in vivo in mice, in the absence of enzyme induction (Lasker *et al.*, 1984). The basis of the stimulation is still not clear. One hypothesis is that these compounds can stimulate oxidation reactions by increasing the affinity of NADPH-P450 reductase with individual P450s (Huang *et al.*, 1981). Alternatively, in some cases evidence for allosteric interactions has been obtained (Schwab *et al.*, 1988). The underlying mechanism has not been defined, although a model steroid has been shown to enhance catalytic activity by increasing substrate affinity (Johnson *et al.*, 1988).

P450 3A4 is involved in the activation of aflatoxin $B_1$ ($AFB_1$) and in many liver samples is the principal catalyst (Shimada & Guengerich, 1989). Two reactions are catalyzed by the reaction: oxidation at the 8,9-double bond yields a reactive epoxide [that can be tapped as a glutathione (GSH) conjugate in the presence of excess GSH S-transferase], and hydroxylation at the 3α-position yields aflatoxin $Q_1$ ($AFQ_1$) (Raney *et al.*, 1992b). Conney and his associates had demonstrated the enhancement of mutagenicity of the $AFB_1$ by α-napthoflavone (α-NF, 7,8-benzoflavone) in human liver microsomes (Buening *et al.*, 1981). We found that many catalytic activities of P450 3A4 were stimulated by α-NF, and the response of this enzyme could account for the patterns seen in microsomes (Shimada & Guengerich, 1989; Shimada *et al.*, 1989). However, some catalytic activities of P450 3A4 were refractory to α-NF and some were inhibited (Yun *et al.*, 1992). α-NF stimulated $AFB_1$ 8,9-oxidation in human liver microsomes and inhibited 3α-hydroxylation (Fig. 1).

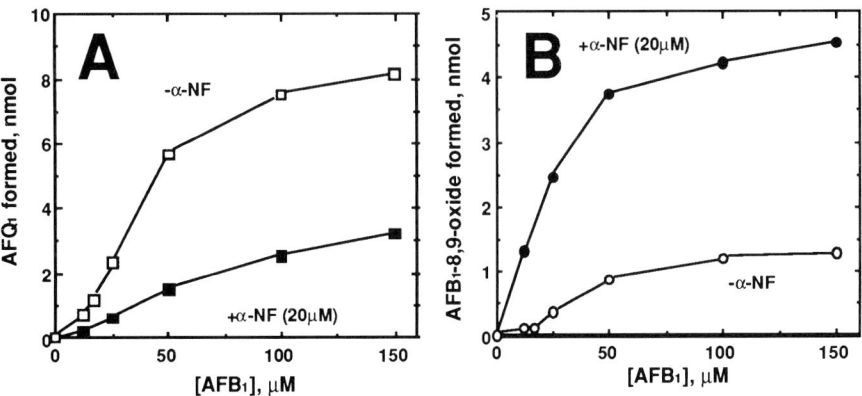

Fig. 1. Plots of AFB$_1$ (A) 3α-hydroxylation and (B) 8,9-epoxidation as a function of AFB$_1$ concentration in human liver microsomes in the absence (open symbols) and presence (closed symbols) of 20 μM α-NF.

Aflatoxin B$_2$ inhibited 8,9-oxidation but not 3α-hydroxylation (Raney *et al.*, 1992b). Conversely, we found some compounds (e.g., the pyridine oxidation product of nifedipine, 2, 6-dimethyl-4-(2'-nitrophenyl)-3,5-pyridine dicarboxylic acid dimethyl ester) could inhibit AFB$_1$ 3α-hydroxylation but not 8,9-epoxidation (Fig. 2).

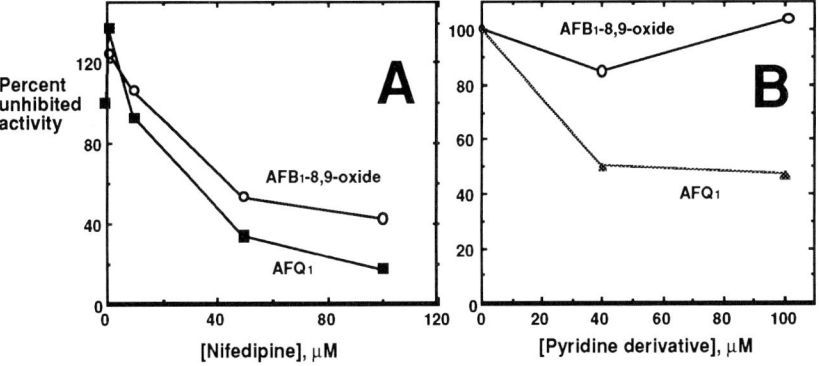

Fig. 2. Effects of (A) α-NF and (B) nifedipine oxidation product on AFB$_1$ 8,9-epoxidation and 3α-hydroxylation in human liver microsomes.

We examined a number of flavones from the list of Conney and his associates (Buening *et al.*, 1981) and found stimulation with flavone and β-naphthoflavone but the most dramatic enhancement with α-NF. Human liver microsomes were examined in detail with low concentrations of the substrate AFB$_1$ and sigmoidal patterns were found when either 3a-hydroxylation or 8,9-epoxidation was plotted against the AFB$_1$ concentration. With both reactions, the pattern appeared to change to a hyperbolic one in the presence of α-NF (Fig. 1).

Further studies on the mechanism of this possibly allosteric behavior of P450 3A4 required a ready source of the purified enzyme and a reconstituted system in which the patterns of Fig. 1 could be reproduced. We expressed an N-terminal-modified P450 3A4 cDNA clone in *Escherichia coli* and developed a facile purification (Gillam *et al.*, 1993). This enzyme was catalytically active in oxidizing nifedipine, testosterone, and AFB$_1$ (both 3α-hydroxylation and 8,9-epoxidation). Optimal catalytic activity required the presence of GSH, which is normally used in trapping AFB$_1$ 8,9-oxide in the reaction. We tested a number of other thiols and found that *N*-acetylcysteine could also support the reaction, while cysteine, dithiothreitol, 2-mercaptoethanol, and others would not. The exact role of the GSH is unclear. We varied the NADPH-P450 reductase concentration and found that both the "$K_m$" and $V_{max}$ were improved. With the addition of 3.25 mM GSH, the $K_m$ for NADPH-P450 reductase dropped from 2.5 (± 0.6) to 0.95 (± 0.4) μM and the $V_{max}$ increased from 0.56 (±0.57) to 9.5 (±1.2) nmol 6β-hydroxytesterone formed min$^{-1}$ (nmol P450)$^{-1}$.

Others have suggested that human P450 1A2 may play a greater role than P4550 3A4 in AFB$_1$ activation However, this was not the case, even when only the intrinsic catalytic activities were considered (Fig. 3).

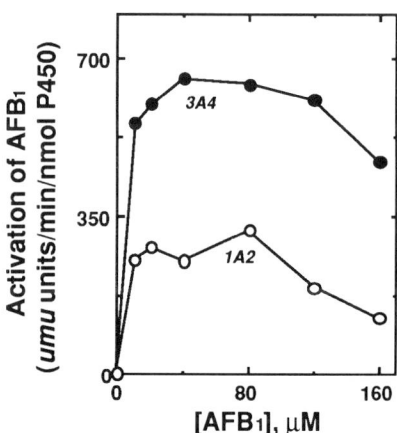

Fig. 3.    Comparison of rates of formation of genotoxic AFB$_1$ products from recombinant P450 1A2 (O) and 3A4 (●) using a *umuC* test system (Shimada *et al.*, 1989).

Preliminary analyses have been done on the AFB$_1$ products formed by P450 1A2. The enzyme appears to form AFQ$_1$ and both the *exo*- and *endo*- isomers of AFB$_1$-8,9-oxide (Raney *et al.*, 1992a) (Fig. 4). The latter is inactive because is does not bind to DNA.

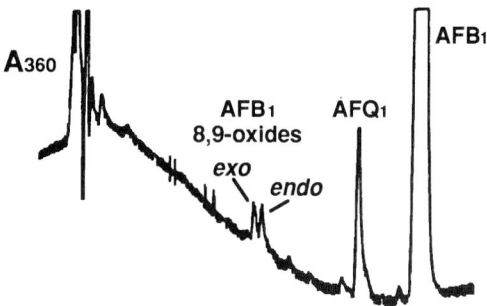

Fig. 4. HPLC analysis of products of oxidation of $AFB_1$ by recombinant human P450 1A2 in the presence of excess mouse GSH S-transferase.

Further analysis of $AFB_1$ oxidation was done with recombinant P450 3A4 in the reconstituted system. Sigmoidal kinetics were seen and, as in the microsomes (Fig. 1) the pattern changed to hyperbolic in the presence of α-NF (Fig. 5).

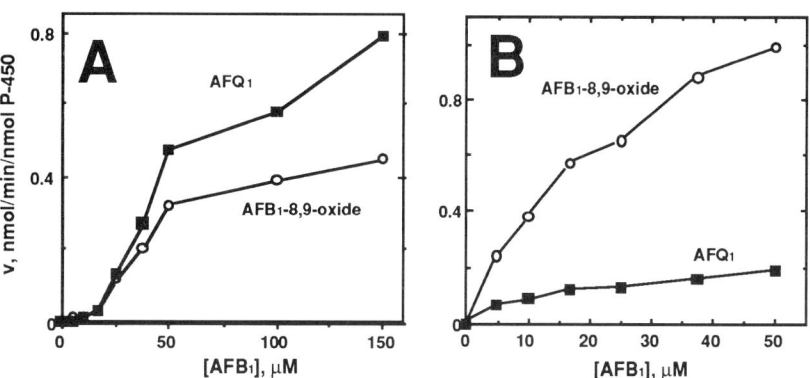

Fig. 5. Plots of $AFB_1$ 3α-hydroxylation (■) and (B) 8,9-epoxidation (○) as a function of $AFB_1$ concentration with a reconstituted system containing recombinant P450 3A4 in the absence (A) and presence (B) of 10 μM α-NF.

Our results suggest that P450 operates in an allosteric mechanism, at least with $AFB_1$ oxidation. Thus, we can hypothesize that there is a binding site nearby the oxidation site of P450 3A4. α-NF can occupy this site to perturb the oxidation site. We also propose that $AFB_1$ itself can occupy this second site to perturb the oxidation site. The availability of a recombinant, purified enzyme should facilitate further examination of this phenomenon. Another hypothesis that we plan to consider is that enhancement of enzyme activity (by α-NF) is due to enhanced interaction of NADPH-P450 reductase with P450 (3A4) (Huang et al., 1981). We have developed a cDNA construct coding for both of the enzymes (Fisher et al., 1992; Shimada et al., 1993) and plan to test the expressed fusion protein for its response to α-NF and other effectors.

# REFERENCES

Buening, M. K., Chang, R. L., Huang, M. R., Fortner, J. G., Wood, A. W. and Conney, A. H. (1981): Activation and inhibition of benzo(a)pyrene and aflatoxin $B_1$ metabolism in human liver microsomes by naturally occurring flavonoids. *Cancer Res.* 41, 67-72.

Cinti, D. L. (1978): Agents activating the liver microsomal mixed function oxidase system. *Pharmac. Ther.* 2, 727-479.

Fisher, C. W., Shet, M. S., Caudle, D. L., Martin-Wixtrom, C. A. and Estabrook, R. W. (1992): High-level expression in *Escherichia coli* of enzymatically active fusion proteins containing the domains of mammalian cytochromes P450 and NADPH-P450 reductase flavoprotein. *Proc. Natl. Acad. Sci. USA* 89, 10817-10821.

Gillam, E. M. J., Baba, T., Kim, B-R., Ohmori, S. and Guengerich, F. P. (1993): Expression of modified human cytochrome P450 3A4 in *Escherichia coli* and purification and reconstitution of the enzyme. *Arch. Biochem. Biophys.* 305, 123-131.

Huang, M. T., Johnson, E. F., Muller-Eberhard, U., Koop, D. R., Coon, M. J. and Conney, A. H. (1981): Specificity in the activation and inhibition by flavonoids of benzo[a]pyrene hydroxylation by cytochrome P-450 isozymes from rabbit liver microsomes. *J. Biol. Chem.* 256, 10897-10901.

Johnson, E. F., Schwab, G. E. and Vickery, L. E. (1988): Positive effectors of the binding of an active site-directed amino steroid to rabbit cytochrome P-450 3c. *J. Biol. Chem.* 263, 17672-17677.

Lasker, J. M., Huang, M. T. and Conney, A. H. (1984): In vitro and in vivo activation of oxidative drug metabolism by flavonoids. *J. Pharmacol. Exp. Ther.* 229, 162-170.

Raney, K. D., Coles, B., Guengerich, F. P. and Harris, T. M. (1992a): The *endo* 8,9-epoxide of aflatoxin $B_1$: a new metabolite. *Chem. Res. Toxicol.* 5, 333-335.

Raney, K. D., Shimada, T., Kim, D-H., Groopman, J. D., Harris, T. M. and Guengerich, F. P. (1992b): Oxidation of aflatoxin $B_1$ and related dihydrofurans by human liver microsomes, significance of aflatoxin $Q_1$ as a detoxication product. *Chem. Res. Toxicol.* 5, 202-210.

Schwab, G. E., Raucy, J. L. and Johnson, E. F. (1988): Modulation of rabbit and human hepatic cytochrome P-450-catalyzed steroid hydroxylations by α-naphthoflavone. *Mol. Pharmacol.* 33, 493-499.

Shimada, T., Gillam, E. M. J., Sandhu, P. and Guengerich, F. P. (1993): Activation of procarcinogens by human cytochrome P450 enzymes expressed in *Escherichia coli*. Simplified bacterial systems for genotoxicity assays. *Cancer Res.*, in press.

Shimada, T. and Guengerich, F. P. (1989): Evidence for cytochrome $P-450_{NF}$, the nifedipine oxidase, being the principal enzyme involved in the bioactivation of aflatoxins in human liver. *Proc. Natl. Acad. Sci. USA* 86, 462-465.

Shimada, T., Iwasaki, M., Martin, M. V. and Guengerich, F. P. (1989): Human liver microsomal cytochrome P-450 enzymes involved in the bioactivation of procarcinogens detected by *umu* gene response in *Salmonella typhimurium* TA1535/pSK1002. *Cancer Res.* 49, 3218-3228.

Wiebel, F. J., Leutz, J. C., Diamond, L. and Gelboin, H. V. (1971): Aryl hydrocarbon (benzo[a]pyrene) hydroxylase in microsomes from rat tissues: differential inhibition and stimulation by benzoflavones and organic solvents. *Arch. Biochem. Biophys.* 144, 78-86.

Yun, C-H., Wood, M., Wood, A. J. J. and Guengerich, F. P. (1992): Identification of the pharmacogenetic determinants of alfentanil metabolism: cytochrome P-450 3A4. An explanation of the variable elimination clearance. *Anesthesiology* 77, 467-474.

# Role of individual human liver P450s and other enzymes in anti-cancer drug metabolism: drug activation and drug resistance mechanisms

Thomas K.H. Chang, Guan Chen, David J. Waxman

*Department of Biological Chemistry and Molecular Pharmacology and Dana-Farber Cancer Institute, Harvard Medical School, Boston, MA 02115, USA*

## INTRODUCTION

Cyclophosphamide (CPA) and ifosphamide (IFA) are alkylating agent anti-cancer prodrugs that require bioactivation to produce pharmacologically active, cytotoxic species (Sladek, 1988). In rat liver, cytochrome P450 (CYP) forms CYP2B1, 2C6 and 2C11 are the major catalysts of CPA 4-hydroxylation (Clarke and Waxman, 1989), whereas these enzymes, together with one or more CYP3A enzymes, catalyze a major fraction of IFA 4-hydroxylation (Weber and Waxman, 1993). Although human P450 has been implicated in the bioactivation of these anti-cancer drugs, the specific P450 forms have not been identified. This information will not only provide insight into the basis for the large interpatient differences in the clinical pharmacokinetics and metabolism of CPA and IFA (Sladek, 1988), but could also be useful for clinicians to predict potential drug-drug interactions in cancer patients, and might lead to rational strategies to enhance drug activation through modulation of liver P450 enzyme levels.

The expression of CPA and IFA therapeutic activity is dependent not only on P450-mediated activation, but is also a function of cellular deactivation of the oxazaphosphorine metabolites catalyzed by two other enzyme systems, aldehyde dehydrogenase (ALDH) and glutathione/glutathione S-transferase (GSH/GST) (Fig. 1). Aldophosphamide, the ring-opened form of 4-hydroxy-CPA, can be inactivated by ALDH-catalyzed oxidation to carboxyphosphamide (Sladek, 1988) or by GSH conjugation to form a hemithioacetal (Lee, 1991). Cellular GSH can also participate in conjugative reactions with the therapeutically significant, DNA-alkylating metabolite phosphoramide mustard and with the cytotoxic, protein-alkylating metabolite acrolein to form non-toxic conjugates. Although there has

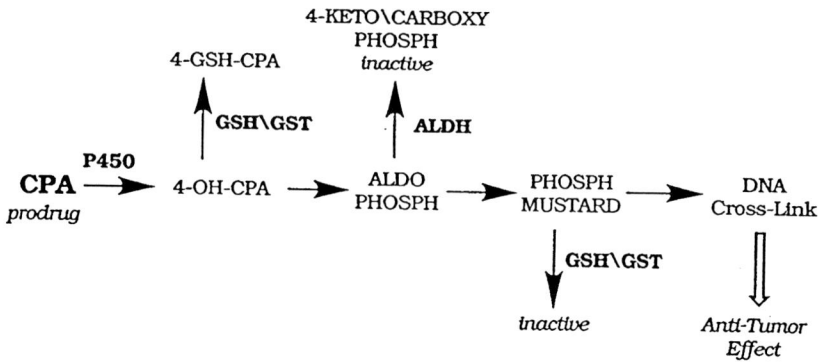

Fig. 1. Cytochrome P450-, ALDH- and GSH/GST-dependent metabolism of CPA.

been no direct demonstration that GST enzymes contribute to these GSH-mediated conjugation in intact cells, cell-free studies indicate that GSTs can enhance the rate of GSH conjugation with CPA and its derivatives (Yuan et al., 1991). GST enzymes, however, are expressed at high levels in many tissues, including tumor, where they may contribute to drug resistance by conjugation of electrophilic chemicals, such as the activated metabolites of CPA (Waxman, 1990). Conceivably, the contribution of these enzymes to CPA inactivation might become even more important in cells that acquired resistance to CPA. Thus, characterization of the resistance phenotype can not only provide important information that may be relevant to drug resistance that emerges clinically, but may also improve our understanding of drug-metabolizing mechanisms that are operative in naive as well as drug-exposed tumor cells.

## OXAZAPHOSPHORINE ACTIVATION BY HUMAN P450 ENZYMES

Considerable inter-sample variation was observed in CPA and IFA activation by our panel of 12 human liver microsomal samples. CPA 4-hydroxylation and IFA 4-hydroxylation varied over a 9-fold and a 4-fold range, respectively. These differences may be related to the induction status of the individual liver donors or to genetic differences relating to expression of the specific liver P450 forms involved in oxazaphosphorine activation. Enzyme kinetic analysis revealed that CPA 4-hydroxylation and IFA 4-hydroxylation by human liver microsomes are best described by a 2-component Michaelis-Menten model, involving both high-affinity ($K_m < 100$ mM) and low affinity ($K_m > 1$ mM) P450 enzymes.

To identify the specific P450 enzymes that are catalytically competent in activating these alkylating anti-cancer agents, we assayed individual cDNA-expressed human P450 enzymes (kindly provided by C. Crespi, Gentest Corp.) for CPA and IFA 4-hydroxylation at 0.25 mM and 2 mM substrate concentrations (Chang et al., 1993). Whereas CYP2C8 and CYP2C9 appeared to be low Km oxazaphorine 4-hydroxylases, CYP2A6, 2B6 and 3A4 were high Km forms, and CYP1A1, 1A2 and 2E1 were inactive (Table 1).

Table 1. Oxazaphosphorine activation by cDNA-expressed human P450.

| P450 | CPA 4-Hydroxylase | | IFA 4-Hydroxylase | |
|---|---|---|---|---|
|  | 0.25 mM | 2 mM | 0.25 mM | 2mM |
|  | (pmol/min/mg protein) | | | |
| 1A1 | <1 | <1 | <1 | <1 |
| 1A2 | <1 | <1 | <1 | <1 |
| 2A6 | 7 | 58 | 3 | 17 |
| 2B6 | 88 | 764 | 11 | 94 |
| 2C8 | 16 | 12 | 19 | 4 |
| 2C9 | 63 | 97 | 31 | 41 |
| 2E1 | <1 | <1 | <1 | <1 |
| 3A4 | <1 | 10 | <1 | 18 |

The above studies establish that CYP2A6, 2B6, 2C8, 2C9 and 3A4 are capable of activating CPA and IFA. However, these data alone do not indicate which of these enzymes make the dominant contribution to drug activation in human liver, where the individual P450 forms are present at varying levels reflecting both differential enzyme expression and interindividual variation due to environmental and/or genetic factors.

Chemical inhibition and immunoinhibition experiments revealed that the CYP2B6 inhibitor orphenadrine, as well as anti-CYP2B IgG, decreased CPA 4-hydroxylation to a greater extent (40-47% inhibition) than IFA 4-hydroxylation (21-28% inhibition) in human liver microsomes (Table 2). By contrast, the CYP3A selective inhibitor triacetyloleandomycin (TAO) and anti-CYP3A IgG substantially reduced microsomal IFA 4-hydroxylation (57-63% inhibition at 2 mM substrate concentration), whereas they had no significant impact on CPA 4-hydroxylation. In contrast, anti-CYP2A IgG inhibited both microsomal CPA and IFA 4-hydroxylation, but by only ~20% (Chang et al., 1993).

Table 2. Chemical and antibody inhibition of CPA and IFA hydroxylation in human liver microsomes.

| Liver Sample HLS9 | CPA 4-Hydroxylase | | IFA 4-Hydroxylase | |
|---|---|---|---|---|
| | 0.25 mM | 2 mM | 0.25 mM | 2 mM |
| | *(Percent Inhibition)* | | | |
| + Orphenadrine | 40 | 47 | 21 | 28 |
| + Anti-CYP2B IgG | 38 | 37 | 15 | 20 |
| + TAO | 6 | 28 | 23 | 57 |
| + Anti-CYP3A IgG | 12 | 0 | 37 | 63 |

These studies directly establish that cytochrome P450 enzymes are major catalysts of CPA and IFA activation in human liver and that a subset of these enzymes are primarily responsible for the activation of these anti-cancer drugs. Although a CYP2B enzyme, likely CYP2B6, was found to make a significant contribution to human liver CPA metabolism, this enzyme had no significant impact on microsomal IFA activation. CYP2B6 is, however, catalytically competent with respect to IFA activation, as revealed by the activity of cDNA-expressed P450. This CYP2B6 activity was ~8-fold lower with IFA than with CPA, a difference that probably accounts for the minor role of liver microsomal CYP2B6 in IFA activation. Whereas CYP3A4, and perhaps other related CYP3A enzymes, are major catalysts of microsomal IFA activation, they make little or no contribution to microsomal CPA hydroxylation. These results are analogous to our recent findings with rat liver microsomes, where a major dexamethasone-inducible rat CYP3A accounts for most of the IFA 4-hydroxylase activity, but it does not contribute to CPA 4-hydroxylation (Weber and Waxman, 1993). Although cDNA-expressed CYP2A6 is catalytically competent in activating CPA and IFA, CYP2A6 plays only a minor role in the activation of these drugs in human liver microsomes, probably due to the low specific content (<10% of the total spectral P450) of CYP2A6 in human liver (Wrighton and Stevens, 1992). The proposed participation by CYP2C8 and CYP2C9, which are low Km oxazaphosphorine 4-hydroxylases, could explain, at least in part, the large inter-individual differences in the clinical pharmacokinetics and metabolism of these drugs (Sladek, 1988) since CYP2C8 and CYP2C9 and closely related human CYP2C enzymes appear to be expressed in a polymorphic fashion in human liver (Wrighton and Stevens, 1992). Overall, these findings suggest ways by which liver P450 inducing agents might be employed in cancer patients to augment CPA and IFA drug activation and to improve therapeutic effectiveness.

## CPA DRUG RESISTANCE PHENOTYPE

Increases in cellular GSH levels and/or GST activity have been observed in several CPA-resistance tumor cell lines. However, the functional significance of these changes with respect to the resistance phenotype has not been established. Our studies in a human breast carcinoma MCF-7 subline selected for resistance to 4-hydroperoxy-CPA (4HC), an activated form of CPA, revealed a 2-fold increase in cellular GSH, in addition to a 2-fold increase in GST activity and ~7-fold increase in ALDH activity in cell extracts. Application of the GSH inhibitor buthionine sulfoximine (BSO) sensitized the cells to 4HC to a similar extent in both the sensitive parental MCF-7 and the drug-resistant MCF/HC cells, indicating that cellular GSH is actively involved in conjugation with 4HC and/or its metabolites in both cell lines. Furthermore, these studies indicate that depletion of cellular GSH can inhibit this process and thereby facilitate formation of cytotoxic metabolites. A similar situation occurred in cells treated with the ALDH inhibitor octanal, although in that case a greater sensitization was observed in the resistant cells. Interestingly, pretreatment with the GST inhibitor ethacrynic acid enhanced 4HC cytotoxicity in the MCF/HC cells but not the parental subline, suggesting that in the resistant cells, the conjugation of cellular GSH with 4HC and/or its metabolites is catalyzed by GST [see Chen & Waxman (manuscript submitted for publication) for further details].

These findings demonstrate that multiple biochemical determinants (ALDH, GSH and GST) can contribute to the expression of 4HC cytotoxicity in tumor cells. The selective enhancing effect of the GST inhibitor ethacrynic acid on 4HC cytotoxicity in the resistant but not in the sensitive cells indicates that in parental MCF-7 cells, GSH conjugation leading to drug inactivation largely occurs without the involvement of GST. By contrast, in the resistant cells, an increase in GST, in addition to the elevated GSH levels, is required for efficient metabolism of 4HC and its metabolites in order for the cells to survive repeated drug treatments and exhibit the resistance phenotype. Since CPA resistance can occur by multiple mechanisms involving ALDH, GSH and GST, application of a multi-modulator approach that targets these resistance determinants may be useful in circumventing resistance that develops in cancer patients.

## CONCLUSIONS

Since a subset of human cytochrome P450 is responsible for CPA and IFA activation, improvements in the therapeutic efficacy of these anti-cancer drugs through modulation of liver drug activation may require clinical strategies that focus on specific P450 enzymes. Furthermore, the findings described in this chapter may have important implications with respect to the underlying mechanisms of anti-cancer drug metabolism and drug resistance, interindividual differences in drug metabolism and efficacy, and the design of novel strategies for optimizing cancer chemotherapy.

## ACKNOWLEDGEMENTS

This work was supported in part by grant CA-49248 from the National Institutes of Health (to D.J.W.). T.K.H.C. is a Canadian Liver Foundation Research Fellow.

## REFERENCES

Chang, T.K.H., Weber, G.F., Crespi, C.L. & Waxman, D.J. (1993): Differential activation of cyclophosphamide and ifosphamide by cytochromes P-450 2B and 3A in human liver microsomes. *Cancer Res.* 53, 5629-5637.

Clarke, L. & Waxman, D.J. (1989): Oxidative metabolism of cyclophosphamide: identification of the hepatic monooxygenase catalysts of drug activation. *Cancer Res.* 49, 2344-2350.

Lee, F.Y.F. (1991): Glutathione diminishes the anti-tumour activity of 4-hydroperoxy-cyclophosphamide by stabilising its spontaneous breakdown to alkylating metabolites. *Br. J. Cancer* 63, 45-50.

Sladek, N.E. (1988): Metabolism of oxazaphosphorines. *Pharmacol. Ther.* 37, 301-355.

Waxman, D.J. (1990): Glutathione S-transferases: role in alkylating agent resistance and possible target for modulation chemotherapy - a review. *Cancer Res.* 50, 6449-6454.

Weber, G.F. & Waxman, D.J. (1993): Activation of the anti-cancer drug ifosphamide by rat liver microsomal P-450 enzymes. *Biochem. Pharmacol.* 45, 1685-1694.

Wrighton, S.A. & Stevens, J.C. (1992): The human hepatic cytochromes P450 involved in drug metabolism. *Crit. Rev. Toxicol.* 22, 1-21.

Yuan, Z.M., Smith, P.B., Brundrett, R.B., Colvin, M., & Fenselau, C. (1991): Glutathione conjugation with phosphoramide mustard and cyclophosphamide. A mechanistic study using tandem mass spectrometry. *Drug Metab. Dispos.* 19, 625-629.

# Comparative analysis of cytochrome P450 3A expression in cultures of rat, rabbit and human hepatocytes

Philip S. Guzelian

Department of Medicine, Section of Medical Toxicology, University of Colorado Health Sciences Center, Denver, CO 80262

Members of the 3A P450s including those forms that are induced by glucocorticoid treatments are found in the livers of all species reported to date including humans (Wrighton, et al., 1985; Watkins, et al., 1985). These enzymes catalyze a remarkable number of oxidation reactions of clinically important drugs (quinidine, warfarin, erythromycin, cyclosporine, midazolam, lidocaine, nifedipine, dapsone), environmental toxins (aldrin, hexachlorinated biphenyls, aflatoxins B1 and G1, benzo(a)pyrene-7,8-diol, and endogenous substances [cortisol, progesterone, testosterone, and dehydroepiandrostenedione-sulfate (DHEA-s)]. The best studied forms of liver 3A in adult rats (3A1), rabbits (3A6) and humans (3A3/4) exhibit structural similarity, while the available data suggest that there may be important differences in their substrate specificities and in their regulatory characteristics. For example, 3A immunoreactive proteins and mRNAs are induced in both rat and rabbit liver by dexamethasone (DEX) whereas the antiglucocorticoid pregnenolone 16$\alpha$-carbonitrile (PCN) induces only in rat, while the macrocyclic antibiotic rifampicin (RIF) induces only in rabbit liver (Wrighton, et al., 1985). To avoid some of the difficulties of experimentation in living animals, many investigators have turned to the system of primary cultures of rat hepatocytes incubated on the reconstituted basement membrane, Matrigel as a substratum. Such culture systems can reliably reproduce the inducible P450 expression seen in living rats (Schuetz, et al., 1988).

Because considerable information on 3A induction has been accumulated in rats and rabbits, it would be useful to compare the effects of treatments on 3A expression in cultures of human hepatocytes with those occurring in hepatocyte cultures prepared from rats or rabbits. To examine the suitability of primary cultures of hepatocytes as a system for evaluating the effects of 3A inducers, we have examined the levels of 3A mRNA in hepatocytes from rat, rabbit, and humans cultured on Matrigel under nearly identical conditions (T. Kocarek, manuscript in preparation). Cells were inoculated on Matrigel-coated dishes in serum-free Waymouth medium, incubated for three days, and then treated for two days with medium containing DEX ($10^{-5}$ M, glucocorticoid), PCN ($10^{-5}$ M, antiglucocorticoid), RIF ($10^{-5}$ M, macrocyclic antibiotic), or PB ($2 \times 10^{-3}$ M, barbiturate). Northern blot analysis, using a cDNA probe that recognizes both 3A1 and 3A2 mRNAs, revealed that treated cultures of rat hepatocytes contained higher amounts of 3A1/2 mRNA when compared to control cultures incubated for three or five days in standard medium (Table 1). Noteworthy was the lack of induction in rat cultures treated with RIF. Analyses performed on cultured hepatocytes prepared from the liver of a male rabbit revealed that DEX, RIF and PB treatments all strongly induced rabbit 3A6 mRNA above the amounts in control cultures (Table 1). The sole exception was PCN which failed to produce any detectable 3A6 mRNA induction (<1% of DEX).

Table 1

## Species Differences in Response to 3A Inducers in Primary Cultures of Hepatocytes on Matrigel

|  | Rat Hepatocytes | Rabbit Hepatocytes |
|---|---|---|
| Treatment | 3A RNA | 3A RNA |
| 0 hr | 5 | 62 |
| 120 hr | < 4 | < 1 |
| Dex | 100 | 100 |
| PCN | 71 | < 1 |
| Rifampicin | 5 | 110 |
| PB | 38 | 108 |

It is well known that the human 3A P450 family is composed of at least four highly homologous genes. To investigate the expression of 3A family members individually in the intact human liver, we prepared oligonucleotides specific for each 3A mRNA and used Northern blot analysis and/or polymerase chain reaction to examine RNA from adult and fetal liver for variation in expression of the 3A forms during development (Schuetz, J.D., et al., 1993). We found that 3A4 (P450NF) mRNA was only expressed postnatally (Table 2). In contrast, 3A7 (HFLa) mRNA, a form previously thought to be confined to fetal liver, was found in all tested samples of fetal liver and in about half of the adult livers. 3A5 (HLp2), an mRNA also found in all the fetal samples, was detected in about 25% of the adults. Two of the livers co-expressed 3A5 and 3A7 mRNA. These results demonstrate polymorphic expression of human liver P450 3A mRNAs and suggest that each form may be under separate regulatory control and may be heterogeneous within the population.

Table 2

## Polymorphic Expression of Human Liver P450 3A mRNAs

| Form | Adult Liver | Fetal Liver |
|---|---|---|
| 3A3/4 | All | None |
| 3A5 | 25% | All |
| 3A7 | 50% | All |

Primary human hepatocyte cultures routinely expressed inducible 3A3/4 immunoreactive protein and 3A mRNA (Schuetz, E. G. et al., 1993). However, we have not detected 3A5 in such cultures. In seeking alternative model systems to examine 3A5 and 3A7, we examined a new human hepatocellular carcinoma cell line, TONG/HCC (kindly provided by Dr. D. Stevenson, LAC/USC School of Medicine, Los Angeles, CA) (Stevenson et al., 1987). Immunoblot analysis revealed that TONG/HCC cells contained a 3A immunoreactive protein that co-migrated with the 52kd protein, 3A5(HLp2)($M_r$=52,000) rather than with 3A3/4(HLp/NF)($M_r$=51,000). The 52kd protein was indistinguishable from 3A5 on immunoblots developed with a form-specific antibody to 3A5 (kindly provided by Dr. S. Wrighton, Eli Lilly Pharmaceuticals, Indianapolis, IN). No other protein members of the 3A family were present in treated or untreated TONG/HCC cells specifically including 3A7 ($M_r$=51,500) (Wrighton et al., 1989). There was no apparent induction of 3A5 protein by any of the 3A inducers tested in TONG/HCC cells. This finding is consistent with the apparent lack of induction of 3A5 liver microsomal protein from treated human subjects (Wrighton et al., 1989). In contrast, Northern blots of TONG/HCC RNA hybridized with a universal 3A cDNA probe (p55b) (170-890bp), revealed 3A mRNAs that were induced by RIF, DEX and PB but not by PCN. Hence, this cell system, in which 3A mRNA but not 3A protein is inducible may provide useful insights into possible defects in 3A protein translation that may be operative *in vivo*. For example, we have found that human fetal livers express both 3A7 and 3A5 mRNAs but only 3A7 protein. Moreover, we have identified some human livers that contain 3A7 mRNA but not 3A7 protein.

A final system we tested was the human hepatoblastoma cell line, HepG2. We had previously observed that HepG2 cells contained a CYP3A immunoreactive protein and a species of mRNA hybridizable to an 3A cDNA probe (Molowa, et al., 1986). With the use of anti-3A antibodies, 3A cDNAs, and 3A3/4, 3A5 and 3A7 specific oligonucleotides as probes we found a single 3A immunoreactive protein in Dex treated HepG2 cells that exhibited an electrophoretic mobility corresponding to that of 3A7 (HFLa) and did not immunoreact with anti-3A5 IgG. HepG2 cells contained only 3A7 protein, a form found in human fetal liver, even after these cultures had been treated with inducers. Treatment of HepG2 cells with dexamethasone, macrolide antibiotics, phenobarbital and phenobarbital-like inducers, or Lovastatin, produced dose-dependent induction of 3A7 immunoreactive protein and of 3A7 mRNA (D. Beach, manuscript in preparation). The profile of inducers of 3A7 mRNA in HepG2 cells more closely resembles those of LM3c, the rabbit ortholog of 3A3/4, rather than 3A1, its rat counterpart. RIF, a non-inducer of 3A1 in rat proved to be a strong inducer of 3A7 mRNA in HepG2 cultures. Moreover, PCN failed to induce 3A7 mRNA in HepG2 cells. Our data in HepG2 demonstrate that *fetal* 3A7 protein and mRNA are inducible by the same agents (Dex, Rif, PB, TAO and Lovastatin) that have been documented to induce the *adult* forms, 3A3/4 in primary human hepatocytes and in vivo. It is clear that the individual 3A genes are co-inducible, but they are not coordinately regulated even in cells derived from a single individual. We conclude that in addition to primary cultures of human hepatocytes, replicating human hepatic lines HepG2 and TONG/HCC are useful for examining regulation of CYP3A genes polymorphically expressed in the living human (Table 3).

Table 3

## Comparisons of Induction of 3A mRNA in Human Liver Cell Cultures

| Characteristic | Human Hepatoblastoma Hep G2 | Human Hepatoma TONG/HCC | Primary Cultures of Human Hepatocytes |
|---|---|---|---|
| Spontaneously Expressed Form | 3A7 protein<br>3A7 mRNA | 3A5 protein<br>3A5 mRNA<br>3A3/4 mRNA | 3A3/4 protein<br>3A3/4 mRNA<br>3A5 mRNA (if present in the donor liver |
| Inducibly Expressed Form | 3A7 protein<br>3A7 mRNA | 3A3/4 mRNA | 3A3/4 protein<br>3A3/4 mRNA |
| Identified Inducers | glucocorticoids, macrolides, phenobarbital lovastatin | glucocorticoids, macrolides, phenobarbital | dexamethasone, phenobarbital, macrolides, lovastatin, clotrimazole, PCN (some cultures only) |
| Mechanism of Induction | Enhanced transcription by dexamethasone (suppressed by protein synthesis inhibition) | Unknown | Unknown |

## REFERENCES

Molowa, D.T., Schuetz, E.G., Wrighton, S.A., Watkins, P.B., Kremers, P., Mendez-Picon, G., Parker, G.A., and Guzelian, P.S. (1986): Complete cDNA sequence of a cytochrome P-450 inducible by glucocorticoids in human liver. *Proc. Natl. Acad. Sci.* 83, 5311-5315.

Schuetz, E. G., Schuetz, J. D., Strom, S. C., Thompson, M. T., Fisher, R. A., Molowa, D. T., Li, D., and Guzelian, P.S. (1993). Regulation of Human Liver Cytochromes P450 in Family 3A in Primary and Continuous Culture of Human Hepatocytes. *Hepatology* (In press)

Schuetz, E.G., Elswick, B., Li, D., Wrighton, S.A., and Guzelian, P.S. (1988). Regulation of gene expression in adult rat hepatocytes cultured on extracellular matrix. *J. Cell Physiol.* 134, 309-323.

Schuetz, J.D., Beach, D. L., and Guzelian, P.S. (1993). Selective expression of human liver cytochrome P450 CYP3A mRNAs in embryonic and adult human liver *Pharmacogenetics* (In press)

Stevenson D, Lin H-H, Tong MJ and Marshall GJ. (1987): Characteristics of a cell line (TONG/HCC) established from a human hepatocellular carcinoma. *Hepatology* 7, 1291-1295.

Watkins, P.B., Wrighton, S.A., Maurel, P., Schuetz, E.G., Mendez-Picon, G., Parker, G.A., and Guzelian, P.S. (1985). Identification of an inducible form of cytochrome P-450 in human liver. *Proc. Natl. Acad. Sci.* 82, 6210-6314.

Wrighton SA, Ring BJ, Watkins PB and Vandenbranden M. (1989). Identification of a Polymorphically Expressed Member of the Human Cytochrome P-450III Family. *Mol Pharmacol*. 36, 97-105.

Wrighton, S.A., Schuetz, E.G., Watkins, P.B., Maurel, P., Barwick, J., Bailey, B.S., Hartle, H.T., Young, B., and Guzelian, P.S. (1985). Demonstration in multiple species of inducible hepatic cytochromes P-450 and their mRNAs related to the glucocorticoid inducible cytochrome P-450 of the rat. *Mol. Pharm.* 28, 312-321.

# Receptor-dependent coregulation of CYP4A genes and peroxisome proliferation

## G. Gordon Gibson

*University of Surrey, Molecular Toxicology Group, School of Biological Sciences, Guildford, Surrey GU2 5XH, England, UK*

## INTRODUCTION

The CYP4A subfamily consists of at least 9 genes encoding enzymes responsible for the hydroxylation of lipid substrates including fatty acids and prostaglandins, particularly at the terminal carbon, or ω, position (Sharma *et al*, 1989a). Many of these genes (Table 1) are regulated by a structurally diverse group of chemicals known as the peroxisome proliferators (PPs) because of their ability to induce peroxisomal fatty acid β-oxidation, and in particular, the rate limiting enzyme of the spiral, acyl CoA oxidase (Gibson, 1993; Moody *et al*, 1991).

**Table 1:** The cytochrome P4504A Gene Subfamily

| Gene/protein Designation | Trivial Name(s) | Species | Tissue | Induction by Peroxisome Proliferators |
|---|---|---|---|---|
| CYP4A1 | P452, LAω | rat | liver, kidney | induced by clofibrate in liver and kidney |
| CYP4A2 | - | rat | liver, kidney | high constitutive expression in kidney, induced by clofibrate in liver |
| CYP4A3 | - | rat | liver, kidney | induced by clofibrate in liver and kidney |
| CYP4A4 | p-2 | rabbit | lung, kidney, liver, uterus, placenta | induced in lung, liver and uterus by progesterone treatment or during pregnancy |
| CYP4A5 | KDB3 | rabbit | kidney, liver small intestine | induced by clofibrate in liver |
| CYP4A6 | R9, KDA6, Ka-1 | rabbit | kidney, liver | induced by clofibrate in liver |
| CYP4A7 | R4, KDB18, Ka-2 | rabbit | kidney | induction by clofibrate |
| CYP4A8 | PP1 | rat | prostate, kidney | ? |
| CYP4A9 | HL14ACON | human | liver | ? |

The latter induction phenomenon is not confined to the enzymes of lipid metabolism but extends to other enzyme systems including UDP - glucuronyl transferase and epoxide hydrolase (Moody *et al*, 1991). The toxicological interest in PPs stems from the observation that PPs cause substantial liver growth, and are non-mutagenic hepatocarcinogens in susceptible species such as the mouse and the rat, species that are also susceptible to CYP4A induction and peroxisomal proliferation.

## STRUCTURAL DIVERSITY OF CYP4A INDUCERS

To date, over 100 chemicals are known to produce peroxisome proliferation and CYP4A induction (Moody *et al*, 1991) and the structural diversity of these inducers is remarkable. These inducers include clofibrate and related oxyisobutyrate hypolipidaemic drugs, phthalate ester plasticisers (di-(2-ethylhexyl)phthalate), perfluoro straight chain saturated fatty acids (perfluorodecanoic acid), naturally occuring steroids (dehydroepiandrosterone), halogenated hydrocarbons (trichloro ethylene, chlorotrifluoroethylene trimer and Tetramer), phenoxy acid herbicides (2,4,5-T), PCBs, tetrazole acetophenone leukotriene antagonists, azole antifungal drugs (bifonazole), high fat diets and the pathophysiological condition of diabetes (Moody *et al*, 1991; Gibson, 1993). Moreover, CYP4A induction by xenobiotics exhibits stereoselectivity in that the S(+)-isomer of a clofibrate structural analogue, 2-(4-p-chlorophenyl-oxy)-2-phenyl ethanoic acid is a better inducer than the R(-)-isomer (Chinje and Gibson, 1991).

At first sight, this structural diversity of CYP4A inducers may appear rather peculiar, but in the majority of cases, the inducers contain a carboxyl acid group or be metabolised to one. Although some exceptions are known, these latter inducers contain functional groups that are bioisosteric with the carboxyl group and may therefore be a unifying hypothesis to explain the structural diversity (see later).

## MECHANISMS OF CYP4A INDUCTION

Induction of the CYP4A genes studied to date is at the level of transcriptional activation, resulting in increased steady state levels of the cognate mRNAs, particularly in the liver, (Hardwick, *et al*, 1987; Sharma *et al*, 1989b), kidney (Kimura *et al*, 1989; Sharma, *et al*, 1989b) and the lung (Matsubara *et al*, 1987).

Although the majority of CYP4A genes are inducible by clofibrate and other PPs in the liver and kidney, the regulation of the CYP4A4 gene in the lung appears distinct. In the lung, the CYP4A4 gene encodes a P450 that does not ω-hydroxylate fatty acids, but has a substrate specificity for prostaglandin ω-hydroxylation (Yamamoto *et al*, 1984; Matsubara *et al*, 1987). The lung CYP4A4 is not inducible by PPs (as in the liver and kidney) but is regulated by progesterone and is elevated in pregnancy (Yamamoto *et al*, 1984; Matsubara *et al*, 1987).

Sex differences in inducibility of CYP4A induction has been reported in the rat, wherein CYP4A1 and CYP4A3 are induced to a greater extent by clofibrate in male rat liver as compared to female, and male-specific CYP4A2 induction in the kidney has also been noted (Sundseth and Waxman, 1992). The precise regulatory mechanisms contributing to this sex-dependency of induction remain to be fully explored, but it is likely that the pattern of growth hormone secretion and testosterone-dependent regulation play a part (Sundseth and Waxman, 1992).

## THE ROLE OF THE PEROXISOME PROLIFERATOR ACTIVATED RECEPTOR (PPAR) IN CYP4A INDUCTION

Recent studies has emphasised the role of the PPAR in transcriptional gene activation of both peroxisomal acyl CoA oxidase and CYP4A6 (Muerhoff *et al*, 1992). The PPAR is a member of the nuclear steroid hormone receptor family and consists of a DNA binding domain and a ligand (PP)-binding domain (Isseman and Green, 1990). The inducer interacts with the ligand binding domain and the binary PPAR-PP complex then binds to response elements in the 5[1] regulatory region of responsive genes such as acyl CoA oxidase and CYP4A6 (Muerhoff *et al*, 1992). The similarity in response elements in the latter two, functionally diverse enzymes, rationalises on a molecular basis, why these two genes are switched on by PPs. Although the majority of the regulatory elements of PP-responsive genes of the PP domain have not been isolated and characterised to date, it is tempting to speculate that all of the PP domain genes contain a similar response element, enabling PPAR interaction, a hypothesis that awaits further investigation.

Several PPARs have been isolated from the mouse (Issemann and Green, 1990) the rat (Gottlicher et al, 1992) and Xenopus (Dreyer et al, 1992). Whether this multiplicity of PPARs, even within one species (Dreger et al, 1992) would rationalise species differences in induction response and help to explain why there are so many structurally diverse CYP4A inducers (see above), still remains an open question and awaits further research.

Although the role of the (mouse) PPAR in regulation of the rabbit CYP4A6 and the rat acyl CoA oxidase genes is reasonably well substantiated, it should be noted that much less information is available on the regulation of the corresponding human genes by the cognate human PPAR. The human PPAR has recently been isolated and sequenced. The human PPAR exhibits 91% sequence similarity in its amino acid sequence to the mouse PPAR, with the DNA binding domains being almost identical (residue 138 is different). Southern blot analysis suggests that the human PPAR consists of only a single gene, although this does not exclude the possibility of the existence of multiple human PPARs that differ from each other only very slightly, and it is salutary to note that restriction fragment length polymorphisms exist. The human PPAR is capable of activating the *rat* palmitoyl CoA oxidase and the *rabbit* CYP4504A6 genes in the presence of either nafenopin, clofibrate or WY-14,643 and dose-response studies indicate that the human and mouse PPARs are approximately equally effective in driving both of the latter genes. However, it must be emphasised that these latter studies still do not fully address the crucial question of human responsiveness to peroxisome proliferators as the regulatory elements of the *human* palmitoyl CoA oxidase and P450 4A genes have not been tested for activation by the newly described human PPAR. In this context, the evidence that human hepatocytes do not respond to CYP4A inducers (Elcombe and Mitchell, 1986) indicates that either human genes do not contain the necessary response elements to the PPAR or that some other element of PPAR activation is absent in the human. Clearly this is a crucial area for further investigation. Using the known PPAR cDNA sequence and hence predicted amino acid sequence, the PPAR ligand-binding domain has recently been examined using a high resolution, molecular modelling approach (Lewis and Lake, 1993). These latter workers have identified a putative peroxisome proliferator binding site consisting of isoleucine-354, lysine-358, phenylalanine-359 and phenylalanine-361. In this model, the carboxylate anion of the peroxisome proliferator (or its bioisostere as occurs in the tetrazole acetophenone peroxisome proliferators) interacts with the ε-amino group of lysine-358, the two phenylalanine residues interact with the aromatic ring of the inducer by π-π stacking and the isoleucine-354 forms a hydrophobic interaction with the cognate portions of the peroxisome proliferator. An interesting facet of this molecular modelling approach is that the interaction energy (i.e. the binding energy and hence the apparent dissociation constant) of the peroxisome proliferator with the PPAR ligand binding domain can be approximated and a good correlation was noted between peroxisome proliferator binding and the ability of the inducers to activate the receptor in either COS-1 cells or in rat hepatocyte cultures (Lewis and Lake, 1993).

The above molecular modelling approach has also been useful in rationalising why the R (-)-isomer of a clofibrate analogue (2-[4-(4-chlorophenyl) benzyloxy]-2-phenyl acetic acid, is an inducer of cytochrome P4504A1 whereas the corresponding S(+)-isomer is only weakly active (Chinje and Gibson, 1991). Molecular modelling studies in our laboratory have demonstrated that the R(-)-isomer interacts with the PPAR much better than its antipode, as reflected in the $\Delta G$ values of -16.9 kcal.mol$^{-1}$ and 12.0kcal.mol$^{-1}$ respectively. Similarly, bifonazole pretreatment of rats results in approximately a 5-fold induction of cytochrome P4504A-dependent fatty acid ω-hydroxylase activity compared to no significant induction of activity with an equimolar dose of the structurally related drug, clotrimazole. Again, using the above molecular modelling approach, the latter specificity for induction is reflected in the more facile docking of bifonazole with the PPAR than clotrimazole, as demonstrated in the more favourable binding values ($\Delta G$) of -37.1 kcal.mol$^{-1}$ and -14.6 kcal.mol$^{-1}$ respectively. (Sabzevari and Gibson, unpublished observations).

An enduring and intriguing aspect of peroxisome proliferator-dependent induction of the cytochrome P4504As is the role of lipids and lipid metabolism. It would appear that the majority of peroxisome proliferators inhibit hepatic fatty acid oxidation and the subsequent accumulation of fatty acids are thought to induce the cytochrome P4504A family. These observations then raise the possbility of a more complex regulation of the CYPP4504As, whereby both exogenous peroxisome proliferators and/or endogenous fatty acids regulate P4504A induction (Fig.1).

Consistent with this hypothesis is the observation that fatty acids can regulate and activate the rat PPAR.

**Figure 1**  Role of Peroxisome Proliferators and Lipids in the PPAR-dependent Regulation of Cytochrome P4504A Genes

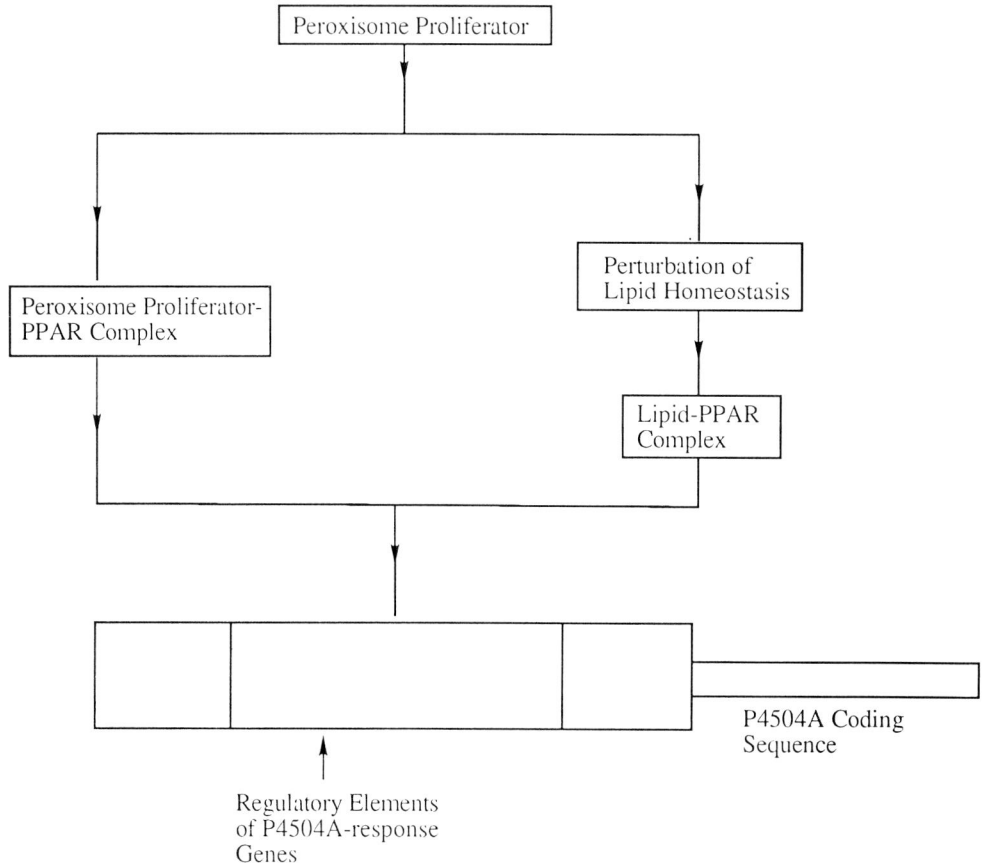

## REFERENCES

Chinje, E. and Gibson, G.G. (1991): Stereochemical selectivity in the induction of cytochrome P450IVA1 (P452) - dependent fatty acid hydroxylation and peroxisome prolierferation. Biochem. Pharmacol. 41, 769-774

Dreyer, C., Krey, G., Keller, H., Givel, F., Helftenbein, G. and Wahli, W. (1992): Control of the peroxisomal β-oxidation pathway by a novel family of nuclear hormone receptors. Cell, 68, 879-888.

Elcombe, C.R. and Mitchell, A.M. (1986): Peroxisome proliferation due to di-(2-ethylhexyl) phthalate (DEHP). Species differences and possible mechanisms. Environ. Health Perspec. 70, 211-219.

Gibson, G.G. (1993): Peroxisome proliferators: paradigms and prospects. Toxicol. Lett. 68, 193-201.

Gottlicher, M., Widmark, E., Li, Q. and Gustafsson, J.A. (1992): Fatty acids activate chimera of the clofibric acid - activated receptor and the glucocorticoid receptor. Proc. Nat. Acad. Sci (USA), 89, 4653-4651.

Issemann, I. and Green, S. (1990): Activation of a member of the steroid hormone receptor superfamily by peroxisome proliferators. Nature, 347, 645-649.

Kimura, S., Hardwick, J.P., Kozak, C.A. and Gonzalez, F.J, (1989): The rat clofibrate-inducible CYP4A subfamily II. cDNA sequence of 4A3, mapping of the CYP4A locus to mouse chromosome 4 and coordinate and tissue-specific regulation of the CYP4A genes. DNA, 8, 517-525.

Lewis, D.F.V. and Lake, B.G. (1993). Interaction of some peroxisome proliferators with the mouse liver peroxisome proliferator activated receptor (PPAR): a molecular modelling and quantitative structure-activity relationship (QSAR) study. Xeno. 23, 79-96.

Matsubara, S., Yamamoto, S., Sogawa, K., Yokotani, N., Fujji-Kuriyama, Y., Haniu, M., Shively, J.E., Gotoh, O., Kusunose, E. and Kusunose, M. (1987). cDNA cloning and inducible expression during pregnancy of the mRNA for rabbit pulmonary prostaglandin w-hydroxylase (cytochrome P450 p-2). J. Biol. Chem. 262, 13366-13371.

Moody, D.E., Reddy, J.K., Lake, B.G., Popp, J.A. and Reese, D.H. (1991). Peroxisome proliferation and nongenotoxic carcinogenesis: Commentary on a symposium. Fund. Appl. Toxicol., 16, 233-248.

Muerhoff, A.S., Griffin, K.J. and Johnson, E.F. (1992). The peroxisome proliferator-activated receptor mediates the induction of CYP4A6, a cytochrome P450 fatty acid w-hydroxylase, by clofibric acid. j. Biol. Chem. 267, 19051-19053.

Reddy, J. and Lalwani, N.D. (1983). Carcinogenesis by peroxisome proliferators: evaluation of the risk of hypolipidaemic drugs and industrial plasticisers to humans. CRC Crit. Rev. Toxicol. 12, 1-58.

Sharma, R., Doig, M.V., Lewis, D.F.V. and Gibson, G.G. (1989a). Role of hepatic and renal cytochrome P4504A1 in the metabolism of lipid substrates. Biochem. Pharmacol. 38, 3621-3629.

Sharma, R., Lake, B.G., Makowski, R., Bradshaw, T., Earnshaw, D., Dale, J.W. and Gibson, G.G. (1989b). Differential induction of peroxisomal and microsomal fatty acid - oxidising enzymes by peroxisome proliferators in rat liver and kidney. Eur. J. Biochem. 184, 69-78.

Sher, T., Yi, H.F., McBride, O.W. and Gonzalez, F.J. (1993). cDNA cloning, chromosomal mapping and functional characterisation of the human peroxisome proliferator activated receptor. Biochem. 32, 5598-5604.

Sundseth, S.S. and Waxman, D.J. (1992). Sex-dependent expression and clofibrate inducibility of cytochrome P4504A fatty acid w-hydroxylation from lung microsomes of rabbits treated with progesterone. J. Biochem. 96, 593-602.

# Phenobarbital induction of P450 cytochromes in rat liver: genetic and endocrine control of a common regulatory mechanism

M. Larsen[1], F. Ikegwuonu[2], P. Brake[1], C. Jefcoate[2]

[1]Environmental Toxicology Center and the [2]Department of Pharmacology, University of Wisconsin, Madison, WI, 53706, USA

## INTRODUCTION

The hepatic xenobiotic drug-metabolizing system, *in vivo* or in appropriately cultured hepatocytes, responds to the presence of many foreign chemicals via a selective increase in Phase I and Phase II enzyme expression. These inductive responses are typically transcriptionally mediated and have been categorized on the basis of the chemical structure of the inducing compound as well as with regard to characteristic gene-specific induction. Phenobarbital (PB) is a prototypical inducer which increases transcription of several forms of cytochrome P450 (2B1, 2B2, 2C6, 2A1, 2A2, 3A1, and 3A2), UDP-glucuronyl transferases, glutathione S-transferases, as well as microsomal epoxide hydrolase (EH), NADPH-dependent cytochrome P450 oxidoreductase, and aldehyde dehydrogenase (Waxman & Azaroff, 1992). A second major category of chemical inducers, typified by 2,3,7,8-tetrachlorodibenzo-p-dioxin (TCDD), mediate the transcriptional induction of generally distinct members of the same superfamilies (cytochromes P-450 1A1 and 1A2, UDP-glucuronyl transferases, glutathione S-transferases, and aldehyde dehydrogenase), while NADPH-dependent cytochrome P450 oxidoreductase and microsomal EH are not affected (Whitlock, 1986). This process is mediated by the Ah-receptor, a cytosolic protein which migrates to responsive genes following binding with the inducer. A third class of steroid inducers, represented by dexamethasone (DEX), demonstrates the receptor-mediated selective induction of several cytochrome P450 forms also stimulated by PB (2B1, 2B2, 3A1, and 3A2) while failing to affect other PB-inducible genes (Gonzalez, 1989).

The PB induction pattern is repeated by a diverse array of chemicals (e.g., non-planar PCBs, benzene derivatives, and organochlorine pesticides), some of which are structurally related to Ah-receptor agonists. A high affinity, non-planal PCB binding protein has recently been identified (Buff & Bründl, 1992). The definitive presence of a PB receptor and the mechanism of PB induction remain to be elucidated. However, He and Fulco (1991) have recently identified PB-responsive cis- and trans-acting DNA regulatory elements that are present in both mammalian and bacterial cytochrome P-450 genes and which bind specific nuclear proteins.

Several forms of cytochrome P450 are preferentially expressed in either male ( 2C11, 2C13, 2A2, 3A2 ) or female ( 2C12 ) rats (Pampori *et al.*, 1991) and are often regulated (2C11 and 2C12) by androgen- and estrogen-induced differences in growth hormone (GH) secretion patterns (Sundseth *et al.*, 1992). Thus, the expression of many of the PB-responsive, xenobiotic-metabolizing enzymes is also under endocrine regulation. GH is responsible for the suppression of the constitutive as well as the induced expression of cytochromes P450 2B1 and 2B2 in a gender-specific manner, with suppression being greater in female than in male rats (Murayama *et al.*, 1991). Thyroxine (T3) mediates a similar suppression of 2B2 (Murayama *et al.*, 1991) while enhancing the expression of NADPH-dependent cytochrome P450 oxidoreductase (Ram & Waxman, 1992).

A genetic linkage of PB inducibility has been demonstrated in Long Evans rats which have been bred as homozygous- responsive (RR) and non-responsive (rr), whereby the rats demonstrate differential responsiveness in aldehyde dehydrogenase induction, leaving CYP 2B1, 2B2, and GST Ya expression unaffected (Dunn et al., 1989). Recently, Lubet et al. (1992) observed a decreased responsiveness in PB-type induction of CYP 2B1, glutathione S-transferases Ya/Yc, and epoxide hydrolase in female lean and obese Zucker rats relative to the F344/NCr and DA strains. A similar, although somewhat diminished, strain sensitivity was observed in the male counterparts. However, the obese phenotype represents a genetic autosomal-recessive animal model of obesity with numerous metabolic and endocrine abnormalities which may affect PB induction. We have identified a gender-selective, endocrine-mediated strain polymorphism of PB-inducible expression in inbred Fischer (F344), Wistar Furth (WF), and Wistar Kyoto (WK) rat strains.

## RESULTS

Immunoblot analysis, using P450-specific IgG, has shown that PB induction of five inducible P450s (2B1, 2B2, 3A1, 2A1, and 2C6) is substantially greater in female F344 than in WF rats, while no difference was observed in the expression of a constitutive form that is not induced by PB (1A2) (Table 1). The data demonstrates that the PB-mediated induction of these P450 forms is strongly suppressed (up to 10-fold) in the WF rat relative to the F344 strain, with the WK rat displaying an intermediate level of responsiveness. Male rats, previously shown to be more effectively induced by PB than their female counterparts (Murayama et al., 1991), failed to exhibit substantial strain-specific differences upon induction. Oligonucleotide and cDNA hybridization analyses revealed a similar pattern of genetic expression following PB administration. Furthermore, equivalent differences in the induction of functional cytochromes were indicated by metabolism of 7-pentoxyresorufin and 12-methyl hydroxylation of dimethylbenz(a)anthracene (DMBA) markers of, respectively, forms 2B1 and 2B2. A similar pattern of inducibility has been obtained with respect to 2B1, 2B2, and 3A1 expression in female F344 and WF rat strains with the structurally dissimilar chemicals, 2-hexanone (HEX), d-limonene (LIM), dieldrin (DIELD), and 2,4,5,2',4',5'-hexachlorobiphenyl (HCB), while DEX demonstrated no strain selectivity of responsiveness (Fig. 1).

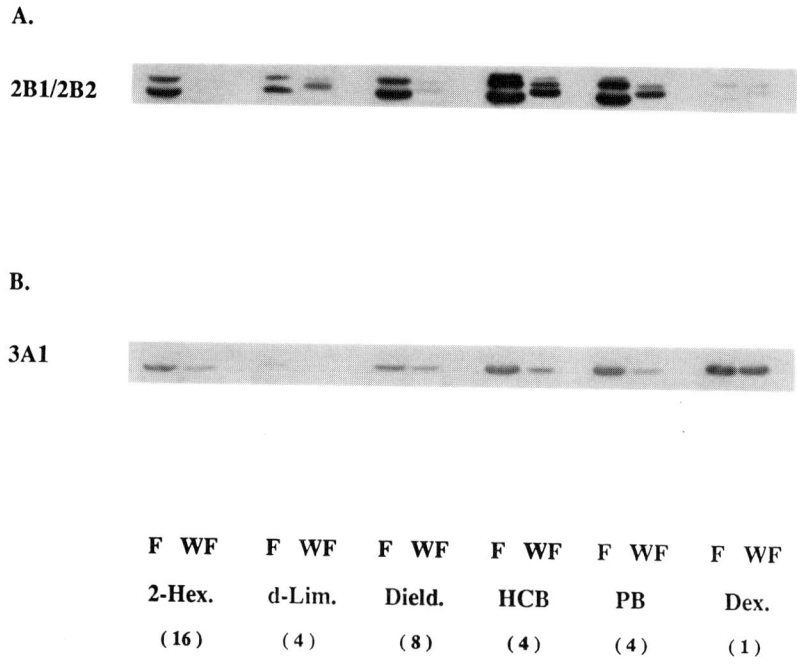

Fig. 1. Immunoblot analysis of P450 2B1/2B2 (A) and 3A1 (B) expression in female rats previously induced with 2-Hexanone (2-HEX.), d-Limonene (d-LIM.), Dieldrin (Dield.), 2,4,5,2',4',5'-Hexachlorobiphenyl (HCB), Phenobarbital (PB), and Dexamethasone (DEX.). Relative microsomal protein loadings are as indicated (μg/lane).

TABLE 1
Normalized Cytochrome P-450 Expression in PB-Induced Female Rats

| STRAIN | 2B1+2B2 | 3A1 | 2C6 | 2A1 | 1A2 | EH[a] |
|---|---|---|---|---|---|---|
| F344 | 100 (2.4)[b] | 100 (5.0) | 100 (1.6) | 100 (4.1) | 100 (13.2) | 100 |
| WF | 12.7 (3.8) | 11.4 (2.1) | 55.5 (11.8) | 28.5 (6.9) | 133.8 (9.0) | 50.4 |
| WK | 25.1 (3.4) | 29.8 (4.7) | 85.9 (15.6) | 59.9 (8.0) | 135.3 (10.0) | 87.5 |

Rats were induced with PB (60mg PB/kg body weight) daily, for three consecutive days, prior to analysis. Immunoblots were quantitated using a scanning laser densitometer and the peak areas normalized relative to the respective F344 strain (as assigned 100).

[a] Additional samples must be analyzed in order to determine statistical variance.
[b] (SEM), representing data from n=4 to n=11 determinations.

The same trend, although involving smaller differences, was observed for the induction of EH by both PB and HCB. Thus, we conclude that a genetic difference between the strains influences the effectiveness of PB and related chemicals as inducers of the cytochrome P450 and possibly EH genes and provides additional evidence for a single receptor-mediated pathway.

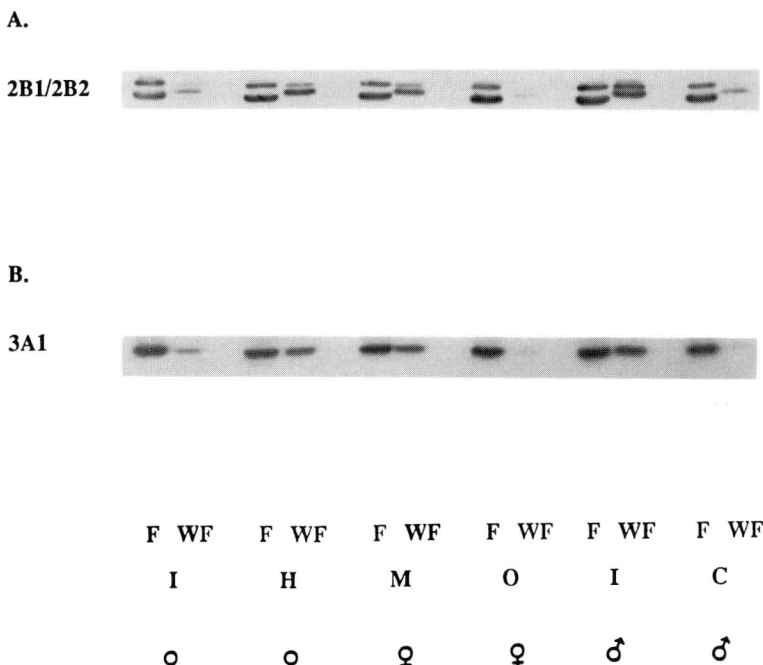

Fig. 2. Immunoblot analysis of P450 2B1/2B2 (A) and 3A1 (B) expression in PB-induced female and male rats. Microsomal fractions (1μg/lane) were analyzed from Intact (I), Hypophysectomized (H), Methimazole-treated (M), Ovariectomized (O), and Castrated (C) rats.

The role of endocrine regulation in the genetic polymorphism of PB-mediated induction was examined in hypophysectomized (20 days prior to PB induction) female rats, such that circulating levels of GH, T3, and sex steroids should be depleted (Fig. 2). While constitutive levels of cytochromes 2B1, 2B2, and 3A1 were raised, the strain differences in PB induction were greatly diminished as the suppression in the WF strain was substantially relieved. The administration of methimazole (0.025% w/v in drinking water for 20 days) to chemically deplete T3 and and indirectly lower GH levels, similarly resulted in an increase in the induced expression of these P450s in the WF rat over levels in the intact animal, while expression in the F344 strain remained constant. Conversely, the genetic differences became increasingly pronounced following ovariectomy (PB induction 20 days post surgery), whereby expression in the F344 strain decreased significantly from that of the intact rat while rendering induction in the F344 unaltered. Short-term castration (20 days) of males, however, yielded only a small decrease in PB induction in the F344 rats, contrasting a large suppression in the WF strain, which only then exhibited responses and strain differences very similar to their female counterparts.

From the data obtained thus far, we propose the model of hormonal regulation of the P450 genes, as shown in Fig. 3. We conclude that there is a general P450 regulatory alteration which suppresses PB-mediated induction of multiple P450s in the WF strain. This suppression mechanism involves pituitary hormone regulation, which is counteracted by estradiol and testosterone. The increased responsiveness of male rats of both strains to PB administration and the castration-mediated suppression of the WF strain suggest that the gender differences may result largely from a greater antagonism by testosterone of the suppressive pituitary hormones (T3 and GH) than by estradiol. This model system should provide an invaluable tool in the delineation of the genetic mechanism(s) responsible for the PB-mediated induction of the Phase I and Phase II xenobiotic-metabolizing enzymes.

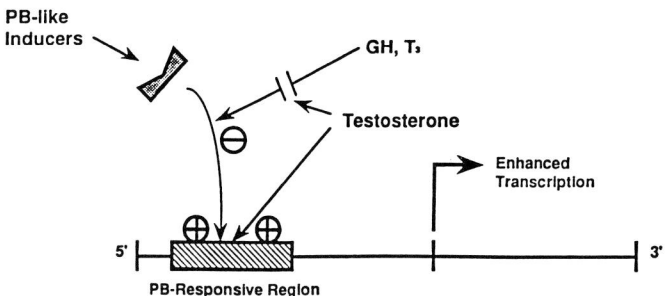

Figure 3. Model for hormonal regulation of phenobarbital (PB) induction of P450 genes. Alternative mechanisms for testosterone stimulation are shown.

**REFERENCES**

Buff, K. & Bründl, A. (1992): Specific binding of polychlorinated biphenyls to rat liver cytosol protein. *Biochem. Pharmacol.* 43, 965-970.

Dunn, T.J., Koleske, A.J. *et al.* (1989): Phenobarbital-inducible aldehyde dehydrogenase in the rat: cDNA sequence and regulation of the mRNA by phenobarbital responsive rats. *J. Biol. Chem.* 264, 13057-13065.

Gonzalez, F.J. (1989): The molecular biology of cytochrome P450s. *Pharmacol. Rev.* 40, 243-288.

He, J.S., and Fulco, A.J. (1991): A barbituate-regulated protein binding to a common sequence in the cytochrome P450 genes of rodents and bacteria. *J. Biol. Chem.* 266, 7864-7869.

Lubet, R.A., Nims, R.W. *et al.* (1992): A markedly diminished pleiotropic response to phenobarbital and structurally-related xenobiotics in Zucker rats in comparison with F344/NCr or DA rats. *Biochem. Pharmacol.* 43, 1079-1087.

Murayama, N., Shimada, M. *et al.* (1991): Difference in the susceptibility of two phenobarbital-inducible forms, P450 2B1 and P450 2B2, to thyroid hormone- and growth hormone-induced suppression in rat liver. *Mol. Pharmacol.* 39, 811-817.

Pampori, N.A., Agrawal A.K. *et al.* (1991): Differential effects of neonatally administered glutamate on the ultradian pattern of circulating growth hormone regulating expression of sex-dependent forms of cytochrome P450. *Biochem. Pharmacol.* 41, 1299-1309.

Ram, P.A., and Waxman, D.J. (1992): Thyroid hormone stimulation of NADPH P450 Reductase expression in liver and extrahepatic tissues: regulation by multiple mechanisms. *J. Biol. Chem.* 267, 3294-3301.

Sundseth, S.S., Alberta, J.A., and Waxman, D.J. (1992): Sex-specific, growth hormone-regulated transcription of the cytochrome P450 2C11 and 2C12 genes. *J. Biol. Chem.* 267, 3907-3914.

Waxman, D.J., and Azaroff, L. (1992): Phenobarbital induction of cytochrome P-450 gene expression. *Biochem. J.* 281, 577-592.

Whitlock, J.P. (1986): The regulation of cytochrome P450 gene expression. *Ann. Rev. Pharmacol. Toxicol.* 26, 333-369.

# Structural and functional characteristics of CYP3A genes related to the development-dependent phenotypic expression and induction by Dexamethasone

Maria Celeste Lechner, Maria Margarida Telhada, Vera Ribeiro, Teresa Maria Pereira, Maria João Gama

*Laboratório de Bioquímica, Instituto Gulbenkian de Ciência, Ap.14, 2781 Oeiras, Portugal*

## INTRODUCTION

Steroid inducible Cytochrome P450's have been long described in rodent liver (Lu *et al*, 1972), named Cytochrome P450 PCN and subsequently purified from Pregnenolone16α-carbonitrile-treated rats (Elshourbagy and Guzelian,1980).
Cloning of two distinct cDNAs from rat liver, the PCN1 and PCN2 (Hardwick *et al*, 1983, Gonzalez *et al*,1986), led to the identification of the Cytochrome P450 sub-family CYP3A, further proven to be widely represented in all mammalian organisms, including man.

CYP3A1 and CYP3A2 are predominantly expressed in the liver . Despite their high degree of structural homology, representing 90% of the mRNA nucleotide sequences, and the overlapping catalytic properties of the encoded enzymes, these two genes display markedly distinct development-dependent and constitutive patterns of expression.
CYP3A1, formerly considered as a non constitutive form, has been later demonstrated to be transiently expressed in the liver at early post-natal stages (Shimada *et al*,1989; Ribeiro and Lechner 1992) and found at residual levels in the adult hepatic tissue, in the absence of stimulation by specific inducers (Nagata *et al*,1990).
CYP3A2 is the major constitutive testosterone 6β-hydroxylase in rat liver. A close relationship between the male hormonal *status* and the concentration of the CYP3A2 mRNA has been demonstrated. Actually, we observed that orchidectomy results in a drastic decrease in CYP3A2 mRNA concentration in the male rat liver, which is almost completely restored by testosterone replacement. Therefore, we postulated that late androgens may regulate the constitutive expression of CYP3A2 gene (Ribeiro and Lechner,1992).

The prototype inducer of CYP3A1 and CYP3A2 is the anti-glucocorticoid PCN, but it has been observed that these forms are markedly induced by the synthetic GC agonist dexamethasone (DEX), as well as by phenobarbital (PB), and by macrolide antibiotics.

Although CYP3A1 and CYP3A2 have been the most intensively studied, the first Southern blot analysis already indicated a larger number of CYP3A genes to exist in the rat genome Accordingly, seven different genes were reported (Kato *et al,* 1992).

The diversity of highly homologous CYP3A genetic entities and their partially overlapping patterns of expression anticipate that these genes are controlled by particularly complex arrays of regulatory elements. Resulting from gene duplication phenomena in evolution, this sub-family genes are expected to contain some common regulatory motifs underlying liver-specific and inducer stimulated transcription which should act in combination with sets of distinct elements controlling the development-dependent and hormone regulated expression of each particular *locus*.

The characterization of the molecular mechanisms involved in the multifactorial regulation of the CYP3A genes has not yet been achieved.

**RESULTS AND DISCUSSION**

We have been interested in the study of CYP3A gene regulation.

Rat genomic libraries were constructed using the EMBL3cosw vector and E.coli NM626 strain. One million recombinants were successively screened with the cDNA probes and specific oligonucleotides, leading to the isolation of CYP3A1 (26Kb) and CYP3A2 (24Kb) rat genomic DNA. Striking structural differences, affecting the 5' regions of these two genes and their flanking promoters, were identified (Telhada *et al*, 1992).

The existence of an unique long intronic sequence in the CYP3A1 gene provided a valuable molecular probe for measuring the activity of this specific gene by run-on-transcription assay in the isolated liver nuclei. This revealed a marked change to occur in the pattern of CYP3A1 gene inductive response to Dexamethasone (DEX), in the adult animals as compared to the immature. Further studies on the time course variations of the gene transcription activities and mRNA accumulation, revealed a sharp and early inductive effect on the CYP3A1 transcription in the immature animals, suggesting that a primary response to DEX occurs, in contrast to the markedly delayed transcription activation that preceeds the DEX-induced accumulation of the mRNA in the adult rat liver (Pereira and Lechner,1994).

Conversely to genes under strict glucocorticoid-receptor (GR)-driven control, requiring ligand activation by agonists, *e.g.* Tyrosine aminotransferase (TAT), CYP3A are activated by both glucocorticoid agonists and antagonists. Yet, both type ligands must be present at one to three orders of magnitude higher concentrations to produce CYP3A inductive response, than in the case of TAT induction by DEX, in the same experimental conditions.

These facts illustrate the complexity and particularities of the physiological mechanisms that govern the expression of CYP3A genes.

Transcription activation of specific genes by DEX has been postulated to occur through two distinct mechanisms which are the binding of the GR complex to GRE in the *locus* controlling region (Chen and Feigelson, 1979), and the action of gene regulatory proteins synthesised *de novo* in response to DEX (Addison and Kurtz, 1986). Based on the observation of the antagonic effects of PCN and other anti-glucocorticoids on TAT and on CYP3A gene expression, it was postulated that CYP 3A1 induction by DEX should not proceed through a classical GR driven mechanism (Schuetz *et al*,1984). The same authors (Burger *et al, 1992*) demonstrated by hepatocytes transfection-transactivation experiments, that a 164-bp-long DNA promoter segment, capable of driving transcription activation by relatively high doses of DEX, does not contain the canonical GRE consensus elements.

By sequencing analysis of the proximal regions upstream to the transcription initiation site of the CYP3A1, and the CYP3A2 genes, and screening for homologies with known *cis*-regulatory elements, we demonstrated the presence of numerous consensus elements that may putatively be involved in the transcriptional control of these genes, namely the tissue-specific expression and the up-regulation by glucocorticoids (Telhada *et al*, 1992). Considering the anticipated complexity of the control mechanisms of CYP3A, that do not correspond to models already established for other genes, we adopted an experimental strategy susceptible of evidentiating the primary targets and interacting regulatory factors that determine the phenotypic expression of each one of the previously cloned genes.

The hierarchy by which transcription factors interact with the native chromatin to accede to the key regulatory elements greatly depends on the supra molecular structure of the DNA at each transcriptionaly active *locus*, characteristic of each differentiated tissue.
Nuclease hypersensitive sites (DHS) have been demonstrated to correlate with the different transcriptional states of mammalian genes, representing "open windows" that provide accessibility of DNA *cis*-regulatory elements to major regulatory factors.

At present, most of the components that control hepatocyte differentiation have been defined. By the other hand, a number of transcription factors have been demonstrated to control the positioning of nucleosomes in model eukaryotic cells, that determine the active transcription of the different genes in a combinatorial control of the critical steps.

GR protein regulates the transcription of numerous liver-specific genes by binding to GRE, disrupting the nucleosome structure and allowing other regulatory proteins to bind, which are essential for the formation of transcription initiation complex (Perlmann,T.,1992; Richard-Foy and Hager, 1987; Cordingley *et al*, 1987).

In an atempt to identify the regulatory elements primarily involved in the developmentaly regulated switch-on of CYP3A genes by glucocorticoids, we analysed the supra molecular organisation of CYP3A1 and CYP3A2 genomic *loci*.
DNase hypersensitive sites have been identified in both active *loci*, sitting in the proximal and in the distal 5' flanking regions, that may contain critical promoter and enhancer elements (Telhada and Lechner, 1994).

Functional transactivation assays are in progress to define the anticipated combinations of regulatory elements and factors that determine the up-regulation of CYP3A1 and CYP3A2 genes by DEX, as a function of the developmental stage, and to establish the analogies and differences with the mechanisms involved in the transcription activation by the anti-glucocorticoids.

# REFERENCES

Addison, W.R., and Kurtz, D.T. (1986): Nucleotide sequences required for the regulation of a rat α2u-globulin gene by glucocorticoids. *Mol.Cell. Biol.* 6, 2334-2346.
Burger,H.J., Schuetz, J.D., Schuetz, E.,and Guzelian,P.S. (1992): Paradoxical transcriptional activation of rat liver cytochrome P450 3A1 by dexamethasone and the antiglucocorticoid pregnenolone 16α–carbonitrile: analysis by transient transfection into primary monolayer cultures of adult rat hepatocytes. *Proc. Natl. Acad. Sci. USA, 89, 2145-2149.*
Chen,C.-L.C., and Feigelson, P. (1979): Cycloheximide inhibition of hormonal induction of α2u-globulin mRNA .*Proc. Natl. Acad. Sci. USA,* 76, 2669-2673.

Cordingley, M.G., Tate Riegel, A. and Hager, G.L. (1987): Steroid-depenent interaction of transcription factors with the inducible promoter of mouse mammary tumor virus in vivo. *Cell*, 48, 261-270.

Elshourbagy, N.A., and Guzelian, P.S. (1980): Separation, purification and characterization of a novel form of hepatic cytochrome P450 from rats treated with pregnenolone-16α-carbonitrile. *J. Biol. Chem.* 255:1279-1285.

Gonzalez, F.J., Song, B.J. and Hardwick, J.P. (1986). Pregnenolone 16α-carbonitrile-inducible P450 gene family: gene conversion and differential regulation. *Mol. Cell. Biol.* 6, 2969-2976.

Hardwick, J.P., Gonzalez, F.J., Kasper, C.B. (1983): Cloning of DNA complementary to cytochrome P450 induced by pregnenolone-16α-carbonitrile. *J. Biol. Chem.* 258:10182-10186.

Kato, R., Nagata, K., Miyata, M. and Yamazoe, Y. (1992): Isolation of CYP3 gene family. *J. Bas. & Clin Physiol. & Pharmacol.*, supplement, 3: 108.

Lu, A.Y.H., Somogyi, A., West, S., Kunzman, R., Conney, A.H. (1972): Pregnenolone -16α-carbonitrile: a new type of inducer of drug-metabolizing enzymes. *Arch. Biochem. Biophys.* 152, 457-462.

Nagata, K., Gonzalez, F.J., Yamazoe, Y. and Kato, R. (1990): Purification and characterization of four catalytically active testosterone 6β-hydroxylase P450s from rat liver microsomes: comparison of a novel form with three structurally and functionally related forms. *J. Biochem.* 107, 718-725.

Pereira, T.M. and Lechner, M.C. (1994): Developmentaly programmed changes in CYP3A1 induction by dexamethasone in male rat liver. Submitted.

Perlmann, T. (1992): Glucocorticoid receptor DNA-binding specificity is increased by the organization of DNA nucleosomes. *Proc. Natl. Acad. Sci. USA*, 89, 3884-3888.

Ribeiro, V.L. and Lechner, M.C. (1992): Cloning and characterization of a novel CYP3A1 allelic variant: analysis of CYP3A1 and CYP3A2 sex-hormone-dependent expression reveals that the CYP3A2 gene is regulated by testosterone. *Arch. Biochem. Biophys.* 293, 147-152.

Richard-Foy, H. and Hager, G.L. (1987): Sequence-specific positioning of nucleosomes over the steroid-inducible MMTV promoter. *EMBO J.*, 6, 2321-2328.

Schuetz, E.G., Wrighton, S.A., Barwick, J.L. and Guzelian, P.S. (1984): Induction of cytochrome P450 by glucocorticoids in rat liver. I. Evidence that glucocorticoids and pregnenolone 16α-carbonitrile regulated *de novo* synthesis of a common form of cytochrome P450 in cultures of adult rat hepatocytes and in the liver *in vivo*. *J. Biol. Chem*, 259, 1999-2006.

Shimada, M., Nagata, K., Murayama, N., Yamazoe, Y. and Kato, R. (1989): Role of growth hormone in modulating the constitutive and phenobarbital-induced levels of two P450 6β (testosterone 6β-hydroxilase) mRNAs in rat livers. *J. Biochem.* 106, 1003-1034

Telhada, M.M. and Lechner, M.C. (1994): Identification of DNase hypersensitive sites at the CYP3A1 and CYP3A2 *locus* in expressing and in non-expressing rat tissues. Submitted.

Telhada, M.M., Pereira, T.M. and Lechner, M.C. (1992): Effect of dexamethasone and phenobarbital on run-on transcription rate and CYP3A mRNA concentration in rat liver. Changes during development. *Arch. Biochem. Biophys.*, 298, 715-725.

## ACKNOWLEDGEMENTS

V. Ribeiro and T.M. Pereira are recipient of JNICT Ph.D. fellowships.

# Expression of xenobiotic-metabolizing cytochrome P450 forms in human adult and fetal liver

Jukka Hakkola[1], Markku Pásanen[2], Raija Purkunen[3], Seppo Saarikoski[3], Olavi Pelkonen[1], Jukka Mäenpää[1], Anders Rane[4], Hannu Raunio[1]

[1]Department of Pharmacology and Toxicology, University of Oulu, 90220 Oulu, Finland. [2]Department of Pharmacology and Toxicology, [3]Department of Obstetrics and Gynaecology, University of Kuopio, 90721 Kuopio, Finland. [4]Department of Clinical Pharmacology, Akademiska Hospital, University of Uppsala, S-751 85 Uppsala, Sweden

## Abstract

Expression of human cytochrome P450 (CYP) genes in human adult and fetal liver were studied using reverse transcriptase-polymerase chain reaction (RT-PCR) method. In adult liver mRNA of CYPs 1A1, 1A2, 2A6/2A7, 2B6/2B7, 2C8-19, 2D6, 2E1, 3A3/3A4 and 3A7 were detected while CYPs 2F1 and 4B1 were absent. In fetal liver mRNA of CYPs 2C8-19, 2D6, 3A3/3A4 and 3A7 were discovered but all other forms studied were undetectable. The results provide a comprehensive qualitative picture of the expression of CYP genes in families CYP1 through CYP4 in human adult and fetal liver.

## INTRODUCTION

The human fetus is capable of metabolizing many xenobiotics relatively actively, usually at a lower rate than the adult (Pelkonen, 1980). Human fetal liver contains multiple but fewer forms of cytochrome P450 as compared with adult liver. There is evidence for the presence of mRNA or protein of members in the CYP1A (Pasanen et al., 1987) and CYP3A (Ladona et al., 1988) subfamilies and CYP2D6 (Treluyer et al., 1991) and CYP17A (Rane et al., 1992) in human fetal liver.

The greatest difficulty in the study of fetal CYP system has been the low levels of individual forms present. RT-PCR allows detection of low quantities of mRNA with great sensitivity and specificity, making it very suitable for analysing human tissues. In this study we used the RT-PCR method to assess P450 gene expression in human adult and fetal liver.

## MATERIALS AND METHODS

**Liver samples.** Six adult livers and 16 fetal livers were used in this study. The adult livers were obtained from the Department of Surgery, University of Oulu. Nine of the fetal livers (U3-U43) were from Akademiska Hospital, Uppsala, Sweden. Seven fetal livers (K4-K13) were from Department of Obstetrics and Gynaecology, University of Kuopio, Finland. All the liver samples were frozen in liquid nitrogen and stored at -70°C until mRNA extraction. The storing times for the Uppsala samples were between 2 and 4 years. The Kuopio samples were used within one year. The use of human tissues was approved by the Ethics Committees of the Universities of Oulu, Kuopio and Uppsala. Messenger RNA was extracted from about 50 mg of tissue with QuickPrep Micro mRNA purification Kit and complementary DNA was synthesized with First-Strand cDNA Synthesis Kit (both from Pharmacia P-L Biochemicals, Uppsala, Sweden).

**PCR.** The PCR reaction contained 1 µl of cDNA (out of 15µl total), 2.5 U DynaZyme DNA polymerase (Finnzymes, Helsinki, Finland), 5 µl 10x DynaZyme reaction buffer, dNTP reaction mix (Pharmacia) at final concentration of 400 µM, 20-50 pmol of each primer, and water to final volume of 50 µl. 25 PCR cycles were performed: 1 min at 94°C, 1 min at 55 °C, 2 min at 72°C. The primers were designed to hybridise only to the desired regions of cDNA. To exclude chances of cross-hybridisation with other sequences, each primer was compared against the EMBL human gene bank with the Genetics Computer Group FASTA program. The primers were also designed to amplify regions containing at least one intron in the gene to exclude contamination of cDNA with genomic DNA. The following primers will detect more than one isoform: CYP2A6/2A7, CYP2B6/2B7, CYP3A3/3A4. The CYP2C(8-19) primers were designed to detect all known human CYP2C cDNAs, i.e. 2C8, 2C9, 2C10, 2C18, 2C19.

## RESULTS

Expression of CYP genes in families *CYP1-4* was studied with RT-PCR method in human adult and fetal liver tissues. A representative amplification result by each pair of primers with adult liver is shown in Figure 1.

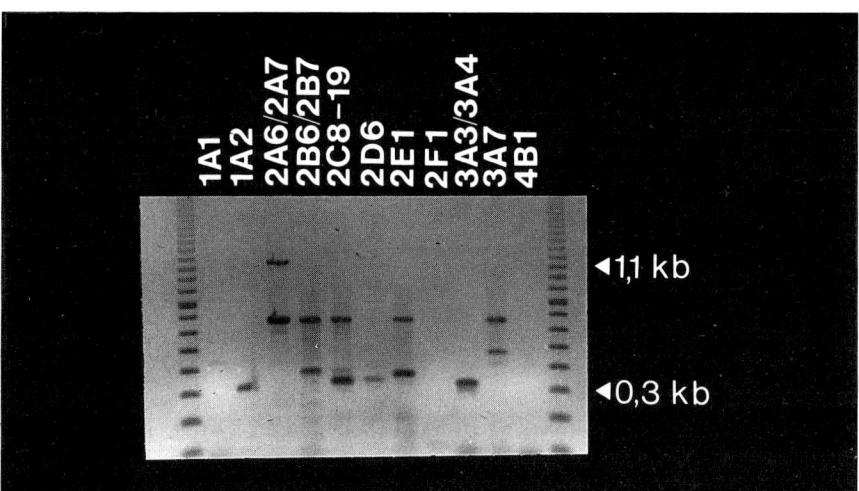

Fig. 1. A representative PCR amplification result by each pair of primers with an adult liver. The PCRs with CYP2A6/2A7, CYP2B6/2B7, CYP2C8-19, CYP2E1 and CYP3A7 were performed as co-amplifications in the same tube with β-actin primers. The 661 bp band in these lanes represents β-actin. The sample lanes are flanked by DNA molecular weight markers (100-base pair ladder)

130

CYP gene expression was studied in 6 adult liver samples (Table 1). The mRNA of CYP2F1 and CYP4B1 was not detected in any of the samples. Low contents of CYP1A1 mRNA were observed in 3/6 adult livers. All other CYPs were readily detected - 2A6/2A7, 2B6/2B7, 2C8-19 and 3A3/3A4 in all samples and 1A2 and 2D6 in 4/6 and 3A7 in 5/6 samples.

TABLE 1. Summary of RT-PCR amplification results

| ADULT LIVER | 1A1 | 1A2 | 2A6/2A7 | 2B6/2B7 | 2C8-19 | 2D6 | 2E1 | 2F1 | 3A3/3A4 | 3A7 | 4B1 |
|---|---|---|---|---|---|---|---|---|---|---|---|
| n=6 | 3/6 | 4/6 | 6/6 | 6/6 | 6/6 | 4/6 | 6/6 | 0/6 | 6/6 | 5/6 | 0/6 |
| FETAL LIVER Uppsala n=9 | 0/9 | 0/9 | 0/9 | 0/9 | 0/9 | 0/9 | 0/9 | 0/9 | 3/9 | 2/9 | 0/9 |
| Kuopio n=7 | 0/7 | 0/7 | 0/4 | 0/7 | 6/7 | 4/4 | 0/3 | 0/7 | 6/7 | 7/7 | 0/3 |

The numbers denote number of positive amplifications/number of samples studied.

RT-PCR of 16 fetal liver samples was used to determine which of the CYP genes expressed in adult are also expressed in fetal liver (Table 1). CYP3A7 and CYP3A3/3A4 yielded highest amounts of amplification products. They were both observed in 9/16 fetal livers. In the Uppsala material (U3-U43, n=9), only members of 3A subfamily were discovered. In the Kuopio material (K4-K13, n=7), also 2D6 (all 4 livers studied) and members in the 2C subfamily (6/7) were detected. mRNA of CYP1A, CYP2A, CYP2B subfamilies and CYP2E1, CYP2F1 and CYP4B1 were constantly undetectable. The fewer forms detected in the Uppsala material can be ascribed to partial degradation of mRNA, obviously due to prostaglandin treatment and longer storage times at -70°C in comparison with the Kuopio material.

## DISCUSSION

Members of the CYP3A subfamily are considered to be the major constituents in human adult and fetal liver (Hulla & Juchau, 1989). Komori et al. (1990) showed by Northern blot analysis that CYP3A7 mRNA could be detected in fetal livers but not in adult livers. In contrast, CYP3A3 and CYP3A4 were detected only in adult liver. With the RT-PCR method used here we were able to detect both CYP3A7 and CYP3A3/3A4 mRNA in both adult and fetal liver.

The presense or absense of CYP1A1 expression in adult liver is still under debate. McKinnon et al. (1991) demonstrated CYP1A1 mRNA in Northern blot analysis in 11 out of 23 human liver samples. Our PCR result in which CYP1A1 mRNA was seen in 3/6 liver samples supports this observation. There is some some evidence for the expression of CYP1A1 in human fetal liver. Omiecinski et al. (1990) detected mRNA of CYP1A1 in fetal liver using PCR and observed an increasing content with increasing fetal age. At the sensitivity level used in this study, we were not able to detect CYP1A1 in any of the fetal liver samples studied. The assay used by Omiecinski et al. was more sensitive since more cycles and Southern blotting of amplification products were used. This facilitates the detection of extremely low contents of mRNA.

Treluyer et al. (1991) detected CYP2D6 mRNA in fetal liver. They also reported low CYP2D6 protein contents in some of the fetal livers. The protein was observed later than RNA, usually only after the 20th gestational week. Here we show that CYP2D6 mRNA is present in at least a subset fetal livers as early as 11 weeks of gestation. There has been very little evidence for the presence of members in the CYP2C subfamily in fetal liver. Ratanasavanh et al. (1991) observed very low quantities of CYP2C with immunohistochemical, immunoblotting and mRNA blotting techniques in some fetal liver samples. Cresteil et al. (1985) and Pons et al. (1991) reported that anti-human CYP2C9 or anti-rat CYP2C11 antibodies did not cross-react with any proteins in fetal liver. Mäenpää et al. (1993) showed that anti-mouse Cyp2c antibody cross-reacted with one to three proteins in the fetal livers. The present study confirms the existence of mRNA of CYP2C subfamily members in fetal liver. Since the primers used for the PCR analysis were designed to hybridise with all known human CYP2C isoforms, the number and specific identity of the CYP2C forms present remain unclear.

CYP2F1 and CYP4B1 have been identified through cDNA cloning in human lung (25,29), but neither one has been detected in human liver. Consistent with this, we could not amplify either CYP2F1 or CYP4B1 transcripts in adult or fetal livers.

It should be emphasized that in this study we used identical PCR conditions for the detection of CYP forms in both adult and fetal liver. This permitted us to make direct comparison of the adult and fetal gene expression. In conclusion, this is the first report to describe a comprehensive panel of xenobiotic-metabolizing CYP expression in fetal liver at early gestation (< 24 weeks).

Supported by the Academy of Finland (Medical Research Council, contract No.1051029) and by the Swedish Medical Research Council (04X-04496).

## REFERENCES

Pelkonen, O. (1980) *Pharmac. Ther.* 10, 261-281.
Pasanen, M., Pelkonen, O., Kauppila, A., Park, S. S., Friedman, F. K., and Gelboin, H. V. (1987) *Dev. Pharmacol. Ther.* 10, 125-132.
Omiecinski, C. J., Redlich, C. A., and Costa, P. (1990) *Cancer Res.* 50, 4315-4321.
Hulla, J. E., and Juchau, M. R. (1989) *Drug Metab. Rev.* 20, 765-799.
Ladona, M. G., Park, S. S., Gelboin, H. V., Hammar, L., and Rane, A. (1988) *Biochem. Pharmac.* 37, 4735-4741.
Treluyer, J-M., Jacqz-Aigrain, E., Alvarez, F., and , Cresteil, T. (1991) *Eur. J. Biochem.* 202, 583-588.
Rane, A., Henningsson, S., Ask., B., and Ladona, M. G. (1992) *J. Steroid. Biochem. Molec. Biol.* 43, 335-341.
Komori, M., Nishio, K., Kitada, M., Shiramatsu, K., Muroya, K., Soma, M., Nagashima, K., and Katamaki, T. (1990) *Biochemistry* 29, 4430-4433.
McKinnon, R. A., Hall, P., Quattrochi, L. C., Tukey, R. H., and McManus, M. E. (1991) *Hepatology* 14, 848-856.
Ratanasavanh, D., Beaune, P., Morel, F., Flinois, J.-P., Guengerich, F. P., and Guillouzo, A. (1991) *Hepatology* 13, 1142-1151.
Cresteil, T., Beaune, P., Kremers, P., Celier, C., Guengerich, F. P., and Leroux, J.-P. (1985) *Eur. J. Biochem.* 151, 345-350.
Pons, C., Dansette, P. M., Amar, C., Jaouen, M., Wolf, C. R., Gregeois, J., Homberg, J. C., and Mansuy, D. (1991) *J. Pharmacol. Exp. Ther.* 259, 1328-1334.
Mäenpää, J., Rane, A., Raunio, H., Honkakoski, P., and Pelkonen, O. (1993) *Biochem. Pharmacol.* 45, 899-907.

# Transcriptional regulation of CYP2B1/B2 gene in rat liver

Govindarajan Padmanaban, Chaitanya Nirodi, Leena Prabhu, Nivedita Ram, Shahana Sultana

*Department of Biochemistry, Indian Institute of Science, Bangalore: 560 012, India*

The near upstream of the 5'-flanking region of the CYPIIB2 gene (upto -178nt) gives rise to two complexes with crude nuclear extracts from rat liver in gel retardation assays. Phenobarbitone (PB) treatment leads to an intensification of the bottom band and this appears to be due to enhanced phosphorylation of the protein involved. Cycloheximide treatment of the animal leads to a time dependent abolition of the top complex, which is due to the turnover of a constitutive protein factor giving rise to the complex. The loss of this complex correlates with the ability of cycloheximide to inhibit basal as well as PB-mediated activation of CYPIIB1/B2 transcription. There is strong intensification of the bottom complex under conditions of cycloheximide treatment, the reason for which is not clear. Okadaic acid mimics PB in giving rise to a similar pattern in gel retardation assays and enhanced transcription of a minigene construct in cell-free transcription of assays. However, okadaic acid is not by itself an inducer of the CYPIIB1/B2 gene and antagonises the action of PB.

---

Earlier studies in this laboratory have led to the identification of the near upstream of the 5' flanking region (upto -178nt) being adequate for basal and induced expression of the CYPIIB1/B2 gene. A positive cis-acting element covering the region -69 to -98nt was also identified (Rangarajan and Padmanaban, 1989; Poornima et al, 1992). The near upstream fragment as well as the positive cis-acting element give two complexes in gel retardation assays with crude nuclear extracts and PB treatment leads to an intensification of the bottom complex. We have also shown earlier that cycloheximide inhibits PB-mediated activation of CYPIIB1/B2 gene transcription (Bhat et al, 1987). In the present study, we have examined the mechanism by which cycloheximide inhibits CYPIIB1/B2 gene transcription and whether PB influences the phosphorylation status of the protein factors binding to the near upstream 5'-flanking region.

MATERIALS AND METHODS

Rats (Wistar strain) weighing 75-80g were injected with saline or phenobarbitone (8 mg/100 g). The animals were usually killed after 6hr and liver nuclei were prepared using citric acid homogenization procedure (Taylor et al, 1973). The nuclei were extracted with 0.4 M NaCl and the extract was used for gel retardation assays. The same extract was subjected to 80% W/V $(NH_4)_2 SO_4$ fractionation and the precipitate was solubilized and used as a cell-free transcription extract. Okadaic acid and cycloheximide were injected at concentrations of 2.5 ug and 1 mg per 100 g body weight respectively. A minigene construct containing -178nt of the 5'-flanking region and the first exon of CYPIIB1/B2 gene was used as a template in cell-free transcription. The cell-free transcripts as well as messenger RNA concentrations in vivo were quantified using RNase protection assay.

RESULTS AND DISCUSSION

Fig.1 gives the gel retardation pattern of the nuclear extracts with [$^{32}$P] labeled -upstream fragment (-178nt). Two complexes are seen and the bottom complex intensifies with nuclear extracts prepared from PB-treated rats. When the extracts are pretreated with calf intestinal alkaline phosphatase, there is a decrease in the intensity of the bottom complex and an increase in that of the top complex.

Fig.1. Effect of Phosphatase pretreatment of liver nuclear extracts on gel retardation complexes

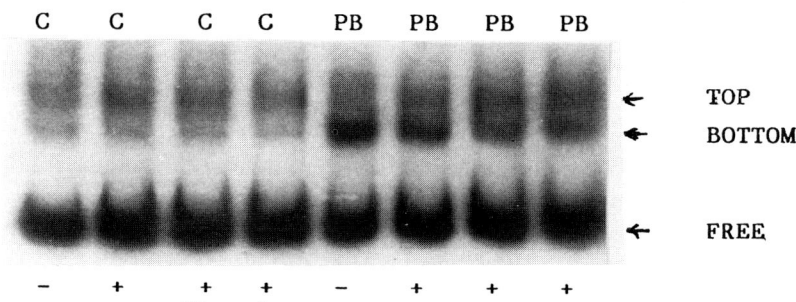

We have earlier shown that cycloheximide treatment of rats prevents the PB-mediated increase in CYPIIB1/B2 gene transcription (Bhat et al, 1987). In the present study, it has been found that the effect of cycloheximide effect is time dependent. One hour after cycloheximide treatment, there is only a marginal effect. However, after 5hr of cycloheximide treatment, there is a decrease in both the basal and PB-induced transcription of the CYPIIB1/B2 gene in whole nuclei. Gel retardation assays with nuclear extracts under these treatment conditions reveal that the top complex shows a time-dependent decrease in intensity and the bottom complex shows a striking intensification. Phosphatase pretreatment of the extracts reveal a significant intensification of the top complex, 1 hr after cycloheximide treatment. However, after 5hr of cycloheximide treatment, there is virtual disappearance of the top complex. The bottom complex shows a marginal decrease after phosphatase treatment (data not presented). These results are interpreted to mean that the top complex is formed by a constitutive fast-turning over protein factor and this interaction is essential for basal transcription of the CYPIIB1/B2 gene.

Since, intensification of the bottom complex in gel retardation assays after PB treatment appears as due to phosphorylation of the protein factor involved, the effect of phosphatase and kinase inhibitors were tested in vivo. The phosphatase inhibitor, Okadaic acid, gave interesting results. Nuclear extracts from okadaic acid treated rats give gel retardation profiles similar to those from PB-treatment with the -178nt upstream fragment and the positive cis-acting element. In cell-free transcription assays using the minigene construct (Fig.2), Okadaic treated extracts are even more efficient than PB-treated extracts.

Fig.2. Cell-free transcription of CYPIIB1/B2 minigene construct using RNase protection assay.

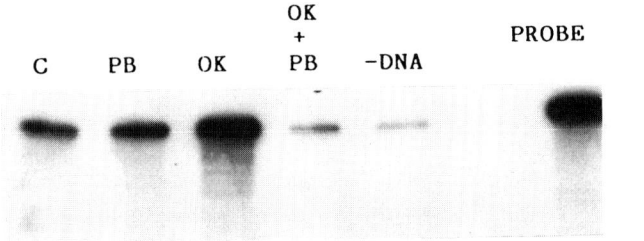

However, okadaic acid does not induce CYPIIB1/B2 mRNA by itself and is a powerful antagonist of PB action in vivo in terms of CYPIIB1/B2 gene transcription as well as interaction of the protein factors with DNA revealed in gel retardation assays. Thus, it appears that a partial mechanism of action of PB may involve inhibition of a protein phosphatase. However, PB action involves certain other steps in the transcription of CYPIIB1/B2 gene not shared by okadaic acid. The antagonistic effect between PB and okadaic acid when given together even at the level of gel retardation profiles indicates that okadaic acid may negatively modulate the phosphorylation and/or synthesis of a protein factor that mediates the action of PB.

ACKNOWLEDGEMENT :

This study was supported by grants from the Department of Science amd Technology, New Delhi.

REFERENCE

Bhat, G.J., Rangarajan, P.N. and Padmanaban, G. (1987). Biochem. Biophys. Res. Commun. 148, 1118-1123.

Poornima Upadhya, Venkateswara Rao, M., Venkateswar, V., Rangarajan, P.N. and Padmanaban, G. (1992). Nucl. Acids Res. 20, 557-562.

Rangarajan, P.N. and Padmanaban, G. (1989). Proc. Natl. Acad. Sci. USA 86, 3963-3967.

Taylor, C.W., Yeoman, L.C., Dashal, I and Bush, H (1973). Exp. Cell Res. 82, 215-226.

# Identification of DNA sequences involved in basal and phenobarbital induced transcription of a CYP2B2 gene

Amal Shervington[1], Ian R. Phillips[2], Elizabeth A. Shephard[1]

[1]Department of Biochemistry and [2]Molecular Biology, University College London, Gower Street, London WC1E 6BT, UK and Department of Biochemistry, Queen Mary and Westfield College, University of London, Mile End Road, London, E1 4NS, UK

## INTRODUCTION

The expression of many members of the CYP1, CYP2, CYP3 and CYP4 families is induced in response to various endogenous and foreign chemicals. Our group is particularly interested in the mechanisms responsible for the increased synthesis in mammalian liver of two members of the CYP2B subfamily, namely CYP2B1 and CYP2B2, in response to phenobarbital treatment (Phillips *et al.*, 1981, 1983). The induction by phenobarbital of CYP2B1/2B2 proteins has been shown to be mediated by an increase in the corresponding mRNAs (Phillips *et al.*, 1981; Shephard *et al.*, 1982), which in turn is due to an increase in the transcription of CYP2B genes (Pike *et al.*, 1985). Owing to the lack of a cell-line capable of supporting high levels of CYP2B gene expression in response to phenobarbital we have used an *in vitro* transcription system to identify DNA sequences that are involved in regulating either the basal transcription of a CYP2B2 gene or the phenobarbital-induced expression of the gene.

## MATERIALS AND METHODS

### Plasmid construction

All plasmids used in the *in vitro* transcription assays were based on the $p(C_2AT)_{19}$ vector, which contains a 380 bp G-free cassette (Sawagodo & Roeder, 1985). Six plasmids were constructed in which the G-free cassette was placed under the control of various 5´ flanking sequences of a CYP2B2 gene: $p[-5600/-1](C_2AT)_{19}$; $p[-2880/-1](C_2AT)_{19}$; $p[-984/-1](C_2AT)_{19}$; $p[-178/-4](C_2AT)_{19}$; $p[-368/-4](C_2AT)_{19}$; and $p[-4/-368](C_2AT)_{19}$. The numbers refer to regions of the CYP2B2 gene with respect to the transcription start site. The last two plasmids are the same except that the CYP2B2 gene sequences are in opposite orientations with respect to the G-free cassette. The plasmid $pML(C_2AT)_{19}$, containing the adenovirus major late promoter, was used as an internal standard in the transcription assays.

### Nuclear extract preparation

Nuclear protein extracts were isolated from the livers of phenobarbital-treated or untreated rats by a slight modification of the procedure of Gorski *et al.* (1986). The homogenization buffer contained a mixture of protease inhibitors (10 µg/ml aprotinin, 1 µg/ml antipain, 1 µg/ml chymostatin and 1 µg/ml leupeptin). Nuclear extracts were frozen on dry-ice and stored in aliquots at -78°C. Protein concentration was measured by the method of Kalb & Bernlohr (1977).

*In vitro* transcription

The procedure was adapted from Sierra (1990). Transcription reactions (20 µl) contained 700 ng of one of the CYP2B2 promoter-p($C_2AT$)$_{19}$ constructs described above and 100 ng of pML($C_2AT$)$_{19}$ as an internal control. In the competition experiments, 400 ng of construct, 100 ng of pML($C_2AT$)$_{19}$, and various amounts of competitor DNA sequence were used. The total DNA content was adjusted to 800 ng by the addition of pUC19. RNA transcripts were analysed by denaturing polyacrylamide gel electrophoresis. After autoradiography, the transcripts were quantified by scanning densitometry with a Bio-Rad Model GS-670 Imaging Densitometer.

RESULTS AND DISCUSSION

To identify the 5´ flanking region of the CYP2B2 gene required for maximal promoter activity each of the six plasmid constructs, p[-5600/-1]($C_2AT$)$_{19}$, p[-2880/-1]($C_2AT$)$_{19}$, p[-984/-1]($C_2AT$)$_{19}$, p[-178/4]($C_2AT$)$_{19}$, p[-368/-4]($C_2AT$)$_{19}$ and p[-4/-368]($C_2AT$)$_{19}$ were transcribed *in vitro* in the presence of nuclear protein extracts from the livers of either phenobarbital-treated or untreated rats. Promoter activity was monitored by quantifying the production of G-free cassette mRNA transcripts. The greatest promoter activity was obtained with the construct containing the CYP2B2 sequences -984/-1. The activity with this construct was 6-fold higher than that obtained with the construct p[-178/-4]($C_2AT$) and two-fold higher than that observed with the construct p[-368/-4]($C_2AT$)$_{19}$. No transcription of the G-free cassette was obtained from the construct p[-4/-368]($C_2AT$)$_{19}$ which carries the CYP2B2 gene sequence in the incorrect orientation with respect to the start site of transcription. The two largest constructs, p[-5600/-1]($C_2AT$)$_{19}$ and p[-2880/-1]($C_2AT$)$_{19}$, were transcribed only weakly.

We have previously identified, by gel retardation and DNase I footprinting, two DNA sequences, located between -199 and -183 and -85 and -31 of the CYP2B2 promoter, which bind proteins that are either activated or enriched in response to phenobarbital treatment. To determine whether these sequences are indeed necessary for CYP2B2 promoter activity the construct p[-984/-1]($C_2AT$)$_{19}$ was transcribed in the presence or absence of each of the sequences. The presence of a 100-fold molar excess of fragment -85 to -31 in the *in vitro* transcription assay reduced the promoter activity of construct p[-984/-1]($C_2AT$)$_{19}$ to 20% of that obtained in the absence of the fragment. A similar excess of double-stranded synthetic oligonucleotide specifying the sequence -199 to -183 of the CYP2B2 promoter decreased the promoter activity of p[-984/-1]($C_2AT$)$_{19}$ to 70% of that observed in the absence of competitor sequence. The results of the *in vitro* transcription competition assays confirmed that sequences between -85 and -31 and -199 and -183 of the CYP2B2 gene bind nuclear proteins that play a role in determining CYP2B2 promoter strength. Computer analysis of these DNA sequences revealed that the more proximal sequence contains a potential binding site for the transcription factor C/EBP whereas the more distal sequence specifies a binding site that differs in only one nucleotide from the consensus binding sequence for the octamer class of POU-domain regulatory proteins.

Because of the lack of a suitable phenobarbital-responsive cell-line we investigated the possibility of using the *in vitro* transcription system to identify sequences responsible for the phenobarbital induction of CYP2B2 gene expression. Plasmid constructs containing a G-free cassette under the control of various regions of the 5´ flanking sequence of a CYP2B2 gene were transcribed in the presence of nuclear protein extracts from the livers of either phenobarbital-treated or untreated rats (Fig. 1). Each reaction contained, as an internal transcription control, the construct pML($C_2AT$)$_{19}$ in which the G-free cassette is under the control of the adenovirus major late promoter. The size of the transcripts produced from the CYP2B2 promoter constructs were between 380 and 376 nucleotides, whereas that derived from pML($C_2AT$)$_{19}$ was 390 nucleotides. To control for differences in transcriptional efficiency between the various nuclear protein extracts, the transcriptional activity of each CYP2B2 promoter construct was expressed as a ratio of the amount of transcript produced from the construct compared with that derived from pML($C_2AT$)$_{19}$. G-free cassette transcription from constructs p[-368/-

4]$(C_2AT)_{19}$ and p[-984/-1]$(C_2AT)_{19}$ was 55 and 32% greater, respectively, with nuclear protein extracts from phenobarbital-treated animals compared with that obtained with extracts from untreated animals. The three other constructs used p[-178/-4]$(C_2AT)_{19}$, p[-5600/-1]$(C_2AT)_{19}$ and p[-2880/-1]$(C_2AT)_{19}$, were not transcribed more efficiently by nuclear protein extracts isolated from phenobarbital-treated rats than by extracts from untreated animals.

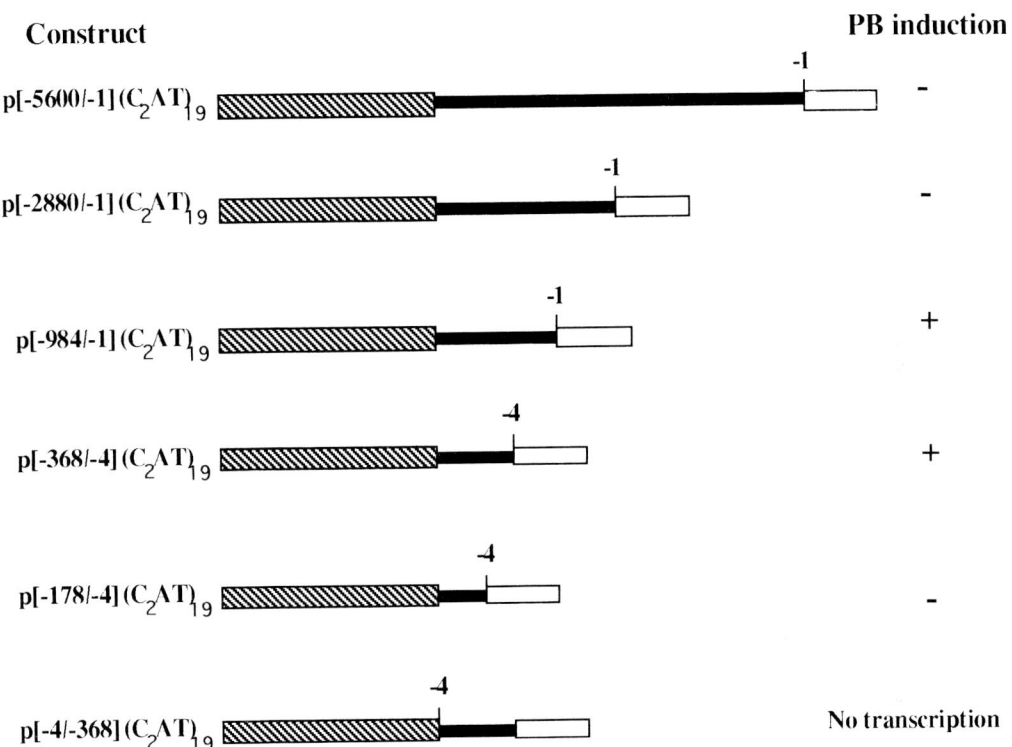

Fig 1: Analysis of CYP2B2 promoter activity by *in vitro* transcription. Various sections of the 5´ flanking region of a CYP2B2 gene were attached to a G-free cassette, $(C_2AT)_{19}$, and the resulting constructs were transcribed in nuclear protein extracts from the livers of phenobarbital-treated or untreated rats. Numbers refer to regions of the CYP2B2 gene (with respect to its transcription start site) and their orientation in the constructs. White boxes represent the G-free cassette. Solid lines represent 5´ flanking sequences of a CYP2B2 gene. Hatched boxes represent pUC13 sequences. The relative trancriptional activity of constructs in extracts from phenobarbital-treated rats, compared with extracts from untreated animals, is indicated by + or -.

The results obtained for construct p[-178/-4]$(C_2AT)_{19}$ suggest that both of the DNA sequences (-85 to -31 and -199 to -183) identified above as being necessary for CYP2B2 promoter activity are required for phenobarbital induction of CYP2B2 gene transcription *in vitro*. The low promoter activity and lack of phenobarbital responsiveness observed with the constructs p[-5600/-1]$(C_2AT)_{19}$ and p[-2880/-1]$(C_2AT)_{19}$ suggest that the CYP2B2 gene may contain, upstream of -984, sequences that repress its transcription *in vitro*.

In conclusion, we have used an *in vitro* transcription assay to identify sequences involved in both the basal expression of a CYP2B2 gene and its induction by phenobarbital. Although the increase in transcription of CYP2B2 genes, obtained in this system is very low compared with that observed *in vivo*, our results demonstrate, that, in the absence of a suitable phenobarbital-responsive cell-line for DNA transfections, *in vitro* transcription may provide a successful alternative for analysis of CYP2B2 promoter function.

ACKNOWLEDGEMENTS

This work is supported by grant from the Cancer Research Campaign, U.K.

REFERENCES

Gorski, K., Carneiro, M. & Schibler, U. (1986): Tissue-specific *in vitro* transcription from the mouse albumin promoter. *Cell* **47**, 467-776.

Kalb, V.F., Jr & Bernlohr, R.W. (1977): A new spectrophotometric assay for protein in cell extracts. *Anal. Biochem.*, **82**, 362-371.

Phillips, I.R., Shephard, E.A., Mitani, F. & Rabin, B.R. (1981): Induction by phenobarbital of the mRNA for a specific variant of rat liver microsomal cytochrome P-450. *Biochem. J.* **196**, 839-851.

Phillips, I.R., Shephard, E.A, Bayney, R.M., Pike, S.F., Rabin, B.R., Heath, R. & Carter, N. (1983): Induction and repression of the major phenobarbital-induced cytochrome P-450 measured by radioimmunoassay. *Biochem J.* **212**, 55-64.

Pike, S.F., Shephard, E.A., Rabin, B.R. & Phillips, I.R. (1985): Induction of cytochrome P-450 by phenobarbital is mediated at the level of transcription. *Biochem.Pharmacol.* **34**, 2489-2494.

Sawadogo, M. & Roeder, G. (1985): Interaction of a gene-specific transcription factor with the adenovirus major late promoter upstream of the TATA box region. *Cell* **43**, 165-175.

Shephard, E.A., Phillips, I.R., Pike, S.F., Ashworth, A., & Rabin, B.R. (1982): Differential effect of phenobarbital and ß-naphthoflavone on the mRNAs coding for cytochrome P450 and NADPH cytochrome P450 reductase. *FEBS Lett.* **150**, 375-380.

Sierra, F. (1990). *A laboratory guide to in vitro transcription*. Basel: Birkhauser Verlag.

# Tissue-specific expression of cytochrome P450arom gene by alternative use of multiple exons 1 and promoters, and a switching of tissue-specific exons 1

Nobuhiro Harada, Shin-Ichiro Honda, Yasuyuki Takagi

*Division of Molecular Genetics, Institute for Comprehensive Medical Science, Fujita Health University, Toyoake, Aichi 470-11, Japan*

## Summary

Extensive screening of aromatase cDNA was carried out in cDNA libraries from various human tissues. The tissue-specific sequences were observed in the 5'-portions of the isolated cDNA clones, corresponding to the region encoded by exon 1. All of them were also found in clones isolated from a human genomic library and mapped between exons 1 and 2, suggesting the presence of multiple exons 1 and promoters in the gene. Reverse transcription-PCR (RT-PCR) analyses of aromatase mRNAs in various tissues revealed that aromatase transcripts are tissue-specifically spliced by alternative use of multiple exons 1. Aromatase mRNA is spliced from 10 exons in most tissues, but from 9 exons in the prostate and from 10 or 11 exons in the placenta. This suggests that tissue-specific regulation of the aromatase gene in various tissues may be explained by alternative use of multiple exons 1 flanked with tissue-specific promoters. The alternative use of multiple exons 1 for liver transcripts was found to change developmentally. Furthermore, switch from an adipose-specific exon 1 to another type of exon 1 was observed in aromatase transcripts of adipose tissues of 3 of 5 breast cancer patients.

## INTRODUCTION

Aromatase, cytochrome P-450$_{AROM}$ or estrogen synthetase, is a unique member of the cytochrome P-450 superfamily. It is localized mainly in the ovary and placenta and catalyzes the conversion of androgen to estrogen, which is a rate-limiting step in estrogen biosynthesis. This enzyme activity has been found in various extra-gonadal tissues such as the brain, liver, skin fibroblasts, and adipose tissue as well as gonadal tissues. It is important that the tissue-specific expression of aromatase is strictly regulated, because estrogen has been suggested to be involved in various physiological functions through not only endocrine, but also autocrine or paracrine actions and to function as a growth- or differentiation-like factor. Indeed, the expression of aromatase has been reported to be regulated tissue-specifically by many kinds of factors.

The human aromatase gene has been isolated (Means et al., 1989; Harada et al., 1990; Toda et al., 1990). The gene consists of 10 exons and spans at least 70 kilobases (kb) (Harada et al., 1990). Furthermore, exon 1 encoding the only 5'-untranslated region is separated from exon 2 by an intron of more than 35 kb. Recent studies demonstrated the presence of multiple exons 1 in several human genes. As analysis of genomic clones of human aromatase suggested that the aromatase gene exists in the human haploid genome as a single copy (Harada et al., 1990), the tissue-specific expression of this gene by alternative utilization of multiple exons 1 and promoters is an attractive idea to explain its complex regulation by various factors.

Recently, the aromatase mRNAs in the ovary and adipose stromal cells were shown to be transcribed from 79 bp and 84 bp upstream, respectively, of the exon 2 identified in placenta (Mahendroo et al., 1991; Means et al., 1991). This finding indicates that the aromatase genes in the ovary and adipose stromal cells utilize a new exon 1 containing the placental exon 2. Furthermore, aromatase mRNA in human skin fibroblasts was shown to be formed by alternative use of a different exon 1 from tissue-specific exons 1 (Harada, 1992).

In this study, to determine how many tissue-specific exons 1 of the human aromatase gene there are and how aromatase transcripts are spliced in various tissues, we isolated aromatase cDNAs from all available cDNA libraries of various human tissues. The results suggested that the human aromatase gene is expressed tissue-specifically by alternative splicing of four major types of exons 1. We also detected tissue-specific expression of several minor forms of aromatase mRNA by alternative splicing, and observed switching of a tissue-specific exon 1 in the adipose tissue of breast cancer patients.

## RESULTS AND DISCUSSION

**Isolation of aromatase cDNAs with unique DNA sequences on the 5'-side from various human tissues.** Extensive screening of cDNA libraries from various tissues revealed the presence of five kinds of unique DNA sequences on the 5'-side of aromatase cDNAs. All the isolated clones had the same DNA sequences in the region encoded by exons 2-10 of the aromatase gene, but had four kinds of unique DNA sequence on the 5'-side region corresponding to exon 1. Furthermore, the sequence observed in the clones depended on the tissue source of the library, indicating tissue-specific expressions of the five kinds of aromatase mRNA. Thus, the aromatase mRNAs in various tissues were tentatively classified into five types: prostate specific mRNA, ovary specific mRNA, brain specific mRNA, fetal liver/skin fibroblast specific mRNA, and placenta specific mRNA. According to this classification, the adipose tissue mainly express low levels of the fetal liver/skin fibroblast specific aromatase mRNA species and the ovary expresses the prostate specific mRNA as well as the ovary specific mRNA. The placenta also expresses a minor species of aromatase mRNA similar in structure to the placenta specific mRNA but with an insertion of an extra fragment of 109 bases(Harada et al., 1993).

**Isolation of alternative exons 1 from a human genomic library.** A human genomic library was screened for the unique sequences found on the 5'-side of aromatase cDNA. All the isolated genomic clones were mapped upstream of exon 2 of the aromatase gene, indicating that the unique sequences found in the aromatase cDNAs were due to alternative splicing of the multiple exons 1 (Fig. 1). The placenta specific exon 1 (exon 1a) and the fetal liver/skin fibroblast specific exon 1 (exon 1b) were located more than 35 and 17 kb, respectively, upstream of placental exon 2, whereas the ovary specific exon 1 (exon 1c) was located within 305 bp upstream of the placental exon 2. Exon 1c was separated by 23 bp from exon 1d and by 102 bp of an intron from placental exon 2. Prostate specific exon 1 (exon 1d) was formed by transcription initiated 79 bp upstream of the placental exon 2 to the 3'-end of exon 2 without splicing at the junction of the placental intron 1 and exon 2. So, in the prostate, aromatase mRNA is encoded by 9 exons of the aromatase gene. On the other hand, the 109 bp sequence observed in the minor species of placental aromatase mRNA was located 23 kb upstream of

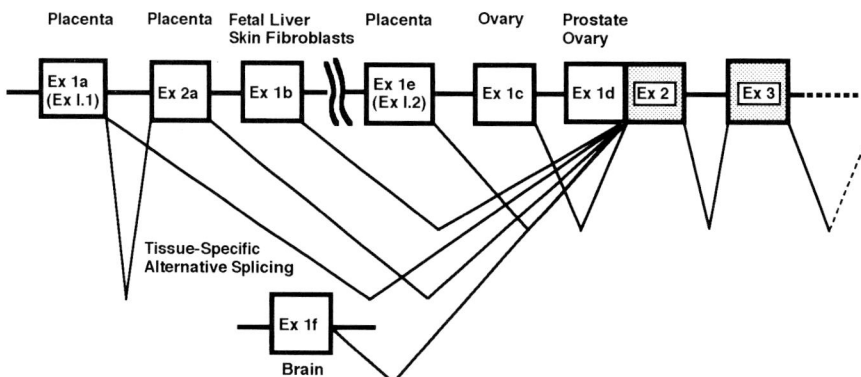

Fig. 1. Schema of alternative utilization of tissue-specific exons 1 and promoters of the human aromatase gene.

placental exon 2. These results indicate that this region functions as a new exon 2 (exon 2a) for the minor species of placental aromatase mRNA and consequently, that the mRNA is formed from 11 exons of the aromatase gene. Exons 1a, 1c, 1d, and 1f, specific for the placenta, ovary, prostate, and brain each have typical TATA box in their 20-30 bp upstream region, whereas exon 1b, specific for fetal liver/skin fibroblasts, has a typical GC box in the upstream region (Harada et al., 1993).

**Analysis of alternative usage of multiple exons 1 in aromatase transcripts of various tissues.** For determination of whether aromatase transcripts are tissue-specifically spliced by alternative usage of multiple exons 1, RNA fractions from various tissues were reverse-transcribed and analyzed by PCR using one of the exon 1-specific primers and an exon 2-specific antisense primer. The major PCR products were formed tissue-specifically, suggesting alternative usage of multiple exons 1 in various tissues and tissue-specific splicing. Apparently, exons 1a, 1b, 1c, 1d and 1f were mainly used for the aromatase mRNAs of placenta, skin fibroblasts and fetal liver, ovary, ovary and prostate, and brain, respectively. Aromatase mRNAs in adipose tissue were transcribed from exon 1b, although their expression levels were low. A major transcript using exon 1b and a minor transcript using exon 1c were observed in fetal liver, whereas in adult liver, the major transcript using exon 1b had completely disappeared and only the minor transcript using exon 1c was observed. The transcript in the ovary seems to use exons 1c and 1d equally for alternative splicing. Leaky levels of minor transcripts were also found in fetal liver, skin fibroblasts, and placenta. These results are summarized in Table 1. At least three kinds of aromatase mRNAs were also found in the ovary, their transcriptions being initiated from positions 79 and 261 bp upstream from the 3'-end of placental exon 2 (Harada et al., 1993).

Table 1. Tissue-specific alternative use of multiple exons 1 of aromatase gene.

| Tissue | Exon 1a | Exon 1b | Exon 1c | Exon 1d | Exon 1f |
|---|---|---|---|---|---|
| Fetal Brain | - | - | - | - | +++ |
| Adult Brain | - | - | - | - | + |
| Fetal Liver | + | ++++ | + | ± | ± |
| Adult Liver | - | - | + | ± | - |
| Ovary | - | - | ++++ | ++++ | - |
| Skin Fibroblasts | - | ++++ | + | + | + |
| Placenta | +++++ | + | + | + | - |
| Adipose Tissue | - | + | - | - | - |
| Prostate | - | - | - | + | - |
| Testis | - | - | - | + | - |

Since the first isolation of an exon 1 used for placental mRNA (Means et al., 1989; Harada et al., 1990; Toda et al., 1990), three other exons 1 have been obtained from mRNAs of placenta (Kilgore et al., 1992), adipose stromal cells (Mahendroo et al., 1991), ovary (Means et al., 1991), and skin fibroblasts (Harada, 1992). The exon 1a in this study is identical in its sequence and transcriptional start site to placental exon 1 (Harada et al., 1990; Toda et al., 1990) or exon I.1 (Means et al., 1989) isolated previously. Although an exon I.2 used for placental mRNA was found in a minor species of aromatase mRNA (Kilgore et al., 1992), we could not isolate it in the present study. Thus, at least six different exons 1 of the human aromatase gene are now known, as shown in Fig. 2. Ovary and adipose stromal cells are reported to utilize an exon (Mahendroo et al., 1991; Means et al., 1991) consisting of placental exon 2 and its 79 or 84 bp upstream region, as a new exon 1. This exon utilized in the ovary (Means et al., 1991) has the same sequence and transcriptional initiation site as exon 1d. However, we could not identify exon 1d transcribed from 84 bp upstream of placental exon 2 in the ovary, prostate, or testis as in adipose stromal cells. Previous findings are inconsistent with the present finding of use of both exons 1c and 1d in the ovary and exon 1b in adipose tissue. As aromatase in the ovary is mainly localized in granulosa cells before ovulation and in lutein cells after ovulation, and its responses to various regulatory factors seem to change during cell differentiation in the ovarian sex cycle (Erickson, 1983), exon 1 of the aromatase gene used may change from one type to another during the ovarian sex cycle. In this study we used pooled aromatase mRNA from the ovaries of four women and so our results may indicate the overall usage of exons 1 in different stages. Similarly, differences between growth conditions or cell types of adipose stromal cells in primary culture and adipose tissue in vivo may result in switch from one type of exon 1 to another.

**Switching of an adipose-specific exon 1b to exons 1c and 1d in adipose tissues from breast cancer patients.** Breast adipose tissues in breast cancer patients often show elevated expressions of aromatase. Therefore we examined the alternative use of exons 1 for aromatase transcripts in adipose tissues from healthy controls and breast cancer patients showing elevated levels of aromatase mRNA (Fig. 2). Aromatase transcripts from healthy controls showed tissue-specific use of exon 1b as seen in Table 1, whereas those from 3 of 5 breast cancer patients examined showed a switch

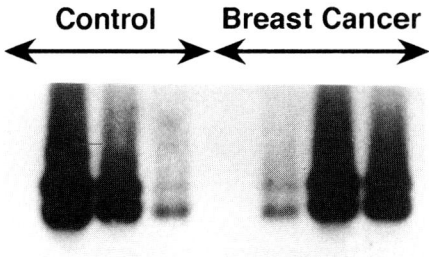

Fig. 2. RT-PCR analysis of switching of tissue-specific exons 1 in breast adipose tissue. RNA fractions form adipose tissues of controls and breast cancer patients were analyzed by RT-PCR using primers specific for exons 1a, 1b, 1c, or 1d. RT-PCR products were subjected to Southern blotting analysis.

from exon 1b to exons 1c and 1d (Harada et al. 1993). Aromatase is an important etiological factor in some human breast cancers by inducing over-production of estrogens as growth factors, resulting in proliferation of hormone-dependent cancer cells. In menopausal women with breast cancer, breast adipose tissue is a major source of estrogens, and aromatase in breast adipose tissue seems to have a significant association with the presence of tumors (Miller et al., 1991; Lipton et al., 1992). From this finding, we postulate that the regulation of aromatase in breast adipose tissue becomes abnormal during carcinogenesis due to switching of exons 1 and promoters, resulting in over-production of estrogens under control of a new promoter. For confirmation of this hypothesis of switching of exons 1 in breast cancer, we are now examining more clinical cases of breast cancer and analyzing the mechanisms regulating transcription from each of the five exons 1.

## REFERENCES

Erick, G.F. (1983): Primary culture of ovarian cells in serum-free medium as models of hormone-dependent differentiation. *Mol. Cell. Endocrinol.* 29, 21-49.

Harada, N., Yamada, K., Saito, K., Kibe, N., Dohmae, S., & Takagi, Y. (1990): Structural characterization of the human estrogen synthetase (aromatase) gene. *Biochem. Biophys. Res. Commun.* 166, 365-372.

Harada, N. (1992): A unique aromatase (P-450AROM) mRNA formed by alternative use of tissue-specific exons 1 in human skin fibroblasts. *Biochem. Biophys. Res. Commun.* 189, 1001-1007.

Harada, N., Utsumi, T., & Takagi, Y. (1993) Tissue-specific expression of the human aromatase cytochrome P-450 gene by alternative use of multiple exons 1 and promoters, and switching of tissue-specific exons 1 in carcinogenesis. *Proc. Natl. Acad. Sci. U.S.A.* in press.

Kilgore, M.W., Means, G.D., Mendelson, C.R., & Simpson, E.R. (1992): Alternative promotion of aromatase P-450 expression in the human placenta. *Mol. Cell. Endocrinol.* 83, R9-R16.

Lipton, A., Santen, R.J., Santner, S.J., Harvey, H.A., Sanders, S.I., & Matthews, Y.L. (1992): Prognostic value of breast cancer aromatase. *Cancer* 70, 1951-1955.

Mahendroo, M.S., Means, G.D., Mendelson, C., & Simpson, E.R. (1991): Tissue-specific expression of human P-450AROM: the promoter responsible for expression in adipose is different from that utilized in placenta. *J. Biol. Chem.* 266, 11276-11281.

Means, G.D., Mahendroo, M.S., Corbin, C.J., Mathis, J.M., Powell, F.E., Mendelson, C.R., & Simpson, E.R. (1989) *J. Biol. Chem.* 264, 19385-19391.

Means, G.D., Kilgore, M.W., Mahendroo, M.S., Mendelson, C.R., & Simpson, E.R. (1991): Tissue-specific promoters regulate aromatase cytochrome P-450 gene expression in human ovary and fetal tissues. *Mol. Endocrinol.* 5, 2005-2013.

Miller, M.R. (1991): Aromatase activity in breast tissue. *J. Steroid Biochem. Mol. Biol.* 39, 783-790.

Toda, K., Terashima, M., Kawamoto, T., Sumimoto, H., Yokoyama, Y., Kuribayashi, I., Mitsuuchi, Y., Maeda, T., Yamamoto, Y., Sagara, Y., Ikeda, H., & Shizuta, Y. (1990): Structural and functional characterization of human aromatase P-450 gene. *Eur. J. Biochem.* 193, 559-56

# III  CYTOCHROME P450 IN CANCER AND IN IMMUNOTOXYCOLOGY

# Importance of individual enzymes in the control of ultimate carcinogens

Franz Oesch, Barbara Oesch-Bartlomowicz, Hansruedi Glatt, Karl-Ludwig Platt, Michael Arand

*Institute of Toxicology, University of Mainz, Obere Zahlbacher Strasse 67, D-55131, Germany*

A prerequisite for the mutagenic and carcinogenic activity of most chemical mutagens and carcinogens is their metabolic activation. Reactive metabolites are under the control of activating, inactivating and precursor sequestering enzymes, which themselves are under the long-term control of induction and represssion and under the short-term control of posttranslational modification which has received little attention in as far as carcinogen-metabolizing enzymes are concerned but may be especially important since it works fast and may affect the enzymatic activity and the degradation of the enzyme. Activators and inhibitors may modify the enzymatic activity, and the compartmentalization of these enzymes and of the target molecules are crucial determinants for the control of ultimate carcinogens. These enzymes differ widely between toxicological test systems, animal species and man. Correct predictions of genotoxic risks are urgently needed because of the long latency time of genotoxic effects. Therefore, careful consideration of the basic mechanisms responsible for the control of the ultimately active species derived from mutagenic tumor initiators is very important.

INTRODUCTION

Most chemical carcinogens require enzymatic activation to the ultimately carcinogenic species. Of those chemical carcinogens which are present as ultimate carcinogens most can be enzymatically inactivated. Thus, in almost all cases enzymes are important for the control of the ultimately carcinogenic metabolites. These enzymes differ drastically in their levels and in their specificity, some even in their presence or absence between toxicological test systems and animal species compared with man so that for any toxicological risk evaluation these control mechanisms require careful attention.

# CONTROL OF MUTAGENIC TUMOR INITIATORS BY CYTOCHROME P-450-DEPENDENT MONOOXYGENASES

Individual cytochromes P-450 are differentially induced by individual foreign groups of compounds (Wilson et al., 1984; Bücker et al., 1979; Conney, 1982). This may lead to shifts in metabolic routes and thereby to the generation of ultimate carcinogens or to the prevention of their generation. Enzyme induction, however, is generally a slow multistep process. We investigated the phosphorylation of fourteen purified cytochromes P450 by purified protein kinases (Pyerin et al., 1987) in order to elucidate the possible role of protein kinase mediated short term regulation of cytochromes P-450. However, in the compartmentalized situation of the cell the substrate cytochrome P-450 may or may not be available to the corresponding protein kinase. Therefore, hepatocytes were exposed to $^{32}$P-orthophosphate in order to label the intracellular ATP pools. In order to induce the two major phenobarbital-inducible cytochromes P450 2B1 and 2B2 hepatocytes were isolated from the liver of rats which had been pretreated with phenobarbital. These two had been, amongst the 14 investigated isoenzymes, the best substrates of the cAMP-dependent protein kinase (PKA) in the cell-free system of incubating purified cytochromes P450 with purified protein kinases (Pyerin et al., 1987). In absence of stimulation the incorporation of radioactive phosphate into cytochrome P450 isoenzymes (isolated and purified after the incubation with $^{32}$P-orthophosphate) was low. A high degree of incorporation of $^{32}$P-phosphate into cytochrome P450 took place (Table 1) after stimulation by extracellular glucagon or by membrane permeating cAMP-derivatives ($N^6$, $O^{2'}$- dibutyryl-cAMP and 8-thiomethyl-cAMP) (Bartlomowicz et al., 1989a).

Table 1. Stimulation of the phosphorylation of CYP2B1 related proteins by glucagon and cAMP derivatives[a]

| Amount of protein | Treatment of hepatocytes | | |
|---|---|---|---|
| | None | Glucagon | cAMP derivatives |
| 25.0 µg | 10 646 | 18 433[b] | 75 219[b] |
| | | 20 729[b] | 73 416[b] |
| 12.5 µg | 6 344 | 9 978 | 39 842 |
| 6.25 µg | - | - | 15 753 |

[a] Cytochromes P-450 (partially purified, octylamino-Sepharose eluate) obtained from heptocytes incubated in the presence of ($^{32}$P) orthophosphate with glucagon ($10^{-7}$M) or $N^6$, $O^{2'}$-dibutyryl-cAMP (1.4 mM) and 8-thiomethyl-cAMP (0.4 mM), subjected to immunoblotting followed by autoradiography. Values are arbitrary intensity units.

[b] Sample applied in two lanes and each lane scanned.

Autoradiography of gel electrophoretically separated proteins from solubilized microsomes and of purified cytochromes P-450 combined with visualization on Western blots by specific antibodies, showed that 4 cytochromes P450 were selectively phosphorylated, namely CYP2B1 and 2B2 as well as two CYP2B1-related proteins (Bartlomowicz et al., 1989a). One of them was inducible by phenobarbital (Oesch et al., 1989). PKA often act via an Arg-Arg-X-Ser-recognition sequence in the substrate protein (Krebs & Beavo, 1979). The phosphorylation of rabbit CYP2B4 (closely related to rat CYP2B2) occurs on serine 128, within this recognition sequence (Müller et al., 1985). On the basis of other studies (Fujii-Kuriyama et al., 1982) one can forecast that serine 128 is the amino acid which is phosphorylated in rat CYP2B1 and 2B2. This recognition sequence is also present in many members of the CYP gene family 2 of rat liver. The presence of this recognition sequence is not always sufficient for phosphorylation to occur in isolated hepatocytes, e.g. CYP2Cll is not significantly phosphorylated (Koch & Waxman, 1989).

The phosphorylation of CYP2B1 and 2B2 in intact hepatocytes led to a marked decrease in the O-dealkylation of 7-pentylresorufin (Table 2; Bartlomowicz et al., 1989b) representing a selective substrate of CYP2B1 and 2B2. The metabolism of testosterone was regio and stereoselectively decreased in those positions which are attacked by CYP2B1 and 2B2 (Bartlomowicz et al., 1989b). The hydroxylation in position 16ß is catalyzed exclusively by CYP2B1 and 2B2 (Levin et al., 1984; Waxman, 1988). The hydroxylation of testosterone at this position was decreased by about 50 per cent (Table 2) after the treatment of the hepatocytes with the agents leading to phosphorylation of CYP2B1 and 2B2 (cAMP derivatives). The influence of the phosphorylation on the hydroxylation at the 16α-position was "diluted" (decrease of about 30 per cent) (Bartlomowicz et al., 1989b) by those isoenzymes which, in addition to the phosphorylated CYP2B1 and 2B2, also catalyze this reaction, namely CYP2C7 and 2C13 (Levin et al., 1984, Waxman 1988). CYP2B1 and 2B2 also cytalyze the oxidation of the 17ß-OH group of testosterone to a keto group. This reaction is still less specific. The decrease in activity due to phosphorylation was even more "diluted" at this position and did not reach statistical significance (Table 2).

The activation of cyclophosphamide and ifosfamide to mutagens (known to be mediated by CYP2B1) was markedly reduced after pretreatment of the hepatocytes with the membrane-permeating cAMP derivative $N^6, O^{2'}$ dibutyryl-cAMP (dbcAMP). When hepatocytes were incubated for 1 h with dbcAMP in the presence of the phosphodiesterase inhibitor theophylline, cyclophosphamide and ifosfamide activation to mutagens for Salmonella typhimurium TA 1535 were reduced to 51 per cent and 38 per cent of unstimulated controls. After pretreatment of the hepatocytes with dbcAMP for 1.5 h without theophylline and using Salmonella typhimurium TA 100 as target strain a marked reduction was observed in mutagenicity of cyclophosphamide (35 per cent compared with unstimulated controls). Continued presence of the CYP2B1 and 2B2 inducer phenobarbital in the medium increased the mutagenicity of cyclophosphamide and led to an even more marked reduction of mutagenicity by pretreatment of the hepatocytes with dbcAMP and theophylline. The mutagenicity of the ifosfamide metabolite ifosfamide mustard which does not require metabolic activation by cytochrome P450 was indistinguishable after incubation with dbcAMP from that after incubation with unstimulated hepatocytes. Also the metabolic formation of cytotoxic metabolites from cyclophosphamide and

Table 2. Isoenzyme-selective decrease of monooxygenase activities in hepatocytes after treatment with cAMP derivatives[a]

| Treatment of hepatocytes | 7-Pentylresorufin depentylation (pmol/mg protein/min) | Testosterone metabolites[b] (nmol/mg protein/min) | |
|---|---|---|---|
| | | 16α-OH | 16β-OH |
| None | 2565 ± 81 | 2.27 ± 0.20 | 0.81 ± 0.10 |
| cAMP derivatives[c] | 1680 ± 52[d] (65) | 1.56 ± 0.14[d] (68) | 0.45 ± 0.07[d] (56) |

[a] Microsomes from hepatocytes of phenobarbital-treated rats were incubated with 7-pentylresorufin or testosterone as described in Bartlomowicz et al., 1989b. The dealkylation of 7-pentylresorufin was assayed spectrophotofluorimetrically. Testosterone metabolites were resolved and quantitated by HPLC. Values: means ± S.E.M. (in parantheses: percent of control).
[b] Abbreviations: hydroxylated testosterone metabolites.
[c] 1mM N$^6$, O$^{2'}$-dibutyryl-cAMP and 0.5 mM 8-thiomethyl-cAMP.
[d] $p < 0.05$ compared with corresponding control; no significant differences in oxidation rates at the following positions: 2α, 2β, 6β, 7α, 17β.

ifosfamide but not that of ifosfamide mustard was markedly decreased by pretreatment of the hepatocytes with dbcAMP and theophylline (Oesch-Bartlomowicz et al., 1990). Thus, the stimulation of PKA in intact cells has important consequences for the control of genotoxic and cytotoxic metabolites derived from cyclophosphamide and ifosfamide.

## CONTROL BY EPOXIDE HYDROLASES AND DIHYDRODIOL DEHYDROGENASE OF MONOFUNCTIONAL EPOXIDES AND OF VICINAL DIOL EPOXIDES DERIVED FROM POLYCYCLIC AROMATIC HYDROCARBONS

The anti-isomer of a non-bay-region diol epoxide, benz(a)anthracene-8,9-dihydrodiol-10,11-epoxide (BA-8,9-diol-10,11-oxide), has a half-life of many hours and is, therefore, useful for metabolic studies. It is often the major DNA-binding and mutagenic species formed from benz(a)anthracene in vivo and in vitro (Wood et al., 1977; Vigny et al., 1980). Enzymatic inactivation of this diol epoxide was obtained with dihydrodiol dehydrogenase but not with microsomal or cytosolic epoxide hydrolase (Glatt, H.R. et al., 1982). Inactivation of the diol epoxide by dihydrodiol dehydrogenase was slower than that of the K-region epoxide by epoxide hydrolases, but would be sufficiently high to substantially affect diol epoxide concentrations in mammalian systems if one assumes that the activities of the investigated enzymes in vivo are comparable to those that are determined in vitro. By the use of purified enzymes and bacterial mutagenicity, the relative role of the three enzymes, microsomal and cytosolic epoxide hydrolase and cytosolic dihydrodiol dehydrogenase in metabolic inactivation was studied, as an indication of their effects on the mutagenicity of BA-8,9-diol-10,11-oxide and benz(a)anthracene 5,6-oxide (BA 5,6-oxide), the K-region epoxide. Test compounds were used at concentrations of the increasing portion of the concentration-mutagenicity curve. As expected from its substrate specificity (Bentley et al., 1976), microsomal epoxide hydrolase readily inactivated BA 5,6-oxide. Even with a 100-fold excess over that sufficient for complete inactivation of the K-region oxide no significant effect on the mutagenicity of BA-8,9-diol 10,11-oxide was obtained. To inactivate BA 5,6-oxide relatively large amounts of cytosolic epoxide hydrolase were required. Neither cytosolic nor microsomal epoxide hydrolase inactivated the diol epoxide. Nevertheless the diol epoxide was inactivated by dihydrodiol dehydrogenase in the presence of $NADP^+$ (Glatt et al., 1982). High concentrations of dihydrodiol dehydrogenase were necessary for inactivation of the diol epoxide. However, the inablility of either $NADP^+$ or of dihydrodiol dehydrogenase alone to inactivate the vincinal diol epoxide and the lack of inactivation of BA 5,6-oxide by dihydrodiol dehydrogenase either with or without $NADP^+$, indicate that the inactivation was a consequence of enzymic activity and not the result of nonspecific binding of the vincinal diol epoxide to protein or to $NADP^+$. A low concentration of microsomal epoxide hydrolase was sufficient for the inactivation of BA 5,6-oxide. By using small quantities of epoxide hydrolases (0.4 unit of purified rat microsomal epoxide hydrolase, equivalent to 1.3 mg of liver, or by 7 units of purified rabbit cytosolic epoxide hydrolase, equivalent to 28 mg of liver) a 50 per cent inactivation of 1 μg BA 5,6-oxide was achieved. However, microsomal epoxide hydrolase equivalent to 200 mg of rabbit liver did not inactivate the vicinal diol epoxide BA-8,9-diol-10,11-oxide, whereas with dihydrodiol dehydrogenase equivalent to 200 mg of liver a 50 per cent inactivation was obtained. The K-region epoxide is inactivated noticeably more efficiently than the vicinal diol epoxide which is in line with the much stronger biological activities in mammalian systems of vicinal diol epoxides compared with K-region epoxides.

In summary, the species and tissue dependent control of mutagenic and carcinogenic metabolites is largely due to the individual pattern of different enzymes and the metabolic cooperation of activating phase I and inactivating phase II carcinogen metabolizing enzymes. *In vivo* their compartmentalization will, in addition, play an important role and factors such as DNA repair and cell proliferation will influence the resulting carcinogenicity profoundly.

Acknowledgements

The authors wish to thank Miss A. Geisel for typing and the Deutsche Forschungsgemeinschaft for financial support.

REFERENCES

Bartlomowicz, B., Waxman, D.J., Utesch, D., Oesch, F., and Friedberg, T. (1989a): Phosphorylation of carcinogen metabolizing enzymes: regulation of the phosphorylation status of the major phenobarbital inducible cytochromes P-450 in hepatocytes. *Carcinogenesis* 10, 225-228.

Bartlomowicz, B., Friedberg, T., Utesch, D., Molitor, E., Platt, K., and Oesch, F. (1989b): Regio- and stereoselective regulation of monooxygenase activities by isoenzyme-selective phosphorylation of cytochrome P-450. *Biochem. Biophys. Res. Commun.* 160, 46-52.

Bentley, P., Schmassmann, H., Sims, P., and Oesch, F. (1976): Epoxides derived from various polycyclic hydrocarbons as substrates of homogeneous and microsome-bound epoxide hydratase. *Eur. J. Biochem.* 69, 97-103.

Bücker, M., Golan, M., Schmassmann, H.U., Glatt, H.R., Stasiecki, P., and Oesch, F. (1979): The epoxide hydratase inducer *trans*-stilbene oxide shifts the metabolic epoxidation of benzo(a)pyrene from the bay- to the K-region and reduces its mutagenicity. *Mol. Pharmacol.* 16, 656-666.

Conney, A.H. (1982): Induction of microsomal enzymes by foreign chemicals and carcinogenesis by polycyclic aromatic hydrocarbons: G.H.A. Clowes Memorial Lecture. *Cancer Res.* 42, 4875-4917.

Fujii-Kuriyama, Y., Mizukami, Y., Kawajiri, K., Sogawa, K., and Muramatsu, M. (1982): Primary structure of cytochrome P-450: coding nucleotide sequence of phenobarbital-inducible cytochrome P-450 cDNA from rat liver. *Proc. Natl. Acad. Sci., USA* 79, 2793-2797.

Glatt, H.R., Cooper,C.S., Grover, P.L., Sims, P., Bentley, P., Merdes, M., Waechter, F., Vogel, K., Guenthner, T.M., and Oesch, F. (1982): Inactivation of a diol-epoxide by dihydrodiol dehydrogenase, but not by two epoxide hydrolases. *Science* 215, 1507-1509.

Koch, J.A. and Waxman, D.J. (1989): Posttranslational modification of hepatic cytochrome P-450. Phosphorylation of phenobarbital-inducible P-450 forms PB-4 (IIB1) and PB-5 (IIB2) in isolted rat hepatocytes and *in vivo*. *Biochemistry* 28, 3145-3152.

Krebs, E.G. and Beavo, J.A. (1979): Phosphorylation of enzymes. *Ann. Rev. Biochem.* 48, 923-959.

Levin, W., Thomas, P.E., Reik, L.M., Wood, A.W., and Ryan, D.E. (1984): In *IUPHAR 9th International Congress of Pharmacology*, eds. W. Paton, J. Mitchell, and P. Turnes, Vol. 3, pp. 203-209. London: Mac Millan Press.

Müller, R., Schmidt, W.E., and Stier, A. (1985): The site of cyclic AMP-dependent protein kinase catalyzed phosphorylation of P-450. *FEBS Lett*. 187, 21-24.

Oesch-Bartlomowicz, B., Vogel, S, Arens, H.-J., and Oesch, F., (1990): Modulation of the control of mutagenic metabolites derived from cyclophosphamide and ifosfamide by stimulation of protein kinase A. *Mutat. Res*. 232, 305-312

Oesch, F., Waxman, D.J., Morrissey, J.J., Honscha, W., Kissel, W., and Friedberg, T. (1989): Antibodies targeted against hypervariable and constant regions of cytochromes P450IIB1 and P450IIB2. *Arch. Biochem. Biophys*. 270, 23-32.

Pyerin, W., Taniguchi, H., Horn, F., Oesch, F., Amelizad, Z., Friedberg, T., and Wolf, C.R. (1987): Isoenzyme-specific phosphorylation of cytochromes P-450 and other drug metabolizing enzymes. *Biochem. Biophys. Res. Commun*. 142, 885-892.

Vigny, P., Kindts, M., Duguesne, M., Cooper, C.S., Grover, P.L., and Sims, P. (1980): Metabolic activation of benz(a)anthracene: fluorescence spectral evidence indicated the involvement of a non-'bay-region'diol epoxide. *Carcinogenesis* 1, 33-41.

Waxman, D.J. (1988): Interactions of hepatic cytochromes P-450 with steroid hormones. Regioselectivity and stereospecificity of steroid hydroxlation and hormonal regulation of rat P-450 enzyme expression. *Biochem. Pharmacol*. 37, 71-84.

Wilson, N.M., Christon, M., Turner, C.R., Wrighton, S.A., and Jefcoate, C.R. (1984): Binding and metabolism of benzo(a)pyrene and 7,12-dimethylbenz(a)anthracene by seven purified forms of cytochrome P-450. *Carcinogenesis* 5, 1475-1483.

Wood, A.W., Chang, R.L., Levin, W., Lehr, R.E., Schaefer-Ridder, M., Karle, J.M., Jerina, D.M. and Conney, A.H. (1977): Mutagenicity and cytotoxicity of the bay region 1,2-epoxides. *Proc. Natl. Acad. Sci. USA* 74, 2746-2750.

# Genetic polymorphism of P450 and human cancer

Kaname Kawajiri, Junko Watanabe, Shin-Ichi Hayashi

*Department of Biochemistry, Saitama Cancer Center Research Institute, 818 Komuro, Ina-machi, Saitama 362, Japan*

A great portion of human cancer is caused by man-made or natural chemical compounds in our environments. Carcinogenic risks from exposures to exogenous chemical carcinogens depend not only on the intrinsic natures and doses of the chemical, but also may depend on the interindividual sensitivities to the carcinogens (Kawajiri and Fujii-Kuriyama, 1991). These may include host factors such as metabolic activation and DNA repair. Some of these host responses may be genetically determined and hence show significant interindividual difference in susceptibility to cancer.

Most chemical carcinogens require metabolic activation by Phase I enzymes, P450s, to their genotoxic intermediates. In some instances, these activated metabolites are subjected to detoxification by conjugation via the various Phase II enzymes. Thus, the coordinate expression and regulation of Phase I and Phase II drug-metabolizing enzymes and their metabolic balance in the cells of target organs may be an important host factors to determine whether exposure to carcinogens results in cancer or not (Nebert, 1991).

Some of the drug-metabolizing enzymes are known to show genetic variability in their activity among individuals, which may be responsible for individual susceptibility to chemical carcinogenesis. Our main focus is to explain genetically determined differences among individuals in susceptibility to lung cancer in terms of genetic polymorphism of drug-metabolizing enzymes, CYP1A1 and GST1. Although several forms of P450s may contribute to occurrence of this cancer, CYP1A1, which is expressed in lung (Shimada et al, 1992) and activates benzo(a)pyrene in cigarette smoke, is the first in line to be responsible for lung cancer. One of the Mu-class of glutathione S-transferase is named GST1, and it detoxificates the metabolites of benzo(a)pyrene including epoxide and hydroxylated forms.

Genetic polymorphisms of CYP1A1 and lung cancer incidence

The polymorphic Msp I site was located at the 264th base downstream from the poly A additional signal (Hayashi et al, 1991). Genotyping of CYP1A1 alleles associated with the presence or absence of the

Msp I site in the 3'-region was carried out by PCR amplification followed by digestion with Msp I. A genotype, designated A, is a predominant homozygote, where the Msp I site is absent at the 3' end. A homozygous rare allele was named genotype C, being derived from one base substitution of thymine with cytosine to form the Msp I site. Genotype B is heterozygous for both alleles. We compared the frequencies of the three genotypes A, B and C between healthy controls and cancer patients (Kawajiri et al, 1990; Nakachi et al, 1991). We isolated lymphocyte DNA of 2500 persons from a cohort of general population and analyzed the DNA polymorphism of 375 randomly selected subjects as healthy controls. Types A, B and C were found 166 (44%), 169 (45%) and 40 (11%) individuals among the healthy controls, respectively. This result gave a good fit to the Hardy-Weinberg equilibrium. On the other hand, the frequency distribution of genotypes among 105 patients with squamous cell carcinoma of the lung, which was closely associated with smoking, was remarkably different from that of healthy controls and genotype C was found in up to 23 % among them. On the other hand, adenocarcinoma of the lung exhibited little or no association with smoking and the frequency distribution was almost identical to that among control. Furthermore, the frequency distributions on the other cancer patients were the same as healthy controls. These results indicated that the individual with genotype C are susceptible to smoking-induced lung cancer than other genotypes.

It is indispensable to investigate whether this Msp I polymorphism is genetically associated with differences in primary structure of the protein or expression of the CYP1A1 gene. PCR direct sequencing of genomic DNA of each genotype, covering all the exons and the 5'-flanking region, revealed a novel point mutation in the coding region of the gene which resulted in different primary structure of proteins (Hayashi et al, 1991). A mutation from adenine to guanine in the exon 7 resulted in replacement of isoleucine (Ile) by valine (Val) at residue 462 near the heme-binding region of CYP1A1. To detect this Ile-Val polymorphism in the population, an allele-specific PCR procedure was developed using two set of primers. The genetic association between these two loci in a general population was examined. The linkage coefficient defined by Hill and Robertson was calculated to be 0.78, showing that these two loci are very closely associated. In order to compare the catalytic activity of these two different CYP1A1 proteins, we constructed and expressed the cDNAs in yeast cells. Both AHH activity and mutagenic activity towards benzo(a)pyrene mediated by expressed CYP1A1 proteins of yeast microsomes were studied. Our results indicate that the valine-type CYP1A1 showed higher AHH activity and mutagenicity, although the enzyme activities were at low levels (Kawajiri et al, 1993b).

## Joint effect of CYP1A1 and GST1 genotyping on lung cancer susceptibility

As already mentioned, the genetically-determined susceptibility to chemically-induced cancers may depend on the metabolic balance of Phase I and Phase II enzymes. From this point of view, we studied the joint effect of CYP1A1 and GST1 genotyping on lung cancer susceptibility (Hayashi et al, 1992). We consider the six genotypes by combining the three genotypes of 1A1 and the two of GST1, and compare the combined genotype frequencies between lung cancer patients and healthy controls (Table 1). GST1(+)

Table 1. Distribution of combined genotypes of CYP1A1 and GST1 genes in lung cancer patients and healthy controls.

| CYP1A1 | Ile/Ile | | Ile/Val | | Val/Val | | Total |
|---|---|---|---|---|---|---|---|
| GST1 | (+) | (−) | (+) | (−) | (+) | (−) | |
| Healthy controls | 127 (35.5) | 106 (29.6) | 55 (15.4) | 53 (14.8) | 9 (2.5) | 8 (2.2) | 358 (100%) |
| Lung cancer | 83 (25.4) | 105 (32.1) | 50 (15.3) | 53 (16.2) | 11 (3.4) | 25 (7.6) | 327 (100) |
| Kreyberg I type | 44 (25.0) | 55 (31.3) | 23 (13.1) | 32 (18.1) | 4 (2.3) | 18 (10.2) | 176 (100) |
| Squamous cell ca. | 24 (23.1) | 37 (35.6) | 13 (12.5) | 17 (16.3) | 2 (1.9) | 11 (10.6) | 104 (100) |
| Undif. cell ca. | 20 (27.8) | 18 (25.0) | 10 (13.9) | 15 (20.8) | 2 (2.8) | 7 (9.7) | 72 (100) |
| Kreyberg II type (Adeno ca.) | 39 (25.8) | 50 (33.2) | 27 (17.9) | 21 (13.9) | 7 (4.6) | 7 (4.6) | 151 (100) |

means the presence of at least one normal allele of GST1 gene, while GST1(−) means homozygous-nulled state. The proportions of GST1(−) genotype among all three genotypes of CYP1A1 were identical in healthy controls, suggesting that these two polymorphisms were genetically independent to each other. Lung cancer patients showed an obviously different distribution over the six genotypes from that in healthy controls, and it is noted that the frequency of valine-homozygote and GST1(−) was 3.5-fold higher in lung cancer patients than in healthy controls. Among 22 patients with Kreyberg I type of lung cancer, whose genotype was valine-homozygous, 18 patients were found to be of GST1(−) genotype.

The relative risk (odds ratio) was calculated from the compared frequencies for each of the combined genotypes, taking the risk of the isoleucine-homozygous and GST1(+) as a baseline of 1.0. Individuals with the combined genotype of valine-homozygous and GST1(−) were at remarkably high risk to lung cancer with an odds ratio of 4.8. It is note worthy that their odds ratio was increased still further to 6.5 in Kreyberg I type carcinomas and to 7.3 in squamous cell carcinoma. On the other hand, the joint effect of the two genotypes resulted in slight elevation of risk to Kreyberg II type of lung cancer.

Genetic difference in susceptibility to lung cancer in relation to cigarette smoking dose

We estimate how lung cancer risk varies among individuals with the different genotypes when cigarette dose is taken into account (Nakachi et al, 1993). We first examined the difference of cigarette dose by genotypes among lung cancer patients. We examined this focusing on the squamous cell carcinoma of the lung, which was mostly closely associated with cigarette smoking. A total of the 85 patients with squamous cell carcinoma were divided into six genotypes, which were composed of the Msp I polymorphism combined with the two genotypes of GST1. We found that the mean cumulative cigarette consumption in each of genotypes A, B and C was higher and lower in combined GST1(+) and GST1(−), respectively, resulting

Table 2. Cigarette consumption of patients in combined genotypes of the CYP1A1 and GST1 genes.

| Genotypes CYP1A1 | GST1 | No. of subjects (frequency) | Mean age ±SD (yr) | Cigarette consumption ±SD ( x 10$^4$) |
|---|---|---|---|---|
| A | (+) | 12(0.141) | 65.1±8.0 | 48.3±30.2 |
| A | (−) | 21(0.247) | 63.4±9.8 | 42.0±21.9 |
| B | (+) | 14(0.165) | 65.9±8.1 | 47.3±14.4 |
| B | (−) | 19(0.224) | 66.9±6.8 | 36.6±18.2 |
| C | (+) | 7(0.082) | 62.4±11.6 | 32.7±23.8 |
| C | (−) | 12(0.141) | 66.8±6.8 | 31.8±7.7 |
| Total | | 85(1.000) | 65.2±8.4 | 40.3±20.5 |

in the lowest cigarette dose of $3.2 \times 10^5$ observed in the combination of susceptible genotype C and GST1(−) (Table 2). The mean cigarette consumption in combined genotyping of the Ile-Val polymorphism and GST1 were also carried out, and the gradation of the cigarette doses in the combined six genotypes was similar to that shown in Table 2. Namely, the patients in combined susceptible genotypes with CYP1A1 and GST1 contracted the carcinoma with lesser cigarette dose than other combined genotypes.

Then relative risk to the cancer was evaluated for combinations of the genotypes and cumulated cigarette dose by means of a case-control study. Controls were individually matched to the patients with respect to sex and age, and they were randomly chosen from 2500 general residents with DNA samples and interview data on cigarette smoking. Two controls were selected for each of the patients within matching conditions. We confirmed that selected controls were exactly in Hardy-Weinberg linkage equilibrium and that no interaction between genotypes and cigarette smoking was found among them.

The different distributions of patients and controls by combined genotypes of Msp I and GST1 polymorphisms allowed us to evaluate the risk to the cancer, taking the mean cigarette amount $3.2 \times 10^5$ among the patients genotype C as a cutoff point. The relative risk was calculated for the combined genotypes, taking the risk of the combined genotype A and GST1(+) at the lower cigarette dose level to be at base line of 1.0. The individuals with genotype C and GST1(−) were extremely high risk of 16-fold at the lower dose compared with those with A and GST1(+). On the other hand, this relative susceptibility of combined genotype C and GST1(−) compared to A and GST1(+) decreased about 2-fold at the higher cigarette dose level. Although the risk of all combined genotypes increased higher dose level, the genetic difference in cancer risk tends to reduce at high dose level where the environmental influence overpowers genetic predispositions.

Figure 1 summarized the genotype-environment interaction in terms of genetic polymorphism of drug-metabolizing enzymes and cigarette dose. Upper graph demonstrate the distribution of genotypes in general population as functions of cigarette dose. Solid and dashed lines show a susceptible and non-susceptible genotypes, respectively. Middle graph shows dose-response relations by the genotypes obtained by a case-control study. A susceptible genotype have a higher risk to cancer than the non-susceptible especially in low dose of cigarette. The observed cigarette distribution among patients by the genotypes is a mathematical product of the cigarette distribution among general population and the dose-response relations. We can thus explain our finding that the patients with susceptible genotype contracted the carcinoma with lesser cigarette dose than other genotypes.

Genetic polymorphism of drug-metabolizing enzymes and p53 mutations

Germ line polymorphisms of other genes involved in multiple steps of carcinogenesis may account for the genetic difference in lung cancer susceptibility. The p53 gene is a candidate because its mutations have been indicated to be important genetic events in lung cancer development. Apart from a rare germ-line mutations in p53 gene associated with Li-Fraumeni syndrome, there are two wild types of p53 molecules in humans, ascribed to amino acid replacement at codon 72 of Arg by Pro in the region of transactivation domain. Then, we examined the relation between Arg-Pro polymorphism of the p53 gene and lung cancer susceptibility (Kawajiri et al, 1993a). We found that a susceptible genotype proline-homozygote showed a risk elevation of 1.7-fold to smoking-induced lung cancer. No etiological association between p53 and CYP1A1 polymorphisms was found (Table 3). Thus, these two risk modification by p53 and CYP1A1 polymorphisms work independently, resulting in an increased attributable risk of "susceptible" genotypes to lung cancer. This is a contrast to the synergistic increase of the risk in susceptible genotype of CYP1A1 and GST1(-).

Overviewing the risk estimate of CYP1A1, GST1 and p53 in genetic susceptibility, CYP1A1 was found to provide a most

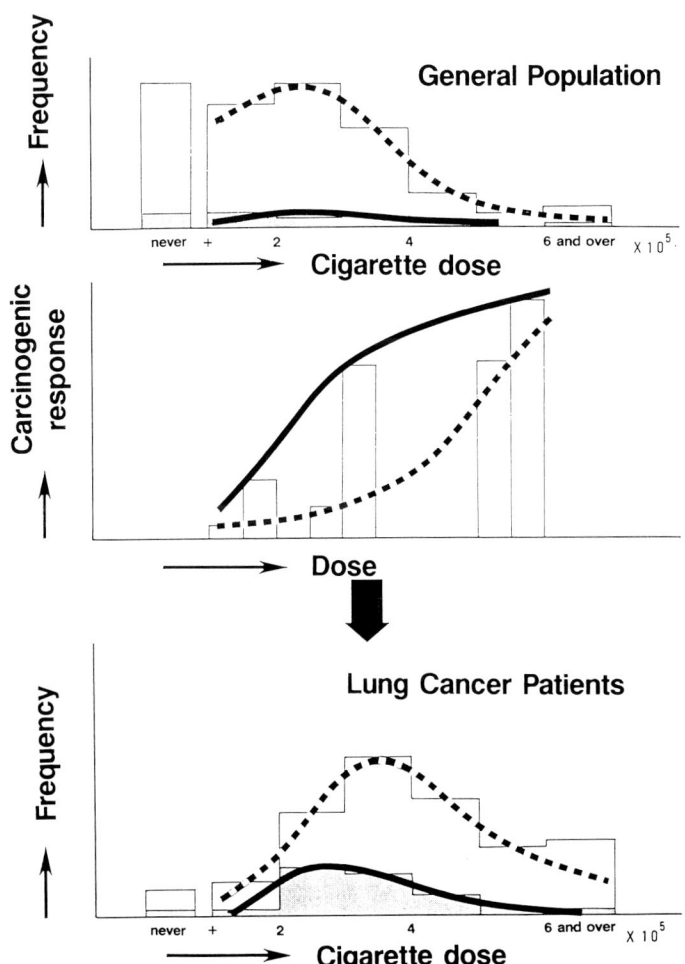

Fig. 1. Genetic difference in dose-response relation determined in lung cancer.

contribution to the interindividual difference in susceptibility to incidence of smoking-induced lung cancer. p53 seems to contribute less to the germ-line susceptibility among populations, but p53 mutation in lung is essential in cancer development as a target gene of chemical carcinogenesis derived from our environments (Suzuki et al, 1992). We then studied the correlation between aberrations of p53 gene in lung cancer and CYP1A1 polymorphism using SSCP analysis (Fig. 2). The p53 alterations in lung cancer tissues were more frequently observed among the patients with a susceptible allele of CYP1A1 gene.

Table 3. O/E ratios by combined genotypes of p53 and CYP1A1 genes in Kreyberg I type of lung cancer.

| CYP1A1 genotypes | p53 genotypes | | | |
|---|---|---|---|---|
| | Arg/Arg | Arg/Pro | Pro/Pro | Total |
| Ile/Ile | 0.88 (42/47.5) | 0.68* (37/54.6) | 1.60* (20/12.5) | 0.86 (99/114.6) |
| Ile/Val | 0.95 (21/22.0) | 0.91 (23/25.3) | 1.89* (11/ 5.8) | 1.03 (55/53.1) |
| Val/Val | 3.14*** (11/ 3.5) | 2.31** ( 9/ 3.9) | 2.22 ( 2/ 0.9) | 2.65*** (22/ 8.3) |
| Total | 1.01 (74/73.0) | 0.82 (69/83.8) | 1.72** (33/19.2) | (176/176) |

*$P<0.05$; **$P<0.01$; ***$P<0.001$

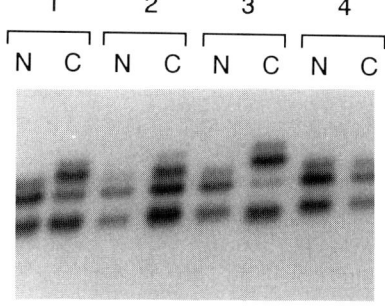

Fig. 2. A representative profile of SSCP analysis of p53 gene. N and C mean normal and cancerous part DNAs, respectively.

## Concluding remarks

Most P450 metabolism results in detoxification of a wide variety of xenobiotics, although certain chemicals are activated in this procedure to electrophilic forms which can damage DNA and on occasion result in carcinogenic transformation of the cells. Since this P450-mediated bioactivation is an initial and obligatory step in chemical carcinogenesis, the interindividual variation in the metabolic activity of P450s may influence subsequent steps, including detoxification by Phase II enzymes and the formation of DNA-carcinogen adducts, and ultimate cancer consequence.

The genetic difference in phenotipic expression and/or structure of P450 genes is a promising candidate to explain the interindividual or interracial difference in cancer susceptibility. Most studies in this area employ a methodology of comparing the frequency of metabolic phenotypes or genotypes of P450 or its gene between cancer patients and controls. In the metabolic phenotype comparison studies, one may be required to pay more attention to the development of methods of determining phenotypes and also to an examination of possible influences of medications on both patients and controls (especially hospital controls). Even a simple case-control comparison in phenotype or genotype frequency needs to examine the selection bias of controls, who must be representative of a general population from which patients occurred. In this sense, hospital controls are, in general, not desirable because they are more or less a biased population. A large size "DNA bank" from a general population may be appropriate as a control pool and follow-up study using the DNA bank can be used to confirm the results of case-control studies (Kawajiri et al, 1993b).

References

Hayashi,S.-I.,Watanabe,J.,and Kawajiri,K. (1992):High susceptibility to lung cancer analyzed in terms of combined genotypes of P450IA1 and Mu-class glutathione S-transferase genes. *Jpn.J.Cancer Res.*,**83**,866-870.

Hayashi,S.-I.,Watanabe,J.,Nakachi,K.,and Kawajiri,K. (1991):Genetic linkage of lung cancer-associated Msp I polymorphisms with amino acid replacement in the heme-binding region of tahe human cytochrome P450IA1 gene. *J.Biochem.*, **110**,407-411.

Kawajiri,K. and Fujii-Kuriyama,Y. (1991):P450 and human cancer. *Jpn.J.Cancer Res.*, **82**,1325-1335.

Kawajiri,K.,Nakachi,K.,Imai,K.,Yoshii,A.,Shinoda,N.,and Watanabe,J. (1990):Identification of genetically high risk individuals to lung cancer by DNA polymorphisms of the cytochrome P450IAI gene. *FEBS Lett.*,**263**,131-133.

Kawajiri,K.,Nakachi,K.,Imai,K.,Watanabe,J.,and Hayashi,S.-I. (1993a):Germ line polymorphisms of p53 and CYP1A1 genes involved in human lung cancer. *Carcinogenesis*,**14**,1085-1089.

Kawajiri,K.,Nakachi,K.,Imai,K.,Watanabe,J.,and Hayashi,S.-I. (1993b):The CYP1A1 gene and cancer susceptibility. *Crit.Rev.Oncol.Hematol.*,**14**,77-87.

Nakachi,K.,Imai,K.,Hayashi,S-I.,and Kawajiri,K. (1993):Polymorphisms of the CYP1A1 and glutathione S-transferase genes associated with susceptibility to lung cancer in relation to cigarette dose in a Japanese population. *Cancer Res.*, **53**,2994-2999.

Nakachi,K.,Imai,K.,Hayashi,S.-I.,Watanabe,J.,and Kawajiri,K. (1991):Genetic susceptibility to squamous cell carcinoma of the lung in relation to cigarette smoking dose. *Cancer Res.*,**51**,5177-5180.

Nebert,D.W.(1991):Role of genetics and drug metabolism in human cancer risk. *Mutat.Res.*,**247**,267-281.

Shimada,T., Yun,C-H., Yamazaki,H., Gautier,J-I., Beaune,P.H.,and Guengerich, F.P. (1992): Characterization of human lung microsomal cytochrome P450 1A1 and its role in oxidation of chemical carcinogens. *Mol.Pharmacol.*,**41**,856-864.

Suzuki,H.,Takahashi,T.,Kuroishi,T.,Suyama,M.,Ariyoshi,Y.,Takahashi,T.,and Ueda,R. (1992): p53 Mutations in non-small cell lung cancer in Japan:Association between mutations and smoking. *Cancer Res.*,**52**,734-736.

# Polymorphisms of cytochrome P450 CYP2D6 as risk factor in carcinogenesis

Urs. A. Meyer

Department of Pharmacology, Biozentrum of the University of Basel, Klingelbergstr. 70, CH-4056 Basel, Switzerland

## Summary

In epidemiological studies genetic polymorphisms of drug metabolism have been variably associated with individual susceptibility to develop cancer. The presumed mechanism of these associations is that the polymorphic enzymes activate or inactivate procarcinogens, e.g., aromatic amines.

The debrisoquine polymorphism affects the metabolism by CYP2D6 of over 30 drugs. 5-10 % of individuals in most populations are poor metabolizers and CYP2D6 is absent in their liver. Twelve different mutations or arrangements of the CYP2D6 gene clusters on chromosome 22 either cause poor, extensive or ultrarapid metabolism of CYP2D6 substrates. Extensive metabolism has been associated in some studies with lung cancer. The association of the debrisoquine metabolizer phenotype or genotype with lung cancer is discussed in regard to the controversial results the different studies have produced.

Exposure to chemical carcinogens is widely accepted as a major etiologic factor in a number of cancers, in particular for smoking and lung cancer, arylamines and bladder cancer and food-derived heterocyclic amines and colorectal cancer. There are clear differences in individual susceptibility consistent with a heritable component to risk. This review is concerned with the role of CYP2D6 and its genetic variability, which may influence individual cancer risk. This enzyme is well known for its polymorphism, the debrisoquine polymorphism, which separates the population into subgroups of different genotypes (and corresponding phenotypes; for review, see Meyer et al., 1992; Idle et al., 1992).

The debrisoquine polymorphism represents one of the best studied examples of the genetic variation of a human cytochrome P450 enzyme. Five to 10 % of individuals in Caucasian populations are so-called poor metabolizers (PMs) and are homozygous for 2 recessive alleles of the gene coding for CYP2D6 (for review, see Meyer et al., 1992). The incidence of the PM phenotype is smaller in Asian populations.

## MOLECULAR MECHANISM OF THE DEBRISOQUINE POLYMORPHISM

In 1984, we accumulated first evidence that the enzyme involved in this polymorphism was a cytochrome P450 (Gut et al, 1984). According to present nomenclature it is called CYP2D6 (Nebert et al., 1991). This enzyme was found absent in livers of poor metabolizers (Zanger et al., 1988). In collaboration with Frank Gonzalez and Shioko Kimura of the National Cancer Institute in Bethesda, USA, the human cDNA for this enzyme was characterized (Gonzalez et al., 1988; for review, see Meyer et al., 1992). Two other genes, designated CYP2D7 and CYP2D8, are located just upstream of CYP2D6 on chromosome 22, but they are defective and not expressed, so-called pseudogenes. Initial analysis of this area of the genome with restriction enzymes revealed several RFLPs (restriction fragment length polymorphisms) which cosegregated in family studies with the poor metabolizer phenotype (Skoda et al., 1988). These RFLPs helped to isolate the variant alleles of the CYP2D6 gene and to identify their mutations (Kagimoto et al., 1990). Over 90% of the mutations of the CYP2D6 gene which cause "poor metabolism" of debrisoquine are now identified (Broly et al., 1991; Daly et al., 1991). The most common mutation (70-80 % of all mutant alleles) is characterized by the mutation "CYP2D6-B", a G to A change at the splice-site recognition sequence between the 3d intron and the 4th exon of CYP2D6. It frequently occurs in association with multiple other mutations of the CYP2D6 gene (Kagimoto et al., 1990). The rare (~5 %) "CYP2D6-A" mutation consists of a single basepair deletion in the 5th exon and the "CYP2D6-D" mutation (10-20 %) is caused by a deletion of the entire CYP2D6 gene (Gaedigk et al., 1991), so that only the 2 pseudogenes CYP2D7 and CYP2D8 remain on the chromosome in this case. The rare "CYP2D6-C" mutation (< 1%) is a deletion of 3 basepairs in the 5th exon (Tyndale et al., 1991). It leads to an enzyme with slightly lower activity than the "wild-type" normal enzyme, but does not cause a "PM phenotype" according to the usual definition (Broly et al., 1993). The major "CYP2D6-B" mutation also occurs in different associations (or linkage disequilibrium) with other mutations in the vicinity of the CYP2D6 gene cluster which cause the above mentioned RFLPs. Thus, the CYP2D6-B" mutation can be found within different haplotypes, for instance on a "normal" XbaI 29kb fragment, associated with XbaI fragments of 16+9 kb size or with a XbaI 44 kb or 42 kb fragment, which harbor 4 D6-like genes (Heim & Meyer, 1992). Knowledge of these mutations (A, B, C) has allowed the development of DNA tests using the polymerase chain reaction and allele-specific (i.e. mutation-specific) DNA amplification. The "D" mutation (gene deletion) has to be identified by restriction analysis. These DNA techniques are used in many laboratories (Heim and Meyer, 1990; Broly et al., 1991; Daly et al., 1991; Dale Smith et al., 1992; Dahl et al., 1992).

Direct genotyping of DNA is important for epidemiological studies, in which phenotyping may be problematic because of concomitant administration of cosubstrates of CYP2D6 or of competitive inhibitors of this enzyme. Caporaso et al. (1992) have recently compared the merits of phenotyping and genotyping to characterize CYP2D6 in population studies.

In summary, most of the mutations of the CYP2D6 gene causing the PM phenotype are identified and simple DNA test to detect them are available. Such a test identifies 90 to 95 % of PM-alleles and in 80 to 90 % of cases correctly predicts the genotype (i.e. identification of both alleles on the chromosome pair) and the phenotype, i.e. if the individual will be in the PM or the EM phenotype subpopulation (Broly et al., 1991, Dahl et al., 1992). However, because of unknown mutations, genotyping does not predict 100 % of phenotypes and by its very nature gives no information, on "how extensive" or "how poor" an individual will be in its metabolic capacity.

Some of this variation has recently been explained by the discovery of additional mutations which cause only a partial deficiency of the CYP2D6 protein (Broly et al., 1993, Yokota et al., 1993) and can explain the so-called intermediate metabolizers. Moreover, Johansson et al. (1992, 1994) have made the fascinating observation of dominantly inherited amplification of a functional CYP2D6 variant in 2 families of so-called "ultrarapid metabolizers" providing a molecular basis for the most extensive metabolizers in the distribution of metabolic ratios.

CYP2D6 PHENOTYPE AND GENOTYPE IN REGARD TO LUNG CANCER

Table 1 summarized the studies in which the frequency of the debrisoquine phenotype, determined by the urinary metabolic ratio after a standard dose of debrisoquine, was determined in lung cancer patients and controls. Critical evaluations of these studies, including the advantages of case control studies over random controls for population studies of multifactorial diseases, have been published by Caporaso et a. (1992) and Roots et al. (1992). Of the "phenotype" studies published, 3 only showed a statistically significant overrepresentation of EMs (or underrepresentation of PMs) in the cancer group. The studies are indeed difficult to compare, they differ markedly in study design, e.g. recruitment of patients, age of the control group, degree of smoking (the genetic risk may only be important at a certain level of exposure), etc. It was hoped that the increasing knowledge of the mutations of CYP2D6 and the development of DNA tests (Heim an Meyer, 1990; Daly et al., 1991, Dale Smith et al., 1992) would allow a better evaluation of the hypothesis of an association of the debrisoquine polymorphism with the risk of lung cancer. DNA tests have the advantage that no drug has to be administered, no urine has to be collected, and the test is not influenced by other drugs or by disease. Table 2 summarizes the published studies using genotyping methodology. Of the 5 studies reported, 3 reveal a significant underrepresentation of individuals with a "PM-genotype" (i.e. 2 mutant alleles of CYP2D6 identified) in the lung cancer group. The 2 studies with non-significant distribution had the lowest percentage of predicted PMs in the control group and this may be related to the study design, in which only one mutant allele of CYP2D6 was tested, reducing the power of prediction to an unknown degree. In the one case-control study where both phenotype and genotype were determined (reviewed by Caporaso et al., 1992), there was considerable misclassification by the genotyping method. The continuous improvement of the genotyping test should allow a more definitive evaluation of the association in the future. Of particular interest is the observation of Agundez et al. (1994), indicating that the CYP2D6-C mutation is 6-times more frequent among Spanish lung cancer patients than in controls. The CYP2D6-C allele is characterized by a 3 bp-deletion in the 5th exon of the gene, which leads to the deletion of Lysine 281 in the CYP2D6 protein and a slightly lower mobility on SDS-PAGE (Tyndale et al., 1991). Moreover, the total amount of protein detected with a monoclonal antibody against normal CYP2D6 was reduced as were the microsomal velocities for debrisoquine, sparteine and bufuralol. On expression in HepG2 cells, CYP2D6-C had a high affinity for these substrates, however. There are several theoretical possibilities why CYP2D6-C may predispose to lung cancer. CYP2D6-C may have a high affinity for certain smoke-derived procarcinogens known to be metabolized by CYP2D6 (Crespi et al., 1991). Another possibility is that there is linkage dysequilibrium of the CYP2D6-C mutations with mutations or genes on

Table 1

## CYP2D6 Phenotype and Lung Cancer

| | Lung Cancer | | Controls | | P |
|---|---|---|---|---|---|
| | N | PMs (%) | N | PMs (%) | |
| Ayesh et al., 1984 | 245 | 1.6 | 234 | 9.0 | 0.0002 |
| Roots et al., 1990 | 400 | 7.0 | 470 | 10.5 | 0.052 |
| Law et al., 1989 | 104 | 1.9 | 104 | 8.7 | 0.029 |
| Duche al., 1991 | 153 | 6.5 | 254 | 7.9 | n.s. |
| Benitez et al., 1991 | 84 (73)[a] | 4.8 (1.4)[a] | 143 | 7.0 | n.s. |
| Caporaso et al., 1990 | 89 | 1.1 | 92 | 13.0 | 0.0015 |
| Speirs et al., 1990 | 82 | 9.8 | - | - | - |
| All studies | 1157 | 4.9 | 1227 | 8.6 | |

[a]  squamons and small cell histology

Table 2

## CYP2D6 Genotype and Lung Cancer

| | Lung Cancer | | Controls | | P | Alleles tested |
|---|---|---|---|---|---|---|
| | N | PMs (%) | N | PMs (%) | | |
| Wolf et al., 1992 | 361 | 3.6 | 720 | 4.3 | n.s. | (B) |
| Kerb et al., 1992 | 115 | 3.5 | 137 | 9.5 | <0.05 | (A,B,D) |
| Hirvonen et al., 1992 | 106 | 0.9 | 122 | 5.7 | <0.05 | (A,B,D) |
| Tefre et al., 1992 | 190 | 9.0 | 220 | 5.0 | n.s. | (B) |
| Agundez et al., 1994 | 89 | 0 | 98 | 7.1 | <0.02 | (A,B,C,D) |
| | | (20) | | (3.0) | <0.0005 | (C alone) |

chromosome 22 which predispose to lung cancer. Several laboratories are in the process of testing if the otherwise rare CYP2D6-C allele occurs at a higher frequency in lung cancer patients also in other populations. This would provide the background for an investigation of the role of this modified enzyme for individual cancer susceptibility.

CONCLUSIONS

The data so far with phenotyping or genotyping methods for the association of lung cancer with the debrisoquine polymorphism have indicated an association of the extensive metabolizer phenotype or the EM genotype with lung cancer in the majority, but not in all studies. The differences in the results may be due to the use of different methodologies or to differences in the genetic and environmental background of the populations studied, which may in addition be exposed to different carcinogens. It has to be kept in mind that cancer is a multifactorial disease and that genotoxicity by chemical carcinogens is only one event in a multistep process involving DNA repair, oncogenes, suppressor genes, etc. Obviously, a multitude of genetic and environmental risk factors and or protective factors determine the ultimate generation of a cancer cell. It is anticipated that future studies will establish a significant but quantitatively variable association of a functionally active CYP2D6 enzyme with lung cancer in the majority of populations.

REFERENCES

Agundez, J.A.G., Martinez, C., Ladero, J., Ledesma, M.C., Ramos, J.M., Martin, R., Rodriquez, A., Jara, C., and Benitez, J. (1994): Debrisoquine oxidation genotype and susceptibility to lung cancer. *Clin. Pharmacol. Ther* (in press).

Ayesh, R., Idle, J.R., Ritchie, J.C., Crothers, M.J., and Hetzel, M.R. (1984): Metabolic oxidation phenotypes as markers for susceptibility to lung cancer. *Nature* 312, 169-170.

Benitez, J., Ladewro, J.M., Jara, C., Carrillo, J.A., Cobaleda, J., Llerena, A., Vargas, E., and Muños, J.J. (1991): Polymorphic oxidation of debrisoquine in lung cancer patients. *Eur. J. Cancer* 27, 158-161.

Broly, F., Gaedigk, A., Heim, M., Eichelbaum, M., Mörike, K., and Meyer, U.A. (1991): Debrisoquine/sparteine hydroxylation genotype and phenotype: Analysis of common mutations and alleles of CYP2D6 in a European Population. *DNA & Cell Biology* 10, 545-558.

Broly, F., and Meyer, U.A. (1993): Debrisoquine oxidation polymorphism: phenotypic consequences of a 3-base-pair deletion in exon 5 of the CYP2D6 gene. Pharmacogenetics 3, 123-130.

Caporaso, N.E., Tucker, M.A., Hoover, R.N., Hayes, R.B., Pickle, L.W., Issaq, H.J., Muschik, G.M., Green-Gallo, L., Buivys, D., Aisner, S., Resau, J.H., Trump, B.F., Tollerud, D., Weston, A., Harris, C.C. (1990): Lung cancer and the debrisoquine metabolic phenotype. J. Natl. Canc. Inst. 82, 1264-1272.

Caporaso, N.E., Shields, P.G., Landi, M.T., Shaw, G.L., Tucker, M.A., Hoover, R., Sugimura, H., Weston, A., and Harris, C.C. (1992): The debrisoquine metabolic phenotype and DNA-based assays: implications of misclassification for the association of lung cancer and the debrisoquine metabolic phenotype. *Environ. Health Perspect.* 98, 101-105.

Crespi, C.L., Penman, B.W., Gelboin, H.V., and Gonzalez, F.J. (1991): A tobacco smoke-derived nitrosamine, 4-(methylnitrosamino)-1-(3-pyridyl)-1-butanone, is activated by multiple human cytochrome P450s including the polymorphic human cytochrome P4502D6. *Carcinogenesis* 12, 1197-1201.

Dahl, M.-L., Johansson, I., Porsmyr Palmertz, M., Ingelman-Sundberg, M., Sjöqvist, F. (1992): Analysis of the CYP2D6 gene in relation to debrisoquine and desipramine hydroxylation in a Swedish population. *Clin. Pharmacol. Ther.* 51, 12-17.

Dale Smith, C.A., Moss, J.E., Gough, A.C., Purr, N.K., and Wolf, C.R. (1992): Molecular genetic analysis of the cytochrome P450-debrisoquine hydroxylase locus and association with cancer susceptibility. *Environ. Health Perspect.* 98, 107-112.

Daly, A.K., Armstrong, M., Monkman, S.C., Idle, M.E., and Idle, J.R. (1991): Genetic and metabolic criteria for the assignment of debrisoquine 4-hydroxylation (cytochrome P4502D6) phenotypes. *Pharmacogenetics* 1, 33-41.

Duche, J.-C., Joanne, C., Barre, J., de Cremoux, H., Dalphin, J.C., Depierre, A., Brochard, P., Tillement, J.P., and Bechtel, P. (1991): Lack of a relationship between the polymorphism of debrisoquine oxidation and lung cancer. *Br. J. clin. Pharmac.* 31, 533-536.

Gaedigk, A., Blum, M., Gaedigk, R., Eichelbaum, M., and Meyer, U.A. (1991): Deletion of the entire cytochrome P450 CYP2D6 gene as a cause of impaired drug metabolism in poor metabolizers of the debrisoquine/sparteine polymorphism. *Am. J. Hum. Genet* 48, 943-950.

Gonzalez F.J., Skoda R.C., Kimura S., Umeno M., Zanger U.M., Nebert D.W., Gelboin H.V., Hardwick J.P., Meyer U.A.. (1988): Characterization of the common genetic defect in humans deficient in debrisoquine metabolism. *Nature* 331, 442-446.

Gut, J., Gasser, R., Dayer, P., Kronbach, T., Catin, T. and Meyer, U.A. (1984): Debrisoquine-type polymorphism of drug oxidation: purification from human liver of a cytochrome P450 isozyme with high activity for bufuralol hydroxylation. *FEBS Lett.* 173, 287-290.

Heim, M., and Meyer, U.A. (1990): Genotyping of poor metabolizers of debrisoquine by allele-specific PCR amplification. *Lancet* 336, 529-532.

Heim, M.H., and Meyer, U.A. (1992): Evolution of a highly polymorphic human cytochrome P450 gene: *CYP2D6. Genomics* 14, 49-58.

Hirvonen, A., Husgafvel.Pursiainen, K., Anttila, S., Karjalainen, A., Sorsa, M., Vainio, H. (1992): Metabolic cytochrome P450 genotypes and assessment of individual susceptibility to lung cancer. *Pharmacogenetics* 2, 259-263.

Idle, J.R., Armstrong, M., Boddy, A.V., Boustead, C., Cholerton, S., Cooper, J., Daly, A.K., Ellis, J., Gregory, W., Hadidi, H., Höfer, C., Holt, J., Leathart, J., McCracken, N., Monkman, S.C., Painter, J.E., Taber, H., Walker, D., and Yule, M. (1992): The pharmacogenetics of chemical carcinogenesis. *Pharmacogenetics* 2, 246-258.

Johansson, I., Dahl, M.-L., Bertilsson, L., Sjöqvist, F., and Ingelman-Sundberg, M. (1992): Gene amplification of an active CYP2D gene as a cause of extremely rapid metabolism of debrisoquine. *J. Basic Clin. Physiol. Pharmacol.* 3, 242.

Johansson, I., Lundqvist, E., Bertilsson, L., Dahl, M.-L., Sjöqvist, F., and Ingelman-Sundberg, M. (1994): Inherited amplification of an active gene in the cytochrome P450 CYP2D locus as a cause of ultrarapid metabolism of debrisoquine. *Proc. Natl. Acad. Sci. USA* (in press)

Kerb, R., Brockmoeller, J., Drakoulis, N., Siegel, J., Roots, I. (1992): CYP2D6 and glutathione S-transferase class mu as host factors of lung cancer susceptibility. *J. Basic Clin. Physiol. Pharmacol.* 3 (suppl.), p. 131.

Kagimoto, M., Heim, M., Kagimoto, K., Zeugin, T., and Meyer, U.A. (1990): Multiple mutations of the human cytochrome P450IID6 gene (CYP2D6) in poor metabolizers of debrisoquine: study of the functional significance of individual mutations by expression of chimeric genes. *J. Biol. Chem.* 265, 17209-1721.

Law, M.R., Hetzel, M.R., and Idle, J.R. (1989): Debrisoquine metabolism and genetic predisposition to lung cancer. *Br. J. Cancer* 59, 686-687.

Meyer, U.A., Skoda, R.C., Zanger, U.M., Heim, M., and Broly, F. (1992): The genetic polymorphism of debrisoquine/sparteine metabolism - molecular mechanism, In: *Pharmacogenetics of Drug Metabolism,* ed. W. Kalow, pp. 609-623. Pergamon Press, New York, Oxford.

Roots, I., Drakoulis, N., Brockmöller, J., Janicke, I., Cuprunov, M., and Ritter, J. (1990): Hydroxylation and acetylation phenotypes as genetic risk factors in certain malignancies. In *Xenobiotic Metabolism and Disposition,* ed. R. Kato, R.W. Estabrook, M.N. Cayen, pp. 499-506. London, New York, Philadelphia: Taylor & Francis.

Roots, I., Drakoulis, N., and Brockmöller, J. (1992): Polymorphic enzymes and cancer risk: concepts, methodology and data review. In *Pharmacogenetics of Drug Metabolism*, ed. W. Kalow, pp. 815-841, Pergamon Press, Inc. New York.

Skoda, R.C., Gonzalez, F.J., Demierre, A. and Meyer, U.A. (1988): Two mutant alleles of the human cytochrome P450db1 gene (P450IID1) associated with genetically deficient metabolism of debrisoquine and other drugs. *Proc. Natl. Acad. Sci. USA* 85, 5240-5243.

Speirs, C.J., Mrray, S., Davies, D.S., Biola Mabadeje, A.F., and Boobis, A.R. (1990): Debrisoquine oxidation phenotype and susceptibility to lung cancer. *Br. J. clin. Pharmac.* 29, 101-109.

Tefre, T., Daly, A., Armstrong, M., Ryberg, D., Hangen, A., Idle, J., Børresen, A.L. (1992): Genotyping of the CYP2D6 gene in lung cancer patients and controls. *J. Basic Clin. Physiol. Pharmacol.* 3 (suppl.), p. 132.

Tyndale, R.F., Aoyama, T., Broly, F., Matsunaga, T., Inaba, T., Kalow, W., Gelboin, H.V., Meyer, U.A., and Gonzalez, F.J. (1991): Identification of a new variant *CYP2D6* allele lacking the codon encoding Lys-281: possible association with the poor metabolizer phenotype. *Pharmacogenetics* 1, 26-32.

Wolf, C.R., Dale Smith, C.A., Gough, A.C., Moss, J., Vallis, K.A., Howard, G., Carey, F.J., Mills, K., McNee, W., Carmichael, J., and Spurr, N.K. (1992): Relationship between the debrisoquine hydroxylase polymorphism and cancer susceptibility. *Carcinogenesis* 13, 1035-1038.

Yokota, H., Tamura, S., Furuya, H., Kimura, S., Watanabe, M., Kanazawa, I., Kondo, I., and Gonzalez, F.J. (1993): Evidence for a new variant CYP2D6 allele CYP2D6J in a Japanese population associated with lower in vivo rates of sparteine metabolism. *Pharmacogenetics* 3, 256-263.

Zanger, U.M., Vilbois, F., Hardwick, J., Meyer, U.A. (1988): Absence of hepatic cytochrome P450buf1 causes genetically deficient debrisoquine oxidation in man. *Biochemistry* 27, 5447-5454.

# CYP1A1-dependent benzo[a]pyrene metabolism activates genes regulated by the human immunodeficiency virus 1 long terminal repeat DNA sequences

Alvaro Puga, Yan Yao

Department of Environmental Health, University of Cincinnati Medical Center, 3223 Eden Avenue, Cincinnati, Ohio 45267-0056, USA

## Summary

We have shown that polycyclic aromatic hydrocarbons (PAHs) such as benzo[a]pyrene (B[a]P) and 2,3,7,8-tetrachlorodibenzo-p-dioxin (TCDD; dioxin) induce the expression of the immediate-early protooncogenes *fos* and *jun* and to activate the DNA binding activity of the transcription factor AP-1, a dimer of the FOS and JUN proteins. To determine whether PAH exposure could induce genes regulated by AP-1, and possibly by other transcription factors, we analyzed the expression of a reporter gene fused to the long terminal repeat (LTR) of the human immunodeficiency virus 1 (HIV-1) in cultured mouse hepatoma cells treated with B[a]P. We find that B[a]P, as well as several CYP1A1-dependent B[a]P metabolites, can induce expression of the reporter gene in a dose-dependent manner. The aromatic hydrocarbon (Ah) receptor negatively regulates LTR-directed gene expression, since wild type cells show low levels of basal expression that are highly derepressed in variant cells lacking a functional Ah receptor. In untreated wild type cells, CYP1A1-dependent metabolism of an endogenous substrate is responsible for maintenance of low basal levels of expression. These results show that B[a]P can activate transcription factors, and suggest a role for PAHs and the *CYP1A1* gene progression of the HIV-1 infection.

## INTRODUCTION

During the last few years it has become increasingly evident that PAHs, such as B[a]P, and halogenated aromatic hydrocarbons, such as TCDD, cause a plethora of apparently unrelated toxic effects without any obvious common denominator other than their capacity to induce expression of the *CYP1A1* gene. In humans, exposure to dioxin and to many other chlorinated phenolic agents causes chloracne, a long-lasting skin disease characterized by the hyperkeratinization of follicular sebocytes (Zugerman, 1990). In rodents, dioxin is one of the strongest tumor promoters known (Poland et al. 1985; Poland and Knutson, 1982), and, during embryogenesis, its administration causes cleft palate and hydronephrosis (Couture et al. 1990). Unlike in the animal, TCDD has no toxic effect in tissue culture cells, but it causes a large elevation of intracellular calcium and induces apoptosis in immature thymocytes (McConkey et al. 1988); (McConkey and Orrenius, 1989). In this regard, the developing immune system is a particularly sensitive target for TCDD, with thymic atrophy being the most common pathological finding (Greenlee and Neal, 1985). B[a]P, on the other hand, is a complete carcinogen, forming mutagenic adducts with DNA, and eliciting a variety of toxic, teratogenic, and carcinogenic responses in exposed animals (Whitlock, 1991; Nebert and Gonzalez, 1987; Nebert, 1989; Landers and Bunce, 1991).

Our current understanding of the regulation of eukaryotic development indicates that growth factors, hormones, neurotransmitters and a variety of extracellular ligands initiate signals that are propagated by a cascade of biochemical events that include increases of ion fluxes across cellular membranes, elevation of intracellular pH, $Ca^{2+}$, and cAMP levels, formation of phosphoinositide metabolites, and activation of protein kinases and phosphatases (Morgan and Curran, 1986). Transduction of these signals to the nucleus induces the expression of a group of immediate-early genes that, among others, encode the protein products of the protooncogenes *fos* and *jun*, components of the transcription factor AP-1. This factor in turn propagates the signal by controlling the expression of genes required for cell proliferation or differentiation (Hunter and Karin, 1992).

Work from our laboratory has shown that treatment of mouse hepatoma cells with PAH inducers of CYP1A1 activity, such as TCDD and B[a]P, causes an increase in the steady state mRNA levels of the protooncogenes c-*fos*, c-*jun*, *jun-B*, and *jun-D*, and the concomitant increase of transcription factor AP-1, reaching levels almost as high as those induced by TPA, a classical phorbol ester tumor promoter (Puga et al. 1992). These results suggested the possibility that genes that have AP-1 binding sites might respond to AP-1 regulation as a result of TCDD or B[a]P treatment. We have tested this hypothesis in mouse hepatoma cells by analyzing the activation of a chloramphenicol acetyltransferase (*cat*) reporter plasmid carrying AP-1 responsive regulatory elements. The chimeric plasmid was constructed by fusion of the *cat* gene sequences to the HIV-1 LTR. We present here results that demonstrating that B[a]P activates CAT expression well over the basal level found in untreated cells.

MATERIALS AND METHODS

The reporter gene that we chose is the bacterial *cat* gene widely used for this type of analyses. The chimeric plasmid pHIVLTRCAT was constructed by fusion of the *cat* gene sequences to the LTR sequences of HIV-1 in pUC8 (a gift of the NIAID Repository of AIDS Reagents). This plasmid was transfected by standard calcium phosphate techniques (Graham and Van der Erb, 1973) into Hepa-1 cells, a mouse hepatoma cell line (Bernard et al. 1973), and into several mutant derivatives of this line, including the *c*2 and *c*37 mutants, that lack a functional Ah receptor or CYP1A1 enzyme, respectively (Hankinson et al. 1985; Kimura et al. 1987), and the CX4 line, a pseudo-wild type derivative of c37, that expresses a functional CYP1A1 enzyme from a stably transfected plasmid vector (Puga et al. 1990; RayChaudhuri et al. 1990). In all experiments we also co-transfected the cells with plasmid pCV-1 (a gift from the NIAID Repository of AIDS Reagents) that expresses HIV-1 TAT under the control of the SV40 early promoter and enhancer. As a negative control we used pSV0CAT, containing a promoterless *cat* gene, and as a positive control we used pSV2CAT, carrying the *cat* gene under the control of the SV40 early promoter and enhancer sequences. To control for variations due to differences in transfection efficiencies, all cultures were co-transfected with plasmid pCMVβgal, which expresses the bacterial β-galactosidase gene under the control of the cytomegalovirus immediate-early enhancer and promoter. Expression of β-galactosidase under regulation by this enhancer is independent of treatment. Twenty-four hours after transfection the cells were placed under low serum (0.1%) conditions and sixteen hours later cells were treated with B[a]P, B[a]P derivatives, or with an equivalent amount of DMSO vehicle. Cell extracts were prepared and expression of CAT and β-galactosidase activities was determined 96 h after transfection. Data were normalized for transfection efficiency by determination of the per cent chloramphenicol (CAM) converted to acetylated forms per unit of β-galactosidase.

Preparation of nuclear extracts and probes for electrophoretic mobility shift analyses (EMSA) was as described previously (Puga et al. 1992).

RESULTS

B[a]P and CYP1A1-dependent B[a]P metabolites activate expression from the HIV-1 LTR

To determine whether genes regulated by the HIV-1 LTR could be induced by B[a]P we treated Hepa-1 cells transfected with pHIVLTRCAT and pCV-1 with increasing doses of B[a]P and measured transient expression of CAT activity. Induction over control levels was clearly evident at a dose of 1 $\mu$M, maximal at 5 - 10 $\mu$M, and started to decline by 20 $\mu$M (Fig. 1). The maximal levels observed correspond to a 14- to 15-fold increase over the basal level of activity observed in cells treated with an equivalent amount of DMSO vehicle. This basal level corresponds to approximately 5% CAM conversion per unit of $\beta$-galactosidase.

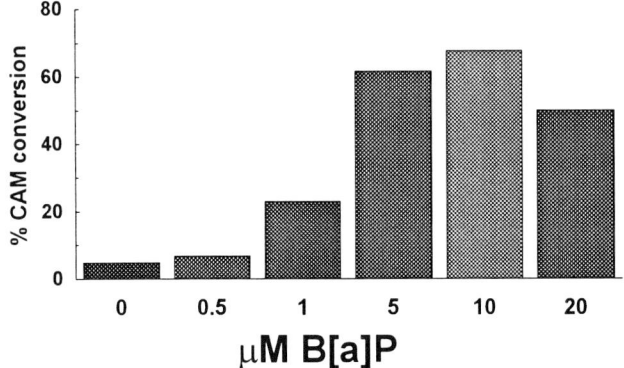

Fig. 1 Dose response curve of the activation of pHIVLTRCAT by B[a]P. Hepa-1 cells were transfected with 10 $\mu$g each of pHIVLTRCAT and pCV-1 and 3 $\mu$g of pCMV$\beta$gal. Transfected cells were treated with the indicated concentrations of B[a]P 48 h prior to extraction. The data for % CAM conversion have been normalized to $\beta$-galactosidase activity in the same extracts.

B[a]P is a procarcinogen that is converted to many reactive intermediates by the sequential action of at least two enzymes: cytochrome P450 CYP1A1 and epoxide hydrolase. To establish whether activation of CAT expression under our experimental conditions was only limited to the parent compound, we tested B[a]P and three of its metabolites, including the ultimate carcinogen, B[a]P-7,8-diol-9,10-epoxide (BPDE), for their capacity to induce CAT expression. All compounds tested were able to induce CAT, although to different extents, possibly a reflection of their different potencies. Treatment with the arene oxide B[a]P-4,5-dihydroepoxide (BPDHE), gave the highest levels of CAT induction; B[a]P was second highest. BPDE and B[a]P-6,12-dione treatments showed lower levels of induction, although significantly higher than those found in cells treated with DMSO vehicle (Fig. 2).

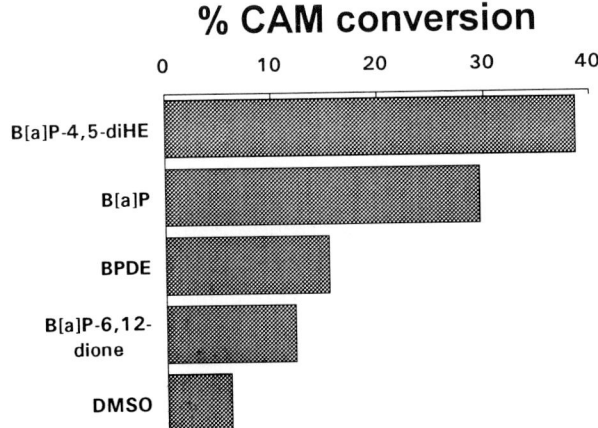

Fig. 2. Activation of CAT activity by B[a]P and by B[a]P metabolites. Cells were transfected as indicated in Fig. 1 and 24 h after transfection they were treated with the compounds shown at a concentration of 20 $\mu$M. Control cells were treated with an equivalent amount of vehicle (DMSO).

CYP1A1-dependent metabolism is required for concerted regulation of CAT expression

Since B[a]P metabolites were capable of activating CAT expression, we argued that CYP1A1-dependent metabolism might be necessary for B[a]P-induced, LTR-directed CAT expression. To test this hypothesis we used the wild type Hepa-1 cell line and two of its variants: the $c37$ cell line, which has two mutations in the structural $Cyp1a1$ gene that render the protein enzymatically inactive (Kimura et al. 1987), and the CX4 cell line, a pseudo-wild type derivative of $c37$, that expresses a transfected expression plasmid containing the functional murine CYP1A1 cDNA (Puga et al. 1990; RayChaudhuri et al. 1990). Expression of the reporter gene was determined in these three cell lines after transfection and treatment with B[a]P, BPDE, or vehicle.

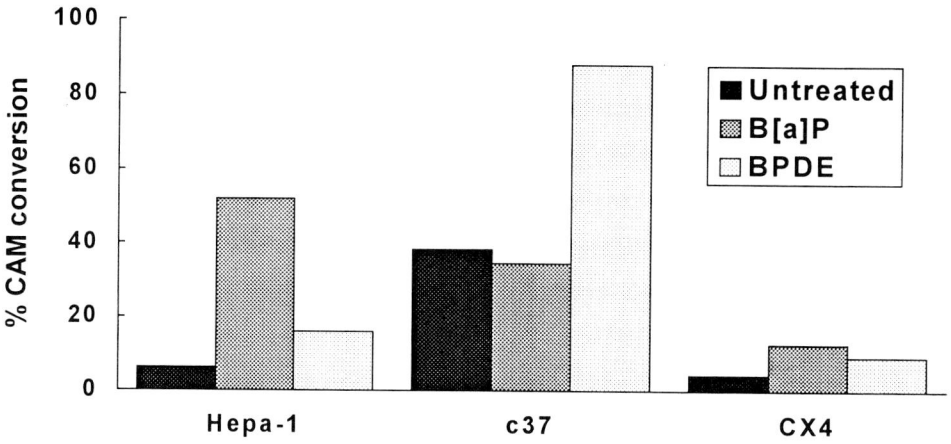

**Fig.3. Activation of CAT expression in Hepa-1 and derivative cell lines after treatment with B[a]P and BPDE.** B[a]P and BPDE were used at 20 $\mu$M. CAT expression was determined 44 h after treatment.

We find that, in untreated wild type cells, there is a low level of expression that is stimulated to a large measure by B[a]P, and to a lesser extent by BPDE. In the mutant $c37$ cell line, CAT expression in untreated cells is highly elevated, reaching levels indistinguishable from those observed in the same cells after B[a]P treatment. Notwithstanding, BPDE activates CAT expression in these cells beyond these derepressed levels. The expression characteristics of the wild type cells are restored by the sole expression of a functional CYP1A1 enzyme in CX4 cells. In this variant line, untreated cells show a low level of expression that increases significantly after B[a]P and BPDE treatments (Fig. 3).

These data indicate that CYP1A1-dependent metabolism is required to modulate at least two different components of LTR-directed expression. On the one hand, in untreated wild type cells, metabolism of an endogenous compound maintains a low basal level of expression that is deregulated in $c37$ cells, resulting in highly derepressed expression in the absence of inducer. On the other hand, metabolic activation of B[a]P is required for its conversion to a compound(s) that induces expression from sequences in the LTR: in the absence of a functional CYP1A1 enzyme in $c37$ cells, only BPDE, the ultimate carcinogenic B[a]P metabolite, and not the parent compound, activates CAT expression.

## The Ah receptor plays an important role in negative regulation of LTR-directed expression

To analyze the role of the Ah receptor in control of gene expression directed by the HIV LTR, we used for our transfection experiments the $c2$ cell line, that has very low levels (<5% of the wild type) of the Ah receptor (Hankinson et al. 1985). We found that, even in untreated cells, CAT expression directed by the HIV-LTR was highly derepressed, reaching levels as high as those found in cells treated with B[a]P (Fig. 4A). This effect seemed to be generalized, since it extended to expression directed by other trans-regulatory sequences; when we used pSV2CAT as the reporter plasmid in these cells, we found a similar extent of deregulation (Fig. 4B). CAT activity levels directed by this plasmid, however, were more than one order of magnitude below those directed by the HIV LTR.

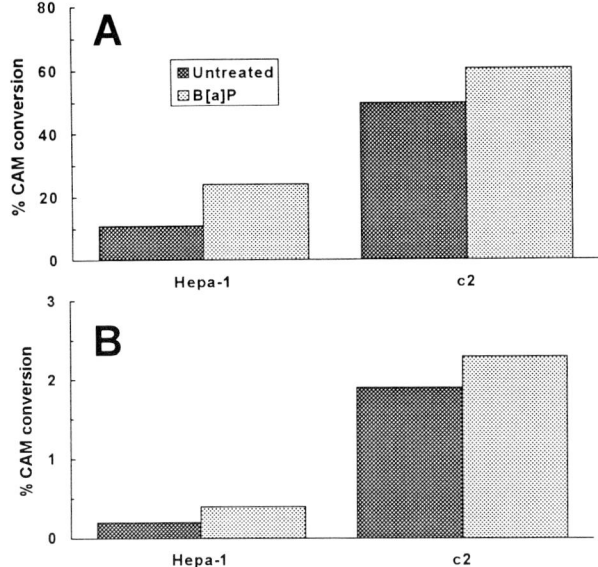

Fig. 4. Deregulation of CAT expression in $c2$ cells, lacking a functional Ah receptor. (A) CAT expression in Hepa-1 and $c2$ cells transfected with pHIVLTRCAT and pCV-1 (TAT). (B) CAT expression in the same cells transfected with pSV2CAT. Notice the scale change for the ordinate in the two panels.

## Activation of AP-1 by B[a]P may be responsible for LTR-directed CAT expression

Several AP-1 binding sites have been mapped in the region of the HIV-1 LTR involved in negative regulation (Lu et al. 1990); (Spandidos et al. 1989). To determine whether AP-1 activation might be involved in the observed effects of B[a]P and BPDE on LTR-directed expression, we prepared nuclear extracts from wild type and from mutant cells treated with B[a]P or BPDE, or left untreated. As a positive control we used TCDD treatment, which we have previously shown to activate AP-1 (Puga et al. 1992). As a measure of AP-1 activation, we used electrophoretic mobility shift analysis with a probe containing a canonical AP-1 binding site (TRE).

AP-1 binding activity in extracts from wild type Hepa-1 cells after with TCDD, B[a]P, or BPDE is 2.5-fold higher than in extracts from untreated controls. The $c37$ cells, deficient in CYP1A1 function, have control levels already 2-fold higher than those in wild type cells, and treatment with TCDD, B[a]P, or BPDE does not affect these levels significantly. In pseudo-wild type CX4 cells, uninduced levels are also high, but treatment with xenobiotic compounds increases these levels to nearly 3-fold greater than those in Hepa-1 cells. By comparison, AP-1 levels in receptorless $c2$ cells are close to 4-fold higher than those in Hepa-1 cells, and are not affected by treatments (Fig. 5). These results suggest that activation of AP-1 is likely to be involved in the induction of CAT expression.

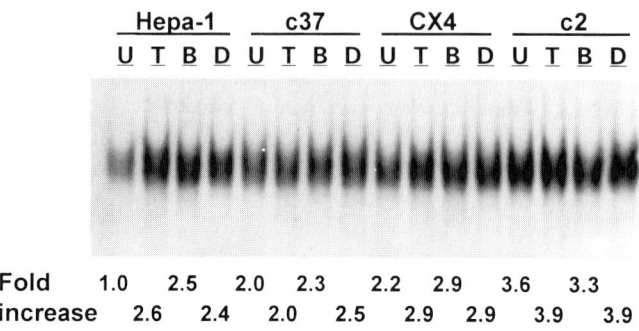

Fig. 5. Activation of AP-1 DNA binding activity in four cell lines by B[a]P and BPDE. EMSA analyses and quantitation were done as described previously (Puga et al. 1992). Treatments were for 4 hr with, U: DMSO vehicle; T: 10 nM TCDD; B: 20 µM B[a]P; D: 20 µM BPDE. *Fold increase* refers to increases in protein-bound probe relative to the value in untreated Hepa-1 cells, set arbitrarily as 1.

Induction of HIV LTR-directed expression by B[a]P is inhibited by NAC and is independent of arachidonic acid metabolism

Recently, an oxidative stress pathway of AP-1 induction has been uncovered (Hunter and Karin, 1992; Devary et al. 1992), and the role of phospholipase A₂ and of arachidonic acid metabolites in the induction of Fos and Jun have been documented (Rao et al. 1993; Fafeur et al. 1991; Haliday et al. 1991). Several cytochromes P450 possess arachidonic acid epoxygenase activity, including the PAH-inducible CYP1A1, CYP1A2, and β-NF$_{AA}$ enzymes, the members of the phenobarbital-inducible CYP2B and CYP2H families, and some members of the CYP2C family (Capdevila et al. 1992; Laethem and Koop, 1992; Kanetoshi et al. 1992). As a consequence, oxidative stress mediated by PAH-inducible cytochrome P450 arachidonic acid epoxygenases may play an important role in the induction of Fos and Jun, in the observed activation of AP-1, and ultimately, in the stimulation of LTR-directed CAT activity by B[a]P. To examine the role of arachidonic acid epoxygenation on CAT activation, we determined the effect of known inhibitors of lipooxygenase and epoxygenase activities on the increase of CAT activity. We used indomethacin as a preferential inhibitor of lipooxygenases, and nordihydroguaiaretic acid (NDGA) as a preferential inhibitor of epoxygenases (Capdevila et al. 1988). In addition, to determine whether thiol-sensitive reactive oxygen species could be involved in the activation of CAT by B[a]P, we used N-acetyl-L-cysteine (NAC) treatment. NAC has been used in other systems to counteract the effect of reactive oxygen species on NFκB and AP-1 activation (Devary et al. 1991; Staal et al. 1990; Roederer et al. 1990; Devary et al. 1992). The results indicate that neither arachidonate cascade inhibitor had an effect on B[a]P activation of LTR-directed CAT expression, whereas NAC partially inhibited (Table 1).

Table 1. Effect of NAC and of inhibitors of arachidonic acid metabolism on the stimulation of CAT activity by B[a]P. Indomethacin (IM) and NDGA were used at 20 µM; NAC at 5 mM. B[a]P was added 24 h after transfection and the inhibitors were added 10 min later.

| Treatment | % CAM Conversion |
|---|---|
| None | 5.3 |
| B[a]P | 45.3 |
| B[a]P + IM | 42.1 |
| B[a]P + NDGA | 37.8 |
| B[a]P + NAC | 27.3 |

DISCUSSION AND CONCLUSIONS

We have shown that B[a]P and several of its metabolites can activate the expression of HIV-1 LTR-directed genes in mouse hepatoma cells. This effect is dependent on the presence of a functional cytochrome P450 CYP1A1 enzyme, yet it is insensitive to the effect of compounds that inhibit the epoxygenation of arachidonic acid, suggesting that oxidative stress from arachidonate metabolites such as the epoxyeicosatrienoic acids and hydroxyeicosatetraenoic acids may not involved in the activation. On the other hand, treatment with NAC reduces the magnitude of the B[a]P-dependent response, indicating that intracellular thiol depletion and thiol-sensitive reactive oxygen species may be an important factor in activation.

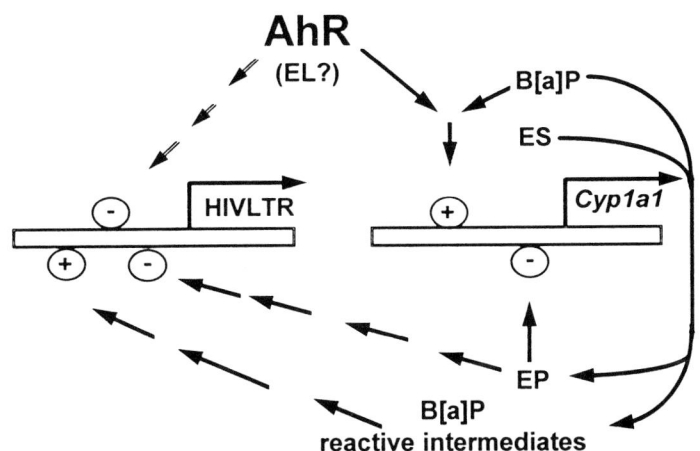

Fig. 6. Proposed model of regulation of the HIV-1 LTR-dependent expression by the Ah receptor and by CYP1A1. *AhR* is the Ah receptor; *EL* is its unknown endogenous ligand; *ES* and *EP* are an endogenous CYP1A1 substrate and its product, respectively. The *positive* and *negative* signs indicate the type of regulation exerted by each compound. Each set of *arrows* with gaps in between indicate that several intermediates steps are likely to intervene.

Three different regulatory activities appear to be exerted over the HIV-1 LTR by PAHs and the Ah receptor. First, based on our results in $c2$ cells, we conclude that the Ah receptor, possibly by interacting with an endogenous ligand, negatively regulates expression from the LTR. Second, based on our results in $c37$ cells, we conclude that CYP1A1-dependent metabolism of an endogenous substrate also down-regulates expression from the LTR. Thirdly, based on the results with B[a]P and its metabolites, we conclude that reactive intermediates of B[a]P can override these two negative regulatory activities and activate expression. This regulatory scheme uses the same components, although with different outcomes, as the model recently reviewed for regulation of *Cyp1a1* and of other genes in the Ah gene battery (Nebert et al. 1993). Both schemes are presented in Fig. 6.

It has been reported that progression of AIDS is accelerated by cigarette smoking (Burns et al. 1991); (Boulos et al. 1990). This epidemiological observation may have a number of molecular explanations, possibly involving the effect of compounds in tobacco smoke on lymphocyte function. It is attractive to speculate that one possible cause for this acceleration is the activation of the LTR by metabolites of B[a]P, one of the major carcinogens found in cigarette smoke, with the concomitant induction of viral RNA synthesis.

ACKNOWLEDGEMENTS

We wish to thank our colleague Daniel W. Nebert for many helpful discussions and criticism. This research was supported by NIH Grants NIEHS P30 ES06096, NIEHS 1R01 ES06321, and NIEHS 1R01 ES06273.

REFERENCES

Bernard, H.P., Darlington, G.J. and Ruddle, F.H. (1973) Expression of liver phenotypes in cultured mouse hepatoma cells: synthesis and secretion of serum albumin. *Dev. Biol.* **35**, 83-96.

Boulos, R., Halsey, N.A., Holt, E., Ruff, A., Brutus, J.R., Quinn, T.C., Adrien, M. and Boulos, C. (1990) HIV-1 in Haitian women 1982-1988. The Cite Soleil/JHU AIDS Project Team. *J. Acquir. Immune. Defic. Syndr.* **3**, 721-728.

Burns, D.N., Kramer, A., Yellin, F., Fuchs, D., Wachter, H., DiGioia, R.A., Sanchez, W.C., Grossman, R.J., Gordin, F.M., Biggar, R.J. and et al (1991) Cigarette smoking: a modifier of human immunodeficiency virus type 1 infection?. *J. Acquir. Immune. Defic. Syndr.* **4**, 76-83.

Capdevila, J., Gil, L., Orellana, M., Marnett, L.J., Mason, J.I., Yadagiri, P. and Falck, J.R. (1988) Inhibitors of cytochrome P-450-dependent arachidonic acid metabolism. *Arch. Biochem. Biophys.* **261**, 257-263.

Capdevila, J., Falck, J.R. and Estabrook, R.W. (1992) Cytochrome P450 and the arachidonate cascade. *FASEB J.* **6**, 731-736.

Couture, L.A., Abbott, B.D. and Birnbaum, L.S. (1990) A critical review of the developmental toxicity and teratogenicity of 2,3,7,8-tetrachlorodibenzo-p-dioxin: recent advances toward understanding the mechanism. *Teratology* **42**, 619-627.

Devary, Y., Gottlieb, R.A., Lau, L.F. and Karin, M. (1991) Rapid and preferential activation of the c-*jun* gene during the mammalian UV response. *Mol. Cell. Biol.* **11**, 2804-2811.

Devary, Y., Gottlieb, R.A., Smeal, T. and Karin, M. (1992) The mammalian ultraviolet response is triggered by activation of Src tyrosine kinases. *Cell* **71**, 1081-1091.

Fafeur, V., Jiang, Z.P. and Böhlen, P. (1991) Signal transduction by bFGF, but not TGF$\beta$1, involves arachidonic acid metabolism in endothelial cells. *J. Cell. Physiol.* **149**, 277-283.

Graham, F.L. and Van der Erb, A.J. (1973) A new technique for the assay of infectivity of human adenovirus 5 DNA. *Virology* **52**, 456-467.

Greenlee, W.F. and Neal, R.A. (1985) The *Ah* receptor: a biochemical and biological perspective.. In: Conn, P.M., (Ed.) *The Receptors*, vol. 2., pp. 89-129. New York: Academic Press,Inc..

Haliday, E.M., Ramesha, C.S. and Ringold, G. (1991) TNF induces c-*fos* via a novel pathway requiring conversion of arachidonic acid toa lipoxygenase metabolite. *EMBO J.* **10**, 109-115.

Hankinson, O., Andersen, R.D., Birren, B., Sander, F., Negishi, M. and Nebert, D.W. (1985) Mutations affecting the regulation of transcription of the cytochrome $P_1$-450 gene in the mouse Hepa-1 cell line. *J. Biol. Chem.* **260**, 1790-1795.

Hunter, T. and Karin, M. (1992) The regulation of transcription by phosphorylation. *Cell* **70**, 375-387.

Kanetoshi, A., Ward, A.M., May, B.K. and Rifkind, A.B. (1992) Immunochemical identity of the 2,3,7,8-tetrachlorodibenzo-*p*-dioxin and $\beta$-naphthoflavone-induced cytochrome P-450 arachidonic acid epoxygenases in chick embryo liver: distinction from the $\Omega$-hydroxylase and the phenobarbital-induced epoxygenase. *Mol. Pharmacol.* **42**, 1020-1026.

Kimura, S., Smith, H.H., Hankinson, O. and Nebert, D.W. (1987) Analysis of two benzo[a]pyrene-resistant mutants of the mouse hepatoma Hepa-1 $P_1$-450 via cDNA expression in yeast. *EMBO J.* **6**, 1929-1933.

Laethem, R.M. and Koop, D.R. (1992) Identification od rabbit cytochromes P450 2C1 and 2C2 as arachidonic acid epoxygenases. *Mol. Pharmacol.* **42**, 958-963.

Landers, J.P. and Bunce, N.J. (1991) The *Ah* receptor and the mechanism of dioxin toxicity. *Biochem. J.* **276**, 273-287.

Lu, Y.C., Touzjian, N., Stenzel, M., Dorfman, T., Sodroski, J.G. and Haseltine, W.A. (1990) Identification of cis-acting repressive sequences within the negative regulatory element of human immunodeficiency virus type 1. *J. Virol.* **64**, 5226-5229.

McConkey, D.J., Hartzell, P., Duddy, S.K., Hakansson, H. and Orrenius, S. (1988) 2,3,7,8-Tetrachlorodibenzo-p-dioxin kills immature thymocytes by $Ca2+$-mediated endonuclease activation. *Science* **242**, 256-259.

McConkey, D.J. and Orrenius, S. (1989) 2,3,7,8-tetrachlorodibenzo-*p*-dioxin kills glucocorticoid-sensitive thymocytes *in vivo*. *Biochem. Biophys. Res. Comm.* **160**, 1003-1008.

Morgan, J.I. and Curran, T. (1986) Role of ion flux in the control of c-fos expression. *Nature* **322**, 552-555.

Nebert, D.W. (1989) The *Ah* locus: genetic differences in toxicity, cancer, mutation, and birth defects.. *Crit. Rev. Toxicol.* **20**, 153-174.

Nebert, D.W., Puga, A. and Vasiliou, V. (1993) Role of the Ah receptor and dioxin inducible [*Ah*] gene battery in toxicity, cancer, and signal transduction. *Ann. New York Acad. Sci.* **685**, 624-640.

Nebert, D.W. and Gonzalez, F.J. (1987) P450 genes. Structure, evolution and regulation. *Annu. Rev. Biochem.* **56**, 945-993.

Poland, A., Knutson, J. and Glover, E. (1985) Studies on the mechanism of action of halogenated aromatic hydrocarbons. *Clin. Physiol. Biochem.* **3**, 147-154.

Poland, A. and Knutson, J.C. (1982) 2,3,7,8,-tetrachlorodibenzo-p-dioxin and related halogenated aromatic hydrocarbons: examination of the mechanisms of toxicity. *Annu. Rev. Pharmacol. Toxicol.* **22**, 517-554.

Puga, A., RayChaudhuri, B., Salata, K., Zhang, Y.-H. and Nebert, D.W. (1990) Stable expression of mouse Cyp1a-1 and human CYP1A-2 cDNAs transfected into mouse hepatoma cells lacking detectable P450 enzyme activity. *DNA and Cell Biol.* **9**, 425-436.

Puga, A., Nebert, D.W. and Carrier, F. (1992) Dioxin induces expression of c-*fos* and c-*jun* proto-oncogenes and a large increase in transcription factor AP-1. *DNA and Cell Biol.* **11**, 269-281.

Rao, G.N., Lasségue, B., Griendling, K.K., Alexander, R.W. and Berk, B.C. (1993) Hydrogen peroxide-induced c-fos expression is mediated by arachidonic acid release: role of protein kinase C . *Nucleic Acids Res.* **21**, 1259-1263.

RayChaudhuri, B., Nebert, D.W. and Puga, A. (1990) The murine Cyp1a-1 gene negatively autoregulates its own transcription and that of other members of the aromatic hydrocarbon-responsive [Ah] gene battery. *Mol. Endocrinol.* **4**, 1773-1781.

Roederer, M., Staal, F.J.T., Raju, P.A., Ela, S.W. and Herzenberg, L.A. (1990) Cytokine-stimulated human immunodeficiency virus replication is inhibited by N-acetyl-L-cysteine. *Proc. Natl. Acad. Sci. USA* **87**, 4884-4888.

Spandidos, D.A., Yiagnisis, M. and Pintzas, A. (1989) Human immunodeficiency virus long terminal repeat responds to transformation by the mutant T24 H-ras1 oncogene and it contains multiple AP-1 binding TPA-inducible consensus sequence elements. *Anticancer Res.* **9**, 383-386.

Staal, F.J.T., Roederer, M. and Herzenberg, L.A. (1990) Intracellular thiols regulate activation of nuclear factor kB and transcription of human immunodeficiency virus. *Proc. Natl. Acad. Sci. USA* **87**, 9943-9947.

Whitlock, J.P., Jr. (1991) Genetic and molecular aspects of 2,3,7,8-tetrachlorodibenzo-p-dioxin action. *Annu. Rev. Pharmacol. Toxicol.* **30**, 251-277.

Zugerman, C. (1990) Chloracne. Clinical manifestations and etiology. *Dermatol. Clin.* **8**, 209-213.

Peptides shown to bind specifically cytochrome P450 2B4 antibodies to a considerable extent have been treated as linear antigenic determinants of this protein; they are listed in the table 1.

Table I. Linear antigenic determinants of cytochrome P450 2B4 found by PEPSCAN.[a]

| # | Peptide | Antiserum binding | | | # | Peptide | Antiserum binding | | |
|---|---------|-------|-------|-------|---|---------|-------|-------|-------|
|   |         | 1:200 | 1:400 | 1:800 |   |         | 1:200 | 1:400 | 1:800 |
| 1 | 25-30 | + | - | - | 36 | 223-228 | + | + | - |
| 2 | 28-33 | + | - | - | 37 | 224-229 | + | + | - |
| 3 | 39-44 | + | - | - | 38 | 225-230 | + | - | - |
| 4 | 40-45 | + | + | - | 39 | 228-233 | +++ | ++ | - |
| 5 | 62-67 | + | - | - | 40 | 229-234 | + | + | - |
| 6 | 67-72 | + | - | - | 41 | 230-235 | + | + | + |
| 7 | 79-84 | + | - | - | 42 | 231-236 | + | + | + |
| 8 | 80-85 | + | - | - | 43 | 242-247 | ++ | + | - |
| 9 | 81-86 | + | - | - | 44 | 243-248 | + | + | - |
| 10 | 91-96 | + | - | - | 45 | 251-256 | + | - | - |
| 11 | 92-97 | + | - | - | 46 | 254-259 | ++ | - | - |
| 12 | 93-98 | ++ | + | + | 47 | 258-263 | + | + | + |
| 13 | 94-99 | + | + | - | 48 | 259-264 | + | + | + |
| 14 | 104-109 | + | + | + | 49 | 261-267 | + | + | + |
| 15 | 105-110 | +++ | + | + | 50 | 279-284 | + | - | - |
| 16 | 106-111 | + | + | + | 51 | 280-285 | ++ | - | - |
| 17 | 107-112 | ++ | + | + | 52 | 281-286 | ++ | - | - |
| 18 | 108-113 | ++ | + | + | 53 | 282-287 | ++ | - | - |
| 19 | 109-114 | + | + | + | 54 | 283-288 | + | + | - |
| 20 | 112-117 | + | + | - | 55 | 284-289 | + | + | + |
| 21 | 117-122 | + | - | - | 56 | 294-299 | + | - | - |
| 22 | 122-127 | + | - | - | 57 | 354-359 | + | + | + |
| 23 | 134-139 | + | - | - | 58 | 374-379 | + | + | + |
| 24 | 143-148 | + | - | - | 59 | 393-398 | +++ | - | - |
| 25 | 181-186 | ++ | - | - | 60 | 405-410 | + | + | + |
| 26 | 182-187 | ++ | - | - | 61 | 407-412 | +++ | ++ | ++ |
| 27 | 183-188 | + | - | - | 62 | 408-413 | + | + | + |
| 28 | 186-191 | ++ | - | - | 63 | 417-422 | + | + | + |
| 29 | 212-217 | + | + | - | 64 | 422-427 | ++ | + | + |
| 30 | 213-218 | + | + | - | 65 | 425-430 | + | + | + |
| 31 | 218-223 | + | + | + | 66 | 467-472 | + | + | + |
| 32 | 219-224 | + | + | + | 67 | 468-473 | +++ | +++ | +++ |
| 33 | 220-225 | + | + | + | 68 | 469-474 | +++ | +++ | +++ |
| 34 | 221-226 | ++ | + | + | 69 | 470-475 | +++ | +++ | +++ |
| 35 | 222-227 | + | - | - | 70 | 481-486 | + | + | + |

[a] +++ - $OD_{405} > 1.0$; ++ - $1.0 > OD_{405} > 0.7$; + - $0.7 > OD_{405} > 0.5$

The antigenic determinants are not uniformly spread on the P450 2B4 molecule. Four clusters with dense packing of antigenic determinants are observed as fig.2 shows: (1) region of No. 60-150 amino acid residues including a part of the predicted Rossmann domain (residues No. 32-120) and the C-helix close to it (residues No. 119-132), (2) region of No. 210-300 residues which includes the G-helix and a part of I-helix, regions (3) preceding (residues No. 393-430) and (4) following (residues No. 467-486) the conservative L-helix. One can mention that B-epitopes are concentrated mainly in the regions known to represent the most variable parts of the cytochrome P450 sequences (Imai et al., 1988): No.88-120, No.195-251,

No.270-295 and the C-terminal domain, the only exception being the highly variable N-terminal domain which is devoid of antigenic determinants. Summarily, about 43 per cent of the 2B4 amino acid residues compose the linear epitopes of this protein. The sorting of the antigenic determinants of 2B4 revealed by PEPSCAN has been achieved by their search in the sequences of cytochromes P450 from the CPD, and its results are summarized in the fig.2. Several kinds of 2B4 epitopes have been estimated according to their presence in molecules of different cytochromes P450. 36 antigenic determinants can be regarded as individual ones found only in cytochromes P450 2B4 and 2B5, the products of 2 possible allele genes with 98% sequence identity. Antibodies which bind to these peptides are expected not to cross-react with any other members of cytochrome P450 superfamily. Moreover, the epitopes characteristic only for 2B4 have been found; it means that there exist antibodies which can distinguish the two cytochromes P450 coded by allele genes with high sequence identity. The individual epitopes are concentrated in the highly variable regions mentioned above. Besides that, group-specific antigenic determinants common to the whole 2B subfamily are estimated; they constitute about 30% of the whole number of epitopes. Three of such peptides - fragments No. 143-148 and 261-266 can be considered as CYP2 family-specific epitopes since they are present in cytochromes P450 belonging to different subfamilies of CYP2 family.

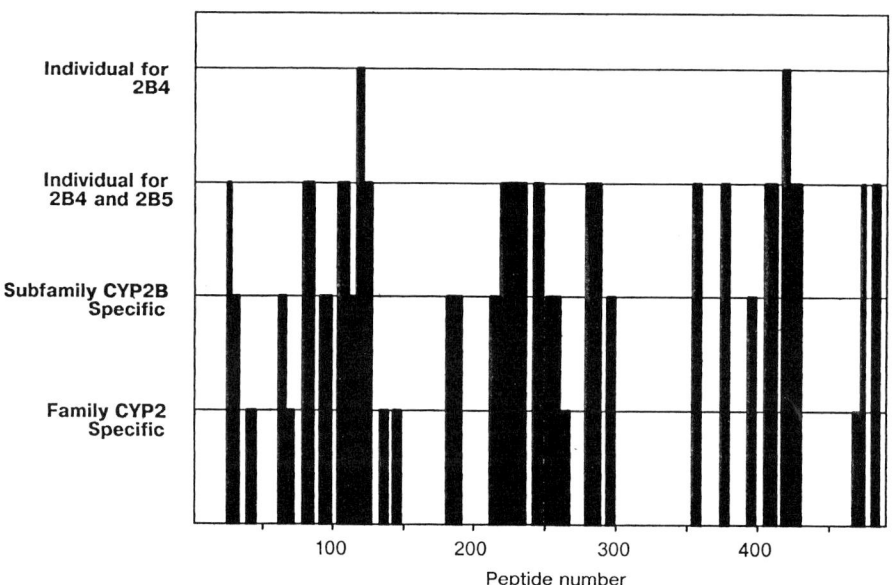

Fig.2. Distribution of sites with antigenic activity (sorted according to their group specificity) in the cytochrome P450 2B4 molecule.

According to our data the great antigenic homology is observed between the cytochromes P450 2B4 and 2B5, the moderate one - between 2B4 and other members of 2B subfamily; the weak homology is found between 2B4 and cytochromes 2G1 (7 common epitopes) and 2H1, 2H2 (5 common epitopes). Very weak antigenic homology exists between cytochromes 2B4 and other proteins from CYP2 family, and no homology is observed between 2B4 and cytochromes P450 belonging to the families other than CYP2. The information on the epitopes of cytochrome P450 2B4, namely on their structure, relative immunogenicity and presence in other cytochrome P450 molecules will help to successfully produce antipeptide antibodies specific to this protein and, vice versa, subfamily-specific ones. More profound studies of antigenic homology of different cytochromes P450 may underlie their further classification, in addition to the ones now existing (Nebert et al., 1991, Gotoh, 1992).

## REFERENCES

Archakov, A.I. and Bachmanova, G.I. (1990) <u>Cytochrome P450 and Active Oxygen</u>, Suppl. Taylor & Francis, London-New York-Philadelphia.

Ching, W.M., Wychowski, C., Beach, M.J., et al. (1992): <u>Proc.Natl.Acad.Sci. USA</u>, 89, 3190-3194.

Das, M.K. and Lindstrom, J. (1991): <u>Biochemistry</u> 30, 2470-2477.

Geysen, M.H., Tainer, J.A., Rodda, S.J., Mason, H.A., Getzoff, E.D. and Lerner, R.A. (1987): <u>Science</u>, 235, 1184-1190.

Grebenshchikova, O.G., Beryozov, A.T., Kolesanova, E.F. and Archakov, A.I. (1992): In <u>Cytochrome P450: Biochemistry and Biophysics</u> (Archakov, A.I., Bachmanova, G.I., Eds.), pp.686-691, INCO-TNC, Moscow, Russia.

Gotoh. H. (1992): In <u>Cytochrome P450: Biochemistry and Biophysics</u> (Archakov, A.I., Bachmanova, G.I., Eds.), pp.686-691, INCO-TNC, Moscow, Russia.

Imai, Y., Komori, M. and Sato, R. (1988): <u>Biochemistry</u>, 27, 80-88.

Nebert, D.W., Nelson, D.R., Coon, M.J. et al. (1991): <u>DNA Cell Biol.</u>, 10, 1-13.

Wang, J.-G., Jansen, R.W., Brown, E.A. and Lemon, S.M. (1990): <u>J.Virol.</u>, 64, 1108-1116.

# Specificity of covalent binding and autoantibodies in drug-induced immunoallergic hepatotoxicity

S. Lecoeur[1], D. Challine[1], J.C. Gautier[1], F. Ballet[2], Ph. Beaune[1]

[1]INSERM U.75, CHU Necker, 156, rue de Vaugirard, 75730 Paris Cedex 15, France.
[2]Rhône-Poulenc Rorer, 3, digue d'Alfortville, 94140 Alfortville, France

## Introduction

Drug-induced hepatotoxicity can be divided in two classes: that occurring systematically (exemplified by paracetamol overdose), and that involving idiosyncratic toxicity by direct mechanism or through an abnormal immune response. The mechanisms of idiosyncratic hepatotoxicity are not well elucidated; upon metabolization especially by P450, drugs produce reactive metabolites able to alkylate cell macromolecules. These reactive metabolites are involved in the first step of the postulated mechanism of autoimmune hepatitis (Beaune et al. 1987):

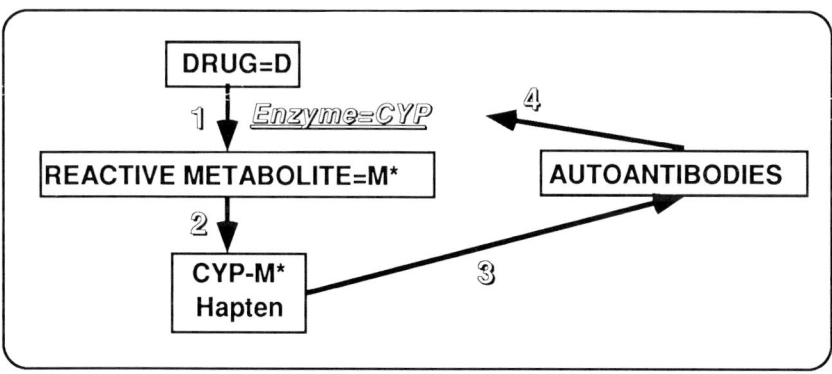

**Step 1**: formation of reactive metabolites mainly by cytochrome P450.
**Step 2**: covalent binding of metabolite(s) on the protein generating them; the complex protein-metabolite behaves as a neoantigen.
**Step 3**: immunological response.
**Step 4**: the autoantibodies recognize the P450 (alkylated and/or native) which produced the reactive metabolite.
Steps 1, 2, and 4 have been demonstrated for dihydralazine and tienilic acid-induced hepatitis: it was shown that the P450 responsible for the production of reactive metabolites (P450 1A and 2C respectively) was specifically alkylated by these metabolites, among liver microsomal proteins. The aim of this work was to uncover a possible relationship between the specificity of covalent binding of drug metabolites to liver proteins and the kind of hepatotoxicity which resulted, namely direct or immunoallergic.
We studied covalent binding of reactive metabolites from:
-two drugs toxic via a direct mechanism: paracetamol and chloroform.
-two immunotoxic drugs: tienilic acid and dihydralazine.

## Methods

### Quantitative covalent binding:
Human liver microsomes were incubated with the radiochemical ($^{14}$C tienilic acid, $^{14}$C chloroform and $^{3}$H paracetamol), and a NADPH-generating system in phosphate buffer. Aliquots of incubation mixtures were loaded on to glass-fiber filter disks. Filters were washed, dried and counted (Dansette et al. 1991). Results were obtained with the radiochemical used at increasing concentrations ($10^{-7}$ to $10^{-4}$ M), and with or without NADPH-generating system. Km and Vmax were graphically determined from double reciprocal plots. Quantitative covalent binding of tienilic acid (TA) on yeast microsomes was performed with TA 0,1 mM and yeast corresponding to 50 pmoles P450.

### Yeast and bacteria preparation:
Human P450 2C9 and 2C18 cDNA coding sequence (Brian et al. 1989, Romkes et al. 1991) were amplified by PCR and inserted in the yeast expression vector V8 (Pompon. 1988). This vector was transfected in yeast *S. cerevisiae.* Human P450 2C9 cDNA was also inserted into bacterial expression vector pGex and was produced in *E.Coli* JM101. Yeast microsomes and bacterial fusion protein were used to characterize the anti-LKM2 antibodies. These autoantibodies appeared in the sera of patients suffering from tienilic acid-induced hepatitis.

### Qualitative covalent binding:
Preparation of samples: Human liver microsomes (1 nmole P450) or yeast microsomes (50 pmoles of P450) were incubated with radiochemical, NADPH-generating system in phosphate buffer. Proteins of the incubation mixture were precipitated by addition of trichloroacetic acid and washed by various solvents. After centrifugation, the pellet was dissolved in 0,5 ml SDS 1%.

Electrophoresis and Western blotting: After migration on SDS-PAGE, proteins were electrotransferred to a nitrocellulose sheet. The sheet was cut perpendicular to the direction of the migration and counted. One part of the sheet was kept for immunoblots and probed with anti-human P450 1A1-2, 2C, 2E1, 3A4. The radioactive count of each band enabled us to obtain a covalent binding profile which was compared with the location of P450 by antibodies.

## Results

The kinetic constants measured on human liver microsomes corresponded to low apparent Km for tienilic acid and paracetamol (respectively 7 and 9 µM) and higher Km for chloroform (250 µM). The covalent binding evaluated at 10 µM (in the range of Km for tienilic acid and paracetamol) and 1mM (corresponding to saturation for the three compounds) showed higher values for tienilic acid than for the two others compounds.

The covalent binding of tienilic acid to yeast and human microsomes showed that P450 2C9 was the main P450 producing tienilic acid-reactive metabolites.

|  | Covalent binding (pmol/min/nmol P450) |
|---|---|
| Human liver | 98 |
| Control | 0 |
| P450 1A1 | 77 |
| P450 1A2 | 44 |
| P450 2C9 | 340 |
| P450 2D6 | 6 |
| P450 3A4 | 0 |
| P450 2C18 | 42 |

The incubation of the compounds with human liver microsomes gave us different covalent binding profiles: very specific for tienilic acid-metabolites, non-specific for paracetamol and chloroform metabolites:

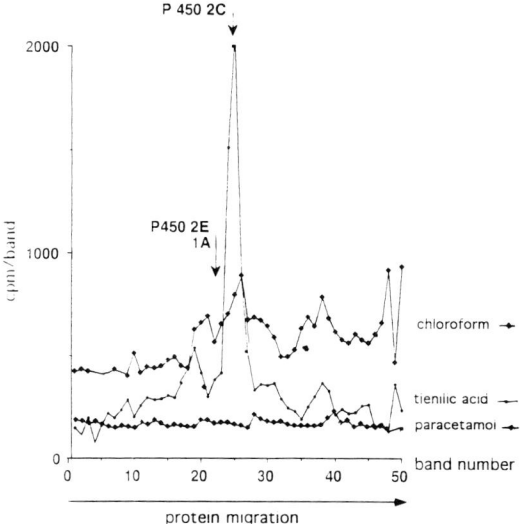

Tienilic acid metabolites were covalently bound to the P450 generating them. This experiment was confirmed by incubating tienilic acid with microsomes of yeast expressing isolated P450: only P450 2C9 gave covalent binding, and we could detect a radioactive peak corresponding to the metabolites bound to this P450.

All the sera from patients treated with tienilic acid and who suffered from autoimmune hepatitis recognized P450 2C9 expressed in the yeast, and this very specifically (they did not recognize others P450s expressed in the yeast or in the bacteria).

## Discusssion

In previous work (Beaune et al. 1988) covalent binding of TA was shown to be inhibited by anti-LKM2 and anti-human P450 2C, indicating that TA was metabolized into reactive metabolite(s) by a P450 2C subfamily member, according to steps 1 and 2 of the scheme 1. Our work attempted to precise the main target of reactive metabolites and autoantibodies.

Data on Km and Vmax showed that TA at a low drug concentration was able to generate many more bound metabolites than the other two hepatotoxic drugs tested: this let us to think that the amount of complex protein-reactive metabolite (neoantigen) formed may play a role in triggering of the immune response. Furthermore, covalent binding of TA metabolite(s) was concentrated to one microsomal protein (one major radioactive peak after electrophoresis), in contrast to covalent binding of paracetamol and chloroform which was spread over several microsomal proteins. As TA, dihydralazine-reactive metabolites exhibited specific covalent binding on the P450 generating them (Bourdi et al. personnal communication).

Specific recognition by autoantibodies of the P450 which metabolized the drug was found in the two cases of drug-induced autoimmune hepatitis: anti-LKM2 recognized P450 2C9 in case of TA-induced hepatitis, and anti-LM recognized P450 1A2 in case of dihydralazine-induced hepatitis. In both cases, specificity of the antibodies was hight because anti-LKM2 did not recognize P450 2C18, and anti-LM did not recognize P450 1A1 (Bourdi et al. 1992). The results presented here indicate that P450 2C9 is a specific target for TA-reactive metabolites and autoantibodies anti-LKM2.

Present results remain to be confirmed by testing TA covalent binding to other P450 2C and by recognition of these P450 2C by anti-LKM2. Other hepatotoxic drugs should be tested before concluding, but the use of quantitative and qualitative covalent binding data should help us to determine the immunotoxic potential of hepatotoxic drugs. Studies on homogenates and cytosolic systems are also necessary, since we studied only covalent binding of drugs to microsomal proteins. An animal model of immunoallergic hepatitis would also help us to elucidate the mechanism of triggering of autoimmune drug-induced hepatitis.

## References.

Beaune P., Dansette P., Mansuy D., Kiffel L., Finck M., Amar C., Leroux J.P., Homberg J.C. (1987). Human antiendoplasmic reticulum autoantibodies appearing in a drug induced hepatitis are directed against a human liver cytochrome that hydroxylates the drug. *Proc. Natl. Acad. Sci. U.S.A.* 84, 551-555.

Beaune P., Dansette P., Mansuy D., Kiffel L., Amar C., Leroux J.P., Homberg J.C. (1988). Autoantibodies to cytochrome P450 during autoimmune hepatitis. In *Liver cells and drugs*, Guillouzo A., editor, pp175-180. Paris: John Libbey Eurotext Ltd.

Bourdi M., Gautier J.C., Mircheva J., Larrey D., Guillouzo A., Andre C., Belloc C., Beaune P. (1992). Anti-liver microsomes autoantibodies and dihydralazine induced hepatitis: specificity of autoantibodies and inductive capacity of the drug. *Mol. Pharmacol.* 42, 280-285.

Brian W.R., Srivastava P. K., Umbenhauer D. R., Lloyd R. S., Guengerich F. P. (1989). Expression of a Human Liver Cytochrome P-450 Protein with Tolbutamide Hydroxylase Activity in *Saccharomyces cerevisiae*. *Biochemistry*. 28, 4993-4999.

Dansette P., Amar C., Valadon P., Pons C., Beaune P., Mansuy D. (1991). Hydroxylation and formation of electrophilic metabolites of tienilic acid and its isomer by human liver microsomes: catalysis by a cytochrome P-450 IIC different from that responsible for mephenytoin hydroxylation. *Biochem. Pharmacol.* 41, 553-560.

Pompon D. (1988). cDNA and functional expression in yeast *Saccharomyces cerevisiae* of β-naphtoflavone-induced rabbit liver P-450 LM4 and LM6. *Eur. J. Biochem.* 177, 285-293.

Romkes M., Faletto M.B., Blaisdell J.A., Raucy J.L., Goldstein J.A. (1991). Cloning and Expression of Complementary DNAs for Multiple Members of the Human Cytochrome P450 IIC Subfamily. *Biochemistry*. 30, 3247-3255.

## Acknowledgements

This work was supported by a grant from the European Economic Community and Rhône-Poulenc Rorer (Bioavenir).

# Metabolic activation of the thiophene ring as a primary event in immunoallergic hepatitis to tienilic acid

Patrick M. Dansette[1], M. Pilar Lopez-Garcia[1,4], Claudine Amar[1], Catherine Pons[1], Philippe Valadon[1], Eric Bonierbale[1], Philippe H. Beaune[2], Jean-Claude Homberg[3], Daniel Mansuy[1]

[1]Laboratoire de chimie et biochimie pharmacologiques et toxicologiques, URA 400 CNRS, Université René-Descartes, 45, rue des Saints-Pères, 75270 Paris Cedex 06, France. [2]CHU Necker, INSERM U.75. [3]CHU Saint-Antoine. [4]Departmento de Bioquímica y Biología Molecular, Facultad de Farmacia, Universidad de Valencia, Spain

Tienilic acid (TA) is a uricosuric drug marketed in France from 1976 until 1992 and in USA from 1979 until 1980, which has caused rare cases of cytolytic hepatitis (1 for 10000 patients in France). In all proven cases investigated, a very caracteristic antimicrosomal antibody was found; since it reacts with rat liver and kidney slices it was called antiLKM2 (anti Liver Kidney Microsomes type 2) (Homberg et al., 1984). The major metabolite in human and rat urine was 5-hydroxytienilic acid (5OHTA), more than 50% of the dose being excreted in human urine in 24h and a small amount of the carbinol derivative (reduced ketone) (Mansuy et al., 1984). Half of the sera of the hepatitis patients (antiLKM2) and none of those of long term patients recognized in immunotransfer one specific protein of human liver microsomes, a cytochrome P450 of family 2C (Cyp 2). Moreover all the antiLKM2 sera tested inhibited very well 3 reactions mediated by Cyp 2C, TA 5-hydroxylation (Beaune et al., 1987) and the activation TA and its 3-isomer (TAI) to reactive metabolites able to bind to microsomal proteins (Dansette et al., 1991). They also inhibited two activities of Cyp 2C : (S)-Mephenytoin 4-hydroxylase (Meier & Meyer, 1987) and tolbutamide hydroxylase (Lopez-Garcia et al., 1993). Thus antiLKM2 are directed against mostly Cyp 2C9, and also react with rat Cyp 2C11 (Pons et al., 1992, see also Lecoeur et al., *this volume*).

Tienilic Acid (TA)

Tienilic Acid Isomer (TAI)

Tienilic acid (TA) and its isomer (TAI) (a minor impurity in some of the early production batches) are activated by cytochrome P450 into reactive metabolites which bind covalently with microsomal proteins. It is possible to decrease this covalent binding with diverse nucleophiles and in the case of TAI this decrease correspond to the trapping of adducts which can be isolated. Using mercaptoethanol as the nucleophile several successive adducts have been isolated and fully identified (Valadon P., *Thesis 1992*). The identification of the first adduct as a 2,5-dihydro-thiophene sulfoxide demonstrate that the reactive metabolite of TAI is the corresponding unstable thiophene sulfoxide <u>A</u> (Mansuy et al. 1991). More recently isolation of the major urinary metabolite of thiophene as a dihydrothiophene sulfoxide mercapturic acid (Dansette et al., 1992) shows that thiophene sulfoxides are probably more general reactive metabolites and thus tienilic acid could also be activated in a similar manner.

The covalent fixation of TA and TAI probably involves their thiophene ring, thus carrier proteins (BSA or ß-lactoglobulin) were attached thru a spacer arm to the thiophene ring of TA and were used to raise antibodies in rabbits, and to establish an ELISA assay. These antibodies (anti-TA) recognized both the covalent binding of TA or TAI to microsomes, and competition studies in ELISA showed that the ketodichloro-phenoxyacetic moiety was necessary for recognition (Valadon P., *Thesis 1992*). Moreover these antibodies could be used in immunotransfer.


Kinetics of the 5-hydroxylation of Tienilic acid in human microsomes were not linear and carefull examination showed progressive inactivation of the enzymatic system. Similar studies using Cyp 2C9 or Cyp 2C10 expressed in yeast demonstrated a catalysis dependent loss of activity with TA, whereas this cytochromes P450 were stable to metabolic products of TAI (Lopez-Garcia et al., 1993b). Trapping of TA or TAI metabolites by glutathione (GSH) using Cyp 2C9 shows that it is posible to decrease TAI binding to very low level (< 0.2 nmol/nmol P450), but TA binding only to a third of that in absence of GSH or about 1 nmol/nmol P450 as shown on Figure 1. When the yeast microsomes were analyzed by immuno-revelation with the anti-TA, only one band corresponding to Cyp 2C9 was enlighted by antiLKM2; in absence of GSH at least two other bands were evident. Thus part of the reactive metabolite can escape from the active site and react at other protein sites or with excess nucleophile. To explain the inactivation of the enzyme, we propose that the active metabolite, a thiophene sulfoxide, reacts with a nucleophile in the active site leading to an adduct which after aromatization is irreversibly bound. If the sulfoxide reacts with water this leads after aromatization to 5OHTA. From the kinetic data there is one inactivation event for 12 5OHTA formed.

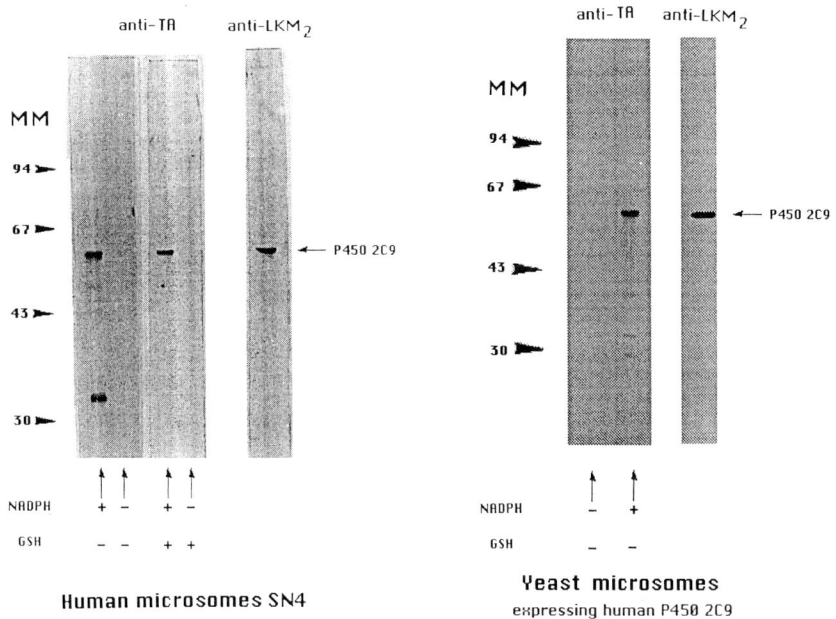

These results suggest that the primary event in TA activation is its oxidation by cytochrome P450 into a very reactive thiophene sulfoxide which reacts at the active site of Cyp 2C9/10, causing its inactivation and probably triggering its disposal by the hepatocytes. Possible exocytosis of a whole or degraded protein or immunogenic peptides presented to the immune system could be at the origin of the autoimmune response. Sera from patients treated with TA for long periods, with or without hepatitis all contain antibodies recognizing ß-lactoglobulin-TA in an ELISA test. This is indicative of the presence of an antigen haptenized with TA, triggering also an immune response against the drug. However the cytolytic hepatitis and autoimmune reaction is very rare in patients treated with TA, thus another individual component is certainly needed for the immune reaction, depending either of the individual immune system or of a molecular mimicry between Cyp 2C9/10 peptides and another viral or bacterial antigen.

It is noteworthy that other organ specific autoimmune diseases are restricted to an estonishingly narrow immune repertoire which is available in only a very small part of the population (Chazenblack et al., 1993). It is also not known if antiLKM2 are pathogenic or only a marker of the disease. Epitope mapping and determination of the target peptide(s) modified by TA are underway.

## REFERENCES

Beaune, P. H., Dansette, P. M., Mansuy, D., Kiffel, L., Finck, M., Amar, C., Leroux, J. P., & Homberg, J. C. (1987) : Human anti-endoplasmic reticulum autoantibodies appearing in a drug induced hepatitis are directed against a human liver cytochrome P-450 that hydroxylates the drug. *Proc. Natl. Acad. Sci. U.S.A. 84*, 551-555.

Chazenback G.D., Portolano S., Russo D., Hutchison S.C., Rapoport B. & McLachlan S. (1993) : Human organ-specific autoimmune disease : Molecular cloning and expression of an autoantibody repertoire for a major autoantigen reveals an antigenic immunodominant region and restricted immunoglobulin gene usage in the target organ. *J. Clin. Invest., 92*, 62-74.

Dansette, P. M., Amar, C., Smith, C., Pons, C., & Mansuy, D. (1990) : Oxidative activation of the thiophene ring by hepatic enzymes, Hydroxylation and formation of electrophilic metabolites during metabolism of tienilic acid and its isomer by rat liver microsomes. *Biochem. Pharmacol. 39*, 911-918.

Dansette, P. M., Amar, C., Valadon, P., Pons, C., Beaune, P. H., & Mansuy, D. (1991) : Hydroxylation and formation of electrophilic metabolites of tienilic acid and its isomer by human liver microsomes : catalysis by a P450IIC different from that responsible for mephenytoin hydroxylation. *Biochem. Pharmacol. 41*, 553-560.

Dansette, P. M., Thang, D. C., Elamri, H., & Mansuy, D. (1992). Evidence for Thiophene-S-Oxide as a Primary Reactive Metabolite of Thiophene Invivo - Formation of a Dihydrothiophene Sulfoxide Mercapturic Acid. *Biochem Biophys Res Commun, 186*, 1624-1630.

Homberg, J. C., André, C., & Abuaf, N. (1984) : A new anti-liver-kidney-microsome antibody (anti-LKM2) in tienilic acid-induced hepatitis. *Clin. Exp. Immunol. 55*, 561-570.

Homberg J.C., Abuaf N., Elmy-Khalil S., Biour N., Poupon R., Islam S., Darnis F., Levy V.G., Opolon P., Beaugrand M., Toulet J., Danan G. & Benhamou J.P. (1985) : Drug induced hepatitis associated with anti-cytoplasmic organelles autoantibodies. *Hepatology, USA*, 5 : 722-727.

Lecoeur S., Challine D., Gautier J.C., Ballet F. & Beaune P. : Specificity of covalent binding and autoantibodies in drug induced immunoallergic hepatotoxicity. *This volume*

López-Garcia, M. P., Dansette, P. M., Valadon, P., Amar, C., Beaune, P. H., Guengerich, F. P., & Mansuy, D. (1993) : Human liver P450s expressed in yeast as tools for rective metabolite formation studies : oxidative activation of tienilic acid by P450 2C9 and P450 2C10. *Eur. J. Biochem.. 213*, 223-232.

López-Garcia, M. P., Dansette, P. M. & Mansuy, D. (1993b) : Thiophene derivatives as new mechanism-based inhibitors of cytochromes P450 : inactivation of yeast expressed human liver cytochrome P450 2C9 by tienilic acid. *(this volume)*

Mansuy D., Dansette P., Foures C., Jaouen M., Moinet G. & Bayer N. (1984) : Metabolic hydroxylation of the thiophene ring : isolation of 5-hydroxytienilic acid as the major urinary metabolite in rat and man. *Biochem. Pharmacol., 33* : 1429-1435.

Mansuy, D., Valadon, P., Erdelmeier, I., López-Garcia, M. P., Amar, C., Girault, J.P., & Dansette, P. M. (1991) : Thiophene-S-oxides as new reactive metabolites : formation by cytochrome P450-dependent oxidation and reaction with nucleophiles. *J. Am. Chem. Soc. 113*, 7825-7826.

Pons C., Dansette P.M., Amar C., Jaouen M., Wolf C.R., Grégeois J., Homberg J.C. & Mansuy D. (1992) : Detection of human hepatitis anti-LKM2 autoantibodies on rat liver sections is predominantly due to reactivity with rat liver cytochrome P450 IIC11. *J. Pharmacol. Exper. Ther., 259*, 1328-1334.

# Anti-cytochrome P450 antibodies in patients with aromatic anticonvulsant-induced hypersensitivity reactions

J.S. Leeder[1], Vicki A. Cook[1], Robert J. Riley[2]

[1]Division of Clinical Pharmacology and Toxicology, The Hospital for Sick Children, Toronto, Ontario M5G 1X8, Canada. [2]Fisons Pharmaceuticals, Department of Biochemistry, Drug Metabolism Section, Bakewell Road, Loughborough, Leics., UK

Described as "hypersensitivity" reactions, systemic idiosyncratic reactions to phenytoin, phenobarbital and carbamazepine are characterized by the delayed onset (days to months after initiation of therapy) of febrile illness accompanied by skin rash, lymphadenopathy and possibly secondary organ involvement including hepatitis, nephritis, exfoliative dermatitis and hematologic or immune abnormalities (Shear and Spielberg, 1988). The clinical manifestations of these reactions are consistent with an immunological origin, and it has been proposed that an imbalance between the activation and detoxification of the parent compounds may lead to increased production of reactive, potentially cytotoxic intermediates capable of interacting covalently with cellular macromolecules (Riley et al., 1988). These covalent adducts (haptens) could serve as potential immunogens intitating an immune response against drug-modified or native proteins. Consistent with this, we previously identified anti-microsomal antibodies in the sera of patients experiencing hypersensitivity reactions to phenytoin, phenobarbital and carbamazepine directed against rat CYP3A and 2C proteins (Leeder et al., 1992) and a similarly regulated human P450, possibly CYP3A3/4 or CYP2C8 (Riley et al., 1993). We present additional data on anti-P450 antibody production with important implications for the pathogenesis of these reactions.

METHODS

Heparinized whole blood was obtained from individuals being evaluated for possible anticonvulsant hypersensitivity reactions by means of *in vitro* drug rechallenge (Shear and Spielberg, 1988). Plasma or sera from 77 such patients and 30 controls (no history of drug exposure) were also screened for the presence of anti-microsomal antibodies by western blotting using as an antigen source liver microsomes from rats or mice pretreated with phenobarbital or dexamethasone. Anti-microsomal antibody-positive sera were further assessed using purified CYPs 2B1, 2B2, 2C6, 2C11, 3A1 and 3A2 generously provided by Mr. Wayne Levin (Hoffmann-LaRoche, Nutley, NJ). Bound antibody was visualized with enhanced chemiluminescence detection (Amersham Canada Ltd.) and peroxidase-conjugated anti-human immunoglobulin (Ig) antibodies: sheep anti-total Ig (Amersham); goat anti-IgG, IgA and IgM (Jackson Laboratories); mouse anti-IgG$_1$, IgG$_2$, IgG$_3$, and IgG$_4$ (WHO clones HP6069, HP6014, HP6047 and HP6025, respectively; Zymed); rabbit anti-IgD (Dakopatts); and goat anti-IgE (Caltag).

RESULTS AND DISCUSSION

The results of the initial screening against hepatic microsomes are presented in Table 1. Antibodies recognizing a 53 kD microsomal protein were identified in 24 patients and 4 controls. Eighty per cent (19/24) patients with anti-microsomal antibodies demonstrated decreased capacity to detoxify reactive

Table 1

## CYP2D6 Phenotype and Lung Cancer

| | Lung Cancer | | Controls | | P |
|---|---|---|---|---|---|
| | N | PMs (%) | N | PMs (%) | |
| Ayesh et al., 1984 | 245 | 1.6 | 234 | 9.0 | 0.0002 |
| Roots et al., 1990 | 400 | 7.0 | 470 | 10.5 | 0.052 |
| Law et al., 1989 | 104 | 1.9 | 104 | 8.7 | 0.029 |
| Duche al., 1991 | 153 | 6.5 | 254 | 7.9 | n.s. |
| Benitez et al., 1991 | 84 (73)[a] | 4.8 (1.4)[a] | 143 | 7.0 | n.s. |
| Caporaso et al., 1990 | 89 | 1.1 | 92 | 13.0 | 0.0015 |
| Speirs et al., 1990 | 82 | 9.8 | - | - | - |
| All studies | 1157 | 4.9 | 1227 | 8.6 | |

a    squamons and small cell histology

Table 2

## CYP2D6 Genotype and Lung Cancer

|  | Lung Cancer | | Controls | | P | Alleles tested |
|---|---|---|---|---|---|---|
|  | N | PMs (%) | N | PMs (%) | | |
| Wolf et al., 1992 | 361 | 3.6 | 720 | 4.3 | n.s. | (B) |
| Kerb et al., 1992 | 115 | 3.5 | 137 | 9.5 | <0.05 | (A,B,D) |
| Hirvonen et al., 1992 | 106 | 0.9 | 122 | 5.7 | <0.05 | (A,B,D) |
| Tefre et al., 1992 | 190 | 9.0 | 220 | 5.0 | n.s. | (B) |
| Agundez et al., 1994 | 89 | 0 | 98 | 7.1 | <0.02 | (A,B,C,D) |
|  |  | (20) |  | (3.0) | <0.0005 | (C alone) |

Cumulatively, the presence of anti-P450 antibodies in the sera of control subjects, the absence of a detectable anti-P450 IgM response and the possibility of pre-existing IgG antibodies suggests that these anti-P450 antibodies may have been a response to another antigen, possibly of infectious origin. Anti-CYP2D6 antibodies have been detected in the sera of patients with non-drug related, type II autoimmune hepatitis. Epitope mapping studies have identified homologies between mapped amino acid segments and herpes simplex virus type 1 and hepatitis C virus (HCV) proteins suggesting that mimicry between viral antigens and CYP2D6 may occur (Manns *et al.*, 1991). Although the relationship between HCV infection and type II autoimmune hepatitis remains to be clearly elucidated (Vergani *et al.*, 1993), a viral etiology in the pathogenesis of this autoimmune disorder is possible.

Sera from halothane hepatitis patients contain antibodies which recognize various trifluoroacetyl (TFA)-protein adducts which arise from oxidative halothane metabolism. Gut *et al.* (1993) have proposed that mimicry may occur between lipoic acid, the prosthetic group of the E2 subunit of the mitochondrial pyruvate dehydrogenase complex (PDC), and TFA-protein adducts. Lipoyl-containing PDC-E2 are major antigens recognized by autoantibodies in primary biliary cirrhosis. Burroughs *et al.* (1992) have proposed a novel hypothesis for the pathogenesis of this autoimmune disorder based on homologies between HLA-DR-$\alpha$ and human and *E. coli* PDC-E2. Whether this hypothesis also applies to halothane hepatitis is presently not known.

We favour the hypothesis that an infectious episode prior to initiation of anticonvulsant therapy predisposes an individual to a hypersensitivity reaction. Imbalances in activation/detoxification pathways may also influence susceptibility inasmuch as they may lead to increased amounts of drug-modified P450 or fragments thereof which could mimic the viral antigen. Anticonvulsant-induced dysregulatory effects on the immune system (Aarli *et al.*, 1980) may also contribute to an altered response to drug-derived neoantigens or self-antigens. Identification of the epitope recognized by patient antibodies, its relationship to viral antigens, if any, and the nature of the host immune response to P450 antigens will lead to a better understanding of the pathogenesis of anticonvulsant hypersensitivity reactions.

*Supported by MRC Canada and a PMAC-Health Research Foundation/MRC Career Award (JSL).*

REFERENCES

Aarli, J. A. and Fontana, A. (1980): Immunological aspects of epilepsy. *Epilepsia* 21, 451-457.

Burroughs, A. K., Butler, P., Sternberg, M. J. E. *et al.* (1992): Molecular mimicry in liver disease. *Nature* 358, 377-378.

Forrester, L. M., *et al.* (1992): Relative expression of cytochrome P450 isoenzymes in human liver and association with the metabolism of drugs and xenobiotics. *Biochem. J.* 281, 359-368.

Gut, J., Christen, U. and Huwyler, J. (1993): Mechanisms of halothane toxicity: Novel insights. *Pharmacol. Ther.* 58, 133-155.

Leeder, J. S., Riley, R. J., Cook, V. A. *et al.* (1992): Human anti-cytochrome P450 antibodies in aromatic anticonvulsant-induced hypersensitivity reactions. *J. Pharmacol. Exp. Ther.* 263, 360-367.

Manns, M. P., Griffen, K. J., Sullivan, K. F. *et al.* (1991): LKM-1 autoantibodies recognize a short linear sequence in P450IID6, a cytochrome P450 monooxygenase. *J. Clin. Invest.* 88, 1370-1378.

Pons, C., Dansette, P. M., Amar, C. *et al.* (1991): Detection of human hepatitis anti-liver kidney microsomes (LKM2) autoantibodies on rat liver sections is predominantly due to reactivity with rat liver P-450 IIC11. *J. Pharmacol. Exp. Ther.* 259, 1328-1334.

Riley, R. J., Maggs, J. L., Lambert, C. *et al.* (1988): An in vitro study of the microsomal metabolism and cellular toxicity of phenytoin, sorbinil and mianserin. *Br. J. Clin. Pharmacol.* 26, 577-588.

Riley, R. J., Smith, G., Wolf, C. R. *et al.* (1993): Human anti-endoplasmic reticulum antibodies produced in aromatic anticonvulsant hypersensitivity reactions recognise rodent CYP3A proteins and a similarly regulated human P450 enzyme(s). *Biochem. Biophys. Res. Commun.* 191, 32-40.

Shear, N. H. and Spielberg, S. P. (1988): Anticonvulsant hypersensitivity syndrome. In vitro assessment of risk. *J. Clin. Invest.* 82, 1826-1832.

Vergani, D. and Mieli-Vergani, G. (1993): Type II autoimmune hepatitis. What is the role of the hepatitis C virus. *Gastroenterology* 104, 1870-1873.

# Presence of cytochrome P450 on human hepatocyte plasma membrane. Recognition by several autoantibodies

Jacqueline Loeper[1], Véronique Descatoire[1], Michèle Maurice[2], Philippe Beaune[3], Gérard Feldmann[2], Frederic P. Guengerich[4], Dominique Pessayre[1]

[1]INSERM U.24, Hôpital Beaujon, 92110 Clichy, France. [2]INSERM U.327, CHU-Bichat, 75018 Paris, France. [3]INSERM U.75, CHU-Necker-Enfants Malades, 75014 Paris, France. [4]Laboratory of Biochemistry, Vanderbilt University, TN, USA

Autoantibodies against cytochrome P-450 are found in several forms of autoimmune hepatitis. Some children with type II autoimmune hepatitis develop anti-liver/kidney microsome type 1 (anti-LKM$_1$) autoantibodies. These anti-LKM$_1$ autoantibodies are directed against human liver P-450 2D6 (Zanger et al, 1988; Kiffel et al, 1989; Manns et al, 1989). Anti-LKM$_2$ autoantibodies are present in patients with ticrynafen (tienilic acid)-induced hepatitis. They are directed against a human liver P-450 of the 2C family, wich transforms tienilic acid into a reactive metabolite (Beaune et al, 1987). Another type of autoantibodies only reacting with liver microsome (anti-LM) are found in patients with dihydralazine-induced hepatitis.and are directed against human liver P-450 1A2 (Bourdi et al, 1990). Anti-P-450 1A2 autoantibodies have been also found in an idiopathic form of autoimmune hepatitis. The pathological consequences of such anti-cytochrome P-450 autoantibodies remain unknown. For an immune response against cytochrome P-450 to possibly damage the hepatocytes, it must be assume that cytochrome P-450 is present on the surface of the hepatocytes. This has been reported in rat hepatocyte plasma membranes (Stasiecki and Oesch, 1980; Loeper et al, 1990). Here, we report the presence of several forms of human liver P-450 (1A2, 2C, 2D6, 2E1 and 3A4) in the plasma membrane of human hepatocyte (Loeper et al, 1993). And the recognition of some epitopes of cytochrome P-450, exposed on the outer surface of the plasma membrane of human hepatocytes, by anti-P-450 autoantibodies as anti-LKM$_1$, anti-LKM$_2$, and anti-LM (Loeper et al, 1993).

To demonstrate this presence,we performed, first, biochemical assays with a very pure plasma membrane fraction practically devoid of microsomal contamination. Plasma membranes were prepared with a technique based on the electrostatic attachment of isolated hepatocytes to polyethyleimine-coated polyacrylamide beads. Polyglutamate was added to cover bare sites left between attached hepatocytes.

Human hepatocytes were obtained from liver of adult organ donor, after reduction in size, when liver had to be transplanted in children. They were isolated by collagenase perfusion and purified on isotonic Percoll (Loeper et al, 1993). This method to prepare plasma membranes yield a very pure fraction, enriched in Na$^+$/K$^+$ ATPase and in 5'-

Hirvonen, A., Husgafvel.Pursiainen, K., Anttila, S., Karjalainen, A., Sorsa, M., Vainio, H. (1992): Metabolic cytochrome P450 genotypes and assessment of individual susceptibility to lung cancer. *Pharmacogenetics* 2, 259-263.

Idle, J.R., Armstrong, M., Boddy, A.V., Boustead, C., Cholerton, S., Cooper, J., Daly, A.K., Ellis, J., Gregory, W., Hadidi, H., Höfer, C., Holt, J., Leathart, J., McCracken, N., Monkman, S.C., Painter, J.E., Taber, H., Walker, D., and Yule, M. (1992): The pharmacogenetics of chemical carcinogenesis. *Pharmacogenetics* 2, 246-258.

Johansson, I., Dahl, M.-L., Bertilsson, L., Sjöqvist, F., and Ingelman-Sundberg, M. (1992): Gene amplification of an active CYP2D gene as a cause of extremely rapid metabolism of debrisoquine. *J. Basic Clin. Physiol. Pharmacol.* 3, 242.

Johansson, I., Lundqvist, E., Bertilsson, L., Dahl, M.-L., Sjöqvist, F., and Ingelman-Sundberg, M. (1994): Inherited amplification of an active gene in the cytochrome P450 CYP2D locus as a cause of ultrarapid metabolism of debrisoquine. *Proc. Natl. Acad. Sci. USA* (in press)

Kerb, R., Brockmoeller, J., Drakoulis, N., Siegel, J., Roots, I. (1992): CYP2D6 and glutathione S-transferase class mu as host factors of lung cancer susceptibility. *J. Basic Clin. Physiol. Pharmacol.* 3 (suppl.), p. 131.

Kagimoto, M., Heim, M., Kagimoto, K., Zeugin, T., and Meyer, U.A. (1990): Multiple mutations of the human cytochrome P450IID6 gene (CYP2D6) in poor metabolizers of debrisoquine: study of the functional significance of individual mutations by expression of chimeric genes. *J. Biol. Chem.* 265, 17209-1721.

Law, M.R., Hetzel, M.R., and Idle, J.R. (1989): Debrisoquine metabolism and genetic predisposition to lung cancer. *Br. J. Cancer* 59, 686-687.

Meyer, U.A., Skoda, R.C., Zanger, U.M., Heim, M., and Broly, F. (1992): The genetic polymorphism of debrisoquine/sparteine metabolism - molecular mechanism, In: *Pharmacogenetics of Drug Metabolism,* ed. W. Kalow, pp. 609-623. Pergamon Press, New York, Oxford.

Roots, I., Drakoulis, N., Brockmöller, J., Janicke, I., Cuprunov, M., and Ritter, J. (1990): Hydroxylation and acetylation phenotypes as genetic risk factors in certain malignancies. In *Xenobiotic Metabolism and Disposition*, ed. R. Kato, R.W. Estabrook, M.N. Cayen, pp. 499-506. London, New York, Philadelphia: Taylor & Francis.

Roots, I., Drakoulis, N., and Brockmöller, J. (1992): Polymorphic enzymes and cancer risk: concepts, methodology and data review. In *Pharmacogenetics of Drug Metabolism*, ed. W. Kalow, pp. 815-841, Pergamon Press, Inc. New York.

Skoda, R.C., Gonzalez, F.J., Demierre, A. and Meyer, U.A. (1988): Two mutant alleles of the human cytochrome P450db1 gene (P450IID1) associated with genetically deficient metabolism of debrisoquine and other drugs. *Proc. Natl. Acad. Sci. USA* 85, 5240-5243.

Speirs, C.J., Mrray, S., Davies, D.S., Biola Mabadeje, A.F., and Boobis, A.R. (1990): Debrisoquine oxidation phenotype and susceptibility to lung cancer. *Br. J. clin. Pharmac.* 29, 101-109.

Tefre, T., Daly, A., Armstrong, M., Ryberg, D., Hangen, A., Idle, J., Børresen, A.L. (1992): Genotyping of the CYP2D6 gene in lung cancer patients and controls. *J. Basic Clin. Physiol. Pharmacol.* 3 (suppl.), p. 132.

Tyndale, R.F., Aoyama, T., Broly, F., Matsunaga, T., Inaba, T., Kalow, W., Gelboin, H.V., Meyer, U.A., and Gonzalez, F.J. (1991): Identification of a new variant *CYP2D6* allele lacking the codon encoding Lys-281: possible association with the poor metabolizer phenotype. *Pharmacogenetics* 1, 26-32.

Wolf, C.R., Dale Smith, C.A., Gough, A.C., Moss, J., Vallis, K.A., Howard, G., Carey, F.J., Mills, K., McNee, W., Carmichael, J., and Spurr, N.K. (1992): Relationship between the debrisoquine hydroxylase polymorphism and cancer susceptibility. *Carcinogenesis* 13, 1035-1038.

Yokota, H., Tamura, S., Furuya, H., Kimura, S., Watanabe, M., Kanazawa, I., Kondo, I., and Gonzalez, F.J. (1993): Evidence for a new variant CYP2D6 allele CYP2D6J in a Japanese population associated with lower in vivo rates of sparteine metabolism. *Pharmacogenetics* 3, 256-263.

Zanger, U.M., Vilbois, F., Hardwick, J., Meyer, U.A. (1988): Absence of hepatic cytochrome P450buf1 causes genetically deficient debrisoquine oxidation in man. *Biochemistry* 27, 5447-5454.

Zanger, U.M., Hauri, A.P., Loeper, J., Homberg, JC., and Meyer, U.A. (1988): Antibodies against human cytochrome P-450 db1 in autoimmune hepatitis type II. *Proc. Natl. Acad. Sci. USA.* 85,8256-8260.

# Activation of promutagens by rat CYP P450 expressing V79 cells

A.S. Rodrigues[1], I. Duarte Silva[1], M.J. Monteiro[1], H. Caria[1], A. Laires[1,2], T. Chaveca[1], J. Rueff[1]

[1]Department of Genetics, Faculty of Medical Sciences, New University of Lisbon, R. da Junqueira, 96, P-1300 Lisbon, Portugal. [2]Faculty of Sciences and Technology, New University of Lisbon, P-2825 Monte da Caparica, Portugal

## INTRODUCTION

It is well established that the presence of numerous chemicals, natural or man made, in the environmental represents a threat to human genetic health. The probability that some of these chemicals are genotoxic, demands the use of methods to evaluate their genotoxicity. The use of short term tests has greatly helped in identifying potentially genotoxic chemicals to which man is exposed. An important drawback of most established cell lines used in short term toxicology testing is their lack of metabolic capabilities, including those of cytochromes P450 (Turchi et al., 1992). This fact has led to the use of exogenous metabolic activation systems to metabolise promutagens, either with whole cells or the rodent post mitochondrial fraction, S9. The use of different metabolizing systems in short term tests will forcibly influence the overall outcome for a promutagen (Langenbach et al., 1992). Moreover, the use of complex metabolizing systems does not permit mechanistic studies as to what metabolic pathway is important in the activation of a xenobiotic. It is desirable, therefore, to use metabolically competent cells as indicators for genotoxicity, thus using single step screening for mutagens/carcinogens. Molecular cloning techniques have allowed the development of cell lines with defined enzyme expression, with the advantage of having known metabolic phenotypes. A number of V79 cell lines have been genetically engineered to express specific rat cytochromes P450, as well as human cytochromes P450. The present study aimed to evaluate cell lines developed by Doehmer et al.(1992) (Dogra et al., 1990; Wölfel et al., 1991) expressing rat cytochrome P450 1A2 (XemdMZ) for their capacity to activate the carcinogenic aromatic amines 2-aminoanthracene (2-AA), 2-aminofluorene (2-AF), 2-acetylaminofluorene (2-AAF), the heterocyclic aromatic amine 2-amino 3-methylimidazo [4,5-$f$] quinoline (IQ) and the noncarcinogen 4-acetylaminofluorene (4-AAF). The end points were chromossomal aberrations and sister chromatid exchanges. Since acetyltransferase activity is known to be important in the activation of aromatic amines we also studied a cell line expressing rat cytP450 1A2 plus endogenous acetyltransferase activity (Xemd-NH) for the activation of the same aromatic amines. Additionally we studied the V79 cell lines expressing rat cytochrome P450 1A1 (Xem-2) and 1A2 (Xemd-MZ) apart from the parental V79 cell line, for their capacity to metabolize the flavonoid quercetin, using chromossomal aberrations as the endpoint.

## METHODS

Genetic engineering of the V79 Chinese hamster cells has been described elsewhere (Doehmer et al.,1988; Dogra et al.,1990; Wolfel et al.,1991).. Full lenght cDNAs were combined with the expression vector psv2 under the control of the SV40 early promotor.These PSV450 were co-tranfected with plasmid pdBPV-MTneo (342-2) which carries the neomycin phosphotransferase gene, conferring resis-

Our current understanding of the regulation of eukaryotic development indicates that growth factors, hormones, neurotransmitters and a variety of extracellular ligands initiate signals that are propagated by a cascade of biochemical events that include increases of ion fluxes across cellular membranes, elevation of intracellular pH, $Ca^{2+}$, and cAMP levels, formation of phosphoinositide metabolites, and activation of protein kinases and phosphatases (Morgan and Curran, 1986). Transduction of these signals to the nucleus induces the expression of a group of immediate-early genes that, among others, encode the protein products of the protooncogenes *fos* and *jun*, components of the transcription factor AP-1. This factor in turn propagates the signal by controlling the expression of genes required for cell proliferation or differentiation (Hunter and Karin, 1992).

Work from our laboratory has shown that treatment of mouse hepatoma cells with PAH inducers of CYP1A1 activity, such as TCDD and B[a]P, causes an increase in the steady state mRNA levels of the protooncogenes c-*fos*, c-*jun*, *jun-B*, and *jun-D*, and the concomitant increase of transcription factor AP-1, reaching levels almost as high as those induced by TPA, a classical phorbol ester tumor promoter (Puga et al. 1992). These results suggested the possibility that genes that have AP-1 binding sites might respond to AP-1 regulation as a result of TCDD or B[a]P treatment. We have tested this hypothesis in mouse hepatoma cells by analyzing the activation of a chloramphenicol acetyltransferase (*cat*) reporter plasmid carrying AP-1 responsive regulatory elements. The chimeric plasmid was constructed by fusion of the *cat* gene sequences to the HIV-1 LTR. We present here results that demonstrating that B[a]P activates CAT expression well over the basal level found in untreated cells.

## MATERIALS AND METHODS

The reporter gene that we chose is the bacterial *cat* gene widely used for this type of analyses. The chimeric plasmid pHIVLTRCAT was constructed by fusion of the *cat* gene sequences to the LTR sequences of HIV-1 in pUC8 (a gift of the NIAID Repository of AIDS Reagents). This plasmid was transfected by standard calcium phosphate techniques (Graham and Van der Erb, 1973) into Hepa-1 cells, a mouse hepatoma cell line (Bernard et al. 1973), and into several mutant derivatives of this line, including the c2 and c37 mutants, that lack a functional Ah receptor or CYP1A1 enzyme, respectively (Hankinson et al. 1985; Kimura et al. 1987), and the CX4 line, a pseudo-wild type derivative of c37, that expresses a functional CYP1A1 enzyme from a stably transfected plasmid vector (Puga et al. 1990; RayChaudhuri et al. 1990). In all experiments we also co-transfected the cells with plasmid pCV-1 (a gift from the NIAID Repository of AIDS Reagents) that expresses HIV-1 TAT under the control of the SV40 early promoter and enhancer. As a negative control we used pSV0CAT, containing a promoterless *cat* gene, and as a positive control we used pSV2CAT, carrying the *cat* gene under the control of the SV40 early promoter and enhancer sequences. To control for variations due to differences in transfection efficiencies, all cultures were co-transfected with plasmid pCMVβgal, which expresses the bacterial β-galactosidase gene under the control of the cytomegalovirus immediate-early enhancer and promoter. Expression of β-galactosidase under regulation by this enhancer is independent of treatment. Twenty-four hours after transfection the cells were placed under low serum (0.1%) conditions and sixteen hours later cells were treated with B[a]P, B[a]P derivatives, or with an equivalent amount of DMSO vehicle. Cell extracts were prepared and expression of CAT and β-galactosidase activities was determined 96 h after transfection. Data were normalized for transfection efficiency by determination of the per cent chloramphenicol (CAM) converted to acetylated forms per unit of β-galactosidase.

Preparation of nuclear extracts and probes for electrophoretic mobility shift analyses (EMSA) was as described previously (Puga et al. 1992).

RESULTS

B[a]P and CYP1A1-dependent B[a]P metabolites activate expression from the HIV-1 LTR

To determine whether genes regulated by the HIV-1 LTR could be induced by B[a]P we treated Hepa-1 cells transfected with pHIVLTRCAT and pCV-1 with increasing doses of B[a]P and measured transient expression of CAT activity. Induction over control levels was clearly evident at a dose of 1 μM, maximal at 5 - 10 μM, and started to decline by 20 μM (Fig.1). The maximal levels observed correspond to a 14- to 15-fold increase over the basal level of activity observed in cells treated with an equivalent amount of DMSO vehicle. This basal level corresponds to approximately 5% CAM conversion per unit of β-galactosidase.

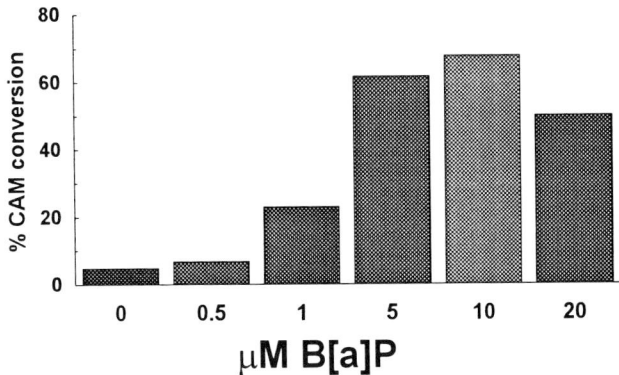

Fig. 1 Dose response curve of the activation of pHIVLTRCAT by B[a]P. Hepa-1 cells were transfected with 10 μg each of pHIVLTRCAT and pCV-1 and 3 μg of pCMVβgal. Transfected cells were treated with the indicated concentrations of B[a]P 48 h prior to extraction. The data for % CAM conversion have been normalized to β-galactosidase activity in the same extracts.

B[a]P is a procarcinogen that is converted to many reactive intermediates by the sequential action of at least two enzymes: cytochrome P450 CYP1A1 and epoxide hydrolase. To establish whether activation of CAT expression under our experimental conditions was only limited to the parent compound, we tested B[a]P and three of its metabolites, including the ultimate carcinogen, B[a]P-7,8-diol-9,10-epoxide (BPDE), for their capacity to induce CAT expression. All compounds tested were able to induce CAT, although to different extents, possibly a reflection of their different potencies. Treatment with the arene oxide B[a]P-4,5-dihydroepoxide (BPDHE), gave the highest levels of CAT induction; B[a]P was second highest. BPDE and B[a]P-6,12-dione treatments showed lower levels of induction, although significantly higher than those found in cells treated with DMSO vehicle (Fig. 2).

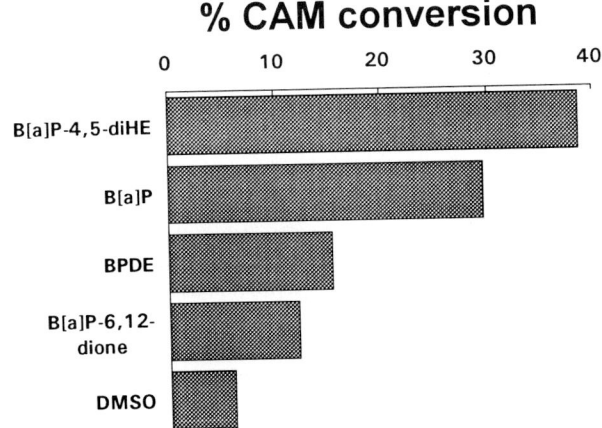

Fig. 2. Activation of CAT activity by B[a]P and by B[a]P metabolites. Cells were transfected as indicated in Fig. 1 and 24 h after transfection they were treated with the compounds shown at a concentration of 20 μM. Control cells were treated with an equivalent amount of vehicle (DMSO).

| [Q] µM | V79 | Xem-2 CYP 1A1 | Xemd-MZ CYP 1A2 |
|---|---|---|---|
| 0 | 0.015 ± 0.007 | 0.03 ± 0 | 0.04 ± 0.007 |
| 8.9 | 0.03 ± 0.028 | 0.02 ± 0 | 0.04 ± 0.028 |
| 32.5 | 0.07 ± 0.021 | 0.06 ± 0.014 | 0.12 ± 0.035 |
| 44.3 | 0.265 ± 0.007 | 0.12 ± 0.007 | 0.30 ± 0.049 |
| 65 | 0.21 ± 0.042 | 0.20 ± 0.07 | |

**Table 3**. CAs obtained with Quercetin, in the cell lines V79, Xem-2 (expressing CYP 1A1) and Xemd-MZ (expressing CYP 1A2). Cells were incubated for 3h in the presence of quercetin, then washed and grown in fresh medium for a further 16h.

## Acknowledgements

A.S.Rodrigues has a Doctoral fellowship by JNICT. This work is supported by CEC (STEP Programme) and the CIENCIA Programme.

## References

Crespi,C.L., F.J.Gonzalez,D.T.Steinel, T.R.Turner,H.V.Gelboin, B.W.Penman and R.Langenbach (1991) A metabolically competent human cell line expressing five cDNA encoding procarcinogen activating enzymes: application to mutagenicity testing, Chem.Res.Toxicol., 4,566-572.

Darroudi,F. and A.T.Natarajan (1993) Metabolic activation of chemicals to mutagenic carcinogens by human hepatoma microsomal extracts in Chinese hamster ovary cell (*in vitro*), Mutagenesis,8,11-15.

Davies,R.L., Crespi,C.L., Rudo,K., Turner,T.R., and Langenbach,R. (1989) Development of a human cell line by selection and drug-metabolizing gene transfection with increased capacity to activate promutagens. Carcinogenesis, 10, 885-891.

Doehmer,J., S.Dogra., T.Friedberg, S.Monier, M.Adesnik, H.R.Glatt, and F.Oesch (1988) Stable expression of rat cytochrome P-450IIB1 cDNA in Chinese hamster cells (V79) and metabolic activation of aflatoxin B1, Proc.Natl.Acad.Sci. (U.S.A.)., 85,5769-5773.

Doehmer,J., Wolfel,C., Dogra,S., Doehmer,C., Seidel,A., Platt,K.L., Oesch,F. and Glatt, H.R. (1992), Applications of stable V79-derived cell lines expressing rat cytochromes P4501A1,1A2 and 2B1. Xenobiotica, 22,1093-1099.

Dogra,S., Doehmer,J., Glatt,H.R., Molders,H., Siegert,P., Friedberg,T., Seidel,A. and Oesch,F. (1990) Stable expression of rat cytochrome P-450 1A1 cDNA in V79 Chinese Hamster Cells and their use in Mutagenicity Testing, Mol.Pharmacol., 37,608-613.

Ellard,S., Y.Mohammed, S.Dogra, C.Wolfel, J.Doehmer and J.M.Parry (1991) The use of genetically engineered V79 Chinese hamster cultures expressing rat liver CYPIA1, IA2 and 2B1 cDNAs in micronucleus assays, Mutagenesis,6,461-470.

Langenbach,R., P.Blaise Smith and C.Crespi (1992) Recombinant DNA approaches for the development of metabolic systems used in in vitro toxicology, Mutation Res., 277,251-275.

Rueff,J., A.Laires, J.Gaspar, H.Borba and A.Rodrigues (1991) Oxygen species and the genotoxicity of quercetin, Mutation Res., 265,75-81.

Turchi,G., A. Nardone and F.Palitti, (1992) Application of an epithelial liver cell line, metabolically competent, for mutation studies of promutagens, Mutation Res.,271,79-88.

Ueno,I.,M.Kohno,K.haraikawa and I.Hirono (1984) Interaction between quercetin and superoxide radicals. Reduction of the quercetin mutagenicity, J.Pharm.Dyn., 7,798-803.

Wolfel,C., Platt,K-L., Dogra,S., Glatt,H.R., Wachter,F. and Doehmer,J. (1991), Stable expression of Rat Cytochrome P4501A2 and Hydroxylation of 17β-Estradiol and 2-Aminofluorene in V79 Chinese Hamster Cells. Mol. Carcinogenesis, 4,489-498.

# P450 and azole tolerance in fungi

Steven L. Kelly[1], David C. Lamb[1], Brian C. Baldwin[2], Diane E. Kelly[1]

[1]Krebs Institute for Biomolecular Research, Department of Molecular Biology and Biotechnology, Sheffield University, Sheffield UK. [2]Zeneca Agrochemicals, Jealotts Hill Research Station, Bracknell, Berks, UK

## Introduction

Azole antifungal compounds inhibit the cytochrome P450 responsible for sterol 14α-demethylation, a step in ergosterol biosynthesis, and *Saccharomyces cerevisiae* cytochrome P450 mutants with defective sterol 14α-demethylase have been observed to be resistant to azole antifungals (Kenna et al.,1989). However, these mutants are also defective in sterol $\Delta^{5,6}$ desaturase (*erg3*) and the latter mutation is required for the viability of strains defective in P450 (Taylor et al.,1983). A biochemical block to growth and sterol biosynthesis produced by azole antifungal inhibition of sterol 14α-demethylation may also be overcome by an *erg3* mutation (Watson et al.,1989). The basis of the resistance is related to the formation of 14α-methylergosta-8,24(28)-dien-3B,6α-diol (14-methyl-3,6-diol) after treatment of wild-type yeast which in *erg3* mutants is inhibited, instead causing increased levels of its precursor 14α-methylfecosterol (Fig. 1). The latter sterol is capable of supporting growth in mutants derived by classical genetic means, but clear interpretation of such data is difficult as there is a low level of leakiness in the mutants.

Gene disruption of *CYP51A1* encoding *S. cerevisiae* sterol 14α-demethylase is lethal, except when cells are cultured anaerobically with an ergosterol and fatty acid supplement. Such mutants can give rise to slow growing colonies capable of aerobic growth by suppressor mutation (Kalb et al.,1987). The sterol profiles of such strains were addressed here, together with the effect on azole tolerance when the P450 gene disruption was complemented by a plasmid expressing *Candida albicans CYP51A1*. These studies indicate sterol $\Delta^{5,6}$ desaturase is solely responsible for the resistance phenotype in P450 mutant strains.

## Results and discussion

### Sterols associated with P450 gene disruption

The gene disruption of a diploid strain of *S.cerevisiae*, using *URA3* for the selection of transformants, produced ergosterol containing strains. Following sporulation of the diploid transformants the tetrads exhibited a 2:2 segregation of ergosterol containing, aerobically growing cultures (eg strain *DK1*) and ergosterol requiring, obligate anaerobe sister cultures e.g. strain *DK3*. With continued

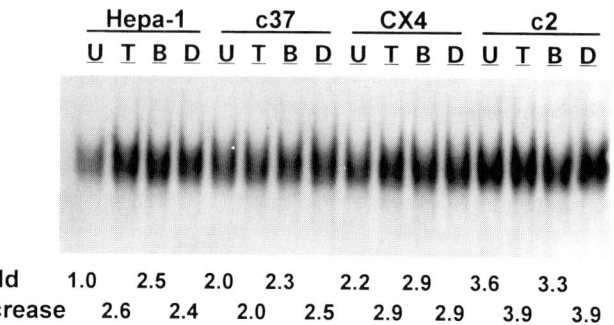

Fig. 5. Activation of AP-1 DNA binding activity in four cell lines by B[a]P and BPDE. EMSA analyses and quantitation were done as described previously (Puga et al. 1992). Treatments were for 4 hr with, U: DMSO vehicle; T: 10 nM TCDD; B: 20 μM B[a]P; D: 20 μM BPDE. *Fold increase* refers to increases in protein-bound probe relative to the value in untreated Hepa-1 cells, set arbitrarily as 1.

## Induction of HIV LTR-directed expression by B[a]P is inhibited by NAC and is independent of arachidonic acid metabolism

Recently, an oxidative stress pathway of AP-1 induction has been uncovered (Hunter and Karin, 1992; Devary et al. 1992), and the role of phospholipase $A_2$ and of arachidonic acid metabolites in the induction of Fos and Jun have been documented (Rao et al. 1993; Fafeur et al. 1991; Haliday et al. 1991). Several cytochromes P450 possess arachidonic acid epoxygenase activity, including the PAH-inducible CYP1A1, CYP1A2, and β-$NF_{AA}$ enzymes, the members of the phenobarbital-inducible CYP2B and CYP2H families, and some members of the CYP2C family (Capdevila et al. 1992; Laethem and Koop, 1992; Kanetoshi et al. 1992). As a consequence, oxidative stress mediated by PAH-inducible cytochrome P450 arachidonic acid epoxygenases may play an important role in the induction of Fos and Jun, in the observed activation of AP-1, and ultimately, in the stimulation of LTR-directed CAT activity by B[a]P. To examine the role of arachidonic acid epoxygenation on CAT activation, we determined the effect of known inhibitors of lipooxygenase and epoxygenase activities on the increase of CAT activity. We used indomethacin as a preferential inhibitor of lipooxygenases, and nordihydroguaiaretic acid (NDGA) as a preferential inhibitor of epoxygenases (Capdevila et al. 1988). In addition, to determine whether thiol-sensitive reactive oxygen species could be involved in the activation of CAT by B[a]P, we used N-acetyl-L-cysteine (NAC) treatment. NAC has been used in other systems to counteract the effect of reactive oxygen species on NFκB and AP-1 activation (Devary et al. 1991; Staal et al. 1990; Roederer et al. 1990; Devary et al. 1992). The results indicate that neither arachidonate cascade inhibitor had an effect on B[a]P activation of LTR-directed CAT expression, whereas NAC partially inhibited (Table 1).

Table 1. Effect of NAC and of inhibitors of arachidonic acid metabolism on the stimulation of CAT activity by B[a]P. Indomethacin (IM) and NDGA were used at 20 μM; NAC at 5 mM. B[a]P was added 24 h after transfection and the inhibitors were added 10 min later.

| Treatment | % CAM Conversion |
| --- | --- |
| None | 5.3 |
| B[a]P | 45.3 |
| B[a]P + IM | 42.1 |
| B[a]P + NDGA | 37.8 |
| B[a]P + NAC | 27.3 |

## DISCUSSION AND CONCLUSIONS

We have shown that B[a]P and several of its metabolites can activate the expression of HIV-1 LTR-directed genes in mouse hepatoma cells. This effect is dependent on the presence of a functional cytochrome P450 CYP1A1 enzyme, yet it is insensitive to the effect of compounds that inhibit the epoxygenation of arachidonic acid, suggesting that oxidative stress from arachidonate metabolites such as the epoxyeicosatrienoic acids and hydroxyeicosatetraenoic acids may not involved in the activation. On the other hand, treatment with NAC reduces the magnitude of the B[a]P-dependent response, indicating that intracellular thiol depletion and thiol-sensitive reactive oxygen species may be an important factor in activation.

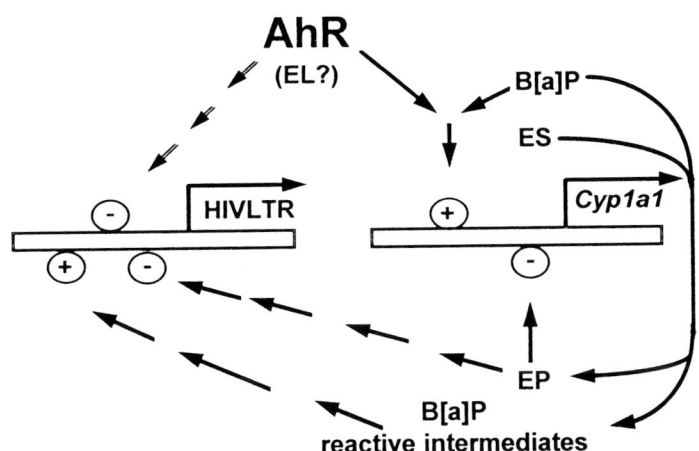

Fig. 6. Proposed model of regulation of the HIV-1 LTR-dependent expression by the Ah receptor and by CYP1A1. *AhR* is the Ah receptor; *EL* is its unknown endogenous ligand; *ES* and *EP* are an endogenous CYP1A1 substrate and its product, respectively. The *positive* and *negative* signs indicate the type of regulation exerted by each compound. Each set of *arrows* with gaps in between indicate that several intermediates steps are likely to intervene.

Three different regulatory activities appear to be exerted over the HIV-1 LTR by PAHs and the Ah receptor. First, based on our results in *c2* cells, we conclude that the Ah receptor, possibly by interacting with an endogenous ligand, negatively regulates expression from the LTR. Second, based on our results in *c37* cells, we conclude that CYP1A1-dependent metabolism of an endogenous substrate also down-regulates expression from the LTR. Thirdly, based on the results with B[a]P and its metabolites, we conclude that reactive intermediates of B[a]P can override these two negative regulatory activities and activate expression. This regulatory scheme uses the same components, although with different outcomes, as the model recently reviewed for regulation of *Cyp1a1* and of other genes in the Ah gene battery (Nebert et al. 1993). Both schemes are presented in Fig. 6.

It has been reported that progression of AIDS is accelerated by cigarette smoking (Burns et al. 1991); (Boulos et al. 1990). This epidemiological observation may have a number of molecular explanations, possibly involving the effect of compounds in tobacco smoke on lymphocyte function. It is attractive to speculate that one possible cause for this acceleration is the activation of the LTR by metabolites of B[a]P, one of the major carcinogens found in cigarette smoke, with the concomitant induction of viral RNA synthesis.

C4-demethylation reactions, which involve a 3-keto intermediate. A similar situation with obtusifolione levels reaching a maximum under fluconazole treatment when cell growth arrest occurs is also observed in *Cryptococcus neoformans*. The mechanisms whereby cells overcome ketosteroid formation is clearly important as different routes to resistance are likely to occur than when 14-methyl-3,6-diol is present at high levels.

Examination of a series of *C. neoformans* isolates from AIDS patients which had failed fluconazole therapy resulted in the identification of some isolates where therapeutic faliure was not associated with *in vitro* resistance. A series of four others exhibited ten-fold resistance in minimum inhibitory concentration assays over a series of wild-type strains and slightly reduced azole content per cell. However, the azole content was still in excess of the P450 content and was unlikely to be the basis of resistance. *In vitro* ergosterol biosynthesis studies in the resistant strains produced $IC_{50}$ values ten-fold in excess of wild-type values. This indicates that for these resistant strains, which are subject to ketosteroid formation on azole treatment, that alterations in the target P450 have occurred. Further studies on the P450 proteins and genes are in progress.

Table 3. Minimum inhibitory concentration and $IC_{50}$ values for *in vitro* ergosterol biosynthesis for a wild-type and fluconazole resistant strains of *C. neoformans*.

|  | MIC(M) | $IC_{50}$(M) |
| --- | --- | --- |
| wild-type | $1 \times 10^{-6}$ | $5 \times 10^{-7}$ |
| R715 | $2 \times 10^{-5}$ | $5 \times 10^{-6}$ |
| R716 | $1 \times 10^{-5}$ | $2 \times 10^{-6}$ |
| R717 | $2 \times 10^{-5}$ | $5 \times 10^{-6}$ |
| R718 | $2 \times 10^{-5}$ | $4 \times 10^{-6}$ |

**Acknowledgements:** This rearch was supported in part by AFRC and The Wellcome Trust. DCL is a SERC CASE student with Zeneca Agrochemicals.

## References

Kalb V.F., Woods C.W., Turi T.G., Dey C.R., Sutter T.R. and Loper J.C. (1987) Primary structure of P450 lanosterol demethylase gene of *Saccharomyces cerevisiae*. DNA 6;528-537.
Kenna S., Bligh H.F.J., Watson P.F. and Kelly S.L. (1989) Genetic and physiological analysis of azole sensitivity in *Saccharomyces cerevisiae*. J. Med. Vet.Mycol. 27;397-406.
Quail M.A., Moore D., Goosey M., Arnoldi A. and Kelly S.L. (1993) Ketoconazole mediated growth inhibition in *Botrytis cinerea* and *Saccaromyces cerevisiae*. Phytochemistry 32;273-280.
Rodriguez R.J., Low C., Bottema C.D.K. and Parks L.W. (1985) Multiple functions for sterol in yeast. Biochim.Biophys.Acta. 837;336-343.
Taylor F.R., Rodrigues R.J. and Parks L.W. (1983) Requirement for a second sterol biosynthetic mutation for viability of a sterol C14 demethylation defect in *Saccharomyces cerevisiae*. J.Bacteriol 155;64-68.
Watson P.F., Rose M.E., Ellis S.W., England H.E. and Kelly S.L. (1989) Defective sterol C5-6 desaturation and azole resistance. A new hypothesis on the mode of action of azole antifungals. Biochem.Biophys.Res.Commun. 164(3); 1170-1175.

# Cytochrome P4502E1-dependent metabolism as a biomarker for genetic susceptibility to chemically-induced cancers: a pilot study

Sharan L. Campleman[1], Stanley J. Tamaki[1], Betty-Ann Hoener[2], Martyn T. Smith[1]

[1]School of Public Health, University of California, Berkeley, CA 94720, USA. [2]School of Pharmacy, University of California, San Francisco, CA 94143, USA

## Summary

Cytochrome P450 2E1 (CYP2E1) metabolizes a wide range of xenobiotics, including the drug chlorzoxazone. Chlorzoxazone phenotyping was used to assess CYP2E1 metabolic status in a group of 17 subjects. DNA samples from this small group were used to develop a polymerase chain reaction (PCR)-based method for rapid, genetic analysis of *CYP2E1*. First, PCR assays were developed for the two previously described polymorphisms, the *RsaI/PstI* regulatory 5'-promoter region and the *DraI* restriction site of intron 6 polymorphisms. Only a single rare *RsaI/PstI* allele (c2) was identified from a heterozygous (c1c2) individual for an allele frequency of 0.029. The rare *DraI* allele (C) was identified in 4 individuals (3 heterozygous, DC, 1 homozygous, CC) for an allele frequency of 0.147. Additionally, PCR conditions targeting all *CYP2E1* coding regions have been developed in order to identify other polymorphic regions possibly associated with metabolic variation. To date, a unique nested-asymmetric PCR-based sequence analysis of exons 3 through 9 has not identified any new polymorphisms.

## INTRODUCTION

Cytochrome P4502E1 metabolizes a wide range of xenobiotics, including the human leukemogen benzene (Koop, 1992). The specific role that CYP2E1 metabolism plays in individual susceptibility to benzene, and other chemically induced disease remains unclear. If variability in CYP2E1 metabolism increases or decreases an individual's risk of developing disease following chemical exposure, then CYP2E1 metabolic status may be used as a biological marker of genetic susceptibility. With the recent demonstration that CYP2E1 almost exclusively mediates the 6-hydroxylation of the muscle relaxant chlorzoxazone, a possible noninvasive probe of intrinsic CYP2E1 activity has been identified (Peter *et al.*, 1990). By utilizing such chlorzoxazone phenotyping to identify slow and rapid CYP2E1 metabolizers, and, by analyzing the *CYP2E1* gene for sequence variation, associations between specific *CYP2E1* polymorphisms and metabolic status may be developed. Such a genetic characterization of CYP2E1 metabolic status may allow the design of polymorphism specific assays suitable for understanding the genetic susceptibility affecting chemical carcinogenesis.

# Cytochrome P450 2B4 antigenic determinants revealed by pepscan

Ekatherine F. Kolesanova, Serge A. Kozin, Alexander I. Archakov

*Institute of the Biomedical Chemistry, Russian Academy of Medical Sciences, 10, Pogodinskaya str., Moscow, 119832, Russia*

## Abstract

Overlapping hexapeptides covering the whole sequence of the cytochrome P450 2B4 have been synthesized on the solid supports and tested by ELISA using the polyclonal antiserum against cytochrome 2B4. 70 hexapeptide fragments have been found to interact specifically with the antiserum, i.e. to possess antigenic activity. The mapped linear epitopes occupy about 43% of the whole sequence of 2B4. They presumably form clusters in the regions of No.60-150, 210-300, 390-430 and 465-486 amino acid residues. The use of cytochrome P450 DataBase has allowed to classify the revealed antigenic determinants into absolutely specific for 2B4, specific only for 2B4 and 2B5, characteristic for 2B subfamily and widely distributed in family 2.

## INTRODUCTION

Antibodies can provide invaluable information on the immunochemical relatedness of different P450s, on screening cDNA expression libraries, on measuring levels of individual cytochromes and on their functioning via antibody inhibition experiments. But many of these tasks demand monospecific antibodies which can be obtained by rather complex procedures. To our mind, the best solution of this problem is to reveal at first the antigenic determinants with the help of the peptide scanning method (PEPSCAN) worked out by Geysen et al. (1987). This method gives an exhaustive information about linear antigenic determinants of proteins as it has been shown in the case of myohemerythrin (Geysen et al., 1987), alpha-subunit of acetylcholine receptor (Das & Lindstrom, 1991), hepatitis delta virus antigen (Wang et al., 1990) and nucleocapsid proteins of hepatitis C virus (Ching et al., 1992). Once the peptide epitopes are revealed, they can be used as conjugates to produce monospecific antisera or as ligands attached to any chromatography medium for the purification of these antibodies from the polyclonal antisera. In view of the existence of sites with both concervative and variable sequences in numeral members of the cytochrome P450 superfamily, both group-specific and individual epitopes may be found pointing out to the possible ways for the production of mono- and group-specific antibodies.

# REFERENCES

Hoener, B. (personal communication), September, 1993.

Hayashi, S., Watanabe, J., and Kawajiri, K. (1991): Genetic polymorphisms in the 5'- flanking region change transcriptional regulation of the human cytochrome P450IIE1 gene. *J. Biochem.* 110: 559 - 565.

Hirvonen, A., Husgafvel-Pursiainen, K., Anttila, S., Karjalainen, A., and Vainio, H. (1993): The human *CYP2E1* gene and lung cancer: *DraI* and *RsaI* restriction fragment length polymorphisms in a Finnish study population. *Carcinogenesis* 14: 85 - 88.

Koop, D.R. (1992) Oxidative and reductive metabolism by cytochrome P4502E1. *FASEB J.* 6: 24 - 30.

Peter, R., Bocker, R., Bequne, P.H., Iwasaki, M., Guengerich, F.P., and Yang, C.S. (1990): Hydroxylation of chlorzoxazone as a specific probe for human liver cytochrome P450II21. *Chem. Res. Toxicol.* 3:566 - 573.

Uematsu, F., Kikuchi, J., Motomiyz, M., Abe, T., Sagami, I., Ohmachi, T., Wakui, A., Kanamaru, R., and Watanabe, M. (1991): Association between restriction fragment length polymorphism of the human cytochrome P450IIE1 gene and susceptibility to lung cancer. *Jpn.J.Cancer Res.* 82: 254 - 256.

Umeno, M., McBride, O.W., Yang, C., Gelboin, H.V., and Gonzalez, F. (1988): Human ethanol-inducible P450IIE1, chormosome mapping, and cDNA-directed expression. *Biochemistry* 27: 9006 - 9013.

# Comparison of the principal drug-metabolizing enzyme systems in human tumors and corresponding peritumoral tissues: importance for sensitivity and/or resistance to anticancer drugs

Guy G. Chabot, Liliane Massaad, Caroline Toussaint, Nicolas Albin, Orlando Parise, Vincent Ribrag, François Janot, A. Gouyette

*Laboratory of Pharmacotoxicology and Pharmacogenetics (CNRS URA 147 and INSERM U.140), Institute Gustave-Roussy, 94805 Villejuif, France*

## Summary

To further understand the sensitivity or resistance of human tumors to chemotherapy, the principal drug-metabolizing enzyme systems were assessed in tumors and peritumoral specimens of colon, head/neck, breast, and non-small cell lung (NSCL) cancers. The systems studied included: cytochromes P-450 (1A1/1A2, 2B1/B2, 2C8-10, 2E1, 3A4), epoxide hydrolase, glutathione S-transferases (GST-$\alpha$, -$\mu$, -$\pi$) (immunoblotting); UDP-glucuronosyltransferase (UDPGT), ß-glucuronidase, sulfotransferase, sulfatase, GST activity (CDNB-GST), and glutathione (GSH) (determined by spectral assays). Concerning the cytochromes P-450, only colon (CYP3A4) and lung (CYP1A1/A2) specimens presented detectable levels, although at significantly lower levels in tumors compared to peritumoral tissues. The following changes were also observed in tumors compared to peritumoral tissues: increased GST activity in breast (increased GST-$\mu$ and -$\pi$); GSH content was increased in colon and head/neck tumors; a significant decrease in UDPGT and an increase in both ß-glucuronidase and sulfatase in breast tumors; epoxide hydrolase was decreased in head/neck and lung tumors. A comparison between the different human tumors revealed that each tumor type presented a qualitatively or quantitatively unique enzymatic profile. In conclusion, several differences were observed between human tumors and peritumoral tissues that could be the target of chemotherapeutic interventions aimed at improving anticancer drugs selectivity. Moreover, the distinct enzymatic profile of the human tumors studied suggests that each tumor type should be considered as a unique entity that would respond differently to anticancer drugs employing these drug-metabolizing enzyme pathways.

## INTRODUCTION

Although the principal drug-metabolizing enzyme systems play a major role in tissue responses to the toxic effects of many drugs or xenobiotics, we still know very little about these enzyme systems in human tumors. In an attempt to better understand the response of human tumors to the drugs used in cancer chemotherapy, we assessed the principal drug-metabolizing enzyme systems in human tumors and peritumoral specimens of colon, head/neck, breast, and non-small cell lung (NSCL) cancers. The phase I and phase II drug-metabolizing enzyme systems assayed are known to either activate and/or favor the excretion of drugs in general.

No.270-295 and the C-terminal domain, the only exception being the highly variable N-terminal domain which is devoid of antigenic determinants. Summarily, about 43 per cent of the 2B4 amino acid residues compose the linear epitopes of this protein. The sorting of the antigenic determinants of 2B4 revealed by PEPSCAN has been achieved by their search in the sequences of cytochromes P450 from the CPD, and its results are summarized in the fig.2. Several kinds of 2B4 epitopes have been estimated according to their presence in molecules of different cytochromes P450. 36 antigenic determinants can be regarded as individual ones found only in cytochromes P450 2B4 and 2B5, the products of 2 possible allele genes with 98% sequence identity. Antibodies which bind to these peptides are expected not to cross-react with any other members of cytochrome P450 superfamily. Moreover, the epitopes characteristic only for 2B4 have been found; it means that there exist antibodies which can distinguish the two cytochromes P450 coded by allele genes with high sequence identity. The individual epitopes are concentrated in the highly variable regions mentioned above. Besides that, group-specific antigenic determinants common to the whole 2B subfamily are estimated; they constitute about 30% of the whole number of epitopes. Three of such peptides - fragments No. 143-148 and 261-266 can be considered as CYP2 family-specific epitopes since they are present in cytochromes P450 belonging to different subfamilies of CYP2 family.

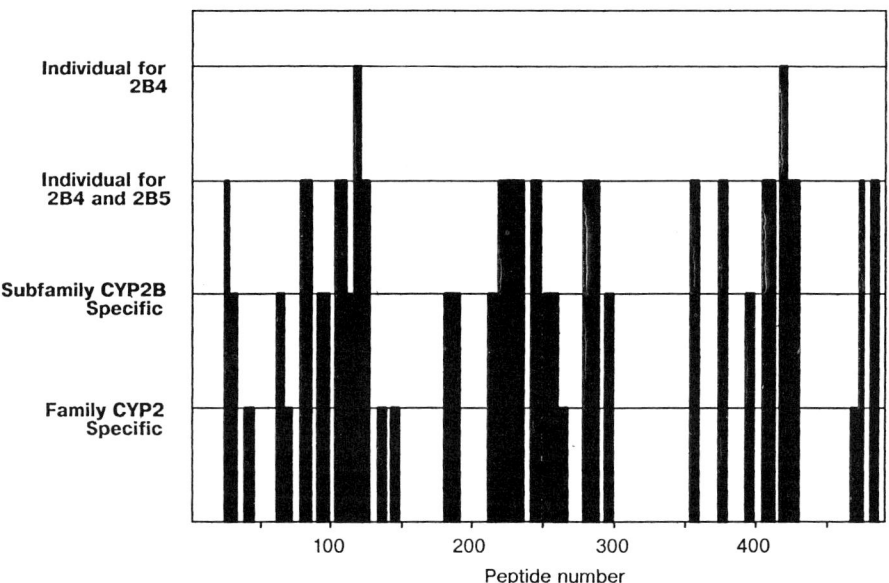

Fig.2. Distribution of sites with antigenic activity (sorted according to their group specificity) in the cytochrome P450 2B4 molecule.

According to our data the great antigenic homology is observed between the cytochromes P450 2B4 and 2B5, the moderate one - between 2B4 and other members of 2B subfamily; the weak homology is found between 2B4 and cytochromes 2G1 (7 common epitopes) and 2H1, 2H2 (5 common epitopes). Very weak antigenic homology exists between cytochromes 2B4 and other proteins from CYP2 family, and no homology is observed between 2B4 and cytochromes P450 belonging to the families other than CYP2. The information on the epitopes of cytochrome P450 2B4, namely on their structure, relative immunogenicity and presence in other cytochrome P450 molecules will help to successfully produce antipeptide antibodies specific to this protein and, vice versa, subfamily-specific ones. More profound studies of antigenic homology of different cytochromes P450 may underlie their further classification, in addition to the ones now existing (Nebert et al., 1991, Gotoh, 1992).

REFERENCES

Archakov, A.I. and Bachmanova, G.I. (1990) Cytochrome P450 and Active Oxygen, Suppl. Taylor & Francis, London-New York-Philadelphia.

Ching, W.M., Wychowski, C., Beach, M.J., et al. (1992): Proc.Natl.Acad.Sci. USA, 89, 3190-3194.

Das, M.K. and Lindstrom, J. (1991): Biochemistry 30, 2470-2477.

Geysen, M.H., Tainer, J.A., Rodda, S.J., Mason, H.A., Getzoff, E.D. and Lerner, R.A. (1987): Science, 235, 1184-1190.

Grebenshchikova, O.G., Beryozov, A.T., Kolesanova, E.F. and Archakov, A.I. (1992): In Cytochrome P450: Biochemistry and Biophysics (Archakov, A.I., Bachmanova, G.I., Eds.), pp.686-691, INCO-TNC, Moscow, Russia.

Gotoh. H. (1992): In Cytochrome P450: Biochemistry and Biophysics (Archakov, A.I., Bachmanova, G.I., Eds.), pp.686-691, INCO-TNC, Moscow, Russia.

Imai, Y., Komori, M. and Sato, R. (1988): Biochemistry, 27, 80-88.

Nebert, D.W., Nelson, D.R., Coon, M.J. et al. (1991): DNA Cell Biol., 10, 1-13.

Wang, J.-G., Jansen, R.W., Brown, E.A. and Lemon, S.M. (1990): J.Virol., 64, 1108-1116.

Among the qualitative differences between tumors were the detection of cytochromes P-450, only in two tumor types (colon and lung), but the isoenzyme detected in these tumors were different (CYP3A4: colon; CYP 1A1/1A2, lung). Qualitative differences also included the GST isoenzymes, since GST-$\alpha$ was not detected in breast tumors, and GST-$\mu$ was not detected in colon tumors.

Important quantitative differences were observed for most other enzyme systems, with differences ranging from 1-fold to as high as 80-fold for UDP-GT, for example. Therefore, each tumor type was observed to possess a distinct enzymatic profile.

**DISCUSSION**

In an attempt to better understand human tumor sensitivity or resistance to anticancer drugs, the main drug-metabolizing enzyme systems were assessed in both tumoral and non-tumoral adjacent peritumoral tissues of colon, head/neck, breast, and non-small cell lung cancers. The drug-metabolizing enzyme systems studied are known to markedly influence the sensitivity of organs or tissues to the cytotoxicity and/or carcinogenicity of drugs or other xenobiotics (Guengerich, 1988). These enzyme systems could also determine tumor response to anticancer agents in either detoxication or toxific effects of these drugs (Powis and Prough, 1987).

Cytochromes P-450, a multigene superfamily of microsomal hemoproteins, play a key role in the biotransformation of a wide variety of xenobiotics and are involved either in detoxication or toxic effects. Not only the P-450's are also involved in the biotransformation of certain anticancer agents, but they are sometimes required for the expression of anticancer activity, *e.g.* in the activation of cyclophosphamide. In the present studies, cytochromes P-450 were detected only in colon and lung tumors, although these enzyme levels were decreased in tumoral tissues compared to peritumoral specimens. The decreased content, or the non detectability of the principal cytochromes P-450 in breast and head/neck tissues, would indicate that tumors in general would be deficient in performing P-450-dependent oxidative reactions such as hydroxylation, epoxidation, or dealkylation, that may toxify or detoxify certain drugs. If drugs are not metabolized by other enzymes, *e.g.*, conjugating enzymes, the retention of drugs within tumors would be likely, thus favoring possible cytotoxic effects of drugs.

The glutathione system, another major enzymatic pathway involved in the detoxication of many exogenous compounds, was also studied in tumoral and peritumoral tissues. The GST isoenzymes levels in tissues are important because these enzymes play a central role in the detoxication of many toxic electrophilic compounds, including cytotoxic drugs and carcinogens. Given the great diversity of their functions, several authors have postulated that the variability in the expression of these enzymes could be a factor in the susceptibility of various tissues to toxins and carcinogens, and suggested that they may be used as potential prognostic markers in carcinogenesis (Wang and Tew, 1985; Mannervik, 1985). Increased GST activity and increased tumoral GSH content were both associated with resistance to cytotoxic agents (Mannervik, 1985). The cytosolic GST enzymes in human have been divided into three distinct groups which are commonly referred to as basic ($\alpha$), neutral ($\mu$) or acidic ($\pi$) transferases, according to their isoelectric points. In agreement with other reports (Di Ilio *et al.*, 1985; Singh *et al.*, 1990), breast tumors had higher GST activity and levels of GST-$\pi$ than their corresponding peritumoral tissues. Moreover, GST activity in tumoral tissues was also well correlated with GST-$\pi$ content. Recent studies have shown that the activity of the anionic GST isoenzyme (GST-$\pi$) is increased in both rat hyperplastic nodules and adriamycin-resistant MCF-7 cells (Cowan *et al.*, 1986; Batist *et al.*, 1986), and that the human form of anionic GST, *i.e.* GST-$\pi$, is increased in multidrug-resistant MCF-7 cells when compared with the parental cell line (Griffin, 1982; Hamilton *et al.*, 1985).

In our studies, GST-α was not detected in breast tissues, and GST-μ was not detected in colon tissues. These findings are in accordance with a Shea et al. (1990) who also showed the absence of GST-α in human breast cancer. These data suggest that GST-α may not play a predominant role in the detoxication of organic hydroperoxides in breast tissues. For GST-μ, which is known to be genetically polymorphic in humans, it was not detected in colon tissues, although the number of observations was small (N=4). This GST isoenzyme is not expressed in 40-50% of the human population (Di Ilio et al., 1985).

Glutathione was increased in both colon tumors and head/neck tumors compared to peritumoral tissues. This increased GSH in tumors could contribute to their resistance to anticancer drugs such as certain alkylating agents. Hamilton et al. (1985) showed an augmentation of adriamycin, melphalan and cisplatin cytotoxicity in drug-resistant ovarian cancer cell lines by buthionine sulfoximine-mediated glutathione depletion. Buthionine sulfoximine (BSO) is a specific inhibitor of the enzyme gamma-glutamyl cysteine synthetase, which is a key enzyme in the synthesis of GSH. These results provide possibility of modulating or circumventing resistance to clinically useful drugs, thus providing more effective treatment modalities using a nontoxic "modulator" in combination with a standard dose of antitumor agents so as to enhance the effect of the cytostatic drugs without producing increased damage to normal tissues.

Little information is presently available concerning role of the glucuronide and sulfate pathways in human tumors. Our data showed significant differences between tumoral and peritumoral tissues for these pathways in breast tumors only. For the glucuronide system, UDP-GT was decreased, whereas the corresponding hydrolytic enzyme ß-glucuronidase was increased in tumoral *versus* peritumoral breast tissues. We can hypothesize that this interplay of conjugating and hydrolytic actions could favor drug retention in tumoral tissues, since the low synthesis of the glucuronide conjugate associated to the high ß-glucuronidase activity could lead to drug retention. Glucuronidation is a pathway employed by many anticancer drugs such as doxorubicin analogues, mitoxantrone and 5-fluorouracil (Mross et al., 1988; Wolf et al., 1986; Sommadossi et al., 1985). Also noteworthy was the high sulfatase activity in breast tumors compared to peritumoral tissues: this hydrolytic activity could also favor drug retention in tumor by hydrolyzing the sulfate conjugates. Consequently, the glucuronide and sulfate pathways could favor a greater drug retention, and perhaps an increased sensitivity of breast tumoral tissues to drugs metabolized *via* these metabolic routes.

Epoxide hydrolase is an enzyme thought to play a protective role with respect to metabolically produced epoxides from xenobiotics involved in carcinogenesis and cytotoxicity (Oesch, 1973). In our studies, epoxide hydrolase was significantly decreased in head/neck and lung tumors which are tumors linked to alcohol and tobacco usage.

Finally, the comparison of four different solid tumors evaluated in our laboratory with regard to their main drug-metabolizing enzyme systems clearly demonstrates that each human tumor type possesses a unique profile.

In conclusion, several differences were observed between human tumors and peritumoral tissues for many drug-metabolizing enzyme systems. These enzyme systems could be the target of chemotherapeutic interventions aimed at improving anticancer drugs selectivity. Moreover, the distinct enzymatic profile of the human tumors studied suggests that each tumor type is a unique entity that would respond differently to anticancer drugs employing these metabolic pathways.

## Methods

### Quantitative covalent binding:
Human liver microsomes were incubated with the radiochemical ($^{14}$C tienilic acid, $^{14}$C chloroform and $^{3}$H paracetamol), and a NADPH-generating system in phosphate buffer. Aliquots of incubation mixtures were loaded on to glass-fiber filter disks. Filters were washed, dried and counted (Dansette et al. 1991). Results were obtained with the radiochemical used at increasing concentrations ($10^{-7}$ to $10^{-4}$ M), and with or without NADPH-generating system. Km and Vmax were graphically determined from double reciprocal plots. Quantitative covalent binding of tienilic acid (TA) on yeast microsomes was performed with TA 0,1 mM and yeast corresponding to 50 pmoles P450.

### Yeast and bacteria preparation:
Human P450 2C9 and 2C18 cDNA coding sequence (Brian et al. 1989, Romkes et al. 1991) were amplified by PCR and inserted in the yeast expression vector V8 (Pompon. 1988). This vector was transfected in yeast *S. cerevisiae*. Human P450 2C9 cDNA was also inserted into bacterial expression vector pGex and was produced in *E.Coli* JM101. Yeast microsomes and bacterial fusion protein were used to characterize the anti-LKM2 antibodies. These autoantibodies appeared in the sera of patients suffering from tienilic acid-induced hepatitis.

### Qualitative covalent binding:
<u>Preparation of samples</u>: Human liver microsomes (1 nmole P450) or yeast microsomes (50 pmoles of P450) were incubated with radiochemical, NADPH-generating system in phosphate buffer. Proteins of the incubation mixture were precipitated by addition of trichloroacetic acid and washed by various solvents. After centrifugation, the pellet was dissolved in 0,5 ml SDS 1%.

<u>Electrophoresis and Western blotting</u>: After migration on SDS-PAGE, proteins were electrotransferred to a nitrocellulose sheet. The sheet was cut perpendicular to the direction of the migration and counted. One part of the sheet was kept for immunoblots and probed with anti-human P450 1A1-2, 2C, 2E1, 3A4. The radioactive count of each band enabled us to obtain a covalent binding profile which was compared with the location of P450 by antibodies.

## Results

The kinetic constants measured on human liver microsomes corresponded to low apparent Km for tienilic acid and paracetamol (respectively 7 and 9 µM) and higher Km for chloroform (250 µM). The covalent binding evaluated at 10 µM (in the range of Km for tienilic acid and paracetamol) and 1mM (corresponding to saturation for the three compounds) showed higher values for tienilic acid than for the two others compounds.

The covalent binding of tienilic acid to yeast and human microsomes showed that P450 2C9 was the main P450 producing tienilic acid-reactive metabolites.

|  | Covalent binding (pmol/min/nmol P450) |
|---|---|
| Human liver | 98 |
| Control | 0 |
| P450 1A1 | 77 |
| P450 1A2 | 44 |
| P450 2C9 | 340 |
| P450 2D6 | 6 |
| P450 3A4 | 0 |
| P450 2C18 | 42 |

The incubation of the compounds with human liver microsomes gave us different covalent binding profiles: very specific for tienilic acid-metabolites, non-specific for paracetamol and chloroform metabolites:

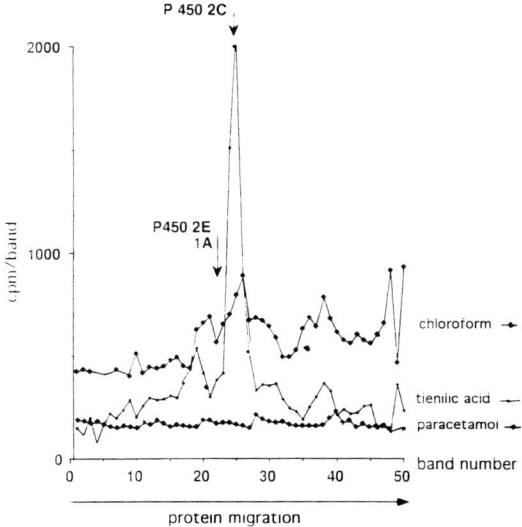

Tienilic acid metabolites were covalently bound to the P450 generating them. This experiment was confirmed by incubating tienilic acid with microsomes of yeast expressing isolated P450: only P450 2C9 gave covalent binding, and we could detect a radioactive peak corresponding to the metabolites bound to this P450.

All the sera from patients treated with tienilic acid and who suffered from autoimmune hepatitis recognized P450 2C9 expressed in the yeast, and this very specifically (they did not recognize others P450s expressed in the yeast or in the bacteria).

## Discusssion

In previous work (Beaune et al. 1988) covalent binding of TA was shown to be inhibited by anti-LKM2 and anti-human P450 2C, indicating that TA was metabolized into reactive metabolite(s) by a P450 2C subfamily member, according to steps 1 and 2 of the scheme 1. Our work attempted to precise the main target of reactive metabolites and autoantibodies.
Data on Km and Vmax showed that TA at a low drug concentration was able to generate many more bound metabolites than the other two hepatotoxic drugs tested: this let us to think that the amount of complex protein-reactive metabolite (neoantigen) formed may play a role in triggering of the immune response. Furthermore, covalent binding of TA metabolite(s) was concentrated to one microsomal protein (one major radioactive peak after electrophoresis), in contrast to covalent binding of paracetamol and chloroform which was spread over several microsomal proteins. As TA, dihydralazine-reactive metabolites exhibited specific covalent binding on the P450 generating them (Bourdi et al. personnal communication).
Specific recognition by autoantibodies of the P450 which metabolized the drug was found in the two cases of drug-induced autoimmune hepatitis: anti-LKM2 recognized P450 2C9 in case of TA-induced hepatitis, and anti-LM recognized P450 1A2 in case of dihydralazine-induced hepatitis. In both cases, specificity of the antibodies was hight

# Molecular engineering steroid hydroxylase activity of mammalian P450s

Masahiko Negishi[1], Masahiko Iwasaki[1], Tatsuya Sueyoshi[1], Thomas A. Darden[2], Lee G. Pedersen[2]

[1]Laboratory of Reproductive and Developmental Toxicology and [2]Molecular Toxicology, National Institute of Environmental Health Sciences, National Institutes of Health, Research Triangle Park, North Carolina 27709, USA

Recent studies of the structure-activity relationships of P450s, using site-directed mutagenesis and heterologous expression systems, have shown that the substrate-heme pocket appears to be geometrically flexible. The pocket structure can be modified by a minor amino acid substitution to properly accommodate various substrates, which leads to inherently versatile P450 activity. As a result, the P450 system can be engineered to increase catalytic specificity or even confer novel activity. Engineered P450s can be expressed in mammalian, yeast and bacterial cells for various purposes - examples include testing toxicity and genetic damage caused by environmental chemicals or newly developed drugs, and developing "bioreactors" to produce existing as well as new chemicals at industrial levels. The large number of P450 enzymes provides us with the opportunity to use P450s in many applied and commercial fields.

Tienilic acid (TA) and its isomer (TAI) (a minor impurity in some of the early production batches) are activated by cytochrome P450 into reactive metabolites which bind covalently with microsomal proteins. It is possible to decrease this covalent binding with diverse nucleophiles and in the case of TAI this decrease correspond to the trapping of adducts which can be isolated. Using mercaptoethanol as the nucleophile several successive adducts have been isolated and fully identified (Valadon P., *Thesis 1992*). The identification of the first adduct as a 2,5-dihydro-thiophene sulfoxide demonstrate that the reactive metabolite of TAI is the corresponding unstable thiophene sulfoxide A (Mansuy et al. 1991). More recently isolation of the major urinary metabolite of thiophene as a dihydrothiophene sulfoxide mercapturic acid (Dansette et al., 1992) shows that thiophene sulfoxides are probably more general reactive metabolites and thus tienilic acid could also be activated in a similar manner.

The covalent fixation of TA and TAI probably involves their thiophene ring, thus carrier proteins (BSA or ß-lactoglobulin) were attached thru a spacer arm to the thiophene ring of TA and were used to raise antibodies in rabbits, and to establish an ELISA assay. These antibodies (anti-TA) recognized both the covalent binding of TA or TAI to microsomes, and competition studies in ELISA showed that the ketodichloro-phenoxyacetic moiety was necessary for recognition (Valadon P., *Thesis 1992*). Moreover these antibodies could be used in immunotransfer.


Kinetics of the 5-hydroxylation of Tienilic acid in human microsomes were not linear and carefull examination showed progressive inactivation of the enzymatic system. Similar studies using Cyp 2C9 or Cyp 2C10 expressed in yeast demonstrated a catalysis dependent loss of activity with TA, whereas this cytochromes P450 were stable to metabolic products of TAI (Lopez-Garcia et al., 1993b). Trapping of TA or TAI metabolites by glutathione (GSH) using Cyp 2C9 shows that it is posible to decrease TAI binding to very low level (< 0.2 nmol/nmol P450), but TA binding only to a third of that in absence of GSH or about 1 nmol/nmol P450 as shown on Figure 1. When the yeast microsomes were analyzed by immuno-revelation with the anti-TA, only one band corresponding to Cyp 2C9 was enlighted by antiLKM2; in absence of GSH at least two other bands were evident. Thus part of the reactive metabolite can escape from the active site and react at other protein sites or with excess nucleophile. To explain the inactivation of the enzyme, we propose that the active metabolite, a thiophene sulfoxide, reacts with a nucleophile in the active site leading to an adduct which after aromatization is irreversibly bound. If the sulfoxide reacts with water this leads after aromatization to 5OHTA. From the kinetic data there is one inactivation event for 12 5OHTA formed.

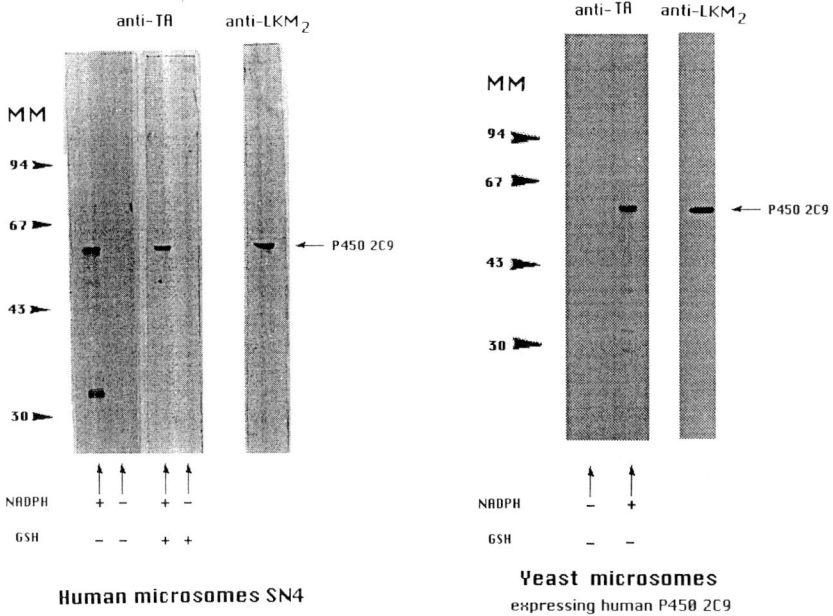

Human microsomes SN4

Yeast microsomes expressing human P450 2C9

These results suggest that the primary event in TA activation is its oxidation by cytochrome P450 into a very reactive thiophene sulfoxide which reacts at the active site of Cyp 2C9/10, causing its inactivation and probably triggering its disposal by the hepatocytes. Possible exocytosis of a whole or degraded protein or immunogenic peptides presented to the immune system could be at the origin of the autoimmune response. Sera from patients treated with TA for long periods, with or without hepatitis all contain antibodies recognizing ß-lactoglobulin-TA in an ELISA test. This is indicative of the presence of an antigen haptenized with TA, triggering also an immune response against the drug. However the cytolytic hepatitis and autoimmune reaction is very rare in patients treated with TA, thus another individual component is certainly needed for the immune reaction, depending either of the individual immune system or of a molecular mimicry between Cyp 2C9/10 peptides and another viral or bacterial antigen.

communication). These 15α–hydroxylase activities are a catalytic characteristics of mouse P450 2a-4 (Harada and Negishi, 1988). Despite of the fact that the rat and mouse P450s are only 54 per cent identity in their amino-acid sequences, the steroid-substrate specificity of rat P450 can be altered to that of mouse P450 by the only two amino-acid mutations at the corresponding positions. The results indicate strongly that the geometry of the substrate-heme pocket can be determined by the types of few residues at the critical positions.

The hypothesis that the pocket is geometerically flexible and the geometry can be determined by a minor mutation is further supported by the following findings. Rabbit P450 2C1 confers progesterone 21-hydroxylase activity depending on the identity of residue at position 113 (Kronbach and Johnson, 1991). A single amino acid mutation at position 365 confers on P450 2C3 high progesterone 6β-hydroxylase activity (Hsu et al., 1993). Uno and Imai (1992) demonstrated that P450 2C2 (laurate ω−1 hydroxylase) acquires novel testosterone 16β-hydroxylase activity by replacing the C-terminal 28 amino acids with the corresponding sequence from P450 2C14 (testosterone 16α-hydroxylase). It appears, therefore, that the corresponding residues for the rat and rabbit system to positions 117, 209, 365 and 481 in P450 2a-4 and -5 are common determinants for definition of the steroid hydroxylase specificity of P450s within the subfamily 2.

Current studies show that the residues at these critical positions alter not only the steroid hydroxylase activity but also activity for metabolizing xenobioics. For example, human P450 2C9 has phenytoin hydroxylase activities altered by mutation of Leu at position 359 to Ile (Veronese et al., 1993), whereas the corresponding residue at position 380 in rat P450 2D1 determines its bufuralol hydroxylase activity (Matsunaga, et al., 1990). Moreover, conservative substitutions such as Val to Ala , Leu to Ile, Ser to Thr, are sufficient to alter the specificity of P450 activity indicating that side-chain size plays a key role in defining the specificity of P450 activity. These important amino acid residues, therefore, must be located in the heme pocket for contact with substrate. Many researchers have previously examined this possibility using the 3D-structure of bacterial P450 101A as the model and have thereby provided a structurual basis for these observed specificities (Poulos, 1991).

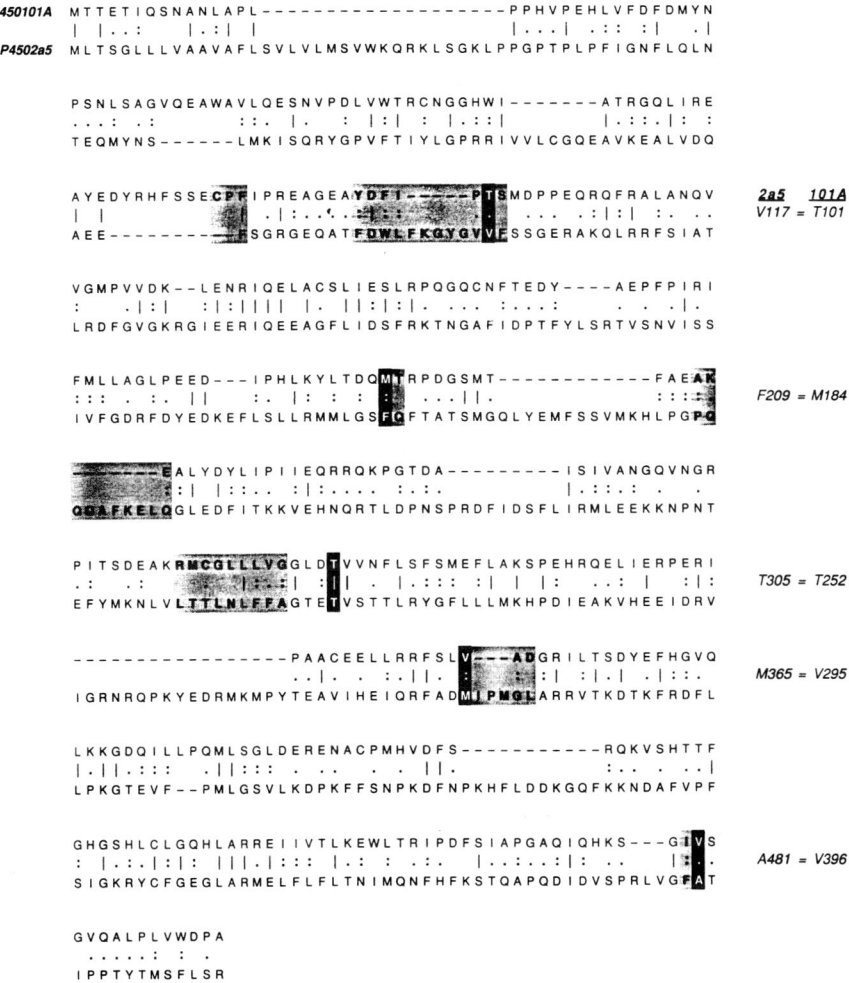

Figure 2. Amino-acid sequence alignment between mouse P450 2a-5 and bacterial P450 101A. The substrate-binding sites of the bacterial P450 and their corresponding regions of the mouse P450 are boxed. The four critical residues and the conserved threonine are darkened and also shown in the right-side of the alignment.

Gotoh (1992) aligned the substrate-binding residues in bacterial P450 101A to mammalian P450s within the CYP 2 subfamily, and proposed the six putative substrate recognition sites (SRS) in the mammalian P450s. Importantly, the SRSs include all of the critical residues identified by previous site-directed mutagenesis studies. Recently, Korzekwa and Jones (1993) have reported a similar alignment which agrees with the Gotoh's SRSs. We also reported the bacterial amino-acid residues which correspond to each of the key residues of mouse P450s (Iwasaki et al., 1993). We have used several different programs with several constraints to align the amino acid sequence of mouse P450 2a-5 with that of bacterial P450 101A. We show one of our alignments in Figure 2. The alignment was accomplished using the global methodology of Needleman and Wunsch (1970) by adopting the following constraints: a gap initiation weight of 3.0, a gap extension length weight of 0.10, and a constraint against gaps in the long helix I. Moreover, we employed a requirement that the alpha carbons of residues 117, 209, 365 and 481 were mapped to residues in the bacterial P450 that are less than a certain distance from the 6th axial position of the heme: 18Å, 15Å, 17Å and 15Å for residues 117, 209, 365 and 481, respectively. Then the mouse residues 117, 209, 365 and 481 corresponded to bacterial Thr-101, Met-184, Val-295 and Val-396, respectively. These bacterial residues are located in the substrate-binding sites of P450 101A. The homology-alignment studies, therefore, have provided a topological basis for the specificity of mouse P450 activity.

To obtain the structural basis for the substrate specificity that is determined by the types of the critical residues, we modeled a steroid-binding orientation in the substrate-heme pocket of P450 101A (Iwasaki et al., 1993). For this purpose, the corticosterone molecule was docked approximately perpendicular to the heme plane so as to direct the 11β-hydroxyl toward Met-184 (corresponding to residue-209 in the mouse P450). Energy-minimization verified that the pocket is large enough to accommodate the corticosterone molecule in this orientation. In this binding model, the oxygen of the 11β-OH is 7Å from the β-carbon of Met-184. Significantly, the C15 position (hydroxylation site) of the steroid molecule is 2.9Å from the activating oxygen in the 6th axial position of the heme. Moreover, Thr-101 (corresponding to residue-117 in the mouse P450) and Val-396 (corresponding to residue-481 in the mouse P450) are close and directed toward the C3 and C11 positions, respectively. This steroid-binding model serves to explain the roles of the residues in defining the steroid-substrate specificity. Moreover, the model provides the structural basis for the steroid-substrate specificity of mammalian P450 and should be a useful tool for redesigning P450 activity.

We are now confident that P450s can be engineered to novel enzymes capable of producing new products (Estabrook, 1989). For instance, in searching for new glucocorticoids, mutant P450 2a-4/5 enzymes could reasonably be designed and a bioreactor subsequently constructed to provide hydroxylated corticosterones and hydrocortisones for basic and clinical studies.

# REFERENCES

Estabrook, R. W. (1989): Future developments with the cytochrome P450: Promises for the apllied chemical and medical sciences. Biochem. Soc. Trans. 18, 34-36.

Gotoh, O. (1992): Substrate recognition sites in cytochrome P450 family 2 (CYP2) proteins inferred from comparative analysis of amino acid and coding nucleotide sequences. J. Biol. Chem. 267, 83-90.

Harada, N. and Negishi, M. (1988): Substrate specificities of cytochrome P450 s, C-P450 16α and P450 15α, and contribution to steroid hydroxylase activities in mouse liver microsomes. Biochem. Pharmacol. 37, 4778-4780.

Harpert, J. R., and He, Y-A. (1993): Engineering of cytochrome P450 2B1 specificity: Conversion of an androgen 16β-hydroxylase to a 15α–hydroxylase. J. Biol. Chem. 268, 4453-4457.

Hsu, M-H., Griffin, K. J., Wang, Y., Kemper, B., and Johnson, E. F. (1993): A single amino acid substitution confers progesterone 6β-hydroxylase activity to rabbit cytochrome P450 2C3. J. Biol. Chem. 268, 6939-6944.

Iwasaki, M., Darden, T., Pedersen, L., Davis, D. G., Juvonen, R.O.,

Sueyoshi, T., and Negishi, M. (1993): Engineering mouse P450coh to Novel corticosterone 15a-hydroxylase and modeling steroid-binding orientation in the substrate-heme pocket. J. Biol. Chem. 268, 759-762.

Iwasaki, M., Lindberg, RLP., Juvonen, R. O., and Negishi, M. (1993): Site-directed mutagenesis of mouse steroid 7α-hydroxylase (cytochrome P4507α); role of residue-209 in determining steroid-cytochrome P-450 interaction. Biochem. J. 291, 569-573.

Lindberg RLP and Negishi, M. (1989): Alteration of mouse cytochrome P450coh substrate specificity by mutation of a single amino-acid residue. Nature 339, 632-634.

Matunaga, E., Zeugin, T., Zanger, U. M., Aoyama, T., Meyer, U. A., and Gonzalez, F. J. (1990): Sequence requirements for cytochrome P-450IID1 catalytic activity: A single amino acid change (Ile380Phe) specifically decreases Vmax of the enzyme for bufuralol but not debrisoquine hydroxylation. J. Biol. Chem. 265, 17197-17201.

Needleman, S. B. and Wunsch, C. D. (1970): A general method applicable to search for similarities in the amino acid sequence of the protein. J. Mol. Biol. 48, 443-453.

Negishi, M., Iwasaki, M., Juvonen, R. O., and Aida, K. (1992): Alteration of the substrate specificity of mouse P450 2A P450s by the identity of residue-209: Steoid-binding site and orientation. J. Steroid Biochem. Molec. Biol. 43, 1031-1036.

Poulos, T. L. (1991): Modeling of mammalian P450s on the basis of P450cam X-ray structure. Methods Enzymol. 206, 11-30.

Uno, T. and Imai, Y. (1992): Further studies on chimeric P450 2C2/2C14 having testosterone 16α-hydroxylase activity which is absent in the parental P450s. J. Biochem. 112, 155-162.

Veronese, M. E., Doecke, C. J., Mackenzie, P. I., McManus, M. E., Miners, J. O., Rees, D. P., Gasser, R., Meyer, U. A., and Birkit, D. J. (1993): Site-directed mutation studies of human cytochrome P-450 isoenzymes in the CYP2C subfamily. Biochem, J. 289, 533-538.

# Co-expression of mammalian, plant or yeast P450s and P450 reductases in *Saccharomyces cerevisiae* as cloning and bioconversion tools

Denis Pompon[1], Gilles Truan[1], Aouatef Bellamine[1], Michaël Kazmaier[2], Philippe Urban[1]

[1] Centre de génétique moléculaire, UPR 2420 du CNRS associée à l'université Pierre-et-Marie-Curie, 91198 Gif-sur-Yvette Cedex. [2] Centre de Recherche ORSAN, 16, avenue de la Baltique, 91953 Courtaboeuf Cedex, France

*Saccharomyces cerevisiæ* is one of the most popular host for heterologous expression of P450s (Urban et al., 1990; Guengerich et al., 1991). However, until recently, the expression levels of mammalian P450s produced in yeast were rather low as compared to results obtained in *Escherichia coli* for modified P450s carrying the signal sequence of bovine P450 17α-hydroxylase (Barnes et al., 1991). Nevertheless the presence in yeast, in contrast to E. coli, of a significant amount of endogenous P450 reductase led to a system self-sufficient for activity without need for the in vitro addition of any extra component. Yeast cells exhibit very little or no endogenous xenobiotic metabolizing activities and are almost devoid of spectrally detectable endogenous P450, provided that suitable strain and culture conditions be used. Therefore whole cells or microsomes prepared from P450-expressing yeast became a material of choice for activity screening purposes or structure-function studies. However, comparison of the P450 turnover numbers in yeast microsomal fraction with values observed in a reconstituted system suggested that a rather poor coupling efficiency of the yeast P450 reductase with eukaryotic P450s strongly limits the activity. Finally, the limited amount of P450 produced from small-scale cultures was an additional difficulty, particularly for spectroscopic studies or crytallogenesis attempts. This paper reviews the new strategies we have developed to maximise both the level of expression and the turnover numbers leading to the concept of "humanized" and "vegetalized" yeast strains.

MAXIMIZING P450 EXPRESSION LEVELS

The simplest and most widely used yeast expression system is constituted by a an expression cassette containing a cDNA-derived P450 open reading frame placed downstream a constitutive yeast promoter. The cassette is held on a multicopy plasmid carrying a suitable auxotrophy complementation marker. Although simple and reliable, such a system involves culture in a minimal media that does not generally allow to reach high culture density in batch condition. Moreover, constitutive overexpression of heterologous protein causes a negative selection leading to a decrease in the plasmid copy number. Dramatic improvement of the P450 yield per liter of culture can be achieved using stabilised plasmid bearing inducible expression cassettes, in cells cultivated in a rich and non-selective medium. For example, host-vector system can be constructed using the properties that a plasmid bearing the URA3 gene (which encodes orotate decarboxylase) is strictly maintained in a *ura3, fur1* double mutant, even in the presence of exogenous uracil (See Fig.1). The *FUR1* gene, which encodes the uracil phosphoribosyl-transferase is strictly required for exogenous uracil utilisation (Kern et al., 1990).

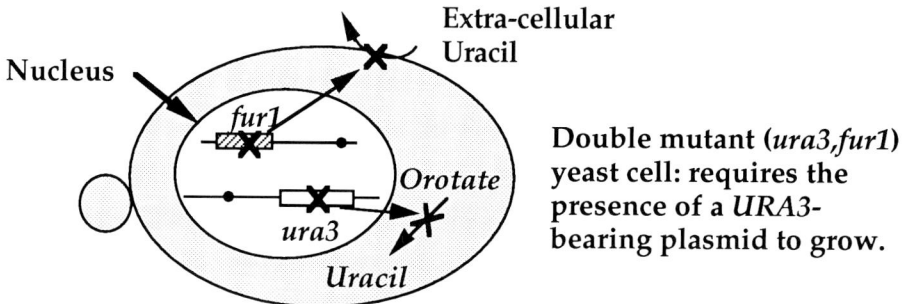

Plasmid stabilisation in *ura3, fur1* double mutant

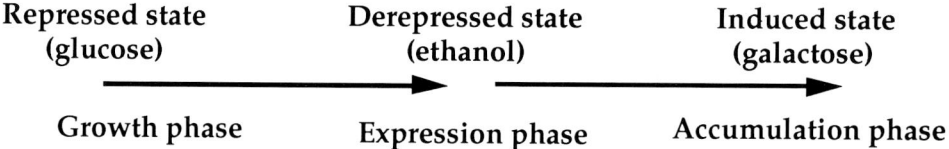

Three-step induction system

**Fig.1** Improving yeast expression system.

The use of such strains and of a three steps glucose-ethanol-galactose induction procedure, which allows to fully separate the biomass growth phase from the expression phase, allows to reach cell density up to two order of magnitude higher than basic expression system. P450 yield exceeding 500 nmoles/liter and specific content of 200-400 pmoles/mg microsomal protein can thus be achieved. The values compare fairly well with the best results obtained in *E. coli*..

**IMPROVING P450 ACTIVITIES BY OVER-EXPRESSION OF P450 REDUCTASES.**

Besides high yields of production, bioconversion and activity screening purposes require high turnover numbers. Although, the endogenous P450 reductase (CPR) content in yeast microsomes is, when expressed as cytochrome c reductase activity, similar to values found in liver, the heterologous nature of the reductase-P450 couple led to a rather inefficient coupling. This results in microsomal fraction containing high P450 levels but exhibiting rather low activity. A first solution considered was to significantly increase the endogenous CPR activity by over-expression either from a multi-copy vector (Urban et al., 1990; Murakami et al., 1990) or from a yeast strain engineered at the genomic DNA level. W(R) strain which overexpresses about 30-fold the reductase was thus constructed by replacing the natural promoter of the endogenous *CPR* gene by an artificial galactose-inducible one (Truan et al., 1993). Similar approaches involving the substitution of both the promoter and the open reading frame can be used alternatively to express either heterologous P450 reductase originating for example from plant (WAT11, WAT22) and human (W(hR)), or to express any other P450-associated gene like human cytochrome b5, yielding W(B) strain, or phase II enzyme like human epoxide hydrolase, yielding W(E) strain (Fig. 2).

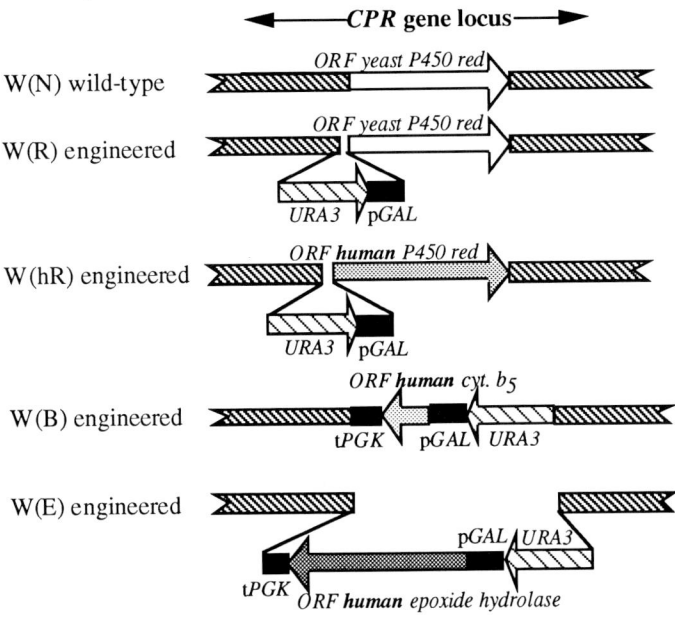

**Fig.2** Structures of the *CPR* locus in the wild-type and engineered yeast strains.

W(R) and related strains expressing various reductases present a conditional CPR expression. When grown on glucose, W(R) cells do not express detectable levels of P450 reductase; thus behaving as a *CPR*-disrupted strain. However, when grown in galactose-containing medium, the yeast CPR is over-expressed and reaches a specific microsomal activity 30-fold higher than that measured in wild-type strain in approximately 10 hours. This increased P450 reductase activity was found to dramatically enhance the heterologous P450 turnover numbers, although to various extents depending on the P450 isoform (Truan et al., 1993; Urban et al., 1994). For instance, human and mouse CYP1A1s exhibit respectively a 5- and a 9-fold increased turnover numbers for EROD activity when compared to values determined in wild-type yeast. In contrast, human CYP3A4 and Jerusalem artichoke CYP73 exhibit respectively a 62- and a 25-fold higher turnover numbers for, respectively, testosterone 6b-hydroxylase and cinnamate 4-hydroxylase activities. These results demonstrate that over-expression of yeast P450 reductase activity permits to significantly enhance the turnover numbers of mammalian and plant P450s produced in yeast. High reductase level can compensate for the rather inefficient heterologous coupling. However, the strong overexpression of the yeast reductase results in the toxic generation of higher amounts of oxygen radicals possibly resulting in membrane peroxidation and direct P450 damages leading to P420 conversion. Moreover some unwanted biotransformations can result from direct redox reactions involving the reductase when the system is used for drug metabolism prediction. Substitution of the yeast reductase to reconstitute homologous reductase-P450 systems was consequently tested, assuming that lower amounts of homologous reductase would be required to support similar P450 activity.

## DESIGNING AN UNIVERSAL P450 REDUCTASES CLONING TOOL.

Although several mammalian including human and fungal cDNA sequences have been cloned, plant CPR sequences were still unavailable in the early 1992. The limited sequence similarity (about 35-40 %) making difficult the use of low stringency hybridisation techniques, this led us to design an universal cloning tool for CPR cDNAs. The method is based on a functional test involving the coupling between an heterologous reductase and the endogenous yeast lanosterol P450 demethylase. The selection of a cDNA expression library in yeast is based on the dependence of the ketoconazole resistance (a well known lanosterol demethylase inhibitor) on the level of P450 reductase activity (endogenous or exogenous ) present in yeast cells (fig. 3).

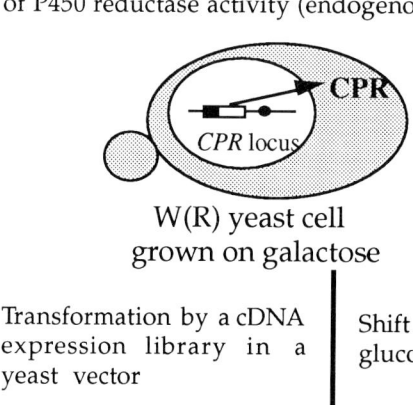

W(R) yeast cell grown on galactose

*CPR* gene is placed under the control of *GAL10-CYC1* artificial promoter.

Phenotype:
Low ketoconazole resistance in glucose
High ketoconazole resistance in galactose

Transformation by a cDNA expression library in a yeast vector

Shift from galactose to glucose as carbon source

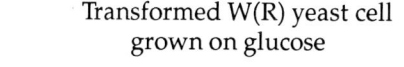

Transformed W(R) yeast cell grown on glucose

**Two cases for the resulting transformants**

1) The introduced heterologous gene codes for protein unrelated to the P450 mechanism of action
>>> Low ketoconazole resistance phenotype remains.

2) The heterologous gene encodes a P450 reductase
>>> Restored resistance to ketoconazole even in a glucose-containing medium.

Fig.3 Universal cloning procedure for P450 reductase-encoding genes.

The method involves the transformation of W(R) cells grown on galactose with a cDNA library placed downstream the constitutive *PGK* promoter in a yeast multicopy vector. When transformed W(R) cells are shifted in a glucose-containing medium, the genomic engineered *CPR* gene transcription is turned off and endogenous P450 reductase activity rapidly decreases and disappears. This results in a concomitant decrease in CYP51 activity and in a sharp increase in the ketoconazole sensibility. The use of a conditional CPR mutant is required because of the very poor transformation efficiency of reductase-deficient strains which is not compatible with a cDNA library screening. Plating of cells onto a glucose-containing selective medium in the presence of ketoconazole achieved the selection procedure. Only transformant cells harbouring a functional expression unit for a cDNA encoding a CPR or a P450 related redox enzyme (vide infra) can grow in such conditions. Two genes encoding *A. thaliana* CPRs were cloned by using the illustrated method. After a reformatting step using PCR, their coding sequences were integrated into the yeast genome at the endogenous *CPR* locus (Fig. 2), yielding WAT11 and WAT22 yeast strains (Mignotte et al., 1992). The same selection procedure, but using for transformation a yeast genomic library on a high copy number plasmid, allowed us to clone an other electron carrier (yeast cytochrome b5) involved in the control of P450 activity (Truan et al. ,1993b).

## REPLACING THE YEAST P450-REDUCTASE

Previous results clearly indicate that heterologous reductases can substitute for the yeast one in supporting the endogenous P450 dependent sterol biosynthesis and eventual others unknown physiological functions. Considering also the potential advantages of the expression of a reductase belonging to the same species as the heterologous P450 of interest, we decided to fully substitute the endogenous yeast *CPR* gene by an overexpressed heterologous one. Formally, production of human enzyme as the sole P450 reductase in yeast requires to substitute the coding sequence of the yeast CPR gene by the human one. For simplicity, this was achieved by transforming a strain disrupted for the endogenous *CPR* gene by a plasmid carrying a double expression cassettes, one coding for human CYP1A1 and the second for human P450 reductase (Eugster et al., 1992). Our strategy differs by the clean substitution in the yeast genomic DNA of the endogenous CPR gene by a galactose inducible artificial gene coding for the human CPR (figure 2). To reach sufficient expression level with a single copy system, we figured out that elimination of an hair pin loop structure standing in the 5'-end of human CPR cDNA was sufficient to significantly improve human CPR expression in yeast. This more versatile strategy yields W(hR) "humanized" strain which is perfectly viable although not expressing endogenous yeast reductase and offers a humanized reductase environment for any P450 expressed from a plasmid-based system (Urban et al., 1993). The same strategy was performed to parallely develop "vegetalized" yeast strains.

In W(hR) cells, the microsomal specific P450 reductase activity, measured by the NADPH-dependent cytochrome *c* reduction, was found to be a third of that measured in wild-type yeast cells (35 versus 120 nmol cyt.c reduced per min per mg protein). Despite that, the bioconversion rate of W(hR) cells transformed by human or mouse CYP1A1s was found as high as that measured with W(R) cells which over-express the yeast P450 reductase. In W(hR) yeast microsomes, the plant CYP73 catalyses the transformation of cinnamate in coumarate at a rate 2-fold higher than that in wild-type yeast. These preliminary results indicate that the human P450 reductase expressed in yeast is much more efficient than the yeast enzyme in supporting the activity of mammalian and even plant P450s, even if the cytochrome *c* reductase activity is lower.

## DESIGNING HUMANIZED OR VEGETALIZED YEAST STRAINS.

Yeast strains genomicaly engineered to express human or plant reductases being constructed, the next step performed was to build coexpression systems including also cytochrome b5 and important phase II enzyme such as the microsomal epoxide hydrolase. The principle is based on an extension of the preceding concepts (Fig.4) which involves multiple genomic integrations in diploides strains (Gautier et al, 1993, Truan et al., 1993, Urban et al. ,1993). This patented procedure allows easily to build up strains coexpressing a complete human redox environment and when required selected phase II enzymes. A complete set of strains optimised for the different requirements of various classes and origins of P450s was thus constructed.

The 'humanized' and 'vegetalized' strains offers significant advantages over any else coexpression approaches including fusion protein or co-expression plasmids. They allow the expression of virtually any human or plant P450 in an optimised environment without the need for particular vector construction. The only requirement is the transformation by a suitable P450 expression vector in the best-adapted "humanized" or 'vegetalized' strains.

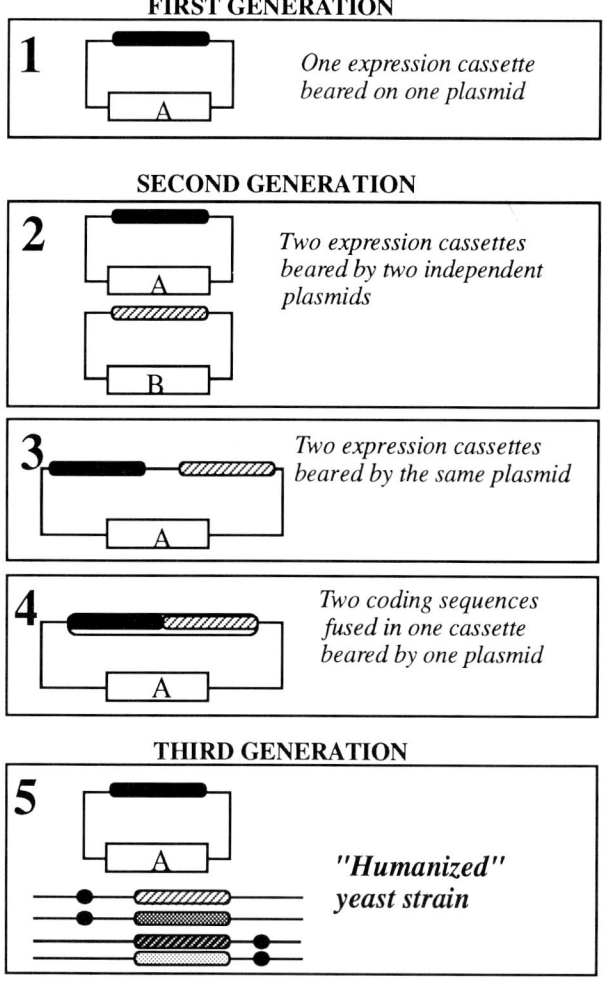

Fig.4 Different available strategies for coexpression.

The multi-integration strategy can be extended very easily to the coexpression of 2x**n**+1 heterologous enzymes, where **n** is the number of targeted diploid loci in the yeast genome. We consider that such an extension opens the field to a 'third generation' of humanized or vegetalized yeast strains, able to simulate complex metabolite pathways or to constitute industrial heterologous biosynthesis tools.

## REFERENCES

Barnes, H.J., Arlotto, M.P., & Waterman, M.R. (1991): Expression and enzymatic activity of recombinant cytochrome P450 17a-hydroxylase in Escherichia coli. *Proc. Natl. Acad. Sci. USA* 88, 5597-5601.

Eugster, H.P., Bärtsch, S., Würgler, F.E., & Sengstag, C. (1992): Functional co-expression of human oxidoreductase and cytochrome P450 1A1 in Saccharomyces cerevisiæ results in increased EROD activity. *Biochem. Biophys. Res. Commun.* 185, 641-647.

Gautier, J.C., Gautier, J.C., Urban, P., Beaune, P., & Pompon, D. (1993): Engineered yeast cells as model to study coupling between human metabolizing enzymes. *Eur. J. Biochem.* 211, 63-72.

Guengerich, F.P., Brian, W.R., Sari, M.A., & Ross, J.T. (1991): Expression of mammalian cytochrome P450 enzymes using yeast-based vectors. *Methods Enzymol.* 206, 130-145.

Kern, L., de Montigny, J., Jund, R., & Lacroute, F. (1990): The FUR1 gene of Saccharomyces cerevisiæ: cloning, structure and expression of wild-type and mutant alleles. *Gene* 88, 149-157.

Mignotte, C., Kazmaier, M., Lacroute, F., & Pompon, D. (1992) Patent pending.

Murakami, H., Yabusaki, Y., Sakaki, T., Shibata, M., & Ohkawa, H. (1990): Expression of cloned yeast NADPH-cytochrome P450 reductase gene in Saccharomyces cerevisiæ. *J. Biochem. (Tokyo)* 108, 859-865.

Shibata, M., Sakaki, T., Yabusaki, Y., Murakami, H., & Ohkawa, H. (1990): Genetically engineered P450 monooxygenases: construction of bovine P450 c17 / yeast reductase fused enzyme.*DNA Cell Biol.* 9, 27-36.

Truan, G., Cullin, C., Reisdorf, P, Urban, P., & Pompon, D. (1993a): Enhanced in vivo monooxygenase activties of mammalian P450s in engineered yeast cells producing high levels of NADPH-P450 reductase and cytochrome b5. *Gene*, 125, 49-55.

Truan G., Epinat J.C., Rougeulle C., Cullin C., Pompon D.(1993b): Cloning and characterisationof a yeast cytochrome b5 gene which suppresses ketoconazole hypersensitivity in NADPH P450 reductase disrupted strain. Submitted to Gene.

Urban, P., Cullin, C., & Pompon, D. (1990): Maximizing the expression of mammalian cytochrome P450 monooxygenase activities in yeast cells. *Biochimie* 72, 463-472.

Urban, P., Truan, G., Gautier, J.C., & Pompon, D. (1993): Xenobiotic metabolism in humanized yeast: engineered yeast cells producing human NADPH-cytochrome P450 reductase, cytochrome b5, epoxide hydrolase and P450s. *Biochem. Soc. Transac.* 21, 1028-1033.

Urban, P., Perret, A., Truan, G., & Pompon, D. (1993): Advantages of the human over yeast NADPH-P450 reductase on supporting human P450 activities in yeast. *This volume*.

Urban, P., Truan, G., Bellamine, A., Lainé, R., Gautier, J.C., & Pompon, D. (1994): A critical review on the scenery of engineered yeasts simulating human drug metabolism. *Drug Metab. Drug Interac.*, in press.

# Expression in *E. coli* of enzymatically active fusion proteins containing cytochrome P450 and cytochrome P450 reductase

Charles W. Fisher, Manju S. Shet, Priscilla L. Holmans, Ronald W. Estabrook

*Department of Biochemistry, University of Texas Southwestern Medical Center, 5323, Harry Hines Boulevard, Dallas, Texas, 75235-9038 USA*

## INTRODUCTION

The successful expression of eukaryotic cytochrome P450s in *E. coli* has been achieved in the past few years with increasingly higher levels of recombinant protein generated. Several expression vectors and growth conditions have been utilized with relatively low levels of expression (Li & Chiang 1991, Winters & Cederbaum 1992, Wada et. al 1991). The highest levels of expression have been obtained with the use of modified amino terminal sequences and the pCWori+ vector (Barnes et al., 1991, Fisher et al. 1992a).

CYP 103 (P450 BM3), isolated from *Bacillus megaterium*, is a naturally occurring fusion protein containing the domains of P450 and P450 reductase. This soluble fusion protein has a turnover number of 1,200 (Narhi & Fulco 1982). In contrast, membrane bound cytochrome P450 proteins typically catalyze reactions with turnover numbers significantly less than 20. This suggests that the rate limiting step in cytochrome P450 catalysis may be the interaction of the flavoprotein reductase and cytochrome P450. Murakami et al. (1982) genetically engineered a fusion of the cDNA for rat liver P450c (CYP1A1) with the cDNA of rat NADPH-P450 reductase to create a P450 fusion protein which they expressed in yeast. These constructs had turnover numbers as high as 206 nmol substrate/min/nmol enzyme.

We have utilized the expression of cytochrome P450 in *E. coli* in order to create similar fusions of different cytochrome P450s with cytochrome P450 reductase (Fisher et. al 1992b). These fusions have allowed us to purify large quantities of cytochrome P450 as fusion proteins. The purified fusion proteins have been examined with regard to their kinetic properties. The stimulatory effect of cytochrome $b_5$ on the reactions catalyzed by these fusion proteins was found to differ with respect to the form of cytochrome P450 examined (Fisher et. al (1992b).

Human cytochrome P450 3A4, which is one of the most abundant forms of cytochrome P450 in the human liver, acts on a wide variety of drugs and steroids. The 3A family of P450 has proven extremely enigmatic in its activity. Reconstitution experiments have shown a requirement for an array additives including cytochrome $b_5$, lipids, and detergents and in one instance glutathione (Gillam et al. 1993). We describe here the properties of a fusion containing human cytochrome P450 3A4 linked to rat NADPH P450 reductase (rF450[mHum3A4/mRatOR]L1). In addition we describe modifications to the linker region of the orginal fusion protein (rF450[mBov17A/mRatOR]L1) which have been expressed. A construct encoding an additional biotinylated protein domain is also described.

**MATERIALS AND METHODS**

A construct expressing bovine P450 17α-hydroxylase (CYP17A) in the vector pCWori+ (pCWmod17) was obtained from the laboratory of Dr. Michael Waterman (Barnes et al. 1992). The plasmid containing the coding sequence for rat liver NADPH reductase (pOR263) was obtained from Dr. Charles Kasper (Shen et al. 1989). A λgt11 clone containing the cDNA encoding human cytochrome P450 3A4 (NF-25) was obtained from Dr. Fred Guengerich (Beaune et al. 1986). The sequence for E. coli biotin carboxyl carrier protein (BCCP, Li & Cronan 1992) was amplified from genomic DNA by standard PCR methods and cloned into the EcoR V restriction site of pBS SKII+ (Stratagene). Mutagenesis of DNA sequences was performed by PCR using primers incorporating the desired sequences and restriction sites. The DNA fragments incorporating the introduced mutations were subcloned into PBS SKII+ and the desired products removed by restriction enzyme digestion. Constructs were transformed into E. coli DH5α (BRL). The transformed cells were grown in TB medium at 28°C, then induced with IPTG and grown for 48 hours. Membranes were prepared from E. coli following sonication of the washed cells (Fisher et al. 1992b).

**RESULTS AND DISCUSSION**

A construct consisting of human P450 3A4 fused to rat liver NADPH-P450 reductase was made by excising the bovine 17α-hydroxylase coding sequence and replacing it with that of the modified P450 3A4 (Fig. 1). The fusion protein was expressed in DH5α and purified as described previously for bovine 17A and rat 4A1 fusion proteins (Fisher et al. 1992b). The purified fusion protein was tested for enzymatic activity using a variety of substrates. We have found (Shet et al. 1993) that the 6β-hydroxylation of testosterone requires cytochrome $b_5$, phospholipids and detergent (0.05% CHAPS). Similarly the oxidation of nifedipine requires the addition of cytochrome $b_5$, lipid and detergent. However, studies on the N-demethylation of erythromycin and benzphetamine showed an inhibition by cytochrome $b_5$ without the need for lipids or detergent. We found no requirement for glutathione in reconstituting either activity.

The human P450 3A4 fusion protein described here is highly active and easily purified. Some properties of the system may have been modified by the creation of a fusion protein. The strength of using a fusion construct lies with the ability to quickly generate a single enzyme which can produce large quantities of metabolites.

We are investigating several alternative means of increasing the enzymatic activity of future constructs. These include: 1. Introduction of additional reductase domains; 2. Modifications of linker sequence; 3. New functional domains; 4. Modifications to the reductase or P450 sequence.

Linker regions of various types have been produced in the fusion proteins created by workers at Sumitomo Chemical Co (Ohkawa et al. 1989, Sakaki et al. 1990, Shibata et al. 1990, Yabusaki et al. 1988, 1990). Recent studies on protein linker regions have suggested several theories on the most effective combinations (Argos 1990). To investigate the role of the linker region in the interaction of the two domains of rF450[mBov17A/mRatOR]L1 a series of linker constructs were generated (Fig. 2). One series was based upon the sequence observed at the junction of the P450 and P450 reductase domains in BM-3 in which multiple lysine residues exist (LYS-LINKER). Argos (1990) suggested a motif to be used linking protein domains based upon observations of naturally occurring linkage domains in

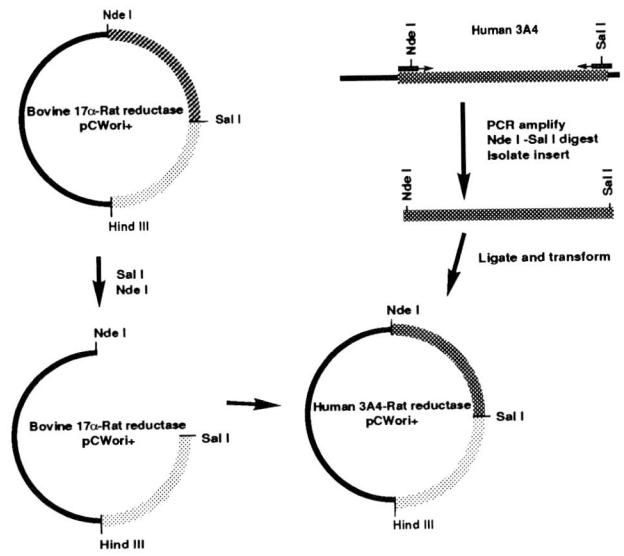

A.

| | |
|---|---|
| Native sequence | MALIPDLAMETWLLLAVSLVLLY |
| Modified sequence | MALLLAVFLVLLY |

B.

**Fig. 1.** A. Outline of the construction for the expression of the human cytochrome P450 3A4 fusion using the pCWori+ vector system. B. Modifications to the amino acid sequence of the amino terminal region of the P450 3A4 to enhance expression. The native sequence is shown above the modified sequence. The amino acids for which the coding sequence has been adapted from the modified bovine 17a-hydroxylase sequence are underlined.

proteins of known crystal structure. A sequence rich in serine, glycine and threonine was constructed based upon these suggestions (STG-LINKER). A final series was based upon the known high mobility of glycine containing peptide chains and the rigid structure created by multiple prolines (PRO/GLY-LINKER). The objective of these constructs is to examine the extent that interaction of the reductase with the P450 can be enhanced.

The plasmid encoding rF450[mBov17A/mRatOR]L1 was digested with *Sal* I and ligated with annealed oligonucleotides. The plasmid constructs incorporating the oligonucleotide linkers were found by colony hybridization with radiolabelled oligonucleotides and then sequenced to identify the orientation of the introduced sequences. In many cases multiples of the introduced oligonucleotides were identified. *E. coli* transformants containing the modified fusion protein expression plasmids were grown and induced with IPTG as described for previous constructs. Samples of cells were taken for Western blot, spectral and enzymatic analysis. All linkers examined to date expressed an intact fusion protein at levels comparable to that observed with the original fusion protein. Even linkers containing multiple positive charges, e.g. +30, expressed functional P450. Membranes prepared from *E. coli* expressing these constructs were assayed enzymatically for the metabolism of progesterone (Fig. 3). A positive effect for some of the constructions was seen when compared with the original ST linker.

```
PRO/GLY-LINKER
    Ser Arg Gly Gly Gly Gly Gly Ser Thr
    Ser Arg Pro Pro Pro Pro Pro Ser Thr

STG-LINKER
    Ser Thr Gly Ser Gly Ser Gly Ser Thr
    Ser Ser Arg Ser Arg Ser Pro Ser Thr

LYS-LINKER
    Ser Lys Lys Lys Lys Lys Ser Thr
    Ser Thr Ser Ser Ser Ser Ser Thr
```

**Fig. 2.** Linker domains adapted from naturally occurring domains. The coding sequences were designed to be contain ends complementary to the Sal I restriction site and to encode an alternate amino acid sequence when inserted in the reverse orientation.

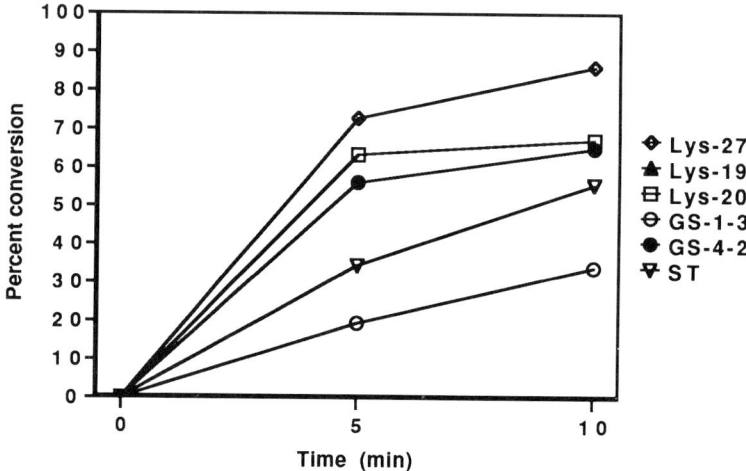

**Fig. 3.** The rates of progesterone metabolism by the bovine 17 fusion proteins containing different linker regions. Reactions were performed at 37°C with E. coli membranes containing the expressed fusion proteins. The reaction mixtures contained 1 nmol/ml of the P450 fusion protein, 10 μM progesterone and a NADPH regenerating system. Linker regions compared were: ST (ST); Lys-27 ($SK_5STS_5TS_5TS_5TS_5T$); Lys-19 ($SK_5SK_5SK_5STS_5T$); Lys-20 ($SK_5SK_5STS_5T$); GS-1-3 ($SRG_5SRG_5ST$); GS-4-2 ($SRG_5ST$).

Further efforts have been made to develop a fusion protein which can be immobilized. This immobilization will also allow the rapid extraction of products and permit one to examine the effects of limited oxygen concentration on the function of P450. To achieve this goal we have chosen to create constructs which attach the BCCP from E. coli by fusing this domain to the carboxy terminus of the reductase (Cronan 1990, Fig. 4). This domain is biotinylated *in vivo* by *E. coli* and generates a protein which can be attached to avidin-containing Sepharose with an affinity approaching that of a covalent bond. The plasmid encoding rF450[mBov17A/mRatOR]L1 was modified for this construction as follows. The 3' end of the coding sequence for the reductase domain was modified to incorporate an *Spe* I site in place of the stop codon. The entire BCCP protein was amplified with *Spe* I sites incorporated into the sequence flanking the coding region. Finally to aid in purification an additional domain encoding 6 histidines was incorporated into the 3' end of the amplified DNA. This modified protein was expressed in *E. coli* as an intact fusion protein that was spectrally active and could be detected by avidin peroxidase by Western blot analysis (Fig. 5).

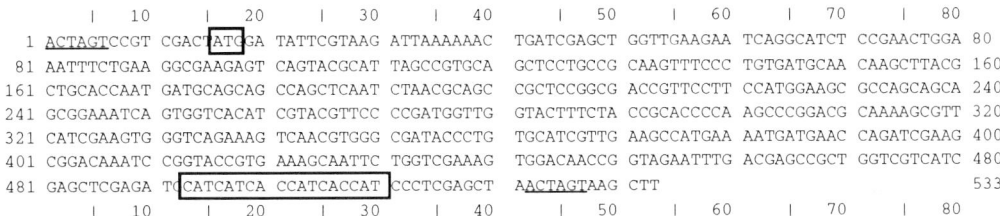

```
         |   10      |   20      |   30      |   40      |   50      |   60      |   70      |   80
   1 ACTAGTCCGT CGACTATGGA TATTCGTAAG ATTAAAAAAC TGATCGAGCT GGTTGAAGAA TCAGGCATCT CCGAACTGGA  80
  81 AATTTCTGAA GGCGAAGAGT CAGTACGCAT TAGCCGTGCA GCTCCTGCCG CAAGTTTCCC TGTGATGCAA CAAGCTTACG 160
 161 CTGCACCAAT GATGCAGCAG CCAGCTCAAT CTAACGCAGC CGCTCCGGCG ACCGTTCCTT CCATGGAAGC GCCAGCAGCA 240
 241 GCGGAAATCA GTGGTCACAT CGTACGTTCC CCGATGGTTG GTACTTTCTA CCGCACCCCA AGCCCGGACG CAAAAGCGTT 320
 321 CATCGAAGTG GGTCAGAAAG TCAACGTGGG CGATACCCTG TGCATCGTTG AAGCCATGAA AATGATGAAC CAGATCGAAG 400
 401 CGGACAAATC CGGTACCGTG AAAGCAATTC TGGTCGAAAG TGGACAACCG GTAGAATTTG ACGAGCCGCT GGTCGTCATC 480
 481 GAGCTCGAGA TCCATCATCA CCATCACCAT CCCTCGAGCT AACTAGTAAG CTT                              533
         |   10      |   20      |   30      |   40      |   50      |   60      |   70      |   80
```

**Fig. 4.** Sequence of the PCR amplified BCCP domain including the modifications to the coding sequence introduced by the use primers containing modified sequences. The *Spe* I sites are underlined and the start codon of BCCP and the region encoding the six histines on the carboxy terminus of the protein are boxed.

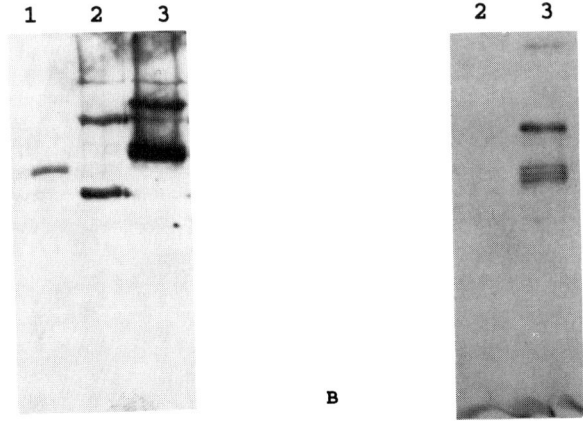

**Fig. 5.** Detection of biotinylated cytochrome P450 fusion protein. A. Western blot of whole cells expressing fusions of bovine P450 17 probed with a polyclonal antibody to rabbit cytochrome P450 reductase (OXYgene) detected with goat anti-rabbit peroxidase conjugate and chemiluminescent substrate (Amersham). B. Western blot of whole cells probed with avidin-peroxidase. Lanes are; 1. Purified recombinant rat liver reductase. 2. *E. coli* cells expressing the fusion of bovine P450 17α-hydroxylase and cytochrome P450 reductase. 3. *E. coli* cells expressing the fusion of bovine P450 17α hydroxylase and cytochrome P450 reductase with the additional BCCP domain.

## CONCLUSION

We have shown that cytochrome P450 can be expressed at high levels in a form that can be easily purified. Fusions of cytochrome P450 with the reductase allow the evaluation of enzymatic properties of the P450 species. It is now possible to choose a DNA sequence encoding one of the many different forms of cytochrome P450 and create a functional unit that can be used for specific oxidative applications. This capability is particularly important for examination of cytochrome P450 species which cannot be easily obtained and/or produce low amounts of important metabolites. Future applications for these fusion proteins will involve the rational design and construction of multifunctional proteins of practical value. Practical uses of the fusion proteins will be be in the area of chemical synthesis via large scale production of metabolites.

*Supported by a grant from the N.I.H. (GM 16488)

## REFERENCES

Argos, P. (1990) An investigation of oligopeptides linking domains in protein tertiary structures and possible candidates for general gene fusion. *J. Mol. Biol.* 211, 943-958.

Barnes, H.J., Arlotto, M.P. & Waterman, M.R. (1991): Expression and enzymatic activity of recombinant cytochrome P450 17α-hydroxylase in *Escherichia coli*. *Proc. Natl. Acad. Sci.* USA 88, 5597-5601.

Beaune, P.H., Umbenhauer, D.R., Bork, R.W., Lloyd, R.S. & Guengerich, F.P. (1986): Isolation and sequence determination of a cDNA clone related to human cytochrome P-450 nifedipine oxidase. *Proc. Natl. Acad. Sci.* USA 83, 8064-8068.

Cronan, J.E. (1990): Biotination of proteins in vivo. *J. Biol. Chem.* 265, 10327-10333.

Fisher, C.W., Caudle, D.L., Martin-Wixtrom, C., Quattrochi, L.C., Tukey, R.H., Waterman, M.R. & Estabrook, R.W. (1992a): High-level expression of functional human cytochrome P450 1A2 in Escherichia coli. *FASEB J.* 6, 759-764.

Fisher, C.W., Shet, M.J., Caudle, D.L., Martin-Wixtrom, C. & Estabrook, R.W. (1992b): High-level expression in E. coli of a enzymatically active fusion proteins containing the domains of mammalian cytochrome P450s and NADPH-P450 reductase flavoprotein. *Proc. Natl. Acad. Sci.* USA 89, 10817-10821.

Gillam, E.M.J., Baba, T., Kim, B.-R., Ohmori, S. & Guengerich, F.P. (1993): Expression of modified human cytochrome P450 3A4 in *Escherichia coli* and purification and reconstitution of the enzyme. *Arch. Biochem. Biophys.* 305, 123-131.

Li, Y.C. & Chiang, J.Y.L. (1991): The expression of a catalytically active cholesterol 7α-hydroxylase cytochrome P450 in *Eschericia coli*. *J. Biol. Chem.* 266, 19186-19191.

Li, S.-J. & Cronan, J.E. (1992): The gene encoding the biotin carboxylase subunit of *Esherichia coli* acetyl-CoA carboxylase. *J. Biol. Chem.* 267, 855-863.

Murakami, H., Yabusaki, Y., Sakaki, T., Shibata, M. & Ohkawa, H. (1987): A genetically engineered P450 monooxygenasee: Construction of the functional fused enzyme between rat cytochrome P450c and NADPH-cytochrome P450 reductase. *DNA* 6, 189-197.

Narhi, L.O. & Fulco, A.J. (1982): Characterization of a catalytically self-sufficient 119,000-dalton cytochrome P-450 monooxygenase induced by barbituates in Bacillus megaterium. *J. Biol. Chem.* 257, 2147-2150.

Ohkwa, H., Yabusaki, Y., Sakaki, T., Murakami, H. & Shibata, M. (1989): Expression of cytochrome P-450 and NADPH-cytochrome P-450 reductase genes in yeast. In *Xenobiotic Metabolism and Disposition*, Eds. Kato, R., Estabrook, R.W. & Cayen, M.N., New York.

Sakaki, T., Shibata, M., Yabusaki, Y., Murakami, H. & Ohkawa, H. (1990): Expression of bovine cytochrome P450C21 and its fused enzymes with yeast

NADPH-cytochrome P450 reductase in *Saccharomyces cerevisiae*. *DNA Cell Biol.* 9, 603-614.

Shen, A., Porter, T.D., Wilson, T.E. & Kasper, C.B. (1989) Structural analysis of the FMN binding domain of NADPH-cytochrome P-450 oxidoreductase by site-directed mutagenesis. *J. Biol. Chem.* 264, 7584-7589.

Shet, M.S., Fisher, C.W., Holmans, P.L. and Estabrook, R.W. (1993): Human cytochrome P450 3A4: enzymatic properties of a purified recombinant fusion protein containing NADPH-P450 reductase. *Proc. Natl. Acad. Sci.* USA, In press.

Shibata, M., Sakaki, T., Yabusaki, Y., Murakami, H. & Ohkawa, H. (1990): Genetically engineered P450 monooxygenases: Construction of bovine P450c17/yeast reductase fused enzymes. *DNA* 9, 27-36.

Wada, A., Mathew, P.A., Barnes, H.J., Sanders, D., Estabrook, R.A. & Waterman, M.R. (1991): Expression of functional bovine cholesterol side chain cleavage cytochrome P450 (P450scc) in *Escherichia coli*. *Arch. Biochem. Biophys.* 290, 376-380.

Winters, D.K. & Cederbaum, A.I. (1992): Expression of a catalytically active human cytochrome P-4502E1 in *Escherichia coli*. *Biochim. Biophys. Acta* 1156, 43-49.

Yabusaki, Y., Murakami, H., Sakaki, T., Shibata, M. & Ohkawa, H. (1988): Genetically engineered modification of P450 monooxygenases: Functional analysis of the amino-terminal hydrophobic region and hinge region of the P450/reductase fused enzyme. *DNA* 7, 701-711.

Yabusaki, Y., Sakaki, T., Murakami, H., Shibata, M. & Ohkawa, H. (1990): Expression of engineered P450/reductase fused enzymes in yeast. *Proc. of the VIIIth International Symposium on Microsomes and Drug Oxidations*.

# The role of P450s in insect-plant interactions

Hataichanoke Prapaipong[1], Chien-Fu Hung[2], May R. Berenbaum[2], Mary A. Schuler[1,3]

[1]Department of Biochemistry, [2]Department of Entomology, [3]Department of Plant Biology, University of Illinois, Urbana, Illinois 61801, USA

In comparison with mammalian P450s, insect P450s are poorly understood at the genetic level, despite the importance of these proteins in the detoxification of insecticides and hostplant secondary metabolites, and in the synthesis of endogenous substrates such as hormones and pheromones. Biochemical analysis of P. polyxenes and several closely related *Papilio* species has revealed that hostplant utilization patterns within this genus are associated not only with changes in P450 metabolic activities but also with changes in the inducibility of specific P450s. In particular, P450s mediate detoxification of furanocoumarins, a group of hostplant defensive compounds, in the black swallowtail (*Papilio polyxenes*). The larvae of this specialist insect are restricted to feeding on one or two plant families, the Umbelliferae and Rutaceae, which contain high levels of furanocoumarins. These naturally occurring plant metabolites (Fig. 1) are highly toxic to a wide variety of organisms because

Fig. 1     xanthotoxin     bergapten
          angelicin      sphondin

when photoactivated they react directly and irreversibly with pyrimidine bases in DNA (Berenbaum, 1991). *In vivo* analysis has demonstrated that this insect species is capable of metabolizing xanthotoxin, a linear furanocoumarin, at levels up to ten times higher than polyphagous lepidopteran species which rarely encounter furanocoumarins in their diet (Berenbaum et al., 1990). *P. polyxenes* larvae are also capable of metabolizing angular furanocoumarins such as angelicin and sphondin. In principle, this increased ability to tolerate plant toxins could involve changes in either regulatory elements or structural components of a P450 target gene.

*In vitro* analysis of the metabolic activities in isolated *P. polyxenes* larval microsomes initially demonstrated that the linear furanocoumarin, xanthotoxin, and its isomer, bergapten, can be metabolized with approximately equal efficiency and that angular furancoumarins, such as angelicin and sphondin, are metabolized at significantly lower efficiencies (Cohen, 1991; Berenbaum and Zangerl, 1993). As measured in isolated microsomes, the ability to metabolize xanthotoxin is induced up to 8-fold by a three day exposure of fifth instar larvae to 0.5% xanthotoxin and approximately 2.6-fold by a similar exposure to bergapten (Berenbaum et al., 1990). Related species such as *P. glaucus*, a generalist feeder which only rarely encounters furanocoumarins in its diet displays, not surprisingly, very low constitutive levels of xanthotoxin metabolism (Cohen, 1991). In this species also, supplementation of the larval diet with xanthotoxin increases xanthotoxin metabolism at least five-fold. The general correlation between their metabolic activities, their inducibilities and the frequency with which each of these species encounters furancoumarins originally suggested that the transcriptional activities of these genes might be dependent on one or more of the substrates which are metabolized by these P450s.

Cloning of *P. polyxenes* cDNAs from xanthotoxin-induced larvae has allowed us to identify two P450 cDNAs, both designated *CYP6B1*, which are more than 98% identical at the amino acid level (Cohen et al., 1992). Baculovirus-mediated expression of the two cloned *CYP6B1* cDNAs in two different lepidopteran cell lines, *Spodoptera frugiperda* Sf9 and *Trichoplusia ni* Tn5, has demonstrated that both of these *CYP6B1* isozymes metabolize substantial amounts of the linear furanocoumarins, xanthotoxin and bergapten, but not angular furanocoumarins, such as angelicin and sphondin (Ma et al., 1994). These reactivities, which can only be accurately defined in this type of heterologous expression system, suggest that *P. polyxenes* express at least two selective furanocoumarin-metabolic P450s: CYP6B1, which metabolizes a discrete set of linear furanocoumarins, and another P450, which metabolizes angular furancoumarins more efficiently than does CYP6B1.

The biological activities of *P. polyxenes* larvae against linear and angular furanocoumarins have now been monitored in response to individual furanocoumarins to determine the level of crossactivation *in vivo*. Because of genetic variation in *P. polyxenes* populations, reactivities against different furancoumarins have been compared in total cell homogenates derived from the midguts of individual fifth instar larvae exposed for one day to 0.01 mmol/g of fresh weight tissue of xanthotoxin, bergapten or angelicin. Metabolism rates for two different insects subjected to each of these treatments are shown in Fig. 2. These results indicate that control larvae which have

Fig. 2. Induction of furanocoumarin metabolism in black swallowtail larvae

| Inducer | xanthotoxin | Metabolism bergapten | (p mol/ min/ mg protein) angelicin | sphondin |
|---|---|---|---|---|
| xanthotoxin | 9.76 + 0.91 | 31.27 + 0.78 | 1.71 + 1.02 | 12.45 + 0.93 |
|  | 18.54 + 4.45 | 24.22 + 2.13 | 2.80 + 0.44 | 12.83 + 2.56 |
| bergapten | 12.51 + 0.28 | 16.91 + 1.04 | 0.53 + 0.04 | 10.01 + 0.06 |
|  | 9.02 + 0.94 | 19.32 + 1.52 | 2.74 + 0.35 | 7.58 + 0.86 |
| angelicin | 19.79 + 2.16 | 27.02 + 6.29 | 1.72 + 1.21 | 17.14 + 0.03 |
|  | 12.21 + 1.31 | 36.20 + 2.49 | 1.31 + 0 | 21.07 + 3.86 |
| control | 1.62 + 1.01 | 10.51 + 0.47 | 0.48 + 0.23 | 9.42 + 0.67 |
|  | 2.61 + 0.74 | 9.57 + 0.84 | 0.61 + 0 | 9.58 + 2.37 |

Means + SE of two experiments assayed in duplicate.

ingested parsley, a hostplant with extremely low levels of furanocoumarins (Cohen et al., 1989) have low constitutive levels of xanthotoxin and angelicin metabolism and slightly higher levels of bergapten and sphondin metabolism. Metabolisms of xanthotoxin and bergapten are significantly enhanced by exposure to these linear furancoumarins and, surprisingly, by exposure to angelicin, an angular furanocoumarin. In some insects, all three of these inducers increase metabolism of angelicin but never to the level observed for xanthotoxin and bergapten. Metabolism of sphondin, another angular furancoumarin, is induced by angelicin but not by either of the linear furancoumarins. These results have indicated that regulated expression of these insect P450s is extremely complex. The angular furanocoumarin, angelicin, is capable of crossactivating expression of P450s which metabolize linear furanocoumarins as well as angelicin and the other angular furanocoumarin, sphondin.

Northern analyses using the cloned *CYP6B1* cDNA as probe at high stringency have indicated that CYP6B1 mRNA is transcriptionally induced in *P. polyxenes* larvae in response to xanthotoxin (Cohen et al., 1992). To determine whether other furanocoumarins induce transcription of CYP6B1 in *P. polyxenes* larvae, fifth instar larvae were exposed to 0.01 mmol/g fresh weight tissue of either a linear furanocoumarin (xanthotoxin, bergapten, or psoralen) or an angular furanocumarin (angelicin or sphondin) for one day.

Northern analysis (Fig. 3) indicates that, at this level of comparison which detects CYP6B1 and closely related mRNAs, the three linear furanocoumarins, xanthotoxin, bergapten, and psoralen, are capable of inducing accumulation of mRNAs complementary to *CYP6B1* cDNA. The magnitude of induction differs, with xanthotoxin and psoralen inducing an 18-fold increase and bergapten inducing a 10-fold increase in these mRNAs. At this level of stringency, angelicin, the angular furanocoumarin, appears to induce expression of CYP6B1-like mRNAs to the same level as xanthotoxin.

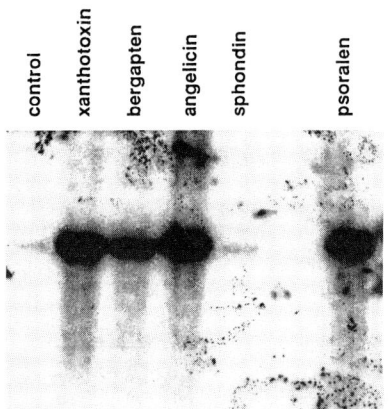

Fig. 3

Because of the potential presence of other closely related P450 genes which may also be induced in response to furanocoumarins, the magnitude and specificity of *CYP6B1* transcriptional regulation is difficult to establish at the level of Northern analysis. To more specifically define the transcriptional inducers and induction patterns of the *CYP6B1* gene, we have cloned and characterized a full-length gene from *P. polyxenes*. Alignment of the full length gene sequence with that of the *CYP6B1* cDNA indicated that the *CYP6B1* gene contains a single intron of 471 bases at amino acid position 445 just beyond the heme axial ligand present in all P450s (Fig. 4). The two allelic

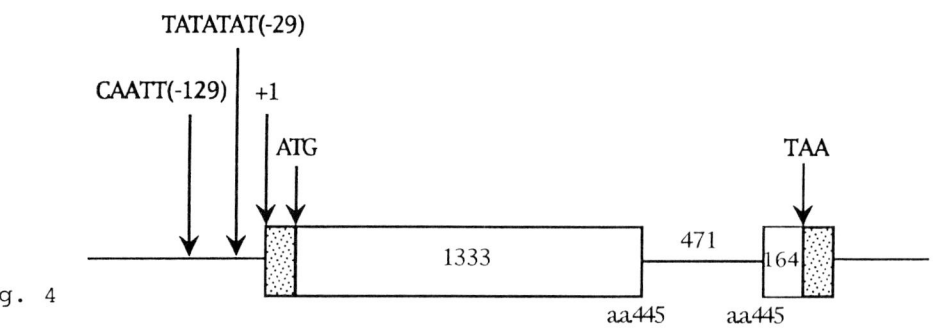

Fig. 4

variants of *CYP6B1* cDNA previously isolated differ in 1.81% (9/498) of their amino acids and 1.87% (28/1494) of their coding nucleotides (Cohen et al., 1992). The *CYP6B1* gene sequence reported here is 99.53% identical in its coding nucleotides to CYP6B1.1 and 98.59% identical to CYP6B1.2. The amino acid variations resulting from these differences occur only at the positions and in the same permutations as in the two cDNAs. Of the nine amino positions in which the two cDNAs vary, six positions in the *CYP6B1* gene correspond to those in CYP6B1.1 and three correspond to those in CYP6B1.2. Variations in the 3' nontranslated region, which is not constrained by protein structure, are minimal, with no differences observed between the genomic DNA and CYP6B1.1 cDNA sequences.

The precise initiation site for the CYP6B1 mRNA has been identified by alignment of a DNA sequencing ladder with primer extension products generated with mRNA from xanthotoxin-induced larvae. The initiation site is situated 29 and 129 nucleotides downstream from putative TATA (TATATAT) and CAAT (CAATT) boxes, respectively, and 10 nucleotides upstream from the 5' end of the *CYP6B1* cDNAs previously characterized (Cohen et al., 1992).

The exact level to which the *CYP6B1* gene is induced by furanocoumarins and the spectrum of transcriptional inducers which operate in *P. polyxenes* larvae has been analyzed by RNase protection using an antisense probe which spans the first 340 nucleotides of the *CYP6B1* gene. In these assays, fifth instar larvae were again exposed for one day to 0.01 mmol/g fresh weight tissue of xanthotoxin, bergapten or angelicin, the angular furanocoumarin. The radioactivities in each of the protected fragments were quantified by phosphorimagery and compared by linear regression analyses. These results indicate that the low levels of *CYP6B1* transcript exist in control larvae (Fig. 5).

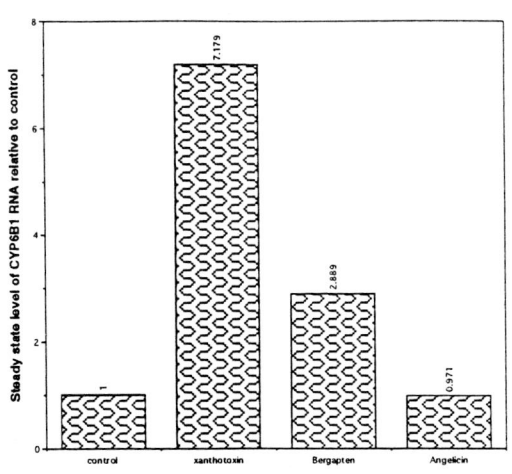

Fig. 5

Transcription of the *CYP6B1* gene is induced 8-fold over the control level by a 24 hr. exposure to xanthotoxin and 2.6-fold by exposure to bergapten. In contrast with the Northern results, *CYP6B1* transcription is not induced to any appreciable extent by angelicin.

Clearly, a hierarchy exists in the transcriptional regulation of the *CYP6B1* gene by furanocoumarins. Two linear furanocoumarins, xanthotoxin and bergapten, that are capable of being metabolized by the *CYP6B1* gene product (Ma et al., 1994) induce its transcription, although to different levels. An angular furanocoumarin, angelicin, which is not metabolized to any significant extent by CYP6B1 (Ma et al., 1994) does not induce transcription of this gene. As suggested by our original analysis of CYP6B1 activities in baculovirus-infected cells, these findings suggest that a second P450 in *P. polyxenes* is responsible for the metabolism of angular furanocoumarins. Comparison of the Northern and RNase protection analyses suggests that the various furanocoumarins have different abilities to induce the two loci. The *CYP6B1* gene is specifically induced in *P. polyxenes* larvae in response to xanthotoxin and bergapten, two linear furanocoumarins which are actively metabolized by this isozyme. It is not induced to any extent by angelicin, an angular furanocoumarin which is metabolized poorly by the CYP6B1 gene product. The high level of xanthotoxin metabolic activity and Northern signal produced in response to angelicin compared with the failure of angelicin to induce *CYP6B1* expression strongly suggest that another gene closely related to *CYP6B1* is induced in response to angelicin. The data we have obtained are consistent with the notion that the product of this second locus metabolizes xanthotoxin and bergapten, and also the angular furanocoumarins, angelicin and sphondin. It is not yet clear whether this second locus can be induced by xanthotoxin and bergapten.

These findings support the presence of a second P450 in *P. polyxenes* which is responsible for the metabolism of angular furanocoumarins. Clearly, transcriptional regulation by furanocoumarins represents a significant adaptive strategy which modulates resistance to plant furanocoumarins. We have now established biochemical and molecular data to support this hypothesis and have begun to detail the organization of this gene family, which is of critical importance in defining insect-hostplant interactions.

To determine if the 5′ flanking sequences upstream from the classic TATA box (-29) and CAAT box (-129) retain promoter activity in a heterologous system, we subcloned nucleotides -838 to +22 of the *CYP6B1* gene upstream from chloramphenicol acetyltransferase (CAT) coding sequences. Transfection experiments with the resulting 838CYP-CAT construct have indicated that the 800 nucleotide long *CYP6B1* promoter fragment contains basal transcription elements which are functional in Sf9 cells derived from *Spodoptera frugiperda*, a lepidopteran species distantly related to *P. polyxenes*. In addition, we have demonstrated that transcription from this promoter fragment can be induced by xanthotoxin in transfected Sf9 cells. These studies provide the basis for future analysis of the mechanisms modulating transcriptional regulation of this insect P450 gene family.


References

Berenbaum, M.R., Cohen, M.B. & Schuler, M.A. (1990): Cytochrome P-450 in plant-insect interactions: Induction and deductions. In *Molecular Insect Science*, eds. H.H. Hagendorn et al., pp. 257-262. New York: Plenum Press.

Berenbaum, M.R. (1991): Coumarins. In *Herbivores: Their Interaction with Secondary Plant Metabolites*, eds. G. Rosenthal & M. Berenbaum, pp. 221-249. New York: Academic Press.

Berenbaum, M.R. & Zangerl, A.R. (1993): Furanocoumarin metabolism in *Papilio polyxenes*: biochemistry, genetic variability and ecological significance. Oecologia, in press.

Cohen, M.B. (1991): Characterization of a hostplant-inducible cytochrome P450 from the black swallowtail caterpillar, *Papilio polyxenes* (Lepidoptera Papilionidae). Ph.D. dissertation, University of Illinois, Urbana-Champaign.

Cohen, M.B., Berenbaum, M.R. & Schuler, M.A. (1989): Induction of cytochrome P450-mediated metabolism of xanthotoxin in the black swallowtail. *J. Chem. Ecol.* 15, 2347-2355.

Cohen, M.B., Schuler, M.A. & Berenbaum, M.R. (1992): A host-inducible cytochrome P450 from a host-specific caterpillar: molecular cloning and evolution. *Proc. Natl. Acad. Sci. USA* 89, 10920-10924.

Ma, R., Cohen, M.B., Berenbaum, M.R. & Schuler, M.A. (1994): Black swallowtail (*Papilio polyxenes*) alleles encode cytochrome P450s that selectively metabolize linear furanocoumarins. Submitted.

# From basic cytochrome P450 research to applications in drug development

Anthony Y.H. Lu

*Department of Drug Metabolism, Merck Research Laboratories, Rahway, New Jersey 07065, USA*

In the last thirty years, tremendous progress has been made in cytochrome P450 research. In the '60s, elegant studies by Omura and Sato (1964) established the hemoprotein nature of cytochrome P450. Differential effects on the metabolism of various substrates in the *in vivo* induction studies in animals provided the first indication for the existence of multiple forms of cytochrome P450. Solubilization, resolution, and reconstitution of the liver microsomal cytochrome P450 system opened the door for detailed biochemical studies for functional components and for addressing the question on cytochrome P450 multiplicity. The '70s highlighted the purification and characterization of numerous cytochrome P450 species from animals, the discovery of the Ah receptor, and the extensive mechanistic studies to define the events involved in the cytochrome P450 catalytic cycle. In the '80s, cytochrome P450 research was dominated by molecular biology technology as scientists began to unravel the secret of gene structure, function, and regulation. Hundreds of cytochrome P450 cDNAs were cloned and numerous P450 proteins were produced in various expression systems for metabolism, toxicity, structure, and function studies. Important progress was also made on the purification, characterization, and cloning of human cytochrome P450s, and the mechanism of regulation of human cytochrome P450 genes. Knowledge from the cytochrome P450 basic research is vital in the '90s for the application in plant science, metabolism, drug discovery and development, and other related fields.

## SEARCHING FOR FUNCTIONAL COMPONENTS

I have been very fortunate to have the opportunity to participate in cytochrome P450 research and to witness many of the key discoveries in this field over the years. My good fortune started in 1966 when I joined Jud Coon's laboratory at the University of Michigan as a Postdoctoral Fellow. Jud has been an admirable mentor and a good friend ever since. His high scientific and moral standard has made an important impact on my scientific career.

At the time I joined Jud's laboratory, Julian Peterson and others were investigating the enzyme components involved in the ω-hydroxylation of hydrocarbons and fatty acids in *Pseudomonas oleovorans*. Using low ionic strength buffer containing EDTA, Peterson succeeded in solubilizing a pink protein from the bacterial cells, purified and identified this electron carrier as rubredoxin, a non-heme iron protein (Peterson et al., 1966). Jud suggested that I study the liver microsomal fatty acid ω-hydroxylase system, hoping perhaps that a similar method can be used to solubilize the mammalian enzyme component. Indeed, the first attempt to solubilize the ω-hydroxylase was to suspend the rabbit microsomal preparation in dilute Tris buffer containing EDTA, followed by centrifugation. Not surprisingly, no functional components were solubilized by this method and all ω-hydroxylase activity was recovered in the microsomal fraction. We learned very quickly that the bacterial and mammalian ω-hydroxylases are two very different systems.

Although at that time we were unaware of the involvement of cytochrome P450 in microsomal hydroxylation, we assumed that the ω-hydroxylase was a multicomponent enzyme system based on the findings of the *Pseudomonas* ω-hydroxylase work. Our strategy was to assay ω-hydroxylation activity at every step following solubilization and fractionation and always to combine all fractions to ensure that no functional components were left out. After trying various methods, the liver microsomal ω-hydroxylation system was solubilized and resolved into three fractions containing cytochrome P450, NADPH-cytochrome P450 reductase, and a lipid later identified as phosphatidylcholine (Lu and Coon, 1968; Lu et al., 1969a; Strobel et al., 1970). All three fractions were required for the ω-hydroxylation of lauric acid. Subsequently, the reconstituted cytochrome P450 system was shown to also metabolize a variety of drugs and steroids (Lu et al., 1969b; Lu et al., 1972).

FROM ONE CYTOCHROME P450 TO MANY

Following my postdoctoral training, I had another good fortune in joining the laboratory of Allan Conney, then at Hoffmann-LaRoche. Allan and his colleagues had previously demonstrated that liver microsomes prepared from phenobarbital- or 3-methylcholanthrene-treated rats have different substrate specificity in the metabolism of drugs and steroids (Conney, 1967). By isolating the three components from rats treated with either phenobarbital or 3-methylcholanthrene and assaying their catalytic activities in the reconstituted system, we demonstrated that substrate specificity of liver microsomes from rats treated with different inducers is determined primarily by the cytochrome P450 component, rather than by the reductase or the lipid (Lu et al., 1971). These studies prompted us to initiate the purification of various cytochrome P450s in rats. In a very productive collaboration with Wayne Levin, a number of cytochrome P450 forms were purified from rats for metabolism, toxicity, and mechanistic studies, and these studies paved the way for the discovery of many more cytochrome P450s.

# FROM DEUTERIUM ISOTOPE EFFECTS TO METABOLIC SWITCHING

The arrival of Gerald Miwa in my laboratory in 1975 added a different dimension to my research direction on cytochrome P450. Because of Gerald's interest and talent in reaction mechanism, we initiated a series of studies to investigate the mechanism of cytochrome P450 reactions using deuterium isotope as a probe. This long collaboration is not only scientifically very fruitful, but also personally very rewarding in establishing our long-lasting friendship.

One of the most interesting studies carried out in this series was the oxidative O-deethylation of 7-ethoxycoumarin by rat cytochrome P450 1A1 and 2B1 (Miwa et al., 1984; Harada et al., 1984). Substitution of the hydrogens on the α-carbon of the ethoxy side chain of 7-ethoxycoumarin resulted in an observed deuterium isotope effect of 2 for P450 1A1 and 4 for 2B1. The intrinsic isotope effects for the O-deethylation of 7-ethoxycoumarin were 13-14 for both enzymes indicating the masking of the expression of the intrinsic isotope effect by other rate factors. Careful analysis of the stoichiometric data by Nobuhiro Harada and Gerald Miwa indicated that deuterated 7-ethoxycoumarin has no effect on rat cytochrome P450 1A1 dependent NADPH oxidation, oxygen and 7-ethoxycoumarin consumption, and $H_2O_2$ formation. Thus, despite a very large intrinsic isotope effect and a significant observed isotope effect on the O-deethylation of 7-ethoxycoumarin, the oxidase and the overall monooxygenase activities of rat 1A1 are not altered. Furthermore, the decreased rate in O-deethylation was associated with a pronounced increase in 6-hydroxylation on the aromatic ring. These results suggest that 7-ethyoxycoumarin can bind to the active site of rat 1A1 in at least two different orientations and that the active oxygen species, once formed, is committed to catalysis.

In addition to its value in probing the mechanism of cytochrome P450 catalyzed reactions, deuterium substitution could provide an alternative approach to improve the design of potential therapeutic agents. For example, deuterium substitution could potentially slow down the major P450 mediated inactivation pathway of a therapeutic agent and thus increase the half-life of the drug. The biological activity of a deuterated drug could be improved if the isotope effect of the inactivation pathway is associated with the increased formation of a minor metabolite which is biologically active. Finally, deuterium substitution could potentially decrease the toxicity of a therapeutic agent if the major metabolic pathway is associated with the toxicity of the drug.

# FROM BASIC RESEARCH TO DRUG DISCOVERY AND DEVELOPMENT

## Rational Drug Design

Major advances in biochemistry, molecular biology, x-ray crystallography, and computer graphics in recent years have made it possible to optimize drug design, based on enzyme or receptor structure, to gain maximum *in vitro* potency. However, it is not unusual to find that compounds with potent *in vitro* activity show little or no activity in the *in vivo* animal studies, quite often due to a lack of absorption or extensive metabolism by drug-metabolizing enzymes, particularly cytochrome P450. Using various tissues and recombinant cytochrome P450s from animals and humans, structure-metabolism relationships can be established to search for more metabolically stable compounds. Thus, it has become increasingly important to use metabolism and pharmacokinetic data to help design and select more desirable drug candidates.

Human Clinical Studies

Significant progress in basic research on human cytochrome P450 (Boobis and Davies, 1984; Gonzales et al., 1991; Guengerich, 1989; Wrighton and Stevens, 1992) has provided valuable tools to address certain critical issues on therapeutic agents prior to or during clinical studies. Using antibodies, specific inhibitors, and stably expressed human cytochrome P450s, it is now possible to identify the major human cytochrome P450(s) responsible for the metabolism of therapeutic agents before clinical study. For example, Regina Wang, Kamlesh Vyas, and other collaborators (Wang *et al.* 1991) identified cytochrome P450 3A as the major enzyme involved in the oxidative metabolism of lovastatin, a potent cholesterol-lowering agent. Identification of cytochrome P450(s) in the metabolism of therapeutic agents in humans could potentially contribute to our understanding on the fate of the drug and the predication of important interactions. For example, the rates and metabolic profiles of drug candidates by human P450 could be established; potential drug-drug interactions and the possible involvement of polymorphic P450 in metabolism can be predicted; the induction and inhibition of human P450s by the therapeutic agents in chronic drug treatment can be evaluated. Understanding the role of specific P450 in metabolism could also help to evaluate the mechanism regarding idiosyncratic reactions in humans. Thus, it has become increasingly feasible to adopt a more rational strategy to incorporate knowledge in cytochrome P450 and metabolism research for the development of better and safer therapeutic agents.

ACKNOWLEDGMENT

I would like to thank Terry Rafferty for her assistance in preparing this manuscript.

REFERENCES

Boobis, A.R. and Davies, D.S. (1984): Human cytochromes P450. *Xenobiotica* 14, 151-185.
Conney, A.H. (1967): Pharmacological implications of microsomal enzyme induction. *Pharmacol. Rev.* 19, 317-366.
Gonzales, F.J., Crespi, C.L. and Gelboin, H.V. (1991): cDNA-expressed human cytochrome P450s, a new age of molecular toxicology and human risk assessment. *Mut. Res.* 247, 113-127.
Guengerich, F.P. (1989): Characterization of human microsomal cytochrome P450 enzymes. *Annu. Rev. Pharmacol. Toxicol.* 29, 241-264.

Harada, N., Miwa, G.T., Walsh, J.S., and Lu, A.Y.H. (1984): Kinetic isotope effects on cytochrome P450-catalyzed oxidation reactions. Evidence for the irreversible formation of an activated oxygen intermediate of cytochrome P448. *J. Biol. Chem.* 259, 3005, 3010.

Lu, A.Y.H. and Coon, M.J. (1968): Role of hemoprotein P450 in fatty acid ω-hydroxylation in a soluble enzyme system from liver microsomes. *J. Biol. Chem.* 243, 1331-1332.

Lu, A.Y.H., Junk, K.W. and Coon, M.J. (1969a): Resolution of the cytochrome P450-containing ω-hydroxylation system of liver microsomes into three components. *J. Biol. Chem.* 244, 3714-3721.

Lu, A.Y.H., Strobel, H.W. and Coon, M.J. (1969b): Hydroxylation of benzphetamine and other drugs by a solubilized form of cytochrome P450 from liver microsomes. Lipid requirement for drug demethylation. *Biochem. Biophys. Res. Commun.* 36, 545-551.

Lu, A.Y.H., Kuntzman, R., West, S. and Conney, A.H. (1971): Reconstituted liver microsomal enzyme system that hydroxylates drugs, other foreign compounds and endogenous substrates. I. Determination of substrate specificity by the cytochrome P450 and P448 fractions. *Biochem. Biophys. Res. Commun.* 42, 1200-1206.

Lu, A.Y.H., Kuntzman, R., West, S., Jacobson, M. and Conney, A.H. (1972): Reconstituted liver microsomal enzyme system that hydroxylates drugs, other foreign compounds and endogenous substrates. II. Role of the cytochrome P450 and P448 fractions in drug and steroid hydroxylations. *J. Biol. Chem.* 247, 1727-1734.

Miwa, G.T., Walsh, J.S. and Lu, A.Y.H. (1984): Kinetic isotope effects on cytochrome P450-catalyzed oxidation reactions. The oxidative O-dealkylation of 7-ethoxycoumarin. *J. Biol. Chem.* 259, 3000-3004.

Omura, T. and Sato R. (1964): The carbon Monoxide-binding pigment of liver microsomes. I. Evidence for its hemoprotein nature. *J. Biol. Chem.* 239, 2370-2378.

Peterson, J.A., Basu, D. and Coon, M.J. (1966): Enzymatic ω-oxidation. I. Electron carriers in fatty acid and hydrocarbon hydroxylation. *J. Biol. Chem.* 241, 5162-5164.

Wang, R.W., Kari, P.H., Lu, A.Y.H., Thomas, P.E., Guengerich, F.P. and Vyas, K.P. (1991): Biotransformation of lovastatin. IV. Identification of cytochrome 3A proteins as the major enzymes responsible for the oxidative metabolism of lovastatin in rat and human liver microsomes. *Arch. Biochem. Biophys.* 290, 355-361.

Wrighton, S.A. and Stevens, J.C. (1992): The human hepatic cytochrome P450 involved in drug metabolism. *Crit. Rev. Toxicol.* 22, 1-21.

# V | BIOSYNTHESIS AND MEMBRANE TOPOLOGY OF MICROSOMAL P450. FUNCTIONAL IMPLICATIONS OF P450 STRUCTURE

# Biosynthesis and membrane topology of microsomal cytochrome P450

Masao Sakaguchi, Katsuyoshi Mihara, Tsuneo Omura

*Department of Molecular Biology, Graduate School of Medical Science, Kyushu University, Fukuoka, Fukuoka 812, Japan*

Two types of cytochrome P450 are found in animal cells, microsomal P450s and mitochondrial P450s, and they are both tightly associated with the membranes, microsomal membrane and mitochondrial inner membrane, respectively. Binding to the membranes seems to be essential for the function of P450s in eukaryotic cells.

Microsomal P450s are synthesized by the membrane-bound ribosomes of rough-surfaced endoplasmic reticulum (ER), and co-translationally integrated into the membrane (Sakaguchi, et al., 1984). The conformation of the membrane-bound P450 molecules, which is essential for the correct insertion of the heme prosthetic group to the newly synthesized molecules, must be formed at the step of their membrane integration, and the conformation seems to be dependent upon the membrane-binding segment at the amino terminal portion of the molecules. When this portion is deleted and the modified P450s are expressed in eukaryotic cells, they are not properly incorporated into the membrane and are usually rapidly degraded in the cells. The amino terminal portion of microsomal P450s seems to contain the information necessary for the correct folding of the membrane-bound P450 molecules in addition to that for the membrane targeting and binding.

<u>Conserved Structural Motifs in the Amino Terminal Portions of Microsomal P450s</u>
When we compare the primary structures of the amino-terminal portions of various microsomal P450s, we find four highly conserved structural motifs. They are 1) a negatively charged amino acid at the amino terminus, 2) a long stretch of highly hydrophobic amino acids, 3) several positively charged amino acids after the hydrophobic stretch, and 4) a proline rich region (PR) after the positively charged amino acid residues. Some microsomal P450s lack the negatively charged amino acid at the amino terminus, and some have a peptide

extension consisting of relatively hydrophilic amino acids before the hydrophobic stretch. Figure 1 shows the amino acid sequence of the amino terminal portion of a microsomal P450, P450(2C11) or P450(M-1) (Yoshioka, et al. 1987), which is a major P450 species in the liver microsomes of male rats. The four conserved structural motifs are indicated in the Figure by the symbols and the underlines.

```
        -                    +   + +
MDPVLVLVLTLSSLLLLSLWRQSFGRGKLPPGPTPLPIIGNT..
━━━━━━━━━━━━━━━━━━━━━━         \\\\\\\\\\\\\\\\\\\\\
     Hydrophobic stretch                Proline rich region
```

Fig. 1. Amino teminal portion of rat P450(2C11). Charged amino acid residues are indicated by - and + symbols. The hydrophobic stretch and the proline rich region are underlined.

Signal-Anchor Function of the Amino Terminal Sequence

The signal-anchor function of the amino terminal portion of microsomal P450s has been studied in detail (Sakaguchi, et al. 1987, 1992, 1993, Szeczesna-Skorupa, et al. 1988, 1989, Sato, et al., 1990). The hydrophobic segment in the amino teminal portion is recognized by the signal recognition particle (SRP), and functions as the signal for targeting the P450 peptides to ER membrane. The negatively charged amino acid, aspartic acid or glutamic acid, at the amino terminus is important in determining the function of the signal-anchor sequence (Szeczesna-Skorupa, 1988, Sato, et al., 1990). While the signal sequence (S) of the secretory proteins and the signal-anchor sequence (SAII) of the type II membrane proteins, which usually have a positively charged amino acid at the amino terminus, induce the translocation of the following peptides across the ER membrane, the P450-type signal-anchor sequence (SAI) inserts itself into the membrane but does not allow the membrane translocation of the following peptide (Fig. 2). Replacement of the negatively charged amino acid residue with a positively charged one converted the SAI of a microsomal P450 to SAII (Szeczesna-Skorupa et al. 1989).

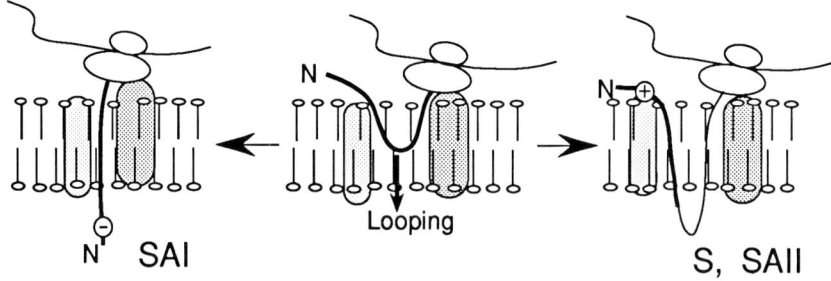

Fig. 2. Functions of Signal (S) and Signal-Anchor (SAI, SAII) Sequences in the Insertion of Peptides into ER Membrane.

Our studies have shown that the functions of the signal-anchor sequences are determined by the balance between the charge at the amino terminus and the length of the following hydrophobic segment (Sato, et al., 1990). As shown in Fig. 2, the hydrophobic segment is pulled into the hydrophobic interior of the membrane, whereas the positive charge at the amino terminus of SAII is attached to the negatively charged polar heads of the membrane phospholipids resisting to being pulled in. The negatively charged amino terminus of SAI may be more easily translocated across the membrane. We can thus expect reverse orientations of SAI and SAII in the ER membrane; the former is $N_{lumen}/C_{cytoplasm}$ and the latter is $N_{cytoplasm}/C_{lumen}$.

In the case of SAI, the signal-anchor sequence functions both as the membrane insertion signal and as the stop-transfer sequence (Sakaguchi, et al., 1987). The latter is a hydrophobic sequence which interrupts the translocation of peptides across the membrane. We analyzed the structural requirements for the stop-transfer sequence, and confirmed that the positively charged amino acid residues immediately after the hydrophobic segments potentiates the stop-transfer efficiency (Kuroiwa, et al. 1990, 1991). It is likely that the positively charged amino acids after the hydrophobic segment in the amino terminal portion of microsomal P450s also contribute to the signal-anchor function by potentiating the stop-transfer efficiency of the hydrophobic segment.

<u>Orientation of P450-type Signal-Anchor Sequence in the Membrane.</u>
The analysis of the membrane integration of type I signal-anchor sequences indicated $N_{lumen}/C_{cytoplasm}$ orientation of the amino terminal portion of microsomal P450s (Sakaguchi, et al., 1987, Szczesna-Skorupa, et al., 1988, Sato, et al., 1990), but direct confirmation of the luminal location of the amino terminus of microsomal P450s has been difficult (Sakaguchi, et al. 1993).

When a short amino acid sequence containing a potential N-glycosylation site was attached to the amino terminus of microsomal P450s, the site was glycosylated when the modified P450s were translated in vitro in the presence of rough microsomes or expressed in cultured mammalian cells (Monier, et al., 1987, Szeczesna-Skorupa and Kemper, 1993). The glycosylation of the introduced N-terminal glycosylation site indicated that the amino terminal portions of the modified P450s were translocated across the ER membrane into the lumen.

Human aromatase P450, P450 19 or P450(arom), has two potential N-glycosylation sites, one is Asn-12 before the hydrophobic segment of the signal-anchor sequence and the other is Asn-180 in the middle of the molecule (Harada, 1988), and the presence of sugars in the purified preparations was reported (Sethumadhavan, et al. 1991). We purified P450(arom) from human placenta and found it glycosylated with a high mannose-type sugar chain. We translated human P450(arom) cDNA in vitro in the presence of dog pancreas rough microsomes, and the newly synthesized P450(arom) incorporated into the microsomes was also glycosyalated with a high mannose-type sugar chain. The glycosylation at the amino teminal potential glycosylation site was confirmed by introducing mutations into each of the two potential glycosylation sites and examining the glycosylation of the mutated P450(arom)s by the in vitro system. P450(arom) is the first natural P450 with which the luminal location of the amino terminus was confirmed by glycosylation assay (Fig. 3).

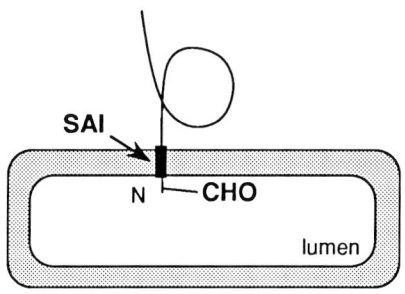

Fig. 3. Orientation of the Signal-Anchor Sequence of P450(arom) in the ER Membrane. The amino terminus and the attached sugar chain are indicated by N and CHO, respectively.

Proline-rich Region and the Conformation of Microsomal P450s
The proline-rich region (PR) is present in all microsomal P450s, following the signal-anchor sequence in the amino terminal portion, and the PR motif is particularly well conserved among the P450s belonging to the families 1, 2 and 3. The conserved motif is PPGPXPXPXXG.

```
1A1     MPSVYGFPAFTSATELLLAVTTFCLGFWVVRVTRTWVPKGLKSPPGPWGLPFIGHVLTLGKN
1A2      MAFSQYISLAPELLLATAIFCLVFWVLRGTRTQVPKGLKSPPGPWGLPFIGHMLTLGKN
2A1            MLDTGLLLVVILASLSVMLLVSLWQQKIRGRLPPGPTPLPFIGNYLQLNTK
2A2            MLDTGLLLVVILASLSVMFLVSLWQQKIRERLPPGPTPLPFIGNYLQLNMK
2A3            MLASGLLLVASVAFLSVLVLMSVWKQRKLSGKLPPGPTPLPFIGNYLQLNTE
2B1           MEPTILLLLALLVGFLLLLVRGHPKSRGNFPPGPRPLPLLGNLLQLDRG
2B2           MEPSILLLLALLVGFLLLLVRGHPKSRGNFPPGPRPLPLLGNLLQLDRG
2B3           MDTSVLLLLAVLLSFLLFLVRGHAKVHGHLPPGPRPLPLLGNLLQMDRG
2C6            MDLVMLLVLTLTCLILLSIWRQSSGRGKLPPGPIPLPIIGNIFQLNVK
2C7            MDLVTFLVLTLSSSLILLSLWRQSSRRRKLPPGPTPLPIIGNFLQIDVK
2C11           MDPVLVLVLTLSSSLLLLSLWRQSFGRGKLPPGPTPLPIIGNTLQIYMK
2C12           MDPFVVLVLSLSFLLLLYLWRPSPGRGKLPPGPTPLPIFGNFLQIDMK
2C13           MDPVVVLLLSLFFLLFLSLWRPSSGRGKLPPGPTPLPIIGNFFQVDMK
2D1         MELLNGTGLWSMAIFTVIFILLVDLMHRRHRWTSRYPPGPVPWPVLGNLLQVDLS
2D2         MGLLIGDDLWAVVIFTAIFLLLVDLVHRHKFWTAHYPPGPVPLPGLGNLLQVDFE
2D3         MELLAGTGLWPMAIFTVIFILLVDLMHRRQRWTSRYPPGPVPWPVLGNLLQVDLC
2D4         MRMPTGSELWPIAIFTIIFLLLVDLMHRRQRWTSRYPPGPVPWPVLGNLLQIDFQ
2D5         MELLNGTGLWPMAIFTVIFILLVDLMHRHQRWTSRYPPGPVPWPVLGNLLQVDPS
2E1            MAVLGITIALLVWVATLLVISIWKKIYNSWNLPPGPFPLPILGNIFQLDLK
2G1           MALGGAFSIFMTLCLSCLLILIAWKRTSRGGKLPPGPTPIPFLGNLLQVRID
3A1         MDLLSALTLETWVLLAVVLVLLYGFGTRTHGLFKKQGIPGPKPLPFFGTVLNYYMG
3A2         MDLLSALTLETWVLLAVILVLLYRLGTHRHGIFKKQGIPGPKPLPFLGTVLNYYKG
```

Fig. 4. Alingnments of the Proline-rich Regions of Rat Microsomal P450s belonging to Families 1, 2 and 3 (Yamazaki, et al. 1993). Conserved prolines and glycines in the proline-rich region are shadowed. Hydrophobic segments are underlined.

We examined the functional significance of the proline residues in this region by systematically altering these residues of P450(2C11). One, two and three proline residues out of the five in the region were exchanged with alanine residues, and the mutated proteins as well as wild type were expressed in *Schizosaccharomyces pombe* cells. The expressed wild type P450(2C11) showed a typical CO difference spectrum and was enzymatically active, whereas all the mutated proteins showed no P450 spectrum, suggesting that the substitution of the proline residues in PR resulted in the defect of proper heme incorporation (Yamazaki, et al., 1993). Proline and glycine are α-helix breakers. The proline residues in the PR seem to be crucial in the formation of the correct conformation of microsomal P450 molecules. The importance of successive proline residues in the carboxy-terminal region of P450(2C2) and P450(2C14) has recently been reported (Uno, et al., 1993).

Retention of Microsomal P450s to ER
Microsome-type P450s are found only in the ER membrane in eukaryotic cells. Since there is a bulk flow of proteins from ER to plasma membrane via Golgi apparatus in animal cells including the hepatocytes, microsomal P450s must be retained to ER by some mechanism without being transported to Golgi apparatus and further.
We have examined the possible contributions of the amino terminal conserved structural motifs to the retention of microsome-type P450s to the ER membrane by attaching various portions of the microsomal P450 molecule to the carboxy-terminus of several reporter proteins which are normally transported from ER to Golgi apparatus and further. Our tentative results obtained with cultured animal cells indicated the importance of the transmembrane hydrophobic segment in the ER retention of P450 molecules to ER.

The four conserved structural motifs in the amino-terminal portion of microsomal P450s are essential for the biogenesis of the P450s. The conserved motifs are needed for the ER-targeting and membrane integration of newly synthesized P450 molecules, and also for the formation of the correct membrane topology and conformation of the P450s. The microsomal localization of the P450s is also dependent on the amino terminal conserved structural motifs.

REFERENCES
Harada,N. (1988): Cloning of a complete cDNA encoding human aromatase: immunological identification and sequence analysis. Biochem.Biophys.Res.Commun., 156:725-732.
Kuroiwa,T., Sakaguchi,M., Mihara,K., and Omura,T.(1990): Structural Requirements for interruption of protein translocation across rough endoplasmic reticulum membrane. J.Biochem., 108: 829-834.
Kuroiwa,T., Sakaguchi,M., Mihara,K., and Omura,T.(1991): Systematic analysis of stop-transfer sequence for microsomal membrane. J.Biol.Chem., 266: 9251-9255.
Monier,S., Van Luc,P., Kreibich,G., Sabatini,D.D., and Adesnik,M. (1987): Signals for the incorporation and orientation of cytochrome P450 in the endoplas-

mic reticulum membrane. J.Cell Biol., 107: 457-470.

Sakaguchi,M., Mihara,K., and Sato,R. (1984): Signal recognition particle is required for co-translational insertion of cytochrome P-450 into microsomal membranes. Proc.Natl.Acad.Sci.USA., 81: 3361-3364.

Sakaguchi,M., Mihara,K., and Sato,R.(1987): A short amino-terminal segment of microsomal cytochrome P-450 functions both as an insertion signal and as a stop-transfer sequence. EMBO J., 6: 2425-2432.

Sakaguchi,M. and Omura,T. (1993): Topology and biogenesis of microsomal cytochrome P-450s. In Frontiers in Biotrnasformation, Volume 8, eds. K.Ruckpaul and H.Rein, pp. 59-73. Berlin, Akademie-Verlag.

Sakaguchi,M., Tomiyoshi,R., Kuroiwa,T., Mihara,K., and Omura,T.(1992): Functions of signal and signal-anchor sequences are determined by the balance between the hydrophobic segment and the N-terminal charge. Proc.Natl.Acad.Sci.USA., 89: 16-19.

Sato,T., Sakaguchi,M., Mihara,K., and Omura,T.(1990): The amino-terminal structure that determine topological orientation of cytochrome P-450 in microsomal membrane. EMBO J., 9: 2391-2397.

Shimozawa,O., Sakaguchi,M., Ogawa,H., Harada,N., Mihara,K., and Omura,T.(1993) Core glycosylation of cytochrome P-450(arom). Evidence for localization of N-terminus of microsomal cytochrome P-450 in the lumen. J.Biol.Chem., 268:(in press).

Sethumadhavan,K., Bellino,F.L., and Thotakura,N.R. (1991): Estrogen synthetase (aromatase). The cytochrome P-450 component of the human placental enzyme is a glycoprotein. Mol.Cell.Endocrinol., 78:25-32.

Szczesna-Skorupa,E., Browne,N., Mead,D., and Kemper,B. (1988): Positive charges at the NH2 terminus convert the membrane-anchor signal peptide of cytochrome P-450 to a secretory signal peptide. Proc.Natl.Acad.Sci.USA., 85: 738-742.

Szczesna-Skorupa,E., and Kemper,B. (1989): NH2-terminal substitution of basic amino acids induce translocation across the microsomal membrane and glycosylation of rabbit cytochrome P450 IIC2. J.Cell Biol., 108: 1237-1243.

Szczesna-Skorupa,E., Straub,P., and Kemper,B.(1993): Deletion of a conserved tetrapeptide, PPGP, in P450 2C2 results in loss of enzymatic activity without a change in its cellular location. Arch.Biochem.Biophys., 304: 170-175.

Szczesna-Skorupa,E. and Kemper,B. (1993): An N-terminal glycosylation signal on cytochrome P450 is restircted to the endoplasmic reticulum in a luminal orientation. J.Biol.Chem., 268: 1757-1762.

Uno,T., Imai,Y., Nakamura,M., Okamoto,N., and Fukuda,T.(1993): Importance of successive prolines in the carboxy-terminal region of P450 2C2 and P450 2C14 for the hydroxylase activities. J.Biochem., 114: 363-369.

Yamazaki,S., Sato,K., Suhara,K., Sakaguchi,M., Mihara,K., and Omura,T.(1993): Importance of the proline-rich region following signal-anchor sequence in the formation of correct conformation of microsmal cytochrome P-450s. J.Biochem., 114: (in press).

Yoshioka,H., Morohashi,K., Sogawa,K., Miyata,T., Kawajiri,K., Hirose,T., Inayama,S. Fujii-Kuriyama,Y. and Omura,T.(1987): Structural analysis and specific expression of micrsoomal cytochrome P-450(M-1) mRNA in male rat livers. J.Biol.Chem., 262: 1706-1711.

# Structure and function in P450s: are bacterial P450s good models for eukaryotic P450s?

Julian A. Peterson[1], Johann Deisenhofer[2], Charles Hasemann[2], K.G. Ravichandran[2], Sekhar S. Boddupalli[2], Sandra Graham-Lorence[1]

[1]Department of Biochemistry. [2]Howard Hughes Medical Institute and Department of Biochemistry, The University of Texas Southwestern, Medical Center at Dallas, Dallas, TX 75235, USA

Approximately thirty years ago, the discovery that the carbon monoxide inhibition of steroid oxidation by adrenal extracts was reversed by light of 450 nm was a hallmark event in the studies of P450s (Estabrook, et al, 1963). The elucidation of the structure, function, and mechanism of this unique class of proteins which had earlier been named cytochrome P450 by Omura and Sato (1961), has been the focus of an intense effort. Over the intervening years there have been many milestone discoveries each of which have spurred greater and more widespread interest in this class of enzymes by an extraordinarily wide ranging group of scientists. Because of this diversity, there are many different definitions of what a P450 is. So that there is a common basis for the present discussion, we will define what constitutes a "P450" based on structure and function. First and foremost, a P450 is a hemoprotein which has as one of its axial ligands a cysteinyl sulfur atom. The carbon monoxide complex of the ferrous form of hemoproteins with a cysteinyl ligand maximally absorb light of 450 nm. Second, P450s catalyze the activation of molecular oxygen and the insertion of one of the atoms of molecular oxygen into some compound, usually a hydrophobic organic compound with the formation of a hydroxyl group, epoxide, or oxide (monooxygenation). Finally, with the determination of the nucleotide sequences of the genes encoding more than 200 unique proteins which meet the first two requirements, it has been recognized that most of these proteins belong to a gene superfamily (Nelson, et al, 1993). Although, there are proteins such as NO synthase (Marletta, 1993), chloroperoxidase (Nuell, et al, 1988), thromboxane synthase (Ullrich, and Hecker, 1990), allene oxide synthase (Song, Funk, and Brash, 1993), etc. which meet one or more of these criteria, we will limit our discussion of structure and function to those proteins which satisfy all three. In addition, we will propose that the proteins which meet these criteria will all have many structural features in common while differing in many important respects reflecting their unique substrate specificities and electron donor partners.

To date, the soluble, bacterial protein, P450cam has provided most of the structural, functional and mechanistic guideposts for both prokaryotic and eukaryotic P450s (Poulos, 1991). The validity of the extrapolations from P450cam to other P450 systems is open to question because of the low amino acid sequence identity and homology between P450cam and other P450s, and the type of redox partners required (i.e. an iron-sulfur protein and a reductase in *Class I* P450s, as in P450cam, or only a reductase in *Class II* P450s). P450cam has been used primarily because it was the best characterized P450, and because the structure of P450cam at atomic resolution was determined by Poulos' research group in 1987 (Poulos, Finzel, and Howard, 1987). The isolation, purification and characterization by Fulco's group at UCLA of P450BM-3 (a soluble, self-sufficient,

*Class II* bacterial fatty acid monooxygenase) (Narhi and Fulco, 1986), and the subsequent crystallization (11) and structural determination of its hemeprotein domain (P450BM-P) at atomic resolution by our research group (Ravichandran, *et al*, 1993) has aided us in the structural analysis of this superfamily of proteins. In addition, we have recently completed the determination of the atomic structure of P450terp (Hasemann, *et al*, 1994) (a *Class I* P450 like P450cam (Peterson, *et al*, 1992)). Thus, we can now begin to inquire as to which of the extrapolations from P450cam to other P450s may be valid regarding structure and function.

As we were completing the determination of the atomic structure of P450BM-P at 2.0 Å resolution (Ravichandran, *et al*, 1993), we were struck by the dramatic overall similarity of this structure to that of the previously determined P450cam. Ribbon diagrams of these two proteins are extremely difficult to tell apart without detailed study which implies that their secondary structural elements are conserved. Fig. 1 is a topology drawing illustrating the secondary structural elements in each of the P450s which have been analyzed to date. As noted by Poulos (Poulos, *et al*, 1985), the structures are composed of two distinct domains linked by three short pieces of random coil. The presence of these structural elements in each of the P450s is striking given the very low sequence homology between these three proteins and indicates that similar structural elements will be present in all P450s. However, there are significant variations among these P450s in the lengths of both the α–helices and strands of β-sheets. In addition each of the P450s has variations on the secondary structure which may reflect the specificity of the electron donor partner as well as substrate binding. Examples of each of these variations will be presented and discussed below.

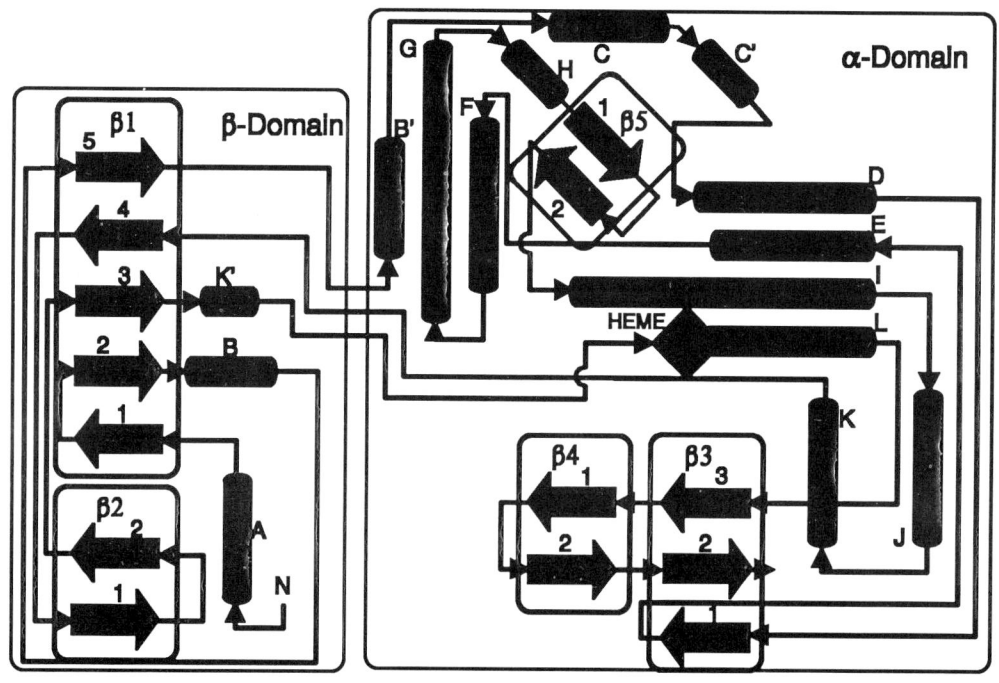

Fig. 1. *Topology drawing for the P450 superfamily of proteins.*

A closer examination of the structures has shown that less than half of the Cα backbone carbon atoms of P450cam, P450terp, and P450BM-P can be overlain in three dimensions with an RMS deviation of approximately 2 A. Thus, we can begin to define those elements which probably represent the "core" of P450s. As can be seen from Fig. 2, which is a plot of the deviation of the Cα backbone atoms of P450cam and P450BM-P, there are certain secondary structural elements which are within 2 A RMS deviation in both proteins. Chief among these elements are the carboxy-terminal half of the I-helix, and the L, D, and E helices. These helices make up a four-helix bundle with three parallel and one antiparallel (E) helix (Presnell and Cohen, 1989). The heme is held between the residues immediately amino-terminal to the L-helix and the I-helix as shown in Fig. 3. Other structural elements which constitute the core of these P450s include the J and K helices and most of the β1 sheet. Thus, although the overall structural topology of these P450s is similar, the details of the tertiary structure are quite different in some parts of the molecules.

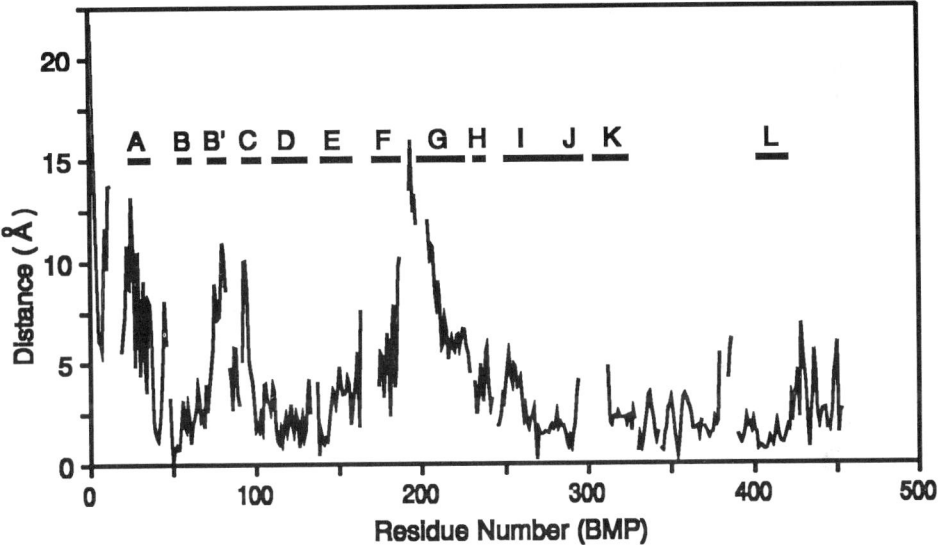

*Fig. 2. Plot of the deviation of the Cα backbone atoms between P450cam and P450BM-P.* The Y-axis of this figure is a measure of the deviation in Angstroms between the two structures in three-dimensional space. The X-axis is the sequence numbers of the optimally aligned sequences (Ravichandran, 1993). The secondary structural elements of P450BM-P are indicated by the horizontal bars and letters in the top part of the figure.

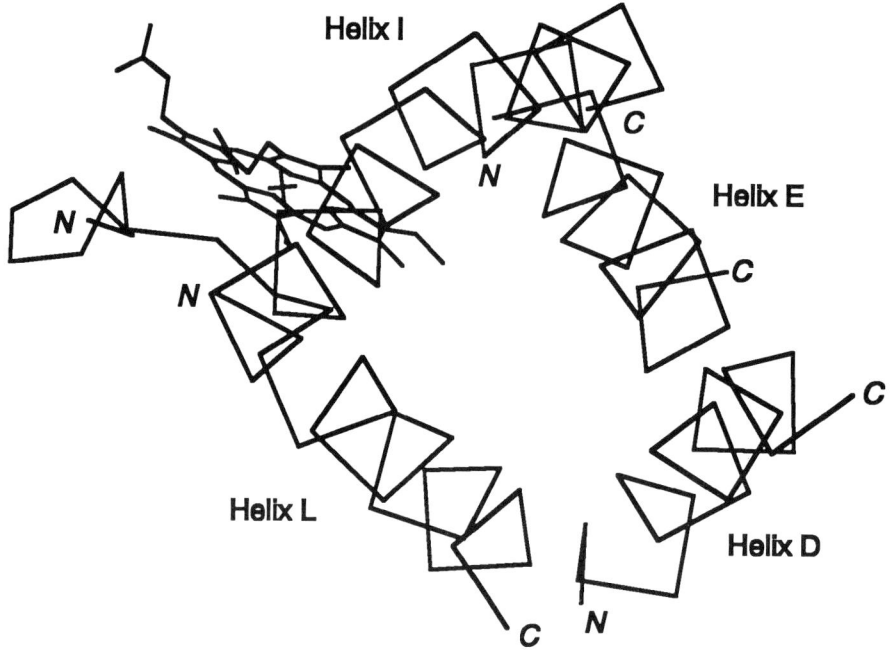

*Fig. 3. Alpha-carbon backbone representation of the 4-α-helical bundle in the core of P450s.*

On examination of the structure of P450BM-P, we were struck by the presence of a large hydrophobic funnel shaped opening (Fig 4A) extending from the surface of the molecule down to the active site heme residue (Ravichandran, 1993). This funnel shaped channel is approximately 20 A long and 10 A in diameter and there is only a single charged residue in this channel, R47, which is located at the mouth of the opening. We have proposed that an electrostatic interaction occurs between the guanidinium group of this arginine and the carboxyl group of the fatty acid substrate thereby orienting the substrate in the substrate access channel with the carbon atoms to be hydroxylated positioned near the heme. An additional feature of the structure was a hydrophobic patch on the surface adjacent to the substrate access channel (Fig. 4B). We have proposed that hydrophobic substrates such as fatty acids are first docked on this solvent exposed hydrophobic patch and then subsequently transferred by hydrophobic effect into the substrate access channel. It should be noted that the residue numbers of the amino acids in the docking region are low and thus near the amino terminus. Fulco's group recognized early on in their studies of P450BM-3 that there were lysine residues near the amino terminus of P450BM-3 that were sensitive to tryptic cleavage in the absence of a fatty acid substrate and that when cleaved, the enzyme would not undergo the low to high spin conversion characteristic of substrate binding (Narhi and Fulco, 1987).

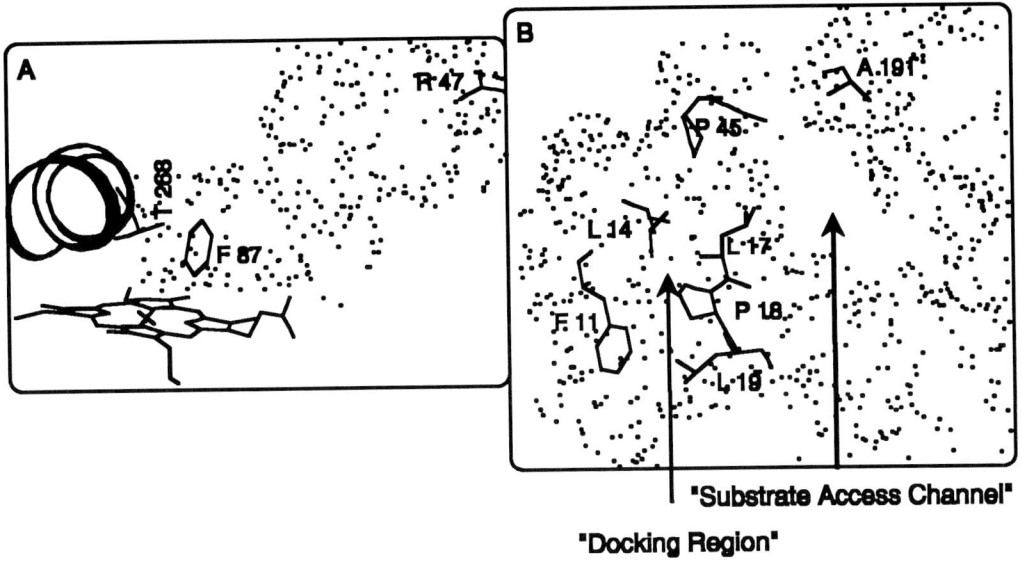

*Fig. 4. Connally surface of the "substrate access channel" and the "docking region" for substrates in P450BM-P.* In part A of this figure, the Connally surface is shown by dots for the access of substrates in P450BM-P from the solvent which would be found in the upper right portion of this figure to the heme oxygen binding site found in the lower left portion of this panel. The I-helix is represented by the series of circles on the left. In part B of this figure is shown the proposed hydrophobic substrate docking region which is exposed on the surface of this molecule near the substrate access channel. The amino acid residues are indicated by their single letter code and sequence number.

With the publication of the structure of P450cam, many investigators were troubled by the apparent lack of a readily discernible substrate access channel. Poulos' group proposed that the flexible loops on the surface of P450cam could open and close with thermal motion letting the substrate in and product out of the enzyme (Poulos, 1991). These residues were on the "top" of the molecule and a considerable distance from the amino-terminus. This troubled those individuals who were aligning sequences of P450s and comparing them to P450cam because if the membrane anchoring leader sequence in these P450s was on the membrane than the hydrophobic substrates of eukaryotic P450s would have to pass from the hydrophobic membrane into solvent and thus into the access channel to the active site. This scenario presumes that the P450s all have a similar shape and that they are anchored to the membrane via a hydrophobic amino terminal helix. With the structure of P450BM-P and our proposed pathway for access of fatty acids into the active site of this enzyme, we would extend the structural analysis of others as shown in Fig. 5 and include the fluorescence energy transfer measurements of Centeno and Gutierrez-Merino which indicate that the heme of eukaryotic P450s is approximately 50 Å from the surface of the membrane. Anchoring a eukaryotic P450 to a membrane via a hydrophobic leader sequence, as illustrated in this cartoon, would position the hydrophobic docking region of P450BM-3 adjacent to the surface of the membrane. The transfer of hydrophobic compounds from the lipid bilayer of the membrane to the active site heme probably would thus be facilitated.

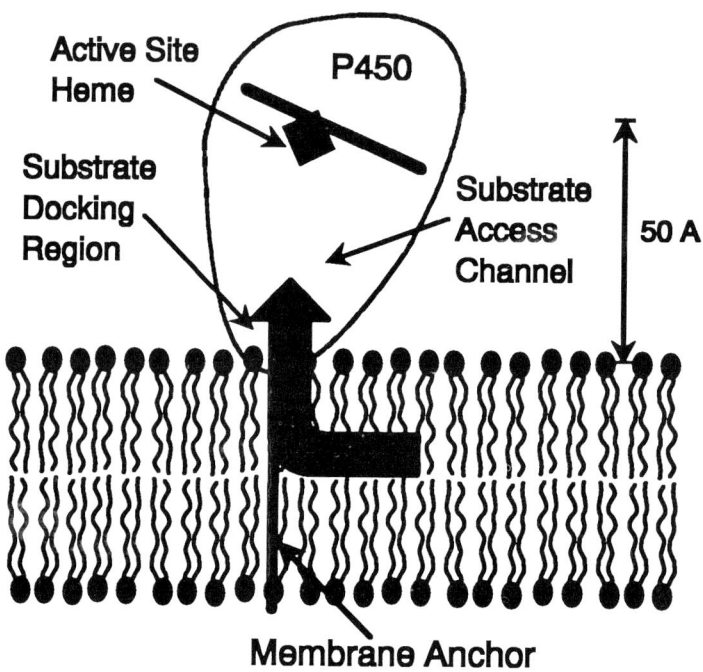

Fig. 5. *A cartoon illustrating how eukaryotic P450 might be anchored to the lipid bilayer of membranes.*

In summary the structures of these P450s will not predict the precise details of the structure of other P450s but they have enabled us to identify those features which constitute the central core of these widely divergent proteins. We have extended our definition of a P450 to now include the structurally conserved four-α-helix bundle, the heme binding region, and the β-sheet region. The other regions which control the substrate and redox donor specificity will vary from protein to protein.

**REFERENCES**

Estabrook, R. W., Cooper, D. Y., and Rosenthal, O. (1963): *Biochem. Z.* **338**, 741-755.
Omura, T., and Sato, R. (1961): *J. Biol. Chem.* **237**, 1375-1376.
Nelson, D. R., Kamataki, T., Waxman, D. J., Guengerich, F. P., Estabrook, R. W., Feyereisen, R., Gonzalez, F. J., Coon, M. J., Gunsalus, I. C., Gotoh, O., Okuda, K., and Nebert, D. W. (1993): *DNA Cell Biol.* **12**, 1-51.
Marletta, M. A., (1993): *J. Biol. Chem.* **268**, 12231-12234.
Nuell, M. A., Fang, G. H., Axley, M. J., Kenigsberg, P., and Hager, L. P. (1988): *J. Bacteriol.* **170**, 1007-1011.

Ullrich, V., and Hecker, M. (1990): *Adv. Prostaglandin Thromboxane Leukot. Res.* **20**, 95-101.
Song, W. C., Funk, C. D., and Brash, A. R. (1993): *Proc. Natl. Acad. Sci. USA* **90**, 8519-8523.
Poulos, T. L. (1991): *Methods Enzymol.* **206**, 11-30.
Poulos, T. L., Finzel, B. C., and Howard, A. J. (1987): *J. Mol. Biol.* **195**, 786-800.
Narhi, L. O., and Fulco, A. J. (1987): *J. Biol. Chem.* **262**, 6683-6690.
Boddupalli, S. S., Hasemann, C. A., Ravichandran, K. G., Lu, J.-Y., Goldsmith, E. J., Deisenhofer, J., and Peterson, J. A. (1992): *Proc. Natl. Acad. Sci. U.S.A.* **89**, 5567-5571.
Ravichandran, K. G., Boddupalli, S. S., Hasemann, C. A., Peterson, J. A., and Deisenhofer, J. (1993): *Science* **261**, 731-736.
Hasemann, C. A., Ravichandran, K. G., PETERSON, J. A., and Deisenhofer, J. (1994): *J. Mol. Biol.* (in press).
Peterson, J. A., Lu, J.-Y., Geisselsoder, J., Graham-Lorence, S., Carmona, C., Witney, F., and Lorence, M. C. (1992): *J. Biol. Chem.* **267**, 14193-14203.
Poulos, T. L., Finzel, B. C., Gunsalus, I. C., Wagner, G. C., and Kraut, J. (1985): *J. Biol. Chem.* **260**, 16122-16130
Presnell, S. C., and Cohen, F. E., (1989): *Proc. Natl. Acad. Sci. USA* **86**, 6592-6596.
Narhi, L. O., and Fulco, A. J. (1987): *J. Biol. Chem.* **262**, 6683-6690.
Centeno, F., and Gutierrez-Merino, C., (1992): *Biochemistry* **31**, 8473-8481.

# Structural and functional mapping of P450 proteins inferred from multiple sequence alignment

Osamu Gotoh

*Department of Biochemistry, Saitama Cancer Center Research Institute, Ina-machi, Saitama 362, Japan*

A multiple sequence alignment of more than 200 complete cytochrome P450 protein sequences was obtained by a two-step method. First, sequences within each family (or a few close families) were aligned with a newly developed method of a combination of progressive and randomized iterative strategies. The average hydrophobicities, secondary-structure forming propensities, and amino-acid compositions at individual alignment positions were calculated to generate a "profile" of each family. The second step of alignment was carried out to maximize the similarity in the profiles of the families using again the randomized iterative strategy. Based on the alignment and the known X-ray structures of P450cam (CYP101) and P450BM3 (CYP102), the locations of various functional and structural units are inferred and mapped on individual P450 sequences. The six putative substrate recognition sites (SRSs) assigned in this way agreed very well with experimental observations obtained with protein engineering techniques. High correlations were found between these SRSs and the regions with high amino-acid replacement rates, indicating that cytochrome P450s, as a whole, gained a wider spectrum of substrate specificities by varying the sequence and structure of their substrate recognition sites while keeping their basic tertiary structure unchanged.

## INTRODUCTION

More than 200 complete cytochrome P450 protein sequences have so far been published and categorized (Nelson et al., 1993). A wealth of information about structural and functional components of individual forms of P450 proteins have been drawn from theses sequence data (Nelson & Strobel, 1989; Gotoh & Fujii-Kuriyama, 1989; Gotoh, 1992) through various kinds of analyses including sequence alignment, structural prediction, and comparison with the known three-dimensional (3D) structure of P450cam (Poulos et al., 1987). Probably the most important outcome of these analyses was the establishment of the concept that all P450 proteins have essentially the same 3D structure in spite of the extensive sequence divergences between members, difference in subcellular localizations, or difference in electron transfer systems involved. Undoubtedly, multiple sequence alignment played a central role for this outcome.

I have previously proposed a structural and functional map of CYP2 family proteins which was obtained from a group-to-group alignment between CYP2 and bacterial type (P450cam-like) P450s

(Gotoh, 1992). The rationale behind the approach was that a group-to-group alignment between distant families is much more reliable than an alignment between individual members of the families. The predicted functional sites, especially substrate recognition sites, have been supported by several experimental observations. All this suggests that detailed structural and functional mapping of all the P450 proteins might be possible if we get a reliable multiple sequence alignment of all members. Multiple sequence alignment, however, remains, for computer science, problematic, belonging to a class of tough problems which elude precise solution. Various efforts are in progress to reduce computation time and to improve reliability. The randomized iterative strategy developed by Berger and Munson (1991) and Gotoh (1993a) is a most successful method for these purposes. By introducing generalized profiles, I have further improved the method so that multiple groups of sequences are aligned efficiently (Gotoh, 1993c). By using this new method, I have obtained an alignment of a total of 210 P450 sequences divided into 19 groups, each of which has been aligned beforehand with a combination of progressive and randomized iterative methods (Gotoh, 1993a). The quality of the results was assessed by comparing the alignment between P450cam and P450BM3 with that obtained recently by Ravichandran et al. (1993) based on 3D structures of P450cam and P450BM3.

## METHODS

Amino acid or coding nucleotide sequences were collected from public databases or directly obtained from the literature. P450 sequences within each family (Nelson et al., 1993) are aligned first by the conventional progressive method, and then refined with the randomized iterative strategy (Gotoh, 1993a). Several close families are aligned further by the same strategy to generate a group alignment. For example, all P450cam-like bacterial P450s (CYP101, 103-109, 111-112) are aligned at this stage, while P450BM3 (residue number from 1 to 460) comprises a single group. Nineteen group alignments are thus obtained, and they were transformed into profile vectors consisting of the average hydrophobicities (Kyte & Doolittle, 1982), secondary-structure forming propensities (Gibrat et al., 1987), and amino-acid compositions at individual alignment positions. The final alignment of all members was obtained again by the use of randomized iterative strategy, but this time the similarities in profiles of the groups were maximized.

## RESULTS AND DISCUSSION

The alignment of 14 bacterial type P450s (Fig[1]. 1) shows an excellent compatibility with the known tertiary structure of P450cam (Poulos et al., 1987). All but B' helix of the 13 helices and the six β structures in P450cam appear to be retained in all of the proteins as judged from the predicted secondary structures and the distribution of gaps. All the sites corresponding to the residues of P450cam that interact with the heme are strictly conserved, although the 'invariant threonine' in the distal helix appears to be more variable than previously thought. In contrast, the putative substrate binding/recognition sites inferred from the alignment and the structure of P450cam are highly variable and contain many gaps, except for the site in the I helix. These findings are fully consistent with the

---

**Fig. 1.** Alignment of bacterial type P450s. The helices (=), sheets (`), and turns (>><<) found in the P450cam 3D structure are indicated above the alignment. Completely and marginally conserved sites are indicated by the capital and lowercase one-letter amino acid codes, respectively. The sites exclusively occupied by large aliphatic (@), hydrophobic (o), positive (+), negative (_), polar (j), or small (.) amino acids are also indicated. Substrate-contact sites and heme-contact residues are shown by triangles and rhombi, respectively.

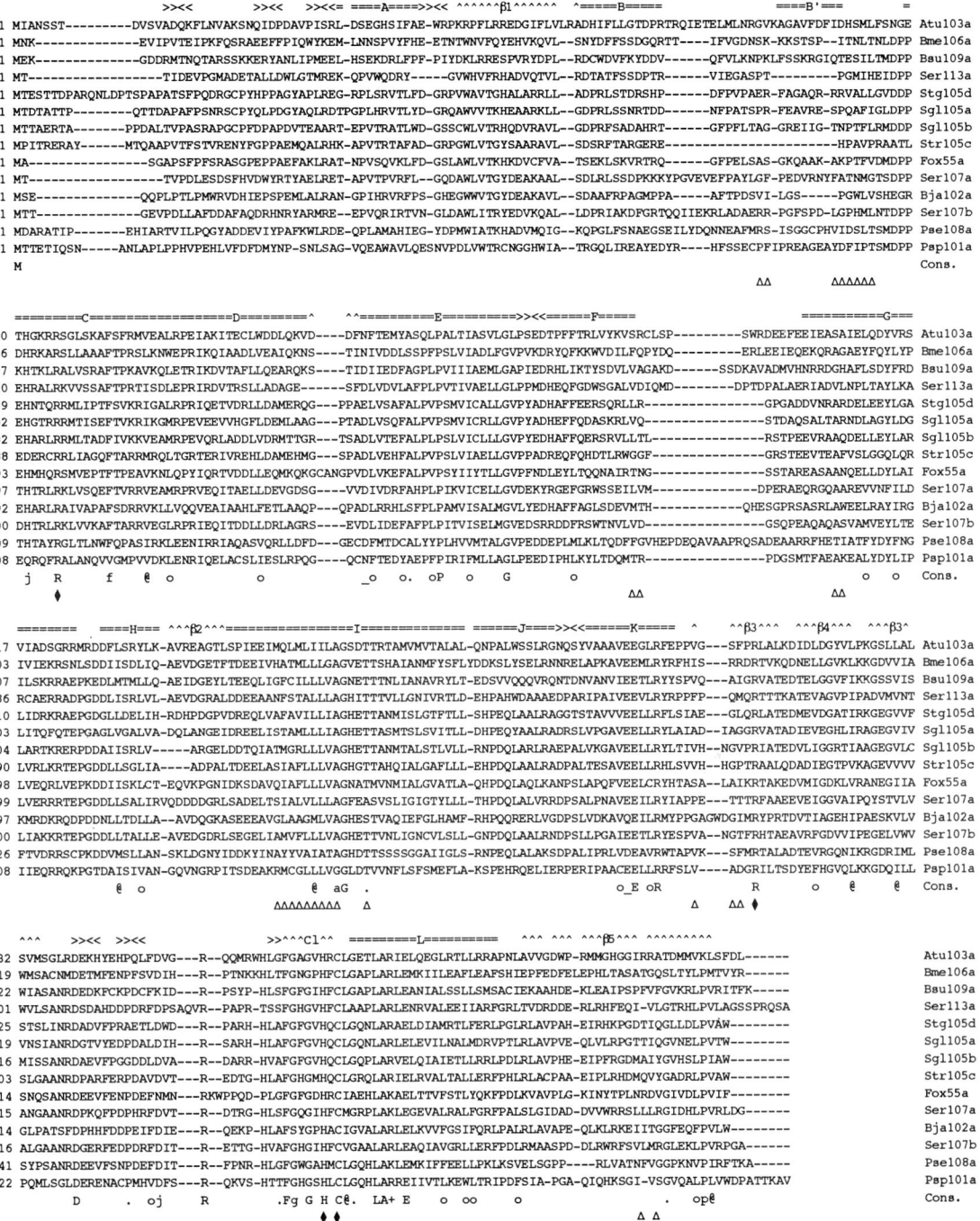

hypothesis that P450 proteins have evolved to widen their catalytic specificities by changing the sequence and structure of substrate-binding/recognition sites but keeping their basic tertiary structure unchanged.

By virtue of the multiple profile comparisons, the reliability in alignment between families is conceivably improved. To confirm this supposition, the P450cam and P450BM3 sequences were extracted from the final alignment, and compared with the structural alignment recently proposed by Ravichandran et al. (1993) (Fig. 2). About 64% of comparable sites in the two alignments agreed precisely (the N-terminal 21 sites were omitted for this comparison). This percentage jumped up to 97%, if we allowed for displacements up to four residues. By contrast, only about a 60% match by this criterion was attained when the two sequences were aligned pairwise by the standard method. P450cam and P450BM3 have evolved on the different linages after the first major divergence of P450 superfamily that probably took place more than 1.5 billion years ago (Gotoh, 1993b). Hence, the result of the test gives us an idea about the lower limit of the quality of the alignment of all P450 members.

Based on a group-to-group alignment between CYP2 and bacteria type P450s, I have mapped six putative substrate recognition sites (SRS1-6) on CYP2 proteins (Gotoh, 1992). Since many experimental findings about substrate recognition sites of enzymes of this family are accumulating, we can directly assess the prediction by comparing the experimentally identified substrate recognition sites with those inferred from the alignment. A surprisingly good correlation has been found between the experimental and inferred substrate recognition sites in CYP2 family proteins (Halpert & He, 1993; Hsu et al., 1993; Veronese et al., 1993; Iwasaki et al., 1993). The distribution of the conserved and variable regions in CYP2 sequences also well correlates with that observed in bacterial type P450s, and hence reinforce the hypothesis about the evolution of P450 proteins as discussed above. Although experimental observations about other families are still scarce, the distributions of sequence variability in CYP3 and CYP4 families are similar to that of CYP2 (Fig. 3), implying similar evolutionary and structural constraints imposed on these drug-metabolizing P450 families. Moreover, a few studies on P450s that are involved in steroidogenesis support our mapping of substrate recognition sites in this

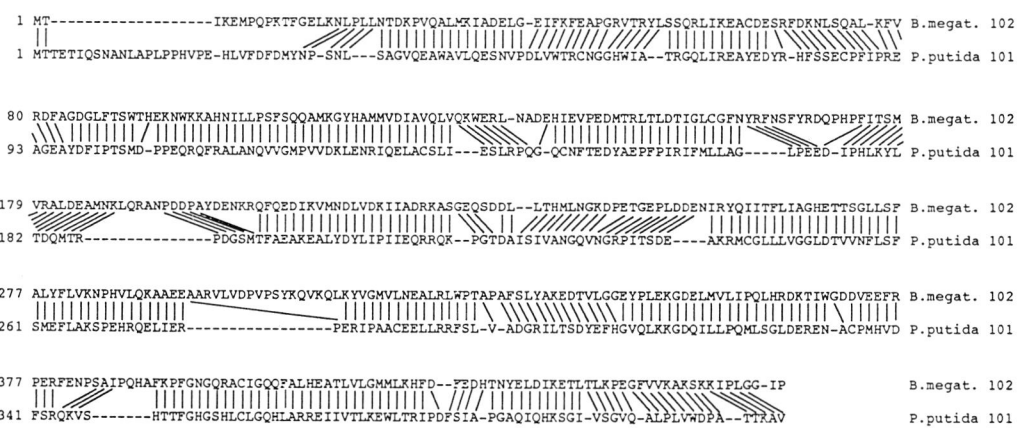

Fig. 2. Alignment of P450cam and P450BM3 sequences extracted from the alignment of all members of P450 proteins. Vertical lines indicate the corresponding residues suggested by Ravichandran et al. (1993) based on the 3D structures of the two proteins.

class of P450s (Matsukawa et al., 1993).

Although the number of available P450 sequences has grown more than five fold, and we can now map more accurate substrate recognition sites in various P450s, our knowledge about the sequence–structure relationships common to all P450s have been essentially unchanged since the time of our first analysis of multiple P450 sequences (Gotoh & Fujii–Kuriyama, 1989). In particular, the functional roles of two well–conserved regions, the K–helix and the aromatic regions, are still obscure. Further experiments and more detailed modelling of eukaryotic P450s will be necessary to solve these problems.

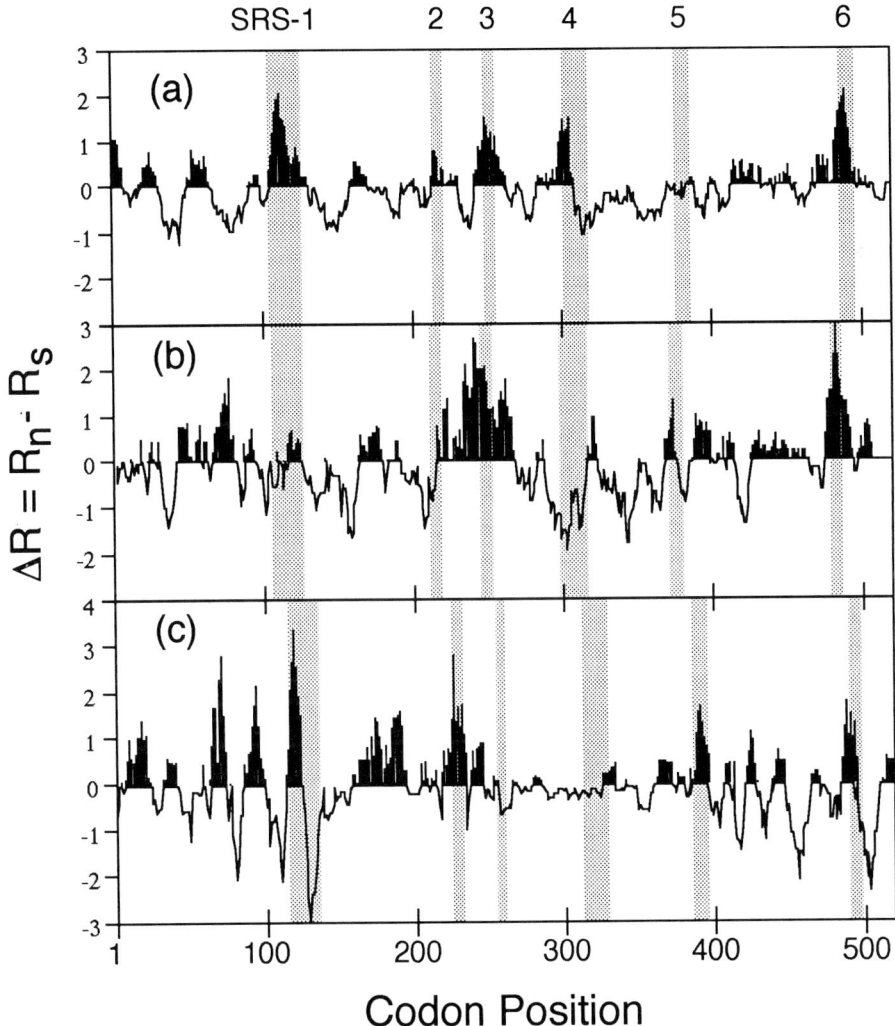

**Fig. 3.** Local variations in difference between rates of nonsynonymous and synonymous nucleotide substitutions. Calculations were made individually for CYP2 (a), CYP3 (b) and CYP4 (c) family members as described by Gotoh (1992). Shaded areas indicate the six putative substrate recognition sites.

This work was partly supported by a Grant-in-Aid for Scientific Research from the Ministry of Education, Science and Culture of Japan.

REFERENCES

Berger,M.P. & Munson,P.J. (1991): A novel randomized iterative strategy for aligning multiple protein sequences. *Comput. Applic. Biosci.* **7**, 479-484.

Gibrat,J.-F., Garnier,J. & Robson,B. (1987): Further developments of protein secondary structure prediction using information theory. New parameters and consideration of residue pairs. *J. Mol. Biol.* **198**, 425-443.

Gotoh,O. (1992): Substrate recognition sites in cytochrome P450 family 2 (CYP2) proteins inferred from comparative analyses of amino acid and coding nucleotide sequences. *J. Biol. Chem.* **267**, 83-90.

Gotoh,O. (1993a): Optimal alignment between groups of sequences and its application to multiple sequence alignment. *Comput. Applic. Biosci.* **9**, 361-370.

Gotoh,O. (1993b): Evolution and differentiation of P-450 genes. in *Cytochrome P-450*, ed. T.Omura,, Y.Ishimura, & Y.Fujii-Kuriyama, pp. 255-272. Tokyo: Kodansha and Weinheim: VCH.

Gotoh,O. (1993c): Further improvement in methods of group-to-group sequence alignment with generalized profile operations. in preparation.

Gotoh,O. & Fujii-Kuriyama,Y. (1989): Evolution, structure, and gene regulation of cytochrome P-450. in *Frontiers in Biotransformation*, ed. K.Ruckpaul & H.Rein, Vol. 1, pp. 95-243. Berlin: Akademie Verlag.

Nelson,D.R., Kamataki,T., Waxman,D.J., Guengerich,P., Estabrook,R.W., Feyereisen,R., Gonzalez,F.J., Coon,M.J., Gunsalus,I.C., Gotoh,O., Okuda,K. & Nebert,D.W. (1993): The P450 superfamily: Update on new sequences, gene mapping, accession numbers, early trivial names of enzymes, and nomenclature. *DNA Cell Biol.* **12**, 1-51.

Nelson,D.R. & Strobel,H.W. (1989): Secondary structure prediction of 52 membrane-bound cytochromes P450 shows a strong structural similarity to P450cam. *Biochemistry* **28**, 656-660.

Halpert,J.R. & He,Y. (1993): Engineering of cytochrome P450 2B1 specificity. Conversion of an androgen 16β-hydroxylase to a 15α-hydroxylase. *J. Bio. Chem.* **268**, 4453-4457

Hsu,M.-H., Griffin,K.J., Wang,Y., Kemper,B., & Johnson,E.F. (1993): A single amino acid substitution confers progesterone 6β-hydroxylase activity to rabbit cytochrome P450 2C3. *J. Biol. Chem.* **268**, 6939-6944.

Iwasaki,M., Darden,T.A., Pedersen,L.G., Davis,D.G., Juvonen,R.O., Sueyoshi,T. & Negishi,M. (1993): Engineering mouse P450coh to a novel corticosterone 15α-hydroxylase and modeling steroid-binding orientation in the substrate pocket. *J. Biol. Chem.* **268**, 759-762.

Kyte,J. & Doolittle,R.F. (1982): A simple method for displaying the hydropathic character of a protein. *J. Mol. Biol.* **157**, 105-132.

Matsukawa,N., Nonaka,Y., Higaki,J. ,Nagano,M., Mikami,H., Ogihara,T. & Okamoto,T. (1993): Dahl's salt-resistant normotensive rat has mutations in cytochrome P450(11β), but salt-sensitive hypertensive rat does not. *J. Biol Chem.* **268**, 9117-9121.

Poulos,T.L., Finzel,B.C. & Howard,A.J. (1987): High-resolution crystal structure of cytochrome P450cam. *J. Mol. Biol.* **195**, 687-700.

Ravichandran,K.G., Boddupalli,S.S., Hasemann,C.A., Peterson,J.A., & Deisenhofer,J. (1993): Crystal structure of hemoprotein domain of P450BM-3, a prototype for microsomal P450's. *Science* **261**, 731-736.

Veronese,M.E., Doecke,C.J., Mackenzie,P.I., McManus,M.E., Miners,J.O., Rees,D.L.P., Gasser,R., Meyer,U.A. & Birkett,D.J. (1993): Site-directed mutation studies of human liver cytochrome P-450 isoenzymes in the CYP2C subfamily. *Biochem. J.* **289**, 533-538.

# Role of the NH$_2$-terminal region of microsomal P450s in catalysis and membrane binding

Steven J. Pernecky, Valerie S. Fujita, Lorelle L. Bestervelt, Minor J. Coon

*Department of Biological Chemistry, University of Michigan Medical School, Ann Arbor, Michigan, USA 48109-0606*

Based upon the results of immunochemical, molecular genetic, and proteolytic studies in several laboratories, the microsomal P450 cytochromes were proposed to be anchored to the endoplasmic reticulum by no more than one or possibly two transmembrane segments located in the NH$_2$-terminal region, with the NH$_2$-terminus in the lumen or cytosol, respectively (Nelson and Strobel, 1988; Black, 1992). On the other hand, the cytosolic location of bacterial cytochromes P450$_{cam}$ and P450$_{BM-3}$ is presumed to be due to the absence of a highly hydrophobic NH$_2$-terminal region. If cellular localization were dependent upon the hydrophobicity in the NH$_2$-terminal region, then deletion of the signal peptide from the microsomal P450 cytochromes would be expected to lead to their accumulation in the cytosol. However, removal of the NH$_2$-terminal region from P450 cytochromes generally does not prevent their insertion into membranes, as observed by Larson *et al.* (1991a) for expression in *Escherichia coli* of rabbit 2E1 lacking amino acids 3-29 and in other laboratories for P450s expressed in bacteria, yeast, or mammalian cells (Gillam *et al.*, 1993; Sakaki *et al.* 1988; Cullin, 1992; Clark and Waterman, 1991). Larson *et al.* (1991b) found that purified 2E1 (Δ3-29) was indistinguishable from the full-length form in its rate of oxygenation of ethanol, *N*-nitrosodiethylamine, aniline, and *p*-nitrophenol and in its interaction with the other components of the enzyme system. Other investigators have found that catalytic activity of truncated forms of other P450 cytochromes is not affected or is decreased, depending upon the P450 examined and the deletion made. In a further attempt to learn the extent to which the NH$_2$-terminal region influences incorporation of the microsomal P450 cytochromes into the membrane, Pernecky *et al.* (1993) constructed two 2E1 (Δ3-29) derivatives in which positive charges were inserted into this region and, alternatively, removed the first two hydrophobic regions to make 2E1 (Δ3-48). P450 2B4 forms lacking amino acids 2-20 or 2-27 were also made for comparative purposes, and the subcellular localization of all shortened 2B4 and 2E1 variants was determined after expression in *E. coli* ; the results of these experiments are summarized in this report. In additional experiments also presented in this study, chimeric P450 2B4 and 2E1 constructs were made by replacement of the 17 NH$_2$-terminal amino acids of 2E1 (Δ2-29) or 2B4 (Δ2-27) with the 19 NH$_2$-terminal residues of P450$_{BM-3}$. Finally, the catalytic proficiency of 2B4 (Δ2-27) was determined following its expression as a fusion protein with a thrombin recognition sequence and glutathione S-transferase, purification of the expressed protein by GSH-Sepharose chromatography, and liberation of the cytochrome from the fusion protein with thrombin.

MATERIALS AND METHODS

Truncated forms of liver microsomal P450 cytochromes 2B4 and 2E1 were constructed in the pJL expression plasmid as previously described by Pernecky *et al.* (1993). The 17 NH$_2$-terminal amino acids of 2E1 (Δ3-29) or 2B4 (Δ2-27) were replaced with the 19 NH$_2$-terminal residues of P450$_{BM-3}$ (Ruettinger *et al.*, 1989) with the use of a double-stranded linker:

```
CATGACAATTAAAGAAATGCCTCAGCCAAAAACGTTTGGAGAGCTTAAAAATTTACCGCTTCTGC
    GTTAATTTCTTTACGGAGTCGGTTTTTGCAAACCTCTCGAATTTTTAAATGGCGAAG
```

The polymerase chain reaction was used to generate a fragment in 2E1 that encompassed nucleotides 136 through 440 (Khani *et al.*, 1988), with the introduction of a *Pst*I site in the 5' end (C to G at nucleotide 138), and to amplify a fragment in 2B4 that comprised nucleotides 130 through 439 (Gasser *et al.*, 1988). The DNA linker and the *Pst*I/*Bam*HI-digested PCR fragment generated from 2E1 or the *Pst*I/*Bsm*I-digested PCR fragment generated from 2B4 were ligated with *Nco*I/*Bam*HI-digested pJL containing 2E1 or *Nco*I/*Bsm*I-digested pJL containing 2B4, respectively. An expression plasmid for a fusion protein of glutathione S-transferase and 2B4 (Δ2-27) was obtained by cloning *Not*I/*Eco*RI-digested 2B4 in pBluescript into similarly digested pGEX-KN (Hakes *et al.*, 1992) and by subsequent ligation of a double-stranded DNA linker, which contained a portion of the thrombin cleavage sequence, with the *Not*I and *Xma*III sites of similarly digested 2B4 in pGEX-KN. The thrombin recognition site in this construct was LeuValProArgMetAla (Chang, 1985), which was cleaved with thrombin (16 units/nmol P450 for 2h at 22 °C) to release 2B4(Δ2-27) with an NH$_2$-terminal Met. *E. coli* strain MV1304 was transformed with the expression plasmids, P450 expression was induced by isopropyl β-D-thiogalactoside, cells were grown with shaking at 37°C and harvested after 4 h, and subcellular fractions were prepared as described by Porter and Larson (1991). Quantitative immunoblots were examined as summarized by Pernecky *et al.* (1993), and catalytic activity of the P450 cytochromes expressed in *E. coli* was determined in a reconstituted system with NADPH-cytochrome P450 reductase and synthetic phospholipid. Ferrous carbonyl difference spectra of P450 forms expressed in *E. coli* membranes were determined after solubilization with β-D-octylglucoside and column chromatography on S-Sepharose to remove spectrally interfering bacterial cytochromes *o* and *d*, or after solubilization with low concentrations of Tergitol NP-10 as described by Pernecky *et al.* (1993).

## RESULTS AND DISCUSSION

A highly hydrophobic signal peptide is required for insertion of proteins into the inner membrane of *E. coli* by a mechanism that is highly similar to the signal recognition particle-dependent insertion of proteins into the endoplasmic reticulum. Consistent with this scheme are recent findings by Larson *et al.* (1991b) and Pernecky *et al.* (1993), which showed that about 75% of full-length 2E1 or 2B4 is located in the membrane fraction when expressed in *E. coli*. The microsomal P450 cytochromes expressed in *E. coli* had a CO-reduced difference spectrum identical to that obtained for the same enzymes isolated from the liver microsomes of adult rabbits. Interestingly, when a derivative of 2E1 lacking amino acids 3-29 was expressed in *E. coli*, the enzyme was also found to have a predominantly membranous location, indicating that the signal peptide is not the sole membrane anchor for the enzyme. The $NH_2$-terminal amino acid of 2E1 ($\Delta$3-29) purified from *E. coli* membranes or cytosol was found by sequence analysis to be Ala, as was observed for the full-length 2E1 isolated from liver microsomes of acetone-treated rabbits. The truncated cytochrome could not be removed by washing *E. coli* membranes with 0.1 M $Na_2CO_3$, pH 11, a treatment conventionally used to remove proteins that are peripherally associated with the membrane. Thus, anchoring of 2E1 ($\Delta$3-29) to the membrane is apparently not attributable to a loose attachment or occlusion of the truncated protein. Moreover, a specific attachment of the substrate-binding site of the shortened form to membrane components was also apparently ruled out, since 2E1($\Delta$3-29) could not be removed from the membrane by a carbonate wash followed by treatment with membrane constituents such as choline, serine, *myo*-inositol, or glycerol (100 mM), or laurate or dilauroylglyceryl-3-phosphorylcholine (100 μg/ml), or with the 2E1 inhibitor 4-methylpyrazole (0.5 mM) or 2E1 substrate *n*-butanol (40 mM). When the highly hydrophobic $NH_2$-terminal region of 2B4 was removed by deletion of amino acids 2-20 or 2-27, most of the enzyme was located in the cytosol in contrast to what was observed with 2E1 ($\Delta$3-39). The fraction of expressed cytochrome located in the membrane was resistant to extraction with 0.1 M $Na_2CO_3$, pH 11. The retention of the membrane-binding properties of both shortened P450 cytochromes suggests that there are protein segments other than the signal peptide that are determinants of membrane binding, but the extent to which these regions contribute to subcellular localization in *E. coli* differs somewhat between 2B4 and 2E1.

As shown in Fig. 1, the $NH_2$-terminal region of 2E1 ($\Delta$3-29) is still quite hydrophobic, which is a fundamental property of $NH_2$-terminal regions of proteins that are translocated across *E. coli* inner membrane (Saier *et al.*, 1989). To determine whether the hydrophobicity in the $NH_2$-terminal region contributes to the localization of 2E1 ($\Delta$3-29) in the membrane fraction of bacteria, a positive charge was inserted into the $NH_2$-terminal region by substitution of Arg for Gly at position 8, and in another experiment a second positive charge was inserted and a hydrophobic residue was removed by subsitution of Lys for Trp at position 3 (Pernecky *et al.*, 1993). Such a modification in the hydrophobic $NH_2$-terminal signal sequence of λ receptor, an *E. coli* outer membrane protein, has been shown by Emr *et al.* (1980) to lead to its accumulation in the cytosol. An alternative approach to reduction in the hydrophobicity of the $NH_2$-terminal region of 2E1 ($\Delta$3-29) was to delete the $NH_2$-

terminal segment to obtain 2E1 (Δ3-48), which has an $NH_2$-terminal 21-amino acid sequence (MetAlaAspLeuLysAspIleProLysSerPheGlyArgLeuAlaGluArgPheGlyPro) that contains 3 fewer hydrophobic residues than the $NH_2$-terminal region of 2E1 (Δ3-29) and has 7 charged residues. The approach used in the present study to reduce the hydrophobicity in the $NH_2$-terminal region of 2E1 (Δ3-29) was to replace the 17 $NH_2$-terminal amino acids with the 19 $NH_2$-terminal amino acids of the soluble, bacterial $P450_{BM-3}$. The construction of the chimeric protein was based upon alignment of the $NH_2$-terminal regions of 2E1 (Δ3-29) and $P450_{BM-3}$, which revealed a homologous segment characterized by the sequence LeuLeu(Gln/Asn)XAsp, where X represents a nonhomologous residue, followed by Lys in one of the two succeeding amino acid positions (Fig. 1). The conserved sequence in the $NH_2$-terminal region of the bacterial cytochrome and 2E1 (Δ3-29) was also apparent in the $NH_2$-terminal region of 2B4 (Δ2-27) as shown in Fig. 1; a chimeric 2B4 cytochrome bearing the $NH_2$-terminal region of $P450_{BM-3}$ was also constructed by replacing the 17 $NH_2$-terminal amino acids of 2B4 (Δ2-27) with the 19 $NH_2$-terminal amino acids of $P450_{BM-3}$.

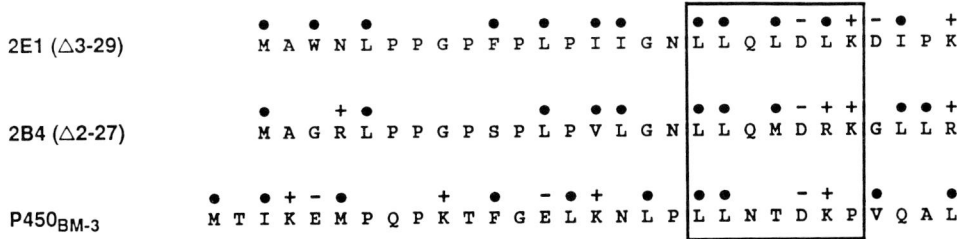

Fig. 1. $NH_2$-terminal sequence alignment of 2E1 (Δ3-29), 2B4 (Δ2-27) and $P450_{BM-3}$. The boxed region indicates a conserved sequence, with the hydrophobic residues shown as filled circles and the ionic residues with the appropriate charge.


None of the expressed forms of P450 2E1 or 2B4 was degraded as judged by immunoblot analysis, and all of the membrane-bound forms had typical CO-reduced difference spectra except 2E1 (Δ3-48), which could not be resolved from spectrally interfering bacterial cytochromes after S-Sepharose column chromatography. As shown in Fig. 2, when the hydrophobicity of the $NH_2$-terminal region of 2E1 (Δ3-29) was lowered by replacement of Gly by Arg at position 8, the fraction of expressed cytochrome localized in the membrane fraction was lower relative to 2E1 (Δ3-29), and an even lower percentage of membrane binding occurred with 2E1 (Δ3-29) when a second positive charge was inserted into the $NH_2$-terminal region. Most striking was the finding that the BM-3:2E1 chimera, which contained the highly hydrophilic $NH_2$-terminal region of $P450_{BM-3}$, was predominantly localized in the cytosol with only about 20% of the expressed protein located in the membrane. In fact, a direct correlation was observed between the hydrophobicity of the $NH_2$-terminal region of the truncated 2E1 forms and the percentage of expressed protein that was membrane-bound. Interestingly, Kaiser et al. (1987) have found a similar connection between hydrophobicity of the $NH_2$-terminal region and subcellular localization in yeast with proteins that are either secreted or localized in the cytosol. Collectively, these data indicate that deletion of the $NH_2$-terminal region of 2E1 and a reduction in the hydrophobicity of the new $NH_2$-terminal segment are required to obtain a predominantly cytosolic form. On the other hand, when the hydrophobic $NH_2$-terminal sequence of 2E1 ($\Delta$3-29) was removed to yield 2E1 ($\Delta$3-48), which substantially lowered the hydrophobicity of the $NH_2$-terminal sequence, 45% of the expressed protein was inserted into the membrane. However, the structural integrity of 2E1 ($\Delta$3-48) has yet to be determined by examination of the CO-reduced difference spectrum, since, unlike the other truncated forms of 2E1, it is not resolved from spectrally interfering bacterial cytochromes by S-Sepharose column chromatography.

As noted earlier, somewhat different from the results obtained with 2E1 ($\Delta$3-29) were those obtained with 2B4 ($\Delta$2-27), which was predominantly localized in bacterial cytosol (Fig. 2). The 27% that remained avidly bound to the membrane fraction could not be extracted with a carbonate wash. The $NH_2$-terminal region of 2B4 ($\Delta$2-27), like that of 2E1 ($\Delta$3-29), is hydrophobic and thus possesses a distinguishing characteristic of a signal peptide, which could explain why one-third of the expressed shortened 2B4 enzyme was located in the membrane. In contrast to what was observed with the BM-3:2E1 chimera, however, when the $NH_2$-terminal region of 2B4 ($\Delta$2-27) was replaced with the relatively hydrophilic $NH_2$-terminal region from P450$_{BM-3}$, there was no significant decline in the fraction of expressed protein that was membrane-bound even though there was a substantial decline in hydrophobicity in the $NH_2$-terminal region. The results of the experiments with 2E1 and 2B4 indicate that the first hydrophobic region in microsomal P450 cytochromes is apparently more critical for membrane localization of 2B4 than it is for 2E1, and that the second hydrophobic sequence in 2E1 as well as other portions of both cytochromes may be secondary determinants of membrane binding.

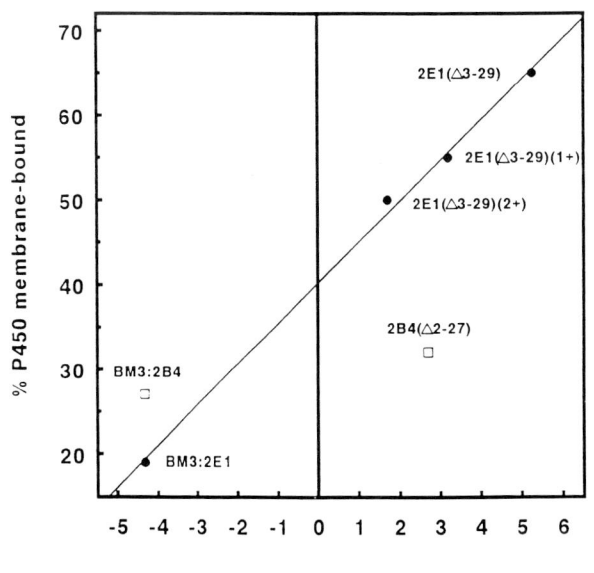

Fig. 2. Relationship of the hydropathicity of the $NH_2$-terminal region and membrane binding of shortened 2B4 and 2E1 variants. The hydropathicity of the 20 $NH_2$-terminal amino acids of 2B4 ($\Delta$2-27), 2E1 ($\Delta$3-29) variants, and P450$_{BM-3}$ chimeras was calculated by the method of Kyte and Doolittle (1985) with the SOAP program of PC/GENE. The amount of P450 expressed in the membrane is indicated as a percentage of the total present in cytosol and membrane, and is given as the average of at least duplicate determinations.

The content of expressed P450 cytochromes in *E. coli* membranes is difficult to determine due to the interference of the CO-reduced difference spectrum by bacterial cytochromes *o* and *d*. Thus, a mutant strain of *E. coli* (RG129) that did not contain these cytochromes (Au *et al.*, 1985) was transformed with pJL expression plasmids containing 2E1, 2E1 (Δ3-29), or BM-3:2E1. Immunoreactive protein that comigrated with 2E1 standard was detected in the membrane fraction of RG129 cells harboring each of these plasmids after anaerobic incubation in the presence of IPTG. The membranes isolated from these cells were then treated with β-D-octylglucopyranoside and subjected to ultracentrifugation to remove insoluble components. The CO-reduced difference spectrum of the detergent-solubilized membranes had an absorption band at about 424 nm in the case of expressed 2E1, an absorption peak at 418 nm and one of substantially lesser intensity at about 450 nm with 2E1 (Δ3-29), and no spectrally detectable species between 390 and 510 nm for BM-3:2E1. Thus, the expressed proteins were not degraded, but unfortunately little, if any, P450 holoenzyme was evident even though the 2E1 stabilizing agent, 4-methylpyrazole, was present throughout cell growth and membrane preparation. Chromatographic separation of the expressed P450 from the bacterial cytochromes is required to accurately examine the catalytic proficiency of the expressed enzymes.

The removal of the first hydrophobic region of 2E1 was previously found by Larson *et al.* (1991b) not to affect the catalytic activity of the purified cytochrome with a number of typical 2E1 substrates, including ethanol, *N*-nitrosodiethylamine, aniline, and *p*-nitrophenol, nor did it affect the interactions of 2E1 with NADPH-cytochrome P450 reductase or alter the effect of lipid and cytochrome $b_5$ on catalytic activity. It was of great interest to determine the catalytic proficiency of the truncated 2B4 forms in light of the observed changes in subcellular distribution following removal and/or modification of the hydrophobic $NH_2$-terminal region of 2B4, although preliminary attempts to purify shortened 2B4 cytochromes from *E. coli* membranes or cytosol were unsuccessful due to their low level of expression. Thus, the cDNA for 2B4 (Δ2-27) was cloned downstream from the glutathione S-transferase and a thrombin recognition sequence in a commercially available expression vector modified by Hakes *et al.* (1992) to allow for efficient cleavage of the expressed fusion protein with thrombin immediately before the Met of 2B4 (Δ2-27) following GSH-Sepharose column chromatography. A preparation of 2B4 (Δ2-27) obtained in this manner from *E. coli* lysate and liberated from glutathione S-transferase with thrombin was found to have the same catalytic activity as 2B4 purified from liver microsomes of phenobarbital-treated rabbits in the NADPH-dependent N-demethylation of *d*-benzphetamine (1 mM). Other investigators have demonstrated the retention of full catalytic activity of P450 3A4 lacking $NH_2$-terminal amino acids 3-12 (Gillam *et al.*, 1993) and of benzo(*a*)pyrene 3-hydroxylase activity of rat 1A1 lacking the 30 $NH_2$-terminal amino acids (Sakaki *et al.*, 1985). However, the same truncated 1A1 cytochrome has 25% less activity with 7-ethoxycoumarin as substrate, and removal of amino acids 2-24 in cholesterol 7α hydroxylase expressed in *E. coli* is associated with a loss of 60% of the catalytic activity, which is not attributable to conversion of the expressed P450s to P420. On the other hand, the complete loss of activity observed in the bovine 17α hydroxylase lacking amino acids 2-17 is due to expression of a cytochrome that is predominantly P420. Thus, the observed catalytic activities of different P450 cytochromes may depend upon the substrate and the particular truncated P450 cytochrome examined and its stability. Moreover, catalytic activity also apparently depends upon the extent of the $NH_2$-terminal deletion, since Cullin (1992) has observed that the catalytic activity of mouse 1A1 expressed in yeast membranes decreases without conversion to P420 as more extensive deletions are made. The overall conclusion derived from our current studies of subcellular distribution and catalytic activity is that the first hydrophobic segment comprising the signal peptide may vary in the extent of its participation in membrane localization, but it is not required for catalytic activity of 2E1 or 2B4. Further studies are being directed toward understanding the effects of $NH_2$-terminal deletions and modifications of hydrophobicity on the catalytic activity and aggregation state of 2E1 and 2B4.

This research was supported by Grant DK-10339 from the National Institutes of Health and Grant AA-06221 from the National Institute on Alcohol Abuse and Alcoholism. S.J.P. was a Postdoctoral Trainee of the National Institute of Environmental Health Sciences, Grant ES-07062.

REFERENCES

Au, D.C.-T., Lorence, R.M., and Gennis, R.B. (1985): Isolation and characterization of an *Escherichia coli* mutant lacking cytochrome *o* terminal oxidase. *J. Bacteriol.* 161, 123-127.

Black, S.D. (1992): Membrane topology of the mammalian P450 cytochromes. *FASEB J.* 6, 680-685.

Chang, J.-Y. (1985): Thrombin specificity: requirement for apolar amino acids adjacent to the thrombin cleavage site of polypeptide substrate. *Eur. J. Biochem.* 151, 217-224.

Clark, B.J., and Waterman, M.R. (1991): The hydrophobic amino-terminal sequence of bovine 17α-hydroxylase is required for the expression of a functional hemeprotein in COS 1 cells. *J. Biol. Chem.* 266, 5898-5904.

Cullin, C. (1992): Two distinct sequences control the targeting and anchoring of the mouse P450 1A1 into the yeast endoplasmic reticulum membrane. *Biochem. Biophys. Res. Commun.* 184, 1490-1495.

Emr, S.D., Hedgpeth, J., Clément, J.-M., Silhavy, T.J., and Hofnung, M. (1980): Sequence analysis of mutations that prevent export of λ receptor, an *Escherichia coli* outer membrane protein. *Nature (London)* 285, 82-85.

Gasser, R., Negishi, M., and Philpot, R.M. (1988): Primary structures of multiple forms of cytochrome P-450 isozyme 2 derived from rabbit pulmonary and hepatic cDNAs. *Mol. Pharmacol.* 32, 22-30.

Gillam, E.M.J., Baba, T., Kim, B.-R., Ohmori, S., and Guengerich, F.P. (1993): Expression of modified human cytochrome P450 3A4 in *Escherichia coli* and purification and reconstitution of the enzyme. *Arch. Biochem. Biophys.* 305, 123-131.

Hakes, D.J., and Dixon, J.E. (1992): New vectors for high level expression of recombinant proteins in bacteria. *Anal. Biochem.* 202, 293-298.

Kaiser, C.A., Preuss, D., Grisafi, P., and Botstein,D. (1987): Many random sequences functionally replace the secretion signal sequence of yeast invertase. *Science* 235, 312-317.

Khani, S.C., Porter, T.D., Fujita, V.S., and Coon, M.J. (1988): Organization and differential expression of two highly similar genes in the rabbit alcohol-inducible cytochrome P-450 subfamily. *J. Biol. Chem.* 263, 7170-7175.

Kyte, J., and Doolittle, R.F. (1982): A simple method for displaying the hydropathic character of a protein. *J. Mol. Biol.* 157, 105-132.

Larson, J.R., Coon, M.J., and Porter, T.D. (1991a): Alcohol-inducible cytochrome P-450IIE1 lacking $NH_2$-terminal segment retains catalytic activity and is membrane-bound when expressed in *Escherichia coli*. *J. Biol. Chem.* 266, 7321-7324.

Larson, J.R., Coon, M.J., and Porter, T.D. (1991b): Purification and properties of a shortened form of cytochrome P-450 2E1: deletion of the $NH_2$-terminal membrane-insertion signal peptide does not alter the catalytic activities. *Proc. Natl. Acad. Sci. USA* 88, 9141-9145.

Li, Y.C., and Chiang, J.Y.L. (1991): The expression of a catalytically active cholesterol 7α-hydroxylase cytochrome P450 in *Escherichia coli*. *J. Biol. Chem.* 266, 19186-19191.

Nelson, D.R., and Strobel, H.W. (1988): On the membrane topology of vertebrate cytochrome P-450 proteins. *J. Biol. Chem.* 263, 6038-6050.

Pernecky, S.J., Larson, J.R., Philpot, R.M., and Coon, M.J. (1993): Expression of truncated forms of liver microsomal P450 cytochromes 2B4 and 2E1 in *Escherichia coli*: Influence of $NH_2$-terminal region on localization in cytosol and membranes. *Proc. Natl. Acad. Sci. USA* 90, 2651-2655.

Porter, T.D., and Larson, J.R. (1991): Expression of mammalian P450s in *Escherichia coli*. *Methods Enzymol.* 206, 108-116.

Ruettinger, R.T., Wen, L.-P., and Fulco, A.J. (1989): Coding nucleotide, 5' regulatory, and deduced amino acid sequences of P-450$_{BM-3}$, a single peptide cytochrome P-450:NADPH-P-450 Reductase from *Bacillus megaterium*. *J. Biol. Chem.* 264, 10987-10995.

Saier, M.H., Werner, P.K., and Müller, M. (1989): Insertion of proteins into bacterial membranes: mechanism, characteristics, and comparisons with the eukaryotic process. *Microbiol. Rev.* 53, 333-366.

Sakaki, T., Oeda, K, Miyoshi, M., and Ohkawa, H. (1985): Characterization of rat cytochrome P-450$_{MC}$ synthesized in *Saccharamoyces cerevisiae*. *J. Biochem.* 98, 167-175.

# Structure-function analysis of eukaryotic P450s expressed in *E. coli*

Irina A. Pikuleva, Christopher M. Jenkins, Michael R. Waterman

Vanderbilt University School of Medicine, Department of Biochemistry, 607 Light Hall, Nashville, Tennessee 37232-0146, USA

## INTRODUCTION

A variety of different heterologous expression systems have been applied to the study of P450 structure/function relationships including mammalian cells and microorganisms (Table 1). Following the lead of Larson *et al.* (1991) and Li and Chiang (1991) we have investigated the utility of bacteria as an expression system for eukaryotic P450s. Bacteria prove to be a particularly facile system

TABLE 1: Expression Systems for P450s

| | |
|---|---|
| Yeast | AHH Cells |
| COS | V-79 Cells |
| Vaccinia | *E. coli* |
| Baculovirus | |

for this purpose for two reasons; very high levels of expression of functional forms of P450 can be achieved and this system is particularly amenable to site-directed mutagenesis. However, the bacterial expression system demonstrates certain characteristics quite distinct from animal cells. In this paper we will describe how this system has been used for P450 structure/function analysis by site-directed mutagenesis and in one respect how the system differs from heterologous expression in animal cells.

## Overexpression of P450s in Bacteria and their Purification

We have succeeded in achieving high levels of expression of both microsomal (Barnes, *et al.*, 1991) and mitochondrial (Wada *et al.*, 1991) P450s in *E. coli*. For reasons that we do not yet understand, it has been necessary to utilize different expression vectors for these two types of P450. Also the maximal expression achieved for microsomal P450s is greater (up to 1 $\mu$mole/liter bacterial culture) than for mitochondrial P450s (250 nmole/liter bacterial culture). Both of these types of P450 are found to be integral proteins in the bacterial inner membrane. Interestingly, and quite unexpectedly, microsomal P450s are active in the intact *E. coli*, while mitochondrial

P450s are not. The generally held view that mitochondria have arisen from bacteria had led us to predict that the opposite might be true. Nevertheless our results indicate that there are no proteins in the strains of *E. coli* that we have investigated which can replace ferredoxin or ferredoxin reductase in supporting mitochondrial P450 activity while these bacteria do contain an enzyme system which can replace NADPH cytochrome P450 reductase. However, this latter activity in intact bacteria is quite low so that maximal activity of recombinant P450s expressed in *E. coli* is obtained by isolating bacterial membranes and reconstituting their activities using purified NADPH cytochrome P450 reductase.

High levels of expression of microsomal and mitochondrial P450s in bacteria permit purification of substantial amounts of these enzymes by standard purification procedures (Wada, et al., 1992; Gillam et al., 1992). In addition, however, recombinant technologies permit novel purification schemes to be developed as well. For example, the human form of P450c17, an enzyme which is normally very difficult to obtain in part because of the lack of ready availability of human tissues, has been purified to homogeneity by metal ion affinity chromatography following addition of a metal ion binding site (in this case 4 histidine residues) to the polypeptide chain (Imai et al., 1993). This modification appears to have no effect on the enzymatic activities associated with P450c17.

Thus methods have been established in this laboratory and by others which permit expression of both microsomal and mitochondrial P450s in *E. coli*. The enzymes can be obtained in relatively high levels by a variety of purification schemes. Consequently P450s which have previously been difficult to obtain, either because they come from tissues which are difficult to obtain (*i.e.*,. human) or because they are expressed at low levels, are now readily accessible via recombinant technology and heterologous expression in bacteria.

## Site-Directed Mutagenesis, Bacterial Expression and Structure/Functional Analysis

The best examples to date of the use of bacterial expression in the study of P450 structure/function come from studies on the mitochondrial P450, cholesterol side chain cleavage cytochrome P450 (P450scc). This is the key enzyme in steroidogenic pathways catalyzing the conversion of cholesterol to pregnenolone. In one case, site-directed mutagenesis has been used to examine the interaction of P450scc with its electron donor adrenodoxin (Wada et al., 1992) and in a second series of studies site-directed mutagenesis has been used to study substrate binding at the active site (Pikuleva et al., 1994).

Based on chemical modification studies and protein sequence alignment, it was predicted that two specific lysine residues in bovine P450scc might be important in binding adrenodoxin. Mutation of these residues (lys 338 and lys 342) to neutral or acidic amino acids was found to have a profound effect on the Ks of adrenodoxin binding (Table 2) without affecting the enzymatic properties (pregnenolone production from 25-hydroxycholesterol). To carry out these studies it was necessary to purify recombinant P450scc from *E. coli* and to measure optically the binding of adrenodoxin to P450scc by titration. The high level of expression of P450scc in bacteria make these experiments possible.

TABLE 2: Estimated Binding Constants of Adrenodoxin to P450scc

| | |
|---|---|
| wild type | 0.23 µM |
| Lys 338 Ala | 60 µM |
| Lys 338 Gln | 74 µM |
| Lys 338 Thr | 130 µM |
| Lys 338 Glu | 130 µM |
| Lys 342 Glu | 130 µM |
| Lys 342 Gln | 35 µM |

Furthermore, in a more recent study, site-directed mutagenesis has been used to probe the active site of P450scc. Tyrosine residues 93 and 94 align closely with Y96 of P450cam and thus are predicted to play a role in substrate binding in P450scc. These particular tyrosine residues are highly conserved among P450scc from several species, further supporting this notion. One of the serious problems associated with investigation of protein structure/function relationships by site-directed mutagenesis was encountered in these particular studies, that being the effect of amino acid alterations on protein stability. Thus the single mutants (Y93S; Y93A; Y94S; Y94A) were not very stable and only limited quantities of the P450 form of these P450scc mutants could be obtained during purification. However, a functional double mutant (Y93S, Y94S) was obtained in sufficient quantities for optical measurement of substrate binding constants by titration. This double mutant bound cholesterol 5-times less well than did the wild type enzyme, while it bound 22-hydroxycholesterol, the first intermediate in the conversion of cholesterol to pregnenolone, with the same affinity as the wild type enzyme. It is concluded that one or both of these tyrosine residues is important for cholesterol binding but is not involved in 22-hydroxycholesterol binding. Studies of enzymatic properties of the single mutants indicate that the key residue is Y93. Once again, the relatively high level of expression of P450scc in bacteria made this study possible.

We predict that bacterial expression systems will be useful for study of structure/function relationships in many different forms of P450. However, the fact that many key mutants may not be stable provides a cautionary note to the success of such efforts. Also it is important to recognize that bacterial expression of mutant P450s can provide results which are not obtained in animal cells. For example, using COS cells it has been clearly demonstrated that removal of the signal anchor sequence from bovine P450c17, a microsomal P450, profoundly alters the folding pathway of this protein leading to a nonfunctional enzyme (Clark and Waterman, 1991, 1992). This truncated form of P450c17 is associated with the COS cell endoplasmic reticulum, yet has no enzymatic activity. Heme binding is abnormal in this truncated protein. However, when this same truncated P450c17 is expressed in *E. coli* large quantities of a functional enzyme are obtained, this enzyme being associated with the bacterial membrane (Sagara et al, 1993). Thus bacteria permit the folding of this truncated form of P450 into a functional enzyme while mammalian cells can not. While bacteria can produce large quantities of functional P450s; they have certain abilities which are not found in animal cells.

## Bacteria Can Support Microsomal P450 Activities

As noted previously, microsomal P450s expressed in *E. coli* are active in the intact microorganism, as the result of an unknown cytosolic reductase system. Partial purification of this system has indicated that this reductase consists of two components. Both factors contain flavoproteins. The component which interacts with the P450 has been purified to homogeneity and by N-terminal sequence analysis, optical properties and molecular size is found to be *E. coli* flavodoxin. Perhaps the other component is flavodoxin reductase, but further purification is necessary to establish this point for certain. Flavodoxin is an FMN containing protein which has previously been predicted by sequence alignment to be a bacterial analog of the FMN binding domain of NADPH cytochrome P450 reductase (Porter and Kasper, 1986). Thus it would seem that a spin-off of studies of heterologous expression of P450s in bacteria is the identification of a small FMN-containing protein which binds microsomal P450s. Perhaps this protein will be useful in localizing the residues of microsomal P450s which interact with the microsomal P450 reductase.

## CONCLUSION

Expression of P450s in bacteria provides an abundant source of functional enzyme (both microsomal and mitochondrial) which can be used for structure/function analysis. Thus by employing recombinant technology mutant forms of specific P450s can be produced which permit evaluation of reductase interactions and the active site. The major advantage of this expression system for these purposes is the high level of expression and relative ease of purification of recombinant forms of P450.

## ACKNOWLEDGEMENTS

These studies were supported by NIH grants GM37942 and ES00267. C.M.J. is supported by NIH ES07028.

## REFERENCES

Barnes, H. J., Arlotto, M. P., and Waterman, M. R. (1991): Expression and enzymatic activity of recombinant cytochrome P450 17α-hydroxylase in *Escherichia coli*. Proc. Natl. Acad. Sci. **88**, 5597-5601.

Clark, B. J., and Waterman, M. R. (1991): The hydrophobic amino-terminal sequence of bovine 17α-hydroxylase is required for the expression of a functional hemoprotein in COS 1 cells. J. Biol. Chem. **266**, 5896-5904.

Clark, B. J., and Waterman, M. R. (1992): Functional expression of bovine 17α-hydroxylase in COS 1 cells is dependent upon the presence of an amino-terminal signal anchor sequence. J. Biol. Chem. **267**, 24568-24574.

Gillam, E. M. J., Baba, T., Bok-Ryang, K., Ohmori, S., and Guengerich, F. P. (1993): Expression of modified human cytochrome P450 3A4 in *Escherichia coli* and purification and reconstitution of the enzyme. *Arch. Biochem. Biophys.* **305** 123-131.

Imai, T., Globerman, H., Gertner, J. M., Kagawa, N., and Waterman, M. R. (1993): Expression and purification of functional human 17α-hydroxylase/17,20-lyase deficiency. *J. Biol. Chem.* **268**, 19681-19689.

Larson, J. R., Coon, M. J., and Porter, T. D. (1991): Alcohol-inducible cytochrome P450IIE1 lacking the hydrophobic $NH_2$-terminal segment retains catalytic activity and is membrane-bound when expressed in *Escherichia coli*. *J.Biol. Chem.* **266**, 7321-7324.

Pikuleva, I., Kagawa, N., and Waterman, M. R. (1994): Evidence for the involvement of tyrosine 93 in the interaction of cholesterol with bovine cholesterol side chain cleavage cytochrome P450 (P450scc), Submitted for publication.

Porter, T. D., and Kasper, C. B. (1986): NADPH-Cytochrome P-450 oxidoreductase: flavin mononucleotide and flavin adenine dinucleotide domains evolved from different flavoproteins. *Biochemistry* **25**, 1682-1687.

Sagara, Y., Barnes, H. J., and Waterman, J. R. (1993): Expression in *Escherichia coli* of functional cytochrome P450c17 lacking its hydrophobic amino-terminal signal anchor. *Arch. Biochem. Biophys.* **304**, 272-278.

Wada, A., Mathew, P. A., Barnes, H. J., Sanders, D., Estabrook, R. W., and Waterman, M. R. (1991): Expression of functional bovine cholesterol side chain cleavage cytochrome P450 (P450scc) in *Escherichia coli*. *Arch. Biochem. Biophys.* **290**, 376-380.

Wada, A., and Waterman, M. R. (1992): Identification by site-directed mutagenesis of two lysine residues in cholesterol side chain cleavage cytochrome P450 that are essential for adrenodoxin binding. *J. Biol. Chem.* **267**, 22877-22882.

# Proton and electron transfer mechanism in dioxygen activation by cytochrome P450cam

Hideo Shimada[1], Ryu Makino[2], Masashi Unno[1], Tadao Horiuchi[3], Yuzuru Ishimura[1]

[1]Departement of Biochemistry, School of Medicine, Keio University, Shinanomachi 35, Shinjuku-ku, Tokyo 160. [2]Molecular Biochemistry, Faculty of Science, Himeji Institute of Technology, Harima Science Park 1479-1, Hyogo 678-12. [3]Department of Bioengineering, Faculty of Engineering, Soka University, Hachioji, Tokyo 192, Japan

## Introduction:

In 1985, Poulos and his coworkers determined the x-ray crystal structure at 2.6 Å-resolution of a ferric substrate-bound form of cytochrome P450cam (P450cam), the terminal monooxygenase for the *d*-camphor hydroxylating multienzyme system of *P. putida* (Poulos *et al.*, 1985). This served as the first demonstration of the three-dimensional structure of a cytochrome P450. The structures of ferric substrate-free and ferrous CO-bound forms were subsequently demonstrated (Poulos *et al.*, 1986; Raag and Poulos, 1989). Since then, a number of studies have been carried out aiming to elucidate the structure-function relationship in this enzyme by utilizing various means including molecular spectroscopic techniques and site-directed mutagenesis. Among them, we and others (Imai *et al.*, 1989; Martinis *et al.*, 1989) have found that threonine252 at the distal side of the heme in P450cam is essential for the dioxygen activation in this monooxygenase reaction: Replacement of this amino acid with others converted the enzyme from a monooxygenase to an oxidase. In this presentation, we discuss proton and electron transfer mechanism in the dioxygen activation by this cytochrome P450.

## Processes of oxygen activation:

The reaction catalyzed by P450cam is:
$$d\text{-Camphor} + NADH + H^+ + O_2 \rightarrow 5\text{-}exo\text{-Hydroxycamphor} + NAD^+ + H_2O$$
The catalytic sequence has been shown to involve several reaction intermediates (Gunsalus *et al.*, 1974; Sligar and Murray, 1986). In the resting state, the monooxygenase is in a low-spin ferric state ($Fe^{3+}$: low-spin), which is converted to a high-spin ferric form ($Fe^{3+}$: high-spin) upon binding of *d*-camphor. Subsequent one-electron reduction of the ferric form yields a high-spin ferrous derivative ($Fe^{2+}$: high spin), which binds with $O_2$ to form an oxy-ferrous intermediate ($Fe^{2+}O_2$) (Ishimura *et al.*, 1971; Tyson *et al.*, 1972) or CO to form the ferrous CO-adduct. Then the oxy-ferrous intermediate is decomposed into the products, *i.e.* 5-*exo*-hydroxycamphor, $H_2O$ and the ferric P450cam by receiving an additional electron; the one-electron reduction of the oxy-ferrous P450cam has been proposed to give a ferric peroxo compound of the heme ($Fe^{3+}O_2^{2-}$) followed by a rapid cleavage of the O-O bond in this intermediate to give $H_2O$ and an oxo-ferryl porphyrin π-cation radical ($Fe^{4+}O\text{-}Por^{+\bullet}$) (Imai *et al.*, 1989). Finally oxygen transfer from the ferryl compound to *d*-camphor generates

the reaction products. It should be noted that two protons are necessary to achieve the O-O bond scission yielding H$_2$O and the oxo compound.

## Electron transfer from putidaredoxin to P450cam:

Two electrons from NADH, which are consumed during the above mentioned reactions, are transferred to P450cam *via* putidaredoxin reductase (an FAD-containing enzyme) and putidaredoxin (Pdx; an Fe$_2$S$_2$Cys$_4$ iron-sulfur protein), where Pdx functions as a one-electron carrier between the reductase and P450cam; both the first and the second electrons are transferred by this iron-sulfur protein to the ferric and the oxygenated forms of P450cam, respectively. Then the binding site of Pdx on P450cam has been proposed to be the same with that of cytochrome $b_5$ ($b_5$) (Stayton et al., 1990); the association reactions of $b_5$ and Pdx with P450cam were competitive with each other. The binding site contains several basic amino acid residues such as Arg72, Arg112, Lys344 and Arg364 (see Fig. 1), and, when glutamine or glutamic acid was substituted for either Arg72 or Lys344, the affinity of P450cam for Pdx decreased by 20~60 % without changing the maximum rate of NADH oxidation (Stayton and Sligar, 1990). Thus the basic amino acids in the region are important in the interaction of P450cam and Pdx.

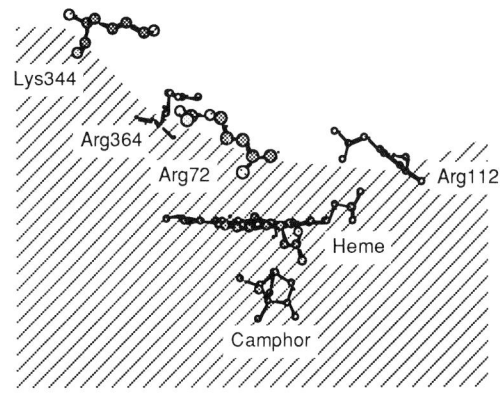

Fig. 1. Schematic representation of the proposed redox partner binding site of cytochrome P450cam. The binding site contains four basic amino acid residues Arg72, Arg112, Lys344 and Lys364. The heme and substrate, $d$-camphor, are also shown.

Recently Koga et al. isolated a Cys112-mutant of P450cam from clones obtained by an *in vitro* random mutagenesis of its gene (Koga et al., 1993). The ferric camphor-bound form of this Cys-enzyme exhibited a low affinity toward Pdx, being consistent with the above view that 112Arg is an important component in the Pdx binding site. When the rate of reduction of the ferric Cys112-mutant by reduced Pdx was studied in the presence of an excess amount of the Pdx, the rate was as small as 1/400 of that of the original P450cam (0.1 *vs.* 40 sec$^{-1}$ at 20° C). These results suggest that the complex formed between the Cys-enzyme and Pdx was fairly inactive with respect to the electron transfer. Furthermore, the replacement of Arg112 by Met or Tyr by site-directed mutagenesis resulted in a very low reduction rate of their heme by reduced Pdx (1/500 ~ 1/1000). On the other hand, the Lys-mutant showed a comparable reduction rate by Pdx to that of the native enzyme in conformity with the view that the basic amino acid residue at the position is important in the interaction of P450cam and Pdx. Catalytic activities of these and other mutants are summarized in Table 1. As seen, oxygen consumed by each mutant enzyme, though small in quantity, was efficiently utilized for the monooxygenase reaction, indicating that the oxygenated intermediates of these mutant enzymes underwent normal O-O bond scission upon receiving the second electron. The very low catalytic activities observed with these mutant enzymes were therefore caused by the decreases in the rates of electron transfer from Pdx to P450cam, but not by other reasons.

Table 1. Oxygen Consumption Rate and the Amount of Product Formed by the Native and 112-, 251-, and 366-Mutant Cytochromes P450cam[1)]

|  | Rate of Oxygen Consumption | Product Formed per $O_2$ Consumed | |
| --- | --- | --- | --- |
|  | ($\mu$M / min / $\mu$M heme) | 5-OH-CAM[2)] in % | $H_2O_2$ in % |
| 112-Arg (native) | 1350 | 97 | 3 |
| 112-Lys | 200 | 88 | 12 |
| 112-Cys | 1.7 | 80 | 15 |
| 112-Met | 1.3 | 69 | 16 |
| 112-Tyr | 0.8 | 50 | 30 |
| 251-Asp (native) |  |  |  |
| 251-Ala | 3 | 90 | 11 |
| 251-Gly | 21 | 99 | 2 |
| 366-Glu (native) |  |  |  |
| 366-Met | 940 | 85 | 12 |

1) The catalytic activities of the native cytochrome P450cam and its mutants were determined in a reconstituted system comprized of putidaredoxin reductase, putidaredoxin and itself as described by Imai et al. (Imai et al. 1989).
2) 5-OH-CAM: 5-exo-hydroxycamphor.

Another interesting feature of Arg112 residue in the binding site is that it forms a hydrogen bond with one of the propionate sidechain of the heme embedded in the protein (Fig. 2). As stated above, the effects of the amino acid replacement at this position markedly affected the rate of the reduction of its heme by Pdx. The finding suggests that the hydrogen bond between the propionate and Arg may also be involved in the electron transfer. Then an Asp251-residue locating next to Thr252 at the distal side of the heme in P450cam was also found to be important in the electron transfer. The residue was essential for the reduction of the oxygenated intermediate by the second electron, but not required for the reduction of the ferric form by the first electron (Koga et al., 1989; Shimada et al., 1990; Gerber and Sligar 1992). These

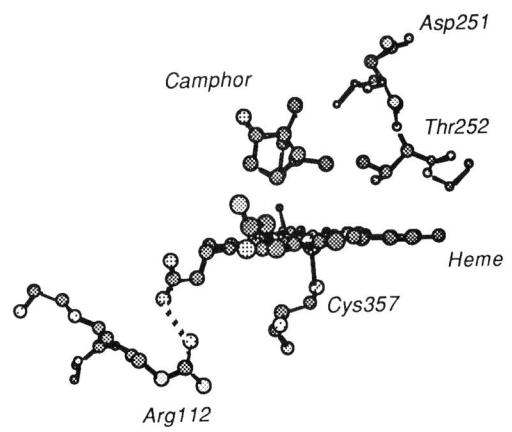

Fig. 2. Heme environment of cytochrome P450cam. The heme, substrate (d-camphor) and some selected amino acid residues are shown. Arg112 is hydrogen bonded with a heme propionate group.

observations suggest that the electron transfer mechani-sms or pathways for the first and second electrons are different.

## Catalyst of O-O bond cleavage reaction:

As mentioned already, the hydroxy amino acid Thr252 at the active site of P450cam plays a crucial role in the activation of dioxygen. Imai *et al.* and Martinis *et al.* have shown that Thr252 is essential for the cleavage of O-O bond in the oxygenated intermediate upon receiving the second electron (Imai *et al.*, 1989; Martinis *et al.*, 1989). Table 2 lists the camphor hydroxylating activities of the native and the 252-mutants determined in the reconstituted assay system described earlier (Shimada *et al.*, 1990; Makino and Shimada, 1993). As seen, the native and mutant enzymes showed a variety of oxygen and NADH consuming activities. In all cases, however, the ratio of the two activities were 1.0. In other words, the stoichiometry between the oxygen consumption and NADH oxidation is 1 : 1. Among them, almost all the oxygen consumed by the native enzyme with Thr252 was utilized to produce the hydroxylated product, 5-*exo*-hydroxycamphor, indicating a highly efficient O-O bond cleavage reaction. This Thr can be replaced by Ser or Asn without significant decreases in the O-O bond cleavage efficiency. On the other hand, the majority of oxygen consumed by the other mutants without such a functional group as -OH and -$NH_2$ was converted to $H_2O_2$. This $H_2O_2$ was produced by the decomposition of the peroxo or hydroperoxo intermediates which was formed by the reduction of the oxygenated intermediate: No O-O bond scission is necessary for the production of $H_2O_2$. For these reasons, the -OH group in Thr252 has been proposed to function as a catalyst in the O-O bond cleavage step of the P450cam-catalyzed hydroxylation reaction (Imai *et al.*, 1989; Martinis *et al.*, 1989; Shimada *et al.*, 1990). Then the importance of this Thr residue emerges also from the amino acid sequences of cytochromes P450. A Thr-residue at the corresponding position was highly conserved among over 100 P450 species (Gotoh and Fujii-Kuriyama, 1989; Ouzounis and Melvin, 1991). In fact, Imai and Nakamura have found that Thr301 locating presumably at the active site of laurate ω-1 hydroxylase and testosterone 16α-hydroxylase is essential for the hydroxylation (Imai and Nakamura, 1988). The substitution of His for Thr301 in the two mammalian P450s made their hydroxylating activities sluggish.

Table 2. Oxygen Consumption and NADH Oxidation Rates and the Amount of Product Formed by the Native and 252-Mutant Cytochromes P450cam[1]

|  | Rate (µM / min / µM heme) | | Product Formed per $O_2$ Consumed | |
|---|---|---|---|---|
|  | Oxygen | NADH | 5-OH-CAM[2] in % | $H_2O_2$ |
| 252-Thr (native) | 1350 | 1380 | 97 | 3 |
| 252-Ala | 1150 | 1180 | 5 | 89 |
| 252-Gly | 1090 | 1090 | 3 | 88 |
| 252-Ser | 830 | 830 | 85 | 15 |
| 252-Cys | 690 | 700 | 7 | 86 |
| 252-Asn[3] | 420 | 440 | 57 | 12 |
| 252-Pro | 100 | 100 | 10 | 77 |

1) The catalytic activities of the native and mutant cytochromes P450cam were determined in a reconstituted system composed of putidaredoxin reductase, putidaredoxin and itself as described by Imai *et al.* (Imai *et al.*, 1989).
2) 5-OH-CAM: 5-*exo*-hydroxycamphor.
3) The catalytic activity was determined in the presence of glycerol.

*Proton relay system:*

The hydroxyl group of the Thr-residue when functions as a proton donor requires another proton to restore the initial state after the catalysis. Thus question arises as to the proton donor and its delivery system. Two plausible pathways are envisaged for the delivery of proton(s) in P450cam. One is the hydrogen bond network extending from the -OH group of Thr252 to the carboxyl group of Glu366 *via* three internally ordered $H_2O$ molecules. The terminal Glu366 in this network is on L-helix, to which the external water is accessible (Poulos *et al.*, 1987; Raag and Poulos, 1991). Another candidate for the pathway also consists of a hydrogen bond network beginning from Thr252 *via* $H_2O$ and Asp251 in this order terminating in Lys178/Arg186 of F-helix, to which external water is accessible (Makino *et al.*, 1991; Gerber and Sligar 1992).

The former hydrogen bond network, Thr252-3$H_2O$-Glu366 is visible in the crystal structure of P450cam (Poulos *et al.*, 1987). On the other hand, only a half portion of the second network, *i.e.* Asp251 to Lys178/Arg186 is visible in the crystal structure, and the remaining half, Thr252-$H_2O$-Asp251, is not observable. The invisible portion of the network has been proposed by Makino *et al.* based on their measurements of redox potential of the native and 251- and 252-mutant P450cam over a wide range of pH (Makino *et al.*, 1991). The midpoint potential of the native enzyme increased at an acidic region of pH, suggesting the presence of a dissociable proton of p$K$a = 6.3. This pH-dependent change disappeared upon either the 251- or 252-mutation to, for example, alanine. In a structural model which explains these observations, the carboxyl group of Asp251 has been proposed to have p$K$a = 6.3 and to be hydrogen bonded with Thr252 *via* water. A relatively high p$K$a of the carboxyl group of Asp251 is probably brought about by the local environments, where two basic amino acid residues, Lys178 and Arg186, are located close to the carboxyl group of Asp251 (Poulos *et al.*, 1985). The network, Thr252-$H_2O$-Asp251, has been supported by the findings on the effect of pH on the infrared C-O stretch frequency of the ferrous CO-form of P450cam (Makino *et al.*, 1991; Shimada *et al.*, 1991); the infrared C-O stretch frequency shifted from 1941 cm$^{-1}$ to 1940 cm$^{-1}$ by changing the solvent from $H_2O$ to $D_2O$. The shift only by 1 cm$^{-1}$ may seem so small, but the 252-mutant enzyme such as 252Ala, for example, did not show any shift in the frequency under identical experimental conditions. From these results, therefore, they have proposed that the bound CO is hydrogen bonded to the -OH group of Thr252 in P450cam (Shimada *et al.*, 1991). If it applies to the oxygenated intermediate, a hydrogen bond network could also be formed between the bound oxygen to Asp251 *via* Thr252 and water.

In an attempt to identify the proton delivery system operating in P450cam, we firstly changed Glu366 in the Thr252-3$H_2O$-Glu366 network to Met, which does not form a hydrogen bond with water. Thus this substitution probably breaks the hydrogen bond network. As seen in Table 1, however, the mutant enzyme showed no significant change in any reaction step in the camphor hydroxylation reaction cycle suggesting that Thr252-3$H_2O$-Glu366 pathway is not operative in the proton delivery. On the other hand, substitution of Asp251 by Ala, Gly or Asn markedly lowered the reduction rate of the oxy-ferrous intermediate to 1/100 ~ 1/500 of that of the native enzyme, resulting in a significantly low O-O bond cleavage activity as seen in Table 2 (Koga *et al.*, 1989; Shimada *et al.*, 1990; Gerber and Sligar, 1992). On these basis, we propose that Thr252-$H_2O$-Asp251-(Lys178/Arg186) pathway could be the hydrogen bond network which provide proton(s) to the oxy-ferrous intermediate of P450cam.

Gerber and Sligar (Gerber and Sligar, 1992) have proposed a similar proton network, 252Thr-251Asp-Lys178,Arg186-Asp182, where Asp251 served as a proton shuttle between Thr252 and Lys178,Arg186-Asp182 network.

Recently, Ravichandran *et al.* reported the crystal structure of the hemoprotein moiety of cytochrome P450BM-3 in its substrate-unbound low spin ferric form (Ravichandran *et al.*, 1993). At the distal side of the heme, there is a conserved Thr268 residue and also a conserved acidic residue, Glu267, which is solvent exposed. In addition to these amino acid residue, an ordered $H_2O$ molecule was found close to both of them. Accordingly, they have proposed that Thr268, $H_2O$ and Glu267 form a proton delivery system. Then, as seen in Fig. 3, the proton delivery systems in cytochromes P450cam and P450BM-3 share the same array of a hydrogen bond network: the conserved Thr, $H_2O$ molecule and a conserved acidic amino acid residue.

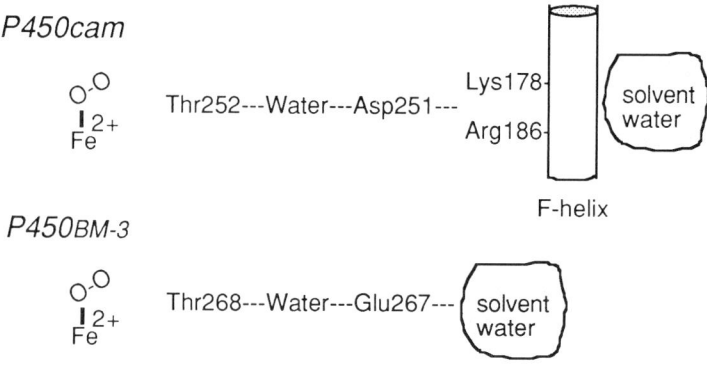

Fig. 3. The proposed proton relay system of cytochromes P450cam and P450BM-3.

*Acknowledgments:*

This work was supported in part by Grant-in-Aids from the Ministry of Education, Science and Culture of Japan, through Special Coordination Funds of the Science and Technology Agency of the Japanese Government, and by grants from Keio University.

*References:*

    Gerber, N. C., and Sligar, S. G. (1992): Catalytic mechanism of cytochrome P450cam: Evidence for a distal charge relay. *J. Am. Chem. Soc.* **114**, 8742-743.
    Gotoh, O., and Fujii-Kuriyama, Y. (1989) Evolution, Structure, and Gene Regulation of Cytochrome P450. In *Frontiers in Biotransformation* (Ruckpaul, K., and Rein, H., eds) Vol 1, pp. 195-243, Akademie-Verlag, Berlin.
    Gunsalus, I. C., Meeks, J. R., Lipscomb, J. D., Debrunner, P., and Munck, E. (1974): Bacterial monooxygenases- The P450 cytochrome system. In

*Molecular Mechanisms of Oxygen Activation* (Hayaishi, O., ed) pp. 559-613, Academic Press, New York.

Imai, M., Shimada, H., Watanabe, Y., Matsushima-Hibiya, Y., Makino, R., Koga, H., Horiuchi, T., and Ishimura, Y. (1989): Uncoupling of the cytochrome P-450cam reaction by a single mutation, threonine-252 to alanine or valine: A possible role of the hydroxy amino acid in oxygen activation. *Proc. Natl. Acad. Sci. U.S.A.* **86**, 7823-7827.

Imai, Y., and Nakamura, M. (1998): The importance of threonine-301 from cytochrome P-450 ( laurate ($\omega$-1)-hydroxylase and testosterone 16$\alpha$-hydroxylase) in substrate binding as demonstrated by site-directed mutagenesis. *FEBS letter* **234**, 313-315.

Ishimura, Y., Ullrich, V., and Peterson J. A. (1971): Oxygenated cytochrome P-450 and its possible role in enzymatic hydroxylation. *Biochem. Biophys. Res. Commun.*, **42**, 140-146.

Koga, H., Makino, R., Yaoi, T., Tsujimura, M., Sagara, Y., Shimada, H., Ishimura, Y., and Horiuchi, T. (1989): Structure and function relationship in cytochrome P450cam (IV): Roles of Asp251 in oxygen activation. *Seikagaku*, 61, 998.

Koga, H., Sagara, Y., Yaoi, T., Tsujimura, M., Nakamura, K., Sekimizu, K., Makino, R., Shimada, H., Ishimura, Y., Yura, K., Go, M., Ikeguchi, M., and Horiuchi, T. (1993): Essential role of the Arg112 residue of cytochrome P450cam for electron transfer from reduced putidaredoxin. *FEBS Letter* **331**, 109-113.

Makino, R., Shimada, H., Okada, Y., Ishimura, Y., and Horiuchi, T. (1991): Analysis of proton delivery system in cytochrome P450cam by employing site-directed mutagenesis. *Seikagaku*, 63, 840.

Makino, R., and Shimada, H. (1993): Structure of cytochrome P-450. In *Cytochrome P-450 Second Edition* (Omura, T., Ishimura, Y., and Fujii-Kuriyama, Y., eds) pp. 17- 30, Kodansha, Tokyo and VCH, Weinheim, New York, Basel, Cambridge.

Martinis, S. A., Atkins, W. M., Stayton, P. S., and Sligar, S. G. (1989): A conserved residue of cytochrome P-450 is involved in heme-oxygen stability and activation. *J. Am. Chem. Soc.* **111**, 9252-9253.

Ouzounis, C. A., and Melvin, W. T. (1991): Primary and secondary structural patterns in eukaryotic cytochrome P-450 families correspond to structures of the helix-rich domain of *Pseudomonas putida* cytochrome P-450cam: Indication for a similar overall topology. *Eur. J. Biochem.* **198**, 307-315.

Poulos, T. L., Finzel, B. C., Gunsalus, I. C., Wagner, G. C., and Kraut, J. (1985): The 2.6-Å crystal structure of *Pseudomonas putida* cytochrome P-450. *J. Biol. Chem.* **260**, 16122-16130.

Poulos, T. L., Finzel., B. C., and Howard, A. J. (1986): Crystal structure of substrate-free *Pseudomonas putida* cytochrome P-450. *Biochemistry*, **25**, 5314-5322.

Poulos, T. L., Finzel, B. C., and Howard, A. J. (1987): High resolution crystal structure of cytochrome P450cam. *J. Mol. Biol.* **95**, 687-700.

Raag, R., and Poulos, T. L. (1989): Crystal structure of the carbon monoxide-substrate-cytochrome P-450cam ternary complex. *Biochemistry*, **28**, 7586-7592.

Raag, R., and Poulos, T. L. (1991): Crystal structure of cytochrome P-450cam complexed with camphene, thiocamphor, and adamantane: Factors controlling P-450 substrate hydroxylation. *Biochemistry,* **30**, 2674-2684.

Shimada, H., Makino, R., Imai, M., Horiuchi, T., and Ishimura, Y. (1990): Mechanism of oxygen activation by cytochrome P-450cam. In *International Symposium on Oxygenases and Oxygen Activation* (Nozaki, M., Yamamoto, S., and Ishimura, Y. eds) pp. 133-136, Yamada Science Foundation.

Shimada, H., Makino, R., Imai, M., Okada, Y., Ishimura, Y., Horiuchi, T., Aichun Dong, and Caughey, W. S. (1991): Effects of amino acid substitutions on infrared C-O stretch spectra of carbonyl cytochrome P450cam. *Seikagaku*, 61, 998.

Sligar, S. G., and Murray, R. I. (1986): Cytochrome P-450cam and other bacterial P-450 enzymes. In *Cytochrome P-450* (Ortiz de Montellano, P. R. ed) pp. 429-503, Plenum Press, New York.

Tyson, C. A., Lipscomb, J. D., and Gunsalus, I. C. (1972): The roles of putidaredoxin and P450cam in methylene hydroxylation. *J. Biol. Chem.* **247**, 5777-5784.

Ravichandran, K. G., Boddupalli, S. S., Hasemann, C. A., Peterson, J. A., and Deisenhofer, J. (1993): Crystal structure of hemoprotein domain of P450BM-3, a prototype for microsomal P450's. *Science*, **261**, 731-736.

# High level expression of integral membrane proteins induces proliferation of the endoplasmic reticulum

R. Menzel, E. Kärgel, C. Wolff, F. Vogel, W.-H. Schunck

Max-Delbrück-Centre for Molecular Medicine, Robert-Rössle-Str. 10, D-13125 Berlin-Buch, Germany

### Summary

Proliferation of the endoplasmic reticulum (ER) can be induced experimentally in *Saccharomyces cerevisiae* by high-level expression of ER-resident integral membrane proteins. In case of cytochrome P45052A3 from the alkane-assimilating yeast *Candida maltosa* the preconditions for induction of ER-proliferation were studied more in detail. Different parts of the membrane anchor region were used to construct several fusions with the cytoplasmic form of yeast invertase as reporter protein.

The property of cytochrome P45052A3 to induce ER-proliferation was found to be transferable with the same sequences that confer targeting to and retention within the ER-membrane.

### INTRODUCTION

The endoplasmic reticulum (ER) of eukaryotic cells is characterized by its capability to adjust its morphology (surface area, volume and subcompartmentation) to different metabolic and secretory functions: Differentiation of lymphozytes and increased immunoglobulin secretion correlates with amplification of rough ER (Shohat et al., 1973). Phenobarbital induction of cytochrome P450 (P450) in rat hepatocytes is associated with an increase in smooth ER (Stäubli et al., 1969). The overproduction of the integral ER-protein 3-hydroxy-3-methylglutaryl CoA reductase in *Saccharomyces cerevisiae* cells induced the proliferation of "karmellae", paired membranes surrounding the nucleus (Wright et al., 1988).
Previous work in our laboratory resulted in cloning and sequencing of cDNAs encoding two different P450-forms (P45052A3 and P450A4, Schunck et al., 1991) and the NADPH-P450-reductase (Kärgel et al., unpublished results) of *Candida maltosa*. These enzymes are integral membrane proteins of the ER (Vogel et al., 1992). As their mammalian counterparts these enzymes contain amino-terminal signal anchor sequences whereas their catalytic domains are exposed to the cytoplasm. Heterologous overexpression in *S. cerevisiae* of P45052A3 and 52A4 resulted in an ultrastructural pecularity of the host organism, a massive proliferation of the ER bearing the newly synthesized proteins (Schunck et al., 1991).

The aim of our present study was to include further membrane and secretory proteins in our investigations of the phenomenon of ER-proliferation and to analyze that topogenic sequences in the P450 primary structure responsible for targeting to and retention within the ER as well as for the induction of ER-proliferation.

MATERIALS AND METHODS

The structural genes encoding the following proteins were expressed in S. cerevisiae:
- alkane-hydroxylating P45052A3, C. maltosa (Schunck et al., 1989)
- fatty acid-hydroxylating P45052A4, C. maltosa (Schunck et al., 1991)
- NADPH-P450-reductase, C. maltosa (Kärgel et al., unpublished)
- lanosterol demethylating P45051, S. cerevisiae (kindly provided by D. Sanglard, Swiss Federal Institut of Technology, Dept. Biotechnology, Zürich, Switzerland))
- invertase, S. cerevisiae (SUC2 coding region both for the secretory and the cytoplasmic form was amplified by PCR from genomic DNA)

The host/vector system used consisted of the S. cerevisiae strain GRF18 (his 3-11, his 3-15, leu 2-3, leu 2-112, can$^r$, kindly provided by D. Sanglard) and the plasmid YEp51 (kindly provided by J.R. Broach, Dept. Biology, Princeton University, Princeton, NJ/USA). The inserted structural genes were expressed under control of the galactose inducible GAL10 promotor.

The intracellular localization of the expressed proteins was achieved by protein A/gold-immunolabeling of ultrathin frozen sections of formaldehyde/glutaraldehyde fixed cells (Vogel et al. 1992). Polyclonal antibodies against the P450-proteins, reductase and cytoplasmic invertase were raised in rabbits and affinity purified.

RESULTS AND DISCUSSION

Expression in S. cerevisiae of the integral membrane proteins P45052A3, P45052A4, P45051, and NADPH-P450-reductase resulted in their correct localization in the ER as shown by immunoelectron microscopy. In each case, a massive ER-proliferation occured. Depending on the protein expressed, distinctly organized stacks of paired membranes and other morphological types of extended ER appeared. Overexpression of the NADPH-P450-reductase caused a proliferation of karmellae like structures.

Overexpressing the secretory form of invertase accumulation in the ER lumen and distensions of this compartment were observed. In contrast, high-level expression of a cytoplasmic protein (invertase without its signal sequence) did not induce ER-proliferation. The newly synthesized invertase is spread over the entire cytoplasm.

ER-proliferation as induced by high-level expression of integral membrane proteins seems to include a coordinated synthesis of further ER resident proteins. Thus, a co-induction of the ER-luminal protein KAR2 protein was observed in cells overexpressing P45052A3, P45052A4, and NADPH-P450-reductase.

Based on these findings we suggest that high-level expression of integral membrane proteins in S. cerevisiae may provide a suitable experimental tool to study basic mechanisms of ER biogenesis.

To analyze those parts of the P450 sequence responsible for membrane anchoring, several fusions were constructed using the cytosolic form of yeast invertase as a reporter protein (Tab. 1). A chimeric protein containing only the hydrophilic sequence preceding the first hydrophobic segment of P45052A3 remained in the cytoplasm. P45052A3 derived sequences including the first

## Tab. 1
## Identification of topogenic sequences of two cytochromes P-450

| Construct (encoded protein) | Membrane-integration | Glycosylation | ER-Proliferation |
|---|---|---|---|
| SUC2 [a] | − | +++ | (−) |
| SUC2 | − | − | − |
| P-450Cm1 | + | − | + |
| ΔN(2-47)Cm1 | − | − | − |
| Cm1(1-16)/SUC2 | − | − | − |
| Cm1(1-34)/SUC2 | (+) [b] | ++ | + |
| Cm1(1-37)/SUC2 | + | + | + |
| Cm1(1-44)/SUC2 | + | − | + |
| Cm1(1-79)/SUC2 | + | − | + |
| SUC2 [a]/Cm1(2-523) | + | + | + |
| P-450Cm2 | + | − | + |
| Cm2(1-44)/SUC2 | (+) [b] | ++ | + |
| Cm2(1-86)/SUC2 | + | (+) | + |

[a] secretory form of invertase
[b] protein is partially removed by alkali treatment

hydrophobic segment turned out to be sufficient for ER-targeting and insertion. Fused to invertase they redirected the cytosolic reporter protein to the ER as shown by immunoelectron microscopy and subcellular distribution of invertase activity.

The membrane orientation of the first hydrophobic segment of P45052A3 was dependend on the charge pattern of the flanking hydrophilic sequences as indicated by differences in glycosylation of the reporter protein (Tab. 1). For a correct retention of P45052A3 in the ER the first 44 amino acids including the first hydrophobic segment and the following hydrophilic sequence were required. A chimeric protein consisting of the complete invertase with its own signal peptide and P45052A3 was retained in the ER and did not reach the golgi as indicated by core glycosylation of the invertase part.

The property of P45052A3 and P45052A4 to induce ER-proliferation was found to be transplanted with the same N-terminal sequences that confer targeting to and retention within the ER (Tab. 1).


REFERENCES


Schunck, W.-H., Kärgel, E., Gross, B., Wiedmann, B., Mauersberger, S., Köpke, K., Kießling, U., Strauss, M., Gaestel, M., and Müller, H.-G. (1989): Molecular cloning and characterization of the primary structure of the alkane hydroxylating cytochrome P450 from the yeast *Candida maltosa*. *Biochem. Biophys. Res. Commun.* 161, 843-850

Schunck, W.-H., Vogel, F., Gross, B., Kärgel, E., Mauersberger, S., Köpke, K., Gengnagel, C., and Müller, H.-G. (1991): Comparison of two cytochromes p450 from *Candida maltosa*: primary structures, substrate specificities and effects of their expression in *Saccharomyces cerevisiae* on the proliferation of the endoplasmic reticulum. *Eur. J. Cell Biol.* 55, 336-345

Stäubli, W., Hess, H., and Weibel, E.R. (1969): Correlated morphometric and biochemical studies on the liver cell. *J. Cell Biol.* 42, 92-111

Shohat, M., Janossy, G., and Dourmashkin, R.R. (1973): Development of rough endoplasmic reticulum in mouse splenic lymphocytes stimulated by mitogens. *Eur. J. Immunol.* 3, 680-687

Wright, R., Basson, M., D'Ari, L., and Rine, J. (1988): Increased amounts of HMG-CoA reductase induce "Karmellae": A proliferation of stacked membrane pairs surrounding the nucleus. *J. Cell Biol.* 107, 101-114

Vogel, F., Gengnagel, C., Kärgel, E. Müller, H.-G., and Schunck, W.-H. (1992): Immunocytochemical localization of alkane-inducible cytochrome P450 and its NADPH-dependent reductase in the yeast *Candida maltosa*. *Eur. J. Cell Biol.* 57, 285-291

# Membrane topology and complex formation of cytochrome P450SCC in proteoliposomes studied by delayed fluorescence depolarization

Dieter Schwarz[1], Alexey Chernogolov[2], Volker Krüger[3], Sergey Usanov[2], Anton Stier[3]

[1]Max Delbrück Center for Molecular Medicine, D-13122 Berlin – Buch, Germany. [2]Institute of Bioorganic Chemistry, Belorussian Academy of Sciences, 220141 Minsk, Belarus. [3]Max Planck Institute of Biophysical Chemistry, 37077 Göttingen, Germany

## INTRODUCTION

An understanding of the molecular mechanism of steroid biosynthesis requires direct information about the molecular dynamics and interaction of cytochrome P450SCC (CYP45011A1) (P450SCC) which catalyses the first and rate-limiting step – the side chain cleavage of cholesterol to produce pregnenolone (Lambeth, 1990). Some of the fundamental problems have not been resolved including the quaternary structure and membrane topology of P450SCC and its electron transfer interactions in the mitochondrial membrane. On the other hand increasing evidence has been accumulated indicating that microsomal P450's form oligomers of a defined size (Greinert et al., 1982; Schwarz et al., 1990), which are anchored in the membrane by a N-terminal short hydrophobic segment of the protomer with a luminal orientation of the N-terminus (for reviews see: Black, 1992; Szczesna-Skorupa & Kemper, 1993).

In membrane systems only few studies were performed to approach the problem of topology, protein interaction and dynamics of P450SCC though it has been manifested that the phospholipid environment in general and cardiolipin specifically play an important role (Lambeth, 1990; Seybert, 1990; Ikushiro et al., 1992). In natural membrane systems the dynamics and interactions of mitochondrial P450's were analyzed directly by rotational diffusion (Ohta et al., 1990 and 1991). Recently experimental evidence has been gained for P450SCC in DOPC vesicles to show an exceptionally high mobility and P450SCC-induced vesicle aggregation resulting in P450SCC transfer between vesicles (Dhariwal et al., 1991).

We follow the line to measure P450SCC rotation in reconstituted phospholipid vesicle systems providing a membrane of defined lipid and protein composition which enables to analyse intermolecular interactions of the distinct P450's separately. The present paper reports application of time-resolved delayed fluorescence anisotropy to probe the rotational motion of P450SCC in lipid vesicles of different lipid composition, especially in dependence on the CL content and on the electron transfer partners adrenodoxin (AD) and adrenodoxin reductase (AR). In combination with electron microscopy and antibody crosslinking experiments we obtained less ambiguous information concerning membrane topology and electron transfer complexes. The results indicate a significant different way of membrane integration for mitochondrial P450 compared to microsomal ones probably as consequence of the lack of a hydrophobic amino-terminal anchoring.

## MATERIALS AND METHODS

<u>Selective labeling and incorporation into liposomes.</u> P450SCC was covalently and specifically labeled with the high yield delayed fluorescence chromophore diiodofluorescein iodoacetamide (DIFIA) at sulfhydryl groups of Cys-264. Selective labeling was proved by limited proteolysis followed by HPLC separation of the labeled peptides and sequence analysis to determine the location of the labeled amino acids. For rotational diffusion labeling at a degree of 0.5 mol DIFIA/mol P450SCC was used which has nearly no effect on the native structure of P450SCC (see this vol.: A.A.Chernogolov et al. "Study of the membrane topology....."). After labeling P450SCC was incorporated into unilamellar phospholipid vesicles of different lipid composition by octylglucoside dialysis/adsorption. Successful incorporation of P450SCC was proved by gel filtration of the vesicles and

measurements of the enzymatic activity in a reconstituted vesicle system (see this vol.: D.Schwarz et al. "Reconstitution.....").

Fluorescence anisotropy measurements. Rotational diffusion of P450SCC was determined by measuring the rotation-modulated decay of the delayed fluorescence emission anisotropy of the DIFIA-labeled P450SCC. The principle of the technique and improved instrumentation are described in details elsewhere (Greinert et al., 1982; Stier et al., 1991; Schwarz et al., 1993). Fluorescence was excited by an excimer-pumped dye laser at 495 nm. Usually 100 000 sweeps (laser shots) were accumulated in about 20 min per experiment. Measurements were made in standard buffer (50 mM K-phosphate, 100 mM KCl, pH 7.3, 0.5 mM EDTA, 0.1 mM DTT) containing 20 % glycerol at 20 °C if not otherwise stated. The use of only 'large P450SCC liposomes' (average diameter ≥ 150 nm) after vesicle fractionation proved no significant contribution of vesicle tumbling to the measured fluorescence anisotropy decays.

Analysis. Multiexponential functions were fitted to the experimental fluorescence anisotropy decays

$$r(t) = (I_{//}(t) - I_{\perp}(t)) / (I_{//}(t) + 2 I_{\perp}(t)) \qquad (1)$$

($I_{//}(t)$ and $I_{\perp}(t)$ .....delayed fluorescence emission with parallel and perpendicular polarization) by a modified non-linear 'damped least-squares' Marquardt-Levenberg algorithm (Wynne & Wormell, 1963). Analysis of r(t) was based on the model of uniaxial rotation of P450SCC about the membrane normal (Kawato & Kinosita, 1981) because it gave in most cases good fits. According to this

$$r(t) / r(0) = A_1 e^{-t/\Phi_1} + A_2 e^{-t/\Phi_2} + A_3, \qquad (2)$$

assuming one single rotating species; with $\Phi_1/\Phi_2 = 4$ an (axial) rotational relaxation time $\Phi_{//} = \Phi_1$ is obtained. r(0) is the initial anisotropy at t = 0 and the normalized amplitudes $A_i$ (i=1,2,3) depend only on the angle $\Theta$ between the axis of rotation and the transition moments for emission and absorption of the chromophore. $A_3$ is the residual anisotropy $r_\infty / r(0)$. For uniaxial rotation the angle $\Theta$ can be evaluated from

$$A_1 / A_2 = 4 \cot^2 \Theta \qquad (3)$$

or from

$$A_3 = 1/4 (3 \cos^2\Theta - 1)^2, \qquad (4)$$

when all P450 molecules are rotating. Otherwise $A_3$ contains a contribution from an immobile population of P450, $f_{im}$. This fraction can be calculated according to

$$A_3 = f_{im} + (1 - f_{im}) A_3^{theo} \qquad (5)$$

with $A_3^{theo}$ as theoretical value determined by (4) using the $\Theta$ value derived from (3).

RESULTS AND DISCUSSION

Lipid effects. The rotational diffusion of P450SCC was examined in phospholipid vesicles of different lipid composition especially in dependence on the CL content (Table). The protein rotates in all cases around axis perpendicular to the membrane surface; the mobility of P450SCC was significantly dependent on the CL concentration. In PC/PE/CL (2:2:1) vesicles having a lipid composition similar to the inner membrane of bovine adrenal mitochondria (Cheng & Kimura, 1983) almost all P450 are rotating with a relatively high mobility corresponding to a $\Phi_{//} = 53$ μs (Fig.1.). With a $\Theta$ of about 70° it follows that the immobile fraction of P450 is between 0 and 3%. In PC/PE/CL (1:1:1) proteoliposomes with a considerably increased fraction of CL P450 rotates slower, $\Theta$ is about 66° and the immobile population increases to about 9 %. In the absence of any CL only 67 – 75 % of the P450 could be reconstituted as mobile fraction. In addition to the last finding the incorporation of P450 into the membrane was incomplete (about 10% of P450 could not properly reconstituted). Parallely to the increase in $A_3$ larger CL contents induce a significant enhancement of r(0) from 0.11 to 0.14.

Membrane topology. Examination of the morphology of the vesicle preparations by freeze-fracturing proved mainly spherical, round and unilamellar vesicles. P450SCC particles which probably represent small aggregates because of their size (6–10 nm) were found to be tightly associated with the outer surface of the liposomes. P450SCC-liposomes show a CL dependent tendency to aggregate. Vesicles with a CL content of about 11 mol%

**Table:** Decay parameters of the time-resolved delayed fluorescence anisotropy of cytochrome P450SCC in proteoliposomes with L/P=5 [a] at 20°C in standard buffer containing 20% glycerol

| P450SCC-Proteoliposomes | $\phi_{//}$ (μs) | $A_1$ | $A_2$ | $A_3$ | $\Theta$ | r(0) | $f_{im}$ |
|---|---|---|---|---|---|---|---|
| PC/PE/CL (2:2:1) | 53.0 | 0.316 | 0.599 | 0.085 | 70.0 | 0.108 | 0.010 |
| +Sucrose (43%) | 69.0 | 0.329 | 0.362 | 0.309 | 64.5 | 0.116 | 0.273 |
| PC/PE/CL (1:1:1) | 68.9 | 0.362 | 0.487 | 0.151 | 66.7 | 0.138 | 0.087 |
| PC/PE/CL (4:4:1) | 62.7 | 0.394 | 0.519 | 0.087 | 66.5 | 0.108 | 0.020 |
| PC/PE (2:1) | 71.2 | 0.304 | 0.390 | 0.306 | 66.2 | 0.086 | 0.258 |

[a] All ratios are weight ratios, i.e. w/w. Values of the parameters are averages of usually 5–8 experiments including different vesicle preparations.

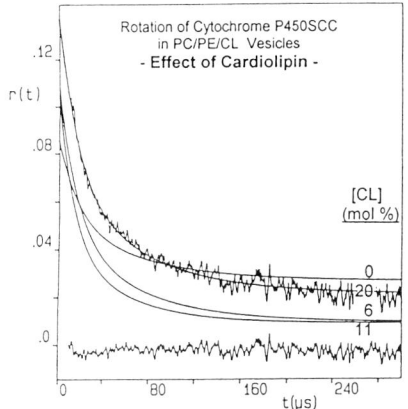
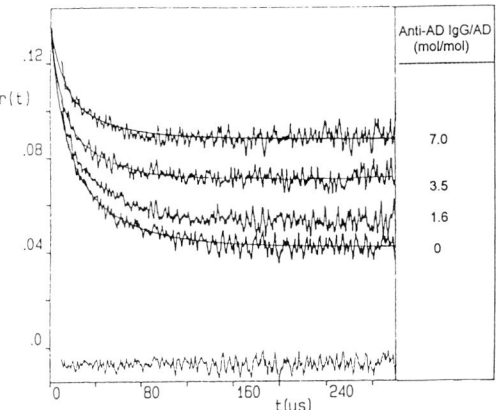

**Fig. 1** Decays of the anisotropy of P450SCC in PC/PE/CL vesicles. Besides one, the experimental curves were omitted because of clarity. Smooth: Calculated (fitted) curves. Lower trace: residuals for the fit of the 20 mol%-curve.

**Fig. 3** Effect of anti-AD IgG on the rotation of P450SCC in PC/PE/CL(2:2:1) vesicles in the presence of AD. Measurements were made in 52 %(w/v) sucrose buffer solution. Lower trace: residuals for the fit of the lowest decay curve.

are nearly not aggregated, whereas vesicles containing about 20 mol% CL mainly associate into groups of vesicles with a predominant localization of P450SCC between the vesicle surfaces. However, the most important result of the freeze-fracturing experiments was the absence of intramembrane particles on fracture faces indicating no large bilayer spanning segments of the P450SCC polypeptide chain. This finding is in striking contrast to results obtained from microsomal cytochrome P450's incorporated into liposomal membranes ( LM2 (CYP4502B4) and LM4 (CYP4501A1) ) (Schwarz et al., 1990).

The results from rotational diffusion and freeze-fracturing indicate a CL dependent insertion and/or conformation of P450SCC in PC/PE/CL vesicles. At least two different states can be discriminated: (i) al lower CL content (mitochondrial-like lipid composition) P450SCC has a relatively high rotational mobility and flexibility (at least of its segment to which DIFIA is bound, the Cys-264), all P450 are mobile and the P450SCC-induced vesicle aggregation is small, and (ii) higher CL incorporation into the vesicles reduces the mobility ( $\Phi_{//}$ ) and flexibility (r(0), initial anisotropy) and leads to partial immobilization of P450SCC (deeper immersion into the bilayer ?) - an characteristic vesicle aggregation occurs probably caused by simultaneous interaction of P450SCC with two vesicles. The vesicle coupling is a specific process which is inhibited by anti-P450SCC IgG (not shown).

Taking these results together we suggest P450SCC oligomers which are tightly, but not deeply, integrated in the bilayer of lipid vesicles; they probably are embedded only within the outer monolayer protruding with its major part to the aqueous phase surrounding the membrane (Fig.2). This model is supported by the finding that all P450SCC in PC/PE/CL vesicles could be enzymatically reduced by addition of the electron transfer components and NADPH (see this volume: D.Schwarz "Reconstitution....."). Assuming such a topology one can predict $\Phi_{//}$ values between 57 and 158 µs from (Greinert et al., 1982)

$$\Phi_{//} = 4\pi\eta a / (kT) \cdot ( h/a + 8/(3\pi) )$$

assuming typical values for the bilayer viscosity $\eta$ between 1 and 2.5 cP (Peters § Cherry, 1982) , a radius a=5.7 nm and a depths of immersion into the bilayer h = 1 nm for a hypothetical hexamer with a shape as depictured in Fig. 2. The dimensions for the monomeric subunits are approximated by the size of crystallized bacterial P450 ( Poulos et al., 1985 ). These estimated values are in reasonable agreement with the observed ones taking into account the errors of the experiment and the curve fitting. The results are in contrast to recent immunochemical studies according to which in mitoplasts a transmembrane (bilayer penetrating) P450SCC was proposed ( Usanov et al., 1990 ). However, the mitochondrial P450SCC lack any hydrophobic amino-terminus (Edwards et al., 1989) that might function as a membrane anchor in contrast to microsomal P450's. Wether the different topology observed for mitoplasts and our proteoliposomes is in fact caused by differences in properties of natural and model membranes ( for instance the lower protein content of the liposomes and the use of lipids differing in their degree of unsaturation from the mitochondrial ones ) remains to be solved by future experiments.

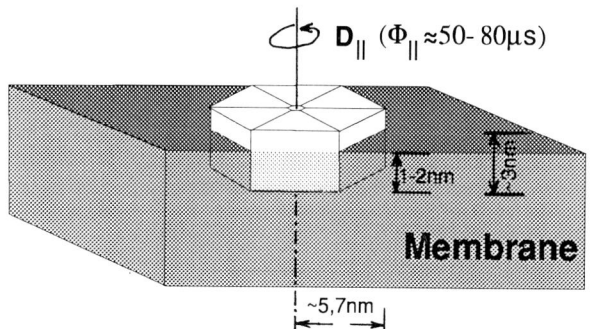

**Fig. 2** Model of the topology and organisation of the mitochondrial cytochrome P450SCC in phospholipid vesicles.
An oligomeric (probably hexameric) aggregate is immersed not too deeply in the outer monolayer of the bilayer membrane and rotates with a relatively high mobility around an axis parallel to the membrane normal.

Complex formation. To study interactions of P450 SCC with its electron transfer partners the effect of AD (externally added) and AR (coreconstituted with P450SCC) on the P450SCC rotation was investigated. A specific complexation was tried to visualize by combination of rotational diffusion and antibody crosslinking. Addition of AD to P450SCC vesicles does not lead to any significant changes in the fluorescence decays. This is not surprising because it can be expected that the binding of such a relatively small protein like AD should not in any case influence the rotation of the P450SCC aggregate. However, a specific effect was observed after addition of anti-AD IgG in the presence of AD (Fig.3). Increasing amounts of antibodies monotonically enhance the immobilized fraction of P450SCC. This indicates an immobilization of P450SCC caused by crosslinking of AD by anti-AD IgG. The absence of any effect after addition of a comparable amount of non-immuno IgG excludes any unspecific influence on P450SCC rotation by antibodies. These results directly prove the existence of binary P450SCC-AD complexes in liposomal membranes.

On the other hand coreconstitution of AR with P450SCC and addition of anti-AR IgG to (P450SCC+AR)-liposomes in the presence and absence of AD do not influence the anisotropic decays of P450SCC in PC/PE/CL vesicles (not shown) indicating both the absence of P450SCC-AR complexes and ternary complexes under inclusion of all three proteins. Taking together these results support the 'shuttle'mechanism of electron transfer from AR to P450 with AD as mobile electron shuttle (Lambeth et al., 1979).

This study was supported by the German Research Foundation DFG (Schw 471/1-1).

REFERENCES
Black, S.D. (1992): *FASEB J.* 6, 680-685.
Cheng, B. & Kimura, T. (1983): *Lipids* 18, 577-584.
Dhariwal, M.S., Kowluru, R.A., and Jefcoate, C.R. (1991): *Biochemistry* 30, 4940-4949.
Edwards, R.J., Murray, B.P., Boobis, A.R., and Davis, D.S. (1989): *Biochemistry* 28, 3762-3770.
Greinert, R. Finch, S.A.E., and Stier, A. (1982): *Xenobiotica* 12, 717-726.
Ikushiro, S., Komimani, S., and Takemori, S. (1992): *J. Biol. Chem.* 267, 1464-1469.
Kawato, S. & Kinosita, K. (1981): *Biophys. J.* 36, 277-296.
Lambeth, J.D., Seybert, D.W., and Kamin, H. (1979): *J. Biol. Chem.* 254, 7255-7264.
Lambeth, J.D. (1990): In *Frontiers in Biotransformation, Vol.3*, eds. K. Ruckpaul & H. Rein, pp. 58-100. Berlin: Akademie Verlag.
Ohta, Y., Mitani, F., Ishimura, Y., Yanagibashi, K., Kawamura, M.,and Kawato, S. (1990):*J.Biochem.*107, 97-104.
Ohta, Y., Yanagibashi, K. Hara, T. Kawamura, M., and Kawato, S. (1991): *J. Biochem.* 109, 594-599.
Peters, R. & Cherry, R.J. (1982): *Proc. Natl. Acad. Sci USA* 79, 4317-4321.
Poulos, T.L., Finzel, B.C., Gunsalus, I.C., Wagner, G.C., and Kraut, J. (1985): *J. Biol. Chem.* 260, 16122-16130.
Schwarz, D., Pirrwitz, J., Meyer, H.W., Coon, M.J., and Ruckpaul, K. (1990): *Biochem. Biophys. Res. Commun.* 171, 175-181.
Schwarz, D., Krüger, V., Chernogolov, A.A., Usanov, S.A., and Stier, A.(1993): *Biochem. Biophys. Res.Commun.* 195, 889-896.
Seybert, D.W. (1990): *Arch. Biochem. Biophys.* 279, 188-194.
Stier, A., Krüger, V., Eisbein, T., and Finch, S.A.E. (1991): In *Molecular Aspects of Monooxygenases and Bioactivation of Toxic Compounds*, eds. E. Arinc et al., pp 93-13, New York: Plenum Press.
Szczesna-Skorupa, E. & Kemper, B. (1993): *J. Biol. Chem.* 268, 1757-1762.
Usanov, S.A., Chernogolov, A.A., and Chashchin, V.L. (1990): *FEBS Letters* 275, 33-35.
Wynne, C.G. & Wormell, P.M.J.H. (1963): *Applied Optics* 2, 1233-1239.

# Core glycosylation of cytochrome P450(arom): evidence for localization of N-terminus of microsomal cytochrome P450 in the lumen

Okura Shimozawa[1], Masao Sakaguchi[1], Hisamitsu Ogawa[2], Nobuhiro Harada[2], Katsuyoshi Mihara[1], Tsuneo Omura[1]

[1]Department of Molecular Biology, Graduate School of Medical Science, Kyushu University, Maidashi, Higashi-ku, Fukuoka 812. [2]Institute for Comprehensive Medical Science, Fujita Health University, Toyoake, Aichi 470-11, Japan

## Summary

The core glycosylation of P-450(arom) was examined with two heterologous expression systems, cultured insect cells and *in vitro* translation system. The P-450(arom) protein expressed in the insect cells was glycosylated, and the sugar chain was sensitive to endoglycosidase H (Endo H). It was also glycosylated when translated with wheat germ cell-free system in the presence of rough microsomal membrane, and the sugar chain could be removed by Endo H treatment. We replaced each of the two asparagine residues (Asn-12 and Asn-180) with alanine by site directed mutagenesis and examined the glycosylation of the two mutant proteins in the cell-free system. The core glycosylation did not occur when the Asn-12 residue was mutated, whereas the mutant protein with modified Asn-180 residue was glycosylated. These results demonstrated that the potential glycosylation site (Asn-12) in the N-terminal portion of P-450(arom) is the site of glycosylation. We conclude that the N-terminus of P-450(arom) is translocated across the ER membrane to be glycosylated at the luminal side.

INTRODUCTION  Microsomal cytochrome P-450s form multigene families, and are the integral membrane proteins localized in the endoplasmic reticulum (ER). It is co-translationally integrated into the membrane, and the integration process is mediated by signal recognition particle (Sakaguchi, *et al.*, 1984). The N-terminal short hydrophobic segment of the P-450 molecules is the type I signal-anchor sequence, which possesses a combined topogenic functions of an insertion signal and a stop-translocation function (Monier, *et al.*, 1988; Sakaguchi, *et al.*, 1987; Szczesna-Skorupa, *et al.*, 1988). We previously demonstrated that the unique topogenic function of the signal-anchor sequence was determined by the balance between the N-terminal charge and the following hydrophobic segment, so that the sequence is integrated into the membrane via protein translocation machinery whereas the following portion of the P-450 molecule is not translocated across the membrane (Sakaguchi, *et al.*, 1992; Sato, *et al.*, 1990).

The membrane topology of microsomal P-450s has been studied by various methods (Sakaguchi, *et al.*, 1993). The microsomal P-450 is anchored to the membrane by the N-terminal portion, which is the only hydrophobic segment in the transmembrane configuration, and the major portion of the molecule is located on the cytoplasmic surface of the membrane (Sakaguchi, *et al.*, 1993). However, the orientation of the N-terminal signal-anchor segment in the membrane has not been definitely elucidated with natural P-450s.

P-450(arom) is responsible for the conversion of androgens to estrogens (Harada, 1988a; Harada, 1988b). Human P-450(arom) has two potential glycosylation sites and the purified preparation has been reported to contain a sugar chain which could be cleaved by endoglycosidase F (Sethumadhavan, *et al.*, 1991). Thus P-450(arom) would provide a natural instance of core glycosylation of microsomal P-450s.

RESULTS AND DISCUSSION
*Core glycosylation of P-450(arom) in vivo and in vitro*    When P-450(arom) purified from human placenta was treated with Endo H and analyzed by SDS-PAGE, the mobility was shifted by about 3 kDa as compared with the untreated preparation, demonstrating that the sugar chain of P-450(arom) possesses an ER-type simple core structure. P-450(arom) which was expressed in insect Sf21 cells with recombinant baculovirus system was glycosylated in the cultured cells. These observations indicate that the sugar chain of the P-450(arom) was not modified by the enzymes within the medial Golgi compartment in the cells. This is consistent with the histochemical observation that P-450(PB) (P-450 2B1) was observed only in the ER membrane and not in the other exocytotic membrane compartments even after fully induced by phenobarbital treatment.

To confirm the core glycosylation of P-450(arom) in the ER, P-450(arom) was synthesized with wheat germ cell-free system in the presence of rough microsomal membrane (RM) from dog pancreas. The *in vitro* synthesized P-450(arom) showed a single band in the absence of RM, whereas another higher molecular weight band was observed in the presence of RM. The high molecular weight band disappeared upon Endo H treatment, demonstrating that the larger product observed in the presence of RM was core glycosylated.

*Determination of glycosylation site of P-450(arom)*    There are two potential N-linked glycosylation sites (Asn-X-Thr/Ser) in P-450(arom), Asn-12 and Asn-180. As shown in Fig. 1, either Asn-12 or Asn-180 of the two potential glycosylation sites was exchanged with alanine residue to yield two mutated P-450(arom)s, pNAla or pCAla. The mutated P-450(arom)s were expressed *in vitro* and the glycosylation of the expressed proteins by RM was examined.

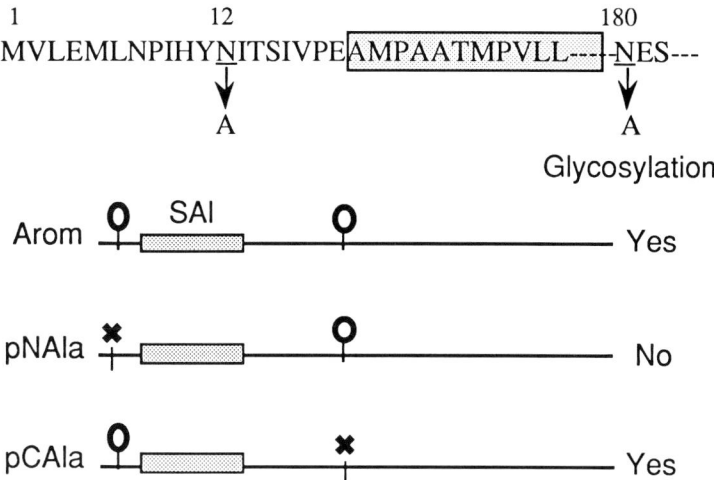

Fig. 1. Construction of mutant P-450(arom)s with mutations in the potential glycosylation sites. Upper panel indicates the two potential glycosylation sites in P-450(arom) molecule. Lower panel shows schematically the constructed mutants whose names are indicated on the left side. The crosses indicate exchange of asparagine residue to alanine residue whereas the open circles indicate the potential glycosylation sites.

The results of core glycosylation assay are indicated on the right side of the constructs. SAI represents type I signal-anchor sequence. When they were synthesized in the absence of RM, single protein band was observed in both cases. pNAla protein was not glycosylated even when translated in the presence of RM, whereas pCAla protein was core glycosylated with RM and the glycosylated product was sensitive to Endo H treatment as in the case of the wild type P-450(arom). These observations clearly demonstrated that Asn-12 in the N-terminal portion of P-450(arom) was glycosylated by RM.

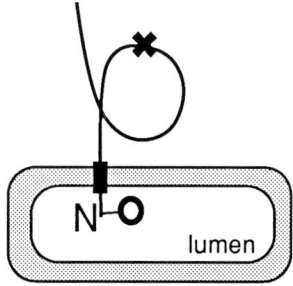

Fig. 2. Orientation of the N-terminal portion of P-450(arom). Open circle indicates sugar chain, whereas the cross indicates the unglycosylated potential glycosylation site.

Peptide segments containing a glycosylation site fused to the N-terminus of P-450s were glycosylated *in vitro* and *in vivo*, indicating that such artificially added sequences could be translocated across the membrane into the lumen by the type I signal-anchor sequence unless the introduced sequences affect the signal-anchor function (Monier, *et al.*, 1988; Szczesna-Skorupa, *et al.*, 1993). Our present study on the glycosylation of P-450(arom) clearly demonstrated that the N-terminus of a natural microsomal P-450 is translocated across the ER membrane and glycosylated as schematically shown in Fig. 2.

## REFERENCES

Harada, N. (1988a): Cloning of a complete cDNA encoding human aromatase: immunological identification and sequence analysis. *Biochem. Biophys. Res. Commun.*, **156**, 725-732.

Harada, N. (1988b): Novel properties of human placental aromatase as cytochrome P-450: purification and characterization of a unique form of aromatase. *J. Biochem.*, **103**, 106-113.

Monier, S., Luc, P.V., Kreibich, G., Sabatini, D.D., and Adesnik, M. (1988): Signals for the incorporation and orientation of cytochrome P450 in the endoplasmic reticulum membrane. *J. Cell Biol.*, **107**, 457-470.

Sakaguchi, M., Mihara, K., and Sato, R. (1984): Signal recognition particle is required for co-translational insertion of cytochrome P-450 into microsomal membrane. *Proc. Natl. Acad. Sci. USA*, **81**, 3361-3364.

Sakaguchi, M., Mihara, K., and Sato, R. (1987): A short amino-terminal segment of microsomal cytochrome P-450 functions both as an insertion signal and as a stop-transfer sequence. *EMBO J.*, **6**, 2425-2431.

Sakaguchi, M., and Omura, T. (1993): Topology and biosynthesis of microsomal P-450s. in *Frontiers in Biotransformation.* (Ruckpaul,K. and Rein,H. ed.) Vol. 8, pp. 59-73 Akademie Verlag, Berlin, Germany.

Sakaguchi, M., Tomiyoshi, R., Kuroiwa, T., Mihara, K., and Omura, T. (1992): Functions of signal and signal-anchor sequences are determined by the balance between the hydrophobic segment and the N-terminal charge. *Proc. Natl. Acad. Sci. USA*, **89**, 16-19.

Sato, T., Sakaguchi, M., Mihara, K., and Omura, T. (1990): The amino-terminal structures that determine topological orientation of cytochrome P-450 in microsomal membrane. *EMBO J.*, **9**, 2391-2397.

Sethumadhavan, K., Bellino, F.L., and Thotakura, N.R. (1991): Estrogen synthetase (aromatase). The cytochrome P-450 component of the human placental enzyme is a glycoprotein. *Mol. Cell. Endocrinol.*, **78**, 25-32.

Szczesna-Skorupa, E., Browne, N., Mead, D., and Kemper, B. (1988): Positive charges at the $NH_2$ terminus convert the membrane-anchor signal peptide of cytochrome P-450 to a secretory signal peptide. *Proc. Natl. Acad. Sci. USA*, **85**, 738-742.

Szczesna-Skorupa, E., and Kemper, B. (1993): An N-terminal glycosylation signal on cytochrome P450 is restricted to the endoplasmic reticulum in a luminal orientation. *J. Biol. Chem.*, **268**, 1757-1762.

# Systematic analysis of N-terminal transmembrane segments of microsomal P450s

Kirill N. Degtyarenko, Alexander I. Archakov

*Laboratory of Physico-Chemical Methods, Institute of Biomedical Chemistry, Pogodinskaya 10, Moscow 119832, Russia*

Microsomal P450s are the group of integral membrane proteins which have N-terminal hydrophobic segment that spans the membrane (Nelson & Strobel, 1988, 1989; Tretiakov *et al.*, 1989). The remainder of the molecule comprises mainly hydrophilic globule that appears to be structurally similar to water-soluble P450s of bacterial origin and does not contain transmembrane segments. Multiple alignment studies of amino acid (aa) sequences revealed obvious homology for major part of P450s from different families while their N-terminal domains show high variability. We showed previously that microsomal P450s have no common motif for transmembrane segment (TMS) (Archakov *et al.* 1992). On the other hand, this segment should serve the same function in different P450s. In this work, we have performed systematic analysis of TMS and its flanks of microsomal P450s. The very common building scheme for analyzed N-terminal domains comprises hydrophilic segment, TMS, charged segment, and proline-rich segment.

## MATERIALS AND METHODS

Amino acid and nucleotide sequences of microsomal P450s were extracted from SWISSPROT and EMBL databases, respectively. N-terminal fragments (1-60 aa) of protein sequences were tested for the stretch of uncharged residues. Multiple sequence alignment within P450 subfamilies and consensus sequence construction have been performed using MULTALIN program (Corpet, 1988). Codon usage for P450 protein coding sequences was calculated with CDUSAGE program from PC/GENE package (IntelliGenetics, Inc.).

## RESULTS AND DISCUSSION

*Vertebrate microsomal P450s.* CYP1A subfamily contains proteins with TMS preceded by hydrophilic segment of 13-16 aa long. All the TMS contain conserved cysteine residue. In contrast to the most of P450 proteins, TMS of trout 1A1 does not contain even the doublets of identical apolar amino acid residues. Most of CYP1A proteins (except dog, rabbit and trout 1A1) show characteristic motif of charged segment, **RxxRxxxxKxxK**. Trout 1A1 contains two negative charges in this segment. TMS in most of CYP2B proteins usually contain two long polyleucine stretches. Charged segment shows the regular motif +x+x+x+x+. TMS of 2D subfamily can be subdivided into long (21 aa) and short (15 aa) ones. Charged segment usually carries stretch of 3-4 positive charged residues. Proteins of 2G, 2H and partly 2C subfamilies show characteristic motif, **CLSC** or **CISC**, in the TMS that should be considered as an example of homology. CYP2C4, 2C5 and 2C16 contain unusual cysteine doublet in the TMS. Most of P450s from 1, 2, and 17 families (except 2B, 2D, 2F, 2J, and trout 1A1) show conserved tryptophan residue at the C-terminus of TMS. Yet CYP21 proteins contain tryptophan-rich block inside the charged segment. CYP3A proteins show the TMS which can be categorized into the medium (17 aa), short (13 aa), and long (19 aa) ones. It is unusual that CYP4B1 proteins do not contain polyleucine clusters within TMS whereas the following charged segment comprises **LLL** sequence. On the contrary, TMS of the most CYP4A proteins contain long (5-6 aa) polyleucine stretch. Rat 4F1 protein contains TMS of 28 aa long. CYP7 proteins contain TMS followed by stretch of three arginine residues; human CYP7 shows unique **CCC** sequence in TMS. Human and mouse CYP19 proteins (aromatases) show TMS preceded by hydrophilic segment of 17-18 aa long. Short of single glutamic acid residue, the charged segment is absent. Rat and chicken 19 contain sole positively charged residue after TMS.

*Yeast microsomal P450s.* There is known three yeast P450 gene families, 51, 52, and 56. In contrast to most of microsomal P450s, CYP51 and CYP52 proteins do not contain proline-rich sequence after TMS and have various charged segments. CYP56, vice versa, contains proline-rich sequence immediately after the putative TMS, and shows sole positive charge within this sequence. 51 family is presented by *Saccharomyces cerevisiae*, *Candida albicans*, and *Candida tropicalis* species. CYP51 proteins have N-terminal hydrophilic segment of 22-30 aa long followed by TMS. Proteins from 52A subfamily contain highly variable N-terminal sequences. All these proteins possess hydrophilic sequence of varying length [from 7 (52A7) to 28 (52A1) residues] preceding the TMS and the long charged segment after TMS. Sanglard *et al.* (1992) point to putative hydrophobic segments of 20 aa in *C. tropicalis* 52A1 and 52A2. However, 52A2 contains lysine and aspartic acid residues in this TMS as well as *C. maltosa* 52A5. It can be proposed that the effect of charges is neutralized through the salt bridge formation between positively and negatively charged amino acids in α-helix positions (*i*, *i*+3). Some yeast P450s contain unusual stretches of hydrophobic aliphatic and aromatic residues, for example **III** (*C. tropicalis* 51 [23-25], *C. maltosa* 52A5 [26-28]), **IIII** (*S. cerevisiae* 51 [34-37]), **VVV** (*C. tropicalis* 52A2 [16-18]), **YFY** (*C. maltosa* 52A5 [34-36], *C. tropicalis* 52B1 [22-24]), **YWY** (*C. tropicalis* 52B1 [14-16]), **WYYF** (*C. tropicalis* 52A1 [29-32]), **YWYY** (*C. tropicalis* 52A7 [8-11]), **YWYF** (*C. maltosa* 52A10 and 52A11 [10-13]). Only few yeast P450s contain proline doublets just after TMS (*S. cerevisiae* 51 [56-57], *C. tropicalis* 52A6 [48-49], 52B1 [46-47]), while in CYP56 proline doublet [25-26] is localized at the C-terminus of hydrophobic segment [6-27]. Several yeast P450s show unusually short hydrophobic segments. TMS of *C. tropicalis* 52A8 and *C. maltosa* 52A11 are of 12 aa long. *C. tropicalis* 52B1 has TMS of 14 aa long (it is surprising that this TMS does not contain leucine residues at all). TMS of 52C subfamily proteins are of 12 (52C1) and 15 (52C2) aa long. *C. maltosa* 52D1 contains TMS of 13 aa long.

*Other microsomal P450s.* The rest of microsomal P450s constitute a very heterogeneous assembly that includes proteins of insect (CYP4C, 6A, 6B), plant (CYP71, 72, 73, 74) and fungal (CYP53) origin. Insect and avocado P450s contain the "classical" TMS of 18-20 aa long whereas *Aspergillus niger* CYP53 shows unusually long TMS (27 aa). Jerusalem artichoke (*Helianthus tuberosus*) CYP73 contains TMS of 16 aa long which, however, can be elongated up to 22 aa assuming the salt bridge formation between $E_7$ and $K_8$. Analogously, 15 aa long TMS of Madagascar periwinkle (*Catharanthus roseus*) CYP72 can be longer due to ion pair $R_{26}$-$D_{29}$. In the last case TMS would be rich in tryptophan. The sequence preceding this TMS contains regular motif **M-M-M-**; single proline residue is located between TMS and charged segment. Flaxseed allene oxide synthase (CYP74) contains 21 aa long TMS enriched by polar residues and followed by sole positively charged residue. Obviously, there is deficiency of sequence information for any conclusions concerning the P450s from these kingdoms of organisms in general; however, extant sequences show that common scheme of membranous organization can be accepted for all known microsomal P450s.

*Leucine residues abundance in the TMS.* Multiple sequence comparisons of microsomal P450s using different scoring matrices revealed little or no homology of TMS from different P450 families and even subfamilies. We suppose that it can reflect either extreme variability or independent origin of these segments during evolution. Anyway, there is no common motif for TMS and therefore structure of common ancestor cannot be proposed. Obviously, only physico-chemical properties of amino acid residues should be decisive in forming of TMS. However, there are strong predominance of leucine residues over the other apolar residues in TMS. This fact can be explained by the abundance of the leucine codons in the universal genetic code. Assuming the equal frequency for each codon corresponding to any hydrophobic amino acid, one should conclude that appearance of leucine-rich TMS as a selected stochastic sequence is more likely than, for example, methionine- or tryptophan-rich ones. However, the codons have different frequencies in different genes and in different genomes (Kurland, 1991), making such approach inapplicable. On the other hand, statistical analysis shows that leucine is a most frequent residue in all P450 sequences except several bacterial P450s (104, 105C, 108). The most frequent leucine codon in the majority of DNA encoding microsomal P450s is **CTG**. In yeast P450s, leucine is the most frequent residue as well, although yeast *CYP* genes have adopted the different "codon usage strategy". Indeed, there are the most abundant leucine codons **TTA** and/or **TTG**, whereas **CT**x codons are rare or absent. However, such a codon usage pattern is very similar to overall codon usage of yeast genes (Sharp *et al.*, 1988; Wada *et al.*, 1990). Using observed frequencies of amino acids, structure of "statistical" TMS can be obtained by simulation in the absence of charged residues. Nevertheless, this model cannot explain why there are no predominance of leucine residues in TMS of yeast P450s.

***What is typical for TMS of microsomal P450s?*** Sato *et al.* (1990) pointed out two prominent differences between signal-anchor sequences of microsomal P450s and eukaryotic secretory signal sequences. First, P450s lack the positively charged amino acid preceding hydrophobic core sequence that is typical for secretory signal sequences; second, the TMS of microsomal P450s are longer than those of secretory signal sequences. Not all the microsomal P450s, however, share both these features. First, several TMS are preceded just by positive charge (CYP2J, 4A, 4B, 51, 52A, 56). Second, there exist TMS of 12-14 aa long (e.g., porcine CYP21A1 [1-12 aa], *C. maltosa* CYP52C2 [4-15 aa], rat CYP3A2, mouse CYP3A11 [11-23 aa]) that is compatible with length of typical hydrophobic core of secretory signal sequences (von Heijne, 1985).

Kuroiwa *et al.* (1991) on the basis of experiments have concluded that the membrane fragments are shorter the higher average hydrophobicity is required for efficient stop-translocation. The second observation was that the positively charged residues downstream the hydrophobic stretch are preferred for membrane anchoring although they are not obligatory. Our analysis shows that TMS of microsomal P450s tend to follow this rule. Indeed, very short TMS (rat CYP3A2, mouse CYP3A11, porcine CYP21A1, *C. tropicalis* CYP52A8 and 52B1, *C. maltosa* CYP52A11, 52C2 and 52D1) are necessarily followed by several (3 or more) positive charges; TMS followed by sole (*S. cerevisiae* CYP56) or no (CYP19) positively charged residues are of sufficient length.

Kuroiwa *et al.* (1991) proposed that more hydrophobic segments are necessary for the stop-translocation than for membrane anchoring function. Von Heijne (1985) points out that there is no preference for leucine residues in TMS of membranous proteins as against in the hydrophobic segments of signal sequences. Thus, it can be supposed that leucine residues abundance in TMS of P450s reflects its relatedness to ancestral signal sequence. However, the proposal of Kuroiwa *et al.* (1991) was based only on the study of polyleucine stretches and therefore should be verified by further experiments with different artificial hydrophobic sequences.

Gavel *et al.* (1991) analyzing proteins from different membrane systems showed that there is strict predominance of arginine and lysine residues in non-translocated versus translocated segments. This "positive-inside rule" holds for integral proteins from the bacterial inner membrane, eukaryotic plasma membrane and thylakoid membrane of chloroplasts. Our analysis shows that N-terminal domain of microsomal P450s commonly follows a "positive-inside rule" as well. Thus, it can be supposed that abundance of positive charges adjacent to hydrophobic core prevents the translocation of polypeptide chain through the membrane. This also can be explained by amphipathic nature of **R** and **K** residues which tend to lie on the lipid-water border.

The issue of the secondary structure of transmembrane segments is still unresolved. Kuroiwa *et al.* (1990) supposed that $\alpha$-helix is not obligatory conformation for the stop-transfer and membrane anchoring functions, and TMS can span the membrane as a $\beta$-strand. However, it would be electrostatically unfavorable for hydrophobic peptide to adopt nonhelical conformation in membrane environment due to uncompensated carbonyl and amino groups. Crystallographic studies revealed that transmembrane domains of several bacterial porins consist of a antiparallel β-barrel (Pauptit et al. 1991), yet single-spanning membrane proteins with TMS in an extended conformation are unknown. Some putative TMS of P450s contain many residues known as "helix-breakers" (G, I, T, V) for water-soluble proteins. However, recent experiments showed that hydrophobic peptides rich in these residues maintain α-helical conformation in membranous environment (Li & Deber, 1992).

The data presented above show that former idea of the typical TMS of the microsomal P450s as the large leucine-rich hydrophobic stretches calls for revision. In fact, only part the of vertebrate microsomal P450s has TMS following this rule.

## REFERENCES

Archakov, A.I., Degtyarenko, K.N., and Lisitsa, A. (1992) Common motifs in microsomal cytochrome P450 N-terminal membrane fragments. *J. Basic Clin. Physiol. Pharmacol.* **3**, Suppl., 97-98.

Corpet, F. (1988): Multiple sequence alignment with hierarchical clustering. *Nucleic Acids Res.* **16**, 10881-10890.

Gavel, Y., Steppuhn, J., Herrmann, R., and von Heijne, G. (1991): The "positive-inside rule" applies to thylakoid membrane proteins. *FEBS Lett.* **282**, 41-46.

Kurland, C.G. (1991): Codon bias and gene expression. *FEBS Lett.* **285**, 165-169.

Kuroiwa, T., Sakaguchi, M., Mihara, K., and Omura, T. (1990): Structural requirements for interruption of protein translocation across rough endoplasmic reticulum membrane. *J. Biochem.* **108**, 829-834.

Kuroiwa, T., Sakaguchi, M., Mihara, K., and Omura, T. (1991): Systematic analysis of stop-transfer sequence for microsomal membrane. *J. Biol. Chem.* **266**, 9251-9255.

Li, S.-C., and Deber, C.M. (1992): Glycine and β-branched residues support and modulate peptide helicity in membrane environments. *FEBS Lett.* **311**, 217-220.

Nelson, D.R., and Strobel, H.W. (1988): On the membrane topology of vertebrate cytochrome P-450 proteins. *J. Biol. Chem.* **263**, 6038-6050.

Nelson, D.R., and Strobel, H.W. (1989): Secondary structure prediction of 52 membrane-bound cytochromes P450 shows a strong structural similarity to P450cam. *Biochemistry* **28**, 656-660.

Pauptit, R.A., Schirmer, T., Jansonius, J.N., Rosenbush, J.P., Parker, M.W., Tucker, A.D., Tsernoglou, D., Weiss, M.S., and Schulz, G.E. (1991): A common channel-forming motif in evolutionary distant porins. *J. Struct. Biol.* **107**, 136-145.

Sakaguchi, M., Tomiyoshi, R., Kuroiwa, T., Mihara, K., and Omura, T. (1992): Functions of signal and signal-anchor sequences are determined by the balance between the hydrophobic segment and the N-terminal charge. *Proc. Natl. Acad. Sci. USA* **89**, 16-19.

Sanglard, D., Seghezzi, W., and Fiechter, A. (1992): Probing the membrane topology of yeast cytochrome P450s with an *in vivo* assay. In *Cytochrome P-450: Biochemistry and Biophysics*, eds. A.I. Archakov and G.I. Bachmanova, pp. 428-432. Moscow: INCO-TNC.

Sato, T., Sakaguchi, M., Mihara, K., and Omura, T. (1990): The amino-terminal structures that determine topological orientation of cytochrome P-450 in microsomal membrane. *EMBO J.* **9**, 2391-2397.

Sharp, P.M., Cowe, E., Higgins, D.G., Shields, D.C., Wolfe, K.H., and Wright, F. (1988): Codon usage patterns in *Escherichia coli, Bacillus subtilis, Saccharomyces cerevisiae, Schizosaccharomyces pombe, Drosophila melanogaster* and *Homo sapiens*; a review of the considerable within-species diversity. *Nucleic Acids Res.* **16**, 8207-8211.

Tretiakov, V.E., Degtyarenko, K.N., Uvarov, V.Yu., and Archakov, A.I. (1989): Secondary structure and membrane topology of cytochrome P450s. *Arch. Biochem. Biophys.* **275**, 429-439.

von Heijne, G. (1985): Structural and thermodynamic aspects of the transfer of proteins into and across membranes. In *Current Topics in Membranes and Transport*, eds. P. Knauf and J.S.Cook, Vol. 24, pp. 151-179. New York: Academic Press.

Wada, K.-n., Aota, S.-i., Tsuchiya, R., Ishibashi, F., Gojobori, T., and Ikemura, T. (1990): Codon usage tabulated from the GenBank genetic sequence data. *Nucleic Acids Res.* **18**, Suppl., 2367-2411.

# Membrane topology of P450c21

Li-Chung Hsu, Bon-Chu Chung

Institute of Molecular Biology, Academia Sinica, Taipei, Taiwan ROC

## Summary

We have used an in vitro translation system to study the membrane topology of P450c21 whose deficiency causes congenital adrenal hyperplasia. Wild type and mutant P450c21 proteins synthesized in vitro or in vivo showed the same pattern of enzymatic activity, demonstrating the feasibility to study the structure of P450c21 in vitro. P450c21 was integrated into membranes through an N-terminal α-helix while the rest of the protein was at the cytoplasmic side. A series of N-terminal deletion mutations demonstrate that the N-terminal hydrophobic domain is responsible for protein stability, membrane integration, and maintenance of protein structure.

## Introduction

Steroid 21-hydroxylase (P450c21) catalyzes the conversion of progesterone and 17-hydroxyprogesterone into deoxycorticosterone and 11-deoxycortisol, respectively. Deficiency in 21-hydroxylase results in impaired steroid synthesis and accounts for more than 90% of a common genetic disease, congenital adrenal hyperplasia (CAH). The enzyme deficiency can be attributed to mutations in the CYP21B gene which encodes P450c21. These mutations include gene deletions, splicing errors, and point mutations (For review, see ref. 1). Many of the point mutations are located in the coding region of the CYP21B gene, resulting in formation of proteins with impaired enzymatic activities. For example, the Asn-172 and Trp-356 mutant enzymes have severely impaired activities and are associated with the simple virilizing form of the disease (2). The Leu-281 and Leu-30 mutant enzymes have slightly higher residual activities and are associated with the milder, nonclassical type of the disease (3, 4). We characterized the membrane topology of P450c21 in relation to the disease.

## Materials and Methods

In vitro transcription coupled to in vitro translation/ protein translocation in the presence of membranes were used to test for the membrane integration properties of the protein. We have generated mutants with N-terminal truncations such as Δ4, Δ7, Δ10, Δ12, Δ23, and Δ52. These mutant cDNAs were transfected into Rat-1 cells to test for their enzymatic activity. Pulse-chase experiments were performed with $^{35}$S-Met labeling to assay the degradation rate of the P450c21 proteins.

## Results and Discussions

In vitro translation was used as the major method to test the membrane topology of the protein. The in vitro translation products were found to be enzymatic active after they were reconstituted with other components of the electron transport chain (Fig. 1). The Δ4 mutant, which lacks amino acids 2-4, still retained 70% of the wild type activity, a situation similar to the in vivo activity. Therefore, this in vitro system could be used to study the structure of these proteins.

The amount of membrane integration of each mutant protein was tested by alkaline extraction followed by ultracentrifugation. The membrane protein will sediment down to the pellet, while soluble protein will stay in the supernatant. While Δ4 and Δ7 proteins still retain normal membrane integration, Δ12, Δ12, and Δ23 were only partially integrated into the membrane, and Δ52 was completely soluble. This decrease in membrane attachment parallels their enzymatic function. The mutants have progressively lower enzymatic activities as the deletions increase. In addition, mutant proteins are degraded much faster indicating they have abnormal structure which triggers faster degradation (5).

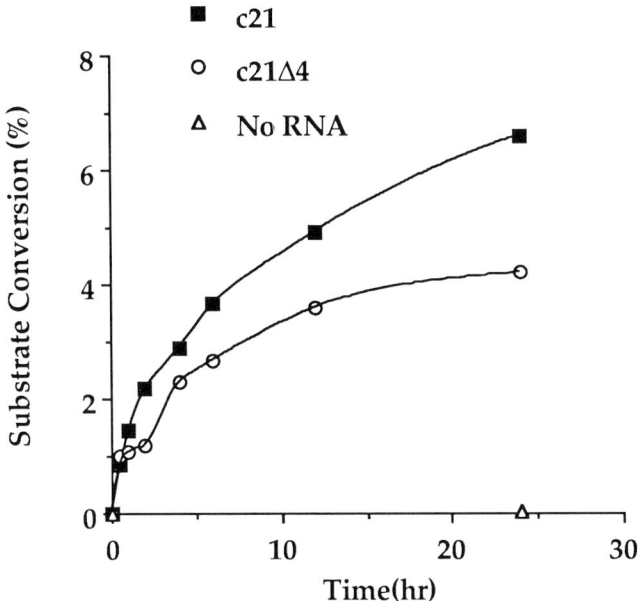

**Fig. 1. Function of normal P450c21 and its Δ4 mutant synthesized in vitro.** Six microliters of the in vitro translation product from RNA derived from P450c21, Δ4, or nothing was incubated with substrate $^{14}C$-17α-hydroxyprogesterone in 500 μl of reaction buffer at 37°C. At the indicated time points, an aliquot of the reaction was stopped and the product was separated from the substrate by TLC. The amount of product formation was plotted versus time.

## References

1. Miller, W. L., and Morel, Y. (1989) Annu. Rev. Genet. 23, 371-393

2. Chiou, S.-H., Hu, M.-C., and Chung, B.-c. (1990) J. Biol. Chem. 265, 3549-3552

3. Speiser, P. W., New, M. I., and White, P. C. (1988) N. Engl. J. Med. 319:19-23

4. Tusie-Luna, M.-T., Speiser, P. W., Dumic, M., New, M. I., and White, P. C. (1991) Mol. Endocrinol. 5, 685-692

5. Hsu, L.-C., Hu, M.-C., Cheng, S.-C., Lu, J.-C., and Chung, B.-c. (1993) J. Biol. Chem. 268, 14682-14686

# Study of the membrane topology and functional importance of the central region of cytochrome P-450scc (CYP11A1) molecule by selective chemical modification of Cys[264]

Alexey Chernogolov[1], Sergey Usanov[1], Regine Kraft[2], Dieter Schwarz[2]

[1]Institute of Bioorganic Chemistry, Academy of Sciences of Belarus, Zhodinskaya st. 5/2, 220141 Minsk, Republic of Belarus. [2]Max-Delbrück-Center for Molecular Medicine, Robert-Roessle st. 10, D-13122 Berlin, Federal Republic of Germany

## INTRODUCTION

Cytochrome P450scc is localized in the inner membrane of mitochondria of steroidogenic tissues including adrenal cortex and responsible for a key enzymatic step of the steroid biosynthesis – pregnenolone formation from cholesterol. For the multi-step hydroxylation of one cholesterol molecule, P450scc requieres six electrons which are sequentially transported from NADPH to P450scc via adrenodoxin reductase and adrenodoxin. The membrane topology of P450scc and protein-protein interrelations during catalysis are open problems. In particularly, the question is how AD which riched with residues of glutamic acid and thus has negative charge is able to interact with the membrane-embedded molecule of P450scc.
P450scc consists of two large domains F1 and F2 formed by N- and C- terminal sequences of the molecule and connected in the trypsin-sensitive region Arg$^{250}$-Asn$^{257}$ via an interdomain hinge. Both F1 and F2 domains were shown to penetrate the membrane, the interdomain hinge being exposed to matrix space of mitochondria (Usanov et al., 1990). Tryptic fragment F2 includes both of the cysteine residues of P450scc – Cys$^{264}$ and Cys$^{422}$. The latter was identified as fifth ligand of the heme group (Tsubaki et al., 1986). The functional importance of Cys$^{264}$ located near the trypsin-sensitive hinge is unclear. In the present work the functional role of the central region of P450scc sequence which includes the trypsin-sensitive interdomain hinge necessary for the specific cholesterol side chain activity and the membrane topology of the region in proteoliposomes was investigated using selective modification of Cys$^{264}$ diiodofluorescein iodoacetamide (a SH-specific fluorescent reagent).

## MATERIALS AND METHODS

DIFIA was synthesized according to Greinert et al. (1982) and kindly provided by Prof. A. Stier (Max Planck Institute for Biophysical Chemistry, Goettingen, Germany). P450scc, AD and AR were purified from bovine adrenocortical mitochondria as described by Usanov et al. (1989).
The labelling of soluble P450scc was performed in 50 mM potassium phosphate buffer, pH 7.2, containing 0.15% sodium cholate, 0.5 M NaCl, 0.5 mM EDTA, 10% glycerol and 0.2% methanol. P450scc and DIFIA were mixed in the final volume of 5 ml and incubated in a two-step incubation procedure: 10 hours at 4 °C and then 1 hour at 23°C. P450scc incorporated into liposomes according to Schwarz et al. (1993) was labelled in 50 mM PB, pH 7.4, with 0.1 M NaCl, 0.5 mM EDTA, 0.1 mM DTT and 20% glycerol. The suspension of liposomes was incubated with DIFIA of 8 hours at 4 °C and 30 min at 23 °C. The reaction mixtures were run on a Sephadex G-25M column (0.4x2.5 cm) and used for the proteolytic and activity experiments. Protein-free liposomes were used as control under the same conditions in order to check possible non-specific incorporation of the label into membranes.

---

**Abbreviations:** P450scc, AD and AR, bovine adrenocortical cytochrome P450scc, adrenodoxin and adrenodoxin reductase; DIFIA, diiodofluorescein iodoacetamide; DIFIA-P450scc, P450scc labelled with DIFIA. PC, phosphatydyl choline; PE, phosphatydyl ethanolamine; CL, cardiolipin; CHL, cholesterol; PG, pregnenolone; DTT, 1,4-dithioteit; SDS-PAGE, polyacrylamide gel electrophoresis in the presence of sodium dodecylsulfate; PB, sodium phosphate buffer.

Differential spectral titration of P450scc with AD was performed according to (Kido & Kimura, 1979), in 50 mM PB, pH 7.4, containing 25 μM CHL and 0.03% Tween 20 at 23 °C. Cholesterol side chain cleavage activity of P450scc was determined in the reconstituted system as described earlier Usanov et al. (1989) followed by treatment of the reaction mixture with cholesterol oxidase to convert CHL and PG into 3-one-4-en steroids (Sugano et al, 1989). To study the trypsinolysis of DIFIA-P450scc and DIFIA-P450scc liposomes, the ones were diluted with 50 mM PB, pH 7.4, containing 0.1 M NaCl and 0.1 mM EDTA up to a P450scc concentration of 3 μM. The suspension was incubated with trypsin at a molar ratio P450scc:trypsin of 50:1 or 20:1 for 30-60 min at 23°C and then soybean trypsin inhibitor was added up to a molar ratio inhibitor:trypsin of 4:1. The samples were analysed by SDS-PAGE followed by immunoblotting in the presence of antibodies against P450scc and its fragments F1 and F2. To detect DIFIA-P450scc and its fragments after proteolysis, gels were scanned using a Cromato-VUE Transilluminator TM-36 (UVP, USA) as an UV-light source. Proteolysis of DIFIA-P450scc with protease from Staphylococcus aureus was performed in 0.1 M ammonium bicarbonate buffer, pH 8.3.

## RESULTS AND DISCUSSION

A cysteine-specific fluorescent reagent DIFIA was used for the selective chemical modification of P450scc. The specific character of the labelling procedure was proved by subjecting a sample of the labelled protein with a molar ratio DIFIA:P450scc of 1:1 to proteolytic digestion with protease from Staphylococcus aureus. The separated peptides obtained by HPLC were sequenced and revealed as main peptide fraction which containing DIFIA as chromophore (absorbance at 517 nm) in the sequence: Phe-Arg-Asn-Tyr-Pro-Gly-Ile-Leu-Tyr-Cys(*)-Leu-Leu-Lys-Ser-Glu. This peptide corresponds to the P450scc sequence $Phe^{255}-Glu^{269}$ (Morohashi et al., 1984; Chashchin et al., 1985). No other labelled peptides were found, particularly not those containing the second cysteine residue of P450scc, $Cys^{422}$. Therefore, it was concluded that the central region of P450scc containing $Cys^{264}$ is accessible for the chemical modification under the conditions applied.

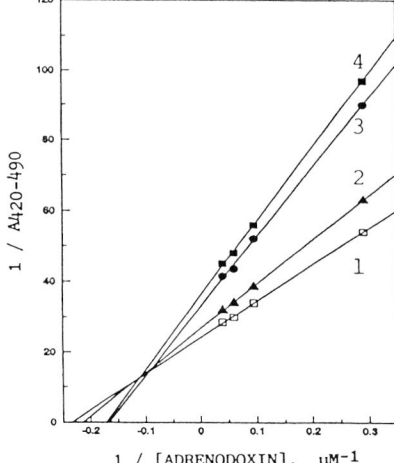

**Fig. 1.** Spectrophotometrical titration of native P450scc (1) and DIFIA-P450scc with a molar ratio 0.45:1 (2), 0.75:1 (3), and 1:1 (4). Concentration of P450scc was 2.3 μM. AD was sequentially added up to the final concentrations 3.4, 10.4, 17.3 and 24.2 μM.

The labelling influenced the interaction of the hemeprotein with AD (Fig. 1). The degree of labelling depends on the ratio DIFIA:P-450scc. The apparent Km of the interaction of 1:1 labelled P450scc with AD under the conditions applied is increased from 3.8 to 6.0 μM, Vmax of the process being simultaneously decreased 1.6-fold. The labelling of P450scc with DIFIA also affected the cholesterol side chain cleavage activity of the hemeprotein in the reconstituted system. The inhibitory effect depends on both the amount of DIFIA bound to P450scc and the ratio AD:P450scc. A 10-fold excess of AD almost prevents the effect, but at a ratio AD:P450scc of 5:1 and DIFIA:P450scc of 1:1 the residual activity of DIFIA-P450scc amounts to only 53% as compared to that of the native hemeprotein. For further characterization the functional implication of the region nearest to $Cys^{264}$, the cholesterol side chain activity in the reconstituted system containing AR, AD and proteoliposomes was measured. As follows from Table 1, the activity of P450scc after incorporation into phospholipid vesicles is maintained in the presence of externally added AR and AD, the liposomal P450scc being able to transform CHL into PG more effectively (81% stimulation) than in aqueous solution. The treatment of P450scc-liposomes with DIFIA inhibited the activity by 38%. On the other hand, the liposomes containing P450scc preliminary labelled with a molar ratio DIFIA:P450scc of 1:1 retained only 32% of the activity as compared with P450scc-liposomes.

Table 1. Enzymatic activity of membrane-bound P450scc and DIFIA-P450scc*.

| Preparation | Activity, nmol of PG/nmol of P450scc/min) | % |
| --- | --- | --- |
| P450scc–liposomes | 4.73 | 181.0 |
| P450scc–liposomes labelled with DIFIA (1:0.51) | 2.94 | 112.6 |
| DIFIA–P450scc (1:1) liposomes | 1.53 | 58.6 |
| DIFIA–P450scc (1:1) in solution | 1.23 | 47.1 |
| P450scc (native) in solution | 2.61 | 100.0 |

* P450scc and DIFIA-P450scc were incorporated into liposomes consisting of PC:PE:CL:CHL with a weight ratio 2:2:1:0.6. The reconstituted system contained in a final volume of 1 ml: 0.5 μM P450scc, 4 μM AD, 0.28 μM AR, 100 μM CHL and 360 μM of NADPH in 50 mM PB, pH 7.4, with 0.01% Tween 20.

To investigate the membrane topology of P450scc, P450scc-containing liposomes were treated with DIFIA. As seen in Fig. 2, the liposomal P450scc can be specifically labelled by DIFIA likewise. However, a final ratio of DIFIA:P450scc accounts to no more than 0.5–0.6. Neither the increase of incubation time nor a higher excess of DIFIA were able to enhance the amount of DIFIA bound to P450scc. P450scc in solution labelled under the same conditions revealed a ratio of labelling of 1:1.

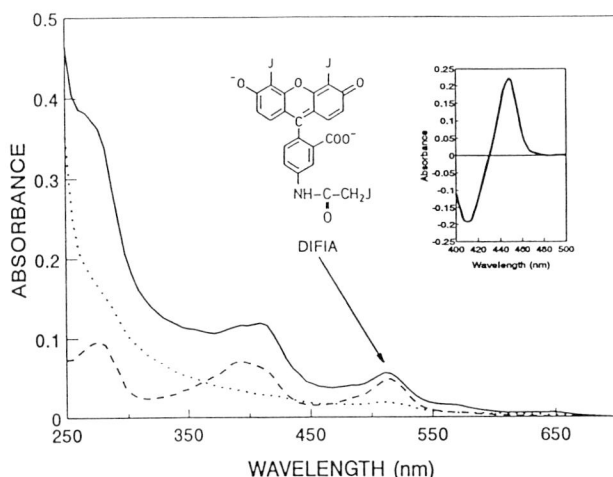

Fig. 2. Labelling of membrane-bound P450scc with DIFIA. P450scc was incorporated into the phospholipid membrane and the proteoliposomes were treated with DIFIA ( ———— ). For comparison: ( – – – – ) native P450scc labelled with DIFIA in solution under the same conditions, ( ········ ) protein-free liposomes treated with DIFIA (control). Insertions: chemical structure of DIFIA and CO-difference spectra of the DIFIA-labelled P450scc–liposomes.

The treatment of DIFIA-P450scc with trypsin led to the formation of two main peptide fragments with molecular weights ranging between 29000 and 26000 Da (Fig. 3, bands 6 and 7). These values correspond to the fragments F1 (Ile$^1$-Arg$^{250}$) and F2 (Asn$^{257}$-Ala$^{481}$) (Fig. 3, bands 1 and 2) which were found for soluble P450scc after trypsinolysis. Indeed, the electrophoretic separation of the reaction mixture followed by immunoblotting in the presence of antipeptide antibodies against F1 and F2 proved the presence both of these fragments and F2 proved to be labelled with DIFIA. Limited proteolysis with trypsin of proteoliposomes which contained untreated native P450scc led to proteolytic digestion of the protein comparable to that of soluble P450scc. Labelling of the P450scc-proteoliposomes with DIFIA with subsequent limited trypsinolysis does also not change the electrophoretic pattern of the tryptic peptides. If DIFIA-P450scc was incorporated into the phospholipid vesicles and subjected to trypsinolysis, the result was similar to that described above (Fig. 3, bands 8 and 9). Further using of immunoblotting in the presence of antibodies against F1 and F2 fragments likewise revealed the label to occur only in F2. Thus, the results obtained suggest that native and DIFIA-P450scc are accessible to trypsin in solution and in proteoliposomes likewise, i.e. in the region Arg$^{250}$-Asn$^{257}$. Therefore, the labelling of P450scc with DIFIA does not influence the position of the trypsin-sensitive interdomain loop.

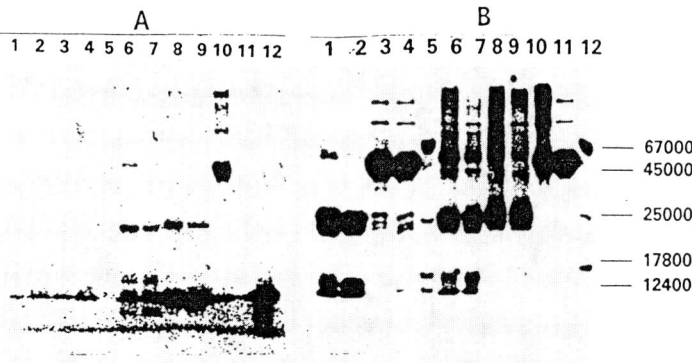

**Fig. 3** Limited proteolysis of soluble and membrane-bound DIFIA-P450scc with trypsin. **A** — SDS-PAGE under UV-light, **B** — the same slab stained with Coomassie BB R-250. 1 and 2 — P450scc treated with trypsin at 50:1 and 20:1 molar ratio; 3 and 4 — control (P450scc preincubated with soubean trypsin inhibitor and treated with trypsin under the conditions of 1 and 2); 5 — standard protein mixture 6 and 7 — DIFIA-P450scc treated with trypsin under the conditions of 1 and 2; 8 and 9 — DIFIA-P450scc-contained proteoliposomes treated with trypsin under the same conditions; 10 — DIFIA-P450scc; 11 — P450scc; 12 — standard protein mixture + pure DIFIA.

The results obtained in the present work suggest that the interdomain hinge in the central part of the P450scc sequence containing $Cys^{264}$ might be involved directly in the interaction of hemeprotein with AD on the surface of the phospholipid membrane. However, on the base of the experiments, it cannot be decided whether $Cys^{264}$ is directly involved in the interaction with AD or the inhibition of binding of AD is caused by a steric hindrance effect of the relatively bulky DIFIA molecule. Taken together, the experiments provide evidence that: i) $Cys^{264}$ of P450scc could be specifically labelled with the chromophore DIFIA under preservation of its native structure ii) $Cys^{264}$ is located at or near the AD binding area and iii) this region is located outside the membrane in proteoliposomes and therefore accessible from the aqueous solution surrounding the membrane. It should be noted that the central region of the bovine P450scc sequence contains $Lys^{267}$ and $Lys^{272}$ which have been previously identified by chemical modification of P450scc with succinic anhydride as being involved in AD binding via charge ion pairing (Adamovich et al., 1989). On the other hand, P450scc sequence $Leu^{266}-Leu^{272}$ was found to correspond to one of the antigenic determinants of P450scc molecule. Antibodies against P450scc as well as F2 fragment were proved to cause a strong inhibitory effect on the cholesterol side chain activity of P450scc (Chernogolov et al., 1991). These results support the high flexibility of the region around Cys264, that provides the possibilities for the interaction between P450scc and AD.

This work was supported by German Research Foundation (DFG, Schw 471/1-1) and Boehringer Ingelheim Fonds.

**Acknowledgments:** We thank Prof. Dr. A. Stier (Max Planck Institute for Biophysical Chemistry, Goettingen) for providing us with fluorescence label DIFIA.

# REFERENCES

Adamovich T.B., Pikuleva I.A., Chashchin V.L. and Usanov S.A. (1989): *Biochim. Biophys. Acta* 996, 247–253.
Chashchin, V.L., Lapko, V.N., Adamovich, T.B., Lapko, A.G., Kuprina, N.C., Kirillova, H.M., Berikbayeva, T.M., Akhrem, A.A., Zolotarev, A.S. (1985): *Bioorganicheskaya Khimia (USSR)* 11, 1048–1067.
Chernogolov A.A., Adamovich T.B. and Usanov S.A. (1992): In: *Cytochrome P-450: Biochemistry and Biophysics,* eds. Archakov A.I. and Bachmanova G.I., pp. 57–59, Moscow: INCO-TNC.
Greinert, R., Finch, S.A.E. and Stier, A. (1982): *Xenobiotica* 12, 717–726.
Kido, T. & Kimura, T. (1979): *J. Biol. Chem.* 254, 11806–11815.
Morohashi, K., Sogawa, K., Omura, T. and Fujii-Kuriuama, Y. (1984): *Proc. Natl. Acad. Sci. USA* 81, 4645–4653.
Schwarz D., Krüger V., Chernogolov A.A., Usanov S.A. & Stier A. (1993): *Biochem. Biophys. Res. Communs* 195, 889–895.
Sugano, S., Morishima, N., Ikeda, N. and Horie, S. (1989): *Anal. Biochem.* 182, 327–333.
Tsubaki M., Tomita S., Tsuneoka Y. and Ichikawa Y. (1986): *Biochim. Biophys. Acta* 870, 564–574.
Usanov S.A., Chernogolov A.A. and Chashchin V.L. (1989): *FEBS Lett.* 255, 125–128.
Usanov S.A., Chernogolov A.A. and Chashchin V.L. (1990): *FEBS Lett.* 275, 33–35.

# VI DETERMINANTS OF COMPONENT INTERACTIONS IN P450 SYSTEMS. ELECTRON TRANSFER AND OXYGEN ACTIVATION

# The importance of one-electron transfers in the mechanism of cytochrome P450

Ronald E. White

*Department of Metabolism and Pharmacokinetics, The Bristol-Myers Squibb Pharmaceutical Research Institute, PO Box 4000, Princeton, New Jersey, 08543-4000, USA*

Cytochrome P450 is an enzyme that moves in one-electron steps. This paper will discuss why it is a one-electron enzyme and the biochemical price that must be paid in order to use a radical mechanism (White, 1991). It will also provide an overview of the last 30 years of P450 mechanistic work to show how we know what we know presently.

## WHAT ARE ONE-ELECTRON TRANSFERS?

The term "one-electron transfer" refers to the movement of either an electron or a hydrogen atom from one functional group to another. The transfer may be intra- or inter-molecular. Thus, several types of reactions are included: *one-electron oxidations*, such as the removal of a single electron from an amino nitrogen atom; *one-electron reductions*, such as reduction of ferric iron to ferrous or the reduction of an alkyl halide to produce a free radical; *hydrogen abstractions*; and the *addition of various free radicals to double bonds*. These are all radical reactions, but the more general term one-electron transfers reminds us that many of the steps in the P450 mechanism are radical steps, not only the attack on substrate.

In discussing the occurrence of radical steps in the P450 mechanism, we will show why on the basis of FIRST PRINCIPLES that P450 must use a radical mechanism. Next we will discuss the EXPERIMENTAL EVIDENCE to support that conclusion. Finally, we will examine the various CONSEQUENCES of the need to generate radicals in the P450 active site.

## FIRST PRINCIPLES

### What Pathways are Possible from the Point of View of the Substrate?
In the overall conversion of $R-CH_3$ to $R-CH_2OH$, a C-H bond had to be broken and a new C-O bond had to be formed. There are five ways this can happen.

The FIRST way is to remove the hydrogen as a proton to generate a carbanion. The carbanion could then react with $O_2$. Carbanions are known to be formed in some enzyme reactions.

However, for most substrates that are hydroxylated by P450, an extremely strong base would be required to remove the proton. It would be difficult to protect either the strong base or the

1. $R\text{-}CH_2\text{-}H \rightarrow R\text{-}CH_2^- + H^+$      (Carbanion)

2. $R\text{-}CH_2\text{-}H \rightarrow R\text{-}CH_2^+ + H^-$      (Carbocation)

3. $R\text{-}CH_2\text{-}H \rightarrow R\text{-}CH_2\bullet + H\bullet$      (Carbon Radical)

4. $R\text{-}CH_2\text{-}H \rightarrow R\text{—}H_2C\cdots H$ (with O bridging)      (Oxene Insertion)

5. $R\text{-}CH_2\text{-}H \rightarrow$ [X—O⋯C(H)(H)(R)⋯H⋯A]      ($S_N^2$ Displacement)

carbanion from water and yet allow substrates to freely move into the active site.

The SECOND way is to remove a hydride ion, generating a carbocation. This would require a powerful Lewis acid that probably could not exist in water. If the carbocation *could* be generated, it would immediately react with water to generate $R\text{-}CH_2\text{-}OH$, the desired product. However, this cannot be the pathway used by P450, since we know from $^{18}O$ labeling that the O in $R\text{-}CH_2\text{-}OH$ comes from $O_2$, not from $H_2O$.

The THIRD way is by hydrogen abstraction. Many types of radicals compatible with water are able to abstract hydrogen atoms, especially oxy radicals. Because the carbon radical formed is planar, the P450 active site would need to hold the carbon radical in place in order to carry out stereospecific hydroxylations. Conversely, a carbon radical mechanism allows the possibility for *loss* of stereochemistry if the P450 active site is unable to prevent the radical from moving, as seems to occur with some substrates.

The FOURTH way is concerted insertion of a singlet oxene into the C-H bond, analogous to insertion of a singlet carbene. This insertion in inherently stereospecific and retains the original stereochemical configuration of the substrate. Since some substrates such as norbornane and ethylbenzene *do* undergo racemization, then the oxene insertion is ruled out on stereochemical grounds.

The LAST way is an $S_N2$-type nucleophilic displacement. Backside displacement of hydride by a bare oxygen anion is conceivable in the P450 active site since hydride transfers are common in biochemistry. The $S_N2$ displacement is inherently stereospecific; furthermore, it results in net *inversion* of configuration. Since P450 retains the original stereochemistry with some substrates such as camphor, the mechanism cannot be $S_N2$.

This simple analysis reveals that only the radical mechanism (hydrogen abstraction) fits *all* the substrate reaction criteria.

What Pathways are Possible from the Point of View of Oxygen?
Since hydrogen abstraction is the likely mode of attack on substrate, what oxygen species are reactive enough to perform hydrogen abstraction? While $O_2$ can attack some allylic, benzylic

and other reactive C-H bonds, it is not able to attack an unactivated C-H bond such as the terminal methyl of a fatty acid. A more reactive form of oxygen must be generated in the active site, which requires some degree of reduction of $O_2$. The only two reduced oxygen forms capable of hydrogen abstraction are oxygen atoms and hydroxyl radicals.

What Pathways are Possible from the Point of View of the Enzyme?
The general observations that limit the mechanistic possibilities are as follows: the oxygen in product comes from $O_2$; the $O_2$ binds at $Fe^{2+}$ so oxygen activation must involve the iron in some way; the P-450-$O_2$ complex can be isolated in some cases and is known not to be able to attack a substrate by itself. Thus, as mentioned above, it must be reductively activated, probably to the level of peroxide since two reducing equivalents are required. On the other hand, we know from chemical studies that iron-peroxide complexes always decompose by splitting the O-O bond. We expect P450 to do the same. Thus, these simple considerations show that while the reactive oxygen intermediate must involve the iron, it cannot be the ferrous-$O_2$ complex or a peroxide complex. The only alternative is that it contains only a *single* oxygen directly bonded to iron. In view of the expected reaction characteristics of oxygen species (see above), we can guess that the intermediate would be a hydroxyl radical or oxygen atom bound to iron. These same intermediates are also observed in the related peroxidase enzymes and in chemical model systems.

## EXPERIMENTAL VERIFICATION OF A RADICAL MECHANISM

Now we will examine some of the best evidence that P450 really *does* follow a radical mechanism. Most of the evidence is indirect. With a variety of substrates, the reaction products are those that are expected from radical attack on the substrate. For example, methylcyclohexane contains primary, secondary and tertiary positions for hydroxylation in a small molecule that probably has rapid movement in the active site. When P450 hydroxylates methylcyclohexane, a mixture of all possible alcohols is formed (White *et al.*, 1979). However, the mixture is not random. When statistically adjusted for the number of each type of hydrogen present, the product distribution reveals a strong preference for secondary and tertiary positions, similar that of radicals such as tBuO•, Cl• and triplet oxygen atoms. P450 is quite different from insertions by singlet carbenes and singlet oxygen atoms, which show almost no preference. The carbanion and $S_N2$ mechanisms would be expected to show the opposite trend, a preference for *primary* positions.

With specially designed substrates, one can demonstrate that the P450 mechanism is not inherently stereospecific. In the experiment illustrated in Fig. 1, ethylbenzene that had been stereospecifically labeled with deuterium in either the *R* or *S* configurations was hydroxylated by P450 (White *et al.*, 1986). Regardless of whether the substrate was *R* or *S*, a mixture of four alcohols (*R* and *S* configurations, with and without deuterium) was produced, as shown in the boxes. Thus, the original stereochemistry of ethylbenzene was scrambled during hydroxylation by P450.

In our interpretation of this result, initial hydrogen or deuterium abstraction yields 1-phenylethyl radical, containing either deuterium or hydrogen as a stereochemical marker. Immediate collapse of the carbon and $Fe^{IV}$-OH(D) radicals produces *R*-1-Phenylethyl alcohol with retention of configuration. However, in 25-40% of the hydroxylations, the 1-phenylethyl radical presents the opposite face for recombination, producing *S*-1-phenylethanol, with inversion of configuration. This experiment proves that the mechanism allows loss of stereochemistry, as would be true with radical, carbocation and carbanion mechanisms, but not oxene insertion or $S_N2$.

Very large deuterium isotope effects can be deduced for the enzymatic step where the C-H bond is broken. When substrates containing both hydrogen and deuterium in similar positions are hydroxylated by P450, the enzyme shows a strong preference to remove

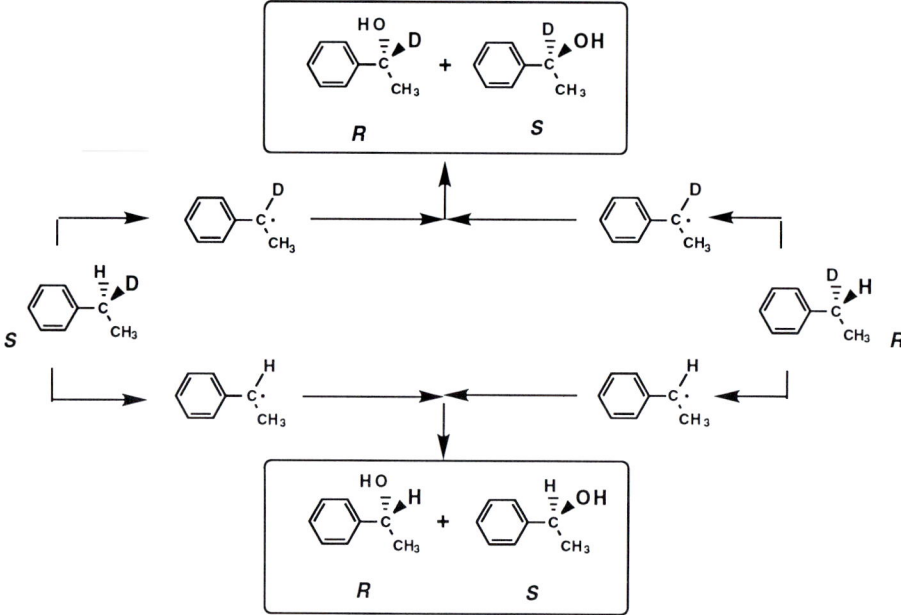

Fig. 1. Stereochemical scrambling by cytochrome P450

hydrogen instead of deuterium. Analysis of products from such an experiment allows one to calculate the intrinsic deuterium isotope effect for hydrogen removal by the P450 reactive oxygen intermediate. Very large isotope effects, in the range 11 to 19, have been repeatedly found (Groves et al., 1978; Harada et al., 1984; White et al., 1986). These large isotope effects are found in chemical reactions involving radical hydrogen abstractions and are expected on the basis of kinetic theory. We do not expect such large isotope effects from the other four mechanisms that were proposed above.

The covalent modification of porphyrin by certain suicide substrates is best explained by a radical process (e.g., terminal alkenes, Fig. 2). In the interpretation of Ortiz de Montellano et al. (1983a), initial radical addition to one end of the double bond produces a transient carbon radical intermediate. Usually this intermediate collapses to the expected epoxide product. But sometimes, the carbon radical attaches to one of the porphyrin nitrogen atoms instead. The N-alkylated "green porphyrin" no longer binds iron. This elegant work allows us to "visualize" the one-electron reactions occurring in these oxidations. These data are representative of a large body of evidence that, taken together, strongly confirm that P450 is operating through a radical mechanism.

## CONSEQUENCES OF A RADICAL MECHANISM

Now that we have established that P450 accomplishes hydroxylations with a radical mechanism, let us examine the implications. As we have seen, P450 must generate the equivalent of a hydroxyl radical or oxygen atom and this requires reduction of $O_2$. In either case, two reducing equivalents must be used. Both species are capable of attacking an

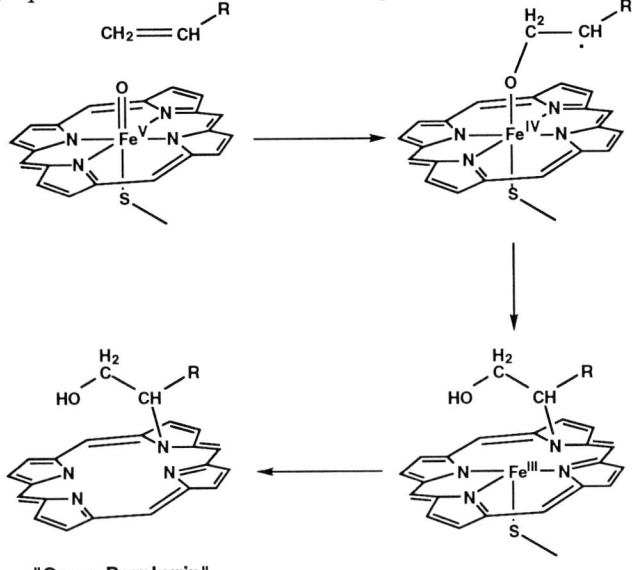

"Green Porphyrin"
Fig. 2. Covalent modification of porphyrin by substrate-derived radicals

unactivated substrate as shown in these equations. When we sum the chemical equations for the reductive and oxidative half-reactions, we get the well-known equation for a monooxygenase.

$$O_2 + 2\,e^- + 2\,H^+ \rightarrow O + H_2O$$
$$\underline{O + RH \rightarrow ROH}$$
$$O_2 + 2\,e^- + 2\,H^+ + RH \rightarrow ROH + H_2O$$

Thus, the requirement for a radical mechanism means that valuable chemical energy in the form of NADPH must be expended. The second consequence is a corollary of the first. Since a heme cannot be reduced by NADPH, another protein is needed to "translate" the obligate two-electron donor for the obligate one-electron acceptor. All P450s require an associated reductase, either as a separate protein or a second domain in a fusion protein. Prokaryotic forms need *two* proteins. Thus, because of the P450 radical mechanism, a cell is required to maintain and regulate additional genes and synthesize additional proteins.

However, the most serious consequence of the P450 mechanism is that reactions often go wrong, especially with xenobiotic substrates, because of the difficulty of controlling reactive radicals. These aborted reactions include the release of free radicals, radical substrate fragmentation, and suicide inactivation of the heme, as will be discussed in a later section.

## P450 MECHANISTIC CYCLE

Figure 3 shows the complete mechanistic cycle for P450 as we understand it today. We will use this familiar diagram to consolidate the concepts presented in the preceding sections. Steps 2, 3, 4, 6 and 7 involve one-electron transfers, and are therefore radical reactions, even though they may not be usually thought of in that way.

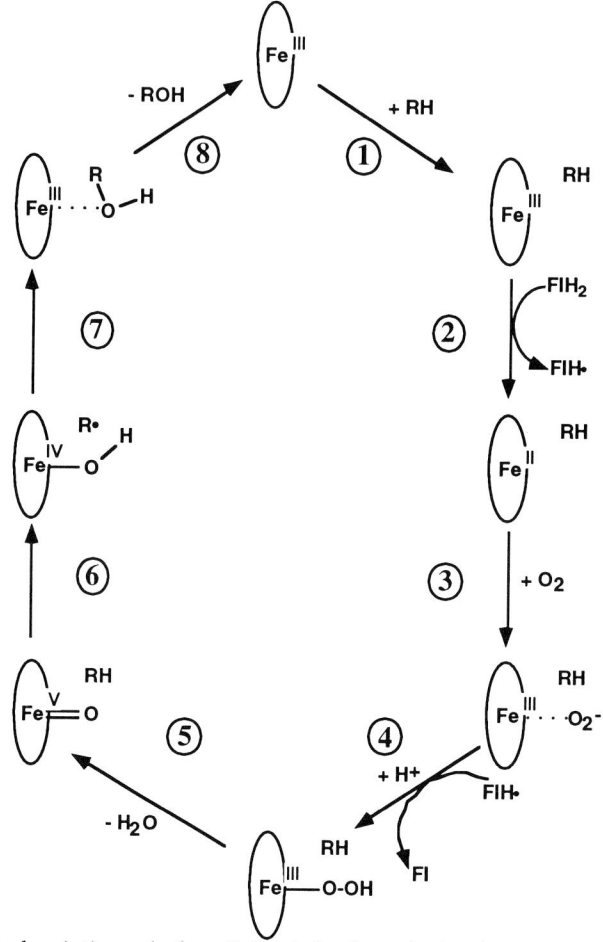

Fig. 3. Mechanistic cycle for aliphatic hydroxylation by cytochrome P450

First Electron Transfer
The first one-electron transfer (Step 2) occurs immediately after the substrate binding (Step 1). One clearly sees the radical nature of the reduction of iron(III) to iron(II) by viewing it from the perspective of the reductase: iron(III) oxidizes the flavoprotein (FlH$_2$) to a stable semiquinone free radical (FlH•). When oxygen concentrations are low, the reduced P450 is able to donate its one electron to an alternate electron acceptor. Because of the low reduction potential of P450, a variety of compounds are reducible. For example, the general anesthetic halothane can be reduced with loss of bromide ion to produce a carbon radical that may be released into solution to initiate a chain reaction of lipid peroxidation (Van Dyke et al., 1988).

Release of Superoxide Radicals, Hydroxyl Radicals and Hydrogen Peroxide
When oxygen *is* present, the ternary P450-substrate-oxy complex is formed (Step 3), but if the P450 reductase is slow to deliver the second electron, then oxygen can escape and take the one electron with it, in the form of superoxide radical. Superoxide released into solution is mainly disproportionated to O$_2$ and hydrogen peroxide, but it can also reduce cellular metal ions such as iron(III) and copper(II). Ferrous or cuprous ions readily react with the hydrogen

peroxide concurrently produced to generate free hydroxyl radicals in solution. The generation of these hydroxyl radicals is a serious consequence of the need for P450 to utilize a radical mechanism. However, let us allow that the oxy-complex is successfully reduced to the iron-peroxide complex (Step 4). If this complex is not efficiently converted to the reactive oxygen intermediate, then there is the potential for the loosely bound hydrogen peroxide to escape, wasting the NADPH.

## Formation of $Fe^V=O$, Hydrogen Abstraction and Oxygen Rebound

However, usually the peroxide complex is successfully converted to the iron-oxo intermediate (Step 5). As shown here and as generally accepted, this is *not* a radical process. But the iron-oxo complex *is* a radical, which we can conveniently view as a triplet oxygen atom coordinated to ferric iron. While more correct resonance forms can be written, the P450 FeO intermediate reacts *as if* it were a triplet oxygen atom. The two-step mechanism is hydrogen abstraction (Step 6) followed immediately be recombination of the two radicals to produce a stable spin-paired product (Step 7).

$$Fe^{III}\ \dot{O}\cdot\ +\ RH\ \longrightarrow\ [Fe^{III}\ \dot{O}H\ +\ R\cdot]\ \longrightarrow\ Fe^{III}\ +\ HO\text{-}R$$

This process was given the name "oxygen rebound" by Groves and McClusky (1976). The mechanistic cycle is completed by the release of the final product alcohol (Step 8).

## Abnormal Reactions

Unfortunately, sometimes the second step, recombination, does not occur properly. Instead, the radical R• undergoes some other reaction. This is definitely the most serious consequence of generating radicals in the active site. Following are several examples of the aborted reactions that can occur.

The first example involves initial oxidation at nitrogen leading to β-scission and ring-opening of cyclopropylamines (Fig. 4). Notice that the carbon radical is formed at a distance from FeO, so that recombination is improbable. Instead, it adds to the heme. In fact, Guengerich *et al.* (1984) provided strong evidence for an initial one-electron oxidation by correlating the oxidation potentials of cyclopropyl analogs with their ability to inactivate P450.

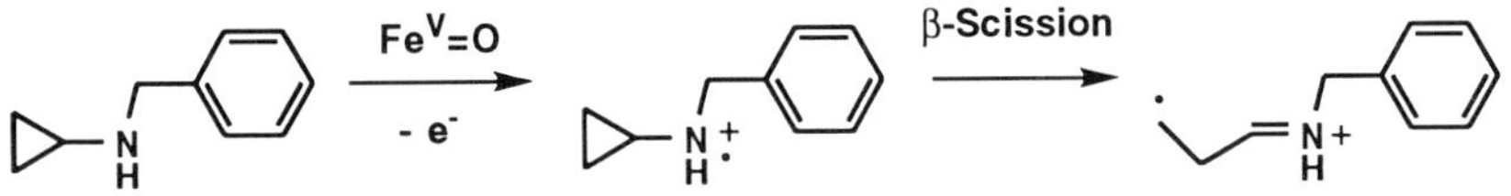

Fig. 4. Radical β-scission of cyclopropylamines

In another example, 4-alkyldihydropyridines are one-electron oxidized to aminium cation radicals that can fragment and release a carbon radical (Guengerich and Böcker, 1988). Again, because of the distance from the FeO, the radical has little chance of recombining to form a stable alcohol. Instead, it may diffuse away from the active site and react with cellular components.

A similar example involves the oxidation of alkyl hydrazines (Ortiz de Montellano *et al.*, 1983b). Again, the intermediate aminium radical fragments to release a reactive carbon radical that usually attacks heme instead of recombining. Finally, the analogous fragmentation of nitrosoamines during oxidation is interesting because the radical released is a stable compound, nitric oxide (Yang *et al.*, 1987).

## SUMMARY

The only feasible way to hydroxylate aliphatic regions of P450 substrates is by initial hydrogen abstraction, requiring that P450 generate a reactive intermediate capable of hydrogen abstraction. The only oxygen species suitable for hydrogen abstraction are HO• and O and their analogs $Fe^{IV}$-OH and $Fe^V$=O, all requiring reduction of $O_2$. Thus, NADPH must be supplied and biosynthesis of at least one reductase protein is required in addition to the P450 protein. Since free radicals are difficult to control, deviations from the expected reaction course are frequent. These deviations result in inefficient usage of NADPH, inactivation of P450, and release of reactive radicals that can have toxicological consequences. The ultimate price paid for the ability of P450 to hydroxylate molecules includes not only the investment of chemical energy and the need to make extra electron-transfer proteins, but also some loss of control of the outcome of oxidative events.

## REFERENCES

Groves, J.T., McClusky, G.A. (1976) Aliphatic hydroxylation via oxygen rebound. Oxygen transfer catalyzed by iron. *J. Am. Chem. Soc.* **98**, 859-861

Groves, J.T., McClusky, G.A., White, R.E., and Coon, M.J. (1978): Aliphatic hydroxylation by highly purified liver microsomal cytochrome P-450. Evidence for a carbon radical intermediate. *Biochem. Biophys. Res. Commun.* **81**, 154-160.

Guengerich, F.P., Willard, R.J., Shea, J.P., Richards, L.E., and Macdonald, T.L. (1984): Mechanism-based inactivation of cytochrome P-450 by heteroatom-substituted cyclopropanes and formation of ring-opened products. *J. Am. Chem. Soc.* **106**, 6446-6447.

Guengerich, F.P. and Böcker, R.H. (1988): Cytochrome P-450-catalyzed dehydrogenation of 1,4-dihydropyridines. *J. Biol. Chem.* **263**, 8168-8175.

Harada, N., Miwa, G.T., Walsh, J.S., and Lu, A.Y.H. (1984): Kinetic isotope effects on cytochrome P-450-catalyzed oxidation reactions. *J. Biol. Chem.* **259**, 3005-3010.

Ortiz de Montellano, P.R., Mangold, B.L., Wheeler, C., Kunze, K.L., and Reich, N.O. (1983a): Stereochemistry of cytochrome P-450-catalyzed epoxidation and prosthetic heme alkylation. *J. Biol. Chem.* **258**, 4208-4213.

Ortiz de Montellano, P.R., Augusto, O., Viola, F., and Kunze, K.L. (1983b): Carbon radicals in the metabolism of alkylhydrazines. *J. Biol. Chem.* **258**, 8623-8629.

Van Dyke, R.A., Baker, M.T., Jansson, I., and Schenkman, J. (1988): Reductive metabolism of halothane by purified cytochrome P-450. *Biochem. Pharmacol.* **37**, 2357-2361.

Wade, D., Yang, C.S., Metral, C.J., Roman, J.M., Hrabie, J.A., Riggs, C.W., Anjo, T., Keefer, L.K., and Mico, B.A. (1987): Deuterium isotope effect on denitrosation and demethylation of N-nitrosodimethylamine by rat liver microsomes. *Cancer Res.* **47**, 3373-3377.

White, R.E., Groves, J.T., and McClusky, G.A. (1979): Electronic and steric factors in regioselective hydroxylations catalyzed by purified cytochrome P-450. *Acta Biol. Med. Germ* **38**, 475-482.

White, R.E., Miller, J.P., Favreau, L.P., and Bhattacharyya, A. (1986) Stereochemical dynamics of aliphatic hydroxylation by cytochrome P-450. *J. Am. Chem. Soc.* **108**, 6024-6031.

White, R.E. (1991): The Involvement of Free Radicals in the Mechanisms of Monooxygenases. *Pharmacol. Therapeut.* **49**, 21-42.

# Studies of the interactions of microsomal cytochrome P450 reductase with cytochromes P450

Henry W. Strobel, Sijiu Shen

*Department of Biochemistry and Molecular Biology, The University of Texas, Medical School at Houston, PO Box 20708, Houston, Texas 77225, USA*

INTRODUCTION

Sequence analysis confirms evidence from a number of studies by various workers indicating that cytochrome P450 reductase and the various isoforms of cytochrome P450 are attached to the endoplasmic reticulum by portions of the protein sequence which are deeply embedded in or traverse the membrane. Included in the evidence leading to this conclusion are solubilization studies showing the requirement of detergent for removal of the catalytically active full-length proteins from the membrane environment, immunochemical studies of the membrane-bound enzymes revealing portions of the proteins to be accessible to antibodies while other portions are not and proteolysis studies correlating catalytic activities with structure (Black, et al., 1979;, Gum and Strobel, 1979, Nelson and Strobel, 1988). Thus, hydrophobic forces have been identified in the attachment of the component enzymes of the mixed function oxidase system to the membrane. For the reductase component, the membrane binding peptide has been isolated, purified and characterized by Black et al., (1979) and by Gum and Strobel (1981). Black et al. (1979) suggested that the isolated peptide may play a role in substrate specificity of the cytochrome P450 while Gum and Strobel (1981) demonstrated that while the membrane binding peptide was absolutely necessary to reconstitute native catalytic activity of the microsomal proteins in a reconstituted assay system they found no evidence for a specificity-conferring role under a variety of substrate and cytochrome P450 isoform combinations. On the other hand, cytochrome P450 and cytochrome P450 reductase self-associate and associate with each other in multimeric arrays (French et al., 1980). Wile these interactions may be due to interactions of the hydrophobic membrane binding portions of the various proteins, a second mode of attraction electrostatic interaction might also be important in these combinations.

Evidence for electrostatic charge pairing in the interaction of cytochrome P450 reductase with cytochromes P450 was summarized in an earlier report (Strobel et al., 1989). Some of the evidence is based on numerous demonstrations that increased salt concentration can damp the interaction of cytochrome P450 reductase with cytochrome P450 as determined by its effect on substrate hydroxylation activities. In order to pursue charge pairing as an interaction modality of cytochrome P450 and cytochrome P450 reductase Nadler and Strobel (1988) utilized protein modification techniques to examine the effect of charge alteration of the reductase molecule on its ability to interact with cytochromes P450. These workers showed that modification of positive charges (i.e. lysine groups) actually increased the ability of the reductase to interact with cytochromes P450 as evidenced in a lowered $K_m$ value for cytochromes P450 whereas modification of negatively-charged residues (carboxylate residues of glutamate and aspartate) markedly reduced interaction by raising the $K_m$ value for cytochromes P450. Modification of some reductase carboxylate residues with ethyldimethylamino propyl carbodiimide in the presence of several different nucleophiles allowed neutralization of negative charges on the reductase or maintenance of the charge and extending the size of the substituent group. Loss of the capacity for charge pairing in each case reduced the interaction of reductase with cytochromes P450.

While there are 101 carboxylate resides in cytochrome P450 reductase, modification of only a few of these elicits a large diminution in the interaction of reductase and cytochromes P450. The identity of the critical carboxylates was studied by Nadler and Strobel (1991) using radiolabelled nucleophile followed by proteolysis of the labelled protein, resolution of the radiolabelled peptides using high performance liquid chromatography and sequence analysis of the purified labelled peptide. In this manner, the carboxylate residues in the peptide derived from residue 109-130 of rat liver cytochrome P450 reductase were identified as involved in electrostatic interaction with cytochromes P450. This base was extended with studies examining the cytochromes P450 for the presence of charge pair partners and with studies using antibodies to perturb charge pairing sites. The results of these studies are reported here.

METHODS

The methods utilized for modification of amino acid residues in cytochrome P450 are those previously described (Nadler and Strobel, 1988; Shen and Strobel, 1992; Shen and Strobel, 1993).

RESULTS AND DISCUSSION

Complimentary charge pair partners of cytochrome P450 for the negatively-charged carboxylate residues previously identified for reductase were anticipated to be the positively charged residues lysine and arginine. Thus, the question of charge pairing was pursued by modification of lysine or arginine residues. The lysine charges were modified by neutralizing the ε-amino group of lysine residues with acetic anhydride. By regulating the concentration of acetic anhydride and the time of incubation, the number of lysines modified could be varied and the reaction extent measured by determining the remaining free lysines with fluorescamine. The ability of the modified cytochromes P450 to catalyze substrate hydroxylation could then be determined as a function of lysine groups modified. The effect of modification of lysine residues on cytochrome P450-dependent substrate hydroxylation as shown in Fig. 1.

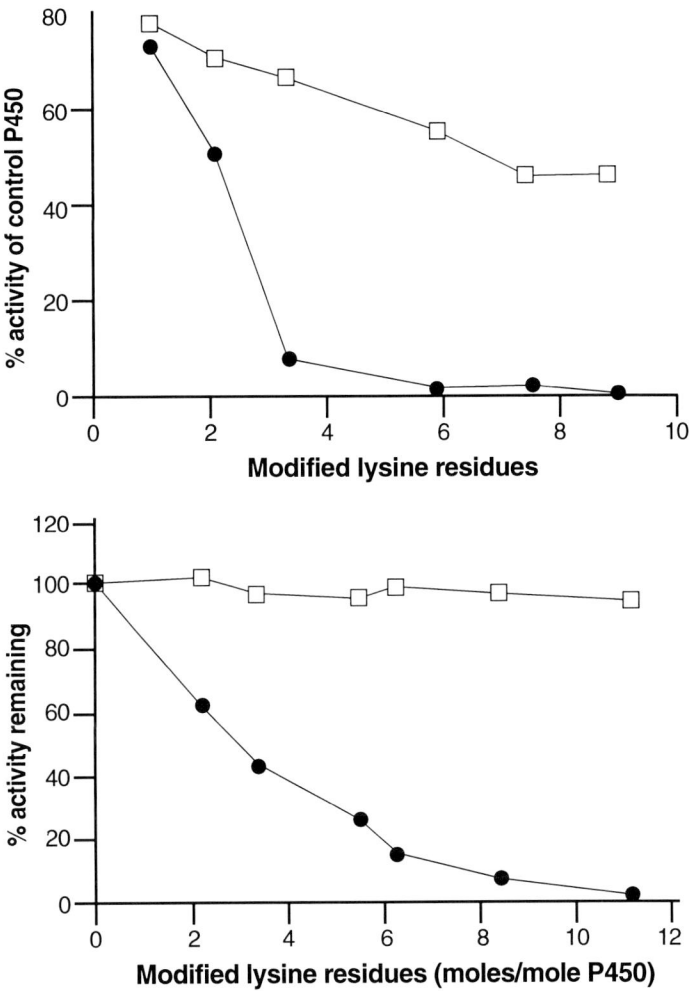

Fig. 1. Effects of cytochrome P450 lysine modification on cytochrome P450 1A1 catalyzed 7-ethoxycoumarin hydroxylation activity (upper panel) and in cytochrome P4502B1 catalyzed benphetamine N-demethylation activity (lower panel). In each panel, cytochrome P450 supported activity is indicated by the closed circles and cumene hydroperoxide supported activity is indicated by the open squares.

A dramatic difference in the response of substrate hydroxylation activity to acetic anhydride modification of cytochrome P450 lysine residues is seen in both panels between reductase-supported and cumene hydroperoxide-supported cytochrome P450 mediated hydroxylation activity. In both panels, the modification of two lysine residues equivalents per mole of cytochrome P450 lower the reductase-supported activity almost to the 50% level, but has either little or no effect on the cumene hydroperoxide-supported activity. This dramatic difference suggests that the lysine residues most sensitive to acetylation on a relative scale are those cytochrome P450 residues interacting with reductase since the cumene hydroperoxoide eliminates the need for reductase to catalyze substrate hydroxylation. Increasing the number of lysine residue equivalents modified eliminates reductase supported substrate hydroxylation (at 6 residues modified/mole P450) while reducing cumene hydroperoxide supported 7-ethoxycoumarin hydroxylation by only 40% and benzphetamine hydroxylation not at all. These data are consistent with a primary effect of lysine modification on reductase/P450 interaction sites and a secondary, less sensitive effect at the active site. Using radiolabelled acetic anhydride, lysines modified at lower levels of acetic anhydride, can be identified by labelling, proteolysis, separation of peptides and sequencing of peptides previously used in our laboratory for the identification of reductase carboxyl containing residues (Nadler and Strobel, 1991). Lysine residues 97, 271, 279 and 407 were identified in the manner for cytochrome P4501A1 and lysines 251, 384, 422, 433 and 473 for cytochrome P4502B1. The residues identified by our studies are correlated in Table 1 with residues implicated in interaction with reductase or, in the case of $P450_{cam}$, with putidaredoxin.

Table 1. Alignment position of positively charged amino acids of cytochrome P450 indicated in interaction with cytochrome P450 reductase.

| Form | Alignment Position | | | | | |
|---|---|---|---|---|---|---|
|  | 110 | 311 | | 466 | 506 | 519 |
| 2B4 |  | K251 |  | K384 |  |  |
| 1A2 | K94 |  |  |  | K440 | K453 |
| 1A1 | K97 | K271 | K279 | K407 |  |  |
| 2B1 |  | K251 |  | K384 | K422 | K433 | K473 |
| cam | R72 |  |  |  | K344 |  |  |

Included in this table are the results of Bernhardt et al. (1988) for P4502B4, the results of Shimizu et al. (1991) for P4501A2 and Stayton and Sligar (1991) for $P450_{cam}$. The position numbers at the top of the table are those assigned by Nelson and Strobel (1988) alignment. In this correlation there are five homologous positions where amino acid residues in at least two of the five isozymes have been identified by some technique of protein modification or by site specific mutagenesis as being important in interaction. These alignment positions are 110, 311, 466, 506 and 519. Position 110 is near the N-terminus of the protein position 311 nearer the middle and positions 466, 506 and 519 near the C-terminal end. At the least, these are candidates for the putative sites of interaction with the reductase. All five do not light up in any form tested so far. Perhaps the pattern is variability dependent on the individual form where two or more likely three residues (from our inhibition studies) from among the five candidates participate in interaction.

Having identified putative sites on the cytochromes P450 and cytochrome P450 reductase, we were faced with the task of further defining those now identified sites as sites critical to productive interaction of reductase and P450. In order to address this question, we chose to prepare antibodies to specific peptides in order to assess their effects on cytochrome P450 interaction with cytochrome P450 reductase. Further, we chose to synthesize peptides from the rat reductase region which Nadler and Strobel (1991) had shown to be involved in interaction with cytochromes P450 and the reductase region implicated by Nisimoto (1986) as being involved in cytochrome c interaction with reductase. These peptide sequences are shown for rat liver reductase as well as rabbit and human in Table 2.

## Putative P450 binding sites on reductase

| | | |
|---|---|---|
| SH2 | rat (109) | S A D P E E Y D L A |
| | rabbit | A A D P E E Y D L A |
| | human | S A D P E E Y D L A |
| | | |
| SH3 | rat (114) | L A D L S S L P E I D K S |
| | rabbit | L A D L S S L P E I N N A |
| | human | L A D L S S L P E I D N A |
| | | |
| SH8 | rat (204) | G L G D D D G N L E E D F I T |
| | rabbit | G M G D D D A N L E E D F I T |
| | human | G L G D D D G N L E E D F I T |

The SH2 peptide contains the N-terminal portion of sequence 109-130 while SH2 contains the C-terminal portion of the peptide identified by Nadler and Strobel (1991) SH8 peptide is the region beginning at 204 identified by Nisimoto (1986). SH8 peptide was synthesized chemically coupled with keyhole limpet hemocyanin and inoculated into rabbits. From the resultant serum was purified antibody 2A by a peptide sepharose 4B affinity column. Peptide SH3 elicited antibody 3B while peptide SH8 gave rise to antibody 8A. The peptides have in common that they have clustered carboxyl groups and are highly hydrophilic. The antibodies all have quite high titers and high specificity. Antibodies 2A and 8A bind equally well to rabbit and human reductase, but no binding was detectable between 3B and either rabbit or human reductase, the last two residues being different from the rat sequence. The data of Fig. 2 show the inhibition of cytochrome P4501A1 dependent 7-ethoxycoumarin hydroxylation by the three antipeptide antibodies.

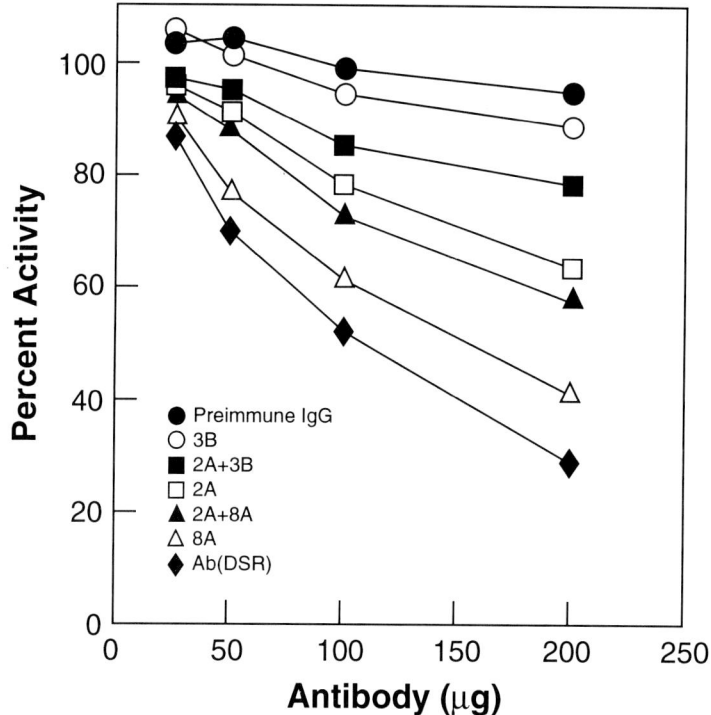

Fig. 2 Antipeptide antibody inhibition of 7-ethoxycoumarin hydroxylation by a cytochrome P4501A1 containing reconstituted system.

Differing amounts of the antipeptide antibodies were incubated with rat liver reductase and then reconstituted with cytochrome P4501A1 and lipid. At 200 µg of antibody 8A and 2A show 60% and 38% inhibition of activity, respectively. Antibody 3B does not have any significant effect on hydroxylation activity. Equal mixtures of 2A and 8A have no additive effect compared on a total antibody basis. A similar pattern of inhibition is seen with P4502B1 as shown in Fig. 3.

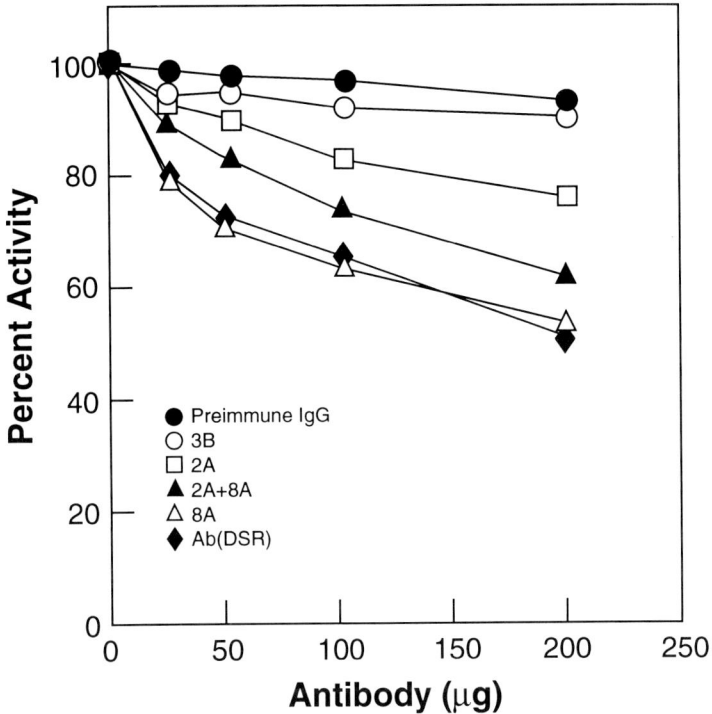

Fig 3. Antipeptide antibody inhibition of pentoxyresorufin dealkylation activity by a cytochrome P4502B1 containing reconstituted system.

Antipeptide antibodies 8A and 2A also inhibit cytochrome P4502B1 catalyzed pentoxyresorufin dealkylation activity. At 200 µg, 8A and 2A inhibit activity by 58 and 26%, respectively. A 40% inhibition is seen when equal amounts of 8A and 2A are mixed at 200 µg final concentration of antibody. Again antibody 3B has little effect.

These data suggest that both regions 109-120 and 204-220 of rat reductase may be involved in the interaction with cytochromes P450. On the other hand, antibody 3B to region 120-130 does not provide evidence to suggest that this region is involved. No differences could be detected between P4501A1 or P4502B1 supported activities in terms of the inhibition of their respective interactions with reductase, though P4502B1 was less sensitive to all antibodies than P4501A1. These data are consistent with a role for those regions identified as sites of electrostatic charge interaction. Any more precise description of roles for particular residues requires further study.

ACKNOWLEDGEMENTS

The authors acknowledge Ms. Anne Bernhardt, Mr. Christopher Boehme and Ms. Laura Bankey for their skillful technical assistance. This work was supported by grant CA53191 from the National Cancer Institute DHHW.

REFERENCES

Bernhardt, R. et al. (1988): Electrostatic interaction between cytochrome P450LM$_2$ and NADPH-cytochrome P450 reductase. *Biomed. Biochim. Acta.* 47, 581-592.

Black, S.D. et al. (1979): Role of hydrophobic polypeptide in the N-terminal region of NADPH-cytochrome P450 reductase in complex formation with P450LM. *Biochem. Biophys. Res. Commun.* 911, 1528-1535.

French, J.S. et al. (1980): Interactions of cytochrome P450, NADPH-cytochrome P450 reductase, phospholipid and substrate in the reconstituted liver microsomal system. *J. Biol. Chem.* 255, 4112-4119.

Gum, J.R. & Strobel, H.W.(1979): Purified NADPH-cytochrome P450 reductase: interaction with hepatic microsomes and phospholipid vesicles. *J. Biol. chem.* 254, 4177-4185.

Gum, J.R., & Strobel, H.W. (1981): Isolation of the membrane-binding peptide of NADPH-cytochrome P450 reductase characterization of the peptide and its role in the interaction of reductase with cytochrome P450. *J. Biol. Chem.* 256, 7478-7486.

Nadler, S.G., & Strobel, H.W. (1988): Role of electrostatic interactions in the reaction of NADPH-cytochrome P450 reductase with cytochromes P450. *Arch. Biochem. Biophys.* 261, 418-429.

Nadler, S.G., & Strobel, H.W. (1991): Identification and characterization of an NADPH-cytochrome P450 reductase derived peptide involved in binding to cytochrome P450. *Arch. Biochem. Biophys.* 290, 277-284.

Nelson, D.R., & Strobel, H.W. (1988): On the membrane topology of vertebrate cytochrome P450 proteins. *J. Biol. Chem.* 263, 6038-6050.

Nisimoto, Y. (1986): Localization of cytochrome C-binding domain on NADPH-cytochrome P450 reductase. *J. Biol. Chem.* 266, 14232-14239.

Shen, S.J. & Strobel, H.W. (1992): The role of cytochrome P450 lysine residues in the interaction between cytochrome P450 1A1 and NADPH cytochrome P450 reductase. *Arch. Biochem. Biophys.* 294, 83-90.

Shen, S.J. & Strobel, H.W. (1993): Role of lysine and arginine residues of cytochrome P450 in the interaction between cytochrome P4502B1 and NADPH cytochrome P450 reductase. *Arch. Biochem. Biophys.* 304, 257-265.

Shimizu, T. et al. (1991): Probing the role of lysine and arginines in the catalytic function of cytochrome P450 by site-directed mutagenesis. *J. Biol. Chem.* 266, 3372-3375.

Stayton, P.S. & Sligar, S.G. (1990): The cytochrome P450$_{cam}$ binding surface as defined by site-directed mutagenesis and electrostatic modeling. *Biochemistry* 29, 7381-7386.

Strobel, H.W. et al., (1989):Cytochrome P450: cytochrome P450 reductase interactions. *Drug Metabol. Rev.* 20, 519-533.

# On the nature of the NADPH-cytochrome P450 reductase-cytochrome P450 electron transfer complex

Andrei I. Voznesensky, John B. Schenkman

Department of Pharmacology, University of Connecticut Health Center, Farmington, CT 06030-1505, USA

Microsomal cytochrome P450 (P450) is a family of very diverse monooxygenases responsible for the metabolism of a variety of exogenous and endogenous compounds. The monooxygenase reaction requires a source of 2 electrons, that is satisfied by NADPH-cytochrome P450 reductase (reductase). Stoichiometry studies indicated that a transient 1:1 electron transfer complex between reductase and P450 occurs (Miwa & Lu, 1984). However, the forces that hold the two proteins together in the electron transfer complex are not fully defined. Many studies have been carried out on a variety of different electron transfer pairs, both physiologic and non-physiologic. Such studies have demonstrated formation of electron transfer complexes and their stabilization to be facilitated by electrostatic interaction of complementary charged residues on the two proteins. For example, the redox interaction between cytochrome c and cytochrome $b_5$ has been shown to require stabilization by the highly conserved lysyl residues on the cytochrome c (Ng et al., 1977); interaction of cytochrome $b_5$ and NADH-cytochrome $b_5$ reductase involves lysyl residues on the reductase (Loverde & Strittmatter, 1968) and carboxyl residues on the cytochrome $b_5$ (Dailey & Strittmatter, 1979). The interaction between methemoglobin and the catalytic fragment of cytochrome $b_5$ likewise is by complementary charge pairing (Mauk & Mauk, 1982).

In contrast with such studies indicative of complementary charge-pairing, at least in some instances, charge-pairing can be shown to interfere with electron transfer between redox centers. Electron transfer between yeast cytochrome c peroxidase and horse heart cytochrome c was shown to be diminished by complementary charge-pairing; it was suggested that complementary charge pairing holds the two proteins in nonproductive complex, and that a loosening of this complex, by increasing ionic strength, facilitated electron transfer, resulting in an increased $V_{max}$ and $K_m$ (Hazzard et al., 1988). Similarly, it was found that in cytochrome $b_5$ and cytochrome c interactions, a number of spurious complexes were formed, as monitored by EDC (1-ethyl-3(3-dimethylaminopropyl) carbodiimide) cross-linking (Mauk & Mauk, 1989). In the present study we made an attempt to evaluate the role of electrostatic interactions in the formation of the reductase-P450 complex for a number of P450 forms.

**Chemical modifications.** Studies to determine the role of charged protein residues in protein-protein interactions have had mixed success when applied to proteins of the P450 monooxygenase system. In accordance with the suggestion that cytochrome P450 uses its amino residues for redox interactions, neutralization of carboxyl residues on NADPH-cytochrome P450 reductase by methylamidation strongly inhibited reduction of cytochrome P450 and substrate turnover in the reconstituted system (Tamburini & Schenkman, 1986a). The kinetic parameters of the interaction were more informative; the $V_{max}$ of ethylmorphine turnover was diminished by the methylamidation, but the $K_m$ for the reductase was unaltered, suggesting that overall alteration of the protein structure is responsible for the effect. Further, when the interaction between reductase and P450 was monitored spectrally, it was seen that the dissociation constant for the binding was unaltered by the methylamidation of reductase carboxyl groups (Tamburini & Schenkman, 1986a). The conclusion made was that the reductase carboxyl residues are called into action at some step subsequent to complex formation, and possibly hydrophobic interactions drive the complexation. Nevertheless, subsequent studies have suggested complementary charge-pairing as necessary for redox interaction between cytochrome P450 and NADPH-cytochrome P450 reductase. Thus, inhibition of substrate turnover as the result of binding of FITC (fluorescein isothiocyanate) to the amino-terminal amino group of cytochrome P450 was taken as evidence for electrostatic interaction between P450 and reductase (Bernhardt et al., 1984). More recently, neutralization of about 3 lysyl residues on rat CYP1A1 were found to decrease turnover of 7-ethoxycoumarin by the reconstituted system as much as 95%; in contrast, using cumene hydroperoxide instead of NADPH and NADPH-cytochrome P450 reductase as the source of active oxygen only showed a 30% inhibition (Shen & Strobel, 1992). The authors concluded that this was an indication of involvement of lysyl residues on the P450 in interaction with the reductase. A similar study was also performed with another form of rat cytochrome P450,CYP2B1, where acetylation of up to 8.5 lysyl residues of the hemoprotein caused 95% inhibition of benzphetamine demethylase activity in the reductase-reconstituted system, but no inhibition of the cumene hydroperoxide supported reaction (Shen & Strobel, 1993).

**Covalent cross-linking.** The water soluble carbodiimides, such as EDC, activate carboxy groups on the protein surface and catalyze covalent cross-linking of the charge-pairing protein complexes. EDC has proven to be a useful tool for determination of complementary charge pairing interactions between proteins. Thus treatment with EDC covalently linked adrenodoxin and adrenodoxin reductase (Lambeth et al., 1984) and cytochrome P450 and cytochrome $b_5$ (Tamburini & Schenkman, 1987) in functionally active complex. Such complexes were also useful for determination of formation of non-optimal electrostatic complexes, as between yeast cytochrome c peroxidase and horse cytochrome c (Hazzard, et al., 1988). Cytochrome c was also covalently linked to the NADPH-cytochrome P450 reductase (Nisimoto, 1986) and competition between cytochrome $b_5$ and cytochrome c was suggested to indicate a single electron donor site on the reductase (Nisimoto & Otsuka-Murakami, 1988). In an attempt to determine the nature of interaction between cytochrome P450 and NADPH-cytochrome P450 reductase we tried to trap complementary charge pairs in a covalent amide linkage using EDC. We found what appeared to be traces of such heterodimer were found (Tamburini et al., 1986), although we could trap more than 30% of cytochrome $b_5$ and cytochrome P450 in a functionally active covalent complex. Subsequent investigation of the parameters of complexation with EDC indicated that despite all attempts, no complex between cytochrome P450 and NADPH-cytochrome P450 reductase could be generated (Nisimoto & Edmondson, 1992; Voznesensky & Schenkman, 1992a). Inability of EDC to covalently link P450 and reductase suggested that the complex of the two proteins is not formed by complementary charge-pairing.

**Electrostatic Effects.** Since we were unable to demonstrate complementary charge pairing between NADPH-cytochrome P450 reductase and cytochrome P450 using EDC as a catalyst, we turned our attention to the influence of ionic strength on the interaction of reductase and P450 as a further test of such interactions. Electrostatic neutralization of charges on the protein surface by increasing ionic strength is the most gentle way to diminish protein-protein electrostatic interactions. The decrease is gradual as opposed to "all or none" influence of chemical modification or site-directed mutagenesis. The basis of the effect is the decrease in the activity coefficients of the ions with increasing ionic content of the medium. When ionic strength is increased charges on the protein surface are also neutralized because from an electrostatic point of view there is no difference between salt ions in solution and charged amino acid residues. Generally, strong inhibition of complex formation by increasing ionic strength is taken as evidence of electrostatic interaction in the complex. Thus, the association constant of the cytochrome $b_5$-methemoglobin complex decreased with increasing ionic content of the medium (Mauk & Mauk, 1982). The interaction between adrenodoxin and adrenodoxin reductase was likewise shown to be electrostatic in nature, with high ionic strength weakening the association (Lambeth et al., 1979). In contrast, studies on NADPH-cytochrome P450 reductase indicated its ability to reduce cytochrome c was increased at high ionic strength (Bilimoria & Kamin, 1973; Phillips & Langdon, 1962), as was its ability to reduce cytochrome $b_5$ (Bilimoria & Kamin, 1973).

Using rabbit CYP2B4 and reductase we observed a stimulation of the rate of reduction of P450, and a major increase in the fast phase rate constant of reduction at 100mM sodium phosphate buffer relative to 10mM buffer (Voznesensky & Schenkman, 1992a). In agreement with the literature, a similar stimulation in the rate of cytochrome c reduction was seen, as well as a stimulation in ferricyanide reduction. The conclusion drawn from our studies was that at low ionic strength electrostatic interaction interferes with the formation of a functional redox complex. In contrast was the stimulation in rate of electron transfer between cytochrome c peroxidase and cytochrome c (Hazzard, et al., 1988), where loosening of a non-productive complex was the explanation for the enhanced activity at elevated ionic strength. With NADPH-cytochrome P450 reductase and cytochrome P450 the $K_m$ for reductase was actually 26-fold lower at elevated ionic strength (Voznesensky & Schenkman, 1992a). The absence of EDC-induced reductase-P450 cross-linking argues against the formation of a non-productive complex and speaks in favor of charge repulsion between the hemoprotein and the flavoprotein (Voznesensky & Schenkman, 1992a). As a test of the hypothesis, an uncharged electron acceptor, 1,4-benzoquinone, was tried. In support of the suggestion, the reduction of this acceptor was unaffected by ionic strength.

Increasing salt concentrations doesn't provide an indication as to the nature of the interacting residues on either protein, since the salt will affect both cations and anions. However, it is possible to use polyanionic or polycationic peptides as the probes of the interacting residues. It was shown that when polylysine was complexed with Clostridium pasteurianum flavodoxin the ability to transfer reducing equivalents to positively charged acceptors was diminished, while the ability to reduce negatively charged acceptors was enhanced (Cheddar & Tollin, 1990), the polylysine binds to anion-dense regions, essentially decreasing potential charge repulsions. When we utilized polylysine in this manner, the rates of reduction of cytochrome P450 and ferricyanide were markedly enhanced, while reduction of cytochrome c and of 1,4-benzoquinone were unaffected.

Rabbit NADPH-cytochrome P450 reductase has a high proportion of charged residues, 87 carboxylate groups and 68 cationic residues (K+R). CYP2B4 likewise is highly charged, containing 53 carboxylate residues and 58 cationic residues. Consequently it is not surprising that its interactions with its redox acceptors are influenced by the conductance of the medium. In contrast to the studies where chemical modification of reductase anionic residues was found to inhibit substrate turnover without affecting affinity of the reductase for the P450, i.e. to influence the step subsequent to complex formation (Tamburini & Schenkman, 1986a) increased ionic strength stimulated reductase-P450 association due to neutralization of electrostatic interactions (Voznesensky & Schenkman, 1992a). Addition of polyols to the medium has an effect opposite to that of increased ionic strength. Glycerol and ethylene glycol have dielectric constants of 43 and 38 respectively vs 78 for water. Their addition decreases the dielectric constant of the medium, thereby promoting ion-pairing. Conductivity, which can serve as a measure of ionic properties of the medium, is increased by increasing ionic strength and is decreased by subsequent addition of the polyols.

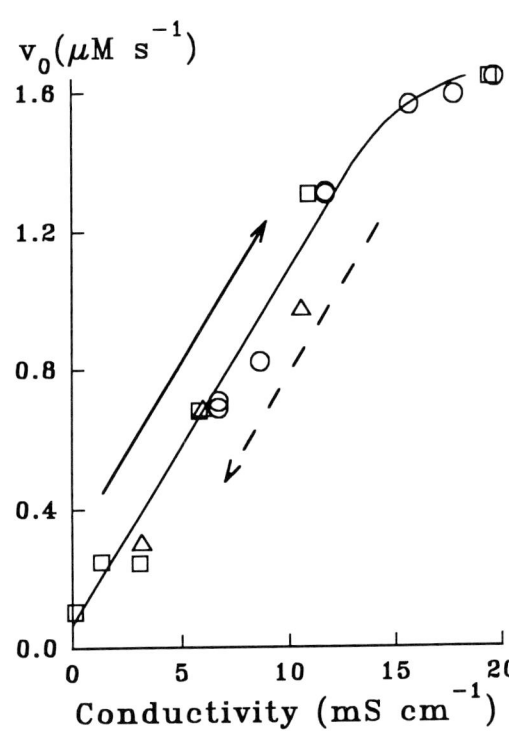

**Fig. 1.** The influence of glycerol and ethylene glycol on the conductivity of the medium and on the reduction of cytochrome P450.

Fig. 1 demonstrates the influence of polyols on the reduction of cytochrome P450. As the ionic strength is increased the rate of cytochrome P450 2B4 reduction rises (solid arrow), peaking at 10mS

(100mM sodium phosphate, □). When increasing concentrations of glycerol (▲) or ethylene glycol (○) were added to a 100mM buffer the rate of P450 reduction decreased along with decreasing conductivity (dashed arrow). Benzphetamine turnover responded in a similar manner to the presence of glycerol or ethylene glycol in the medium (Voznesensky & Schenkman, 1992b). In contrast, neither alteration of ionic strength nor polyol addition affected the rate of electron transfer from NADPH-cytochrome P450 reductase to an uncharged acceptor 1,4 benzoquinone (Voznesensky & Schenkman, 1992b). A plot of the apparent $K_m$ of CYP2B4 for reductase as a function of conductivity in the medium is shown in Fig. 2. The highest value was in the presence of 10mM buffer and 20% glycerol (▼), followed by 100mM buffer plus 25% glycerol (■), and 100mM buffer plus 20% glycerol (▲), and 100mM buffer plus 5% glycerol (●). The data are in agreement with the suggested inhibitory role of electrostatic forces in the reductase-P450 functional complex formation.

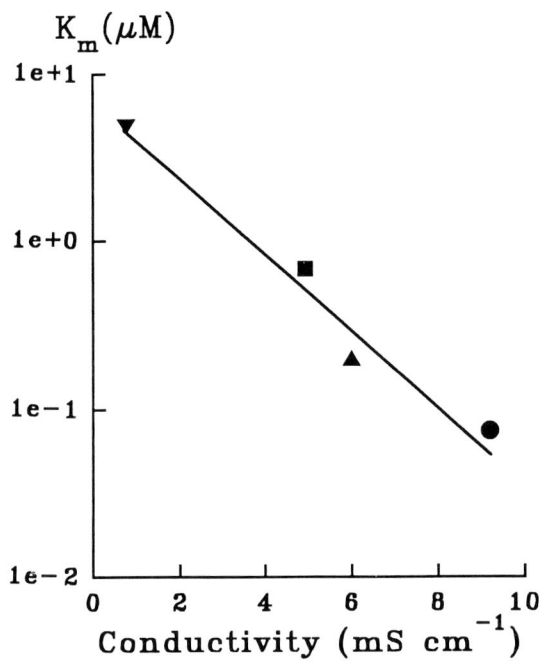

**Fig. 2.** Relationship between the $K_m$ of CYP2B4 for reductase and the conductance of the medium. Modified from Voznesensky & Schenkman, 1992b

Stimulation of reductase-P450 interactions by increased ionic strength was not a unique property of the CYP2B4. When we compared the rate of reduction of CYP1A2, CYP2C6, CYP2C12, CYP2C13 and CYP2E1 in 5mM phosphate buffer to that in 100mM buffer in all cases a stimulation was observed, but the extent of the stimulation was different. The highest increase was seen with CYP2C13 and CYP2E1 (6-fold). A 3-fold increase in the rate of reduction was observed with CYP2C12 and reduction of CYP2C6 was increased only by 50%. The fact that reduction of all four P450 enzymes was stimulated by ionic strength suggests that electrostatic forces inhibit all of them in their interaction with the reductase. The different extents of stimulation suggests that the local distribution of charged residues on the protein surfaces involved in the interaction with the reductase is different for the various forms of P450.

If increased ionic strength stimulates reductase-P450 interactions by lowering the activity coefficients of the ions there should be a relationship between the association constant of the two proteins and the activity coefficient. A relationship between the association constant of the two proteins ($K_a$) and the log of ionic activity coefficient ($\log(y)$) in a given medium is

$$\log(K_a/K_{ao}) = -2mn\log(y)$$

where m and n are the numbers of charges on respective proteins affecting the complex formation, and $K_{a_0}$ is the association constant at zero ionic strength. When reductase-CYP1A2 association constants were plotted in the above coordinates a direct relationship was observed (Fig.3). The best fit line had a positive slope (13.27) indicating weaker association with increasing electrostatic interaction. The magnitude of the slope, which is equal to the product of the number of interacting charges on the two proteins, suggests that repulsion between 3 charges on one protein and 2 charges on the other is influencing the complex formation.

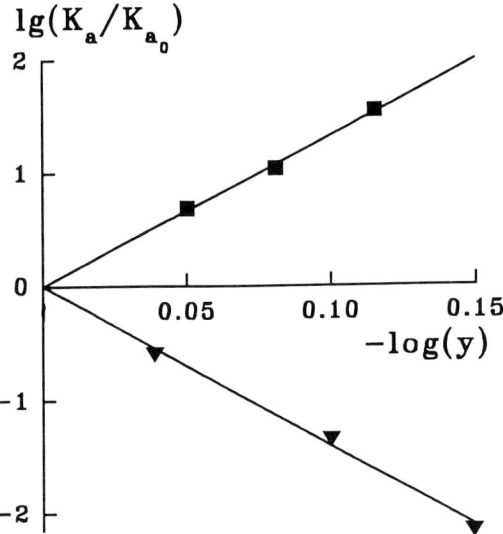

**Fig. 3.** Dependence of the reductase-CYP1A2 (■) association constants and the CYP2C11-cytochrome $b_5$ (▼) association constants on the ionic activity coefficients.

When the cytochrome $b_5$ - CYP2C11 association constants (Tamburini & Schenkman, 1986b) were plotted in the same coordinates a linear relationship was observed, but unlike reductase-CYP1A2 association best fit line had a negative slope. Previously we showed that cytochrome $b_5$ interacts with cytochrome P450 by complementary charge-pairing (Tamburini & Schenkman, 1987) and the plot in fig. 3 is in agreement with this conclusion.

If the net effect of the interaction of charged residues on the reductase and the P450 is disruptive to the complex formation what force is responsible for bringing the two protein together? We suggest that this force is the interaction of the complementary hydrophilic and hydrophobic patches on the two proteins. Formation of protein-protein complexes by the interaction of complementary hydrophobic patches was shown for a number of multisubunit proteins (Korn & Burnett, 1991). The concept of reductase-P450 complex formation by interactions of complementary hydrophobic patches agrees well with the flexible nature of the P450 monooxygenase system. One reductase supplies electrons to an array of functionally diverse and structurally different P450 forms as well as non-P450 acceptors. The formation of the complex by complementary charge pairing interactions would require charged amino acids to be positioned precisely in the same orientation in a large number of different proteins, imposing considerable constraints on their structure. Protein-protein recognition can be more easily achieved through hydrophobic patch complementarity, because it allows high structural variability as long as the general patch motif is preserved. Existence of conserved hydrophobic domains in P450 (Nelson & Strobel, 1988) supports such a possibility. The role of the observed charge repulsion between P450 and reductase may be to allow proteins to stay in the complex only long enough for the electron transfer to occur.

# REFERENCES

Bernhardt, R., Makower, A., Janig, G.-R., & Ruckpaul, K. (1984). Selective chemical modification of a functionally linked lysine in cytochrome P-450 LM2. Biochim. Biophys. Acta, 785, 186-190.

Bilimoria, M., & Kamin, H. (1973). The effect of high salt concentrations upon cytochrome c, cytochrome B5, and iron-EDTA reductase activities of liver microsomal NADPH-cytochrome C reductase. Annals of the New York Academy of Sciences, 212, 428-448.

Cheddar, G., & Tollin, G. (1990). Electrostatic effects on the spectral and redox properties of Clostridium pasteurianum flavodoxin: effects of salt concentration and polylysine. Arch. Biochem. Biophys., 278, 265-268.

Dailey, H., & Strittmatter, P. (1979). Modification and identification of cytochrome b5 carboxyl groups involved in protein-protein interaction with cytochrome b5 reductase. J. Biol. Chem., 254, 5388-5396.

Hazzard, J., Moench, S., Erman, J., Satterlee, J., & Tollin, G. (1988). Kinetics of intracomplex electron transfer and of reduction of the components of covalent and noncovalent complexes of cytochrome c and cytochrome c peroxidase by free flavin semiquinones. Biochemistry, 27, 2002-2008.

Korn, A., & Burnett, R. (1991). Distribution and complementarity of hydropathy in multisubunit proteins. Proteins: structure, function, and genetics, 9, 37-55.

Lambeth, J., Geren, L., & Millett, F. (1984). Adrenodoxin interaction with adrenodoxin reductase and cytochrome P-450scc. Cross-linking of protein complexes and effects of adrenodoxin modification by 1-ethyl-3-(3-dimethylaminopropyl)carbodiimide. J. Biol. Chem., 259, 10025-10029.

Lambeth, J., Seybert, D., & Kamin, H. (1979). Ionic effects on adrenal steroidogenic electron transport. The role of adrenodoxin as a electron shuttle. J. Biol. Chem., 254, 7255-7264.

Loverde, A., & Strittmatter, P. (1968). The role of lysyl residues in the sturcture and reactivity of cytochrome b5 reductase. J. Biol. Chem., 243, 5779-5787.

Mauk, M., & Mauk, A. (1982). Interaction between cytochrome b5 and human methemoglobin. Biochemistry, 21, 4730-4734.

Mauk, M., & Mauk, A. (1989). Crosslinking of cytochrome c and cytochrome b5 with a water- soluble carbodiimide. Reaction conditions, product analysis and critique of the technique. Eur. J. Biochem, 186, 473-486.

Miwa, G., & Lu, A. (1984). The association of cytochrome P-450 and NADPH-cytochrome P-450 reductase in phospholipid membranes. Arch. Biochem. Biophys., 234, 161-166.

Nelson, D., & Strobel, H. (1988). On the membrane topology of vertebrate cytochrome P-450 proteins. J. Biol. Chem., 263, 6038-6050.

Ng, S., Smith, M., Smith, H., & Millett, F. (1977). Effect of modification of individual cytochrome c lysines on the reaction with cytochrome b5. Biochemistry, 16, 4975-4978.

Nisimoto, Y. (1986). Localization of cytochrome c-binding domain on NADPH-cytochrome P-450 reductase. J. Biol. Chem., 261, 14232-14239.

Nisimoto, Y., & Edmondson, D. (1992). Effect of KCl on the Interactions Between NADPH - Cytochrome-P-450 Reductase and Either Cytochrome-c, Cytochrome-b5 or Cytochrome-P-450 in Octyl Glucoside

Micelles. Eur. J. Biochem., 204, 1075-1082.

Nisimoto, Y., & Otsuka-Murakami, H. (1988). Cytochrome b5, cytochrome c, and cytochrome P-450 interactions with NADPH-cytochrome P-450 reductase in phospholipid vesicles. Biochemistry, 27, 5869-5876.

Phillips, A., & Langdon, R. (1962). Hepatic triphosphopyridine nucleotide-cytochrome c reductase: isolation, characterization, and kinetic studies. J. Biol. Chem., 237, 2652-2660.

Shen, S., & Strobel, H. (1992). The Role of Cytochrome-P450 Lysine Residues in the Interaction Between Cytochrome-P450IA1 and NADPH-Cytochrome-P450 Reductase. Arch. Biochem. Biophys., 294, 83-90.

Shen, S., & Strobel, H. (1993). Role of Lysine and Arginine Residues of Cytochrome- P450 in the Interaction Between Cytochrome-P4502B1 and NADPH-Cytochrome P450 Reductase. Arch. Biochem. Biophys., 304, 257-265.

Tamburini, P., MacFarquhar, S., & Schenkman, J. (1986). Evidence of Binary Complex Formation Between Cytochrome P-450, Cytochrome b5 and NADPH-cytochrome P-450 reductase of Hepatic Microsomes. Biochem. Biophys. Research Commun., 134, 519-526.

Tamburini, P., & Schenkman, J. (1986a). Differences in the mechanism of functional interaction between NADPH-cytochrome P-450 reductase and its redox partners. Mol. Pharm., 30, 178-185.

Tamburini, P., & Schenkman, J. (1986b). Mechanism of interaction between cytochromes P-450 RLM5 and b5: evidence for an electrostatic mechanism involving cytochrome b5 heme propionate groups. Arch. Biochem. Biophys., 245, 512-522.

Tamburini, P., & Schenkman, J. (1987). Purification to homogeneity and enzymological characterization of a functional covalent complex composed of cytochromes P-450 isozyme 2 and b5 from rabbit liver. Proc. Natl. Acad. Sci, 84, 11-15.

Voznesensky, A., & Schenkman, J. (1992a). The Cytochrome-P450 2B4-NADPH Cytochrome-P450 Reductase Electron Transfer Complex Is Not Formed by Charge-Pairing. J. Biol. Chem., 267, 14669-14676.

Voznesensky, A., & Schenkman, J. (1992b). Inhibition of Cytochrome-P450 Reductase by Polyols Has an Electrostatic Nature. Eur. J. Biochem., 210, 741-746.

# Electron pathways in adrenal mitochondrial cytochrome P450 systems: relative rates of leakage and hydroxylation

Revital Rapoport, Moshe Raikhinstein, David Sklan, Israel Hanukoglu

Department of Hormone Research, Weizmann Institute of Science, Rehovot, 76100, and The Faculty of Agriculture, Hebrew University, Rehovot, 76100, Israel

## INTRODUCTION

Mitochondrial P450 systems are located on the matrix side of the inner mitochondrial membrane and catalyze crucial steps in the synthesis of steroid hormones, and many other steroidal compounds (Lambeth, 1990; Hanukoglu, 1992). The hydroxylation reactions catalyzed by these P450s require electrons which are transferred from NADPH to P450 by an electron transport chain that consists of a flavoprotein (adrenodoxin reductase), and an iron-sulfur protein (adrenodoxin). This electron transfer may be "coupled" or "leaky". In a coupled system all electrons from NADPH are used in substrate hydroxylation reactions, whereas in a leaky system some electrons escape to other acceptors, such as $O_2$ (Fig. 1). Among the microsomal cytochromes P450 some are highly leaky (Archakov and Bachmanova, 1990; Ding et al., 1991; Hornsby, 1989). In the present study we determined the sites and rates of electron leakage reactions in two mitochondrial P450 systems reconstituted with purified enzymes (Hanukoglu et al., 1993).

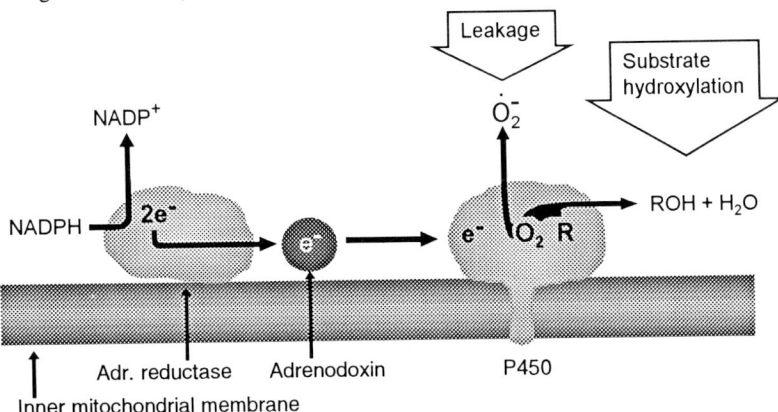

Fig. 1. Schematic representation of the possible pathways of electron flow in mitochondrial P450 systems. In systems reconstituted without P450, adrenodoxin is the major source of superoxide formation. In the absence of other side reactions, the rate of NADPH oxidation should equal the rate of superoxide formation (electron leakage) plus the rate of substrate hydroxylation.

## MATERIALS AND METHODS

Adrenodoxin reductase, adrenodoxin and the mitochondrial P450scc and P45011β were purified from bovine adrenal cortex (Hanukoglu et al., 1981). NADPH oxidation, cholesterol side chain cleavage and 11β-hydroxylation reactions were assayed as described (Hanukoglu et al., 1993). Superoxide generated by the tested systems was converted to $H_2O_2$ which was assayed after catalase mediated conversion to formaldehyde using a new fluorimetric method more accurate and sensitive than the previous assay used in our recent study (Hanukoglu et al., 1993).

## RESULTS AND DISCUSSION

*Electron leakage from the adrenodoxin reductase-adrenodoxin system*
Under aerobic conditions adrenodoxin reductase alone oxidizes NADPH only at a relatively low rate (Fig. 2; Chu and Kimura, 1973; Sugiyama et al., 1979). Addition of adrenodoxin greatly stimulates the rate of NADPH oxidation. Simultaneous monitoring of NADPH oxidation and adrenodoxin redox state revealed three distinct phases (Fig. 2):

1. The initial rapid phase of NADPH oxidation corresponded exactly to the time required for the reduction of adrenodoxin. The number of electrons donated by NADPH during this phase matched the number of adrenodoxin molecules reduced (adrenodoxin is a single electron acceptor).

2. In the second phase of NADPH oxidation (from 0.5 to 4 min in Fig. 2) all adrenodoxin molecules remained reduced. During this "steady-state phase" of electron transfer, adrenodoxin molecules were continuously being reduced and oxidized, transferring the electrons to $O_2$. Thus, in this phase adrenodoxin functioned as an automatic "electron gun", constantly being loaded and fired, until electron supply (NADPH) was exhausted.

3. After NADPH depletion (at 4 min in Fig. 2), reduced adrenodoxin molecules oxidized, but during this auto-oxidation phase, the rate of electron leakage from adrenodoxin$_{red}$ was nearly ten fold slower than that in the steady-state phase. The different rates indicate that the pathway of electron transfer and/or molecular interactions of adrenodoxin with $O_2$ are also different in the two phases (Hanukoglu et al., 1993).

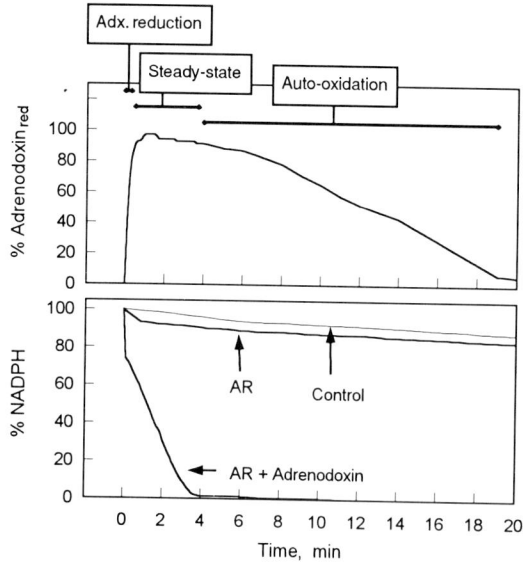

Fig. 2. NADPH oxidation by adrenodoxin reductase and adrenodoxin, and phases of electron leakage from adrenodoxin. Reaction conditions: 10 mM Hepes, pH 7.2, 50 mM KCl, 48 μM NADPH, 0.3 μM adrenodoxin reductase (AR), and 11.8 μM adrenodoxin. The boxes mark the three phases of electron transfer. The percentage of reduced adrenodoxin (adrenodoxin$_{red}$), and NADPH oxidation were determined by absorbance at 414 nm, and 340 nm respectively. The reaction was initiated by the addition of reductase.

Kinetic analyses of adrenodoxin dependence of NADPH oxidation during the steady-state phase showed a good fit to Michaelis-Menten kinetics (Hanukoglu et al., 1993). In contrast, the auto-oxidation phase of adrenodoxin behaved as a non-enzymatic process, fitting best to pseudo first order reaction kinetics (with respect to reduced adrenodoxin). The rate of electron transfer from reduced adrenodoxin during the auto-oxidation phase was much slower than the turnover number during the steady state phase. The rate constant derived from this linear fit was highly dependent on temperature. The activation energy calculated from the Arrhenius plot was 24.2 kcal/mole. This value is higher than activation energies of most enzymatic reactions which are usually less than 20 kcal/mole (Eisenberg and Crothers, 1979).

During the steady-state phase, adrenodoxin$_{red}$ may reduce $O_2$ while bound to reductase or in unbound form. The results indicated that binding of adrenodoxin to reductase does not block its reduction of $O_2$. In contrast, in similar experiments reductase inhibited P450 catalytic activity by competing with P450 for binding to adrenodoxin (Hanukoglu et al., 1981) as the sites of binding for these two proteins apparently overlap (Tuls et al., 1987; Coghlan and Vickery, 1991). The possibility that $O_2$ is reduced by the reductase$_{red}$-adrenodoxin$_{red}$ complex is also consistent with the redox potentials. The redox potential of adrenodoxin in unbound form (-290 mV) is shifted to -360 mV upon its binding to reductase (Lambeth et al., 1979), while the redox potential for the $O_2/O_2^-$ couple is -330 mV (Ilan et al., 1976). Thus, the complex would have a more favorable potential to reduce $O_2$ than unbound adrenodoxin$_{red}$.

*Electron leakage in the presence of P450*

In a P450 system reconstituted with purified enzymes the total electron flow can be quantitated by monitoring NADPH oxidation at 340 nm. Measurement of both NADPH oxidation and superoxide converted to $H_2O_2$ showed that the rates of these two reactions are essentially identical in the presence of P450scc or P450$11\beta$ but in the absence of their substrates (Fig. 3). These findings showed that, in absence of a substrate all electrons from NADPH are channeled to superoxide formation, and that side reactions of electrons with other molecules, such as proteins in solution, are essentially negligible under the assay conditions.

The presence of P450scc enhanced NADPH oxidation above the rate observed with reductase-adrenodoxin (Fig. 3). Kinetic analyses demonstrated that the stimulation of leakage by P450scc was not a simple additive effect over that of adrenodoxin and that at low adrenodoxin$_{red}$ concentrations electrons leaked mainly through P450scc. P450scc was more active than adrenodoxin in promoting electron leakage to $O_2$, and with maximal effect at lower concentrations: 0.3 $\mu$M vs. > 7 $\mu$M for adrenodoxin (Hanukoglu et al., 1993). In adrenal mitochondria the concentrations of P450scc and

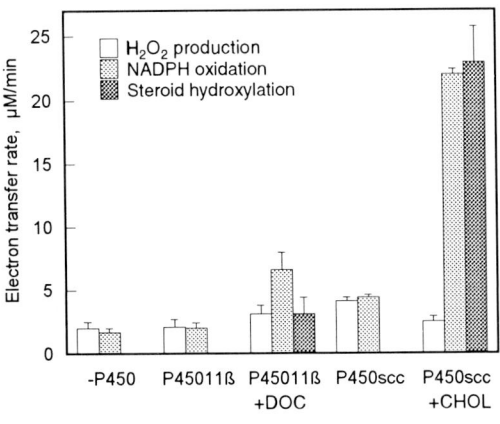

Fig. 3. The rates of electron leakage ($H_2O_2$ production) and steroid hydroxylation catalyzed by P450scc and P450$11\beta$. Reaction conditions: 10 mM Hepes, pH 7.2, 100 mM KCl, 60 $\mu$M NADPH, 0.1 $\mu$M adrenodoxin reductase, and 9 $\mu$M adrenodoxin. P450scc (0.3 $\mu$M) reactions included 0.3% Tween 20, and 200 $\mu$M cholesterol (CHOL) at 37°C. P450$11\beta$ (0.16 $\mu$M) reactions included 50 $\mu$M deoxycorticosterone (DOC) and were carried out at 30°C to avoid enzyme inactivation. In all reactions 200 U/ml catalase, 10 U/ml SOD, and 6% methanol were included for the assay of $H_2O_2$.

adrenodoxin are similar (Hanukoglu and Hanukoglu, 1986). In system reconstituted with this molar ratio of the two proteins, electrons appeared to leak mainly through P450scc and not through adrenodoxin as depicted in Fig. 1.

Previously, substrate free P450scc was thought to be not readily reducible by reduced adrenodoxin, based on the observation that under anaerobic conditions its reduction potential (-412 mV) is lower than that of adrenodoxin (-290 mV) (Lambeth et al., 1982). However, experimental observations showed that substrate free P450scc can be reduced easily under aerobic conditions (Hanukoglu et al. 1993). A possible explanation for this apparent discrepancy is that the binding of $O_2$ and adrenodoxin increases the redox potential of P450scc. Many microsomal P450s have been observed to function as NADPH oxidases reducing oxygen in the absence of a substrate, and producing superoxide anion, hydrogen peroxide or water (Archakov and Bachmanova, 1990; Hornsby, 1989). NADPH oxidase activity of some of these microsomal P450s (e.g. Ding et al., 1991), is much higher than that of the mitochondrial P450s studied here.

*Electron leakage in the presence of P450 and substrate*

The presence of substrate enhanced the rate of NADPH oxidation consistent with the turnover rates of P450scc and P45011β systems. However, these two P450s differed in the rate of leakage during substrate metabolism. In the P450scc system with cholesterol as substrate, the leakage was reduced significantly and represented only about 10% of the total NADPH oxidation (last set of columns in Fig. 3). In contrast, in the P45011β system with deoxycorticosterone as substrate the leakage represented about 50% of the total NADPH oxidation (Fig. 3). These results indicate that the P45011β is more "leaky" than the P450scc, i.e., during 11β-hydroxylation a greater percentage of the electrons is channeled to superoxide formation (Martsev et al., 1984).

The tighter degree of coupling in the P450scc system during cholesterol conversion to pregnenolone, may be explained by the following observations: (a) cholesterol enhances the binding of adrenodoxin to P450scc (both display mutual positive cooperativity, and shift the heme Fe(III) into the high spin state), (b) cholesterol binding increases the redox potential of P450scc, and (c) the binding of cholesterol and its hydroxylated derivatives stabilize the oxyferro complex of P450scc (Hanukoglu and Jefcoate, 1980; Lambeth and Pember, 1983, Tuckey and Kamin, 1982). One possible explanation for the substrate dependent inhibition of leakage from P450scc is that, as observed in the P450cam structure (Poulos et al., 1985), the substrate may fully occupy the active site, excluding water and decreasing the polarity of the microenvironment of the $FeO_2$ complex, and consequently inhibiting the release of $O_2^-$. In other P450 systems, various substrates and substrate analogs show wide variation in their effects on the coupling of the electron transfer process (Archakov and Bachmanova, 1990; Hornsby, 1989).

*Is NADPH availability to the mitochondrial P450 system regulated?*

The present findings raise several questions regarding the function of the P450 system within the mitochondria: Is NADPH constantly available to the P450 systems even in absence of the substrate cholesterol? If so, is adrenodoxin constantly kept in a reduced state, and does it or P450 "leak" electrons? If they leak electrons then what types of cellular mechanisms operate to control the damaging effects of the oxygen radicals?

Previous observations in cultured cells showed that in contrast to P450scc, P45011β undergoes rapid steroid dependent inactivation in adrenocortical cells (Hanukoglu et al., 1990; Hornsby, 1989). This was suggested to result from oxygen radicals generated by the P450 itself (Hornsby, 1989). The present data that P45011β is much more leaky than P450scc, provide evidence that the studies with purified enzymes reflect electron transfer and leakage reactions in living cells.

Based on the current findings the hypothesis could be raised that, in order to minimize electron leakage, NADPH synthesis or availability may be coregulated with cholesterol availability in

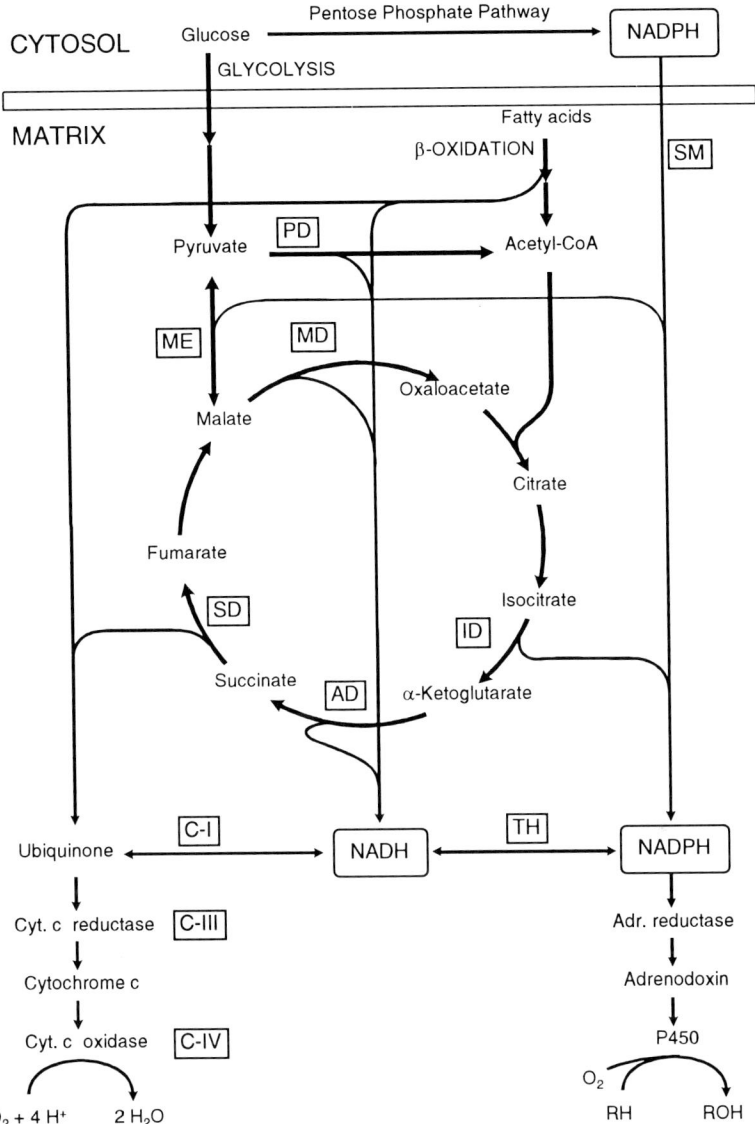

Fig. 4. Routes of NADPH synthesis in the mitochondria of steroidogenic cells. Abbreviations: SM, shuttle mechanisms of NADPH transport into mitochondria; AD, α-ketoglutarate dehydrogenase; ID, isocitrate dehydrogenase; MD, malate dehydrogenase; ME, malic enzyme; PD, pyruvate dehydrogenase; SD, succinate dehydrogenase; TH, transhydrogenase; C-I, C-III, and C-IV, complexes of oxidative phosphorylation system.

steroidogenic cells. The delivery of cholesterol to P450scc is regulated by trophic hormones in steroidogenic tissues and represents the rate-limiting step in acute stimulation of steroidogenesis (Jefcoate et al., 1992). Several early studies provided evidence that ACTH also regulates availability of NAPDH (Haynes, 1975). Yet, our current understanding of the regulation of NADPH production in adrenocortical cells is very meager.

NADPH that supplies electrons to the mitochondrial P450 systems may be generated by several alternative routes (Fig. 4) (McNamara and Jefcoate, 1990; Simpson et al., 1969). The synthesis of NADPH may be regulated in short-term at the level of the activity of the enzymes involved in the routes of NADPH biosynthesis. Our recent studies indicated that NADPH biosynthesis may also be regulated at the level of the expression of the enzymes on the pathways of NADPH production. Our differential screening of an adrenal cortex cDNA library for ACTH inducible genes surprisingly led to the isolation of a group of cDNAs for mitochondrial genes encoding subunits of oxidative phosphorylation system enzymes (Raikhinstein and Hanukoglu, 1993). Northern blot analyses using these cDNAs revealed that the mRNA encoding NADH dehydrogenase subunit 3 was most highly induced (nearly 40 fold). This enzyme system participates in succinate dependent pathway of reducing equivalent supply to the P450scc system (McNamara and Jefcoate, 1990). Thus, its dramatic induction by ACTH may be related to its role in NADPH generation, in addition to oxidative phosphorylation (Fig. 4).

The questions raised by the present studies emphasize a significant gap in our current knowledge of regulation of reducing equivalent supply, and require further study of the regulation of NADPH synthesis and availability to the mitochondrial P450 systems in steroidogenic cells.

## ACKNOWLEDGEMENTS

This research was supported in part by grants from Israel Cancer Research Fund, Association Suisse Pour Favoriser Les Recherches Contre Le Cancer en Israel, Friends of the Israel Cancer Association, and the Kimmelman Center for Biomolecular Structure and Assembly at the Weizmann Institute of Science.

## REFERENCES

Archakov, A.I. & Bachmanova, G.I. (1990) *Cytochrome P-450 and active oxygen.* Taylor & Francis, Hants, U.K.

Chu, J.W. & Kimura, T. (1973) Studies on adrenal steroid hydroxylases. Complex formation of the hydroxylase components. *J. Biol. Chem.* 248, 5183-5187.

Coghlan, V.M. & Vickery, L.E. (1991) Site-specific mutations in human ferredoxin that affect binding to ferredoxin reductase and cytochrome P450scc. *J. Biol. Chem.* 266, 18606-18612.

Ding, X., Pernecky, S.J. & Coon, M.J. (1991) Purification and characterization of cytochrome P450 2E2 from hepatic microsomes of neonatal rabbits. *Arch. Biochem. Biophys.* 291, 270-276.

Eisenberg, D. & Crothers, D. (1979) *Physical chemistry with applications to the life sciences.* The Benjamin/Cummings Publ. Co., Menlo Park, Calif

Hanukoglu, I. (1992) Steroidogenic enzymes: structure, function and regulation of expression. *J. Steroid Biochem. Mol. Biol.* 43, 779-804.

Hanukoglu, I., Feuchtwanger, R. & Hanukoglu, A. (1990) Mechanism of ACTH and cAMP induction of mitochondrial cytochrome P450 system enzymes in adrenal cortex cells. *J. Biol. Chem.* 265, 20602-20608.

Hanukoglu, I. & Hanukoglu, Z. (1986) Stoichiometry of mitochondrial cytochromes P-450, adrenodoxin and adrenodoxin reductase in adrenal cortex and corpus luteum: Implications for membrane organization and gene regulation. *Eur. J. Biochem.* 157, 27-31.

Hanukoglu, I. & Jefcoate, C.R. (1980) Mitochondrial cytochrome P-450scc: Mechanism of electron transport by adrenodoxin. *J. Biol. Chem.* 255, 3057-3061.

Hanukoglu, I., Privalle, C.T. & Jefcoate, C.R. (1981) Mechanism of ionic activation of mitochondrial cytochromes P-450scc and P-45011β. *J. Biol. Chem.* 256, 4329-4335.

Hanukoglu, I., Rapoport, R., Weiner, L. & Sklan, D. Electron leakage from the mitochondrial NADPH - adrenodoxin reductase - adrenodoxin - P450scc (cholesterol side chain cleavage) system. *Arch. Biochem. Biophys.* 305, 489-498, 1993.

Hanukoglu, I., Spitsberg, V., Bumpus, J.A., Dus, K.M. & Jefcoate, C.R. (1981) Adrenal mitochondrial cytochrome P-450scc: Cholesterol and adrenodoxin interactions at equilibrium and during turnover. *J. Biol. Chem.* 256, 4321-4328.

Haynes, R.C. (1975) Theories on the mode of action of ACTH in stimulating secretory activity of the adrenal cortex. In: *Handbook of Physiology-Endocrinology* vol. 6, Am. Physiol. Soc., Washington, D.C., pp. 69-76.

Hornsby, P.J. (1989) Steroid and xenobiotic effects on the adrenal cortex: Mediation by oxidative and other mechanisms. *Free Radicals Biol. Med.* 103-115.

Ilan, Y.A., Czapski, G. & Meisel, D. (1976) The one electron transfer redox potentials of free radicals: The oxygen-superoxide system. *Biochim. Biophys. Acta* 430, 209-224

Jefcoate, C.R., McNamara, B.C., Artemenko, I. & Yamazaki T. (1992) Regulation of cholesterol movement to mitochondrial cytochrome P450scc in steroid hormone synthesis. *J. Steroid Biochem. Mol. Biol.* 43, 751-767.

Lambeth, J.D. (1990) Enzymology of mitochondrial side-chain cleavage by cytochrome P-450scc *Frontiers in Biotransformation* 3, 58-100.

Lambeth, J.D. & Pember, S.O. (1983) Cytochrome P-450scc-adrenodoxin complex. Reduction properties of the substrate-associated cytochrome and relation of the reduction states of heme and iron-sulfur centers to association of the proteins. *J. Biol. Chem.* 258, 5596-5602.

Lambeth, J.D., Seybert, D.W., Lancaster, J.R., Salerno, J.C. & Kamin, H. (1982) Steroidogenic electron transport in adrenal cortex mitochondria. *Mol. Cell. Biochem.* 45, 13-31.

Lambeth, J.D., Seybert, D.W. & Kamin, H. (1979) Ionic effects on adrenal steroidogenic electron transport. The role of adrenodoxin as an electron shuttle. *J. Biol. Chem.* 254, 7255-7264.

Martsev, S.P., Chashchin, B.L. & Akhrem, A.A. (1984) Reconstitution and study of multienzyme steroid-11β hydroxylating system. *Biokhimiya* 50, 243-257.

McNamara, B.C. and Jefcoate, C.R. (1990) Heterogeneous pools of cholesterol side chain cleavage activity in adrenal mitochondria from ACTH-treated rats: differential responses to different reducing precursors. *Mol. Cell. Endocrinol.* 73, 123-134.

Raikhinstein, M. & Hanukoglu, I. Mitochondrial genome encoded RNAs: differential regulation by ACTH in bovine adrenocortical cells. *Proc. Natl. Acad. Sci. USA.* 90, 10509-10513, 1993.

Simpson, E.R., Cooper, D.Y. & Estabrook R.W. (1969) Metabolic events associated with steroid hydroxylation by the adrenal cortex. *Recent Progr. Horm. Res.* 25, 523-562.

Sugiyama, T., Miura, R. & Yamano, T. (1979) Differences between the reactivities of two pyridine nucleotides in the rapid reduction process and the reoxidation process of adrenodoxin reductase. *J. Biochem.* 86, 213-223.

Tuls, J., Geren, L., Lambeth, J.D. & Millett, F. (1987) The use of a specific fluorescence probe to study the interaction of adrenodoxin with adrenodoxin reductase and cytochrome P-450scc. *J. Biol. Chem.* 262, 10020-10025.

# Reaction mechanism of P450 dependent steroidogenesis: regulation of catalytic activity of bovine adrenal P450 (11β)

Shigeki Takemori, Shiro Kominami, Shin-Ichi Ikushiro, Takeshi Yamazaki, Daisuke Harada

*Faculty of Integrated Arts and Sciences, Hiroshima University, Higashihiroshima 724, Japan*

## Summary

In the adrenal cortex, a mineralocorticoid, aldosterone, is produced in the zona glomerulosa but not in the zonae fasciculata-reticularis. Based on the results from immunoinhibition and cross-linking, the zone specificity of aldosterone synthesis in bovine adrenal cortex would appear to quite likely arise from the differences of interactions between P-450(11β) and P-450(SCC) in mitochondria in different cortical zones. The reactions of corticosterone and aldosterone synthesis from deoxycorticosterone were studied kinetically using P-450(11β)-proteoliposomes and Tween 20-solubilized P-450(11β). The catalytic properties of the latter were virtually the same as those of proteoliposomes containing P-450(11β) and P-450(SCC). In the reactions of P-450(11β) complexed with P-450(SCC) in proteoliposomes and Tween 20-solubilized P-450(11β), corticosterone as an intermediate easily leaves from the catalytic site of P-450(11β), consequently hindering three successive monooxygenation reactions for deoxycortricosterone to form aldosterone as the final product.

# INTRODUCTION

Adrenal steroid hormones are produced in different cortical compartments. The zona glomerulosa (ZG) produces mineralocorticoid(aldosterone) and zonae fasciculata-reticularis (ZFR), glucocorticoid (corticosterone or cortisol) and androgen. In bovine adrenal, two forms of P-450(11β) differing in molecular weight have been found. Both are distributed in three zones and are capable of catalyzing the terminal steps in aldosterone and corticosterone synthesis from deoxycorticosterone (DOC) (Ogishima et al., 1989). Intact mitochondria isolated from bovine adrenal ZFR are not capable of producing aldosterone, while those from ZG are. Cholate extracts from mitochondria of ZFR produce aldosterone almost at the same rate as those from ZG (Ohnishi et al., 1988; Ikushiro et al., 1991). ZG-specific aldosterone synthesis in bovine adrenal would thus not be due to differences in P-450 species in the mitochondria of each zone, as observed in the adrenals of mouse (Domalik et al., 1991), rat (Ogishima et al., 1989) and human (Ogishima et al., 1991). A certain local factor in mitochondria would certainly be the cause of selective suppression of aldosterone synthetic activity inherent to P-450(11β) in bovine adrenal ZFR. When using P-450(11β)proteoliposomes (Ikushiro et al., 1989), the addition of another mitochondrial P-450, P-450(SCC), was observed to enhance 11β-hydroxylase activity but suppress aldosterone synthetic activity (Ikushiro et al., 1992). Kinetic analysis of this effect indicated the formation of an equimolar complex between P-450(11β) and P-450(SCC) on liposomal membranes.

To determine whether specific interactions between P-450(11β) and P-450(SCC) are physiologically involved in the regulation of aldosterone synthesis, the strength of each interaction between the two P-450 proteins was assessed in inner mitochondrial membranes from ZG and ZFR by immunoinhibition and cross-linking. The effect of P-450(SCC) on aldosterone synthetic activity of P-450(11β) was similar to that of Tween 20 toward P-450(11β); Tween 20-solubilized P-450(11β) (detergent P-450(11β)) showed very high activity for 11β-hydroxylation but was incapable of producing aldosterone. Using P-450(11β)proteoliposomes and detergent P-450(11β), the relationship between 11β-hydroxylation and aldosterone synthesis activity of P-450(11β) was examined kinetically.

## RESULTS AND DISCUSSION

### Interactions between P-450(11β) and P-450(SCC) in inner mitochondrial membranes from ZG and ZFR of bovine adrenal cortex

To examine the effects of endogenous P-450(SCC) on corticosterone and aldosterone synthetic activity of P-450(11β) toward DOC, each inner mitochondrial membrane from ZG and ZFR was treated with anti-P-450(SCC) IgG. As shown in Fig.1, this treatment reduced the production of corticosterone from DOC in inner mitochondrial membranes from ZFR but not in those from ZG. In inner mitochondrial membranes from ZFR, the treatment caused significant stimulation of aldosterone production. Interactions between P-450(11β) and P-450(SCC) in mitochondrial membranes would thus appear stronger in ZFR than in ZG and the aldosterone synthetic activity of P-450(11β) in mitochondrial membranes of ZFR may possibly be suppressed more by interaction with P-450(SCC) than by interactions in ZG.

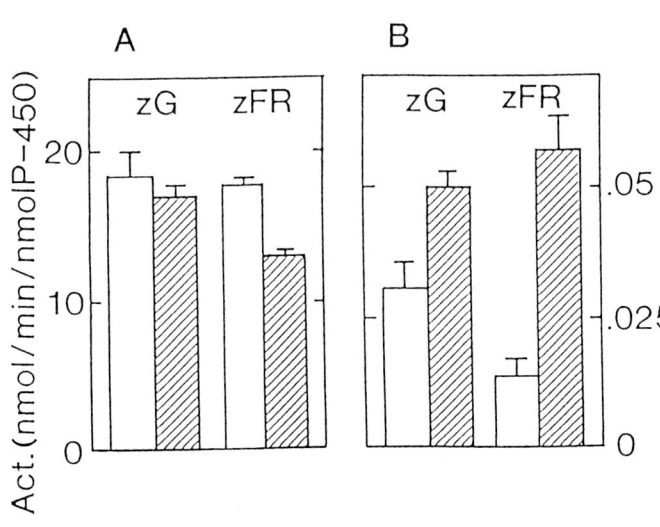

Fig. 1 Effects of anti-P450(SCC) IgG on the production of corticosterone (A) and aldosterone (B) from DOC in inner mitochondrial membrane from ZG and ZFR of bovine adrenal cortex.
Inner mitochondrial membrane (0.1 mg protein) from ZG and ZFR were incubated without (open bar) or with (closed bar) 0.5 mg of anti-P-450(SCC) IgG in 0.2 ml potassium phosphate buffer (50 mM, pH 7.1) for 10 h at 0°C. The mixture was added to 50 nmol DOC, 12 nmol adrenodoxin, 0.2 nmol adrenodoxin reductase and 100 nmol NADPH. Reaction was allowed to proceed for 15 min at 37 °C.

Interactions between P-450(11β) and P-450(SCC) in inner mitochondrial membranes were examined by chemical cross-linking, as shown in Fig.2. The heterobifunctional, cleavable, cross-linking reagent used was sulfosuccinimidyl(4-azidophenyldithio)propionate (sulfo-SADP) (Alston et al., 1991). Following their cross-linking in inner mitochondrial membranes, P-450s were solubilized and isolated with Sepharose 4B-bound anti-P-450(SCC) IgG. This was followed by SDS-PAGE in the presence of β-mercaptoethanol, with consequent cleavage of the disulfide in cross-linker to yield the individual components of the cross-linked complex. Cross-linked P-450s were then identified by immunoblotting with antibodies against P-450(SCC) and P-450(11β). Figure 3, lane 1 shows immunoblotting of inner mitochondrial membranes from ZG and ZFR, where two bands corresponding to P-450(SCC) and P-450(11β) are present. Lane 2 shows control immunopurification in the absence of cross-linking. A single band corresponding to P-450(SCC) from each membrane was observed. After cross-linking and immunopurification, P-450(11β) as well as P-450(SCC) appeared by cleavage with β-mercaptoethanol (lane 3). The relative intensity of P-450(11β) band in ZFR exceeded that in ZG, suggesting strong association between the two P-450 proteins in ZFR.

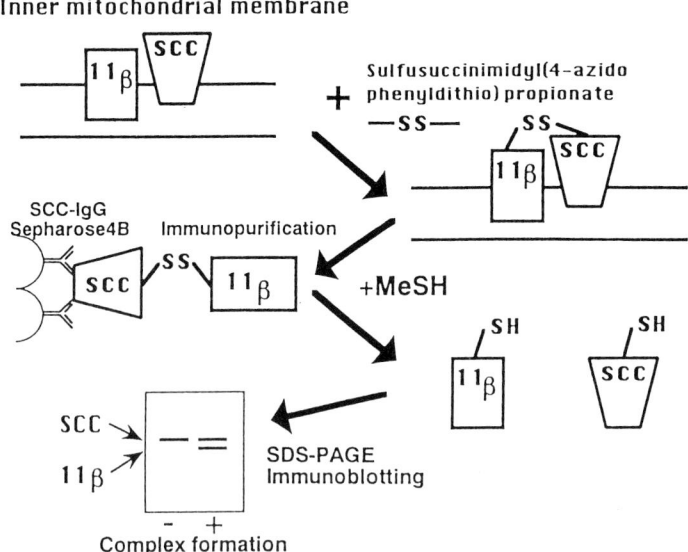

Fig.2  Procedure for the chemical cross-linking experiment.

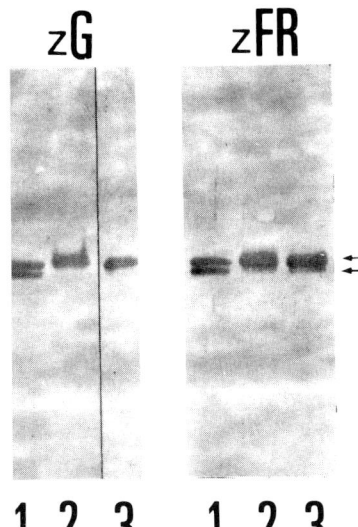

Fig. 3 Immunoblot analysis of proteins immunopurified from cross-linked inner mitochondrial membranes from ZG and ZFR.

Lane 1 contains 1 μg (ZG) or 0.5 μg (ZFR) of inner mitochondrial membrane proteins. Lane 2 contains the protein immunopurified from inner mitochondrial membranes not treated with sulfo-SADP. Lane 3 is the same as lane 2 except that the membrane have been cross-linked with sulfo-SADP. The upper and lower arrows indicate P-450(SCC) and P-450(11β), respectively.

## Substrate interactions with P-450(11β) proteoliposomes in the presence or absence of P-450(SCC) and with detergent P-450(11β)

All experiments were performed at 5 °C since P-450(11β) became labile following solubilization with Tween 20. On adding DOC or corticosterone to P-450(11β)proteoliposomes, type-1 difference spectra were obtained, whose amplitudes were proportional to the amount of substrate-P-450(11β) complex. Spectral titration of P-450(11β) was not possible in the presence of P-450(SCC) since DOC and corticosterone caused spectral change in P-450(SCC). In detergent P-450(11β), a type-1 difference spectrum was detected for DOC, but which was negligible for corticosterone. Based on the results of spectral titration, the apparent dissociation constant of P-450(11β) proteoliposomes was 0.054 μM for DOC and 13 μM for corticosterone and that of detergent P-450(11β), 4.0 μM for DOC and >2500 μM for corticosterone. Corticosterone thus leaves P-450(11β) in Tween 20 more easily than in liposomes. The maximum activity (nmol/min/nmol P-450) of P-450(11β)proteoliposomes was about 2.5 times higher for conversion from DOC to corticosterone (Vmax: 0.24) than to aldosterone (Vmax: 0.11). In detergent P-450(11β), Vmax for conversion from DOC to corticosterone was 1.1, this being higher than that of P-450(11β)proteoliposomes. No aldosterone formation, however, could be detected.

In proteoliposomes containing P-450(11β) and P-450(SCC), Vmax for conversion from DOC to corticosterone was 1.1 but there was no activity for aldosterone production, as also observed for detergent P-450(11β). In P-450(11β)proteoliposomes, corticosterone competitively inhibited the production of corticosterone (Ki: 16 μM) and aldosterone (Ki: 13 μM) from DOC. In detergent P-450(11β), corticosterone at 0.5 mM failed to exert any inhibitory effect on corticosterone production from DOC. No inhibitory effect of corticosterone could be detected on proteoliposomes containing P-450(11β) and P-450(SCC).

By radioisotope dilution, aldosterone production in P-450(11β)proteoliposomes was observed to occur through successive reactions without the intermediate, corticosterone, leaving P-450(11β) (Ikushiro et al., 1991). The easy dissociation of corticosterone from the active site of P-450(11β) would possibly hinder the successive reactions. In a detergent solubilized state, P-450(11β) has weaker ability to bind with corticosterone. This is the reason why detergent P-450(11β) cannot produce aldosterone from DOC and has higher 11β-hydroxylase activity for producing corticostrerone. The kinetic parameters of P-450(11β) complexed with P-450(SCC) in membranes resemble those of detergent P-450(11β), suggesting essentially the same catalytic state of P-450(11β) in both forms. In membranes, P-450(11β) complexed with P-450(SCC) alters the binding affinity of corticosterone and modulates successive reactions for DOC. Thus, interaction between P-450(11β) and P-450(SCC) in mitochondrial membranes would be required for regulating aldosterone synthesis in bovine adrenal cortex (Fig. 4).

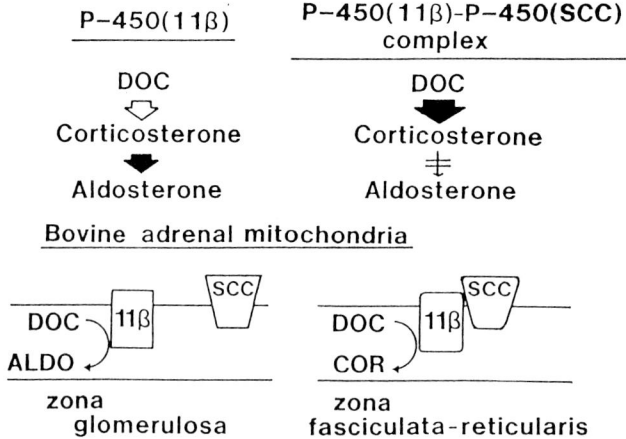

Fig. 4  Control mechanism of the catalytic activity of bovine adrenal P-450(11β).

# REFERENCES

Alston, K., Robinson, R.C., Park, S.S., Gelboin, H.V. & Friedman, F.K. (1991): Interactions among cytochromes P-450 in the endoplasmic reticulum: Detection of chemically cross-linked complexes with monoclonal antibodies. *J. Biol. Chem.* 266, 735-739.

Domalik, L.J., Chaplin, D.D., Kirkman, M.S., Wu, R.C., Liu, W., Howard, T.A., Seldin, M.F. & Parker, K.L. (1991): Different isozymes of mouse 11β-hydroxylase produce mineralocorticoids and glucocorticoids. *Mol. Endocrinol.* 5, 1853-1861.

Ikushiro, S., Kominami, S. & Takemori, S. (1989): Adrenal cytochrome P-450(11β) proteoliposomes catalyzing aldosterone synthesis: Preparation and characterization. *Biochim. Biophys. Acta* 984, 50-56.

Ikushiro, S., Kominami, S., Yamazaki, T. & Takemori, S. (1991): Control Mechanism of P-450(11β) dependent catalytic activity in bovine adrenal mitochondria. In *Cytochrome P-450: Biochemistry and Biophysics,* ed. A.L. Archakov & G.I. Bachmanova, pp. 95-100. Moscow: INCO-TNC.

Ikushiro, S., Kominami, S. & Takemori, S. (1992): Adrenal P-450(SCC) modulates activity of P-450(11β) in liposomal and mitochondrial membranes:Implication of P-450(SCC) in zone specificity of aldosterone biosynthesis in bovine adrenal. *J. Biol. Chem.* 267, 1464-1469.

Ogishima, T., Mitani, F. & Ishimura, Y. (1989): Isolation of two distinct cytochromes P-450(11β) with aldosterone synthase activity from bovine adrenocortical mitochondria. *J. Biochem.* 105, 497-499.

Ogishima, T., Mitani, F. & Ishimura, Y. (1989): Isolation of aldosterone synthase cytochrome P-450 from zona glomerulosa mitochondria of rat adrenal cortex. *J. Biol. Chem.* 264, 10935-10938.

Ogishima, T., Shibata, H., Shimada, H., Mitani, F., Suzuki, H., Saruta, T. & Ishimura, Y. (1991): Aldosterone synthase cytochrome P-450 expressed in the adrenals of patients with primary aldosteronism. *J. Biol. Chem.* 266, 10731-10734.

Ohnishi, T., Wada, A., Lauber, M., Yamano, T. & Okamoto, M. (1989): Aldosterone biosynthesis in mitochondria of isolated zones of adrenal cortex. *J. Steroid Biochem.* 31, 73-81.

# Electron transfer associated dioxygen activation in P450 systems

Stephen G. Sligar, John Aikens, Nancy Gerber, Mark McLean,
Kenneth Suslick, David Benson

*Departments of Biochemistry and Chemisty, University of Illinois, Urbana, Illinois 61801, USA*

The microbial derived cytochrome P-450cam has served as an excellent vehicle for studying the molecular mechanisms of electron transport and dioxygen activation in this important class of heme oxygenases. The reaction cycle of cytochrome P-450 has remained essentially unchanged for the last decade with regard to the stable states easily observable by optical and resonance specroscopies. Recently specific insight into the identity and reactivity of the hypothetical intermediates has become available due to advances in rapid reaction methodology and the study of site directed mutants of the P-450 heme component. Perhaps most interesting are the steps of dioxygen and substrate activation which follow the input of the second reducing equivalent to the quasi-stable ferrous oxyenated adduct. A major difficulty has been the slowness of this second electron transfer process, occuring at roughly 20 sec$^{-1}$. Early studies in the Gunsalus laboratory (Pederson et al, 1977) showed that the rate for oxidation of the iron-sulfuer protein donor in a diprotein complex with P-450cam was identical to the rate for appearance of the regenerated ferric protein, suggesting that the actual electron transfer rate was limiting in the overall turnover of the ferrous dioxygen adduct, explaining why no further iron-porphyrin intermediates are seen in the normal catalytic cycle. Two major areas of investigation have opened up the detailed bioinorganic chemistry of oxygen activation for direct analysis. The first is our incorporation, via genetic engineering, of a specific attachement site on the surface of cytochrome P-450cam and the derivitization of this site with a ruthenium bipyridal chelate which allows rapid input of reducing eqiivalents into the iron prophyrin system. The second is the use of isotope effect methodologies to probe the transition state in dioxygen bond scission which direcly allows definition of competing mechanisms for proton donation to the intermediate structures that have been proposed from site directed mutagenesis studies. We begin the discussion by presenting specific chemical models for the involvement of distal residues in the pocket of cytochrome P-450cam in the catalytic cycle.

A critical role for an alcohol containing side chain at position 252 in the P-450cam sequence was determined several years ago (Martinis, et. al, 1989; Imai et. al, 1989). Clearly evident through studies of the temperature dependence of the one electron autoxidation rate (Martinis et. al, 1989) was a hydrogen bonding role between Thr252 and the distal oxygen of the ferrous dioxygen complex:

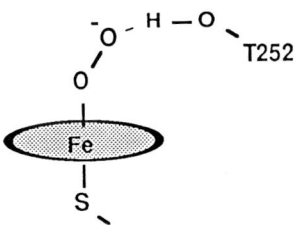

With removal of this hydrogen bond, the reaction cycle is essentially completely uncoupled, with reducing equivalents appearing as hydrogen peroxide.

The second residue playing a critical role in the activation of dioxygen is that of Asp251 in the P-450cam sequence nomenclature (Gerber et. al, 1992; Imai, et. al, 1992). Of the numerous substitutions at this position, it appears that the Asp251Asn is the most stable and, due to its near isosteric relationship to the aspartate side chain, perhaps the least pertubative of the critical acitve site structure. The Asp251Asn varient appears quite different in its chemical effect than substitutions at Thr252. WHereas removal of the Thr252 alcohol side chain yields an enzyme which is "dead" in product formation with the release of all reducing equivalents as hydrogen peroxide at essentially the wild-typen NADH oxidation rate, the Asp251Asn varient is almost completely coupled in product formation but with a drastically reduced turnover rate. The rate of NADH oxidation by this mutant is so slow that one must be careful in kinetic analysis since shunt processes ignored for the wild-type protein become important in analyzing variants at position 251. It was postulated that Asp251 served as a general proton donor in the catalytic cycle of P-450cam, perhaps by delivering a proton to the incipient alkoxide Thr252 side chain as it served to protonate the distal oxygen atom following dioxygen bond scission:

Although the mechanism above explains the overall effects of mutagenesis of these key P-450cam residues, we were anxious to develop methods to prove or disprove specific mechaisms that have been postulated. In particular, we wondered whether it would be possible to distinguish between direct protonation of a peroxo-iron complex via solvent or perhaps through a protein stabilized water channel (Raag, et. al, 1991) which involves Asp251 and Thr252 in a specific distal charge relay mechanism for proton delivery. Additionally, we wished to examine whethter state of the art transient spectroscopic methods could detect further intermediates in the P-450cam catalytic cycle in both wild-type and the slow Asp251Asn mutant. Finally, we are anxious to develop rapid means for injecting reducing equivalents into the iron-prophyrin-oxygen system to examine new potential intermediates that would appear when the constraint of the low putidaredoxin - P-450 electron transfer rate was removed.

In order to address these questions we have applied kinetic solvent isotope effect, KSIE (Schowen, Schowen 1977, 1978) methodology, wherein deuterium oxide in place of protium oxide, to dissect the mechanism of oxygen activation and determine the source and involvement of protons in the reaction. The study of KSIEs in cytochrome P-450cam requires a kinetic study of the individual steps in the catalytic cycle beginning with substrate binding and proceeding through first electron transfer, oxygen binding, second electron transfer and product release. The goal is to identify which steps are isotopically sensitive to solvent substitution ($D_2O$ for $H_2O$). These first three steps displayed solvent isotope effects of $0.97 \pm 0.1$, $1.03 \pm 0.04$, and $1.00 \pm 0.03$ respecively. The second electron transfer reaction which is coupled to O-O bond scission and product formation is the only step in the catalytic cycle that displayed a significant solvent isotope effect, $1.8 \pm 0.05$ using single turnover experiments. This large isotope effect observed is important since it links proton transfer to the unusually slow electron transfer from putidaredoxin (Pd) to P-450cam (Sligar and Gunsalus, 1979) and committment to catalysis. The lack of a measurable solvent isotope effect on the first electron transfer reaction suggests, but does not prove, that the observed KSIE in second electron transfer is not the result of an isotope effect in the binding of putidaaredoxin to P-450cam. We verified that slow exchange of hydrogenic sites within the protein were not responsible for the observed results. A more detailed study of KSIEs was performed by determinng the overal proton inventory (Schowen 1977) to determine the mechanism of proton involvement in oxygen activation. The figure below shows the nonlinear polynomial fit to the measured KSIE as a function of the percentage $D_2O$ in the reaction mixture.

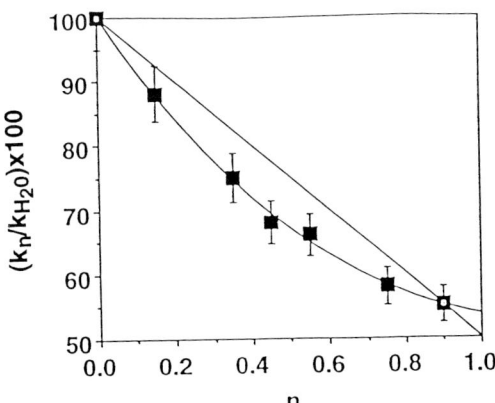

The curvature of the fitted line is consistent with multiple proton involvement in the transition state of the reaction, and taken in concert with the magnitude of the KSIE, provides an opportunity to develop a mechanistic hypothesis of oxygen activation. Using functional group fractionation factors taken from the literature we calculated the expected magnitude of the KSIE for different models of oxygen activation proposed to involve multiple protons. Only one mechanism is consistent with our experimental results. Clearly indicated is the involvement of a charge relay between the reduced oxygenated complex and a hydrogen bond network composed of the Thr252 hydroxyl, Asp251 carboxyl and the amines of Lys178 and/or Arg186 as discussed earlier. The charge relay model argues that as second electron transfer occurs, charge density builds on the bound distal oxygen in the complex which is stabilized by a hydrogen bonding interaction with Thr252. In the transition state the Thr252-peroxy hydrogen bonded complex receives a proton from Asp251 which is a strong donor due to electrostatic interactions with Lys178 and/or Arg186. In this treatment, Thr252 can be thought of as a protonated oxoium-like species which would be expected to yield an overall isotope effect of approximately 2.1 given a fractionation factor of 0.69 (Schowen 1972). Other possible mechanisms that involve direct protonation of the bound peroxy complex with Asp251 or via a water channel mediated process are unlikely as these mechanistic models would yield KSIE's significantly different from the experimental value observed.

Further insight into the mechaisms of dioxygen activation will come from a direct obsrevation of the intermediate steps in the catalytic cycle following the dioxygen adduct. Two separate approaches are being used. The first makes use of site sirected mutants to alter the relative rate constants of the catalytic steps such that new intermediates can be observed. For example, disruption of the proton donor network illustrated previously slows the dioxygen bond scission step such that a new steady state turnover spectra is obtained:

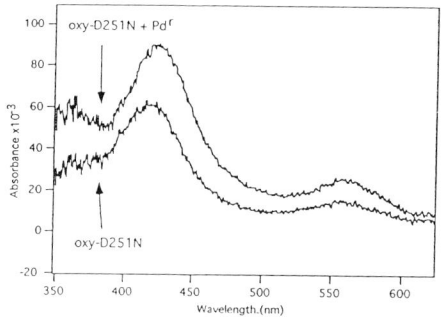

A second method is to speed up electron transfer processes such that they no longer are rate determining and hence new iron-dioxygen, iron-peroxo or iron-oxo states may become populated. In this regard we have built on the pioneering work of the Durham, Millett, (Durham, et al, 1990) and Gray laboatoaries (Winkler & Gray, 1992) in using covalently attached ruthenium chelates. Particulary useful in this regard has been the development of ruthenium complexes which can be targeted to uniquly incorporated cysteine residues on the surface of the P-450 molecule (Stayton et al,; Willie, et. al, 1992). In order to engineer a photoacitvatble redox center into the P-450 molecule, we conducted extensivie modeling work to define the most suitable pathway for electron transfer into the cytochrome. The graph below presents the results of calculations based on the "pathway's" model of Berretan and Onuchic (Bertran & Onuchic 1989) and which displays a ratio of electron transfer couplings between a direct through bond mechanisms and through space tunneling (Moser, et al, 1992) for each atom on the surface of vartious metalloproteins to the heme center. Unlike other proteins, P-450cam has a subset of atoms which display a high degree "wiredness" from a surface attachment site to the heme active center.

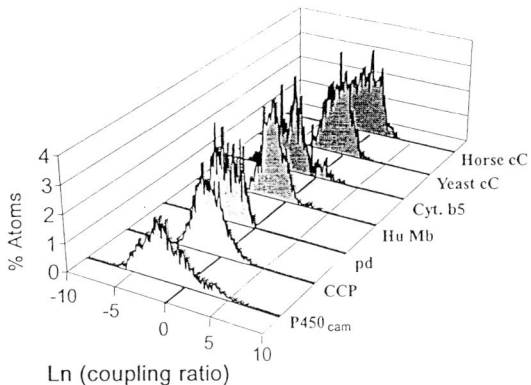

Using this information, together with careful consideration of electrostatic potential surfaces and steric hinderance of the P-450cam molecular surface, allowed us to predict novel attachment sites for ruthenium trisbipyridyl complexes. Most exciting is our engineering of a unique cysteine residue at position 113 (Ala113Cys) which was generated in a background of Cys334Ser + Cys136Ser which removes the wild-type reactive sulfhydryl. Transient optical, Raman and ESR studies are now being used to precisely define the structure of these subsequent intermediates states.

Our work is supported through NIH grants GM33775 and GM31756.

REFERENCES

Bertran, D. N. & Onuchic, J. N. (1989): Electron Transfer: From Model Compounds to Proteins. *Advances in Chemistry Series* 228, 71-89.
Durham, B., Pan, L. P., Haum, S., Long, J., & Millet, F. (1990): Electron Transfer Kinetics of Singly Labeled Ruthenium (II) Polypyridine Cytochrome c Derivatives, *Advances in Chemistry Series* 226, 181-193.
Gerber, N. C. & Sligar, S. G. (1992): Catalytic Mechanism of Cytochrome P-450: Evidence for a Distal Charge Relay. *J. Am. Chem. Soc.* 114, 8742-8743.
Imai, M., Shimada, H., Watanabe, Y., Matsushima-Hibiya, Y., Makino, R., Koga, H., Horiuchi, T., & Ishimura, Y. (1989): Uncoupling of the Cytochrome P-450can Monooxygenase Reaction by a Single Mutation, Threonine-252 to Alanine or Valine: A Possible Role of the Hydroxy Amino Acid in Oxygen Activation. *Proc. Natl. Acad. Sci. U.S.A.* 86, 7823-7827.
Martinis, S. A., Atkins, W. M., Stayton, P. S., & Sligar, S. G. (1989): A Conserved Residue of Cytochrome P-450 is Involved in Oxygen Activation. *Journal of the Americal Chemical Society* 111, 9252-9253.
Moser, C. C., Keske, J. M., Warncke, K., Farid, R. S., &Dutton, P. L. (1992): Nature of Biological Electron Transfer. *Nature* 355, 796-802.
Pederson, T. C., Austin, R. H., & Gunsalus, I. C. (1977): Redox and Ligand Dynamics in P-450 cam Putidaredoxin Complexes. Pergamon Press, Oxford.
Schowen, R. L. (1972): Mechanistic Deductions from Solvent Isotope Effects. *Prog. Phys. Org. Chem.* 9, 275-332.
Schowen, K. B. J. In Transitio States of Biochemical Processes; R. D. Grandour and R. L. Schowen, Ed.; Plenum Press: New York, 1978, 225-283.
Schowen, R. L. In Isotope Effects on Enzyme-Catalyzed Reactions; W. W. Cleland, M. H. O'Leary and D. B. Northrop, Ed.; University Park Press: Baltimore, 1977; 64-99.

Raag, R., Martinis, S. A., Sligar, S. G., & Poulos, T. L. (1991): Crystal Structure of the Cytochrome P-450cam Active Site Mutant Thr252Ala. *Biochemistry* 30, 11420-11429.

Sligar, S. G., & Gunsalus, I. C. (1979): Proton Coupling in the Cytochrome P-450 Spin and Redox Equilibria. *Biochemistry* 18, 2290-2295.

Tabushi, I. (1988): Reductive Dioxygen Activation by Use of Artificial P450 Systems. *Coordination Chemistry Reviews* 86, 1-42.

Willie, A., Stayton, P. S., Sligar, S. G., Durham, B., & Millet, F. (1992): Genetic Engineering of Redox Donor Sites: Measurement of Intracomplex Electron Transfer between Ruthenium-65-Cytochrome b5 and Cytochrome c. *Biochemistry* 31, 7237-7242.

Winkler, J., & Gray, H. B. (1992): Electron Transfer in Ruthenium Modified Proteins. *Chemical Reviews* 92, 369-379.

# Conformational control of interprotein electron transfer

F. Millett[1], B. Durham[1], A. Willie[1], S. Hahm[1], L. Geren[1], R. Liu[1], P. Stayton[2], M. McLean[2], M. Miller[4], S. Sligar[2], G. Peilak[3], J. Kraut[4]

[1]Department of Chemistry, University of Arkansas, Fayetteville, AR 72701. [2]Department of Biochemistry, University of Illinois, Urbana, IL. [3]Department of Chemistry, University of North Carolina, Chapel Hill, NC. [4]Department of Chemistry, University of California, San Diego, CA, USA

## INTRODUCTION

The reaction between two electron transfer proteins generally involves three distinct steps: a) formation of a 1:1 substrate complex between the two proteins, b) electron transfer within the 1:1 complex, and c) dissociation of the product complex. Depending on conditions, any one of these three steps may be rate-limiting. A number of different techniques have recently been developed to study step b), the actual rate of electron transfer within the 1:1 bound complex. These include pulse radiolysis (McLendon and Miller, 1985) and flash photolysis utilizing flavins (Hazzard et al., 1987, 1988A,B) or zinc-substituted heme proteins (McLendon and Miller, 1985; Wallin et al., 1991; Everest et al., 1991). We have recently introduced a new method to study interprotein electron transfer that utilizes a trisbipyridine ruthenium complex attached to one of the proteins (Pan et al., 1988, 1990; Durham et al., 1989; Geren et al., 1991; Hahm et al., 1992). The Ru(II) complex can be photoexited with a short laser flash to form the metal-to-ligand charge-transfer state, Ru(II*), which is a strong reducing agent that can transfer an electron to the redox center of the covalently attached protein on a nanosecond time scale (Durham et al., 1989). The system is then poised to study intracomplex electron transfer between redox centers on a microsecond time scale. This method has a number of advantages. First, the pathway for reduction of the first redox center by Ru(II*) is well-defined, highly selective, and rapid compared to subsequent electron transfer steps. Second, the position of the ruthenium group can be designed so as not to interfere in the intracomplex electron transfer reaction between the two proteins. Third, the excited state redox chemistry of ruthenium complexes is highly flexible, allowing the properties to be tuned to the specific application needed. The new technique has been applied to measure intracomplex electron transfer reactions between cytochrome c and a number of physiological partners, including cytochrome $b_5$ (Willie et al., 1992, 1993), cytochrome c peroxidase (Geren et al., 1991; Hahm et al., 1992), cytochrome oxidase (Pan et al., 1993), cytochrome $c_1$ (Heacock et al., 1993), as well as the non-physiological partner plastocyanin (Pan et al., 1990).

We have developed a number of different strategies to covalently attach ruthenium groups to proteins. First, a novel two step method was developed to attach ruthenium trisbipyridine to lysine amino groups on cytochrome c (Pan et al., 1988). Over fifteen different singly labeled derivatives of cytochrome c have been purified and characterized. Second, a two-step method was developed to attach the ruthenium (bisbipyridine)(imidazole) group to histidine residues on proteins (Durham et al., 1990). Third, a new sulfhydryl-selective ruthenium reagent has been developed to attach the ruthenium trisbipyridine group to cysteine residues on proteins (Geren et al., 1989). The cysteine residue can be introduced into the desired location on the surface of the protein by site-directed mutagenesis techniques.

## THE CYTOCHROME $b_5$ - CYTOCHROME c COMPLEX

We have developed a general method for the *de novo* design and synthesis of a ruthenium-labeled redox protein that is optimized for the measurement of interprotein electron transfer. A mutant of cytochrome $b_5$ was prepared with a single cysteine at residue 65 (Stayton et al., 1988). The new sulfhydryl-selective reagent (Geren et al., 1991) was used to attach ruthenium to this cysteine residue, forming Ru-65-cyt $b_5$. The placement of the ruthenium complex at Cys 65 was designed to promote rapid electron transfer to the heme group of cytochrome $b_5$ but not interfere with electron transfer to cytochrome c. There is a direct 12-covalent-bond pathway for electron transfer between the ruthenium complex and the heme group of Ru-65-cyt $b_5$. We have recently measured the rate constant for photoinduced electron transfer between the Ru(II*) and the heme group Fe(III) to be $k_1 = 1.4 \pm .3 \times 10^7$ s$^{-1}$ and the rate of the thermal back reaction from the heme Fe(II) to Ru(III) to be $k_2 = 6.0 \pm .7 \times 10^6$ s$^{-1}$ (Scott et al., 1993).

The reaction between cytochrome $b_5$ and cytochrome c is a very favorable system for investigating fundamental questions about biological electron transfer. Using computer graphics techniques, Salemme (1976) proposed a model for the 1:1 complex between cytochrome $b_5$ and cytochrome c based on the X-ray crystal structures of the individual proteins (Takano and Dickerson, 1981; Argos and Mathews, 1975). This complex is stabilized by charge interactions between cytochrome c lysines 13, 27, 72, and 79, and cytochrome $b_5$ carboxylates on Glu 48, Glu 44, Asp 60, and the exposed heme propinate, respectively. The heme groups of the two proteins are nearly coplanar in the complex, with their edges separated by 8.4 Å. The general features of this model have been verified using a number of different experimental approaches (Willie et al., 1992, 1993).

Photoexcitation of a solution containing Ru-65-cyt $b_5$ and horse cytochrome c in low ionic strength buffer with a 450 nm laser flash resulted in rapid electron transfer from Ru(II*) to heme Fe(III) in Ru-65-cyt $b_5$, followed by electron transfer to the heme Fe(III) in cytochrome c (Fig. 1). Aniline was used as a sacrificial electron donor, D, to reduce Ru(III) to Ru(II) and prevent the back reaction between Ru(III) and heme Fe(II). The transient at 556.5 nm was biphasic, with a fast phase rate constant of $4 \pm 1 \times 10^5$ s$^{-1}$ and a slow phase rate constant of $3 \pm 1 \times 10^4$ s$^{-1}$. The two rate constants were independent of concentration, consistent with electron transfer within a 1:1 complex between Ru-65-cyt $b_5$ and cytochrome c with a dissociation constant of less than 5 μM. The rate constants decreased and the transients became monophasic as the ionic strength was increased, indicating dissociation of the complex. The ionic strength dependence of the second-order rate constant for the reaction of Ru-65-cyt $b_5$ with cytochrome c was nearly the same as for native bovine cytochrome $b_5$ measured by stopped-flow (Eltis et al., 1991), indicating that there are no significant conformational differences between the two proteins.

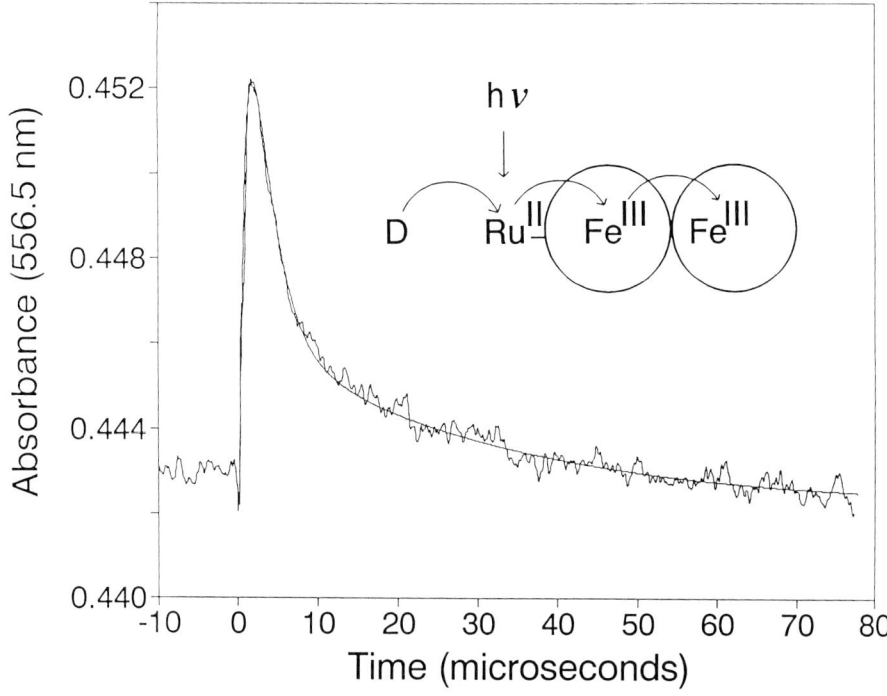

Fig. 1. Laser-induced electron transfer kinetics in a solution containing 20 μM Ru-65-cyt $b_5$, 26 μM cytochrome c, 1 mM sodium phosphate, pH 7, 10 mM aniline. Reproduced with permission from Willie et al., 1993.

The reactions of Ru-65-cyt $b_5$ with position 82 variants of yeast iso-1-cyt c have also been studied, as shown in Tables I and II from Willie et al. (1993).

Table I. Rate constants for electron transfer between Ru-65-cyt $b_5$ and cytochrome $c$ variants at low ionic strength. The buffer contained 2 mM N-phenyl glycine and 1 mM sodium phosphate.

| Cytochrome $c$ | $k_a$ (s$^{-1}$) | f | $k_b$ (s$^{-1}$) |
|---|---|---|---|
| horse | 4.0 x 10$^5$ | 0.80 | 3.0 x 10$^4$ |
| iso-1 | 1.0 x 10$^5$ | 0.70 | 1.8 x 10$^4$ |
| C102T | 1.1 x 10$^5$ | 0.60 | 2.0 x 10$^4$ |
| F82Y;C102T | 4.2 x 10$^4$ | 0.50 | 5.0 x 10$^3$ |
| F82G;C102T | 1.9 x 10$^5$ | 0.60 | 3.5 x 10$^3$ |
| F82L;C102T | 1.9 x 10$^5$ | 0.67 | 2.4 x 10$^3$ |
| F82I;C102T | 2.0 x 10$^5$ | 0.58 | 2.0 x 10$^3$ |

Table II. Rate constants and dissociation constants for the reactions of Ru-65-cyt $b_5$ with cytochrome $c$ variants at 50 mM ionic strength. The buffer contained 2.5 mM sodium phosphate, 5 mM N-phenyl glycine and 40 mM NaCl.

| Cytochrome $c$ | $k_c$ (s$^{-1}$) | K ($\mu$M) |
|---|---|---|
| iso-1 | 3.8 x 10$^4$ | 14 |
| F82Y;C102T | 3.1 x 10$^4$ | 16 |
| F82G:C102T | 3.9 x 10$^4$ | 13 |
| F82L;C102T | 5.0 x 10$^4$ | 12 |
| F82I;C102T | 4.5 x 10$^4$ | 11 |

The biphasic kinetics observed for the reaction of Ru-65-cyt $b_5$ with all of the variants of cytochrome $c$ at low ionic strength suggests that there are at least two different conformational forms of the complex. A mechanism for the kinetics is shown in Scheme I, where f and s stand for the "fast" and "slow" forms of the complex. The rate constant $k_a$ for the fast phase of the F82G;C102T variant is two-fold greater than that of wild-type iso-1-cytochrome $c$ at low ionic strength, while the slow phase rate constant $k_b$ is five-fold smaller (Table I). As the ionic strength increases, the fast phase rate constant decreases and the slow phase rate constant increases until a single phase is present at 50 mM ionic strength (Table II). This behavior is consistent with Scheme I if both the dissociation and interconversion rate constants increase with increasing ionic strength to satisfy the rapid equilibrium assumption at 50 mM ionic strength. The very dramatic increase in the rate constant of the slow phase with increasing ionic strength suggests that this phase might be controlled by conformational gating, with $k_b$ = $k_7$. These studies show that an aromatic side chain at residue 82 of cytochrome $c$ is not needed for rapid electron transfer with cytochrome $b_5$. This argues against the proposal of Wendoloski et al. (1987) that the phenyl group of Phe-82 acts as an electron transfer bridge between the two hemes in the complex.

$$b_5^o\text{-}c_f^o \xrightarrow{h\nu} b_5^r\text{-}c_f^o \xrightarrow{k_{etf}} b_5^o\text{-}c_f^r$$

$$k_3 \uparrow \downarrow k_4 \qquad\qquad k_7 \uparrow \downarrow k_8 \qquad\qquad \uparrow \downarrow$$

$$b_5^o\text{-}c_s^o \xrightarrow{h\nu} b_5^r\text{-}c_s^o \xrightarrow{k_{ets}} b_5^o\text{-}c_s^r$$

$$k_1 \uparrow \downarrow k_2 \qquad\qquad k_5 \uparrow \downarrow k_6 \qquad\qquad \uparrow \downarrow$$

$$b_5^o + c^o \xrightarrow{h\nu} b_5^r + c^o \qquad\qquad b_5^o + c^r$$

Scheme I

# THE CYTOCHROME c - CYTOCHROME c PEROXIDASE COMPLEX

The cytochrome c - cytochrome c peroxidase complex is an important paradigm for investigating fundamental questions about biological electron transfer. Poulos and Kraut (1980) proposed a hypothetical model for the 1:1 complex between tuna cytochrome c and cytochrome c peroxidase that was based on the crystal structures of the individual proteins. This model complex is stabilized by charge-pair interactions between lysines 13, 27, 72, 86, and 87 surrounding the heme crevice of cytochrome c and the carboxylate groups on Asp 34, 37, 79 and 216 on cytochrome c peroxidase. Very recently, Pelletier and Kraut (1992) determined the three-dimensional structure of a 1:1 complex between yeast cytochrome c peroxidase and yeast iso-1-cytochrome c crystallized at high ionic strength (150 mM NaCl, pH 7). The binding domain is different than that proposed in the Poulos-Kraut model, and hydrophobic and van der Waals interactions are particularly important in stabilizing the complex. Most importantly, the structure revealed a potential electron transfer pathway extending from the exposed heme methyl group CBC of cytochrome c through cytochrome c peroxidase residues Ala-194, Ala-193, and Gly-192 to the indole group of Trp-191, which is in van der Waals contact with the heme group. In the mechanism for cytochrome c peroxidase, the resting ferric state, CcP, is oxidized by hydrogen peroxide to CMPI, which contains an oxyferryl heme Fe(IV) and a radical on the indole ring of Trp 191 (Mauro et al., 1988; Sivaraja et al., 1989; Fishel et al., 1991). CMPI is then sequentially reduced to CMPII and CcP in two one-electron reactions each involving ferrocytochrome c (Kim et al., 1990). Coulson et al. (1971) first identified two forms of the singly oxidized state, CMPII(IV,R) containing the oxyferryl heme Fe(IV), and CMPII(III,R$^\bullet$) containing the radical. A number of different rapid kinetics techniques have been used to study the reaction between cytochrome c and CMPI. Summers and Erman (1988) and Hazzard et al. (1987, 1988A-C; 1991) reported that cytochrome c initially reduced the heme Fe(IV) site in CMPI at low ionic strength using stopped-flow and flavin flash photolysis techniques, respectively. In contrast, Geren et al., (1991) and Hahm et al. (1992) found that five different ruthenium-labeled horse and yeast cytochrome c derivatives reacted first with the radical in CMPI and then with the Fe(IV) in CMPII at pH 7 according to Scheme II:

$$\text{CMPI(IV,R}^\bullet\text{)} + \text{cyt c}^{2+} \rightarrow \text{CMPII(IV,R)} + \text{cyt c}^{3+} \quad (1)$$
$$\text{CMPII(IV,R)} + \text{cyt c}^{2+} \rightarrow \text{CcP(III,R)} + \text{cyt c}^{3+} \quad (2)$$

Scheme II

The rate constant for the intracomplex reaction with the radical at low ionic strength was very large, 55,000 s$^{-1}$ for the cytochrome c derivative labeled at lysine 27. Hahm et al. (1993) have recently found that native horse cytochrome c also obeys this mechanism at high ionic strength and pH 7, and have developed a stopped-flow protocol to independently measure the second-order rate constants of reactions (1) and (2).

We have recently investitated the reactions of cytochrome c with a wide range of cytochrome c peroxidase mutants. These include mutants of charged residues that are potentially involved in the interaction with cytochrome c (such as Asp-32, Glu-33, Asp-34, Asp-35, and Glu-290), mutants of groups that might be involved in the electron transfer pathway (such as Ala-193, His-181, Tyr-39, Tyr-42, Tyr-229, and Trp-223), and mutants that could affect the equilibrium between the radical site at Trp-191 and the oxyferryl heme Fe(IV) site (such as Met-230). Kinetic studies with these mutants support the Pelletier-Kraut pathway for electron transfer (Fig. 2).

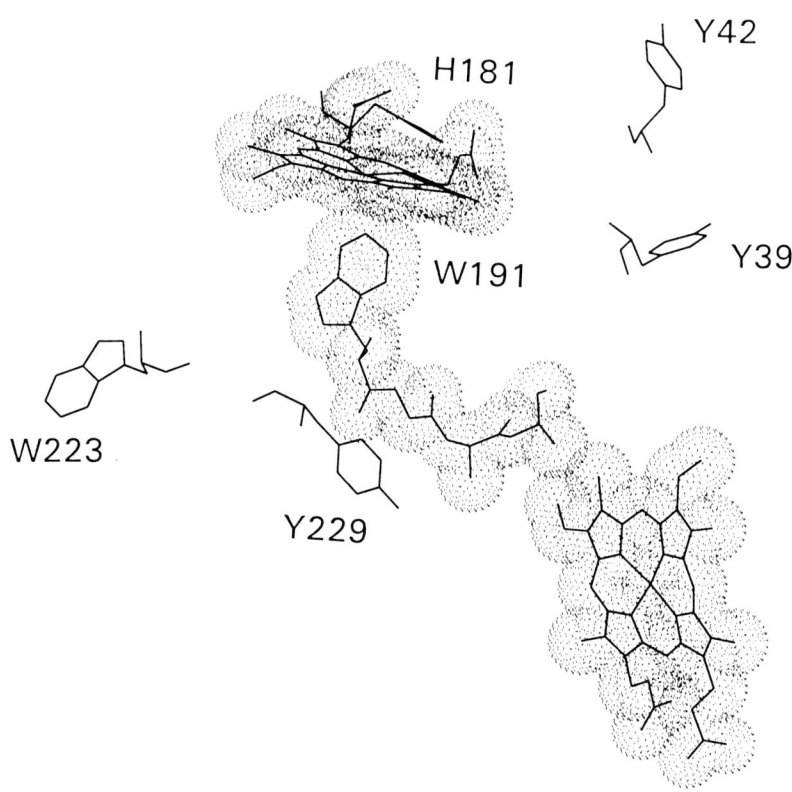

Fig. 2. X-ray crystal structure of the complex between yeast cytochrome c peroxidase and yeast iso-1-cytochrome c (Pelletier and Kraut, 1992). The heme groups and residues Trp-191, Gly-192, Ala-193, and Ala-194 that are proposed to be involved in the electron transfer pathway are enclosed in Connolly surfaces.

Acknowledgments: This work was supported by NIH grants GM20488 (F. M and B. D.), GM33775 (S. S.) and GM31756 (S. S.).

# References

Argos, P., & Mathews, F. S. (1975) J. Biol. Chem. 250 747-751.

Coulson, A. J. W., Erman, J. E., and Yonetani, T. (1971) J. Biol. Chem. 246 917-924

Durham, B., Pan, L. P., Long, J., and Millett, F. (1989) Biochemistry 28 8659-8665

Durham, B., Pan, L. P., Hahm, S., Long, J., and Millett, F. (1990) Adv. Chem. 226 181-196.

Eltis, L. D., Herbert, R. G., Barker, P. D., Mauk, A. G., & Northrup, S. H. (1991) Biochemistry 30 3663-3674.

Everest, A. M., Wallin, S. A., Stemp, E. D. A., Nocek, J. B., Mauk, A. G., and Hoffman, B. M. (1991) J. Am. Chem. Soc. 113 4337-4338

Fishel, L. A., Farum, M. F., Mauro, J. M., Miller, M. A., Kraut, J., Liu, Y., Tan, X., & Scholes, C. P. (1991) Biochemistry 30 1986-1996.

Geren, L. M., Hahm, S., Durham, B., and Millett, F. (1991) Biochemistry 30 9450-9457

Hahm, S., Durham, B., & Millett, F. (1992) Biochemistry 31 3472-3477.

Hahm, S., Geren, L., Durham, B., and Millett, F. (1993) J. Am. Chem. Soc. 115 3372-3373.

Hazzard, J.T., Poulos, T., and Tollin, G. (1987) Biochemistry 26 2836-2848

Hazzard, J. T., McLendon, G., Cusanovich, M. A., and Tollin, G. (1988A) Biochem. Biophys. Res. Comm. 151 429-434

Hazzard, J.T., Moench, S.J., Erman, J. E., Satterlee, J.D. and Tollin, G. (1988B) Biochemistry 27 2002-2008

Hazzard, J.T., Mclendon, G., Cusanovich, M.A., Das, G., Sherman, F., and Tollin, G. (1988C) Biochemistry 27 4445-4451.

Hazzard, J. T., and Tollin, G. (1991) J. Am. Chem. Soc. 113 8956-8957.

Heacock, D., Liu, R.-Q., Yu, C.A., Yu, L., Durham, B., and Millett, F. (1993) J. Biol. Chem. (in press).

Kim, K. L., Kang, D. S., Vitello, L. B. and Erman, J. E. (1990) Biochemistry 29 9150-9159

Mauro, J. M., Fishel, L. A., Hazzard, J. T., Meyer, T. E., Tollin, G., Cusanovich, M. A., and Kraut, J. (1988) Biochemistry 27 6243-6256

McLendon, G., & Miller, J. R. (1985) J. Am. Chem. Soc. 107 7811-7816.

Pan, L. P., Durham, B., Wolinska, J., and Millett, F. (1988) Biochemistry 27 7180-7184

Pan, L. T., Frame, M., Durham, B., Davis, D., & Millett, F. (1990) Biochemistry 29 3231-3236.

Pan, L. P., Hibdon, S., Liu, R.-Q., Durham, B., and Millett, F. (1993) Biochemistry 32 8492-8498.

Pelletier, H. and Kraut, J. (1992) Science 258 1748-1755.

Poulos, T. L., and Kraut, J. (1980) J. Biol. Chem. 255 10322-10330

Salemme, F. R. (1976) J. Mol. Biol. 102 563-568.

Scott, J. R., Willie, A., McLean, M., Stayton, P. S., Sligar, S. G., Durham, B., & Millett, F. (1993) J. Am. Chem. Soc. (in press)

Sivaraja, M., Goodin, D. B., Smith, M., And Hoffman, B. M. (1989) Science 245 738-740

Stayton, P. S., Fisher, M. T., & Sligar, S. G. (1988) J. Biol. Chem. 263 13544-13548.

Summers, F., E., and Erman, J. E. (1988) J. Biol. Chem. 263 14,267-14,275

Takano, T, and Dickerson, R. E. (1981) J. Mol. Biol. 153 79.

Wallin, S. A., Stemp, E. D. A., Everest, A. M., Nocek, J. M., Netzel,, T. L., and Hoffman, B. M. (1991) J. Am. Chem. Soc. 113 1842-1844

Wendoloski, J. J., Matthew, J. B., Weber, P. C. & Salemme, F. (1987) Science 238 794-797.

Willie, A., Stayton, P. S., Sligar, S. G. Durham, B., & Millett, F. (1992) Biochemistry 31 7237-7242.

Willie, A., McLean, M., Liu, R.-Q., Hilgen-Willis, S., Sanuders, A., Pielak, G. J., Sligar, S. G., Durham, B., and Millett, F. (1993) Biochemistry 32 7519-7525.

# Studies on electron transfer pathways in cytochrome P450 systems

R. Bernhardt[1], V. Beckert[1], H. Uhlmann[1], S.G. Sligar[2]

[1]Max Delbrück Center for Molecular Medicine, D-13125 Berlin-Buch, Robert-Roessle Str. 10, FR Germany. [2]School of Chemical Sciences, University of Illinois, Urbana, IL, 61801-3792, USA

Cytochrome P450 protein docking and electron transfer processes require clarification (Sligar et al., 1974; Bernhardt et al., 1988; Stayton et al., 1989). More than 200 amino acid primary structures differing in substrate pocket, reactivities, and two classes of reduced pyridine nucleotide redox energy coupling enzymes are known, which roughly fall into two classes (Nelson et al., 1993).

P450LM2 (CYP2B4) (EC 1.14.14.1) represents the microsomal monooxygenases, with a NADPH reducible flavoprotein (EC 1.6.2.4) containing FAD/FMN as electron donor, whereas the mitochondrial and most procaryote monooxygenases, including P450cam (CYP101), require two redox proteins: an iron sulfur redoxin that connects a flavoprotein, containing a single FAD, to the heme protein.

In previous studies we (Bernhardt et al., 1983, 1984, 1988) applied the protein modification reagent FITC to the microsomal P450LM2 and reported a reaction at M1 and K384, leading to an interference with electron transfer from the microsomal reductase. Furthermore the presence of the reductase protected both M1 and K384 from chemical modification, suggesting the location of both residues within the reductase binding site (Bernhardt et al., 1988). From these data and the observation that carboxyl groups of reductase, on the other hand, are involved in binding of P450LM2 (Bernhardt et al., 1987), we concluded that salt bridges play an essential role in mutual recognition between P450 and its electron donor (Bernhardt et al., 1984, 1988).

Since then the existence of salt bridges has been shown between evolutionary distinct cytochromes P450 and their electron donors (TABLE 1). Interestingly, there is not only the mechanism of recognition between P450 and the electron donor conserved, but there seems to exist a conserved binding site on evolutionary distinct P450s for their respective electron donor as shown by heterologous reconstitution of CYP2B4 activity using the bacterial electron donors putidaredoxin (Pd) and putidaredoxin reductase (PdR) or linredoxin and linredoxin reductase instead of microsomal reductase (Bernhardt & Gunsalus, 1992). This result makes the bacterial P450cam, which 3D-structure became available nearly a decade ago

(Poulos et al., 1985), a good model for studying the mechanism of electron transfer in P450s and for drawing from this conclusions to other P450 systems.

X-ray crystallographic data have shown that the heme in P450cam is deeply embedded into the protein being not accessible from the surface (Poulos et al., 1985). Thus, no direct pathway from the prosthetic group of the electron donor to the prosthetic group of P450cam seems to be available for the electron to be transfered, but a pathway via protein groups of at least P450cam is necessary. The role of the protein medium in directly modulating biological electron transfer processes has been the subject of intriguing but relatively few experimental and theoretical investigations.

TABLE 1  Salt bridges between amino groups of P450 and carboxyl groups of of the electron donor

| P450 | electron donor | references |
|---|---|---|
| chemical modification | | |
| 2B4 | microsomal reductase | Bernhardt et al., 1984 |
| 1A1 | microsomal reductase | Shen & Strobel, 1992 |
| 2B1 | microsomal reductase | Shen & Strobel, 1993 |
| 11A1 | adrenodoxin | Tuls et al., 1989 |
|  |  | Tsubaki et al., 1989 |
| cross-linking experiments | | |
| 2B4 | microsomal reductase | Tamburini et al., 1986 |
|  |  | Bernhardt et al., 1987 |
| 11A1 | adrenodoxin | Hara & Miyata, 1990 |
| site-directed mutagenesis | | |
| 101 | putidaredoxin | Stayton & Sligar, 1990 |
| 1A2 | microsomal reductase | Shimizu et al., 1991 |
| 11A1 | human ferredoxin, adrenodoxin | Coghlan & Vickery, 1991 |
|  |  | Wada & Waterman, 1992 |

At present there are three main models how the electron transfer process proceeds in biological systems. One model is assuming aromatic residues to play a key role in the transmission of the electron (cf. Veitch & Williams 1992). There are only very few experimental data trying to check this model. So far cytochrome c seems to be one of the few examples where the hypothesis of involvement of aromatic side chains into electron transfer has been experimentally tested (Liang et al., 1988; Everest et al., 1991; Ingln et al., 1991; Willie et al., in press). Further studies using other proteins seem to be necessary to elucidate the function of aromates in this process unambigously. The second mechanism proposes an uniform one-dimensional square tunneling barrier model for electron tunneling with $ß=1.4$ $A^{-1}$ and according to Moser et al. (1992) describes a broad range of natural and synthetic electron transfer pathways. Finally, a specific through bond pathway model for electron transfer was developed, which accounts for unique covalent, hydrogen bonded, and Van der Waals contacts linking donor and acceptor in a protein. This model gives a

consistent description of electron transfer rates in ruthenated proteins (cytochrome c, myoglobin, cytochrome b5) (Beratan et al. 1990, 1991, 1992; Onuchic & Beratan, 1990).
There is so far very little known on the role of aromatic residues for P450-dependent catalysis. In order to ascertain their function in the electron transfer within P450cam or in the conformational gating process, we have mutated Tyr 75 and Phe 81, located between the recognition site for the electron donor, Pd, and the active site, heme, and determined fundamental rates and equilibria of the cycle. The data obtained were compared to calculated pathways using the model of Beratan and coworkers (1990, 1991). In addition, the unique tyrosine of bovine adrenodoxin (Adx), Y82, which has been suggested to be involved in electron transfer in the mitochondrial steroid hydroxylases in previous studies (Taniguchi & Kimura, 1975), was replaced by phenylalanine, serine and leucine and the effect of the mutations has been investigated.

## MATERIALS AND METHODS

Most of the enzymes used were obtained from Boehringer Mannheim. *Taq* DNA polymerase was from Amersham-Buchler KG. The steroids used were from Sigma. Other chemicals were purchased from commercial sources.
*Site-directed mutagenesis* - The P450cam single mutants Y75S, F81I and the double mutants Y75F/F81Y and Y75S/F81I as well as mutants Y82F, Y82S and Y82L of Adx were obtained using PCR. The PCR primers were synthesized by BioTez GmbH. PCR was carried out according to Sambrook et al. (1989) and Uhlmann et al. (1992) under optimal conditions for primer annealing. Ligation, transformation, plasmid preparation, and dideoxy sequencing of the plasmids were performed using standard protocols (Sambrook et al. 1989).
*Protein purification* - P450cam and the P450 mutants were purified from recombinant Escherichia coli cells according to the protocol previously described by Gunsalus & Wagner (1978) with modifications according to Martinis (1990). Pd and PdR are purified from recombinant E. coli cells according to Gunsalus & Wagner (1978) and Davies et al. (1990). Adx and its mutants were purified from E. coli according to Beckert et al. (1993); AdR, CYP11A1 and CYP11B1 were isolated from bovine adrenals as previously described (Akhrem et al., 1979).
*Enzyme assays* - The activities of the P450cam and Adx mutants were measured in reconstituted systems as described (Gunsalus & Wagner, 1978; Beckert et al., 1993).
*Stopped-flow kinetic measurements* were performed essentially as described in Davies & Sligar (1992). Solutions and proteins used in the kinetic experiments were thoroughly evacuated by gradually applying a vacuum of about 1 torr until no bubbles were evident for at least 5 min and then purged with argon. All experiments were carried out in an anaerobic chamber. To remove residual oxygen from the reaction mixtures an oxygen-scavenging system, consisting of 2 mM glucose, 0.1 mg/ml glucose oxidase and 3000 units/ml catalase, was added. Five to six traces (0.2 ml total volume per shot) were acquired per concentration of Pd.
*Electron-tunneling pathway calculations* - To survey P450cam for electron tunneling pathways the numerical algorithm for the calculation of long-range electron transfer pathways developed by Beratan & Coworkers (Beratan et al., 1990; Onuchic & Beratan, 1990) was used. The calculations were performed with the software package PATHWAYS II.

## RESULTS AND DISCUSSION

There are so far only a few attempts to study the potential role of aromatic residues of cytochromes P450 in electron transfer (Gotoh & Fujii-Kuriyama, 1989; Furuya et al., 1989), in contrast to studies on the role of aromatic residues in substrate binding and specificity (Poulos et al., 1985; Furuya et al., 1989; Lindberg & Negishi, 1989). Our experimental results and data from others suggest that some of the most conserved aromatic residues in the P450 superfamily seem to have a function for the structural integrity of the proteins rather than in electron transfer. This concerns F81 of P450cam, the replacement of which leads to the formation of the enzymatically inactive P420 (TABLE 2), although it is located 20.6 A away from the heme in a hydrophobic patch, together with Y78 and Y305.

TABLE 2    Effect of Y75 replacement of P450cam

| Mutant | CO-difference spectrum |
|---|---|
| Y75S | P450 |
| F81I | P420 |
| Y75F, F81Y | P420 |
| Y75S, F81I | P420 |

In contrast, replacement of Y75 by serine leads to an active P450cam (TABLE 2). As expected, this replacement did not cause any changes in product pattern, stoichiometry of reducing equivalents used to product formed or D (+) camphor and Pd binding (data not shown), indicating that the conformation of the protein has not been significantly changed by the replacement. The activity of substrate conversion, however, decreased by about 30 %, due to a decrease in Vmax. The diminuation of the activity has been shown to arise from hindrance of first electron introduction as traced by a decrease in $k_{obs}$ from 60.8 $sec^{-1}$ for the wild type protein to 50.0 $sec^{-1}$ for the Y75S mutant. This alteration is relatively small and thus does not indicate on an essential role of Y75 in the intracomplex electron transfer between P450 and Pd. However, the replacement of Y75 seems to modulate the electron transfer rate. This could be due to small conformational changes in the intervening (involved in electron transfer) sequence, which are induced by the replacement.

Possible pathways of the electron from the surface of the protein to the heme can be calculated using the model of Beratan and coworkers (Beratan et al., 1990, 1991). To check whether Y75 is close to the intervening sequence involved in electron transfer, these pathways were calculated using the PATHWAYS II software package. The most coupled pathway is depicted in Fig. 1. It can be seen that the pathway leads from the glutamine 360 side chain through the backbone of Q 360, followed by a 3.4 A through-space jump to the sulfur of C 357 and this way directly to the heme iron. Y75 is located within about 10 A of this pathway. Thus, its replacement may cause disturbances of this pathway region so that especially the through-space jump could be affected. Another possible explanation of the effect of Y75S replacement on the electron transfer comes from a more dynamic view on the P450 structure and electron transfer and suggests different pathways to work at about the same time, in dependence on the structural motions of the protein. Y75 could be involved in one of the less favorable pathways.

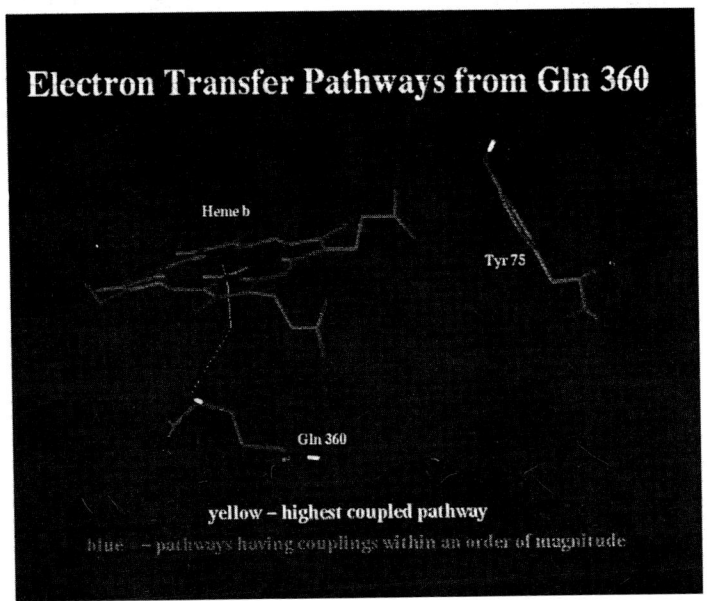

Fig. 1

Electron tunneling pathway from the strongest coupled residue of P450$_{cam}$, Gln360, to the heme. Data were calculated according to beratan et al. (1990, 1991).

The direct involvement of an aromatic residue in the electron transfer process in P450 systems was further investigated using mitochondrial Adx. In previous studies to elucidate the function of Y82 mainly chemical modification was used. Modification of bovine Adx with tetranitromethane lead to a decrease in cytochrome c reduction to 19 % of the control for the nitrotyrosine- and to 7 % for the aminotyrosine-derivative. From these data a participation of Y82 in electron transfer was suggested (Taniguchi and Kimura, 1975).
To check this the residue in our studies was replaced by phenylalanine, serine and leucine and the effect of these mutations has been studied with respect to intra- or intermolecular electron transfer by detailed structural and functional characterization of the mutant proteins. As concluded from unchanged absorption, CD and EPR spectra as well as redox potentials, introduction of phenylalanine, leucine or serine instead of tyrosine into position 82 did not lead to any changes

of the microenvironment of the /2Fe-2S/ cluster. Thus, Y82 is obviously not located in the immediate iron-sulfur cluster environment. This location is consistent with data from $^1$H-NMR studies, suggesting that Y82 does not interact with the iron-sulfur center although it is at most 11-13 Å away from it (Greenfield et al., 1989). However, although Y82 is not located in the immediate vicinity of the iron-sulfur cluster, an involvement in electron transfer cannot be excluded on the basis of these data only. Thus, the participation of Y82 of Adx in intra- or intermolecular electron transfer from AdR to P450 was tested by analyzing the influence of replacement of this amino acid on the efficiencies of cytochrome c reduction and P450-dependent substrate conversion.

No influence of the replacement of Y82 on $V_{max}$ in the cytochrome c reduction has been observed (TABLE 3). Furthermore, the P45011A1-dependent conversion of cholesterol to pregnenolone and the P45011B1-dependent formation of corticosterone from deoxycorticosterone did not show any significant changes of the $V_{max}$ values when using Y82 mutants instead of wild type Adx as electron

TABLE 3   Influence of Y82 replacement of efficiencies of cytochrome c, P450 11A1 and P450 11B1-dependent reactions

| Adx | $V_{max}$-values | | |
|---|---|---|---|
| | Cyt c reduct. assay (nmol Cyt c/min) | 11ß-hydrox. activity (min$^{-1}$) | cholesterol scc-act. (min$^{-1}$) |
| wild type | 24.5 ± 0.6 | 15.8 ± 1.5 | 5.1 ± 0.1 |
| Y82F | 25.5 ± 0.5 | 17.3 ± 0.5 | 5.2 ± 0.1 |
| Y82S | 25.8 ± 1.0 | 16.6 ± 0.5 | 4.8 ± 0.2 |
| Y82L | 25.7 ± 0.5 | 15.7 ± 1.0 | 4.9 ± 0.2 |

mediator (TABLE 3). These results indicate that Y82 does not play a significant role in intra- or intermolecular electron transfer.

Taken together, these experimental results show that tyrosine residues, at least Y75 of P450cam and Y82 of bovine Adx, are not involved in intra- or intermolecular electron transfer in cytochrome P450 systems. Instead, it seems to be probable that the electron is transferred via distinct pathways which can be calculated using the PATHWAY II program developed by Beratan and coworkers (1990, 1991). Further studies, especially the cocrystallization of electron transfer complexes of P450 systems, are necessary to experimentally confirm these theoretical predictions.

*Acknowledgments* - We thank D. Benson for preparing the computer graphics using PathwayII. The work is supported by DFG grants Be 1343/1-2 and 1343/2-1. H.U. is recipient of a grant from the Boeringer Ingelheim Fonds.

# REFERENCES

Akhrem, A.A., Lapko, V.N., Lapko, A.G., Shkumatov, V.M., Chashchin, V.L. (1979): Acta biol. med. germ. 38, 257-274.
Beckert, V., Dettmer, R., Bernhardt, R.; J. Biol. Chem., in press.
Beratan, D. N., Betts, B. E., Onuchic, J. N. (1991): Science 252, 1285-1288.
Beratan, D. N., Onuchic, J. N., Betts, B. E., Bowler, Gray, H. B. (1990): J. Am. Chem. Soc. 112, 7915-7921.
Beratan, D. N., Onuchic, J. N., Winkler, J. R., Gray, H. B. (1992) Science 258, 1740-1741.
Bernhardt, R., Ngoc Dao, N. T., Stiel, H., Schwarze, W. Friedrich, J., Jänig, G.-R., Ruckpaul, K. (1983): Biochem. Biophys. Acta 745, 140-148.
Bernhardt, R., Makower, A., Jänig, G.-R., Ruckpaul, K. (1984): Biochem. Biophys. Acta 785, 186-190.
Bernhardt, R., Pommerening, K., Ruckpaul, K. (1987): Biochem. Int. 14, 823-832.
Bernhardt, R., Kraft, R., Otto, A., Ruckpaul, K. (1988): Biomed. Biochim. Acta 47, 581-592.
Coghlan, V.M., Vickery, L.E. (1991): J. Biol. Cehm. 266, 18606-18612.
Davies, M. D., Qin, L., Beck, J. L., Suslick, K. S., Koga, H., Horiuchi, T., Sligar, S. G. (1989): J. Am. Chem. Soc. 112, 7396-7398.
Davies, M. D., Sligar, S. G. (1992): Biochemistry 31, 11383-11389.
Everest, A. M., Wallin, S. A., Stemp, E. D., Nocek, J. M., Mauk, A. G., Hoffman, B. (1991): J. Am. Chem. Soc. 113, 4337-4338.
Furuya, H., Shimizu, T., Hirano K., Fujii-Kuriyama, Y. (1989): Biochemistry 28, 6848-6857.
Gotoh, O., Fujii-Kuriyama, Y. (1989): in Frontiers in Biotransformation (Ruckpaul, K., Rein, H., Eds.) pp 195-243, Akademie-Verlag, Berlin.
Greenfield, N.J., Wu, X., Jordan, F. (1989): Biochim. Biophys. Acta 995, 246-254.
Gunsalus, I. C., Wagner, G. C. (1978): Methods Enzymol. 52, 166-188.
Hara, T., Chiyata, T. (1990): Abstrtacts of the VIIIth International Symposium on Microsomes and Drug Oxidations, Stockholm, 25.-29.6.90.
Ingln, S. C., Guillernette, J. G., Johnson, J. A., Smith, M. (1991): Protein Eng. 4, 569-574.
Liang, N., Mauk, A.G., Pielak, G. J., Johnson, J. A., Smith, M., Hoffman, B. M. (1988): Science 240, 311-313.
Lindberg, R.L.P., Negishi, M. (1989): Nature 339, 632-634.
Martinis, S. A. (1990): PhD Thesis, University of Illinois, Urbana.
Moser, C.C., Keske, J.M., Warneke, Farid, R.S., Dutton, P.L. (1992): Nature 355, 796-802.
Onuchic, J.N., Beratan, D.N. (1990): J. Chem. Phys. 92, 722-733.
Poulos, T. L., Finzel, B., Gunsalus, I. C., Wagner, G. C., Kraut, J. (1985): J. Biol. Chem. 260, 16122-16130.
Sambrook, J., Fritsch, E.F., Maniatis, T. (1989): Molecular Cloning: A Laboratory Manual, 2nd Ed. Cold Spring Harbo Laboratory, Cold Spring Harbor, NY.
Shen, S.J., Strobel, H.W. (1993): Arch. Biochem. Biophys. 304, 257-265.
Shen, S.J., Strobel, H.W. (1992): Arch. Biochem. Biophys. 294, 83-90.

Shimizu, T., Tateishi, T., Hatano, M., Fujii-Kuriyama, Y. (1991): J. Biol. Chem. 266, 3372-3375.
Sligar, S. G., Debrunner, P. G., Lipscomb, J. D., Namvedt, M. J., Gunsalus, I. C. (1974): Proc. Natl. Acad. Sci. U.S.A 71, 3906-3910.
Stayton, P. S., Poulos, T. L., Sligar, S. G. (1989): Biochemistry 28, 8201-8205.
Stayton, P. S., Sligar, S. G. (1990): Biochemistry 29, 7381-7386.
Tamburini, P.P., MacFarquhar, S., Schenkman, J.B. (1986): Biochem. Biophys. Res. Commun. 134, 519-526.
Taniguchi, T., Kimura, T. (1975): Biochemistry 14, 5573-5578.
Tuls, J., Geren, L., Millett, F. (1989): J. Biol. Chem. 264, 16421-16425.
Tsubaki, M., Iwamoto, Y., Hiwatashi, A., Ichikawa, Y. (1989): Biochemistry 28, 6899-6907.
Uhlmann, H., Beckert, V., Schwarz, D., Bernhardt, R. (1992): Biochem. Biophys. Res. Commun. 188, 1131-1138.
Veitch, N.C., Williams, R.J.P. (1992): in frontiers in Biotransformation (Ruckpaul, K., Rein, H., Eds.), Akademie-Verlag, pp 279-320.
Wada, A., Waterman, M.R. (1992): J. Biol. Chem. 267, 22877-22882.
Willie, A., McLean, M., Liu, R. Q., Pielak, G. J., Sligar, S. G., Durham, B., Millett, F.; J. Am. Chem.in press.

# Electron transfer and protein-protein interactions in soluble reconstituted liver monooxygenase systems*

Galina I. Bachmanova, Irina P. Kanaeva, Irina F. Sevrukova, Olga V. Nikityuk, Natalia V. Stepanova, Tatiana V. Knushko, Yakov M. Koen, Alexander I. Archakov

Institute of Biomedical Chemistry, Pogodinskaya str. 10, Moscow, Russia

## Summary

The kinetic parameters of NADPH-dependent cytochrome P450-reductase and monooxygenase reactions of two reconstituted soluble systems (RSS), consisting of rabbit liver NADPH-cytochrome P-450 reductase (R), cytochromes P-450 2B4 (2B4) or P-450 1A2 (1A2) and Emulgen 913 were compared. The monomers of 2B4 and 1A2 are obtained at protein to detergent molar ratios 1:100 and 1:3200, respectively. Monomers of 2B4 and R were active in NADPH-dependent N-demethylation reactions. On the contrary, the monomers of 1A2 and R were inactive in 7-ethoxyresorufin deethylation reaction (7-EROD). Maximal 7-EROD activity was observed in RSS with 1A2 pentamers and R dimers. In RSS, 2B4 monomers reduction did not depend on initiation reaction mode, while 1A2 reduction parameters were higher when the reaction was initiated by NADPH. The $K_d$ value for the complex of monomeric R and 2B4 was 6.4 μM based on kinetic data (Kanaeva et al., 1992b), while $K_d$ value for the complex of dimeric R and pentameric 1A2 was 1 μM based on spectral data (Sevrukova et al., 1993). Using cross-linking agent carbodiimide (EDC) the binary equimolar complex of R and 1A2, R and cytochrome b5 (b5), 2B4 and b5, 1A2 and b5 but not R and 2B4 was found electrophoretically. When tryptic catalytical fragments of R and b5 were used, there was no complex formation with their partners as well as reduction and monooxygenation reactions did not occur. The study of the effect of medium ionic strength indicated that electrostatic forces play a role in flavo- and hemoprotein interaction. Thus, the results obtained provide evidence for the important role of hydrophobic membrane domains in protein-protein interaction. It means that hydrophobic tails of R and b5 are of important role in interaction of both ones with 2B4 and 1A2.

*This work has been supported by grant from Russian Foundation of Fundamental Research (93-04-6227)

INTRODUCTION

The possibility of reconstitution of microsomal monooxygenase systems in solution from R and 2B4 or 1A2 in the presence of Emulgen 913 was shown by Kanaeva et al. (1992a,b) and Sevrukova et al. (1993). These RSS were able to catalyze hemoprotein reduction and oxygenation reactions with different rates depending on detergent concentrations. The components of RSS with 2B4 interacted by random collisions in accordance with mass action law with maximal activity at the 1:1 R and 2B4 stoichiometry (Kanaeva et al., 1992b). On the contrary, the system containing monomers of 1A2 and R possessed very low activity (Sevrukova et al., 1993). The maximum activity was observed for RSS with 1A2 pentamers and R dimers. The role of hydrophobic tails of R and b5 was investigated in respect of their involving in interaction with 2B4 and 1A2. Their absolute necessity for complex formation with their partners was shown.

MATERIALS AND METHODS

Phenobarbital- and methylcholanthrene-treated rabbit liver microsomes (MC), 2B4, 1A2, R, b5, were obtained and monomerization of 2B4 and R was carried out as described earlier (Kanaeva et al., 1992a,b; Sevrukova et al., 1993). Incubation of oligomers of isolated 1A2 with Emulgen 913 was carried out as follows: to 5 nmole of each protein in 20-30 µl 100 mM K-phosphate buffer, pH 7.5, was added such volume of 2% (w/v) Emulgen 913 so after the following dilution the necessary concentration of the detergent was obtained. After incubation for 5-10 min at room temperature, the concentration of the proteins was brought to 5 µM with buffer and the protein-detergent mixture was incubated at 4° C for another 24 h. Gel filtration studies of 1A2 and R and their mixture were performed as described previously (Sevrukova et al., 1993). The determination of NADPH-dependent reduction of 2B4 or 1A2 and their activities, the treatment of proteins by EDC have been performed as described previously (Kanaeva et al., 1992a,b; Sevrukova et al., 1993). Purified tryptic fragments of R and b5 were obtained from rabbit liver microsomes after treatment with trypsin (Omura and Takesue, 1970).

RESULTS AND DISCUSSION

Two liver monooxygenase soluble systems containing R and 2B4 or 1A2 in the presence of nonionic detergent Emulgen 913 were reconstituted. Different action of detergent on aggregate states of 2B4 and 1A2 by gel filtration method was found (Table 1). 2B4 monomers were obtained at 0.25 g/l Emulgen concentration (Kanaeva et al., 1992 a). At the same time, 70% of 1A2 was monomerized at high detergent concentration - 8 g/l (Sevrukova et al., 1993). In RSS with 2B4 the maximum catalytic activity was observed when R and 2B4 existed as monomers (Kanaeva et al., 1992b), and in the case of RSS with 1A2 - when R dimers and 1A2 pentamers were presented at Emulgen concentration of 0.1 g/l (Sevrukova et al., 1993). It was shown that RSS containing R and 1A2 monomers possessed low catalytic activity (150 and 10 pmole of resorufin per nmole of 1A2 per min for pentamers and monomers, respectively). The study of binding spectra of 1A2 aggregates, pentamers and monomers with 7-ethoxyresorufin (7-ER), R, and b5 indicated that 1A2 was able to bind of the substrate and both partners (data not shown) at every case (Sevrukova et al., 1993). The spectral $K_d$ values for complex of 1A2 with 7-ER and with b5 did not depend on 1A2 aggregate state. Monomeric 2B4 has been shown also to form complexes with substrate (Kanaeva et al., 1992a) and b5 (Kanaeva et al., 1992b). In RSS, 2B4 and 1A2 have different affinity to R. Contrary to 1A2, the $K_d$ value for the complex of monomeric 2B4 with R was calculated from kinetic data, but not from spectral ones and was equal to 6.4 μM (Kanaeva et al., 1992b). The spectral $K_d$ value for the complex of 1A2 with R was minimum for 1A2 pentamers and was equal to 1 μM. At the same time with 1A2 monomers the $K_d$ value for the complex of 1A2 with R was 3 and 6 times higher compared those with aggregates and pentamers, respectively. It means that high concentration of detergent may interfere to the interaction of R and 1A2.

TABLE I.

The effect of Emulgen 913 concentration on aggregate states of 2B4 and 1A2

| Emulgen 913 g/l | 2B4 | | 1A2 | | |
|---|---|---|---|---|---|
| | Mol. weight kDa | Subunits[a] number | Mol. weight kDa | Subunits[a] number | % |
| 0 | 700 | 11 | 2500 | 40 | 35 |
| | | | 500 | 8 | 65 |
| 0.05 | 270 | 4 | n.d.[b] | | |
| 0.1 | n.d. | | 2500 | 40 | 5 |
| | | | 300 | 5 | 95 |
| 0.25 | 65 | 1 | 2500 | 40 | 5 |
| | | | 270 | 4 | 95 |
| 8 | n.d. | | 2500 | 40 | 30 |
| | | | 62 | 1 | 70 |

[a] Calculated from the molecular weights of 65 kDa for 2B4 and from 62 kDa for 1A2.
[b] n.d. - not determined

The significant differences in NADPH-dependent reduction of 2B4 and 1A2 were observed. In the case of RSS with monomeric 2B4, the kinetic parameters of reduction reaction did not depend on the mode of mixing of the components (Kanaeva et al., 1992b), while in 1A2-containing system its reduction depended on initiation reaction mode and was maximum when initiating of reaction by NADPH (Sevrukova et al., 1993) (Table II). In both RSS in the absence of substrate the reactions obeyed monophasic kinetics. In RSS with 1A2 the rate constants were maximum at Emulgen concentration of 0.1 g/l in the absence of 7-ER. In both RSS the addition of substrate turned the monophasic kinetics into biphasic, increasing the rate constants and reduction levels. In the case of 1A2 reduction when the reaction was initiated by R, the kinetics remained monophasic (Sevrukova et al., 1993). In RSS with 1A2, the dependence of the kinetic parameters on the order of reagent mixing, indicates that the rate of the first electron transfer in the fast phase is limited by the protein interaction.

TABLE II.

Kinetic parameters of NADPH-dependent reduction of 1A2 in reconstituted system in the presence of 7-ER.

| System | Initiation by NADPH | | | | Initiation by R | |
|---|---|---|---|---|---|---|
| | $k_f$[a] $s^{-1}$ | $k_s$[a] $s^{-1}$ | Reduction level for 1 min (% of initial level[b]) | $F_1$[c] | $k$ $s^{-1}$ | Reduction level for 1 min (% of initial level) |
| Aggregates | $0.97 \pm 0.05$ | $0.005 \pm 0.0001$ | 12 | 0.14 | $0.01 \pm 0.004$ | 10 |
| Pentamers | $0.67 \pm 0.17$ | $0.001 \pm 0.0001$ | 47 | 0.60 | $0.18 \pm 0.080$ | 30 |
| Monomers | $0.37 \pm 0.08$ | $0.001 \pm 0.0003$ | 15 | 0.38 | $0.11 \pm 0.060$ | 16 |

Incubation mixture (1.5 ml) contained 100 mM K-phosphate buffer, pH 7.6, with various Emulgen 913 concentrations, 1 mM NADPH, 0.65 μM 1A2 and 0.65 μM R, 2 μM 7-ER. The anaerobic system included 15 mM D-glucose, 90 U/ml glucose oxidase and 2500 U/ml catalase. The samples were bubbled with CO for 1 min. T=30°C.
[a] $k_f$ and $k_s$ - 1A2 reduction rate constants of fast and slow phases respectively.
[b] Initial level (100%) was the value of $\Delta A_{450-490}$ of 0.65 μM 1A2.
[c] $F_1$ - The proportion of the fast phase.

To elucidate the role of electrostatic forces in R and 2B4 interaction, the 2B4 reduction and N-demethylation of benzphetamine in RSS and MC were studied (Table III). It is seen that in MC the raising of ionic strength of medium from 25 to 100 mM led to the increase in the values of the fast phase rate constants of reduction, and rate constants of demethylation reaction and did not alter the slow phase rate constants of reduction. The same dependence of 2B4 reduction parameters on ionic strength was observed in RSS. At the highest ionic strength, e.g. 500 mM K-phosphate buffer, reduction rates declined in RSS but rate constants of benzphetamine oxidation did not alter. Our data are with accordance with the results of Voznesensky and Schenkman (5), who also studied on the influence of ionic strength on reduction and demethylation reaction in RSS with phospholipid.

TABLE III.
The influence of ionic strength on kinetic parameters of NADPH-dependent reductase and oxygenase reactions in MC and RSS with 2B4 in the presence of benzphetamine.

| System | ionic strength, mM | $k_f^a$ min$^{-1}$ | $k_s^a$ min$^{-1}$ | $k^b$ min$^{-1}$ |
|---|---|---|---|---|
| Microsomes | 25 | 16.6 | 2.16 | 5.2 |
| | 50 | 22.6 | 2.22 | 7.1 |
| | 100 | 48.7 | 1.68 | 8.6 |
| | 500 | 60.9 | 1.26 | 5.8 |
| RSS | 25 | 73.8 | 1.8 | 14.1 |
| | 50 | 87.0 | 1.26 | 15.1 |
| | 100 | 113.4 | 1.74 | 14.3 |
| | 500 | 49.2 | 4.00 | 12.9 |

RSS for 2B4 reduction contained different concentrations of K-phosphate buffer, pH 7.5, with 0.25 g/l Emulgen 913, 0.5 μM 2B4, 0.5 μM R and 2 mM benzphetamine; anaerobic conditions (see Table II). RSS for N-demethylation reaction contained different concentrations of K-phosphate buffer, pH 7,5, 0.25 g/l Emulgen 913, 2 mM benzphetamine, 1 μM 2B4 and 1 μM R. With MC, incubation mixture contained buffer without detergent, 1 μM cytochrome P-450. T=30°C.
[a] $k_f$, $k_s$ - 2B4 reduction rate constants of fast and slow phases respectively.
[b] rate constants of benzphetamine N-demethylation.

All of these data suggested that the functional electron transfer complex between flavo- and hemoprotein is not stabilized by electrostatic forces and their interaction not due to complementary charged protein amino acid residues. To elucidate the role of hydrophobic interaction for electron transfer and protein-protein interaction the role of hydrophobic tails of R and b5 for their complex formation with 2B4 and 1A2 was investigated. Table IV shows that in all cases the using of tryptic fragments of R or b5 instead of the whole R or b5 as partners for 2B4 and 1A2 was ineffective in reaction of electron transfer as well as in monooxygenase reaction and spectral visible complex formation.

TABLE IV.
The reductase and oxygenase activities in RSS contained whole or tryptic fragments of R or b5 and 2B4 or 1A2, and spectrophotometrically available complex partner formation

| Composition of system | Oxygenation k, min$^{-1}$ | Reduction k, min$^{-1}$ | Binding spectra |
|---|---|---|---|
| 2B4: | | | |
| 2B4 + d-R | 13.0 | 0.57 | - |
| 2B4 + t-R | 0.04 | 0.0 | - |
| 2B4 + d-R + d-b5 | 19.0 | n.d. | n.d. |
| 2B4 + d-R + t-b5 | 12.0 | n.d. | n.d. |
| 2B4 + t-R + d-b5 | 0.05 | n.d. | n.d. |
| 2B4 + t-R + t-b5 | 0.04 | n.d. | n.d. |
| 1A2: | | | |
| 1A2 + d-R | 0.15 | 0.085 | + |
| 1A2 + t-R | 0.003 | 0.0 | - |

Footnotes: Oxygenation reactions in RSS with 2B4 were performed with benzphetamine; in RSS with 1A2 - with 7-ER.
n.d. - measurements were not performed;
"-" or "+" - the appearance or absence of binding spectra.

The same situation was observed when the complex formation was registered by SDS-electrophoresis after cross-linking of protein partners by EDC (Table V).
EDC treatment of the mixture of monomers or aggregates of R and 2B4 did not reveal bands that could be attributed to equimolar R-2B4 complex. It is surprising that monomeric R can transfer electrons to 2B4 and this monomeric system catalyzes monooxygenase reactions very efficiently (Kanaeva et al., 1992b). R-2B4 complex formation was observed by spectrophotometrically using aggregates of R and 2B4. These results allowed us to suggest the absence of charged pairs in the sites of R and 2B4 interaction.
However, the appearance of the band corresponding to equimolar 96-kDa R-b5 complex and the band corresponding to the equimolar 66-kDa molecular weight 2B4-b5 complex was observed (Kanaeva et al., 1992b). On the contrary 2B4, 1A2-R

equimolar complex (130 kDa) was fixed upon treatment of proteins with EDC (Table V). The complex formation between 1A2 and other redox partner b5 (70 kDa) was also observed.

TABLE V.

Study on complex formation between partners of monooxygenase system by using of EDC (SDS-PAGE analysis)

| Component | Complex, kDa | | | |
|---|---|---|---|---|
| | 2B4 | 1A2 | d-b5 | t-b5 |
| d-R | - | 130 (1:1) | 96 (1:1) | - |
| t-R | - | - | - | - |
| d-b5 | 66 (1:1) | 70 (1:1) | - | n.d. |
| t-b5 | - | - | n.d. | - |

Footnotes: Proteins' ratio in complexes are indicated in branches;
"-" complex formation was not observed;
n.d. - measurements were not performed.

In all pair combinations of R, 2B4, 1A2 and b5 no complex formation were observed when we used tryptic catalytically active fragments of R (t-R) and b5 (t-b5). These data are in full agreement with results above given in Table IV, which showed that the hydroxylase activity in RSS consisting of 2B4 was not supported by t-R and t-b5. Thus, the results obtained provide evidence for the main role of hydrophobic tails of R and b5 in protein-protein interaction.

REFERENCES

Kanaeva, I.P., Dedinskii, I.R., Skotselyas, E.D., Krainev, A.G., Guleva, I.V., Sevrukova, I.F., Koen, Y.M., Kuznetsova, G.P., Bachmanova, G.I., and Archakov, A.I. (1992): Arch. Biochem. Biophys. 298: 395-402.

Kanaeva, I.P., Nikityuk, O.V., Davydov, D.R., Dedinskii, I.R., Koen, Y.M., Kuznetsova, G.P., Skotselyas, E.D., Bachmanova, G.I., and Archakov. A.I. (1992):Arch. Biochem. Biophys. 297: 403-412.

Omura, T., and Takesue, S. (1970): J. Biochem. 67: 249-257.

Sevrukova, I.F., Kanaeva, I.P., Koen, Y.M., Samenkova, N.F., Bachmanova, G.I., and Archakov, A.I. (1992): Arch. Biochem. Biophys., in press.

Voznesensky, A.I., and Schenkman, J.B. (1992): J. Biol. Chem. 267:14669-14676.

# Thermodynamic and kinetic studies of the process of camphor entry into cytochrome P450cam

Gaston Hui Bon Hoa, Eric Deprez

INSERM-INRA U.310, Institut de Biologie Physico-Chimique, 13, rue Pierre-et-Marie-Curie, 75005, Paris, France

## INTRODUCTION:

Cytochrome P-450cam from *Pseudomonas putida* catalyses the hydroxylation of camphor to 5-exo-hydroxycamphor. The substrate recognition in the active site is highly regio- and stereospecific, as probed by site directed mutagenesis experiments (Atkins & Sligar, 1989). However little is known about the intriguing question of how the camphor enters and leaves the active site of this enzyme because the active site is buried in the interior of the enzyme and is isolated from the solvent without any obvious channels that are large enough for camphor molecule to pass through (Poulos et al., 1987). The crystal structure of the protein determined by T. Poulos et al., indicates that several salt-bridges in particular Arg186-Asp251 salt bridge which connects the F-G loop to the I helix (Poulos et al., 1987), could be implicated in the control of the channel opening required for the access of the substrate. To elucidate the problems, we have used a two step binding model of camphor as probed by previous T-jump and P-jump experiments (Ficher & Sligar, 1987, Hui Bon Hoa et al., 1991) in order to test the possibility of any electrostatic control of both the diffusion step of the hydrophobic substrate into the heme pocket of cytochrome P-450cam and its relaxational binding step inside the pocket. Spectroscopic titration and kinetic experiments were investigated as a function of ionic strength, alteration of the dielectric constant of the solvent, and temperature. Data obtained on wild type enzyme and enzyme with point mutation (D251N) are compared and discussed.

## MATERIALS & METHODS:

Cytochrome P-450cam and mutant D251N in which Asp251 charged residu is replaced by a neutral Asn251 residu, were generated and purified in the laboratory of professor S.G.Sligar in collaboration with Nancy Gerber, as described previously (Atkins & Sligar, 1988, Gerber, 1993). Substrate-free protein was obtained by passage of the protein though a Sephadex G-25 fine column equilibrated at 4 ° C with the buffer used for the experiments (Tris-HCl 100 mM pH 7). d-camphor was purchased from Aldrich Chemical Co. Sodium chloride from Sigma was used to increase ionic strength effects. Ethylene Glycol from Carlo Erba, was mixed to water with a range of 0 to 50 % (V/V) in order to modulate the overall dielectric constant of the solvent. The protonic activity of the mixed buffer was nearly identical (7 to 7.2) (Douzou et al., 1976). Optical titrations of the binding of substrate and spin transition were recorded on a Uvikon 940 spectrophotometer equipped with a cryostat Huber HS40 for temperature control of the sample. The overall association constant $K_{ass}$ of the reaction was obtained by Eadie-Hofstee plots of $K_{ass}$ = P-450 $os_{hs}$ / P-450 $o_{ls}$ . s . Spin equilibrium constant $K_2$ = P-450 $os_{hs}$ / P-450 $os_{ls}$ at saturation of substrate was determined using a computer program improved by C.Jung (Jung et al., 1991). The association constant $K_1$ = P-450 $os_{ls}$ / P-450 $o_{ls}$ . s of the first step of the entry of camphor was calculated according to the well established relationship $K_1 = K_{ass} / (1 + K_2)$, (Ficher & Sligar, 1987, Sligar, 1976). Enthalpy and Entropy parameters were calculated from the Van't Hoff plots of the titration curves. Camphor dissociation rate constant $k_{off}$, were measured using a stopped-flow from Photo Physics, by trapping the free ferric cytochrome P-450 with metyrapone to form the irreversible metyrapone-ferric cytochrome P-450 complex. The kinetics were followed at 422 nm, after mixing a solution of 2 $10^{-2}$ M metyrapone with a solution of camphor bound ferric cytochrome P-450 (2 $10^{-6}$ M).

**RESULTS & DISCUSSION:**

Salt effects affect differently the association constant ($K_1$) of the diffusion step and the relaxation binding ($K_2$) step (Hui Bon Hoa et al., 1991). Potassium cations ($K^+$) induce a sharp increase of $K_2$ owing to their specific binding properties on cytochrome P-450cam, influencing the correct orientation of the substrate for the regio- and stereo-selective reaction, (Poulos et al., 1987, Di Primo et al., 1990). In the contrary sodium monovalent cations have much less effects on the second step ($K_2$) and are good candidate to modulate any electrostatic gates controlling the access of the camphor. In the presence of 100 mM tris buffer pH 7, 20 °c, both the overall camphor association constant $K_{ass}$(overall) of Wt and D251N enzymes increase respectively from $10^5$ M and 1.4 $10^5$ M till 4 $10^5$ M at 400 mM $Na^+$ (Fig. 1). The association constants $K_1$ of the first step which are calculated from the following relationship $K_1 = K_{ass}$(overall) / (1+$K_2$), increase recpectively from 2 and 2.9 to 3.4 $10^4$ M. (Fig. 2). Salt effects and point mutation increase

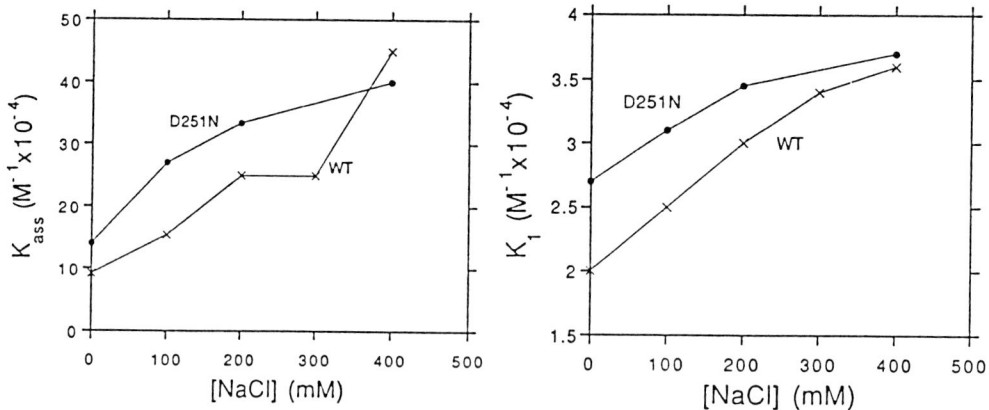

Fig. 1 : Overall association constant (Kass) of camphor versus sodium concentration.

Fig. 2 : Association constant (K1) of the diffusion step versus sodium concentration

the association constant of the first step. However the small increase of $K_1$ of D251N could be due to non-specific effects perturbing some hydrogen bonding of N251 with K178. Lowering the dielectric constant of the medium to 64.5 by addition of 50 % ($^V/_V$) of ethylene glycol induces an opposite behavior of the binding constants for Wt enzyme. Both $K_{ass}$(overall) and $K_1$ decrease respectively to $1.8\ 10^4$ M and $2\ 10^3$ M, indicating some strengthen of ionic interactions on the access channel. Addition of salts ($Na^+$) can recover the binding constants.

Enthalpies of the overall camphor binding reaction (Kass) are similar for both Wt and D251N mutant, while the enthapy of the last relaxation step (K2) is 2.4 time smaller for the D251N mutant compared to the Wt. Interesting differences are found in the first step of camphor entry, as shown in Fig. 3 and 4. $\Delta H°_1$ is 3.5 kcal / mol. for Wt with a break of the Van't Hoff plots, and 7 kcal / mol. for D251N protein. An increase of sodium concentration to 100 mM, mainly affects the enthalpy and entropy of the first step of Wt enzyme: $\Delta H°_1$ increases by a factor of 1.7 (from 3.5 to 6 kcal.mol$^{-1}$) and $\Delta S°_1$ by a factor of 1.28 (from 32 to 41 cal.mol$^{-1}$.K$^{-1}$) while these thermodynamic parameters remain unchanged for D251N. However point mutation and neutralization of charges by salts on

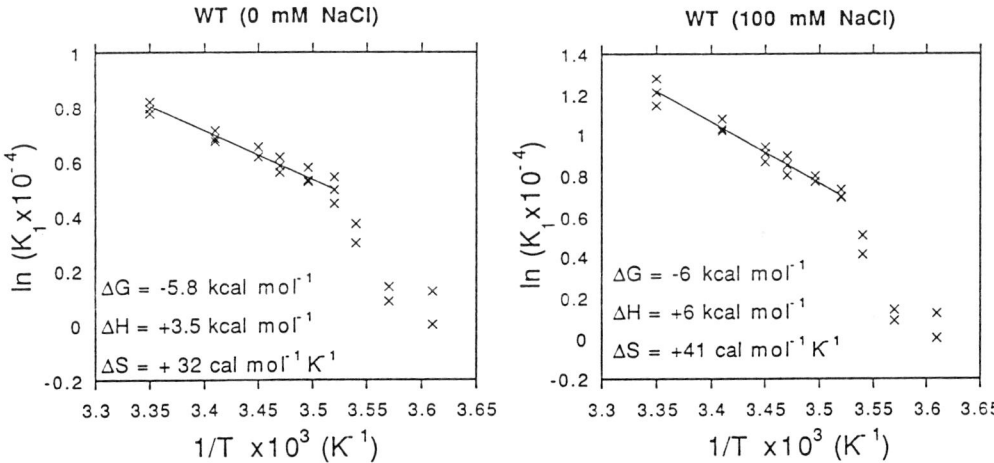

Fig. 3 : Van't Hoff plot of the binding constant (K1) of the first step at 0 mM NaCl, for wild type cytochrome P-450.

Fig. 4 : Van't Hoff plot of the binding constant (K1) of the first step at 100 mM NaCl, for wild cytochrome P-450.

Wt enzyme have similar effects : $\Delta\Delta H°_1$ and $\Delta\Delta S°_1$ (D251N - Wt) are respectively equal to +3.5 kcal.mol$^{-1}$ and +12 cal.mol$^{-1}$.K$^{-1}$. These results suggest important contributions of conformational changes of the protein that occur as substrate pass to and from the active site, on the overall thermodynamic parameters.

Stopped-flow kinetic measurements of the camphor dissociation rate constants ($k_{off}$) as a function of salts showed complex biphasic decrease of $k_{off}$ from 130 to 1.5 s$^{-1}$. The decrease is faster (particularly in the first exponential phase) when potassium cations are used as modulators, suggesting that $k_{off}$ is mainly controlled by the conformation of Tyr96 which increases the binding affinity of camphor in the active site. In the presence of sodium monovalent cations the decrease of $k_{off}$ rate constant is less pronounced; however the measured $k_{off}$ are overall dissociation rate constants of the entire reaction, P-jump experiments (Hui Bon Hoa et al., 1991) showed that the rate limiting step should be the low spin relaxation (dissociation) rate constants of the second step; $k_{-2}$ were shown to decrease sharply with $K^+$ (cation binding site occupied) and presumably with high concentration of $Na^+$. So the measured overall $k_{off}$ reflect the behavior of the rate limiting relaxation step and can hide the expected increase of $k_{on}$ of the first step: at 100 mM $Na^+$, $K_{ass}$ = 1.53 $10^5$ $M^{-1}$, $k_{off}$ = 76 $s^{-1}$, and the calculated overall $k_{on}$ = 11.6 $10^6$ $M^{-1}$ $s^{-1}$ is identical that at zero salt. At 400 mM $Na^+$, $K_{ass}$ = 4.55 $10^5$ $M^{-1}$, $k_{off}$ = 32 $s^{-1}$, and $k_{on}$ calculated = 14.6 $10^6$ $M^{-1}$ $s^{-1}$, a small increase of the overall $k_{on}$ rate. Ionic strength had practically no effect on the dissociation activation enthalpies $\Delta H^*_{-1}$ for wild type as well as for D251N enzymes.

## CONCLUSION:

Two-step camphor binding model was used to analyse the effects of ionic strength and dielectric constant of the medium on the interactions of cytochrome P-450cam with its substrate. It was shown that the first step of the entry of the hydrophobic camphor molecule is controlled by electrostatic interactions. Point mutation of a charged residu Asp251 to a neutral Asn251 residu which abolished one of the salt bridges (Asp251-Arg186) connecting the F-G loop to I helix, increased the camphor equilibrium binding constant of the first step ($K_1$) and minimized the ionic strength effect. This salt linkage was proposed to couple substrate access / binding to catalysis through control of substrate entry (Poulos et al.,1987). However the remaining ionic effect could suggest the contribution of others salt linkages. Thermodynamic parameters obtained, especially the enthalpy and entropy of the first step, are composite parameters. They could reflect some compensation effects between change of electrostatic interaction term and conformational change of the protein during the binding step. Kinetic experiments using stopped-flow apparatus cannot separate the rate constants of the two steps. The measured off rate reflected mainly the relaxation low spin rate limiting step ($k_{-2}$). Other techniques such as P-Jump or T-Jump working at non saturating substrate concentrations are possibly helpful for further characterization of the dynamics of the salt bridges and their implications on the control of the camphor access channels.

## ACKNOLEDGEMENTS:

We gratfully acknowledgeNancy Gerber for the generous gift of D251N mutant. We thank Dominique Pantaloni and Laurent Blanchouen for technical assistance in the use of Photo Physics stopped-flow. This work was supported by the Institut National de la Santé et de la Recherche Médicale and by the Institut National de la Recherche Agronomique.

## REFERENCES:

Atkins, W, M. and Sligar, S. G., (1989) : Molecular Recognition in Cytochrome P-450: Alteration of Regioselective Alkane Hydroxylation via Protein Engineering. *J. Am. Chem. Soc.*, *11*, 2715 - 2717.

Alteration of Regioselective Alkane Hydroxylation via Protein Engineering. *J. Am. Chem. Soc.*, *11*, 2715 - 2717.

Atkins, W.M., and Sligar, S.G., (1988) : The Roles of Active Site Hydrogen Bonding in Cytochrome P-450cam as Revealed by Sited-directed Mutagenesis. *J. Biol. Chem.*, *263*, 18842 -18849.

Di Primo, C., Hui Bon Hoa, G., Douzou, P. and Sligar, S. G., (1990) : Mutagenesis of a Single Hydrogen Bond in Cytochrome P-450 Alters Cation Binding and Heme Solvation. *J. Biol. Chem.*, *265*, 5361-5363.

Douzou, P., Hui Bon Hoa, G., Maurel, P., and Travers, F., (1976) : Physical Chemical Data for Mixed Solvents used in Low Temperature Biochemistry. *Handbook of Biochemistry and Molecular Biology,* 3rd Edition, Physical Chemical data, *1*, 522-529.

Ficher, M. T. and Sligar, S. G., (1987) : Temperature Jump Relaxation Kinetics of the P-450cam Spin Equilibrium. *Biochemistry*, *26*, 4797 - 4803.

Gerber, N., (1993) : *Ph. D.* , University of Illinois, Urbana, U.S.A.

Hui Bon Hoa, G., Di Primo, C. and Deprez,E., (1991) : Conformational Dynamics of Cytochrome P-450cam as Related to the Binding of the Substrate. In *7th International Conference on Biochemistry and Biophysics of Cyt. P-450: Structure and Function, Biotechnological and Ecological Aspects,* Moscou, July 28 - August 2.

Jung, C., Ristau, O., & Rein, H., (1991) : The High-Spin / Low-Spin Equilibrium in Cytochrome P-450 : a New Method for Determination of the High-Spin Content. *Biochimica and Biophysica Acta*, *1076*, 130-136.

Poulos, T. L., Finzel, B. C., Howard, A. J., (1987) : High-Resolution Crystal Structure of Cytochrome P-450cam. *J. Mol. Biol.*, *195*, 687-700.

Sligar, S. G., (1976) : Coupling of Spin, Substrate and Redox Equilibria in Cytochrome P-450. *Biochemistry*, *15*, 5399 - 5406.

# Cytochrome P450: topology and catalysis

Paul R. Ortiz de Montellano[1], Noboru Shirane[1], Zhihua Sui[1], Julia Fruetel[1], Julian A. Peterson[2], James J. De Voss[1]

[1]Department of Pharmaceutical Chemistry, School of Pharmacy, University of California, San Francisco, CA 94143-0446. [2]Department of Biochemistry, University of Texas Southwestern Medical Center, Dallas, Texas 75235-9038, USA

INTRODUCTION

Cytochrome P450 enzymes are distinguished from other enzymes in that their positive catalytic action is primarily focused on activation of molecular oxygen to a reactive oxidizing species. Although not directly demonstrated for cytochrome P450, it is believed that a ferryl ($Fe^{IV}=O$) species coupled with a porphyrin or protein radical is the dominant oxidizing species that is formed (Ortiz de Montellano, 1986). This ferryl species is sufficiently reactive that subsequent reaction with the substrate, whether it result in hydroxylation, epoxidation, heteroatom oxidation, or some other transformation, requires no direct assistance by active site amino acid residues. The exquisite interplay of forces that lowers the energy of the transition state in the catalytic action of most enzymes, resulting in catalysis, is called into play in cytochrome P450 for oxygen activation but not substrate oxidation. This view of cytochrome P450 catalysis is supported by the ability of model iron porphyrins to catalyze most of the transformations facilitated by cytochrome P450 enzymes (Meunier, 1992), by the relatively slow turnover numbers for cytochrome P450 monooxygenases, and by the absence of identifiable "catalytic" residues in the active sites of cytochrome P450 enzymes for which crystal structures are available (Poulos et al., 1987; Ravichandran et al., 1993). The influence of the enzyme on the reaction of the ferryl oxygen with the substrate appears to be limited to sequestration of the substrate in a cavity of appropriate polarity within which it is oriented with respect to the ferryl oxygen by steric, hydrogen bonding, and (occasionally) electrostatic, interactions. This orientation can be relatively stringent, resulting in regiospecific substrate oxidation, or loose enough to allow different sites on the substrate to be competitively oxidized by the ferryl oxygen. Although active site residues may be directly involved in substrate oxidation in exceptional cases, the outcome of cytochrome P450 catalysis generally appears to be determined by substrate orientation rather than by facilitation of the oxygen transfer by protein residues.

As part of a general study of the origins of cytochrome P450 specificity, we are investigating the factors that control the catalytic action of cytochromes $P450_{cam}$ (CYP101) and $P450_{BM-3}$ (CYP102), the two cytochrome P450 enzymes for which crystal structures are available (Poulos et al., 1987; Ravichandran et al., 1993).

## CYTOCHROME $P450_{cam}$ AS A XENOBIOTIC METABOLIZING ENZYME

Camphor, the normal substrate for cytochrome $P450_{cam}$, is specifically hydroxylated at the 5-exo position (Sligar & Murray, 1986):

$$\text{camphor} \xrightarrow{P450_{cam}} \text{5-exo-hydroxycamphor}$$

The crystal structure of the enzyme-camphor complex indicates that this specificity is due to non-bonding steric interactions, most notably with Val-247 and Val-295, as well as a hydrogen bond between the carbonyl group of camphor and Tyr-86 of the enzyme (Poulos et al., 1987). In accord with this conclusion, site specific mutation of the tyrosine to a phenylalanine (Atkins & Sligar, 1989), or replacement of the carbonyl oxygen of the substrate with a non-hydrogen bonding sulfur (Atkins & Sligar, 1988), results in less regiospecific hydroxylation of the substrate. It was first thought that cytochrome $P450_{cam}$ was a substrate-specific enzyme akin to the sterol biosynthetic enzymes and would only turn over camphor-related structures. Our demonstration that cytochrome $P450_{cam}$ readily oxidizes styrenes (Fruetel et al. 1992), and subsequent studies with other substrates (e.g., Jones et al., 1992), have demonstrated that this is not the case. Styrene is oxidized to styrene oxide and small amounts of phenylacetaldehyde and phenyl-hydroxylated products (Fruetel et al., 1992). Benzaldehyde is simultaneously formed via a $H_2O_2$-dependent mechanism associated with the uncoupled formation of $H_2O_2$ (see below). Cis- and trans-β-methylstyrene are oxidized to the corresponding epoxides and allylic alcohols as well as traces of phenylacetone and ring-hydroxylated products. The reaction stoichiometry indicates, however, that styrene oxidation is much less efficient than camphor oxidation (Fruetel et al., 1992). Thus, whereas each mole of oxygen and NADH consumed yields one mole of 5-exo-hydroxycamphor, less than 10% of the oxidation equivalents provided by NADH and oxygen are used to oxidize styrene and even less to oxidize the β-methylstyrenes. The bulk of the NADH is utilized in the presence of styrene to reduce molecular oxygen to $H_2O_2$. These results suggest that substrates that fit into the active site of cytochrome $P450_{cam}$ can be oxidized by the enzyme, but that substrate binding must suppress uncoupled reduction of oxygen for efficient oxidation to occur. The factors that control uncoupled enzyme turnover remain obscure, but it is likely that exclusion of water from the active site, possibly from specific regions of the active site, is a critical factor (Raag et al, 1991). If so, the extent to which the substrate fills the active site may be important in determining oxidation efficiency. It is interesting, in this context, that most membrane-bound cytochrome P450 enzymes oxidize substrates of different sizes with comparable efficiency.

In studies currently in progress, we are investigating the degree to which "fit" within the active site is a sufficient criterion for acceptability as a cytochrome P450 substrate. Although efforts are being made to clarify the substrate specificities of membrane bound enzymes by defining their active site volume and shape, it is not yet clear to what extent the substrate specificity of an enzyme can be defined from such criteria even when a crystal structure of the enzyme is available. To address this question, we have used the program DOCK to identify compounds that fit the active site of the enzyme. DOCK is an interactive program that produces a positive image of the active site cavity and then is able to search structural data bases for compounds that fit within the given volume (Desjarlais et al., 1988). Ten compounds have been chosen from the high scoring (best fit) structures identified by DOCK from the fine chemical index. The choice excluded compounds that were ionic or not readily available commercially. Preliminary results on oxygen consumption and metabolite formation indicate that seven of the ten compounds are bound and stimulate catalytic turnover, although the turnover in some cases results primarily in uncoupled oxygen reduction (unpublished results). Three of the compounds neither stimulate oxygen consumption nor give detectable metabolites and therefore appear not to be substrates. These preliminary results suggest that fit is a reasonable but not absolute criterion for substrate acceptability.

## CYTOCHROME P450$_{BM-3}$: A STRUCTURAL DETERMINANT OF REGIOSPECIFICITY

Cytochrome P450 catalysis is most clearly represented by hydrocarbon hydroxylation reactions because such reactions are rarely catalyzed by other types of hemoproteins. A substantial body of evidence indicates that carbon hydroxylation involves abstraction of the hydrogen by the ferryl oxygen to yield a carbon radical, which recombines with the ferryl-bound hydroxyl moiety to give the hydroxylated product (Ortiz de Montellano, 1986). The recombination rate is very rapid because it is essentially the recombination of a carbon radical with the equivalent of an iron-bound hydroxyl radical (P = porphyrin):

$$P^{+\cdot}(Fe^{IV}=O) + R_3C-H \rightarrow P(Fe^{III} OH\cdot) + R_3C\cdot \rightarrow P(Fe^{III}) + R_3C-OH$$

As suggested by this mechanism, the susceptibility of C-H bonds to cytochrome P450-catalyzed hydroxylation depends on their bond strength, which in turn depends on the relative stabilities of the carbon radicals generated when the hydrogen radical is abstracted (Ortiz de Montellano, 1986). Because a primary carbon radical is considerably less stable than a secondary or tertiary carbon radical, the C-H bonds of the terminal methyl of a hydrocarbon chain are stronger than the C-H bonds of the methylene groups of the chain (Egger & Cocks, 1973). This difference in bond strength underlies the well known preference for the formation of ω-1- over ω-hydroxylated products. This is as true for the oxidation of hydrocarbons by metalloporphyrins as it is for cytochrome P450 enzymes. Indeed, ω-hydroxylation is a minor hydroxylation event in the oxidation of heptane by even the best available porphyrin model for an ω-hydroxylase (Cook et al., 1986).

The fatty acid ω-hydroxylases overcome the preference for ω-1 over ω-hydroxylation and catalyze the latter over the former by factors as high as 20:1 (Table 1). We have provided evidence that this is achieved by constraining the substrate so that only the methyl is accessible to the ferryl oxygen, but the details of the mechanism by which ω-hydroxylases overcome the preference for in-chain hydroxylation remain to be defined (CaJacob et al, 1988). Cytochrome P450$_{BM-3}$ represents the alternative situation in which the preference for in-chain hydroxylation is reinforced to the complete exclusion of ω-hydroxylation (Table 1) (Ho & Fulco, 1976; Ruettinger & Fulco, 1981; Boddupalli et al., 1992). The mechanism employed in reinforcing the intrinsic bias against ω-oxidation is amenable to detailed study because the crystal structure of cytochrome P450$_{BM-3}$ has recently been determined (Ravichandran et al., 1993). Cytochrome P450$_{BM-3}$ thus provides an excellent situation in which to explore the effect of structure on substrate regiospecificity.

If the reason for ineffective ω-hydroxylation is the higher bond energy of terminal methyl than in-chain methylene C-H bonds, replacement of the terminal methyl by a double bond should result in epoxide formation because double bonds are easily oxidized. Indeed, cytochrome P450 enzymes readily oxidize terminal double bonds to epoxides and greatly favor epoxidation over ω-2 (allylic) hydroxylation. Thus, cytochrome P450 4A1, the lauric acid ω-hydroxylase, oxidizes 12-dodecenoic acid almost exclusively to the corresponding epoxide (Table 2) (Cajacob et al., 1988), and cytochrome P450 2B1, a relatively non-specific enzyme, oxidizes octene to the epoxide and ω-2 hydroxylated products in a 98:2 ratio (White et al., 1986; Ortiz de Montellano et al., 1983). These results clearly show that oxidation of the terminal double bond is favored over ω-2 oxidation if the double bond is sterically accessible. In this context, the results obtained with cytochrome P450$_{BM-3}$ are highly significant (Shirane et al., 1993). This enzyme oxidizes 12-tridecenoic acid predominantly to the ω-2 hydroxylated product with the epoxide accounting for only 3% of the product (Table 2). Cytochrome P450$_{BM-3}$ similarly oxidizes 95% of 17-octadecenoic acid to the ω-2 hydroxylated product and only 5% to the epoxide (Table 2).

In order to rationalize the virtual inability of cytochrome P450$_{BM-3}$ to oxidize a reactive π-bond occupying the ω/ω-1 positions, one must postulate that the site in question is not sterically accessible to the ferryl oxygen. We therefore deduce from these results that the

Table 1. Regiospecificity in the hydroxylation of hydrocarbon chains by cytochrome P450 enzymes.

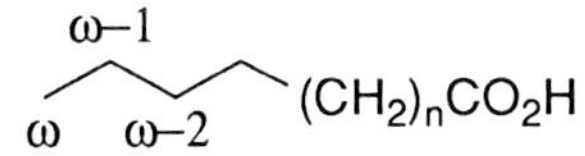

$$\text{ω} \quad \text{ω-2} \overset{\text{ω-1}}{\frown} (CH_2)_nCO_2H$$

| Enzyme | n | ω | ω-1 | ω-2 | ω-3 |
|---|---|---|---|---|---|
| $P450_{BM-3}$ | 7 | 0 | 36 | 30 | 34 |
|  | 8 | 0 | 17 | 65 | 18 |
|  | 13 | 0 | 39 | 47 | 14 |
| P450 2B1 | 7 | 12 | 88 |  |  |
| P450 4A1 | 7 | 95 | 5 |  |  |

Figure 2. Epoxidation versus allylic hydroxylation of ω/ω-1 unsaturated fatty acids

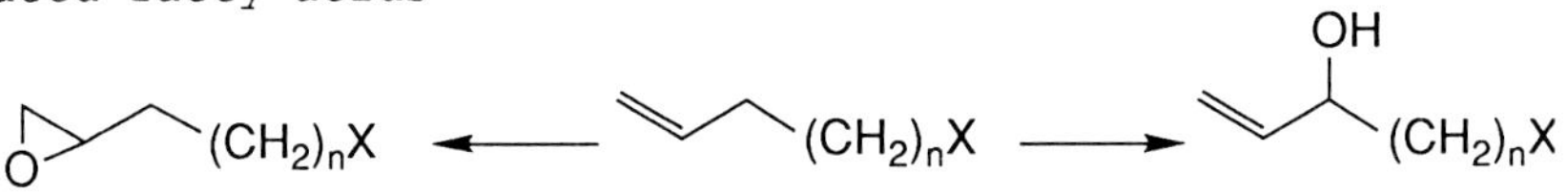

| Enzyme | n | X | Epoxide | ω-2 Hydroxyl |
|---|---|---|---|---|
| $P450_{BM-3}$ | 9 | $CO_2H$ | 3 | 97 |
|  | 14 | $CO_2H$ | 5 | 95 |
| P450 2B1 | 4 | $CH_3$ | 98 | 2 |
| P450 4A1 | 8 | $CO_2H$ | >98 | <2 |

substrate is bound in the active site with the methyl group sequestered in a lipophilic cavity that keeps it away from the reactive oxygen (Shirane et al, 1993). Support for this conclusion is provided by the fact that oxidation of a mixture of *cis*- and *trans*-16-octadecenoic acids, substrates with the double bond at the ω-1/ω-2 positions, give the epoxide and ω-3 hydroxylated products in a ratio of 1:4.6 (unpublished results) without detectable formation of the ω-hydroxylated product:

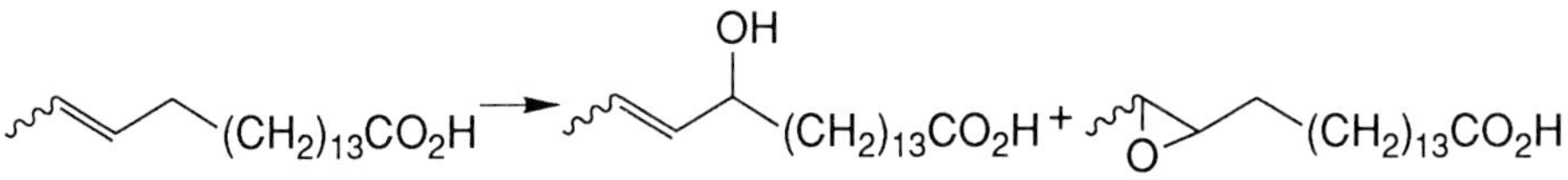

Epoxidation is thus much more important with this substrate than it is when the double bond is located at the terminal position. The exact degree of discrimination between the double bond and the allylic position requires quantification of the products obtained individually from the *cis*- and *trans*-isomers of the double bond, but it is clear that the double bond is not protected in this substrate. The formation of both epoxide and allylically hydroxylated products is consistent with the earlier finding that fatty acids with a double bond in the middle of the chain are subject to both epoxidation and hydroxylation (Ruettinger & Fulco, 1981).

The crystal structure of cytochrome $P450_{BM-3}$ shows that the heme is at the bottom of a broad channel that extends to the surface of the protein (Ravichandran et al. 1993). It is likely that fatty acids bind in the channel with the carboxyl group near the mouth of the channel and the terminal methyl near the heme moiety. The crystal structure shows that Phe-87 is so placed in the active site that it forms a lipophilic pocket in the vicinity of the heme group. The present results are readily explained if the terminal methyl group is bound within this pocket by hydrophobic forces, protecting it from oxidation while positioning the $\omega-1$, $\omega-2$, and $\omega-3$ methylene protons in the vicinity of the reactive oxygen species.

*Acknowledgment*: This work was supported by National Institutes of Health Grant GM25515

REFERENCES

Atkins, W.M., and Sligar, S.G. (1988): The roles of active site hydrogen bonding in cytochrome P-450*cam* as revealed by site specific mutagenesis. *J. Biol. Chem.* 263: 18842-18849.

Atkins, W.M., and Sligar, S.G. (1989): Molecular recognition in cytochrome P450: alteration of regioselective alkane hydroxylation via protein engineering. *J. Am. Chem. Soc.* 111: 2715-2717.

Boddupalli, S.S., Pramanik, B.C., Slaughter, C.A., Estabrook, R.W., and Peterson, J. A. (1992): Fatty acid monooxygenation by $P450_{BM-3}$: product identification and proposed mechanisms for the sequential hydroxylation reactions. *Arch. Biochem. Biophys*. 292: 20-28.

CaJacob, C.A., Chan, W., Shephard, E., and Ortiz de Montellano, P.R. (1988): The catalytic site of rat hepatic lauric acid $\omega$-hydroxylase. Protein vs prosthetic heme alkylation in the $\omega$-hydroxylation of acetylenic fatty acids. *J. Biol. Chem.* 263: 18640-18649.

Cook, B.R., Reinert, T.J., and Suslick, K.S. (1986): Shape selective alkane hydroxylation by metalloporphyrin catalysts. *J. Amer. Chem. Soc.* 108: 7281-7286.

DesJarlais, R.L., Sheridan, R.P., Seibel, G.L., Dixon, J.S., Kuntz, I.D. and Venkataraghavan, R. (1988): Using shape complementarity as an initial screen in designing ligands for a receptor binding site of known three-dimensional structure. *J. Med. Chem.* 31: 722-729.

Egger, K.W., and Cocks, A.T. (1973): Homopolar and heteropolar bond dissociation energies and heats of formation of radicals and ions in the gas phase. I. Data on organic molecules. *Helv. Chim. Acta.* 56: 1516-1536.

Fruetel, J.A., Collins, J.R., Camper, D.L., Loew, G.H., and Ortiz de Montellano, P.R., (1992): Calculated and experimental absolute stereochemistries of the styrene and β-methylstyrene epoxides formed by cytochrome P450$_{cam}$. *J. Am. Chem. Soc.* 114: 6987-6993.

Ho, P.P., and Fulco, A.J. (1976): Involvement of a single hydroxylase species in the hydroxylation of palmitate at the omega-1, omega-2, and omega-3 positions by a preparation from *Bacillus megaterium*. *Biochim. Biophys. Acta* 431: 249-256.

Jones, J.P., Trager, W.F., and Carlson, T.J. (1992): The binding and regioselectivity of reaction of (R) and (S)-nicotine with cytochrome P-450*cam*: parallel experimental and theoretical studies. *J. Am. Chem. Soc.* 115: 381-387.

Meunier, B. (1992): Metalloporphyrins as versatile catalysts for oxidation reactions and oxidative DNA cleavage. *Chem. Rev.* 92: 1411-1456.

Ortiz de Montellano, P. R. (1986): Oxygen activation and transfer. In *Cytochrome P450: structure, mechanism, and biochemistry*, ed. P.R. Ortiz de Montellano, pp. 217-271. New York: Plenum Press.

Ortiz de Montellano, P.R., Mangold, B.L.K., Wheeler, C., Kunze, K.L., and Reich, N.O. (1983): The cytochrome P-450 active site. Regiospecificity of the prosthetic heme alkylation by olefins and acetylenes. *J. Biol. Chem.* 258: 4202-4207.

Poulos, T.L., Finzel, B.C., and Howard, A.J. (1987): High-resolution crystal structure of cytochrome P450$_{cam}$. *J. Mol. Biol.* 195: 687-700.

Raag, R., Martinis, S.A., Sligar, S.G., and Poulos, T.L. (1991): Crystal structure of the cytochrome P-450$_{cam}$ active site mutant Thr252Ala. *Biochemistry* 30: 11420-11429.

Ravichandran, K.G., Boddupalli, S.S., Hasemann, C.A., Peterson, J.A., and Deisenhofer, J. (1993): Crystal structure of the hemoprotein domain of P450$_{BM-3}$, a prototype for microsomal P450's. *Science* 261: 731-736.

Ruettinger, R.T., and Fulco, A.J. (1981): Epoxidation of unsaturated fatty acids by a soluble cytochrome P-450-dependent system from *Bacillus megaterium*. *J. Biol. Chem.* 256: 5728-5734.

Shirane, N., Sui, Z., Peterson, J.A., and Ortiz de Montellano, P.R. (1993) Cytochrome P450$_{BM-3}$ (CYP102): regiospecificity of oxidation of ω-unsaturated fatty acids and mechanism-based inactivation. *Biochemistry*: in press.

Sligar, S.G., Murray, R.I. (1986): Cytochrome P450$_{cam}$ and other bacterial cytochrome P450 enzymes. In *Cytochrome P450: structure, mechanism, and biochemistry*; ed. P.R. Ortiz de Montellano, pp. 429-503. New York: Plenum Press.

White, I.N.H., Green, M.L., Bailey, E., and Farmer, P.B. (1986): Metabolic activation of olefins. Conversion of 1-octene to a putative reactive intermediate 1-octen-3-one: an alternative pathway to epoxidation. *Biochem. Pharmacol.* 35: 1569-1575.

# Conclusive evidence of a quaternary cluster model for cholesterol side chain cleavage reaction catalyzed by cytochrome P-450scc

Takayuki Hara, Mikako Takeshima

*Department of Biochemistry, Graduate School of Health and Nutrition Sciences, Nakamura Gakuen College, 5-7-1 Befu, Jonan-ku, Fukuoka 814-01, Japan*

## Introduction

Mitochondrial steroid hydroxylase system consists of NADPH-adrenodoxin reductase (AR), adrenodoxin (AD), and cytochromes P-450 in steroidogenic tissues (Kimura, 1981). Cholesterol side chain cleavage (SCC) reaction catalyzed by mitochondrial cytochrome P-450scc (P-450scc) is the rate-limiting step of steroid hormone biosynthesis and regulated by the action of ACTH (Kimura, 1981). Elucidation of the reaction mechanism has been attempted by many investigators to understand the regulation mechanism of steroid hormone biosynthesis. Lambeth *et al.* (1979) have proposed the shuttle mechanism in which AD functions as a mobile electron carrier between AR and P-450scc. In this model, AD first binds to AR and forms a complex with AR. Electrons are donated from NADPH to flavin moiety of AR, and then transferred to the iron-sulfur of AD. Reduced AD dissociates from AR, then binds to P-450scc to reduce P-450scc. Oxidized AD again dissociates from P-450scc, and a second cycle starts. This model is based on the following lines of evidence:

1. Titration with AD in SCC reaction indicated that 1 more mol of AD is necessary for the maximum activity (Seybert *et al.*, 1979; Hanukoglu & Jefcoate, 1980).
2. AR can donate electrons from NADPH to excess AD, indicating that reduced AD is easily dissociated from AR and a second AD can bind to AR (Lambeth *et al.*, 1979).
3. The effective concentration of ions on steroid hydroxylation reactions is about 10-times higher than that on the cytochrome *c* reduction, suggesting that the dissociation step of AD is involved in the former reaction (Lambeth *et al.*, 1979).

In spite of these extensive studies, a cluster model has been proposed by several groups. Kido and Kimura (1979) reported that a stable complex among AR, AD, and P-450scc is formed in the presence of cholesterol and phospholipid. This model has been supported by Usanov *et al.* (1985) from their chemical modification studies. We reported that a cross-linked complex between AR and AD prepared with 1-ethyl-3-(3-dimethylaminopropylcarbodiimide) is active in NADPH-cytochrome *c* reduction, but that 1 more mol of AD is necessary for P-450scc reduction reaction (Hara & Kimura, 1989ab). From these results, we have proposed a quaternary cluster model for SCC reaction, which has been considered as a candidate of the mechanism for SCC reaction (Seybert *et al.*, 1979).

In this study, we present conclusive evidence of the quaternary cluster model for SCC reaction, in which 1 mol each of AD binds to AR and P-450scc.

## Materials and Methods

Purification of AR, AD, and P-450scc from bovine adrenal cortex mitochondria was carried out as described (Hara & Kimura, 1989ab). Preparation of P-450scc-containing liposomes (PC/PE/CL=4/4/1, mol/mol/mol) was carried out as follows: the flask containing 13.3 μmol each of PC and PE, 3.3 μmol of CL, and 12.9 μmol of cholesterol was solubilized with 50 mM Hepes buffer (pH 7.5) containing 20 % glycerol, 0.15 M KCl, 1 mM dithiothreitol, 1 mM EDTA, and 2 % cholic acid by sonication. To the flask, 30 nmol of P-450scc was added, and then shaken gently at $4^\circ$ C for 24 h. The solution was dialyzed against 50 mM Hepes buffer (pH 7.5) containing 20 % glycerol, 0.15 M KCl, 1 mM dithiothreitol, and 1 mM EDTA extensively, and then centrifuged at 30,000 rpm for 20 min to remove the precipitate. SCC activity was measured as described (Takeshima & Hara, 1991). High density lipoprotein (HDL) was purified as described (Takeshima & Hara, 1991). HDL-Sepharose was prepared by the same method as described (Hara & Kimura, 1989a).

## Results and Discussion

<u>Titration with AD in SCC activity</u>: We determined SCC activity in a reconstituted system using P-450scc-containing liposome which contains 30 mol % of cholesterol as a model of inner membrane of mitochondria. As shown in Fig.1, the titration with AD in SCC activity showed a clear Michaelis-Meten type curve with the stoichiometry of AR/AD/P-450scc=1/2/1. A "lag phase" of titration with AD at less than AD/P-450scc=1 which has been typically observed by previous workers (Hanukoglu & Jefcoate, 1980) was seen only when cholesterol dispersed with the detergent system was used as the substrate (data not shown). These results suggest that the data of reaction mechanism using cholesterol-detergent system is not relevant to the natural environment of P-450scc.

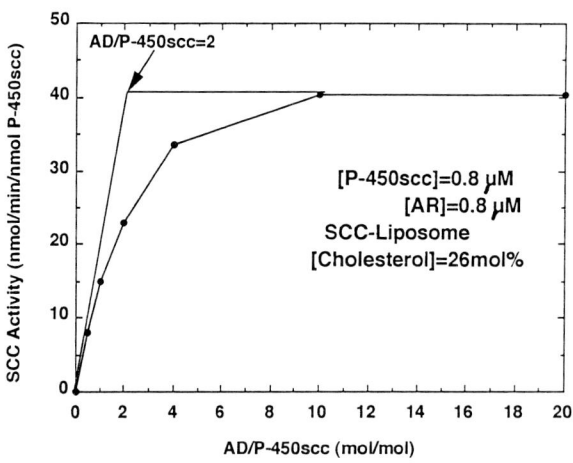

**Fig. 1 Titration with AD in SCC activity**

The reaction mixtures contained 30 mM K-phosphate buffer (pH 6.6), 0.23 mM liposomal cholesterol, 0.8 μM each of AR and liposomal P-450scc, and various amounts of AD as indicated in the figure. The reaction tube was incubated at $37^\circ$ C for 1 min after addition of 0.5 mM NADPH.

<u>Titration with AR in SCC activity</u>: Fig. 2 shows the titration with AR in SCC activity in a reconstituted system using DOPC-liposomal cholesterol. Surprisingly, SCC activity was diminished when AR/P-450scc was increased. The maximum activity point was shifted to the higher ratio of AR/P-450scc when AD/P-450scc was increased from 2 to 20. Hanukoglu and Jefcoate (1980) have already done the similar experiments. However, they used the experimental condition of AD/P-450scc=27, and they titrated with AR up to AR/P-450scc=5. Under these conditions, they must have missed the inhibitory effect starting from AR/P-450scc=20 as shown in Fig. 2 as the similar conditions by them. This novel observation is

not due to the salt effect nor to the liposomal cholesterol effect, because AR dissolved in the buffer had no effect on SCC activity at all and this inhibitory effect was more clearly seen when the cholesterol-detergent system was used as the substrate (data not shown). These results suggest that AD bound to P-450scc is deprived by AR when AR is increased and AD bound to P-450scc is necessary for SCC reaction.

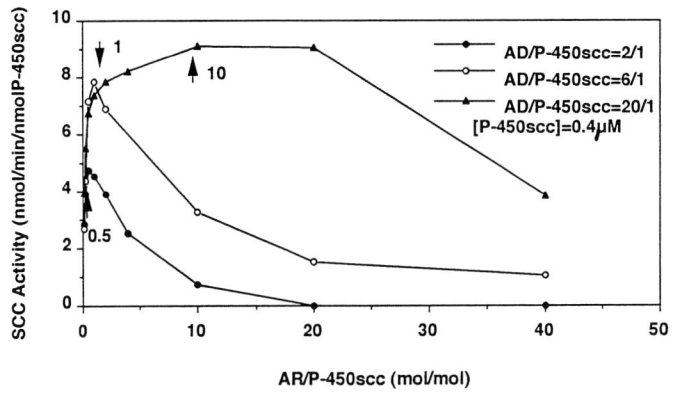

**Fig. 2 Titration with AR in SCC activity**

The reaction mixtures contained 30 mM K-phosphate buffer (pH 6.6), 0.2 mM liposomal cholesterol, 0.4 µM P-450scc, and various amounts of AR and AD as indicated in the figure. The incubation time was 5 min.

Determination of AD bound to P-450scc using HDL-Sepharose method: To test the above possibility, we applied HDL-Sepharose method to determine AD bound to P-450scc. We found that P-450scc was incorporated into HDL vesicles spontaneously, and AR and AD did not bind to HDL at all. If the reaction mixtures contain AR, AD, and P-450scc together with HDL-Sepharose, AD which is bound to P-450scc is sedimented together with P-450scc-HDL-sepharose. Free AD or AD which is bound to AR should remain in the supernatant. Fig. 3 shows the determination of AD in the supernatant by Western blotting analysis. When AD and P-450scc (0.8 µM each) were present, about 60 % of AD was bound to P-450scc. In the presence of NADPH, when AR was added to the mixtures and increased, AD in the supernatant was increased to the plateau level at AR/P-450scc=10. Under the same conditions using HDL-Sepharose, the SCC activity was comparable with that in Fig. 2 at AR/P-450scc=1, from which the activity was decreased sharply, then diminished at AR/P-450scc=40 (data not shown).

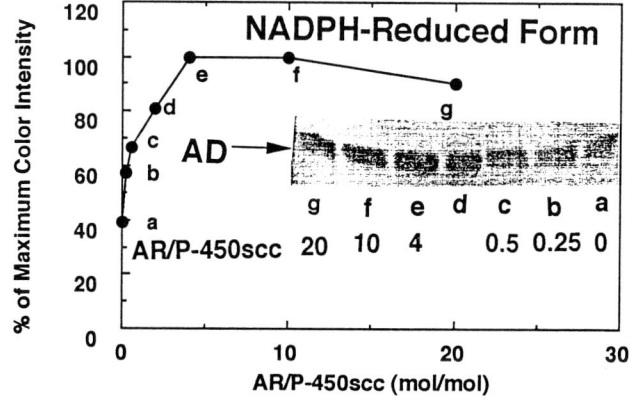

**Fig. 3 Effects of increasing amount of AR on the binding capacity of AD to P-450scc**

The reaction mixtures contained 30 mM K-phosphate buffer (pH 6.6), 0.2 mM free cholesterol of HDL-Sepharose, 0.8 µM each of P-450scc and AD, and various amounts of AR as indicated in the figure. The tubes were incubated with 0.5 mM NADPH at 37° C for 5 min, and then centrifuged at 2,000 rpm for 5 min. AD amounts of the supernatant were determined by Western blotting analysis.

If this mechanism is correct, the observed SCC activity is assumed by the following equation:

$$V_{scc} = C[ARAD][ADP\text{-}450scc] = R \times \frac{X}{K_1+X} \times S \times \frac{X}{K_2+X} \quad (1)$$

where K1 and K2 are dissociation constants between AD and AR, and AD and P-450scc, respectively. R and S are the total concentrations of AR and P-450scc. X is the concentration of free AD. If this equation is applied to the computer analysis and we assume that K1 and K2 are 1 nM and 100 nM, respectively, Vscc especially fitted very well to the curves of Fig. 2 under the conditions of AD/P-450scc=2 and 6 (data not shown, Dr. Kominami personal communication). From these results, we conclude that this model is correct, if the K1 and K2 values are not so greatly changed under oxidation-reduction states. According to the data of Fig. 3 and the data of the analysis under oxidized state (data not shown), K1 value does not seem to change greatly under oxidation-reduction states. This supports our previous findings that AD does not need to dissociate from AR during the catalysis (Hara & Kimura, 1989a).

The molar ratio of AR/AD/P-450 existing in bovine adrenal cortex mitochondria is known to be 1/10/10. If the ratio of P-450scc/P-450 11β is 4, the molar ratio of AR/AD/P-450scc=1/10/8. The main support for the shuttle mechanism is based on the existence of excess AD relative to AR. From our present results, excess AD seems to be necessary to divide AD into AR and P-450scc to maintain AD bound to P-450scc.

**References**

Hanukoglu, I. & Jefcoate, C. R. (1980): Mitochondrial cytochrome P-450scc. Mechanism of electron transport by adrenodoxin. *J. Biol. Chem.* 255, 3057-3061.

Hara, T. & Kimura, T. (1989a): Purification and catalytic properties of a cross-linked complex between adrenodoxin reductase and adrenodoxin. *J. Biochem.* 105, 594-600.

Hara, T. & Kimura, T. (1989b): Active complex between adrenodoxin reductase and adrenodoxin in the cytochrome P-450scc reduction reaction. *J. Biochem.* 105, 601-605.

Kido, T. & Kimura, T. (1979): The formation of binary and ternary complexes of cytochrome P-450scc with adrenodoxin and adrenodoxin reductase-adrenodoxin complex. The implication in ACTH function. *J. Biol. Chem.* 254, 11806-11815.

Kimura, T. (1981): ACTH stimulation on cholesterol side chain cleavage activity of adrenocortical mitochondria: Transfer of the stimuli from plasma membrane to mitochondria. *Mol. Cell. Biochem.* 36, 105-122.

Lambeth, J. D., Seybert, D. W., & Kamin, H. (1979): Ionic effects on adrenal steroidogenic electron transport. The role of adrenodoxin as an electron shuttle. *J. Biol. Chem.* 254, 7255-7264.

Seybert, D. W., Lambeth, J. D., & Kamin, H. (1979): The participation of a second molecule of adrenodoxin in cytochrome P-450-catalyzed 11β hydroxylation. *J. Biol. Chem.* 253, 8355-8358.

Takeshima, T. & Hara, T. (1991): High density lipoprotein cholesterol as a mechanistic probe for the side chain cleavage reaction. *Biochem. Biophys. Res. Commun.* 179, 161-169

Usanov, S. A., Turko, I. V., Chashchin, V. L., & Akhrem, A. A. (1985): Cross-linking studies of steroidogenic electron transfer: Covalent complex of adrenodoxin reductase with adrenodoxin. *Biochim. Biophys. Acta* 832, 288-296.

# Structural features of the bovine mitochondrial electron transfer system CYP 11A1

Ralf-Jürgen Kuban, Andreas Marg, Klaus Ruckpaul

*Max-Delbrück-Centrum of Molecular Medicine, Robert-Rössle-Straße 10 D-13122 Berlin, Federal Republic of Germany*

## Introduction

The general aim of this investigation is to determine the 3d-structure of the essential components of the steroid hydroxylating mitochondrial electron transfer system: *adrenodoxin-reductase (AR), adrenodoxin (Adx)* and *CYP 11A1*. Prerequisites are (1) highly purified homogenous proteins, (2) crystallizability and stability of the proteins, (3) crystallographic characterization of the protein components (determination of lattice constants and space groups).

Furthermore, heavy metal derivatives have to be prepared; the enzymatic stability and the structural integrity of the derivatives to be determined. Synchrotronic radiation experiments have to be performed as alternative possibility in comparison to X-ray structure analysis.

The present study is aimed at crystallographically characterizing the protein components for pursuing detailed structure analysis.

## Methods and Material

The 3 essential components of the steroid hydroxylase were isolated and purified from bovine adrenals according to established methods [AR: Hiwatashi et al., 1976; Nonaka et al., 1985; Suhara et al., 1982. Adx: Sakihama et al., 1988. CYP 11A1: Chashchin et al., 1984].

The concentration of *adrenodoxin* was determined spectrophotometrically by using an extinction coefficient of 9,800 $M^{-1} * cm^{-1}$ at 414 nm. Despite employment of HPLC separation techniques the Adx preparation proved heterogeneous by mass spectrometric control. In the present study we used the main fraction which was characterized by an absorption ratio A 415 nm/A 276 nm of 0.91 and a molecular mass of 13,675 ± 2 Da corresponding to an adrenodoxin consisting of 125 amino acids (full length bovine adrenodoxin consists of 128 amino acids).

The concentration of *adrenodoxin reductase* was obtained spectrophotometrically (extinction coefficient = 11,300 $M^{-1} * cm^{-1}$) at 450 nm. The final AR preparation showed an absorption ratio $A_{272\,nm} / A_{450\,nm}$ of 7.7. The homogeneity of the main fraction of AR was proved by mass spectrometry and revealed a value of 51,390 ± 150 Da in good agreement with a calculated $M_r$ for a molecule with FAD without sugar residues.

The concentration of *CYP 11A1* was determined as CO-complex by means of an extinction coefficient of 91,000 $M^{-1}$ at 450 nm. The high- and low-spin content were calculated from the absorption at 393 nm ($\varepsilon = 91,000$ $M^{-1} * cm^{-1}$) and 418 nm ($\varepsilon = 110,000$ $M^{-1} * cm^{-1}$) respectively.

Mass spectra were recorded either on a Finnigan-MAT triple stage quadrupole mass spectrometer (adrenodoxin) or a MALDI-TOF mass spectrometer from Kratos (Manchester, UK) and Shimadzu (Tokyo) based on depolarization-flight in time.

## Results and Discussion

**Adrenodoxin**
Brown crystals of bovine adrenodoxin which have been grown at 4°C or 15°C (0.8 * 0.3 * 0.2 mm in size) were mounted in thin-walled glass capillaries. Diffraction photographs were taken with a Nonius Fast diffractometer using $CuK_\alpha$ (radiation $\lambda$ =1.54 Å ) generated by a rotating anode source run at 40 kV, 56 mA, with focal size 1.0 * 0.2 mm². The reflection pattern indicates a hexagonal crystal lattice with unit cell dimensions a = b = 172.50 (9) Å , c = 183.49 (9) Å, $\alpha = \beta = 90$ °, $\gamma = 120$°. The crystals diffract with a resolution to 4 Å. The space group could not be determined because we were not able to record ($hk0$) and ($k0l$) precession photographs with the coaxial protein cooling device. ($00l$) reflections with $l$ odd are either absent or weak  (F< 4$\sigma$) and it is difficult to conclude whether they are systematically absent or characteristically weak due to non-crystalline graphic translation elements between molecules. Assuming 144 molecules in the unit cell, equivalent to 12 molecules per asymmetric unit in the hexagonal unit cell results in a $V_m$ value of 2.28 Å³/ Da. The $V_m$ value is very similar to the value of 2.20 Å³ /Da calculated for a [2Fe-2S] ferredoxin (Ogawa et al. 1977).  The number of 12 molecules in the asymmetric unit does permit only with difficulties to solve the structure of the molecule. Consequently structure analysis requires crystals in another modification with a smaller number of molecules in the asymmetric unit.

**Adrenodoxin reductase**
Yellow crystals of bovine adrenodoxin reductase were grown to hedgehog-like bunches within 16 days at 0°C. After mounting crystals with dimensions of 0.6 * 0.5 * 0.2 mm³ one three-dimensional data set has been collected by means of  an Enraf Nonius FAST detector system. The latter was mounted on a FR 571 rotating anode operating at 45 kV and 65 mA with a focal spot of  0.2 * 2.0 mm². Data were collected in oscillations of 0.1° with a crystal detector distance of 75 mm and a detector swing angle of 5° using $CuK_\alpha$ radiation ($\lambda$= 1.5418 Å). Unit cell parameters and cell orientation were refined by least squares methods on the basis of 198 reflections up to 4.0 Å using the online version of the MADNES program.
AR crystallizes as a dimer with 2 * 460 amino acids in the asymmetric unit. The density of a crystal was determined to 1.27 g * cm$^{-3}$ by use of the method of the hydrostatic balance which is in good agreement with a value of the X-ray density. The reflection pattern indicates a monoclinic crystal lattice with unit cell dimensions a = 85.9 Å, b = 62.6 Å, c = 128.6 Å and $\beta$= 99.8 °. The volume of the unit cell is 682,020 Å³. The systematic lack of (hkl) reflections were observed for h+k=2n+1. Thus the crystals are monoclinic and belong to the space group C2. Assuming 8 molecules in the unit cell, equivalent to 2 molecules per asymmetric unit in the monoclinic unit cell results in a $V_m$value of 1.67 ...1.69 Å³/ Da.
According to Nonaka et al. (1985) the cell volume is 295,909 Å³. From this data which is based on a molecular weight of 110.5 kDa of  the asymmetric unit, a X-ray density of 1.25 g* cm$^{-3}$ has been calculated. The molar volume of the residues $V_m$ is not in the normal range for protein crystals  (Matthews, 1977).
Two reasons may explain the differences between the X-ray density and  the molar volume of residues in the paper of Nonaka et al. (1985) as compared to the parameters in this paper. On the one hand the volume of the unit cell according to Nonaka et al. (1985) is too small and on

the other the molecular mass of AR based on amino acid residues derived from cDNA sequencing including FAD is not higher than 52.5 kDa.

**Table 1: Unit cell volume, molecular weight, density and molar volume of adrenodoxin reductase**

| Parameter | according to Nonaka et al. (1985) with FAD (54.0 kd) | according to Nonaka et al. (1985) (55.2 kd) | this paper without FAD | this paper with FAD |
|---|---|---|---|---|
| Formation | rosette | rosette | hedgehog | hedgehog |
| Volume of unit cell [Å$^3$] | 296000 | 296000 | 682020 | 682020 |
| Number of molecules in the unit cell | 4 | 4 | 8 | 8 |
| Volume of one molecule [Å$^3$] | 74000 | 74000 | 85253 | 85253 |
| Molecular weight of residues Mr [d] | 54000 | 55250 | 50299 | 51083 |
| X-ray density for Mr [g* cm-$^3$] | 1.22 | 1.25 | 0.99 | 1.00 |
| Molar volume of residues Vm [Å$^3$/d] | 1.37 | 1.34 | 1.69 | 1.67 |
| Vol. of solvents in a unit cell [vol %] | 10 | 8 | 27 | 26 |
| Crystal density [g* cm-$^3$] | 1.25 | 1.25 | 1.27 | 1.27 |

The vectors of the primitive and the C-faced centered unit cell belong to the same crystal class with a correlation between their lattice vectors but differ in the space group symmetry. The amounts and directions of the both **a**- and **b**-vectors are nearly but not quite equal and the **c**-vector of the C-face-centered unit cell is very equal to **a** + 2**c** of primitive unit cell according to Nonaka et al. (1985). The data obtained in this paper provides a basis for obtaining precise results by X-ray crystal structure analysis.

**Adrenodoxin / adrenodoxin reductase complex**

Due to the complex character of the steroid hydroxylase consisting of 3 essential components it is considered important not only to elucidate the structure of each isolated component but also that of complexes between them. Therefore we carried out first experiments aimed at crystallizing complexes of Adx either with AR or CYP 11A1. As necessary requirement we used homogenous preparations of Adx consisting of 115 amino acids which were isolated from trypsinized Adx and homogenous AR. Suhara et al. (1982) described AR as being separable into 2 fractions differing in their content of sugar residues. The separability proved reproducible. But despite careful and broad analyses we were not able to find any molecular differences between both fractions (spectral behaviour, enzymatic activity, determination of N- and C-termini, sugar residues, mass spectrometry, crystallizability). Due to its little higher stability at storage and radiation, fraction 2 was cross-linked with Adx consisting of 115 amino acids (by means of 1-ethyl-3(3-dimethylaminopropyl)carbodiimide; ECD) and the resulting AR/Adx-complex isolated and analysed. Mass spectrometry revealed a resonable molecular mass of 62,990 ± 50 Da, the complex proved enzymatically active with regard to cytochrome c reductase activity (Hara et al., 1989). First crystals of the cross-linked AR/Adx-complex were obtained. After dissolution of the crystals SDS-PAGE clearly evidenced an AR/Adx complex.

Similar experiments were performed with the CYP 11A1/Adx-complex. The complex was cross-linked by means of ECD and chromatographically isolated after removal of the uncomplexed proteins. The molecular mass was determined with ~ 64 kDA by SDS-PAGE. Experiments to get crystals from the CYP 11A1/Adx complexes are under study.

**Acknowledgment**

The authors (R.K.and K.R.) are grateful to WIP for a grant. This study was further supported by DFG (grant YE2/D8 in Sfb 312) and by Fonds der Chemischen Industrie. The authors gratefully acknowledge continuous interest and helpful advice of Prof. Dr. W. Saenger. We also wish to thank Mrs. S. Just and Mr. R.Dettmer for skilful isolation and purification of the proteins and Dr. Franke for providing the mass spectrometer for the measurements.

**References**

1. Chashchin, V.L., Vasilievsky,.V.I. Shkumatov, V.M., and Akhrem, A.A.: The domain structure of the cholestreol side-chain cleavage cytochrome P-450 from bovine adrenocortical mitochondria. Biochim. Biophys.Acta 787, 27 - 38.
2. Hara, T., and Kimura, T. (1989): Purification and catalytic properties of a cross-linked complex between adrenodoxin reductase and adrenodoxin. J. Biochem (Tokyo) 105, 594 - 600.
3. Hiwatashi, A., Ichikawa, Y., Yamano,T., and Maruya, N. (1976): Properties of crystalline reduced nicotinamide dinucleotide phosphate adrenodoxin reductase from bovine adrenocortical mitochondria. II. Essential histidyl and cysteinyl residues at the NADPH binding site of NADPH-adrenodoxin reductase. Biochemistry 15, 3091-3096.
4. Kuban, R.-J., Marg, A., Resch, M., and Ruckpaul, K. (1993): Crystallization of bovine adrenodoxin-reductase in a new unit cell and its crystallographic characterization. J. Mol. Biol. 234, 245 - 248.
5. Marg, A., Kuban, R.-J., Behlke, J., Dettmer, R., and Ruckpaul, K. (1992): Crystallization and X-ray examination of bovine adrenodoxin. J. Mol. Biol. 227, 945 - 947.
6. Matthews, B.W. (1977): In *The Proteins* (Neurath, H. & Hill, R.L., eds.) pp. 403 - 590, Academy Press, New York.
7. Nonaka, Y., Aibara, S., Sugiyama, T., Yamano, T., and Morita, T.(1985): A crystallographic investiagtion on NADPH-adrenodoxin oxidoreductase. J.Biochem. (Tokyo) 98, 257 - 260.
8. Ogawa, K., Tsukihara, T., Tahara, H., Katsube, Y., Matsu-Ura, Y. Tanaka, N., Kakudo, M., Wada, K., and Masubara, H. (1977): Location of the iron-sulfur cluster in *Spirulina platensis* ferredoxin by X-ray analysis. J. Biochem.(Tokyo) 81, 529 - 531.
9. Sakihama, N., Hiwatashi.A., Miyatake, A., Shin, M., and Ichikawa, Y.(1088): Isolation and purification of nature bovine adrenocortical ferredoxin with an elongated carboxyl end. Arch. Biochem. Biophys. 264, 23 - 29.
10. Suhara, K., Nakayama, K., Takikawa, O., and Katagiri, M. (1982): Two forms of adrenodoxin reductase from mitochondria of bovine adrenal cortex. Eur. J. Biochem. 125, 659 - 664.

# An integrated, phase-theoretic approach to cytochrome P450c17 (CYP17) catalysis:
## Probabilities of hydroxyprogesterone stabilization vs decomposition depend on reductase "arrival" frequencies in stochastic relation to intermediate "departure" frequencies at the active site

W. Nikolaus Kühn-Velten

*Department of Obstetrics and Gynecology and Institute for Physiological Chemistry, Laboratory of Biochemical Endocrinology, Heinrich-Heine-Universität, Moorenstrasse 5, D-40225 Düsseldorf, Germany*

INTRODUCTION:
From a spatial, concentration-oriented towards a temporal, frequency-oriented view of CYP17 catalysis

Cytochrome P450c17 (CYP17), the second rate-limiting enzyme at the bifurcation of steroid hormone synthesis in the gonads and adrenals (Kühn-Velten, 1993), has been characterized as an enzyme protein with a bifunctional catalytic site. Its bifunctionality has been conventionally identified in terms of isolated enzyme activities (EC 1.14.99.9: steroid-17-monooxygenase, and EC 4.1.2.30: 17-hydroxyprogesterone aldolase, or steroid-17,20-lyase), but recent work has suggested that a more reasonable definition of CYP17 enzyme function should discriminate between substrate (e.g. progesterone) fractions undergoing either incomplete (or "abortive") transformation (i.e. 17-hydroxyprogesterone is released) or complete (or "productive") transformation (i.e. the intermediate is retained and androstenedione is produced). Alternatively, a primary catalytic event (first electron-pair transfer and oxygen activation cycle producing either hydroxyprogesterone as the intermediate from progesterone, or producing androstenedione from *exogenous* hydroxyprogesterone) should be differentiated from a secondary catalytic event (second coordinated electron-pair transfer and oxygen activation cycle which processes the enzyme-bound, *endogenous* intermediate towards androstenedione as the final product) (Kühn-Velten et al., 1991; Kühn-Velten & Lessmann, 1992). Both reaction cycles require the coordinated release of two electrons which are provided (directly or indirectly via cytochrome $b_5$) by the NADPH-ferrihemoprotein reductase (cytochrome reductase; EC 1.6.2.4).- Though an individual CYP17 molecule performs only one of both functions at a given time, depending on the type of ligand (exogenous or endogenous) bound within its catalytic site, the actual ratio of both functions catalyzed by a population of CYP17 enzyme molecules in their natural membrane environment varies to a considerable degree. This phenomenon can be adequately simulated *in vitro*, but up to now, detailed phenomenological descriptions failed to provide an unequivocal explanation for those changes of catalytic coupling efficiency.

A multitude of parameters has been demonstrated to affect CYP17 catalysis, including membrane fluidity, membrane lipid/protein concentration ratio, substrate accumulation in the membrane compartment, the continuous transition from 3-dimensional via 2-dimensional (intracellular membrane compartment) to nearly 1-dimensional (substrate-access channel) substrate diffusion, concentration ratio between CYP17 and the cytochrome reductase, possible facilitation by cytochrome $b_5$ of a 2nd $e^-$ transfer for an individual CYP17 reduction cycle, and intermediate retention within the CYP17 active site. It can be anticipated that such a variety of modulating parameters will make it extremely difficult to formulate a reaction mechanism based upon classical enzyme kinetics which relies primarily on concentration-dependent terms, since relative diffusion rates and collision probabilities can not adequately be described within the conventional system.

To achieve a prediction of the variability of intermediate processivity by CYP17, the present work proposes a novel, phase-theoretic approach to CYP17 catalysis. This strategy, which replaces the conventional ratios of concentration terms by ratios of frequencies of certain events on a common time-scale, is derived from previous attempts to apply stochastic models such as queuing theory to catalytic processes (Bartholomay, 1964). From a simplified outline of CYP17 catalysis (Fig.1), it can be derived that the mutually dependent and crucial parameters within this novel concept are the CYP17:reductase collision frequencies (or reductase arrival times) and the intermediate retention periods (or intermediate departure times). Since the formation of an androgen molecule from progesterone involves two catalytic cycles, each requiring productive CYP17:reductase collision, the working hypothesis was formulated that in this system, androgen formation will be favoured whenever the collision frequency is (stochastically) higher than the mean intermediate retention capability.

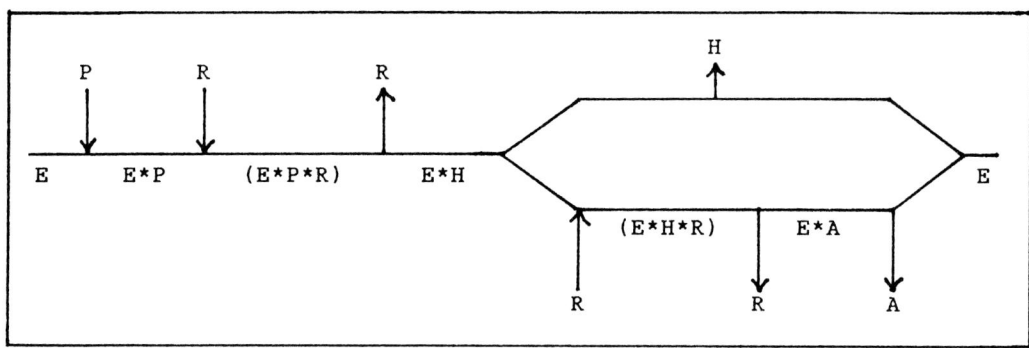

Fig.1. Simplified scheme of CYP17 (E) catalysis according to Cleland. Interaction with steroids (P = progesterone; H = hydroxyprogesterone; A = androstenedione) and with cytochrome reductase (R) are presented; electron-pair transfer is considered as a quasi-one-step process on this time-scale, possible effects of cytochrome $b_5$ are thus neglected, and the oxygen activation cycle is omitted for clarity. The central complexes are presented in parentheses while the transitory complexes are not. Androstenedione is only produced if the enzyme*intermediate complex receives a second $e^-$ pair through productive collision with a second reductase molecule within a sufficiently short time span (intermediate decomposition; otherwise, stabilization of the putatively labilized intermediate (yielding free hydroxyprogesterone) occurs.

## METHODS:
### Re-analysis of previous experiments involving variation of fractional saturation of CYP17 and variation of available electron flux rates

Initial-rate kinetic analyses of CYP17 with progesterone as its substrate have been performed using microsomal membrane suspensions from rat testis homogenates under various conditions: Firstly, progesterone concentrations were varied from 36 nM to 15 µM in the presence of 130 µM NADPH (Kühn-Velten et al., 1991). Secondly, ketoconazole, estradiol or promegestone were included as non-substrate-type competitive inhibitors of progesterone binding and turnover (Kühn-Velten et al., 1991; Kühn-Velten & Lessmann, 1992). Thirdly, exogenous 17-hydroxyprogesterone was added as a substrate-type competitive inhibitor. Finally, cytochrome c was included as a competitive inhibitor of electron transfer from the reductase to CYP17, resulting in a non-competitive inhibition of overall progesterone metabolism (Kühn-Velten & Vogt, submitted). In all cases, progesterone conversion to hydroxyprogesterone, androstenedione and testosterone was quantified; for the present study, the catalytic processivity was calculated which is defined as the ratio of the rates of secondary / primary catalytic events (cleavage rate / hydroxylation rate, or rate of androgen formation / rate of hydroxyprogesterone plus androgen formation, or rate of in-intermediate processing / rate of intermediate formation).

## RESULTS AND DISCUSSION:
### Prediction of CYP17 catalytic processivity and comparison with other steroidogenic P450 enzymes

It has repeatedly been shown that the catalytic processivity of CYP17 decreases continuously with increasing progesterone concentrations. A linear relation can be obtained if the processivity is referred to Y, the fractional saturation of the enzyme with the substrate. At a given substrate concentration, the processivity is enhanced in the presence of non-substrate-type competitors (though the absolute product formation rates are reduced), whilst addition of hydroxyprogesterone reduces the processivity of CYP17. Cytochrome c leads to a marked decrease of CYP17 processivity (Kühn-Velten et al., 1991; Kühn-Velten & Lessmann, 1992; Kühn-Velten & Vogt, submitted).

For all experiments evaluated so far, a linear relation exists between CYP17 processivity on the one hand (y) and the available electron transfer rate which is available for the 2nd CYP17 catalytic (cleavage) cycle (x; dimension: nM * min$^{-1}$ * nM CYP17$^{-1}$). This relation is described by the formula

$$y = +0.0839x + 0.196 \quad (r = +0.863; n = 136)$$

and is no longer dependent on the fractional saturation of CYP17. The available electron transfer rate corresponds to the collision frequency between the fraction of CYP17 species occupied by the metabolic intermediate and those reductase molecules which are not engaged in serving other CYP17 molecules for the first catalytic cycle (or other electron acceptors such as cytochrome c). Therefore, high substrate concentrations reduce CYP17 processivity since a relatively large fraction of reductase provides electrons for the hydroxylation cycles with the consequence that the transfer of a second electron pair through CYP17 collision with a reduced reductase becomes less probable. With high progesterone concentrations, this double-collision probability within a certain time limit becomes less than the probability that the CYP17-bound steroid intermediate stabilizes itself into free hydroxyprogesterone at the cost of androgen formation. Non-substrate-type competitive inhibitors reduce the fractional saturation of the enzyme with its substrate; the fraction of reductase molecules operating for the primary catalytic events is likewise reduced, and the probability of a second collision providing electrons for intermediate processing increases. If exogenous hydroxyprogesterone is added as a competitive inhibitor, its conversion to androstenedione employs a certain fraction of reductase molecules; therefore, the probability of a sufficiently short delay between CYP17 * progesterone : reductase collisions is reduced, and the androgen formation probability decreases in favour of hydroxyprogesterone ejection. Experiments performed in the presence of cytochrome c confirm previous assumptions that the second cycle of CYP17 catalysis is more sensitive to variation of electron transfer rates than the first one. Thus, the crucial parameter determining the efficiency of CYP17 catalysis is the fraction of the overall CYP17 reducing activity, corresponding to effective collision probabilities, that is available for the second (cleavage) catalytic cycle of CYP17 within a certain time span. Androgen formation rates by rat testicular CYP17 can therefore be interpreted as a sensor for protein-protein collision frequency in relation to intermediate lifetime within an active site: Intermediary hydroxyprogesterone decomposition into androstenedione indicates that the time difference between CYP17:reductase collisions was less than the intermediate retention time. On the other hand, the relatively low affinity of hydroxyprogesterone to CYP17, if compared with reaction intermediates of other multifunctional steroidogenic cytochromes P450 (e.g. CYP11A), can be reconsidered as a relatively low probability to be efficiently retained and processed by the active site of testicular CYP17. In conclusion, the results presented herewith allow for the first time a prediction of hormone secretion rates by a specialized multifunctional enzyme on the basis of an introduction of temporal aspects into the evaluation of cytochrome P450 catalysis.

REFERENCES:

Bartholomay, A.F. (1964): The general catalytic queue process. In *Stochastic Models in Medicine and Biology*, ed. J.Gurland, pp. 101-142. Madison: The University of Wisconsin Press.
Kühn-Velten, W.N. (1993): Cytochrome P450c17: Regulation of gene expression and enzyme function at the bifurcation in steroid hormone synthesis. In *Handbook of Experimental Pharmacology Vol.105: Cytochrome P450*, ed. J.B.Schenkman & H.Greim, pp.667-676. Berlin: Springer.
Kühn-Velten, W.N., Bunse, T. & Förster, M.E.C. (1991): Enzyme kinetic and inhibition analyses of cytochrome P450XVII, a protein with a bifunctional catalytic site. Quantification of effective substrate concentrations at the active site and their significance for intrinsic control of the hydroxylase/lyase reaction sequence. *J.Biol. Chem.* 266, 6291-6301.
Kühn-Velten, W.N. & Lessmann, M. (1992): Ketoconazole inhibition of the bifunctional cytochrome P450c17 does not affect androgen formation from the endogenous lyase substrate. The catalytic site remains refractory in the course of intermediary hydroxyprogesterone processing. *Biochem.Pharmacol.* 44, 2371-2378.

# Kinetic studies on a genetically engineered fused enzyme between rat cytochrome P4501A1 and yeast NADPH-P450 reductase

Toshiyuki Sakaki[1], Shiro Kominami[2], Shigeki Takemori[2], Hideo Ohkawa[3], Megumi Akiyoshi-Shibata[1], Yoshiyasu Yabusaki[1]

[1]Biotechnology Laboratory, Takarazuka Research Center, Sumitomo Chemical Co., Ltd., Takarazuka, Hyogo 665, Japan. [2]Faculty of Integrated Arts and Sciences, Hiroshima University, Hiroshima 730, Japan. [3]Department of Plant Protection, Kobe University, Kobe, Hyogo 657, Japan

## Summary

An expression plasmid for a fused enzyme between rat P4501A1 and yeast P450 reductase was constructed and introduced into *Saccharomyces cerevisiae* cells. The microsomal fraction prepared from the recombinant yeast cells was subjected to kinetic studies on zoxazolamine 6-hydroxylation. The apparent Km and $V_{max}$ values for the hydroxylation by the fused enzyme were 0.38 mM and 0.42 $sec^{-1}$, respectively. The rate constant for reduction of the fused enzyme with NADPH in the presence of 1 mM zoxazolamine was larger than 50 $sec^{-1}$, indicating that electrons are rapidly transferred from NADPH through FAD and FMN to the heme iron of the fused enzyme. The rate constant $k_{on}$ for the substrate binding to the fused enzyme was 25 $mM^{-1} \cdot sec^{-1}$ which is not much different from that of non-fused P4501A1. These results together with spectral data measured during the hydroxylation in the steady state suggest that the rate-limiting step of the reaction by the fused enzyme might be the release of the product.

## INTRODUCTION

We have constructed genetically engineered fused enzymes between P450 and NADPH-P450 reductase, and expressed them in yeast. (Murakami et al., 1987, Yabusaki et al., 1988, Shibata et al., 1990, Sakaki et al., 1990). The optimized structure of the fused enzyme resembles the naturally occurring soluble fused enzyme P450BM-3 (Ruettinger et al., 1989) except for the presence of an amino terminal membrane anchor in our fused enzymes.

Here, we constructed an fused enzyme between rat P4501A1 and yeast P450 reductase and analyzed its reaction mechanism in yeast microsomes.

MATERIALS AND METHODS

Stopped flow analysis was performed using a dual wavelength stopped flow device as described previously (Kominami et al., 1992).
Spectra during steady state zoxazolamine 6-hydroxylation were measured as follows. The reaction mixture of the microsomal fraction containing P4501A1 or the fused enzyme and the substrate with NADH was poured into both sample and reference cuvettes. After base line correction, NADPH was added to the sample cuvette, and the difference spectrum was measured after 3 min at 10 °C.

RESULTS AND DISCUSSION

P450 and NADPH-P450 reductase contents were determined in the microsomal fractions prepared from the recombinant yeast strains, AH22/pAMC1 (MC)(Oeda et al., 1985), AH22/pAMR2 (MR)(Murakami et al., 1990) and AH22/pAFCR1 (FCR) (Fig. 1). The ratio of P4501A1 to the reductase was estimated to be about 1 : 0.03 (mol/mol) in MC microsomes and 1 : 0.6 in MR microsomes.

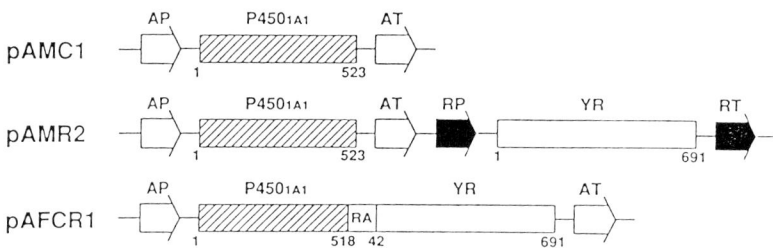

Fig. 1 Schematic representation of the expression plasmids: pAMC1 for rat P4501A1, pAMR2 for P4501A1 with overproduction of yeast P450 reductase (YR), and pAFCR1 for the fused enzyme between P4501A1 and YR. A and R indicate amino acid residues derived from the synthesized DNA linkers. Numbers indicate amino acid numbers of the corresponding enzymes. AP and AT ; ADH promoter and terminator, RP and RT ; yeast P450 reductase promoter and terminator

There are not much difference in Km values of P4501A1 in MC and MR microsomes, and the fused enzyme in FCR microsomes (0.32, 0.33 and 0.38 mM). The $V_{max}$ value of P4501A1 in MR microsomes was 0.29 $sec^{-1}$, 20 times larger than the $V_{max}$ in MC microsomes (0.015 $sec^{-1}$). The $V_{max}$ value of the fused enzyme was calculated to be 0.42 $sec^{-1}$, this being the highest among the three microsomes.
Figure 2 shows stopped flow analysis on the reduction of

heme iron by NADPH in the presence of zoxazolamine. The rate constants k in MC and MR microsomes were estimated to be 0.025 and 0.40 sec$^{-1}$, respectively. Comparison of the k and $V_{max}$ values suggests that first electron transfer could be the rate-limiting step of zoxazolamine 6-hydroxylation in MC and MR microsomes. On the other hand, we could not measure any NADPH-dependent reduction rate of the fused enzyme in FCR microsomes, since the reaction had been completed within the dead time of the device ( < 20 msec). On the other hand, we could not measure any NADPH-dependent reduction rate of the fused enzyme in FCR microsomes, since the reaction had been completed in the dead time of the device ( < 20 msec).

Stopped flow analysis on substrate-binding showed the $k_{on}$ values for P4501A1 and the fused enzyme to be 20 and 25 mM$^{-1}$·sec$^{-1}$, respectively. These data clearly indicate that substrate-binding is not the rate-limiting step of zoxazolamine 6-hydroxylation in the MC, MR and FCR microsomes.

Fig. 2 Typical kinetic traces of the reduction of P4501A1 and its fused enzyme in yeast microsomes.

The reduction of P450 in AH22/pAMC1 (A), AH22/pAMR2 (B), and AH22/pAFCR1 (C) microsomes was measured in the presence of CO and 1 mM zoxazolamine. The microsomal solution was mixed rapidly with an equal volume of NADPH solution and the time course of increase in the absorbance difference between 450 and 490 nm was measured with a dual wavelength device. Solid lines in (A) and (B) show the simulated curves. In (C), M and S indicate the initiation and termination of the mixing, respectively.

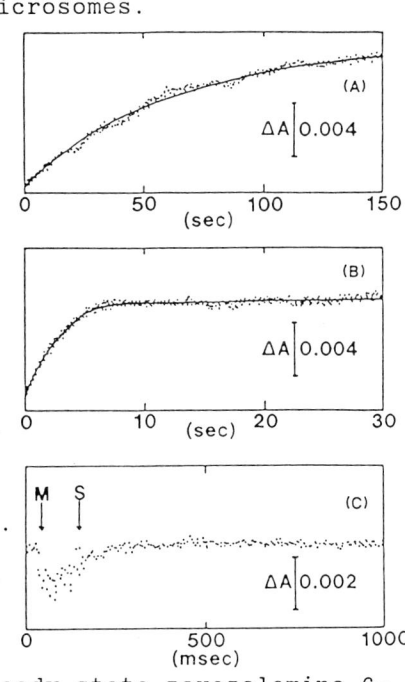

The difference spectra during steady state zoxazolamine 6-hydroxylation were quite different, while the P4501A1 content in MR microsomes is nearly identical to the content in MC microsomes. Both spectra have a trough around 460-480nm, but the trough in the MR microsome spectrum is much deeper than that observed in the MC microsome spectrum. These features are very simlar to the difference spectra of the reduced vs. oxidized form of P450 reductase. Thus, the difference in the spectra between MR and MC microsomes can be clearly explained by the contents of P450 reductase. In addition, these spectra suggested that P4501A1 during steady state zoxazolamine 6-hydroxylation in MC and MR microsomes is in the ferric high spin state. The difference spectrum of FCR microsomes during steady state zoxazolamine 6-hydroxylation was nearly the same as that of MR microsomes, suggesting that most of the heme iron of the fused enzyme during the steady state is also in the ferric high spin state. Since the first electron transfer is not rate limiting, most of the fused enzyme seems to be in the $Fe^{3+}$-SOH form where the product remains in the substrate binding pocket.

In conclusion, the rate limiting step in zoxazolamine 6-hydroxylation by the fused enzyme may be the release of the product. On the other hand, the rate limiting step of the reaction in MC microsomes is suggested to be the first electron transfer. In the case of MR microsomes, both steps might operate on a similar time scale.

## REFERENCES

Kominami,S., Ogawa,N., Morimune,R., De-Ying,H., & Takemori,S. (1992): The role of cytochrome $b_5$ in adrenal microsomal steroidogenesis. *J. Steroid Biochem. Molec. Biol.* 42, 57-64

Murakami,H., Yabusaki,Y., Sakaki,T., Shibata,M., & Ohkawa, H. (1987): A genetically engineered P450 monooxygenase: Construction of the functional fused enzyme between rat cytochrome P450c and NADPH-cytochrome P450 reductase. *DNA* 6, 189-197

Oeda,K., Sakaki,T., & Ohkawa,H. (1985): Expression of rat liver cytochrome P450MC cDNA in *Saccharomyces cerevisiae*. *DNA* 4. 203-210

Ruettinger,R.T., Wen, L.P., & Fulco, A.J. (1989): Coding nucleotide, 5' regulatory, and deduced amino acid sequences of P-450BM-3, a single peptide cytochrome P-450:NADPH-P-450 reductase from *Bacillus megaterium*. *J. Biol. Chem.* 264, 10987-10995

Sakaki,T., Shibata,M., Yabusaki,Y., Murakami,H., & Ohkawa, H. (1990): Expression of bovine cytochrome P450c21 and its fused enzymes with yeast NADPH-cytochrome P450 reductase in *Saccharomyces cerevisiae*. *DNA Cell Biol.* 9, 603-61

Shibata,M., Sakaki,T., Yabusaki,Y., Murakami,H., & Ohkawa, H. (1990): Genetically engineered P450 monooxygenases: Construction of bovine P450c17 yeast reductase fused enzymes. *DNA Cell Biol.* 9, 27-36

Yabusaki,Y., Murakami,H., Sakaki,T., Shibata,M., & Ohkawa,H. (1988): Genetically engineered modification of P450 monooxygenases: Functional analysis of the amino-terminal hydrophobic region and hinge region of the P450/reductase fused enzyme. *DNA* 7, 701-711

# Overexpression of yeast cytochrome $b_5$ gene can supress ketoconazole hypersensitivity of a NADPH-P450 reductase deficient strain

Gilles Truan[1], Jean-Charles Epinat[2], Claire Rougeulle[2], Christophe Cullin[1], Denis Pompon[1]

[1]Centre de Génétique Moléculaire du Centre National de la Recherche Scientifique, 91198 Gif-sur-Yvette Cedex, France. [2]Unité de Biologie Moléculaire de l'Expression Génique, Institut Pasteur, 25, rue du Docteur-Roux, 75015 Paris, France

Lanosterol-C14-demethylase is an essential P450 enzyme of the sterol pathway in yeast and a target for antifungal compounds such as ketoconazole. Absence of the NADPH-P450 reductase (Yred) decreases 20-fold the ketoconazole resistance. We have identified a multicopy suppressor of the ketoconazole hypersensitivity: the yeast cytochrome $b_5$ gene (*YB5*). Yeast cytochrome $b_5$ (Yb5) is a 120 aminoacid (aa) protein that shares about 30% sequence identity with mammalian cytochromes $b_5$ (b5). Disruption of the Yb5 gene is viable demonstrating that Yb5 is not an essential protein. Nevertheless, *YB5*-disrupted strain shows modifications of the redox environment in microsomal fractions.

## INTRODUCTION

In yeast *S. cerevisiae*, CYP51 (14DM) is a key enzyme in ergosterol biosynthesis thus rendering *CYP51* disruption lethal in aerobiosis (Kalb, V.F. et al., 1987). The 14DM is the target of various antifungal drugs, such as ketoconazole which inhibits the 14-demethylation step and can arrest cell growth by depletion of the ergosterol pool (Vanden Bossche, H. et al., 1990; Yoshida, Y. & Aoyama, Y., 1987). In contrast, disruption of yeast NADPH-P450 reductase-encoding gene (*YRED*) is not lethal but increases ketoconazole sensitivity (Sutter, T.R. & Loper, J.C., 1989). Therefore, the presence of another electron carrier was hypothesised. We describe here the cloning and characterisation of a suppressor of the ketoconazole hypersensitivity of a *YRED* disrupted strain.

## CLONING OF THE *YB5*

WRΔ is a strain that carries a null allele of the *YRED* gene (*YRED* disrupted by the *TRP1* gene). Ketoconazole at a concentration of 2 µg per ml of culture medium fully inhibits the growth of WRΔ whereas the parent strain is resistant to 30-40 µg of the inhibitor per ml of culture medium. WRΔ strain was transformed with a multicopy genomic library. Transformants were selected, pooled and sprayed onto minimal medium plates containing increasing concentrations of ketoconazole. Different classes of clones were made by pooling resistant clones at each concentration of ketoconazole. Plasmids from each class of resistant clones were analysed by restriction mapping. Restriction analysis of plasmid DNA extracted from classes resistant to 4 µg/ml of ketoconazole revealed the presence of a plasmid with a restriction pattern not corresponding to that of the YRED restriction pattern. The plasmid was isolated and named pDA.

## IDENTIFICATION OF THE YB5 AND SEQUENCING

Eight progressive deletions of pDA were constructed (Fig. 1). After WRΔ transformation, only the genomic fragments resulting from the first two deletions (1 and 2) gave rise to ketoconazole resistant clones. Thus the suppressor function was localised between deletion 2 and 3. Sequencing was performed on pDA, deletions of pDA and subclones of the region surrounding the deletions 1 to 5. An open reading frame (ORF) of 360 bp was found, encoding a putative protein of 13,287 Da.

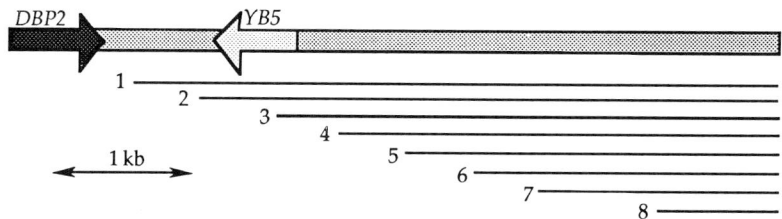

Fig. 1: Deletions of pDA and mapping of YB5. DBP2 is the 3' terminus of a gene which was mapped on chromosome XIII.

Sequence comparison of the predicted protein with the EMBL data bank using the FASTA program revealed a significant identity with other b5s (30% sequence identity with mammalian proteins). Hydropathy plot of the predicted protein showed a highly hydrophobic stretch at the C-terminal end of the sequence. This is in agreement with the knowledge that the C-terminal end of microsomal b5 is the membrane anchor (Mitoma, J.-Y. & Ito, A., 1992; Vergères, G. & Waskell, L., 1992). Alignments of the Yb5 protein sequence with other known b5s is shown on Fig. 2. Two highly conserved regions, surrounding the His37 and His64 residues of Yb5, can be aligned with sequences surrounding the heme iron binding histidines observed by crystallographic studies of a proteolytic fragment of bovine b5 (Mathews, F.S. et al., 1972).

```
Hum    .....MAEQSDEAVKYYTLEEIQKHNHSKSTWLILHHKVYDLTKFLEEHPGGEEVLREQAGGDATENFEDVG    70
Rab    .....MAAQSDKDVKYYTLEEIKKHNHSKSTWLILHHKVYDLTKFLEEHPGGEEVLREQAGGDATENFEDVG    70
Rat    ......AEQSDKDVKYYTLEEIQKHKDSKSTWVILHHKVYDLTKFLEEHPGGEEVLREQAGGDATENFEDVG    69
Hor    .......EDASKAVKYYTLEEIKKHNHSKSTWLILHHKVYDLTKFLEDHPGGEEVLREQAGGDATENFEDIG    68
Bov    .....MAEESSKAVKYYTLEEIQKHNNSKSTWLILHYKVYDLTKFLEEHPGGEEVLREQAGGDATENFEDVG    70
Pig    ......AEQSDKAVKYYTLEEIQKHNSKSTWLILHHKVYDLTKFLEEHPGGEEVLREQAGGDATENFEDVG    69
Chi    MVGSSEAGGEAWRGRYYRLEEVQKHNNSQSTWIIVHHRIYDITKFLDEHPGGEEVLREQAGGDATENFEDVG    75
Cau    ........MASEKKVLGFEEVSQHNKTKDCWLIISGKVYDVTPFMDDHPGGDEVLLSSTGKDATNDFEDVG    66
Yea    ...........MPKVYSYQEVAEHNGPQNFWIIIDDKVYDVSQFKDEHPGGDEIIMDLGGQDATESFVDIG    63
Con                 E...H......W.I.....YD...F...HPGG.E......G.DAT..F.D.G

Hum    HSTDAREMSKTFIIGELH....PDDRPKLNKPPETLITTIDSSSSWWTNWVIPAISAVAVALMYRLYMAED   134
Rab    HSTDARELSKTFIIGELH....PDDRSKLSKPMETLITTVDSNSSWWTNWVIPAISALIVALMYRLYMADD   134
Rat    HSTDARELSKTYIIGELH....PDDRSKIAKPSETLITTVESNSSWWTNWVIPAISALVVALMYRLYMAED   133
Hor    HSTDARELSKTFIIGELH....PDDRSKIAKPVETLITTVDSNSSWWTNWVIPAISAVVVALMYRIYTAED   132
Bov    HSTDARELSKTFIIGELH....PDDRSKITKPSESIITTIDSNPSWWTNWLIPAISALFVALIYHLYTSEN   134
Pig    HSTDARELSKTFIIGELH....PDDRSKIAKPSETLITTVESNSSWWTNWVIPAISALVVSLMYHFYTSEN   133
Chi    HSTDARALSETFIIGELH....PDDRPKLQKPAETLITTVQSNSSSWSNWVIPAIAAIIVALMYRSYMSE.   138
Cau    HSDTARDMMEKYYIGEIDSSTVPATRTYVAPVQPAYNQDKTPEFMIKILQFLVPILILGLALVVRQYTKKE   134
Yea    HSDEALRLLKGLYIGDVDK...TSERVSVEKVSTSENQSKGSGTLVVIL....AILMLGVAYYLLNE....   120
Con    HS..A........IG........R........................I.................
```

Fig. 2: Multiple alignment of Yb5 among other b5s. Hum: human; Rab: rabbit; Rat: rat; Bov: bovine; Hor: horse; Pig: pig; Chi: chicken; Cau: cauliflower; Yea: yeast; Con: consensus.

## *YB5* -DISRUPTED STRAIN PHENOTYPE

An heterozygous strain was constructed for *YB5* gene (*YB5* disruption by the *HIS3* gene). After sporulation, tetrads were dissected and analysed for spores viability and for histidine prototrophy. Surprisingly, there was no difference of viability among the four spores, demonstrating that disruption of *YB5* is not lethal. Furthermore, mutant spores (disrupted for *YB5*, His$^+$ phenotype) do not have lengthened generation time nor modification of ketoconazole sensitivity compared to wild-type strain. Each of the four spores was analysed for their b5 and cytochrome P450 content as shown on Fig. 3.

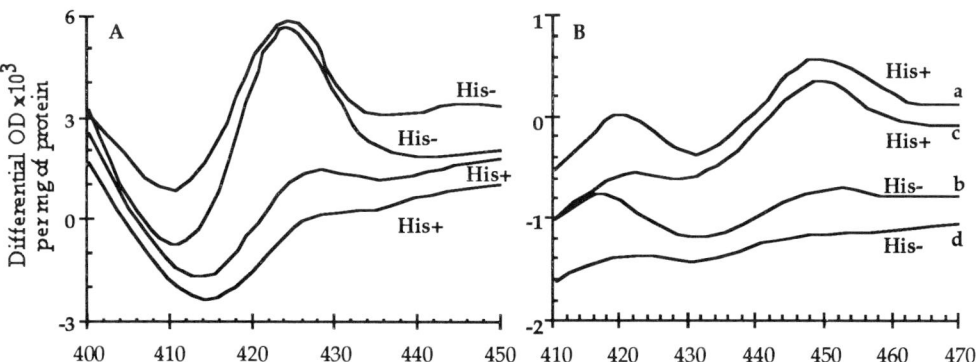

Fig. 3: b5 and P450 spectra of the four spores from a tetrad. A: reduced versus oxidised differential spectra and B: reduced carbon monoxide difference spectra of the microsomal fractions from each spore. Each spectrum was normalised by dividing the differential absorbance by the protein content.

# CONCLUSION

We have cloned a yeast gene encoding microsomal cytochrome $b_5$, by suppression of the ketoconazole sensitivity of a NADPH-P450 reductase deficient strain. It is well known that b5 can play certain roles in the sterol biosynthesis pathway (Aoyama, Y. et al., 1981; Reddy, V.V.R. et al., 1977). Nevertheless, the cloning of YB5 as a multicopy suppressor of the ketoconazole hypersensitivity of a YRED disrupted strain was surprising. The role of Yb5 as the sole electron donor for 14DM raise questions that will de addressed in the future. The nearly total absence of phenotype for the YB5 disrupted strain is intriguing, suggesting that other b5 isoforms could exist. Further studies, particularly at the level of 14DM activity in different strains, will be carried out to analyse the role of Yb5 in the microsomal redox environment.

# REFERENCES

Aoyama, Y., Yoshida, Y., Sato, R., Susani, M. and Ruis, H. (1981): Involvment of cytochrome $b_5$ and a cyanide-sensitive monooxygenase in the 4-demethylation of 4,4-dimethyl-zymosterol by yeast microsomes. *Biochimica et Biophysica Acta* 663, 194-202.

Kalb, V.F., Woods, C.W., Turi, T.G., Dey, C.R., Sutter, T.R. and Loper, J.C. (1987): Primary structure of the P450 lanosterol demethylase gene from *Saccharomyces cerevisiae*. *DNA* 6, 529-537.

Mathews, F.S., Argos, P. and Levine, M. (1972): The structure of cytochrome $b_5$ at 2.0 angstroms. *Cold Spring Harbor Symp. Quant. Biol.* 36, 387-395.

Mitoma, J.-Y. and Ito, A. (1992): The carboxy-terminal 10 aminoacid residues of cytochrome $b_5$ are necessary for its targetting to the endoplasmic reticulum. *EMBO J.* 11, 4197-4203.

Reddy, V.V.R., Kupfer, D. and Caspi, E. (1977): Mechanism of C-5 double bond introduction in the biosynthesis of cholesterol by rat liver microsomes. *J. Biol. Chem.* 252, 2797-2801.

Sutter, T.R. and Loper, J.C. (1989): Disruption of the *Saccharomyces cerevisiae* gene for NADPH-cytochrome P450 reductase causes increased sensitivity to ketoconazole. *Biochem. Biophys. Res. Commun.* 160, 1257-1266.

Vanden Bossche, H., Marichal, P., Gorrens, J., Bellens, D., Moereels, H. and Janssen, P.A. (1990): Mutation in cytochrome P-450-dependent 14 alpha-demethylase results in decreased affinity for azole antifungals. *Biochemical Society Transactions* 18, 56-59.

Vergères, G. and Waskell, L. (1992): Expression of cytochrome $b_5$ in yeast and characterisation of mutants of the membrane anchoring domain. *J. Biol. Chem.* 267, 12583-12591.

Yoshida, Y. and Aoyama, Y. (1987): Interaction of azole antifungal agents with cytochrome P-45014DM purified from Saccharomyces cerevisiae microsomes. *Biochemical Pharmacology* 36, 229-235.

# Structure-function interrelationships in mitochondrial and microsomal cytochrome P-450-dependent monooxygenases

S.A. Usanov[1], M. Iwasaki[2], M. Negishi

[1]Institute of Bioorganic Chemistry, Academy of Sciences of Belarus, Minsk, Belarus. [2]National Institute of Environmental Health Sciences, NIH, Research Triangle Park, N.C. 27709, USA

## INTRODUCTION

The cytochrome P-450 (P-450) enzymes are a superfamily of monooxygenases comprising about 180 individual proteins (Nelson et al., 1993). Two types of P-450-dependent monooxygenases localized in mitochondria and microsomes of different organs and tissues as well as bacteria are responsible for the extreme diversity in hydroxylase activity. The mitochondrial P-450 require FAD-containing reductase and iron sulphur protein for their activity whereas microsomal one need only a reductase containing both FAD and FMN. The interaction between the components of the two types of P-450 systems as well as regulation of their specificity and coupling of electron transfer during monooxygenation are still not clear. P-450-dependent monooxygenases both from mitochondria (steroid hydroxylases from adrenocortical mitochondria) and microsomes (steroid 7a-, 15a-, and 16a-hydroxylases and coumarin 7-hydroxylase) are involved in stereoselective hydroxylation of steroids.
Recently, specific interaction between $b_5$ and P-450scc from bovine adrenal cortex was demonstrated (Usanov & Chaschin, 1991). Cytochrome $b_5$ stimulates cholesterol side chain cleavage being unable to substitute adrenodoxin (Ad) in the reconstituted system. The aim of the present work is to understand the interaction between the components of the two types of P-450 systems via heterologous reconstitution of activity with respect to steroids.

## METHODS

Site-directed mutagenesis and expression. cDNAs encoding wild type of P-$450_{7a}$, P-$450_{15a}$ and P-450coh were cloned in pAAH5 vector and expressed in *Saccharomyces cerevisiae* AH22 cells as described (Iwasaki et al., 1991 and 1992; Junonen et al., 1991 and 1992). Construction of P-$450_{7a}$ mutants was carried out using a kit supplied by Promega (Iwasaki et al., 1993). The mutated P-$450_{7a}$ were ligated to the yeast expression vector pAAH5 and expressed in *Saccharomyces cerevisiae* AH22 cells as previously described (Juvonen et al., 1991).
Protein purification. P-$450_{7a}$, P-$450_{15a}$ and P-450coh and mutants have been purified from the microsomes of the recombinant yeast by the combination of chromatography on aminooctyl Sepharose 4B and hydroxylapatite columns

according previously published methods (Iwasaki et al., 1991 and 1992; Juvonen et al., 1991 and 1992). NADPH-P-450 reductase and $b_5$ were co-purified during purification P-450coh from the pyrazole-treated mouse by the combination of affinity chromatography on 2',5'-ADP-Sepharose 4B and ion-exchange chromatography, respectively.

P-450scc, P-450$_{11}$, adrenodoxin reductase (AdR) and Ad have been purified from the bovine adrenocortical mitochondria as previously described (Usanov et al., 1984). Antibodies to AdR and Ad were prepared as previously described (Usanov et al., 1989).

Determination of steroid hydroxylase activity. Reconstituted system consisted of purified P-450 (100 pmol), NADPH-P-450 reductase (30 pmol) or AdR and Ad, NADPH (0.5 mM), $MgCl_2$ (5 mM), dilauroylphosphatidylcholine (6.0 nmol), and [4-$^{14}$C]testosterone (100 uM) in 0.5 ml of 50 mM Tris-HCl buffer, pH 7.5. The steroid metabolites were extracted with methylene chloride, separated by thin layer chromatography and exposed to the X-ray film. The spots containing the metabolites were scraped from plates and counted using liquid scintillation counting.

## RESULTS AND DISCUSSION

The mouse 2A subfamily - steroid 7a- and 15a-hydroxylases (2A4) and coumarin 7-hydroxylase (2A5) - provide an excellent model to study structure-function relationships between the components of the two types of P-450 systems since the subfamily members share high amino acid sequence similarities, but exhibit different hydroxylase activity. P-450$_{7a}$ from mouse liver is 70% and 71% identical in its amino acid sequence to the P-450$_{15a}$ and P-450coh, respectively (Negishi et al., 1992). Mouse P-450$_{7a}$ exhibits the high amino acid sequence identity (88.4%) to the rat testosterone 7a-hydroxylase P-4502A1 (Matsunaga et al., 1990). In contrast to P-450coh and P-450$_{15a}$ which contain the Phe and Leu at position 209, respectively, that play the critical role in determining the steroid hydroxylase activity of P-450, mouse P-450$_{7a}$ has Asn at position 209 (Iwasaki et al., 1993). Despite the diversity, the amino acid region around the residues 209 of P-450$_{7a}$, P-450coh and P-450$_{15a}$ should conserve a similar function to interact with the steroid molecule. The members of 2A subfamily share only 23% of amino acid similarity with P-450scc.

From the members of mouse 2A subfamily, only P-450$_{7a}$ was able to catalyze testosterone 7a-hydroxylation in the reconstituted system containing the components of mitochondrial cytochrome P-450-dependent monooxygenases - AdR and Ad (Fig. 1).

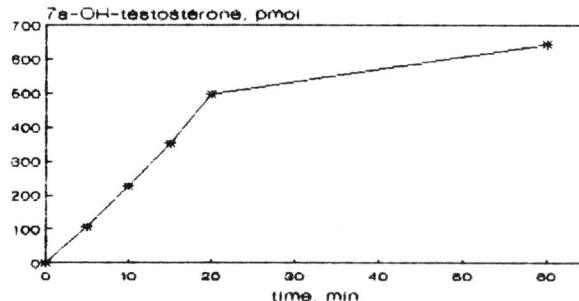

Fig. 1. Kinetic of testosterone hydroxylation at 7a-position in heterologously reconstituted system containing recombinant P-450$_{7a}$ (0.1 nmol), AdR (1.3 nmol) and Ad (4.65 nmol)

The formation of 7a-hydroxytestosterone is linear over 20 min and decreases thereafter. This indicates that microsomal P-450$_{7a}$ is able to accept electrons from the AdR and Ad - the mitochondrial electron transfer servicing proteins and selectively hydroxylate testosterone at 7a-position as P-450$_{7a}$ does in the presence of NADPH-P-450 reductase. However, heterologous reconstitution needs 10-50-fold excess of mitochondrial electron transfer proteins. The efficiency of the heterologous system with respect to testosterone is 50-100-fold less then in the presence of natural electron donor - NADPH-P-450-reductase.

Mitochondrial P-450scc and P-450$_{11}$ were found to be unable to hydroxylate neither cholesterol nor 11-deoxysteroids in the presence of microsomal flavoprotein (data not shown).

Subfamily 2A members P-450$_{7a}$ and P-450$_{15a}$ specifically catalyze testosterone 7a- and 15a-hydroxylations, respectively. In spite of their divergent catalytic activities, the structures of these P-450 are very similar. Nevertheless, neither P-450$_{15a}$ having the similar reaction product in reverse ratio nor P-450coh couldn't catalyze testosterone and coumarin hydroxylation in heterologous system, respectively. P-450$_{7a}$ and 450$_{15a}$ being able to bind the same substrate - testosterone - differ in their abilities to be stimulated by b$_5$ which is thought to be connected with the absence of Arg at position 129 in P-450$_{15a}$ - the key residue responsible for the interaction with b$_5$ (Iwasaki et al., 1992).

Besides 7a-hydroxy-, 15a-hydroxytestosterone is usually formed in heterologous system despite of the fact that P-450$_{15a}$ is inactive in the heterologous system. This fact indicates that P-450$_{7a}$ and 450$_{15a}$ being able to bind the same substrate - testosterone - differ in abilities to except electrons from Ad.

The reaction is obligatory with respect to P-450$_{7a}$ and increases with the increase of P-450$_{7a}$ concentration up to 100-200 pmol per tube. AdR and Ad are also obligatory for the heterologous reconstitution and maximal rate is reached at 3.0 and 4.5 nmol of AdR and Ad, respectively. Exception of any protein from the reaction mixture results in the loss of activity. Addition of the specific IgG against AdR or Ad inhibits the activity up to 60 and 80%, respectively (not shown).

P-450$_{7a}$ mutants (Gln207Gly, Asn209Leu, Lys209Gln) were used to understand the role of these amino acid residues in the monooxygenation. Mutants Gln207Gly and Asn209Leu which are less active then wild type P-450$_{7a}$ in hydroxylation of testosterone at 7a-position proved to be also less effective in the heterologous system, while Lys209Gln mutant of P-450$_{7a}$ having the same activity as wild type was also active in heterologous system.

The results obtained in the present work indicate that ability to be reconstituted in the presence of AdR and Ad is tightly correlates with 7a-hydroxylase activity of P-450$_{7a}$. Neither P-450$_{15a}$ that able to to selectively bind testosterone but unable to interact with b$_5$ nor P-450coh which is sensitive to b$_5$ but very loosely binds testosterone are unable to serve as efficient catalysts in heterologous system. There are at least two requirements for testosterone hydroxylation in heterologous system: (i) P-450 should be able to specifically bind steroid molecule; (ii) P-450 should be able to interact and except electrons from b$_5$ or Ad. The activities of P-450coh and P-450$_{7a}$ having Arg at the position 129 were found to stimulated by b$_5$ indicating that Arg$^{129}$ plays a key role in the process of interaction with b$_5$ and b$_5$ does not effect or even inhibit reactions catalyzed by P-450$_{15a}$ that Ser at position 129 (Juvonen et al., 1992). P-450coh also catalyzes testosterone 7a- and 15a-hydroxylations, but 50- and 100-fold lesser levels, when they were compared with the activities of wild type P-450$_{7a}$ and P-450$_{15a}$, respectively.

7a-Hydroxylation activity of P-450coh is increased by cytochrome b$_5$ when residue 209 is presented by Phe, Asn or Ser. 15a-Hydroxylation is stimulated by cytochrome b$_5$ when residue 209 is Phe, Ser, Lys and to a lesser extent when Asn (Juvonen et al., 1992). Recently, we have shown that b$_5$ and Ad compete for the same binding site on the P-450scc molecule. That means that Arg$^{129}$ may be the potential Ad-binding site on the P-450$_{7a}$.

In summary, the data presented in this work indicate that there is no strictly conserved mechanism of interaction between the components of microsomal and mitochondrial monooxygenases and highly homologous P-450 from 2A subfamily differ in their ability to accept electrons from Ad and this ability is tightly connected with regioselectivity of steroid hydroxylation.

Acknowledgements. This work was supported by Fogarty International Center of the National Institutes of Health, Public Health Service, U.S.A. is a recipient of CEEHS Fellowship.

## REFERENCES

Bernhardt R., Gunsalus I.C. (1992): Reconstitution of cytochrome P-4502B4 (LM2) activity with camphor and linalool monooxygenase electron transfer. Biochem. Biophys. Res. Commmuns. 187:310-317

Juvonen R.O., Iwasaki M., Negishi M. (1991): Structural Function of Residue-209 in Coumarin 7-Hydroxylase (P-450coh). Enzyme kinetic studies and site-directed mutagenasis. - J. Biol. Chem. 66.5.16431-16435

Juvonen R.O., Iwasaki M., Negishi M. (1992) Roles of residues 129 and 209 in the interaction by cytochrome $b_5$ of hydroxylase activity in mouse 2A P-450s. Biochemistry. 31. 46. 11519-11523.

Matsunaga T., Nomoto M., Kozak A., Gonzalez F.J. (1990): Structure and in vitro transcription of the rat CYP2A1 and CYP2A2 genes regional localization of the CYP2A gene subfamily on mouse chromosomes. 29. 5. 1329-1341.

Negishi M., Iwasaki M., Juvonen R.O., Aida K. (1992): Alteration of the substrate specificity of mouse 2A P450s by the identity of residue-209: steroid-binding site and orientation. J. Steroid Biochem. Molec. Biol. 43: 1031-1036

Nelson D.R., Kamataki T., Waxman D.J., Guengerich F.P., Estabrook R.W., Feyereisen R., Gonzalez F.J., Coon M.J., Gunsalus I.C., Gotoh O., Okuda K., Nebert D. (1993): The P450 superfamily: update on new sequences, gene mapping, accession numbers, early trivial names of enzymes, and nomenclature. DNA and Cell Biol. 12: 1-51.

Iwasaki M., Juvonen R., Lindberg R., Negishi M. (1991): Alteration of High and Low Spin Equilibrium by a Single Mutation of Amino Acid 209 in Mouse Cytochromes P450. - J. Biol. Chem. 1991. 266. 6. 3380-3382.

Iwasaki M., Lindberg R., Juvonen R., Negishi M. (1993): Site directed mutagenesis of mouse steroid 7a-hydroxylase (cytochrome P-$450_{7a}$): role of residue-209 in determining steroid-P-450 interaction. Biochem. J., in press

Usanov S.A., Pikuleva I.A., Chashchin V.L., Akhrem A.A. (1984): Chemical modification of adrenocortical cytochrome P-450scc with tetranitromethane. Biochim. Biophys. Acta. 790. 3. 259-267.

Usanov S.A., Chernogolov A.A., Chashchin V.L. (1989): Inhibitory domain-specific antibodies to cytochrome P-450scc. FEBS Letts. 255. 1. 125-128

Usanov S.A., Chashchin V.L. Interaction of cytochrome P-450scc with cytochrome $b_5$. - FEBS Letts. 1991. 278.2. 279-282

Cytochrome P450. 8th International Conference. Ed. M.C. Lechner. John Libbey Eurotext, Paris © 1994, pp. 441-444.

# Direct evidence of the electron releasing character of the cysteinate proximal ligand of cytochrome P450 from X-ray absorption near edge-spectroscopy

John H. Dawson[1], Hongbin Isaac Liu[2], Masanori Sono[1], Saloumeh Kadkhodayan[1], Britt Hedman[3], Keith O. Hodgson[2,3]

[1]Department of Chemistry and Biochemistry, University of South Carolina, Columbia, SC 29208 USA. [2]Department of Chemistry and [3]Stanford Synchrotron Radiation Laboratory, Stanford University, Stanford, CA 94305 USA

## INTRODUCTION

The cytochromes P-450 are found throughout the plant and animal kingdom (Sato & Omura, 1978) and are noted for their ability to insert one atom of dioxygen into unactivated C-H bonds (eq. 1) (Dawson, 1988). Because catalytic oxygen atom insertion from dioxygen into hydrocarbons is not a well-precedented reaction, there has been substantial interest in the mechanism of action of P-450. The name P-450 was derived from the unusually red-shifted Soret absorption band near 450 nm for the ferrous-CO derivative of the enzyme. This property has been clearly shown to depend on the presence of an anionic sulfur donor atom from the cysteine proximal heme iron ligand in P-450 (Scheme 1) (Dawson & Sono, 1987). However, the role that ligand may play in the mechanism of action of the enzyme is less well established.

$$R\text{-}H + O_2 + 2H^+ + 2e^- \longrightarrow R\text{-}OH + H_2O \qquad (1)$$

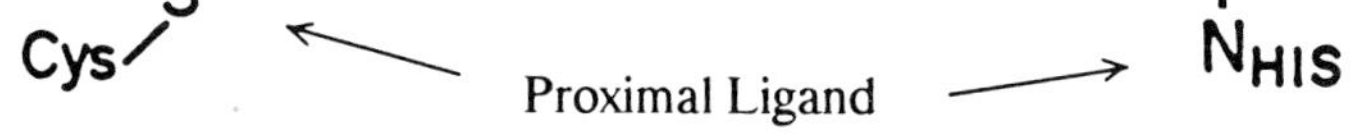

Scheme 1. The Heme Iron Coordination Structures of Ferric Cytochrome P-450 (left) and Myoglobin (right). The porphyrin is abbreviated as a parallelogram of pyrrole nitrogens.

The myoglobins are monomeric heme-containing dioxygen carriers found in the vertebrate muscle tissue (Buchler, 1978). The proximal ligand to the heme iron of the myoglobins is a neutral nitrogen donor atom from the imidazole ring side chain of histidine (Scheme 1). The ability of the ferrous myoglobins and related tetrameric hemoglobins to bind dioxygen reversibly, as required for respiration to occur, has led to intense scrutiny of these proteins as well.

X-ray absorption spectroscopy using synchrotron radiation has proven to be a particularly useful technique because it directly probes the properties of the central metal in complex biological systems (Penner-Hahn & Hodgson, 1989; Andersson & Dawson, 1990). The local coordination structures of the metal sites for numerous heme iron proteins including four P-450 states have been studied by extended X-ray absorption fine structure (EXAFS) spectroscopy (Andersson & Dawson, 1990). The X-ray crystal structure of three of these four P-450 states have been reported (Raag & Poulos, 1992). The metal-ligand bond distances reported with these two methods are in close agreement (Andersson & Dawson, 1990). Using this method, it is possible to determine $Fe-N_{pyrrole}$ and $Fe-X_{axial}$ bond distances to an accuracy of 0.01 to 0.02 Å. However, as usually applied, EXAFS is insensitive to the three-dimensional arrangement of atoms. On the other hand, the X-ray absorption near edge spectroscopy contains information about the oxidation state, electronic structure, and coordination geometry of the heme iron such as Fe-ligand bond angles and changes in site symmetry due to small atomic displacement around iron. X-ray absorption edge spectra are easier to measure than EXAFS spectra and can be obtained on less concentrated samples. Comparison of the edge spectra of P-450 and myoglobin will increase our understanding of the electronic properties of their metal centers.

We have systematically examined parallel low-spin ferric and ferrous derivatives of P-450 and myoglobin in order to carefully compare the energy of the X-ray absorption edge under conditions where the oxidation, spin, and coordination states of the pair are kept constant. In this way, the specific effect of the thiolate ligand that is present in P-450 on the properties of the central heme iron can be gauged. It has been proposed that the cysteinate proximal ligand of P-450 plays a key mechanistic role as a strong internal electron donor to facilitate cleavage of the O-O bond of the putative iron-peroxide intermediate to generate the active hydroxylation catalyst (Dawson et al., 1976; Dawson, 1988). The shifts in the energy of the X-ray absorption edge reported herein for P-450 derivatives relative to those for parallel myoglobin ligand adducts provide the most direct evidence to date for the specific influence of the cysteinate ligand on the properties and possibly on the reactivities of the heme iron center of cytochrome P-450.

## RESULTS AND DISCUSSION

I. *Low-Spin Ferric State.* The rationale of the present study is that the energy of the X-ray absorption near edge for two heme protein derivatives examined under conditions of constant oxidation, spin and coordination state will be influenced by the electron donor properties of their respective axial ligands. To test this hypothesis, it is first necessary to show that the edge energy is the same for parallel P-450 and myoglobin complexes having the same set of axial ligands (proximal thiolate/exogenous imidazole for P-405, *vice versa* for myoglobin). Several imidazole adducts of ferric P-450 and thiolate complexes of ferric myoglobin have been compared; the data for one such comparison is reported in Table I. The energies of the X-ray absorption near edge is the same, within experimental error, for each protein under the conditions where they have the same ligand set (as well as the same oxidation, spin and coordination state).

With the initial control experiment completed, we next examined the effect of changing the exogenous ligand of myoglobin from imidazole to thiolate. Consistent edge shifts were observed (see Table I for one such result) when imidazole-ligated myoglobin (*bis*-imidazole ligand set) is compared with thiolate-ligated myoglobin (thiolate/imidazole ligand set). The energy difference for the absorption edge of these two myoglobins is $1.07 \pm 0.13$ eV. The absorption edge of 1-propanethiolate-ligated ferric myoglobin is shifted in the same direction (*i.e.*, to lower energy) as occurs upon reduction of the ferric metal center (Penner-Hahn & Hodgson, 1989). Thus, the presence of a thiolate ligand has the same effect on the energy of the X-ray absorption edge as lowering the oxidation state. The thiolate ligand therefore appears to increase the electron density at the central heme iron relative to imidazole ligated heme centers.

Next, we compared the near edge spectra of 1-methylimidazole-bound ferric P-450 and myoglobin. As in the previous experiment, this is a comparison of thiolate/imidazole versus *bis*-imidazole ligand sets. Once again, there is a shift in the energy of the edge (Table I, $0.98 \pm 0.17$ eV) in the direction of ferrous state for both imidazole-ligated P-450 when compared with imidazole-ligated myoglobin. The difference in the edge energy for the imidazole adducts of ferric P-450 and myoglobin is close to that observed for thiolate-ligated *vs.* imidazole-ligated ferric myoglobin. The X-ray absorption edge spectra are clearly sensitive to the presence of the thiolate ligand in P-450 and the shift in the energy of the edge suggests that the thiolate ligand is more electron-donating relative to the imidazole of myoglobin. This is consistent with the proposed role of the thiolate proximal ligand of P-450 as a strong electron donor to promote the cleavage of the bound peroxide in the putative ferric-peroxide intermediate to generate the putative ferryl-oxo "active oxygen" state of the cycle (Dawson *et al.*, 1976; Dawson, 1988).

Table 1. Relative X-Ray Absorption Near Edge Energy Shift.

| Sample Comparison | Edge Shift |
| --- | --- |
| Ferric-1-propanethiolate myoglobin *vs.* ferric-1-methylimidazole P-450-CAM | No shift |
| Ferric-1-methylimidazole myoglobin *vs.* ferric-1-propanethiolate myoglobin | $1.07 \pm 0.13$ eV |
| Ferric-1-methylimidazole myoglobin *vs.* ferric-1-methylimidazole P-450-CAM | $0.98 \pm 0.17$ eV |
| Ferrous-CO myoglobin *vs.* ferrous-CO P-450-CAM | $1.62 \pm 0.09$ eV |

II. *Low-Spin Ferrous State.* Finally, the spectra of low-spin ferrous-CO myoglobin and P-450 were studied. This comparison is between thiolate/CO and imidazole/CO ligand sets. Once again, the edge energy observed for ferrous-CO P-450 is at lower energy than that of ferrous-CO myoglobin (Table I, 1.62 ± 0.09 eV). As above, we interpret this result as indicating that the thiolate axial ligand of P-450 is more electron releasing than the imidazole ligand of myoglobin.

CONCLUSION

Direct evidence has been obtained through the use of X-ray absorption near edge spectroscopy to show that a thiolate axial ligand is more electron releasing to the central heme iron than an imidazole axial ligand. This provides support for the supposition that the thiolate ligand of cytochrome P-450 is capable of serving as a strong electron donor to the proposed ferric peroxide intermediate and thereby facilitating the cleavage of the peroxide O-O bond to yield the putative ferryl-oxo "active oxygen" state.

*Acknowledgements*: Support provided to JHD and KOH by the National Science Foundation. The Stanford Synchrotron Radiation Laboratory is supported by the Department of Energy (BES and OHER) and in part by the National Institutes of Health NCRR BRTP program.

REFERENCES

Andersson, L.A., and Dawson, J.H. (1990): EXAFS Spectroscopy of Heme-Containing Oxygenases and Peroxidases. *Struct. Bonding* 74: 1-40.

Buchler, J.W. (1978): Hemoglobin - An Inspiration for Research in Coordination Chemistry. *Angew Chem. Int. Ed. Engl.* 17: 407-423.

Dawson, J.H. (1988): Probing Structure-Function Relations in Heme-Containing Oxygenases and Peroxidases. *Science* 240: 433-439

Dawson, J.H., Holm, R.H., Trudell, J.R., Barth, G., Linder, R.E., Bunnenberg, E., Djerassi, C., and Tang, S.C. (1976): Oxidized Cytochrome P-450. Magnetic Circular Dichroism Evidence for Thiolate Ligation in the Substrate Bound Form. Implications for the Catalytic Mechanism. *J. Am. Chem. Soc.* 98: 3707-3709.

Dawson, J.H., and Sono, M. (1987): Cytochrome P-450 and Chloroperoxidase: Thiolate-Ligated Heme Enzymes. Spectroscopic Determination of Their Active Site Structures and Mechanistic Implications of Thiolate Ligation. *Chem. Rev.* 87: 1255-1276

Penner-Hahn, J.E., and Hodgson, K.O. (1989): X-Ray Absorption Spectroscopy of Iron Porphyrins. In *Iron Porphyrins Part III*, Eds. A.B.P. Lever and H.B. Gray, pp. 235-304. New York: VCH Publishers.

Raag, R., and Poulos, T. L. (1992): X-Ray Crystallographic Structural Studies of Cytochrome P-450-CAM: Factors Controlling Substrate Metabolism. *Frontiers in Biotransformation* 7: 1-43.

Sato, R., and Omura, T., Eds. (1978): *Cytochrome P-450*. New York: Academic Press.

# Electron transfer in cytochrome P450 containing semi-artificial enzymes

Valentin Yu. Uvarov, Victoria V. Shumyantseva, Elena A. Bykhovskaya, Ludmila N. Kolyada, Alexander I. Archakov

Institute of Biomedical Chemistry, Pogodinskaya 10, Moscow 119832 Russia

## Summary

To create a semi-artificial monomolecular oxygenase system FAD or FMN was covalently bound to cytochrome P450 2B4 as electrondonor centers and bleomycin to NADPH-cytochrome P450 reductase as a generator of active oxygen species. The most catalytically active was the conjugate of cytochrome P450 with FMN, able to initiate the reactions of dimethylaniline and aminopyrine demethylation as well as the reaction of aniline p-hydroxylation. The conjugate of cytochrome P450 with FAD oxidized these substrates with a much slower rate. The bleomycin-reductase complex was capable of dimethylaniline and aminopyrine demethylation but failed to oxidize aniline.

## INTRODUCTION

The microsomal chain of electron transfer is a membraneous system consisting of at least two protein molecules: cytochrome P450 and its NADPH-dependent reductase. The activation of the oxygen molecule necessary for insertion of one of its atoms in the substrate molecule, requres a successive transfer of two electrones from NADPH through FAD and FMN of the reductase to the heme iron of hemoprotein. Besides, there are naturally occurring systems, combining in one protein molecule the flavin and the monooxygenase centers. They are found both in prokaryotes (cytochrome P450BM3) (Ruettinger et al., 1989) and in eukaryotes (nitrooxydsyntase) (White et al., 1992). Lately, attempts have been made to obtain artificial analogs of such systems. Thus, a chimeric protein was described obtainable by gene engineering technique and comprising reductase and cytochrome P450 (Fisher et al., 1992). An attempt to obtain an analog of such a system by means of a covalent binding of flavin to hemoglobin was also described (Kukubo et al., 1987). The aim of the present study consists in creation a hydroxylase system by means of a covalent binding of artificial electron-donor (flavins) or electron-acceptor (bleomycin) centers with cytochrome P450 or its reductase, respectively.

## EXPERIMENTAL

The micrisomal fraction was isolated from the liver of male New Zealand rabbits treated with 0.1% (w/v) sodium phenobarbital in drinking water for 1 week (Karuzina et al., 1979). Cytochrome P4502B4 (P450) was isolated according to the method of Imai et al (1980) NADPH-cytochrome P450 reductase (FP) was isolated according to the method Kanaeva et al (1985). The reductase (specific activity 40-43 μmol cytochrome c min$^{-1}$ mg$^{-1}$ at 30°C, specific content 12-13.5 nmol mg protein$^{-1}$) and P450 2B4 (specific content 17-18 nmol mg protein$^{-1}$, $A_{276}/A_{417}$ = 0.5 - 1) showed a single band on SDS-PAGE. NADPH-dependent N-demethylation of aminopyrine and demethylaniline and p-hydroxylation of aniline were estimated by measuring the amounts of formaldehyde and p-aminophenol formed, respectively (Kanaeva et al., 1992). NADPH-cytochrome c reductase activity was determined from the cytochrome c reduction rate at 30°C, based on the extinction coefficient of reduced cytochrome c, 21.1 mM$^{-1}$cm$^{-1}$ for 550 nm (Digman et al., 1977). Ferricyanide reductase activity was determined from the $K_3[Fe(CN)_6]$ reduction rate at 30°C, based on the extinction coefficient of reduced product, 1.02 mM$^{-1}$cm$^{-1}$ for 420 nm. The concentration of FAD or FMN was determined based on a molar extinction coefficient of 11300 mM$^{-1}$cm$^{-1}$. The iron content was determined by atomic absorption spectroscopy with applying Perkin-Elmer 3000 spectrometer (USA). The FMN-containing flavocytochrome was carried out by the use of water-soluble carbodiimide (1-ethyl-3-(3-dimethylaminopropyl)-carbodiimide, EDC). The flavocytochrome obtained had 3 ± 1 molecules of the covalently bound FMN. FAD-containing flavocytochromes were obtained by two different methods. The former was based on FAD phosphorylation and its subsequent binding with protein using EDC. The other method for obtaining of FAD-containing flavocytochromes was based on FAD binding with cytochrome P450 by activation of FAD with carbonyldiimidazole. The flavocytochrome obtained had 6 ± 1 molecule of bound FAD. Coupling of bleomycin (Blm) with reductase was carried out using EDC; the unreacted Blm was removed by repeated washing with 10 mM Tris-HCl buffer, pH 6.5, using an Amicon cell with PM-30 membrane. As a result, 16 ± 2 molecules of Fe bounded with bleomycin were obtained.

## RESULTS AND DISCUSSION

Addition of 1000-fold excessive FAD to cytochrome P450 did not allow us to register any of the oxygenase reactions. Addition of the same amount of FMN to the hemoprotein stimulated both the demethylation and p-hydroxylation reaction processes. The reaction rate was 3-7% compared to the reaction rate in microsomes (Table 1). The covalent binding of FMN with cytochrome P450 using carbodiimide allowed us to obtain the conjugate which brought about the reactions of NADPH dependent demethylation and p-hydroxylation at rates which were only 30-40% lower than those in the microsomal system. That's why for the ferther detailed studies we have chosen the conjugate of cytochrome P450 and FMN. Data given in Table 1 show that such a semi-artificial system is capable of catalysing both NADPH- and NADH-dependent processes, while the oxidation rates of aminopyrine and aniline were even higher than in microsomes. At the same time, the rate of dimethylaniline demethylation was lower in comparison with microsomes. Of interest was to clear up the influence of lipid environment on the rate of catalytic processes in such a semi-artificial enzyme. It is evident of the rates of aminopyrine and NADH-dependent dimethylaniline demethylation, rates hadn't changed following the P450-FMN conjugate insertion in liposomes, than other processes slowed up.

To demonstrate the possibility of creating a semi-artificial protein with oxygenase activity on the basis of the NADPH-dependent flavoprotein, the latter was coupled, by use of EDC, with the bleomycin molecule. This antibiotic is similar in its properties to the active center of cytochrome P450. As seen from Table 1, the enzyme, containing 16 covalently bound bleomycin molecules per mole of flavoprotein, is capable of catalyzing the demethylation reactions. The binding of a lesser number of electron-acceptor groups was not accompanied by the appearance of the oxygenase activity. It is of interest that the rate of aminopyrine oxidation in this system was higher than that of dimethylaniline. Conjugates

Table 1. Catalytic activity of semi-artificial enzymes

| System | $k_{cat} \times 10^3$, $s^{-1}$ for reaction of: | | | | | |
|---|---|---|---|---|---|---|
| | N-demethylation | | | | p-hydroxylation | |
| | aminopyrine | | dimethylaniline | | aniline | |
| | NADPH | NADH | NADPH | NADH | NADPH | NADH |
| Microsomes | 60 ± 5 | 12 ± 1 | 600 ± 40 | 100 ± 10 | 150 ± 10 | 2 ± 1 |
| P450 + FMN (1 : 1000) | 2 ± 0.1 | – | 20 ± 1 | – | 10 ± 1 | – |
| P450-FAD (1 : 6) | | | | | | |
| a | 4 ± 0.2 | – | 25 ± 1 | – | 7 ± 0.5 | – |
| b | 3 ± 0.2 | – | 20 ± 1 | – | 5 ± 0.3 | – |
| P450-FMN (1 : 3) | 40 ± 2 | 20 ± 2 | 320 ± 20 | 40 ± 5 | 100 ± 7 | 40 ± 5 |
| Proteoliposomes P450-FMN (1 : 3) | 40 ± 5 | 16 ± 2 | 170 ± 30 | 40 ± 4 | 30 ± 5 | 20 ± 3 |
| FP-Blm (1 : 16) | 6 ± 0.3 | – | 3 ± 0.1 | – | 0 | – |
| | 100 ± 5* | – | 50 ± 2* | – | – | – |

Note. Abbreviations used: a – phosphorylated FAD is coupled by EDC; b – FAD is coupled by carbodiimide. The apparent catalytic constants were calculated from the maximal rates of the corresponding reactions: – per mole of active centers; * – per mole of the enzyme.

of P450 with FMN or FAD oxidized dimethylaniline at a higher rate than aminopyrine. It is should also be noted that when the conjugate of the reductase with bleomycin was tested, the demethylation (per mole of protein) was twice as rapid as in microsomes.

Thus, the covalent binding of FMN ( FAD ) with the cytochrome P450 or that of bleomycin with cytochrome P450 reductase allowed us to obtain the monomolecular systems with oxygenase activity.

## REFERENCES

Dignam, J.D. and Strobel, H.M. (1977): NADHP-cytochrome P-450 reductase from rat liver: purification by affinity chromatography and characterization. Biochemistry 16, 1116-1123.

Fisher, Ch.W., Shet, M.S., Caudle, D.L., Martin-Wixtrom, C.A. and Estabrook,R.W. (1992): High-level expression in Escherichia coli of enzymatically active fusion proteins containing the domains of mammalian cytochromes P450 and NADPH-P450 reductase flavoprotein. Proc. Natl. Acad. Sci. USA 89, 10817-10821.

Imai, I., Hashimoto,Y.C., Sakake, H., Girardin, A. and Sato, R. .(1980): Multiple forms of cytochrome P 4 50 purified from liver microsomesof 3-methycholantrene and phenobarbital treated rabbits. J. Biochem. 88, 489-503.

Kanaeva, I.P., Scotselas, E.D., Kuznetsova, G.P., Antonova, G.N., Bachmanova, G.I. and Archakov A.I. (1985): Reconstitution of microsomal monooxygenase cytochrome P450-containing liver system with detergents in solution. Biokhimia (rus) 50, 1382-1388.

Kanaeva, I.P., Nikituk, O.V., Davydov, D.R., Dedinskii, I.R., Koen, Ya. M., Kuznetsova, G,P. Skotselas, E.D., Bachmanova, G.I. and Archakov, A.I.(1992): Comparative stady of monomeric reconstituted and membrane microsomal monooxygenase systems of the rabbit liver. Arch. Biochem. Biophys. 298, 403-412.

Karuzina, I.I., Bachmanova, G.I., Mengazetdinov, D.E., Myasoedova, K.N., Zhihareva, V.D., Kuznetsova, G.P. and Archakov, A.I. (1979): Biokhimia (rus) 44, 1049-1057.

Kukubo, T., Sassa, S. and Kaiser, E.T.(1987): Flavohemoglobin: A semisynthetic hydroxylase acting in the absence of reductase. J. Am. Chem. Soc. 109, 606-607.

Ruettinger, R.T., Wen, L.-P. and Fulko, A,J. (1989): coding nucleotide, 5' regulatory and deduced amino acid sequences of $P-450_{BM-3}$, a single peptide cytochrome P-450: NADPH-P-450 reductase from Bacillus megaterium. J. Biol. Chem. 264, 10987-10995.

White, K.A. and Marletta, M.A. (1992): Nitric oxide syntase is a cytochrome P-450 type hemoprotein. Biochemistry 31, 6627-6631.

Electron transfer in cytochrome P-450 containing semi-artificial enzymes.
Valentin Yu.Uvarov, Victoria V.Shumyantseva, Elena A.Bykhovskaya, Ludmila N.Kolyada and Alexander I.Archakov
Institute of Biomedical Chemistry, Pogodinskaya 10, Moscow 119832 Russia
Correspondence address: V.Yu.Uvarov, Institute of Biomedical Chemistry, Pogodinskaya 10, Moscow 119832, Russia
TEL.(095) 2463374;
FAX (095) 2450857;
Electronic Mail: inst@ibmh.msk.su

# Segment directed mutagenesis of human cytochrome P450 1A2 by PCR and effect on the catalytic activity and substrate selectivity

Romuald Lainé, Philippe Urban, Denis Pompon

*Centre de Génétique Moléculaire du Centre National de la Recherche Scientifique, 91198 Gif-sur-Yvette Cedex, France*

The superfamily of P450 isoenzymes of more than 220 known members offers almost unlimited opportunities for the investigation of structure-function relationships [1].
Each P450 usually metabolizes several substrates, yielding for each substrate different products. The substrate can be the same for very different P450 enzymes, but very close P450 enzymes (in terms of sequence identities) may not metabolize the same substrates [2].
It is the case for the two human isoenzymes P450s 1A1 and 1A2 (70% homology). Only the former is able to activate procarcinogenic polycyclic hydrocarbons like benzo(a)pyrene into mutagenic compounds.

We are interested by Gotoh's notion of the Substrate Recognition Site (SRS) [3]. A comparaison of the amino acid sequence between the CYP1 family and the CYP2 family of SRS1 allows one to identify a segment which may be involved in the determination substrate selectivity in the CYP1 family.

The B' helix of P450 cam and P450 BM-3 are contained in the SRS1. Alignment of proteins in the CYP2 family show that 2 positions are more degenerate that the others. One of these position is very conserved in the CYP1 family.(Figure 1).
If this region is also a SRS for the CYP1 family, a mutation of a very conserved amino acid must be critical for the enzyme function.
In the human cytochrome P450 1A2 we have replaced isoleucine 117 by threonine.

We have compared the activity for wild type (1A2 WT) and the mutant (I117T) with the usual substrates of the CYP1 family, Etoxyresorufin and Metoxyresorufin.(Figure 2).

## Figure 1.
The Substrate Recognition Site (SRS) of the CYP1 family. The position of isoleucine 117 is in bold.

```
                              B'cam
           80              [---------------]         100
cam        H F S S E C . P F I P R E A G E A Y . D F I P
BM-3       R F D K N L . S Q A L K F V R D F A G D G L F
           66          [---------------------------]    83
                         B' BM-3
Hum 1A2    D F K G R P D L Y T S T L I . T D G . Q S L T
Rat 1A2    D F K G R P D L Y S F T L I . T N G . K S M T
Rab 1A2    D F K G R P D L Y S S S F I . T E G . Q S M T
Mou 1A2    D F K G R P D L Y S F T L I . T N G . K S M T
Hum 1A1    D F K G R P D L Y T F T L I . S N G . Q S M S
Rat 1A1    D F K G R P D L Y S F T L I . A N G . Q S M T
Rab 1A1    D F K G R P D L Y S F S F V . T K G . Q S M I
Mou 1A1    D F K G R P D L Y S F T L I . T N G . K S M T
Rtr 1A1    D F A G R P D L Y S F K F I . N D G . K S L A
```

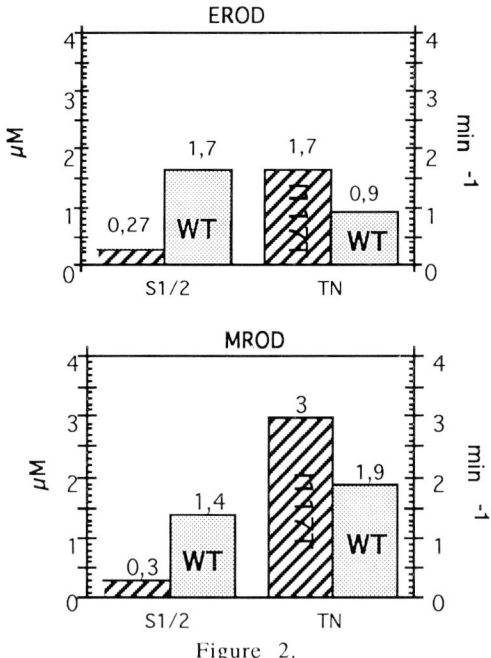

Figure 2.

We notice a decrease in the S1/2, but the TN is approximately the same. S1/2 : half saturation constant. TN : Turnover Number.

If the mutant has a good EROD and MROD activity, we noticed that the mutant had no acetanilid hydroxylase activity and no phenacetin-O-deethylase activity.(data not shown).
We shown that in fact the mutant is unable to bind this two substrates.(Figure 3).

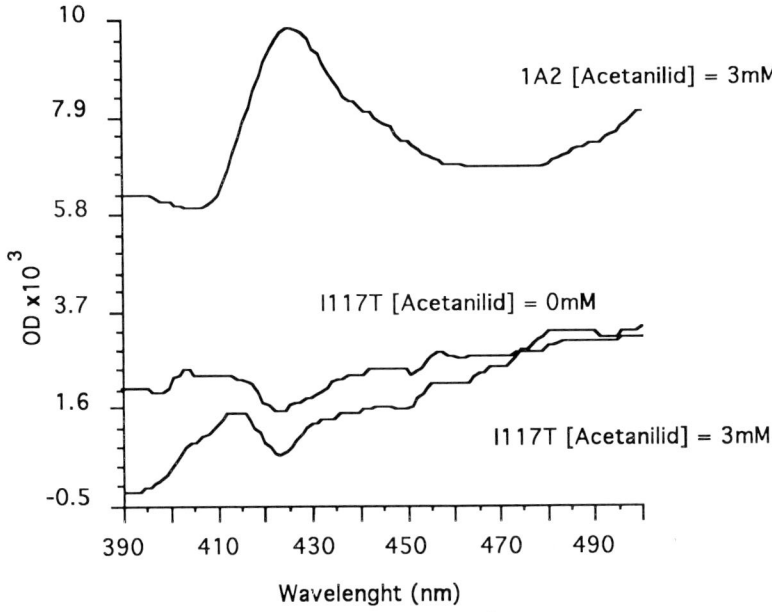

Figure 3
Comparaison of the binding spectra of 1A2 and I117T at the same concentration of P450. The mutated cytochrome P450 1A2 (I117T) cannot bind acetanilid.

## CONCLUSION

We have characterized a mutation at the position 117 that affects the binding of substrates.
This mutation does not affect the EROD and MROD turnover number. The segment that contains isoleucine 117 is involved in the Substrate Recognition Site (SRS).
Is this segment a part of an hydrophobic channel involved in the control of substrate's binding although not directly implicated in the catalytic activity?

# Specific inhibition of the last steps of the aldosterone biosynthesis by 18-vinylprogesterone in bovine adrenocortical cells

G. Defaye[1], A. Piffeteau[2], C. Delorme[2], A. Marquet[2]

[1]INSERM U.244, Département de Biologie Moléculaire et Structurale, 85X F-38041 Grenoble. [2]Laboratoire de Chimie Organique Biologique, URA CNRS 493, Université P.-et-M.-Curie, F-75252 Paris Cedex 05, France

## Summary

Last steps of aldosterone as well as cortisol biosynthesis implie cytochrome P-450$_{11\beta}$. The effect of 18-vinylprogesterone (18-VP), a mechanism-based specific inhibitor of cytochrome P-450$_{11\beta}$, on the different enzymatic steps involved in the biosynthesis of cortisol and aldosterone were studied. The results clearly demonstrate that 18-VP inhibited only the last enzymatic steps and more specifically the 18- than the 11-hydroxylation.

## Introduction

Aldosterone, a potent natural occuring mineralocorticoid is produced in the glomerulosa zone of the adrenal cortex. It is synthesized, from cholesterol through a pathway including, pregnenolone, progesterone, 11-deoxycorticosterone, corticosterone and 18-hydroxycorticosterone as intermediate metabolites. We have developed enzyme inhibitors of aldosterone biosynthesis by synthesizing progesterone derivatives, modified at the 18-methyl group : 18-vinylprogesterone (18-VP) and 18-ethynylprogesterone (18-EP) (Viger & al, 1989). Indeed, the last steps of the aldosterone biosynthesis involve a 18-hydroxylation of corticosterone and the transformation of 18-hydroxycorticosterone to aldosterone. These steps, and the 11-hydroxylation, are catalyzed by cytochrome P-450$_{11\beta}$. In the bovine, two different cytochromes P-450$_{11\beta}$ encoded by two different genes have been isolated, both are found in the glomerula as well as in the fasciculata-reticularis zones, and have the same catalytic properties (Müller, 1993 ; Hashimoto & al, 1989 ; Morohashi & al, 1990). Recently, a 18-ethynyl derivative of deoxycorticosterone (18-EDOC) has been described and its action on calf adrenal cells studied (Yamakita & al, 1991).

This paper present studies of the action of 18-VP on the cortisol and aldosterone productions of bovine adrenocortical cells. The results obtained confirm that 18-VP inhibits the last steps of the aldosterone biosynthesis and more specifically the 18- than the 11-hydroxylation.

## Methods

### Adrenocortical cells
Bovine adrenocortical (BAC) cells were prepared by successive digestion of fresh adrenal glands. Cells were grown in Hams'F12 medium supplemented with insulin (5 µg/ml), transferrin (5 µg/ml) and selenous acid (5 ng/ml) (ITS medium).

### Incubation of the cells
Cells were used at day 5 of culture. The incubations were performed at 37°C, in Ham's F12 medium (Gibco), during 2h in the presence of 18-VP (0-10 µM). For stimulation of steroidogenesis a concentration of 10 nM of ACTH and 0.3 µM of AII was used. Cortisol production was measured in the medium by RIA. For the enzymatic activities, the cells were incubated, in the same conditions, with the apropriate substrate. The steroids were extracted from the medium with chloroform. The organic extract was analyzed by thin layer chromatography on silica gel plates (F-254, Merck). The spots corresponding to the substrate and the product were scrapped off the plates and counted for radioactivity.

### Enzymatic activities
side chain cleavage activity was determined by the measure of the transformation of 25-hydroxycholesterol (15µM) to pregnenolone in the presence of trilostane (2 µM) as a blocker of Δ5-3ß-hydroxysteroid dehydrogenase/isomerase and SU 10603 as a blocker of 17-hydroxylase. Pregnenolone was quantified in the medium by RIA.

The 3ß-hydroxysteroid-dehydrogenase/isomerase activity was measured by incubation of [$^3$H]-dehydroepiandrosterone (DHEA) (100 µM) in the presence of metyrapone (2 µM) as a blocker of 11-hydroxylase.; DHEA and androstenedione were separated on TLC using chloroform-acetone (9:1).

The 17α-hydroxylase, 11ß-hydroxylase or 21-hydroxylase activities were assayed by measuring the transformation of, [$^3$H]-pregnenolone (50 µM) to [$^3$H]-17α-hydroxypregnenolone in the presence of trilostane (2 µM), [$^3$H]-deoxycortisol (100 µM) to [$^3$H]-cortisol, or [$^3$H]-17-hydroxy-progesterone (50 µM) to [$^3$H]-deoxycortisol in the presence of metyrapone (2 µM) as substrates respectively ; pregnenolone, 17-hydroxypregnenolone, 17-hydroxyprogesterone, cortisol and deoxycortisol were separated on TLC using chloroform-ethyl acetate (1:1, v/v).

The 18-hydroxylation of corticosterone was measured with [$^3$H]-corticosterone (10 µM). The products: corticosterone, 18-hydroxycorticosterone and aldosterone were separated by HPLC (silica RP-18 (5µm) column and a mixture methanol/water (53%) at 1 ml/mn flow rate). The elution was collected and the radioactivity corresponding to each fraction was measured.

The aldosterone formation was performed in the same conditions with corticosterone (10 mM); aldosterone produced was measured by radioimmunoassay ; aldosterone anti-serum was a generous gift from N. Cittanova (Paris, France).

## Results and discussion

The cortisol secretion of BAC cells, either basal, or stimulated by ACTH or Angiotensine II (A II), was inhibited by 18-VP at a concentration of 10 µM (Table 1).

### Table 1 Effect of 18-VP on basal and hormone activated cortisol production

| compound added | cortisol production (ng/$10^6$cells/h) 18-VP (µM) | |
|---|---|---|
| | 0 | 10 |
| no | 28 ± 4.7 | 5.2 ± 1 |
| ACTH (10 nM) | 416 ± 47 | 79 ± 3 |
| Angiotensin II (0.3 µM) | 296 ± 12 | 50 ± 5 |

The cortisol production, stimulated by ACTH (10 nM), was measured in the presence of various concentrations of 18-VP (0-10 µM) ; it is inhibited in a dose dependent manner with a maximum inhibition at 5 µM (Fig. 1).

We then explored each enzymatic step of the pathway leading from cholesterol to cortisol to determine which was affected by the action of 18-VP. The different activities examined were : the side chain cleavage of cholesterol, the 17- and 21-hydroxylases as well as the 3β-hydroxysteroid dehydrogenase/isomerase activities (Table 2). None of these enzymatic activities were affected by 18-VP. Together these results suggested that the inhibition of cortisol production by the BAC cells was due to a selectively decrease of the last step, namely the 11-hydroxylation of deoxycortisol.

**Table 2 Effect of 18-VP on the individual enzymatic activities**

| | nmoles produced/$10^6$ cells/h | | | |
|---|---|---|---|---|
| 18-VP (μM) | 0 | 1 | 5 | 10 |
| pregnenolone*** (from cholesterol 50 μM) | 0.8 ± 0.1 | | 0.85 ± 0.2 | |
| 17-OH-pregnenolone* (from pregnenolone 50μM) | 13 ± 2 | 14 ± 0.5 | 11.2 ± 1.5 | 12.8± 1 |
| deoxycortisol** (from 17-OH-progesterone 100μM) | 21 ± 1 | 22 ± 2 | 19 ± 2 | 18 ± 2 |
| Androstenedione** (from DHEA 100μM) | 13.8 ± 0.7 | | | 12.3 ± 0.5 |

* in the presence of trilostane (2 μM) ; ** metyrapone (2 μM) ; *** trilostane and SU 10603

In the bovine adrenal cortex, 11-hydroxylation and aldosterone production are all achieved by cytochrome P-450$_{11β}$. Consequently, the cells were incubated with deoxycortisol to measure the production of cortisol (11-hydroxylase), and corticosterone to measure that of 18-hydroxycorticosterone and aldosterone (18-hydroxylase and aldosterone synthase) respectively.

The dose dependence of the inhibition by 18-VP of these activities are reported on Fig 2. The aldosterone synthase and the 18-hydroxylase were inhibited with an $I_{50}$ = 0.05 ± 0.01 μM whereas the 11-hydroxylase was inhibited with an $I_{50}$ = 0.3 ± 0.1 μM. In fact, at a concentration of 0.1 μM of 18-VP, the aldosterone formation, from corticosterone, was inhibited by 90%, 18-hydroxycorticosterone by 60% and cortisol from deoxycortisol only by 35%.

Fig. 2

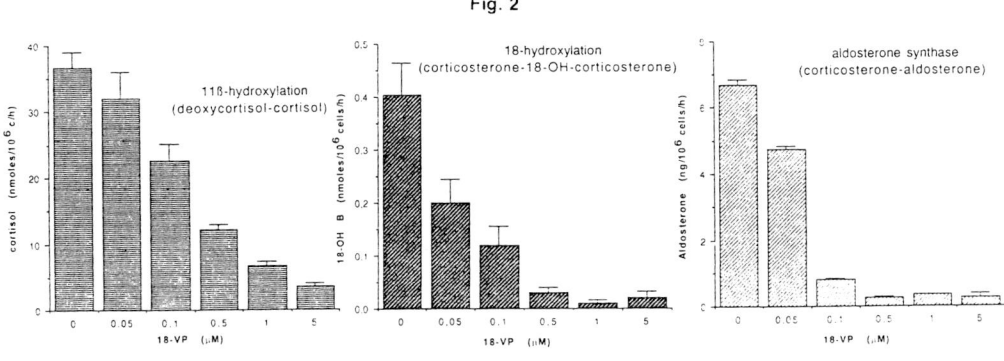

At last, we compared the action of 18-VP and 18-EP on the cortisol production. These two compounds were shown (Viger & al, 1989) to completely inhibit the aldosterone production from corticosterone at 0.8 and 8 µM respectively in rat adrenal cell-free extracts. The results on Fig. 3 confirm the previous observations ; at a concentration of inhibitor of 1 µM, there is a strong decrease of 11β-hydroxylation with 18-VP (only 20% of the total activity) whereas there is no effect with 18-EP. The inhibition of 11-hydroxylation is evident at 5 µM of 18-EP.

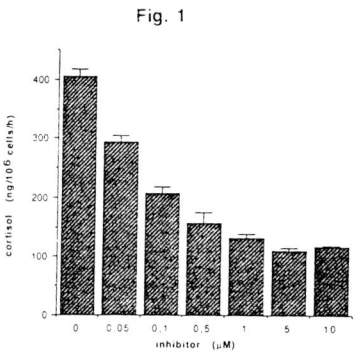

Fig. 1

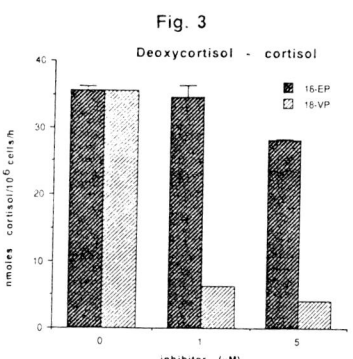

Fig. 3

Taken together, these observations have shown that 18-VP is an inhibitor of the different enzymatic activities of cytochrome P-450$_{11\beta}$, and not an inhibitor of the earlier steps of the aldosterone biosynthesis. 18-VP inhibits more specifically the 18- than the 11-hydroxylation, in agreement with, *in vitro*, measurements (Delorme & al). 18-EP, which structure is very closed to 18-EDOC (Yamakita & al, 1991) has a different behavior, and is a poor inhibitor of 11-hydroxylation.

**References**

Delorme, C., Piffeteau, A., Viger, A. and Marquet, A. (1993) : Bovine cytochrome P-450$_{11\beta}$ : dual activity at the same active site examined by inhibition with 18-unsaturated progesterone derivatives, submitted

Hashimoto T, Morohashi, K. and Kimura, T (1989) : Cloning and characterization of bovine cytochrome P-450(11β) genes J.Biochem 105 : 676:-679

Kirita, S and Morohashi, K. (1988) : Expression of two kinds of cytochrome P-450(11β) mRNA in bovine adrenal cortex J. Biochem 104 : 683-686

Morohashi, K, Nonaka, Y, Kirita, S, Hatano, O, Takakusu, A, Okamoto, M and Omura Y (1990) : Enzymatic activities of P-450(11β)s expressed by two cDNAs in COS-7 cells J. Biochem 107 : 637-640

Müller, J (1993) : Final steps of Aldosterone biosynthesis : molecular solution of a physiological problem J. Steroid Biochem. Molec. Biol. 45 : 153-159

Viger, A., Coustal, S., Perard, S., Piffeteau, A. and Marquet, A. (1989) : 18-Substituted progesterone derivatives as inhibitors of aldosterone biosynthesis J. Steroid Biochem. 33 : 119-124

Yamakita, N, Chiou, S, and Gomez-Sanchez, C.E (1991) : Inhibition of aldosterone biosynthesis by 18-Ethynyldeoxycorticosterone Endocrinology 129 : 2361-2366

# Mechanism based inactivation of nitric oxide synthase, a P450 like enzyme, by xenobiotics

Yoichi Osawa, Julio C. Davila, Carl A. Meyer, Mikiya Nakatsuka

Laboratory of Chemical Pharmacology, National Heart, Lung, and Blood Institute, National Institutes of Health, Bethesda, MD 20892, USA

Nitric oxide synthase (NOS) was recently shown to be a cytochrome P450 like enzyme that catalyzes the metabolism of L-arginine to L-citrulline and nitric oxide. These findings prompted us to examine whether NOS could be inactivated by xenobiotics through generation of reactive intermediates. We have found that phencyclidine (PCP), an arylcyclohexylamine used as a drug of abuse, irreversibly inactivated brain NOS in a time dependent manner. The inactivation was dependent on NADPH and the concentration of PCP and was prevented by the addition of L-arginine, but not D-arginine, in the reaction mixture. The kinetics of inactivation could be described by the apparent dissociation constant for the initial reversible complex ($K_i$) and the pseudo first order inactivation constant ($k_{inact}$) of 4.9mM and 0.3min$^{-1}$, respectively. We have also found that BrCCl$_3$, but not CCl$_4$, caused a time- and cofactor-dependent inactivation of NOS activity. Since the liver microsomal P450 cytochromes are known to be inactivated by tetrahalomethanes and PCP in a metabolism based manner, these studies indicate the similarity of NOS to the microsomal P450 cytochromes in this respect. It also appears that NOS, like microsomal hepatic P450, can be inactivated by both oxidative and reductive bioactivation reactions. The lower activation energy needed to reductively cleave the C-Br versus the C-Cl bond is consistent with these results. Owing to the numerous physiological roles ascribed to nitric oxide, the inactivation of NOS by xenobiotics, such as drugs, may be an important mechanism for eliciting toxicity.

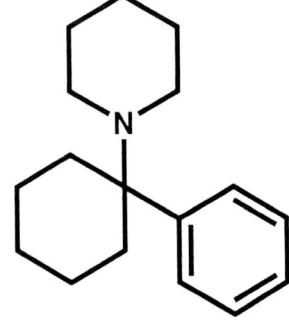

N$^G$-Monomethyl-L-arginine       Phencyclidine

Fig. 1. Structures of some suicide inhibitors of NOS

## INTRODUCTION

Nitric oxide synthase (NOS), the enzyme responsible for the conversion of L-arginine to L-citrulline and nitric oxide was recently shown to be a P450-like enzyme, containing a heme prosthetic group (White and Marletta, 1992; Stuehr and Ikeda-Saito, 1992; McMillan et al., 1992). Unlike the liver microsomal P450 cytochromes, which require a flavoprotein reductase for activity, NOS is a fully competent protein having FAD and FMN prosthetic groups in addition to the heme (Marletta, 1993). Currently, the regulation of NOS is of intense interest owing to the variety of important biological roles ascribed to nitric oxide (for review see (Moncada et al., 1991)), which have been in large part defined by pharmacological methods with the use of substrate analogs that inhibit NOS. One of the most extensively used inhibitors, N-monomethyl-L-arginine (Fig. 1), has recently been shown to be a mechanism-based inactivator of the NOS isolated from macrophages (Olken et al., 1991). Recently phencyclidine (Fig. 1), a pyschomimetic drug of abuse unrelated in structure to arginine, was shown to inactivate brain NOS in a mechanism-based manner (Osawa and Davila, 1993). We report here that $BrCCl_3$, a hepatotoxic and neurotoxic agent, as well as phencyclidine inactivates brain NOS with the kinetics of a suicide inactivator. Since both of these compounds are known to inactivate liver microsomal P450 cytochromes in a suicide manner and since $BrCCl_3$ has been used in model systems to characterize the covalent alterations that can occur with hemoproteins (Osawa and Pohl, 1989), these agents may be useful probes in studies on the mechanism of action of NOS. In addition, these results suggest that xenobiotics could act to inhibit NOS activity to elicit pharmacological or toxicological effects.

## MATERIALS AND METHODS

PCP was purchased from Research Biochemicals Inc. (Natick, MA). $BrCCl_3$ was purchased from Aldrich (Milwaukee, WI) and was purified by distillation prior to use. Rat brain cytosol was prepared from male Wistar rats as previously described (Osawa and Davila, 1993). PCP or $BrCCl_3$ was placed in a reaction mixture containing NADPH (1 mM) and brain cytosol (6.8 mg of protein/ml), in a total volume of 1.5 ml of 40 mM potassium phosphate, pH 7.4. This mixture was incubated at 37°C in a Dubunoff shaker and at 5-min intervals 250 µl aliquots were removed and placed in a second reaction mixture or "assay mixture" for determination of NOS activity by the oxidation of oxyhemoglobin as described previously (Knowles et al., 1990). The assay mixture contained NADPH (1 mM), $CaCl_2$ (1 mM), L-arginine (1 mM), and bovine oxyhemoglobin (1.6 µM) in a total volume of 1.0 ml of 40 mM potassium phosphate, pH 7.4. In some experiments, NOS activity was measured by released $^{14}C$-citrulline. In this case 100 µl of the initial reaction mixtures were placed in an assay mixture described above except that $^{14}C$-arginine (NEN, Boston, MA, final concentration of 30 µM at 20 mCi/mmol)

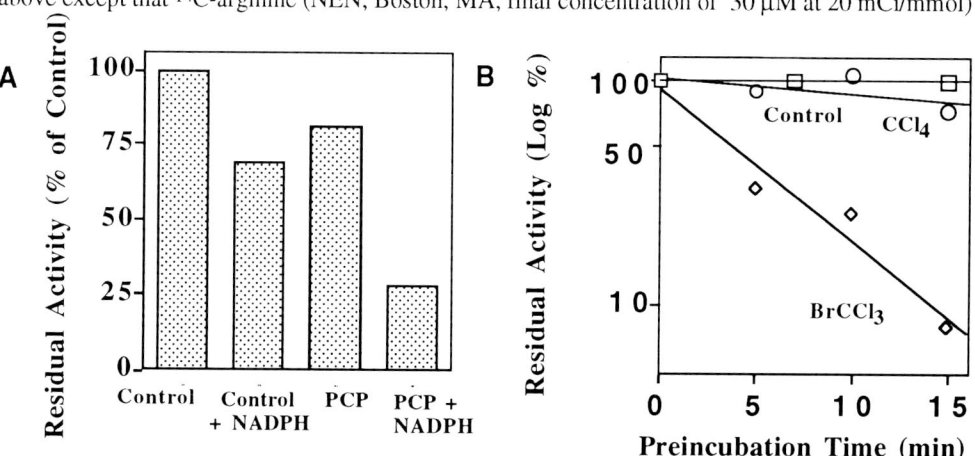

Fig. 2. Effect of PCP (Panel A) or $BrCCl_3$ or $CCl_4$ (Panel B) on NOS activity. Panel A, activities were measured by $^{14}C$-citrulline assay; Panel B, activities were measured by oxyhemoglobin assay. The method is as described in "Materials and Methods".

replaced unlabeled arginine and the total volume was 250 µl. The reaction was stopped after 10 min by addition of 0.1 vol of 500 mM citric acid, pH 2.5 (Solvent B). An aliquot of this mixture (250 µl) was placed onto an HPLC column (BioGel SP-5-PW, 7.5 x 75mm, 10 µm, BioRAD, Hercules, CA) that was equilibrated with a mixture of solvent A (500mM sodium citrate, pH 8.5) and solvent B and solvent C (water) in a ratio of 4:16:80 at a flow rate of 1 ml/min. At 2 min the ratio of solvents A:B:C was changed to 20:80:0. The retention time of citrulline and arginine were 3.6 and 8.2 min, respectively. Radioactivity was quantified by an on-line detector (Radiomatic FloOne, Packard Instrument Co., Meriden, CT). The logarithmic percentage of residual NOS activity was plotted against time.

## RESULTS AND DISCUSSION

Phencyclidine was shown to be a suicide inactivator of NOS in brain cytosol with the use of the oxyhemoglobin assay (Osawa and Davila, 1993). This inactivation could also be observed when a more specific NOS assay measuring the amount of $^{14}C$-citrulline produced was used (Fig. 2 Panel A). Approximately 75% inactivation was observed after 20 min in reaction mixtures treated with PCP and NADPH, in comparison to that of controls. PCP has also been shown to be a mechanism-based inactivating agent for the major phenobarbital inducible isoform of P450 (CYP2B4) (Osawa and Coon, 1989). The mechanism of this inactivation involves a loss of Soret absorbance due to the heme. A similar mechanism may also be involved in the inactivation of NOS.

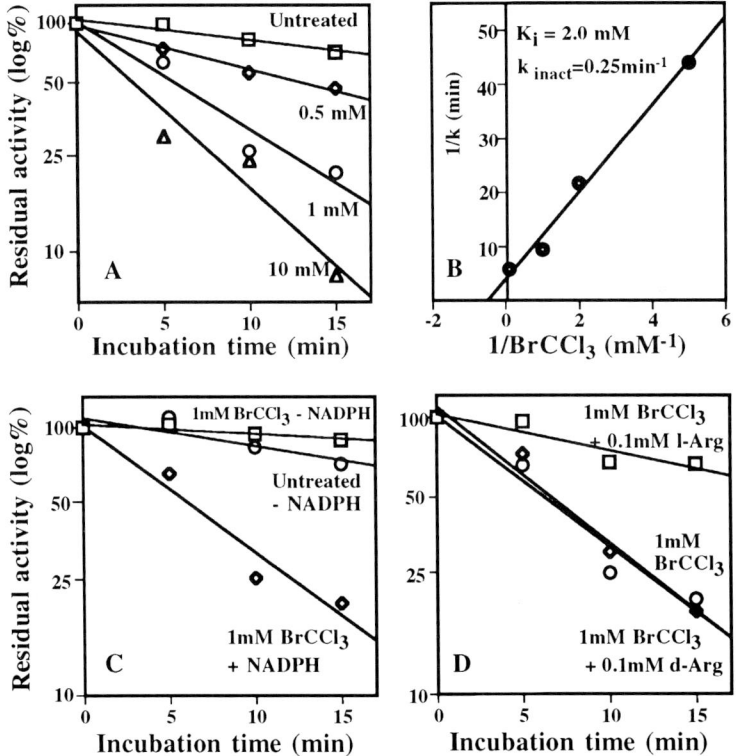

Fig. 3. Suicide inhibition of brain NOS by $BrCCl_3$. Activities were measured by oxyhemoglobin assay as described in "Materials and Methods". (A) Concentration and time-dependent inhibition of NOS. $BrCCl_3$ concentrations were: [], 0mM; ◊, 0.5mM; O, 1mM; Δ, 10mM. (B) Replot of the data from panel A to determine kinetic constants. (C) NADPH-dependence of inhibition. [], 1mM $BrCCl_3$ without NADPH; ◊, 1mM $BrCCl_3$ and NADPH; O, untreated without NADPH. (D) Effect of substrate on inhibition of NOS. O, 1mM $BrCCl_3$; [], 1mM $BrCCl_3$ and 0.1mM L-arginine; ◊, 1mM $BrCCl_3$ and 0.1mM D-arginine.

BrCCl$_3$, but not CCl$_4$, was shown to irreversibly inactivate NOS (Fig. 2 Panel B). This inactivation fulfills the kinetic criteria for suicide inactivation (Fig. 3), similar to that found for PCP. The inactivation is dependent on time and the concentration of BrCCl$_3$ (Fig. 3 Panel A). The kinetics of inactivation could be described by the apparent dissociation constant for the initial reversible complex ($K_i$) and the pseudo first order inactivation constant ($k_{inact}$) of 2.0 mM and 0.25 min$^{-1}$, respectively (Fig 3 Panel B). The inactivation depends on the presence of NADPH in the preincubation (Fig 3 Panel C). Furthermore, the addition of substrate, L-arginine, protected it from inactivation, whereas D-arginine had no effect (Fig. 3 Panel D). Analogous reactions with myoglobin and hemoglobin with BrCCl$_3$ results in regiospecific ring I vinyl modified heme adducts (Osawa et al., 1989; Osawa et al., 1990; Kindt et al., 1992; Osawa and Pohl, 1989). It remains to be determined if such adducts form with NOS. Three mechanisms of inactivation of P450 cytochromes have been described (for review see Osawa and Pohl, 1989). They involve covalent alteration of the heme, the protein, and the crosslinking of heme to protein. In the case of NOS similar reactions may occur, but, other mechanisms such as alteration of the flavin, tetrahydrobiopterin, or calmodulin binding site have to be considered as well.

REFERENCES

Kindt, J.T., Woods, A., Martin, B.M., Cotter, R.J., and Osawa, Y. (1992): Covalent alteration of the prosthetic heme of human hemoglobin by BrCCl$_3$ : Cross-linking of heme to cysteine residue 93. *J. Biol. Chem.* 267, 8739-8743.

Knowles, R.G., Merrett, M., Salter, M., and Moncada, S. (1990): Differential induction of brain, lung and liver nitric oxide synthase by endotoxin in the rat. *Biochem. J.* 270, 833-836.

Marletta, M.A. (1993): Nitric oxide synthase structure and mechanism. *J. Biol. Chem.* 268, 12231-12234.

McMillan, K., Bredt, D.S., Hirsch, D.J., Snyder, S.H., Clark, J.E., and Masters, B.S.S. (1992): Cloned, expressed rat cerebellar nitric oxide synthase contains stoichiometric amounts of heme, which binds carbon monoxide. *Proc. Natl. Acad. Sci. U. S. A.* 89, 11141-11145.

Moncada, S., Palmer, R.M.J., and Higgs, E.A. (1991): Nitric oxide: Physiology, pathophysiology, and pharmacology. *Pharm. Rev.* 43, 109-142.

Olken, N.M., Rusche, K.M., Richards, M.K., and Marletta, M.A. (1991): Inactivation of macrophage nitric oxide synthase activity by N$^G$-methyl-L-arginine. *Biochem. Biophys. Res. Commun.* 177, 828-833.

Osawa, Y and Coon, M.J. (1989): Selective mechanism-based inactivation of the major phenobarbital-inducible P-450 cytochrome from rabbit liver by phencyclidine and its oxidation product, the iminium compound

Osawa, Y., Highet, R.J., Murphy, C.M., Cotter, R.J., and Pohl, L.R. (1989): Formation of heme-derived products by the reaction of ferrous deoxymyoglobin with BrCCl$_3$. *J. Am. Chem. Soc.* 111, 4462-4467.

Osawa, Y. and Pohl, L.R. (1989): Covalent bonding of the prosthetic heme to protein: A potential mechanism for the suicide inactivation or activation of hemoproteins. *Chem. Res. Toxicol.* 2, 131-141.

Osawa, Y., Martin, B.M., Griffin, P.R., Yates, J.R., Shabanowitz, J., Hunt, D.F., Murphy, A.C., Chen, L., Cotter, R.J., and Pohl, L.R. (1990): Metabolism-based covalent bonding of the heme prosthetic group to its apoprotein during the reductive debromination of BrCCl$_3$ by myoglobin. *J. Biol. Chem.* 265, 10340-10346.

Osawa, Y. and Davila, J.C. (1993): Phencyclidine, a psychotomimetic agent and drug of abuse, is a suicide inhibitor of brain nitric oxide synthase. *Biochem. Biophys. Res. Commun.* 194, 1435-1439.

Stuehr, D.J. and Ikeda-Saito, M. (1992): Spectral characterization of brain and macrophage nitric oxide synthases. *J. Biol. Chem.* 267, 20547-20550.

White, K.A. and Marletta, M.A. (1992): Nitric oxide synthase is a cytochrome P-450 type hemoprotein.*Biochemistry* 31, 6627-6631.

ACKNOWLEDGMENTS: We are grateful to Drs. Lance R. Pohl and James R. Gillette for critically reviewing this manuscript.

# Molecular recognition in cytochrome P450: control of uncoupling reactions *via* site-directed mutagenesis

Paul J. Loida, Stephen G. Sligar

*Department of Biochemistry, University of Illinois, Urbana, Illinois, USA*

The pathway for utilization of pyridine nucleotide derived reducing equivalents in the cytochrome P-450 monoxygenase systems has three major branch points (Atkins & Sligar, 1988; Fig. 1). The first is a partitioning between autoxidation of a ferrous, oxygenated heme adduct and input of the second reducing equivalent required for monoxygenase stochiometry (Sligar et al, 1974; Ishimura et al, 1971). The second is between dioxygen bond scission and release of two electron reduced $O_2$ as hydrogen peroxide. The third is between substrate hydrogen abstraction initiated by a putative higher valent iron-oxo species and reduction of this intermediate by two additional electrons to produce water in an overall oxidase stochiometry (Atkins & Sligar, 1987; Gorsky et al., 1984). In order to elucidate the aspects of molecular recognition of a substrate-P-450 complex which effect these individual branch points in the catalytic cycle, we have measured the NADH derived reducing equivalents recovered in hydroxylated substrate, hydrogen peroxide (Fig. 2A) and water (Fig. 2B) for a series of active site mutants designed to alter the coupling of ethylbenzene hydroxylation (Loida & Sligar, 1993). For all substrates investigated, the direct release of superoxide at the first branch point never competes with second electron input. The partitioning of reducing equivalents at the second and third branch point is effected by the site-directed mutations that introduce Leu, Ile, Met and Phe residues into the active site (Fig. 3). The increased commitment to catalysis observed for all mutants suggests that active site hydration is important in the uncoupling to form hydrogen peroxide at the second branch point. By adding steric bulk to the active site at a variety of different locations the liberation of hydrogen peroxide is consistently decreased, and the effect is largest when the active site hydrophobicity is maximized. A strong correlation is observed between water production at the third branch point and the location and size of the amino acid side chain in the substrate binding pocket. Larger hydrophobic side chains introduced in the upper regions of the binding pocket (Fig. 4; Tiers 2 and 3) increase the ratio of hydroxylated product to water production by two to four fold relative to wild-type, while similar substitutions in residues near the heme plane (Fig. 4; Tier 1) result in diminished 1-phenyl-ethanol production. Overall, the partitioning between hydroxylation and oxidase activities at the iron-oxo species varies by over 65% due to the location of non-polar substituents engineered into the active site. Substrate access to the oxidizing species is the key determinant in tight coupling at third reaction branch point and is rationally modulated by changing the position and steric packing volume of nonpolar amino acid substitutions in the binding pocket. The orientation of substrate in the buried binding pocket and active site solvation of the heme iron represent general mechanisms of control of the reaction specificity at the oxidase pathway and peroxide shunt for the P-450 family of hemoproteins.

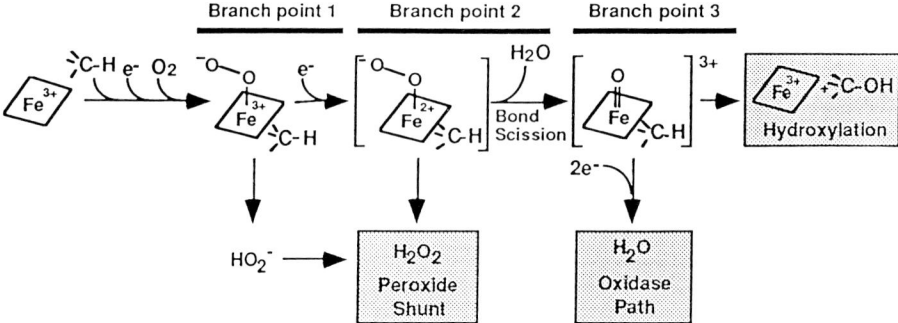

Figure 1. Schematic of the cytochrome P-450 reaction illustrates the unproductive pathways that give rise to the side products hydrogen peroxide and water. The one electron reduced oxy-P-450 intermediate marks the first branch point at which autoxidation or second electron reduction occurs. The two electron reduced peroxy-P-450 species either undergoes dioxygen bond cleavage or peroxide release at the second branch point. At the third fork the putative iron-oxo intermediate is partitioned between substrate hydroxylation and additional reduction via the oxidase pathway to form water.

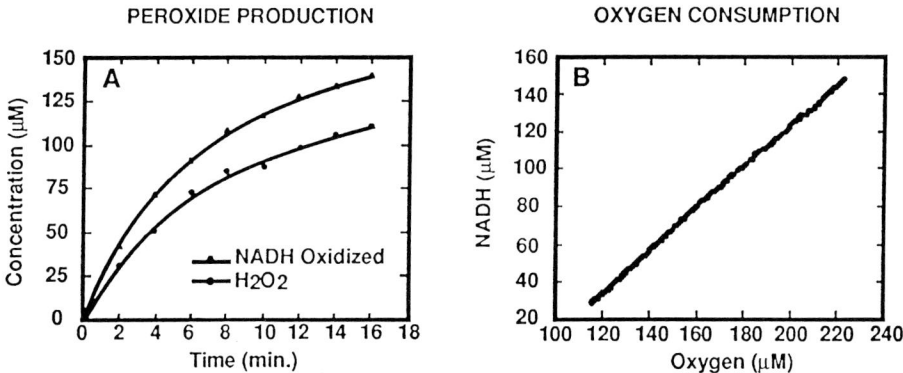

Figure 2. (A.) Hydrogen peroxide formation relative to NADH consumption. The liberation of hydrogen peroxide in the presence of ethylbenzene was determined using an iron-thiocyanate colorimetric assay and NADH oxidation was monitored spectrophotometrically at 340 nm. Key: NADH is denoted by triangles and hydrogen peroxide is shown in circles. Reaction conditions were typically 2 μM P-450cam, 5 μM putidaredoxin, 1 μM putidaredoxin reductase, 50 mM Tris-HCl, pH 7.4, 50 mM KCl, and 500 μM substrate. (B.) NADH oxidation was determined simultaneously with oxygen consumption by monitoring the reaction spectrophotometrically at 340 nm and potentiometrically with a Clark electrode. The degree of oxidase activity and the quantity of water formed in the reaction can be calculated from the slope of the line which is the ratio of NADH/oxygen consumed.

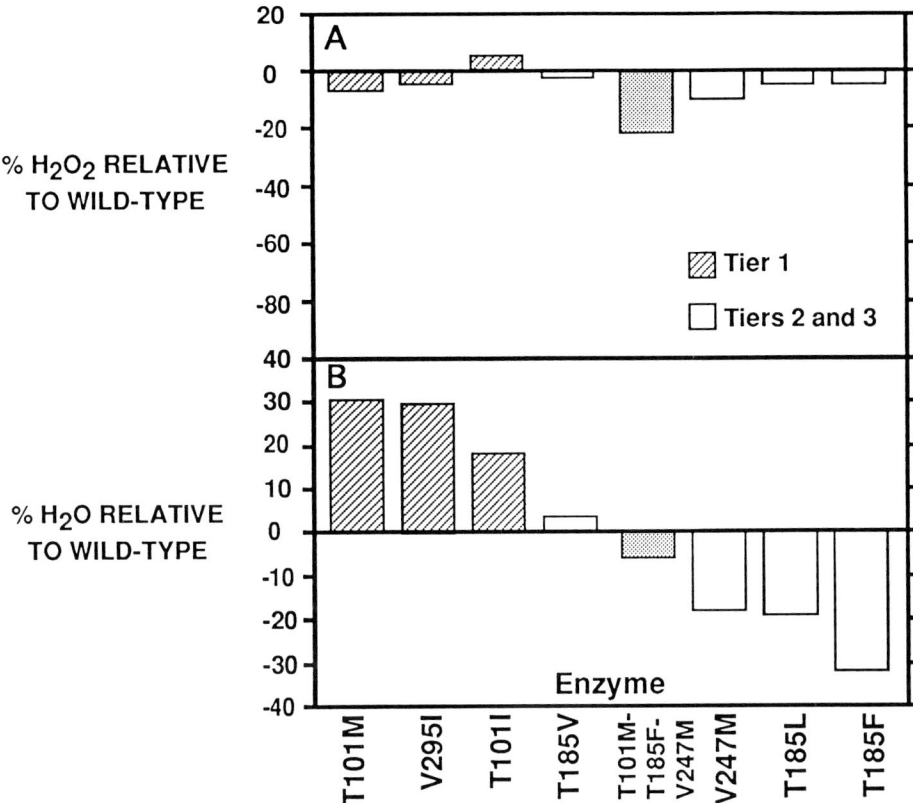

Figure 3. Reaction specificity differences between wild-type cytochrome P-450cam and a series of active site mutants in the hydroxylation of ethylbenzene. The large hydrophobic side chains of Ile, Leu, Met and Phe were substituted for Val and Thr at four positions in the binding pocket. Residues in Tier 1 are located near the heme and are shown in hatched bars, and residues in Tiers 2 and 3 are distant from the heme and are indicated by open bars. (A.) Each bar represents the difference in the partitioning of oxygen to peroxide at the second branch point for the mutants minus wild-type P-450cam. The hydrogen peroxide is expressed as a percentage of the total oxygen consumed in the reaction. (B.) Each bar represents the difference in water formed at the oxidase pathway relative to wild-type P-450cam. The water is expressed as a percentage of the total oxygen equivalents passing through the third reaction branch point. For active site residues in Tiers 2 and 3 there is a negative correlation between side chain packing volume and the production of water. This relationship is reversed for mutations in Tier 1.

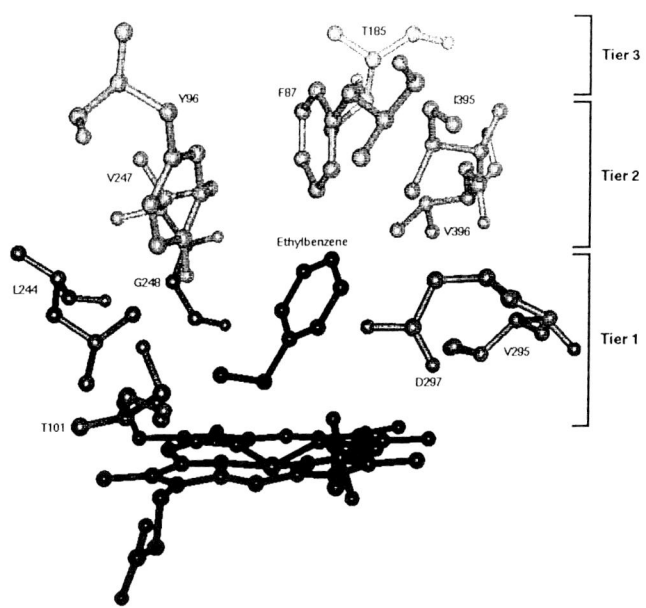

Figure 4. (A.) Side-on view of the cytochrome P-450$_{cam}$ active site with a model of ethylbenzene oriented for benylic hydroxylation. Residues in Tier 1, shown in dark gray, are positioned near to the heme and include T101, L244, V295, G248 and D297. Tiers 2 and 3, shaded light gray, form the upper region of the binding pocket and are comprised of residues F87, Y96, T185, V247, I395 and V396.

REFERENCES

Atkins, W. M., & Sligar, S. G. (1987) Metabolic Switching in Cytochrome P-450cam: Deuterium Isotope Effects on Regiospecificity and the Monooxygenase/ Oxidase Ratio. *J. Am. Chem. Soc. 109*, 3754-3760

Atkins, W. M., & Sligar, S. G. (1988) Deuterium Isotope Effects in Norcamphor Metabolism by Cytochrome P-450cam: Kinetic Evidence for the Two-Electron Reduction of a High-Valent Iron-Oxo Intermediate. *Biochemistry 27, 1610-1616..*

Gorsky, L. D., Koop, D. R., & Coon, M. J. (1984) On the Stoichiometry of the Oxidase and Monooxygenase Reactions Catalyzed by Liver Microsomal Cytochrome P-450: Products of Oxygen Reduction. *J. Biol. Chem. 259*, 6812-6817.

Ishimura, Y., Ullrich, V., & Peterson, J. A. (1971) Oxygenated Cytochrome P-450 and its Possible Role in Enzymic Hydroxylation. *Bioch. and Biophys. Res. Comm. 42*, 140-146.

Loida, P. J., & Sligar, S. G. (1993) Engineering cytochrome P-450cam to increase the stereospecificity and coupling of aliphatic hydroxylation. *Protein Eng. 2*, 207-212.

Sligar, S. G., Lipscomb, J. D., Debrunner, P. G., & Gunsalus, I. C. (1974) Superoxide Anion Production by the Autooxidation of Cytochrome P450cam. *Bioch. Biophys. Res. Comm. 61*, 290-296.

# Metabolism of two dopaminergic ergot derivatives in genetically engineered V79-cells expressing CYP450-enzymes

Hille Gieschen, Michael Hildebrand, Birgit Salomon

Institute of Pharmacokinetics, Schering AG, D-13342 Berlin, FRG

## Abstract

V79 cells expressing rat and human CYP4501A1 and 1A2 were used to characterize and identify biotransformation steps of the ergot derivatives lisuride and terguride. Lisuride was monodealkylated at the diethylurea moiety by CYP1A1 of both species, whereas with terguride this reaction was only catalyzed by human CYP1A1. Rate of monodeethylation of lisuride was similar to the dealkylation rate of 7-ethoxyresorufin, while the affinity of human 1A1 to terguride was less. The consecutive dideethylated metabolite known from in vivo investigations was not formed by any enzyme. Rat 1A2 degradated terguride to a second unknown metabolite, rat 1A1 and human 1A2 showed overlapping activity with parental V79 cells. In summary these investigations show that biodegradation of chemically highly similar compounds involves neither the same CYP450 enzymes in one nor in different species. Slight molecular modifications might deeply influence substrate properties. V79 cells expressing CYP450 enzymes can contribute to a deeper insight and understanding of complex biodegradation processes.

## INTRODUCTION

Studies on biotransformation are an important part of preclinical characterization of drugs. Rapid advance in biotechnology resulted in an additional promising tool for metabolic research, permanent cell lines which express defined enzymes of various species. In the wide field of putative applications ranges from investigations of substrate properties, interactions at the enzyme level, structure-biotransformation correlations to production of single enzyme mediated metabolites. In order to test some of these possibilities two structurally closely related ergot derivatives, lisuride and terguride, were tested in V79 cell lines expressing CYP450 enzymes, which were shown to be metabolically active in case of caffeine (Fuhr et al, 1992), phenacetine (Jensen et al., 1993) and 17β-estradiol (Woelfel et al., 1991). Lisuride, a selective D2-agonist, and its transdihydro-derivative terguride with partial antagonistic effects at the CNS-dopamine receptor, are used as prolactin-inhibitory drugs. Both are known to be extensively metabolized in animals and man. An initial important biodegradation pathway is dealkylation of the diethylurea moiety. In a set of V79 cell lines with rat and human CYP450 enzymes the substrate properties of both compounds should be screened and possible metabolites identified to attribute metabolic reactions to define enzymes, to compare CYP450 activity in both species and to check the feasibility and the relevance of information obtained by this novel approach.

## MATERIALS AND METHODS

Compounds: (*Position of $^{14}$C-label)

Lisuride

Terguride

Cell lines:
V79 cells expressing rat 1A1, rat 1A2, human 1A1, human 1A2 and the parent V79 cells were provided by PD Dr. Doehmer, Institute for Toxicology, Munich, F.R.G.

Metabolic patterns in V79 cell lines (Part A):
The metabolism studies were performed with whole cells. The V79 cells were maintained in DMEM, supplemented with 5% FCS and incubated at 37°C in 5% $CO_2$-athmosphere in 95% humidity. $2 \times 10^6$ cells were seeded in 75 cm$^2$ flasks, incubated for 24 hours in 10 ml medium. After medium exchange 0.3 $\mu$M of radiolabeled $^{14}$C-substrate was added and incubated for 72 hours.

Time dependent degradation of ergot derivatives (Part B):
The time course was estimated throughout cell culture incubated in 6-well-tissue culture clusters. Two wells were used per timepoint. $2 \times 10^5$ cells were seeded in one well, incubated for 24 hours in 2 ml medium. After medium exchange 20 nM of radiolabeled $^{14}$C-substrate was added and incubated for 0, 6, 24, 48 and 72 hours. After sampling the supernatant, cells were counted, homogenised and after ultracentrifugation (100000 g, 60min) resuspended in Hepes-buffer (50mM, pH 7.6) to a cytosole free membrane proteinpellet (CfP). The protein content was determined by BCA-method.

Analysis:
The supernatants of samples obtained in parts A and B were extracted with acetone/methanol (3/1; v/v). After removal of the liquid phase by $N_2$ the residues were solved in eluate and analyzed by HPLC.

*HPLC-Conditions*
Part A: For separation of metabolites a Spherisorb ODS II (125 x 4.6 mm, 5 $\mu$m) with precolumn was used as stationary phase. The mobile phase consisted of eluate A (acetonitrile) and eluate B (ammoniumcarbonate 0.01 M) at a flow rate of 1.0 ml/min. The chromatographic run increased linear from 20 % A to 100% A in 30 min.
Part B: As stationary phase a Lichrospher 60 RP select B-column, (120 x 3.0 mm, 5 $\mu$m) with precolummn was used. The mobile phase consisted of 50 % methanol, 25 % $H_2O$, 25 % $Na_2HPO_4$ (1/15 M) at a flow rate of 1.0 ml/min.
For metabolite detection in both parts fluorescence detection (ext. 282 nm, em. 370 nm) and continuous flow radiodetection (RAMONA, raytest) was used.

*Evaluation*
The HPLC-chromatograms were recorded both as fluorescence and on-line radioactivity detection. Metabolic profiles were always compared for transformed versus parent V79 cell lines.

## RESULTS

Lisuride was monodeethylated by rat and human CYP1A1 and almost stable against both CYP1A2. Parent V79-cells showed negligible metabolic activity. The monodeethylated metabolite was identified by cochromatography and photodiodearray-detection. This compound is known as a main metabolite from other in-vitro (rat hepatocytes, liver-perfusion) and in-vivo investigations (Huempel et al., 1989). The consecutive main metabolite dideethyllisuride was not formed by any of the tested enzymes. Metabolic pattern obtained by fluorescence and $^{14}C$-detection were similar (Fig. 1).

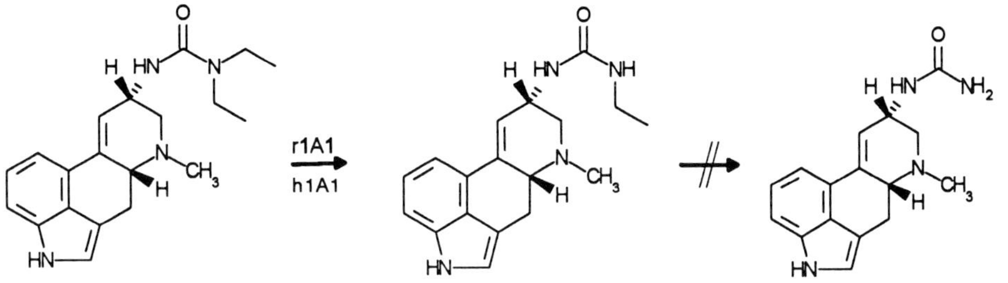


Lisuride Monodeethyllisuride Dideethyllisuride

*Fig. 1. Metabolic pathway of lisuride after incubation with rat and human CYP 1A1 and 1A2*

Terguride was monodeethylated by human CYP1A1, but only to a minor extent by rat CYP1A1 (Fig. 2). The parent V79 cells formed another unknown metabolite, which was also found at a larger amount after incubation with rat CYP1A1 and human CYP1A2. Interestingly this metabolite was only detectable by radiochromatography. With rat CYP1A2 a second unknown, but fluorescing metabolite was observed.

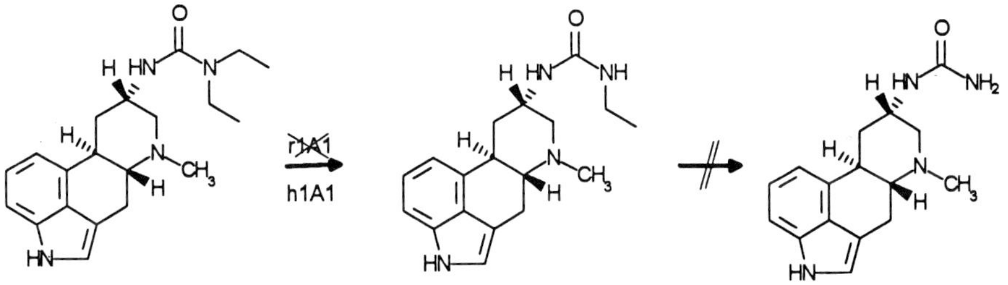


Terguride Monodeethylterguride Dideethylterguride

*Fig. 2. Known metabolic pathway of terguride after incubation with rat and human CYP 1A1 and 1A2*

Interesting is that terguride with a very similar structure to lisuride is also deethylated by human CYP1A1, but rat CYP1A1 forms a different metabolite that also is found with human CYP1A2 and in the parent V79-cells.

Turnover:

The turnover rate of lisuride after 24 h incubation with rat and human CYP1A1 was similar to that obtained for 7-ethoxyresorufin with these enzymes, though ergot incubation was done in cell culture and the resorufin-activity was determined out of a CfP. In contrast terguride turnover with the human CYP1A1 was lower than lisuride.

| CYP450 | Deethyllisuride pmol/min/mg | Deethylterguride pmol/min/mg | 7-Ethoxyresorufin pmol/min/mg |
|---|---|---|---|
| rat 1A1 | 9.8 | - | 8.3 |
| human 1A1 | 21.2 | 2.23 | 18.8 |

## DISCUSSION

Despite high structural analogy lisuride and terguride exhibited species and enzyme specific difference concerning rate and extent of metabolic degradation by human and rat CYP 1A1. Monodeethylation of lisuride was catalyzed by CYP 1A1 from both species, whereas only human CYP 1A1 transformed terguride by the same pathway. Additionally lisuride has a higher affinity to human CYP1A1 than terguride. It remains to investigate enzyme kinetics by Michaelis-Menten to determine maximum of turnover. Metabolic activity was limited to single dealkylation, although a second identical reaction to a dideethlylated metabolite is known to occur in vivo. Another ergot derivative CQA 206-291 was shown to be monodeethylated at the sulfonamide moiety by CYP3A4 in human microsomes (Ball et al., 1992). Evaluating these results it can be concluded that a combination of CYP450 enzymes seems to be responsible for this reaction. Soon we will be able to investigate our ergot derivatives with V79 cells expressing human CYP3A4, which are genetically engineered in-house. It will be possible to clarify rate and extent of monodeethylation by these different CYP450 enzymes. As all investigations performed up to now only revealed single dealkylation of CYP450, it remains to be clarified, which enzymes catalyse the in vivo relevant dideethylation.

Our results suggest that slight molecular modifications might deeply influence substrate properties. The assumption that chemically highly similar compounds have to be metabolized by the same CYP450 in one species has to be verified by direct investigation. V79 cells expressing CYP450 enzymes can contribute to a deeper insight and understanding of complex biodegradation processes.

## ACKNOWLEDGEMENTS

The authors wish to thank Dr. J. Doehmer (Institute of toxicology, University of Munich, F.R.G.) for providing V79 cells and advisory contribution for model adaptation. The research work was part of a scientific project sponsored by the Schering Research Foundation, Berlin, F.R.G.

## REFERENCES

Ball et al. (1992): Characterisation of the Cytochrome P450 gene family responsible for the N-Dealkylation of the ergot alkaloid CQA 206-291 in Humans. *Drug Metabolism and Disposition* 20 No1: 56-63

Fuhr et al. (1992): Biotransformation of caffeine and theophylline in mammalian cell lines genetically engineered for expression of single Cytochrome P450 isoforms. *Biochem Pharmacol.* 43: 225-235

Huempel et al. (1989): Studies on the biotransformation of Ionazolac, bromerguride, lisuride and terguride in laboratory animals and their hepatocytes. *Xenobiotica* 19, No. 4: 361-377

Jensen et al. (1993): Metabolism of phenacetin in V79 Chinese hamster cell culture expressing liver Cytochrome P450 1A2 compared to isolated hepatocytes. *Biochem Pharmacol.* 45, 1171-1173

Woelfel et al. (1991), Stable expression of rat Cytochrome P450 1A2 cDNA and hydroxylation of 17β-estradiol and 2-aminofluorene in V79 Chinese hamster cells. *Mol Carcinogen* 4: 489-498

# Mutants affecting substrate recognition and substrate binding in P450BM-P

S. Graham-Lorence, D. Sanders, J.A. Peterson

Department of Biochemistry, University of Texas Southwestern Medical Center, Dallas, Texas, USA

## INTRODUCTION

Cytochromes P450 are a super family of heme-containing proteins which in combination with a redox partner (e.g. an iron-sulfur protein and/or a reductase) monooxygenate hydrophobic organic compounds. There are two classes of cytochromes P450: *class I* which requires both an iron-sulfur protein and an FAD-containing NAD(P)H reductase, and is found generally in bacteria and in mitochondria of eukaryotes; and *class II* which requires an FAD/FMN-containing NADPH P450 reductase and is generally found in the endoplasmic reticulum of eukaryotic organisms. One exception to this categorization is the soluble P450-BM3 which is a *class II* P450 found in *Bacillus megaterium* and is a fusion protein with its FAD/FMN-containing redox partner. Because this *class II* P450 is soluble and readily isolated in large quantities and because of its fusion with its redox partner, P450-BM3 is a good model for *class II* P450s both structurally and functionally.

Recently, the three-dimensional structure of cytochrome P450-BM3 has been determined (1). It has a structure similar to that of cytochrome P450cam whose structure was determined in 1987 by Poulos and co-workers (2). However, the structures are sufficiently different that one can not use P450cam as a template to build P450-BM3. A large portion of the structural differences are in those regions involved in redox-partner binding, and in substrate recognition and substrate orientation into the heme pocket/active site. To better understand some of the subtleties of structure in those regions involved in substrate recognition and binding, initially we have chosen to make mutations in two residues: R47 and F87.

## METHODS AND RESULTS

### Rationale of Mutants

R47 is located at the mouth of the access channel leading to the heme pocket and F87 is located in the active site/heme pocket over the meso edge between the C and D pyrrole rings. Since R47 is the only charged residue in the channel, we believe it may interact with the carboxyl group of fatty acids; therefore, we have asked the question "How does mutagenesis of R47 affect the selectivity, and thus, activity and amount of product formation of P450BM-P?" To answer this question, we have mutated R47 to Ala and to Glu. R47A removes the charge and reduces the size of the sidechain making the channel more hydrophobic and possibly less selective, and R47E changes the charge from basic to acidic possibly inhibiting binding of fatty acids.

The other residue F87 is located directly over the heme and may be involved in steric steering of the substrate, and thus we ask "How does mutagenesis of residue F87 affect substrate binding in the heme pocket, and therefore, the activity and products formed?". To test the role of this residue, we have mutated this residue to Val and Tyr. F87V will remove the steric bulk from the pocket possibly increasing the number of products formed or changing the ratio of product formation. Additionally, in sequence alignments with omega hydroxylases, this residue is a valine or leucine. The other mutant, F87Y, while maintaining the steric bulk, will make the pocket more polar.

## Turnover Rates with Palmitate and Arachidonate

Turnover rates of palmitate were determined in the wildtype and mutant proteins by measuring the rate of NADPH consumption in a reconstituted system as shown in Table 1 using a 1:20 ratio of BM3 NADPH-reductase-domain to BM3 P450-domain and in the presence of 500 uM palmitate (3). In the R47 mutants, R47A had essentially the same rate as wildtype, while R47E had consumed NADPH at approximately 60% of wildtype. In the F87 mutants, F87Y had the same rate of NADPH consumption as wildtype, but the rate for F87V was twice that of wildtype. (The background NADPH consumption rate for the wildtype and the mutants was approximately 0.2 uM/min/uM P450.)

Generally, the rates of turnover for arachidonate were approximately twice those for palmitate in both the wildtype and mutant proteins except for F87Y. That is, of the other mutant proteins, R47A had a similar rate to WT, while F87V had a higher rate, and R47E had a slower rate of metabolism for both compounds; however, in F87Y, the rate was the same for both substrates.

## TABLE 1

### Rate of NADPH Consumption with Palmitate and Arachidonate
(uM/min/uM P450)

| Protein | Palmitate | Arachidonate | Background |
|---|---|---|---|
| Wildtype | 3.4 +/-0.5 | 8.3 +/-0.6 | 0.196 +/-.003 |
| F87V | 6.8 +/-0.4 | 11.6 +/-0.7 | 0.251 +/-.009 |
| F87Y | 3.3 +/-0.2 | 3.0 +/-0.5 | 0.23 +/-.01 |
| R47A | 3.0 +/-0.6 | 8.6 +/-0.8 | 0.203 +/-.007 |
| R47E | 2.1 +/-0.6 | 5.4 +/-0.6 | 0.205 +/-.006 |

## Percent Product Formation and Percent Coupling with Palmitate

Initial experiments on total product formation were done using 650 uM palmitate containing $^{14}$C-palmitate, followed by separation of products from substrate by HPLC using the methods of Okita *et al.* (4). In the F87 mutants, F87V, where the NADPH turnover rate was twice that of WT, formed hydroxylated product approximately half that of WT indicating that the oxidation of NADPH was coupled to hydroxylated product formation about 25% of the time. In the other F87 mutant - F87Y, where the NADPH turnover was similar to WT, the amount of polar product formed was less than 7% as compared to WT which is essentially the background rate, *i.e.*, no significant product formation could be seen in this mutant. In the R47 mutants, R47E had a similar coupling rate as F87V although its rate of NADPH turnover was much slower, and R47A whose rate of NADPH consumption was not significantly different than wildtype was coupled approximately 40% of the time. In the future, we will look at the ratio of products formed.

## DISCUSSION

We have chosen two residues for our initial mutagenesis experiments: R47 which is at the mouth of the access channel and F87 which is located directly above the gamma-meso edge of the heme. R47 may bind to the carboxyl group of the fatty acid either to anchor the fatty acid at the top of the pocket or as a mechanism to recognize the fatty acid. F87 may serve to orient substrates in the pocket.

Of the R47 mutants, changing the Arg to Ala (R47A) did not significantly change the rate of NADPH turnover; however, it increased the amount of uncoupling. This probably results from appropriate recognition of substrate due to the hydrophobic interactions, but frequent inappropriate positioning of the terminal carbons in the active site due to "slipping" of the substrate up or down in the access channel/heme pocket. R47E changed the charge on this residue which decreased the NADPH turnover, and increased the amount of uncoupling. This most likely is due to both a decreased recognition of the substrate at the access channel and inappropriate positioning of the terminal carbon into the active site.

Of the F87 mutants, F87V showed an increased turnover rate probably because of the increased accessibility of the substrate into the active site/heme pocket; however, the lack of steering by removal of the bulky phenolic ring along with the increased room in the pocket - possibly filled by water - may have caused the increased amount of uncoupling. F87Y which maintained the steric bulk but introduced a polar group, did not alter the turnover with palmitate, but was almost completely uncoupled. We believe this to be due to the increased polarity in the active site. More specifically, the Tyr may interact with the water ligand on the heme iron causing uncoupling of the reaction. This mutant also shows very little shift from low spin to high spin in the presence of excess substrate. Interestingly, the F87Y mutant had a slower NADPH consumption rate with arachidonate as compared to WT, i.e. its rate with palmitate and arachidonate were suspiciously similar.

From the results of the F87Y mutant, one might postulate that the NADPH rate which is directly related to the rate of reductase binding to the P450 and reduction of the P450 may be dependent on a conformational change of the P450 induced by substrate binding. This would explain the comparable rates of NADPH turnover for both palmitate and arachidonate in this mutant, but the absence of product formation. This concept is supported by the fact that there were two conformationally different molecules found in the crystal structure of P450BM-P (1) such that the access channel of one molecule was more open than the other. This theory will be further checked in the future by determining the binding constant for palmitate to this mutant.

In summary, it appears that R47 may play a role in substrate recognition and binding of fatty acids, although the hydrophobic effect may be the primary determinant in recognition and binding. Additionally, F87 does appear to play an important role in substrate steering in the active site. These mutants will be further tested with other substrates.

## References

1. Ravichandran, K.G., Boddupalli, S.S., Haseman, C.A., Peterson, J.A., and Deisenhofer, J. (1993). Science **261**, 731-736.
2. Poulos, T.L., Finzel, B.C., and Howard, A.J. (1987). J Mol Biol **195**, 687-700.
3. Boddupalli, S.S., Oster, T., Estabrook, R.W., and Peterson, J.A. (1992). J Biol Chem **267**, 10375-10379.
4. Okita, R.T., Clark, J.E., Okita, J.R., and Masters, B.S.S. (1991). Methods in Enzymology, vol. 206, 432-441.

# VII | CYTOCHROME P450. BIOCHEMISTRY, BIOPHYSICS AND MOLECULAR BIOLOGY

# The use of synthetic peptides for the investigation of components interaction in monooxygenase rabbit liver system

A.M. Kritsky, I. P. Kanaeva, D.R. Davydov, N.V. Stepanova, G.I. Bachmanova

*Institute of Biomedical Chemistry, Russian Academy of Medical Sciences, Pogodinskaya 10, Moscow, 119832, Russia*

Local homology of NADPH−cytochrome P450 reductase and cytochrome b5 amino acid sequences allowed to presume structural similarity between the cytochrome c binding centers in cytochrome b5 and NADPH−cytochrome P450 reductase (Davydov et al.,1992). On the basis of this presumption the peptide LEEHPGGEE was synthesized which simulates the part of the cytochrome b5 binding center with cytochrome c and cytochrome P450. The peptide's influence on the NADPH−dependent benzphetamine N-demethylation reaction in monomeric soluble reconstituted system (MRS), containing NADPH− cytochrome P450 reductase, cytochrome P450 2B4 (2B4) and cytochrome b5 was studied. The action of this peptide on NAD(P)H-cytochrome c reductase activities was studied in microsomes as well. It was shown, that it inhibits these reactions, but the process is not highly specific.

Electrostatic interactions have been proposed to play important role in the formation of active cytochrome P450 complexes with its redox partners. These interactions might involve positively charged regions on the cytochrome P450 molecule and negatively charged sites in the amino acid sequence of NADPH−cytochrome P450 reductase (Bernhardt et al., 1988; Strobel et al., 1989). In microsomal monooxygenase system this interaction has been presumed to be based on the interaction of aspartic or glutamic acid rich regions of the flavoprotein and cytochrome b5 with positively charged lysine or arginine rich zones in partner molecule (Poulos and Mauk, 1983; Mauk et al., 1986; Holloway and Mantsh, 1988).

The purpose of this work is to clear up whether the interaction mentioned involves only the local charge of a limited (about ten residues) peptide sequence in the binding site. In this case its charge recognition will be sufficient to provide for protein interaction and overcome spatial restrictions caused by polypeptide folding. The alternative could be a more complex situation when parts of the folded polypeptide molecule, not adjacent to binding site, also contribute to its charge pattern and binding capacity.

Among the ways to verify this hypothesis, synthesis of a peptide which amino acid sequence reproduces the binding site of cytochrome b5 to cytochrome P450 and cytochrome c was proposed. Like the cytochrome b5 binding site, the peptide bears four negative charges alternating with a non charged region (Davydov et al,1992). If electrostatic forces caused by this charged region prevail in recognition process, the peptide would stick up to the partner's binding domain much stronger than some peptide a with different charge distribution can do it. In other case, we had to conclude that electrostatic interactions caused by this local zone of the cytochrome b5 molecule do not play main role in protein recognition.

**Materials and Methods**

Peptide synthesis. The peptide LEEHPGGEE was synthesized on an automatic peptide synthesizer Applied Biosystems 431A using Fmoc−strategy. Commercial amino acid derivatives, reagents and solvents of the highest available purity were used. The amino acid composition of the reversed phase HPLC purified peptide was: Glu − 3.95 (4), Pro−0.98 (1), Gly−2.00 (2), Leu−0.9 (1), His−1.08 (1). Phenobarbital−induced rabbit liver microsomes, 2B4, and reductase preparation, monomerization of the proteins as well as NADPH−dependent reduction of microsomal cytochrome P450 and benzphetamine demethylation rate measurements were carried out according to methods, described previously (Kanaeva, et al., 1992a; Kanaeva, et al., 1992b).

**Results and Discussion**

Addition of the LEEHPGGEE peptide to microsomes influenced the NADH−cytochrome c reductase and NADPH−cytochrome c reductase activity only to a small extent. At the peptide concentration varying from 33 μM to 133 μM, i.e. to 0.6 − 2.6 molecules per binding site of cytochrome c, both reductase systems retained about 80 per cent of the activity. Even at the 667 μM concentration the loss of activity did not exceed 20 per cent for NADH−cytochrome c reductase and 40 per cent for the NADPH−cytochrome reductase (Tables 1 and 2).

**Table 1.** Influence of the peptide on NADH−cytochrome c reductase activity in microsomes. The incubation mixture contained 100 mM K−phosphate buffer, pH 7.5, 50 μM cytochrome c, 330 μM NaCN, 200 μM NADH, 0.5 μM of microsomal cytochrome P450. $\lambda = 550$ nm. $T = 30^{\circ}$ C.

|  | Peptide, μM | | | | | |
| --- | --- | --- | --- | --- | --- | --- |
|  | 0 | 33 | 67 | 133 | 333 | 667 |
| Activity, % of the control | 100 | 90 | 80 | 80 | 80 | 80 |

**Table 2.** Influence of the peptide on NADPH−cytochrome c reductase activity of NADPH−cytochrome P450 reductase. The incubation mixture contained 100 mM K−phosphate buffer, pH 7.7, 50 μM cytochrome c, 330 μM NaCN, 100 μM NAPDH, 0.82 μM of NADPH−cytochrome P450 reductase. $\lambda = 550$ nm. $T = 30^{\circ}$ C.

|  | Peptide, μM | | | |
| --- | --- | --- | --- | --- |
|  | 0 | 133 | 334 | 667 |
| Activity, % of the control | 100 | 80 | 80 | 60 |

Such small loss of activity after the addition of the LEEHPGGEE peptide shows that the peptide could not seriously affect redox partner interaction between the components in monooxygenase system.

Apparent rate of the NADPH−dependent cytochrome P450 reduction in monooxygenase system includes two kinetic components with $k_1$ and $k_2$ which can be separated by monitoring the reaction in the presence of benzphetamine which adding enhances mostly the fast ($k_1$) component. It has been shown, that addition of the LEEHPGGEE peptide practically did not influence the velocity of the $k_2$ reaction. The $k_1$ value slightly decreased in the presence of the peptide. These results indicate that the peptide might shield some sites on the cytochrome P450 molecule from reductase and cytochrome b5 binding. This binding, however, was not of high specificity (Table 3).

**Table 3.** Effect of the peptide on NADPH−dependent reduction of microsomal cytochrome P 450. The incubation mixture contained 100 mM K−phosphate buffer, pH 7.5, 0.5 μM of microsomal cytochrome P450, 1mM NADPH, 1mM peptide, saturating concentration of carbon monoxide. T=30° C.

| Substrate | without peptide | | plus peptide | |
|---|---|---|---|---|
| | $k_1$ (s$^{-1}$) | $k_2$ (s$^{-1}$) | $k_1$ (s$^{-1}$) | $k_2$ (s$^{-1}$) |
| without benzphetamine | 0.128 | 0.021 | 0.129 | 0.020 |
| in the presence of benzphetamine | 0.300 | 0.025 | 0.247 | 0.024 |

The effect of the LEEHPGGEE peptide on the benzphetamine demethylation rate in MRS was studied. As an alternative peptide lacking specific negative charge pattern, we used YAGFLR, kindly presented by Dr. V.F.Pozdnev. Both the peptides produced almost the same effect on the rate of demethylation in MRS, some small difference in the activity of two peptides could be observed with low, non−saturating concentrations of LEEHPGGEE (Table 4).

**Table 4.** Effect of the peptide on the benzphetamine demethylation rate in MRS. The incubation mixture contained 100 mM K−phosphate buffer, pH 7.5, 0.25 g/liter Emulgen 913, 2 mM NADPH, 2 mM benzphetamine, 1 μM reductase, 2B4 and b5 monomers. Incubation was carried out for 5 min. at 30° C. YAGFLR peptide was used as control one.

| Concentration of peptide, μM | % of the benzphetamine demethylation rate in the absence of peptide | | |
|---|---|---|---|
| | LEEHPGGEE + 2B4 + reductase | LEEHPGGEE + 2B4 + reductase + b5 | YAGFLR + 2B4 + reductase |
| 0 | 100 | 100 | 100 |
| 500 | 80 | 85 | 90 |
| 1000 | 70 | 75 | 85 |
| 2000 | 55 | 60 | 55 |

The data presented conflict with presumed specific competition of the LEEHPGGEE with cytochrome b5 and reductase for binding to cytochrome P450 and cytochrome c. In this connection the suggestion on critical significance of a relatively small (about 10 residues) peptide region in formation of the partner complex between these proteins needs at least some revision. In the following experiments such factors as overall polypeptide folding should be considered. In contrast to a small isolated peptide, the folded protein molecule may form some specific charge profile by exposing together the residues located distally from each other in its primary structure. In this case synthetic analogs for the binding region can be realistically designed only from a detailed 3D model of the studied protein.

**References**

1. Bernhardt, R., Kraft, R., Otto, A., Ruckpaul, K. (1988), Biomed. Biochim. Acta 47, 581−592.
2. Davydov, D.R., Darovsky, B.V., Dedinsky, I.R., Kanaeva, I.P., Bachmanova, G.I., Blinov, V.M., Archakov, A.I. (1992), Arch. Biochem. Biophys. 297, 304−313.
3. Holloway, P.V. and Mantsh, H.H. (1988), Biochemistry 27, 7991−7993.
4. Kanaeva, I.P., Dedinskii, I.R., Skotselyas, E.D., Krainev, A.G., Guleva, I.V., Sevryukova, I.F., Koen, Y.M., Kuznetsova, G.P., Bachmanova, G.I., Archakov, A.I. (1992), Arch. Biochem. Biophys. 298, 395−402.
5. Kanaeva, I.P., Nikityuk, O.V., Davydov, D.R., Dedinskii, I.R., Koen, Y.M., Kuznetsova, G.P., Skotselyas, E.D., Bachmanova, G.I., Archakov, A.I. (1992), Arch. Biochem. Biophys. 298, 403−412.
6. Mauk, M.R., Mauk, A.G., Mathew, J.B., Weber, P.C. (1986), Biochemistry 25, 7085−7091.
7. Poulos, T.L., and Mauk, A.G. (1983), J. Biol. Chem. 258, 7369−7375.
8. Strobel, H.W., Nadler, S.G., Nelson, D.R. (1989), Drug. Metab. Dispos. 20, 519−534.

# Multiple forms of NADPH-cytochrome P450 reductase in higher plants

Agnès Lesot, Catherine Bègue-Kirn, Marie-Paule Hasenfratz, Francis Durst, Irène Benveniste

*CNRS, Institut de biologie moléculaire des plantes, Laboratoire d'enzymologie cellulaire et moléculaire, 28, rue Goethe, 67083 Strasbourg Cedex, France*

## Introduction

Plant cytochrome P450 monooxygenases are implicated in important physiological pathways leading to the biosynthesis of lignins and pigments, sterols, cutins and suberins, hormones, phytoalexins, alkaloids, and in the metabolism of xenobiotics (Durst and Benveniste, 1993).
Microsomal monooxygenation activities need the transfer of reducing equivalents from NADPH to the cytochrome P450 via a flavoprotein, the NADPH-cytochrome P450 reductase. In animals and microorganisms, it is generally admitted that all the isoforms of cytochrome P450 are reduced by a unique form of NADPH-cytochrome P450 reductase. In contrast, recently we have described a polymorphism of the NADPH-cytochrome P450 reductase at the protein level, in Jerusalem artichoke tuber microsomes (Benveniste *et al.*, 1991). This observation is here extended to the microsomal reductase from *Vicia sativa* seedlings and from several other higher plants. The molecular bases of this multiplicity are investigated and a differential expression of the reductases in Jerusalem artichoke tuber after wounding is described.

## Materials and Methods

Preparation of microsomes, purification of NADPH-cytochrome P450 reductase and measurement of NADPH-cytochrome c reductase activity are performed according to Benveniste *et al.*, 1991.
Characterization of polyclonal antibodies raised against Jerusalem artichoke NADPH-cytochrome P450 reductase and Western blot conditions are described in Benveniste *et al.*, 1989.
Total RNAs are purified and characterized as described in Lesot *et al.*, 1990.
Cloning, isolation and sequencing of three partial cDNAs encoding NADPH-cytochrome P450 reductases from Jerusalem artichoke tuber are described in Hasenfratz (1992).
Radioactive labelling of cDNA probes and hybridization conditions in Northern blots with different RNAs purified from *Helianthus tuberosus* tuber are reported in Lesot (1993).

# Results

### Purification of three NADPH- cytochrome P450 reductases from *Helianthus tuberosus* tuber and *Vicia sativa* seedlings

Solubilization of microsomal proteins from Jerusalem artichoke tuber and vetsch seedlings by a non-ionic detergent ( Emulgen 911) and chromatography on DEAE-Trisacryl M and 2'5' ADP Sepharose purified specifically three proteins which comigrates nearly at the same position on an 7.5% SDS-polyacrylamide gel.(Fig. 1)
The molecular weights of Jerusalem artichoke reductases are 80 000, 82 000 and 84 000.

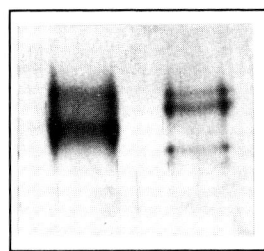

Fig. 1. : Comparison on 7.5% SDS-PAGE of the purified NADPH-cytochrome P450 reductases from Jerusalem artichoke (left) and *Vicia sativa* (right) after chromatography on DEAE-Trisacryl and 2'5' ADP-Sepharose.

Polyclonal antibodies against the three Jerusalem artichoke reductases recognized the three *Vicia sativa* reductases.

Further chromatography of the three proteins from *Helianthus tuberosus* and *Vicia sativa* on Concanavaline A-Sepharose showed the binding of the highest MW form on the lectin column and specific elution by α- methylmannoside, indicating the glycoprotein nature of this form of reductase.

We have shown for the Jerusalem artichoke reductases that all three forms allow functional reconstitution of monooxygenase activity in the presence of purified cytochrome P450 (cinnamic acid hydroxylase) and phospholipids.

### Induction of NADPH-cytochrome c reductase activity by wounding and aeration of the tuber tissues in water

The microsomal NADPH-cytochrome c reductase activity was rapidly and strongly enhanced by wounding and aging of the tuber tissues. The activity, very low in quiescent tissues was increased after a lag phase of only 45 minutes. The maximal stimulation (20 times) was reached 24 hours after wounding and then the activity declined rapidly.

### Differential induction of the expression of NADPH-cytochrome P450 reductases in response to wounding and aging of the tuber tissues on water.

Western blot analysis on microsomal proteins after different times of aging of the tuber tissues, with polyclonal antibodies against the purified three tuber reductases, revealed a differential expression of the flavoproteins (Fig. 2).

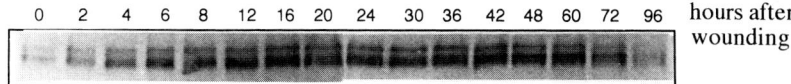
hours after wounding

Fig. 2. : Western blot analysis on microsomal proteins from Jerusalem artichoke tuber tissues at different times after wounding and aging, with polyclonal antibodies against the three tuber reductases.

The 80 kDa protein was absent in the quiescent tissues, which contained only low amounts of the 82 and 84 kDa. Two hours after wounding, the 80 kDa form was already present and its quantity increased as the NADPH-cytochrome c reductase activity. The amount of the 82 and 84 kDa were also enhanced by wounding. A good correlation between activity and quantity of the reductases was observed during the stimulation phase. Therefore, the activity stimulation could be the result of a *de novo* synthesis of the three isoforms. After the maximum, the activity decreased rapidly, whereas the amount of enzymes lowered more slowly. Inactivation seemed to occur more rapidly than protein degradation.

Evolution of the transcripts of NADPH-cytochrome P450 reductases after wounding and aging of Jerusalem artichoke tuber.

Northern blot analysis on total RNAs from tuber tissues extracted after different times of aging, with a cDNA probe specific of an *Helianthus tuberosus* reductase (HTR1 described in Hasenfratz 1992), is represented on Fig. 3.

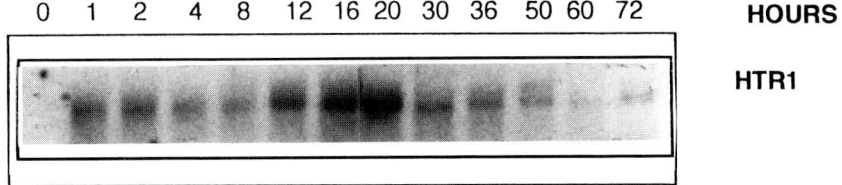

Fig. 3. : Northern blot analysis on total RNAs from Jerusalem artichoke tuber tissues extracted at different times after wounding and aging, with a radioactive *Helianthus tuberosus* reductase cDNA probe (HTR1).

No specific transcript was detectable in the non-induced tissues. After one hour wounding, a transcript was clearly revealed. The intensity of hybridization increased until 20 hours and then decreased rapidly. It is note worthy that the HTR1 probe revealed two populations of RNAs (well visible from 12 to 50 hours after wounding), which could code for the different forms of reductase. The strongest hybridization signal revealed a transcript of 2.4 kb, the weaker one at 2.6 kb.

An heterogenous hybridization pattern (2 populations near 2.4 kb) has been also observed on a Northern blot with poly $A^+$ RNAs from *Vicia sativa* seedlings and an homologous reductase cDNA probe (VSR1, access number Z26252).

## Conclusion

The stimulation of NADPH-cytochrome P450 reductase activity by wounding and aging of Jerusalem artichoke tuber tissues was well correlated with the increase of reductase protein and of specific transcripts. This stimulation is probably the result of *de novo* synthesis of the enzyme in response to transcriptional activation by wounding and aeration of the tissues. At the protein level, a differential expression was observed : the 80 kDa form, which is absent in quiescent tuber, and the 82 and 84 kDa forms, which are slightly represented in these tissues, are rapidly increased after wounding.

The results also clearly show the existence of at least two transcripts for the Jerusalem artichoke reductases, indicating that the polymorphism at the protein level was not only due to post-translational modifications.
This observation is in excellent agreement with the isolation of three partial cDNAs, which show more than 10% divergence.
Another example of NADPH-cytochrome P450 reductase polymorphism at the protein and transcript level was furnished by *Vicia sativa* seedlings. Furthermore two full-length cDNAs are isolated and their sequences are investigated.
Finally, two full-length cDNAs coding for the NADPH-cytochrome P450 reductase from *Arabidopsis thaliana* presenting only 65% amino-acid positional identity were isolated (D. Pompon, personal communication).

Therefore, it is an original situation that the NADPH-cytochrome P450 reductases in higher plants could be encoded by a multigenic family, in comparison with the reductases from other kingdoms (microorganisms ans animals) (Shen and Kasper, 1993).

**Acknowledgements** : We would like to thank Marie-France Castaldi for her excellent technical assistance.

## REFERENCES

Benveniste, I., Lesot, A., Hasenfratz, M.P., and Durst, F. (1989) Immunochemical characterization of NADPH-cytochrome P450 reductase from Jerusalem artichoke and other higher plants. Biochem. J. 259 847-853.

Benveniste, I., Lesot, A., Hasenfratz, M.P., Kocks, G., and Durst, F. (1991). Multiple forms of NADPH-cytochrome P450 reductase in higher plants. Biochem. Biophys. Res. Commun. 177 105-112.

Durst, F., and Benveniste, I. (1993). Cytochrome P450 in plants. In Handbook of Experimental Pharmacology - Cytochrome P450, eds Schenkman J.B. and Greim H. (Springer Verlag) vol. 105 pp 293-310.

Hasenfratz, M.P. (1992). Clonage de la NADPH - cytochrome P450 réductase et d'une protéine "calnexine like" chez *Helianthus tuberosus*. Thèse d'Université - Strasbourg.

Lesot, A. (1993) Polymorphisme et régulation de l'expression des NADPH-cytochrome P450 réductases (CPR) d'*Helianthus tuberosus*. Thèse d'Etat - Strasbourg.

Lesot, A., Benveniste, I., Hasenfratz, M.P., and Durst, F. (1990). Induction of NADPH-cytochrome P450 (L) reductase in wounded tissues from *Helianthus tuberosus* tubers. Plant Cell. Physiol. 31 1177-1182.

Shen A.L., and Kasper, C.B. (1993). Protein and gene structure and regulation of NADPH - cytochrome P450 oxido reductase. In Handbook of Experimental Pharmacology - Cytochrome P450, eds Schenkman J.B. and Greim H. (Springer Verlag) Vol 105 pp 35-60.

# Strain- and tissue-specific expression of rat CYP2B

Alevtina Yu. Grishanova, Lyudmila F. Gulyaeva, Eugeny V. Petchkovski, Vyacheslav V. Lyakhovich

*Institute of Molecular Pathology and Ecological Biochemistry, Russian Academy of Medical Sciences, Novosibirsk 630117, Russia*

We studied the expression of CYP2B in liver of different rat strains (Wistar, Sprague-Dawly, non-insulin diabetic (NID)) treated with phenobarbital-type inducers Aroclor 1254 (Ar), triphenyldioxane (TPD) and perfluorodecalin (PFD) at the posttranscriptional level as well as of CYP2B in liver and lung of Wistar rats treated with Aroclor 1254 both at the transcriptional and posttranscriptional levels.
Remarkable inter-strain differences were shown in 7-penthoxyresorufin-O-depenthylase activity associated with CYP2B1 which can be interpreted as a reflection of an interstrain variability of CYP2B1 genes.
Aroclor 1254 treatment of Wistar rats compared to PB treatment significantly changes the distribution of CYP2B forms: both in liver and in lung besides CYP2B1 also CYP2B2 and the third CYP2B immunoreactive protein are detected. With the use of dot-blot analysis with cDNA to CYP2B1/B2 the increase of the mRNA level was detected in lung of Aroclor-treated rats compared to the PB-treated one.

## I. INTRODUCTION

Induction of genetically controlled CYPs by chemicals is tissue-specific, and it was suggested that the multiplicity of CYPs is a factor determining the pharmacological susceptibility to drugs and carcinogenic compounds. Among these isozymes there are the phenobarbital-inducible (PB) forms, products of CYP2B genes. Although the liver is generally an organ with the highest drug-metabolizing activity, P-450s and their associated monooxygenase activities are also present in other tissues, including kidney, lung, skin, small intestine etc. The regulation of the expression of CYP allows consequently an important determination of the metabolic fate of drugs as well as a large number of chemicals of different structures. Existence of tissue-specific and interstrain differences complicates the pattern of CYP regulation.

## II. MATERIALS AND METHODS

Male Wistar, Sprague-Dawly, non-insulin diabetic (NID) rats weighing 160-180g were used in our experiments. The animals were treated with phenobarbital (80mg/kg of body weight during 4 days), polychlorinated biphenyls Aroclor 1254 (300mg/kg), triphenyldioxane (10mg/kg) and 10perfluorodecaline emulsion (10 ml/kg body weight during 3 days) [Mishin et al,1989]. Phenobarbital was dissolved in saline, whereas Aroclor 1254 and triphenyldioxane were dissolved in corn oil and administered by intraperitoneal injection. Control rats were given 0.5ml of corn oil or saline by the same routes. Liver microsomes were obtained by a conventional method of differential centrifugation. The protein content was measured by the method of Lowry et al [1951]. The rate of penthoxyresorufin-O-dealkylation was assayed by the fluorescence method using resorufin as standard [Burke et al,1985]. Western blot experiments were performed as described previously [Towbin et al,1979], using monoclonal antibodies (clone 12C10) against CYP2B1/B2 [Grishanova and Lyakhovich, 1992]. Isolation of liver and lung mRNA were performed as described by Chomczynski and Sacchi [1987].

## III. RESULTS AND DISCUSSION

We studied the CYP2B catalytic activity in liver microsomes of different rat strains treated by a set of inducers causing the increase of the CYP2B subfamily. Wistar, Sprague-Dawley and NID (non insulin dependent diabet) rats were chosen. CYP2B1 is active in the metabolism of some compounds (penthoxyresorufin, testosterone, androstenedione). CYP2B2 does not metabolize these substrates. We showed remarkable interstrain differences in the penthoxyresorufin metabolism (Fig.1). Wistar rats treated with phenobarbital are less active in this metabolism whereas the the catalytic activity of CYP2B1 in liver of phenobarbital-treated Sprague-Dawley and NID rats was higher. No differences were observed during Aroclor 1254 induction. Triphenyldioxan which is an inducer of the CYP2B subfamily [Mishin et al, 1990] caused significant increase of the penthoxyresorufin-O-dealkylase activity in NID rats.

Interesting results were obtained during induction of perfluorodecalin. In this case the penthoxyresorufin-O-dealkylase activity was not induced in Wistar rats, in contrast to Sprague-Dawley rats, which showed the increase of this activity.

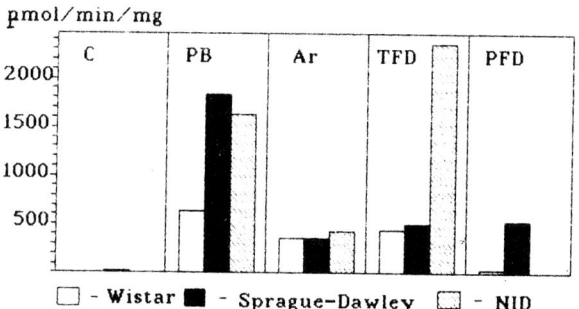

Figure 1: Interstrain differences of penthoxyresorufin-O-dealkylase activity during different type of inductions in liver of control (C), phenobarbital-, Aroclor 1254-, triphenyldioxane-, and perfluorodecalin-treated rats.

The differences observed, resulting in variations of the rate of penthoxyresorufin metabolism can be interpreted as a reflection of an inter-strain variability of CYP2B genes. In fact, several examples of such gene polymorphism among animal strains have been described for CYP genes other than CYP2B1 [Gonzalez, 1989].

Fig. 2 presents the results of Western-blot analysis using monospecific antibodies against cytochrome CYP2B1/B2. Treatment of rats with phenobarbital - classical inducer of CYP2B causes synthesis of CYP2B1 and 2B2 in liver while the latter is not observed in lung. Marcus et al [1990] have also shown that CYP2B1, rather than CYP2B2, mRNA is present in lung. In contrast to the response in liver, phenobarbital treatment did not induce the lev-

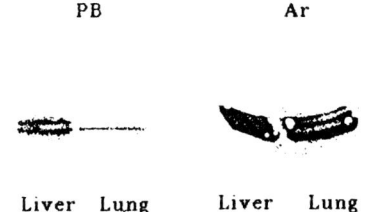

Figure 2: Immunostained Western blot of liver and lung microsomes of phenobarbital and Aroclor 1254-treated Wistar rats. Proteins were transferred from the acrylamide gel to a sheet of nitrocellulose later incubated with monoclonal antibodies clone 12C10 [Grishanova and Lyakhovich, 1992] and treated as in [Towbin et al, 1979].

els of CYP2B1 in the lung. In our experiments we detected CYP2B1 in lung of phenobarbital-induced animals. Treatment of rats with Aroclor 1254 significantly changes the distribution of CYP forms: both in liver and in lung besides two cytochromes CYP2B one more protein is detected which probably belongs to the same family. This conclusion relies on the following. First, it is known that the IIB subfamily contains about 10 genes [Atchison and Adesnic, 1986] from which besides CYP2B1 and 2B2 one more CYP is constitutively expressed referred to as CYP2B3 [Affolter et al, 1986]. However, our unknown protein can not be this one since it is not induced by xenobiotics. Second, the observed protein is with high probability from the CYP2B subfamily since in immunoblot

we used monoclonal antibodies recognizing only homologous proteins of this subfamily. Besides that, of importance is induction by Aroclor 1254 of all members of the CYP2B subfamily in lung unknown before. Thus, the nature of induction of microsomal CYPs is even more complicated, i.e. it is inducer type dependent in addition to tissue-specificity. Dependence of expression of CYP genes in liver and lung of mouse embryo on the inducer (methylcholanthrene and beta-naphtoflavone) is described in literature [Miller et al, 1989].
Synthesis of these CYPs can be due to transcriptional or posttranscriptional changes of CYP2B genes. To clarify this we made an attempt to determine the mRNA level in Aroclor- and phenobarbital -induced rat lung and liver. Induction of CYP2B1 and 2B2 mRNA was assessed by quantitative dot blot using a cDNA which recognizes both CYPs [Ravishankar and Padmanaban, 1985]. Fig. 3 shows the levels of CYP2B mRNA in liver and lung of Wistar rats treated with phenobarbital or Aroclor 1254. Control animals did not receive the mentioned inducers. The relation of mRNA in liver and lung during phenobarbital and Aroclor 1254 induction was different. We detected slight increase of the mRNA level in lung of Aroclor-treated preparations. Induction of CYP2B mRNA is probably due to activation of other CYP2B genes in good agreement with the picture of Western-blot analysis.

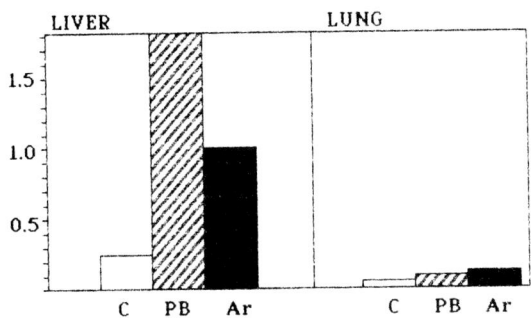

Figure 3: Graphic representation of quantitative dot-blots of liver and lung P-450 IIB1/IIB2 mRNA from control (C), phenobarbital- (PB) and Aroclor 1254-treated (AR) rats.

In summary, there is a rather complicated picture of the regulation of the P-450 IIB subfamily, which plays a very important role in detoxication processes and drug metabolism. Expression of these isozymes depends on the tissue, inducer type and strain of rats.

## REFERENCES

Affolter M. et al, (1986): cDNA clones for liver cytochrome P-450s from individual Aroclor -treated rats: constitutive expression of a new P-450 gene related to phenobarbital-inducible forms. DNA 5: 209-218.
Atchison M., and Adesnik M. (1986): Gene conversion in a cytochrome P-450 gene family. Proc. Natl. Acad. Sci. USA 83: 2300-2304.
Burke M.D. et al, (1985): Ethoxy-, penthoxy- and benzyloxyphenoxazones and homologues: a series of substrates to distinguish between different induced cytochromes P-450. Biochem Pharmacol. 34:3337-3345.
Chomczynski P., and Sacchi N. (1987): Anal. Biochem. 162:156-159.
Gonzalez F.J. (1989): The molecular biology of cytochrome P-450 s. Pharmacol. Rev. 40: 243-288.
Grishanova A.Y., and Lyakhovich V.V. (1992) Using of antibodies to cytochrome P-450 library in xenobiotic metabolism studies. In " Cytochrome P-450: Biochemistry and Biophysics". Proceedings of the 7th International Conference, Archakov A.I., and Bachmanova G.I. (eds): Moscow: pp 525-527.
Lowry O.H. et al, (1951): Protein measurement with the Folin phenol reagent. J. Biol. Chem. 193:265-275.
Marcus C.B. et al, (1990): Selective expression of cytochrome P-450 isozymes by 4 -n-alkyl-methylenedioxybenzenes in rat lung cells. Arch. Biochem. Biophys. 277: 17-25.
Miller M.S. et al, (1989): Differential induction of fetal mouse liver and lung cytochromes P-450 by beta-naphthoflavone and 3-methylcholanthrene. Carcinogenesis 10: 875-891.
Mishin V.M. et al, (1990): Comparison of triphenyldioxane bis-(dichlorpyridinloxy)-benzene and phenobarbital inducing activities on the liver monooxygenase. Biokhimia. Russia 55: 29-36.
Mishin V.M. et al, (1989): The phenobarbital-type induction of rat liver microsomal monooxygenases by perfluorodecalin. Chem.-Biol. Interactions 72: 143-155.
Ravishankar H., and Padmanaban G. (1985): Regulation of cytochrome P-450 gene expression. J. Biol. Chem. 260: 1588-1592.
Towbin H., Staehelin T., and Gordon J. (1979): Electrophoretic transfer of proteins from polyacrylamide gels to nitrocellulose sheets - procedure and some applications. Proc. Natl. Acad. Sci. USA 76: 4350-4354.

# The studies of cytochrome P450scc in liposomes by means of limited trypsinolysis

Alexander V. Krivosheev, Oleg N. Murashko, Sergey A. Usanov

Institute of Bioorganic Chemistry, Academy of Sciences of Belarus, 5/2 Zhodinskaya St., 220141 Minsk, Belarus

INTRODUCTION

The investigations of molecular characteristics of highly purified cytochrome P-450scc in aqueous phase by means of limited trypsinolysis resulted in estimation of a domain structure of P-450scc (Chashchin et al., 1984). To continue the studies of P-450's molecular organization we applied this approach for understanding structure-functional behavior of cytochrome's molecule incorporated in artificial lipid membranes. The first stage of this work dealt with the studies of P-450-containing liposomes treated with trypsin and followed by SDS PAGE and immunochemical analysis.

METHODS

The incorporation of highly purified bovine adrenal cortical cytochrome P-450scc into egg phosphatidylcholine vesicles was carried out by a cholate-filtration method according to Yamakura et al., 1981, except of using 25 mM Na-phosphate buffer (pH 7.4) instead of 50 mM HEPES buffer. The lipid:protein ratio was 200:1 (mol/mol). The integrity of liposomes was detected by reduction of ferricyanide trapped inside the vesicles with electrons transferred from NADPH by NADPH-adrenodoxinreductase. The trypsin was added to liposomes at the 1:25 ratio to a protein content (w/w) at room temperature and the liposomes were treated for 60 minutes or less. The reaction was stopped by 3 fold excess of soybean trypsin inhibitor. The Western blotting procedure was run with antibodies to P-450scc's fragments raised in rabbits.

RESULTS AND DISCUSSION

The present paper describes the results of the studies of artificially reconstituted system which are essential for learning the topographic properties of P-450scc molecules in the liposomal structure that mimics the native membrane. Since the inner mitochondrial membrane contains of more than 30 per cent of phosphatidylcholine (Demel et al., 1989) and was shown to be the most suitable lipid for vesicles construction (Seybert et al., 1979) (Yamakura et al., 1981) we used this lipid for our experiments. The incorporation of P-450scc into vesicles revealed more than 90 per cent of the total protein amount. However the experiments on stability showed initial rapid degradation within first 3 hours of storage at 4°C under argon resulted in formation of up to 60 per cent of P-420 as judged from CO-bound Na-dithionite-reduced difference spectra of P-450 in liposomes that corresponded to the previous data (Yamakura et al., 1981) and increased slightly to 68 per cent during the 48-hours interval. The experiments were therefore carried out in 30 min after the liposomes were eluted from gel-filtration column. It should be pointed out that on the whole time span starting at 2hr point we registered the constant amount of P-450scc reduced by reconstituted enzyme system. The addition of dithionite showed about 80 per cent of all P-450 form to be accessible for interaction with adrenodoxin and its reductase. This data could be an evidence of a different character of P-450's incorporation in the membrane, thus exhibiting two pools of cytochrome molecules: 1-with high lability and so subjected to degradative influence of lipid-peroxidation reactions etc.; 2-more steadily oriented and showing the ability to be reduced enzymatically.

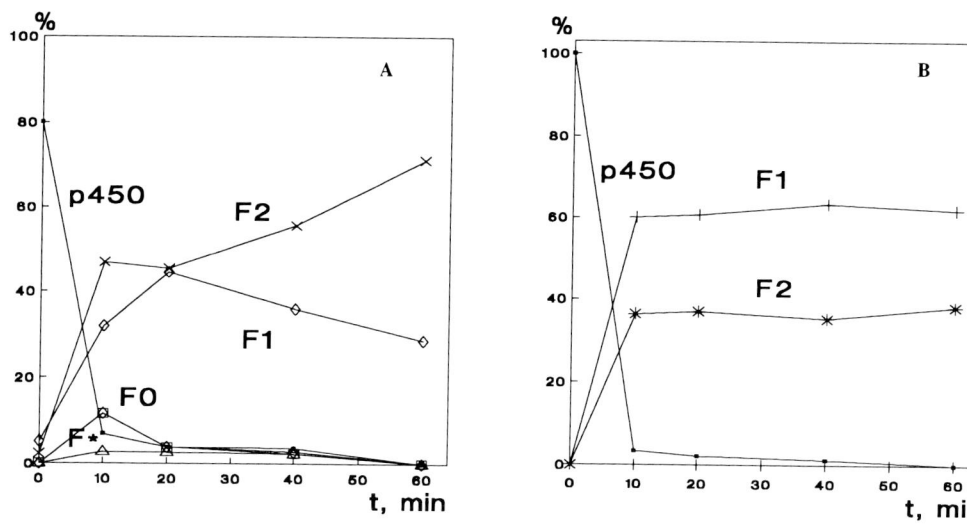

Fig.1 Comparison of P-450scc fragments formation during trypsin treatment in liposomes (a) and in solution (b).

The limited trypsinolysis of P-450scc-containing liposomes made it possible to predict the localization of the two main domains. According to SDS-PAGE data processed by means of densitometry the amount of N-terminal domain F1 was initially lower than that of C-terminal F2 (Fig.1a) contrary to the control series carried out in solution (Fig.1b).

The formation of fragments in control series without trypsin did not exceed 6 per cent of total protein amount. The percentage of F1 decreased in time as well as F2 amount remained approximately constant making its relative quantity higher in time course. In addition, the specific band of Mr 37,800 appeared within first 40 minutes of tryptic digestion and diminished by 60 minutes. The formation of both F0 (Mr 40,000) and F* (Mr 37,800) shows that besides the usual tryptic available site at Arg92 for P-420 form (Chashchin et al., 1986) the second site appears. According to the difference in F0 and F* weights and assuming the average amino acid weight to be 130 one can compute the length of a polypeptide chain to consist of 17-18 amino acids from Arg92 towards C-terminus. The Lys109-Lys110-Asp111 site

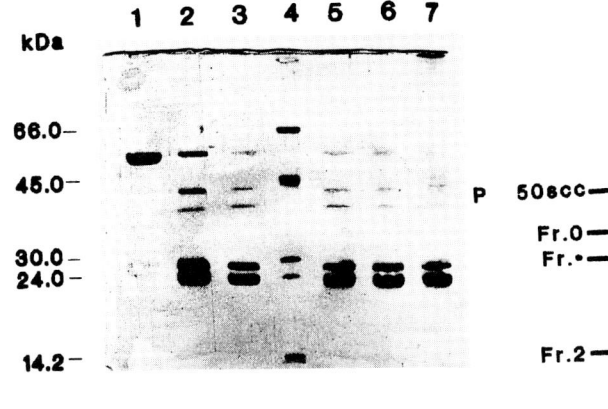

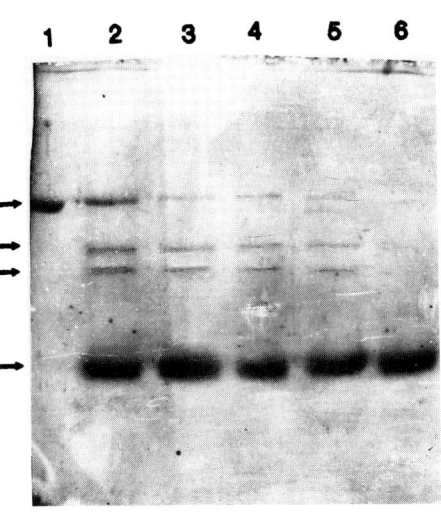

Fig.2 12 % SDS PAGE of P-450scc-containing liposomes treated with trypsin for 0 (1), 10 (2), 20 (3), 30 (5), 40 (6) & 60 (7) minutes. Lane 4 - standards.

Fig.3 Western blotting with anti-F2 antibodies for P-450scc-containing liposomes treated with trypsin for 0 (1), 10 (2), 20 (3), 30 (4), 40 (5) & 60 (6) minutes.

could be concerned as a probable one for F* formation during tryptic cleavage. Another evidence for that comes from Western-blotting data with anti-F2 antibodies (Fig.3, 4). It is definitely seen that the F* fragment contains F2 antigen determinants and thus is recognized by the correspondent antibodies; anti-F1 did not bind the fragment mentioned (data not shown). The higher stability of F2 to trypsin influence allows suggesting its localization to be either intramembrane or spanning through and exposed to the inner space of the liposome as the F1 been faced to the outer side of the liposome and therefore digested greatly during proteolysis.

The data presented could be concerned as a starting point for elucidation of structure-functional characteristics of cytochrome P-450scc both in artificial vesicles and in natural membranes as well as an approach for protein-protein interactions studies in the P-450's system.

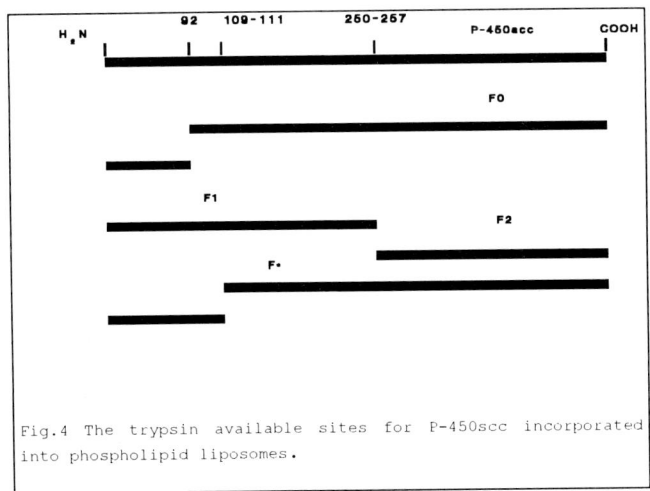

Fig.4 The trypsin available sites for P-450scc incorporated into phospholipid liposomes.

# REFERENCES

Chashchin, V.L., Vasilevsky, V.M., Shkumatov, V.M., Lapko, V.N., Adamovich T.B., Berikbaeva T.M. & Akhrem, A.A. (1984): The domain structure of the cholesterol side-chain cleavage cytochrome P-450 from bovine adrenocortical mitochondria. Localization of haem group and domains in the polypeptide chain. *Biochim. et Biophys. Acta* 791, 375-383.

Chashchin, V.L.,, Lapko, V.N., Lapko, A.G., Kuprina, N.S. & Akhrem, A.A. (1986): Primary structure of the cholesterol side-chain cleavage cytochrome P-450 from bovine adrenocortical mitochondria and some aspects of its functioning on a structural level. *Biochim. et Biophys. Acta* 871, 217-223.

Demel, R.A., Jord, W., Lambrechts, H., Van Damme, H., Hovius, R. & Den Krujiff, B. (1989): Differential interactions of apo- and holocytochrome c with acidic membrane lipids in model systems and the implication for their import into mitochondria. *J. Biol. Chem.* 264, 3988-3997.

Seybert, D.W., Lancaster, J.R., Lambeth, Jr.J. & Kamin, H. (1979):Participation of the membrane in the side chain cleavage of cholesterol. Reconstitution of cytochrome P-450scc into phospholipid vesicles. *J. Biol. Chem.* 254, 12088-12098.

Yamakura, F., Kido, T & Kimura,T. (1981): Characterization of cytochrome P-450scc-containing liposomes. *Biochim. et Biophys. Acta.* 649, 343-354.

# Computer modeling of cytochrome P450cam and $b_5$ interaction

Alexis S. Ivanov, Vladlen S. Skvortsov, Yury A. Lyulkin, Andrey B. Rumyantsev

*Laboratory of Computers in Biochemistry, Institute of Biomedical Chemistry, Pogodinskaya str. 10, Moscow, 119832, Russia*

## Summary

The interaction of cytochromes P450cam and $b_5$ was investigated by methods of computer molecular modeling and docking procedure. 3-D coordinates of cytochrome P450cam and polar domain of cytochrome $b_5$ were got from the PDB. The original computer program MOLWIN for protein structure investigation was used. It allows to obtain 3-D images of proteins (up to 32,000 atoms) both as wire and as VDW presentation with different coloring according to the hydrophobicity, charge distribution, secondary structure, etc. Membrane-binding nonpolar segment which anchors cytochrome $b_5$ to the membrane was drawn by ALCHEMY II program in a "loop" conformation (residues 104-126). This model consists of two $\alpha$-helixes (residues 104-113, 117-126) and $\beta$-turn (residues 114-116) in the middle. The molecular mechanics program PCMODEL (ver.4) was used both for energy minimization (by MMX method) and for dot cloud surfaces calculations. The model of the whole cytochrome $b_5$ was designed by combination of PDB structure of polar domain, computer model of membrane anchor as well as computer models of adjacent to anchor domains (residues 87-103, 127-133). The docking of cytochrome P450cam and both polar domain and total cytochrome $b_5$ model was done by original program DOCKMASTER. The dependence of molecular complex numbers via distance between hems as well as via complex dissociation energy allows to find the main binding sites for cytochromes P450cam and $b_5$.

INTRODUCTION

Cytochrome $b_5$ consists of two structurally independent domains, a catalytic (polar) segment, and membrane-binding (nonpolar) anchor (Ozols, 1989). 3-D structure of polar segment is well known (PDB, 1993) while the membrane anchor structure is not. It was shown that this protein part does not span membrane bilayer (Ozols, 1989). The interaction of water soluble part of cytochrome $b_5$ with cytochrome P450cam was shown by Stayton et al., 1989 and the molecular complex model was designed. In this study we report the computer modeling of full 3-D structure of bovine cytochrome $b_5$ and computer simulation of its interaction with cytochrome P450cam.

MATERIALS AND METHODS

Computer molecular modeling.
ALCHEMY II (Tripos) was used for building of 3-D molecular structures and its energy minimization. Molecular surface area calculations were done by PCMODEL v.4 (Serena Software). The final molecular energy minimization was done by MMX force field method. PC/GENE 6.6 (IntelliGenetics) was utilized for protein sequence and secondary structure analysis. Cytochrome $b_5$ (bovine) sequences were got from CD-ROM database ENTREZ v.4.0 (Nat.Centre Biotech.Information).

Correlation Docking.
Molecular interaction modeling was conducted by geometrical docking and by selection of sample conformations based on their potential energy evaluation. Conformation of each molecule was constant during the docking procedure. The molecular complex was modeling by changing the position and space orientation of smaller molecule while bigger molecule remained fixed. Thus each experimental docking result was characterized by seven values. Three of them represented a vector of mass center displacement of the movable molecule. Three others represented the molecular rotation around the mass center. The last one was an evaluation parameter. The geometrical docking was based on lattice model. The molecule was presented by 3-D cubic lattice (Katchalski-Katzir, 1992). The value of lattice knots was equal to: 0 - when the knot was out of molecule; 1 - when it was on the molecular surface; high positive - when the knot was in the fixed molecule and low negative - when it is in a movable molecule. The evaluation parameter was described by correlation function:

$$C_{a,b,c} = \sum_{i=1}^{N} \sum_{j=1}^{N} \sum_{k=1}^{N} A_{ijk} \cdot B_{ijk}$$

where $A_{ijk}$ is the value of lattice knot (ijk) for a fixed molecule. $B_{ijk}$ is the same data for a movable molecule. An original computer program based on the modified algorithm of correlation docking was designed. The number of experimental geometrical docking results is great (dozens of thousands). Clusterization of these data in 6-dimension choice space decreased their number up to several dozens. After clusterization of molecular complexes their potential energy was minimized and the final conformation was chosen. A special program based on the algorithm of accelerated Monte Carlo method with annealing and force bias method (Kotelyanskii, 1992) was designed. As a target function the potential OPLS (included Lennard-Jones potential and electrostatic interaction) was used (Jorgensen et al., 1988). The energetic function is:

$$U_{a,b} = \sum_{i}^{for\ A} \sum_{j}^{for\ B} \left( \frac{q_i q_j e^2}{R_{ij}} + \frac{A_{ij}}{R_{ij}^{12}} - \frac{B_{ij}}{R_{ij}^{6}} \right)$$

where $R_{ij}$ is the distance between i-atom of molecule A and j-atom of molecule B, e is the electron charge, $q_i$ and $q_j$ are the coefficients of atom partial charges, $A_{ij}$ and $B_{ij}$ are the constants for Lennard-Jones potential. The last one depends on types of atoms that form the given couple. The resulting conformations should have the minimal potential energy.

RESULTS

Membrane-binding segment which anchors cytochrome $b_5$ (bovine) to the membrane was drawn in a "loop" conformation (residues 104-126):

```
GELHPDDRSK ITKPSESIIT TIDSNPSWWT NWLIPAISAL FVALIYHLYT SEN
   | 90         100  |      110       120     |  130
```

The model consists of two α-helixes (residues 104-113, 117-126) and β-turn (residues 114-116) in the middle. The loop-like conformation of membrane anchor for rat microsomal cytochrome $b_5$ was shown earlier by Ozols, 1989.

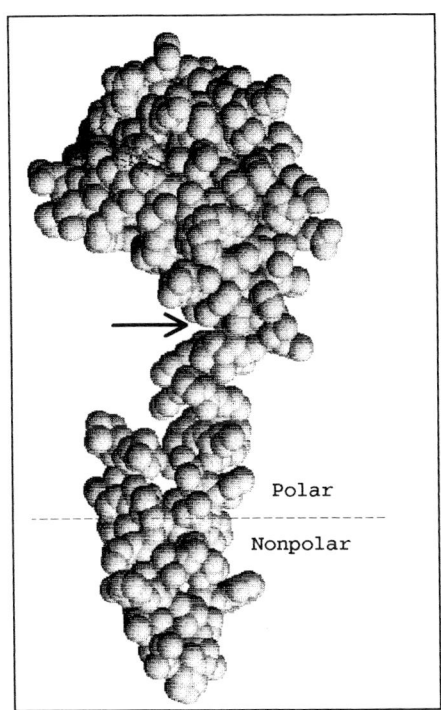

Fig. 1. Computer model of the complete cytochrome $b_5$.
The connection point between water soluble part (PDB) and membrane anchor part (computer modeling) is shown by arrow.

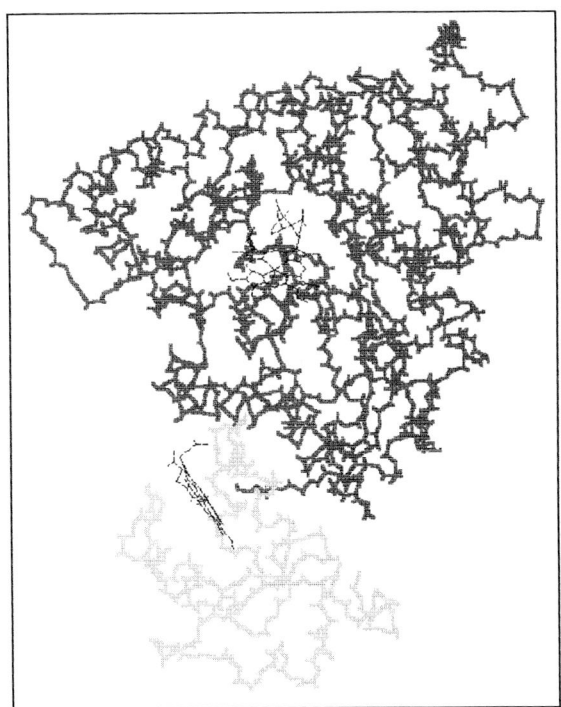

Fig. 2. The interaction of cytochromes P450cam and $b_5$.
(Docking computer simulation).

Molecular mechanics methods were used both for energy minimization (by MMX force field) and for dot cloud surfaces calculations. The model of the whole cytochrome $b_5$ was designed by combination of PDB structure of polar domain, computer model of membrane anchor as well as computer models of adjacent to anchor domains (residues 87-103, 127-133). The computer model of full cytochrome $b_5$ in VDW mode is shown in Fig. 1. The docking of cytochrome P450cam and both polar domain and complete cytochrome $b_5$ model was done. The dependence of molecular complex numbers via distance between hems as well as via complex dissociation energy allows to find the main binding sites for cytochromes P450cam and $b_5$. The list of amino acid residues which are involved in the interaction of cytochromes P450cam and $b_5$ is shown in Table 1.

Table 1. Amino acid residues which are involved in the interaction.

| Cytochrome P450cam | Cytochrome $b_5$ |
|---|---|
| 121-Met | 56-Glu |
| 125-Asp | 59-Glu |
| 128-Glu | 60-Asp |
| 278-Pro | 61-Val |
| 279-Glu | 62-Gly |
| 285-Pro | 64-Ser |
| 365-Arg | 65Thr |
|  | 66-Asp |

LITERATURE

Jorgensen, W.L., Tirado-Rives, J. (1988): The OPLS Potential Functions for Proteins. Energy Minimizations for Crystals of Cyclic Peptides and Crambin. *J.Am.Chem.Soc.* 110, 1666-1671.

Katchalski-Katzir E., Shariv I., et all. (1992): Molecular surface recognition: Determination of geometric fit between proteins and their ligands by correlation techniques. *PNAS USA*, 89, 2195-2199.

Kotelyanskii, M.J., Suter, U.W. (1992): A dynamic Monte Carlo method suitable for molecular simulations. *J.Chem.Phys.* 96, 5383-5388.

Ozols, J. (1989): Structure of cytochrome b5 and its topology in microsomal membrane. *BBA* 997, 121-130.

Protein Data Bank, (January 1993 Release), Chemistry Department, Building 555 Brookhaven National Laboratory Upton, NY 11973 USA.

Stayton, P.S., Poulos, T.L., and Sligar, S.G. (1989): Putidaredoxin Competitively Inhibits Cytochrome b5 - Cytochrome P-450cam Association: A Proposed Molecular Model for a Cytochrome P-450cam Electron-Transfer Complex. *Biochemestry* 28, 8201-8206.

# Wild mice as bioindicators for pollution monitoring

Amrit Bhatia, Karl Mazzucco

Institute of Tumorbiology-Cancer Research, University of Vienna, Borschkegasse 8a, A-1090 Vienna, Austria

A number of studies were recently conducted to establish the induction of hepatic cytochromes P450 in wild rodents as a suitable parameter for monitoring terrestrial environmental pollution (Lubet et al., 1992; Payne, et al., 1987; Peakall 1992; Schrenk et al., 1991; Simmons & McKee, 1992). These investigators observed a good correlation between the induction of hepatic cytochrome P-450 in wild rodents and the concentration of various pollutants in the soil sample, in the liver, or in the whole body. No attempt has been made to use feral mice of the genus Apodemus for such studies. The study presented here was to find out the suitability of *Apodemus flavicollis,* the yellow necked mouse, for such a purpose.

## MATERIALS AND METHODS

All the sites selected for the studies were mixed wood forests, three industrialized urban sites and a nonindustrialized rural one serving as reference:
1. Rural site:
    Pulkau (PUL), a village about 90 km north-west of Vienna.
2. Urban sites:
    a) Institute of Wildlife Research & Ecology at Wilhelminenberg, Vienna (IWT).
    b) Danubia river bank near Albern Port Vienna (ALB).
    c) Siedersgraben in Weidling, Klosterneuburg (KLB).

Mice were trapped alive during the months of July to September 1992 using Longworth aluminium traps. Traps with fresh baits were placed in the fields in the evening, and captured mice were collected early in the morning and taken to the laboratory. The mice were killed by cervical dislocation, weighed, and the freshly removed livers were cut into small slices, placed in 2 ml of 0.05 M Tris pH 7.4 containg 20% glycerol and frozen at $-80^{\circ}C$. The microsomes were prepared using ultracentrifuge as described by Gibson & Skett (1986). These were frozen at $-80^{\circ}C$ in aliquots required for analysis. Protein content was determined by the method of Lowry (Lowry et al., 1951) using BSA as a standard. Total cytochrome P450 and b5 contents were measured as given by Gibson & Skett (1986). Benzyloxyresorufin-O-deethylase (BROD), pentoxyresorufin-O-deethylase (PROD) and ethoxyresorufin-O-deethylase (EROD) were determined as described by Lubet and co-workers (1985) with minor modifications. Incubation was carried out in 2 ml volume for 30 minutes at 37°C and terminated by adding 2 ml methanol (Rodman et al., 1989; Kennedy et al., 1993). Fluorescence was measured at Aminco SPF 500 spectrofluorimeter using $\lambda_{ex}$ at 530 nm and $\lambda_{em}$ at 585 nm. Resorufin was used as a standard. Ethoxycoumarin-O-deethylase (ECOD) was measured as given by Prough et al. (1978). Incubation was carried out in 2 ml volume at 37°C for 30 minutes and stopped by adding 2 ml glycine buffer pH 10.3. Fluorescence was measured at $\lambda_{ex}$ 365 nm and $\lambda_{em}$ 455 nm. 7-hydroxycoumarin was used as a standard. All data were subjected to two way sample analysis of variance (ANOVA). Both male and female animals were compared separately. The differences were considered significant at p<0.05 and highly significant at P<0.01.

## RESULTS

None of the groups showed any significant difference in body weight, liver weight, liver to body weight ratio, and total hepatic microsomal protein content. The data for cytochrome b5 and P450 contents are shown in Fig. 1 and 2 respectively. These were increased in all three female urban groups, whereas only one of the male urban groups, IWT, showed significantly higher values than the control.

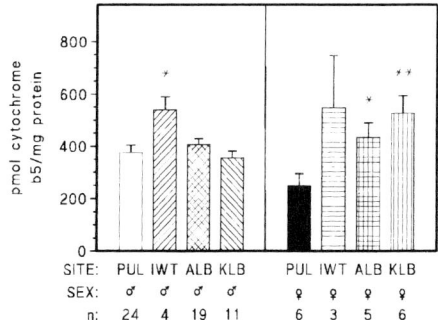

Fig.1 Hepatic cytochrome b5 level in A. flavicollis from one rural (PUL) & 3 urban areas (IWT, ALB, KLB).
The data displayed is group mean (±S.E).
Statistical significance, ANOVA test, * (p<0.05), ** (p<0.01).

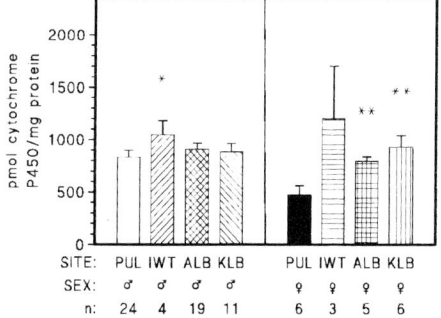

FIG.2 Hepatic cytochrome P-450 level in A. flavicollis from one rural (PUL) & 3 urban areas (IWT, ALB, KLB).
The data displayed is group mean (±S.E).
Statistical significance, ANOVA test, * (p<0.05), ** (p<0.01).

The variations in catalytic activities of ECOD and alkoxyresorufin O-dealkylase in both sexes of the four groups are shown in Figs. 3-6. For the ECOD activity no difference was observed among the groups studied, the exception being one of the urban male groups, ALB, where slightly reduced values were observed. Significantly increased levels of BROD, PROD, and EROD (Figs. 4-6) were observed in animals trapped from the KLB area. These values were also two to three fold higher in IWT as well as ALB animals in comparison to the rural area PUL. The substrate specificities for EROD, BROD and PROD have been established and confirmed by a number of authors (Nims et al., 1992;

Lubet et al., 1990). The measure of EROD activity by using ethoxyresorufin as a substrate is an indirect assay for P450IA, whereas BROD and PROD activities, with the use of benzyloxyresorufin or pentoxyresorufin as respective substrates, are considered indirect assays for P450IIB. The differences among the four populations between PROD or BROD activities are to be attributed to different levels of CYPIIB enzymes.

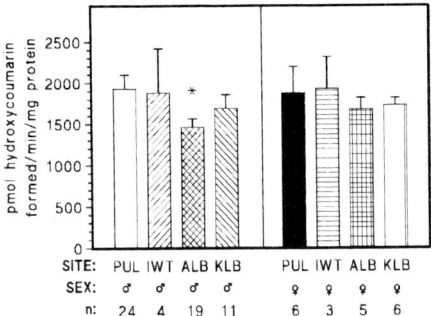

Fig.3 Hepatic cytochrome P-450 ethoxycoumrin O-deethylase activity in A. flavicollis from one rural (PUL) & 3 urban areas (IWT, ALB, KLB).
The data displayed is group mean (±S.E).
Statistical significance, ANOVA test, * (p<0.05), ** (p<0.01).

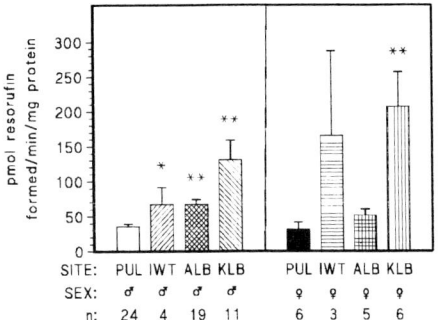

Fig.4 Hepatic cytochrome P-450 benzyloxyresorufin O-deethylase activity in A. flavicollis from one rural (PUL) & 3 urban areas (IWT, ALB, KLB).
The data displayed is group mean (±S.E).
Statistical significance, ANOVA test, * (p<0.05), ** (p<0.01).

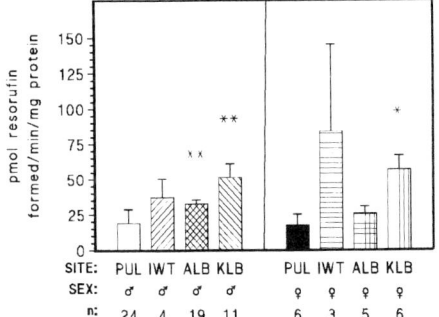

Fig.5 Hepatic cytochrome P-450 pentoxyresorufin O-deethylase activity in A. flavicollis from one rural (PUL) & three urban areas (IWT, ALB, KLB).
The data displayed is group mean (±S.E).
Statistical significance, ANOVA test, * (p<0.05), ** (p<0.01).

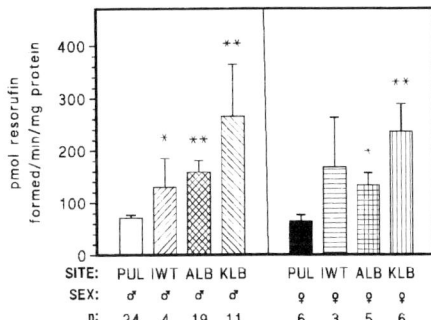

Fig.6 Hepatic cytochrome P-450 ethoxyresorufin O-deethylase activity in A. flavicollis from one rural (PUL) & 3 urban areas (IWT, ALB, KLB).
The data displayed is group mean (±S.E).
Statistical significance, ANOVA test, * (p<0.05), ** (p<0.01).

## DISCUSSION AND CONCLUSION

Pollution of urban environment is caused by emissions of industries, thermal electric power plants, auto engine exhausts and domestic fires, etc. (Schlatter and Poiger, 1989). The higher alkoxyresorufin activities of hepatic microsomes in the urban mice in comparison to the rural group reflect thus higher pollution very clearly. In conclusion, we can say that among the three urban sites biomonitored using A. flavicollis as an indicator animal, KLB seems to be the most polluted one, while ALB and IWT take the second and third place respectively in comparison to PUL as a reference site. Our results confirm that feral mice like *A. flavicollis,* which can be easily caught in the vicinity of human beings, are very well suited for bioindication of the terrestrial pollution load.

## ACKNOWLEDGEMENT

The authors extend their thanks and appreciation to the following:
(a) Rolf Schulte-Hermann, Wolfram Parzefall, Waltraud Gerdenitsch & Karin Bhatia for helpful discussions in preparation of this manuscript.
(b) Gerhard Lepschy, Friedrich Tobil and Xenia Werk for their excellent technical assistance.
(c) Helmuth Götz, Institute of Zoology, University of Agriculture, Vienna, for identification of animals.
(d) Ilse Wawra and Erich Klansek, Institute of Wildlife Research & Ecology, Vienna, for trapping mice at Wilhelminenberg.

## REFERENCES

Gibson G.G and Skett P. (1986): Introduction to Drug Metabolism. Chapman and Hall. pp. 237-284.

Kennedy, S.W., Lorenzen, A., James, C.A., and Collins, B.T. (1993): Ethoxyresorufin-O-deethylase & Porphyrin Analysis in Chicken Embryo Hepatocyte Cultures with a Fluorescence Multiwell Plate Reader. Analytical Biochemistry, 211, 102-112.

Lowry O.H., Rosebrough N.J., Farr A.L., Randall R.J. (1951): Protein Measurement with Folin Phenol Reagent. J Biol. Chem., 193, 265-275.

Lubet R.A., Nims, R.W., Mayer, R.t, Cameron, J.W., Schechtman, L.M. (1985): Measurement of Cytochrome P-450 dependent Dealkylation of Alkoxyphenoxazones in Hepatic S9s and Hepatocyte Homogenates: Effects of Dicumarol. Mutation Research, 142, 127-131.

Lubet, R.A., Syi, J.-L., Nelson, J.O., and Nims, R.W. (1990): Induction of Hepatic Cytochrome P-450 mediated Alkoxyresorufin O-dealkylase Activities in Different Species by Prototype P450 Inducers. Chem. Biol. Interact., 75, 325-339

Lubet, R.A., Nims, R.W., Beebe, L.E., Fox, S.D., Issaq, H.J., and McBee, K. (1992): Induction of Hepatic CYP1A Activity as a Biomarker for Environmental Exposure to Aroclor® 1254 in Feral Rodents. Arch. Environ. Contam. Toxicol,. 22, 339-344.

Nims R.W., Beebe L.E., Dragnev K.H., Thomas P.E., Fox S.D., Issaq H.J., Jones C.R., and Lubet R.A. (1992): Induction of Hepatic CYP1A in Male F344/NCr Rats by Dietary Exposure to Aroclor 1254: Examination of Immunochemical, RNA, Catalytic, and Pharmacokinetic Endpoints. Environmental Research, 59, 447-466.

Payne, J.F., Fancey, L.L, Rahimtulla, A.D., Porter, E.L. (1987): Review and Perspective on the Use of Mixed-function Oxygenase Enzymes in Biological Monitoring. Comp. Biochem. Physiol., 86C, 233-245.

Peakall, D. (1992): Animal Biomarkers as Pollution Indicator. Chapmann & Hall. London . New York . Tokyo . Melbourne . Madras.

Prough, R.A., Burke, M.D., and Mayer, R.T. (1978): Direct Fluorimeteric Methods for Measuring Mixed Function Oxidase Activity. Methods in Enzymology, 52, 372-377.

Rodman, L.E., Shedlofsky, S.I., Swim, A.T. and Robertson L.W. (1989): Effect of Polychlorinated Biphenyls on Cytochrome P450 Induction in the Chick Embryo Hepatocyte Culture. Arch. Biochem. Biophys., 275 (1), 252-262.

Schlatter, Ch. and Poiger, H. (1989): Chlorierte Dibenzodioxine und Dibenzofurane (PCDDs/PCDFs)- Belastung und gesundheitliche Beurteilung. UWSF-Z. Umweltchem. Ökotox., 2, 11-17.

Schrenk, D., Lipp, H.P., Brunner, H., Wiesmüller, T., Hagenmayer, H., and Bock, K.W. (1991): Induction of Hepatic P450-dependent Monoxygenase in feral Mice from a PCDD/PCDF-contaminated area. Chemosphere, 22 (11) 1011-1018.

Simmons, G.J. and McKee, M.J. (1992): Alkoxyresorufin Metabolism in White-footed Mice at Relevent Environmental Concentrations of Aroclor 1254. Fundamental and Applied Toxicology, 19, 446-452.

# Expression of functional human microsomal epoxide hydrolase in *Escherichia coli*

Andrea Gaedigk, Hidefumi Nakamura, Denis M. Grant

Division of Clinical Pharmacology and Toxicology, The Hospital for Sick Children, Toronto, Ontario M5G 1X8, Canada

## INTRODUCTION

Microsomal epoxide hydrolase (**mEH**; EC 3.3.2.3) is one of a small number of functionally related enzymes that catalyse the hydrolysis of aliphatic and aromatic epoxides to *trans*-dihydrodiols. mEH shows catalytic selectivity for the hydrolysis of *cis*-disubstituted oxiranes such as *cis*-stilbene oxide, and especially for arene oxides formed from polycyclic and other aromatic hydrocarbons, such as benzo[a]pyrene, via cytochrome P450-mediated oxidation. The human enzyme is a monomeric protein with a molecular mass of 53 kDa, and it resides predominantly in the endoplasmic reticulum of a variety of cell types, although evidence from kinetic, immunochemical and hybridization experiments suggests that it is present at highest levels in liver hepatocytes (Guenthner, 1990).

mEH plays a vital role in the modulation of toxicity arising subsequent to P450-mediated epoxide formation from drugs and chemical carcinogens, since such epoxides are often highly chemically reactive electrophiles. As shown in Fig. 1, arene oxides formed from aromatic compounds by P450 may either spontaneously decompose to non-toxic phenols and be excreted as such (or as their glucuronide and sulfate conjugates), or be enzymatically detoxified to *trans*-dihydrodiols by mEH or to glutathione conjugates by glutathione transferase (GST). Depending upon the balance of their intrinsic chemical reactivities and their ability to act as substrates for mEH or GST, certain arene oxides generated *in vivo* may thus be capable of binding covalently to nucleophilic sites on cellular macromolecules (Nelson and Pearson, 1990; Park et al., 1992), leading to direct cytotoxicity, genotoxicity and/or the initiation of immunologic responses (Park and Kitteringham, 1990).

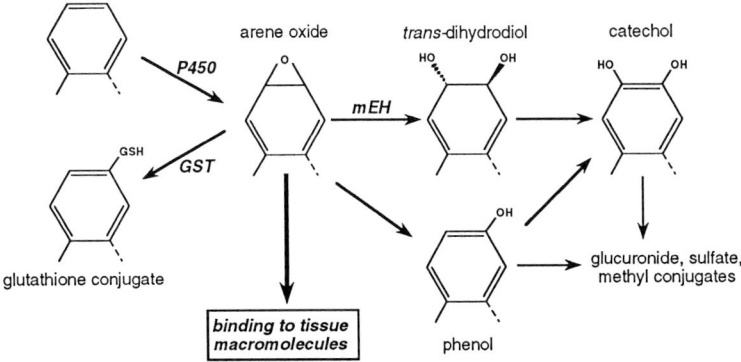

Fig. 1: Pathways of arene oxide formation and detoxication

Our interest in mEH arose from our investigations into the mechanisms underlying the occurrence of inherited idiosyncratic toxicity associated with therapy using the aromatic anticonvulsants phenytoin, phenobarbital and carbamazepine (Spielberg, 1984). Although such adverse drug reactions (ADRs) during anticonvulsant therapy are relatively rare they are potentially life-threatening, manifesting with symptoms including skin rash and fever, and resulting in damage to organs such as liver, bone marrow, kidney, lung and heart (Thomsick, 1983; Shear and Spielberg, 1988). Since the biotransformation of each of these drugs is thought to proceed significantly via P450-mediated formation of reactive arene oxide intermediates (Lertratanangkoon and Horning, 1982; Moustafa et al., 1983, 1990), it has been proposed that the drug toxicity observed in genetically predisposed individuals is related to a functional defect in mEH that impairs its ability to efficiently detoxify these metabolites. Indeed, a substantial amount of indirect evidence gathered from *in vitro* lymphocyte toxicity tests and from enzyme inhibitor studies has provided support for this notion (Spielberg et al., 1981; Gerson et al., 1983; Shear and Spielberg, 1988; Pirmohamed et al., 1992).

In order to more directly test the hypothesis that mEH defects are involved in the pathogenesis of anticonvulsant ADRs, we recently undertook a detailed biochemical and molecular genetic comparison of mEH genes, transcripts and proteins from patients who had presented with anticonvulsant toxicity and from control subjects in whom no *in vivo* or *in vitro* evidence of drug hypersensitivity could be demonstrated (Gaedigk et al., manuscript submitted). The results of these studies demonstrated that 1) no alterations in mEH protein content or overall transcript levels could be demonstrated in peripheral lymphocytes (a convenient target tissue demonstrating increased susceptibility to reactive anticonvulsant metabolites) from affected patients compared to those of controls; 2) no major gene rearrangements disrupting the human mEH gene were detected in genomic DNA from susceptible patients; and 3) although a number of point mutations could be detected within exons of the human mEH gene in a sample of 26 individuals, no single mutation correlated with the occurrence of anticonvulsant ADRs. Thus it seems that the mechanism underlying the occurrence of anticonvulsant-induced toxicity is more complex than had previously been presumed. For instance, it is possible that multiple defects in mEH occur in the population, some altering enzyme properties and others affecting transcript processing or tissue-selective expression. Alternatively, defects in other pathways of arene oxide formation or elimination (such as GST) may contribute, solely or in combination with mEH variation, in producing a susceptible phenotype.

As part of our continuing studies to investigate the role of variations in mEH function in the pathogenesis of these and other chemical-induced toxicities, we have undertaken the development of systems for the heterologous expression of functional human mEH in *E. coli*. One of the first applications of such a system would be to use site-directed mutagenesis to introduce three specific amino acid changes within the mEH protein that we have already predicted to exist in certain individuals, based upon our analysis of mEH genes in the human population. We will then be able to determine whether these structural alterations in the mEH protein affect its function, and thus whether these could have any functional significance for toxicity induced by arene oxide metabolites.

## METHODS

Vector Construction. A full-length cDNA encoding human mEH (Skoda et al., 1988) was kindly provided by U.A. Meyer, University of Basel. The mEH protein coding region was amplified from the cDNA template using oligonucleotide primers designed to produce a PCR product with the initiation ATG at the extreme 5' terminus and the termination TGA at the 3' end. This fragment was treated with Klenow fragment and T4 polynucleotide kinase, then blunt-end ligated into the *BamHI*-digested, S1 nuclease-treated prokaryotic expresson vector pTrcHIS (Invitrogen). This tac promoter-based plasmid, bearing the *bla* gene conferring ampicillin resistance, is designed to produce a fusion protein between the insert protein and an amino-terminal peptide fragment containing both a histidine hexamer tag (for affinity purification of the recombinant product on nickel-agarose columns) and an enterokinase proteolytic cleavage site (for subsequent cleavage of the affinity tag to produce the native protein). The resulting recombinant plasmid, designated as pEHex, was transformed into the *E. coli* strain TOP10.

Expression Conditions. Preliminary optimization tests produced the following expression protocol which was followed in most experiments. Plasmid-bearing TOP10 cells were restarted from a fresh overnight bacterial culture in SOB medium plus 50 mg/ml ampicillin to an $OD_{600}$ of 0.5, IPTG was added to a final concentration of 0.4 mM, and growth was continued for 6 hr at 37°C. Cultures were centrifuged at 4000 x g for 10 min, cell pellets were resuspended in sonication buffer and sonicated on ice with a probe sonicator (3 x 15 sec, 60% full power). Lysates were centrifuged for 10 min at 9000 x g, and the supernatant fraction was used directly for enzyme activity measurements or further centrifuged for 60 min at 100,000 x g to isolate a bacterial 'microsomal' fraction. In experiments to determine the effect of detergent addition on recovery of enzyme activity, detergent was added to sonicated cells, mixed for 30 min at 4°C and centrifugation was then performed as described above.

Enzyme Assay. A specific HPLC assay modified from (Eaton and Stapleton, 1989) was used to monitor dihydrodiol formation from the mEH-selective substrate benzo[a]pyrene-4,5-dihydroepoxide (BaPO). Incubations, in a final volume of 250 µl, contained 50 mM Tris-HCl, pH 7.4, 0.1 mM EDTA, enzyme source (25 µg protein) and BaPO (80 µM). Reactions (37°C for 30-45 min) were started with the addition of 10 µl of 2 mM BaPO in DMSO and terminated by the addition of 250 µl of acetonitrile. Following overnight incubation at 4°C, precipitated protein was removed by centrifugation, the supernatant was injected onto a reversed-phase HPLC column, eluted with a solvent consisting of acetonitrile/triethylamine/acetic acid/$H_2O$ (58:0.05:1:42, v/v/v/v) at a flow rate of 1.0 ml/min and peaks were detected by UV absorbance at 275 nm. Under these conditions the retention times of BaPO and its diol metabolite were 15 min and 3.8 min, respectively. The standard curve for BaPO-diol was linear from 2.5 to at least 500 pmoles with a correlation coefficient of 0.999, and the detection limit for BaPO-diol formation rate under these conditions was about 0.4 pmoles/min/mg.

## RESULTS AND DISCUSSION

Using the system described above we were able to express functional human mEH in *E. coli*. The enzyme was catalytically active with the amino terminal histidine affinity purification tag attached, and activity in the 100,000 x g pellet ('microsomal') fraction of sonicated bacterial cells (typically 300 pmoles/min/mg) was 10-fold higher than that detected in the corresponding supernatant fraction. The enzyme activity is roughly 30-fold lower (per mg protein) than that observed in human or rat liver microsomes, but about 6-fold higher than that in human placental microsomes. Western blot analysis using a polyclonal antiserum specific for mEH (kindly provided by U.A. Meyer, University of Basel) demonstrated the presence of a single immunoreactive protein of 53 kDa in both human liver and placenta, but three bands of immunoreactivity in preparations from bacterial cells. The most slowly migrating band likely corresponds to the fusion protein of mEH with the 31 amino acid affinity tag, while the most rapidly migrating band is coincident with that seen in human liver and probably consists of native mEH which has been produced by downstream initiation at the start ATG of the mEH coding sequence. Addition of either Lubrol PX (0.5%) or CHAPS (10 mM) following cell sonication produced a 40% increase in the quantity of catalytically active enzyme recovered during centrifugation procedures. On the other hand, addition of a cocktail of trace elements to culture media or the protease inhibitor PMSF or lysozyme to cell homogenization buffers had no effect on recoverable enzyme. The expressed fusion protein was quite stable; incubation at 37°C for 24 hrs led to a 10% decrease in catalytic function.

This system will be used to construct and express mutant forms of human mEH corresponding to allelic variants that we have detected in the course of our studies of anticonvulsant hypersensitivity. We are also attempting to improve the yield of recombinant enzyme. After the completion of these studies, Bell and Kasper (1993) reported the expression of rat mEH in *E. coli* to a level of at least 5% of the total bacterial protein. One difference in their system from that reported here is the addition of an OmpA signal peptide to the amino terminus, directing expression of native protein to the periplasmic space, instead of an affinity purification tag. We are presently constructing vectors that will allow us to determine whether the OmpA signal sequence will help to improve the yield of human mEH in *E. coli*.

## REFERENCES

Bell, P.A. and Kasper, C.B. (1993): Expression of rat microsomal epoxide hydrolase in *Escherichia coli*. *J Biol Chem* 268: 14011-14017.

Eaton, D.L. and Stapleton, P.L. (1989): Simultaneous determination of cytosolic glutathione S-transferase and microsomal epoxide hydrolase activity toward benzo[a]pyrene-4,5-oxide by high-performance liquid chromatography. *Anal Biochem* 178: 153-158.

Gerson, W.T., Fine, D.G., Spielberg, S.P. and Sensenbrenner, L.L. (1983): Anticonvulsant-induced aplastic anemia: increased susceptibility to toxic drug metabolites in vitro. *Blood* 61: 889-893.

Guenthner, T.M., Epoxide hydrolases. In: G. J. Mulder, ed. Conjugation reactions in drug metabolism. London: Taylor & Francis, 1990: 365-404.

Lertratanangkoon, K, and Horning, M.G. (1982): Metabolism of carbamazepine. *Drug Metab Disp* 10: 1-10.

Moustafa, M.A., Claesen, M., Adline, J., Vandervorst, D. and Poupaert, J.H. (1983) : Evidence for an arene-3,4-oxide as a metabolic intermediate in the *meta-* and *para-*hydroxylation of phenytoin in the dog. *Drug Metab Disp* 11: 574-580.

Moustafa, M.A., el-Emam, A.A., Subbagh, H.I. and el-Din, M,K, (1990) : Comparative study on the para-metabolic oxidation of phenytoin and decadeuteriophenytoin. *Arzneim Forsch* 40: 1076-1078.

Nelson, S.D. and Pearson, P.G. (1990) : Covalent and noncovalent interactions in acute lethal cell injury caused by chemicals. *Annu Rev Pharmacol Toxicol* 30: 169-195.

Park, B.K. and Kitteringham, N.R. (1990) : Drug-protein conjugation and its immunological consequences. *Drug Metab Rev* 22: 87-144.

Park, B.K., Pirmohamed, M. and Kitteringham, N.R. (1992) : Idiosyncratic drug reactions: a mechanistic evaluation of risk factors. *Br J Clin Pharmac* 34: 377-395.

Pirmohamed, M., Kitteringham, N.R., Guenthner, T.M., Breckenridge, A.M. and Park, B.K. (1992) : An investigation of the formation of cytotoxic, protein-reactive and stable metabolites from carbamazepine *in vitro*. *Biochem Pharmacol* 43: 1675-1682.

Shear, N.H. and Spielberg, S.P. (1988) : Anticonvulsant hypersensitivity syndrome: in vitro assessment of risk. *J Clin Invest* 82: 1826-1832.

Skoda, R.C., Demierre, A., McBride, O.W., Gonzalez, F.J. and Meyer, U.A. (1988) : Human microsomal xenobiotic epoxide hydrolase: cDNA sequence, cDNA-directed expression in COS-1 cells, and chromosomal localization. *J Biol Chem* 263: 1549-1554.

Spielberg, S.P. (1984) : *In vitro* assessment of pharmacogenetic susceptibility to toxic drug metabolites in humans. *Fed Proc* 43: 2308-2313.

Spielberg, S.P., Gordon, G.B., Blake, D.A., Mellits, E.D. and Bross, D.S. (1981) : Anticonvulsant toxicity in vitro: possible role of arene oxides. *J Pharmacol Exp Ther* 217: 386-389.

Thomsick, R.S. (1983) : The phenytoin syndrome. *Cutis* 32: 535-541.

# Regulation of CYP2B gene expression in the intestinal mucosa

Andrew P. Elia[1], Elizabeth A. Shephard[1], Ian R. Phillips[2]

[1]Department of Biochemistry and Molecular Biology, University College London, Gower Street, London WC1E 6BT, UK. [2]Department of Biochemistry, Queen Mary and Westfield College, Mile End Road, University of London, London E1 4NS, UK

## INTRODUCTION

The intestinal mucosa is exposed to a large variety of ingested naturally occurring and synthetic chemicals including drugs, pollutants and chemical carcinogens. The tissue is known to contain enzymes capable of metabolising potentially harmful compounds and thus may act as the primary site of metabolism of many such chemicals (Chhabra & Eastin, 1984). One of these enzyme systems that plays an important role in the detoxification of foreign compounds is the cytochrome P450 (CYP)-mediated mono-oxygenase.

It is well established that members of the CYP2B subfamily are induced in the liver by the barbiturate phenobarbital (Phillips *et al.*, 1983a). More recently, it has been demonstrated that members of this subfamily are also induced in the small intestine by this compound (Traber *et al.*, 1988). However, whereas in the liver both CYP2B1 and CYP2B2 are induced by phenobarbital (Omiecinski, 1986), in the small intestine only CYP2B1 is induced by this drug (Traber *et al.*, 1990).

Picrotoxin, a convulsant used therapeutically as an antidote for barbiturate poisoning, has been shown to be a potent inducer, in the liver, of CYP2B protein (Yamada *et al.*, 1993). We have investigated the effects of both picrotoxin and phenobarbital on the expression of CYP2B genes in various sections of the small intestine.

## MATERIALS AND METHODS

Male Sprague-Dawley rats weighing 180-200 g were used in these studies. Animals were treated with phenobarbital (Phillips *et al.*, 1983a) and picrotoxin (Yamada *et al.*, 1993) as described previously. The small intestine was excised, washed with ice-cold 0.15 M NaCl, opened longitudinally and again washed in ice-cold saline. The intestinal mucosa was removed by scraping with a microscope slide and was frozen immediately on dry ice.

Total RNA was isolated from intestinal mucosa and from liver using RNAzol B (Biogenesis) according to the supplier's recommendations. Total RNA was analysed by northern blot hybridization with a rat cDNA clone coding for CYP2B2 as described previously (Phillips *et al.,* 1983b). The final wash was in 150 mM sodium chloride/15 mM sodium citrate/0.1% SDS at 55ºC (Phillips *et al.,* 1983b). Under these conditions the cDNA recognises both CYP2B1 and CYP2B2 mRNAs. The autoradiograms were scanned with an image densitometer (BIORAD, model GS-670).

## RESULTS AND DISCUSSION

We have investigated the expression of mRNAs encoding members of the CYP2B subfamily in the intestinal mucosa and liver of untreated rats and those treated with either phenobarbital or picrotoxin. The expression of CYP2Bs along the small intestine was investigated by dissecting the tissue into five sections each of about 15 cm in length.

CYP2B mRNAs were not detected in any of the five sections isolated from the small intestine of untreated animals. In phenobarbital-treated animals, CYP2B mRNAs were detected in the four proximal sections (Fig. 1). Maximal induction was obtained in the most proximal section and expression decreased towards the distal end of the tissue. The expression of CYP2B mRNAs in

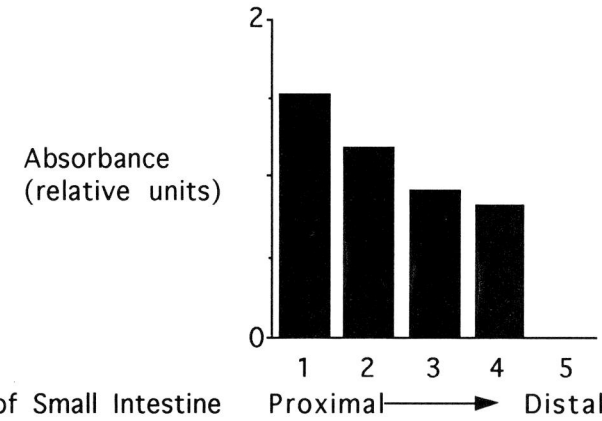

**Fig. 1.** Expression of CYP2B mRNAs in five regions of the small intestine. RNA was isolated from phenobarbital-treated animals and analysed by northern blot hybridization with a rat CYP2B cDNA probe.

section 1 was 2-fold greater than that in section 4 of the small intestine. CYP2B mRNAs were induced more than 50-fold by phenobarbital in the livers of these animals (Fig. 2).

The expression of CYP2Bs in the small intestine was also induced by picrotoxin. As was the case with phenobarbital, the induction was confined to the four sections most proximal of the small intestine and maximal induction was obtained in section 1 of this tissue (data not shown). In these animals picrotoxin induced CYP2B mRNAs about 43-fold in the liver (Fig. 2).

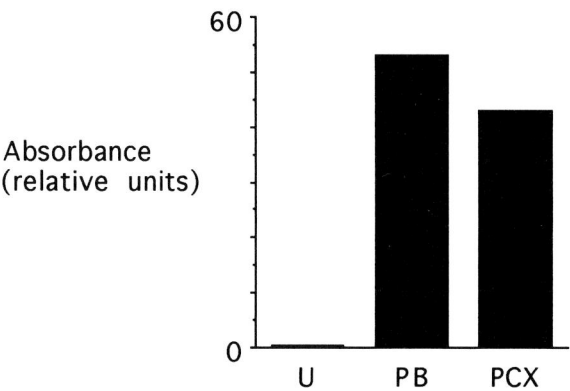

Fig. 2. Expression of CYP2B mRNAs in the livers of untreated (U) animals and those treated with either phenobarbital (PB) or picrotoxin (PCX). RNA was analysed by northern blot hybridization with a rat CYP2B cDNA probe.

ACKNOWLEDGEMENTS

A.P.E. is a recipient of a postgraduate studentship from the Science and Engineering Research Council, U.K.. The work was supported by a grant from the Cancer Research Campaign.

REFERENCES

Chhabra, R.S. & Eastin, W.C. (1984): Intestinal absorption and metabolism of xenobiotics in laboratory animals. In *Intestinal Toxicology*, ed. C.M. Schiller, pp. 145-160. New York: Raven Press.
Omiecinski, C.J. (1986): Tissue-specific expression of rat mRNAs homologous to cytochromes P-450b and P-450e. *Nucl. Acids Res.* 14, 1525-1539.
Phillips, I.R., Shephard, E.A., Bayney, R.M., Pike, S.F., Rabin, B.R., Heath, R. & Carter, N. (1983a): Induction and repression of the major phenobarbital-induced cytochrome P-450 measured by radioimmunoassay. *Biochem. J.* 212, 55-64.
Phillips, I.R., Shephard E.A., Ashworth A. & Rabin B.R. (1983b): Cloning and sequence analysis of a rat liver cDNA coding for a phenobarbital-inducible

microheterogeneous cytochrome P-450 variant: regulation of its messenger level by xenobiotics. *Gene* 24, 41-52.

Traber, P.G., Chianale, J., Florence, R., Kim, K., Wojcik, E. & Gumucio, J.J. (1988): Expression of cytochrome P-450b and P-450e genes in small intestinal mucosa of rats following treatment with phenobarbital, polyhalogenated biphenyls, and organochlorine pesticides. *J. Biol. Chem.* 263, 9449-9455.

Traber, P.G., Wang, W., McDonnell, M. & Gumucio, J.J. (1990): P450IIB gene expression in rat small intestine: cloning of intestinal P450IIB1 mRNA using the polymerase chain reaction and transcriptional regulation of induction. *Mol .Pharmacol.* 37, 810-819.

Yamada, H., Fujisaki, H., Kaneko, H., Ishii, Y., Hamaguchi, T. & Oguri, K. (1993): Picrotoxin as a potent inducer of rat hepatic cytochrome P450, CYP2B1 and CYP2B2. *Biochem. Phamacol.* 45,1783-1789.

# Clofibrate decreases CYP2A5 in male DBA/2N mice

Anneli Kojo[1], Sari Laitinen[1], Pertti Pellinen[1], Risto Juvonen[1], Hannu Raunio[2], Olavi Pelkonen[2], Markku Pasanen[1,2]

[1]Department of Pharmacology and Toxicology, University of Kuopio, PO Box 1627, SF-70211 Kuopio, Finland. [2]Department of Pharmacology and Toxicology, University of Oulu, SF-90220 Oulu, Finland

## INTRODUCTION

The peroxisome proliferators are a structurally diverse class of chemicals which have been shown to induce several peroxisomal (*e.g.* acyl-CoA oxidase, catalase) and microsomal (CYP4A1) enzyme activities. Unlike other P450 proteins CYP4A1 metabolizes fatty acids but not xenobiotics, *e.g.* benzphetamine, 7-ethoxyresorufin or testosterone (see Milton *et al.*, 1990). Consequently, the emphasis of peroxisome proliferators is focused on CYP4A1 rather than the CYP subfamilies 1 through 3, the principal xenobiotic metabolizing P450 families.

Coumarin is 7-hydroxylated by a specific CYP2A5 coded by *Cyp2a-5* gene in mouse and *CYP2A-6* in man (Negishi *et al.*, 1989). The expression of CYP2A5 has been shown to increase in mice after treatment with phenobarbital-type inducers (Juvonen *et al.*, 1985, Raunio *et al.*, 1988) or in connection with chemically-induced liver damage and in neoplastic lesions (see Pellinen *et al.*, 1993). Since the peroxisome proliferators have been associated with an increased incidence of hepatocellular carcinomas in rodents we wanted to examine the effects of the peroxisome proliferator clofibrate on the expression of mouse hepatic CYP2A5 after a single dose and one month treatment.

## MATERIALS AND METHODS

<u>Animals and microsomal preparations.</u> Male DBA/2N mice (6-8 weeks) were given clofibrate (200 mg/kg) dissolved in corn oil as a single *i.p.* injection. The control animals were dosed with the same volume of corn oil. For the one month treatment, clofibrate (500 ppm) was given with the food. The animals were killed and the livers were divided into two parts. One part was frozen in liquid nitrogen

microsomes.

Analytical methods. Preparation of microsomes and monooxygenase activities were determined using protocols described elsewhere (Honkakoski & Lang, 1989). Western blotting, preparation of total RNA and slot blotting were performed according to Honkakoski *et al.* (1992).

Statistical analysis. The values are means (±S.D., n=6-8) and a statistical significance was determined by Mann-Whitney U-test.

RESULTS

Exposure of mice to clofibrate resulted in a significant decrease (50-60 %) in the hepatic 7-hydroxylation of coumarin after one day and one month treatment (Fig. 1). Treatment with clofibrate caused slight changes in testosterone 15α-hydroxylase activity: an increase after one day and a decrease after one month treatment (Fig.1).

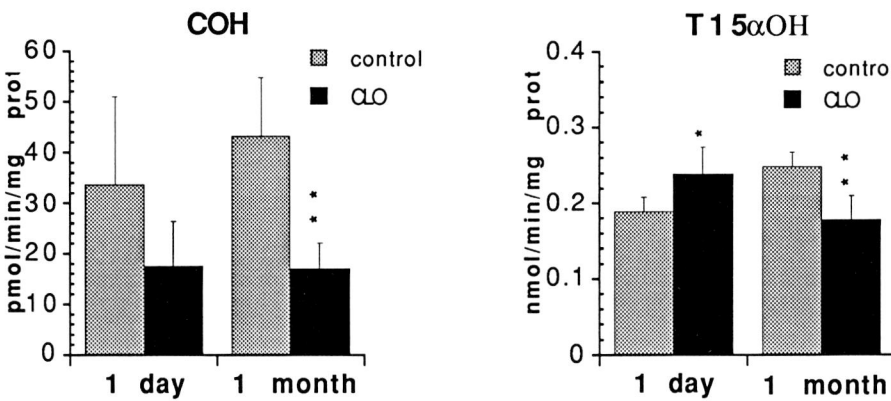

Fig. 1. The effect of clofibrate treatment (one day and one month) on coumarin 7-hydroxylase (COH) and testosterone 15α-hydroxylase (T15αOH) activities in the livers of male DBA/2N male mice. Data are means ± SD of 6-8 animals. * = p<0.05, ** = p<0.01.

The effect of clofibrate on other monooxygenase activities are summarized in Table 1. Total cytochrome P450 content and the activities tested: ethoxyresorufin O-dealkylase, pentoxyresorufin O-dealkylase and benzphetamine N-demethylase were not altered significantly by the treatments.

Table 1. The effect of one day and one month treatment with clofibrate (CLO) on total P450 content and ethoxyresorufin O-dealkylase (EROD), benzphetamine N-demethylase (BZDM) and pentoxyresorufin O-dealkylase (PROD) activities in the livers of male DBA/2N mice.

|  |  | P450[a] | EROD | BZDM | PROD |
|---|---|---|---|---|---|
| control | 1 day | $0.382 \pm 0.080$ | $0.264 \pm 0.047$ | $1.89 \pm 0.56$ | $8.18 \pm 1.13$ |
| CLO | " | $0.333 \pm 0.023$ | $0.308 \pm 0.047$ | $2.45 \pm 0.55$ | $8.64 \pm 0.99$ |
| control | 1 month | $0.521 \pm 0.090$ | $0.357 \pm 0.064$ | $3.35 \pm 0.56$ | $13.69 \pm 1.53$ |
| CLO | " | $0.471 \pm 0.069$ | $0.305 \pm 0.026$ | $2.88 \pm 0.35$ | $11.90 \pm 1.24$ |

[a] Total P450 content is expressed as nmol/mg protein, EROD and BZDM activities as nmol/min/mg protein, and PROD activity as pmol/min/mg protein. Data are means ± SD from 6-8 animals.

Western blot analysis employing polyclonal antibody against mouse CYP2A5 showed no significant changes in protein content after clofibrate treatment (Table 2). Slot blot analysis showed that CYP2A5 mRNA content decreased by 50 % both after one day and one month of treatment (Table 2.). The extent of the decrease in COH activity and mRNA content were very similar.

Table 2. Specific content of CYP2A5 protein, and relative level of CYP2A5 mRNA in the livers of clofibrate (CLO) treated and control DBA/2N mice.

|  |  | CYP2A5 protein pmol/mg prot | CYP2A5 mRNA relative level |
|---|---|---|---|
| control | 1 day | 64.6 | 1 |
| CLO | " | 75.9 | 0.40 |
| control | 1 month | 60.1 | 1 |
| CLO | " | 52.8 | 0.47 |

DISCUSSION

Here we report a strong inhibition of COH activity and CYP2A5 mRNA levels in mouse liver after CLO treatment. COH activity is catalyzed by pyrazole-inducible CYP2A5 in mouse liver (Juvonen *et al.*, 1985) which has a 98.3% homology with CYP2A4 responsible for T15αOH activity, which was not decreased after CLO treament (Negishi *et al.*, 1989, Lindberg *et al.*, 1989). The present results demonstrate that the T15αOH activity does not follow the decrease in COH activity after CLO treatment indicating that the two closely related P450 enzymes are differently regulated in their response to CLO. Exposure of mice to CLO had no effect on other xenobiotic monooxygenases mainly detecting the expression of CYP1A, 2B and 3A gene subfamilies although other investigators (Wortelboer *et al.*, 1991) reported the enhancement of several P450s after CLO treatment. Our results do not agree with those of Sharma *et al.* (1988) who showed reduced EROD and BZDM activities in rats after CLO treatment.

Dietary lipids, especially linoleic acid, have been shown to increase hepatic P450 levels (McDanell *et al.*, 1992, Yoo *et al.*, 1992). CYP2A5 seems to be exceptional in its response to corn oil. It has been reported to decrease after corn oil administration (Honkakoski *et al.*, 1992). We have also shown a decrease in COH activity by corn oil with only minor changes in several other monooxygenases (data not shown). Therefore, it is possible that changes in the fatty acyl composition of microsomal lipid may affect CYP2A5. Changes in fatty acyl composition alter membrane fluidity which in turn can influence the conformation, mobility and function of membrane bound proteins, including P450 proteins. However, this cannot alone explain the observed decrease in the mRNA content.

## REFERENCES

Juvonen, R.O., Kaipainen, P.K. & Lang, M.A. (1985): Selective induction of coumarin 7-hydroxylase by pyrazole in $D_2$ mice. *Eur. J. Biochem.* 152, 3-8.

Honkakoski, P. & Lang, M.A. (1989): Mouse liver phenobarbital-inducible P450 system: purification, characterization, and differential inducibility of four cytochrome P450 isozymes from the D2 mouse. *Arch. Biochem. Biophys.* 273, 42-57.

Honkakoski, P., Auriola, S. & Lang, M.A. (1992): Distinct induction profiles of three phenobarbital-responsive mouse liver cytochrome P450 isozymes. *Biochem. Pharmacol.* 43, 2121-2128.

Lindberg, R., Burkhart, B., Ichikawa, T. & Negishi, M. (1989): The structure and characterization of type I $P-450_{15\alpha}$ gene as major steroid 15α-hydroxylase and its comparison with type II $P-450_{15\alpha}$ gene. *J. Biol. Chem.* 264, 6465-6471.

McDanell, R.E., Beales, D., Henderson, L. & Sethi, J.K. (1992): Effect of dietary fat in the *in vitro* hepatotoxicity of paracetamol. *Biochem. Pharmacol.* 44, 1303-1306.

Milton, M.N., Elcombe, C.R. & Gibson, G.G. (1990): On the mechanism of induction of microsomal cytochrome P450IVA1 and peroxisome proliferation in rat liver by clofibrate. *Biochem. Pharmacol.* 40, 2727-2732.

Negishi, M., Lindberg, R., Burkhart, B., Ichikawa, T., Honkakoski, P. & Lang, M.A. (1989): Mouse steroid 15α-hydroxylase gene family: identification of Type II P45015α as coumarin 7-hydroxylase. *Biochemistry.* 28, 4169-4172.

Pellinen, P., Stenbäck, F., Rautio, A., Pelkonen, O., Lang, M. & Pasanen, M. (1993). Response of mouse liver coumarin 7-hydroxylase activity to hepatotoxins: dependence on strain and agent and comparison to other monooxygenases. *Naunyn-Schmiedeberg's Arch. Pharmacol.* in press.

Raunio, H., Kojo, A., Juvonen, R., Honkakoski, P., Järvinen, P., Lang, M.A., Vähäkangas, K., Gelboin, H.V., Park, S-S. & Pelkonen, O. (1988): Mouse hepatic cytochrome P-450 isozyme induction by 1,4-bis[2-(3,5-dichloropyridyloxy)]benzene, pyrazole, and phenobarbital. *Biochem. Pharmacol.* 37, 4141-4147.

Sharma, R., Lake, B.G., Foster, J. & Gibson, G.G. (1988): Microsomal cytochrome P-452 induction and peroxisome proliferation by hypolipidaemic agents in rat liver. *Biochem. Pharmacol.* 37, 1193-1201.

Wortelboer, H.M., de Kruif, C.A., van Iersel, A.A.J., Falke, H.E., Noordhoek, J. & Blaauboer, B.J. (1991): Comparison of cytochrome P450 isoenzyme profiles in rat liver and hepatocyte cultures. *Biochem. Pharmacol.* 42, 381-390.

Yoo, J-S.H., Smith, T.J., Ning, S.M., Lee, M-J., Thomas, P.E. & Yang, C.S. (1992): Modulation of the levels of cytochromes P450 in rat liver and lung by dietary lipids. *Biochem. Pharmacol.* 43, 2535-2542.

Cytochrome P450. 8th International Conference. Ed. M.C. Lechner. John Libbey Eurotext, Paris © 1994, pp. 515-518.

# Exchange of N- or C- terminal segments between human P4501A1 and 1A2 leads respectively to high stability and high activity chimeras

Aouatef Bellamine, Denis Pompon

*Centre de Génétique Moléculaire, Laboratoire Propre 2420 du CNRS associé à l'Université Paris-VI, Gif-sur-Yvette, France*

Cytochrome P450s 1A subfamily metabolize mainly aromatic molecules like polycyclic hydrocarbons. In human organism, the two members of this family, although sharing relatively high sequence homology (73%), display significantly different substrate specificities.

Here we report the construction of a chimeric cDNA library obtained by homeologous recombination, between 1A1 and 1A2 P450 cDNAs, in *Saccharomyces cerevisea.* Two chimeric cytochrome P450 coding sequences were selected on their enhanced ethoxyresorufin-*O*-deethylase activity in hole cells. Chimeric P450s sequences are represented in figure 1.

**FIG1**

**Chimera S1-2**

```
NH2MLFP--ISMSATEFLLASVIFCLVFWVMRASRPQVPKGLKNPPGPWGWPLI      1A1
NH2malsqsvpf....l....a........lkgl..r......s..e......l..v..  1A2

  MLTLGKNPHLALSRMSQQYGDVLQIRIGSTPVVVLSGLDTIRQALVRQGDDFk....  1A1
  .................r.............l..r...............QGRPD   1A2
```

**Chimera S7-1**

```
>>> FLFLAILLQrv.....l.....m...........c...f.mql--rs COOH  1A1
>>> .........QLEFSVPPGVKVDLTPIYGLTMKHARCEHVQARRFSIN COOH  1A2
```

The first chimera S1-2 contains the 264 bp 5'-end of *CYP1A1* ORF fused in frame with the last 1284 bp of *CYP1A2* ORF, and, so differs from 1A2 by the substitution of the 90 N-terminal residues of 1A2 with the corresponding residues of 1A1 representing a total of 22 aminoacid changes. The second chimera S7-1 contains the 472 first aminoacid residues of 1A1 followed by the 44 last residues of 1A2. Thus S7-1 chimera differs from 1A1 by only 13 aminoacid changes.

Chimeric and parental P450s expressed as hemoprotein content were determined on the basis of the reduced carbon monoxide difference spectrum (Omura and Sato, 1964). CPR (cytochrome P450 reductase) specific contents (Table 1) appear rather constant whatever the P450 expressed. In contrast, the P450 contents were found consistently very different for chimeras and parental enzymes. Very surprisingly, the S1-2 production level was found to be 9-fold higher than those of 1A1 and 1A2. On the other hand, the production level of S7-1 appeared to be 3-fold lower than the parental ones. In contrast, determination of the P450 content for chimera S7-1 indicated that this enzyme exists in microsomes at a 3-fold lower content than parental 1A1 and 1A2, moreover an unusual high level of P420 form is present for S7-1. These observations and results suggest that the enhanced turnover numbers likely account for the high activity of this chimera observed on living cells. Northern blot analysis shows similar mRNA levels for the different forms (Table 1). These results indicate that differences in P450 expression levels do not appear to be correlated with transcriptional or mRNA stability effects.

**Table 1**

| Strain | Expressed P450 | Microsomal CPR activity | Microsomal contents (in pmoles per mg protein) | | CPR to P450 molar ratio | Specific mRNA level |
| --- | --- | --- | --- | --- | --- | --- |
| | | | CPR | P450 | | |
| W (R) | 1A1 | 1800 ± 40 | 500 | 30 | 17 | 1 |
| W (R) | S7-1 | 2710 ± 440 | 750 | 10 | 75 | 2 |
| W (R) | 1A2 | 2310 ± 44 | 640 | 30 | 21 | 1 |
| W (R) | S1-2 | 2220 ± 420 | 610 | 280 | 2 | 0.7 |
| W (N) | control | 130 | 40 | - | - | - |
| W (R) | control | 2100 | 580 | - | - | - |

To understand molecular bases of S1-2 overproduction, the protein stability was evaluated within the microsomal membranes. Experiments are based on P450 denaturation in the presence of the strong chaotropic reagent, guanidinium thiocyanide. Denaturation was first estimated by determination of the residual P450 EROD activity after preincubation of yeast microsomal membranes for a fixed time in the presence of increasing concentrations of reagent. As shown in figure 2A, P450s 1A1 and S7/1 share the same stability pattern, wether P450s 1A2 and S1/2 seem to have radically different behaviours.

**FIG 2**

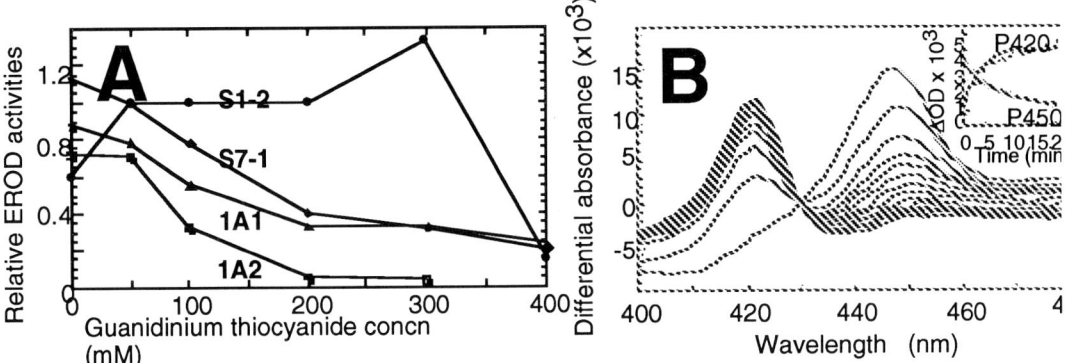

Moreover, S1-2 denaturation occured abruptly for denaturant concentrations lying in the range of 350-400 mM, in comparison with the progressive denaturation observed for 1A1 and 1A2. This suggests that a highly cooperative process could trigger the conversion of S1-2. To discriminate between effects affecting only S1-2 or effects involving associated redox enzymes (CPR), kinetics of the S1-2 denaturation reaction were performed. The conversion of the P450 species of S1-2 to the inactive P420 species was monitored at a critical guanididium thiocyanide concentration determined by the previous experiments. Results presented in Fig. 2B were further analysed using the second order derivate of the spectra to calculate P450 and P420 contents after various denaturation times. Experimental values thus calculated were plotted against time and the theoretical trace was analysed by multiexponential fitting. Experimental rate of the decrease in P450 was found to follow a simple monoexponential law characteristic of a first-order reaction ($t_{1/2}$ = **3 min**), consistent with denaturation occuring in a single limiting step. Moreover, P420 accumulation follows the same kinetic law suggesting a quantitative conversion of P450 to P420 form. Such a result suggests that a single limiting molecular events triggers the S1-2 denaturation in contrast to the other P450 forms. Owing that S1-2 chimera is build only from 1A1 and 1A2 sequences, we propose that the native folding state of S1-2 is locked by some specific molecular interaction(s) shared by residues belonging to the 1A1 and 1A2 moieties of S1-2.

Analysis of some enzymatic properties of 1A1, 1A2 and chimeric proteins was performed to evaluate whether significant perturbation of the active site folding could be detected. Four typical substrates: 7-ethoxyresorufin, 7-methoxyresorufin, phenacetin and benzo[a]pyrene were used. Figure 3A shows that, at saturating substrate concentration, S7-1, which is mainly 1A1, retains the benzo[a]pyrene stereospecificity preference of 1A1; while S1-2, which is mainly 1A2, does not metabolise benzo[a]pyrene as is the case for its parent 1A2. The overall substrate specificities were found to be very similar for 1A1 and S7-1 in one hand, and for 1A2 and S1-2 in the other hand. Nevertheless absolute values of all turnover numbers are 3-times increased in S7-1 as compared to the ones measured for 1A1. On the other hand, the turnover numbers are decreased in a similar extent for S1-2 as compared to that of 1A2. Ethoxyresorufin half-saturation concentration (Fig.3 B, solid rectangles) seems to be slightly increased (approx. 2-fold) for S7-1 as compared to the one for 1A1, going from 170 nM to 70 nM respectively. Similarly, the ethoxyresorufin half-saturation concentration for S1-2 is also increased 2-times compared to the one for 1A2, going from 0.54 µM to 0.27 µM respectively. Nevertheless for a substrate of tightly related structure, 7-methoxyresorufin, the affinity effects were found to be inverted (Fig. 3, open rectangles). Moreover, no significant difference was observed for phenacetin and benzo[a]pyrene half-saturation concentrations between chimeric and parental P450s (Fig. 3A, gray and hatched rectangles). Therefore, no clear perturbation of the substrate affinities is found associated with the aminoacid sequence substitutions considered.

FIG3

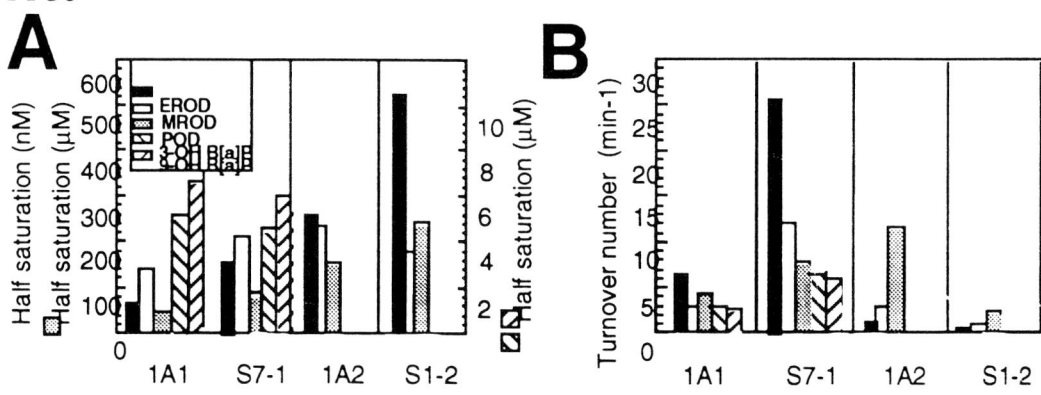

# Assessment of CYP2A6 and CYP3A4 activities *in vivo* in different diseases in man

Arja Rautio[1], Erja Salmela[1], Pentti Arvela[1], Olavi Pelkonen[1], Eero A. Sotaniemi[2]

[1]Departments of Pharmacology and Toxicology, [2]Internal Medicine, University of Oulu, SF-90220 Oulu, and Deaconess Institute of Oulu, Finland

## Summary

The effect of liver condition on coumarin metabolism in vivo was investigated in young healthy volunteers and in patients with liver cirrhosis and epilepsy. The formation rate of 7-hydroxy metabolite (7-hydroxycoumarin, 7OHC) by coumarin 7-hydroxylase (CYP2A6) did not change after alcohol ingestion or after treatment with rifampisine or cimetidine for 6 six days in healthy volunteers. Subjects with cirrhosis had delayed 7OHC excretion and epileptics had increased formation rate. The indicator of CYP3A4 function, the formation rate of the lidocaine metabolite monoethylglycinexylide (MEGX), was lowered both in subjects with liver cirrhosis and epilepsy. In epileptic patients therapy with drugs competing for CYP3A4 may delay MEGX formation.

## INTRODUCTION

The main pathway in the metabolism of coumarin in man is 7-hydroxylation (Ritschel et al. 1977), catalyzed by a single isozyme, CYP2A6 (Raunio et al. 1988). After hydroxylation 7OHC is glucuroniced and excreted into urine. The formation of monoethylglycinexylidide (MEGX) after i.v. infusion of lidocaine reflects CYP3A4 activity (Oellerich et al. 1987). Both drugs have been suggested to be useful as indicators of liver function in clinical practice.

This study was undertaken to investigate the effect of liver condition on the formation rate of 7OHC in healthy volunteers and in patients with liver cirrhosis and epilepsy. The effect of an acute alcohol exposure, and treatment with cimetidine (microsomal enzyme-inhibiting drug) and rifampisine (microsomal enzyme-inducing drug) on the 7-hydroxylation rate of coumarin was investigated in healthy volunteers.

## MATERIALS AND METHODS

### Protocol for in vivo studies

The study protocol was approved by the Ethics Committee of the Medical Faculty of University of Oulu. The drug metabolism tests were done after an overnight fast. Venalot[R], a capsule (containing 5 mg coumarin and 25 mg rutosides, Schaper & Brummer, Ringelheim, Germany) was given orally with 200 ml water. The subjects had breakfast 2 hours after the administration of the test drug. Urine was voided before and 2, 4 and 24 h after capsule administration. The samples were kept frozen (-20°C) until analyzed by the HPLC method of Rautio et al. (1992). Lidocaine (1 mg/kg, Astra, Södertelje, Sweden) was given iv bolus during 1 minute injection before the coumarin test, and blood samples for monoethylglycinexylidide (MEGX) determinations were drawn before the injection and 15 min after it (Oellerich et al. 1987).

**Volunteers.** Twenty medical students (7 men and 13 women) participated in this study. Their body mass index (BMI, kg/m2) was 21.5 + 2.0 and their age was 22.2 + 1.2 years. Nine of them used contraceptives and one continuously drugs for arthiritis rheumatoides. They were non-smokers and were advised to avoid alcohol for at least two weeks before the tests.

**Treatment.** Healthy volunteers received rifampisine at a dose of 450 mg/day (Rimapen®, Orion Co, Helsinki, Finland, n=9), or cimetidine 600 mg/day (Tagamet®, Lääkefarmos Co, Turku, Finland, n=5), or placebo (n=6) divided into three doses for 6 days. Coumarin tests were done before and after the treatment. The last dose of cimetidine, rifampisine and placebo were taken at the morning of the test day 2 h before coumarin administration.

Three weeks later the volunteers participated in a social dinner, as used previously (Sotaniemi et al. 1981). Briefly, they had a full dinner with drinks starting at 7 p.m., and the last drink was given at 2 a.m. Next morning (at 8 a.m.) the coumarin test was performed.

**Patients.** Patients with alcoholic liver cirrhosis (n=12), epilepsy (n=12) and age and sex matched controls (n=12) participated in this part of the study (Table II).

## RESULTS

Healthy volunteers excreted within the first 4 h about 95 % of the formed 7OHC. The total amount of 7OHC formed (70.7-79.8% of the given dose) was at the same level in all the treated groups and after the drinking party and did not differ from controls (Table I).

Table I. The urinary excretion of 7OHC in healthy volunteers after a drinking party and after treatment with cimetidine, rifampisine or placebo for 6 days.

| Treatment | 7OHC excretion, % of given dose | | |
|---|---|---|---|
| | 0-2 h | 2-4 h | 4-24 h |
| Before drinking party (n=20) | 61.6±15.4 | 11.0±8.1 | 3.8±3.6 |
| After drinking party (n=20) | 61.8±14.6 | 8.9±5.7 | - |
| Rifampisine 450 mg/day (n=9) | 61.4±6.9 | 9.3±2.8 | - |
| Cimetidine 600 mg/day (n=5) | 62.6±20.4 | 8.3±4.9 | - |
| Placebo 6 tablets/day (n=6) | 68.5±8.5 | 11.3±6.2 | - |

Data are means±SD

7OHC excretion at 2 h was delayed in patients with cirrhosis and increased formation rate was seen in subjects with epilepsy at the 2-4 h sample as compared to controls. MEGX formation was decreased both in patients cirrhosis and in subjects with epilepsy. (Table II)

Table II. The clinical data and urinary 7-hydroxycoumarin (7OHC) excretion in patients with liver cirrhosis and epilepsy and age and sex-matched controls.

| Subjects | Age yr | Sex F/M | BMI kg/m2 | 7OHC excretion, % of given dose | | | MEGX ng/ml |
|---|---|---|---|---|---|---|---|
| | | | | 0-2 h | 2-4 h | 4-24 h | |
| Liver cirrhosis (n=12) | 52±9 | 4/8 | 27.5±3.9 | 18.0±10.3a | 13.0±5.6c | 21.5±12.8a | 5.6±3.0a |
| Epilepsy (n=12) | 53±11 | 6/6 | 23.8±1.8 | 69.513.2± | 11.3±6.7b | 0.0 | 41.0±13.1a |
| Controls (n=12) | 53±10 | 5/7 | 26.3±3.8 | 59.2±13.6 | 21.7±8.9 | 3.5±7.4 | 69.0±12.0 |

BMI = body mass index
Data mean±SD
p-values compared with control subjects a p<0.001, b p<0.01, c p<0.05

## DISCUSSION

Knowledge of the hepatic drug metabolizing ability is important in many clinical conditions, and easy and reliable tests are needed in the follow-up of the liver function. CYP2A6 is the sole catalyst of coumarin 7-hydroxylation (Mäenpää et al. 1993), and as seen in this study, acute alcohol ingestion, or treatment with enzyme inhibitors or inducers did not change the 7OHC formation rate in healthy volunteers.

Our data demonstrate that patients with liver cirrhosis metabolized coumarin and lidocaine at a slow rate. It is known that many drugs, such as erythromycin, steroids and calcium antagonists, are metabolized by the same CYP3A isoforms as lidocaine, and the rate of MEGX formation is decreased. Epileptic patients treated with carbamazepine/phenobarbital showed poor ability to metabolize lidocaine, probably because antiepileptic drugs interfered with the metabolism of lidocaine.

Simultaneous assessment of CYP2A6 and CYP3A4 activities gives better information about the liver function than either test alone, especially in situations where drugs metabolized by CYP3A isoforms are used by the subjects.

## REFERENCES

Mäenpää J, Sigusch H, Raunio H, Syngelmä T, Vuorela P, Vuorela H, and Pelkonen O. (1993): Differential inhibition of coumarin 7-hydroxylase activity in mouse and human liver microsomes. Biochem. Pharmacol. 45: 1035-1042.

Oellerich M, Raude E, Brudelski M, Schulz M, Schmidt FW, Ringe B, Lamesch P, Pichlmayr R, Raith H, Schuruhn M, Wrenger M, and Wittekind Ch. (1987): Monoethylglycinexylidide formation kinetics: A novel approach to assessment of liver function. J. Clin. Chem. Clin. Biochem. 25: 845-853.

Raunio H, Syngelmä T, Pasanen M, Juvonen R, Honkakoski P, Kairaluoma MA, Sotaniemi EA, Lang M, and Pelkonen O. (1988): Immunochemical and catalytical studies on hepatic coumarin 7-hydroxylase in man, rat and mouse. Biochem. Pharmacol. 37: 4141-4147.

Rautio A, Kraul H, Kojo A, Salmela E, and Pelkonen O. (1992): Interindividual variability of coumarin 7-hydroxylation in healthy volunteers. Pharmacogenetics 2: 227-233.

Ritschel WA, Brady ME, Tan HSI, Hoffman KA, Yiu IM, and Grummich KW. (1977): Pharmacokinetics of coumarin and its 7-hydroxymetabolites upon intravenous and peroral administration of coumarin in man. Eur. J. Clin. Pharmacol. 12: 457-461.

Sotaniemi EA, Anttila M, Rautio A, Stengård J, Saukko P, and Järvensivu P. (1981): Propranolol and sotalol metabolism after a drinking party. Clin. Pharmacol. Ther. 29: 705-710.

# Putative active site model for CYP2C9 (tolbutamide hydroxylase)

B.C. Jones[1], G. Hawksworth[3], V. Horne[2], A. Newlands[3], M. Tute[2], D.A. Smith[1]

[1]Departments of Drug Metabolism. [2]Computational Chemistry, Pfizer Central Research, Sandwich, Kent CT10 2NF, UK. [3]Department of Medicine and Therapeutics, University of Aberdeen, Aberdeen, AB9 2ZD, UK

INTRODUCTION

The cytochrome P450s are a superfamily of enzymes responsible for the metabolism of a large number of drug molecules. Binding of substrates to P450s has traditionally been viewed as relying solely on hydrophobic interactions. However, recent understanding of the various isozymes indicate that other binding interactions can be of importance giving the enzyme high regio- and even stereo-selectivity. For instance CYP2D6 relies on an ion-pair interaction as a major binding factor (Strobl et al., 1993) whilst metabolism by CYP3A4 seems to occur as a result of chemical lability rather than a specific binding interaction (Smith and Jones, 1992).

There are now a number of substrates reported which have routes of metabolism ascribed to CYP2C9 (tolbutamide hydroxylase) (Smith and Jones, 1992). These substrates are chemically diverse but examination of the structures suggest that all the substrates have the capability to act as either hydrogen bond donors or acceptors. We have therefore rationalised the substrate specificity of the isoform in terms of a hydrogen bond donor model by the use of molecular modelling.

CYP2C9 SUBSTRATES/INHIBITOR

A number of the substrates with differing chemical lability at the site of metabolism were modelled including phenytoin (aromatic hydroxylation), warfarin (aromatic hydroxylation), ibuprofen (aliphatic hydroxylation), naproxen (O-demethylation), diclofenac (aliphatic hydroxylation), Δ1-tetrahydrocannabinol (THC) (aliphatic hydroxylation), tolbutamide (benzylic hydroxylation) and the potent inhibitor sulphaphenazole ($IC_{50}$ of 0.6μM Doecke et al. 1991). The structures and sites of CYP2C9-mediated metabolism of these compounds are shown in Fig 1.

MODELLING

Initially a hydrogen bond donor site and site of metabolism were identified for phenytoin which was then used as a template since phenytoin is the most rigid of the substrates. Computer-assisted molecular modelling studies were carried out using a Silicon Graphics Iris work station. The structures of the substrates/inhibitor were obtained from the Cambridge Crystal database or built interactively using the SYBYL program. Using the RMS fitting routine within SYBYL the sites of metabolism and potential hydrogen bonding sites for each of the substrates were overlayed with the corresponding sites in phenytoin. From these overlays the putative active site model was constructed

Fig 1.

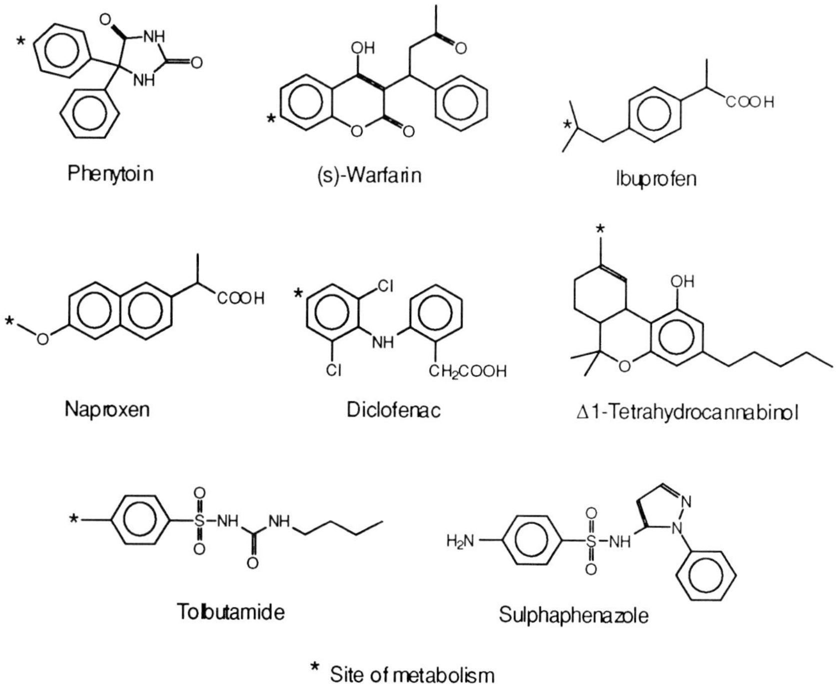

[Structures of Phenytoin, (s)-Warfarin, Ibuprofen, Naproxen, Diclofenac, Δ1-Tetrahydrocannabinol, Tolbutamide, Sulphaphenazole]

\* Site of metabolism

RESULTS

All the substrates produced good overlays with the sites of metabolism and hydrogen donation on phenytoin. By analysis of each of the overlays it was possible to define an active site template involving the distance (a) from the site of metabolism (Y) to the heteroatom supporting the hydrogen bond (X) and the corresponding angle between this vector and the hydrogen bond as shown in Fig 2.

Fig 2.

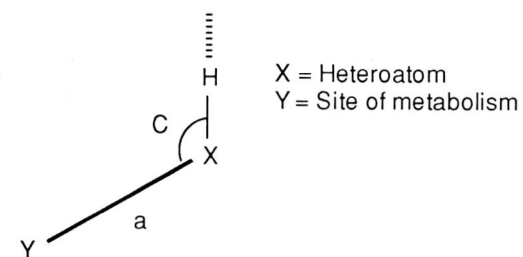

X = Heteroatom
Y = Site of metabolism

Assuming that sulphaphenazole inhibits CYP2C9 by a ligand interaction between a nitrogen lone pair and the haem then there are two possible orientations that the inhibitor could adopt, either the nitrogen of the aniline group or the nitrogen of the imidazole interacts with the haem. This model suggests the interaction with the aniline group is most likely since this orientation gives the best overlap with the hydrogen bonding sites and the sites of metabolism/interaction.

The distance a, and the angle C were measured for each of the 8 compounds of interest (Table 1):

Table 1

| Substrate | a (Å) | C (°) |
|---|---|---|
| Phenytoin | 5.9 | 142 |
| (s)-Warfarin | 6.7 | 172 |
| Ibuprofen | 6.5 | 132 |
| Naproxen | 8.1 | 120 |
| Diclofenac | 7.1 | 124 |
| Δ1-THC | 4.9 | 139 |
| Tolbutamide | 6.9 | 105 |
| Sulphaphenazole | 6.7 | 113 |

CONCLUSIONS

All the substrates with routes of metabolism attributable to CYP2C9 have hydrogen bond donating groups. The hydrogen bond donating groups and sites of metabolism on each of the substrates have been overlayed with those of phenytoin to produce a putative template of the active site of CYP2C9. The dimensions of the model are defined by a, the distance between the site of metabolism and the hydrogen bond donor heteroatom and C, the angle between this and the hydrogen bond. The mean dimensions (± S.D.) for the 8 compounds (a = 6.6±1Å, C = 133±21°) illustrates the degree of overlap achieved.

REFERENCES

Doecke, C.J., Veronese, M.E., Pond, S.M., Miners, J.O., Birkett, D.J. & McManus, M.E. (1991): Relationship between phenytoin and tolbutamide hydroxylations in human liver microsomes. Br. J. Clin. Pharmacol. 31 125-130

Smith, D. A. & Jones, B. C. (1992): Speculations on the substrate structure-activity relationship (SSAR) of cytochrome P450 enzymes. Biochem. Pharmacol. 44 2089-2098.

Strobl, G. R., von Kruedener, S., Stockigt, J., Guengerich, F. P. & Wolf, T. (1993): Development of a pharmacophore for inhibition of human liver cytochrome P4502D6: molecular modelling and inhibition studies. J. Med. Chem. 36 1136-1145

# Expression of a bovine P450c17 cDNA in the yeast *Saccharomyces cerevisiae*

Bruno Dumas[1], Gilles Cauet[2], Eric Degryse[2], Roberto Spagnoli[1], Tilman Achstetter[2]

[1]ROUSSEL-UCLAF, Biotechnology Department, 111, route de Noisy, F-93230 Romainville, France. [2]TRANSGENE SA, Yeast Department, 11, rue de Molsheim, F-67082 Strasbourg Cedex, France

## INTRODUCTION

Steroidogenesis in higher eukaryotic cells is a complex process guided by a series of enzymes, most of them are members of the cytochrome P450 family (for a review, see Miller, 1988). These specialized cytochromes carry out a number of hydroxylation reactions, they localize to mitochondrial or microsomal membranes. The microsomal P450 17α-hydroxylase (E.C.1.14.99.9; P450c17) catalyzes the hydroxylation of progesterone into 17α-hydroxyprogesterone as well as pregnenolone into 17α-hydroxypregnenolone. Depending on the species the $C_{17,20}$-lyase activity of P450c17 leads also to conversion of 17α-hydroxypregnenolone and 17α-hydroxyprogesterone into dehydroepiandrosterone and androstenedione, respectively (Fevold et al., 1989). Reducing equivalents originate from NADPH via a membrane-bound flavoprotein oxidoreductase (NADPH:cytochrome P450 reductase, EC 1.6.2.4; P450red). P450c17 cDNAs have been successfully expressed in *E. coli* (Barnes et al., 1991), in the yeast *Saccharomyces cerevisiae* (Sakaki et al., 1993), as well as in mammalian cells (Clark & Waterman, 1992).

## RESULTS

A P450c17 cDNA was cloned from a bovine adrenal cortex cDNA library by classical screening. It was found to be identical to the published sequence (Zuber et al., 1986). The P450c17 cDNA was inserted in a 2µm-based *E. coli*-yeast shuttle plasmid, where its expression is controlled by the yeast *CYC1* promoter (pTG10131). Cells from a transformed clone of yeast strain W303-1A (W303/pTG10131) were disrupted with glass beads. Cell debris and mitochondria were removed from the crude extract by centrifugation. Microsomal and cytosolic fractions, respectively were obtained after a 150,000 x g centrifugation step of the clear supernatant. Proteins of these fractions were analyzed by Western blot using an anti bovine P450c17 antibody. As shown in Fig. 1 the W303/pTG10131-derived microsomal fraction contains an antigen of 57 kDa (Barnes et al., 1991), the same apparent molecular weight (MW) as the bovine P450c17 produced in *E. coli*. No signal is seen in the cytosolic fraction demonstrating the particulate nature of the antigen when produced in yeast.

The microsomal fraction of W303/pTG10131 was submitted to a spectral analysis. A reduced CO-difference spectrum exhibits a maximum at 450 nm indicative for the presence of a hemoprotein (Fig. 2). The concentration of recombinant bovine P450c17 holoenzyme was estimated at 0.3 % of microsomal proteins.

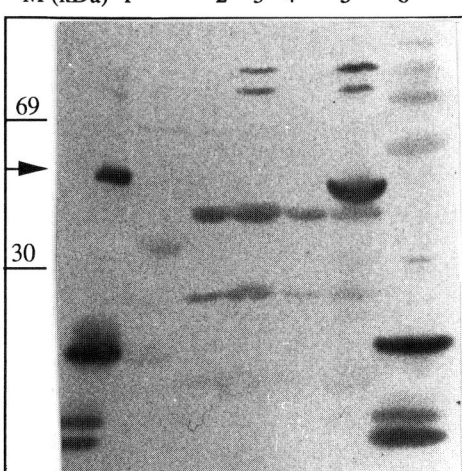

**Fig. 1.** Bovine P450c17 made in yeast. Microsomal (mic) and cytosolic (cyt) fractions of W303/pTG10131 were analyzed by Western blotting using a rabbit anti bovine P450c17 antiserum (kindly provided by Dr. M. Waterman). Subcellular fractions from untransformed W303 cells as well as membranes of an untransformed *E. coli* strain or an *E. coli* strain expressing a bovine P450c17 cDNA (Barnes *et al.*, 1991) served as negative and positive control respectively. M = MW markers; Lane 1: membranes of *E. coli* expressing a bovine P450c17 cDNA; Lane 2: W303 cyt; Lane 3: W303 mic; Lane 4: W303/pTG10131 cyt; Lane 5: W303/pTG10131 mic; Lane 6: membranes of an untransformed *E. coli* strain.

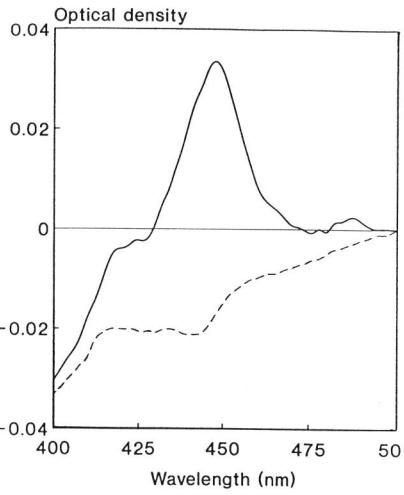

**Fig. 2.** Reduced CO-difference spectrum of a microsomal preparation of W303/pTG10131. The microsomes were reduced with dithionite and the sample cuvette was flushed with CO prior to analysis. Protein concentration was 5.4 mg/ml (W303/pTG10131) and 4.1 mg/ml (W303). ——— W303/pTG10131; - - - W303.

In order to correlate an enzymatic activity with the positive signal obtained in the spectral analysis the 17α-hydroxylase and the $C_{17,20}$-lyase activities were measured both *in vivo* and *in vitro*. Yeast cells were incubated with either [$^3$H]-pregnenolone or [$^3$H]-progesterone (10 μM, 2 x 10$^7$ cells/ml) for 8 h at 37°C. Pregnenolone and progesterone were almost completely converted into their respective 17α-hydroxy derivatives, but neither dehydroepiandrosterone nor androstenedione as products of the $C_{17,20}$-lyase activity could be detected. The enzymatic test demonstrates that recombinant bovine P450c17 can couple with the endogenous yeast P450red to catalyze the 17α-hydroxylation of pregnenolone and progesterone (data not shown).

The lyase activity of P450c17 was suggested to depend on the availability of reducing equivalents and thus on the amount of P450red (Yanagibashi & Hall, 1986). In order to determine the influence of the concentration of P450red on the lyase activity, a sample of the microsomal fraction of W303/pTG10131 was incubated with [$^3$H]-pregnenolone and [$^3$H]-progesterone (2.5 μM) for 30 min at 37°C and with increasing amounts of purified rat liver P450red (Fig. 3). Apparently, endogenous (yeast) P450red is limiting since the hydroxylase activity of the recombinant bovine P450c17 is stimulated upon addition of purified rat P450red. In agreement with the literature (Fevold *et al.*, 1989), no lyase activity is observed when 17α-hydroxyprogesterone is produced. A low conversion of 17α-hydroxypregnenolone into dehydroepiandrosterone is detected only in the presence of a large excess of purified rat liver reductase over the P450c17. Kinetic parameters were also determined on microsomes of W303/pTG10131 in the presence of saturating amounts of purified rat liver P450red. Like the pig enzyme (Nakajin *et al.*, 1984), the bovine P450c17 has a higher affinity for pregnenolone ( 0.2 ± 0.008 μM) than for progesterone (1.4 ± 0.3 μM).

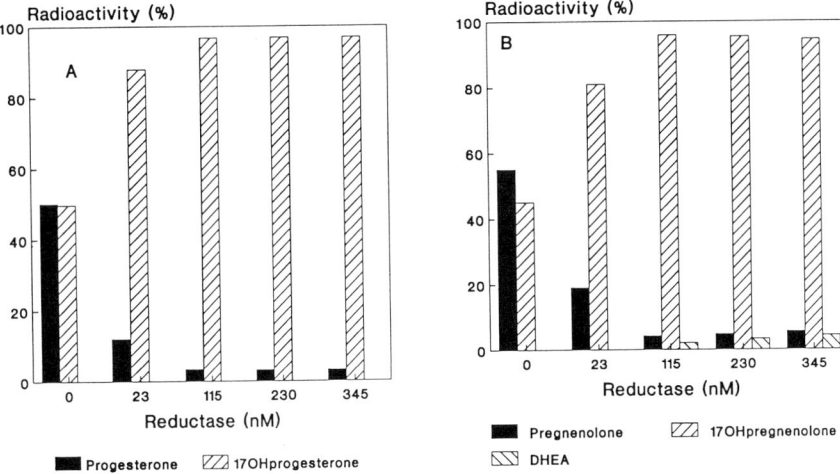

**Fig. 3.** 17α-hydroxylase and $C_{17,20}$-lyase activities in a yeast microsomal fraction (W303/pTG10131) as a function of the concentration of rat liver P450red. Active P450c17 in the test was estimated at less than 5 nM. Incubation time was 30 min at 37°C with [³H]-progesterone (A) or [³H]-pregnenolone (B). 17OH, 17α-hydroxyl; DHEA, dehydroepiandrosterone.

*In vivo* conversion of progesterone was analyzed with growing cells in a medium supplemented with 100 mg/l of progesterone. Figure 4 shows that progesterone is rapidly converted into 17α-hydroxyprogesterone. A side product appears after 30 h of incubation which has been identified as 17α-hydroxy,20α-dihydroprogesterone (data not shown). This compound is formed independent of the presence of the P450c17 cDNA indicating the ability of yeast to reduce 17α-hydroxyprogesterone. This reducing activity was detected in all the yeast strains tested.

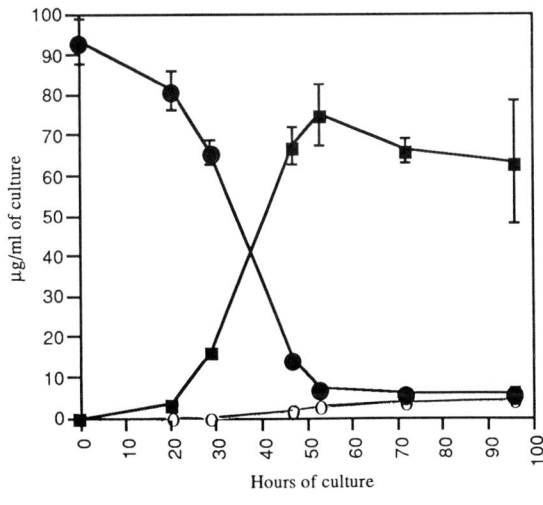

**Fig. 4.** *In vivo* conversion of progesterone to 17α-hydroxyprogesterone by W303/pTG10131. Substrate and product were extracted with dichloromethane from the culture broth and were quantitated by subsequent RP-HPLC analysis. - ● - , progesterone [μg/l]; - ■ - , 17α-hydroxyprogesterone [μg/ml]; - ○ - , 17α-hydroxy, 20α-dihydroprogesterone [μg/ml].

## CONCLUSIONS

Expression in yeast of a bovine cDNA encoding P450c17 allows the production of a protein of 57 kDa apparent MW which crossreacts with an antiserum raised against bovine P450c17. The protein localizes to membranes enriched in a microsomal preparation from a yeast transformant. A typical CO difference spectrum indicates the assembly of the protein in a functional form. Spectral analysis data are corroborated by detection of *in vitro* conversion activity of isolated microsomes and *in vivo* conversion of transformed yeast cells. Whereas no $C_{17,20}$-lyase could be detected *in vivo*, a large excess of rat P450red added induces a weak lyase activity *in vitro*. In the *in vivo* conversion experiment an endogenous yeast activity was detected which leads to reduction of the ketone in position 20 of 17α-hydroxyprogesterone. This reducing activity is dependent on the presence of a hydroxyl group in position 17, as no reduction occurred when untransformed cells were incubated with progesterone.

## ACKNOWLEDGEMENTS

The authors are grateful to M. Janes and H. Kolbe for fruitful discussion, to L. Laruelle, C. Ledoux and M. Dietrich for expert technical assistance, and to Gist-brocades for providing the cDNA and start-up technology for the project. The project was supported by the French Ministery of Research and Technology (grant No. 92T0426).

## REFERENCES

Barnes, H.J., Arlotto, M.P., & Waterman M.R. (1991): Expression and enzymatic actvity of recombinant cytochrome P450 17α-hydroxylase in *Escherichia coli*. *Proc. Natl. Acad. Sci. USA* 88, 5597-5601.

Clark, B.J., & Waterman, M.R. (1992): Functional expression of bovine 17α-hydroxylase in COS 1 cells is dependent upon the presence of an amino-terminal signal anchor sequence. *J. Biol. Chem.* 267, 24568-24574.

Fevold, H.R., Lorence, M.C., McCarthy, J.L., Trant, J.M., Kagimoto, M., Waterman, M. R., & Mason, J.I. (1989): Rat $P450_{17\alpha}$ from testis: characterization of a full-length cDNA encoding a unique steroid hydroxylase capable of catalyzing both $\Delta^4$- and $\Delta^5$-steroid-17,20-lyase reactions. *Mol. Endocrinol.* 3, 968-975.

Miller, W.L. (1988): Molecular biology of steroid hormone synthesis. *Endocrine Rev.* 9, 295-318.

Nakajin, S., Shinoda, M., Haniu, M., Shively, J.E., & Hall, P. (1984): C21 steroid side chain cleavage enzyme from porcine adrenal microsomes. *J. Biol. Chem.* 259, 3971-3976.

Sakaki, T., Shibata, M., Yabusaki, Y., Murakami, H., & Ohkawa, H. (1989): Expression of bovine P450c17 cDNA in *Saccharomyces cerevisae*. *DNA* 8, 409-418.

Yanagibashi, K., & Hall, P.F. (1986): Role of electron transport in the regulation of the lyase activity of C-21 side-chain cleavage from porcine adrenal and testicular microsomes. *J. Biol. Chem.*, 261, 8429-8433.

Zuber, M.X., John, M.E., Okamura, T., Simpson, E. R., & Waterman, M.R. (1986): Bovine adrenocortical cytochrome $P450_{17\alpha}$: regulation of gene expression by ACTH and elucidation of primary sequence. *J. Biol. Chem.* 261, 2475-2482.

# Conformational dynamics of cytochrome P450$_{cam}$ as monitored by photoacoustic calorimetry

Carmelo Di Primo[1], Gaston Hui Bon Hoa[1], Eric Deprez[1], Pierre Douzou[1], Stephen G. Sligar[2]

[1]INSERM-INRA U.310, 13, rue Pierre-et-Marie-Curie, 75005 Paris, France. [2]Department of Biochemistry, University of Illinois, 600 S Mathews, Urbana, IL 61801, USA

*Introduction.* One of the major challenge in cytochrome P-450$_{cam}$ catalysis is to understand how the substrate accesses the buried active site. The three-dimensional X-ray structure (Poulos et al., 1985, 1986, 1987) does not show an apparent access channel for the substrate molecule but associated thermal factors suggest that the protein probably undergoes large conformational changes during binding of metabolizable substrates.

The recent development of the time-resolved photoacoustic calorimetry (Rothberg et al., 1982; Peters et al., 1991) allows monitoring protein dynamics such as those that might be coupled to ligand or substrate egress from the active site. Small changes in volume and enthalpy can be detected in the nanosecond to microsecond time scale, in aqueous medium.

These motions were triggered by the photoinduced dissociation of CO from the ferrous heme. We present in this communication the quantitation of structural changes of cytochrome P-450$_{cam}$ in the presence and absence of camphor.

*Materials and Methods.* The instrument developed in our laboratory (Di Primo et al., 1993) uses a Laser Science nitrogen-pumped dye laser operating at 536 nm with a 3-ns pulse width at a 2.5 Hz repetition rate. An output energy of 8 µJ was chosen to be within the linear relationship between the acoustic signal of the CO-bound protein and the energy of the laser. The acoustic waves generated by the pulses were detected by a 1-MHz piezoelectric transducer. Its signal was amplified then recorded by a Lecroy 940 oscilloscope.

The instrument was calibrated by using a compound which converts the light absorbed only as heat deposited in the medium (no conformational changes). The oxidized form of cytochrome P-450$_{cam}$ can be used for the calibration since no conformational changes are expected upon excitation by the laser.

Our analysis of the photoacoustic signals follows the pioneering works of the Peters group (Peters et al., 1991). The data obtained in the presence of camphor were analyzed assuming a two-step model where the first step is faster than the time response of the instrument and the second step slow enough to be temporally resolved.

*Results and Discussion.* The photoacoustic spectra of cytochrome P-450$_{cam}$, at 17°C, for different binding states of the heme pocket are presented in Figure 1. The oxidized form of the

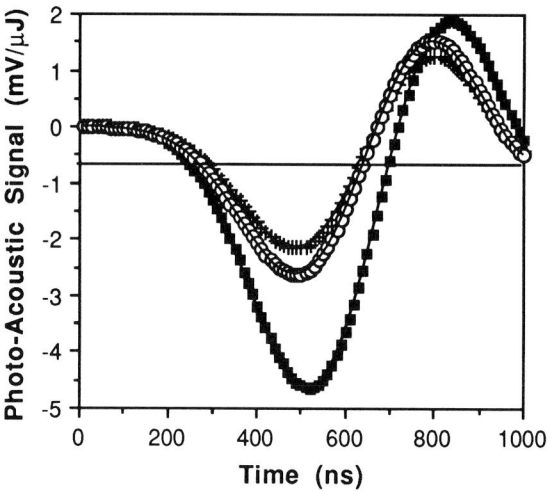

Fig. 1 : Photoacoutic spectra of cytochrome P-450$_{cam}$ at 17°C. (+) Cytochrome P-450$_{cam}$ oxidized (calibration compound), (o) ferrous cytochrome P-450$_{cam}$-CO, (■) ferrous cytochrome P-450$_{cam}$-CO-camphor.

protein was used to calibrate the instrument. The signal corresponds to the fastest response detectable as only heat is deposited in the medium. The theory of photoacoustic calorimetry predicts that if chemistry occurs faster than the time scale of the instrument response, the reaction will not be temporally resolved but the instrument will still be sensitive to changes in enthalpy and volume within the system.

The amplitude of the signal for ferrous CO-bound cytochrome P-450$_{cam}$ substrate-free is higher than that generated with the calibration compound. The signal is not time shifted. This clearly means that all photochemical events in the substrate-free protein are occuring faster than the time scale of the instrument response. In this case, no kinetic information is available. The volume and enthalpy changes deduced from the data are 1.8 ± 0.1 ml/mol and 2.2 ± 0.9 kcal/mol, respectively.

When camphor is present, the signal increases in amplitude and is significantly shifted in time. The parameters deduced from the experimental data, analyzed according to a two-step model are $\Delta V_1$ = 4.4 ± 0.4 ml/mol, $\Delta H_1$ = 12.4 ±3.1 kcal/mol and $\Delta V_2$ = 5.9 ± 0.4 ml/mol, $\Delta H_2$ = -27.9 ± 3.2 kcal/mol for the first and second step, respectively. The lifetime of the intermediate formed is 130 ns at 17°C.

The exact meaning of the thermodynamic parameter is not yet clear. Shimada and co-workers (Shimada et al., 1979) who first reported the photodissociation of CO from cytochrome P-450$_{cam}$ found that 6% and 100% of the CO photodissociated diffused out of the protein for the substrate-free and -bound ferrous CO-bound complexes, respectively, in good agreement with our determination. The amplitude of the signal generated without camphor is half the amplitude of the signal obtained with the substrate. This result is surprising considering that only 6% of the CO molecules photodissociated would leave the protein. Clearly other contributions are being measured.

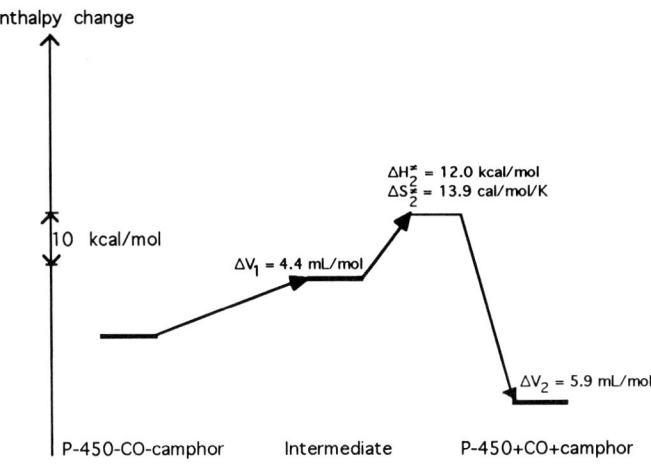

Fig. 2 : Energy landscape of the photoinduced dissociation of CO from cytochrome P-450$_{cam}$ in the presence of camphor.

The question is to understand why in the presence of camphor the signal is time shifted relative to the calibration compound signal. In our experiments an intermediate is formed within the time scale of the instrument with 130-ns lifetime and larger changes in enthalpy and volume. Does this reflect a migration process of CO and/or camphor through the protein matrix or a relaxation process after CO and camphor have left the active site as also suggested by Raman studies (Wells et al., 1992)? The slow and large dynamics expected to open a channel for the substrate could then explain the origin of the phase shift and the larger amplitude of the signals.

In Figure 2 we show an energy landscape for the tertiary complex ferrous CO-cytochrome P-450$_{cam}$ substrate-bound as derived from the experimental data. In summary increase in the amplitude of the signals for ferrous CO-bound cytochrome P-450$_{cam}$ substrate-free and -bound compared to the calibration compound involves structural reorganization of the protein during the photoinduced dissociation of CO. Camphor clearly affects the dynamics of the protein. A reaction not temporally resolved becomes slow enough to be detected as illustrated by the phase shift of the photoacoustic signal.

*References.*
Di Primo, C., Hui Bon Hoa, G., Deprez, E., Douzou, P. & Sligar, S.G. (1993) : Conformational dynamics of cytochrome P-450$_{cam}$ as monitored by photoacoustic calorimetry. *Biochemistry* 32, 3671-3676.
Peters, K.S., Watson, T. & Marr, K. (1991) : Time-resolved photoacoustic calorimetry. A study of myoglobin and rhodopsin. *Ann. Rev. Biophys. Biophys. Chem.* 20, 343-362.
Poulos, T.A., Finzel, B.C. & Howard, A.J. (1986) : Crystal structure of substrate-free *Pseudomonas putida* cytochrome P-450$_{cam}$. *Biochemistry* 25, 5314-5322.
Poulos, T.A., Finzel, B.C. & Howard, A.J. (1987) : High resolution crystal structure of cytochrome P-450$_{cam}$. *J. Mol. Biol.* 195, 687-700.
Poulos, T.A., Finzel, B.C., Gunsalus, I.C., Wagner, G.C. & Kraut, J. (1985) : The 2.6

Angstrom crystal structure of *Pseudomonas putida* cytochrome P-450. *J. Biol. Chem.* 260, 16122-16130.

Rothberg, L.J., Simon, J.D., Berstein, M. & Peters, K.S (1983) : Pulsed laser photo-acoustic calorimetry of metastable species. *J. Am. Chem. Soc.* 105, 3464-3468.

Shimada, H., Iizuka, T., Ueno, R. & Ishimura, Y. (1979) : Correlation between the quantum yields of photodissociation and CO stretching frenquencies of carbon monoxide hemeproteins. *FEBS Letters* 98, 290-294.

Wells, A.V., Pusheng, L., Champion, P.M., Martinis, S.A. & Sligar, S.G. (1990) : Resonance raman investigations of *E. coli*-expressed *P. putida* cytochrome P-450 and P-420. *Biochemistry* 31, 4384-4393.

# The molecular basis of aromatase deficiency in an adult female due to mutations in the CYP19 gene

Carolyn R. Fisher[1], Yuji Ito[1], Felix A. Conte[2], Melvin M. Grumbach[2], Evan R. Simpson[1]

[1]The Green Center for Reproductive Biology Sciences, UT Southwestern, Medical Center, 5323 Harry Hines Blvd, Dallas, TX 75235-9051. [2]Department of Pediatrics, University of California at San Francisco, San Francisco, CA 94143, USA

The biosynthesis of estrogens from androgens is catalyzed by an enzyme complex known as aromatase, whose activity results in aromatization of the A ring of androgens to form the phenolic A ring characteristic of estrogens, with the concomitant loss of the $C_{19}$ angular methyl group (Goto & Fishman, 1977; Cole & Robinson, 1988). This enzyme complex is located in the endoplasmic reticulum of estrogen-producing cells and consists of two components. The first is a form of cytochrome P450 named aromatase cytochrome P450(P450arom), the product of the CYP19 gene. The second component, NADPH-cytochrome P450 reductase, transfers reducing equivalents to P450arom. A single species of P450arom catalyzes the conversion of the three substrates, testosterone, androstenedione, and 16α-hydroxydehydroisoandrosterone to the respective $C_{18}$ estrogens (Corbin, et al., 1988).

In the human, CYP19 is expressed in a variety of cells and tissues including ovarian granulosa cells, testicular Sertoli and Leydig cells, placenta, adipose, and brain (Mendelson, et al., 1989). We have presented evidence that the tissue-specific expression of the CYP19 gene is regulated, in part, by the use of tissue-specific promoters through mechanisms involving alternative splicing of untranslated 5' exons (Mahendroo, et al., 1991). These untranslated exons are spliced into a common 3'-splice junction upstream of the start of translation, thus the coding region is unaffected and the protein product in each tissue site is identical (Mahendroo, et al., 1993). The overall length of the gene is at least 75 kb, and it is composed of 9 coding exons and at least 4 untranslated 5' exons.

In spite of the size and complexity of the CYP19 gene, only one definitively characterized case of placental aromatase deficiency has been reported, although deficiencies of most of the other steroidogenic forms of cytochrome P450 have been well characterized. Shozu, et al. (1991) reported a case of female pseudohermaphrodism secondary to placental aromatase deficiency in a Japanese infant, and biochemical and molecular genetic studies were conducted by Harada, et al. (1992). Examination of genomic DNA from the patient revealed that the consensus 5'-splice acceptor sequence was mutated from GT to GC resulting in the use of a cryptic splice acceptor sequence further upstream in intron 6. This resulted in the incorporation of 87 bases from intron 6, which were translated into an additional 29 amino acids present in the mature polypeptide (Harada, et al., 1992).

Our patient is an 18 year old 46, XX female who presented at birth with mild virilization. At age 14 she exhibited primary amenorrhea, failure of breast development, and her plasma estrogen level was extremely low (E2 < 5 pg/ml). Plasma FSH and LH levels were markedly elevated. Sonography and MRI revealed multiple bilateral 4-6 cm ovarian cysts. Genomic DNA was extracted from the patient's ovarian fibroblasts and exons II through X encompassing the coding region of the CYP19 gene were amplified and sequenced directly. Sequencing revealed two single base mutations in exon X at bp1303(C→T) and bp1310(G→A) (Fig. 1). Subcloning and sequencing of each allele confirmed that the patient is a compound heterozygote for these mutations (data not shown). Further restriction enzyme digestion and direct sequencing experiments demonstrated that the patient inherited the C1303→T mutation from her mother (Fig. 1). Her father is deceased and there are no other siblings.

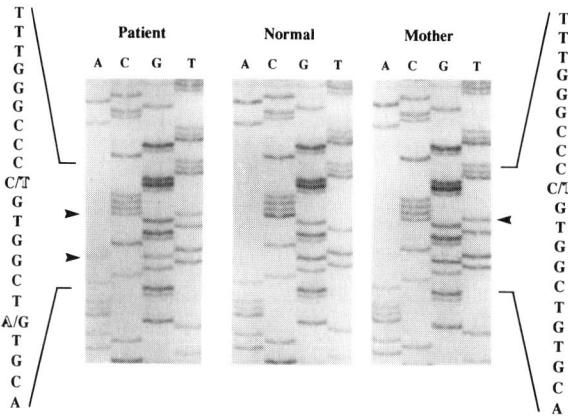

Fig. 1. Nucleotide sequences of the region of exon X of the CYP19 gene in the patient showing single base changes at bp1303 (C→T) and bp1310(G→A). The patient is heterozygous for these mutations as judged by the fact that bands corresponding to both C and T and G and A could be detected in the sense strand at bp1303 and 1310, respectively. In the normal CYP19 gene, only C and G are present at these positions whereas the maternal DNA has only a single base change at bp1303(C→T).

These mutations result in codon changes of R435C and C437Y, respectively. Transient expression experiments showed that the R435C mutant protein had approximately 1.1% of the activity of wild-type aromatase, whereas the C437Y protein was totally inactive (Table 1). Western blot analysis of total cellular protein showed similar levels of protein expression compared to normal cDNA (data not shown).

Table 1. Transient expression of human cytochrome P450arom mutant or control cDNA in COS-1 cells.

| cDNA construct | Activity (pmol/mg protein/2 hrs) |
|---|---|
| Wild-type | 127.0 ± 8.2 (100%) |
| R435C | 1.4 ± 0.1 (1.1%) |
| C437Y | 0.05 ± 0.01 (0%) |
| Vector | 0.02 ± 0.05 |

pCMV5 plasmids containing wild-type or mutant cDNA inserts or no insert were transfected into COS-1 cells by electroporation, and aromatase activity was assayed by the incorporation of tritium into [$^3$H] water from [$1\beta$-$^3$H] androstenedione. The data presented are mean ± SD of triplicate results.

These missense mutations reside within the heme-binding region of the protein. Since cysteine437 is the conserved cysteine which comprises the 5th coordinating ligand of the heme iron, the gene product of the C437Y allele would be expected to be inactive. Arginine435 is also a highly conserved residue within the heme-binding region among mammalian P450s with the exception of cholesterol $7\alpha$-hydroxylase cytochrome P450 in which it is a threonine (Noshiro & Okida, 1990). Only one other documented mutation (R448H) of this arginine has been reported, in $11\beta$-hydroxylase deficiency. Using a transient transfection assay, Curnow, et al. (1993) confirmed that the R448H mutation reduces $11\beta$-hydroxylase activity to undetectable levels. Therefore, a mutation at this position in P450arom would also be expected to result in substantial loss of enzymatic activity, which indeed was the case.

In spite of the possibility that some cases of aromatase deficiency may have been overlooked, it appears to be a rare condition. Previously it was believed that the complete absence of P450arom activity was potentially lethal, perhaps because of a need for estrogens at the time of implantation of the blastocyst. However, in spite of this, it is now apparent that individuals with mutations in P450arom do exist, and based on the structure of the gene it is possible to predict not only defects due to mutations in the coding region, but also tissue-specific defects caused by mutations in specific promoter regions as well as in splice junctions.

In conclusion, we identified two novel mutations (R435C and C437Y) in the CYP19 gene responsible for aromatase deficiency in an adult female with sexual infantilism, primary amenorrhea, ambiguous external genitalia, and polycystic ovaries. Moreover, we confirmed that mutation of the conserved cysteine, C437Y, totally inactivates P450arom. On the other hand the mutation, R435C, gives rise to a protein with only marginal activity. We have also demonstrated the heritability of one of these two alleles. Therefore, we suggest that some cases of infertility in women may be accounted for by mutations in the CYP19 gene.

# REFERENCES

Cole, P.A., and Robinson, C.H.(1988) A peroxide model reaction for placental aromatase. *J. Am. Chem. Soc.* 110:1284-1285.

Corbin, C.J., Graham-Lorence, S., McPhaul, M., Mason, J.I., Mendelson, C.R., and Simpson, E.R. (1988) Isolation of a full length cDNA insert encoding human aromatase system cytochrome P450 and its expression in non-steroidogenic cells. *Proc. Nat. Acad. Sci. USA* 85:8948-8952.

Curnow, K.M., Slutsker, L., Vitek, J., Milkelson, M., Cole, T., Speicer, P.W., New, M.I., White, P.C., and Pascoe, L. (1993) Mutations in the CYP11B1 gene causing congenital adrenal hyperplasia and hypertension cluster in Exons 6, 7, and 8. *Proc. Nat. Acad. Sci. USA* 90:4552-4556.

Goto, J. and Fishman, J. (1977) Participation of a non-enzymatic transformation in the biosynthesis of estrogens from androgens. *Science* 195:80-81.

Harada, N., Ogawa, H., Shozu, M., Yamada, K., Suhara, K., Nishida, E., and Takagi, Y. (1992) Biochemical and molecular genetic analyses on placental aromatase (P450arom) deficiency. *J. Biol. Chem.* 267:4781-4785.

Harada, N., Ogawa, H., Shozu, M., and Yamada, K. (1992) Genetic studies to characterize the origin of the mutation in placental aromatase deficiency. *Am. J. Human. Gen.* 51:666-672.

Mahendroo, M.S., Means, G.D., Mendelson, C.R., and Simpson, E.R. (1991) Tissue-specific expression of human P450arom: the promoter responsible for expression in adipose is different from that utilized in placenta. *J. Biol. Chem.* 266:11276-11281.

Mahendroo, M.S., Mendelson, C.R., and Simpson, E.R. (1993) Tissue-specific and hormonally controlled alternative promoters regulate aromatase cytochrome P450 gene expression in human adipose tissue. *J. Biol. Chem.* 268:19463-19470.

Mendelson, C.R., Corbin, C.J., Means, G.D., Mahendroo, M., Graham-Lorence, S. Merril, J.C., and Simpson, E.R. (1989) The aromatase cytochrome P450 gene and its regulation. *Serono Symposia Publications* 21:5-19.

Noshiro, M. and Okida, K. (1990) Molecular cloning and sequence analysis of cDNA encoding human cholesterol 7α-hydroxylase. *FEBS Lett.* 268:137-140.

Shozu, M., Akasofu, K., Harada, T., and Kubota, Y. (1991) A new cause of female pseudohermaphroditism: Placental aromatase deficiency. *J. Clin. Endocrinol. Metab.* 72:560-566.

# The loss of cytochromes P450 (CYPs) in rat liver cell culture is triggered during hepatocyte isolation and again during the first 4 hours of culture

Charles R.W. Padgham, Xiu-Jun Wang, Alan J. Paine

DH Department of Toxicology, St Bartholomew's Hospital Medical College, London EC1A 7ED, UK

## Abstract

When hepatocytes are placed into culture total cytochrome P450 and CYP mRNAs decline rapidly. In order to understand and study this loss the levels of CYP mRNAs and the expression of two liver specific transcription factor mRNAs have been quantified during the first 24 hours of culture. Hepatocytes were isolated from 250-290gm male CD rats (anaesthetized with 100 mg pentobarbitone/kg:i.p.) followed by perfusion of the liver with 0.025% (w/v) collagenase H (Padgham & Paine 1993). Comparsion of the abundance of CYP mRNAs in the caudate lobe of the liver prior to addition of collagenase to that found in the freshly isolated hepatocytes demonstrated that the relatively brief (1.5hr) hepatocyte isolation and washing procedure caused small but statistically significant ($p<0.01$) increases in the abundance of CYP 1A2, 2B1/2, 3A1/2 and 4A1 mRNAs. These are probably due to the use of pentobarbitone as the anaesthetic. However such a mechanism may not account for the loss of the CYP2C11 mRNA.

## INTRODUCTION

Liver cell culture is a popular model system to study mechanisms of hepatotoxicity under defined conditions *in vitro*. However a major limitation to the use of hepatocyte cultures in the study of xenobiotic metabolism/toxicity is their rapid loss of cytochrome P450 (CYP) content. This is due to the inability to synthesise the cytochrome and may be related to the loss of specific CYP mRNAs (Padgham & Paine 1993). The stimuli that underly these losses are unknown. In order to determine the effects of both hepatocyte culture and the isolation procedure upon cytochrome P450 expression we have quantified CYP, C/EBPα / β and HNF1 mRNAs expression during the isolation procedure and during the first few hours of culture.

## METHODS

Rat hepatocytes were isolated from 250-290g male CD rats (anaesthetized with 100 mg pentobarbitone/kg:i.p.). Prior to collagenase perfusion the caudate liver lobe was ligated and removed, its total RNA being isolated with RNAzol B (Biogenesis). Perfusion of the liver with 0.025% (w/v) collagenase H was performed and hepatocytes were isolated (Padgham & Paine 1993). Isolated hepatocytes were cultured in Williams E medium containing $10^{-6}$M insulin, $10^{-4}$M hydrocortisone, 60μg/ml gentamicin and maintained for upto 24 hours at 37°C under these conditions (Padgham & Paine 1993). At hourly intervals for the first six hours and after 24 hours of culture total RNA was isolated from the hepatocytes and analysed by northern blotting (Padgham & Paine 1993). Labeled cDNA and oligonucleotide probes to CYP 1A1, 1A2, 2A1/2, 2B1/2, 2C11, 2E1, 3A1/2, 4A1 (Padgham & Paine 1993) and the liver transcription factors C/EBPα (Landschulz et al 1988), C/EBPβ (Descombes et al 1990) and HNF1 (Chouard et al 1990) were used to quantify the abundance of their corresponding mRNAs by filter hybridisation (Padgham & Paine 1993). The abundance of mRNA in each sample was normalised to to the total amount of ethidium bromide stainable RNA (Padgham & Paine 1993).

## RESULTS AND DISCUSION

**Alteration of CYP, C/EBP and HNF1 mRNA levels during the hepatocyte isolation procedure.**

During the perfusion and isolation of hepatocytes small changes occurred in the expression of certain CYP mRNAs as evidenced by the values for the liver lobe being different from that determined in the isolated cell suspension used for the subsequent culture of hepatocytes. CYP 1A2, 2B1/2, 2E1, 3A1 and 4A1 mRNAs are elevated during the cell isolation procedure by about 2-3 fold Table 1. The other CYP mRNAs remain unchanged or decrease. The major constitutive CYP *in vivo*, CYP2C11 decreases rapidly during hepatocyte isolation. Expression of liver specific transcription factors C/EBP α,β and HNF1 were unaltered during isolation. These findings suggest that the expression of these CYPs are altered during the relatively brief (1.5hr) isolation procedure as well as indicating that the individual CYPs studied respond differently. These effects could be due to the loss of specific regulatory factors present in either the intact liver or blood (e.g. glucocorticoids, growth hormone, transcription factors) or to the presence/induction of such factors in the perfusate/medium used to isolate the hepatocytes. As the expression of three liver specific transcription factors appear to be unaffected by this isolation procedure it seems unlikely that they are involved in the changes of CYP expression. However other mechanisms may act upon these factors (e.g. phosphorylation/dephosphorylation, thiol group oxidation/reduction, interactions with other factors) increasing or decreasing their activities, thus altering CYP expression without any effect upon their own expression. It is also possible that induction by the anaesthetic used for surgery namely pentobarbitone may be be the cause of these effects. Pentobarbitone being a barbituate like phenobarbitone (PB) may well induce the PB inducible CYPs i.e. CYP2Bs and CYP3A1 leading to these effects. It is of interest to note that CYP1A2, 2B and 3A1 mRNAs are induced by glucocorticoids and that the medium used during washing contains hydrocortisone. Evidently more studies as to what is happening during cell isolation are required, as these early effects seemingly influence CYP expression throughout the culture period and therefore play a role in CYP expression in culture and in particular in the loss of the constitutive CYP2C11 which comprises about 50% of the total cytochrome P450 *in vivo* (Steward *et al* 1985).

**Table 1 Alteration of CYP, C/EBPα/β and HNF1 mRNA levels during the hepatocyte isolation procedure.**

| CYP | Fold change in mRNA abundance in freshly isolated hepatocytes relative to the value determined in the intact caudate liver lobe. |
|---|---|
| 1A1 | n.d. |
| 1A2 | 2.4 ± 0.9* (n=9) |
| 2A subfamily | 1.1 |
| 2B1/2 | 2.5 ± 0.9* (n=10) |
| 2C11 | 0.4 ± 0.1* (n=4) |
| 2E1 | 1.8 |
| 3A1/2 | 3.5 ± 1.1* (n=10) |
| 4A1 | 3.3 ± 0.2* (n=3) |
| C/EBPα | 1.1 ± 0.1 (n=4) |
| C/EBPβ | 1.1 |
| HNF1 | 1.0 ± 0.3 (n=4) |

Fold change in CYP, C/EBP α / β and HNF1 mRNA abundance in freshly isolated hepatocytes relative to the value determined in the intact caudate liver lobe. n.d. = not detectable, * = significantly different ($p<0.01$).

**Effect of hepatocyte culture on the abundance of CYP, C/EBPα / β and HNF1 mRNAs.**

In hepatocytes cultured upto 24 hours there is a general decline in CYP, C/EBP and HNF1 mRNAs apart from CYP 1A1 and 3A1/2 which increase Table 2. In the case of CYP1A1 the levels rise from being undetectable in isolated hepatocytes prior to culture becoming detectable after 1 hour of culture. Further induction of CYP1A1 occurs, the abundance being elevated 8 fold by 24 hours of culture. As previously reported this induction of CYP1A1 is transcriptional in origin (Padgham & Paine 1993), CYP1A1 not being constitutively expressed in rat hepatocytes (Gonzalez 1990) and the presence of a glucocorticoid is necessary (data not shown). CYP3A1 mRNA in comparison is induced by nearly 2 fold under control conditions throughout the period of culture above its constitutive levels. As this CYP is glucocorticoid inducible it appears likely that this elevation is due to glucocorticoid induction and in fact if hydrocortisone is omitted from the culture medium this induction is decreased (data not shown).

These results suggest that there is a general decline in CYP expression except for CYPs 1A1 and 3A1/2 which increase. The necessity of hydrocortisone for CYP1A1 and 3A1/2 induction illustrates the importance of media constituents in CYP expression and the differential regulation of CYPs reflecting their different regulatory mechanisms. CYPs are downregulated at different rates and they may thus be under control of different mechanisms. A relationship between the loss of CYP expression and loss of expression of liver specific transcription factors is implied, but whether this loss is the cause or a parallel effect remains to be tested. Transfection of cultured hepatocytes with expression vectors containing the complete coding sequence of these factors should indicate if in fact they are the effecting agents involved in CYP loss.

**Table 2 Effect of hepatocyte culture on the abundance of CYP, C/EBP and HNF1 mRNAs.**

| CYP | Fold change in mRNA abundance in hepatocytes relative to the value determined in freshly isolated hepatocytes. Time in culture (hours):- | | | | | |
|---|---|---|---|---|---|---|
|  | 1 | 2 | 3 | 4 | 6 | 24 |
| 1A1 | 1.0 | 4.1 | 6.6 | 6.0 | - | 8.0 |
| 1A2 | 0.8 | 0.7 | 0.7 | 0.2 | 0.2 | - |
| 2B1/2 | 0.7 | 0.8 | 0.7 | 0.1 | 0.1 | 0.1 |
| 2C11 | 0.4 | 0.4 | 0.5 | 0.04 | 0.02 | n.d. |
| 2E1 | 1.1 | 0.7 | 0.9 | 0.4 | n.d. | n.d. |
| 3A1/2 | 0.4 | 0.7 | 0.9 | 1.5 | 1.5 | 2.0 |
| 4A1 | 0.6 | 0.5 | 0.4 | 0.3 | 0.1 | 0.1 |
| C/EBPα | 0.8 | 0.7 | 0.7 | 0.4 | 0.2 | 0.1 |
| C/EBPβ | 0.6 | 0.8 | 0.9 | 0.9 | 0.6 | 0.5 |
| HNF1 | 0.7 | 0.7 | 0.6 | 0.6 | 0.2 | 0.5 |

n.d. = not detectable

Fold change in CYP, C/EBP α/β and HNF1 mRNA abundance in hepatocytes cultured for 1-24 hours relative to the value determined in freshly isolated hepatocytes.

In conclusion these studies show that the isolation procedure alters CYP mRNA expression leading to an increase of CYP1A2, CYP2B, CYP2E1, CYP3A, CYP4A1 mRNAs and a decrease in CYP2C11 mRNA, whereas the expression of the liver specific transcription factors C/EBPα/β, HNF1 mRNA is unaltered. In culture the expression of CYP mRNAs is differentially regulated, however most rapidly decrease, whilst CYP1A1 and CYP3A1 mRNAs are induced, hydrocortisone being involved in this induction. The expression of the liver specific transcription factors C/EBPs and HNF1 decline in parallel with CYPs and may play a role in the loss of CYPs in culture.

## REFERENCES

Padgham C.R.W. and Paine A.J. Altered expression of cytochrome P450 mRNAs, and potentially of other transcripts encoding key hepatic functions, are triggered during the isolation of rat hepatocytes. (1993) *Biochem. J.* **289**, 621-624.

Landschulz W.H., Johnson P.F., Adashi E.Y., Graves B.J. and Mcknight S. Isolation of a recombinant copy of the gene encoding C/EBP. (1988) *Genes and Development* **2**, 786-800.

Descombes P., Chojkier M., Lichtsteiner S., Favey E. and Schibler U. LAP a novel member of the C/EBP gene family, encodes a liver enriched transcriptional activator protein. (1990) *Genes and Development* **4**, 1541-1551.

Chouard T., Blumenfeld M., Bach I., Vandekerckhove J., Cereghini S. and Yaniv M. A distal dimerisation domain is essential for DNA binding by the atypical HNF1 homeodomain. (1990) *Nucleic Acid Research* **18**, 5853-5863.

Steward A.R., Dannan G.A., Guzelian P.S. and Guengerich F.P. Changes in the concentration of seven forms of cytochrome P450 in primary cultures of adult rat hepatocytes. (1985) *Molecular Pharmacology* **27**, 125-132.

Gonzalez FJ. (1990) Molecular genetics of the P-450 superfamily. *Pharmac. Ther.* **45**, 1-38.


**Acknowledgement.** Charles Padgham is supported by a grant from the Animals (Scientific Procedures) Committee of the U.K. Home Office.

# Conformational states and substates of cytochrome P450$_{cam}$ – insight in protein dynamics and folding

Christiane Jung, Wolfgang Pfeil, Karla Köpke, Heike Schulze, Otto Ristau

Max-Delbrück-Centre for Molecular Medicine, 13122 Berlin-Buch, FRG

## INTRODUCTION

The crystal structure of cytochrome P450$_{cam}$ does not show dramatic differences in the overall conformation between the substrate-bound and the substrate-free protein (Poulos et al., 1987). However, the overall stability and the dynamics of the active site structure is strongly influenced by substrate binding. To understand this phenomenon we analyzed the thermal unfolding of the protein and local conformational changes in the heme pocket.

## MATERIALS & METHODS

Cytochrome P450$_{cam}$ from Pseudomonas putida (CYP101, P450$_{cam}$), expressed in Escherichia coli, was purified as described by Jung et al., 1992. Scanning calorimetric measurements were performed on the MicroCAL MC-2D scanning calorimeter (MicroCal Inc., Northampton, MA) with the DA-2 data acquisition system with a scan rate of 1 K/min. The protein concentration ranged from 1.2 to 5 mg/ml. Thermal unfolding was also followed by optical spectroscopy in the Soret band with a heat scan rate of 0.5 K/min.
Local conformational changes in the heme pocket were monitored using the stretching mode of the carbon monoxide ligand in reduced P450$_{cam}$ on the Mattson Sirius 100 with 2 cm-1 resolution. Temperature and pressure dependent infrared measurements were carried out as described by Scholl (1991). The infrared samples were prepared as given by Jung et al. (1992). All infrared spectra were fitted with Voigt or log-normal distribution line shapes. Thermodynamic parameters were determined from the temperature and pressure dependent population of the CO-conformational substates by a global least square fit analysis with a self-written software package (Jung et al., 1994). For the assignment of structural domains the hydrophobicity scales of Chothia (1976) and of Rose et al.(1985) and the coordinates of the crystal structure (Brookhaven Data Bank, entry code 2CPP, Poulos et al., 1987) were used.

## RESULTS

Thermal unfolding of P450$_{cam}$ is an nonreversible process indicating a complex transition behavior. Aggregation of the unfolded protein can be avoided by blocking the sixth heme iron ligand position by cyanide (Pfeil et al., 1993). Deconvolution of the scanning calorimetric curves of the cyanide complex of oxidized P450$_{cam}$ reveals three cooperative unfolding units with transition temperatures at 41.9°C, 47.8°C and 54.3°C for the substrate-free protein and at 48.5°C, 57.8°C and 61.7°C for the camphor-bound protein (Fig. 1, Tab. 1). The unfolding curve in the absence of substrate followed in the heme absorption shows a very broad transition region between 42 and 47°C, in the presence of camphor, however, a sharp transition at 57.3°C.

For a structural assignment of the unfolding units we searched for hydrophobic cores which are relatively separated from each other by hydrophilic amino acid residues. Three structural domains (Fig. 2) with following amino acid numbers were found:
domain I: residues 1-96 (ß1, A, B, B') + 295-338 (ß3, ß4)
domain II: residues 127-169 (ß5, D, E) + 251-294 (J, K, part of I) + 339-414 (ß5, L)
domain III: residues 97-126 (C) + 170-250 (ß2, F, G, H, part of I).

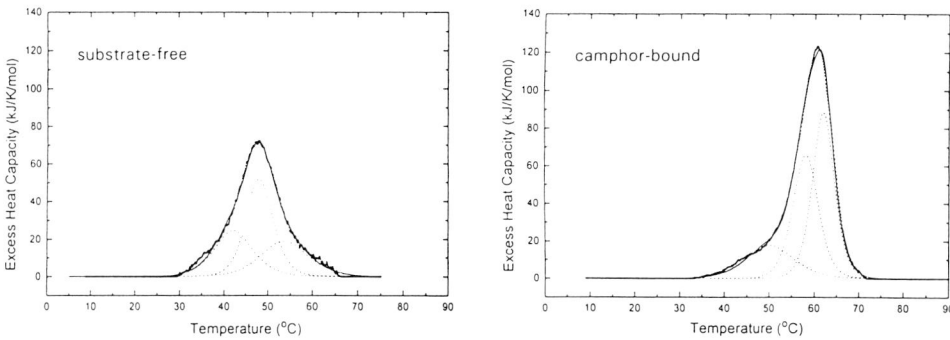

Fig. 1 Deconvoluted scanning calorimetric curves

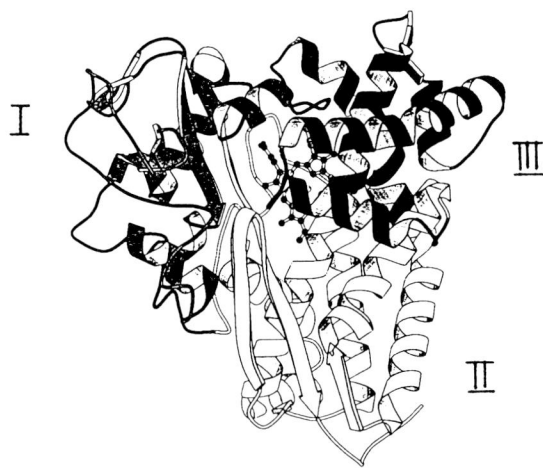

Fig. 2 Three-dimensional structure of Cytochrome P450$_{cam}$ indicating the domains

The heme is located in the contact region of all three domains and does not belong to a particular domain. Domain II and III are not so strongly separated by hydrophilic amino acids as in the interface to domain I. Nevertheless, there are several hydrophilic amino acids even between the protein parts II + III which allows to define an interface crossing the I helix in the region of Asp 251 and Thr252 where an anomalous hydrogen bonding pattern is found in the crystal structure (Poulos et al., 1987). In the substrate-

bound protein camphor is in contact only with domains I and III. The heme pocket is formed by the interfacing amino acid residues of all three domains.
Strong structural disorder in the heme pocket is found for the substrate-free protein reflected in a broad infrared spectrum of several overlapping bands for the CO stretching mode ( bands: 1(1915cm$^{-1}$), 2(1930cm$^{-1}$), 3&4(ensemble of two bands at 1946cm$^{-1}$), 5(1953cm$^{-1}$). Camphor binding reduces the structural disorder, only one infrared band is observed which splits in two bands (1(1942cm$^{-1}$, 2(1935cm$^{-1}$)) at low temperatures or high pressures (Jung & Marlow, 1987, Jung et al., 1992,1994). Tab. 1 summarizes the thermodynamic parameters for the different local conformations of the heme pocket.

## DISCUSSION

Conformational changes in proteins take place within an energetical hierarchy of states (Frauenfelder et al., 1991). The native conformation exists in an equilibrium of several, probably distinct conformational substates whose thermodynamics can be probed by spectroscopic methods in dependence on temperature and pressure. The denatured protein is an ensemble of many disordered conformations. It is not clear whether there is a relation between the flexibility and a high degree of structural substate multiplicity in the native overall conformation and the stability of the protein.

Tab. 1   Thermodynamic parameters for thermal unfolding and conformational substates

| native state | $T_{trs}$ | $\Delta H$ | $\Delta S$ | $\Delta V$ | $\Delta G$ |
|---|---|---|---|---|---|
| substrate-free | | | | | |
| substate   1 | -34.8 | 16.1 | 67.6 | 7.7 | -3.7 |
| 2 | -70.2 | 15.4 | 75.6 | 8.8 | -6.8 |
| 3&4 | -64.2 | 24.9 | 118.9 | 4.8 | -9.9 |
| camphor-bound | | | | | |
| substate   1 | -76.2 | 41.5 | 210.2 | 64.0 | -20.0 |

| unfolding: | substrate-free | | camphor-bound | |
|---|---|---|---|---|
| | $T_{trs}$ | $\Delta H$ | $T_{trs}$ | $\Delta H$ |
| heme absorp. | 42-47 | 185.0 | 57.3 | 353.0 |
| unit   1 | 41.9 | 296.0 | 57.8 | 480.0 |
| 2 | 47.8 | 405.0 | 48.5 | 243.0 |
| 3 | 54.3 | 279.0 | 61.7 | 586.0 |

$\Delta H, \Delta G$ in kJ/mol; $\Delta S$ in J/K mol; $\Delta V$ in ml/mol; $T_{trs}$ in °C; 20°C, 0.1 MPa

Three cooperative unfolding units have been experimentally found by scanning calorimetry for the P450$_{cam}$ cyanide complex and three structural domains can be seen using hydrophobicity criteria. At the first glance one would assign the three unfolding units to the three structural domains. However, the following arguments let suppose another explaination. In substrate-free P450 the transition temperature of unfolding unit 1 at about 42°C matches the value at 41°C observed by optical spectroscopy in the heme absorption region. The heme group, however, is located in the contact region of three domains and is not buried within a particular domain. Therefore, unfolding unit 1 may correspond to loosening the interaction between the three domains. That would explains the effect of camphor binding. In camphor-bound P450 the transition temperature of unfolding unit 1 is increased to 57.8°C with the respective transition temperature at 57.3°C obtained by spectroscopic measurements. Unfolding unit 2 seems to be not strongly influenced by substrate binding. The transition temperatures are very narrow at 47.8°C (free) and 48.5°C (bound) suggesting an assignment to domain II which has no contact to camphor. Unfolding unit 3 is also influenced by camphor binding and may be related in some, still unclear way to domains I and III.

The local heme pocket conformation is strongly disordered in the absence of the substrate. The transitions from the substates 1, 2 and 5 to the substate ensemble 3&4 is entropically driven (Tab. 1) which may indicate hydration changes. The input thermal energy for the temperature range studied [ $R(T_{293°C} - T_{200°C})$, about 0.8 kJ/mol] is only a minimal part of the enthalpy difference between the substates. In the camphor-bound protein the transition from substate 2 to substate 1 is even more endothermic and more strongly entropically induced. The volume differences between the substate conformations in substrate-free P450 are rather small (4-9 ml/mol) which may indicate only rearragement of water molecules in the heme pocket during substate transitions. However, in the presence of camphor the volume difference is rather big (64ml/mol) which let suppose an influx of water into the heme pocket. Di Primo et al. (1992) have shown by pressure-induced $P420_{cam}$ formation that a big inactivation volume is observed if the water content in the heme pocket is low.

Summarizing the results, we conclude that the heme pocket in substrate-free $P450_{cam}$ is rather flexible and able to facilitate rearrangements of internal water molecules which might be necessary for thermal unfolding which starts in the heme pocket. Camphor binding renders the hydration of the heme pocket more difficult which results in a shift of the unfolding unit 1 to higher temperatures.


## ACCKOWLEDGEMENT

C.J. thanks Professor Hans Frauenfelder for the possibility to run the infrared measurements in his laboratory in the Physics Department of the University of Illinois at Urbana-Champaign.


## REFERENCES

Chothia, C. (1976): The nature of the accessible and buried surfaces in proteins. J. Mol. Biol. 105, 1-14.

Di Primo, C., Hui Bon Hoa, G. & Douzou, P. (1992): Heme-pocket-hydration change during the inactivation of cytochrome $P-450_{cam}$ by hydrostatic pressure. Eur. J. Biochem. 209, 583-588.

Jung, C. & Marlow, F. (1987): Dynamic behavior of the active site structure in bacterial cytochrome P450. Studia Biophysica 120, 241-251.

Jung, C., Hui Bon Hoa, G., Schröder, K.-L., Simon, M. & Doucet, J.P. (1992): Substrate analogue induced changes of the CO-stretching mode in the cytochrome $P450_{cam}$-carbon monoxide complex. Biochemistry 31, 12855-12862.

Jung, C., Scholl, R., Ristau, O., Frauenfelder, O. & Sligar, S.G. (1994): in preparation.

Frauenfelder, H., Sligar, S.G. & Wolynes, P.G. (1991): The energy landscapes and motions of proteins. Science 254, 1598-1603.

Pfeil, W., Nölting, B.O. & Jung, C. (1993): Apocytochrome $P450_{cam}$ is a native protein with some intermediate-like properties. Biochemistry 32, 8856-8862.

Poulos, T.L., Finzel, B.C., & Howard, A.J. (1987): High-resolution crystal structure of cytochrome $P-450_{cam}$. J. Mol. Biol. 195, 687-700.

Rose, G.D., Geselowitz, A.R., Glenn, J.L., Lee, R.H. & Zehfus, M.H. (1985): Hydrobicity of amino acid residues in globular proteins. Science 229, 834-838.

Scholl, R. (1991): Relaxation dynamics in heme proteins. PhD thesis, University of Illinois, Urbana.

# Hepatic cytochrome P450 profile in BB rats with spontaneous insulin-dependent diabetes mellitus

Christopher R. Barnett[1], Peter R. Flatt[1], Adrian J. Bone[2], Costas Ioannides[3]

[1]Department of Biological and Biomedical Sciences, University of Ulster, Coleraine, N. Ireland. [2]Department of Pharmacy, University of Brighton, East Sussex. [3]Division of Toxicology, School of Biological Sciences, University of Surrey, Guildford, Surrey, UK

INTRODUCTION

The effect of streptozotocin-induced diabetes mellitus on hepatic cytochrome P450 expression has been extensively studied during recent years (Barnett et al., 1990a,b,c; 1992). These models of insulin-dependent diabetes mellitus (IDDM) compare favourably with the uncontrolled disease in humans with respect to certain pathophysiological parameters. However, the animals do not succumb to the severe hyperketonaemia, hyperglycaemia and electrolyte imbalances observed in spontaneously diabetic individuals (Rifkin and Porte, 1990). The spontaneously diabetic BB wistar rat has many similarities with human IDDM. It occurs in a non-obese laboratory rat derived from a non-inbred Wistar line. Both sexes are affected with the onset of the disease occurring after the time of sexual maturation. Both genetic and immune processes are involved in the aetiology but their precise nature remains to be defined (Like et al., 1982). The clinically overt syndrome occurs over a period of a few hours to a few days and an intense insulitis is observed. This is accompanied by destruction of the pancreatic B cells. Within 7 - 21 days following the appearance of glycosuria the B-cells are completely destroyed. The investigation of hepatic cytochrome P450 modulation in the diabetic BB rat has only been investigated recently. Initial studies indicate that alterations in hepatic cytochrome P450 expression are manifest in these animals as a consequence of diabetes (Bellward et al., 1988; Favreau and Schenkman, 1988). In the present study the effects of duration of diabetes on hepatic cytochrome P450 expression was investigated using the spontaneously diabetic BB rat.

METHODS

Twelve diabetic BB/S and twelve diabetes-resistant BB/S rats were used. Three groups were studied corresponding to duration of diabetes of 4±1 weeks, 12±2 weeks and 24±3 weeks. Each group was age and sex matched with the diabetes-resistant BB/S rats. Diabetic animals received daily insulin injection ranging from 1.81 IU - 3.6 IU insulin calculated on the basis of body weight and glycated haemoglobin. Insulin treatment was withdrawn for five days before the animals were killed by cervical dislocation and the livers excised. Terminal blood samples were obtained and microsomal fractions prepared from the livers of all animals. Microsomal parameters were investigated and selective cytochrome P450 catalyzed reactions employed to investigate cytochrome P450 expression. These included; pentoxyresorufin-O-dealkylase (CYP2B), ethoxyresorufin-O-dealkylase (CYP1A), p-nitrophenol hydroxylation (CYP2E), lauric acid hydroxylation (CYP4A) and erythromycin N-demethylase (CYP3A).

RESULTS

After withdrawal of insulin therapy the diabetic BB rats rapidly developed the characteristic signs of uncontrolled IDDM including polyuria and polydipsia although hyperphagia was not apparent. Analysis of plasma parameters indicated that all diabetic animals were significantly (P<0.001) hyperglycaemic and hyperketonaemic compared to their age matched non-diabetic controls (Table 1.).The levels of microsomal protein and total carbon-monoxide discernible cytochrome P450 was not altered in any of the diabetic groups. Similarly, the activity of NADPH-dependent cytochrome P450 reductase was not affected by diabetes indicating that the flow of electrons to cytochrome P450 was not perturbed (Table 2.). The levels of cytochrome $b_5$, although elevated in all diabetic groups, was only statistically significant in the 12 and 24 week diabetic animals (Table 2.). Dealkylations of both ethoxy- and pentoxyresorufin were elevated in all diabetic animals (Table 3.). Similarly p-nitrophenol and lauric acid hydroxylations and the demethylation of ethylmorphine were all significantly elevated in the diabetic animals (Table 3). As the duration of diabetes increased the inter-animal variability for these reactions increased resulting in less significant increases at 24 weeks compared to that at 4 weeks of diabetes duration.

Table 1.  Plasma glucose and ketone body concentrations

| Group | Duration of diabetes | Plasma glucose (mM) | 3-hydroxybutyrate + Acetoacetate (mM) |
|---|---|---|---|
| Diabetic | 4 weeks | 28.2±4.1*** | 11.4±0.5*** |
| Control |  | 6.8±0.54 | 0.51±0.1 |
| Diabetic | 12 weeks | 26.6±5.0*** | 11.0±2.3*** |
| Control |  | 7.1±0.8 | 0.43±0.08 |
| Diabetic | 24 weeks | 32.0±5.8*** | 8.4±1.2*** |
| Control |  | 5.9±0.9 | 0.49±0.2 |

Data represent Mean ± SEM
***P<0.001 compared to respective controls

## DISCUSSION

Withdrawal of insulin therapy from the diabetic animals was associated with the characteristic signs of uncontrolled IDDM. The animals became lethargic and displayed polyuria and polydipsia, although there was pronounced inter-animal variation in the onset of the symptoms. Investigation of plasma parameters revealed that all the diabetic animals were significantly hyperglycaemic and hyperketonaemic (Table 1.) and that duration of the disease had no significant effect on these parameters.

Table 2. Microsomal parameters for diabetic BB rats

| GROUP | Protein (mg/g liver) | Cyt. P450 (nmol/mg Pt) | Cyt. $b_5$ (nmol/mg Pt) | NADPH cyt. P450 reductase (nmol/min/mg Pt) |
|---|---|---|---|---|
| Diabetic (4 wks) | 38.1±1.2 | 0.34±0.10 | 0.91±0.19 | 15.0±2.4 |
| Control | 37.1±0.9 | 0.40±0.07 | 0.63±0.05 | 18.1±1.8 |
| Diabetic (12 wks) | 30.3±0.4 | 0.43±0.05 | 0.82±0.16** | 20.9±3.1 |
| Control | 29.6±0.7 | 0.38±0.10 | 0.53±0.04 | 25.1±2.4 |
| Diabetic (24 wks) | 35.1±1.2 | 0.52±0.06 | 0.73±0.10** | 15.0±2.1 |
| Control | 32.0±0.4 | 0.35±0.07 | 0.53±0.05 | 17.3±1.0 |

Data represent Mean ± SEM     mg Pt = mg protein
**$P<0.01$ compared to respective controls

Cytochrome P450 investigation revealed that as the duration of diabetes increased, the extent of increase in the activities of CYP3A, CYP2E and CYP4A proteins became less pronounced. The effects of duration of the disease on the activities of CYP1A and CYP2B were less apparent with elevation of these activities only reaching significance at 12 weeks of IDDM duration (Table 3.). These results suggest that in the BB rat the duration of the disease may influence the level of induction of the cytochrome P450 proteins. It is possible that these effects may be due to an adaptation process. As insulin treatment cannot mimic the normal secretory responses of the pancreatic B cell to nutrients, there will be periods of both hyper- and hypoglycaemia occurring between insulin injections with associated fluctuations in ketone body concentrations. As the duration of the disease progresses the animals may become more refractory to the inducing effects of the circulating ketone body and hormonal imbalances which accompany the disease. We have observed similar effects of diabetes duration on cytochrome P450 proteins with streptozotocin-induced diabetic animals (Barnett et al., 1993).

Table 3. Hepatic cytochrome P450 activities in diabetic BB rat

GROUP

| | Diabetic (4 wks) | Control | Diabetic (12 wks) | Control | Diabetic (24 wks) | Control |
|---|---|---|---|---|---|---|
| EROD | 18±5.5 | 10±2.1 | 24±4** | 13±2.1 | 18±1.2 | 14±2 |
| PROD | 6.1±2.1 | 3.1±2 | 7.4±2** | 3±0.7 | 9.1±6 | 3.4±1.6 |
| EM | 22±4*** | 10±2 | 17±2.1** | 9.4±1.1 | 14.8±2* | 11±0.8 |
| p-NP | 1.7±0.1** | 1.0±0.2 | 1.8±0.3** | 0.8±0.1 | 1.3±0.3* | 0.76±0.2 |
| LA | 11±3*** | 3.1±0.8 | 18±5*** | 2.9±1.1 | 8.7±2** | 3.2±1.5 |

Data represent Mean ± SEM ***$P<0.001$, **$P<0.01$, *$P<0.05$ compared to respective controls.
EROD - Ethoxyresorufin O-deethylase (pmol/min/mg protein)
PROD - Pentoxyresorufin O-depentylase (pmol/min/mg protein)
EM - Ethylmorphine N-demethylase (nmol/min/mg protein)
p-NP - p-Nitrophenol hydroxylase (nmol/min/mg protein)
LA - Lauric acid hydroxylase (nmol/min/mg protein)

REFERENCES

Barnett, C. R., Flatt, P. R. and Ioannides, C. (1992). In: Food, Nutrition and Chemical Toxicity, Ed D. V. Parke, C. Ioannides and R. Walker. Smith Gordon and Co Ltd. London p163-172

Barnett, C. R., Flatt, P. R. and Iannides, C. (1990a). Induction of hepatic microsomal P450 I and IIB proteins by hyperketonaemia Biochemcial Pharmacology 40 393-397

Barnett, C. R., Gibson, G. G., Wolf, C. R., Flatt, P. R. and Ioannides, C. (1990b). Induction of cytochrome P450III and P450IV family proteins in streptozotocin-inducd diabetes. Biochemical Journal 268 765-769

Barnett, C. R., Flatt, P. R. and Ioannides, C. (1990c). Hyperketonaemia markedly modulates the metabolic activation of chemical carcinogens. Chemico-Biological Int. 74 281- 289

Barnett, C. R., Flatt, P. R. and Ioannides, C. (1993). Modulation of the rat hepatic cytochrome P450 composition by long-term streptozotocin-induced insulin-dependent diabetes. Journal of Biochemical Toxicology In press

Bellward, G. D., Chang, T., Rodrigues, B., McNeill, J. H., Maines, S., Ryan, D. E., Levin, W. and Thomas, P. E. (1988). Hepatic cytochrome P450j induction in spontaneously diabetic BB rat. Molecular Pharmacology 33 140-147

Favreau, L. V. and Schenkman, J. B. (1988). Cytochrome P450 alterations in the BB/Wor spontaneously diabetic rat. Biochemical Pharmacology 37 3505-3509

Rifkin and Porte. Diabetes Mellitus: Theory and Practice, 4th Edn., Elsevier, New York.

# 2,3,7,8-tetrachlorodibenzo-p-dioxin *versus* 3-methylcholanthrene. Ah receptor binding, transformation and induction of CYP1A1

David S. Riddick[1], Patricia A. Harper[2], Allan B. Okey[1]

[1]Department of Pharmacology, University of Toronto, Toronto, Canada M5S 1A8. [2]Division of Clinical Pharmacology, and Toxicology, Hospital for Sick Children, Toronto, Canada M5G 1X8

## INTRODUCTION

The aromatic hydrocarbon (Ah) receptor mediates induction of CYP1A1 by halogenated aromatic hydrocarbons such as 2,3,7,8-tetrachlorodibenzo-p-dioxin (TCDD) and polyclyclic aromatic hydrocarbons such as 3-methylcholanthrene (MC) (Okey, 1990; Whitlock, 1990; Landers & Bunce, 1991). Following binding of ligand to the Ah receptor in the cytoplasm, the ligand-receptor complex undergoes a process of transformation to a form that is tightly associated with the nucleus (Okey *et al.*, 1979). The nuclear form of the receptor interacts with specific *cis*-acting dioxin-responsive enhancers (DRE) located upstream of the *CYP1A1* gene (Denison *et al.*, 1988), leading to stimulated transcription (Israel & Whitlock, 1984) and enhanced expression of CYP1A1 mRNA and protein (Israel & Whitlock, 1983).

The affinity with which a ligand binds to the Ah receptor is a major determinant of that ligand's potency as a CYP1A1 inducer (Safe, 1986). Although such structure-activity relationships hold well within individual classes of Ah receptor ligand (*e.g.* within the dibenzofuran class, within the polychlorinated biphenyl class, *etc.*), there are major discrepancies in inter-class comparisons. For example, the affinity with which rat or mouse cytosolic Ah receptors bind the nonhalogenated ligand, MC, *in vitro* is nearly equal to the affinity with which the receptor binds TCDD (Okey & Vella, 1982). However, in rat liver *in vivo*, TCDD is about 30,000-times more potent than MC as an inducer of CYP1A1-mediated aryl hydrocarbon hydroxylase (AHH) activity (Poland & Glover, 1974). Thus, a factor or factors other than binding affinity must account for the much greater biological potency of TCDD relative to MC. We have hypothesized that TCDD and MC differ in potency and/or efficacy for stimulating events in the CYP1A1 induction pathway downstream from initial binding to the cytosolic Ah receptor.

In the present investigation, we compared systematically TCDD and MC at several key stages in the CYP1A1 induction mechanism in the mouse hepatoma cell line, Hepa-1, in order to determine if differences at any key step(s) may account quantitatively for the large difference in biological potency observed for these compounds. Specifically, we have compared TCDD and MC at the following mechanistic steps: (a) binding to cytosolic Ah receptor, (b) nuclear translocation of the ligand-Ah receptor complex, (c) *in vitro* transformation of the Ah receptor to its DRE-binding form, (d) affinity of the ligand-Ah receptor complex for a DRE sequence, (e) induction of CYP1A1 mRNA, and (f) induction of AHH activity. We also addressed the importance of MC metabolism as a factor that contributes to this compound's reduced biological potency relative to TCDD.

## MATERIALS AND METHODS

Cell culture. The mouse hepatoma cell line, Hepa-1c1c9, was grown in monolayer culture in α-minimal essential medium supplemented with 10% fetal bovine serum.

Radioligand binding. Hepa-1 cytosol was incubated with various concentrations of [$^3$H]TCDD or [$^3$H]MC in the absence or presence of a 100-fold molar excess of nonradioactive 2,3,7,8-tetrachlorodibenzofuran (TCDF) for 1 h at 4°C. Samples were analyzed by sucrose density gradient centrifugation (Tsui & Okey, 1981). We also measured the ability of various concentrations of nonradioactive TCDD or MC to compete with 1 nM [$^3$H]TCDD for binding to the cytosolic Ah receptor.

AHH activity. Hepa-1 monolayers were treated with various concentrations of [$^3$H]TCDD or MC. At defined time points, AHH activity was determined in whole cell suspensions by a fluorometric method (Nebert & Gelboin, 1968). A modification of this assay was used to examine the effect of MC exposure on the ability of Hepa-1 cells to convert [$^3$H]MC to alkali-extractable metabolites.

Nuclear translocation. Hepa-1 monolayers were treated with ~1 nM [$^3$H]TCDD in the absence or presence of various concentrations of nonradioactive TCDD or MC for 2 h. Nuclear extracts were prepared (Miller et al., 1983) and analyzed by sucrose density gradient centrifugation.

Gel retardation. Hepa-1 cytosol was incubated with various concentrations of [$^3$H]TCDD or [$^3$H]MC for 4h at 30°C, and interaction of the *in vitro* transformed Ah receptor-ligand complex with a $^{32}$P-labeled DRE oligonucleotide was examined on nondenaturing polyacrylamide gels (Denison & Yao, 1991). Saturation experiments were also performed to determine the apparent affinity of the ligand-Ah receptor complex for the DRE sequence.

CYP1A1 mRNA. Hepa-1 monolayers were treated with various concentrations of [$^3$H]TCDD or MC for 2 h, and then total RNA was isolated (Chomczynski & Sacchi, 1987). CYP1A1 mRNA was measured by slot-blot hybridization analysis using a $^{32}$P-labeled mouse *Cyp1a-1* cDNA provided by Dr. R.H. Tukey (University of California, San Diego) (Tukey & Okino, 1991). CYP1A1 mRNA signals were normalized for actin mRNA levels.

## RESULTS AND DISCUSSION

Saturation analysis of radioligand-binding experiments was used to determine the apparent $K_d$ and $B_{max}$ for the interaction of [$^3$H]TCDD and [$^3$H]MC with cytosolic Ah receptor. Hepa-1 cytosol contained the same number of Ah receptor sites for both radioligands and the Ah receptor bound [$^3$H]TCDD with an affinity that was only ~3-fold greater than that for [$^3$H]MC. Similarly, competitive binding studies demonstrated that the apparent affinity of Ah receptor for TCDD was ~4-fold greater than that for MC.

Following 14 h of ligand exposure, TCDD was found to be ~1,000-fold more potent than MC as an inducer of AHH activity in Hepa-1 cells. However, at an earlier time-point (*i.e.* 4 h), TCDD was only ~10-fold more potent than MC as an AHH inducer.

In order to understand the mechanistic basis for the large difference in biological potency between TCDD and MC, we first examined the ability of nonradioactive TCDD or MC to compete for the nuclear uptake of the [$^3$H]TCDD-Ah receptor complex. After 2 h of exposure, TCDD was ~25-fold more potent than MC as a competitor for this process.

Under *in vitro* conditions, TCDD and MC were equipotent for transforming the cytosolic Ah receptor to its DRE-binding form. In addition, the TCDD-Ah receptor complex and the MC-Ah receptor complex bound to the DRE sequence with similar apparent affinity.

Following 2 h of ligand exposure, TCDD was shown to be ~10-fold more potent than MC as an inducer of CYP1A1 mRNA.

These data are summarized schematically in Fig.1, in which numbers are used to indicate the quantitative differences in affinity/potency between TCDD and MC at each key step in the CYP1A1 induction pathway. For events assessed *in vitro* (*e.g.* Ah receptor binding, Ah receptor transformation and DNA binding), TCDD and MC differed in affinity/potency by only 0 to 4-fold. For events measured in intact cells at early time-points (*e.g.* nuclear translocation of Ah receptor and CYP1A1 mRNA induction at 2 h, and AHH induction at 4 h), TCDD and MC differed in potency by ~10 to 25-fold. For events assessed in intact cells at late time-points (*e.g.* AHH induction at 14 h), TCDD was ~1,000-fold more potent than MC.

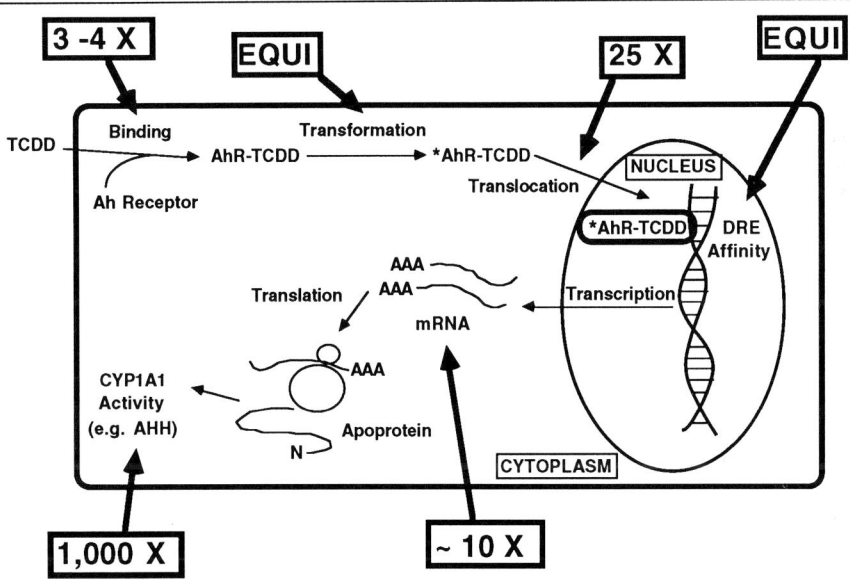

Fig. 1. Summary of the quantitative affinity/potency differences between TCDD and MC at key steps in the CYP1A1 induction pathway in Hepa-1 cells. Boxed bold numbers indicate the degree to which TCDD displays greater affinity/potency compared to MC at each step. (Abbreviations: EQUI = equal potency or affinity; AhR = Ah receptor).

This analysis suggested that TCDD and MC do not differ dramatically in their abilities to activate processes in the CYP1A1 induction pathway, but instead the greater rate of metabolism of MC may limit its biological potency in intact cells. We have three sources of evidence to support this hypothesis. First, AHH induction caused by MC is transient compared to the persistent induction caused by TCDD. Second, we observed a time-dependent decrease in the potency of MC as an AHH inducer. Third, exposure of Hepa-1 cells to MC increases the ability of these cells to convert [$^3$H]MC to alkali-extractable metabolites. This study demonstrates that *in vivo* pharmacokinetic differences between ligands of different chemical nature may, in some instances, render inaccurate or invalid predictions that are based strictly on receptor binding or transformation experiments conducted *in vitro*.

Acknowledgments. This work was supported by grants from the Medical Research Council of Canada (A.B.O.) and the National Cancer Institute of Canada (A.B.O. and P.A.H.), and a post-doctoral fellowship from the Medical Research Council of Canada (D.S.R.).

**REFERENCES**

Chomczynski, P. & Sacchi, N. (1987): Single-step method of RNA isolation by acid guanidinium thiocyanate-phenol-chloroform extraction. *Anal. Biochem.* 162, 156-159.

Denison, M.S., Fisher, J.M. & Whitlock, J.P., Jr. (1988): The DNA recognition site for the dioxin-Ah receptor complex. Nucleotide sequence and functional analysis. *J. Biol. Chem.* 263, 17221-17224.

Denison, M.S. & Yao, E.F. (1991): Characterization of the interaction of transformed rat hepatic cytosolic Ah receptor with a dioxin responsive transcriptional enhancer. *Arch. Biochem. Biophys.* 284, 158-166.

Israel, D.I. & Whitlock, J.P., Jr. (1983): Induction of mRNA specific for cytochrome $P_1$-450 in wild type and variant mouse hepatoma cells. *J. Biol. Chem.* 258, 10390-10394.

Israel, D.I. & Whitlock, J.P., Jr. (1984): Regulation of cytochrome $P_1$-450 gene transcription by 2,3,7,8-tetrachlorodibenzo-*p*-dioxin in wild type and variant mouse hepatoma cells. *J. Biol. Chem.* 259, 5400-5402.

Landers, J.P. & Bunce, N.J. (1991): The Ah receptor and the mechanism of dioxin toxicity. *Biochem. J.* 276, 273-287.

Miller, A.G., Israel, D. & Whitlock, J.P., Jr. (1983): Biochemical and genetic analysis of variant mouse hepatoma cells defective in the induction of benzo(a)pyrene-metabolizing enzyme activity. *J. Biol. Chem.* 258, 3523-3527.

Nebert, D.W. & Gelboin, H.V. (1968): Substrate-inducible microsomal aryl hydroxylase in mammalian cell culture. I. Assay and properties of induced enzyme. *J. Biol. Chem.* 243, 6242-6249.

Okey, A.B. (1990): Enzyme induction in the cytochrome P-450 system. *Pharmacol. Ther.* 45, 241-298.

Okey, A.B., Bondy, G.P., Mason, M.E., Kahl, G.F., Eisen, H.J., Guenthner, T.M. & Nebert, D.W. (1979): Regulatory gene product of the *Ah* locus. Characterization of the cytosolic inducer-receptor complex and evidence for its nuclear translocation. *J. Biol. Chem.* 254, 11636-11648.

Okey, A.B. & Vella, L.M. (1982): Binding of 3-methylcholanthrene and 2,3,7,8-tetrachlorodibenzo-*p*-dioxin to a common *Ah* receptor site in mouse and rat hepatic cytosols. *Eur. J. Biochem.* 127, 39-47.

Poland, A. & Glover, E. (1974): Comparison of 2,3,7,8-tetrachlorodibenzo-*p*-dioxin, a potent inducer of aryl hydrocarbon hydroxylase, with 3-methylcholanthrene. *Mol. Pharmacol.* 10, 349-359.

Safe, S.H. (1986): Comparative toxicology and mechanism of polychlorinated dibenzo-*p*-dioxins and dibenzofurans. *Annu. Rev. Pharmacol. Toxicol.* 26, 371-399.

Tsui, H.W. & Okey, A.B. (1981): Rapid vertical tube rotor gradient assay for binding of 2,3,7,8-tetrachlorodibenzo-*p*-dioxin to the *Ah* receptor. *Can. J. Physiol. Pharmacol.* 59, 927-931.

Tukey, R.H. & Okino, S.T. (1991): Quantitation of related gene products by nuclear run-on and Northern blot analysis. *Methods Enzymol.* 206, 284-290.

Whitlock, J.P., Jr. (1990): Genetic and molecular aspects of 2,3,7,8-tetrachlorodibenzo-p-dioxin action. *Annu. Rev. Pharmacol. Toxicol.* 30, 251-277.

# Reconstitution of cytochrome P450scc by octylglucoside dialysis/adsorption into (large) unilamellar phospholipid vesicle. Structural and functional characterization

Dieter Schwarz[1], Alexey Chernogolov[2], Klaus Gast, Walter Richter[3]

[1]Max Delbrück Center for Molecular Medicine, D-13122 Berlin – Buch, Germany. [2]Institute of Bioorganic Chemistry, Belorussian Academy of Sciences, 220141 Minsk, Belarus. [3]Institute of Ultrastructural Research, Friedrich-Schiller-University, 07740 Jena, Germany

## INTRODUCTION

It has been manifested that phospholipid environment in general and cardiolipin (CL) particularly play an important role for the structural, functional and topological properties of cytochrome P450SCC (P450SCC, CYP45011A1). Reconstitution into a membrane-like environment results in significant activation depending on both the method of reconstitution and on the lipid composition of the membrane (for review see: Lambeth, 1990, and citations therein). Successful incorporation has been achieved by both cholate dialysis or gel filtration (Hall et al., 1979; Yamakura et al., 1981) and by adding the enzyme directly to lipid vesicles preformed by sonication (Seybert et al., 1979) or detergent dialysis (Ikushiro et al., 1992; Dhariwal & Jefcoate, 1989), respectively.

We have utilized reconstitution by octylglucoside (OG, from Calbiochem, USA) dialysis/adsorption of detergent-lipid-P450SCC mixed micelles for several reasons, mainly

(i) to prepare larger unilamellar proteoliposomes, particularly convenient for rotational diffusion and freeze-fracturing studies (following the line of successful incorporation of microsomal P450's (Schwarz et al., 1984),

(ii) because OG could be removed effectively and quickly and to a minimal residual level to overcome possible problems with residual levels of OG as reported by Dhariwal (1989),

(iii) to exclude possible problems caused by more or less aggregation of the cytochrome preparations which may arise by using preformed vesicles (Pember et al., 1983), and

(iv) to detect different modes of membrane insertion, for instance intramembrane particles indicating transmembrane, bilayer spanning segments of the P450SCC peptide chain which may not occur using preformed vesicles.

## PREPARATION

P450SCC could not reconstituted successfully by OG dialysis under conditions which resulted in proper incorporation of microsomal P450's (Schwarz et al., 1984). In the presence of higher OG concentrations P450SCC showed a strong tendency to precipitate and to degrade to P420. We found it optimal to use OG concentrations ≤ 0.43 % at a molar detergent/protein ratio of about 3300 and a detergent/lipid ratio of about 10. This guarantees both monomerization and no precipitation and/or degradation. P450SCC was incorporated into vesicles by coreconstitution of lipid(s) and protein(s) combining OG dialysis with simultaneous adsorption of the detergent by polysterene beads for an effective detergent removal (Schwarz et al., 1988). The following is a typical protocol for the reconstitution of P450SCC into unilamellar vesicles with a lipid composition nearly approximating that of bovine adrenal mitochondria (Cheng § Kimura, 1983):

10 mg of lipid ( standard mixture was egg PC, egg PE (both from Lipid Products, UK), and bovine heart CL (from Serva, Germany) : PC/PE/CL (2:2:1,w/w/w) ) were dried and subsequently solubilized by an OG-buffer solution ( standard buffer: 50 mM potassium phosphate, 100 mM KCl, pH 7.3, 0.5 mM EDTA, 0.1 mM DTT containing 20 % (v/v) glycerol ). This suspension was incubated with P450SCC 45 min at room temperature in a final volume of 9 ml ( final concentrations: 0.43 % OG, 1.1 mg/ml lipid, 4 µM P450, lipid/protein (L/P) ratio of 5 (w/w) )

followed by 48 hours dialysis at 4 °C in the presence of detergent adsorbing BioBeads SM-2 ( Biorad, USA) to remove the OG effectively. After dialysis the vesicle suspension was gel filtrated using Sephacryl S-1000 ( Pharmacia, Sweden ) (i) to reduce the residual OG content further, (ii) to separate a possible part of non-incorporated P450, and (iii) to fractionate the vesicles according to size. To reduce vesicle tumbling in the rotational diffusion experiments only fractions with large vesicles were collected and concentrated using Centricon-100 microconcentrators ( Amicon, USA ) to a final P450 concentration of about 1-2 µM (Fig. 1). P450SCC concentrations were determined according to Omura & Sato (1964).

Vesicles prepared this way were stable during storage at 4 °C under an athmosphere of $N_2$ for at least one week. During this time no lipid degradation and denaturation of P450SCC could be detected; the P450SCC concentration decreased only about 22 %. The vesicles were stable in relation to their size and size distribution (homogeneity) as checked by dynamic light scattering (DLS); no aggregation and/or fusion took place.

## STRUCTURAL CHARACTERIZATION

Gel filtration. Fig. 1 shows a typical elution profile after Sephacryl S-1000 gel filtration of PC/PE/CL (1:1:1). Almost all lipid ( ≥90 %) and 55-70 % of P450SCC elute together in a relatively broad peak between the void volume $V_o$ and the total volume $V_t$ of the column. The coincidence of the fractions containing P450SCC ($A_{416}$) and lipid ($[^{14}C]PC$) and the enzymatic activity measured in these fractions (see below) indicate that P450SCC is associated with the vesicles. During the reconstitution no degradation to P420 could be detected. Almost no P450 ( ≤3%) elutes at a position where free P450SCC in aqueous solution (without lipid) elutes. Without CL this fraction is enhanced, for instance in PC/PE(2:1) vesicles up to 10 %. Fig.1 shows a significant shift in the profiles for lipid and protein, respectively: the larger vesicles which elute at or shortly after $V_o$ had a lower protein content (L/P=25-30) whereas the smaller vesicles were characterized by a larger one (L/P ≤5).

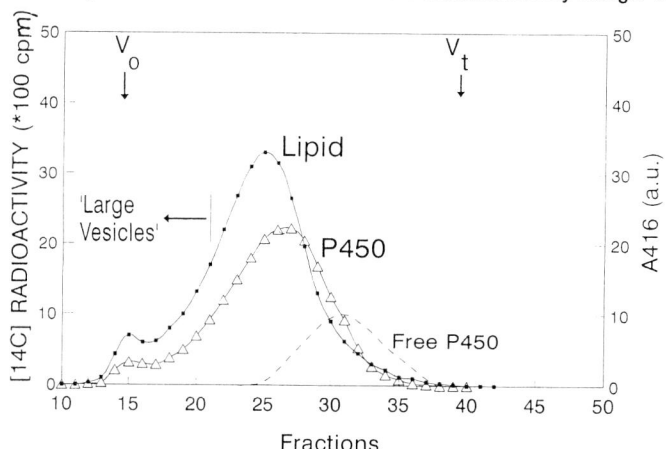

**Fig. 1** Elution profile after Sephacryl S-1000 gelchromatography of P450SCC liposomes ( PC/PE/CL(1:1:1, w/w/w), L/P=10, w/w).

From Fig. 1 it can be seen that a discrimination of free, non-incorporated P450SCC and small P450SCC vesicles by Sephacryl S-1000 gel filtration could not be achieved. However, this gel is well suited to fractionate the vesicles according to size (Reynolds et al.,1983). Thus 'large vesicles' (average diameter ≥ 150 nm) were separated and collected corresponding to fractions ≤ No.21). To do this the column was calibrated in relation to particle size by using phospholipid vesicles of different size. The sizes were determined by DLS. In agreement with electron microscopic analyses of the size distribution DLS studies showed that P450SCC liposomes consist of mainly two populations: larger ones with an average diameter of about 150 nm ( ≈27 %) and smaller ones with a diameter around 60 nm (≈73 %).

Freeze-fracturing. Electron micrographs showed mainly spherical and round vesicles with typical smooth fracture faces indicating the unilamellar nature of the membrane. PC/PE/CL vesicles (without P450SCC) were not aggregated and characterized by a Gaussian size distribution whereas proteoliposomes showed a broad size distribution consisting of two main peaks. P450SCC vesicles had a tendency to aggregate in dependence on the CL content. In PC/PE/CL(2:2:1) liposomes they are mainly not aggregated showing tightly associated protein particles at the outer vesicle surface (Fig. 2 A). On the other side PC/PE/CL(1:1:1) vesicles predominantly associate into small groups with the particles mainly located between the vesicles (Fig. 2 B). These results indicate a specific P450SCC-induced vesicle aggregation because addition of anti-P450SCC IgG inhibits the aggregation (not shown). The P450SCC particles represent most probably aggregates because of their size

(6–10 nm). We never have seen any intramembrane particles on both the concave and convex fracture faces in contrast to OG-reconstituted microsomal P450 liposomes (LM2 and LM4, CYP4502B4 and CYP4501A1) (Schwarz et al., 1990).

In general cholate-reconstituted P450SCC vesicles prepared by the same method are characterized by smaller average diameters, particularly the portion of larger liposomes was smaller. However in relation to the other morphological properties they showed a qualitatively similar behaviour.

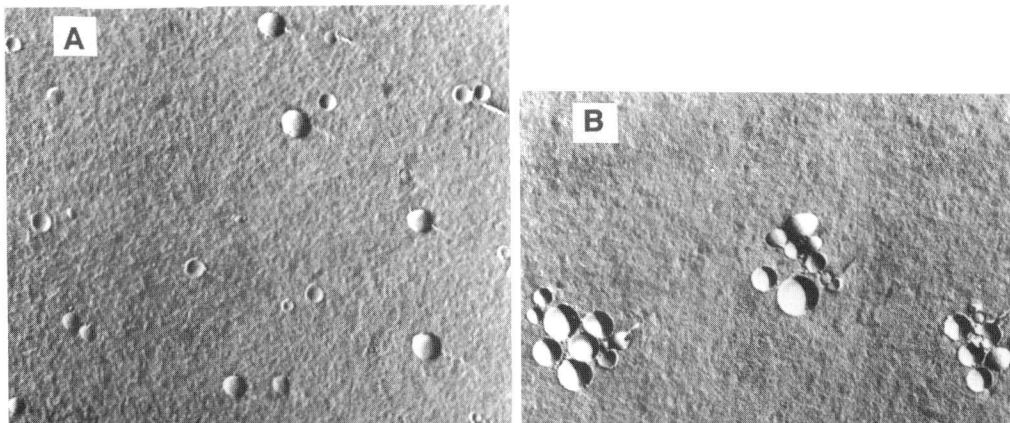

**Fig. 2** Freeze-fracturing electron micrographs of octylglucoside reconstituted PC/PE/CL vesicles. A: Cytochrome P450SCC liposomes ( PC/PE/CL(2:2:1) ), B: Cytochrome P450SCC liposomes ( PC/PE/CL(1:1:1) ). Conditions: L/P=5, lipid:0.86 mg/ml, [P450SCC]=3.3 µmol/l, 4 °C, standard buffer with 20 % glycerol, magnification: 59400 x.

Residual octylglucoside. To determine the residual amount of detergent in the vesicle preparation an appropriate amount of [$^{14}$C]OG ( about 200 000 cpm ) was added to the initial solution of the lipid. The amount of the remaining OG after 24 h dialysis/adsorption was determined to be about 0.1 % of what was added initially, corresponding to 1 OG molecule per 100 molecules of lipid. By the following gel filtration step using Sephacryl S-1000 this value could be further diminished to final level of about 0.06 %, corresponding to about 1 mol OG/160 mol of lipid. These results show that OG, as it is already known from the preparation of (pure) liposomes (without protein) can be removed as quickly to such a small residual concentration as no other detergent under comparable experimental conditions ( in the presence of glycerol and at 4 °C !). Particularly cholate, which was used earlier (Yamakura et al.,1981) to prepare P450SCC liposomes can be removed only by a rate approximately two times slower than OG.

The characterization of the vesicles corresponding to the remaining detergent concentration is in particular important because recently Dhariwal § Jefcoate (1989) reported a high stimulation by low concentrations of OG or certain fatty acids exclusively in large DOPC vesicles of the P450SCC activity. The cholesterol side chain cleavage activity in large vesicles was remarkably increased by addition of OG corresponding to >2 µM. The concentration required for remarkable stimulation by OG represents a ratio of D/L of only 0.3-1.5 %(w/w) corresponding to 0.8-4 mol OG/100 mol of lipid. The residual OG content that has been determined by [$^{14}$C]OG in our PC/PE/CL vesicles ( OG/L = 1/160 (mol/mol)) corresponds to 0.6 OG molecules per 100 lipid molecules, thus lying well under the 'stimulating level'. This was strongly supported by the effect of extensively prolonged dialysis, that did not change the enzymatic activity of our (large) P450SCC proteoliposomes.

## FUNCTIONAL CHARACTERIZATION
Reduction. To analyze the orientation of P450SCC in the liposomal membrane the reducibility by external addition of the physiological electron transfer partners was examined. After addition of a sufficient amount of AD, AR and NADPH in the presence of CO nearly all P450SCC could be converted to the P450SCC-CO complex. Additon of dithionite did not lead to further enhancement of the proportion of the reduced P450-CO complex (Fig. 3, upper curve). The experiments indicate a totally asymmetric distribution of P450SCC: almost all P450SCC are located in/at the outer side of the bilayer membrane since the reducible fraction should represent the molecules embedded in the outer monolayer of the membrane. This asymmetric incorporation of P450SCC in

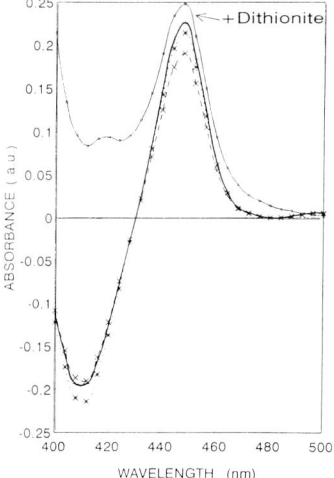

**Fig. 3** Enzymatic reduction of P450SCC in PC/PE/CL vesicles. 0.5 nmol liposomal P450SCC were reduced by 5 nmol AD, 0.5 nmol AR and 12.5 nmol NADPH in 1 ml standard buffer. After reduction the suspension was saturated with CO before the spectra were recorded at 2 (-×-×), 10 (·*··*) and 20(+–+) min after addition of NADPH. Lastly a few grains of dithionite were added.

unilamellar PC/PE/CL vesicles has a strong impact on the interpretation of the rotational diffusion experiments and is in accordance with our electron microscopic studies which indicate mainly surface location of P450SCC at/in the outer vesicle monolayer. By comparison the proportion of P450 located in the outer monolayer of cholate-reconstituted DOPC-vesicles was found to be about 70 % (Yamakura et al., 1981).

Activity. Enzymatic activity was examined in a reconstituted system consisting of P450SCC vesicles, AR, AD and NADPH as cholesterol side chain cleavage activity to produce pregnenolone. Cholesterol and pregnenolone could be separated and analyzed by reversed phase HPLC using a method of Sugano et al.(1989) with some modifications. P450SCC liposomes were found functionally active, for instance the activity of P450SCC in PC/PE/CL/Cholesterol (2:2:1:0.6) vesicles (i.e. 20 mol% cholesterol) was significantly enhanced from 2.6 in aqueous solution (100 %) to 4.7 nmol pregnenolone/nmol P450SCC/min (180 %) in the liposomal state (for further discussion and conditions see this vol.: A.A.Chernogolov et al. "Study of the membrane topology.....").

This study was supported by the German Research Foundation DFG (Schw 471/1-1).

REFERENCES
Cheng, B. & Kimura, T. (1983): *Lipids* 18, 577-584.
Dhariwal, M.S. & Jefcoate, C.R. (1989): *Biochemistry* 28, 8397-8402.
Dhariwal, M.S., Kowluru, R.A., and Jefcoate, C.R. (1991): *Biochemistry* 30, 4940-4949.
Hall, P.F., Watanuke, M., and Hamkalo, B.A. (1979): *J. Biol. Chem.* 255, 3057-3061.
Ikushiro, S., Komimani, S., and Takemori, S. (1992): *J. Biol. Chem.* 267, 1464-1469.
Lambeth, J.D., Seybert, D.W., and Kamin, H. (1979): *J. Biol. Chem.* 254, 7255-7264.
Lambeth, J.D. (1990): In *Frontiers in Biotransformation, Vol.3*, eds. K. Ruckpaul & H. Rein, pp. 58-100. Berlin: Akademie Verlag.
Omura, T. & Sato, R. (1964): *J. Biol. Chem.* 239, 2370-2378.
Pember, S.O., Powell, G.L., and Lambeth, J.D. (1983): *J. Biol. Chem.* 258, 3198-3206.
Reynolds, J.A., Nasaki, Y., and Tanford, C. (1983): *Anal. Biochem.* 130, 471-474.
Schwarz, D., Gast, K., Meyer, H.W., Lachmann, U., Coon, M.J., and Ruckpaul, K. (1984): *Biochem. Biophys. Res. Commun.* 121, 118-125.
Schwarz, D., Zirwer, D., Gast, K., Meyer, H.W. and Lachmann, U. (1988): *Biomed. Biochim. Acta* 47, 609-621.
Schwarz, D., Pirrwitz, J., Meyer, H.W., Coon, M.J., and Ruckpaul, K. (1990): *Biochem. Biophys. Res. Commun.* 171, 175-181.
Schwarz, D., Krüger, V., Chernogolov, A.A., Usanov, S.A., and Stier, A.(1993): *Biochem. Biophys. Res.Commun.* 195, 889-896.
Seybert, D.W., Lancaster, Jr., J.R., Lambeth, J.D., and Kamin, H. (1979): *J. Biol. Chem.* 254, 12088-12098.
Seybert, D.W. (1990): *Arch. Biochem. Biophys.* 279, 188-194.
Sugano, S., Morishima, N., Ikeda, H., and Horie, S. (1989): *Anal. Biochem.* 182, 327-333.
Yamakura, F., Kido, T., and Kimura, T. (1981): *Biochim. Biophys. Acta* 649, 343-354.

# Pressure-induced transitions in cytochrome P450 IIB4 (LM2): evidence of the conformational inhomogeneity in the oligomers in solution and in proteoliposomes

Dmitri R. Davydov, Gaston Hui Bon Hoa

*Institut de biologie physico-chimique, INSERM U.310, 13, rue Pierre-et-Marie-Curie, 75005, Paris, France*

In the last few years pressure-induced transitions in protein structure were successfully used to study the changes in the heme moiety of cytochromes P-450cam and P-450lin during substrate binding. Recently we have reported the first results on the pressure-induced transitions in rabbit liver microsomal cytochrome P-450 2B4 reduced CO-complex (Davydov, et al., 1992). An application of high hydrostatic pressure was shown to induce conversion of 2B4 ($Fe^{2+}$)-CO into inactive P-420 form. However, only about 65% of the oligomeric 2B4 could be converted into P-420. The remaining portion of the protein was stable up to 6 kbar. This was shown both in solution and 2B4-containing proteoliposomes. After monomerization of hemoprotein oligomers by 0.2% Triton N-101 such inhomogeneity of 2B4 pool vanished. Monomers could be completely converted to P-420 [1]. These results were interpreted as an evidence of the existence of the static splitting of 2B4 pool in the oligomers between two conformers with different pressure sensitivity (Davydov, et al., 1992). Such splitting was proven to exist in solution as well as in membranes.

The present study was undertaken to investigate pressure-induced transitions in oxidized P-450 LM2 to further test this hypothesis and in to examine possible deference in the spin equilibria and substrate-binding properties of the apparent P-450 conformers.

## MATERIALS AND METHODS

<u>Cytochrome P-450 2B4 (LM2) and NADPH-cytochrome P-450 reductase</u> were isolated from phenobarbital-pretreated rabbits (Imai, 1987). Experiments were performed at 25 °C in 0.1 M K-Hepes buffer, pH 7.4 containing 1 mM EDTA and 1 mM dithiotrietol. Proteoliposomes were prepared from 2:1:1 mixture of PC:PE:PS by Sephadex LH-20 gel filtration (Mishin, et al., 1979)

<u>High-Pressure system.</u> Computer-controlled optical pressure system, capable to generate a pressure of 6200 bar, has been described previously (Hui Bon Hoa, et al., 1982).

<u>Processing of the spectra.</u> To interpret the results of our experiments in terms of pressure-induced changes in the concentrations of 2B4 species (ls, hs, and P-420) a special mathematical method based on factor analysis technique was developed. This method is designed to resolve the optical changes in multicomponent system into components reflecting separately each particular simultaneous processes, taking place in the system. It was also used for the algebraic correction of the spectra to compensate the turbidity of the proteoliposomes and to suppress the changes in the turbidity, which is due to the partial denaturation of the protein at high pressures. Extinction standards of 2B4 species were deduced from independent experiments including titration by benzphetamine and ethylbenzene, spin state temperature dependence and osmotic shock-promoted inactivation studies.

<u>Analysis of pressure-induced transitions.</u> The fitting of the concentrations of 2B4 species versus pressure was based on the classical equation for the first order reversible reaction A $\rightleftarrows$ B:

---

[1]Present address: Institute of Medical and Biological Chemistry, Acad. Med. Sci. USSR, Pogodinskaya ul., 10, 119832, Moscow, Russia.

$$[A] = \frac{C_0}{1 + K_{eq}(0) \cdot e^{-p\Delta V^\circ/RT}} = \frac{C_0}{1 + e^{(p_{1/2}-p)\Delta V^\circ/RT}} \quad (1)$$

where $K_{eq}(p)$ and $K_{eq}(0)$ are the equilibrium constant of the reaction at pressure p and at zero pressure respectively, $p_{1/2}$ is the pressure at which $K_{eq}=1$ ("half pressure" of the conversion), $\Delta V^\circ$ is molar reaction volume, associated with the conversion at p=0, and $C_0=[A]+[B]$. Computer program based on Marquardt and Neilder-Mead optimization algorithms was used to fit the data.

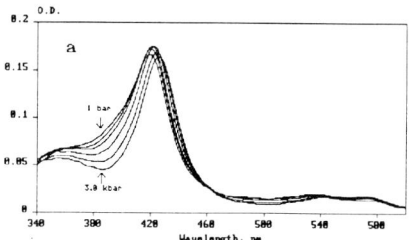

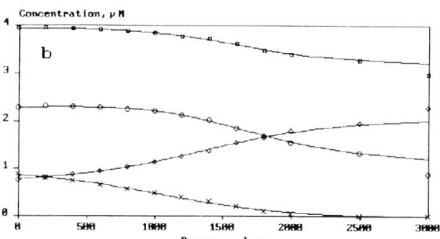

Fig.1. Pressure-induced changes in cytochrome P-450 2B4 in solution, 2 mM benzphetamine. Conditions: 4 µM 2B4 in 100 mM Na-Hepes buffer, pH 7.4, 1 mM EDTA, 1 mM dithiothreitol, 2 mM benzphetamine, 25 °C. Optical pathlength 5 mm. Spectra were measured at 1 bar and 0.6, 1.0, 1.6, 2.5, 3.0 kbar. a: Absolute spectra. b: changes in the concentrations of 2B4 species: P-450$_{hs}$ (x), P-450$_{ls}$ (o), P-420 ($\diamond$) and total hemoprotein ($\square$). Solid lines correspond to the results of data fitting.

## RESULTS

Pressure-induced transitions in oligomers of P-450 2B4 were found to be complex. Two processes of inactivation (P-450→P-420 and heme loss) were overlapped here with pressure-induced spin shift (fig 1). An appropriate results of the fitting of these inactivation processes could be obtained only from the assumption that the fraction of P-450, which is exposed to P-450→P-420 conversion, does not undergoes heme loss and vise versa. The fraction of P-450 exposed to P-450→P-420 conversion was amounted to be about 65% of initial 2B4 concentration. Amplitude of heme loss depends on the concentration of substrate, obeying hyperbolic (Michaelis-Menten) equation. Maximal amplitude of heme loss of about 30% was observed in the absence of substrate.

While heme loss process was irreversible, large part of pressure-generated P-420 reconverts into P-450 after decompression. To reach the maximal reconversion level about 2 hours of incubation at normal pressure are required. When decompressed after maximum pressure of 1.3 kbar, where about no heme loss was observed, 2B4 immediately turns back to the initial state. However, decompression after exposition of 2B4 to 3 kbar results only in partial reconversion of pressure-generated P-420 to P-450. The most interesting result here is the finding that the spin state of P-450 after decompression was not the same as that determined before application of the pressure. While 27% of the P-450 hemoprotein was initially represented by the high spin state, this portion amounted only to 10% after decompression. This finding might be interpreted as an evidence that the fractions of 2B4 exposed to P-450→P-420 conversion and heme loss have different positions of the spin equilibrium.

Based on these results we are inferring that two processes of pressure-induced inactivation (P-450→P-420 conversion and heme loss) are applied here to two different conformers of the hemoprotein representing about 65 and 35 per cent of its initial concentration respectively. No interconversions between these two states were detected. We suppose such 2:1 conformational splitting of 2B4 to be engendered by some peculiarities of organization of 2B4 oligomers (hexamers), which were shown to be organized as dimers of trimers (Tsuprun, et al., 1986).

*Pressure-induced spin transitions* of 2B4 versus benzphetamine concentration were studied. The following model of substrate binding and spin transition was used to interpret the data:

$$\begin{array}{ccc} P_l & \xrightleftharpoons{K_h} & P_h \\ K_s \updownarrow & & \updownarrow \\ PS_l & \xrightleftharpoons{K_{hs}} & PS_h \end{array}$$

$$K_s = \frac{([PS_l]+[PS_h])}{([P_l]+[P_h])\cdot[S]} ; \quad K_h = \frac{[P_l]}{[P_h]} ; \quad K_{hs} = \frac{[PS_l]}{[PS_h]} \quad (2)$$

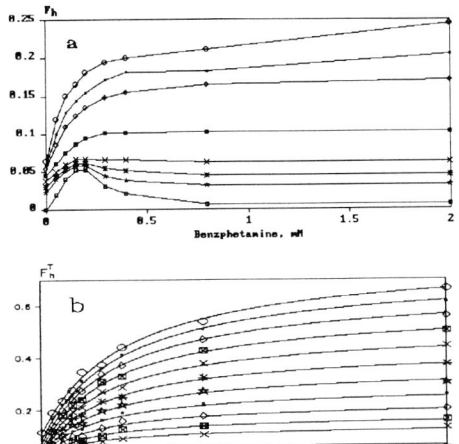

Fig.2. Spin state of 2B4 versus benzphetamine concentration at various pressures. a: Fraction of high spin form in the total P-450 ($F_h$). Curves obtained at 1 bar (o) and 0.4, 0.8, 1.2, 1.6, 1.8, 2.2, 2.5 ( ) kbar are shown. The points are connected by straight lines. b: Fraction of the high spin form in apparent heme-losing conformer ($F_h^T$). Curves obtained at 1 bar (o) and 0.2, 0.4, 0.6, 0.8, 1.0, 1.2, 1.4, 1.6, 1.8, 2.0, 2.5 (□) kbar are shown. Solid lines show the results of data fitting and correspond to the parameters shown in Table 1.

Table 1. Equilibrium constants and reaction volumes of the benzphetamine binding and spin shift transitions of P-450 2B4.

| Transition [a] | Equilibrium constant at zero pressure | $\Delta V°$ [b] ml/mol |
|---|---|---|
| $P450_l \rightleftharpoons P450_h$ | $K_h$ = 8.38 | -13.3 |
| $(P450-S)_l \rightleftharpoons (P450-S)_h$ | $K_{hs}$ = 0.28 | -36.2 |
| $P450 + S \rightleftharpoons P450-S$ | $K_s$ = 0.42 mM | 9.2 |

[a] The designations used here correspond to those in scheme (2);
[b] The sign of $\Delta V°$ corresponds to hs→ls spin shift and substrate complex dissociation transitions.

The spin state of 2B4 may be expressed by the fraction of the high-spin state ($F_h$) in the total P-450. The dependence of $F_h$ on benzphetamine concentration at different pressures is shown in fig. 2a. It could be seen that at the pressures higher than 1.7 kbar this dependency show a clear maximum at about 0.2 mM benzphetamine. At lower pressures these curves exhibit an apparent increase of the affinity of 2B4 for the substrate with increasing pressure. Such behavior of the hemoprotein was unexpected and seemed confusing.

In trying to explain this peculiarity we have supposed fraction of 2B4 exposed to P-450→P-420 transition to be represented by low spin only and excluded from spin equilibrium transitions. The dependence of high spin fraction in apparent heme-losing conformer, $F^T_h$, on benzphetamine concentration at different pressures is shown in fig. 2b. Apparent irregularities mentioned above disappeared. The results of global fitting of this data set to the equation reflecting pressure dependence of system (2) are shown in fig 2b by solid lines. Square coefficient of correlation for this fitting was greater 0.997. Therefore, we can deduce, that the proposition made is correct and that only the heme-losing conformer is exposed to the spin equilibrium system (2).

Parameters of substrate binding and spin transitions found by this fitting are shown in Table 1. The binding of the substrate itself is associated with a negligible reaction volume. Although, spin equilibrium constants at zero pressure for substrate-free and benzphetamine-bound hemoprotein differ by a factor of about 30. Thus, for 2B4, pressure-induced high-to-low spin shift is caused not by the dissociation of the substrate complex, as proposed for P-450CAM (Marden and Hui Bon Hoa, 1987), but by the spin equilibrium shift of both substrate-bound and substrate-free hemoprotein.

*Pressure-induced spectral changes in cytochrome P-450 monomers* in the presence of 0.2% Triton N-101, were found to be much simpler than those in oligomers. Only one pressure induced process which corresponds to P-450 inactivation by heme loss was found here. There was also some appearance of low amounts P-420 state during experiment. However, both disappearance of hemoprotein and production of P-420 reveal the same pressure-related parameters ($\Delta V°$ = 116 ml/mol; $P_\frac{1}{2}$ = 1.5 kbar) and were attributed to be consequences of single inactivation process involving disappearance of 6 molecules of P-450 and production of 1 molecule of P-420 instead. About 85% of 2B4 was exposed to such inactivation. Thus, in the presence of 0.2% Triton N-101 the pool of 2B4 monomers is considered to be nearly homogeneous.

For <u>cytochrome P-450 2B4 incorporated into proteoliposomes</u> pressure-related behavior was in general the same as for 2B4 oligomers in solution. P-450→P-420 transition and irreversible heme loss were characterized here by the same value of $P_{\frac{1}{2}}$ (1.6 kbar), but $\Delta V°$ of heme loss (125 ml/mol) was about 2 times higher that of P-450→P-420 transition (70 ml/mol). Amplitudes of both processes were accounted to be about 35% of the initial 2B4 content; the rest of the hemoprotein was stable up to 6 kbar. No dependence of these parameters on the concentration of benzphetamine was detected. Complete reconversion of pressure-generated P-420 back to P-450 was observed immediately after decompression. It was found, that pressure-related behavior of the sample, which was already pressurized (sample after decompression), was notably different from that of intact preparation. No further heme loss was detected during second application of high pressure, only reversible P-450→P-420 conversion and spin shift were observed. These results also could be interpreted to be an evidence of existence of few oligomerization-engendered conformers of 2B4 in the membrane.

<u>Incorporation of NADPH-cytochrome P-450 reductase</u> into 2B4-containing proteoliposomes (1:1 to hemoprotein) results in very prominent changes in the hemoprotein pressure-related behavior. About no heme loss was observed here. Some P-450→P-420 transition was detected only at very high pressures (above 2.5 kbar). The whole pool of the hemoprotein appears to be exposed to pressure-sensitive spin equilibrium. The same was observed also for P-450 3A4 in yeast microsomes containing high concentrations of NADPH-cytochrome P-450 reductase. These observation could be easily explained from the point of view of our hypothesis taking into account that P-450 oligomers in membrane appear to be dissociated in the presence of equimolar NADPH-cytochrome P-450 reductase (Gut, et al., 1986).

## CONCLUSIONS.

1. The pool of cytochrome P-450 2B4 (LM2) in oligomers in solution and in proteoliposomes are divided into two fractions, which have different pressure-related behavior and, therefore, are represented by distinct conformers of the hemoprotein.
2. The splitting of the pool of P-450 2B4 between these conformers is due in the proportion near to 2:1 and this ratio seems to be constant. No interconversion processes between these conformers were detected. This inhomogeneity of the pool of P-450 2B4 appears to be a result of it's supramolecular organization in the oligomers in solution and in the proteoliposomes as it vanishes at the solubilization of the oligomers by the detergent.
3. Our data is consistent with the hypothesis that there is no true overall spin equilibrium in LM2 oligomers. Only fraction of P-450, which undergoes pressure-induced heme loss (35%) is exposed to spin equilibrium. The second conformer, which is sensitive to pressure induced P-450→P-420 transitions is represented by low-spin state only.
4. Pressure-induced spin shift in cytochrome P-450 LM2 is not associated with the dissociation of substrate, since value of $K_s$ has a very weak dependence on pressure. This shift is caused by spin transitions of both substrate-free and substrate-bound protein.
5. Incorporation of equimolar NADPH-cytochrome P-450 reductase into 2B4-containing proteoliposomes stabilizes the hemoprotein. Only pressure-induced spin shift and some P-450→P-420 transition at high pressures were found here. All pressure-induced transitions here were completely reversible immediately after decompression.

<u>Acknowledgments.</u> The authors gratefully acknowledge that the present investigation became possible by a long-term fellowship given by INSERM to Dr. D.R. Davydov.

*References.*

Davydov,D.R., Knyushko,T.V., Hui Bon Hoa,G.(1993) <u>Biochem.Biophys.Res.Commun. 188</u>: 216-221.
Gut, J., Cherry, R.J., Kawato, S. (1982) <u>J.Biol.Chem. 257</u>: 7030-7036.
Hui Bon Hoa, G., Douzou, P., Dahan, N., and Balny, C. (1982) <u>Anal. Biochem. 120</u>:125-145.
Imai, Y. (1987) In: <u>Cytochrome P-450</u>, eds. R. Sato and T. Omura,pp.37 - 46. Tokyo: Kodansha Ltd.
Marden, M.C., and Hui Bon Hoa, G. (1987) <u>Arch. Biochem. Biophys. 253</u>: 100-107.
Mishin, V.M., Grishanova, A.Yu., and Lyakhovich, V.V. (1979) <u>FEBS Lett.104</u>: 300-302.
Tsuprun, V.L., Myasoedova, K.N., Berndt, P., et al. (1986) <u>FEBS Lett. 205</u>: 35 - 40.

# The effect of phenobarbitone and β-naphthoflavone on selected biochemical liver parameters following subchronic administration to female beagle dogs

E. Molitor[1], H. Thomas[1], J. Weymann[2], F. Waechter[1]

[1]Toxicology Services, Cell Biology, Ciba-Geigy Ltd., Basle, Switzerland. [2]Department of Biochemistry, Knoll AG, Ludwigshafen, FRG

## INTRODUCTION

In toxicity studies, Beagle dogs are frequently used as a non-rodent species. Compared to the huge number of data elaborated in rats and mice, only limited information about the effects of foreign compounds on dog liver drug metabolizing enzymes is available from the literature. This investigation was intended to characterize the liver effects of the model inducers phenobarbitone (PB) and ß-naphthoflavone (NF) in the female Beagle dog after intraperitoneal administration of 10 and 20 mg/kg body weight/day for 7 days, respectively.

## MATERIALS AND METHODS

Female Beagle dogs were randomly assigned to three groups of three individuals. Control animals were treated by intraperitoneal injection of sesame oil for 7 days. One group each, received a single daily intraperitoneal injection of 20 mg/kg body weight PB or 10 mg/kg body weight NF in sesame oil for 7 days. After the final administration the animals were fasted for 20 hours and then killed under Ketanest/Rompun-combination anesthesia (1 mg/kg body weight each) by exsanguination. Livers were removed, immediately frozen in liquid nitrogen and stored at -80°C until processed. 100 G supernatants were prepared by centrifugation of 20% (w/v) homogenates in 10 mM Tris/HCl buffer, pH 7.5, containing 250 mM sucrose for 15 min.

Microsomal and cytosolic fractions were obtained by centrifugation of a 12'000 g supernatant for 1 hour at 100'000 g. Microsomes were washed once by resuspending the 100'000 g pellets in 50 mM Tris/HCl buffer, pH 7.5, and centrifugation at 100'000 g for 1 hour.

Protein contents of 100 g supernatants, microsomal and cytosolic fractions were determined according to Smith et al. (1985). The microsomal cytochrome P-450 content was determined as described by Omura and Sato (1964).

Enzyme activities were elaborated as follows: cyanide-insensitive peroxisomal ß-oxidation (FAO) as described by Lazarow (1981); microsomal hydroxylation of lauric acid at position 11 (LA-11-OH) and 12 (LA-12-OH) as described by Thomas et al. (1992); microsomal (mEH) and cytosolic (cEH) epoxide hydrolase activities according to Waechter et al. (1988); microsomal ethoxycoumarin O-deethylase activity (ECOD) using the method of Aitio (1978); microsomal 7-ethoxyresorufin- (EROD) and 7-pentoxyresorufin-O-dealkylase (PROD) activities according to the method of Burke et al. (1985); regio- and stereoselective microsomal testosterone hydroxylation as described by Van der Hoeven (1984); microsomal UDP-glucuronosyltransferase (UDPGT) activity using 1-naphthol, morphine and bilirubin as substrates as described by Mackenzie and Hänninen (1980), Miners et al. (1988) and Heirwegh et al. (1972), respectively; cytosolic glutathione S-transferase (GST) activity as described by Habig et al. (1974).

For immunoblot analysis mixtures of equal volumes from microsomal suspensions of all animals per treatment group were subjected to SDS-PAGE, and monoclonal antibodies (purified IgG fractions) against purified rat liver cytochrome P-450 CYP1A, CYP2B, CYP3A and CYP4A proteins were used essentially as described by Waechter et al. (1988).

## RESULTS AND DISCUSSION

**Protein contents and enzyme activities:** Among the investigated parameters neither the subcellular protein contents nor the enzyme activities of LA-12-OH, cEH and GST responded to the treatment with PB or NF. No glucuronosyltransferase activities towards morphine and bilirubin were expressed constitutively or inducibly. Similar to rat and mouse (Okey, 1990; Waxman and Azaroff, 1992), phenobarbitone significantly induced the microsomal cytochrome P-450 content as well as PROD (10-fold), ECOD (3.4-fold), LA-11-OH (3.5-fold), mEH (1.8-fold) and a-naphthol-UDPGT (1.5-fold) activities.

Analogous to PB, treatment with NF caused increased cytochrome P-450 contents (3.1-fold) as well as ECOD (5.7-fold) and α-naphthol-UDPGT (1.7-fold) activities. In contrast to the barbiturate, however, NF induced EROD (9-fold) but not the activities of PROD, LA-11-OH and mEH (Table 1).

**Cytochrome P-450 isoenzyme proteins:** At the cytochrome P-450 protein level, immunoblot analysis with MAbs specific for rat liver CYP1A and CYP3A (Saad et al., 1993) visualized the induction of related cytochromes P-450 in the liver of dogs treated with NF and PB, respectively. No signal was found with MAb be4 specific for rat liver CYP2B1 and CYP2B2. However, MAb clo4, a monoclonal antibody diagnostic for CYP4A proteins in the rat and mouse, yielded a single protein signal in dog liver microsomes irrespective of treatment (Figure 1).

**Microsomal testosterone hydroxylation:** Total microsomal testosterone metabolism was increased in liver microsomes from PB treated dogs (2.5-fold) and, to a lesser extent, in microsomes from NF treated dogs (1.8-fold). With PB, the most prominent changes comprised the appearance of 16β-OHT, which was not detectable in control animals, and a more than 5-fold induction of a major unidentified metabolite U1. A distinct induction of 2β-, 6β- and 15β-OHT, known to be indicative for a CYP3A co-induction by PB in rodents (Sonderfan et al., 1987), was also observed in the dog. With NF, major inductive effects were seen for the 6α-, 6β- and 16α-hydroxylation products as well as for the major unidentified metabolite U2.

To the current knowledge, the dog liver microsomal cytochrome P-450 system of untreated dogs appears to be much less complex than the corresponding monooxygenase system in rodents. It rather seems to be represented mainly by the dominating cytochrome P-450 forms PBD-2 (Duignon et al., 1988; Graves et al., 1990) and PBD-1 (Ciacco and Halpert, 1989). Nevertheless, treatment with PB and NF, two known model inducers in rodents (Okey, 1990) evoked a surprisingly similar induction pattern in dogs as compared to rats. This with respect to enzyme activities and immunochemically detectable cytochrome P-450 isoenzyme contents. Therefore, in conclusion, the currently available set of parameters to phenotype a cytochrome P-450 inducer in rodents, was shown to be also diagnostically sufficient for an inducer phenotyping in dogs.

**TABLE 1:** The Effect of PB and NF on Selected Biochemical Parameters in the Dog Liver.

| Parameter | | CON | PB | NF |
|---|---|---|---|---|
| 100 x g Supernatant Protein | (mg/g liver) | 148 ± 6 | 152 ± 7 | 146 ± 1 |
| Microsomal Protein | (mg/g liver) | 19 ± 1 | 22 ± 4 | 20 ± 4 |
| Cytosolic Protein | (mg/g liver) | 83 ± 2 | 89 ± 5 | 95 ± 17 |
| Cytochrome P-450 | (nmol/g liver) | 6.5 ± 0.7 | 21.2 ± 2.8*** | 20.2 ± 3.0**** |
| EROD | (nmol/min/g liver) | 1.1 ± 0.3 | 1.1 ± 0.5 | 9.9 ± 4.5** |
| PROD | (nmol/min/g liver) | 0.2 ± 0.1 | 2.1 ± 0.5*** | 0.3 ± 0.1 |
| ECOD | (nmol/min/g liver) | 5.8 ± 2.1 | 19.5 ± 5.7** | 33.3 ± 3.2*** |
| LA-11-OH | (nmol/min/g liver) | 3.2 ± 0.1 | 11.0 ± 1.9*** | 3.9 ± 0.4 |
| LA12-OH | (nmol/min/g liver) | 3.8 ± 0.5 | 4.3 ± 0.6 | 4.0 ± 0.6 |
| FAO | (nmol/min/g liver) | 608 ± 141 | 497 ± 126 | 268 ± 96* |
| mEH | (nmol/min/g liver) | 234 ± 52 | 433 ± 125 | 235 ± 49 |
| cEH | (nmol/min/g liver) | 13 ± 1 | 11 ± 2 | 10 ± 2 |
| GST | (μmol/min/g liver) | 63 ± 3 | 70 ± 9 | 53 ± 14 |
| α-Naphthol-UDPGT | (nmol/min/g liver) | 2.9 ± 0.7 | 4.5 ± 0.7* | 5.0 ± 0.5* |
| Morphine-UDPGT | (nmol/min/g liver) | n.d. | n.d. | n.d. |
| Bilirubin-UDPGT | (nmol/min/g liver) | n.d. | n.d. | n.d. |

Values are means ± S.D. from three animals.
Asterisks indicate results different (two sided Dunnett's Test) from control: * $p<0.05$, ** $p<0.01$, *** $p<0.001$.
n.d.: not detectable

**TABLE 2:** The Effect of PB and NF on Dog Liver Microsomal Testosterone Metabolism.

| Metabolites | CON | PB | NF |
|---|---|---|---|
| 6α-OHT | 1.8 ± 0.8 | 2.5 ± 1.1 | 4.7 ± 1.5* |
| 7α-OHT | 0.2 ± 0.3 | 0.2 ± 0.3 | 0.4 ± 0.1 |
| 16ß-OHT | <0.1 | 1.1 ± 0.2** | <0.1 |
| 16α-OHT | 12.6 ± 6.7 | 27.8 ± 20.4 | 33.4 ± 2.9 |
| Androstendione | 4.2 ± 0.5 | 7.9 ± 2.8 | 6.5 ± 2.4 |
| 2α-OHT | <0.1 | <0.1 | <0.1 |
| 6ß-OHT | 7.0 ± 1.8 | 18.2 ± 5.0* | 16.5 ± 7.0* |
| 2ß-OHT | 1.3 ± 0.5 | 4.3 ± 1.0** | 1.0 ± 0.3 |
| 15ß-OHT | 6.4 ± 1.4 | 11.9 ± 8.8 | 1.7 ± 1.5 |
| unidentified U1 | 9.1 ± 3.2 | 49.0 ± 15.7** | 6.5 ± 1.1 |
| unidentified U2 | 12.6 ± 4.8 | 24.4 ± 13.8 | 34.4 ± 9.2* |
| unidentified U3 | 2.1 ± 1.7 | 2.2 ± 1.4 | 2.2 ± 0.6 |
| unidentified U4 | 8.3 ± 0.3 | 7.6 ± 2.9 | 8.7 ± 1.9 |
| total metabolites | 66 ± 18 | 157 ± 56* | 117 ± 22 |

The values, expressed in nmol/min/g liver, are means ± S.D. from three animals
Asterisks indicate results different (two sided Dunnett's Test) from control: * $p<0.05$, ** $p<0.01$, *** $p<0.001$.

# 1,3-butadiene exposure of rats: hemoglobin adducts of 1,2-epoxybutene and cytochrome P450-related changes in styrene metabolism

Eivor Elovaara[1], Siv Osterman-Golkar[2], Tamie Nakajima[3], Harry V. Gelboin[4], Vesa Riihimäki[1], Marja Sorsa[1]

[1]Institute of Occupational Health, 00250 Helsinki, Finland. [2]Department of Radiology, Stockholm University, 10691 Stockholm, Sweden. [3]Shinshu University School of Medicine, Matsumoto, 390 Japan. [4]National Cancer Institute, Bethesda, MD 20892, USA

## Summary

Rats were exposed by inhalation to 1,3-butadiene (**BD**) gas for 5 days (500 ppm, 6h/day) and simultaneously to L-buthionine-[S,R]-sulfoximine (**BSO**) and/or acetone in drinking water. Overt adverse effects (weight loss, decrease in tissue non-protein sulfhydryls, **NPSH**) were observed in the group treated with BD+BSO+acetone. The rate of styrene 7,8-oxidation (at 87 µM styrene concn.) was enhanced 1.5-fold in liver microsomes but less in lung microsomes of rats exposed to BD only. Cytochrome P450 (CYP) 2E1 (by Western blotting) and epoxide hydrolase were clearly induced by BD in the liver. In the lungs, CYP 2E1 was not detected, and CYP 2B1/2, the main pulmonary P450 form, was not induced by any treatment. A 6-fold higher hepatic activity of N-nitrosodimethylamine N-demethylase was observed in BD+acetone treated rats as compared to BD-only treated rats, but no difference was found between these two groups in the 1,2-epoxybutene-hemoglobin adduct levels. BSO, which inhibits glutathione synthesis, lowered the levels of glutathione in the liver, lung and blood, and increased the level of 1,2-epoxybutene-hemoglobin adducts in BD+BSO+acetone treated animals.

## Introduction

Potential carcinogenicity of 1,3-butadiene is the dominant reason to control occupational exposure. Analysis of adducts of 1,2-epoxybutene in peripheral blood offers a tool for the dosimetry of the primary reactive metabolite of BD in man and animals. There are marked species differences in the toxicity, metabolism and exposure-induced body burden of BD. The metabolic clearance of BD is linearly related to the ambient exposure concentration up to about 1000 ppm in rats (ECOTOC, 1993). In order to explore the relationship between BD-induced toxicity, hemoglobin adduct levels and BD metabolism, rats were exposed to BD and simultaneously given BSO and/or acetone in drinking water. Acetone was given to induce CYP 2E1, because this cytochrome P450 is known to catalyze the initial oxidation of BD to 1,2-epoxybutene (Csanady et al., 1992). Subsequent inactivation of 1,2-epoxybutene, which is a potential carcinogenic metabolite of BD, is catalyzed by epoxide hydrolases and glutathione S-transferases. In this study, BSO was given to inhibit the synthesis of glutathione, and thereby detoxification by glutathione S-tranferases.

Table 2. Effects of 1,3-butadiene inhalation exposure on 1,2-epoxybutene-hemoglobin adducts, and tissue non-protein sulfhydryl levels in differently treated rats.

| BUTADIENE CONCN. [ppm] | ADDITIONAL TREATMENT | HEMOGLOBIN ADDUCTS[A,B] [pmol / g globin] | NON-PROTEIN SULFHYDRYLS [µmol / ml or g tissue] | | |
|---|---|---|---|---|---|
| | | | BLOOD | LIVER | LUNG |
| 0 | Water | n.a. | 0.80 ± 0.09 | 5.99 ± 0.17 | 1.17 ± 0.11 |
| 0 | Acetone | n.a. | 0.97 ± 0.06 | 6.11 ± 0.21 | 1.13 ± 0.07 |
| 0 | BSO-Acetone | n.a. | 0.48 ± 0.01 | 2.26 ± 0.28 | 0.58 ± 0.04 |
| 500 | Water | 4657 ± 154 | 0.78 ± 0.04 | 6.01 ± 0.20 | 1.33 ± 0.06 |
| 500 | Acetone | 4634 ± 271 | 0.85 ± 0.08 | 5.34 ± 0.30* | 1.07 ± 0.09 |
| 500 | BSO-Acetone | 5281 ± 815 | 0.45 ± 0.04 | 1.07 ± 0.20* | 0.36 ± 0.03* |

Values are the mean ± SE of 5-6 rats. [A] Preliminary results; in BSO-acetone group only 4 rats.
[B] Hemoglobin adducts are not detected in control rats (n.a = not analysed).
* Differs significantly from controls: $p \leq 0.05$ (Student's t-test).

METHODS

Male Wistar rats (316±31g) were exposed in a dynamic exposure chamber (1m$^3$) to 500 ppm BD (6h/d, 5 days) and subgrouped by additional treatment: (i) BD (ii) BD + acetone, or (iii) BD + BSO + acetone. Acetone (1%) and BSO (25 mM) were given in drinking water throughout the 5-day experiment. Controls were sham exposed to ambient air. Rats were killed and sampled within 1 to 4 hours after the end of the last exposure. Adducts of 1,2-epoxy-3-butene to N-terminal valine in hemoglobin were measured essentially as described by Osterman-Golkar et al. (1991). Analyses of protein, NPSH, N-nitrosodimethylamine N-demethylase (**NDMAD**), 7-ethoxycoumarin O-deethylase (**ECOD**), 7-pentoxyresorufin O-dealkylase (**PROD**) and epoxide hydrolase (**EH**) were assayed by published methods (Elovaara et al., 1990 & 1991). Styrene metabolism was measured with HPLC essentially as described (Nakajima et al., 1993). Monoclonal antibodies against CYP 2E1 and 2B1/2 were used for Western blotting (Elovaara et al., 1991).

RESULTS AND DISCUSSION

The BD exposure elicited only adaptive effects (induction of hepatic enzymes) in control rats. This response to BD was marginally influenced by simultaneous exposure to acetone, whereas the treatment with BSO+acetone clearly aggravated the toxicity by BD; one rat died (after 3 days of BD exposure) and the survivors (5 rats) showed a severe decrease of body weights. The relative weights of the lung and liver were not affected but the relative spleen weight was lowered. None of the treatments elevated alanine aminotransferase levels in plasma.

No enzyme induction was observed in lung microsomes by BD (Table 1). CYP 2B1/2 was the only form detected by Western blotting. The pulmonary PROD activity was higher than the ECOD activity, and the former was also 10 times higher than that in the liver. Acetone treatment did not enhance the rate of styrene 7,8-oxidation in the lungs as it did in the liver. Pulmonary oxidation (activation) of styrene was highest in BD-only treated rats and lowest in BSO+acetone+BD. A

Table 1. Effects of 1,3-butadiene inhalation exposure on the drug-metabolizing enzymes of the liver and the lung in differently treated rats

| BUTADIENE CONCN. [ppm] | ADDITIONAL TREATMENT | STYRENE 7,8-OXIDATION (at 87 µM) | (at 1.7 mM) | ECOD | NDMAD | PROD | EH |
|---|---|---|---|---|---|---|---|
| | | | L I V E R | | | | |
| 0 | Water | 0.26 ± 0.07 | 2.21 ± 0.31 | 0.76 ± 0.12 | 0.34 ± 0.03 | 0.010 ± 0.003 | 4.08 ± 0.98 |
| 0 | Acetone | 0.50 ± 0.05 | 1.87 ± 0.60 | 2.55 ± 0.47 | 2.44 ± 0.58 | 0.011 ± 0.002 | 4.59 ± 0.84 |
| 0 | BSO-Acetone | 0.55 ± 0.13 | 1.07 ± 0.21 | 2.95 ± 0.64 | 2.38 ± 0.61 | 0.012 ± 0.002 | 5.89 ± 1.33 |
| 500 | Water | 0.39 ± 0.06* | 2.39 ± 0.31 | 0.93 ± 0.12* | 0.47 ± 0.13* | 0.012 ± 0.003 | 5.44 ± 0.64* |
| 500 | Acetone | 0.51 ± 0.12 | 1.90 ± 0.24 | 2.36 ± 0.42 | 2.30 ± 0.32 | 0.010 ± 0.002 | 5.39 ± 0.55* |
| 500 | BSO-Acetone | 0.46 ± 0.13 | 1.19 ± 0.43 | 1.80 ± 0.32* | 1.77 ± 0.45* | 0.020 ± 0.004* | 7.27 ± 0.39* |
| | | | L U N G | | | | |
| 0 | Water | 0.66 ± 0.15 | 2.11 ± 0.24 | 0.074 ± 0.025 | | 0.133 ± 0.041 | |
| 0 | Acetone | 0.53 ± 0.18 | 2.22 ± 0.51 | 0.093 ± 0.020 | | 0.119 ± 0.021 | |
| 0 | BSO-Acetone | 0.41 ± 0.12 | 1.92 ± 0.20 | 0.082 ± 0.016 | | 0.136 ± 0.030 | |
| 500 | Water | 0.75 ± 0.11 | 2.18 ± 0.13 | 0.092 ± 0.023 | | 0.117 ± 0.028 | |
| 500 | Acetone | 0.58 ± 0.10 | 2.00 ± 0.23 | 0.094 ± 0.022 | | 0.114 ± 0.017 | |
| 500 | BSO-Acetone | 0.36 ± 0.12 | 1.68 ± 0.25 | 0.069 ± 0.024 | | 0.095 ± 0.022* | |

Values are the mean ± SD of 5-6 rats, and the enzyme activities are given as [nmol/min/mg microsomal protein].
ECOD = 7-ethoxycoumarin O-deethylation;   NDMAD = N-nitrosodimethylamine N-demethylation;
PROD = 7-pentoxyresorufin O-dealkylation;   EH = epoxide hydrolase with (50 mM) styrene oxide as substrate.
* Differs significantly from controls: $p \leq 0.05$ (Student's t-test).

plausible explanation for this is that the combined treatment caused P450 lesions in the lungs of the latter group (Table 1).

BD increased CYP 2E1 expression in rat liver microsomes (shown by Western blots) and also the rate of styrene 7,8-oxidation (1.5 times) when measured at a low styrene concentration. The activities of ECOD, NDMAD and EH were also induced by BD (Table 1). On the other hand, it was shown, by immunoblotting, that BD lowered the amount of 2E1 in BSO+acetone rats, and by enzyme assays, that BD concomitantly decreased the activities of styrene 7,8-oxidation (slightly), ECOD and NDMAD. The activity of EH, which was increased by BD in all groups, was highest in BD+BSO+acetone treated rats. The activity of NDMAD was greatly induced with acetone in all rat livers. NDMAD was 6 times higher in the BD+acetone treated animals than in the BD-only treated animals (Table 1). NDMAD is a marker enzyme for CYP 2E1, which is the major P450 enzyme oxidizing BD to 1,2-epoxybutene in the liver (Csanady et al., 1992). In spite of the marked differences in the hepatic expression of 2E1, there was no difference in the hemoglobin adduct levels between BD+acetone and BD-only exposed rats (Table 2).

The treatment with BSO effectively lowered NPSH levels in blood (by 40%), liver (62%) and lungs (50%) whereas acetone had no effect (Table 1). BD inhalation decreased significantly the NPSHs in liver and lungs of the BSO+acetone treatment group. The 1,2-epoxybutene adduct levels were slightly (but not significantly) increased in BD+BSO+acetone treated rats when compared to those of BD+acetone or BD-only-treated rats, being consistent with the protective role of glutathione in detoxifying hemoglobin-reactive species. To what extent hemoglobin adduct levels reflect DNA adduct levels in rat target tissues (liver, lung) is not known.

ACKNOWLEDGEMENTS. We wish to thank the support from CEC (Contract STEP-CT91-0152).

REFERENCES

Csanady, G.A., Guengerich, F.P. and Bond, J.A. (1992): Comparison of the biotransformation of 1,3-butadiene and its metabolite, butadiene monoepoxide, by hepatic and pulmonary tissues from humans, rats and mice. *Carcinogenesis* 13, 1143-1453.

ECETOC Special Report No.4 (1993): 1,3-Butadiene criteria document. European Centre for Ecotoxicology and Toxicology of Chemicals, B-1160 Brussels, Belgium.

Elovaara, E., Vainio, H. and Aitio, A. (1990): Pulmonary toxicity of inhaled styrene in acetone-, phenobarbital- and 3-methylcholanthrene-treated rats. *Arch. Toxicol.* 64, 365-369.

Elovaara, E., Engström E., Nakajima, T., Park, S.S., Gelboin, H.V. and Vainio, H. (1991): Metabolism of inhaled styrene in acetone-, phenobarbital-, and 3-methylcholanthrene pretreated rats: stimulation and stereochemical effects by induction of cytochromes P450IIE1, P450IIB and P450IA. *Xenobiotica* 21, 651-661.

Nakajima, T., Elovaara, E., Gonzalez, F.P., Gelboin, H.V., Vainio, H. and Aoyama, T. (1993): Characterization of the human cytochrome P450 isozymes responsible for styrene metabolism. In *Butadiene and Styrene: Assessment of Health Hazards*. IARC Scientific Publications No. 127, ed. M.Sorsa, K.Pelkonen, H.Vainio & K.Hemminki. Lyon, International Agency for Research on Cancer, IARC, (in press).

Osterman-Golkar, S., Kautiainen, A., Bergmark E., Håkansson, K. and Mäki-Paakkanen, J. (1991): Hemoglobin adducts and urinary mercapturic acids in rats as biological indicators of butadiene exposure. *Chem-Biol Interactions* 80, 291-302.

# Tritium planigraphy study on cytochrome P450 2B4 structure

Ekaterina V. Apletalina, Valentin Yu. Uvarov, Olga M. Ipatova

*Institute of Biomedical Chemistry, Russian Academy of Medical Sciences, Pogodinskaya 10, Moscow 119832, Russia*

## INTRODUCTION

At present it is not yet possible to determine the tertiary structure of microsomal cytochrome P450 by X-ray analysis. Therefore we used the tritium planigraphy method to study topography of cytochrome P450 2B4. The essence of the method is that protein (or any other organic compound) is treated with thermally activated tritium atoms. Such treatment leads to label incorporation to amino acid residues, the probability of incorporation being dependent on both chemical nature of residue and its spatial arrangement in protein molecule (Baratova et al.,1982; Goldansky et al., 1982). The short path run by reactive atoms in condensed phase (1-3 Å) determines the predominant label localization in the surface-located polypeptide chain's regions that are accessible for non-scattered stream of tritium atoms. That in its turn determines the correlation between intramolecular distribution of tritium label and protein's spatial structure. The tritium planigraphy method was applied with profit to studying the topography of some proteins, for example, TMV protein and bacteriorhodopsin (Goldanskii et al., 1988; Alyonycheva et al., 1987).

## MATERIALS AND METHODS

Cytochrome P450 2B4 purified from liver microsomes of phenobarbital-treated rabbits was kindly provided by Dr. Ya.M. Cohen. Cytochrome P450 2B4-containing proteoliposomes were prepared by incubation of cytochrome with sonicated egg yolk phosphatidylcholine liposomes. For incorporation of radioactive label proteoliposomes were irradiated with tritium atoms with temperature 2000 K as described by Baratova et al.(1982). For this purpose proteoliposomes (enzyme concentration approximately 0.5 mg/ml) in 4.8 ml of 50 mM K-phosphate buffer, pH 7.5, were applied as a thin layer on the inner surface of precooled special vessel. To prevent cytochrome thermal degradation the vessel was kept at temperature of liquid nitrogen all the time during irradiation. On completion of irradiation procedure the preparation was defreezed, lyophilized and subjected to chromatography on a Sephadex G-25 column (1.8 x 30 cm) to remove labile label. Then lipids were removed by adding 2 volumes of chloroform:methanol mixture (1:2, v/v) to proteoliposomes solution, and cytochrome was carboxymethylated under reducing conditions accoding to standard procedure. After that cytochrome was treated sequentially with CNBr and lysylendopeptidase (LEP). Standard procedure was used for CNBr hydrolysis. Cleavage by LEP was done in 50 mM ammonium bicarbonate buffer, pH 9.0, containing 2.3 M Gu HCl, at enzyme:substrate molar ratio 1:25. Reaction proceeded for 22 h at $25^{\circ}C$. Peptide mixture was separated on Ultrasphere ODS column (4.6 x 250 mm, Altex) using a linear gradient of 0-70% acetonitrile (v/v) in 0.1% (aqueous) TFA (v/v) over 90 min at flow rate 1 ml/min. Further HPLC purification of peptides collected involved the same solvents and a more shallow gradient of acetonitrile. Purified peptides were sequenced on Milligen/Biosearch ProSequencer 6600 equipped with on-line HPLC. Recoveries of PTH-derivatives of amino acids were determined by comparison with internal standards. Fractions collected after each cycle of Edman degradation were concentrated and then quantitated by liquid scintillation counting. Based on obtained data the specific radioactivities of amino acids were calculated by assignment of radioactivities to yields of amino acids' PTH-derivatives according to sequence data.

## RESULTS AND DISCUSSION

Using HPLC the quite good separation of peptide mixture obtained by hydrolysis was acieved. By now we have purified and sequenced 5 peptides. They correspond to residues 317-326, 347-373, 374-384, 422-427 and 428-433 of amino acid sequence. For some of amino acids of these peptides the specific radioactivity values were determined. Thus in peptide 317-326 the values were determined for all but the last residue, in peptides 347-373 and 374-384 - for the first 4 and 5 residues, correspondingly, in peptide 422-427 - for residues 423, 424, and 426, and in peptide 428-433 the specific radioactivity values were determined for all amino acid residues. For the remaining residues these values were not found because the poor yields of PTH-derivatives of these amino acids did not allow the valid radioactivity determination.

Fig. 1 represents the profile of accessibility of amino acid residues for tritium. The profile is obtained from data on specific radioactivities of amino acids. As seen, fragments 347-350, 374-378 and 317-325 are well accessible for tritium. This indicates their location on water-exposed surface of cytochrome P450 2B4 molecule. On the other hand, residues 428-432 are much worse labeled as compared with these three fragments, and K433 is not labeled at all. In fragment 422-427 N423 and F426 are also non-radioactive, but E424 is well accessible for tritium. The following interpretation for these data may be given. Two alternatives of spatial arrangement of fragment 428-433 that may explain its low degree of labeling are possible. First, this fragment may be located in a pocket or groove formed by water-exposed surface of cytochrome molecule. Second, this fragment may be located near lipid bilayer. In both cases such spatial arrangement of this peptide would hinder the tritium access to residues. But according to model of tertiary structure of cytochrome P450 2B4 (see below) the second alternative is less preferential. Region 423-426 is apparently packed in such a way that N423 and F426 face the inside of protein globule and E424 is directed to water phase.

The data obtained were compared with cytochrome P450 2B4 tertiary structure model derived by computer-assisted comparison with the known tertiary structure of cytochrome P450cam (Kruger, 1992). Skeleton of cytochrome P450 2B4 was built using the program SYBYL on Graphics Work Station of Evans and Sutherland 350 (the coordinates of atoms are kindly provided by Drs. Stier and Kruger). According to model peptides 317-325, 347-350 and 374-378 are arranged on prominent areas of water-exposed surface of cytochrome which is in agreement with our results. Peptide 428-433 is located in small pocket formed by surface. This is also consistent with the experimental results obtained. The discrepancy concerns the arrangement of peptide 423-426. According to model these residues face to water phase. At present we can not explain this disagreement, but the further investigations on cytochrome P450 2B4 structure by means of tritium labeling ( that are in progress now) are believed to resolved the contradiction and provide the information on topography of other regions of the enzyme.

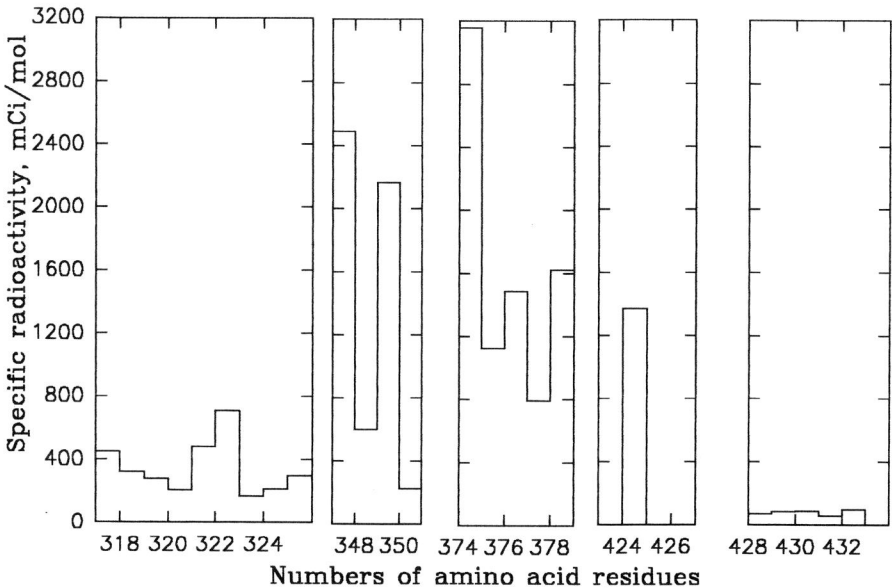

Fig. 1. Intramolecular distribution of tritium label in cytochrome P450 2B4 peptides

REFERENCES

Alyonycheva, T.N., Kuryatov, A.B., Antropova, L.P., Shemyakin, V.V., Neiman, L.A., Tsetlin, V.I., and Ivanov, V.T. (1987): Tritium planigraphy study on the bacteriorhodopsin topography. *Bioorg. khimia* 13, 898-907 (in Russian).

Baratova, L.A., Goldansky, V.I., Rumyantsev, Yu.M., Unukovich, M.S., and Shishkov, A.V. (1982): Study of the three-dimensional structure of proteins by means of tritium labelling. I. Free amino acids as a model of residues in an unfolded polypeptide chain. *Mol. biologiya* 16, 117-122 (in Russian).

Goldansky, V.I., Rumyantsev, Yu.M., Shishkov, A.S., Baratova, L.A., and Belyanova, L.P. (1982): Study of the three-dimensional structure of proteins by means of tritium labelling. II. Intramolecular distribution of tritium in the N-terminal part of myoglobin and the tertiary structure of protein. *Mol. biologiya* 16, 117-122 (in Russian).

Goldanskii, V.I., Kashirin, I.A., Shishkov, A.V., Baratova, L.A., and Grebenshikov, N.I. (1988): The use of thermally activated tritium atoms for structural-biological investigations: the topography of the TMV protein-accessible surface of the virus. *J. Mol. Biol.* 201, 567-574.

Kruger, V. (1992) Fluoreszenzespectroskopische untersuchungen der membrantopologie von cytochrome P450. *Ph. D.-degree thesis* Georg-August-Universitat zu Gottingen, Gottingen, 117 pages.

# Carcinogenic activation of cooked food heterocyclic amines by cytochrome P450 1A2. A theoretical study

Ferran Sanz[1], Juan José Lozano[1], Elena López-de-Briñas[1], Rafael de-la-Torre[2], Francesc Manaut[1]

*Departments of [1]Medical Informatics and [2]Pharmacology and Toxicology, Institut Municipal d'Investigació Mèdica, Universitat Autónoma de Barcelona, Dr. Aiguader 80, E-08003 Barcelona, Spain*

Some heterocyclic amines present in cooked food become carcinogenic after an activation by the cytochrome P-450 1A2 which catalyses the oxidation of the amino group (Shimada, 1989). N-hydroxylamines metabolites appear to be responsible for the observed mutagenicity of these compounds (Wakabayashi, 1992).

In previous works (Sanz, 1993), we described patterns of Molecular Electrostatic Potential (MEP) that are present in the caffeine molecule, in other xanthines and in some quinolones that inhibits the metabolism of caffeine.

Our group has developed and tested several algorithms and computer programs for analyzing MEP distributions, and for computing electrostatic similarity coefficients (Sanz, 1993).

In the present work, the structural and electrostatic features of a series of heterocyclic amines are analyzed and compared among themselves and with those of other substrates of cytochrome P-450 1A2.

Work partially supported by a CESCA grant.

## METHODS

Structural formulae of the considered compounds are shown in Fig. 1.

Biological data were taken from the work of Shimada, 1989 (umu gene expression in presence of human P-450 1A2).

The starting geometries were created with the INSIGHT II module of the BIOSYM software running on a Silicon Graphics workstation. The molecular geometries of considered procarcinogens were optimized at the STO-3G level using the GAUSSIAN 92 package on a CRAY supercomputer.

A 3-21G "single point" of these molecules was calculated in order to obtain the molecular orbital coefficients to be used in the computation of the molecular electrostatic potential.

MEP was computing in a 3-D grid around the molecules using the following expression:

$$V(R) = \Sigma \frac{Z_I}{|R - R_I|} - \int \frac{\delta(r)}{|R-r|} dr$$

where **V(R)** is the MEP value at the point R, $Z_I$ are the nuclear charges, $R_I$ are the nuclear coordinates and $\delta(r)$ is the electron density distribution of the molecule. Electron density distributions can be obtained both from semiempirical or *ab initio* wavefunctions.

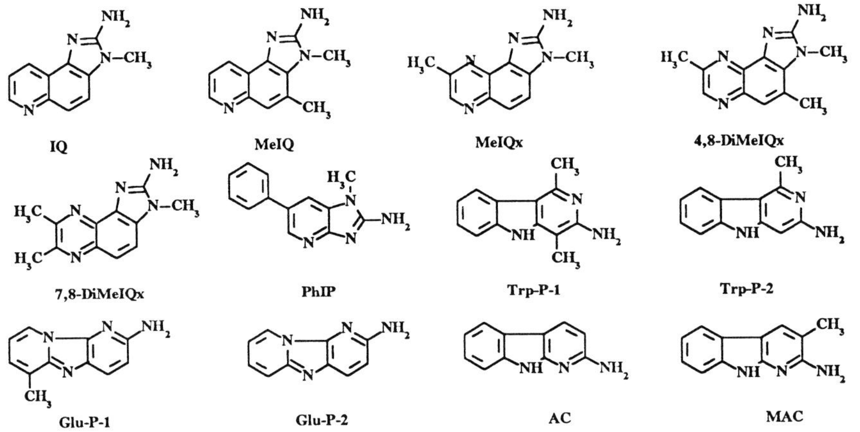

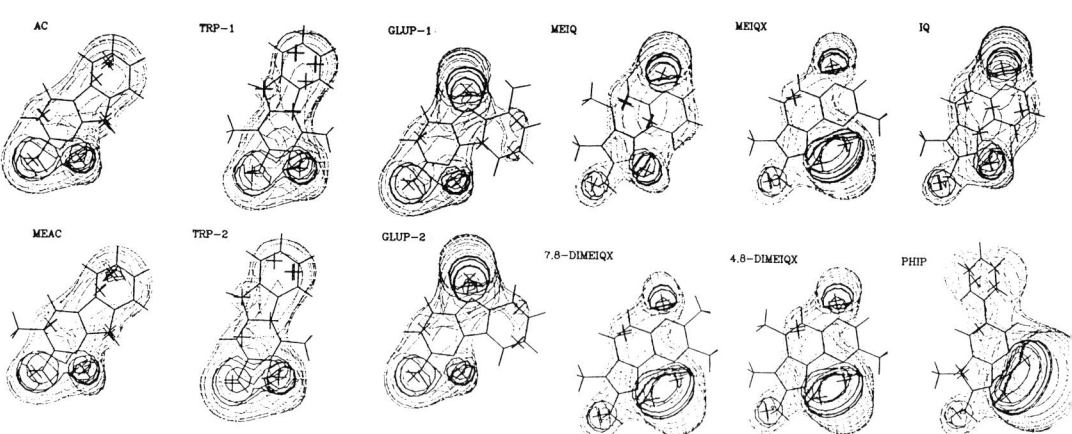

**Fig. 1** Considered Compounds.

Minimum values of MEP and electrostatic similarity coefficients were obtained using the MEPSIM package (4) developed by the authors. Similarity coefficients are defined using the Spearman rank correlation coefficient.

Statistic analysis were carried out using the SSPS package.

**Fig. 2** Isopotential surfaces and MEP minima.

# RESULTS and CONCLUSIONS

All the molecules studied are structurally different but have similar MEP shapes. Fig. 2 shows isopotential surfaces corresponding to -25 Kcal/mol and -10 Kcal/mol. MEP minima around the molecules are depicted by crosses.

A preliminary quantitative study of the similarity of the MEP distributions of the amines of the series studied by Shimada (1) versus the most active one (MEIQ) has been performed.

| COMPOUND | LOG(ACTIVITY) | SIMILARITY |
|----------|---------------|------------|
| MEIQ     | 4.53          | 1.00       |
| IQ       | 4.47          | 0.93       |
| MEIQX    | 3.73          | 0.83       |
| GLU-P1   | 3.49          | 0.73       |
| TRP-P1   | 3.06          | 0.47       |
| TRP-P2   | 2.87          | 0.46       |
| GLU-P2   | 1.95          | 0.72       |

The two variables have a Pearson correlation coefficient of 0.69 ($p=0.087$). The elimination of the compound GLU-P2 produces an improvement of the coefficient to 0.97 ($p<0.002$). It must be investigated if there are other factors different from electrostatics that can explain this low activity.

All the studied molecules have a minimum of MEP near the group to be oxidated by cytochrome P450-1A2. Fig. 3 shows the distances between this MEP minimum and the group to be oxidated. This figure also display the deepness of the minimum.

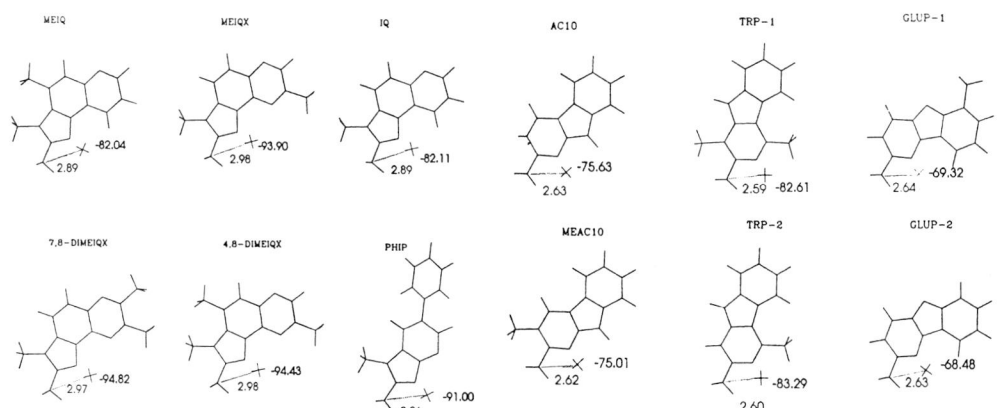

**Fig. 3** Distance and deepness of MEP minimum close to the amine to be oxidated.

The distance (2.5-3 Å) between the MEP minimum and the amino group to be oxidated is the same that is present in other substrates like caffeine, and inhibitors like furafylline and enoxacin (3). Fig. 4 displays the molecules of caffeine, furafylline and enoxacin showing their MEP minima and their group involved in oxidation or inhibition.

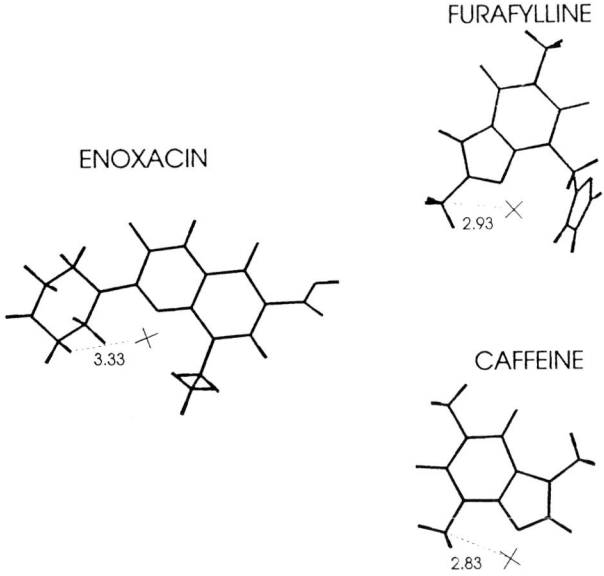

Fig. 4 MEP mimimum of other substrates and inhibitors of P-450 1A2.

REFERENCES

BIOSYM Technologies Inc, 10065 Barnes Canyon Rd, San Diego, CA 92121.
GAUSSIAN 92, Gaussian Inc, Pittsburgh, PA 15213.
Sanz F, López de Briñas E, Rodríguez J, Manaut F.(1993): In: Trends in QSAR and Molecular Modelling 92. pp 193-196. ESCOM. Leiden.
Sanz F, Manaut F, Rodríguez J, Lozoya E, López de Briñas E.(1993): MEPSIM: A computational package for analysis and comparison of molecular electrostatic potentials. J.Comput.-Aided Mol.Design 7, 337-347.
Shimada T, Iwasaki M, Martin M.V, Guengerich F.P. (1989): Human liver Microsomal Cytochrome P-450 Enzymes Involved in the bioactivation of procarcinogens detected by umu gene response in Salmonella typhimurinum TA 1535/pSK1002. Cancer Res 49, 3218-3228.
SPSS, SPSS Inc, Chicago.
Wakabayashi K, Nagao M, Esumi H, Sugimura T (1992): Food-derived mutagens and carcinogens. Cancer Res 52s, 2092-2098.

# Three-dimensional structure of active-site of cytochrome P450 1A2

Francesc Manaut, Elena López-de-Briñas, Juan José Lozano, F. Sanz

Department of Medical Informatics, Institut Municipal d'Investigació Mèdica, Universitat Autònoma de Barcelona, Dr. Aiguader 80, E-08003 Barcelona, Spain

Although the primary sequence of several mammalian cytochromes P-450 has been established, the obtention of 3-D structures of these cytochromes using homology techniques is difficult because the only crystallographic structure described is that of cytochrome P-450cam, and this prokaryote cytochrome has only 10-20% of homology with the mammalian forms.

Our study is focused on building a 3-D model for the active-site of human cytochrome P-450 1A2.

This isoform is involved in the metabolism (demethylations) of caffeine and other methylxanthines (Fuhr, 1992), these demethylations are inhibited by others xanthines (furafylline) (Segura, 1988) and by quinolones (enoxacin) (Valero, 1991). This isoform also is implicated in the bioactivation (N-hydroxylation) of heterocyclic amines from cooked food (MEIQX) to potent mutagenic agents (González, 1990). Other ligands are 7-ethoxyresorufin and phenacetin (O-deethylation).

This work has been partially supported by a CESCA grant.

## METHODS

From the University of Geneva protein sequence data bank we obtained the primary sequences of P-450 1A2 included: human, rat, mouse, rabbit and golden hamster (IntelliGenetics, 1993). We aligned all these sequences.

From the crystallographic structure of P-450cam (Poulus, 1987) we selected a subset of amino acids close to the camphor molecule.

Based on alignments of primary sequences of P-450cam and rat P-450 1A2 proposed by Lewis, 1991; Zvelebil, 1991 and Kawajiri 1984 and from our alignments, the $\alpha$-carbons of active site of human P-450 1A2 were placed in the same spatial position than those of P-450cam.

The caffeine molecule was placed into the active site. We carried out a molecular dynamics analysis at high temperature (600° K) during 40 ps in order to explore a wide conformational space of the model and to study its stability.

Along this dynamics, several characteristic geometries were taken as starting points for a geometrical optimization process of the complex. For each final conformation, we substitute the caffeine molecule by other ligands and a new energy minimization process was carried out.

AMBER force field implemented in the DISCOVER module of BIOSYM was used. During the molecular mechanics and dynamics processes the position of α-carbons and the heme group were kept fixed. The iron atom was removed from the calculations because it is not parametrized in the AMBER force field.

| | | | | | | | | | | |
|---|---|---|---|---|---|---|---|---|---|---|
| P450cam | Pro 86 | Phe 87 | Tyr 96 | Asp 97 | Phe 98 | Ile 99 | Pro 100 | Thr 101 | Met 184 | Thr 185 |
| P4501A2(rat) | Leu 109 | Tyr 110 | Gly 118 | Lys 119 | Ser 120 | Met 121 | Thr 122 | Phe 123 | Arg 241 | Tyr 242 |
| P4501A2(human) | » | » | » | Gln | » | Leu | » | » | » | » |
| P4501A2(rabbit) | » | » | » | Gln | » | Met | » | » | » | » |
| P4501A2(hamster) | » | » | » | Lys | » | Met | » | » | » | » |
| P4501A2(mouse) | » | » | » | Lys | » | Met | » | » | » | » |

| | | | | | | | | | | | | |
|---|---|---|---|---|---|---|---|---|---|---|---|---|
| P450cam | Thr 192 | Phe 193 | Ala 194 | Glu 195 | Leu 244 | Leu 245 | Leu 246 | Val 247 | Gly 248 | Gly 249 | Leu 250 | Asp 251 Thr 252 |
| P4501A2(rat) | Lys 249 | Arg 250 | Phe 251 | Lys 252 | Asp 310 | Ile 311 | Phe 312 | Gly 313 | Ala 314 | Gly 315 | Phe 316 | Glu 317 Thr 318 |
| P4501A2(human) | Gln | » | » | Lys | » | Ile | » | » | » | » | » | Asp » |
| P4501A2(rabbit) | Arg | » | » | Lys | » | Ile | » | » | » | » | » | Asp » |
| P4501A2(hamster) | Lys | » | » | Gly | » | Leu | » | » | » | » | » | Asp » |
| P4501A2(mouse) | Lys | » | » | Phe | » | Ile | » | » | » | » | » | Asp » |

| | | | | | | |
|---|---|---|---|---|---|---|
| P450cam | Leu 294 | Val 295 | Ala 296 | Asp 297 | Ile 395 | Val 396 |
| P4501A2(rat) | Phe 378 | Val 379 | Pro 380 | Phe 381 | Leu 494 | Thr 495 |
| P4501A2(human) | » | Leu | » | » | » | » |
| P4501A2(rabbit) | » | Val | » | » | » | » |
| P4501A2(hamster) | » | Val | » | » | » | » |
| P4501A2(mouse) | » | Val | » | » | » | » |

Fig. 1. Alignment between P450cam and P4501A2.

## RESULTS

The considered active site model contained 29 residues from seven segments as well as the heme group totalizing 572 atoms. The mammalian P-450 1A2 amino acids of the active site obtained by alignment were quite different than those of P-450cam (only 3 amino acids are identical). There are 7 positions in the sequence of the active site were changes between species appear (Fig. 1). These residues are not the closest to the ligand.

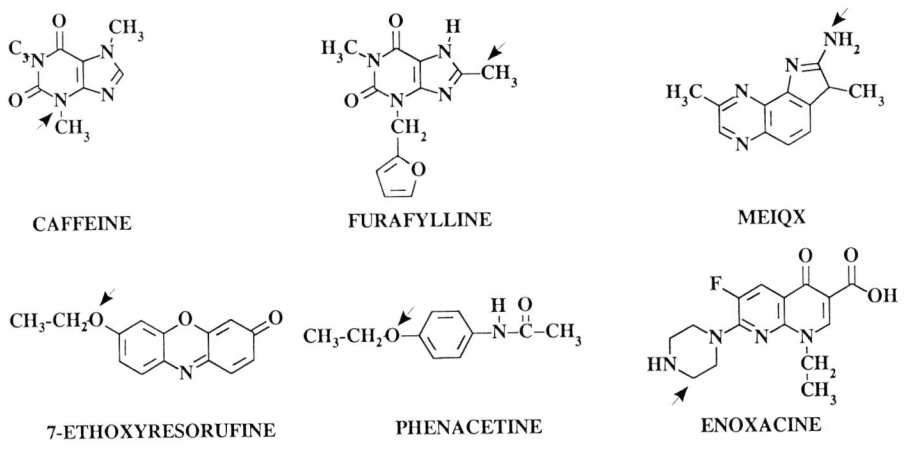

Fig. 2. Ligands considered. Arrows mark groups to be oxidated.

580

Along the dynamics of the caffeine-active site complex, 5 characteristics geometries are considered as starting points for an energy minimization process and 5 local minima of energy were found.

In each final geometry of the complex the caffeine molecule was substituted by MEIQX, furafylline, enoxacin, 7-ethoxyresorufin, and phenacetin (Fig. 2) and a new geometrical optimization process was carried out. Fig. 3 shows a table with the obtained energies and the hydrogen bonds found in each final position. Fig. 4 shows a reduced model of active-site (10 residues) with the amino acids that present hydrogen bonds with at least one ligand. Figure 5 shows three final positions of the ligands tested.

| CAFFEINE | | | | |
|---|---|---|---|---|
| POS.A | POS.B | POS. C | POS. D | POS. E |
| E= -24.78 | E= -36.39 | E= -28.87 | E= -33.29 | E= -34.29 |
| Tyr 242 | No H.Bonds | No H.Bonds | Ser 120 Arg 241 Arg 250 | Arg 241 Asp 310 |

| FURAFYLLYNE | | | | |
|---|---|---|---|---|
| POS. A | POS. B | POS. C | POS. D | POS. E |
| E= -42.27 | E= -49.39 | E= -47.71 | E= -58.10 | E= -56.88 |
| Tyr 242 Asp 310 | No H.Bonds | No H.Bonds | Ser 120 Arg 241 Arg 250 | Arg 241 Asp 310 |

| MEIQX | | | | |
|---|---|---|---|---|
| POS. A | POS. B | POS. C | POS. D | POS. E |
| E= -22.43 | E= -33.66 | E= -31.75 | E= -34.27 | E= -36.22 |
| Asp 310 Hemo 417H | Hemo 417H | Asp 317 | Arg 250 | Hemo 417H |

| ENOXACINE | | | | |
|---|---|---|---|---|
| POS. A | POS. B | POS. C | POS. D | POS. E |
| E= -38.14 | E= -51.40 | E= -40.50 | E= -51.01 | E= -56.72 |
| Gly 118 Tyr 242 Hemo 417H | Arg 250 Hemo 417H | Gly 118 Asp 317 | Arg 241 Arg 250 Hemo 417H | Arg 250 Hemo 417H |

| 7-ETHOXYRESORUFINE | | | | |
|---|---|---|---|---|
| POS. A | POS. B | POS. C | POS. D | POS. E |
| E= -20.70 | E= -34.56 | E= -28.81 | E= -31.14 | E= -37.05 |
| Gly 118 Ser 120 | Arg 250 Arg 250 | No H.Bonds | Gly 118 | No H.Bonds |

| PHENACETINE | | | | |
|---|---|---|---|---|
| POS. A | POS. B | POS. C | POS. D | POS. E |
| E= -42.11 | E= -51.76 | E= -51.85 | E= -52.09 | E= -55.14 |
| Tyr 242 | No H.Bonds | Leu 109 | Ser 120 Arg 241 | Arg 241 Tyr 242 |

**Fig. 3**. Energies and Hydrogen Bonds obtained at the end of each energy minimization process.

Starting position E was the most stable in the case of MEIQX, 7-ethoxyresorufin, phenacetin and enoxacin and the second in the case of caffeine and furafylline. The amino acids involved in hydrogen bonds are different depending on the ligand considered.

A stable model for the active-site was obtained (see Fig. 4). It was stabilized by several intraresidues hydrogen bonds. The stability of the complex is due to hydrophobic interactions between the aromatic or conjugate systems of the ligand and Phe 123 and Tyr 242, and hydrogen bonds with several amino acids depending of the considered ligand.

In all the cases the group to be oxidated appeared close to the center of heme group (Fig. 5).

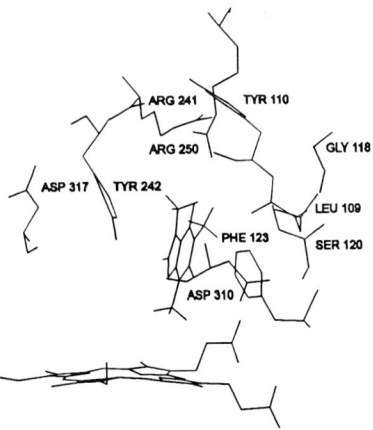

**Fig.4.** Reduced model (10 residues) of the active site of human P450 1A2.

## CONCLUSIONS

A tentative model for active-site of human cytochrome P-450 1A2 has been proposed. This model is in accordance with several published findings:

i) This isoform is not highly specific: can oxidate numerous different substances (Lewis, 1987). The pocket obtained has such a size that can accommodate several ligands. Several hydrogen bonds can be formed depending of chemical structure of ligand.

ii) The active ligands contain fused aromatic or heterocyclic rings resulting in molecular planarity (Lewis, 1987). We found that plane aromatic molecules can form complexes with the two aromatic rings of Phe 123 and Tyr 242. In this complexes the rings can be parallel or perpendicular.

iii) The group to be oxidated appears correctly placed close to the center of heme group in all the ligands.

iv) Residues which have been considered important by other authors (Poulus, 1987), are included in our model.

v) Changes in the residues of the active site between the mammalian species had a small influence on the interaction with the ligand, because they are not the closest to it.

This model describes possible docking positions of ligands within the active site; however, the model does not account for:
i) The ligand recognition site
ii) The oxygen transport mechanism
iii) The electron transfer mechanism
iv) The oxidation mechanism

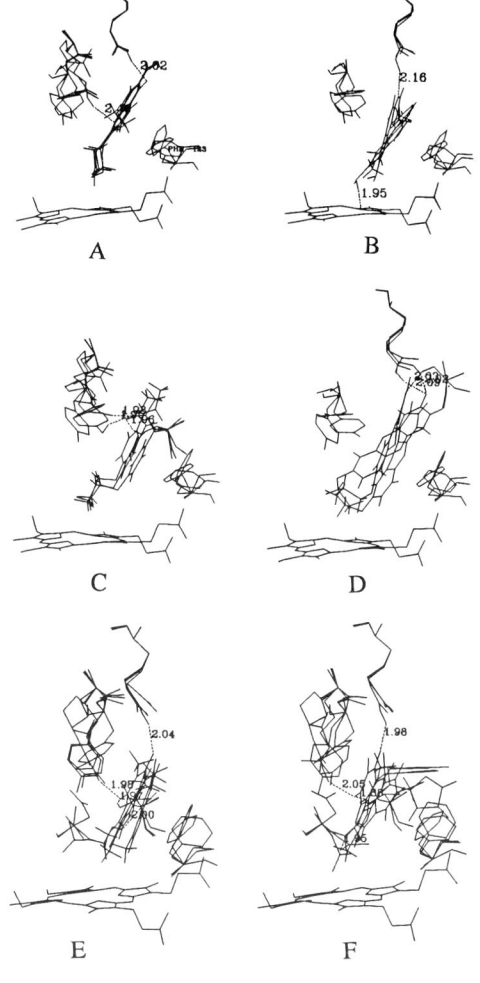

**Fig.5.** Active site with several ligands. A) Enoxacin, B) MEIQX, C) Phenacetin, D) 7-Ethoxyresorufin, E) Caffeine, F) Furafylline

## REFERENCES

Fuhr, U., Doehmer, J., Battula, N., Wölfel, C., Kudla, C., Keita, Y. and Staib, A.H. (1992): Biochem. Pharmacol. 43 : 225-235.
González, F.J., Aoyama, T. and Gelboin H.V. (1990): Mutat Environ B: 77-86.
IntelliGenetics Suite Software, Mountain View, USA.
Kawajiri, K., Gotoh, O., Sogawa, K., Tagashira, Y., Muramatsu, M., Fujii-Kuriyama, Y. (1984): Proc. Natl. Acad. Sci. USA 81: 1649.
Lewis, D.F.V. and Moereels, H. (1992): J.Comp.-Aided Mol.Design, 6: 235-252.
Lewis, D.F.V., Ioannides, C. and Parke, D.V. (1987): Chem.-Biol. Interactions 64: 39-60.
Poulus, T.L., Finzel, B.C. and Howard A.J. (1987): J.Biol. Chem. 195: 687-700.
Segura, J., Roberts, D.J. and Tarrús, E. (1988): . J.Pharm.Pharmacol. 41: 129-131.
Valero, F., de la Torre, R. and Segura, J. (1991): J.Pharm. Pharmacol. 43: 17-21.
Zvelebil, M.J.J.M., Wolf, C.R. and Sternberg M.J.E. (1991): Protein Engineering 4: 271-282.

# Expression of a bovine P450$_{11\beta}$ cDNA in the yeast *Saccharomyces cerevisiae*

Gilles Cauet[1], Bruno Dumas[2], Eric Degryse[1], Roberto Spagnoli[2], Tilman Achstetter[1]

[1]TRANSGENE SA, Yeast Department, 11, rue de Molsheim, F-67082 Strasbourg Cedex, France. [2]ROUSSEL-UCLAF, Biotechnology Department, 111, route de Noisy, F-93230 Romainville, France

## INTRODUCTION

Synthesis of adrenal steroids from cholesterol involves five enzymatic steps, from which four are catalyzed by a class of specific cytochrome P450 enzymes. P450 steroid 11β-hydroxylase (P450$_{11\beta}$; EC1.14.15.4) is an enzyme of the adrenal mitochondria. It is associated with the inner membrane as is the cholesterol side-chain cleavage enzyme (P450scc; EC1.14.15.6) (for a review, see Miller, 1988). It catalyzes a number of steroid hydroxylations generating mineralo- and glucocorticoids through 11β, 18 and even 19 hydroxylations. It receives the necessary electrons from NADPH via adrenodoxin reductase (ADR; EC1.18.1.2) and adrenodoxin (ADX).

Yeast expression systems have been widely employed for expression of cDNAs encoding mammalian microsomal cytochrome P450 enzymes (Guengerich *et al.*, 1991). Yeast is equipped for import of mitochondrial membrane proteins, however, in yeast mitochondria no P450 enzymes have been detected until now. Therefore, production in yeast of heterologous mitochondrial P450s is challenging. Successful expression in yeast mitochondria of mammalian mitochondrial P450$_{LMT25}$ has been reported (Akigoshi-Shibata *et al.*, 1991). Data are presented on the expression of a bovine P450$_{11\beta}$ cDNA in *Saccharomyces cerevisiae*. Subcellular localization and *in vitro* activity of the recombinant enzyme are examined.

## RESULTS

A full length P450$_{11\beta}$ cDNA was cloned from a bovine adrenal cortex cDNA library. Alignment of the protein sequence revealed that this cDNA was identical to the 11β-3 cDNA published by Mathew *et al.* (1990) except for positions 6 and 167 where the sequence translates into an Ala instead of a Val and an Arg instead of a Ser, respectively. The amino acid (aa) changes Ala$_6$→Val and Arg$_{167}$→Ser were described by Mathew *et al.* (1990) and Chua *et al.* (1987), respectively.

The P450$_{11\beta}$ cDNA encodes a precursor (preP450$_{11\beta}$) bearing a N-terminal extension of 24 aas (Mathew *et al.*, 1990). This pre sequence is considered to be necessary and sufficient to target the protein to mitochondria. It is removed during mitochondrial import. Addition of a yeast localization signal should overcome potential problems arising from incompatibilities of the mammalian targeting signal with the yeast mitochondrial protein import machinery. The sequence encoding mature P450$_{11\beta}$ was joined to sequences encoding the yeast cytochrome *c* oxidase subunit VI (COXVI) pre sequence (consisting of 39 aas) (i.e. COXVIpre::mature P450$_{11\beta}$). The chimaeric cDNA was subcloned in the

2μm-based *E.coli*-yeast shuttle plasmid pTG10042 (yielding pTG10145) harboring the yeast *CYC1* promoter which governs the cDNA expression. The plasmid was transferred into the yeast strains TGY73.4, a Leu$^+$ derivative of strain cl3ABYS86 (Schüller & Entian, 1988).

Since proteolytic maturation of the chimaeric protein might be inefficient and/or inaccurate, a fusion protein was produced consisting of the precursor of P450$_{11\beta}$ (i.e. preP450$_{11\beta}$) preceded by the yeast COXVI pre sequence (i.e. COXVIpre::preP450$_{11\beta}$). Thus, this protein has two mitochondrial targeting sequences and it has been previously shown that the use of a duplicate pre sequence can improve import (Galanis et al., 1991). The fusion protein also retains the natural junction of pre sequence and mature P450$_{11\beta}$ allowing for correct cleavage. The corresponding cDNA was subcloned in the same shuttle plasmid pTG10042 (yielding pTG10144).

Subcellular fractions were prepared from transformed yeast cells by differential centrifugation. Samples were submitted to a Western blot analysis using a polyclonal antibody directed against the purified bovine P450$_{11\beta}$ (OXYgene, Dallas, TX, USA). The P450$_{11\beta}$ antigen is detected as a single band in the mitochondrial fraction of the transformed strain (TGY73.4/pTG10145) at an apparent molecular weight (MW) of 51 kDa comigrating with the purified bovine protein indicating that the yeast COXVI pre sequence is efficiently cleaved (Fig. 1). The P450$_{11\beta}$ antigen is estimated to be present at about 0.3 % of total mitochondrial protein.

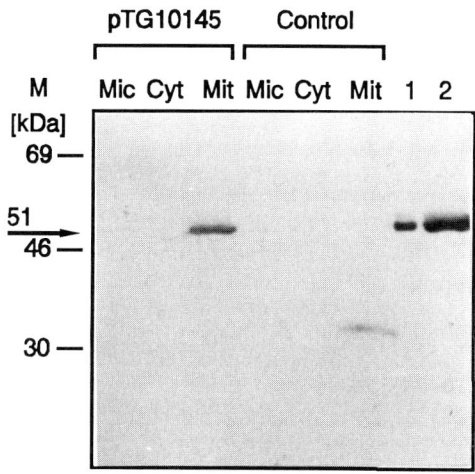

Fig.1. Western Blot of TGY73.4/pTG10145 subcellular fractions (cytosol = Cyt, 105.000 x g supernatant; microsomes = Mic, 105.000 x g pellet., and mitochondria = Mit, 10.000 x g pellet). Subcellular fractions of the untransformed TGY73.4 served as negative control samples. 20 μg of protein per sample were separated by SDS PAGE, transferred to a nitrocellulose membrane and immunodetected with an anti bovine P450$_{11\beta}$ antiserum. M = MW marker; Lane 1: 15 ng of purified bovine P450$_{11\beta}$; Lane 2: 150 ng of purified bovine P450$_{11\beta}$.

A mitochondrial fraction of a strain producing the COXVIpre::preP450$_{11\beta}$ (TGY73.4/pTG10144) shows two bands, one of 51 kDa comigrating with the purified bovine P450$_{11\beta}$ and an additional band with a slightly higher MW which could correspond to an incompletely processed or an unprocessed form of the precursor (data not shown).

The hydroxylase activity of the recombinant bovine P450$_{11\beta}$ was measured *in vitro* with [$^{14}$C]-11-deoxycortisol as substrate. The reconstituted system contained the subcellular fractions of transformed yeast cells, NADPH, purified bovine ADR and ADX, and a NADPH-regenerating system. After a 2 h incubation at 37°C, steroids were extracted and analyzed by reverse-phase (RP-) HPLC. A low but significant activity was found essentially in the sonicated (see below) mitochondrial fraction of TGY73.4 transformed with pTG10145 or pTG10144. The low activity detected in the microsomes results most probably from a contamination by mitochondria (Fig. 2).

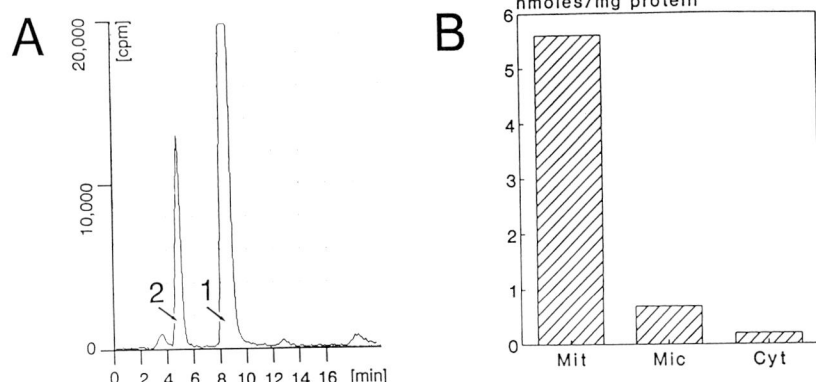

Fig. 2. (A) RP-HPLC analysis of conversion of 11-deoxycortisol (1) into hydrocortisone (2) by sonicated yeast mitochondria (TGY73.4/pTG10145). (B) Distribution of the $P450_{11\beta}$ activity in subcellular fractions of strain TGY73.4/ pTG10145. Subcellular fractions were prepared as in Fig.1. Mitochondria were sonicated before testing. Activity is expressed as nmoles hydrocortisone produced in 2 hrs at 37°C per mg of protein.

Reduced CO-difference spectrum carried out on isolated mitochondria did not show a significant absorption maximum at 450 nm.

It is worth noting that the 11β-hydroxylase activity as well as the fumarase activity (a soluble matrix enzyme) are stimulated by sonicating the mitochondria (TGY73.4/pTG10145). We conclude that sonication, leading to the disruption of the mitochondrial structure, makes the $P450_{11\beta}$ available to the added electron carriers in the assay. The 11β-hydroxylase activity was found associated with the cytochrome c oxidase activity (an inner membrane enzyme) when sonicated mitochondria were fractionated by centrifugation at 105.000 x g for 1 h (Table 1).

| Enzyme | Activity[1] [%] | |
|---|---|---|
| | Pellet | Supernatant |
| 11β-hydroxylase | 68 | 32 |
| Cytochrome c oxidase[2] | 70 | 30 |

Table 1: Distribution of 11β-hydroxylase activities and cytochrome c oxidase between mitochondrial matrix and mitochondrial membranes of TGY73.4/pTG10145.[1] Expressed as % of the sum of activities found in both compartment ;[2] activity was measured according to Mason et al. (1973).

## DISCUSSION

The precursor or the mature form of bovine $P450_{11\beta}$ preceded by the yeast mitochondrial COXVI pre sequence have been successfully produced in yeast. Both chimaeras were cleaved giving rise to a

pre sequence have been successfully produced in yeast. Both chimaeras were cleaved giving rise to a protein of the same apparent MW as the authentic bovine $P450_{11\beta}$. Further work is needed in order to assess the precise cleavage site. In the case of the double pre sequence processing is incomplete giving rise to an additional protein of higher MW. Although no signal at 450 nm was obtained by reduced CO-difference spectrum analysis, mitochondria of transformed yeast cells exhibit 11β-hydroxylase activity towards 11-deoxycortisol in a reconstituted system. A substantial increase in activity is observed after disruption of the mitochondrial structure by sonication suggesting that the recombinant $P450_{11\beta}$ is located inside the mitochondria. Moreover, fractionation of the disrupted mitochondria has shown that the 11β-hydroxylase activity was associated with the mitochondrial membranes together with the cytochrome oxidase activity. This result is indicative for the correct localization of the recombinant $P450_{11\beta}$. The recombinant enzyme has a very low specific activity. This could be the result of the aa difference in position 167. Nevertheless, incorrect proteolytic maturation, a defective heme addition or misfolding cannot be excluded.

## ACKNOWLEDGEMENTS

The authors are grateful to M. Janes and H. Kolbe for fruitful discussions, to L. Laruelle, C. Ledoux and M. Dietrich for expert technical assistance. The project was supported by the French Ministery of Research and Technology (grant No. 92T0426).

## REFERENCES

Akiyoshi-Shibata, M., Usui, E., Sakaki, T., Yabusaki, Y., Noshiro, M., Okuda, K., & Ohkawa, H. (1991): Expression of rat liver vitamin $D_3$ 25-hydroxylase cDNA in *Saccharomyces cerevisiae*. *FEBS Letters* 280, 367-370.

Chua, S.C., Szabo, P., Vitek, A., Grzeschik, K.-H., John, M., & White, P.C. (1987): Cloning of cDNA encoding 11β-hydroxylase (P450c11). *Proc. Natl. Acad. Sci. USA* 84, 7193-7197.

Guengerich, F.P., Brian, W.R., Sari, M.-A. & Ross, J.T. (1991): Expression of mammalian cytochrome P450 enzymes using yeast-based vectors. *Methods. Enzymol.* 206, 130-145.

Mason, T.L., Poyton, R.O., Wharton, D.C., & Schatz, G. (1973): Cytochrome c oxidase from baker's yeast. *J. Biol. Chem.* 248, 1346-1354.

Mathew, P.A., Mason, J.I., Trant, J.M., Sanders, D., & Waterman, M.R. (1990): Amino acid substitutions Phe$^{66}$→Leu and Ser$^{126}$→Pro abolish cortisol and aldosterone synthesis by bovine $P450_{11\beta}$. *J. Biol. Chem.* 265, 20228-20233.

Miller, W.L. (1988): Molecular biology of steroid hormone synthesis. *Endocrine Rev.* 9, 295-318.

Morohashi, K., Yoshioka, .H., Gotoh, O., Okada, Y., Yamamoto, K., Miyata, T., Sogawa, K., Fujii-Kuriyama, & Omura, T. (1987): Molecular cloning and nucleotide sequence of DNA of mitochondrial cytochrome P-450(11β) of bovine adrenal cortex. *J. Biochem. (Tokyo)* 102, 559-568.

Schüller, H.J. & Entian, K.D., 1988: Molecular characterization of yeast regulatory gene *CAT3* necessary for glucose derepression and nuclear localization of its product. *Gene* 67, 247-257.

# Induction by phenobarbital of CYP2B gene expression in primary rat hepatocytes and in rat hepatoma cell line

Giuseppe Ciaramella[1], Mina Edwards[1], Elizabeth A. Shephard[1], Ian R. Phillips[2]

[1]Department of Biochemistry and Molecular Biology, University College London, Gower Street, London WC1E 6BT, UK. [2]Department of Biochemistry, Queen Mary and Westfield College, University of London, Mile End Road, London E1 4NS, UK

INTRODUCTION

Adult rat hepatocytes lose many of their differentiated functions, including the ability to respond to phenobarbital (PB), within twenty four hours of their isolation (Bissell & Guzelian, 1980). Several approaches have been taken to develop culture conditions for optimising the expression of differentiated functions in isolated cultured hepatocytes. These include the use of Matrigel, a reconstituted basement membrane, to coat culture dishes. When hepatocytes are cultured in this way the expression of two cytochromes P450 (CYPs) known as CYP2B1 and 2B2 is increased in response to treatment with phenobarbital (Schuetz *et al.*, 1988; Ben-Ze'ev *et al.*, 1988). Vitrogen coated plates have also been used to culture hepatocytes in the presence of Chee's medium (Waxman *et al.*, 1990). Recently, Sidhu *et al* (1993) have developed a sandwich culture technique in which the hepatocytes are first allowed to attach to type I collagen-coated plates and then overlaid with Matrigel. The last two approaches have obtained some success in supporting induction of CYP2B1/2 genes in response to phenobarbital.

A cell system that responds to phenobarbital and that can be used in DNA transfection experiments would be a valuable asset for studying the mechanisms involved in regulating CYP2B gene expression by this compound. However, it has been reported that Matrigel inhibits the uptake of DNA into hepatocytes (Pasco & Fagan, 1989).

Our approach has been to establish simple culture conditions that permit the induction of CYP2B1/2 mRNAs by phenobarbital and allow the up-take of foreign DNA. We have also developed culture conditions that permit a continuously dividing hepatoma cell-line to respond to phenobarbital.

MATERIALS AND METHODS

Isolation and culture of hepatocytes and hepatoma cells

Hepatocytes were isolated from the livers of adult, male Sprague-Dawley rats weighing 200-250 g by a modification of the two-step collagenase perfusion of Seglen (1976). Rat hepatocytes and the rat hepatoma cell-line FAZA 967 were cultured on Vitrogen-coated plates (Waxman *et al.*, 1990) in the presence of Williams' E medium. FAZA 967 cells were first allowed to propagate in Minimal Essential Medium, supplemeted with 10 % foetal calf serum on standard tissue culture plates (Nunc) and then transferred onto Vitrogen-coated plates. For induction experiments, phenobarbital or picrotoxin was included in the medium at a final concentration of 0.75 mM. For primary hepatocytes, phenobarbital was added to the culture medium 48 hr after cell attachment.

RNase protection assay

Total RNA was isolated from cells pooled from 5-7 culture dishes (60 mm diameter). CYP2B1/2 mRNAs were quantified by RNase protection (Akrawi *et al.*, 1993) using an antisense RNA probe that recognizes both of these mRNAs.

RESULTS AND DISCUSSION

We have tested several published hepatocyte culture conditions for their ability to maintain CYP2B gene expression and its induction in response to phenobarbital. In our hands, the best results were achieved when hepatocytes were cultured on Vitrogen-coated plates (Waxman *et al.*, 1990) in Williams' E medium. Although under these conditions, the expression of CYP2B1/2 falls initially, subsequent treatment of the cells with phenobarbital for 48 hr results in a 60-fold induction of the mRNAs (Fig.1). The extent of induction in the cultured cells compares favourably to that found *in vivo* (Phillips *et al.*, 1983).

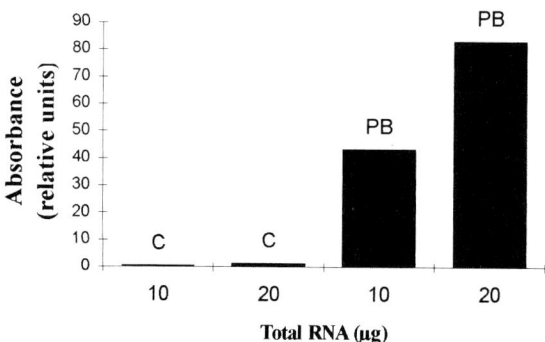

Fig.1 Phenobarbital-mediated induction of CYP2B1/2 mRNAs in primary cultures of hepatocytes. Total RNA was isolated from either untreated (C) or phenobarbital-treated (PB) primary hepatocytes. The abundance of CYP2B1/2 mRNAs was measured by RNase protection.

The expression of CYP2B1/2 mRNAs was also investigated in the continuously replicating rat hepatoma cell-line FAZA 967. Treatment of these cells with phenobarbital for 96 hr resulted in a greater than 30-fold increase in CYP2B1/2 mRNAs (Fig.2). Although the extent of induction in these cells is similar to that obtained in primary cultures of hepatocytes, the overall level of expression of the mRNAs is lower in the hepatoma cells.

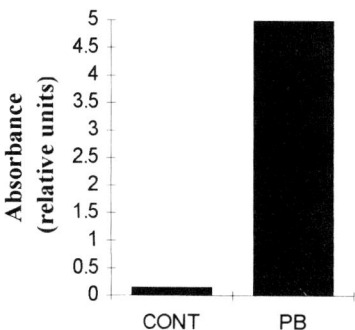

Fig.2 Phenobarbital-mediated induction of CYP2B1/2 mRNAs in FAZA 967 hepatoma cells. The abundance of CYP2B1/2 mRNAs in 20 µg of total RNA isolated from untreated (CONT) or phenobarbital-treated (PB) FAZA 967 cells was measured by RNase protection.

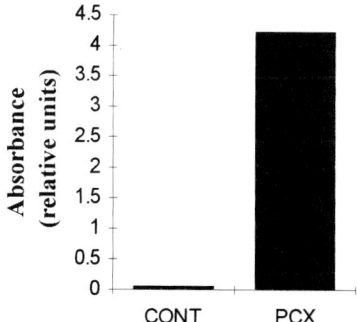

Fig.3 Picrotoxin-mediated induction of CYP2B1/2 mRNAs in FAZA 967 hepatoma cells. The abundance of CYP2B1/2 mRNAs in 20 µg of total RNA isolated from untreated (CONT) or picrotoxin-treated (PCX) FAZA 967 cells was measured by RNase protection.

Recently, it has been shown that CYP2B1/2 proteins are inducible in rat liver by the barbiturate antagonist picrotoxin (Yamada *et al.*, 1993). We have found that this compound induces CYP2B1/2 mRNAs in FAZA 967 hepatoma cells (Fig.3). The extent of induction (70-fold) was greater than that obtained in these cells by phenobarbital (Fig.2).

In conclusion, we have identified culture conditions that support the induction of CYP2B1/2 mRNAs by phenobarbital in primary rat hepatocytes and in a rat hepatoma cell-line. As the extent of induction compares favourably with that observed *in vivo*, these culture systems may prove useful for the investigation of mechanisms responsible for regulating the expression of CYP2B genes.

REFERENCES

Akrawi, M., Rogiers, V., Vandenberghe, Y., Palmer, C. N. A., Vercruysse, A., Shephard, E. A., and Phillips, I. R. (1993): Maintenance and induction in co-cultured rat hepatocytes of components of the cytochrome P450-mediated mono-oxygenase system. *Biochem. Pharmacol.* 45:1583-1591.

Ben-Ze'ev, A., Robinson, G. S., Bucher, N. L. R., and Farmer, S. R. (1988). Cell-cell and cell-matrix interactions differentially regulate the expression of hepatic and cytoskeletal genes in primary cultures of rat hepatocytes. *Proc. Natl. Acad. Sci.* 85: 2161-2165.

Bissell, D. M., and Guzelian, P. S. (1980). Phenotypic stability of adult rat hepatocytes in primary monolayer culture. *Ann. N.Y. Acad. Sci.* 349: 85-98.

Phillips, I. R., Shephard, E. A., Ashworth, A., and Rabin, B. R. (1983). Cloning and sequence analysis of a rat liver cDNA coding for a phenobarbital-inducible microheterogenous cytochrome P-450 variant: regulation of its messenger level by xenobiotics. *Gene* 24: 41-52.

Schuetz, E. G, Li, D., Omiecinski, C. J., Muller-Eberhard, U., Kleinmann, H. K., Elswick, B., and Guzelian, P. S. (1988). Regulation of gene expression in adult rat hepatocytes cultured on a basement matrix. *J. Cell. Physiol.* 134: 309-323.

Sidhu, J. S., Farin, F. M., and Omiecinski, C. J. (1993): Influence of extracellular matrix overlay on phenobarbital-mediated induction of CYP2B1, 2B2, and 3A1 genes in primary adult rat hepatocyte culture. *Arch. Biochem. Biophys.* 301: 103-113.

Waxman, D. J., Morissey, J. J., Naik, S., and Jauregi, H. O. (1990). Phenobarbital induction of cytochromes P-450. *Biochem. J.* 271: 113-119.

Yamada, H., Fujisaki, H., Kaneko, H., Ishii, Y., Hamaguchi, T., and Oguri, K. (1993). Picrotoxin as a potent inducer of rat hepatic cytochrome P450, CYP2B1 and CYP2B2. *Biochem. Pharmacol.* 45: 1783-1789.

# Molecular cloning of cDNA for a renal 27-hydroxylating cytochrome P450

Hans Postlind

Division of Biochemistry, Department of Pharmaceutical Bioscience, University of Uppsala, Box 578, S-751 23 Uppsala, Sweden

## Summary

The aim of the present investigation was to clone cDNA for 25-hydroxyvitamin D3 27-hydroxylase. Several cDNAs were isolated from a pig kidney cDNA library by using a rabbit liver cytochrome P-450$_{27}$ cDNA insert as probe. The longest isolated cDNA clone was 2.2kb and contained a 1506bp open reading frame encoding 502 amino acids and a 268bp 3'-untranslated region. Protein sequence analysis showed that the pig kidney 27-hydroxylase was 70-80% similar to that of the human, rabbit and rat liver mitochondrial sterol 27-hydroxylases, indicating that the enzyme belongs to the CYP 27 family. However, no significant homology was found between the 5'-flanking region of the human gene and the 5'-leader sequence of the pig kidney cDNA or in the possible pig kidney mitochondrial signal sequence compared to other known mitochondrial signal sequences. Surprisingly, a stop codon was found in the correct reading frame of the pig kidney mitochondrial signal sequence. Several hypotheses to explain these findings are discussed.

Vitamin D3 is maintaining calcium homeostasis in animals. However, it must be modified before it exerts any physiological functions. First, it undergoes a 25-hydroxylation in the liver. 25-Hydroxyvitamin D3 is then transported to the kidney where it undergoes further metabolism to 23,25-, 24,25-, 25,27-, or 1α,25-dihydroxyvitamin D3 depending on the serum level of calcium.

27-Hydroxylation of 25-hydroxyvitamin D3 has been recently shown to be catalyzed by immunologically similar mitochondrial cytochromes P-450 in pig kidney and liver. In addition to 27-hydroxylation of 25-hydroxyvitamin D3, both enzymes showed 27-hydroxylase activity towards cholesterol and bile acid intermediates.

Sterol 27-hydroxylase from rabbit and rat liver has been purified and characterized as a mitochondrial member of the cytochrome P-450 family. The enzyme hydroxylates a spectrum of sterol substrates as well as vitamin D3.

The cDNA for the human, rabbit and rat liver sterol 27-hydroxylase has been cloned and shown to belong to the CYP 27 family. At least in the case of rabbit it has also been shown that the mRNA is abundant in several extra-hepatic organs including kidney.

To understand the molecular mechanism of calcium homeostasis it is necessary to elucidate the structure of the genes responsible for modification of vitamin $D_3$ and also to compare the substate specificity of extra-hepatic 27-hydroxylases with those of hepatic sterol 27-hydroxylases.

## MATERIALS AND METHODS

<u>Materials.</u> The librarian$^{Tm}$ cDNA construction system was from Invitrogen. Gigapack II Gold packaging extracts, λZAP II cloning kit, pBluescript II KS⁻ were from Stratagene. Nitrocellulose BA 85 disk membranes used in hybridization experiments were obtained from Schleicher & Schuell. Restriction endonucleases, T4 DNA ligase, megaprime DNA labeling system, d-5'-[α-$^{32}$P]CTP and d-5'-[α-$^{35}$S]ATP were purchased from Amersham. Oligo(dT)-cellulose, T3 and T7 primers were from Pharmacia and the Sequenase Version 2.0 DNA sequencing kit was from United States Biochemicals.

<u>cDNA Cloning.</u> Total RNA was isolated from pig kidney or liver by the guanidine-HCl procedure and poly (A)⁺ RNA was prepared by oligo(dT)-cellulose chromatography. A pig kidney cDNA library was prepared from kidney poly (A)⁺ RNA according to the manufacturer's instruction, ligated into λZAP II arms and packed to reveal the library which then was amplified. The pig kidney library was screened with a $^{32}$P-labeled rabbit liver cytochrome P-450$_{27}$ cDNA (bp 8-1641) by standard methods. Hybridization-positive clones were purified through several rounds of screening. Subcloning into pBluescript SK⁻ was performed by the in vivo excision method.

<u>DNA sequence analysis.</u> Restriction fragments derived from the hybridization-positive clones were isolated and cloned into pBluescript SK⁻ or pBluescript II KS⁻. The DNA sequence was generated directly from the double-stranded plasmids by the chain termination method and compressions were resolved using 7-deaza-dGTP and dITP.

<u>Northern blot analysis</u>. Analysis of cytochrome P-450$_{27}$ transcripts were determined following electrophoresis of poly(A)⁺ RNA on 1% denaturing agarose gels containing formaldehyde. Following transfer of the RNA to nitrocellulose filter, hybridization was carried out at 42°C with mega-prime labeled pig kidney cDNA insert. After hybridization, the filters was washed and then exposed to film in the presence of an intensifying screen for 12h at -70°C.

## RESULTS AND DISCUSSION

Screening of 5 x 10⁵ plaques of a pig kidney λZAP II cDNA library with a $^{32}$P-labeled rabbit liver cytochrome P-450$_{27}$ cDNA insert (bp 8-

1641) resulted in nine positive clones which were purified through several rounds of screening. In vivo excision to form the pBluescript SK⁻ plasmid revealed the plasmid, containing cDNA insert varying in size between 500bp and 2.2kb. Northern blot analysis, revealed mRNA expression in both kidney and liver. In similarity with the rabbit, the intensity of the band in the liver was significantly stronger.

The three largest clones with 1.5kb, 2.2kb and 2.2kb insert were subjected to restriction mapping and nucleotide sequencing. The pig kidney cDNA contains a 1506bp open reading frame encoding 502 amino acids and a 268bp 3'-untranslated region which is in the same range as the 3'-untranslated region for the genes belonging to the CYP 27 family. The highly conserved heme-binding region as well as a possible renodoxin binding region have also been identified. Protein sequence analysis shows that the pig kidney 27-hydroxylase is 70-80% similar to that of the human, rabbit and rat liver mitochondrial sterol 27-hydroxylases, indicating that the enzyme belongs to the CYP 27 family.

The 5'-leader sequence is 392bp long and much longer than for the previously cloned cDNAs belonging to the CYP 27 family. The recently cloned human CYP 27 gene contains nine exons and eight introns. The junctions between the exon-introns in the human gene have all been identified in the pig kidney cDNA. However, no significant homology was found between the 5'-flanking region of the human gene and the 5'-leader sequence in the pig kidney cDNA or in the possible pig kidney mitochondrial signal sequence compared to other known mitochondrial signal sequences. Surprisingly, a stop codon was found in the correct reading frame of the pig kidney mitochondrial signal sequence.

These findings raise many questions. The following speculative hypotheseis may be put forward. First, the isolated pig kidney cDNA is a ligation product between two cDNAs, a mitochondrial cytochrome P-450 likely encoding a pig kidney 27-hydroxylase and an undefined cDNA that appeared during library construction. Second, the stop codon is an error caused by the reversed transcriptase during first strand synthesis due to compressions in the region. Third, the stop codon is caused by a point mutation in the gene of this particular pig and therefore the expression blocked. Fourth, it is a pseudogene, where the stop codon would block translation of the transcribed RNA.

Most likely appears to be the first hypothesis. The high homolgy between the pig kidney cDNA and other cloned cDNAs belonging to the CYP

27 family is consistent throughout the entire mature protein but is lacking in the 5'-leader sequence and in the mitochondrial signal sequence. The rabbit and human liver enzymes show almost identical mitochondrial signal sequences and one would have predicted at least some homology with those and the pig kidney enzyme. Even if the two longest isolated clones are identical, they could be from the same origin since the library was amplified.

The above hypotheses could initially be tested by first strand cDNA synthesis from a pool from pig kidney of purified 5'-ends of poly(A)$^+$ RNA, homopolymer tailing and preparation of target cDNA for subsequent amplification with oligonucleotide primers by the PCR method.

## REFERENCES

Andersson, S., Davis, D.L., Dahlbäck, H., Jörnvall, H. and Russell, D.W. (1989) J.Biol. Chem. 264, 8222-8229.

Bergman, T. and Postlind, H. (1991) Biochem. J. 276, 427-432.

Björkhem, I., Hansson, R., Holmberg, I. and Wikvall, K. (1979) Biochem. Biophys. Res. Commun. 90, 615-622.

Björkhem, I., Holmberg, I., Oftebro, H. and Pedersen, J.I. (1980) J. Biol. Chem. 255, 5244-5249.

Dahlbäck, H. and Wikvall, K. (1988) Biochem. J. 252, 207-213.

Cali, J.J. and Russell, D.W. (1991) J. Biol. Chem. 266, 7774-7778.

DeLuca, H.F. and Schnoes, H.K. (1983) Annu. Rev. Biochem. 52, 411-439

Guo, Y-D., Strugnell, S., Back, D.W. and Jones, G. (1993) Proc. Natl. Acad. Sci. USA 90, 8668-8672.

Leitersdorf, E., Reshef, A., Meiner, V., Levitzki, R., Schwarts Pressman, S., Dann, E.J., Berkman, N., Cali, J.J., Klapholz, L. and Berginer, V.M. (1993) J.Clin. Invest. 91, 2488-2496.

Madhod, T.C. and DeLuca, H.F. (1979) Biochem. J. 184, 491-499

Masumoto, O., Ohyama, Y. and Okuda, K. (1988) J. Biol. Chem. 263, 14256-14260.

Usui, E., Noshiro, M. and Okuda K. (1990) FEBS Lett. 262, 135-138.

# Low temperature behaviour of the high-spin/low-spin equilibrium of cytochrome P450$_{cam}$: the role of the cooling rate and temperature induced changes of the protonic activity

Heike Schulze, Otto Ristau, Christiane Jung

*Max-Delbrück-Centre for Molecular Medicine, Robert Rössle Str. 10, D-13122 Berlin-Buch, Federal Republic of Germany*

## INTRODUCTION

A native protein exists in an equilibrium of a multitude of isoenergetic substates at physiological temperatures. Analysis of the protein structure at cryogenic temperatures shows, that the protein can be frozen or stuck into particular conformational substates. The equilibria between them are influenced by physical and chemical properties of the solution. The question whether the temperature dependent pH changes of the solvent or effects by different cooling rates could influence the spin conversion has been addressed in this work.

## MATERIALS AND METHODS

Isolated and purified cytochrome P450cam from *Pseudomonas Putida* expressed in the Escherichia coli strain TB1 was used for the absorption spectroscopy at the concentration of 5-8 µM and for the EPR measurements at the concentration of 1.1-1.7 mM. For the substrate-free P450cam camphor was removed by dialysis and column chromatography according to Jung et al. (1992). Generally, the measurements were done in aqueous buffered glycerol solutions (60 per cent (w/w) glycerol) containing 0.1M potassium-phosphate buffer pH 7.0. The camphor concentration was 0.44mM for the absorption spectroscopy and 4.0mM for the EPR measurements. Changes of the protonic activity were detected using dye pH-indicators as described at the pH-pK-procedure (Schulze et al.,1993). Absorption spectra were recorded on a double-beam spectrophotometer, Shimadzu UV 2101 PC. Temperature dependent spectra were measured by using the Oxford cryostat DN 704. The temperature was adjusted with a temperature controller DTC-2 with a stability of 0.2K. The cell path length was 1cm. The Oxford cryostat and temperature controller was used for slow cooling the EPR-sample, too. EPR spectra were obtained at 91K on a E300 spectrometer, Bruker, under the following instrumental conditions: modulation amplitude 31.887G, microwave power 50.2mW, modulation frequency 100 kHz. EPR g-values were calculated for the field set (2500-4000 G).

## RESULTS

The spin states of Fe(III) of the camphor-bound and substrate-free cytochrome P450cam was examined in dependence on temperature changes from 298K to 78K and on cooling rates using electronic absorption spectroscopy and EPR spectroscopy.

The heme iron (Fe(III)) of camphor-bound P450cam exists in the well-known high-spin($S=5/2$)/low-spin($S=1/2$) equilibrium which is temperature dependent. At low temperature the equilibrium is shifted to the low-spin state. Surprisingly, it was found, that different amounts of the low-spin state were formed depending on the cooling rate of freezing the sample.

At a very slow cooling rate almost complete conversion from 100 per cent high-spin state at 298K to 5 per cent low-spin state at 200K was detected at the electronic absorption spectrum (Fig.1a). The adjustment of the new equilibrium state after the temperature jump of 5K was controlled by absorbance changes at 417nm. On cooling from 200K to 78K the sample was frozen in at the glass transition temperature of the solution (near 180K). In the temperature range of 180K-78K only a sharpening of the low-spin Soret band was observed. Additionally, EPR studies revealed that after slow cooling only the low-spin state with the g-values of 2.45, 2.26, 1.92 was present (Fig.1b).

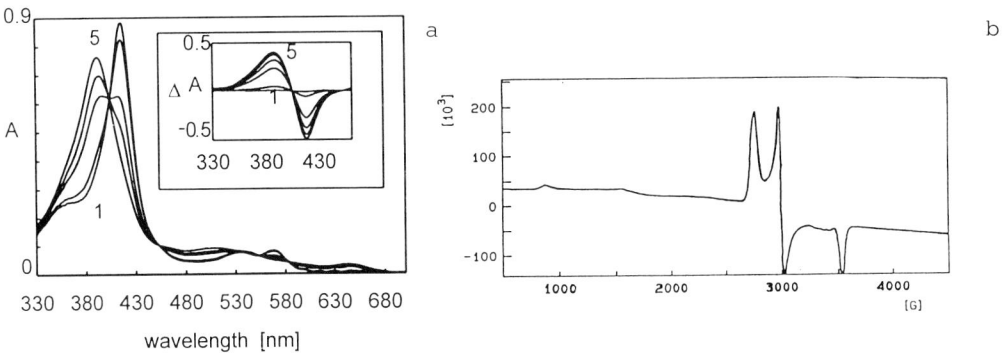

Fig.1. Camphor-bound P450cam; cooling rate 0.3K/min until 200K, then 5 K/min down to 78K; electronic absorption spectra (a) at (1) 210K; (2) 229K; (3) 242K; (4) 250K; (5) 294K; insert: difference spectra to 78K; EPR spectrum (b)

At a very rapid cooling to 78K (the sample was frozen in as quickly as possible after loading the sample in the cryostat (21K/min) ), only 12 per cent of the high-spin state (Soret band at 392nm) is converted to the low-spin state (Soret band at 417nm, Fig.2a). Additionally, at 78K a new spectroscopically detectable species appears, absorbing at about 400nm, as indicating in a shoulder in the second derivative spectrum at 78K (Fig.2a, insert ) and in the temperature difference spectra to 78K on warming-up from 78K to 190K (Schulze, et al. 1993). On warming up from 190-220K the new species disappears and the normal low-spin content increases. Further warming-up to 298K induces the conversion from the low-spin state (417nm) to the high-spin state. The EPR spectrum (Fig.2b) measured according to the commonly used method of immersion of the sample in liquid nitrogen - corresponding to the fastest cooling rate - shows a mixture of the high-spin state (g-values: 7.85, 3,97, 1,78), the normal low-spin state (g-values: 2.45, 2.26, 1.92) and the unidentified low-spin state with g-values of 2.42, 2.26, 1.98 as also detected earlier by Lipscomb (1980).

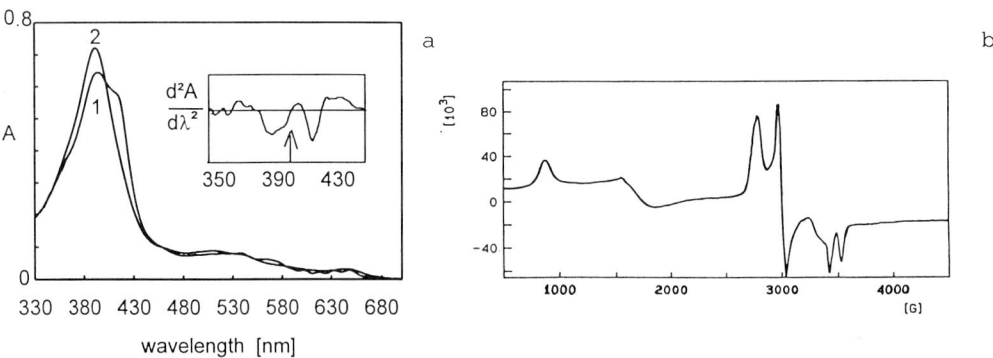

Fig.2. Camphor-bound P450cam; cooling rate at least 21K/min; electronic absorption spectra (a) at (1) 78K; (2) 293K; insert: second derivative of the spectrum at 78K; EPR spectrum (b)

On cooling the solution to 78K changes in the protonic activity value (pa$_H$) additionally occur. The increase of the pa$_H$ amounts to 1.09 units (pa$_H$ 7.0-8.09) from 298K to 78K for potassium-phosphate buffer for the slow cooling rate. On rapid cooling the adjustment of the equilibrium state is kinetically hindered. A pa$_H$-shift of 0.95 units was stucked. It could be shown on pH-titration experiments at 298K (Schulze et al., 1993) that a shift of from 7.0 to 8.09 has no influence on the adjustment of the high-spin/low-spin equilibrium of P450cam as observed by cooling the sample. Furthermore, the new species could also not be detected at pa$_H$ 8.09 at 298K.

Additionally, the described electronic absorption and EPR spectroscopic measurements were done on the substrate-free cytochrome P450cam, too. Both on rapid and slow cooling the sample to 78K the low-spin state of Fe(III) completely remains, detected at 417nm as the Soret band in the absorption spectroscopy (Fig.3a). The EPR spectra (Fig.3b) have the same g-values as for the normal low-spin state of camphor-bound P450cam (2.45, 2.26, 1.92).

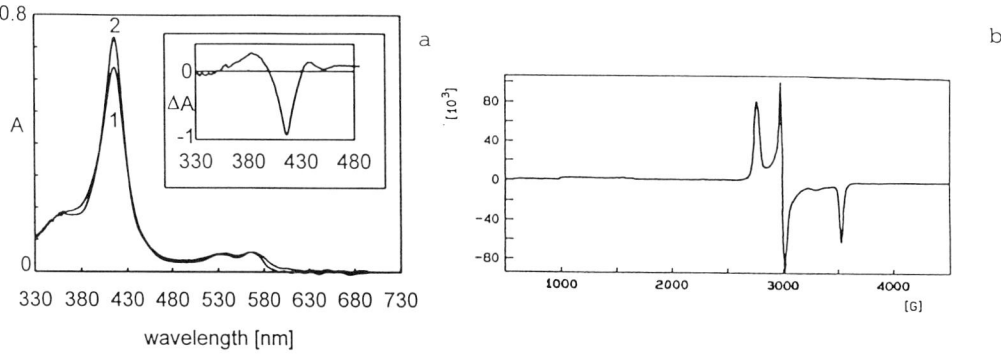

Fig.3. Substrate-free P450cam; slow/fast cooling rate; electronic absorption spectra (a) at (1) 295K; (2) 78K; insert: difference spectra to 78K; EPR spectrum (b)

DISCUSSION

To analyze their conformational substate behaviour proteins are usually frozen in at low temperatures. Temperature dependent changes of the solvent and the effect of the cooling rate has to be considered to understand the conformational changes. Using electronic absorption spectroscopy and EPR measurements of cytochrome P450cam we observed different spin states of the heme iron depending on the cooling rate.

The crystal structure of camphor-bound P450cam shows that the heme iron is pentacoordinated inducing the high-spin state (Poulos et al., 1987). In contrast, the low-spin state in the substrate-free P450cam is caused by the binding of a water cluster to the iron. The identity of the g-values for the low-spin state in substrate-free P450 and in camphor-bound-P450 at slow cooling rate indicate that water is also the sixth ligand in the low-spin state of the camphor-bound protein at low-temperatures. The slow kinetics of the low-spin state formation in camphor-bound P450 supports a ligand binding process. There are two possible pathways how water molecules can move to the iron ligand position. Water can access the heme pocket from the bulk solvent while exchanging with camphor or a water molecule within the protein near the heme pocket is displaced to the iron. In both cases protein fluctuations and geometrical rearrangements of amino acid residues are necessary. We exclude the possibility that camphor is removed from the heme pocket during the low-spin state formation because the infrared spectrum of the CO stretching mode in camphor-bound P450cam at low temperatures does not show the broad structurated infrared signal which is characteristic for substrate-free P450 (Jung & Marlow, 1987). However, Wade (1990) has shown by calculating GRID energy contours that a water binding region exist near Val247 and Asp251. Also Raag, R.& Poulos, T.L. (1991) have seen in the crystal structure a water chain behind the I helix. Additionally, molecular dynamics simulations by Paulsen (1991) show a flexibility of the protein in this region and a conformational switch of the camphor molecule from the normal orientation to an almost perpendicular orientation with maintaining the hydrogen bond to Tyr96. This motion may induce a movement of internal or external water to the heme iron.

For the new spectroscopic species observed at 78K on very fast cooling rate a low-spin producing iron ligand has to be assumed. As revealed by the 175 ps molecular dynamics simulation (Paulsen, 1991) camphor is only in 85% of all possible conformations hydrogen bonded to Tyr96. This may be a requisite for the disorder of the camphor molecule in the heme pocket. We suggest that the camphor ketone group can be oriented toward the heme iron and may function as low-spin ligand. Rapid cooling to 78 K may trap this only weakly populated conformation and cause the spectral species absorbing at 400 nm and showing the g-values at 2.42, 2.26 and 1.98. This conclusion is supported by EPR studies on Fe(TPP)cam synthetic complex having camphor as one iron ligand which shows g-values of 2.37, 2.29 and 1.94 (Tang et al., (1976). Generally, iron porphyrin complexes with oxygen ligands have closer g-values for the x, y, and z-components (2.3-2.4;2.22-2.26;1.95-1.97). Interestingly, two low-spin states were also found for P450scc, especially for complexes with cholesterol analogues containing ketone groups. (Orme-Johnson et al., 1979).

Our observation of the appearance of a new low-spin species and the transition of the high-spin state to the low-spin state in camphor-bound P450cam matches qualitatively the finding by Lipscomb, (1980). However, we could clearly show that the amount of population of a particular spin state at low temperatures is in the first place a function of the cooling rate. Therefore, a pH-dependence of the low-spin species with the g-values at 2.42, 2.26 and 1.98 observed by Lipscomb has to be reexamined under steady conditions.

REFERENCES

Jung, C. (1992): Substrate analogue induced changes of the CO-stretching mode in the cytochrome P450cam- carbon monoxide complex. *Biochemistry* 31, 12855-12862

Jung, C.& Marlow, F. (1987): Dynamic behaviour of the active site structure in bacterial cytochrome P-450. *Studia biophysica* 120, 241-251

Lipscomb, J.D. (1980): Electron paramagnetic resonance detectable states of cytochrome P450cam. *Biochemistry* 19, 3590-3599

Orme-Johnson et al.(1979): Steroid binding properties of beef adrenal cortical cytochrome P450 which catalyses the conversion of cholesterol into pregnenolone. *J. biol Chem.* 254, 2103-2111

Paulsen, M.D. et al. (1991): Analysis of the active site motions form a 175 picosecond molecular dynamics simulation of camphor-bound cytochrome P450cam. *J.Biomol.Structure &Dynamics* 2, 187-203

Poulos, T.L. et al.(1987): High-resolution crystal structure of cytochrome P450cam. *J.Mol.Biol.* 195, 687-700

Raag, R.& Poulos, T.L. (1991): Crystal structures of cytochrome P450cam complexed with camphane, tiocamphor, adamantane; factors controlling P450 substrate hydroxylation. *Biochemistry* 30, 2674-2684

Schulze, H. et al. (1993): The proton activity at cryogenic temperatures - a possible influence on the spin state of the heme iron of cytochrome P450cam in supercooled buffered solutions. *BBA*, in press

Tang, S.C. et al.(1976): Axial ligation modes in iron (III) porphyrins. Models for oxidized reaction states of cytochrome P450 enzymes and the molecular structure of iron (III) protoporphyrin IX dimethylester p-nitrobenzenethiolate. *J Am Chem Soc* 98, 2414-2434

Wade, R.C. (1990): Solvation of the active site of cytochrome P450cam. *J.computer-aided Mol. Design* 4, 199-204

# Induction of cytochrome P450IA1 mRNA in aryl hydrocarbon hydroxylase-deficient mutants of mouse hepatoma line, Hepa-1

Hideaki Kikuchi, Masahiro Usuda, Ikuko Sagami, Shuntaro Ikawa, Minro Watanabe

*Department of Molecular Genetics, Institute of Development, Aging and Cancer, Tohoku University, 4-1 Seiryo-machi, Aoba-ku, Sendai 980, Japan*

## INTRODUCTION

Cytochrome P-450IA represents one of eight known mammalian families in the P-450 gene superfamily (Nebert and Gonzalez, 1987). Two mouse genes, P-450IA1 and P-450IA2, are induced by combustion products such as TCDD (2,3,7,8-tetrachlorodibenzo-p-dioxin) and benzo[a]pyrene. Although the parent benzo[a]pyrene is not directly carcinogenic, the metabolites by the action of P-450IA1 are reactive intermediates shown in many systems to be carcinogenic. It has been suggested that increased levels of P-450IA enzymes are highly correlated with environmental carcinogenesis (Conny, 1982).

From mouse hepatoma Hepa-1 cell cultures treated with benzo[a]pyrene, it is possible to isolate benzo[a]pyrene-resistant mutants. These mutant cells do not induce metabolic activity toward benzo[a]pyrene, and therefore these cells do not generate the cytotoxic intermediates (Hankinson, 1979). The availability of mutant cells permits the use of both genetic and biochemical approaches in analyzing the induction mechanism for the carcinogen-metabolizing P-450IA1 enzyme.

## MATERIALS AND METHODS

<u>Cells</u>  The Hepa-1 cell line, derived from mouse hepatoma (Bernhard et al., 1973), was kindly provided by Dr. Sogawa of Tohoku University. The cells were maintained in Dulbecco's modified Eagle's (DME) medium containing 10% fetal calf serum, 100 µg/ml streptomycin and 100 units/ml penicillin G in humidified 95% air-5% $CO_2$ at 37°C. Benzo[a]pyrene resistant Hepa-1 cells were isolated according to the protocol described by Hankinson (1981) with minor modifications.

<u>Enzyme assays</u>  The AHH activity (Cantfort et al., 1977), and a gel mobility assay (Fujisawa-Sehara et al., 1988) were performed as described previously. Specifically bound [$^3$H] TCDD in cytosol was assayed by adsorption to hydroxylapatite as previously described (Gasiewicz and Neal, 1982; Houser et al., 1986).

<u>PCR reaction and subcloning</u>  Complementary DNA of the mutant cells were synthesized with AMV reverse transcriptase (Life Science Inc., St. Petersburg, FL, USA) (Maniatis et al., 1982). The oligonucleotide primers were 5'-CCACCTAGATCATGCCTTCC-3' (P1S-1), 5'-GCCAACCTCTGCCATCACCT-3' (P1AS-2), 5'-ATCTCGTCAGCAAACTTCAG-3' (P1S-3), 5'-CTGTGTCTAGTTCCTCCTGG-3' (P1AS-4), 5'-GTGGAGCCTCATGTACCTGG-3' (P1S-5) and 5'-

CTAAGCCTGAAGATGCTGAG-3' (P1AS-6). The PCR was performed using 1/10 volume of cDNA reaction mixture and 2.5 units of Taq DNA polymerase (Perkin-Elmer/Cetus, Emeryville, CA, USA) in 100 µl of a reaction mixture (Saiki et al., 1988). PCR products were subcloned into the pCR1000 vector using a TA cloning kit (Invitrogen Corporation, San Diego, CA, USA). The DNA sequence was determined by the dideoxy termination method (Sanger et al., 1977), using a Sequenase Version 2.0 DNA sequencing kit (United States Biochemical Corp., Cleveland, OH, USA).

Other methods  Western blotting and Northern blotting were performed as described in a previous report (Kikuchi et al., 1990).

## RESULTS AND DISCUSSION

Table 1. Comparison of biochemical characteristics of Hepa-1 and mutant cells

| Name of cell | Induction of AHH activity | Induction of CYPIA1-mRNA | Induction of CYPIA1-protein | Activity of Ah receptor |
|---|---|---|---|---|
| Hepa-1 | +++ | +++ | +++ | +++ |
| cl-19 | - | - | - | - |
| cl-21 | - | +++ | - | - |
| cl-32 | - | +++ | + | - |

A large number of clones observed in the benzo[a]pyrene selection dishes were isolated. Of these, 40 clones were then cultured in the presence of benzo[a]pyrene. All showed resistance to benzo[a]pyrene at the dose of 4 µg/ml. AHH activities were assayed in both noninduced and induced cultures of 40 mutant cells. Induced cultures were treated with 2 nM TCDD for 24 h. All 40 mutants had reduced AHH activities under both conditions. Wild type cells showed a 15-fold induction of AHH (73.0 ± 39.9 pmol/min/mg protein) by the treatment of 3-MC against noninduced AHH (4.7 ± 2.1 pmol/min/mg protein). The mean value of noninduced AHH of mutant cells was 4.9 ± 7.3 pmol/min/mg protein and the induced AHH was 7.5 ± 8.0 pmol/min/mg protein (summarized in Table 1).

In the course of this experiment, almost all the clones showed no hybridization signal or very faint signals at the position of P-450IA1 (published elsewhere), but we found two clones cl-21 and cl-32, which exhibited an increased level of P-450IA1-mRNA . Expression of β-actin served as a control in these experiments, and all RNA preparations were shown to hybridize equally with this actin probe .

Western blot analysis of wild-type lysate and mutant-cell lysate revealed P-450IA1 protein of a size identical to that in mouse liver microsomes (56,000 daltons), but the intensity of the band of mutant cells was absent (cl-21) or faint (cl-32).

Sequence comparisons of P-450IA1 cDNA from Hepa-1 wild type, and the cl-21 and cl-32 mutants revealed (1) complete identity of P-450 cDNA between cl-21 or cl-32, except one mutation noted below in each mutant, (2) a single base difference between wild type and cl-21, resulting in an amino acid change from Leu (118) to Arg, and (3) a single-point mutation of cl-32, resulting in an amino acid change from Arg (359) to Trp. To eliminate misincorporation of an incorrect base by Taq DNA polymerase, several independent PCR reactions were performed and sequenced, when the base substitution was found in the amplified DNA sequence.

When intact cells are exposed to Ah receptor ligands, such as TCDD, there is initial binding to a cytosolic form of Ah receptor protein (Landers and Bunce, 1991). After a temperature-dependent transformation step (Okey et al., 1980), the ligand-receptor complex becomes tightly associated with the cell nucleus and appears to acquire the ability to bind to specific DNA regions, thereby leading to enhanced rates of transcription of mRNA coding for P-450IA1. In order to clarify the changed step of the induction mechanism in the mutant cells, cytosolic Ah receptor was assayed by a method utilized hydroxylapatite (Gasiewicz and Neal, 1982). In these experiments, 1.0 nM of [$^3$H]TCDD was used. The cytosol from wild-type cells, Hepa-1, contained a high activity of binding site (70.3 fmol/mg protein). But mutant cells showed a low activity of binding site, cl-21 (1.9 fmol/mg protein) and cl-32 (8.8 fmol/mg protein). The DNA-binding factor specific for xenobiotic responsive elements (XRE) was also assayed by the gel mobility shift method (Fujisawa-Sehara et al., 1988). When the XRE-1 DNA fragment was used as a probe, a specific retarded band was observed only with the nuclear extract from 3-MC-treated Hepa-1 cells. However, this band was extremely faint in the experiment in which the nuclear extract from 3-MC treated mutant cl-21 was used.

In the present paper, we examined the nucleotide sequence of P-450IA1 mRNA of two mutants. Both of the mutant-cell lines had a P-450IA1 protein with an amino acid differing from that of the Hepa-1 wild-type line. These mutants had no cytosolic Ah receptor activity, nor showed nuclear XRE-binding activity; however they showed a level of P-450IA1-mRNA equal to that of wild type when the cells were treated with 3-MC or TCDD. Our results imply that mutant cells can be categorized into a new group and that other control factors participate in the P-450IA1 induction mechanism. The cell lines described herein should be valuable in assessing the contribution of P-450IA1 enzyme to the process of carcinogenesis.

ACKNOWLEDGEMENTS

This research was supported, in part, by a Grant-in-Aid for Cancer Research from the Ministry of Education, Science and Culture, Japan.

REFERENCES

Bernhard, H. P., Darlington, G. J., and Ruddle, F. H. (1973): Expression of liver phenotypes in cultured mouse hepatoma cells: synthesis and secretion of serum albumin. *Dev. Biol.* 35, 83-96.

Cantfort, J. V., Graeve, J. D., and Gielen, J. E. (1977): Radioactive assay for aryl hydrocarbon hydroxylase. Improved method and biological importance. *Biochem. Biophys. Res. Comm.* 79, 505-512.

Conny, A. H. (1982): Induction of microsomal enzymes by foreign chemicals and carcinogenesis by polycyclic aromatic hydrocarbons. *Cancer Res.* 42, 1982.

Fujisawa-Sehara, A., Yamane, M., and Fujii-Kuriyama, Y. (1988): A DNA-binding factor specific for xenobiotic responsive elements of P-450c gene exists as a cryptic form in cytoplasm: Its possible translocation to nucleus. *Proc. Natl. Acad. Sci. USA* 85, 5859-5863.

Gasiewicz, T. A., and Neal, R. A. (1982): The examination and quantitation of tissue cytosolic receptors for 2,3,7,8-tetrachlorodibenzo-p-dioxin using hydroxylapatite. *Anal. Biochem.* 124, 1-11.

Hankinson, O. (1979): Single-step selection of clones of a mouse hepatoma line deficient in aryl hydrocarbon hydroxylase. *Proc. Natl. Acad. Sci. USA* 76, 373-376.

Hankinson, O. (1981): Evidence that benzo[a]pyrene-resistant, aryl hydrocarbon hydroxylase-deficient variants of mouse hepatoma line, Hepa-1, are mutational in origin. *Somat. Cell Genet.* 7, 373-388.

Houser, W. H., Zielinski, R., and Bresnick, E. (1986): Further characterization of the polycyclic aromatic hydrocarbon binding properties of the 4s protein. *Arch. Biochem. Biophys.* 251, 361-368.

Kikuchi, H., Sagami, I., Fujii, H., Ohmachi, T., and Watanabe, M. (1990): Complementary DNA sequence of 3-methylcholanthrene-inducible P-450 from the rat lung. *Tohoku J. Exp. Med.* 160, 323-332.

Landers, J. P., and Bunce, N. J. (1991): The *Ah* receptor and the mechanism of dioxin toxicity. *Biochem. J.* 276, 273-287.

Maniatis, T., Fritsch, F. E., and Sambrook, J. (1982): Analysis of recombinant clones. Southern blotting. In *Molecular Cloning*, pp. 365-401. Cold Spring Harbor: Clod Spring Harbor Laboratory.

Nebert, D. W., and Gonzalez, F. J. (1987): P450 genes: Structure, evolution, and regulation. *Ann. Rev. Biochem.* 56, 945-993.

Okey, A. B., Bondy, G. P., Mason, N. E., Nebert, D. W., Forster-Gibson, C. J., Muncan, J., and Dufresne, M. J. (1980): Temperature-dependent cytosol-to-nucleus translocation of the Ah receptor for 2,3,7,8-tetrachlorodibenzo-p-dioxin in continuous cell culture lines. *J. Biol. Chem.* 255, 11415-11422.

Saiki, R. K., Gelfan, D. H., Stoffel, S., Scharf, S. J., Higuchi, R., Horn, G. T., and Mullis, K. B. (1988): Primer-directed enzymatic amplification of DNA with a thermostable DNA polymerase. *Science* 239, 487-491.

Sanger, F., Nicklen, S., and Coulson, A. R. (1977): DNA sequencing with chain-terminating inhibitors. *Proc. Natl. Acad. Sci. USA* 74, 5463-5467.

# Antigenic determinants of cytochrome P450 superfamily and their classification

Ilya R. Dedinsky, Sergey A. Kozin, Alexander I. Archakov

Institute of Biomedical Chemistry, Pogodinskaya, 10, 119832 Moscow, Russia

## Abstract

A novel algorithm for predicting epitopes was designed to try to reach more accuracy compared to existing ones. The method is based on the set of chemical rules reflecting side-chain interactions of amino acid residues. It was shown that such interactions may be described as motifs and simple algorithm of their using. The comparision with standard epitope prediction algorithms such as Antigenic Index (Jameson and Wolf, 1988) showed that the method developed has less tendency of overprediction. The accuracy of novel method was tested by the comparison of experimental data of scanning P450 2B4 protein by the Pepscan method (Kolesanova et al., 1993). So, this algorithm was used to find antigenic determinants in cytochrome P450 superfamily as in a large set of related proteins. The data obtained were interpreted in the context of Nelson et al. (1993) classification. The common epitopes of subfamilies and families and cross-reacting epitopes were found and described. Then results of search were used to try to build classification basing on antigenic determinants.

## Introduction

Antigenic determinants (epitopes) of proteins play a key role in the interaction with antibodies produced by B-cells. Their set forms an "immunological portrait" of molecule used by the immune system for exact protein identification and starting the immune response. Intensive studies of these sites revealed their general features: small size and discrete boundaries of epitopes; their presence only in a limited number; surface location; sensibility to conformational changes and amino acid substitutions (Atassi, 1984); preference location in beta-turns or loops (Ripoll, 1992). Epitopes may be continuous or discontinuous, and in the latter case they consist of two or more continuous ones (Atassi, 1984; Stern, 1991). On the whole, antigenic site has no clever peculiarities in structure and composition for a simple *a priori* identification without experimental investigation.

Hopp and Woods (1981) first raised the problem of prediction of antigenic determinants from their primary sequence. They used a simple assumption that correlation between properties of individual amino acids (hydrophilicity), averaged along the protein sequence, would reflect the antigenic structure. Later many other amino acid properties, being converted to scales, were used to determine B-cells epitopes: hydropathy (Kyte, Doolittle, 1982); alpha-chain flexibility (Karplus, Shultz, 1985); "antigenicity" (frequence of appearance in epitopes; Welling, 1985); "surface probability" or frequence of appearance on surface (Emini et al., 1985) and many others. Some methods use secondary structure information mostly obtained by other prediction programs, e.g., beta-turn scale of Chou and Fasman (1978). Joint prediction algorithms use superposition of several scales to improve prediction force (antigenic index of Jameson and Wolf, 1988). For the last time these methods were transformed to templates for construction of any combination of any scales (Maksutov, Zagrebelnaya, 1993). However, testing against the set of proteins with the known epitope structure demonstrated that all the methods have got unstable prediction and many of them showed random results (Pellequer, 1991). One of the main reason would be that these models don't take into account the interactions between amino acids, or do it implicitly. The simplest and explicit way to describe the epitope structure basing on interaction of amino acids is to use amino acid patterns and motifs. Such methods successfully used for description and prediction of T-cell epitopes (Rothbard, Taylor, 1988).

Thus, the aim of this work was to develop the method for predicting antigenic determinants with the most accuracy, and to compare its prediction with experimental results from our laboratory (Kolesanova et al., 1993). Then the method proposed was used to predict antigenic determinants of superfamily P450, and to find epitope patterns specific to individual proteins, subfamilies and families.

## System and methods

### The algorithm

We proposed the model basing on the consideration that antigenic determinant is a small and conformationally rigid sequence site, stabilized by amino acids' interactions, which play a dominant role. This features would be the reason of both exact recognition and high affinity of antibodies. Results of experiments with synthetic peptide design correspond well to this consumption (Edmundson, 1991). The main theses of method are:

First, B-cells epitope must be 3-6 residues in length (Geysen et al., 1987a; Novotny et al., 1989; Briggs et al., 1993).

Second, epitope must be conformationally rigid. It, therefore, must contain residues with low side-chain flexibility, such as Pro (due to cyclic structure) and Gly (due to small size). Or, the amino acids must interact within the site by forming polarized hydrogen bond or a salt bridge. Experiments of Edmundson et al. (1991) showed that this may stabilize the peptide conformation, if (1) interacting groups in residues are close to each other, and (2) these groups don't interact with other amino acids containing or surrounding the site. In alpha- and $3_{10}$- helices and in beta-turns, side chains arranged azimutally in $\pm 3$, $\pm 4$ (and $\pm 5$, if two Gly present) positions and has the possibility of interaction.

Then we performed the refinement of these criteria using the learning set of experimentally found epitopes by the Pepscan method, namely myohemerythrin (Geysen et al., 1987a) and hepatitis delta-virus (Wang et al., 1990). This refinement reveals that epitope motifs are either tri-, tetra-, pentapeptides containing proline and/or glycine (PG-motifs), or tetra-, penta-, hexapeptides of complex structure, in which ending residues (terminators) electrostatically interact with each other. Amino acids between N-terminator (the Head) and C-terminator (the Tail) form the Body. The correspondence between Heads and Tails is: **[D,E]** -

[H,K,N,Q,R,S,T], [H,K,R] - [D,E,N,Q,S,T], [N,Q,S,T] - [D,E,K,H,R]. Then, the propagation of method conceptions has got five principles of selection for terminator-containing motifs:

1. All motifs must not contain Cys, Met and Trp. Terminator motifs must not contain Arg and His at the distance of 3 or 4 residues from the Cys residue.
2. Motifs containing Pro must be processed first. Terminator motifs with stronger interaction between charged terminators (Asp, Glu, Arg, His, Lys) must be processed earlier than ones with neutral residues (Asn, Gln, Ser).
3. Compositions of two overlapped terminator motifs of equal size must be processed earlier than simple motifs. Instead, two overlapped motifs of unequal size are unstable and must be rejected.
4. Non-overlapped (simple) terminator motifs containing the pairs of residues **[TN][TN]** or **[FILVY][FILVY]** must be rejected.
5. Epitope is the part of sequence which is not rejected and matched by the Pro-containing or terminator motif.

We have omitted the detailed description of motif system and search procedure to save the space.

Antigenic Index

Antigenic Index according to Jameson and Wolf (1988) was also used to compare the results of prediction with the method proposed. We used the level 0.68 of standard deviation for the cutting of resulting index profile.

Source data and post-processing of predicted epitopes

All the sequences were obtained from the CPD database, version 2.0 (Archakov et al., 1993).

Sequences with predicted epitopes were aligned manually within the subfamilies to refine the set consensus (common) epitopes. Then the search for every consensus epitope was performed through the database. Results of search were used for considering epitope to be specific to the individual protein, subfamily or family.

## **Results and Discussion**

Although there are many available experimentally determined antigenic sites, the learning and testing set contains only data obtained by the Pepscan. This method gives stable results and guarantees that the most of the protein epitopes was skipped. It also precisely defines peptide boundaries, which is important for motif analysis.

Table 1. Classes for $X^2$ criterion calculation.

|  | Epitope | Non-Epitope |
|---|---|---|
| Predicted | A | C |
| Not predicted | B | D |

Comments: Class **A** expresses true positive prediction, **D** - true negative, **B** - underprediction, **C** - overprediction. The $X^2$ value for this scheme was calculated by equation $(|AD-BC|-N/2)^2 \ast N / ((A+C)(B+D)(A+B)(C+D))$, where $N = A+B+C+D$ = protein length. (Pellequer et al., 1991).

We use the $X^2$ criterion (see table 1) to test method reliability, because most of wide-spread criteria give too high results due to dominancy of true negative prediction. Table 2 consists of data for two proteins of learning set (#1 and #2) and one protein (cytochrome P450 2b4, #3), which antigenic determinants were determined by the Pepscan method in our laboratory (Kolesanova et al., 1993) after the prediction. Data in this table show that Motif-based method gives statistically significant results for the level of 0.99 for all three proteins. It seems that method proposed has minimal overprediction. Antigenic Index got statistically significant results for level 0.99 for protein #2, and for level 0.95 for proteins #2 and #3, but it had less accuracy (percentage of correctly predicted residues) and revealing (percentage of revealed residues in epitopes). When the protein length increases, the accuracy of Motif method slowly decreases, but was not less than 62%. The accuracy of Antigenic Index varied from 28% to 61%. Motif method seems to get relatively high revealing ability, and 52% to 85% epitope residues were detected by the algorithm.

Table 2. Precision of methods of epitope prediction.

| Protein and method | Distribution of the amino acids | | | | | | | | $X^2$ value | accuracy, % | revealing, % |
|---|---|---|---|---|---|---|---|---|---|---|---|
| | A | % | B | % | C | % | D | % | | | |
| 1. Myohemerythrin (118 AA) (from learning set) | | | | | | | | | | | |
| AIDX | 9 | 7 | 39 | 33 | 23 | 9 | 47 | 39 | 2.20 | 28 | 18 |
| Motifs | 41 | 34 | 7 | 5 | 7 | 5 | 63 | 53 | **64.0** | 85 | 85 |
| 2. Hepatitis B virus, delta-antigen (214 AA) (from learning set) | | | | | | | | | | | |
| AIDX | 30 | 14 | 48 | 22 | 19 | 8 | 117 | 54 | **15.5** | 61 | 38 |
| Motifs | 63 | 29 | 15 | 7 | 27 | 12 | 109 | 50 | **73.0** | 70 | 80 |
| 3. Cytochrome P450 2B4RAB (491 AA) (from test set) | | | | | | | | | | | |
| AIDX | 79 | 16 | 134 | 27 | 64 | 13 | 214 | 43 | 10.8 | 55 | 37 |
| Motifs | 111 | 22 | 102 | 20 | 68 | 13 | 210 | 42 | **38.6** | 62 | 52 |

Comments: Every class represented by absolute value and percentage of the whole protein. Statistically significant $X^2$ values (higher than 10.83, for level 0.99) are bold. Accuracy (**A**/(**A+C**)) is a part of correctly predicted residues from all predicted residues. Revealing (**A**/(**A+B**)) is a part of correctly predicted residues from all real epitopes of the protein. **AIDX** is Antigenic Index.

The comparision "by sites" (fig. 1) shows that Motif-based method resolves epitope boundaries more precisely than Antigenic Index. So, epitopes **HEIQRL** (354-359), **SSALHD** (393-398), **NGALKRNE** (417-424) were matched almost exactly by Motif method. Epitopes **YGDVFTVYLGS** (62-72) and **PPSYQI** (481-487) were not predicted by only Antigenic Index, and **DFGMGK** (134-139) and **EERIQE** (142-148) not matched by Motifs. Some sites as **SLFFAG** (294-299), **PNTFNPGHF** (405-413) were not found by both methods. There are two correlated overpredictions: **ERVQKE** (322-327) and **SHRP** (334-447). Due to the 6-residues averaging in the Pepscan method, most of epitopes was merged to long zones of high antigenicity, up to 23 amino acids in length. In some cases Antigenic Index predicted very short epitopes, such as **GT(h)R** (229-232), **R** (358), **KR** (433-434), because of profile cutting. It is necessary to note that Motif method has no such a problem, as all profile-based methods. The overprediction of Antigenic Index appears due to the fact that method's scales are oriented to the prediction of molecule surface. Surface, however, doesn't correspond exactly to antigenic area.

To predict antigenic structure of P450 superfamily, we used Motif-based method. All 162 sequences of cytochromes P450 were scanned by the prediction program. The search through database shows that most of the epitopes was specific to individual proteins, but some of then were matched by subfamilies and families (fig. 2). For example, antigenic patterns of cytochrome 2B4 **GHPKAHGRLPPGPSPLPVLG, RLREKYGDVFT, TDAIR, EAFSGRGK, DPIFQG, HFPGT, QSVEKHRATLDPSNPRD, EKDKS, SSEFHH, ERVQKE, GSHRP, HEIQRLGDLIPFGVPH-TVTK, SSALHD, NGALKRNE, EGIAR, DLTPRESG, GNVP, SYQIR** were specific to it only, whereas **NGER** and **KRFD** matched for the majority of CYP2B subfamily proteins. Pattern **RRFS** was found in families 2, 1 and 105. Concerning the bacterial proteins, there were no antigenic homology within subfamilies. Some of their epitopes could be found in mitochondrial and eucariotic microsomal forms, especially in family 2.

We also attempted to find any homology among the epitopes of different P450s from the same organism. No tracks of such homology were detected. The number of common epitopes was not greater than the organismless comparision.

It's interesting to note that the antigenic motif **E![VL][RK]** was found in 98% of cytochromes. Results of searching through the SWISS-PROT database, however, showed that this motif is widely distributed in all proteins (40% matched), although its probability for random sequence is 0.01% per residue (3-5% of proteins).

To find determinants specific to subfamilies, we performed manual alignment within them and revealed its consensus epitopes, described as motifs. The search through the database determined their specificity on P450 superfamily. The motif was considered to be specific, if it has matched more than 80% proteins of this subfamily, and no more than 1-2 proteins of other subfamilies. We tried to compare the data obtained with hierarchical clasterization of P450 superfamily, using the philogenetic tree by Degtyarenko and Archakov (1992). Individual subfamily epitopes (mentioned above), and patterns for branching points were found and marked (fig. 2).

For subfamily 1A, patterns **DLY[ST]** (110-120), **RYLPH** (245-250) is specific to the 92% of sequences.

Subfamily 2A is characterized by epitope **SFLI** (270-273). Specific epitope for subfamily 2B was **EK[YH]GD** (60-67), **RRFS** site (125-129) was found in 2D, 2E, 2G subfamilies. For subfamily 2C there was **HGY!!!K** (84-90), for 2D there were two epitopes, **DQA[RH]** and **STLR** (137-140), for 2E - **FS!GKR** (430-437), for 2F - **NLPRPFQ** (480-486), for 2H - **EK!FK** (105-109). Family 2 members contain many common patterns (**S!GKR, HE[IV]QR, E!LAR**). Epitopes **[ED]AV[RK]** (80-90), **SFLI** (270-273), **QRF![ND]** (360-368), **S!GKR** (432-437), **[ED][GS]LA[RK]** (442-455), being aligned with corresponding sites of other family members, formed zones of strong antigenicity.

Subfamily 3A is characterized by pattern **KETQ** (471-474), 4A - **QHRR** (138-141), 4B - **NIYE** (90-93), 4C - **ELILR** (90-94). **DGRS** (400-410) and **SAGPR** (450-455) are common patterns for subfamilies 4A and 4B, and all the family 4, respectively.

Subfamilies 11A and 11B had specific patterns **SPGD** (55-59) and **DLVLQNYN** (392-400), respectively, and common pattern **DLVL** for family 11. Family 27 had pattern **KPAE** (165-169) and one common epitope **R![AV]LN** (146-155) with family 11.

Thus, the set of antigenic determinants of various -- individual, subfamily and family -- specificity of P450 superfamily was revealed. No common epitopes correlated to the organism, not to taxon, were detected. Information on individual and subfamily- and family-specific epitopes may be used for production of antibodies specific to these protein groups, and for further epitope homology investigation.

```
:         .   10    .    20    .    30    .    40    .    50    .
A    mefsllllaflaglllllfrgHPKAHGRlppgpsplpvlgnllqmDRKGllrsflRLR
M    mefsllllaflaglllllfrGHPKAHGRLPPGPSPLPVLGnllqmdrkgllrsflRLR
E    ......................*********....******................

:  60    .    70    .    80    .    90    .   100    .   110    .
A    EKygdvftvylgsrpvvvlcgtdaiREalvdqaeAFSGRGKiavvdpifqgygvifanGE
M    EKYGDVFTvylGSRPvvvlcgTDAIRealvdqaEAFSGRGKiavvDPIFQGygvifaNGE
E    ..***********......*********....*********....****************

:  120   .   130    .   140    .   150    .   160    .   170    .
A    RWRALRrfslatmrdfGMGKRSVEERIQEEARclvEELRKSKgalldntllfhsitsnii
M    RwralRRFSlatmrdfgmgkrsveeriqeeearclveelrkskgalldntllfhsitsnii
E    ********......******.******....,...........................

:  180   .   190    .   200    .   210    .   220    .   230    .
A    csivfgKRFDYKDpvflrlldlffqsfslissfssqvfelfpgflkhfpGThRqiyrNlq
M    csivfgKRFDykdpvflrlldlffqsfslissfssqvfelfpgflkHFPGThrqiyrnlq
E    .**********.......................************************...

:  240   .   250    .   260    .   270    .   280    .   290    .
A    eintfigqsvEKHRATLDPSNPRDfidvyllrmEKDKSDPSSEFhhqnliltvlslffag
M    eintfigQSVEKHRATLDPSNPRDfidvyllrmEKDKSdpSSEFHHqnliltvlslffag
E    ..*******..******************..........***********....******

:  300   .   310    .   320    .   330    .   340    .   350    .
A    tettsttlrygfllmlkyphvtERVQKEieqvigSHRPPALDDRAKmpytdavihEiqRl
M    tettsttlrygfllmlkyphvtERVQKEieqviGSHRPpalddrakmpytdaviHEIQRL
E    .............................................******

:  360   .   370    .   380    .   390    .   400    .   410    .
A    gdlipfgvphtvTKDTQFrgyvipkntevfpvlssalHDPRYFEtpntfnpghfldangA
M    GDLIPFGVPHTVTKDtqfrGYVIPkntevfpvlSSALHDpryfetpntfnpghfldaNGA
E    ............*******.....******......*********...***

:  420   .   430    .   440    .   450    .   460    .   470    .
A    LKRNegfmpfslgKRiclgegiartelflffttilqnfsiaspvPPEDIDLTPRESgvgn
M    LKRNEgfmPFSLGkriclgEGIARtelflffttilqnfsiasPVPpediDLTPRESGvGN
E    **********...........................................********....

:  480   .   490
A    vppsyqirflar
M    VPpSYQIRflar
E    .******.....
```

Figure 1. The comparision of prediction by Antigenic Index and Motif method on the Cytochrome 2B4 sequence. A - Antigenic Index, M - Motif method, E - Experimental data by Pepscan (Kolesanova et al., 1993). Capital letters are predicted epitope residues, stars marks experimental data. Sites of true positive prediction of epitopes are bold.

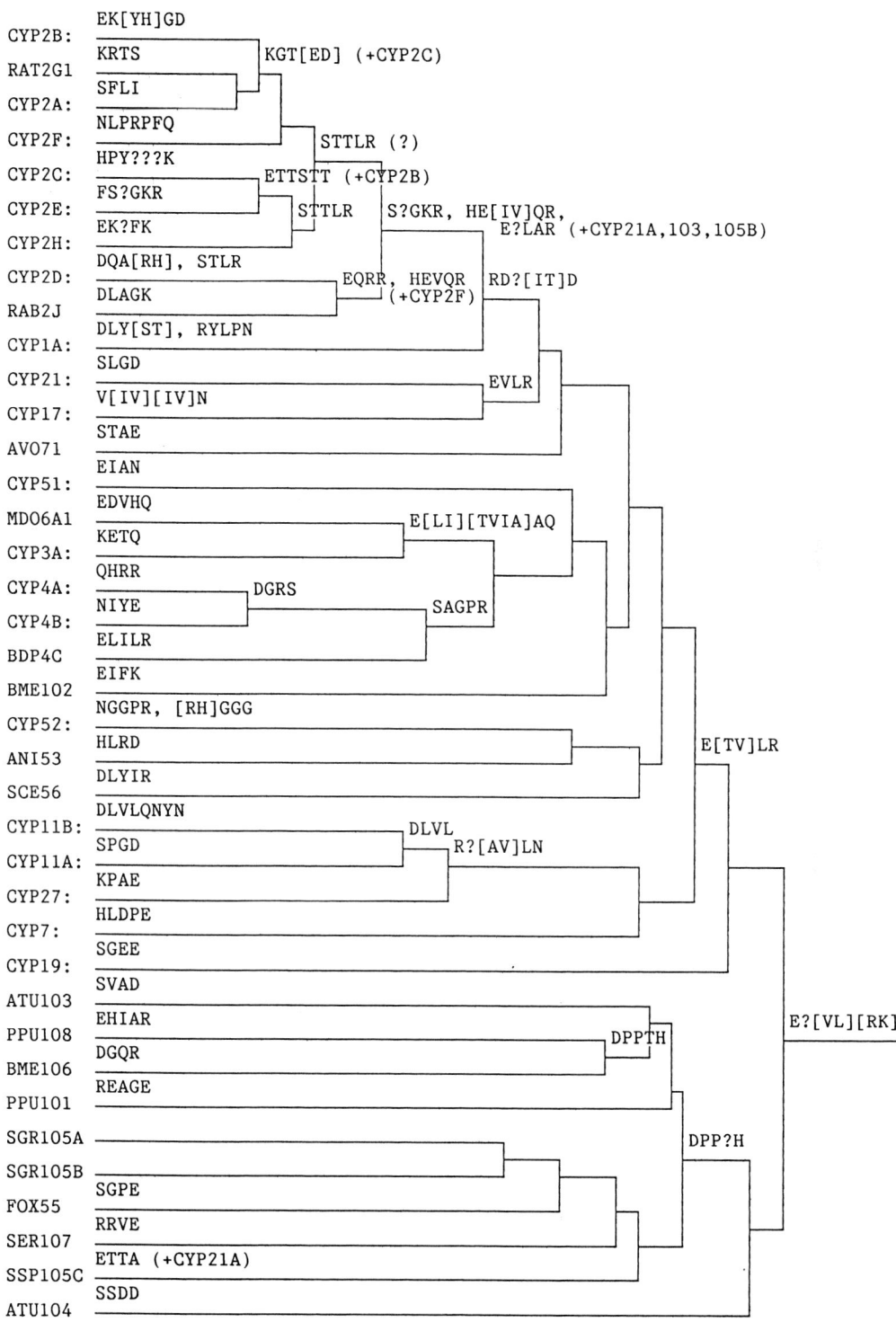

Fig.2 Philogenetic tree (Degtyarenko et al., 1993) with epitope patterns marked. Conflicts in specificity between tree branches listed in paren-

609

## References

Archakov,A.I., Bachmanova,G.I., Sandler,M.K., Tutochkin,I.Yu., Lisitsa A.V. (1991) Cytochrome P450 database and its scientific application. In: "Cytochrome P450: Biochamistry and Biophysics", Moscow (Proc. of the 7th International Conference), 673-679.

Atassi,M.Z. (1984) Antigenic structure of proteins. Their determination has revealed important aspects of immune recognition and generated strategies for synthetic mimicking of protein binding sites. Eur. J. Biochem., 145, 1-20.

Briggs,S., Price,M.R. and Tendler,S.J.B. (1993) Fine specifity of antibody recognition of carcinoma-associated epithelial mucins- antibody binding to synthetic peptide epitopes. Eur. J. Cancer, 29A, 230-237.

Chou,P.Y., Fasman G.D. (1978) Prediction of the secondary structure of proteins from their amino acid sequence. Adv. Enzymol., 47, 45-148.

Corpet,P. (1988) Multiple sequence alignment with hierarchical clustering. Nucleic Acids Res., 16, 10881-10890.

Das,M.K. and Lindstrom,J. (1991) Epitope mapping of antibodies to acetylcholine receptor "a" subunits using peptides synthesized on polypropylene pegs. Biochemistry, 30, 2470-2477.

Degtyarenko,K.N., Archakov,A.I. (1993) Molecular evolution of P450 superfamily and P450-containing monooxygenase systems. FEBS Letters, accepted for publication.

Edmundson,A.B., Harris,D.L., Tribbick,G. and Geysen,H.M. (1991) Binding of peptides to proteins: an exercise in molecular design. In: 1991 Host-guest molecular interactions: from chemistry to biology. Wiley, Chichester (Ciba Foundation Symposium 158), 213-230.

Geysen,H.M., Rodda,S.J., Mason,T.J., Tribbick,G. and Schoofs,P.G. (1987a) Strategies for epitope analysis using peptide synthesis. J.Immunol.Meth., 102, 259-274.

Geysen,H.M., Tainer,J.A., Rodda,S.J., Mason,T.J., Alexander,H., Getzoff,E.D., Lerner,R.A. (1987b) Chemistry of antibody binding to a protein. Science, 235, 1184-1190.

Higgins,D.G. and Sharp,P.M. (1989) Fast and sensitive multiple sequence alignment on a microcomputer. CABIOS, 5, 151-153.

Hopp,T.P. and Wood,K.R. (1981) Prediction of protein antigenic determinants from amino acid sequences. Proc.Natl.Acad.Sci.USA, 78, 3824-3828.

Jameson,B.A. and Wolf,H. (1988) The antigenic index: a novel algorithm for predicting antigenic determinants. CABIOS, 4, 181-186.

Karplus,P.A. and Schulz,G.E. (1985) Prediction of chain flexibility in proteins. Naturwissenschaften, 72, 212-213.

Kolaskar,A.S. and Tongaonkar,P.C. (1990) A semiempirical method for prediction of antigeniccc determinants on protein antigen. FEBS Letters, 276, 172-174.

Kolesanova,E.F., Kozin,S.A., Archakov,A.I. (1993) Epitope mapping of cytochrome P450 2B4. Proceedings of this conference.

Krchnak,V., Mach,O. and Maly,A. (1987) Computer prediction of potential immunogenic determinants from protein amino acid sequence. Analytical Biochemistry, 165, 200-207.

Kyte,J. and Doolitle,R.F. (1982) A simple method for displaying the hydropathic character of a protein. J.Mol.Biol., 157, 105-132.

Maksyutov,A.Z. and Zagrebelnaya,E.S., (1993) ADEPT: a computer program for prediction of protein antigenic determinants. CABIOS, 9, 291-297.

Milich,D.R. (1989) Synthetic T and B cell recognition sites: implications for vaccine development. Advances in Immunology, 45, 195-282.

Novotny,J., Bruccoleri,R.E. and Saul,F.A. (1989) Biochemistry, 28, 4735-4742.

Parker,J.M.R., Guo,D. and Hodges,R.S. (1986) New hydrophilicity scale derived from high-performance liquid chromatography peptide retention date: correlation of predicted surface residues with antigencity and X-ray-derived accesible sites. Biochemistry, 25, 5425-5432.

Pellequer,J.L., Westhof,E. and Van Regenmortel,M.H.V. (1991) Predicting location of continuous epitopes in proteins from their primary structures. Methods in enzymology, 203, 176-201.

Ripoll,D.R. (1992) Conformational study of a peptide epitope shows large preferences for beta-turn conformations. Im.J.Peptide Protein Res., 40, 575-581.

Rothbard,J.B. and Taylor,W.R. (1988) A sequence pattern common to T cell epitopes. EMBO J., 7, 93-100.

Stern,S. (1991) Predicting antigenic sites on proteins. TIBTECH, 9, 163-169.

Taylor,W.R. (1988) Review: pattern matching methods in protein sequence comparasion and structure prediction. Protein Engineering, 2, 77-86.

Wang,J.-G., Jansen,R.W., Brown,E.A. and Lemon,S.M. (1990) Immunogenic domains of Hepatitis delta virus antigen: peptide mapping of epitopes recognized by human and woodchuck antibodies. J.Virol., 64, 1108-1116.

# Vitamin D3 hydroxylases in human skin

Inge Schuster, Helmut Egger, Gerda Herzig, Georg Vorisek

*Sandoz Forschungsinstitut, Department of Dermatology, A-1235 Vienna, Austria*

Skin, in particular epidermis, plays an unique role in the synthesis and metabolism of Vitamin D3 (D3) and is a target organ for resulting D3-metabolites. Exposed to sunlight/UV-light skin is capable of Vitamin D3 (D3)-synthesis from the precursor 7-dehydrocholesterol by a photochemical reaction. Keratinocytes - the main epidermal constituents - possess high activities of D3-hydroxylases. They convert 25OHD3 to calcitriol $(1,25(OH)_2D3)$ - the metabolite with strongest hormonal activity which acts in the manner of steroidal hormones on gene transcription - by the 1-OHase. Local calcitriol upregulates enzymes which cause its catabolism by (multiple) attacks on the C20-27 side chain (Fig.1).

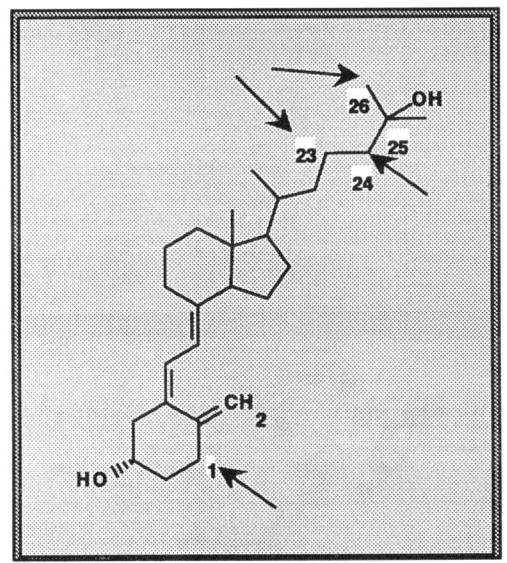

**Fig 1** 25OH-D3: sites of oxidative attack by different epidermal Cyt P450's

All these enzymes belong to the multigene family of cytochrome P-450. However, there has been no positive report on the conversion of D3 to 25-OHD3 in this tissue by a different cytochrome P-450. 25-OHase activity is described to be mainly restricted to the liver (mitochondria, microsomes) and the kidney (microsomes) (DeLuca and Schnoes 1983; Bergman and Postlind, 1990)

Receptors for calcitriol demonstrated in keratinocytes (Horiuchi et al. 1985) point to an autocrine/paracrine role of the hormone in cellular processes; calcitriol has been shown to stimulate the differentiation of keratinocytes. The present paper investigates whether keratinocytes are autonomous in the production of calcitriol or whether they rely on the systemic supply of the precursor 25OHD3. We also have started to explore the possible regulatory role of other D3-metabolites in the maintenance of skin integrity and function. A screening of azole-group (imidazole/triazole) containing compounds capable of a direct link to the heme iron of P-450s has yielded selective inhibitors of distinct D3-OHases. Using these tools we are able to elevate and prolong the cellular levels of otherwise metabolically unstable D3-metabolites which is fundamental to recognize their potential regulatory effects.

## METHODS

Human keratinocytes from breast skin (obtained by surgery) in their second passage and at confluency were used. In order to identify 25-OHase activity we incubated cells with 20nM $^3$H-D3 (Amersham) during 1-24 hrs in 6-multiwell plates (each well 3.3cm diameter) in (serum free) KGM (Clonetics) using a protocol of low and high calcium (0.06, 1.2mM). Moreover, keratinocytes were also overnight-pretreated with 20nM calcitriol (Duphar) which upregulates the catabolism of calcitriol and might also have an influence on the 25-OHase. In parallel, 20nM $^3$H-25D3 was incubated and processed according to Bikle et al. (1986) in order to show the metabolic competence of the used cultures. The uptake of the very lipophilic D3 into keratinocytes was proven in a separate experiment by centrifugation of the cells through a silicon layer which entirely strips off material adhering to the cell wall (Baur et al. 1975). Incubates were extracted with chloroform which dissolves total 25D3 and metabolites of comparable and reduced lipophilic character. Very polar products (resulting from side chain cleavage) may be retained in the aqueous phase and may even be lost by evaporation. Analysis of the extracts was performed by HPLC on silicagel using several (inactive) D3-metabolites as reference compounds (Bikle et al. 1986). After chloroform extraction the formation of individual D3-metabolites was followed by HPLC as described above.

## RESULTS AND DISCUSSION

Independent of the calcium concentration and a preceding pretreatment with calcitriol our incubations showed no evidence for a substantial oxidation of the D3-molecule (FIG 2; radioactivity pattern have been recorded at 20x magnification; small peaks due to impurities from $^3$H-D3; <3%).), although the very lipophilic D3 had entered the cells to a substantial amount (D3 sticking to the outside of the cells was stripped by separation of cells and medium through a silicon layer, data not shown.) In incubations up to 24 hours no peak at the position of the 25D3 standard was formed and no other peak was observed in the HPLC-profile. Since 24-OHase and other catabolic enzymes are present in cells pretreated with calcitriol it had been questioned whether these enzyme(s) could attack D3 to give e.g. 24OHD3. Sequential oxidation of this product by the epidermal 1-OHase could produce 1,24-OH2D3, a molecule with comparable hormonal activity as calcitriol. However, it is obvious from our data that D3 is not recognized as a substrate by enzymes oxidizing at the positions C23 or 24 or 26; moreover, under conditions of high 1-hydroxylase activity no conversion of D3 (attack at position 1α in ring A) is observed likewise.

Therefore, the D3-molecule clearly has to gain an OH-group in the side chain to become fixed on these enzymes. Moreover, no increased radioactivity was found in the aqueous phase indicating that no metabolites of increased polarity had been formed (data not shown. In contrast, $^3$H-25-D3 incubated in parallel as a standard was efficiently metabolized to calcitriol (13.2% within the first hour) either or after calcitriol pretreatment to the cascade of catabolic products emerging from (repeated) C20-27 side chain attack (15% within the first hour) thus indicating the metabolic competence of the cells (FIG 3). These data are in line with recent findings of MacLaughlin et al. (1991) who, however, had not considered a possible regulatory effect of calcium (which some other P-450s require (Reiners et al. 1990)) or of calcitriol pretreatment and had not proven cellular uptake of D3.

Our data strongly support a lacking or very low 25-OHase in human keratinocytes which cannot account for the substrate required for the action of 1-OHase. Whether different epidermal cells (Langerhans cells, melanocytes) could convert D3 to a sufficient extent is still under investigation.

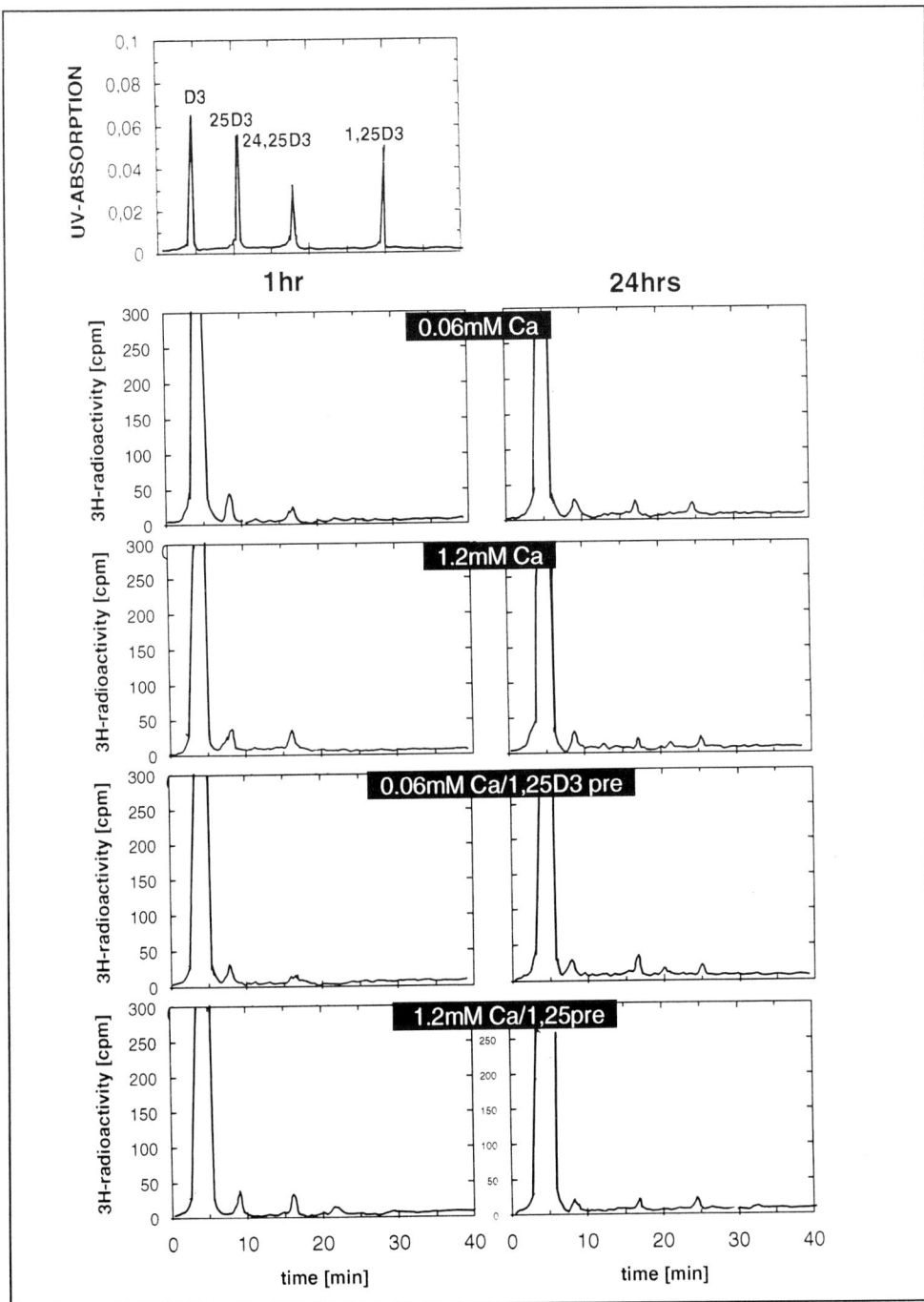

**Fig 2** Incubations (1hr and 24 hrs) of human keratinocytes with ³H-D3, HPLC profiles from extracts.

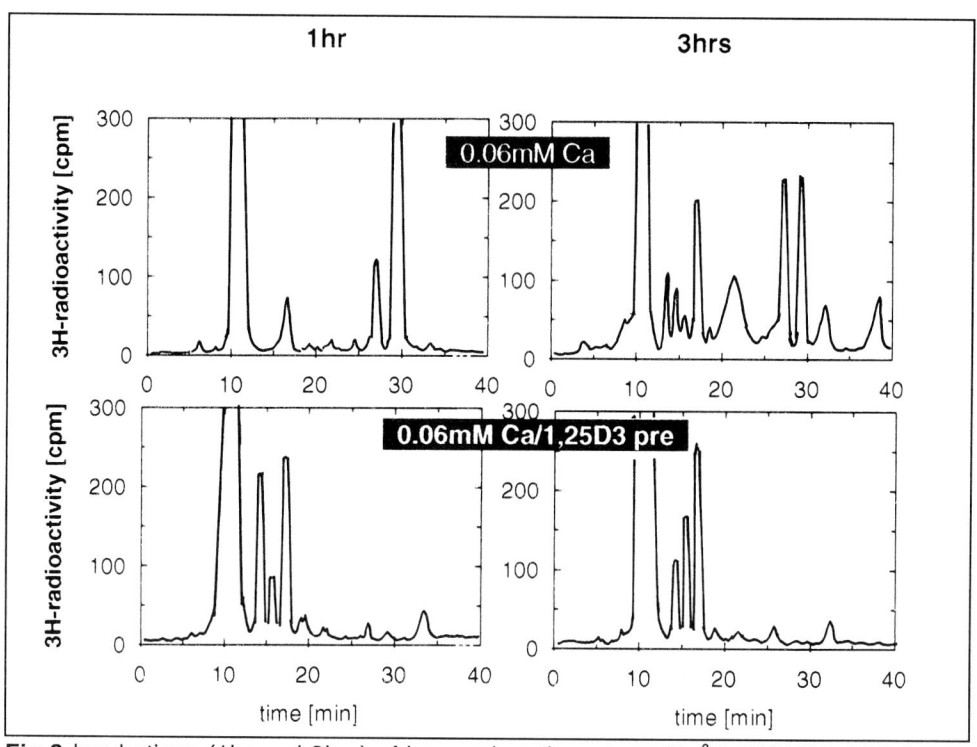

**Fig.3** Incubations (1hr and 3hrs) of human keratinocytes with $^3$H-25OHD3; HPLC profiles from extracts. Conditions as described in Methods.


REFERENCES:

Baur H.,S.Kasperek,E.Pfaff (1975) Criteria of viability in isolated liver cells. Hoppe-Seyler's Z.Physiol.Chem. 356, 827

Bergman T.,H.Postlind (1990). Characterization of pig kidney microsomal P-450 catalysing 25-hydroxylation of vitamin D3 and C-27 steroids. Biochem.J. 270, 345-350

D.D.Bikle , M.K.Nemancik, E.Gee, P.Elias (1986) 1,25-Dihydroxyvitamin D3 production by human keratinocytes: kinetics and regulation. J.Clin.Invest. 78, 557-566

DeLuca H.F.,H.K.Schnoes (1983) Vitamin D: Recent advances Annu.Rev.Biochem. 52, 411-439.

N.Horiuchi,T.L.Clemens,A.L.Schiller,M.F.Holick (1985) Detection and developmental changes of the 1,25(OH)2D3 receptor concentration in mouse skin and intestine. J.Invest.Dermatol.84,461 - 464.

MacLaughlin J.A.,M.A.Castonguay,M.F.Holick (1991) Cultured human keratinocytes cannot metabolize vitamin D3 to 25-OH vitamin D3. FEBS-Letters 282, 409-411

Reiners J.J.Jr,A.R.Cantu,A.Pavone (1990) Modulation of constitutive cytochrome P-450 expression in vivo and in vitro in murine keratinocytes as a function of differentiation and extracellular Ca concentration . PNAS, 87,1825-29

# Two cytochrome P450 aldosterone synthase mRNAs are present in the hamster adrenal

Jean-Guy LeHoux, Hugues Bernard, Lyne Ducharme, Andrée Lefebvre, Jacques LeHoux, Steeve Véronneau

Département de biochimie, Faculté de médecine, Université de Sherbrooke, Sherbrooke, QC, Canada, J1H 5N4

## Summary

We isolated two different cDNAs from a hamster adrenal cDNA library, similar in sequence to the rat $P450_{11\beta}$ and $P450_{c18}$ respectively. The hamster $P450_{c18}$ cDNA, however, was shorter than the rat $P450_{c18}$ cDNA at its 5'-end. From a hamster genomic library we isolated and sequenced the first coding exons of the *CYP11B2* gene. With this information we were subsequently able to obtain, using PCR, a $P450_{c18}$ cDNA similar in length to those of the rat and mouse. Northern analyses were performed on adrenal from hamsters sodium restricted for 0 and 7 days, using two [$^{32}$P]-labeled oligonucleotides specific to $P450_{11\beta}$ and to $P450_{c18}$ respectively. In the zona glomerulosa two mRNA bands developed at 2 kb and 3.5 kb. The intensity of both bands was increased 3- and 5-fold respectively under sodium restriction when compared to controls, but only one mRNA band developed at 2.3 kb for $P450_{11\beta}$. Immunoblotting analyses were performed using an antipeptide antibody that specifically recognizes the hamster $P450_{c18}$ protein and with an antibovine $P450_{11\beta}$. In whole adrenal homogenates, two protein bands were detected in the area of 45 kDa and 43 kDa respectively, the intensity of both bands increased 5-fold when hamsters were maintained on a low sodium intake for 7 days. These results strongly suggest that the two $P450_{c18}$ mRNAs detected by northern analysis, were effectively translated into their respective $P450_{c18}$ protein. In contrast, a principal protein band was shown for $P450_{11\beta}$ and its intensity did not change with sodium restriction.

## INTRODUCTION

The adrenal cortex of most mammalian species, including the human, mouse and rat, express two distinct forms of enzymes which catalyze the terminal steps in the biosynthesis of glucocorticoids and mineralocorticoids. These enzymes, 11β-hydroxylase ($P450_{11\beta}$) and aldosterone synthase ($P450_{c18}$) are the products of *CYP11B1* and *CYP11B2* genes. In the bovine adrenal, however, a single *CYP11B* gene was shown to transform deoxycorticosterone to corticosterone and also corticosterone to aldosterone.

We have previously established that the hamster adrenal synthesizes the mineralocorticoid aldosterone (1) and the glucocorticoid cortisol (2), and that there is different control of the secretion of these two steroids. The regulation of the last steps of aldosterone formation in hamster adrenal has not been studied hitherto. In this study we report the presence of two adrenal $P450_{c18}$ mRNA and two $P450_{c18}$ protein species and the presence of only one mRNA species for $P450_{11\beta}$. The abundance of the two $P450_{c18}$ mRNAs and $P450_{c18}$ proteins was increased by a low sodium intake, whereas those of $P450_{11\beta}$ remained unchanged.

## EXPERIMENTAL

*Animals*: Male Syrian golden hamsters were fed Purina rat chow and tap water *ad libitum* or they were maintained on a sodium deficient diet (< 0.01 mEq $Na^+$/ g) for 7 days.

*RNA analysis by northern blotting*: Total RNA was denatured with glyoxal, electrophoresed in 1% agarose in 0.01 M phosphate buffer pH 6.5 and transferred to GeneScreenPlus. The membrane was prehybridized for 4 h at 42°C and then hybridized for 16 h at 42°C in the presence of specific [$^{32}$P]-oligonucleotides derived from hamster adrenal $P450_{11\beta}$ and $P450_{c18}$ cDNA sequences. The blots were also analyzed with a 28 S [$^{32}$P]-ribosomal probe for the quantitation of mRNA (3).

*Immunoblotting*: Adrenal tissues were homogenized and solubilized as previously described (4) in Laemmli buffer but containing protease inhibitors and iodoacetamide. They were passed through a 26-gauge needle and then boiled for 5 min. Proteins were electrophoresed on a 10% polyacrylamide gel in the presence of sodium dodecyl sulfate, as described by Laemmli (4), and subsequently transferred onto nitrocellulose. P450s were detected using an anti-bovine adrenal $P450_{11\beta}$ antibody (kindly provided by Dr M.R. Waterman) and an anti-peptide antibody specific to the hamster adrenal $P450_{c18}$.

## RESULTS

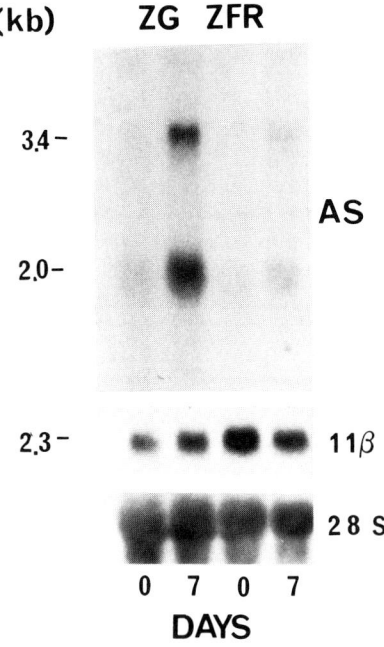

**Fig.1. Northern analysis of hamster adrenal $P450_{c18}$ and $P450_{11\beta}$.**
Total RNA was hybridized with [$^{32}$P]-labeled oligonucleotides. $P450_{c18}$: (AS); $P450_{11\beta}$: (11β). 28 S: [$^{32}$P]-labeled ribosomal probe. ZG: zona glomerulosa; ZFR: zona fasciculata-reticularis.

Fig.1 shows results of a northern blotting analysis performed on hamster adrenal zona glomerulosa (ZG) and zona fasciculata-reticularis (ZFR) preparations. Using the specific [$^{32}$P]-labeled oligonucleotide derived from the hamster P450$_{c18}$ cDNA (5´-ACAGTGGCATCGTGGCAGAACTAATGTCCCAGGGA-3´), two mRNA bands were detected in ZG at 2.0 kb and 3.4 kb. On the same membrane, an oligonucleotide specific to P450$_{11\beta}$ (5´-TGCAAAGTGGTGGCCCACAGTCCTGGAGTGTCATATCACAGCTGGT-3´) hybridized to a single mRNA band at 2.3 kb, which clearly differs from those detected by the P450$_{c18}$ probe. This figure also shows the effects of a low sodium intake for 7 days on P450$_{c18}$ and P450$_{11\beta}$ mRNAs in hamster adrenal ZG and ZFR. The upper 3.4 kb band was less abundant than the lower 2.0 kb band but the intensity of both bands increased with the low sodium intake, suggesting a functional role for these two mRNAs. Some P450$_{c18}$ mRNA was found in the ZFR, and this is not surprising since, in contrast to the rat, it is difficult to cleanly separate the hamster adrenal ZG from the ZFR. P450$_{11\beta}$ mRNA was found in similar quantity in both ZG and ZFR and its level did not change with sodium restriction.

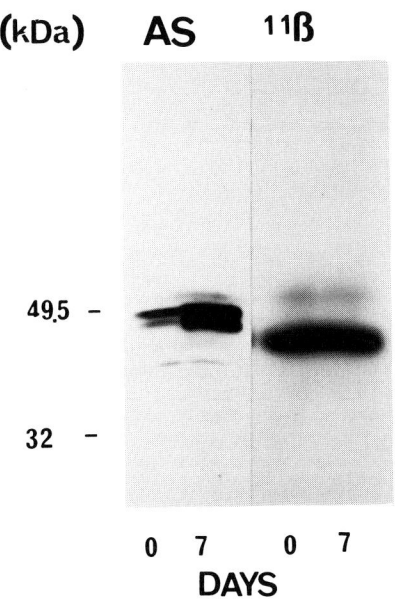

Fig.2. Immunoblotting analysis of hamster adrenal P450$_{c18}$ and P450$_{11\beta}$.

Adrenal tissues were homogenized and solubilized in sodium dodecyl sulfate and then electrophoresed in a 7.5% polyacrylamide gel. The P450$_{c18}$ protein was detected with an antibody directed against a specific peptide sequence deduced from adrenal P450$_{c18}$, whereas P450$_{11\beta}$ protein was detected with an anti-bovine adrenal P450$_{11\beta}$. With the P450$_{c18}$ peptide antibody, two protein bands were detected in the area of 45 kDa and 43 kDa in homogenate preparations of whole adrenal. A sodium restriction for seven days increased the intensity of these two bands compared to similar preparations from hamsters maintained on a normal sodium intake, indicating that P450$_{c18}$ mRNAs were effectively translated into proteins. This figure also shows that the antibovine P450$_{11\beta}$ antibody coupled to a principal protein band and that the low sodium intake did not increase the intensity of the signal.

Plasma aldosterone, corticosterone and cortisol levels were analyzed in plasma of hamster maintained on a low sodium diet for 0 and 7 days. The plasma aldosterone level increased 12-fold over control after 7 days of sodium restriction. The plasma corticosterone level increased 2-fold whereas that of cortisol remained similar to control, indicating a different regulation of the synthesis betwen these corticosteroids.

## DISCUSSION

Northern analysis of total hamster adrenal mRNA or polyA$^+$ mRNA (results not shown) revealed two mRNA bands that hybridized specifically to a specific $P450_{c18}$ oligonucleotide, indicating that both species were polyadenylated. Also, using a specific oligonucleotide, only a single $P450_{c18}$ mRNA species could be demonstrated in the rat thereby indicating a difference between these two closely related animal species. A mRNA band of 2.3 kb hybridized with an oligonucleotide specific to $P450_{11\beta}$ indicating that $P450_{11\beta}$ and $P450_{c18}$ are different entities. We isolated and sequenced a truncated $P450_{c18}$ cDNA from a hamster adrenal cDNA library. This cDNA is shortened at its 5'-end by 380 bp compared to the cDNA generated by PCR. The truncated cDNA, however, posesses an ATG at the beginning of the NH$_2$-terminal, in frame with the full length coding sequence. We have not yet expressed this truncated cDNA, and consequently we cannot say whether it will be translated into an active protein. The low sodium intake selectively affected the levels of both the $P450_{c18}$ mRNAs and the $P450_{c18}$ proteins but not that of $P450_{11\beta}$ mRNA and $P450_{11\beta}$, thereby showing a specific regulation at the final step of aldosterone synthesis which was mediated by sodium restriction. Collectively these results show that the hamster adrenal expresses $P450_{11\beta}$ and $P450_{c18}$. These two P450s are more closely related to murine than to those of human and bovine species. The hamster possesses an additional adrenal $P450_{c18}$ species, inducible by a low sodium intake, whose physiological relevance is yet to be established.

*Acknowledgement*-This work was supported by a grant from the Medical Research Council of Canada.

## REFERENCES

1. Lehoux J.G. and Ducharme L.: The differential regulation of aldosterone output in hamster adrenal by angiotensin$_{\text{II}}$ and adrenocorticotropin. *J. Steroid Biochem. Molec. Biol.* 41 (1992) 809-814.
2. Lehoux J.G., Mason J.I. and Ducharme L.: *In vivo* effects of adrenocorticotropin on hamster adrenal steroidogenic enzymes. *Endocrinology* 131 (1992) 1874-1882.
3. Lehoux J.G., Lefebvre A., Bélisle S. and Bellabarba D.: Effect of ACTH suppression on adrenal 3-hydroxy-3-methylglutaryl coenzyme A reductase mRNA in 4-aminopyrazolopyrimidine-treated rats. *Molec. Cell Endocr.* 69 (1990) 41-49.
4. Laemmli U.K.: Cleavage of structural proteins during the assembly of the head of bacteriophage T$_4$. *Nature (Lond)* 227 (1970) 680-685.

*Cytochrome P450. 8th International Conference*. Ed. M.C. Lechner. John Libbey Eurotext, Paris © 1994, pp. 619-622.

# Purification of the insect P450 enzyme, ecdysone 20-hydroxylase, from the migratory locust, *Locusta migratoria*, by a new affinity method

Jochen Winter, Hartmut Kayser

*Plant Protection Division, Ciba-Geigy AG, R-1093.P.39, CH-4002 Basel, Switzerland*

## Summary

Ecdysone 20-hydroxylases (E20H) are P450 enzymes of microsomal and/or mitochondrial location. To isolate the microsomal enzyme of the locust, *Locusta migratoria*, a new affinity approach was developed. Purification of E20H was followed on the basis of substrate binding spectra. After solubilization with Synperonic NP10 the P450 proteins were first chromatographed on ω-aminooctyl-agarose followed by hydroxyapatite. Gradient elutions with phosphate buffer were performed in both steps. For the third step a new affinity matrix was developed with a triazole-based inhibitor (a presumed general ligand to P450 enzymes) coupled to an activated agarose matrix. Specific elution of E20H in predominantly native form was achieved with its substrate ecydsone at a specific content of 13.1 nmol P450/mg protein. The overall yield of P450 was 4 per cent. Triazole-based affinity chromatography may be applied as a general tool for P450 purification. In combination with specific substrate elution it may provide an efficient new way for the isolation of specific forms of P450 enzymes from various sources.

## INTRODUCTION

Insect development is under control of polyhydroxylated steroid hormones, called ecdysteroids, which are derived from dietary cholesterol through a series of oxidation steps (Rees, 1989; Grieneisen *et al*, 1993). The terminal step in the formation of these steroids is catalyzed by ecdysone 20-hydroxylase (E20H) which is considered as the regulating enzyme for the synthesis of the ultimate moulting hormone, 20-hydroxyecdysone (Fig. 1). Hence this enzyme may provide a new target for selective insect control.

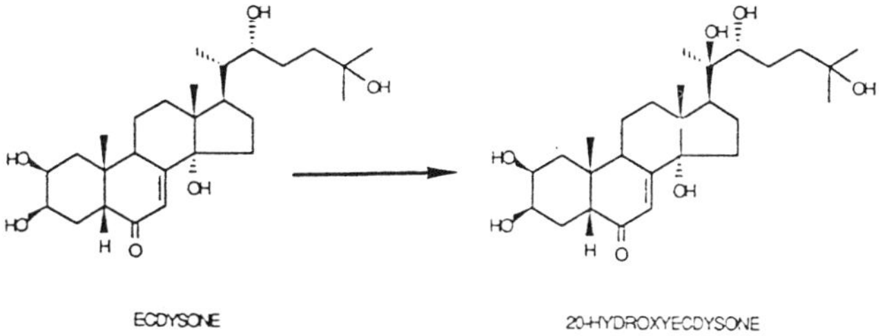


Fig. 1. Hydroxylation of ecdysone to the moulting hormone 20-hydroxyecdysone catalyzed by ecdysone 20-hydroxylase, a microsomal P450 in Malpighian tubules of the migratory locust.

Ecdysone 20-hydroxylases from several insects have been characterized as typical P450-dependent monooxygenases located in microsomes and/or mitochondria dependent on the tissue and species (Weirich, 1989). Several P450 isozymes, catalyzing ecdysone 20-hydroxylation in *in vitro* reconstitution assays, have been obtained from housefly mitochondria (Srivatsan et al., 1990). Isolation of a microsomal form of this key enzyme has not yet been reported. We argued that purification of E20H might be achieved from the migratory locust, *Locusta migratoria*, where the bulk enzyme activity is concentrated in the excretory system, the Malpighian tubules, and strictly localized to the microsomes (Feyereisen and Durst, 1978). However, a successful purification of E20H could never be achieved as most of the methods commonly employed for the isolation of P450 enzymes turned out not to be satisfactory in this case. So, it became clear that a new and specific chromatographic method had to be developed in order to reach the goal.

RESULTS

As E20H reaches peak activity just before a moult Malpighian tubules from 5 day old last instar larvae of *Locusta* were used to prepare microsomes by standard methods of differential centrifugation followed by a further purification by glycerol gradient centrifugation. The microsomes showed a specific content of 0.1 nmol P450 per mg protein and a hydroxylase activity of 0.21 pmol 20-hydroxyecdysone formed per min and per pmol of P450 (typical values). For solubilization of P450 Synperonic NP10 (formerly Renex 690) was selected as it did not lead to denaturation to P420 and provided a maximum (41 per cent) of enzymatic activity after reconstitution.

For chromatography many methods were tried in various combinations. From extensive trials the following procedure turned out to provide optimal results: after solubilization (0.5 per cent Synperonic NP10) and centrifugation (100000g; 1 h) the supernatant was applied to ω-aminooctyl-agarose which was eluted with a gradient of potassium phosphate buffer (pH 7.5; 5-300 mM; 15% glycerol; 0.2% Synperonic NP10). The fractions absorbing at 410 nm were checked by recording absolute spectra (peak at 412 nm) and difference spectra due to substrate binding (type I: peak at 386 nm, trough at 405 nm) and carbon monoxide binding (peak at 448 nm; fully native). Pooled fractions from several runs were, after dialysis against 5 mM potassium phosphate buffer pH 7.5, applied to hydroxyapatite equilibrated with the same buffer. Again, gradient elution was performed and fractions were subjected to spectroscopy as described above. In this run the substrate binding spectra were obtained in the reversed type I form (trough at 385 nm, peak at 412 nm) which is typically obtained with polyhydroxylated ligands. The main fractions of P450, again after dialysis, were then subjected to chromatography on a new affinity matrix for final purification.

As no method commonly used in the isolation of P450 enzymes was successful we tried ligands of E20H to develop an affinity method for this enzyme. Ecdysone, the enzyme's substrate, as a highly functionalized molecule, was not considered useful for this. We screened for a heterocyclic inhibitor of E20H, synthesized inhouse, with an affinity comparable to that of the substrate ($K_m$ = 0.2 μM). The ligand finally chosen was the triazole-based compound, shown in Fig. 2. It was coupled to divinylsulfon-activated agarose (Kem-En-Tec). After blocking of residual groups with dimethylamine and thorough washing the affinity matrix was equilibrated with 5 mM potassium phosphate buffer pH 7.5. The dialyzed pooled eluate of the hydroxyapatite column was slowly pumped onto the triazole column (0.8 x 5 cm) which was then washed with the same buffer until baseline absorption at 410 nm was obtained. Then, exactly one column volume of the 5 mM phosphate buffer supplemented with 20 μM ecdysone was applied to the loaded column and left for 6 h for equilibration. Elution provided a sharp fraction (Fig. 2), as monitored at 410 nm, from which an absolute spectrum of the substrate-bound P450 (peak at 410 nm) and a CO difference spectrum (peak at 447 nm) was obtained demonstrating mainly native P450. With this affinity column a 13-fold purification was achieved with a yield of 14%. Later studies showed that the yield at this step can be considerably increased up to estimated 50% by raising the ecdysone concentration to 100 μM. This means that the large fraction, absorbing at 410 nm, obtained with the buffer wash after affinity elution (Fig. 2) still contains predominantly E20H. Details of this affinity method will be reported elsewhere. An overview of a typical purification is given in Table 1.

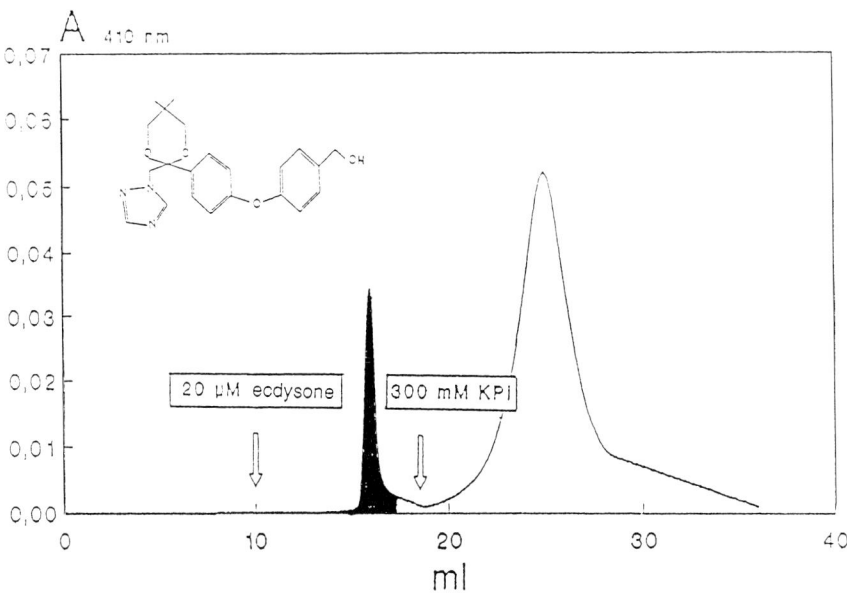

Fig. 2. Elution diagram of the triazole-agarose loaded with microsomal P450 obtained after chromatography on ω-aminooctyl-agarose and hydroxyapatite. Specific elution of E20H (peak marked black) was performed with its substrate ecdysone. The salt wash was later shown to also contain predominantly E20H as evidenced on the basis of substrate binding spectra. Elution of the total amount requires 100 µM ecdysone.

Table 1: Purification of ecdysone 20-hydroxylase from Malpighian tubules of *Locusta migratoria*. Representative numbers for a batch of 1000 locusts are given.

| Step | Protein (mg) | P450 (nmol) | spec. content (nmol P450 per mg protein) | purification (x-fold) | yield of P450 (%) |
|---|---|---|---|---|---|
| microsomes | 40.0 | 4.0 | 0.1 | 1 | 100 |
| ω-aminooctyl-agarose | 5.6 | 2.8 | 0.5 | 5 | 70 |
| hydroxyapatite | 1.2 | 1.2 | 1.0 | 10 | 30 |
| triazole-agarose | 0.013 | 0.17 | 13.1 | 131 | 4 |

SDS-PAGE of the ecdysone eluate showed a prominent band, supposed to represent E20H, indicating a protein with a molecular weight of about 55 kD which is in the range of P450 enzymes. This band was cut out and used for raising antibodies and to obtain partial sequences after proteolytic digestion. E20H obviously represents the only (dominant) P450 in the locust microsomes. It is currently being cloned and sequenced as its cDNA derived from Malpighian tubules.

Besides the triazole compound which, as a N-heterocyclic compound, presumably binds to the heme iron of P450 as a basically group-specific ligand several other compounds as far related mimics of the steroidal substrate of E20H were also coupled to the activated agarose matrix. These were derivatives of stilbestrol and testosterone, respectively. The stilbestrol-based affinity matrix was comparably effective as the triazole column in providing E20H after elution with ecdysone (Winter, 1993).

CONCLUSION

The isolation of a microsomal form of the P450 enzyme ecdysone 20-hydroxylase with high purity has been reported for the first time. This goal was only achieved by the development of a new mode of affinity chromatography for P450 proteins. This methods is a two-fold one: P450 proteins, as a group, are first adsorbed to a matrix-bound triazole inhibitor and then selectively eluted by a substrate (or substrate analogue) of the desired P450. Preliminary tests (not shown) with P450 from mammalian liver microsomes support the view that this concept can also be applied to P450 proteins from other sources after careful adaptations from case to case.

REFERENCES

Feyereisen, R. (1993): Cytochrome P450 in insects. In *Cytochrome P450*, Handbook of Experimental Pharmacology. Vol. 105, ed. G.V.R. Born, P. Cuatrecasas & H. Herken, pp. 311-324. Berlin/Heidelberg: Springer.
Feyereisen, R. & Durst, F. (1978): Ecdysterone biosynthesis: a microsomal cytochrome P-450-linked ecdysone 20-monooxygenase from tissue of the african migratory locust. Eur. J. Biochem. 88, 37-47.
Grieneisen M.L., Warren J.T. & Gilbert L.I. (1993): Early steps in ecdysteroid biosynthesis: evidence for the involvement of cytochrome P-450 enzymes. *Insect Biochem. Molec. Biol.* 23, 13-23.
Rees, H.H. (1989): Pathways of biosynthesis of ecdysone. In *Ecdysone. From chemistry to mode of action*, ed. J. Koolman, pp. 152-160. Stuttgart/New York: Thieme.
Srivatsan, J., Weirich M. & Agosin M. (1990): Cytochrome P-450-catalyzed formation of 20-hydroxy-ecdysone in larval housefly mitochondria. *Biochem. Biophys. Res. Commun.* 166, 1372-1377.
Weirich, G.F. (1989): Enzymes involved in Ecdysone Metabolism. In *Ecdysone. From chemistry to mode of action*, ed. J. Koolman, pp. 174-80. Stuttgart/New York: Thieme.
Winter, J. (1993): Ecdyson-20-Hydroxylase von *Locusta migratoria*: Proteinreinigung eines Cytochrom P450 sowie Klonierung und Sequenzierung eines 1,1 kb-cDNA-Fragmentes. Dissertation Universität Ulm.

# Effect of carrageenan-induced granuloma on the apoprotein content and activities of cytochrome P450 isozymes in liver and nasal mucosa in rats

Jordi Muntane[1,4], Vincenzo Longo[2,4], M. Teresa Mitjavila[1], Pier Giovanni Gervasi[2], Magnus Ingelman-Sundberg[3]

[1]Unitat de Fisiologia, Departament de Bioquímica i Fisiologia, Facultat de Biologia, Av. Diagonal 645, 08028 Barcelona, Spain. [2]Laboratory of Genetics and Biochemical Toxicology, Instituto di Mutagenesi e Differenziamento, CNR, Via Svezia 10, 56124 Pisa, Italy. [3]Department of Physiological Chemistry, Karolinska Institutet, S-10401 Stockholm, Sweden. [4]At the time of the study visiting scientists at the Department of Physiological Chemistry, Karolinska Institutet, S-10401 Stockholm, Sweden

## INTRODUCTION

Cytochromes P450 participate in the metabolism of exogenous and endogenous compounds. These enzymes show a high versatility against their substrates due to their great number in one species and the comparatively low specificity of their action (Nelson *et al.*, 1993). Cytochrome P450 activities are affected by smoke, alcohol, ingestion, malnutrition, age, sex and certain phatophysiological processes (Nebert & Gonzalez, 1982).

Reduction of the cytochrome P450 levels observed during experimental inflammation (Muntané *et al.*, 1993) is correlated to decreases in hepatic drug metabolism during adjuvant-induced arthritis (Cawthorne *et al.*, 1976) during carrageenan-induced inflammation in rats (Ishikawa *et al.*, 1991) and during cancer (Bertini *et al.*, 1992).

It has been emphasized that cytokines released locally by inflammatory cells and/or by Kupffer cells in the liver after stimulation, could be responsible for the changes that take place in hepatocytes during inflammation (Chen *et al.*, 1992). The purpose of the present study was to investigate the effect of an acute and chronic experimental inflammation *in vivo* on the content and the activity of several cytochromes P450 in the liver and nasal mucosa.

## MATERIALS AND METHODS

### Animals and materials

Three groups of four male Sprague-Dawley rats each weighing approximately 225 g were used: one group received saline only (Control) whereas two groups were killed at 1 (I-1 day) and 6 (I-6 days) days after induction of a granuloma per carrageenan, enabling us to discriminate between the acute and chronic phases of the inflammation (Fukuhara & Tsurufuji, 1969).

### Hepatic microsomal assays

Hepatic microsomal fraction was obtained as described by Longo *et al.* (1986), and the protein concentration was measured by the method of Lowry *et al.* (1951). Cytochrome P450 in the samples was separated by SDS-PAGE electrophoresis and quantified by Western blot (Eliasson *et al.*, 1988).

### Evaluation of cytochrome P450 dependent catalytic activities

Ethoxycoumarin O-deethylase (ECOD) was determined according to Aitio (1978). Ethoxyresorufin O-deethylase (EROD) and methoxyresorufin-demethylase (MTRD) were measured by the formation of the

corresponding hydroxy product in a Perkin-Elmer Spectrofluorimeter (Lubet *et al.*, 1985). The activities for the N-demethylation of benzphetamine (BzD) and erytromycin (ErD) were assayed by measuring the formation of formaldehyde (Tu & Yang, 1983). Aniline hydroxylase (AnH) and imipramine demethylase (ImD) activities were measured according to the methods described by Ko *et al.* (1987) and *Fujita et al.* (1989). Testosterone hydroxylase was assayed by a HPLC method described by Platt *et al.* (1989).
Results are expressed as mean ± SEM (four rats per group). Statistical evaluation was carried out using the Student's *t*-test.

## RESULTS

Effect of an acute and chronic inflammation on the apoprotein content of several cytochrome P450 isozymes and related activities

During the acute phase of the carrageenan-induced granuloma, the level of different cytochrome P450 isozymes (Fig. 1A) and related activities (Fig. 1B) were reduced by 30-80 %, depending of the isoenzyme studied. CYP2A1 and CYP2B1 were less affected by inflammation and their level were reduced by 30 % as compared to control whereas the levels of the isozymes CYP1A1, CYP2D, CYP2E1, CYP3A1 and CYP4A and related activities (AnH, ErD and EROD) were reduced by 60 to 80 %. Lowest testosterone hydroxylation activities were mainly catalyzed by CYP2A1 (6α- and 7α-hydroxylation) and CYP2B1 (16α- and 16ß-hydroxylation were reduced during the acute phase by about 50% (data not shown).

During the chronic phase of the inflammation the isozymes (Fig. 1A) and related activities (Fig. 1B) that were mostly affected during the early stages, recovered significantly and, in case of CYP1A1, almost the original levels were reached. By contrast, CYP2A1 and CYP2B1 remained at the same level as during the acute phase.

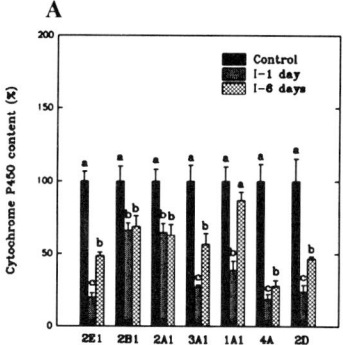
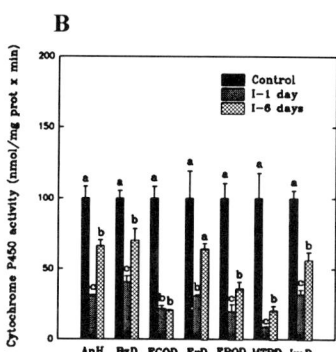

Fig. 1. Levels of cytochrome P450 apoprotein (2E1, 2B1, 2A1, 3A1, 1A1, 4A and 2D) (A) and corresponding monoxygenase activities (B) in hepatic microsomes. All microsomal fractions obtained from Control, I-1 day and I-6 days treated rats. Results are expressed as % of respectively control. Results are the mean ± SEM (each group contained four animals). The mean are given several superscript letters. Groups from the same cytochrome P450 isoenzyme with different superscript letter are statistically different $p<0.05$) from respectively Control group.

Effect of an acute and chronic inflammation on the level of CYP2E1 and CYP2B in nasal mucosa

Analysis of the level of expression of P450 isozymes in microsomes of nasal mucosa revealed a reduction of CYP2E1 to 30% of the control value and that of CYP2B1 to 70% of control, during the acute and chronic phase of inflammation (Fig. 2).

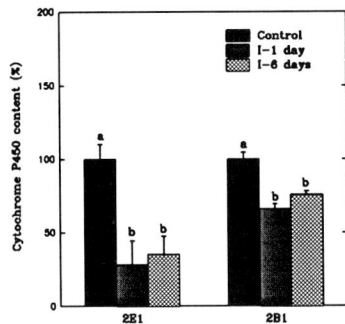

Fig. 2. Levels of cytochrome P450 apoprotein (2E1, 2B1) in nasal mucosa microsomes. Microsomal fractions obtained from Control, I-1 day and I-6 days treated rats. Results are expressed as % of respectively control. Results are the mean ± SEM (each group contained four animals). The mean are given several superscript letters. Groups from the same cytochrome P450 isoenzyme with different superscript letter are statistically different $p<0.05$) from respectively Control group.

## DISCUSSION

Two of the host reactions against infection are the increased synthesis of acute phase proteins and the depression of the hepatic cytochrome P450 content (Muntané et al., 1993). Recent studies show that these effects could be triggered by the injection of several inflammatory mediators implicated in the local and systemic reaction of inflammation as Il-1, Il-6 and TNF (Chen et al., 1992).

The specific mechanisms by which these inflammatory mediators cause decrease in cytochrome P450 expression are not understood. Under our experimental conditions, we could classify the isozymes studied in two groups depending on their sensitivity to inflammation. One group was composed of CYP2A1 and CYP2B1, whose levels were reduced to 70 % of the control level during the acute and chronic inflammation. In the second group, we included CYP1A1, CYP2D, CYP2E1, CYP3A1 and CYP4A, which were expressed to about 20-40 % of the control levels, during the acute phase of inflammation. This group of isozymes recovered their level to between 30 and 85 % of the control values, during the chronic phase of inflammation.

Interestingly, the isozymes most sensitive to inflammation are also the ones most implicated in several pathophysiological processes. In this sense, CYP2E1 exhibit an unusually high rate of oxidase activity with the subsequent formation of reactive oxygen species and thus to initiate lipid peroxidation (Tindberg & Ingelman-Sundberg, 1989). The reduced amount of P450 CYP2E1 (Fig. 1A) could be a crucial point for the survival of the hepatocytes during an inflammatory process, if the inflammatory mediators induce oxidative stress as it has been shown *in vivo* (Muntané et al., 1993) and *in vitro* studies (Adamson & Billings, 1992).

In conclusion, our data suggest that various isozymes of hepatic P450 exhibit different sensitivity towards inflammatory reactions induced by carrageenan. This might have important implications during physiological conditions and for the effectiveness of pharmacotherapy.

## REFERENCES

Adamson, G.M. & Billings, R.E. (1992): Tumor necrosis factor induced oxidative stress in isolated mouse hepatocytes. *Archiv Biochem Biophys* 294, 223-229, 1992.

Aitio, A. (1978): A simple and sensitive assay of 7-ethoxycoumarin eethylation. *Anal Biochem* 85, 488-491.

Bertini, R., Gervasi, P.G., Longo, V. & Ghezzi, P. (1992): Depression of hepatic drug metabolism in endotoxin-treated and sarcoma-bearing mice. *Res Commun Chem Pathol Pharmacol* 76, 223-231.

Cawthorne, M.A., Palmer, E.D. & Green, J. (1976): Adjuvant-induced arthritis and drug-metabolizing enzymes. *Biochem Pharmacol* 25, 2683-2688.

Chen, Y.L., Florentin, I., Batt, A.M., Ferrari, L., Giroud, J.P. & Chauvelot-Moachon, L. (1992): Effects of inteleukin-6 on cytochrome P450-dependent mixed function oxidases in the rat. *Biochem Pharmacol* 44, 137-148.

Eliasson, E., Johansson, I. & Ingelman-Sundberg, M.(1988): Ligand-dependent maintenance of ethanol-inducible cytochrome P450 in primary rat hepatocyte cell cultures. *Biochem Biophys Res Commun* 150, 436-443.

Fujita, S., Morimoto, R., Chiba, M., Kitani, K. & Suzuki, T. (1989): Evaluation of the involvement of a male specific cytochrome P450 isozyme in senescence associated decline of hepatic drug metabolism in male rats. *Biochem Pharmacol* 38, 3925-3931.

Fukuhara, M. & Tsurufuji, S. (1969): The effect of locally injected anti-inflammatory drugs on the carrageenin granuloma in rats. *Biochem Pharmacol* 18, 475-484.

Ishikawa, M., Sasaki, k., Ozaki, M., Watanabe, K., Takayanagi, Y. & Sasaki, K. (1991): Hepatic drug metabolizing activity in rats with carrageenan-induced inflammation. *J Pharmacobio-Dyn* 14, 132-138.

Ko, I.Y., Park, S.S., Song, B.J., Patten, C., Tan, Y., Han, Y.C., Yang, S, & Gelboin, H.V. (1987): Monoclonal antibodies to ethanol-induced rat liver cytochrome P450 that metabolizes aniline and nitrosamines. *Cancer Res* 41, 3101-3109.

Longo, V., Citti, L. & Gervasi, P.G. (1986): Metabolism of diethylnitrosamine by nasal mucosa and hepatic microsomes from hamster and rat: Species specificity of nasal mucosa. *Carcinogenesis* 7, 1323-1328.

Lowry, O.H., Rosebrough, N.J., Farr, A.L. & Randall, R.J. (1951): Protein measurement with the Folin phenol reagent. *J Biol Chem* 193, 265-275.

Lubet, R.A., Mayer, R.T., Cameron, J.W., Nims, R.W., Burke, M.D., Wolff, T. & Guengerich, F.P. (1985): Dealkylation of pentoxyresorufin: A rapid and sensitive assay for measuring induction of cytochrome (s) P450 by phenobarbital and other xenobiotics in the rat. *Arch Biochem Biophys* 238, 43-48.

Muntané, J., Fernandez, Y., Mitjavila, S. & Mitjavila, M.T. (1993): Hepatic changes during a carrageenan-induced graruloma in rats. *Mediators of Inflammation* 2, 79-83.

Muntané, J., Puig-Parellada, P., Fernandez, Y., Mitjavila, S., Mitjavila, M.T. (1993): Antioxidant defenses and its modulation by iron in carrageenan-induced inflammation in rats. *Clin Chim Acta* 214, 185-193.

Nebert, D.W. & Gonzalez, F.J. (1982): P450 genes: structure, evolution and regulation. *Annu Rev Biochem* 56, 945-993.

Nelson, D.R., Kamataki, T., Waxman, F., Guengerich, F.P., Estabrook, R.W., Feyereisen, R., Gonzalez, F.J., Coon, M.J., Gunsalus, I.C., Gotoh, O., Okuda, K. & Nebert, D.W. (1993): The P450 superfamilly: updated on new sequences, gene mapping, accession numbers, early trivial names of enzymes, and nomenclature. *DNA and Cell Biol* 8, 1-51.

Platt, K.L., Molitor, E., Dohmer, J., Dogra, S. & Oesch, F. (1989): Genetically engineered V79 Chinese hamster cell expression of purified cytochrome P450IIB1 monoxygenase activity. *J Biochem Toxicol* 4, 1-6.

Tindberg, N. & Ingelman-Sundberg, M. (1989): Cytochrome P450 and oxygen toxicity. Oxygen-dependent induction of ethanol-inducible cytochrome P450 (IIE1) in rat liver and lung. *Biochemistry* 28, 4499-4504.

Tu, Y.Y. & Yang, C.S. (1983): High-affinity nitrosamine dealkylase system in rat liver microsomes and its induction by fasting. *Cancer Res* 43, 623-629.

# Interaction of ethanol and type I binding substrate n-heptane with liver cytochrome P450 from male rats

Joseph S. Rakoto, Miroslaw M. Szutowski

Department of Toxicology, Institute of Biopharmacy, Warsow Medical School, Banacha 1, 02-097 Warszawa, Poland

Interactions between numerous organic ligands and the oxidized cytochrome P-450 give rise to difference spectra which fall predominantly into three classes: type I with a peak at 385-390 nm, through at 418-427 nm, caused by binding to a hydrophobic part of cytochrome, type II with a peak at 425-430 nm, through at 390-410 nm, believed to be due to, direct binding to the heme iron and reverse type I with a peak at 416-420 nm, through at 385-390 nm, that is characteristic of alcohols and ketones (Remmer et al., 1966; Schenkman et al., 1967; Schenkman et al., 1973).

Alcohols, e.g. 1-butanol competitively inhibits the binding of type II substrates and in female or induced microsomes also the binding of type I substrates (van den Berg et al., 1979a; van den Berg et al., 1979b; Mohn et al., 1981). Recent study (Szutowski & Brzdękiewicz, 1990) involving binding of methanol and n-heptane has shown even more complex spectral interactions. It was, therefore, of interest to further investigate the nature of simultaneous ethanol and n-heptane binding to rat liver cytochrome P-450.

MATERIALS AND METHODS

Male Wistar rats weighed 160-180g were obtained from the Animal Breeding Station, Przeźmierowo (Poznań). They received standard LSM diet and water ad libitum. Some experiments were carried out on microsomes obtained from rats exposed to petroleum fraction (b.p. 90-160°), (34 mg/l) for 6 days, 12 hours daily as described earlier (Szutowski et al., 1981).

Partially purified cytochrome P-450 was prepared from microsomes solubilized by treatment with sodium cholate and deoxycholate, and applied to a column of DEAE-cellulose. A single cytochrom P-450 fraction eluted with KCL and Lubrol PX gradient was concentrated to 2 nmoles/ml by ultrafiltration (Amicon PM30). The sample was free from cytochrome b5 and cytochrome P-420. Finally it was dialyzed against 10 mM phosphate buffer pH 7.7 containing 0.1% sodium cholate, glycerol (30% v/v), 0.05 mM DTT, 0.05 mM EDTA. Cytochrome P-450 concentration was determined by the method of Omura & Sato (1964).

Spectral binding studies were carried out on Shimadzu UV-2101PC spectrophotometer. Nothing was added to the reference cuvette unless otherwise indicated. Zero baseline correction and subtraction of ethanol induced spectrum from subsequent n-heptane titration spectra were applied. Spectral binding was quantitated for types I, I alc and RI by the sum of absolute absorbances A at peak and through wavelengths of the three types of spectra observed. Spectral dissociation constant $K_s$ and $\Delta A_{max}$ values were calculated according to the Michaelis-Menten equation. The respective linear correlation coefficients were calculated by the least square analysis. Ethanol and n-heptane of GC standard grade (POCh) were used.

## RESULTS AND DISCUSSION

Additional type I spectral change

During titration of cytochrome P-450, obtained from rats inhaled with petroleum fraction, with ethanolic solutions of n-heptane a significant increase in $\Delta A_{max}$ of type I spectral change was observed (table 1.).

Table 1.

Characteristic of ethanol and n-heptane-induced spectral change of hepatic cytochrome P-450 from hydrocarbons-inhaled male rats.

| No. | Titration solution | | $A_{max}$[1] | $K_s$ (mM) | $A_{max}-A_{max(nH)}$[2] type I | $A_{max}$[3] type RI |
|---|---|---|---|---|---|---|
| | sample cuvette | reference cuvette | | | | |
| 1 | n-heptane (nH) | - | 19.72 | 18.8 | | |
| 2 | ethanol (Et) | - | - | 124.0 | | 13.0 |
| 3 | 5% nH in Et | - | 31.74 | 0.61 | 12.02 | |
| 4 | 5% nH in Et | 5% H₂0 in Et | 46.42 | 0.72 | | 14.68 |
| 5 | 25% nH in Et | - | 31.67 | 3.1 | 11.95 | |
| 6 | 25% nH in Et | 25% H₂0 in Et | 44.00 | 7.0 | | 12.33 |

[1] $\Delta A_{max}$ 390-422 $\times 10^{-3}$/2 nmoles P-450
[2] $A_{max}$ for pure n-heptane is substracted
[3] $A_{max}$ for ethanol from exp. 2, 4-3, 6-5.

Table 2.

Characteristic of ethanol and n-heptane-induced spectral change of hepatic cytochrome P-450 from untreated male rats.

| No. | substrate sample cuvette[1] | type of spectral change | $A_{max}$[2] | $K_s$ (mM) | $A_{max}-A_{max(nH)}$ |
|---|---|---|---|---|---|
| 1 | ethanol (Et) | RI | 30.1 | 1520 | |
| 2 | 1% nH in Et | I | 21.1 | 0.097 | 1.4 |
| 3 | 5% nH in Et | I | 53.3 | 0.64 | 33.6 |
| 4 | 25% nH in Et | I | 47.4 | 3.63 | 27.7 |
| 5 | 50% nH in Et | I | 26.4 | 10.31 | 6.7 |
| 6 | n-heptane (nH) | I | 19.7 | 38.7 | 0 |

[1] nothing was added into reference cuvette
[2] type I: $A_{max}$ 390-422 $\times 10^{-3}$/2 nmoles P-450,
    type RI: $A_{max}$ 419-390 $\times 10^{-3}$/2 nmoles P-450

If in the reference cuvette simultaneous titration with water-ethanolic solution was done, it was possible to show that the magnitude of type I spectral change increase (exp.3,5) is comparable with the magnitude of ethanol induced reverse type I spectral change (exp.2,4,6). However the type RI spectrum dissapeared during titration with ethanolic solution of n-heptane and conversion of type RI into type I spectrum could explain the phenomena.

The observed increase is characteristic both for cytochrome P-450 samples from inhaled and untreated rats (table 2.). It can be seen, that additional type I spectral change is available in the specified range of n-heptane concentration that has not been fully investigated yet. We also observed similar effect after titration of cytochrome P-450 from inhaled rats with 25 per cent n-heptane in methanol (Szutowski & Brzdękiewicz, 1990). But we could not show this effect with cytochrome P-450 from untreated rats at any n-heptane concentration.

Novel type I spectral change (Type I alc)

Presented below experiments were designed to show the inhibitory effect of ethanol on the formation of n-heptane-induced type I spectral change of hepatic male rat cytochrome P-450. First of all ethanol was added into the sample cuvette and spectral change was recorded. Than cytochrome P-450 was titrated with n-heptane in the same cuvette. The results are presented in fig.1,2 and table 3. Ethanol affects the n-heptane-induced type I spectral change by decreasing the magnitude of spectral change. But the most evident alteration is a novel type I spectral change with a peak at 412 nm and through at 433-437 nm (type I alc). This kind of spectrum has already been observed if regenerated cytochrome P-450 was chosen for spectral studies (Szutowski & Rakoto, 1990). To regenerate cytochrome P-450, samples after titration with ethanol and n-heptane were pooled, applied to a column of DEAE-cellulose, washed to remove substrates and eluted. Only n-heptane induced type I spectral change of this cytochrome. Both ethanol and n-heptane in the presence of ethanol gave type I alc spectral change.

In conclusion, presented spectral properties of cytochrome P-450 molecule suggest that n-heptane and ethanol bind to different binding sites. Binding of ethanol strongly influences the microenvironment of the heme and resulted spectral changes.

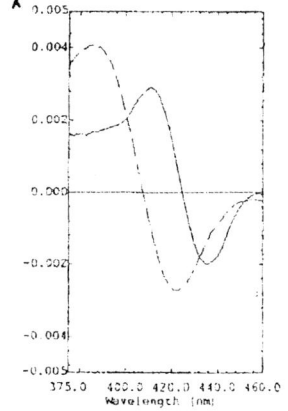

Fig. 1. Spectral changes induced by n-heptane
- - - - - - type I
----------- type I alc

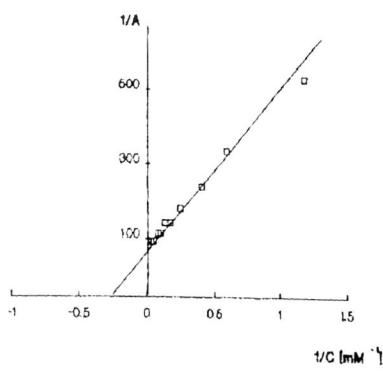

Fig. 2. A reciprocal plot of changes in absorbance at 412 nm relative to 435 nm (type I alc) caused by consecutive additions of n-heptane in the presence of 0.3M ethanol.

Table 3.

Characteristic of n-heptane-induced spectral change of hepatic cytochrome P-450 from untreated male rats in the presence of ethanol.

| No. | Ethanol (M) sample cuvette | Substrate sample cuvette | Type of spectral change[1] | $A_{max}$[2] | $K_s$ (mM) |
|---|---|---|---|---|---|
| 1 | 0 | n-heptane | I | 21.9 | 9.81 |
| 2 | 0.305 | n-heptane | I alc | 13.0 | 4.97 |
| 3 | 0.61 | n-heptane | I alc | 13.8 | 4.16 |
| 4 | 1.5 | n-heptane | I alc | 9.0 | 4.19 |

[1] nothing was added into reference cuvette. Type of spectral change became evident after subtracting of ethanol induced spectrum.
[2] type I : $A_{max}390\text{-}422 \times 10^{-3}/2$ nmol P-450,
type I alc: $A_{max}412\text{-}435 \times 10^{-3}/2$ nmol P-450

Acknowledgments

The authors wish to thank the Stefan Batory Foundation for partial support of the participation in the congress.

REFERENCES

van den Berg, A.P., Noordhoek, J., and Koopman-Kool, E. (1979a): The relation between sex-dependency of typeI binding of ethylmorphine and the 1-butanol-induced spectral change in mouse liver microsomes.*Biochem.Pharmacol.* 28,31-36.

van den Berg, A.P., Noordhoek, J., and Koopman-Kool, E. (1979b): The use of competitive inhibition of substrate-binding to cytochrome P-450 in the determination of spectral dissociation constants for substrates with multiple types of binding, as illustrated with 1-butanol.*Biochem.Pharmacol.* 28,37-42.

Mohn, G., Schmidt, M., Janthur, B.W., and Klemm, G. (1981): Induction of MEOS and binding of ethanol in the liver: microsomes of female mice. In *Industrial and Environmental Xenobiotics*, eds. I. Gut, M. Cikrt and G.L. Plaa. Berlin, Heidelberg, New York: Springer-Verlag.

Omura, T., & Sato, R. (1964): The carbon monoxide-binding pigment of liver microsomes.I.Evidence for its hemoprotein nature.*J.Biol.Chem.* 239,2370-2378.

Remmer, H., Schenkman, J.B., and Estabrook, R.W. (1966): Drug interaction with hepatic microsomal cytochrome.*Mol.Pharmacol.* 2,187-190.

Schenkman, J.B., Cinti, D.L., Moldeus, P.W., and Orrenius, S. (1973): Newer aspects of substrate binding to cytochrome P-450.*Drug Metab.Dispos.* 1,111-120.

Schenkman, J.B., Remmer, H., and Estabrook, R.W. (1967): Spectral studies of drug interaction with hepatic microsomal cytochrome P-450.*Mol.Pharmacol.* 3,113-123.

Szutowski, M.M., Brzeziński, J., Buczkowska, E., and Walecka, E. (1981): The effect of inhaled $C_6$-$C_9$ petroleum fraction on the biotransformation of benzene in the rat. In *Industrial and Environmental Xenobiotics*, eds. I. Gut, M. Cikrt and G.L. Plaa. Berlin, Heidelberg, New York: Springer-Verlag.

Szutowski, M.M. & Brzdękiewicz, D. (1990): Spectral interactions of cytochrome P-450 with n-heptane and methanol.*Folia Med.Cracov.* 31,175-184.

Szutowski, M.M., Rakoto, J., and Ndamira, A.A. (1990): New type of spectral changes of cytochrome P-450 induced by alcohols.*commun.*IVth Congress of Pol.Soc.Toxicology, Jastrzębia Góra. pp.171.

*Cytochrome P450. 8th International Conference*. Ed. M.C. Lechner. John Libbey Eurotext, Paris © 1994, pp. 631-634.

# Coumarin and methoxsalen metabolism by CYP2A6 and CYP2a-5 isoforms in man and mouse

Jukka Mäenpää[1], Risto Juvonen[2], Hannu Raunio[1], Arja Rautio[1], Olavi Pelkonen[1]

[1]*Department of Pharmacology and Toxicology, University of Oulu, 90220 Oulu, Finland.*
[2]*Department of Pharmacology and Toxicology, University of Kuopio, 90721 Kuopio, Finland*

## Summary

Methoxsalen (8-methoxypsoralen) inhibited *in vivo* coumarin metabolism in humans. Methoxsalen was metabolized in human liver microsomes at the rate of 50-100 pmol/mg prot/min. The metabolism was not inhibited by the anti-Cyp2a-5 antibody in human liver microsomes and NIH 3T3 cells stably expressing catalytically active CYP2A6 enzyme did not metabolize methoxsalen. Methoxsalen does not appear to be a substrate of CYP2A6. In pyrazole induced mouse liver microsomes, methoxsalen metabolism was inhibited by the anti-Cyp2a-5 antibody. Cyp2a-5 expressed in the yeast was capable of metabolizing methoxsalen, indicating that methoxsalen is a substrate of Cyp2a-5.

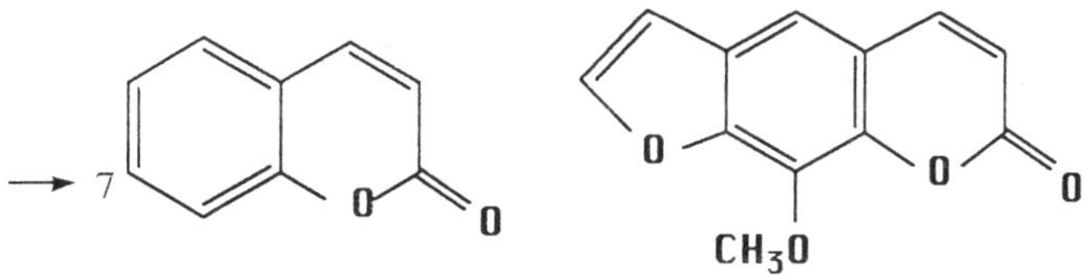


**Coumarin** **Methoxsalen**

**Fig. 1.** Chemical structures of coumarin and methoxsalen. The arrow indicates the carbon atom preferentially hydroxylated in coumarin.

## INTRODUCTION

Methoxsalen (8-methoxypsoralen) is a furanocoumarin derivative (fig. 1) that is widely used as an anti-psoriatic agent in combination with long wavelength ultraviolet light (UVA) radiation (Anderson and Voorhees, 1980). The metabolism of methoxsalen is catalyzed by P450 enzymes in rat and mouse liver microsomes (Mays *et al.* 1987, 1990).

Coumarin (fig. 1) is being actively tested as a drug for lymphedema, various infections and cancer (Egan *et al.*, 1990). In humans, coumarin is 7-hydroxylated effectively *in vitro* and *in vivo* (Raunio *et al.*, 1988; Rautio *et al.*, 1992). Coumarin 7-hydroxylase (COH) activity is catalyzed almost exclusively by the cytochrome P450 isoform CYP2A6 in humans and Cyp2a-5 in mice (Raunio *et al.*, 1988; Yamano *et al.*, 1990). Coumarin is the only clinically used agent that has been shown to have high affinity for CYP2A6.

In a survey of inhibitors of CYP2A6 and Cyp2a-5 (Mäenpää et al., 1993) methoxsalen turned out to be the most potent inhibitor *in vitro*. This study was designed to further elucidate the metabolic interactions of methoxsalen and coumarin in man and mouse and to find out whether methoxsalen is a substrate of CYP2A6 and Cyp2a-5.

MATERIALS AND METHODS

Pretreatment of DBA/2N male mice and the preparation of anti-Cyp2a-5 antibody were done as described (Raunio et al. 1988, Lang et al. 1989). Human liver samples were obtained from renal transplant donors. The microsomal fractions were prepared as described (Mäenpää et al. 1993) and stored at -70°C until assayed. [$^3$H-Methoxyl] methoxsalen (85 Ci/mmol) was from Amersham (U.K.) and Puvaderm[R] (containing 15 mg methoxsalen) tablets were from Leiras (Turku, Finland).

Inhibition of urinary excretion of 7OHC by methoxsalen in humans. Coumarin metabolism *in vivo* in humans was determined as described in detail previously (Rautio et al., 1992). The effect of methoxsalen on 7OHC excretion was studied after at least a six-day period. In the interaction experiment, 45 mg methoxsalen (Puvaderm[R]) was ingested 1 h before coumarin.

Enzyme assays. COH and methoxsalen metabolism were determined as described (Aitio, 1978; Mays et al. 1987).

Methoxsalen and coumarin metabolism by CYP2A6 and Cyp2a-5. The CYP2A6 cDNA was cloned and expressed in NIH 3T3 cells by retrovirus-mediated gene transfer (Salonpää et al., 1993). Cyp2a-5 cDNA was isolated and expressed in the yeast *Saccharomyces cerevisiae* AH22 cells (Iwasaki et al., 1991). The Cyp2a-5 protein produced by these cells was purified and used in reconstitution experiments.

RESULTS

Inhibition of 7OHC excretion by methoxsalen in humans. Methoxsalen inhibited effectively coumarin metabolism in every subject (table 1). The degree of inhibition 8 hours after coumarin administration was $47 \pm 9.2$ % (mean ± S.E.M., range 20-70 %) compared to the amount of 7OHC recovered in the control coumarin test.

Table 1. Inhibition of coumarin metabolism by methoxsalen in five humans *in vivo*.

|  | Control 4 hours | Methoxsalen | Control 8 hours | Methoxsalen |
|---|---|---|---|---|
|  |  | mg 7OHC excreted |  |  |
| Mean ±S.E.M. | $3.67 \pm 0.21$ | $2.02 \pm 0.38$** | $4.03 \pm 0.26$ | $2.15 \pm 0.40$** |

Each subject was given 5 mg coumarin (control) and one week later 45 mg methoxsalen (methoxsalen) one hour before coumarin.
**$P < 0.01$, significantly different vs. control (Student's paired t-test).

Kinetics of COH inhibition. The type of methoxsalen-caused inhibition on COH activity was determined by Lineweaver-Burk analysis in pyrazole induced mouse liver microsomes and human liver microsomes (data not shown). The results indicate that the inhibition is competitive.

Methoxsalen metabolism in mouse and human liver microsomes. Pyrazole and phenobarbital are known to induce COH activity in the mouse liver (Raunio et al. 1988). Pyrazole and phenobarbital increased the metabolism of methoxsalen (1.5 and 3.4-fold above control level, respectively) (table 2). The human liver microsomes also metabolized methoxsalen at a rate of about 30 % of that in control mouse liver microsomes.

The anti-Cyp2a-5 antibody effectively inhibits COH activity in mouse and human liver microsomes (Raunio et al., 1988). The antibody produced a weak (32 %) inhibition on methoxsalen metabolism in pyrazole induced mouse liver microsomes. It had, however, no effect on methoxsalen metabolism in control or phenobarbital induced mouse liver microsomes. The antibody did not inhibit methoxsalen metabolism in human liver microsomes (table 2). The specific activities given denote the disappearance of methoxsalen.

Table 2. The effect of anti-Cyp2a-5 antibody on methoxsalen metabolism in mouse and human liver microsomes.

|  | activity[a] | anti-Cyp2a-5 antibody[b] |
|---|---|---|
| Control | 258 ± 31 | 121 |
| Pyrazole | 376 ± 21 | 68 |
| Phenobarbital | 875 ± 39 | 98 |
| HL[c] (5) | 76 ± 11 | 117 |

[a]pmol/mg protein/min
[b]Percent of control activity.
[c]Human liver microsomes

Methoxsalen and coumarin metabolism by CYP2A6 and Cyp2a-5. The effect of methoxsalen on COH activity was studied in NIH 3T3 cells producing the CYP2A6 enzyme (data not shown). Methoxsalen was a very potent inhibitor of COH activity; 50 µM methoxsalen in the culture medium caused a 99 % reduction in COH activity (IC50 = 0.3 µM). CYP2A6 did not catalyze methoxsalen metabolism (data not shown). In the reconstitution experiment, methoxsalen (1.0 µM) was metabolized by the purified Cyp2a-5 protein to polar metabolites at a rate of 590 pmol/min/nmol P450 (data not shown).

DISCUSSION

This study indicates that methoxsalen is a potent inhibitor of coumarin metabolism in humans *in vivo* and caution should be exerted when methoxsalen is given concurrently with other drugs. Our earlier study showed that methoxsalen efficiently blocks COH activity in human and mouse liver microsomes *in vitro* (Mäenpää et al., 1993). The key question of this work was whether methoxsalen is also accepted as a substrate by CYP2A6. Methoxsalen is clearly metabolized by human liver microsomes, although the rate of the metabolism is lower than in mouse liver microsomes. The competitive nature of the inhibition of coumarin metabolism by methoxsalen suggests that methoxsalen might serve as a substrate of CYP2A6. However, this does not appear to be the case, because anti-Cyp2a-5 antibody had no effect on methoxsalen metabolism in human liver microsomes and CYP2A6 expressing cells could not metabolize methoxsalen. All these data indicate that although methoxsalen is capable of interacting with CYP2A6, it is not a substrate of CYP2A6. The situation is analogous to quinidine, which is a potent inhibitor of CYP2D6, but is metabolized by CYP3A isoforms (Guengerich et al., 1986).

Immunoinhibition studies with anti-Cyp2a-5 antibody suggested that Cyp2a-5 catalyzes methoxsalen metabolism (about 30 %) only in pyrazole induced mouse liver. The reconstitution studies with purified Cyp2a-5 protein indicated that Cyp2a-5 metabolize methoxsalen. These data suggest that predominantly P450 isoforms other than Cyp2a-5 participate in methoxsalen metabolism in control and phenobarbital induced mouse liver.

In summary, although methoxsalen is a potent inhibitor of coumarin metabolism *in vitro* and *in vivo* and methoxsalen and coumarin share structural similarities they are not metabolized by the same P450 isoforms in human liver microsomes. Even though Cyp2a-5 metabolizes methoxsalen in mouse liver microsomes, CYP2A6 does not catalyze methoxsalen metabolism in human liver.

## ACKNOWLEDGEMENTS

We thank Professor Urs Meyer (Department of Pharmacology, Biocenter of the University of Basel, Basel, Switzerland) for providing us a part of the human liver samples used, under the CostNet scheme. The scientific contents of this paper contribute to the goals of the COST B1 Project. We thank Markku Pasanen, Ritva Tauriainen, Päivi Kylli and Liisa Kärki for their contribution to this work. This research was supported by the Academy of Finland (Medical Research Council Contract No. 1051029).

## REFERENCES

Aitio A. (1978): A simple and sensitive assay of 7- ethoxycoumarin deethylation. *Anal. Biochem.* 85: 488-491.

Anderson, T.F. & Voorhees, J. J. (1980): Psoralen photochemotherapy of cutaneous disorders. *Annu. Rev. Pharmacol. Toxicol.* 20: 235-257.

Egan, D., O'Kennedy, R., Moran, E., Cox, D., Prosser, E. & Thornes, R. D. (1990): The pharmacology, metabolism, analysis, and applications of coumarin and coumarin-related compounds. *Drug Metab. Rev.* 22: 503-529.

Guengerich, F. P., Muller-Enoch, D. & Blair, I. A. (1986): Oxidation of quinidine by human liver cytochrome P-450. *Mol. Pharmacol.* 30: 287-295.

Iwasaki, M., Juvonen, R., Lindberg, R. & Negishi, M. (1991): Alteration of high and low spin equilibrium by a single mutation of amino acid 209 in mouse cytochromes P450. *J. Biol. Chem.* 266: 3380-3382.

Lang, M.A., Juvonen, R., Järvinen, P., Honkakoski, P. & Raunio, H. (1989): Mouse liver P450Coh: genetic regulation of the pyrazole-inducible enzyme and comparison with other P450 isozymes. *Arch. Biochem. Biophys.* 271: 139-148.

Mäenpää, J., Sigusch H., Raunio, H., Syngelmä, T., Vuorela, P., Vuorela, H. & Pelkonen,O. (1993): Differential inhibition of coumarin 7-hydroxylase activity in mouse and human liver microsomes. *Biochem. Pharmacol.* 45: 1035-1042.

Mays, D. C., Hecht, S. G., Unger, S. E., Pacula, C. M., Sharp, D. E. & Gerber, N. (1987): Disposition of 8-methoxypsoralen in the rat: Induction of metabolism *in vivo* and *in vitro* and identification of urinary metabolites by thermospray mass spectrometry. *Drug Metab. Dispos.* 15: 318-328.

Mays, D. C., Hillliard, J. B., Wong, D. D., Chambers, M.A., Park, S. S., Gelboin, H. V. & Gerber, N. (1990): Bioactivation of 8-methoxypsoralen and irreversible inactivation of cytochrome P-450 in mouse liver microsomes: Modification by monoclonal antibodies, inhibition of drug metabolism and distribution of covalent adducts. *J. Pharmacol. Exp. Ther.* 254: 720-731.

Raunio, H., Syngelmä, T., Pasanen, M., Juvonen, R., Honkakoski, P., Kairaluoma, M. A., Sotaniemi, E. A., Lang, M. A. & Pelkonen, O. (1988): Immunochemical and catalytical studies on hepatic coumarin 7-hydroxylase in man, rat, and mouse. *Biochem. Pharmacol.* 37: 3889-3895.

Rautio, A., Kraul, H., Kojo, A., Salmela, E. & Pelkonen, O. (1992): Interindividual variability of coumarin 7-hydroxylation in healthy volunteers. *Pharmacogenetics* 2: 227-233.

Salonpää, P., Hakkola, J., Pasanen, M., Pelkonen, O., Vähäkangas, K., Battula, N., Nouso, K. & Raunio, H. (1993): Retrovirus- mediated stable expression of human CYP2A6 in mammalian cells. *Eur. J. Pharmacol. Environ. Toxicol. Pharmacol.* 248: 95-102.

Yamano, S., Tatsuno, J. & Gonzalez, F. J. (1990): The CYP2A3 gene product catalyzes coumarin 7-hydroxylation in human liver microsomes. *Biochemistry* 29: 1322-1329.

# Accelerated caffeine metabolism after omeprazole treatment: measurements in breath, plasma and urine

Karl Ludwig Rost[1], Ivar Roots[2]

[1]Institute of Clinical Pharmacology, Klinikum Steglitz, Free University of Berlin, Hindenburgdamm 30, D-12200 Berlin, FRG. [2]Institute of Clinical Pharmacology, Charité, Humboldt University of Berlin, D-10098 Berlin, FRG

## INTRODUCTION

Omeprazole may serve as a prominent example of a non-toxic compound that is able to induce CYP1A isozymes in man (Rost et al., 1992, 1993a, b). The $^{13}$C-[N3-methyl]-caffeine breath test (CBT) most specifically reflects N3-demethylation of caffeine by CYP1A2 (Butler et al., 1989; Gu et al., 1992). The extent of induction was shown to depend on dose and the genetic trait of S-mephenytoin-hydroxylase (Rost et al., 1992, 1993a, b). Poor metabolizers of S-mephenytoin (PM) are deficient to form one omeprazole metabolite (5´-OH-omeprazole) and show therefore higher and prolonged omeprazole plasma concentrations than extensive metabolizers (EM) (Andersson et al., 1992). PMs who comprise only 3-4% of the Caucasian population (Wedlund et al., 1984) might therefore be more prone to react with CYP1A2 induction. In the present study we investigated, whether the inducing potency of omeprazole can be quantified by plasma clearance of caffeine and the determination of urinary caffeine metabolite ratios, which are based on the urinary excretion of N3-demethylated metabolites. Because the validity of caffeine metabolite ratios to monitor CYP1A2 activity is presently under debate (Kalow & Tang, 1993 for review), three different ratios were investigated.

## METHODS

40 mg omeprazole p.o. were administered for 7 days to six PMs (including one intermediate metabolizer, IM) and twelve EMs. Eight of these EMs took part in a second trial with a 120 mg/d (2 x 60 mg) omeprazole dose. Before and at day 7 of treatment CBT was performed for 8h (Rost et al., 1992) and caffeine plasma kinetics were followed for 24h. Caffeine and its metabolites from 5-8h urine collection were measured by slightly modified rp-HPLC methods for plasma (Campbell et al., 1987) and for urine (Grant et al., 1984) by gradient elution (Rost et al., 1993b). The ratios $CMR_1$ (17X/137X; Kadlubar et al., 1990), $CMR_2$ ([17X+17U]/137X; Butler et al., 1992), and ([1X+1U+AFMU+AAMU]/17U; (Kalow & Tang, 1991).

---

137X, caffeine; 17X, paraxanthine; 17U, 1,7-dimethyluric acid; 1X, 1-methylxanthine; 1U, 1-methyluric acid; AFMU, 5-acetylamino-6-formylamino-3-methyluracil; AAMU, 5-acetylamino-6-amino-3-methyluracil (instead of converting all AFMU to AAMU we measured both AFMU and the non-enzymatic hydrolysis product AAMU).

## RESULTS

Plasma caffeine clearance increased by 31.6 ± 7.3% in EMs after 120 mg omeprazole and correlated by R=0.742 (p=0.004, n=8) with the increase observed in the CBT. The accelerated CYP1A2 activity in the CBT was paralleled by all caffeine urinary metabolite ratios. Table 1 summarizes the results of all study groups. Fig. 1 indicates the individual results of the CBT and three urinary caffeine metabolite ratios. When compared with plasma clearance, good correlations were obtained with all metabolite ratios (Fig. 2). CYP1A2 index showed the highest correlation (r=0.897, p < 0.001, n=13).

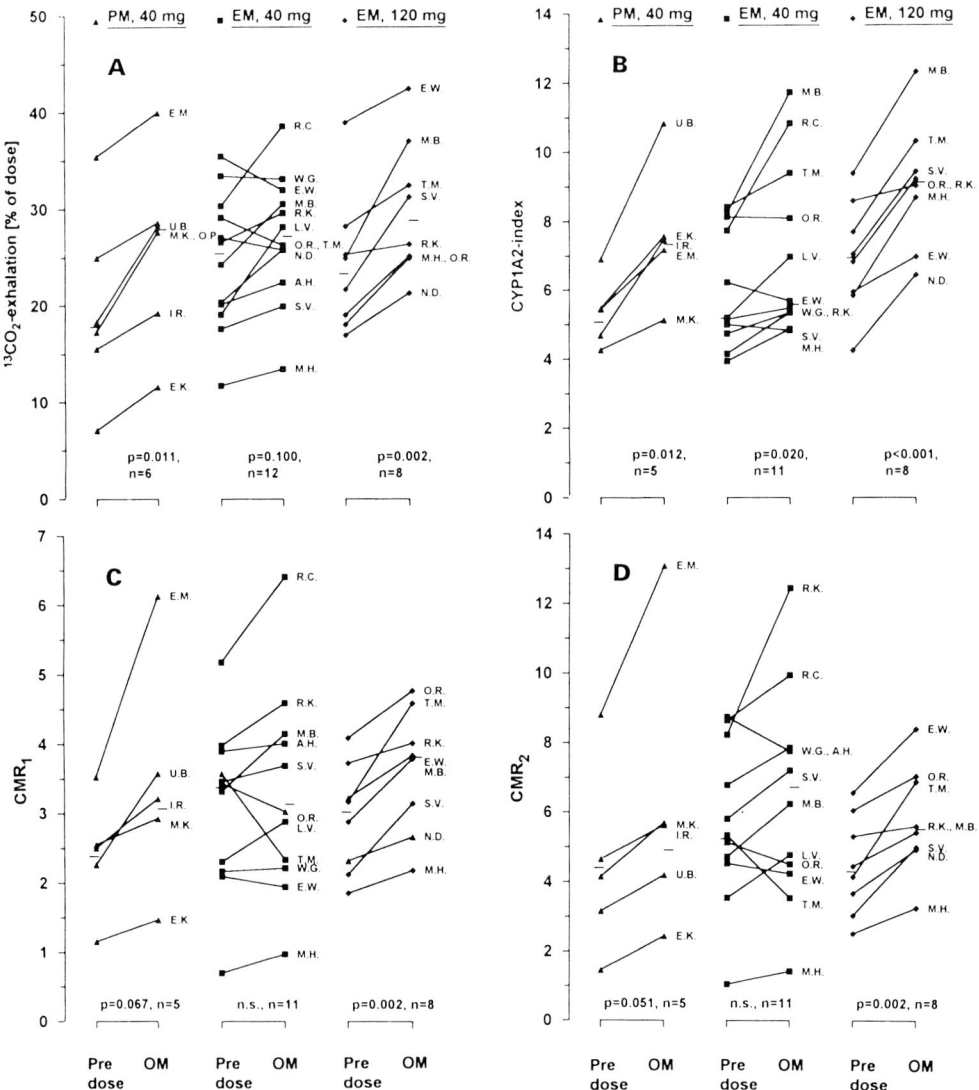

**Fig. 1:** Values of CBT (A), caffeine urinary metabolite ratios (B-D) before (predose) and at the 7th day of omeprazole (OM) treatment. Medians are indicated by a small dash.

## CONCLUSION

The gradual acceleration of caffeine metabolism dependent on omeprazole dose and genetic trait of S-mephenytoin hydroxylase as evidenced by the CBT was reflected by all caffeine urinary metabolite ratios. In this study, CYP1A2 index appeared as superior parameter of CYP1A2 activity, since it most closely correlated with the results of both CBT and plasma clearance. Plasma caffeine clearance also increased to a comparable extent and was well correlated with both CBT and urinary metabolite ratios. Thus, all parameters of caffeine metabolism clearly show the CYP1A2 induction by omeprazole. Clinical relevance, however, might only be expected in PMs or after uncommonly high doses of omeprazole in EMs.

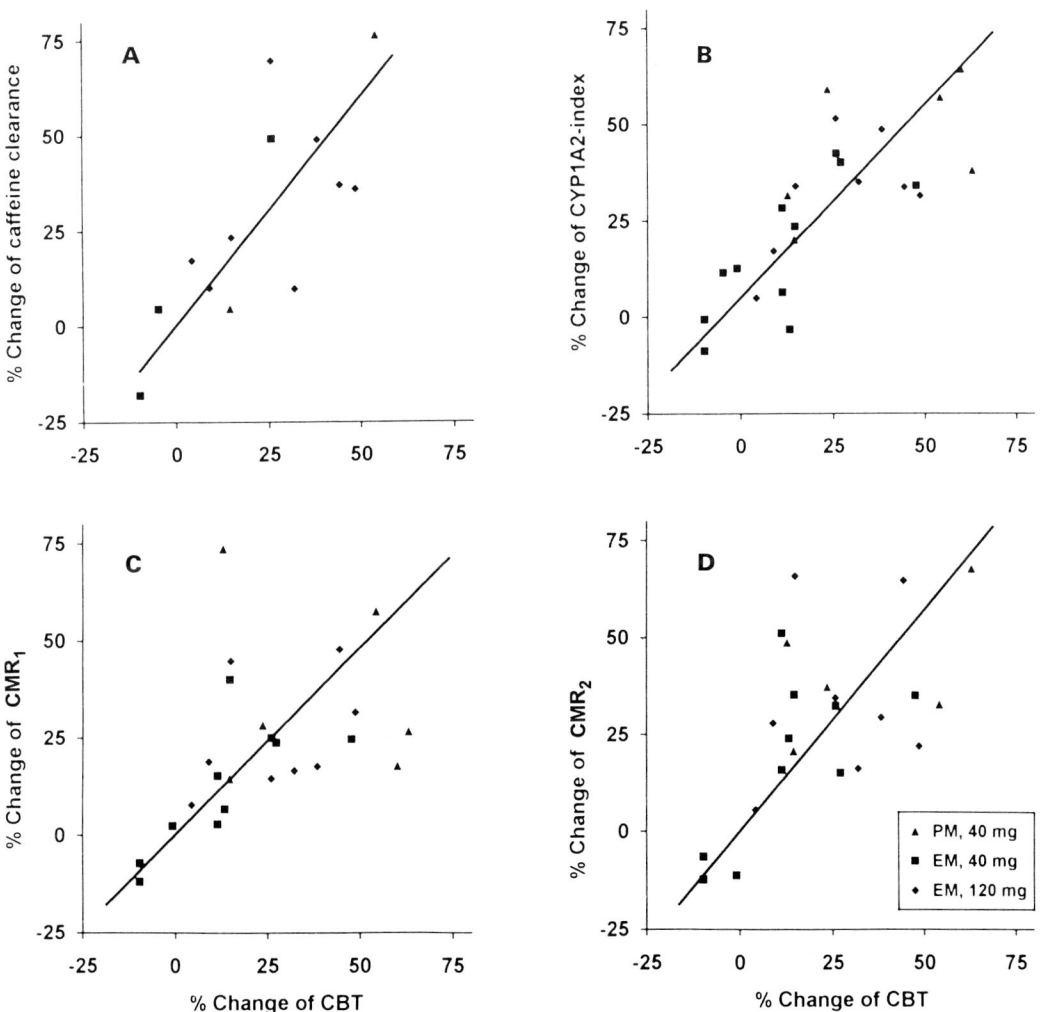

**Fig. 2:** Correlation of relative changes after treatment with omeprazole between CBT and clearance (A), CYP1A2 index (B), $CMR_1$ (C), and $CMR_2$ (D).

Table 1: Acceleration of caffeine metabolism after a 7-day omeprazole treatment in EMs and PM/IMs of S-mephenytoin hydoxylase

|  | EM 40 mg | EM 120 mg | PM/IM 40 mg |
|---|---|---|---|
| $CMR_1$ (5-8h) | 7.9 ± 6.3  n.s. (11) | 25.0 ± 5.2  p=0.002 (8) | 40.2 ± 11.9  p=0.067 (5) |
| $CMR_2$ (5-8h) | 13.2 ± 7.8  n.s. (11) | 33.3 ± 7.7  p=0.002 (8) | 41.5 ± 7.9  p=0.051 (5) |
| CYP1A2 index (5-8h) | 17.0 ± 5.4  p=0.02 (11) | 32.1 ± 5.4  p<0.001 (8) | 41.2 ± 7.5  p=0.012 (5) |
| CBT | 11.5 ± 5.3  n.s. (12) | 27.2 ± 5.8  p=0.002 (8) | 33.6 ± 10.4  p=0.011 (6) |
| Plasma clearance | 12.0 ± 19.8  (3) | 31.6 ± 7.3  p<0.001 (8) | 40.7  (2) |

Values are expressed as % change of predose [means ± SEM, n in parentheses].

### REFERENCES

1. Andersson T., Regårdh C.G., Lou Y.C., Zhang Y., Dahl M.L. & Bertilsson L. (1992): Polymorphic hydroxylation of S-mephenytoin and omeprazole metabolism in caucasian and chinese subjects. *Pharmacogenetics* 2, 25-31.
2. Campbell M.E., Spielberg S.P. & Kalow W. (1987): A urinary metabolite ratio that reflects systemic caffeine clearance. *Clin. Pharmacol. Ther.* 42, 157-165.
3. Butler M.A., Iwasaki M., Guengerich F.P. & Kadlubar F.F. (1989): Human cytochrome P-450$_{PA}$ (P-450IA2), the phenacetin O-deethylase, is primarily responsible for the hepatic 3-demethylation of caffeine and N-oxidation of carcinogenic arylamines. *Proc. Natl. Acad. Sci., USA* 86, 7696-7700.
4. Butler M.A., Lang N.P., Young J.F., Caporaso N.E., Vineis P., Hayes R.B., Teitel C.H., Massengill J.P., Lawsen M.F. & Kadlubar F.F. (1992): Determination of CYP1A2 and NAT2 phenotypes in human populations by analysis of caffeine urinary metabolites. *Pharmacogenetics* 2, 116-127.
5. Grant D.M., Tang B.K. & Kalow W. (1983): Variability in caffeine metabolism. *Clin. Pharmacol. Ther.* 33; 591-602.
6. Gu L., Gonzalez F.J., Kalow W. & Tang B.K. (1992): Biotransformation of caffeine, paraxanthine, theobromine and theophylline by cDNA-expressed human CYP1A2 and CYP2E1. *Pharmacogenetics* 2, 73-77.
7. Kadlubar F.F., Talaska G., Butler M.A., Teitel C.H., Massengill J.P. & Lang N.P. (1990): Determination of carcinogenic arylamine N-oxidation phenotype in humans by analysis of caffeine urinary metabolites. In *Mutation and the environment. Part B: Metabolism, testing methods, and chromosomes*, ed. M.L. Mendelsohn, & R.J. Albertini, pp. 107-114. New York: John Wiley & Sons.
8. Kalow W. & Tang B.K. (1991): Use of caffeine metabolite ratios to explore CYP1A2 and xanthine oxidase activities. *Clin. Pharmacol. Ther.* 50, 508-519.
9. Kalow W. & Tang B.K. (1993): The use of caffeine for enzyme assays: A critical appraisal. *Clin. Pharmacol. Ther.* 53, 503-14.
10. Rost K.L., Brösicke H., Brockmöller J., Scheffler M., Helge H. & Roots I. (1992): Increase of cytochrome P450IA2 activity by omeprazole: Evidence by the $^{13}$C-[N3-methyl]-caffeine breath test in poor and extensive metabolizers of S-mephenytoin. *Clin. Pharmacol. Ther.* 52, 170-180.
11. Rost K.L., Brösicke H., Heinemeyer G. & Roots I. (1993a): Specification of the enzyme inducing effect of omeprazole in man and its dependence on dose and genetic polymorphism. *Submitted*.
12. Rost K.L. & Roots I. (1993b): Accelerated caffeine metabolism after omeprazole treatment is indicated by urinary metabolite ratios - coincidence with plasma clearance and breath test. *Clin. Pharmacol. Ther.* (accepted).
13. Wedlund P.J., Aslanian W.S., McAllister C.B., Wilkinson G.R. & Branch R.A. (1984): Mephenytoin hydroxylation deficiency in caucasians: frequency of a new oxidative drug metabolism polymorphism. *Clin. Pharmacol. Ther.* 36, 773-780.

# The effect of dexamethasone and phenobarbital coadministration on the activity of rat liver P450 system

Katalin Monostory, László Vereczkey

*Department of Biochemical Pharmacology, Central Research Institute for Chemistry, Hungarian Academy of Sciences, PO Box 17, H-1525 Budapest, Hungary*

## Summary

Phenobarbital and dexamethasone are potent inducers of the same cytochrome P450 form, CYP3A, but the mechanism of action is not quite clear. If the mechanism of induction by phenobarbital and dexamethasone is different, additive effect may be observed in the specific activities of CYP3A: ethylmorphine or aminopyrine N-demethylation of liver microsomes from rats treated with phenobarbital and dexamethasone in combination. The results of recent work could not display differences in CYP3A activities between the groups of animals single-administered and coadministered with phenobarbital and dexamethasone. However, p-nitrophenol hydroxylation surprisingly increased threefold in both male and female groups as a result of double induction. Similar changes could be detected in additional studies after treatment with Zixoryn®+dexamethasone. 3-amino-1,2,4-triazole inhibition study shows that CYP2E1 is responsible for 60% of enhanced p-nitrophenol hydroxylase activity, but the residual 40% indicates the participation of other isoenzyme(s).

## INTRODUCTION

The hepatic microsomal hemoproteins called cytochromes P450 (P450) catalyse the oxidative metabolism of many lipophilic chemicals including drugs, pesticides, food additives or environmental pollutants. Some of the individual isoenzyme forms of P450 exhibiting distinct substrate specifities can be induced by these compounds through processes involving stimulated synthesis or stabilization of the mRNAs and cytochrome proteins, but the mechanism of induction is not revealed completely (Okey, 1990). The induction of various P450 can alter metabolic properties of drugs resulting in changes in their pharmacological effects. Combined actions of drugs are quite common phenomena in therapy and can cause some more changes in responses of P450 system.

Two potent inducers that cause the induction of the same P450 form were chosen. If the mechanism of action is different, additive effect may be expected in the concentration of the individual P450 form and its specific activities after treatment with these two compounds in combination comparing to single-treated animal groups.

Phenobarbital (PB), an anticonvulsant drug, increases the transcription of CYP2B genes and also induces CYP3A (Schutz et al., 1986). Zixoryn® (Flumecinol [ZX]), being developed by Chemical Works of Gedeon Richter, Ltd. (Budapest, Hungary) for treatment of icterus neonatorum, has the same inducing capacity as PB (Szeberényi et al., 1978). CYP3A, the most sensitive to glucocorticoids is induced by dexamethasone (DXM), an anti-inflammatory drug (Heuman et al., 1982).

Although increased transcription plays some role in the induction of CYP3A, a strong posttranscriptional influence believed to be message stabilization was found. Our resent study sought to define the responses of P450 system to combined exposure of PB and DXM or ZX and DXM and to compare the results to those obtained in liver microsomes prepared from rats treated with PB, ZX or DXM alone.

## MATERIALS & METHODS

Chemicals: Dexamethasone and phenobarbital were obtained from Sigma (St. Louis, USA); Zixoryn® (flumecinol) and isoniazid were generous gift of Chemical Works of Gedeon Richter, Ltd. (Budapest, Hungary); 3-amino-1,2,4-triazole was purchased from Merck (Darmstadt, Germany).

Animals: Male and female Wistar rats purchased from LATI (Gödöllő, Hungary), were treated with PB (80 mg/kg, i.p.) or DXM (100 mg/kg, p.o.) alone or in combination for 3 consecutive days. In additional studies ZX (80 mg/kg, p.o.) was administered alone or with DXM for 3 days. The liver microsomes were prepared by differential centrifugation as described (van der Hoeven & Coon, 1974).

Assay methods: Published methods were followed in the assays given below: p-nitrophenol hydroxylation (Reinke & Moyer, 1985), ethylmorphine and aminopyrine N-demethylation detected by formaldehyde formation (Nash, 1953), and pentoxyresorufin O-dealkylation (Burke et al., 1985). Studies on inhibition of p-nitrophenol hydroxylase activity were carried out by the method of Koop (1990).

## RESULTS

PB+DXM or ZX+DXM coadministration caused an increase of the rates of ethylmorphine or aminopyrine N-demethylation as well as pentoxyresorufin O-dealkylation in both sexes comparing to untreated groups. Our findings show that similar changes in pentoxyresorufin O-dealkylation and ethylmorphine or aminopyrine N-demethylation activities can be found in rat liver microsomes treated with PB or ZX alone, which agree with published information. DXM administration caused an increase of the ethylmorphine and aminopyrine N-demethylation.

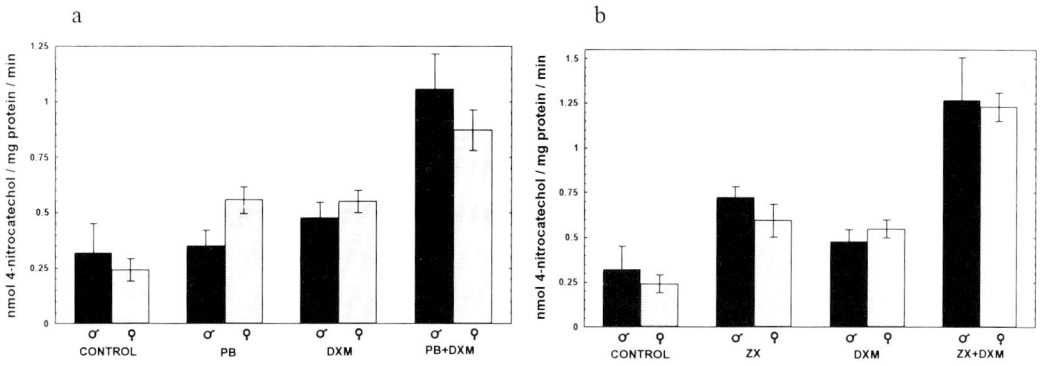

Fig. 1. Effect of treatement with PB and DXM (a) or ZX and DXM (b) alone or in combination on p-nitrophenol hydroxylase activity in male and female rat liver. P-nitrophenol hydroxylation was determined in microsomes with 0.2 mM p-nitrophenol, 1.5 mg/ml hepatic microsomal protein and NADPH generating system. Each value represents the mean ±SD.

P-nitrophenol hydroxylase activity in male rat liver microsomes was not altered after treatment with PB or DXM alone, but their combination surprisingly enhanced it threefold (Fig. 1a). Similar elevation (threefold) could be observed in liver microsomes from female rats induced by PB+DXM. It should be mentioned that p-nitrophenol hydroxylation activity was lifted in female groups single treated with PB or DXM (Fig. 1a).

Additional studies showed that similar changes could be detected in catalytic activities of rat P450s after ZX+DXM coadministration. P-nitrophenol hydroxylase activity of microsomes from ZX+DXM induced male and female rats was increased by fourfold and fivefold, respectively, comparing to untreated animals. However, ZX could elevate p-nitrophenol hydroxylation of liver microsomes by twofold as compared to untreated rats (Fig. 1b).

P-nitrophenol hydroxylation assay is considered to be an indicator of participation of CYP2E1, an ethanol- or isoniazid-inducible isoenzyme form (Koop et al., 1989). Koop (1990) reported that greater than 90% of p-nitrophenol hydroxylation activity of microsomes from ethanol treated rabbits can be inhibited by the nonselective herbicide, 3-amino-1,2,4-triazole. In our experiments 95% of p-nitrophenol hydroxylase activity of microsomes from isoniazid-induced rats was found to be inhibited by 3-amino-1,2,4-triazole and inhibitor concentration for half a maximal inhibition was 3.0 mM.

The unexpected enhancement of p-nitrophenol hydroxylation as a result of double induction by PB+DXM or ZX+DXM can be inhibited by 3-amino-1,2,4-triazole, a specific inhibitor for P4502E1 by about 60% (Fig. 2) and inhibitor concentration for half a maximal inhibition was 3.0 mM.

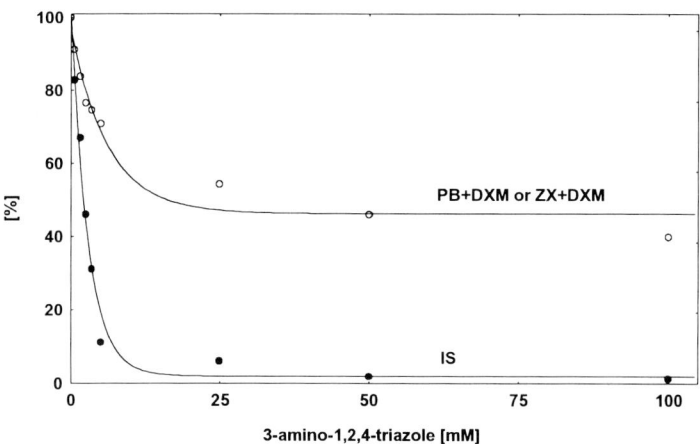

Fig. 2. Effect of 3-amino-1,2,4-triazole on p-nitrophenol hydroxylation in microsomes from PB+DXM-, ZX+DXM- and isoniazid (IS)-treated rats. The incubations contained 1.5 mg of microsomal protein/ml, 0.2 mM p-nitrophenol and NADPH generating system. 3-amino-1,2,4-triazole was added to the incubation mixtures 15 min prior to the addition of p-nitrophenol. The values represents the % of specific activity of the sample without inhibitor.

DISCUSSION

Induction of microsomal monooxigenase system has been extensively investigated. The number of chemicals implicated as inducers is large and is growing further more. Exposure of xenobiotics can alter activity of drug-metabolizing cytochromes P450 through induction of different forms. Combined actions of drugs are quite common phenomena in therapy and can cause some more changes in responses of P450 system. Induction of P450 by xenobiotics is the result of multiple mechanisms: increased transcription of P450 genes, message stabilization, enhanced transport of mRNA from the nucleus to the cytoplasm, elevated P450 protein synthesis or P450 enzyme stabilization. If the mechanism of CYP3A induction caused by PB or DXM is different, additive effect may be found in ethylmorphine or aminopyrine N-demethylase activities of liver microsomes from rats coadministered with PB+DXM. The results of the present study show that such findings cannot be proved; double induction caused similar elevation of CYP3A activities as it can be seen in the groups of rats treated with PB or DXM alone.

However some alterations can be observed in P450 responses after coadministration of PB+DXM: p-nitrophenol hydroxylation increased about threefold in both male and female groups as compared to the activity of microsomes from untreated rats. Microsomes from the animals treated with PB or DXM alone did not display similar increase in p-nitrophenol hydroxylation. Moreover, coinduction with ZX+DXM could produce elevation in p-nitrophenol hydroxylation comparing to untreated group or animals treated with ZX or DXM alone.

To explain our results 3-amino-1,2,4-triazole inhibition studies were carried out. 3-amino-1,2,4-triazole is the specific inhibitor of P4502E1 form (Koop, 1990). 60% of p-nitrophenol hydroxylase activity of microsomes from PB+DXM- and ZX+DXM-treated rats could be inhibited and 40% of it remained. As a result, while CYP2E1 is an effective catalyst of p-nitrophenol hydroxylation, the residual activity remaining in the presence of 3-amino-1,2,4-triazole suggests that there are other isoenzyme(s) which are able to hydroxylate p-nitrophenol.

Acknowledgements - This work has been supported in part by Hungarian Scientific Foundation.

REFERENCES

Burke, M.D., Thompson, S., Elcombe, C.R., Halpert, J., Haaparanta, T., Mayer, R.T. (1985): Ethoxy-, pentoxy- and benzyloxyphenoxazones and homologues: a series of substrates to distinguish between different induced cytochromes P-450. Biochem. Pharmacol. 34, 3337-3345.

Heuman, D.M., Gallagher, E.J., Barwick, J.L., Elshourbagy, N.A. Guzelian, P.S. (1982): Immunochemical evidence for induction of a common form of hepatic cytochrome P-450 in rats treated with pregnenolone-16α-carbonitrile or other steroidal or non-steroidal agents. Mol. Pharmacol. 21, 753-760.

van der Hoeven, T.A. & Coon, M.J. (1974): Preparation and properties of partially purified cytochrome P-450 and reduced nicotinamide adenine dinucleotid phosphate-cytochrome P-450 reductase from rabbit liver microsomes. J. Biol. Chem. 249, 6302-6310.

Koop, D.R., Laethem, C.L., Tierney, D.J. (1989): The utility of p-nitrophenol hydroxylation in P450IIE1 analysis. Drug Metab. Rev. 20, 541-551.

Koop, D.R. (1990): Inhibition of ethanol-inducible cytochrome P450IIE1 by 3-amino-1,2,4-triazole. Chem. Res. Toxicol. 3, 377-383.

Nash, T. (1953): The colorimetric estimation of formaldehyde by means of the Hantzsch reaction. Biochemistry 55, 416-421.

Okey, A.B. (1990): Enzyme induction in the cytochrome P-450 system. Pharmac. Ther. 45, 241-298.

Reinke, L.A. & Moyer, M.J. (1985): P-nitrophenol hydroxylation. Drug Metab. Dispos. 13, 548-552.

Schuetz, E.G., Wrighton, S.A., Safe, S.H., Guzelian, P.S. (1986): Regulation of cytochrome P-450p by phenobarbital and phenobarbital-like inducers in adult rat hepatocytes in primary monolayer culture and in vivo. Biochemistry 25, 1124-1133.

Szeberényi, Sz., Pálosi, É., Szporny, L. (1978): Effects of 3-trifluoromethyl-α-ethylbenzhydrol (RGH-3332), a new enzyme inducer, on the microsomal drug metabolism. Arzneim.-Forsch./Drug Res. 28, 663-668.

# Polymerase chain reaction mediated detection of cytochrome P450 gene in the yeast *Candida apicola*

Katrin Lottermoser[1], Otmar Asperger[1], Wolf-Hagen Schunck[2]

[1]Faculty of Biosciences, University of Leipzig, Talstraße 33, D-04103 Leipzig, Germany.
[2]Max Delbrück Centre of Molecular Medicine, Robert-Rössle-Straße 10, D-13125 Berlin, Germany

## Summary

The yeast *Candida apicola* produces surface-active extracellular glycolipids containing w- and (w-1)-hydroxy fatty acids. Subterminal hydroxylation of fatty acids is probably catalyzed by a novel P450 form. Homologous probes for screening of a genomic library were obtained by PCR with genomic DNA of *C. apicola*. Oligonucleotide primers used in the PCR were directed against the heme-binding region and helix I region of CYP52A3. Highest homology of the first isolated P450 of *C. apicola* was shown to CYP52A11: 42.9 per cent amino acid sequence identity. A multiplicity of P450 in *C. apicola* was indicated by hydroxylation studies and southern hybridisations.

## INTRODUCTION

Recently, several alkane-inducible P450 genes have been cloned from *Candida tropicalis* and *Candida maltosa*. They constitute the family CYP52 within the P450 supergene family [see the references in Nelson et al., 1993]. Some of these P450 forms were shown to catalyze the terminal hydroxylation of n-alkanes or the w-hydroxylation of fatty acids [Schunck et al., 1991; Seghezzi et al., 1991 and 1992]. The alkane-assimilating yeast *C. apicola* was found in our laboratory to express P450 not only during growth on n-alkanes but also during the production of surface-active extracellular glycolipids [Kleber et al., 1990]. These surfactants consist of sophorose and w- or (w-1)-hydroxy fatty acids [Weber et al., 1990]. Subterminal fatty acid hydroxylation would represent a novel function of a P450 form in alkane-assimilating yeasts. A mixed function oxidase that catalyzes the (w-1)-hydroxylation of long-chain fatty acids was already reported to occure in a strain of torulopsis yeast by Heinz et al. (1969). It is the aim of our studies to identify the molecular nature and to characterize the functional properties of such an enzyme. As a first step the present paper describes a PCR-mediated strategy and its successful application in cloning a P450 gene from *C. apicola*.

## MATERIAL AND METHODS

Strain and culture conditions: *C. apicola* IMET43747 was obtained from the Institute of Microbiology and Experimental Therapy (Jena, Germany). It was cultured in a 10-l-fermentor on Reader medium with 50 g/l glucose as carbon source. P450 was induced either by stepwise addition of n-hexane (12 x 0,3 g/l every half hour) during logarithmic growth or by decreasing the culture pH below 3 causing growth limitation.

Microsomes: Microsomal fractions were prepared after cell-disintegration in a Dyno-Mill followed by consecutive centrifugations of the supernatants at 10,000 and 100,000 x g.

Analytical methods: Hydroxylation of n-[1-$^{14}$C]hexadecane and [1-$^{14}$C]fatty acids (Amersham, Buchler & Co KG) was determined as reported by Schunck et al. (1991).

Polymerase chain reaction: Oligonucleotide primers were PHR2 (GACCCAAACAAATTCTTGGA CCACC) and PHI (GGTAGAGATACCACTGC). A 50 µl reaction mixture contained 100 ng total DNA from C. apicola prepared according to Sherman et al. (1986) and 25 pmol of each primer. Taq polymerase (USB) was used according to the manufacturers instructions. 5 cycles of 1 min/94 °C, 3 min/46 °C, 3 min/72 °C and 30 cycles of 1 min/94 °C, 1.5 min/50°C, 2 min(+3 sec)/72 °C were performed. The product was cloned into pUC19 and sequenced.

Isolation of C. apicola P450 genomic DNA clones: A genomic library was constructed by cloning C. apicola genomic DNA Sau3A fragments of 6-9 kb length into the BamHI site of pUC19. A 281 bp AccI/ApaI-fragment of a PCR product was prepared as probe. It was labeled and screened with the ECL random-labeling and detection system (Amersham) under the recommended standard conditions.

Computer analysis: Analyses of DNA sequences and deduced amino acid sequences were performed with the programmes of PCGene (IntelliGenetics, California, USA).

## RESULTS AND DISCUSSION

Alkane and fatty acid hydroxylation by microsomal P450 forms of C. apicola

Using two different cultivation procedures, microsomal P450 systems could be induced in C. apicola that differed significantly in their substrate specificities (Table 1). Microsomes prepared after n-hexane induction catalyzed preferentially the hydroxylation of n-hexadecane and lauric acid. In contrast, microsomal P450 produced during the stationary growth phase showed highest activities with long-chain fatty acids and did not hydroxylate n-hexadecane.

Table 1: Hydroxylation activities of microsomal P450 systems of C. apicola in dependence of inducing conditions[1]

| Substrate | Hydroxylation rates (nmol/nmol P450 x min) | |
|---|---|---|
| | Induction by n-hexane | Growth limitation |
| Hexadecane | 8.0 | n.d.[2] |
| Lauric acid | 15.0 | 4.8 |
| Palmitic acid | 2.0 | 9.2 |
| Oleic acid | n.d.[2] | 15.2 |

[1] see "Materials and Methods",
[2] n.d.: not detectable

PCR amplification of P450 gene fragments

P450 forms of the CYP52 family share a number of highly conserved domains in their amino acid sequences. We used two of them - the heme-binding HR2 region and the distal helix I region - to design oligonucleotide primers for amplification of related gene fragments from the total DNA of C. apicola.

PCR products of the expected size (about 400 bp) were cloned and sequenced. Among them, two 412 bp fragments were found to encode polypeptides having 48.2 % and 46.7 % amino acid identity to the corresponding region of CYP52A3. Southern-blot experiments confirmed that the two fragments amplified were derived from different genes. Moreover, low-stringency hybridization patterns indicated the existence of several further P450-related genes.

Isolation and sequence analysis of a first P450 gene from *C. apicola*

The obtained PCR fragments provided suitable hybridization probes to screen a genomic library of *C. apicola* for P450 genes. One of the isolated clones was further characterized. It contained a DNA fragment of about 5 kb and included the sequence of the probe. Further sequencing revealed the presence of an open reading frame encoding a protein of 519 amino acids. The deduced protein (designated as P450CA1) contained in its C-terminal part a heme-binding HR2 region as characteristic for P450 proteins (Fig. 1).

```
P450CA1      FNGGPRICLGQQFALTEA
CYP52A3      FNGGPRICLGQQFALTEA
CYP4A2       FSGGARNCIGKQFAMNEL
```

Fig. 1: Comparison of amino acid sequences of the heme-binding HR2 region of P450CA1 with CYP52A3 (*C. maltosa*) and CYP4A2 (Rat)

Highest sequence similarities were found to members of the CYP52 family (42.6 per cent to CYP52A3, 44.1 per cent to CYP52A11, 36.1 per cent to CYP52B1, 36.1 per cent to CYP52C2, 41.1 per cent to CYP52D1), indicating that P450CA1 is a new member of this family. A distant relationship seems to exist also to mammalian and bacterial fatty acid-hydroxylating P450 forms (25.2 per cent to CYP4A2 and 24.2 per cent to CYP102).

In conclusion, both the induction of microsomal P450 systems with different substrate specificities and Southern hybridization experiments suggest the presence of multiple P450 genes in the glycolipid-producing yeast *C. apicola*. A first P450 gene has been cloned using a PCR-mediated approach. Based on these results, experiments are in progress to characterize the substrate specificity and regioselectivity of individual P450 forms by means of heterologous expression in *Saccharomyces cerevisiae*.

## ACKNOWLEDGMENT

This study has been supported by the Deutsche Forschungsgemeinschaft.

## REFERENCES

Kleber, H.-P., Asperger, O., Stüwer, O., Stüwer, B., Hommel, R. (1989): Occurrence and Regulation of Cytochrome P450 in *Torulopsis apicola*. In *Cytochrome P450, Biochemistry and Biophysics*, ed. Schuster, I., 169-172. London: Taylor and Francis.

Nelson, D., Kamataki, T., Waxman, D., Guengerich, F., Estabrook, R., Feyereisen, R., Gonzalez, F., Coon, M., Gunsalus, I., Gotoh, O., Okuda, K., Nebert, D. (1993): The P450 superfamily: update on new sequences, gene mapping, accession numbers, early trivial names of enzymes, and nomenclature. *DNA Cell Biol.* 12, 1-51.

Schunck, W.-H., Vogel, F., Gross, B., Kärgel, E., Mauersberger, S., Köpke, K., Gengnagel, C., Müller, H.-G. (1991): Comparison of two cytochromes P450 from *Candida maltosa*: primary structures, substrate specificities and effects of their expression in *Saccharomyces cerevisiae* on the proliferation of the endoplasmic reticulum. *Eur. J. Cell Biol.* 55, 336-345.

Seghezzi, W., Sanglard, D., Fiechter, A. (1991): Characterization of a second alkane-inducible cytochrome P450-encoding gene, *CYP52A3*, from *Candida tropicalis*. *Gene* 106, 51-60.

Seghezzi, W., Meili, C., Ruffiner, R., Kuenzi, R., Sanglard, D., Fiechter, A. (1992): Identification and characterization of additional members of cytochrome P450 multigene family CYP52 of *Candida tropicalis*. *DNA Cell Biol.* 11, 767-780.

Sherman, F., Fink, G., Hicks, J. (1986): Methods in yeast genetics. Cold Spring Harbor Laboratory, 125-126.

Weber, L., Stach, J., Haufe, G., Hommel, R., Kleber, H.-P. (1990): Elucidation of the structure of an unusual cyclic glcyolipid from *Torulopsis apicola*. *Carbohydr. Research* 206, 13-19.

# Chicken cytochrome P450$_{17\alpha}$: cloning of active type E2 and mutated type A2 defective in 17,20-lyase activity

Katsuhiko Nishimori, Hideto Ikushima, Shigeki Mizuno

*Department of Applied Biological Chemistry, Faculty of Agriculture, Tohoku University, Tsutsumidori-Amamiyamachi 1-1, Aoba-Ku, Sendai 981, Japan*

A putative chicken P450$_{17\alpha}$ cDNA was cloned as pLOA0511 by Ono, H. et al.., from cDNA library constructed from mRNA of left ovary of chickens sacrificed in one to three days after hatching. By computer analysis, comparison of the overall amino acid sequence deduced from nucleotide sequence of pLOA0511 with those from mammalian derived P450$_{17\alpha}$s, as from bovine (Zuber et al., 1986), rat (Namiki et al., 1988) and human (Chung et al., 1987), showed 47-48% similarity, and the amino acid sequence of pLOA0511 remarkably exhibited higher similarity (63%) to rainbow trout P450$_{17\alpha}$(Sakai et al.). After insertion of this cDNA into high-level expression vector pcDLSRα296, $10^6$ of cos7 cells were transfected with 20 µg of purified plasmid DNA. This in vivo assay indicated only weak activity, as 0.01~0.02 nmol/hr conversion of progesterone to 17α-hydroxyprogesterone and pregnenolone to 17α-hydroxypregnenolone, under conditions with 1µM of substrates, in contrast to conversion rate of bovine P450$_{17\alpha}$. But the weak enzyme activity from this putative chicken P450$_{17\alpha}$ cDNA also showed the possibility that it can catalyzed 17,20-lyase activity in both Δ4 and Δ5 passways after elongated incubation (48~72 hr, data not shown). This weak activity derived from pLOA0511 was presumed to be due to amino acid substitution(s) in coding region occurred in cDNA synthesis of the clone, or due to critical traits of chicken P450$_{17\alpha}$ in itself.

To seek for the actual cause of such weak activity shown by transfection with pLOA0511, we attempted to newly synthesize chicken P450$_{17\alpha}$ cDNA by RT-PCR method. For this purpose, two sets of primers were used to generate 854 bp N-terminal fragment and 1174 bp C-terminal fragment of the enzyme. As the N-terminal region of ch.P450$_{17\alpha}$ had high GC composition, replacement of Taq DNA polymerase in PCR reaction by vent DNA polymerase (New England Biolabs) led to amplification of designed fragment. Two ch.P450$_{17\alpha}$ expression plasmids, pA2 and pE2, were reconstructed by ligation of N-terminal fragments and C-terminal fragments, derived from individual mRNA of 40-days old chickens' ovaries, and $10^6$ of cos7 cells were transfected with 20 µg of each plasmid to study the P450$_{17\alpha}$ activity of the cDNAs. In this experiment, E2 appeared full activity, at the rate of about 1 nmol conversion of progesterone to 17αOH-progesterone per hour, and slightly less conversion rate for similar reaction with pregnenolone, by transfected cells in

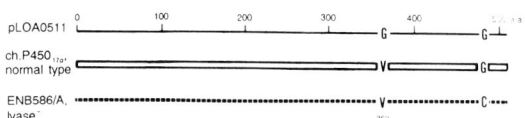

Fig.1. Location of amino acid residues, substitution of which affect the total activity and 17,20-lyase activity of chicken P45017α. Upper line (solid line), middle line (open line) and lower line (hatched line) indicates amino acid sequence of pLOA0511, presumably normal amino acid residues of ch.P45017α deduced from pLOA0511, pE2 and pA2, and amino acid residues of chimeric construct pENB586/A.

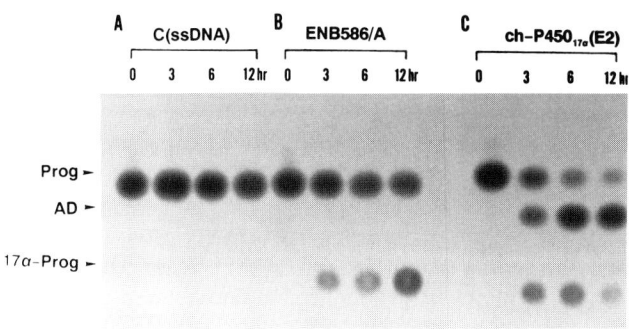

Fig.2. Analysis of products converted from progesterone by chimeric and normal ch.P45017αs. $10^6$ of cos7 cells were transfected with 20 μg of (A), salmon sperm DNA, (B), chimeric pENB586/A, and (C), normal type chicken P45017α (E2). Twenty four hours after transfection, cells were incubated at 37°C with 3.7 kBq of $^{14}$C-progesterone (1 μM), and at the times indicated, the steroids were extracted with dichloromethane and chromatographed on silica gel plate as described. Prog, 17α-prog and AD denotes the position of progesterone, 17αOH-progesterone and androstenedione on the TLC plate, respectively.

10 ml of medium per φ10cm dish, containing 1.0 μM of each substrate. This value are more active to some extent than that of bovine $P450_{17α}$ under control of the same expression vector in the same conditions (data not shown). Remarkable character of ch.$P450_{17α}$ E2 clone is distinct 17,20-lyase activity in both Δ4 and Δ5 passways in contrast with $P450_{17α}$s of bovine (Zuber et al., 1988 and Barnes et al., 1991) and human which lack 17,20-lyase activity in Δ4 passway. On the contrary, A2 appeared relatively moderate 17α-hydroxylase activity as compared with E2 clone, and surprisingly it showed less or no activity of 17,20-lyase activities for both OH-progesterone and OH-pregnenolone.

To further characterize A2 and E2 clones, DNA sequences of both clones were determined, and some amino acid substitutions were detected. To identify the amino acid residues responsible for restoration of $P450_{17α}$ activity observed in E2 clone, and for missing of 17,20-lyase activity in A2 clone, several chimeric composites ligated from each fragment derived from pLOA0511, E2 and A2, were constructed (data not shown). In vivo expression assay of these constructs with cos7 cell clearly demonstrated that conversion of Gly[363] of pLOA0511 to Val in E2(Fig.1) caused the restoration of putatively full activity of normal ch.$P450_{17α}$ (Fig.2 C). On the other hand, the composite construct ENB586/A revealed that conversion of Gly[484] of pLOA0511 and E2 to Cys caused the abolishment of 17,20-lyase activity observed in A2 clone (Fig.2B). Val[363] located at inside of Picado-

Leonard and Miller binding site (Picado-Leonard et al., 1988), was one of conserved amino acid residues observed in mammalian P450$_{17\alpha}$, and Gly$^{484}$ locates at the inside of SRS-6, proposed as substrate recognition sequence for cytochrome P450 family 2, by Gotoh,O. The amino acid substitution occurred in pA2 demonstrates the possibility that distally C-terminal region of ch.P450$_{17\alpha}$ contributes to determine the substrate binding specificity of P450$_{17\alpha}$ or the specificity of 17,20-lyase activity of P450$_{17\alpha}$. The result obtained from this experiments clearly demonstrate that the C-terminal region of P450$_{17\alpha}$s bear notable function in the enzyme activity, as previously reported (Yanase,T et al., 1992). And it also indicates that amino acid residue near the C-terminus of P450$_{17\alpha}$ has similar contribution on the activity of 17,20-lyase to the role of amino acid residues located on Picado-Leonard and Miller binding site, reported by Kitamura,M et al.

We are grateful to O.Gotoh of Saitama Cancer Center Research Institute, Japan and to S.Kominami of Hiroshima University, Japan for helpful and stimulating discussion.

## REFERENCE

Ono, H., Iwasaki, M., Sakamoto, N. and Mizuno, S (1988).:cDNA cloning and sequence analysis of a chicken gene expressed during the gonadal development and homologous to mammalian cytochrome P-450c17. *Gene 66:* 77-85 Barnes,H.J., Arlotto,M.P. and Waterman,M.R.: Expression and enzyme activity of recombinant cytochrome P450$_{17\alpha}$-hydroxylase in *Escherichia coli*.: *Proc.Natl.Acad.ci.USA 88:* 5597-5601

Chung,B.-C., Picado-Leonard,J., Haniu,M., Bienkowski,M., Hall,P.F., Shively,J.E. and Miller,W.L.(1987): Cytochrome P450c17 (steroid 17-alpha-hydroxylase/17,20 lyase): Cloning of human adrenal and testis cDNAs indicates the same gene is expressed in both tissues: *Proc. Natl. Acad. Sci. U.S.A. 84:* 407-411

Gotoh, O. (1992): Substrate recognition sites in cytochrome P450 family 2(CYP2) proteins inferred from comparative analyses of amino acid and coding nucleotide sequences.: *J Biol Chem 267:* 83-90

Kitamura,M., Buczko,E. and Dufau,M.L.(1991): Dissociation of hydroxylase and lyase activities by site-directed mutagenesis of the rat P450$_{17\alpha}$. *Mol. Endocrinol. 5:* 1373- 1380

Namiki,M., Kitamura,M., Buczko,E. and Dufau,M.(1988):Rat testis P-450$_{17\alpha}$: Deduced amino acid sequence, Expression and secondary structural configuration. *Biochem. Biophis. Res. Comm. 157:* 705-712

Picard-Leonard, J. and Miller,W.L.(1988): Homologous sequence in steroidogenic enzyme, steroid receptors and a steroid binding protein suggest a consensus steroid binding sequence.: *Mol. Endocrinol. 2:* 1145-1150

Sakai, N., Tanaka,M., Adachi,S., Miller,W.L. and Nagahama,Y. (1992): Rainbow trout cytochrome P450c17 (17α-ydroxylase/17,20-lyase): cDNA cloning, enzymatic properties and temporal pattern of ovarian P450c17 mRNA expression during oogenesis.: *FEBS letter 30:* 60-64 Yanase,T., Waterman,M., Zachmann,M., Winter,J.S.D., Simpson,E.R. and Kagimoto,M.(1992): Molecular basis of apparent isolated 17,20-lyase deficiency: compound heterozygous mutation in the C-terminal region (Arg(496)→Cys, Gln(461)→stop) actually cause combined 17α-hydroxylase/17,20-lyase deficiency: *Biochem. Biophis. Acta 1139:* 275-279

Zuber,M.X., Simpson,E.R. and Waterman,M.R. (1986):Expression of bovine 17α-hydroxylase cytochrome P-450 cDNA in nonsteroidogenic (cos 1) cells. *Science 234:* 1258-1261

# Expression of aromatase cytochrome P450 in choliocarcinoma cells

Kazuyo Yamada, Seiya Katoh, Nobuhiro Harada, Yasuyuki Takagi

Division of Molecular Genetics, Institute for Comprehensive Medical Science, Fujita Health University, School of Medicine, Toyoake, Aichi 470-11, Japan

## INTRODUCTION

Aromatase catalyzes aromatization of androgens, the rate-limiting step of estrogen synthesis. In human, this activity is found mainly in the ovary and placenta, and tissues such as the brain, liver, adipose tisssues and skin fibrablasts are located as extra-gonadal sites of estrogen production. A number of agents which include cAMP, glucocorticoids, phorbol ester and growth factors are known to affect the expression of aromatase. The regulation by these agents is coordinated in a tissue specific manner. Recent investigations revealed that aromatase mRNA expressed in adipose tissue, ovary and fetal liver are different in untranslated exon I from that of previously reported placenta(Mahendroo et.al 1991, Harada, et.al.1993) The tissue specific usage of different transcriptional initiation sites and Exon I seems to be the underlying molecular bases of complexity of the aromatase gene expression. In this context, for the further investigation of molelular mechanism of the regulation of aromatase in a given organ, it is prerequiste to get a cell model system which exhibits the same type of aromatase expression. In this paper we described some characteristics of aromatase expressed in JEG-3 which is known to retain many differentiated characteristiccs of normal trophoblast.

## RESULT AND DISCUSSION

### Expression of aromatase in JEG-3

Choriocarcinoma Cells JEG-3(ATCC) were cultured in various stimuli for 24 hr and membrane fraction were prepared. Aromatase activity was measured with (19-$^{14}$C)androst-4-ene-3,17-dione(NEN) as a substrate. As shown in Table1, membrane fraction from unstimulated JEG-3 exhibited significant aromatase activity. Stimulation by phorbol 12-myrystate 13-acetate(TPA), forskolin and β-estradiol resulted in 1.6, 3.3 and 1.6 fold increase of aromatase activity, respectively. TPA potentiated the effect of forskolin. RNA fractions were prepared from the cells

Table 1  Induction of aromatase by various reagents

| condition | aromatase activity (pmole/mg protein) | aromatase mRNA (attomol/μgRNA) |
|---|---|---|
| none | 0.61 ± 0.19 | 1.81 ± 0.54 |
| TPA | 0.97 ± 0.24 | 6.76 ± 0.80 |
| forskolin | 2.04 ± 0.51 | 11.49 ± 1.62 |
| dexamethasone | 0.57 ± 0.05 | 3.36 ± 0.17 |
| β-estradiol | 0.96 ± 0 | 2.72 ± 0.63 |
| forskolin+TPA | 4.11 ± 0.63 | 29.1 ± 5.15 |

cultured under the same conditions and the content of aromatase mRNA was determined by reverse transcriptase-polymerase chain reaction method with a fluorescent primer(Harada and Yamada 1992).This method enabled us to quantitate mRNA content over the wide range of mRNA concentration. Stimulation by forskolin and a combination of forskolin and TPA increased the content of aromatase mRNA by 6.3 fold and 16 fold, respectively. Other reagents did not affect the amount of the mRNA significantly. The concentration of aromatase mRNA and the enzyme activity corelated well. In Fig1, time course of the induction of aromatase activity and mRNA are presented. Increase of the activity was detected after 12 h when the increase in the content of aromatase mRNA was also observed. No acute modulation of enzyme activity was observed. The induction of aromatase mRNA by forskolin was completely blocked when cycloheximide was simultaneously added in the medium. Continuous stimulation by forskolin was necessary to maintain high level of aromatase mRNA in this cell line. Withdrawal of forskolin from culture medium resulted rapid loss of aromatase mRNA(Fig2). Cycloheximide (20 μg/ml) faciliated the loss of the mRNA in the absense of forskolin. It also caused loss of the mRNA in the presense of the stimulant.

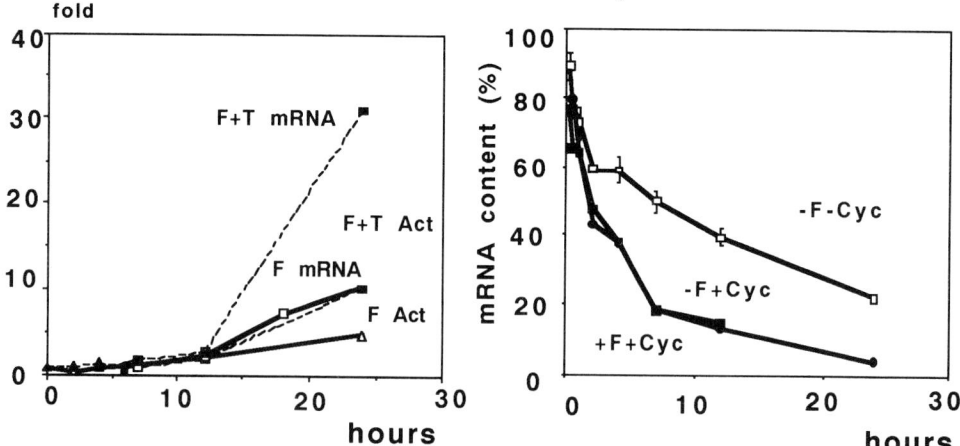

Fig 1  Time course of aromatase induction

Fig 2  Time course of Arom mRNA Decay

The rates of decay caused by cycloheximide were similar in the presense or absense of the stimulant and in basal level expression. These results suggest that degradation of aromatase mRNA is not affected by cAMP and control of transcription rate is primary in the regulation of mRNA content in this cell line.

Structure of aromatase mRNA expressed in JEG-3
For the investigation of transcriptional regulation of aromatase gene, transcription initiation site has to be located in the cell used in expression experiment. It is especially critical for the aromatase gene since multiple initiation sites were discovered in different tissues. Primer extension assay was carried out to compare the transcription initiation site of aromatase mRNA expressed in JEG-3 to that of human placenta. Placental RNA and RNA prepared from JEG-3 gave the same sizes of extension products with a primer complementary to exon I and exon II,respectively, that showed the aromatase mRNA expressed in JEG-3 is placental type.

Expression of aromatase-CAT vectors in JEG-3 cells
To determine the sequence required for aromatase expression The 2.7Kbp fragment and deletion mutant of the aromatase promoter was inserted into *Hind* III site of pSV00CAT 5' of the gene for CAT(2.7KCAT). A series of deletion mutants of this plasmid were also prepared. These vectors were transfected into JEG-3 cells in the presence or absence of forskolin to find DNA sequences involved in basal and cAMP-mediated gene expression in JEG-3. Chloramphenycol acetyl transferase activity was negligble in the cells transfected with pSV00CAT(<0.25%), which lacks a promoter. The CAT vector containing -301bp upstream of transcription start site showed high CAT activity, while the vector containing -212 showed modest CAT activity. Forskolin failed to enhance expression of all chimeric constructs examined. These results indicate that the sequences between -212 to -301 were important for basal expression of aromatase.

Tissue specificicty of the aromatase promotor
Promoter activity of this region was compared with that of thimidine kinase(tk) promotor in three cell lines: JEG-3, a breast tumor cell line MCF-7 and a human fibloblast cell line IMR-90. Membrane fractions from the later two cell lines did not exhibit aromatase activity. Relative CAT activity expressed driven by placental aromatase promoter and tk promoter was in MCF-7 and in IMR-90, respectively. On the contrary the ratio was in JEG-3. The promoter activity located in the 5' flanking region of aromatase gene confines placenta-specific expression.

**References**
Mahendroo, M.S. Means, G.D. Mendeleson, C.R. and Simpson, E.R. (1991) *J.BIol.Chem.266* 11276-11281
Harada,N.(1992) *Biochem.Biophys.Res.Commun.189* 1001-1007
Harada,N. and Yamada,K.(1992) *Endocrinology131* 2306-2312

# Immortalization of hepatocytes *via* a transgenic approach

Kerstin Kramer[1], Giuseppe Ciaramella[2], Mina Edwards[2], Elizabeth A. Shephard[2], Ian R. Phillips[1]

[1]Department of Biochemistry, Queen Mary and Westfield College, Mile End Road, London E1 4NS. [2]Department of Biochemistry and Molecular Biology, University College London, Gower Street, London WC1E 6BT, UK

## INTRODUCTION

Primary cultures of hepatocytes are increasingly being used for a wide range of fundamental and applied pharmaco-toxicological research. However, the value of such cultures as *in vitro* models is dependent on the extent to which the abundance of proteins responsible for foreign compound metabolism reflects that found *in vivo*. Despite recent progress achieved by culturing hepatocytes on biologically-derived matrices (Schuetz *et al.*, 1988, Waxman *et al.*, 1990, Sidhu *et al.*, 1993) or in the presence of liver epithelial cells (Akrawi *et al.*, 1993), the expression of several liver-specific functions, particularly some of the cytochromes P450 involved in xenobiotic metabolism, declines rapidly in primary hepatocyte cultures.

One of the approaches we are taking towards overcoming this problem is to attempt to establish a conditionally immortalized, differentiated hepatocyte cell-line from a transgenic mouse harbouring a temperature-sensitive mutant of the SV40 large T antigen gene under the control of the mouse major histocompatibility complex H-2K$^b$ class I promoter which is inducible by $\gamma$- interferon. (Jat *et al.*, 1991) The permissive temperature for the gene product is $33^0$C; at the non-permissive temperature of $37^0$C it is rapidly degraded. Theoretically, cells isolated from these transgenic mice should be able to divide and be in a dedifferentiated state at the permissive temperature and should be able to differentiate at the non-permissive temperature. If successful, the system will provide a continuously dividing, minimally transformed cell-line whose function mimic those of adult hepatocytes.

## MATERIALS AND METHODS

Isolation and culture of hepatocytes from transgenic mice

Hepatocytes were isolated from the livers of adult transgenic mice weighing 20 - 25g by a modification of the two-step collagenase perfusion technique. The cells were cultured on collagen-coated culture dishes (Corning) in Williams' E medium supplemented with insulin (10 µg/ml), transferrin (10 µg/ ml), hydrocortisone (300 ng/ml) and epidermal growth factor (EGF)(10ng/ml).

The cells were plated at a density of $10^6$ cells per 60 mm (diameter) culture dish and propagated at 33ºC. They reached confluency 48 hrs after isolation. The hepatocytes were passaged at a subconfluent stage using a solution containing 0.05% Trypsin and 0.02% EDTA. The cells were recovered and washed with culture medium containing 2% BSA. The cells were replated in culture medium containing 5% foetal calf serum but lacking EGF and kept at 37ºC for 4 hrs to obtain better attachment. The medium was then replaced with serum-free culture medium containing EGF and the cells were shifted to the permissive temperature of 33ºC. For induction experiments, hepatocytes plated at a density of $10^6$ cells per plate were left to propagate for 24 hrs at 33ºC or 37ºC. β-Naphthoflavone was then added at a final concentration of 10 µM and the cells were harvested 24 hrs later.

RNase protection assay

Total RNA was prepared from cells pooled from two culture dishes (60 mm diameter) according to the method of Chomczynski & Sacchi (1987) with slight modifications as described by Xie & Rothblum (1991). The construct used to generate the antisense probe for the RNase protection assay was prepared by amplification by PCR of the third exon of the mouse CYP1A1 gene and subsequent subcloning of the amplified fragment into pBluescript. The RNase protection assay was caried out as described by Cullingford et al. (in press). For each hybridization, 10 µg of total RNA was used.

RESULTS AND DISCUSSION

We are attempting to establish a system whereby hepatocytes isolated from transgenic mice harbouring thermolabile mutant of the SV40 large T antigen are able to proliferate and form a propagatable cell-line that, if cultured under appropriate conditions should be capable of expressing differentiated liver-specific functions.

After isolation, hepatocytes plated at an original density of $10^6$ cells per plate reach confluency after 48 hrs and form a monolayer which clearly resembles a primary culture of normal hepatocytes plated at a density of $3-4 \times 10^6$ cells per plate. After passaging, cells tend to grow in colonies rather than in a uniform monolayer, but they retain their liver-specific morphology (Fig. 1).

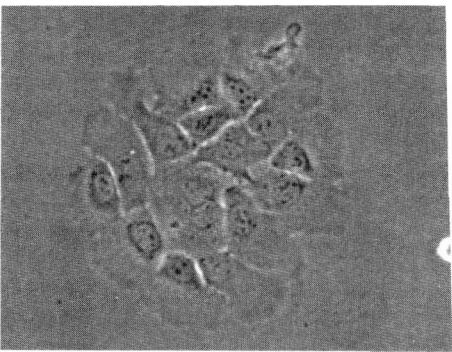

Fig. 1. Immortalized hepatocytes 48 hrs after passaging

The hepatocytes obtained after passaging are mainly mono- or binucleated but some polynucleated cells can be observed. We found that induction of the activity of the promoter controlling the expression of the SV40 large T antigen mutant by γ-interferon is not essential to maintain cell proliferation whereas EGF seems to stimulate cell division. The presence of serum in the culture medium is not required.

The cells have been maintained in culture for several generations by continuous passaging. They retain at $33^0$C or $37^0$C the characteristic morphology associated with differentiated adult hepatocytes and secrete albumin into the culture medium. To assess the extent to which liver-specific functions are maintained in this cell system we have begun to investigate the expression of components of the cytochrome P450-mediated monooxygenase system in freshly isolated cells and in passaged cells cultured at both the permissive and non-permissive temperatures. The results of RNase protection assays showed that CYP1A1 mRNA was undetectable in untreated cells but was strongly induced by β-naphthoflavone. The induction appeared to be greater in cells maintained at the non-permissive temperature (Fig. 2). The extend of induction is maintained after passaging the cells (results not shown).

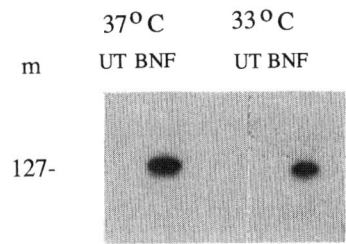

Fig. 2. RNase protection assay using a mouse 1A1 probe hybridized to 10 µg of total RNA from untreated cells (UT) and cells treated with β-naphthoflavone (BNF) and cultured at $33^0$C or $37^0$C , (m) size marker (bp)

Probes have also been constructed for RNase protection assay of mRNAs encoding CYPs 2B, 1A2, 3A and cytochrome b5 of mouse. The expression and inducibility of these mRNAs in the immortalized hepatocytes is currently being investigated.

### Acknowledegments

We thank Dr. P. Jat and Dr. M. Noble of the Ludwig Institute of Cancer Research for access to the transgenic mouse strain. K. K. is a recipient of a Wellcome Trust Toxicology Studentship. G. C. is supported by a studentship from Glaxo Group Research.

# References

Akrawi, M., Rogiers, V., Vandenberghe, Y., Palmer, C. N. A., Vercruysse, A., Shephard, E. A., & Phillips, I. R. (1993) *Biochem. Pharmacol.* 45: 1583-1591.

Chomczynski, P. & Sacchi, N. (1987) *Anal. Biochem.* 162: 156-159

Cullingford, T.E., J.B. Clark & I.R. Phillips (1993) *J. Neurochem.* (in press)

Jat, P.S., M.D. Noble, P. Ataliotis, Y.Tanaka, N. Yannoutsos, L. Larsen & D. Kioussis (1991) *Proc. Natl. Acad. Sci. USA* 88: 5096-5100.

Schuetz, E. G., Li, D., Omiecinski, C. J., Muller-Eberhard, U., Kleinmann, H. K., Elswick, B., Guzelian, P. S. (1988) *J. Cell. Physiol.* 134: 309-323.

Sidhu,J. S., Farin, F. M. & Omiecinski, C. J. (1993) Arch. Biochem.Biophys. 301: 103-113.

Waxman, D. J., Morissey, J. J., Naik, S. & Jauregi, H. O. (1990) *Biochem. J.* 271: 113-119

Xie, W.Q., L.I. Rothblum (1991) *Bio Techniques* 11: 325-327.

# Analyses of promoter regions of rat CYP11B genes involved in mineralo- and glucocorticoid syntheses

Kuniaki Mukai, Hideo Shimada, Yuzuru Ishimura

Department of Biochemistry, School of Medicine, Keio University, Shinjuku-ku, Tokyo 160, Japan

## Summary

Promoter activities of 5'-flanking regions of rat CYP11B genes, which are responsible for the zone-specific syntheses of steroids in the adrenal cortex, have been analyzed by transient transfection experiments using mouse Y1 adrenocortical cells. The 0.5-kb 5'-flanking region of CYP11B1, product of which is expressed in the zonae fasciculata-reticularis, had a 4- and 10-fold higher promoter activity than the corresponding regions of CYP11B2 and -B3, respectively. CYP11B2 expresses in the zona glomerulosa, while the *in vivo* products of CYP11B3 is unknown. The analyses of protein binding site on the promoter region of CYP11B1 with nuclear extracts from Y1 cells indicated that a sequence around at -300 contained a positive regulatory element for the zone-specific expression of CYP11B1.

## INTRODUCTION

Rat CYP11B gene subfamily consists of four forms of structurally related homologs (Mukai et al., 1993). CYP11B1 encodes P45011β which synthesizes corticosterone, the major glucocorticoid in rat, from deoxycorticosterone in the inner zones of adrenal cortex, i.e. the zonae fasciculata-reticularis (zFR), while CYP11B2 encodes P450aldo which catalyzes the formation of aldosterone from the same substrate in the outer zone, the zona glomerulosa (zG) (Ogishima et al., 1989; Lauber & Müller, 1989). CYP11B3 is a gene without a known expression product (Nomura et al., 1993), and CYP11B4 is a pseudogene. By comparison of the nucleotide sequences of the exonic and the 5'-flanking regions, CYP11B1, -B3, and -B4 were found to form a subgroup in the four. Since the 5'-flanking region of CYP11B2 shares a higher homology to those of CYP11B genes of bovine and human than to the above three, the three are unique in CYP11B genes found in animal species (Mukai et al., 1991; Mukai et al., 1993; see the references therein). To investigate the zone-specific expression mechanisms of the CYP11B genes, we examined in this study the promoter activities of the 5'-flanking regions of CYP11B1, -B2, and -B3 using mouse Y1 adrenocortical cells. The promoter region of CYP11B1, the strongest promoter among the three, was analyzed to show the presence of a positive regulatory element for its expression.

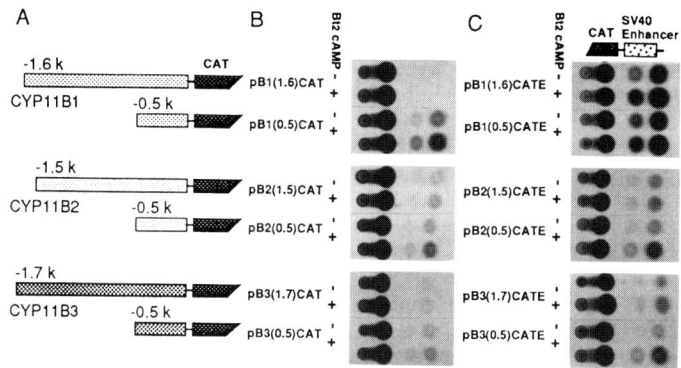

Fig. 1. **Promoter activities of 5'-flanking regions of rat CYP11B genes.** The 5'-flanking regions (1.5 to 1.7 kb and 0.5 kb) of CYP11B1, -B2, and -B3 were each placed in front of CAT gene as shown in panel A. Adrenocortical Y1 cells were transiently transfected by electroporation with 20 μg each fusion CAT construct, which was without (panel B) or with SV40 enhancer (panel C). The cells were cultured for 66 h, and then the cell extracts were prepared. Dibutyryl cAMP (1 mM) was added (+) or not (-) to the medium at 24 h after transfection. The levels of CAT in the cell extracts were assayed using radiolabeled chloramphenicol followed by thin layer chromatography and autoradiography. *Right two spots* in each autoradiogram represent acetylated forms of chloramphenicol.

## EXPERIMENTAL PROCEDURES

<u>Plasmid constructions</u> DNA fragments of 5'-flanking regions of CYP11B1 (1.6 and 0.5 kb), -B2 (1.5 and 0.5 kb), and -B3 (1.7 and 0.5 kb) (Mukai et al., 1993) were placed in front of chloramphenicol acetyltransferase (CAT) gene of pCAT-Basic and pCAT-Enhancer vectors (Promega Corp.). A series of deletion plasmids carrying the 5'-region of CYP11B1 was constructed utilizing restriction enzyme sites in the 0.5-kb 5'-flanking sequence which was cloned in pCAT-Basic (Mukai et al., in preparation).

<u>Transient transfection</u> Mouse Y1 adrenocortical cells (Schimmer, 1985) were transiently transfected with the CAT plasmids by electroporation, and CAT activity of the cell extract was mesured as described (Mukai et al., 1993).

## RESULTS AND DISCUSSION

The promoter activities of 5'-flanking regions of the rat CYP11B genes were analyzed in transient transfection experiments using mouse Y1 adrenocortical cells. Each 5'-flanking region of CYP11B1 (1.6 and 0.5 kb), -B2 (1.5 and 0.5 kb), and -B3 (1.7 and 0.5 kb) was inserted into promoterless CAT plasmid, pCAT-Basic as shown in Fig. 1A. When their promoter activities were examined by the CAT assay, the cells transfected with pB2(1.5)CAT and pB3(1.7)CAT produced very low levels of CAT, while pB1(1.6)CAT showed an undetectable level of CAT activity (Fig. 1B). In the case of the 0.5-kb flanking regions, on the other hand, CAT levels of pB2(0.5)CAT and pB3(0.5)CAT were higher than those of pB2(1.5)CAT and pB3(1.7)CAT by 3-fold, respectively. Interestingly, the 0.5-kb flanking region of CYP11B1 showed the highest level of CAT expression among the 0.5-kb 5'-flanking regions of CYP11B1, -B2, and -B3. These results indicated that the 0.5-kb region of CYP11B1 had the strongest promoter among the 0.5-kb regions of the three genes in Y1 cells and that the DNA sequence from -1.6 to -0.5 kb relative to the CYP11B1 transcription initiation site contained a sequence suppressing the promoter activity within the proximal 0.5-kb region.

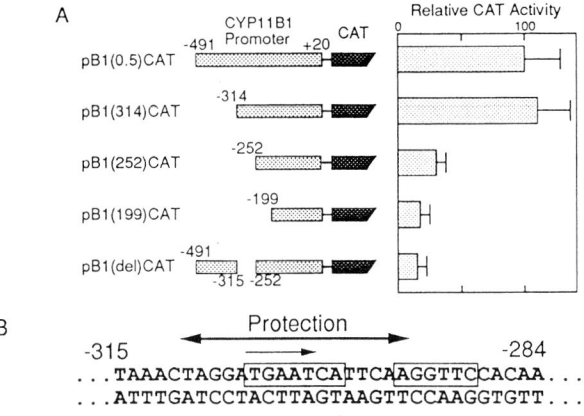

B
```
                       Protection
     -315         ←————————————→           -284
...TAAACTAGGATGAATCATTCAAGGTTCCACAA...
...ATTTGATCCTACTTAGTAAGTTCCAAGGTGTT...
                   ←————————————
```

Fig. 2. **A. Deletion analysis of the promoter region of CYP11B1.** The 0.5-kb promoter region of CYP11B1 in pB1(0.5)CAT was deleted successively from the 5'-end, and the resulting 5'-ends were at -314, -252, and -199. pB1(del)CAT was produced by an internal deletion removing the sequence from -314 to -253. The levels of CAT expression from these plasmids in Y1 cells were analyzed by transient transfection in the absence of dibutyryl cAMP as described in Fig. 1. **B. Nucleotide sequence around a protein binding site in the promoter region of CYP11B1.** The sequence was identified by DNase I protection assay using nuclear extracts from Y1 cells. Numbers indicate nucleotide positions relative to the transcription start site. A pair of arrows indicates palindromic sequences. TGAATCA and AGGTTC in boxes indicate potential binding sites for AP-1-like protein and for a member of steroid hormone receptors, respectively.

We next tested effects of SV40 enhancer on the promoter activities of the three genes. Another set of CAT plasmids, which were identical to the CAT plasmids described above except for the presence of SV40 enhancer, were constructed and analyzed for the activity. As shown in Fig. 1C, SV40 enhancer stimulated all the CAT levels by more than 10-fold. Although the 1.6-kb region of CYP11B1 did not show CAT expression as seen in Fig. 1B, pB1(1.6)CATE carrying the enhancer showed a promoter activity comparable to that of pB1(0.5)CATE. Thus, the suppression of the promoter activity within the 0.5-kb region of CYP11B1 by its upstream sequence from -1.6 to -0.5 kb was overridden by the enhancer. Figures 1B and C also showed that addition of dibutyry cAMP to the culture medium after transfection enhanced the CAT levels from all the CAT constructs by approximately 3-fold. Thus the 5'-regions of the three genes seem to contain DNA sequences conferring response to cAMP. It has been known that adrenocorticotropin stimulates steroid synthesis of Y1 cells through cAMP-mediated processes (Schimmer, 1985).

In the transient transfection experiments, strengths of promoter activities of the 0.5-kb regions of the three genes were in a ratio of 10: 2.5: 1. The very low activity of CYP11B3 was in conformity with the fact that the in vivo expression products of this gene were not detected so far (Nomura et al., 1993). The promoter region of CYP11B1 was four times more active than that of CYP11B2 in this cell line, being consistent with the view that this cell line has properties of the zFR cells. Accordingly, we analyzed the promoter region of CYP11B1 using this cell line as a model of the zFR cells.

To investigate the transcriptional regulation of the CYP11B1 gene, various deletion mutants of the CYP11B1 promoter region were tested in transient transfection experiments. When CAT plasmids carrying the promoter region with their 5'-ends at -491, -314, -252, and -199 relative to the transcription start site were tested, the CAT level of pB1(252)CAT was only one third of pB1(314)CAT (Fig. 2A). The changes in promoter activities by a deletion in the other regions were smaller than that from -314 to -252. The CAT activity of pB1(199)CAT was at the same level of a construct with the 5'-end at -41. An internal deletion from -314 to -253 in the 0.5-kb sequence (pB1(del)CAT) resulted in a decrease in CAT expression to an almost equal amount to that of pB1(199)CAT. Thus, we concluded that the DNA sequence from -314 to -253 contained a positive regulatory element(s) for the expression of CYP11B1.

Protein binding site on the CYP11B1 promoter region was then analyzed by DNase I protection experiments using nuclear extracts from Y1 cells (Mukai et al., in preparation). The sequence from -310 to -294 was protected by the nuclear extract from DNase I digestion (Fig. 2B). This site located in the sequence between -314 and -253, which was shown to contain the positive regulatory sequence (Fig. 2A). The protected sequence contains a pair of palindromic sequences and an AP-1-like protein binding sequence. The AP-1-like sequence is TGAATCA, which differs at the center from the consensus sequence, TGAG/CTCA. In addition, a homolog of a half-site for steroid hormone receptor, AGGTTC, is present at the immediate downstream of the protection. This receptor half-site could have regulatory function in this gene expression. Recently, it was reported that bovine Ad4 binding protein (Honda et al., 1993) and its mouse homolog, steroidogenic factor 1 (Ikeda et al., 1993), both of which are the members of the nuclear steroid hormone receptor superfamily, have an important role in the regulation of steroidogenic CYP gene expression.

The protected sequence from -310 to -294 in the CYP11B1 promoter region was not found in the promoter region of CYP11B2, and was modified in that of CYP11B3. Thus, the results described here suggest that the -300 region of CYP11B1 is a positive regulatory element for the transcription of this gene and is responsible for the zone-specific transcription within the adrenal cortex.

Supported in part by grants from the Ministry of Education, Science and Culture of Japan, from the Takeda Science Foundation, and from Keio University.

# REFERENCES

Honda, S., Morohashi, K., Nomura, M., Takeya, H., Kitajima, M., & Omura, T. (1993): Ad4BP regulating steroidogenic P-450 genes is a member of steroid hormone receptor superfamily. *J. Biol. Chem.* **268**, 7494-7502.

Ikeda,Y., Lala, D. S., Luo, X., Kim, E., Moisan, M.-P., & Parker, K. L. (1993): Characterization of the Mouse *FTZ-F1* gene, which encodes a key regulator of steroid hydroxylase gene expression. *Mol. Endocrinol.* **7**, 852-860.

Lauber, M. & Müller, J. (1989): Purification and characterization of two distinct forms of rat adrenal cytochrome P-450$_{11\beta}$: functional and structural aspects. *Arch. Biochem. Biophys.* **274**, 109-119.

Mukai, K., Imai, M., Shimada, H., Okada, Y., Ogishima, T., & Ishimura, Y. (1991): Structural differences in 5'-flanking regions of rat cytochrome P-450$_{aldo}$ and P-450$_{11\beta}$ genes. *Biochem. Biophys. Res. Commun.* **180**, 1187-1193.

Mukai, K., Imai, M., Shimada, H., & Ishimura, Y. (1993): Isolation and characterization of rat CYP11B genes involved in late steps of mineralo- and glucocorticoid syntheses. *J. Biol. Chem.* **268**, 9130-9137.

Nomura, M., Morohashi, K., Kirita, S., Nonaka, Y., Okamoto, M., Nawata, H., & Omura, T. (1993): Three forms of rat CYP11B genes: 11β-hydroxylase gene, aldosterone synthase gene, and a novel gene. *J. Biochem.* **113**, 144-152.

Ogishima, T., Mitani, F., & Ishimura, Y. (1989): Isolation of aldosterone synthase cytochrome P-450 from zona glomerulosa mitochondria of rat adrenal cortex. *J. Biol. Chem.* **264**, 10935-10938.

Schimmer, B. P. (1985): Isolation of ACTH-resistant Y1 adrenal tumor cells. *Methods Enzymol.* **109**, 350-356.

# Regulatory proteins involved in the tissue-specific expression and induction of a CYP2B2 gene

Levi M. Fernandez[1], Lesley A. Forrest[2], Amal Shervington[2], Elizabeth A. Shephard[2], Ian R. Phillips[1]

[1]Department of Biochemistry, Queen Mary and Westfield College, University of London, Mile End Road, London E1 4NS, UK. [2]Department of Biochemistry and Molecular Biology, University College London, Gower Street, London WC1E 6BT, UK

## INTRODUCTION

Many cytochromes P450 (CYPs) are selectively inducible by the compounds they metabolize. For instance, two members of the CYP2B subfamily, namely CYP2B1 and CYP2B2, are induced in mammalian liver by the anti-epileptic drug phenobarbital (Thomas et al., 1981; Phillips et al., 1981, 1983a). The induction of these proteins is mediated by an increase in the corresponding mRNAs (Phillips et al., 1981, 1983b; Shephard et al., 1982; Gonzalez & Kasper, 1982; Atchison & Adesnik, 1983), which in turn has been shown to be due to an increase in the transcription of CYP2B genes (Hardwick et al., 1983; Atchison & Adesnik, 1983; Pike et al., 1985). Because of the lack of a cell-line capable of maintaining suitable levels of expression and inducibility of CYP2B genes by phenobarbital little is known concerning the mechanisms responsible for regulating the expression of these genes. We have used gel retardation and DNase footprinting to identify, within a novel CYP2B2 gene promoter, potential regulatory sequences via their ability to interact with trans-acting nuclear proteins.

## MATERIALS AND METHODS

Nuclear protein extracts from the livers, brains and kidneys of phenobarbital-treated or untreated rats were prepared essentially as described by Burgoyne et al. (1970) as modified by Hewish & Burgoyne (1973) or according to the method of Gorski et al. (1986). All solutions contained Antipain, Chymostatin and Leupeptin at a concentration of 1 µg/ml each and 1 mM phenylmethanesulphonylfluoride. Restriction fragments of the CYP2B2 gene and double-stranded synthetic oligonucleotides were used as radiolabelled probes and/or unlabelled competitor fragments in gel retardation assays carried out essentially as described by Jose-Estanoyl et al. (1989). When included, unlabelled competitor DNA fragments were added to the reaction mix 10 min before the addition of the radiolabelled DNA probe. Binding reactions were carried out at room temperature for 30 min. Samples were then electrophoresed through a 4% polyacrylamide gel using a buffer containing 40 mM Tris, 5 mM sodium acetate (pH 7.5), 1 mM EDTA. Gels were dried *in vacuo* and autoradiographed at -78°C. DNA-protein complexes were quantified by scanning the autoradiograms with a Bio-Rad imaging densitometer (Model GS-670). DNase footprinting was carried out as described in the SureTrack Footprinting kit manual (Pharmacia P-L Biochemicals).

## RESULTS AND DISCUSSION

Gel retardation assays revealed the presence, between -1390 and +160 of a CYP2B2 gene, of several binding sites for liver nuclear proteins. The majority of these sites bound proteins that were present in equal abundance in extracts from the livers of phenobarbital-treated or untreated rats. However, two fragments of the promoter, -31/-177 and -178/-368, bound more protein with extracts isolated from phenobarbital-treated animals than from those isolated from untreated animals, indicating that these fragments contained binding sites for protein(s) whose abundance or activity was increased in response to phenobarbital.

To localize more precisely the sequences responsible for the increased binding of protein from extracts isolated from phenobarbital-treated animals, fragment -178/-368 was cleaved to produce two smaller subfragments (-178 to -217, and -215 to -368). A single DNA-protein complex, whose abundance was greater with extracts from phenobarbital-treated animals compared with that observed with extracts from untreated animals, was obtained when fragment -178/-217 was used as a probe (Fig. 1A). In contrast, fragment -215/-368 produced a weak pattern of DNA-protein complexes that was qualitatively and quantitatively identical with either extract (data not shown).

The relative abundance, within the two nuclear extracts, of the protein(s) that bound to fragment -178/-217 was determined by incubating the fragment with various amounts of each extract. Image densitometry of the resulting autoradiogram revealed a 4-fold increase in protein binding to fragment -178/-217 with extracts isolated from phenobarbital-treated animals (Fig. 1A). When used as a competitor in the gel retardation assay, a double-stranded synthetic oligonucleotide specifying the sequence between -183 and -199 inhibited the binding of nuclear protein(s) to the -178/-217 fragment.

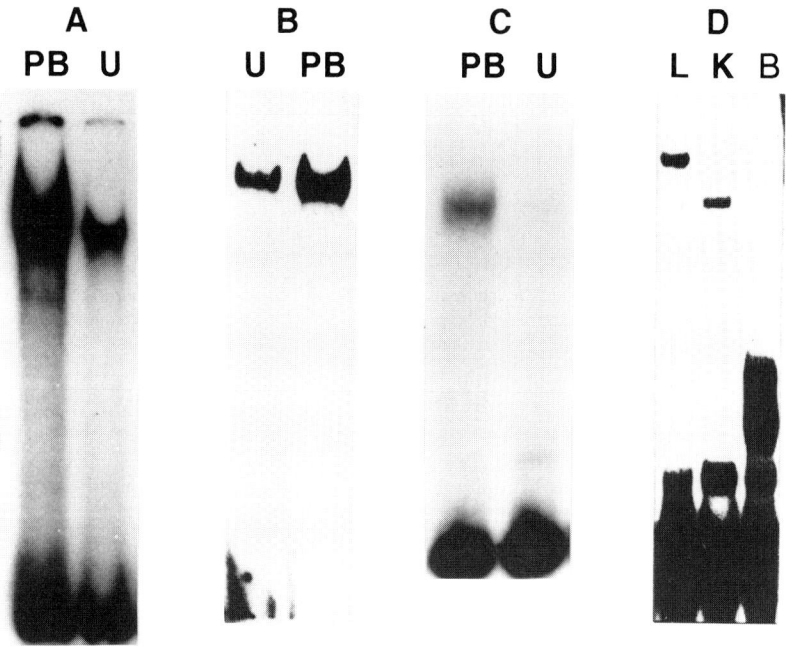

Figure 1 Gel retardation analysis of a CYP2B2 promoter. DNA fragments -178/-217 (A), -183/-199 (B) and -31/-85 (C) were incubated with nuclear extracts isolated from the livers of untreated (U) or phenobarbital-treated (PB) animals. In D the DNA fragment was -178/-217 and the nuclear extracts were isolated from the liver (L), kidney (K) and brain (B) of a phenobarbital-treated animal.

When this 17 bp-oligonucleotide was used as a probe it formed a DNA-protein complex that was about 4-fold more abundant with liver nuclear extracts from phenobarbital- treated rats compared with those from untreated animals (Fig. 1B). Thus, the DNA sequences responsible for binding of a phenobarbital activated or enriched protein to fragment -178/-368 of the promoter are localized in a 17 bp-region between -183 and -199.

In a similar manner, the location, within fragment -31/-177 of the CYP2B2 gene promoter, of the sequence responsible for binding a protein that was enriched or activated in liver nuclear extracts from phenobarbital-treated rats was refined to the region -31/-85 (Fig. 1C). Quantification of the DNA-protein complex obtained using fragment -31/-85 as a probe revealed a 3.8-fold greater abundance with extracts isolated from phenobarbital-treated, compared with untreated, animals.

Fragments -31/-85 and -178/-217, when used as self-competitors in gel retardation assays, were able to abolish completely the binding of protein to their respective radiolabelled probes. However, the binding of protein(s) to fragment -31/-85 was not competed with a 200- fold molar excess of unlabelled fragment -178/-217. Similarly, the binding of protein(s) to fragment -178/-217 was not affected by the presence of a 200-fold molar excess of unlabelled fragment -31/-85. The results of these cross-competition gel retardation assays demonstrate that fragments -31/-85 and -178/-217 interact with different nuclear proteins.

Gel retardation analysis revealed that proteins that bind to the -178/-217 region of the CYP2B2 gene promoter are present also in brain and kidney (Fig. 1D). However, the binding proteins present in these tissues differ from each other and from that found in liver. Thus, the -178/-217 region of the promoter constitutes a binding site for tissue-specific (or enriched) nuclear proteins. The binding protein present in liver is enriched or activated in this tissue, but not in brain or kidney, in response to phenobarbital.

DNase footprinting analysis of the sense strand of a CYP2B2 gene promoter revealed the presence of several protein binding sites. Two of the footprints were more prominent with nuclear protein extracts from phenobarbital-treated, than from untreated, rats. The positions of these footprints corresponded closely to those of the sequences identified by gel retardation as binding proteins enriched or activated in response to phenobarbital. On the template strand of the promoter, a very large footprint, extending from about -250 to -6, was obtained with extracts from phenobarbital-treated rats but not with extracts from untreated animals. The first part of the footprint to develop, between -173 and -198, corresponds closely to the position of one of the sites identified by gel retardation as binding a phenobarbital-enriched protein. Thus, the binding of a protein to this site may be responsible for recruiting other regulatory proteins to adjacent regions of the promoter, hence giving rise to the large footprint.

We have detected, by gel retardation and DNase footprinting, the presence within the CYP2B2 gene promoter of two DNA sequences which bind proteins that are more abundant or more active in liver nuclear extracts from phenobarbital-treated rats. These sequences and the proteins with which they interact may well be involved in mediating the induction of expression of CYP2B2 genes in response to phenobarbital. The -183 to -199 region of the promoter contains an 8 bp-long sequence which differs by only a single nucleotide from the consensus binding site for members of the octamer class of POU domain regulatory proteins (reviewed in Verrijzer & Van der Vliet, 1993). Computer analysis of the -31 to -85 region revealed the presence of a sequence with similarity to a C/EBP binding site (Landschulz *et al.*, 1988). However, the identity of the transcription factors that bind to regions -183 to -199 and -31 to -85 of the promoter remains to be established.

ACKNOWLEDGEMENTS

LMF is the recipient of an Overseas Research Studentship from the Committee of Vice-Chancellors and Principals of the U.K. and thanks the Wellcome Trust for a travel grant to enable her to attend this meeting. The work was supported by a grant from the Cancer Research Campaign, U.K.

# REFERENCES

Atchison, M. & Adesnik, M. (1983): A cytochrome P-450 multigene family : characterisation of a gene activated by phenobarbital administration. *J. Biol. Chem.* 258, 11285-11295.

Burgoyne, L. A., Waqar, M. A. & Atkinson, M. P. (1970): Calcium-dependent priming of DNA synthesis in isolated rat liver nuclei. *Biochem. Biophys. Res. Commun.* 39, 254-258.

Gonzalez, F. J. & Kasper, C. B. (1982): Cloning of DNA complementary to rat liver NADPH-cytochrome c (P-450) oxidoreductase and cytochrome P-450b mRNAs. *J. Biol. Chem.* , 257, 5962-5968.

Gorski, K., Carneiro, M. &Schibler, U. (1986): Tissue-specific *in vitro* transcription from the mouse albumin promoter. *Cell* , 47, 767-776.

Hardwick, J. P., Gonzalez, F. J.& Kasper, C. B. (1983): Transcriptional regulation of rat liver epoxide hydratase, NADPH-cytochrome P-450 reductase, and cytochrome P450b genes by phenobarbital. *J. Biol. Chem.* , 258, 8081-8085.

Hewish, D. R. & Burgoyne, L. A. (1973): The calcium dependent endonuclease activity of isolated nuclear preparations. Relationships between its occurrence and the ocurrence of other classes of enzymes found in nuclear preparations. *Biochem. Biophys. Res. Commun.* , 52, 475 - 481.

Jose-Estanoyl, M., Poliard, A., Foiret, D. & Danan, J.-L. (1989): A common liver-specific factor binds to the rat albumin and α-foetoprotein promoters *in vitro* and acts as a positive trans-acting factor *in vivo*. *Eur. J. Biochem.* 181, 761-766.

Landschulz, W. H., Johnson, P. E.,Adashi, E. Y., Graves, B.J. & McKnight, S. L. (1988): Isolation of a recombinant copy of the gene encoding C/EBP. *Genes Dev.* , 2, 786-800.

Phillips, I. R., Shephard, E. A., Mitani, F. & Rabin, B. R. (1981): Induction by phenobarbital of the mRNA for a specific variant of rat liver microsomal cytochrome P-450. *Biochem. J.* , 196, 839-851.

Phillips, I. R., Shephard, E. A., Bayney, R. M., Pike, S. F., Rabin, B. R., Heath, R. & Carter, N. (1983a): Induction and repression of the major phenobarbital-induced cytochrome P-450 measured by radioimmunoassay. *Biochem. J.* 212, 55 - 64.

Phillips, I. R., Shephard, E. A., Ashworth, A. & Rabin, B. R. (1983b): Cloning and sequence analysis of a rat liver cDNA coding for a phenobarbital-inducible microheterogeneous cytochrome P-450 variant: regulation of its messenger level by xenobiotics. *Gene* , 24, 41-52.

Pike, S. F., Shephard, E. A., Rabin, B. R.& Phillips, I. R. (1985): Induction of cytochrome P-450 by phenobarbital is mediated at the level of transcription. *Biochem. Pharmacol.* 34, 2489-2494.

Shephard, E. A., Phillips, I. R., Pike, S. F., Ashworth, A. & Rabin, B. R. (1982): Differential effect of phenobarbital and β-naphthoflavone on the mRNAs coding for cytochrome P450 and NADPH cytochrome P450 reductase. *FEBS Lett.* , 150, 375-380.

Thomas, P. E., Reik, D. E., Ryan, D. E.& Levin, W. (1981): Regulation of three forms of cytochrome P-450 and epoxide hydrolase in rat liver microsomes: effects of age, sex and induction. *J. Biol. Chem.* , 256, 1044-1052.

Verrijzer, C. P. & Van der Vliet, P. C. (1993): POU domain transcription factors. *Biochim. et Biophys. Acta* , 1173, 1-21.

# Breast cancer aromatase: identification and expression of cytochrome P450 aromatase mRNA in T47D breast cancer cells

Liat Tan[1], Svetlana I. Sadekova[2], Terry Y.-K. Chow[2]

[1]Biochemistry Department, University of Sherbrooke Medical Faculty, Fleurimont, Quebec J1H 5N4. [2]Department of Oncology, Division of Radiation Oncology, Montreal General Hospital, McGill University, Montreal, Que., Canada H3G 1A4

## Summary

*After first establishing through the transformation of [4-$^{14}$C]androstenedione into radioactive estrogens by HPLC-isolation of the latter, that an endogenous steroidal androgen-aromatizing enzyme is indeed present in the ER-positive, breast cancer cell line T47D (1), we analyzed Aromatase cDNA expression in these cells. Total T47D RNA was subjected to Reverse Transcriptase, followed by PCR amplification in the presence of two primers: 5'-AGAAATCTTAAAGAAGATGTCTGG-3' and 5'-GATGGTTTT-GGAAATGCTGA-3'. After 40 cycles, a distinct band of the expected 710 bp size was observed. This DNA product was hybridized with a $^{32}$P-labeled probe: 5'-GCTCTTCTTGAGGATCCCTTTGGAC-3', with a sequence in the region between the two primers. The presence of Aromatase mRNA in T47D cells was confirmed by Northern hybridization analysis. PCR-synthesized fragment corresponding to the 5'-end coding region was extracted, purified and used as a probe in Northern hybridization. Three messengers of, resp., 4.4 kb, 2.4 kb and 600 bp were found. The presence of the 2.4 kb band in non-cancerous tissue, e.g. placenta, is known (2), but that of the 4.4 kb and 600 bp mRNA's is currently not understood. We are presently looking into the possibility of improper RNA processing in breast cancer cells, as suggested earlier by Zhou et al. (3).*

## INTRODUCTION

Although their exact molecular action in the phenotype modification from a normal to a cancerous cell remains to be elucidated, estrogens are generally considered to play a critical role in maintaining the growth of ER-dependent breast cancer cells, once they are formed. In mammalians, the rate-limiting step in estrogen biosynthesis is controlled by P450Aromatase (4). The presence of this enzyme in placental microsomes, ovarian granulose cells, adipose, brain and liver tissue is now well-established (5). Several laboratories have studied the biological function of the aromatizing enzyme system present in breast cancer tissue. Despite some progress that has been achieved in this area, the significance and exact role of *in situ* estrogen biosynthesis in the complex process that ultimately results in the transformation of a healthy breast cell into a malignant one, remains an enduring enigma that has as yet to be clarified. This is even more called for, as thus far no androgen-aromatizing enzyme from a cancerous mammalian source has ever been purified and characterized. Thus, it is not known if any natural aromatase isozyme exists. As of the writing of this article, we are also not aware of the actual isolation in appreciable quantities of an aromatase-active protein through recombinant DNA technology by any laboratory. In a long-term research program focused on the possible involvement of **BCarom** -the locally produced aromatase enzyme in breast cancer cells- in the development and growth of this disease, we have first searched for the endogenous Aromatase-mRNA in the hormone-dependent, human breast cancer cell line T47D. In this communication, we present and discuss the results that we have obtained sofar.

## EXPERIMENTAL

*Reverse Transcriptase-Polymerase Chain Reaction.*
Total RNA (10 µg) was heated at 65°C with upstream primer (sequence indicated below) and subjected to the RT reaction protocol of Gibco-BRL. This was followed by PCR amplification in the presence of two, placental aromatase-derived primers: 5'-AGAAATCTTAAAGAAGATGTCTGG-3' and 5'-GATGGTTTTGGAAATGCTGA-3'. The PCR step (denaturation at 95°C, 1.5 min; annealing at 55°C, 1 min; and extension at 72°C for 3 min) was then carried out for 40 cycles. The DNA product (710 bp) was loaded onto 1% agarose gel, stained with ethium bromide, cut, and purified for use as a probe. To confirm that it was indeed the DNA fragment we expected, it was transferred to Hybon-N membrane and hybridized with a third probe: 5'-GCTCTTCTTGAGGATCCCTTTGGAC-3', which has a sequence derived from the aromatase-coding region between the two primers.

*Northern analysis.*
Total RNA (10-20 µg) was electrophoresed in a 1% agarose gel containing 1xMOPS (4-morpholinepropanesulphonic acid) and 5% formaldehyde (37% w/w). RNA was visualized by staining with ethidum bromide, transferred to Hybon-N membrane, and hybridized with the RT-PCR synthesized, 710 bp fragment, corresponding to the 5'-endterminal of the Aromatase mRNA coding region (2). Prehybridization and hybridization was carried out at 65°C with pBR 322 DNA and $2 \times 10^6$ cpm of 710 bp probe, labeled with random hexamer.

*Measurement of Aromatase activity in T47D cells by radiometric assay.*
The cells were suspended in 50 mM potassium phosphate buffer (pH 7.4), containing 1% EDTA, vortexed briefly and centrifuged for 15 min at 1500g. The assay mixture (final volume 500 µl) contained about 40,000 cpm [$1\beta,2\beta$-$^3$H]androstenedione and a NADPH-generating system. After 5 hr incubation at 37°C, addition of charcoal and centrifugation, 300 µl of the supernatant was counted.

## RESULTS

In contrast to the relatively high aromatase activity found in T47D cells based on extraction, followed by direct counting of the HPLC-purified, $^{14}$C-radioactively labeled estrogens formed (1), the radiometric, tritium-based assay performed with these breast cancer cells gave only a very low activity value (see Table 1). However, we must point out that the first measurement was carried out after 192 hr of steady

**Table 1**. *Endogenous aromatase activity in T47D cells compared to placenta as determined by the incorporation of the tritium label of [$1\beta,2\beta$-$^3$H]Androstenedione into [$^3$H]water.*

|  | Radioactivity (cpm) | | | | |
| --- | --- | --- | --- | --- | --- |
|  | Expt. #1 | Expt #2 | Expt. #3 | Expt. #4 | Mean value |
| Substrate added | 61448 | 61509 | 61538 | 61538 | 61508 |
| Control | 21 | 27 | 45 | 47 | 35 |
| Placenta | 7072 | 7110 | 7164 | 7170 | 7129 (11.59%) |
| T47D cells | 159 | 209 | 276 | 279 | 230 ( 0.37%) |

cell growth in the presence of the radioactive substrate, whereas in the second assay, aromatase activity was measured after a much shorter incubation period. The average amount of radioactive water from four experiments that was produced after 5 hr incubation of the cells, amounted to only 0.37% of the original radioactivity of the [$1\beta,2\beta$-$^3$H]androstenedione substrate. By comparison, with human placental microsome preparations, 11.59% radioactivity was found in the aqueous supernatant after only 20 min incubation. We must assume that at the very low enzyme concentrations in T47D cells, the radiometric aromatase activity assay, which is based on the stereospecific loss of the $1\beta$ and $2\beta$-tritium labels and their incorporation into [$^3$H]water, lacks sensitivity.

The results of Reverse Transcriptase-PCR amplification of Aromatase mRNA, and of Northern blotting of total RNA, extracted from T47D cells, are shown in Figs. 1 and 2, respectively.

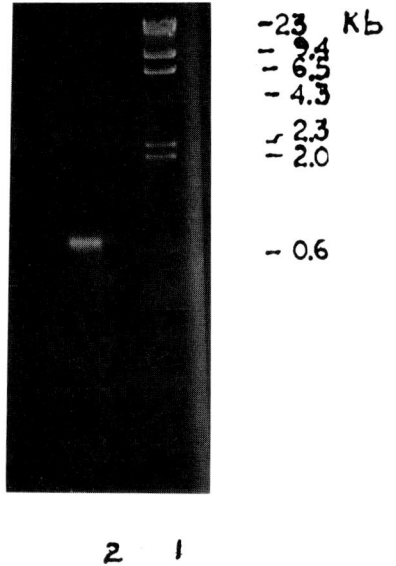 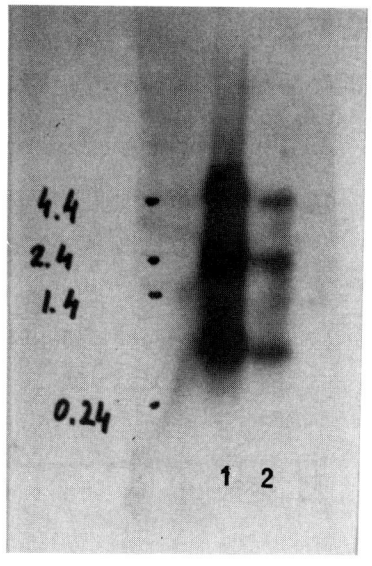

**Figure 1**. BCarom-mRNA detection by RT-PCR.
 1 = λ DNA Hind III digest (Standard)
 2 = 710 bp RT-PCR product from T47D

**Figure 2**. Northern blot hybridization of total T47D RNA.
 Lane 1, 20 µg   Lane 2, 10 µg

## DISCUSSION

Working with a heterogeneous tissue such as the human breast, it would be technically very difficult, if not impossible to distinguish between locally produced aromatase and enzyme, transported from elsewhere. To eliminate this problem, we decided to use for our studies a homogeneous aromatase source, namely T47D breast cancer cells. This cell line was first established from the pleural effusion of a breast cancer patient in Israel (6). In the cytosol of T47D cells, receptors for estradiol, progesterone, and glucocorticoids are present. Consequently, these cancer cells are eminently suitable for our studies. The structure and nature of BCarom, the androgen-aromatizing enzyme that is present in breast cancer tissue, is completely unknown as this enzyme has never been purified from any cancerous source. Thus far, it has been assumed that BCarom is fully identical to the placental aromatase. We have found that there are subtle differences in the $K_m$ and $V_{max}$ values of androstenedione obtained with T47D breast cancer cells and with placental microsomes (Tan, unpublished data). A similar difference in kinetic parameters between these two aromatase sources has also been reported for another important steroid-acting enzyme, namely, Estrone Sulfatase (7). Independent analysis of the cDNA sequence of the placental aromatase by four groups (8-11), all seem to point to the conclusion that there is only a single aromatase protein. However, cDNA sequence analysis does not convey information about any posttranslational modification that may subsequently take place specifically in cancerous breast cells. In a recent elegant and interesting biochemical paper, differentiation of rat ovarian and human placental aromatase has been deduced from tritium isotope distribution analysis with the 2(3)-enol forms of androgen substrates, and also from the formation of distinguishable reaction intermediates (12).

On the basis of our successful use of a Con-A affinity chromatography column, we had in our original 1986 aromatase purification paper already concluded that the human placental aromatase is a glycoprotein (13). Although this was initially disputed at the second Aromatase Conference in Miami (14), our finding has now been definitively confirmed by another laboratory (**15**). In breast cancer cells, BCarom may very well undergo some kind of posttranslational modification which could explain its different kinetic properties. However, until some laboratory is willing to undertake the arduous and possibly nongratifying task of purifying and fully characterizing BCarom, our hypothesis must be considered a conjecture.

We have shown in the preceding that in T47D cells, besides a known 2.4 kb messenger, a larger size 4.4 kb mRNA is also present. We have currently no adequate explanation for the formation of the latter, but we are looking into the possible occurrence of unusual RNA processing in hormone-dependent breast cancer cells.

## ACKNOWLEDGEMENTS.

This work is supported in part by the National Cancer Institute of Canada (TYKC) and by a 1990-91 Medical Research Council of Canada award (to LT). TYKC is a Scholar of the Fonds de Recherches en Santé du Québec.

## REFERENCES

1. Gervais M and Tan L. (1993): 6-Hydroximinoandrostenedione, a New Specific Inhibitor of Estrogen Biosynthesis and its Effect on T47D Human Breast Cancer Cells. Anticancer Res. 13: 383-388.
2. Simpson, E.R., Evans, C.T., Corbin, C.J., Powell, F.E., Ledesma, D. and Mendelson, C.R. (1987): Sequencing of cDNA inserts encoding aromatase cytochrome P-450 (P-450AROM). Mol. Cell. Endocrinol. 52: 267-272.
3. Zhou, D, Pompon, D. and Chen, S. (1990): Stable Expression of Human Aromatase Complementary DNA in Mammalian Cells: A Useful System for Aromatase Inhibitor Screening. Cancer Res. 50, 6949-6954.
4. Tan, L.: Aromatase: Function, Reaction Mechanism and Biological Significance. In Frontiers in Biotransformation, Vol.6, eds. K. Ruckpaul and H. Rein, pp. 63-113. Berlin: Akademie Verlag.
5. Miller, W.R. (1991): Relationship between tumour aromatase activity, tumour characteristics and response to therapy. J. Steroid Biochem. Mol. Biol. 37: 1055-1059.
6. Keydar, J, Chen, L., Karby, S., Weiss, F.R., Delarea, J., Radus, M., Chaitcik, S. and Brenner, H.J. (1979): Establishment and characterization of a cell line of human breast carcinoma origin. Eur. J. Cancer 15: 659-670.
7. Evans, J.T.R., Rowlands, M.G. and Coombes, C.R. (1991): Inhibition of estrone sulfatase enzyme in human placenta and human breast carcinoma. J. Steroid Biochem. Mol. Biol. 39: 493-499.
8. Chen, S., Besman, M., Soarkes, R.S., Zollman, S., Klisak, I., Mohandas, T., Hall, P.F. and Shively, J.E. (1988): Human Aromatase: cDNA Cloning, Southern Blot Analysis, and Assignment of the Gene to Chromosome 15. DNA 7: 27-38.
9. Harada, N. (1988): Cloning of a complete cDNA encoding human aromatase: Immunochemical identification and sequence analysis. Biochem. Biophys. Res. Commun. 156: 725-732.
10. Corbin, C.J., Graham-Lorence, S., McPhaul, M., Mason, J.I., Mendelson, R. and Simpson, E.R. (1988): Isolation of a full-length cDNA insert encoding human aromatase system cytochrome P-450 and its expression in nonsteroidogenic cells. Proc. Natl. Acad. Sci. USA, 85: 8948-8952.
11. Toda, K. Terashima, M., Kawamoto, T., Sumimoto, H., Yokoyama, Y., Kuribayashi, I., Mitsuuchi, Y., Maeda, T., Yamamoto, Y., Sagara, Y., Ikeda, H. and Shizuta, Y. (1990): Structural and functional characterization of human aromatase *P*-450 gene. Eur. J. Biochem. 193: 559-565.
12. Swinney, D.C., Watson, D.M. and So, O-Y. (1993): Accumulation of Intermediates and Isotopically Sensitive Enolization Distinguish between Aromatase (Cytochrome P450 CYP19) from Rat Ovary and Human Placenta. Arch. Biochem. Biophys. 305: 61-67.
13. Tan, L. and Muto, N. (1986): Purification and reconstitution properties of human placental aromatase. A cytochrome P-450-type monooxygenase. Eur. J. Biochem. 156: 243-250.
14. Hall, P.F. (1987) Steroids 50: 321 (in Panel Discussion section).
15. Sethumadhavan, K., Bellino, F.L. and Thotakura, N.R. (1991): Estrogen Synthetase (aromatase). The cytochrome P-450 component of the human placental enzyme is a glycoprotein. Mol. Cell. Endocrinol. 78: 25-32.

# Spectral map analysis of cytochrome P450 isoenzymes

Luc M.H. Koymans, Henri Moereels, Paul J. Lewi, Paul A.J. Janssen

*Janssen Research Foundation, Turnhoutseweg 30, B-2340 Beerse, Belgium*

Most European languages make use of an alphabet of 26 letters to form words that can be combined in numerous ways to create sentences that may contain some kind of message. In a similar way, nature uses the alphabet of life, i.e. the 20 naturally occurring amino acids, in numerous combinations to form amino acid sequences that code for proteins with a particular function. In this way, amino acid sequences can be envisioned as the language of mother nature. Unraveling these hieroglyphics of life should give us some insight into the structure and mode of actions of proteins.

In this respect cytochromes P450 are a thankful object of study as hundreds of amino acid sequences of these enzymes are known to date (Nelson et al., 1993). The P450s are thought to have diverged from a common ancestor sequence more than 3 milliard years ago. The sequence divergence is caused by random substitution in the DNA template of the enzymes. The number of substitutions, accumulated over time, in the nucleotide sequence of DNA is reflected in the dissimilarity with other sequences that evolved parallel from the common anchestor.

The (dis)similarities in amino acid sequence between 280 cytochrome P450 enzymes have been determined pairwise by use of CGEMA and VGAP (Colour Graphics Editor for Multiple Alignment using a variable GAP penalty; Moereels et al., 1990). As a result we obtain a symmetrical table (dimensions 280X280) containing no less than 39060 (280X279/2) amino acid identities (Table 1).

Although this elaborate table contains all the relations between the sequences it is very hard to interpret and lacks a comprehensible representation of the information present. In order to overcome this inconvenience we made use of spectral map analysis (SMA) which is a factorial method of data analysis designed for graphical analysis of contrasts (Lewi, 1989). In this application the result is similar to that of Principal Coordinate Analysis (Gower, 1967).

**Table 1.** Schematic table containing a total of 39060 mutual amino acid identities between 280 P450 sequences

|  | P450 1A1 (rat) | P450 1A1 (human) | P450 1A1 (rabbit) | ...up to... | P450 108 (Ps.spp.) | P450 109 (Bac. subt.) | P450 110 (Ana. spp.) |
|---|---|---|---|---|---|---|---|
| P450 1A1 (rat) |  | 79% | 75% |  | 17% | 20% | 25% |
| P450 1A1 (human) |  |  | 75% |  | 17% | 19% | 24% |
| P450 1A1 (rabbit) |  |  |  |  | 16% | 17% | 25% |
| ... |  |  |  |  |  |  |  |
| up to |  |  |  |  |  |  |  |
| ... |  |  |  |  |  |  |  |
| P450 108 (Ps. spp.) |  |  |  |  |  | 30% | 21% |
| P450 109 (Bac. subt.) |  |  |  |  |  |  | 20% |
| P450 110 (Ana. spp.) |  |  |  |  |  |  |  |

To illustrate the usefulness of SMA we consider the case of a tourist going to the tourist office in Lissabon to obtain information about the geographical position of interesting cities in Portugal. Instead of the expected map, the tourist gets a table of distances between the major cities in Portugal (Table 2).

**Table 2.** Table of interdistances (km) between 10 major cities in Portugal

| | Lisboa | Porto | Coimbra | Elvas | Faro | Beja | Evora | Vila Real | Pinhel |
|---|---|---|---|---|---|---|---|---|---|
| Lisboa | | | | | | | | | |
| Porto | 275 | | | | | | | | |
| Coimbra | 177 | 107 | | | | | | | |
| Elvas | 174 | 282 | 182 | | | | | | |
| Faro | 217 | 465 | 356 | 218 | | | | | |
| Beja | 136 | 356 | 248 | 114 | 111 | | | | |
| Evora | 110 | 294 | 187 | 73 | 172 | 62 | | | |
| Vila Real | 310 | 75 | 134 | 272 | 475 | 364 | 303 | | |
| Pinhel | 289 | 136 | 130 | 210 | 421 | 314 | 255 | 81 | |
| Viseu | 239 | 82 | 66 | 206 | 404 | 293 | 231 | 73 | 72 |

Naturally, the tourist protests and wants his money back. However, the salesman persists and claims that he delivered what the tourist asked for. And the salesman is right: all the information the tourist needs is in the table, only in an incomprehensible way. The tourist can obtain a 'normal' map of Portugal just by analyzing Table 2 using SMA. The result is depicted in Fig. 1.

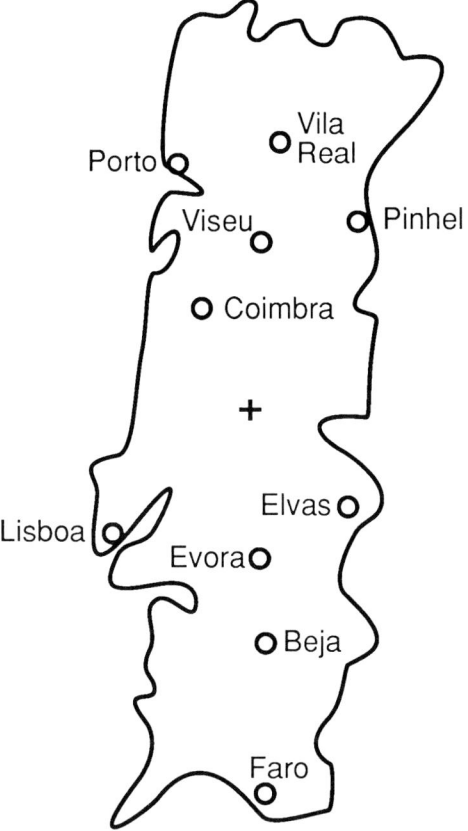

Fig. 1. Graphical result of spectral map analysis of the 10 X 10 matrix of interdistances between 10 major cities in Portugal (Table 2).

In a similar way SMA can be used to visualize the giant table of similarities between the 280 P450 amino acid sequences, which can be regarded as measures of interdistance. The same algorithm for latent variable extraction which we applied to the distances in Table 2 produced the coordinates of the 280 cytochromes P450 along the two principal látent variables. In first instance the "P450-map" contained 280 sequences. To clarify the view, P450 sequences belonging to the same P450 (sub)family were averaged providing that the overall picture of the map was not changed. The resulting "P450-map" contains 35 cytochrome P450 (sub)families and is depicted in Fig. 2. Each point on the "P450-map" in Fig. 2 represents a cytochrome P450 (sub)family. Looking at the "P450-map" it can be observed that several P450 (sub)families appear as isolated points. Cytochromes P450 belonging to family 1A (the average of all known 1A1 and 1A2 sequences) are separated from all the other (sub)families. In the lower right-hand part of the map all mitochondrial P450 families (11A, 11B, 24 and 27) are grouped together. In the left-hand part of the map the xenobiotica-metabolizing P450s of family 2 are localized. In the upper part subfamilies 4A and 4B (metabolism of fatty acids) together with P450 3A are found. The other P450 (sub)families are grouped between these boundaries.

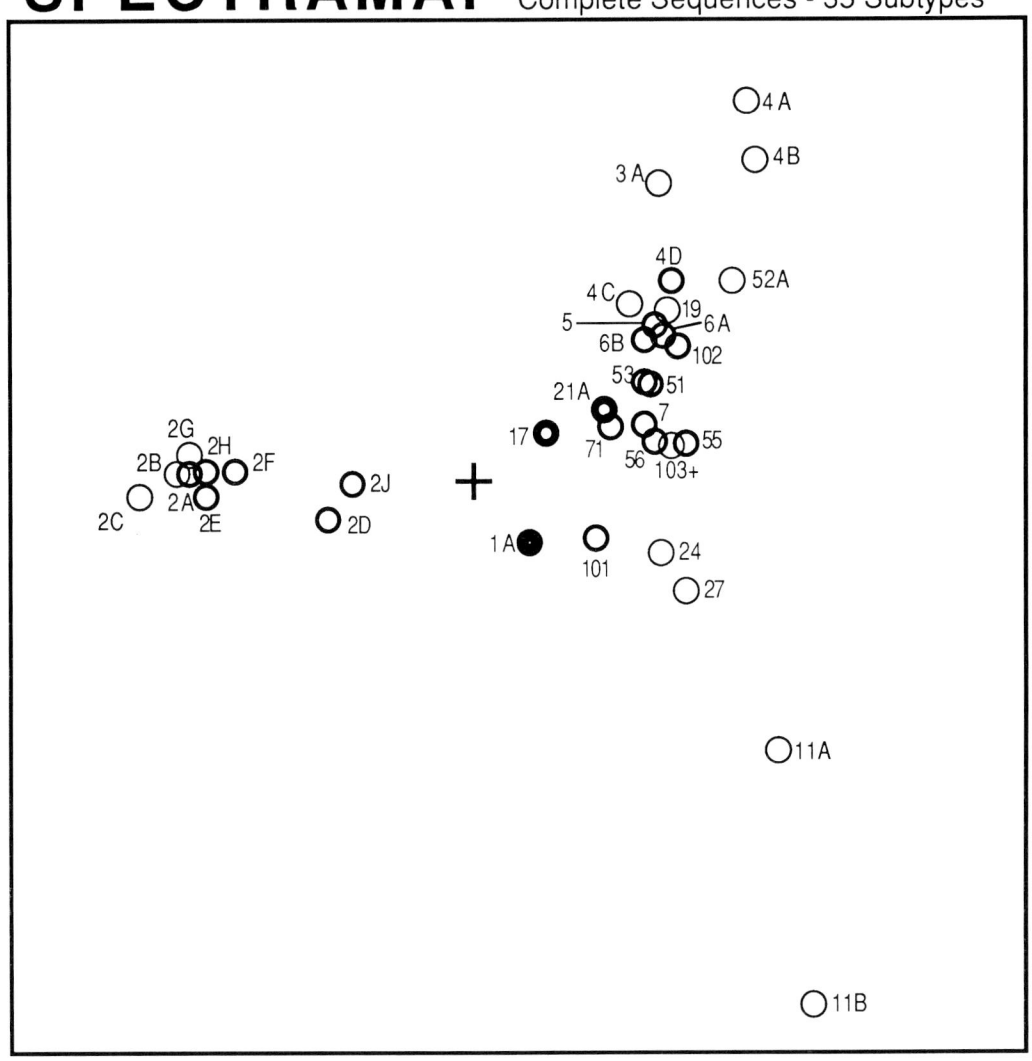

Fig. 2. Graphical result of spectral map analysis of the 280 X 280 matrix of (dis)similarities between 280 cytochrome P450 sequences (Table 1) as determined by the VGAP algorithm (Moereels et al., 1990).

In the "tourist-map" 100% of the total variance can be explained by the two principal latent variables which in fact represent the geographical coordinates (distances in Portugal are short in comparison to the radius of the earth, such that spherical effects can be neglected). In contrast, in the "P450-map" only 42% of the information in Table 1 can be represented by the first three latent variables. The third latent variable (9% of the total variance) is visualized in the "P450-map" in Fig. 2 by the thickness of the points. Points with a fat outline are situated above the plane of the paper and points with a thin outline below. P450 family 1A for example is found far above the plane of the paper and the mitochondrial P450 families 11A, 11B, 24 and 27 are located under the plane.

It can be concluded that spectral map analysis is a suitable method to visualize the relationships between all cytochrome P450 (sub)families. Currently, similar analyses are undertaken for particular regions of the P450 sequences, for example the I-helix and L-helix. Preliminary results on the I-helix suggest that the "P450-I-helix-map" contains some information about the relationship between amino acid sequence and substrate selectivity.

REFERENCES

Gower, J.C. (1967): Multivariate analysis and multidimensional geometry. *The Statistician* 17, 13-28.

Lewi, P.J. (1989): Spectral map analysis: factorial analysis of contrasts, especially from log ratios. *Chemometrics Intelligent Lab. Syst.* 5, 105-116.

Moereels, H., De Bie, L. and Tollenaere, J. (1990): CGEMA and VGAP: a color graphics editor for multiple alignment using a variable gap penalty. Application to the muscarinic acetylcholine receptor. *J. Comp.-Aided Mol. Design* 4, 131-145.

Nelson, D.R., Kamataki, T., Waxman, D.J., Guengerich, F.P., Estabrook, R.W., Feyereisen, R., Gonzales, F.J., Coon, M.J., Gunsalus, I.C., Gotoh, O., Okuda, K. and Nebert, D.W. (1993): The P450 superfamily: update on new sequences, gene mapping, accession numbers, early trivial names of enzymes, and nomenclature. *DNA Cell. Biol.* 12, 1-51.

# Oxidative dealkylation of N,N-dialkylamides

Luis Constantino[1], Jim Iley[2]

[1]Faculdade de Farmácia de Lisboa, Av das Forças Armadas, 1600 Lisboa Portugal.
[2]Chemistry Department, The Open University, Milton Keynes, MK7 6AA, UK

INTRODUCTION

A variety of N,N-dialkylamides have important biological activity. For example N,N-diethyl-3-toluamide is the most effective insect repellent in world-wide use, N,N-diethylnicotinamide is a respiratory stimulant and N,N-dimethylformamide is an industrially important solvent that is toxic on occupational exposure. All these compounds are known to undergo oxidative N-dealkylation by *in vivo* metabolism by the liver. We are currently studying the mechanism of this N-dealkylation reaction using substituted N,N-dialkylbenzamides as our model substrates. The microsomal dealkylation of N,N-dialkylbenzamides follows Michaelis-Menten kinetics, from which $V_{max}$ and $V_{max}/K_m$ constants can be obtained. We have used a combination of substituent, kinetic deuterium isotope effects, N-alkyl structural effects on the kinetic constants and preference of product formation, to probe which of the possible pathways is followed.

MATERIALS AND METHODS

Microsomal incubations and analysis were done according to previous published methods (Constantino *et al.*,1992). AM1 SCF Molecular orbital calculations were done using MOPAC 4.0 QPCE 455 (Indiana University).

RESULTS

The incubation of N,N-dimethylbenzamide with rat liver microsomes produced N-methylbenzamide and formaldehyde. An intermediate has been isolated and identified as N-hydroxymethyl-N-methylbenzamide by comparison (mass spectrometry, HPLC retention time and diode array spectra) with an authentic standard synthesised and characterised in our laboratory. The same pattern of oxidation was present in the microsomal metabolism of all the N,N-dimethylbenzamides studied and results in the formation of unstable N-hydroxymethyl-N-methylbenzamides witch subsequently decompose to the corresponding N-methylbenzamides and formaldehyde.

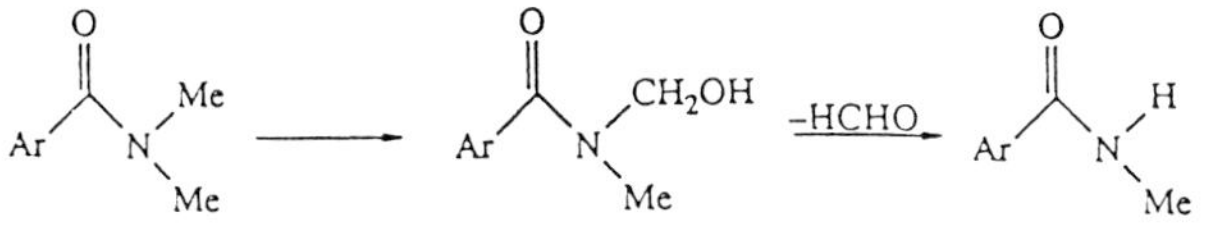

As for the analogous N,N-dimethylamines,(Miwa et al. 1983) two mechanisms that can account for the formation of the N-hydroxyalkyl metabolite are hydrogen atom transfer to form a carbon centred radical or electron transfer from the substrate to form a radical cation .If the carbon centred radical is formed it may lose an electron to give an iminium ion or interact directly with the appropriate haem hydroxy complex to form the product. If the reaction goes via the radical cation, it may lose an hydrogen atom to form an iminium ion, or a proton to form the carbon centred radical.

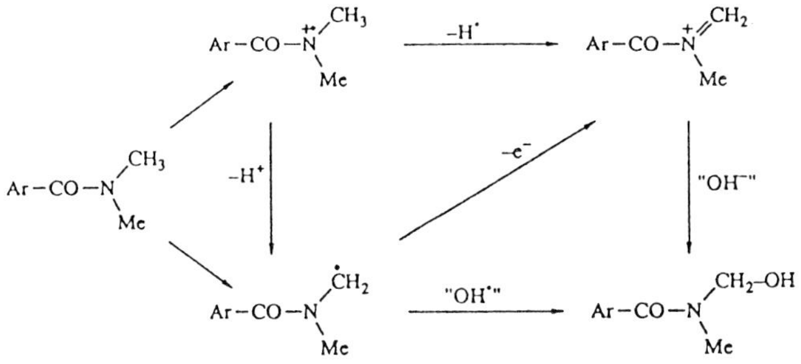

Scheme 1. Pathways for the microsomal oxidation of $N,N$-dimethylbenzamides to $N$-hydroxymethyl-$N$-methylbenzamides.

Values of $V_{max}$ and $V_{max}/K_m$ for different N,N-dimethylbenzamides and for N,N-bis(trideuteriomethyl)benzamide are shown in table 1. Clearly neither of the kinetic constants correlate with either the Hammett σ parameter or the ionisation potential of the amides. Comparation between the kinetic constants of N,N-dimethylbenzamide and N,N-bis(trideuteriomethyl)benzamide reveal that there is no intermolecular isotope effect in the reaction. However in the compound N-methyl-N-trideuteriomethylbenzamide the methyl group is metabolised faster than the $CD_3$ group giving an intramolecular isotope effect of 6 ± 0.3 in the reaction.

Table 1 - Ionisation potentials and kinetic parameters for microsomal oxidation of the benzamides $ArCON(CD_3)_2$

| Ar | $10^3 V_{max}$ (mol/dm³.h) | $10^3 K_m$ (mol/dm³) | $V_{max}/K_m$ (1/h) | IP (e.v.) |
|---|---|---|---|---|
| 4-MeO-$C_6H_4$ | 1.64 ± 0.10 | 1.42 ± 0.24 | 1.15 | 9.26 |
| Ph | 3.31 ± 0.10 | 2.49 ± 0.19 | 1.33 | 9.55 |
| Ph | 3.70 ± 0.26 [a] | 3.84 ± 0.57 [a] | 0.97 [a] | |
| 4-Cl-$C_6H_4$ | 3.55 ± 0.31 | 2.21 ± 0.56 | 1.60 | 9.66 |
| 4-$NO_2$-$C_6H_4$ | 2.05 ± 0.25 | 1.85 ± 0.63 | 1.11 | 10.01 |

In the microsomal oxidation of N-methyl-N-alkylbenzamides only N-hydroxymethyl-N-alkyl intermediates are detected. The N-alkyl group gives directly the dealkylated product. Similarly in the oxidation of dialkylbenzamides only dealkylated products are observed. $V_{max}$ values for total oxidation (demethylation + dealkylation) are generally independent of the alkyl group and are ca. 0.6 x10-3 mol/dm³·h per metabolisable hydrogen atom (except for substrates with branched alkyl groups). $V_{max}/K_m$ is however dependent on the lipophilicity of the substrate (Figure 1)

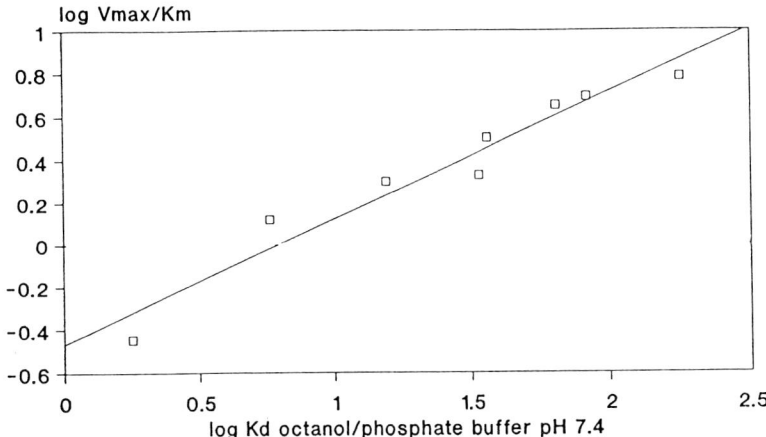

Fig 1. Plot of total dealkylation $V_{max}/K_m$ values versus log Kd octanol/phosphate buffer pH 7.4 for the N-methyl-N-alkylbenzamides.

This result coupled with the lack of intramolecular isotope effect, implies that the C-H bond breaking step is masked by other processes, probably diffusion of the substrate into the enzyme active site. Since our method of analysis can measure independently $V_i$ for dealkylation and demethylation of the same substrate, kinetic constants can be obtained for the amides in which R1 and R2 are different (table 2).

Table 2 - Kinetic parameters for the demethylation and dealkylation of the benzamides $PhCON(R^1)R^2$

| $R^1$ | $R^2$ | Demethylation | | Dealkylation | |
|---|---|---|---|---|---|
| | | $10^3 V_{max}$ (a) | $V_{max}/K_m$ (a) | $10^3 V_{max}$ (a) | $V_{max}/K_m$ (a) |
| Me | Me | 0.55 | 0.21 | 0.55 | 0.21 |
| Me | Et | 0.67 | 0.59 | 0.65 | 0.25 |
| Me | Bu | 0.6 | 1.33 | 0.55 | 1.11 |
| Me | $Pr^i$ | 0.77 | 1.27 | 0.06 | 0.06 |
| Me | $Pr^c$ | 0.60 | 0.73 | 0.00 | 0.00 |
| Me | $Pr^cMe$ | 0.37 | 2.23 | 0.28 | 0.27 |
| Me | Al | 0.18 | 0.90 | 0.89 | 1.57 |

(a) corrected for the number of metabolisable Hidrogen atoms. $V_{max}$ (mol/dm³.h) $V_{max}/K_m$ (1/h)
$Pr^i$ -isopropyl · $Pr^c$ -cyclopropyl, $Pr^cMe$ -cyclopropylmethyl, Al - allyl.

$V_{max}$ for demethylation are essentially constant and are not affected by the nature of the other N-alkyl group but $V_{max}$ for dealkylation are very sensitive to the nature of the alkyl group. Branched alkyl groups exhibit reduced rate of metabolism specially when branching is in the α carbon. $V_{max}/K_m$ values for demethylation correlate well with lipophilicity of the molecule whereas $V_{max}/K_m$ for dealkylations have no clear dependence of steric or lipophilic factors. Theses results

$V_{max}/K_m$ for dealkylations have no clear dependence of steric or lipophilic factors. Theses results may be explained by the presence of a hydrophobic pocket in the enzyme into which the alkyl group sits preferentially presenting the methyl group to haem.

The kinetic deuterium isotope effects suggests a mechanism involving hydrogen atom abstraction and the absence of a substituent effect points to the exclusion of the radical cation as an intermediate. Theoretical calculations show that the ionisation potential of the amides is independent of the alkyl group and whether or not the reaction proceeds via the carbon centred radical or the iminium ion, metabolism of the N-alkyl group should be favoured. Our results show that kinetic factors are usually very important in reaction control. However in two substrates thermodynamic factors may play an important role. In N-methyl-N-cyclopropylbenzamide the cyclopropyl group is not metabolised and in N-methyl-N-allylbenzamide the N-alkyl group is metabolised faster than the N-methyl. The absence of metabolism of the cyclopropyl group cannot be explained in terms of steric effects since for N-methyl-N isopropylbenzamide the N-isopropyl group is metabolized. Rather, it is more likely a result of the difficult formation of a cyclopropyl radical as compared to a methyl radical (Walborsky 1981). For N-methyl-N-allylbenzamide AM1 calculations reveal that the ionisation potential, the $\Delta\Delta H$ for the formation of the radical cation, the $\Delta\Delta H$ for the formation of iminium ion and $\Delta\Delta H$ for the formation of the radical from the methyl group are about the same as for other amides. However $\Delta\Delta HH$ for the formation of the N-allyl radical is much more favoured than alkyl radical formation from the other amides.

The relative rates at which one of the two alkyl groups in an N-methyl-N-alkylbenzamide are metabolised are function not only of the rate constant for the reaction (linked to the stability of the intermediate) but also of its relative concentration ( fraction of the substrate that approaches the active site in the required conformation). Judging from the results of N-methyl-N-alkyl benzamides with saturated acyclic N-alkyl groups, the amide approaches the active site at a conformation that favours metabolism of the smallest group. However in a substrate like N-methyl-N-allylbenzamide the stability of the N-allyl radical could result in a higher rate constant for metabolism of the allyl group. A higher rate constant can result in higher amounts of product in spite of lower availability of the required substrate conformation. In N-methyl-N-cyclopropylbenzamide the low stability of the intermediate toghether with the low availability of the required conformation could result in a very low reaction rate for the reaction. Further work is currently in hand to intercept the carbon centred radical and to investigate the stereochemistry of alkyl group loss.

REFERENCES

Constantino, L., Rosa, E., Iley, J. (1992): The microsomal demethylation of N,N-dimethylbenzamides: Substituent and kinetic deuterium isotope effects. *Biochem. Pharmacol.* **44**, 651-658

Miwa, G.T., Walsh, J.S., Kedderis, G.L., Hollenberg, P.F. (1983): The use of intramolecular isotope effects to distinguish between deprotonation and hydrogen atom abstraction mechanisms in cytochrome P-450 and peroxidase catalysed N-demethylation reactions. *J. Biol. Chem.* **258**, 14445-14449.

Walborsky, H.M. (1981): The cyclopropyl radical.*Tetrahedom*, **37**, 1625-1651.

Acknowledgements - We would like to thank JNICT (Portugal for the award of grant BD 1061/ID to Luis Constantino

# Influence of isoamyl alcohol and ethanol on the oesophageal and hepatic cytochrome P450 metabolism of nitrosamines and their carcinogenicity

L.F. Ribeiro Pinto, Peter F. Swann

CRC Nitrosamine-Induced Cancer Group, Department of Biochemistry and Molecular Biology, University College, London, UK

Throughout the Western world and in Africa alcohol consumption is associated with an increased risk of oesophageal cancer. Although an increase of these tumours is associated with all alcoholic beverages, the greatest risk is associated with certain spirits, notably the apple brandy Calvados which is made and drunk in Normandy, France (Tuyns et al., 1979). Calvados contains an unusually large concentration of long, usually branch, chain alcohols, in particular isoamyl alcohol, derived from the breakdown of aminoacids. It has been suggested ( see Craddock, 1993 for a review) that these alcohols are the causative factor in the association between Calvados and oesophageal cancer. Ethanol and these branched chain alcohols are not carcinogenic thus it seems likely that their influence is actually on some other carcinogenic insult. It has long been suspected that this primary insult is exposure to nitrosamines.

Nitrosamines are the only carcinogens known to produce oesophageal cancer in animals and exposure to these chemicals is ubiquitous, thus making it more difficult to establish a direct link between exposure to these carcinogens and development of human tumours. Nevertheless, a recent study has shown the direct correlation between endogenous formation of nitrosamines and oesophageal cancer in China (Wu et al., 1993), a country which has areas with very high incidences of the disease. Nitrosamines are precarcinogens and they require metabolic activation carried out by cytochrome P450 in order to exert their tumorogenic effect. After P450 mediated hydroxylation at the $\alpha$-carbon, there is formation of an alkyldiazonium ion which has a very short half-life, thus limiting it from travelling through the body and exerting its effect in cells far from the original activation site. The final carcinogen is an alkylating species which will react with macromolecules, in particular with specific sites of the bases in DNA in order to produce tumours. A general feature of nitrosamines is their remarkable organ specificity as carcinogens, which depends on the distribution of the nitrosamine inside the body, the organ distribution of the P450(s) responsible for the metabolic activation of the nitrosamine and the susceptibility of each individual tissue to the DNA damage inflicted by the metabolites of the nitrosamine. So, the carcinogenic potential of a determined nitrosamine depends not only on the exposure to this nitrosamine, but also to the balance of its metabolism among the target organs.

Ethanol has been shown to change the organotropism and the carcinogenicity of a number of nitrosamines (Gibel, 1967; Griciute et al., 1981; Aze et al., 1993), often with an increase of their carcinogenicity towards extrahepatic organs. Swann et al (Swann et al., 1984) have shown that this effect of ethanol on nitrosamine metabolism and carcinogenicity is probably mediated through changes in their pharmacokinetics in a study showing that when moderate amounts of ethanol are given together with N-nitrosodimethylamine and N-nitrosodiethylamine, it produced a dramatic change in the distribution and metabolim of these nitrosamines, resulting in an increase in exposure of extrahepatic organs to them. While the effect of ethanol on N-nitrosodimethylamine was through inhibition of first-pass clearance, the effect on N-nitrosodiethylamine was through *selective* inhibition of its metabolism in the liver, whilst metabolism in the oesophagus was unaffected. This shows that the metabolism of this nitrosamine is carried out by different P450s in these two organs. P450 2E1 is the main enzyme, but not the only one (Heath, 1962), involved in nitrosodiethylamine metabolism in the liver (Yoo *et al.*, 1990), while the P450(s) responsible for the metabolism of this nitrosamine in the oesophagus remains unknown. The high activity of the oesophageal microsomal fraction in metabolizing N-nitrosomethylbenzylamine, a nitrosamine which predominantly produces oesophageal tumours in rats, compared to its metabolism by the hepatic microsomal fraction (Labuc and Archer, 1982) also suggested that different cytochrome P450(s) are responsible for the activation of this nitrosamine in these two organs. The P450(s) responsible for the metabolism of this nitrosamine in the oesophagus is also not known, although in the liver the participation of members of the 2B family seems possible (Yang *et al.*, 1985; Lee *et al.*, 1989). Several other studies examining the influence of ethanol on nitrosamine metabolism and pharmacokinetics (Wiestler *et al.*, 1987; Ludeke *et al.*, 1991) showed similar effects as for N-nitrosodiethylamine and have reinforced the view that different P450s are responsible for the liver metabolism and for the oesophageal metabolism of a number of nitrosamines.

There has been only one previous short report on the effect of isoamyl akcohol on nitrosamine metabolism (Craddock and Henderson, 1991). This showed that isoamyl alcohol, unlike ethanol was able to inhibit the *in vitro* metobolism of N-nitrosomethylbenzylamine by oesophageal and hepatic microsomes. To investigate this further we have looked at the effect of isoamyl alcohol, with and without concurrent administration of ethanol, on the metabolic activation and damage to DNA by nitrosodiethylamine in whole rats. Administration of 5% ethanol increased the ethylation of oesophageal DNA by nitrosodiethylamine by about 2 fold, while decreased the ethylation of liver and kidney DNA when compared to rats which received tap water. Administration of 2.8% isoamyl alcohol in water reduced the ethylation of DNA caused by nitrosodiethylamine by about 30% in all organs, when compared to rats which received the nitrosamine in tap water. However concurrent administration of a mixture containing 5% ethanol and 2.8% isoamyl alcohol unexpectdely produced the same increase in ethylation as had been seen when ethanol was administered alone.

To explain these observations we have investigated the effect of isoamyl alcohol on the metabolism of nitrosodiethylamine *in vitro*. Isoamyl alcohol (1.6 mM) inhibited the *in vitro* oesophageal and hepatic metabolism of nitrosodiethylamine, and although the Ki could not be accurately calculated, it was about the same for both organs (about 0.4 mM). So, there appears to be a lack of selectivity in the inhibitory effect of isoamyl alcohol on the metabolism of nitrosamines in the two organs, and that does not support its involvement in increasing nitrosamine related oesophageal tumours.

As a first step towards identification of tha actual P450(s) involved in the metabolism of nitrosamines in the oesophagus, we have looked at the ability of isoamyl alcohol and ethanol to inhibit various P450 mediated reactions. Both alcohols strongly inhibited N-nitrosodimethylamine demethylase, a reaction catalyzed exclusively by P450 2E1, and had no effect on the deethylation of ethoxyresorufin, a reaction catalyzed exclusively by P450 1A1. Isoamyl alcohol was more efficient than ethanol in inhibiting the depentylation of pentoxyresorufin (P450 2B1/B2), the deethylation of ethoxycoumarin (P450 1A, 2B and others) and the 7-hydroxylation of coumarin (unknown) by liver hepatic microsomes. Ethanol was more efficient than isoamyl alcohol in inhibiting the demethylation of aminopyrene (P450 2B and others). So, there is cleary a difference between these two alcohols in their inhibitory effect on reactions catalyzed by different P450s, although at the moment is not possible to establish any link between the P450s involved in these reactions and the P450(s) responsible for the metabolism of N-nitrosodiethylamine in the oesophagus.

The difference in the metabolism of nitrosamines seen with the oesophagus and the liver does not support any involvement of induction of P450 2E1 by ethanol in the oesophagus for the effect of ethanol on nitrosamine metabolism and carcinogenicity and for its association with a high incidence of oesophageal cancer. However the possible inducing effects of isoamyl alcohol in this are, at present unknown but it has been shown that isoamyl alcohol induces P450 2H1/H2 in cultured chick hepatocytes and increases the induction by ethanol of P450 2B1/B2 in cultured rat hepatocytes (Louis *et al.*, 1993).

This work was generously supported by the Cancer Reseach Campaign.

Aze, Y., Toyoda, K., Furukawa, F., Mitsumori, K. and Takahashi, M. (1993). Ehancing Effect of Ethanol on Esophageal Tumor Development in Rats by Initiation of Diethylnitrosamine. *Carcinogenesis*, **14**: 37-40.

Craddock, V. M. (1993). Cancer of the Oesophagus. Cambridge, Cambridge University Press.

Craddock, V. M. and Henderson, A. R. (1991). Potent Inhibition of Oesophageal Metabolism of N-nitrosomethylbenzylamine, an Oesophageal Carcinogen, by Higher Alcohols Present in Alcoholic Bevarages. Relevance to Human Cancer of N-Nitroso Compounds, Tobacco Smoke and Mycotoxins. Lyon, International Agency for Research on Cancer. 564-567.

Gibel, W. (1967). Experimentelle Untersuchungen zur Syncarcinogenese beim Oesophaguscarzinom. *Arch. Geschwulstforsch.*, **30**: 181-189.

Griciute, L., Castegnaro, M. and Bereziat, J.-C. (1981). Influence of Ethyl Alcohol on Carcinogenesis with N-nitrosodimethylamine. *Carcinogenesis*, **4**: 821-825.

Heath, D. F. (1962). The Decomposition and Toxicity of Dialkylnitrosamines in Rats. *Biochemical Journal*, **85**: 72-91.

Labuc, G. E. and Archer, M. C. (1982). Esophageal and hepatic microsomal metabolism of N-nitrosomethylbenzylamine and N-nitrosodimethylamine in the rat. *Cancer Res.*, **42**: 3181.

Lee, M., Ishizaki, H., Brady, J. F. and Yang, C. S. (1989). Substrate specificity and alkyl group selectivity in the metabolism of N-nitrosodialkylamines. *Cancer Res.*, **49**: 1470.

Louis, C. A., Sinclair, J. F., Wood, S. G., Lambrecht, L. K., Sinclair, P. R. and Smith, E. L. (1993). Synergistic Induction of Cytochrome P-450 by Ethanol and Isopentanol in Cultures of Chick Embryo and Rat Hepatocytes. *Toxycology and Applied Pharmacology*, **118**: 169-176.

Ludeke, B., Meier, T. and Kleihues, P. (1991). Bioactivation of Asymmetric N-dialkylnitrosamines in Rat Tissues Derived from the Ventral Entoderm. Relevance to Human Cancer of N-Nitroso Compounds, Tobacco Smoke and Mycotoxins. Lyon, International Agency for Research on Cancer. 286-293.

Swann, P. F., Coe, M. A. and Mace, R. (1984). Ethanol and dimethylnitrosamine and diethylnitrosamine metabolism and disposition in the rat. Possible relevance to the influence of ethanol on human cancer incidence. *Carcinogen.*, **5**: 1337.

Tuyns, A. J., Pequignot, G. and Abbatucci, J. S. (1979). Oesophageal Cancer and Alcohol Consumption; Importance of Type of Beverage. *International Journal of Cancer*, **23**: 443-447.

Wiestler, O. D., von Deimling, A., von Hofe, E., Schmerold, I., Wiestler, E. and Kleihues, P. (1987). Interorgan Shift of Nitrosamine Metabolism by Dietary Ethanol. *Arch. Toxicol. Suppl.*, **11**: 53-65.

Wu, Y., Chen, J., Ohshima, H., Pignatelli, B., Boreham, J., Campbell, T. C., Peto, R. and Bartsch, H. (1993). Geographic Association Between Urinary Excretion of N-nitroso Compounds and Oesophageal Cancer Mortality in China. *International Lournal of Cancer*, **54**: 713-719.

Yang, C. S., Tu, Y. Y., Koop, D. R. and Coon, M. J. (1985). Metabolism of nitrosamines by purified rabbit liver cytochrome P-450 isozymes. *Cancer Res.*, **45**: 1140.

Yoo, J.-S. H., Ishizaki, H. and Yang, C. S. (1990). Roles of CYP2E1 in the dealkylation and denitrosation of N-nitrosodimethylamine and N-nitrosodiethylamine in rat liver microsomes. *Carcinog.*, **11**: 2239.

# Regulation of the rat liver and lung CYP1A expression by different inducers

Lyudmila F. Gulyaeva, Alevtina Yu. Grishanova, Evgeny V. Petchkovski, Vyacheslav V. Lyakhovich

Institute of Molecular Pathology and Ecological Biochemistry, Novosibirsk 630117, Russia

## Summary

*Using specific oligonucleotide probes for CYP1A1 and 1A2 we have estimated the mRNA level in lung and liver of Wistar rats treated with Aroclor 1254 (AR) and 3-methylcholanthrene (MC). Our results showed that the mRNA level for CYP1A considerably increased in liver of rats treated by these inducers. Significant differences in the expression of CYP1A were revealed in lung. While the mRNA level for CYP1A1 in MC- and AR-lung was rather high, its level for CYP1A2 increased during AR-treatment only. These data are in good accordance with those on the catalytic activity of these P450s in the metabolism of specific substrates. The increase of the methoxyresorufin-O-demethylase (MROD) activity in AR-lung was observed whereas it was not seen in MC-lung. The ethoxyresorufin-O-deethylase (EROD) activity was very high in liver and rather high in lung of AR- and MC-treated rats. The obtained results suggest that the CYP1A expression in liver and lung is regulated at the transcriptional level and is inducer-dependent.*

## Introduction

By the present time the important role of CYP1A in chemical carcinogenesis has been proven. It was shown that the CYP1A subfamily is predominantly inducible by polycyclic aromatic hydrocarbons (PAH). The mechanism of inducer-dependent control of the CYP1A expression is reasonably understood (Fujisava-Sehara et al. 1988). However, the CYP1A1 and CYP1A2 induction or suppression in different tissues and organs remains unclear. During preliminary studies we investigated the tissue-specific induction of CYP1A in rats (Gulyaeva et al., 1992). Using Western blot we have shown that treatment of rats with AR induces CYP1A1 and CYP1A2 in lungs whereas MC induces the former only. Lung CYP1A2 expression was accompanied with the increase of its specific MROD activity. For clarification of the possible mechanism of CYP1A expression by above mentioned xenobiotics we have determined the mRNA level for CYP1A1 and CYP1A2 in rat liver and lung preparations.

## Materials and Methods

To estimate the mRNA level in lung and liver of male Wistar rats we have used the method of Dot blot hybridization of the total mRNA with specific oligonucleotides for both of CYP1A genes mentioned above (Raval et al., 1991). The total mRNA was isolated from liver and lung of rats 22 hours after the treatment by MC- and AR-inducers. The microsomal fraction from liver and lung of individual MC-, AR-treated and untreated rats was prepared by differential ultracentrifugation. EROD and MROD activities were determined by usual methods (Burke et al., 1985).

## Results and Discussion

Our results, presented in Table 1, showed that the mRNA level for these proteins in liver significantly increased (by a factor of 50-100) during the induction both by MC and AR. The content of CYP1A1 mRNA in lung also increased (by a factor of 20-30) compared to untreated rats. A different pattern was observed for CYP1A2 in lung tissues: for MC-induction the mRNA level did not increase, whereas AR-treatment resulted in its ten-fold increase.

Table 1. Activities of microsomal P4501A and mRNA level in rat lung and liver.

|  | Activity* | | mRNA level** | |
|---|---|---|---|---|
|  | ER | MR | 1A1 | 1A2 |
| **LIVER** | | | | |
| Control | 66.5 | 30.0 | <2.0 | 6.4 |
| MC | 7430.0 | 943.0 | 47.0 | 93.0 |
| AR | 5200.0 | 1500.0 | 61.0 | 91.8 |
| **LUNG** | | | | |
| Control | 2.5 | 0.7 | <1.0 | 5.2 |
| MC | 455.0 | 15.5 | 29.0 | 9.0 |
| AR | 533.0 | 350.0 | 16.3 | 42.0 |

* Data expressed in pmol of resorufin/min/mg of protein.

*Autoradiographic images were quantitated using "Hitachi-557" spectrophotometric densitometer, results for each sample were expressed as arbitrary optical density units/microgram of total mRNA. The EROD activity was very high in liver and rather high in lung of AR- and MC-treated rats.

We detected a significant increase of MROD activity for both inducers in liver. In lung this activity significantly increased during AR-treatment only. From the results shown in the Table one can conclude that the data on the protein induction, i.e. catalytic activity agree with those on the mRNA level suggesting that the CYP1A expression in liver and lung is regulated at the transcriptional level. Moreover, this regulation is inducer-dependent which probably is due either to their different interaction with trans-acting factors, or to the different interaction of the inducer-protein complex with cis-acting elements of CYP1A genes.

**References**

Burke, M.D., Thompson, S., Elcombe, C.R., Halpert, G., Haaparanta, T., Mayer, R.T. (1985): Ethoxy-, penthoxy-, benzyloxy- phenaxozones and homologues: a series of substrates to distinguish between different induced cytochrome P-450. Biochem. Pharmacol. 34, 3337-3346.

Gulyaeva, L.F., Grishanova, A.Yu., Lyakhovich, V.V.(1992): Strain- and tissue- dependent induction of cytochrome P-4501A and 2B subfamilies during different type induction. In Proc. 7th Int. Conf. of Cytochrome P-450. eds. A.I. Archakov and G.I. Bachmanova, p.388-390. Moscow: INCO-TNC, Joint Stock Company, Russia.

Fujisava-Sehara, N., Yamane, M., Fujii-Kuriyama, Y. (1988): A DNA-binding factor specific for xenobiotic responsive elements of P450c gene exists as a cryptic form of cytoplasm: its possible translocation to nucleus. In Proc. Nat. Acad. Sci. USA. 85, 5859-5863.

Raval, P., Patrick, L.I., Bresnic, E. (1991): Induction of cytochromes P4501A1 and P4501A2 as determined by solution hybridization. Biochem. Pharmacol. 41, 1719-1723.

# Expression study of CYP genes in drosophila strains resistant or sensitive to insecticides

M. Amichot[1], A. Brun[1], A. Cuany[1], C. Helvig[1], J.P. Salaun[2], F. Durst[2], J.B. Berge

[1]Laboratoire de biologie des invertébrés, INRA, 123, bd F.-Meilland, BP 2078, 06606 Antibes Cedex, France. [2]Institut de biologie moléculaire des plantes, CNRS, 28, rue Goethe, F 67083 Strasbourg Cedex, France

## I/ Introduction

Insect resistance to insecticides is a major problem in Agronomy and Medicine and virtually no insecticide class can avoid this phenomenon. Resistance mechanisms can be classified in four main categories : decreased insecticide penetration, trapping of the insecticide, modification of its target, increased metabolization of the insecticide. Metabolization can involve esterases, gluthatione transferases or Cytochromes P-450. We have focused on this last enzymatic system.

Little is known about molecular biology of the P-450 in insects. Cloned genes belong to two families : CYP4 (*CYP4C1*, Bradfield et al. 1991 ; *CYP4D1*, Gandhi et al.1992 ; *CYP4E1*, Snyder et al. 1983) and CYP6 (*CYP6A1*, Feyereisen et al.1989 ; *CYP6A2*, Waters et al.1992 ; *CYP6B1*, Cohen et al.1992). The last family is found only in insects and has been each time associated with resistance to foreign compounds (insecticides or plant toxins). To get more data on Insect P-450 and on resistance to insecticides, we choosed to work on *Drosophila melanogaster* and study the resistance to DDT.

## II/ Material and methods

### 1/ Drosophila strains and toxicology

Through out this work, Raleigh is the reference strain from which a DDT resistant strain, RDDT, has been selected (Cuany et al. 1990). Resistance factor is over 1,000 for DDT (LD$_{50}$ for Raleigh is $6 \times 10^{-4}$ M whereas 0.1 M treatment don't kill RDDT), 12 for deltamethrin and 9 for parathion.

Toxicology were performed in hemolysis glass tubes using the tarsal contact method. About ten drosophilas are placed in each tube and mortality is scored after a 16 hours period.

### 2/ in vitro metabolizations, inductions

Incubations were performed and metabolites analyzed as published earlier (Cuany et al.1990). Inductors, phenobarbital or clofibrate, are incorporated in the feeding media when melted at final concentrations : 0.1 M for phenobarbital, 0.01 M for clofibrate.

### 3/ Molecular biology techniques

a/ RNA extraction and Northern blots
RNAs were extracted from adults with the use of the guanidine hydrochloride method (Ashburner, 1989). 30 µg of total RNAs were electrophorized in a 1% denaturing agarose gel and transferred on a nitrocellulose membrane. Hybridizations were held at high stringency in 6XSSPE, 5X Denhardt's, 0.5% SDS, 10 mM EDTA at 65°C. Blots were finally washed with 0.2XSSPE, 0.2% SDS at 65°C.
b/ PCR
PCR templates are first strand cDNAs and have been obtained from total RNAs using the 'first strand cDNA synthesis kit' (Pharmacia). PCR were performed using the enzyme sold by Appligene (Illkirch, France) following their specifications.
The sequences of the primers are based upon *CYP4C1* (Bradfield et al., 1991) ones :
(ARD code)
GAY ACN UUY AUG UUY GAR GG is the 5' oligo (aminoacid sequence : DTFMFEG)
AUR CAR UUN CGB GGD CCR GC is the 3' oligo (aminoacid sequence : AGPRNCI). The expected size for the amplified product is 435 mers.

PCR conditions were : 40 cycles 90°C for 1 min. then 51 °C for 1 min. then 70°C for 1.5 min.
                    1 cycle 90°C for 1 min. then 51°C for 1 min. then 70°C for 10 min.
PCR products are analysed on 0.8% agarose gels.
c/ Subcloning, plasmid preparations, DNA sequencing, probes labelling
Blunt end fragments subclonings were realised after klenow polishing of the PCR products in dephosphorylated pUC 18 vector purchased from Appligene. Plasmid preparatfns were as described in Maniatis *et al*, 1982. DNA sequencing were performed with a 'Sequanase' kit (USB). Probes labelling protocole was based upon Feinberg and Vogelstein, 1984.

III/ Results and Discussion
1/ *in vitro* metabolism
These experiments were design to demonstrate the implication of P-450 in the resistance mechanism of RDDT. Results presented in Fig. n°1 concern DDT and lauric acid. All these three substrates show, at least for some sites in the molecule, an increased metabolism due to P-450 (lack of NADPH in the incubation or presence of piperonyl butoxyde blocks the apparition of metabolites). The same kind of results have been obtained with testosterone or 2-22 dihydroxyecdysone (data not shown).

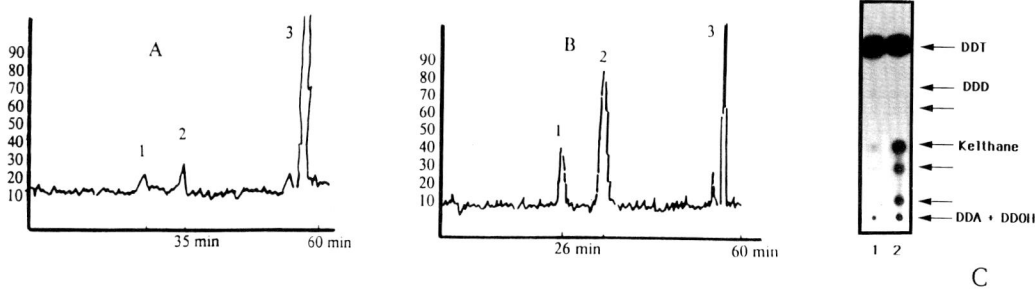

Fig. n°1 : in vitro lauric acid and DDT metabolisms
A and B : RP-HPLC of lauric acid metabolites after incubation with Raleigh (A) or RDDT (B) microsomes. peak 1 : ω-1/ω-2 hydroxylaurate, 2 : ω hydroxylaurate, 3 : lauric acid ; x axis : retention time; y axis : radioactive counts (arbitrary units)
C : autoradiography of a TLC plate of DDT metabolites after incubation with Raleigh (1) or RDDT (2) microsomes.

2/ PCR cloning of CYP4 genes
From the previous results, we decided to clone genes members of the CYP4 family in drosophila. After unsuccessful attempts to clone a gene using *CYP4C1* (kindly provided by Dr. Bradfield) as a probe with low stringency conditions, we designed PCR experiments (conditions are presented in Fig n°2).
We obtained a single band at the expected size (data not shown) and subcloned it. Sequences analysis clearly indicated that we have cloned two group of clones (Fig n°2). It appears that one sequence is identical to *CYP4E1*. The other, although very similar, is clearly different from *CYP4E1*. We tentatively call it *CYP4"F"*.

```
CYP4C1      DTFMFEGHDT  TSAGICWALF  LLGSHPEIQD  KVYEELDHIF  QGSD__RSTT  MRDLADMKYL
CYP4E1      DTFMFEGHDT  TTSGVGFAVY  LLSRHPDEQE  KLFNEQCNVM  GASGLGRDAT  FQEISTMKHL
CYP4"F"     DTFMFEGHDT  TTSGVSFAVY  LLSRHQDEQR  KLFKEQREVM  GNSELGRDAT  FQEISTMKHL
CYP4D1      DTFMFKGHDT  TSSALMFFFY  NIATHPEAQK  KCFEEIRSVV  GNDKSTPV_S  YELLNQLHYV

ERVIKESLRL  FPSVPFIGR_  VLKEDTKIG   DYLVPAGCMM  NLQIYHVHRNQ DQYPNPEAFN  PDNFLPER__
DLFIKEAQRL  YPSVPFIGR_  LTEKDYVID   GDIVPKGTTL  NLGLLMLGYND RVFKDPHKFQ  PERF__ER__
DLFIKEAQRL  YPSVPFIGR_  FTEKDYVID   GDLVPKGTTL  NLGLVMLGYNE KVFKDPHKFR  PERF__EL__
DLCVKETLRM  YPSVPLLGRK  VLE_DCEIN   GKLIPAGTNI  GISPLYLGRRE ELFSEPNIFK  PERF__DVVT

__VAKRHPYA  YVPFSAGPRN  CI
_E_KPGPFE   YVPFSAGPRN  CI
_E_KPGPFE   YVPFSAGPRN  CI
TAE_KLNPYA  YIPFSAGPRN  CI
```

Fig n°2 : deduced aminoacid sequences of CYP4E1 and CYP4"F"
Sequences are aligned with the two other known CYP4 in insects

Homologies between CYP4E1 and CYP4"F" are 75% in nucleotides and 85% in aminoacids. Interestingly, CYP4E1 is more homologous to CYP4C1 (45,6%) than to CYP4D1 (42,6%). On the other hand, CYP4"F" is as homologous to CYP4C1 (44,9%) as to CYP4D1 (45,3%).

### 3/ Transcription analysis of the 4 CYP genes in drosophila

Considering the results obtained after *in vitro* metabolizations, we decided to check the transcription level of all the CYP genes available in the Raleigh strain and in RDDT. CYP6A2 probe is a kind gift from Dr. Ganghuli and we partially reclone *CYP4D1* by PCR. We have cloned the two other probes. Figure n°3 shows that two genes are overexpressed in RDDT : *CYP6A2* and *CYP4"F"*. On the contrary, *CYP4D1* and *CYP4E1* transcriptions remain unaffected by the resistance phenomenon. Several points are noticeable :(1) a CYP6 family gene is overexpressed in the resistant strain as described elsewhere (Feyereisen *et al.*, 1989 ; Waters *et al.*, 1992 ; Cohen *et al.*, 1992),(2) a CYP4 family gene is also overexpressed (this correlates the higher lauric acid metabolization),(3) more than one but not all CYP genes are overexpressed in the resistant strain. The increasement of the expression of *CYP6A2* and *CYP4"F"* is linked to resistance.

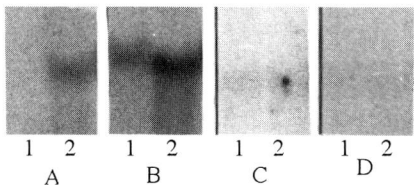

Fig. n°3 : Northern blots with total RNAs from Raleigh (1) or RDDT (2) hybridized with CYP6A2 (A), CYP4"F" (B), CYP4D1 (C) or CYP4E1 (D).

### 4/ Effects of phenobarbital and clofibrate on the expression of the CYP genes

Effects of these inducers (Juchau, 1990 ; Okey, 1990) have been tested by *in vitro* metabolization of lauric acid, northern blots or toxicology.

#### a/ Lauric acid metabolization

Results are reported in table 1. From that table, we can conclude that, surprisingly, clofibrate seems to be a weak inducer of lauric acid hydroxylation in drosophila while phenobarbital is able to induce ω hydroxylation and especially well ω-1/ω-2 hydroxylation. In addition, RDDT has lost the ability to have lauric acid hydroxylations be induced by these molecules.

|  | Raleigh | Raleigh clofibrate | Raleigh phenobarbital | RDDT | RDDT clofibrate | RDDT phenobarbital |
|---|---|---|---|---|---|---|
| ω hydrox. pmole/min/mg of prot | 83850 | 132450 | 350250 | 322500 | 396600 | 349200 |
| ratio induced /non induced | 1 | 1.6 | 4.2 | 1 | 1.2 | 1.1 |
| ω-1/ω-2 hydrox. pmole/min/mg of prot | 14700 | 31650 | 256500 | 91500 | 71100 | 132600 |
| ratio induced /non induced | 1 | 2.1 | 17.5 | 1 | 0.8 | 1.5 |

Table n°1 : Effects of clofibrate and phenobarbital on lauric acid hydroxylations.

#### b/ Northern blots

As RDDT is uninducible, we worked only with Raleigh RNAs. We have got very interesting results as it appears that the phenobarbital inducible genes, *CYP6A2* and *CYP4"F"*, are the ones overexpressed in the resistant RDDT strain (Fig n°4). Thus, we have a link between resistance and phenobarbital regulation. It is to note that clofibrate do not or weakly promote any gene expression and this correlates well the previous results (table n°1). Furthermore, *CYP4E1* and *CYP4"F"*, closely homologous, are not regulated by the same way and thus could have different functions.

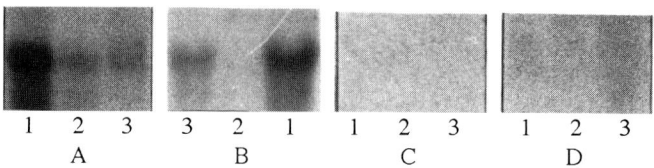

Fig. n°4 : Northern blot analysis of the effects of clofibrate and phenobarbital on CYP genes transcription. Lane 1 : RNAs from phenobarbital treated drosophilas, lane 2 : RNAs from clofibrate treated drosophilas, Lane 3 : RNAs from uninduced drosophilas. A : probed with *CYP6A2*, B : probed with *CYP4"F"*, C : probed with *CYP4E1*, D : probed with *CYP4D1*.

Comparison between *CYP4"F"* transcription in Raleigh, Raleigh induced by phenobarbital and RDDT on one hand and, on the other hand, lauric acid metabolism suggest that *CYP4"F"* could code for an ω-1/ω-2 laurate hydroxylase.

c/ Toxicology

We measured the effects of DDT on Raleigh drosophilas after the inductive treatments. Though clofibrate did not modify DDT $LD_{50}$ ($6 \times 10^{-4}$ M), phenobarbital augmented Raleigh ability to resist to DDT ($LD_{50}=3 \times 10^{-3}$). In addition, phenobarbital treatment induced Raleigh resistance to deltamethrin and to parathion up to RDDT levels. This suggests that RDDT resistance is not only due to overexpression of a P-450 but also to mutations enhancing this P-450 DDT metabolizing ability.

These experiments point out once more a relation between resistance and phenobarbital regulation.

IV Conclusions

This work is the starting point for the elucidation at a molecular level of resistance in the RDDT strain. We have shown that *CYP4"F"* and *CYP6A2* are overexpressed in the resistant strain and these genes are controlled by phenobarbital. Induction experiments and toxicology lead us to think that resistance is a combination of overexpression and mutation(s) in a CYP gene, probably *CYP6A2*. The achievement of this study could allow the identification of phenobarbital regulation process. In addition, we have cloned a new member of the CYP4 family. Analysis of this gene is in progress.

Heterologous expression coupled with directed mutagenesis experiments are planned to determine catalitic specificities of *CYP4"F"* and the alleles of *CYP6A2* from sensitive or resistant strains. Transformation of drosophila will help to determine the role of these genes in resistance.

References :
-Ashburner M., 1989. "Drosophila, a Laboratory Handbook", Cold Spring Harbor Lab., Plainview, NY.
-Bradfield J., Y., *et al.*, 1991. Proc. Natl. Acad., Sci. USA ; **88**, 4558-4562.
-Cohen M., B., *et al.*, 1992. Proc. Natl. Acad., Sci. USA ; **89**, 10920-10924.
-Cuany A. *et al.*, 1990. Pestic. Biochem. Physiol.; **37**, 293-302.
-Feinberg A., P. and Vogelstein B., 1984. Anal. Biochem. ; **137**, 266-267.
-Feyereisen R. *et al.*, 1989. Proc. Natl. Acad. Sci. USA ; **90**, 5643-5647.
-Gandhi R. *et al.*, 1992. DNA and Cell Biol. ; **11**, 397-404.
-Juchau M., R., 1990. Life Sci. ; **47**, 2385-2394.
-Maniatis T. *et al.*, 1982. "Molecular Cloning, A Laboratory Manuel" Cold Spring Harbor Lab., Plainview, NY
-Okey A., B., 1990. Pharmac. Ther. ; **45**, 241-298.
-Snyder M. and Davidson N., 1983. J. Mol. Biol. ; **166**, 101-118.
-Waters L. *et al.*, 1992. Proc. Natl. Acad. Sci. USA ; **89**, 4855-4859.

# Metabolism of selected sex steroids by rat CYP1A1, 1A2 and 2B1 expressed in V79 cell lines

M. Hildebrand, H. Gieschen, B. Salomon

*Institute of Pharmacokinetics, Schering AG, D-13342 Berlin, FRG*

## Abstract

Ethinylestradiol (EE), gestodene (GEST), levonorgestrel (LNG) and testosterone (TEST) were screened in V79 cell lines, expressing rat CYP1A1, 1A2 and 2B1, for metabolic degradation. Apart from TEST, which was known to be hydroxylated in 16α- and 16ß-position by CYP2B1 and metabolized to androstenedione and androsterone in non-transformed cells, no specific degradation product was obtained with the other steroids. V79 cell lines were capable to form glucuronic acid and/or sulfate conjugates as concluded from enzymatic hydrolysis of highly polar radiolabel fractions. In studies on the inhibition of standard resorufin reactions TEST exhibited the highest $IC_{50}$-values, while the three other steroids inhibited especially rat CYP1A1 and 2B1 mediated transformations with $IC_{50}$-values of 7 to 19 μM. Based on the present experiments rat CYP 1A1, 1A2 and 2B1 do not contribute relevantly to the metabolic degradation of the synthetic sex steroids investigated.

## INTRODUCTION

Biotransformation is an important research field in pharmacokinetics to describe the fate of any xenobiotic in the body. In the past these investigations were perfomed in complete biosystems (i.e. animals or man), in isolated organs (e.g. liver perfusion technique), in isolated cells and cell fractions (hepatocytes and microsomes) or with isolated enzymes. The availability of biotechnologically modified expression systems, providing single enzymes, further facilitated specific metabolic questions to be studied on the enzyme level. Due to its negligible endogenous CYP450 contents and its short generation period V79 cells were considered a promising tool to express different CYP450 enzymes, which are known to play a key role in xenobiotic metabolism [Fuhr et al., 1992; Waxman et al., 1989; Wölfel et al., 1991].

Sex steroids are known to be extensively biodegraded by various CYP450 families due to their endobiotic nature. Modified analogues used in drug therapy and as contraceptive agents are likely to be subject to similar metabolic pathways. Additionally inhibitory interactions with CYP450 enzymes have been demonstrated for several steroids [Guengerich et al., 1988; 1990]. Therefore levonorgestrel (LNG), ethinylestradiol (EE), gestodene (GEST) and testosterone (TEST) were screened with V79 cells containing rat CYP1A1, 1A2 and 2B1 to study the contribution of these enzymes to steroid metabolism. Due to the use of metabolically stable radiolabeled materials all possible degradation products should be detectable.

## MATERIALS AND METHODS

Compounds (* position of $^3$H-label)

ETHINYLESTRADIOL

GESTODENE

LEVONORGESTREL

TESTOSTERONE

V79 cell lines

Parent cell line (V79MZ) and cell lines expressing rat CYP1A1, 1A2 and 2B1 were provided by Dr. J. Döhmer (Inst. of Toxicology, University Munich, Germany). Resorufin dealkylase acitivity was 3 - 10 (1A1), 3 - 6 (1A2) and 18 - 38 (2B1) pmol/min/mg protein.

Studies on steroid metabolic degradation
Radiolabeled and non-labeled steroids (3 and 6 µg) were incubated with approx. $2 \times 10^6$ cells in 6-well tissue culture clusters in DMEM with 5% FCS. Incubation was perfomed in a 5% $CO_2$-atmosphere at 95% humidity. At 0, 6, 24, 48 and 72 h medium was sampled and analyzed by HPLC with gradient systems to detect possible metabolites of different polarity. A Hypersil ODS II (250 x 4.6 mm, 5 µm)column was used with a water (A)/methanol (B)-gradient starting at 35% B for 3 min followed by a linear increase to 80% B in 35 min with a flow of 1.5 ml/min at ambient temperature. Detection was performed by on-line LSC and UV at 254 nm (LNG), 246 nm (GEST) and 283 nm (EE). For TEST a Supelcosil LC-18DB (250 x 4.6 mm, 5 µm) column and a linear water/methanol (both solvents with 7.5% THF) gradient with 27% B for 3 min followed by an increase to 60% B in 35 min were used. UV-detection was performed at 254 nm.

Studies on enzyme inhibition
Inhibitory effects were investigated using standard ethoxy-, methoxy- and benzoxy-resorufin dealkylation reactions. V79 cells were grown in Petri dishes in DMEM with 5% FCS. An almost confluent cell layer was harvested and homogenized in a potter. After ultracentrifugation at $10^5$ g for 60 min the protein pellet was resuspended in 50 mM Hepes buffer pH 7.6. All steps were performed at 4°C. Protein content was determined by BCA-method. Resorufin ether dealkylase activity was measured fluorimetrically by a Perkin Elmer LS-50B at 550 nm (excitation) and 580 nm (emission). Resorufin ethers were dissolved in DMSO and used at 0.5 to 1.0 µM. Solutions of steroids were made in methanol and final concentrations ranged from 2 to 100 µM. The pellet suspension and a NADPH-generating system (0.6 mM NADP, 8 mM G6P, 1.4 U G6PDH, 38 mM KCl and 5 mM $MgCl_2$) incubated with the respective resorufin ether and steroid at 37°C. Fluorescence signals were registered continuously. For the determination of inhibition reaction types steroid solutions with 0.5 $IC_{50}$, $IC_{50}$ and 2 $IC_{50}$ and four concentration levels (0.1 to 1.5 µM) of resorufin ethers were used against a methanol control.

Evaluation
Radio- and UV- HPLC-chromatograms of parent and transformed cell lines were compared to detect possible metabolites.
Interaction experiments were evaluated by means of probit analysis to determine $IC_{50}$. Inhibition types were characterized from graphic data analysis by standard methods.

# RESULTS

Non-transformed parent V79 cell lines formed androstenedione and androsterone from TEST. CYP2B1 formed the known metabolites 16α- and 16ß-OH-TEST. None of the other tested rat enzymes formed any specific biodegradation product in detectable amounts with LNG, EE and GEST. In several samples highly polar radiolabel fractions were detected which could be enzymatically hydrolyzed by ß-glucuronidase/arylsulfatase.

In studies on the inhibition of standard resorufin reactions TEST exhibited the highest $IC_{50}$-values, while the three other steroids inhibited especially rat CYP1A1 and 2B1 mediated transformations with $IC_{50}$-values of 7 to 19 µM (Figure 1 and Table 1). Inhibition types were enzyme and substrate specific. GEST and LNG were reversible competitive inhibitors of ethoxyresorufin dealkylation, while EE exhibited a time-dependent irreversible inhibition. In case of CYP2B1 all three compounds were characterized by time dependent inhibition profiles.

Figure 1: Inhibition of rat CYP1A1, 1A2 and 2B1 resorufin dealkylase activity by EE, GEST, LNG and TEST (displayed as percent remaining activity)

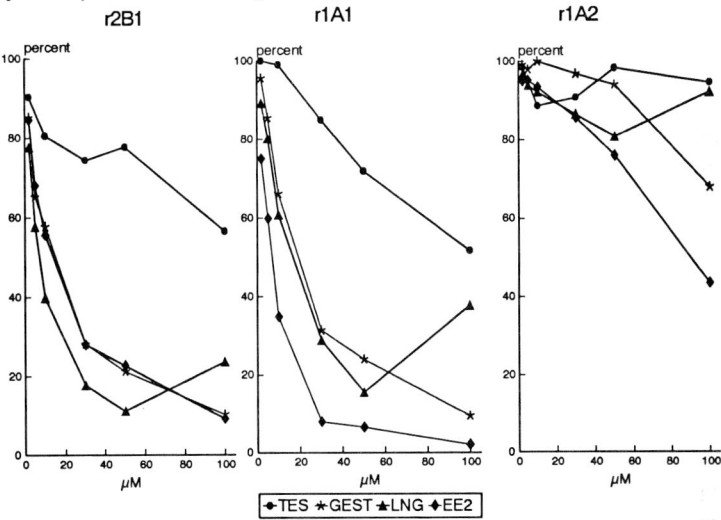

Table 1: $IC_{50}$-values [µM] and inhibition types of steroid interactions with rat CYP1A1, 1A2 and 2B1

| CYP450 | TEST | GEST | LNG | EE |
|---|---|---|---|---|
| r1A1 | 97 | 18 | 15 | 5 |
| r1A2 | >>100 | 154 | 1240 | 97 |
| r2B1 | 290 | 11 | 7 | 11 |

## DISCUSSION

V79 cell lines expressing rat CYP1A1, 1A2 and 2B1 were used to screen several sex steroids for substrate properties. TEST is known to be transformed to its 16α and 16ß-OH derivative by CYP2B1; additionally androstenedione and androsterone are even formed by non-transformed V79 cells. Structurally similar steroids, e.g. GEST and LNG, are not metabolized to a detectable extent by CYP2B1 and the other enzymes tested as well as EE. Interestingly all three steroids are able to inhibit standard resorufin dealkylase reactions mediated by the CYP450s. However, the type of enzyme interaction was highly structure and enzyme dependent. Especially the ethinyl-substitution in position 17 seems to play an important role for inhibitory effects in case of CYP2B1. V79 cell lines are able to perform phase II reactions, i.e. conjugation with glucuronic acid and/or sulfate. The CYP450 enzymes investigated might require different partial structures for metabolic activity. Other rat CYP450 families must play an important role in the multiplicity of metabolic pathways of tested steroids.

## REFERENCES

Fuhr, U., Doehmer, J., Battual, N., Wölfel, C., Kudla, C., Keita, Y., and Staib, H.(1992): Biotransformation of caffeine and theophylline in mammalian cell lines genetically engineered for expression of single cytochrome P450 isoforms. Biochem. Pharmacol. 43: 225-235.

Guengerich, F.P. (1988): Oxidation of 17α-ethynylestradiol by human liver cytochrome P-450. Mol. Pharmacol.22: 500-508.

Guengerich, F.P. (1990): Mechanism-based inactivation of human liver microsomal cytochrome P450IIIA4 by gestodene. Chem. Res. Tox. 3,: 363-371.

Waxman, D.J., Lapenson, D.P., Morrissey, J.J., Park, S.S., Gelboin, H.V, Doehmer, J., and Oesch, F. (1989): Androgen hydroxylation catalysed by a cell line (SD1) that stably expresses rat hepatic cytochrome P-450 PB-4 (IIB1). Biochem. J. 260: 81-85.

Wölfel, C., Platt, K.L., Dogra, S., Glatt, H., Wächter, F., and Doehmer, J.(1991): Stable expression of rat cytochrome P450IA2 cDNA and hydroxylation of 17ß-estradiol and 2-aminofluorene in V79 Chinese hamster cells. Mol. Carcinogen. 4: 489-498.

## ACKNOWLEDGEMENTS

The authors wish to thank Dr. J. Döhmer (Institute of Toxicology, University of Munich, Germany) for providing V79 cell lines and advisory contribution for model adaptation. The research work was part of a scientific project sponsored by Schering Research Foundation, Berlin, Germany.

# Thiophene derivatives as new mechanism-based inhibitors of cytochromes P450: inactivation of yeast expressed human liver cytochrome P450 2C9 by tienilic acid

M. Pilar López-Garcia[1,2], Patrick M. Dansette[1], Daniel Mansuy[1]

[1]Laboratoire de chimie et biochimie pharmacologiques et toxicologiques, URA 400 CNRS, Université René-Descartes, 45, rue des Saints-Pères, 75270 Paris Cedex 06, France.
[2]Departmento de Bioquímica y Biología Molecular, Facultad de Farmacia, Av. Dr Molinar, Universidad de Valencia, Burjasot 46100 Valencia, Spain

Many compounds act as mechanism-based inactivators of P450s (Ortiz de Montellano, 1988). Recently, it has been found that some thiophene derivatives are oxidized by liver P450s into the corresponding very reactive thiophene sulfoxides (Mansuy et al., 1991) which react with nucleophilic groups. Tienilic acid (TA) is an uricosuric diuretic drug which causing immunoallergic hepatitis in about 1 per 10,000 of the patients treated with (Homberg et al., 1984) with the appearance of circulating anti-organelle autoantibodies directed against a protein of human liver endoplasmic reticulum and called anti-LKM$_2$ antibodies. They are specifically directed against a human liver P450 from the 2C subfamily which was also found to be the responsible for the oxidative metabolism of TA in human liver microsomes (Beaune et al., 1987; Dansette et al., 1993; Lecoeur et al., 1993). P450 2C-mediated TA metabolization by human liver microsomes leads to the formation of 5-hydroxy tienilic acid (5-OHTA) as a major stable metabolite and also to electrophilic reactive intermediates which irreversibly bind to liver microsomal proteins. Thus a hypothetical scheme of the molecular mechanism for the development of TA-induced immunoallergic hepatitis involves (i) the oxidation of the thiophene ring of TA into reactive metabolites by a P450 2C enzyme, (ii) the alkylation of this P450 by the reactive metabolite(s), (iii) the appearance of antibodies against this modified protein, and (iiii) the destruction of hepatocytes bearing P450 2C-anti-P450 2C immune-complexes by the immune system. P450 2C9, one of the major P450s in human liver, expressed in yeast under its catalytically active form (Brian et al., 1989a,b; Srivastava et al., 1991) was shown to metabolize TA into 5-OHTA and to electrophilic intermediates that irreversibly bind to microsomal proteins (López-Garcia et al., 1993). A carefull analysis of the time-course evolution of TA 5-hydroxylation by microsomes from yeast expressing P450 2C10 showed a short linear period (ca 5 min) followed by a fast decrease of the reaction rate (Fig. 1). In fact, after a 30 min period no significant hydroxylation of TA could be detected even if more substrate was added to the incubation mixture (data not shown). A very similar decrease as a function of time was observed for the activation of TA into reactive metabolites able to covalently bind to microsomal proteins (Fig. 1B). On the contrary and in the same conditions, the oxidation of tienilic acid isomer (TAI) into reactive metabolites which covalently bind to proteins remained linear as a function of time for at least 20 min (Fig. 1C). The covalent binding of TA metabolites to proteins was considerably reduced in the presence of nucleophiles like glutathione (Fig. 1B). However, a residual level of covalent binding remained unsensitive to the trapping action of 5 mM GSH. The time-dependent decrease of P450 2C10-catalyzed 5-OHTA formation was the same in the presence or absence of 5 mM GSH (Fig. 1A). Identical results were obtained by using 2C9.

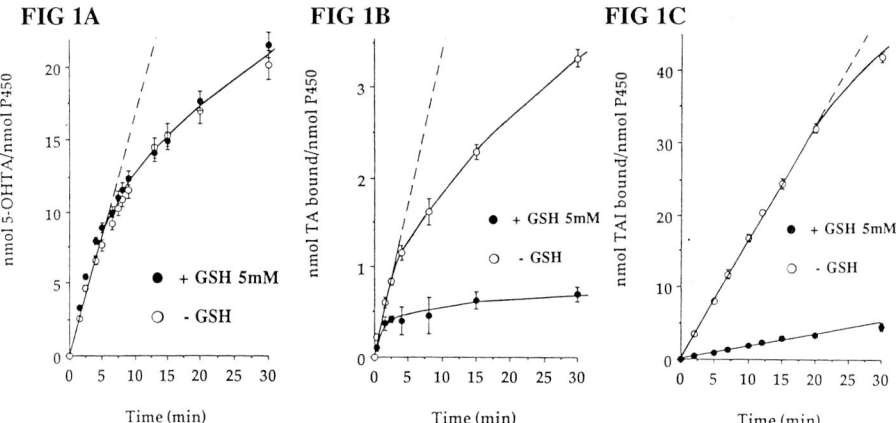

Yeast microsomes were incubated with either TA or TAI in the presence and absence of NADPH (i.e., under catalysis and non-catalysis conditions), and the remaining enzyme activity was measured as a function of time. In the absence of NADPH-generating system (Fig. 2A and 2B), the activity of pAAH5/P450 2C10 microsomes remained constant (100%) for at least 30 min. Similar results were obtained with NADPH-generating system but in the absence of substrate (TA or TAI).

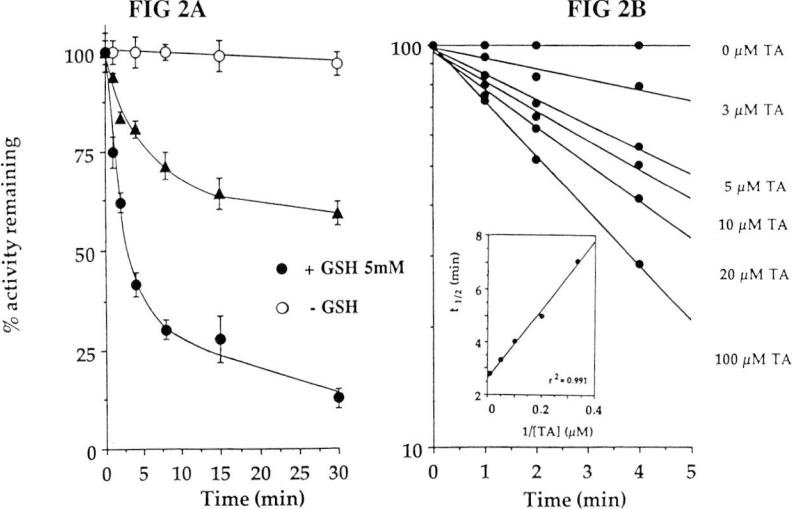

When microsomes were incubated in the presence of NADPH and TAI (100 µM), even in the absence of GSH in order to maximize any inactivating effect, no significant loss of activity could be detected between 0-15 min of incubation (not shown). On the contrary, when microsomes were incubated in the presence of NADPH and TA (● 100 µM or ▲ 18 µM) in the presence of 5 mM GSH (Fig 2A) a dramatic loss of activity was observed. These results demonstrate a catalysis-dependent loss of P450

2C10 (2C9) monooxygenase activity upon TA (but not TAI) oxidation. The general criteria (Walsh, 1982; Walley, 1980) to prove that irreversible loss of activity upon TA oxidation is a mechanism-based process were then studied. For a given time of incubation, loss of activity was found dependent on TA concentration, with saturation kinetics similar to that observed for 5-OHTA formation . The time required for half-maximal inactivation ($t_{1/2}$) at each substrate concentration, as well as the apparent first-order rate constant $k_{inact}$, were calculated from the plot of logarithmic transformation of the remaining activity as a function of time, as depicted in Fig. 2B. From extrapolation to infinite TA concentration, the time required to inactivate half of the enzyme at the maximal rate ($t_{1/2max}$) and the maximal $k_{inact}$ resulted $3.4 \pm 0.8$ min and $3.6 (\pm 0.8) \cdot 10^{-3}$ s$^{-1}$, respectively (n=3). The dissociation constant ($K_I$) was found to be $4.3 \pm 0.6$ µM, a value not significantly different from the $K_m$ for TA 5-hydroxylation determined with several microsomal preparations ($5.9 \pm 2.3$ µM). The *partition ratio*, r ($k_{+3}/k_{+4}$), which represents the number of productive turnovers (leading to 5-OHTA in this case) by inactivation event was derived from Fig. 3, which correlates the % of activity remaining after TA oxidation with the amount of 5-OHTA formed during the same time. A linear relationship was observed whatever the TA concentration or the incubation time. Extrapolation to 0% remaining activity gave a mean r value of $11.6 \pm 0.2$ (n=3).

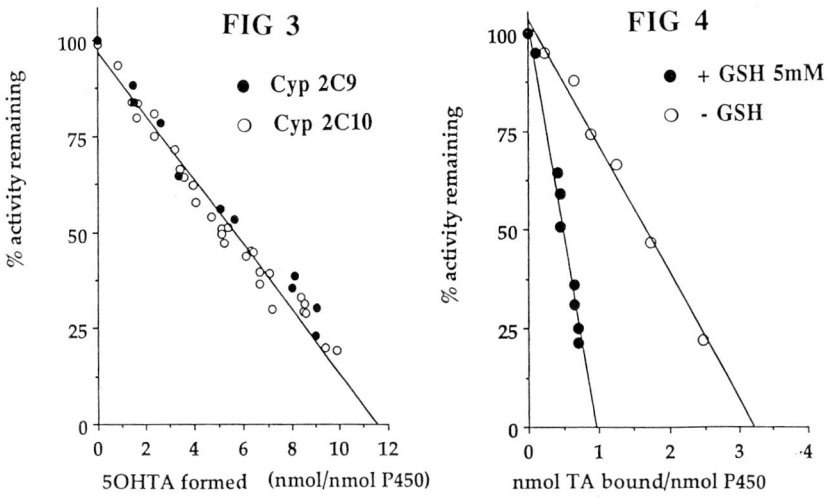

In a mechanism-based process, since binding to the active site always preceeds inactivation, the presence of a competitive inhibitor of the inactivator substrate must protect the enzyme from inactivation (Walsh, 1982). As expected, the presence of tolbutamide, a known substrate for P450 2C9/10 enzymes which has been characterized as a competitive inhibitor for TA 5-hydroxylation (López-Garcia et al., 1993), inhibited both 5-OHTA formation and P450 inactivation . The $t_{1/2}$ of the enzyme increased significantly when tolbutamide was present (from 5.0 to 20.3 min for 5 µM TA and 200 µM tolbutamide), but tolbutamide did not prevent completely the TA-dependent P450 inactivation.

Observations that GSH efficiently decreased the covalent binding of TA electrophilic intermediates to microsomal protein indicate that part of TA reactive metabolites diffuse out from the P450 2C10 (2C9) active site. However a 5 mM GSH concentration which leads to a maximum decrease on the binding had no significant effect neither on the stability of yeast microsomes nor on the rate of P450 2C10 inactivation in the presence of TA and NADPH (identical $t_{1/2}$ for inactivation with and without GSH of $5.2 \pm 0.5$ and $6.5 \pm 0.9$ min, respectively). Extrapolation of the % remaining activity *vs* the amount of covalent binding gave 0.9-1 nmol of TA metabolite(s) covalently bound per nmol P450 for total inactivation and immunoblot with an anti-TA antibody demontrated that only the P450 band was labeled by TA (see Dansette et al., *this volume*). It is also noteworthy that TAI, the isomer of TA does not act as a mechanism-based inhibitor of P450 2C9/10 enzymes although it is metabolically activated to electrophilic intermediates by the same P450 2C that oxidizes TA (López-Garcia et al., 1993) and binds extensively to microsomes (42 nmol/nmol P450 in 30 min). As explained (Dansette et al., *this volume*) TA and TAI are probably oxidized in reactive thiophene sulfoxide which can diffuse out from the P450 active site and react either with microsomal protein with water or GSH present in the medium. The reactive position of TA thiophene sulfoxide is probably position 5, on the contrary that of TAI sulfoxide is the more hindered 2 position, not well positioned for reacting with the nucleophile in the active site. Thus TAI sulfoxide would diffuse out from the enzyme active site and react with GSH or protein nucleophiles in the medium.

This study reports that thiophene derivatives can act as mechanism-based inhibitors of cytochromes P450. The structural properties of TA make this molecule a good suicide substrate of P450 2C9 and 2C10. Its site of covalent binding on the P450 2C9/10 active site is now under study. This should be very helpful for the determination of the topology of the active site region of P450 isoenzyme(s).

## *REFERENCES*

Beaune, P. H., Dansette, P. M., Mansuy, D., Kiffel, L., Finck, M., Amar, C., Leroux, J. P., & Homberg, J. C. (1987) : Human anti-endoplasmic reticulum autoantibodies appearing in a drug induced hepatitis are directed against a human liver cytochrome P-450 that hydroxylates the drug. *Proc. Natl. Acad. Sci. U.S.A. 84*, 551-555.

Brian, W. R., Srivastava, P. K., Umbenhauer, D. R., Lloyd, R. S., & Guengerich, F. P. (1989b) : Expression of a human liver cytochrome P450 protein with tolbutamide hydroxylase activity. *Biochemistry 28*, 4993-4999.

Dansette, P. M., Lopez-Garcia, P., Amar, C., Pons, C., Valadon,P., Bonierbale, E., Beaune, P. H., & Mansuy, D. : Metabolic activation of the thiophene ring as a primary event in immunoallergic hepatitis to Tienilic acid. *This volume*.

Homberg, J. C., André, C., & Abuaf, N. (1984) : A new anti-liver-kidney-microsome antibody (anti-LKM2) in tienilic acid-induced hepatitis. *Clin. Exp. Immunol. 55*, 561-570.

López-Garcia, M. P., Dansette, P. M., Valadon, P., Amar, C., Beaune, P. H., Guengerich, F. P., & Mansuy, D. (1993) : Human liver P450s expressed in yeast as tools for reactive metabolite formation studies : oxidative activation of tienilic acid by P450 2C9 and P450 2C10. *Eur. J. Biochem.. 213*, 223-232.

Mansuy, D., Valadon, P., Erdelmeier, I., López-Garcia, M. P., Amar, C., Girault, J.P., & Dansette, P. M. (1991) : Thiophene-S-oxides as new reactive metabolites : formation by cytochrome P450-dependent oxidation and reaction with nucleophiles. *J. Am. Chem. Soc. 113*, 7825-7826.

Ortiz de Montellano, P. R. (1988) : Suicide substrates for drug metabolizing enzymes : Mechanism and biological conseqences. *Progress in Drug Metab. 11*, 99-148.

Srivastava, P. K., Yun, C., Beaune, P. H., Ged, C., & Guengerich, F. P. (1991) : Separation of human liver microsomal tolbutamide hydroxylase and (S)-mephenytoin hydroxylase cytochromes P450 enzymes. *Mol. Pharmacol. 40*, 69-79.

Walley, S. G. (1980) : Kinetics of suicide substrates. *Biochem. J. 185*, 771-773.

Walsh, C. (1982) : Suicide substrates : Mechanism-based enzyme inactivators.*Tetrahedron 38*, 871-909.

*Cytochrome P450. 8th International Conference.* Ed. M.C. Lechner. John Libbey Eurotext, Paris © 1994, pp. 701-704.

# Cholesterol-binding region of cytochrome P450scc (P-450XIA1): identification and amino acid sequence of the cholesterol binding domain

Maki Tsujita, Yoshiyuki Ichikawa

*Department of Biochemistry, Kagawa Medical School, Miki-cho, Kita-gun, Kagawa 761-07, Japan*

Cytochrome $P\text{-}450_{scc}$ (*P-450 XIA1*) is the terminal enzyme of the side-chain-cleavage-monooxygenase system for cholesterol, and exists in the mitochondrial inner membrane of the adrenocortex. This enzyme catalyzes the first and limiting reaction during the biosynthesis of steroid hormones from cholesterol. It also plays an important role in the production of pregnenolone, a common precursor of various steroid hormones other than the renal steroid hormone: 1α, 25-dihydroxyvitamin $D_3$.

To elucidate the molecular mechanism underlying electron transport, we have investigated the adreno-ferredoxin-binding region of cytochrome $P\text{-}450_{scc}$, and the adreno-ferredoxin and NADPH-binding regions of NADPH-adreno-ferredoxin reductase.

In this study, the cholesterol-binding region of cytochrome $P\text{-}450_{scc}$ was investigated to clarify the binding amino acid sequence for cholesterol, using [14C]methoxychlor, a suicide substrate (Fig. 1) (Tsujita and Ichikawa, 1993).

It is well-known that methoxychlor, which is a pesticide of the DDT class and is used as a substitute for the banned DDT, undergoes metabolic activation by the rat liver microsomal cytochrome *P-450*-monooxygenase system, resulting in covalent binding to microsomal cytochrome *P-450* (Kupfer and Bulger, 1987).

Cytochrome $P\text{-}450_{scc}$ was purified from mitochondria fraction of bovine adrenocortices. The heme content of the purified cytochrome $P\text{-}450_{scc}$ was 18.9 nmol/mg protein. The SDS-PAGE pattern of the purified cytochrome $P\text{-}450_{scc}$ showed a single protein band.

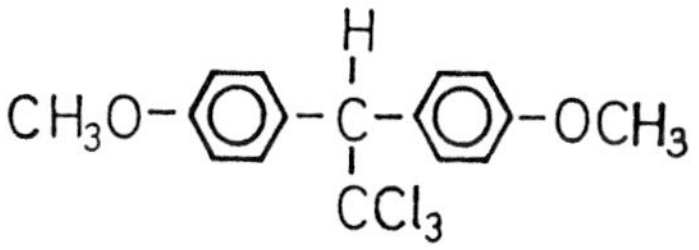

Figure 1. Suicide substrate, methoxychlor, of cytochrome $P\text{-}450_{scc}$

NADPH-ferredoxin reductase was purified from the mitochondrial fraction of bovine adrenocortices. Adreno-ferredoxin was also purified from the mitochondrial fraction of bovine adrenocortices. The NADPH-ferredoxin reductase and adreno-ferredoxin proteins each gave a single protein band on SDS-PAGE.

The methoxychlor indicated the substrate-induced difference spectrum of cytochrome $P\text{-}450_{SCC}$. The methoxychlor-induced difference spectrum, like the cholesterol-induced one, was of typical type I, and the $K_s$ value was 164 $\mu$M. This value was apparently identical with the $K_m$ value of cytochrome $P\text{-}450_{SCC}$ for cholesterol. The intensity of the cholesterol-induced difference spectrum of cytochrome $P\text{-}450_{SCC}$ was decreased competitively by methoxychlor.

The effect of 200 $\mu$M methoxychlor on the activity of the cytochrome $P\text{-}450_{SCC}$-linked monooxygenase system was examined at pH 7.2 and 37°C. During the enzymatic reaction, the enzymatic activity decreased in a time-dependent manner. The enzymatic inactivation caused by methoxychlor exhibited bi-phasic behavior, and the first step of the inactivation proceeded in an apparent first-order manner until apparently 75% of the activity had disappeared.

The activity of the cytochrome $P\text{-}450_{SCC}$-linked monooxygenase system was investigated kinetically with methoxychlor at various concentrations in a competitive manner. The $K_i$ value of methoxychlor was 120 $\mu$M. Unexpectedly, the $V_{max}$ values were the same, although the concentration of active cytochrome $P\text{-}450_{SCC}$ is thought to decrease with the covalent inactivation.

The degree of inactivation of the methoxychlor-bound cytochrome $P\text{-}450_{SCC}$ was considerably different in the absence and presence of cholesterol, although cholesterol is not an antioxidant or a free-radical scavenger. An average of 0.75 nmole methoxychlor per nmole cytochrome $P\text{-}450_{SCC}$ was covalently incorporated on maximal inactivation. When 250 $\mu$M cholesterol was included in the reaction mixture, a total of 200 $\mu$M methoxychlor was incorporated in to the cytochrome $P\text{-}450_{SCC}$ under the same experimental conditions. This protective effect of cholesterol suggests that the incorporation of methoxychlor is responsible for the inactivation of cytochrome $P\text{-}450_{SCC}$. The protective effect of cholesterol as to the methoxychlor inactivation was significantly stronger than that of 20$\alpha$-hydroxycholesterol or deoxycorticosterone.

The methoxychlor-binding region of cytochrome $P\text{-}450_{SCC}$ was determined by comparison of the tryptic peptide mapping profiles on reverse-phase HPLC of cytochrome $P\text{-}450_{SCC}$ bound with [14C]methoxychlor in the absence and presence of cholesterol.

Representative chromatograms on reverse-phase HPLC monitored at 210 nm are presented in Figures 2A and B. The radioactivity associated with each peptide fraction is also indicated in the histograms in Figures 2C and D, i.e., without and with 200 $\mu$M cholesterol. These experiments were performed repeatedly and the same results were obtained as expected under the two sets of experimental conditions.

As shown in Figure 2C, high radiolabeling was observed for two tryptic fractions with retention times of 35 min and 46 min throughout the tryptic peptide maps. No other radioactive fractions were detected. However, the peptide fraction was preferentially labeled in the tryptic peptide fraction with the retention time of 35 min without cholesterol, and the other radiolabeled fraction was confirmed to contain no peptide on amino acid sequence analysis.

The radioactive peptide was rechromatographed and purified on a reverse-phase HPLC column from the crude radioactive peptide fraction by reverse and preparative HPLC. One single large peak of a radioactive peptide was observed and radioactivity was only associated with this peak eluted from the HPLC column with 38% (v/v) acetonitrile/ water. This peptide was subjected to amino acid sequence

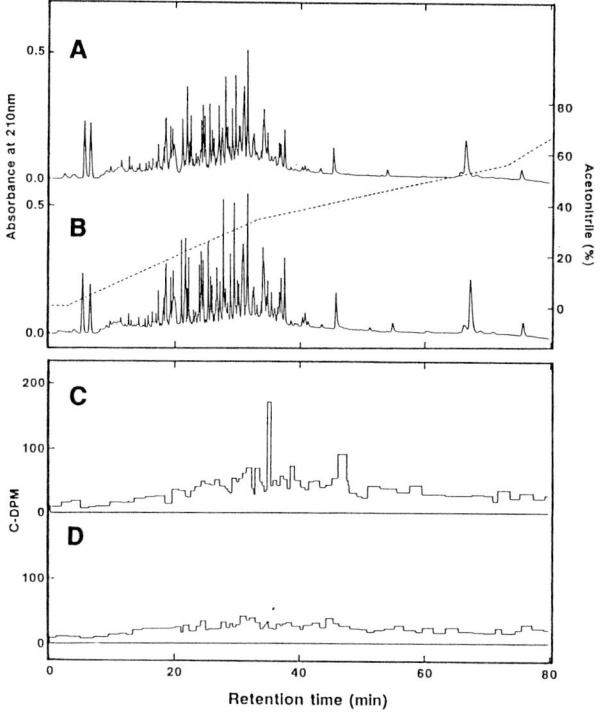

Figure 2. High-pressure liquid chromatographic peptide maps of tryptic digests of cytochrome P-450$_{scc}$ modified with [14C]methoxychlor in the absence and presence of 200 $\mu$M cholesterol.

A, C) A portion (2 nmol) of a 24-hour tryptic digest of cytochrome P-450$_{scc}$ without cholesterol was subjected to RP-HPLC using a linear acetonitrile gradient. Peptide elution was monitored as to the absorbance at 210 nm. The histogram shows the radioactivity associated with each peak.

B, D) A portion (2 nmol) of a 24-hour tryptic digest of cytochrome P-450$_{scc}$ with cholesterol.

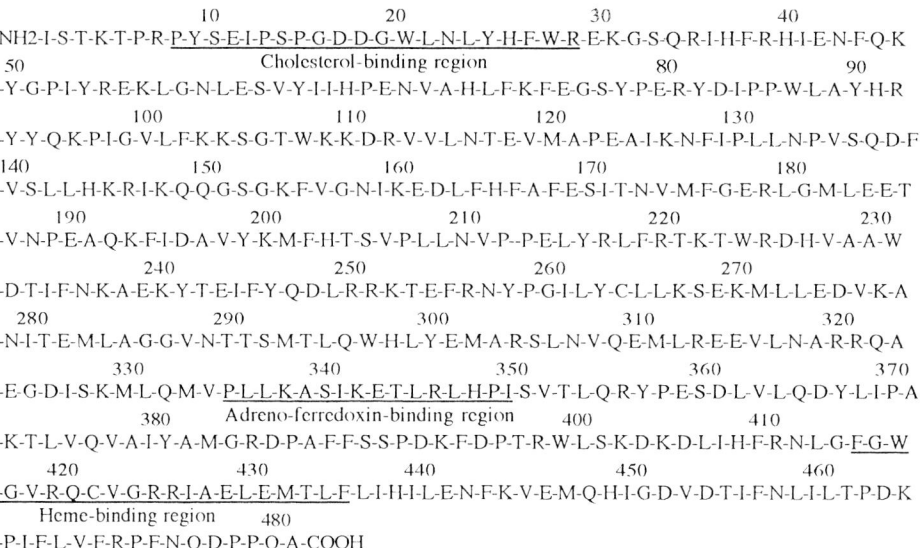

Figure 3. Amino acid sequence of bovine cytochrome P-450scc (Chashchin, 1986). The binding region of cholesterol, adreno-ferredoxin and heme are indicated by underlines.

analysis. Twenty-one amino-acid residues were detected, as follows: Pro-Tyr-Ser-Glu-Ile-Pro-Ser-Pro-Gly-Asp-Asn-Gly-Trp-Leu-Asn-Leu-Tyr-His-Phe-Trp-Arg. The majority of methoxychlor was expected at the 13th or 20th cycle of the peptide analysis, with an extremely low amount at the 13th and 20th cycles, respectively.

The composition deduced from the amino-acid residues of the major radioactive peptide was completely identical with the composition expected from the amino-acid sequence of the peptide from proline-8 to arginine-28 (Figure 3).

## References

Chashchin, V.L., Lapko, V.N., Adamovich, T.B., Lapko, A.G., Kuprina, N.S. and Akhrem, A.A. (1986): Primary structure of the cholesterol side-chain cleavage cytochrome *P*-450 from bovine adrenocortical mitochondria and some aspects of its functioning on a structural level. Biochim. Biophys. Acta 871: 217-223.

Kupfer, D. and Bulger, W.H. (1987): Metabolic activation of pesticides with proestrogenic activity. Federation Proc. 46: 1864-1869.

Tsujita, M. and Ichikawa, Y. (1993): Substrate-binding region of cytochrome *P*-450$_{scc}$ (P450XIA1). Identification and primary structure of the cholesterol binding region in cytochrome P-450$_{scc}$. Biochim. Biophys. Acta 1161: 124-130.

# Inhibition of cytochrome P450 enzyme activity using specific inhibitory antibodies

Malcolm Hatcher, Richard Brown, Michael J. O'Sullivan, Stuart Swinburne

*Amersham International plc, Metabolism and Toxicology, Cardiff Laboratories, Forest Farm, Whitchurch, Cardiff, Wales, UK, CF4 7YT*

INTRODUCTION

Cytochrome P450's are responsible for the oxidative 'phase one' metabolism of a wide variety of endogenous and exogenous compounds. The metabolites from the cytochrome P450 reactions are often further metabolised by 'phase two' enzymes to water soluble conjugates. These polar compounds are generally rapidly excreted. However, cytochrome P450 catalysed reactions can occasionally lead to the formation of highly reactive electrophiles. These activated metabolites can react with normal cellular components leading to potentially toxic, mutagenic and carcinogenic events.

A variety of xenobiotics are known to induce considerable elevations in the levels of a number of cytochrome P450's. This may cause serious side effects during drug treatment. Induction of P450's can lead to an increased rate of drug metabolism with decreased plasma half-life, an alteration in plasma levels of any co-administered drugs and to the increased formation of toxic metabolites. For these reasons it is considered essential to identify any interaction with the cytochrome P450 system during the development and safety evaluation of new drugs.

The induction of specific cytochrome P450's can be provisionally identified using selective enzyme assays. However, due to the complexity of the cytochrome P450 superfamily and the relative lack of specificity of many cytochrome P450 enzyme assays, it is considered desirable to confirm these results by an alternative technique. The use of well characterised specific antibodies is a powerful tool for confirming the identity of induced cytochrome P450's.

We describe the development of optimised antibody inhibition protocols for cytochrome P450's 1A1 and 2B. The inhibitory properties of monoclonal and polyclonal antibodies have also been compared.

METHODS

The fluorimetric assays for cytochrome P450's 1A1 and 2B were based upon ethoxyresorufin O-deethylation (EROD) and pentoxyresorufin O-dealkylation (PROD) respectively. The deethylation of 7-ethoxyresorufin was measured at ambient temperature (22 - 24°C) in quartz curvettes (1cm path lengths) by monitoring the rate of formation of resorufin using a Perkin-Elmer LS50B fluorimeter. Excitation and emission wavelengths of 530 and 585nm, and a slit width of 5nm were employed. The concentrations of each component of the assay was optimised. The final EROD assay conditions were as follows: tris buffer (0.1M, pH7.8), $MgCl_2$ (12.5mM), ethoxyresorufin (2.0$\mu$M) and rat liver microsomes (100$\mu$g) in a volume of 2230$\mu$l. The baseline fluorescence was recorded for approximately 30 seconds and then the reaction was initiated by the addition of NADPH (50$\mu$l, 50$\mu$M) to the cuvette. The increase in fluorescence was monitored for approximately 200 seconds. The rate of resorufin formation was determined by comparison to the fluorescence of a resorufin standard curve. The PROD assay was as described above except that 10$\mu$M pentoxyresorufin was substituted for the ethoxyresorufin.

The EROD and PROD activity of liver microsomes from untreated rats and rats treated with $\beta$-naphthoflavone ($\beta$NF), phenobarbital (PB) and dexamethasome (DEX) were determined. The effect of monoclonal and polyclonal antibodies to cytochrome P450 1A1/2 and 2B on the EROD and PROD activity of the rat liver microsomes was assessed. The monoclonals were used as ascites fluid, the polyclonals as an immunoglobulin fraction. In these inhibition experiments, the antibodies were incubated with the microsomes for 30 minutes at ambient temperature prior to the enzyme assays.

RESULTS

1. Cytochrome P450 1A1
Under the EROD assay conditions described the $\beta$NF microsomes had a specific activity of 3.67nmoles of resorufin formed per minute per mgram of microsomal protein. This represents a 126 fold increase in EROD activity compared to microsomes from untreated rats. Microsomes from rats treated with PB or DEX had increases of 7 and 2.5 fold respectively compared to control microsomes. The effect of monoclonal antibody to cytochrome P450 1A1/2 on the EROD activity of microsomes from $\beta$NF treated rats is shown in Fig. 1. Maximum inhibition (84%) of activity was observed at an antibody to microsomal protein ratio of 2:1. Lower inhibition (67%) occurred with the polyclonal, even at a higher (3:1) antibody to microsomal protein ratio.

2. Cytochrome P450 2B
Microsomes from rats treated with PB showed a large (230 fold) increase in PROD activity. Microsomes from $\beta$NF treated and DEX treated animals had similar levels of PROD activity as microsomes from control animals. Microsomes from PB treated rats had a PROD specific activity of 2.44nmoles resorufin formed per minute per mgram. The PROD activity of the microsomes from the control rats was extremely low.

The effect of monoclonal antibody to cytochrome P450 2B on the PROD activity of PB induced microsomes is shown in Fig. 2. Maximum inhibition (98%) was observed at an antibody to microsomal protein ratio of 0.5:1. Similar inhibition occurred with the polyclonal antibody, but a much higher antibody to microsomal protein ratio (3:1) had to be used.

3. Human Cytochrome P450 1A1
The effect of monoclonal anti-rat cytochrome P450 1A1/2 antibody on the activity of human cytochrome P450 1A1 expressed in a V79 cell line is shown in Fig. 3. The results indicate that the antibody is an effective inhibitor of both human and rat enzymes.

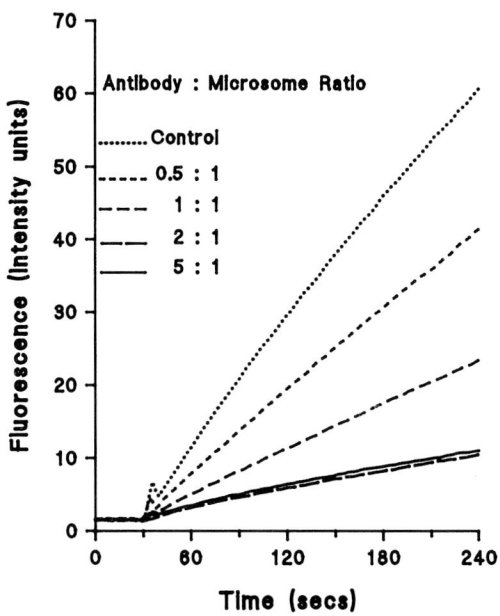

Fig. 1. Antibody inhibition of the EROD activity of βNF microsomes (100μg) using monoclonal anti-CYP 1A1/2

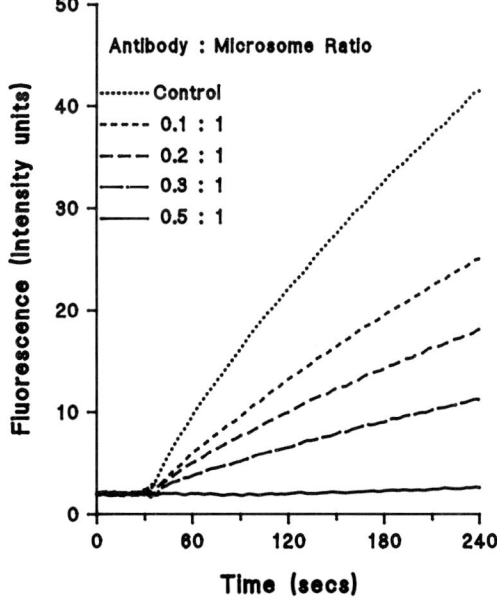

Fig. 2. Antibody inhibition of the PROD activity of PB microsomes (100μg) using monoclonal anti-CYP 2B

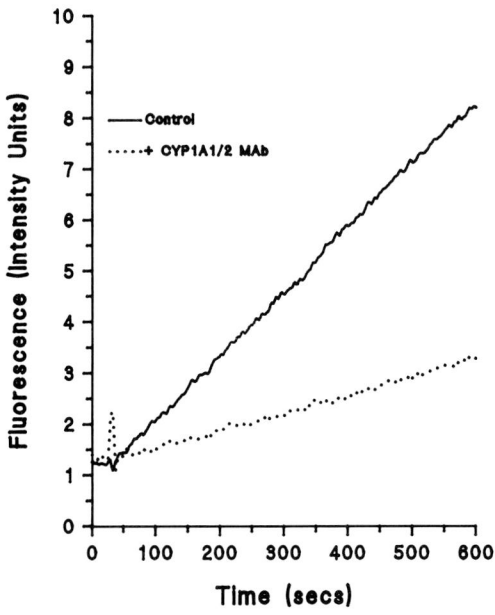

Fig. 3. Antibody inhibition of the EROD activity of human CYP1A1 expressed in V79 Chinese hamster cells

DISCUSSION

We have developed rapid convenient inhibition protocols for rat cytochrome P450's 1A1 and 2B based upon monoclonal antibodies. The cytochrome P450 1A1/2 antibody also inhibits human P450 1A1 activity. Contrary to some previous literature reports, inhibition was achieved even when using low antibody to microsomal protein ratios. Rat cytochrome P450 2B activity can be quantitatively inhibited, but a maximal inhibition of 84% was observed with cytochrome P450 1A1. The polyclonals also inhibited the enzymes but higher antibody to microsomal protein ratios were required. The monoclonal antibodies are very specific. A large excess of anti-cytochrome P450 2B had little effect on the EROD activity of βNF induced microsomes. Similarly, a large excess of anti-cytochrome P450 1A1/2 had little or no effect on the PROD activity of PB induced microsomes. The polyclonals are less specific in this application. Anti-cytochrome P450 1A1/2 at a 10:1 ratio of antibody to microsomal protein inhibited the PROD activity of PB induced micromes by 40%. This is rather surprising as we have previously demonstrated that the polyclonal antibodies are very specific in Western blotting applications. It may represent differences in epitopes on native compared to denatured enzymes.

CONCLUSIONS

We have described rapid, convenient inhibitory protocols for cytochrome P450's 1A1 and 2B based on potent, specific monoclonal antibodies.

# Mammalian cell lines stably expressing bovine adrenal 11β-hydroxylase cDNA

Manfred Husemann, Thomas Petri

Research Laboratories of Schering AG, Müllerstr. 170, D-13342 Berlin, Germany

## Summary

Cytochrome P450 enzymes (P450) catalyse either the biosynthesis of steroids and prostaglandins or are involved in the degradative metabolism of endogenous or xenobiotic lipophilic substrates. The aim of the present study was to construct recombinant animal cell lines functionally expressing a P450 for potential use in steroid converting reactions. The cloned cDNA of the bovine 11β-hydroxylase, an adrenal steroidogenic P450, was inserted into a mammalian expression vector containing the strong viral MPSV promoter and the CMV enhancer. CHO cells were transfected with this plasmid together with a selectable marker gene. Stably transformed clones could be isolated. Genomic cDNA integration and transcription were demonstrated by PCR analysis of the cellular DNA and of reverse transcribed RNA respectively. Functional expression was shown by HPLC analysis of cell culture extracts after incubation with tritium labelled progesterone, a substrate for 11β-hydroxylase.

## INTRODUCTION

A large number of cytochrome P450 enzymes (P450) catalyse the oxidative metabolism of structurally diverse lipophilic substrates.

Four distinct P450 are involved in the biosynthesis of steroid hormones from cholesterol in the adrenal cortex, leading to the production of gluco- and mineralocorticoids as well as adrenal androgens. One of these four P450, the mitochondrial 11β-hydroxylase, is the key enzyme for the synthesis of gluco- and mineralocorticoids, catalysing the 11β-hydroxylation of progesterone, 11-deoxycorticosterone or 11-deoxycortisol. The bovine enzyme also catalyses the 18-hydroxylation as well as 18-aldehyde formation which are essential for aldosterone synthesis. Several subforms of 11β-hydroxylase have been described (Kirita et al., 1988; Mathew et al., 1990).

In the present study one form of the bovine adrenal 11β-hydroxylase (corresponding to P450 (11β)-3 described by Kirita et al., 1988) was stably expressed in CHO cells as a specific means of analysing the catalytic specificity of this enzyme. The recombinant cells were analysed at the level of cDNA integration, transcription and functional expression.

**RESULTS**

The cloned cDNA of the adrenal 11β-hydroxylase - flanked at the 5'-end by a synthetic Kozak site to allow the efficient translation of the mRNA - was inserted into the plasmid pMPSV/CMV (Wirth et al., 1991) giving pMPSV/CMV-11β. This plasmid together with the mouse *dhfr* cDNA containing expression plasmid pUdhfr was transfected into dhfr- CHO cells using the calcium phosphate transfection technique. Dhfr+ clones were selected and further propagated for subsequent analysis of the clones.

Total genomic DNA was isolated from three clones in order to check by PCR whether the dhfr+ cells had integrated the hydroxylase cDNA into their genome. A 0.9 kb fragment spanning an internal 11β-hydroxylase cDNA region could be amplified demonstrating its stable integration. PCR with genomic DNA from non transfected CHO cells did not result in amplification products (negative control) (Fig. 1).

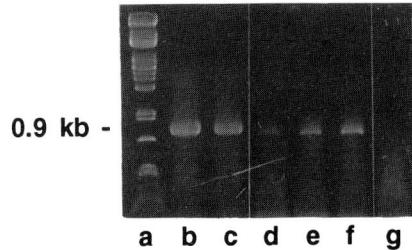

Fig. 1. PCR analysis of genomic DNA of three independent CHO cell clones transfected with pMPSV/CMV-11β. **a** molecular weight marker, **b,c** positive controls, **d-f** CHO cell clones, **g** non transfected CHO cells

To test whether the 11β-hydroxylase cDNA was transcribed into mRNA, total RNA was isolated from same three clones, treated with RNase free DNase, reverse transcribed with MMLV reverse transcriptase and amplified by PCR. A 0.2 kb 3' terminal fragment was obtained using specific primers (Fig. 2). No signal was obtained from RNA which had not been reverse transcribed prior to PCR excluding the presence of contaminating DNA (data not shown).

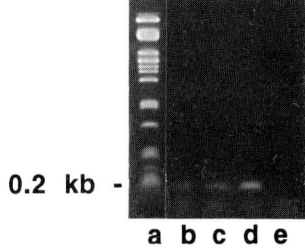

Fig. 2. PCR analysis of reverse transcribed RNA of three independent CHO cell clones transfected with pMPSC/CMV-11β. **a** molecular weight marker, **b-d** CHO cell clones, **e** non transfected CHO cells

To test for the functional activity of the 11β-hydroxylase one of the recombinant CHO cell clones was analysed in comparison to non transfected cells by incubating the cells for 48 h with radiolabelled progesterone. Lipophilic products were extracted and analysed by HPLC. A radioactive product comigrating with cold 11β-hydroxy-progesterone could only be detected in the extract from the recombinant cells indicating that the 11β-hydroxylase was functionally active (Fig. 3). Products hydroxylated in the 18 position could not be detected, which could be due to the known low proportion of the 18-hydroxylase activity.

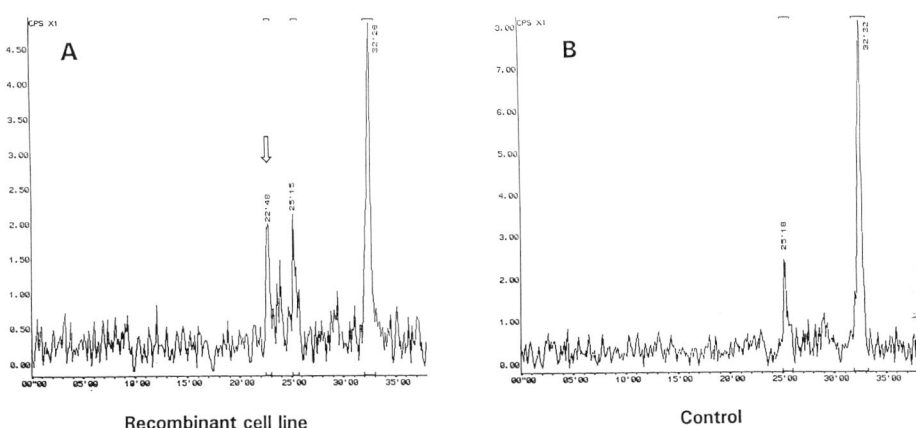

**Fig. 3.** HPLC analysis of a lipophilic extract from pMPSV/CMV-11β transfected (**A**) or non transfected (**B**) CHO cell cultures incubated with radiolabelled progesterone. The cells were incubated for 48 h with $^3$H labelled progesterone (740 Bq, 0.13 ng). Lipophilic products were extracted from the cells and culture medium with acetone/methanol/Triton X100 (750/250/2) and analysed by reverse phase HPLC. The arrow indicates the product comigrating with 11β-hydroxy-progesterone.

## DISCUSSION

The present data show that CHO cells can efficiently and stably express a functional active adrenal mitochondrial P450. In contrast, tansient expression of the bovine 11β-hydroxylase could be demonstrated in COS cells by Mathew et al., 1990. Permanent cell lines however offer the advantage of having a stable and reproducible phenotype. Stable expression of P450 in CHO cells could also be shown by Thompson et al. (1992) for P450 IA2.

Alternatively we have also tried to achieve functional expression of the 11β-hydroxylase in BHK cells using the same expression vector as above in combination with a puromycin selection system. The stable integration of the 11β-hydroxylase cDNA into the BHK cell genome as well as its transcription into mRNA could be demonstrated by PCR (data not shown). However, we could not detect a specific hydroxylation product when the corresponding BHK cell clones were incubated with radiolabelled progesterone. Only endogenous progesterone metabolizing activity of the BHK host cells could be observed which was different from the endogenous activity in untransfected CHO cells.

The missing specific 11β-hydroxylase activity in the otherwise mRNA positive BHK cell clones could be due to a dominant endogenous metabolism or to insufficient activity of adrenodoxin and/or adrenodoxin reductase, enzymes which are necessary for the activity of mitochondrial P450.

## ACKNOWLEDGEMENT

We thank D. Klews and Da. Schmidt for excellent technical assistance, De. Schmidt for synthesizing the oligonucleotides and H. Gieschen and B. Salomon for their help with the HPLC analysis. We also thank W. Boidol for cloning and providing the 11β-hydroxylase cDNA.

## REFERENCES

Kirita, S., Morohashi, K.-i., Hashimoto, T., Yoshioka, H., Fujii-Kuriyama, Y. & Omura, T. (1988): Expression of two kinds of cytochrome P450(11β) mRNA in bovine adrenal cortex. *J. Biochem. 104*: 683-686.

Mathew, P.A., Mason, J.I., Trant, J.M., Sanders, D. & Waterman, M.R. (1990): Amino acid substitutions Phe$^{66}$ → Leu and Ser$^{126}$ → Pro abolish cortisol and aldosterone synthesis by bovine cytochrome P450$_{11β}$. *J. Biol. Chem. 265:* 20228-20233.

Thomson,L.H., Wu, R.W. & Felton, J.S. (1991): Introduction of cytochrome P450IA2 metabolic capability into cell lines genetically matched for DNA repair proficiency / deficiency. *Proc. Natl. Acad. Sci. USA 88* : 3827-3831.

Wirth, M., Schumacher, L. & Hauser, H. (1991): Construction of new expression vectors for mammalian cells using the immediate early enhancer of the human cytomegalovirus to increase expression from heterologous enhancer/promoters. In *Protein Glycosylation: Cellular, Biotechnical and Analytical Aspects*, Vol. 15, ed. H.S. Conradt, pp. 49-52. Weinheim: VCH Publishers.

# Expression of cDNAs encoding human and marmoset CYP2As

Manoj Nanji[1], Philippe Clair[2], Ian R. Phillips[2], Elizabeth A. Shephard[1]

[1]Department of Biochemistry and Molecular Biology, University College London, Gower Street, London WC1E 6BT, UK. [2]Department of Biochemistry, Queen Mary and Westfield College, University of London, Mile End Road, London E1 4NS, UK

INTRODUCTION

Cytochrome P450-dependent mono-oxygenases (CYPs) are a large superfamily of enzymes that catalyze the oxidation of lipophilic chemicals by the insertion of a single atom of molecular oxygen into the substrate. They play an important role in the metabolism of a wide variety of foreign, as well as endogenous, compounds (Gonzalez, 1989). The expression of individual CYPs has become a valuable tool for studying the structure-function relationship of these proteins and their metabolic capacities, and a variety of heterologous expression systems for CYPs have been developed, including yeast (Ching *et al.*, 1991), COS cells (Zuber *et al.*, 1988), vaccinia virus (Liu *et al.*, 1991), bacteria (Fisher *et al.*, 1992) and baculovirus (Assefa *et al.*, 1989). In this paper we report the isolation of a full-length cDNA encoding human CYP2A6 and its expression both in the baculovirus-Sf9 insect cell system and as a fusion protein in *Escherichia coli*.

The marmoset is often used as a primate model for drug metabolism studies. However, very little is known about the functions of marmoset CYPs and how they compare with those of the corresponding CYPs of man. Based on the sequence of human CYP2A6 cDNA, we have designed PCR primers that have enabled us to isolate and characterize a cDNA that encodes a member of the CYP2A subfamily of marmoset.

MATERIALS AND METHODS

Isolation and characterization of cDNA clones

Total RNA isolated from human and marmoset liver was copied into cDNA by reverse transcriptase. The cDNAs were used as templates for PCR amplifications primed by oligonucleotides based on sequences located at the ends of the coding region of human CYP2A6 mRNA. Amplified cDNAs were inserted into pUC19 and their sequence determined by the dideoxy chain-termination method using T7 DNA polymerase and a sequencing kit from Pharmacia P-L Biochemicals Inc.

Expression in *E.coli*

Full-length cDNA encoding human CYP2A6 was inserted into an *Xba*I site of the expression vector pMAL-p.The recombinant plasmid was introduced into *E.coli* JM109 cells and expressed as described by Riggs (1990). In this system the foreign protein is expressed fused to maltose-binding protein via a Factor Xa proteolytic cleavage site. Expression is under the control of a strong, inducible *tac* promoter and *malE* translational initiation signals.

Expression in the baculovirus-Sf9 cell system

CYP2A6 cDNA was inserted into the transfer vector pAcC5. Construction, expression and maintenance of recombinant baculovirus were essentially as described by Summers et al. (1987) and O'Reilly et al. (1992). Recombinant baculovirus was generated by co-lipofection of Sf9 cells with 5µg of recombinant transfer vector and 0.5 µg of linearized AcMNPV DNA (Baculogold, AMS Biotechnologies). For expression of CYP2A6, fresh Sf9 cells were infected with high-titre recombinant virus and cultivated in suspension culture in medium supplemented with 100 µM each of iron citrate and 5-aminolaevulinic acid. For details see Clair et al. (1993).

Identification of expressed proteins

E. coli and Sf9 insect cells containing expression vectors coding for CYP2A6 were lysed in 2x SDS gel loading buffer (4% SDS, 125 mM Tris-HCl (pH 6.7), 30% (v/v) glycerol, 0.002% (w/v) bromophenol blue, and 2% (v/v) β-mercaptoethanol). Cell lysates were electrophoresed through an SDS/polyacrylamide gel and analysed by western blotting with an antibody raised against CYP2A7 isolated from baboon liver.

Coumarin 7-hydroxylase assay

Coumarin 7-hydroxylase activity was measured in a reaction mixture (1ml) containing 50 mM Hepes (pH 7.4), 15mM $MgCl_2$, 50 µM coumarin (Sigma), 500 µM NADPH and a sample of Sf9 microsomal membranes containing 1mg of protein. The reaction mix was incubated at $37^oC$ for various times. The reaction was stopped by addition of 0.5 ml of 6% (w/v) trichloroacetic acid, and precipitated protein was removed by centrifugation. The supernatant was added to 3 ml of 0.8 M Tris/0.8 M glycine (pH 9.0), and 7-hydroxycoumarin was determined fluorometrically (Miles et al., 1990).

Northern blot hybridization analysis

Total RNA was isolated from livers of untreated and phenobarbital-treated marmosets by the method of Chirgwin et al. (1979). The RNA was analyzed by northern blot hybridization with the marmoset CYP2A cDNA sequence isolated as described above.

RESULTS AND DISCUSSION

Full-length cDNA sequences encoding members of the CYP2A subfamily were amplified from total RNA isolated from human and marmoset liver. The human cDNA was identical in sequence to that previously published for CYP2A6 (Yamano et al., 1990). The marmoset cDNA had 90% nucleotide sequence identity with human CYP2A6. Northern blot hybridization analysis of RNA isolated from the livers of untreated or phenobarbital-treated marmosets revealed that the marmoset CYP2A subfamily member was induced 8-fold by phenobarbital.

Human CYP2A6 was expressed in E. coli as a fusion protein with maltose binding protein. The fusion protein (relative molecular mass 91000) was identified by western blotting using an antibody to baboon CYP2A7 (Fig.1A). Its expression was substantially increased after exposure of the E. coli cells to IPTG (Fig.1A, tracks 2 & 3). The fusion protein was unable to catalyse the 7-hydroxylation of coumarin.

CYP2A6 was also expressed using the Baculovirus-Sf9 insect cell system. Immunoblot analysis (with an antibody to baboon CYP2A7) of Sf9 cells infected with recombinant baculovirus revealed a band with a relative molecular mass of 49000 (Fig. 1B, track 4). The antibody did not cross-react with protein extracts from non- infected Sf9 cells nor with extracts from Sf9 cells infected with wild-type baculovirus (Fig. 1B, tracks 2&3). Microsomes isolated from cells infected with wild-type baculovirus were unable to hydroxylate coumarin. However, microsomes isolated from cells infected with the CYP2A6 recombinant virus catalyzed the conversion of coumarin to 7-hydroxycoumarin. The amount of product produced was increased about 3-fold if exogenous NADPH-dependent cytochrome P-450 reductase (either purified from rat liver, or human enzyme expressed in Sf9 cells) was added to the reaction mix. This result demonstrates that the Sf9 cells contain insufficient endogenous NADPH-dependent cytochrome P-450 reductase to cope wuth the high levels of CYP2A6 expression. Experiments are in progress to express, in *E. coli* and insect Sf9 cells, the cDNA encoding a member of the CYP2A subfamily of marmoset.

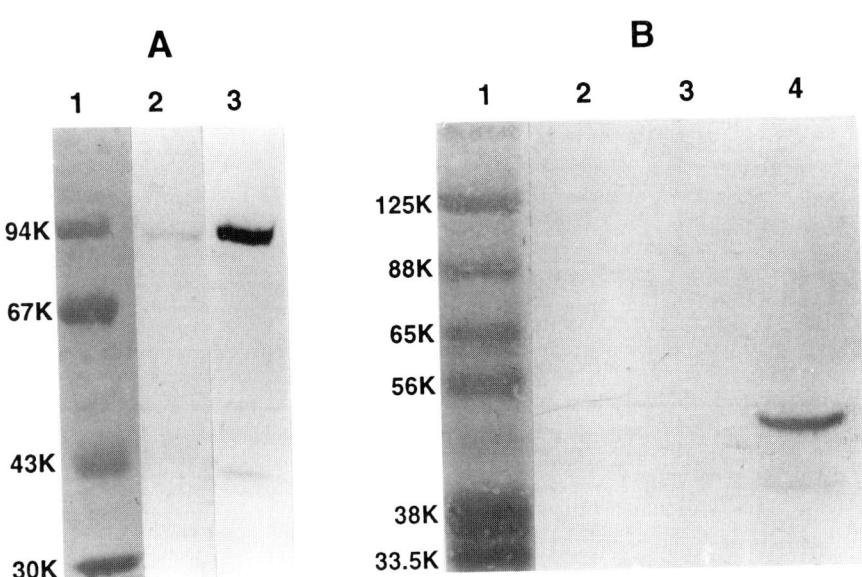

Fig. 1. Western blot analysis with an antibody against baboon CYP2A7 of protein extracts from *E. coli* (A) and Sf9 cells (B).
A. Protein extracts were from uninduced (track 2) or IPTG-treated (track 3) *E. coli* cells that had been transformed with pMAL-p containing sequences encoding CYP2A6. Track 1 shows molecular weight markers.
B. Protein extracts were from non-infected cells (track 2), and cells infected with wild- type baculovirus (track 3) or with CYP2A6 recombinant virus (track 4). Track 1 shows molecular weight markers.

ACKNOWLEDGEMENTS

MN is a recipient of a post-graduate studentship from the Medical Research Council, U.K. The work was supported by a grant from the Cancer Research Campaign.We thank Dr. Claude Bonfils for CYP2A7 antibodies.

REFERENCES

Assefa, A.,Smith, S.J., Nagata, K., Gillette, J., Gelboin, H.V. & Gonzalez, F.J. (1989): Novel exogenous heme dependent expression of mammalian cytochrome P450 using baculovirus. *Arch. Biochem. Biophys.* 274, 481-490.
Ching, M. S., Lennard, M. S., Tucker, G. T., Woods, H. F. & Kelly, S. L.(1991): The expression of human cytochrome 1A1 in the yeast *Saccharomyces cerevisiae*. *Biochem. Pharmacol.* 42, 753-758.
Chirgwin, J. M., Przybyla, A. E., MacDonald, J. R. & Rutter, J. W (1979): Isolation of biologically active ribonucleic acid from sources enriched in ribonuclease. *Biochemistry* 18, 5294- 5299.
Clair, P., Phillips, I. R. & Shephard, E. A. (1993): Expression of components of the cytochrome P-450 mediated mono-oxygenase system using recombinant baculoviruses. *These Proceedings*.

Fisher, C. W., Caudle, D. L., Martin-Wixtrom, C., Quattrochi, L. C., Tukey, R. H., Waterman, M. R. & Estabrook, R. W (1992): High level expression of functional human cytochrome P4501A2 in *Escherichia coli. FASEB. J.* 6, 759-764.

Gonzalez, F. J. (1989): The molecular biology of cytochrome P-450s. *Pharmacol. Rev.* 40, 243-287.

Liu, G., Gelboin, V. & Myers, M. J. (1991): Role of cytochrome P4501A2 in acetanilide 4-hydroxylation as determined with cDNA expression and monoclonal antibodies. *Arch. Biochem. Biophys.* 284, 400-406.

Miles, J. S., McLaren, A. W., Forrester, L. M., Glancey, M. J., Lang, M. A. & Wolf, C. R. (1990): Identification of the human liver cytochrome P-450 responsible for coumarin 7-hydroxylase activity. *Biochem. J.* 267, 365-371.

O'Reilly, D. R., Miller, L. K. & Luckow, A. V. (1992): Baculovirus expression vectors. A laboratory manual. New York: W. H. Freeman and Company.

Summers, M. D. & Smith, G. E. (1987): A Manual of methods for baculovirus vectors and insect cell culture procedures, Texas Agricultural Experiment Station.

Riggs, P. (1990): Expression and purification of maltose-binding protein fusions. Current protocols in molecular biology, suppl. 19, Unit 16.6.

Yamano, S., Tatsuno, J. & Gonzalez, F. J.(1990): The CYP2A3 gene product catalyses coumarin 7-hydroxylation in human liver microsomes. *Biochemistry* 29, 1322-1329.

Zuber, M. X., Mason, J. I., Simpson, E. R. & Waterman, M. R. (1988): Simultaneous transfection of cos-1 cells with mitochondrial and microsomal steroid hydroxylases: incorporation of a steroidogenic pathway into non-steroidogenic cells. *Proc. Natl. Acad. Sci. USA* 85, 699-703.

# Localization and distribution of CYP2B1 in the rat lung by immunocytochemistry and by *in situ* hybridization

Maria H. Figueiredo[1], Vera Ribeiro[2], Domingos Henrique[2], Maria C. Lechner[2], Vasco Bairos[1]

[1]Institute of Histology and Embryology, Faculty of Medicine, 3049 Coimbra Codex, Portugal.
[2]Laboratory of Biochemistry, Gulbenkian Institute of Science, 2781 Oeiras Codex, Portugal

INTRODUCTION

The lung is a complex organ with a great histological and biochemical heterogeneity. Having a large airway surface and extensive capillary network, it becomes a prominent target for a great variety of xenobiotics.
The pulmonary distribution of the enzymes responsible for the detoxication or bioactivation of xenobiotics may vary substantially from one cell type to another. Thus, it is of the utmost importance to identify the metabolic capabilities of each different cellular population. The purpose of the present study was to define the cellular localization and distribution of CYP2B1 in the rat lung, by immunocytochemistry and by *in situ* hybridization.

MATERIAL AND METHODS

Animals - Adult male Wistar rats were used throughout the present investigation.
Immunocytochemistry - The animals were perfused with 0.1 M phosphate buffered saline (PBS) pH=7.4, followed by 4% formaldehyde in PBS. The lungs were further excised and immersed in the same fixative for 9 hours at +4 °C. Paraffin sections were processed for immunocytochemical light microscopy using the avidin-biotin-peroxidase procedure (ABC) with diaminobenzidine tetrahydrochloride as the chromogen (Hsu et al., 1981).
Antibodies against CYP2B1 were raised in young female rabbits according to Kamataki et al. (1976) by using CYP2B1 purified from the liver microsomal fraction of phenobarbital treated rats by affinity chromatography on a tryptamine-CH sepharose 4B column (Lu et al., 1980). The eluted fraction was further submitted to DEAE-cellulose chromatography.
The quantitative distribution of the CYP2B1 in the bronchoalveolar

lining cells was determined by microdensitometrical analysis of the immunostaining intensity expressed in relative integrated optical density.

CYP2B1 mRNA *in situ* hybridization - The detection of CYP2B1 mRNA by *in situ* hybridization was performed on cryostat sections employing an anti-*sense* [$^{35}$S]RNA probe (Harper and Marselle, 1987). This riboprobe was synthesized by transcription of a 0.63 kb CYP2B1 cDNA fragment, subcloned in a pBS vector, using [$\alpha$-$^{35}$S]UTP and T7 polymerase. The probe size was reduced to an average length of 100 ntd by alkaline hydrolysis.

In order to enhance the autoradiographic signal the lung sections were observed with simultaneous transmitted and incident light using the filter block IGS usually employed for immunogold labelling.

RESULTS

Immunocytochemistry

The immunocytochemical studies shown in Figs. 1 to 3, demonstrate that the CYP2B1 is present in the bronchoalveolar lining cells mainly in the Clara cells and type II pneumocytes, but also in ciliated cells. It is clearly observed that the immunostaining is much less intense in the bronchi than in the bronchioles.

The quantitative assessment of the immunocytochemical labelling, Fig.4, carried out with different dilutions of the primary antiserum, revealed notorious differences in the staining intensity among those epithelial cells. For the same dilution of the specific antiserum Clara cells are systematically much more intensely stained than type II pneumocytes or ciliated cells. These data demonstrate that Clara cells contain the greatest concentration of CYP2B1 in the rat lung. Conversely, ciliated cells exhibit a very low content of this haemoprotein, showing a positive immunolabelling only when low dilutions of the primary antiserum are used. Actually, when dilutions higher than 1/1 000 are used, only Clara cells reveal a persistently strong reaction, preferencially located in the apical cytoplasm, Fig. 2. No specific reaction was noticed in the goblet cells, type I pneumocytes or endothelial cells.

The different pattern of staining intensity, observed in the lining epithelium of the proximal and distal regions of the conducting airways, reflects the presence of different cellular populations. Indeed, the bronchi are characterized by a higher proportion of ciliated cells, while the bronchioles are predominantly lined by Clara cells.

*In situ* hybridization

The autoradiographic labelling, obtained after *in situ* hybridization reaction, is characterized by a rather uniform and homogeneous distribution of the silver grains over the epithelial lining cells of the conducting airways. Autoradiographic signal variations between proximal, Fig. 5, and distal conducting airways, Fig. 6, were not detected as it would be expected from the immunocytochemical results. In the respiratory portion, an intense autoradiographic labelling was

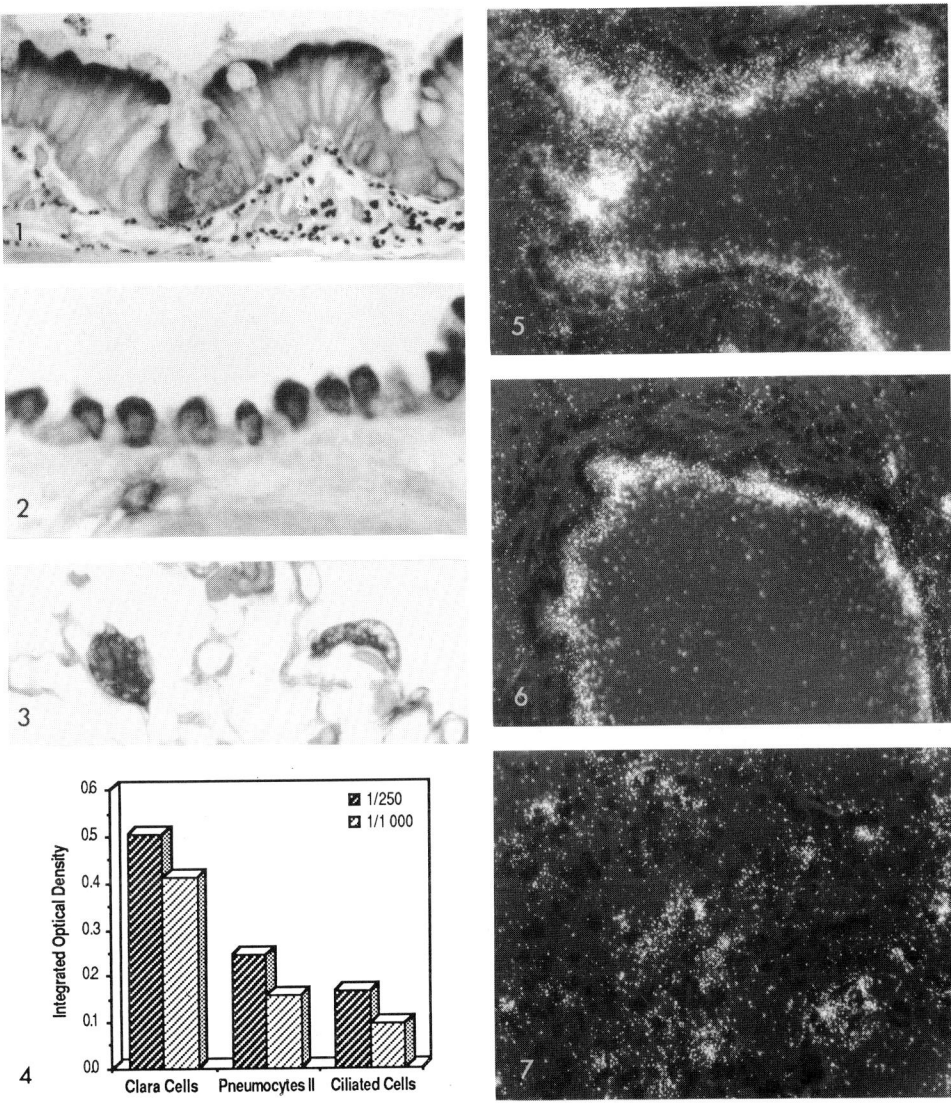

Figs. 1. to 3. Immunocytochemical localization of CYP2B1 in the ciliated cells (Fig. 1. 620 x), in the Clara cells (Fig. 2. 600 x) and in type II pneumocytes (Fig. 3. 900 x).

Fig. 4. Microdensitometrical analysis of the immunocytochemical labelling using 1/250 and 1/1 000 dilutions of the primary antiserum.

Fig. 5. Localization of the CYP2B1 mRNA in the bronchial lining cells by *in situ* hybridization. 260 x

Fig. 6. Photomicrograph of the autorradiographic signal observed in the bronchiolar epithelium. 270 x

Fig. 7. Lung respiratory region showing the autorradiographic labelling of the type II pneumocytes. 135 x

observed over the type II pneumocytes, Fig. 7, while no specific signal was detected over type I pneumocytes or endothelial cells.
Taken together, these observations lead to the conclusion that the CYP2B1 mRNA is present at comparable concentrations in most bronchoalveolar epithelial cells.

DISCUSSION

The results presented herewith demonstrate that the CYP2B1 gene is transcriptionally active in the bronchoalveolar lining cells where this mRNA is found at equivalent concentrations. However, the different concentrations of the protein found in each individual cell type, with a net predominance in the Clara cells, suggest that distinct post-transcriptional regulatory mechanisms modulate the synthesis of this mono-oxygenase in Clara cells, in type II pneumocytes and in the ciliated cells.
The presence of the CYP2B1 in the broncho-alveolar epithelium should represent one of the most important pulmonary defence mechanisms against many xenobiotics, mainly air pollutants. We admit that the presence of an excess of apparently untranslated mRNA in the ciliated cells, as compared to the Clara cells, may correspond to a stock of this messenger, in a latent form, ready to respond to potential environmental agressions.

REFERENCES

Harper, M.E. & Marselle, L.M. (1987): RNA detection and localization in cells and tissue sections by *in situ* hybridization of $^{35}$S labeled RNA probes. *Methods in Enzymology* 151, 539-551.
Hsu , S.M., Raine, L. & Fanger, H. (1981): Use of avidin-biotin--peroxidase complex (ABC) in immunoperoxidase techniques: a comparison between ABC and unlabeled antibody (PAP) procedures. *J. Histochem. Cytochem.* 29, 557-580.
Kamataki, T., Belcher, D.H. & Neal, R.A. (1976): Studies of the metabolism of diethyl p-nitrophenyl phosphothionate (parathion) and benzphetamine using an apparently homogeneous preparation of rat liver cytochrome P-450: effect of a cytochrome P-450 antibody preparation. *Mol. Pharmacol.* 12, 921-932.
Lu, A.Y.H., Morin, J.W. & Warner, M. (1980): Purification of liver microsomal cytochrome P-450 by affinity chromatography. In *Biochemistry Biophysics and Regulation of Cytochrome P-450*, ed. J.-A. Gustafsson, J. Carlstedt-Duke, A. Mode & J. Rafter, Developments in Biochemistry vol. 13, pp. 17-24: New York, Elsevier/North Holland Biomedical Press.

# Site directed mutagenesis on bovine adrenodoxin to target amino acids involved in electron transfer and binding to adrenodoxin reductase and cytochrome P450$_{scc}$

Marie-France Palin, Jurgen Sygusch, Jean-Guy LeHoux

Département de Biochimie, Faculté de Médecine, Université de Sherbrooke, Sherbrooke, Québec, Canada J1H 5N4

## Summary

In order to study which amino acid mediates the specific interaction between adrenodoxin reductase, adrenodoxin and cytochrome P450$_{scc}$, recombinant bovine adrenodoxins was produced in *E. coli* using the expression vector PKK223-3. Site-directed mutagenesis was used for targeting amino acid residues that could play a role either in electron transfer or in the binding of adrenodoxin to its electron transfer partners. Mutations Glu74Ala, Glu74Gln, Asp79Ala and Asp79Asn were made to determine the effect of these changes on the binding of adrenodoxin to adrenodoxin reductase and to P450$_{scc}$. These mutations resulted in a decreased activity and affinity of the expressed protein for both adrenodoxin reductase and P450$_{scc}$.

Mutations Phe64Ala, Phe59Leu, Phe59Ala and Phe43Leu were made to locate residues involved in electron transfer. Mutation Phe64Ala did not affect the activity of the enzyme which was identical to that of the wild type adrenodoxin. Mutation Phe59Ala yielded an apoprotein which did not display any electron transfer activity. However, its affinity for P450$_{scc}$ was preserved. Preliminary studies showed that mutants Phe59Leu and Phe43Leu resulted in a decreased affinity for adrenodoxin reductase as tested by cytochrome *c* reductase assay. Mutants Glu74Ala, Glu74Gln and Phe43Leu yielded soluble adrenodoxin that incorporated the [2Fe-2S] cluster only when expressed at 30°C. The cluster was not incorporated when bacteria were grown at 37°C.

In conclusion, these results demonstrate that amino acid residues Asp79 and Glu74 are essential to the binding of adrenodoxin to P450$_{scc}$ and that residues Asp79, Glu74, Phe59 and Phe43 are crucial for its interaction with adrenodoxin reductase.

## Introduction

Adrenodoxin is a small iron sulfur protein (~14kDa) which transfers reducing equivalents from a NADPH-dependent flavoprotein, adrenodoxin reductase, to the cholesterol side chain cleavage cytochrome P450 (P450$_{scc}$) or to 11β-hydroxylase cytochrome P450 (P450$_{11β}$). Two models were proposed to describe how adrenodoxin transfers electrons from adrenodoxin reductase to cytochrome P450$_{scc}$. One model suggests that adrenodoxin acts as an electron shuttle, dissociating from adrenodoxin reductase before reassociating with $P450_{scc}$. The other model proposes that those three proteins associate together forming a ternary complex. Adrenodoxin is a highly acidic protein that shows two conserved regions of six or more acidic residues. Chemical modification studies on adrenodoxin demonstrated that modification of some acidic residues inhibited its binding to adrenodoxin reductase (Geren & O'Brien, 1984). Although chemical modifications can give us valuable information, multiple and nonstoichiometric labelling often occur, and the bulky labelling reagent could hide important interactions with specific amino acids. In order to overcome these problems we used site-directed mutagenesis approach. Acidic residues as glutamate and aspartate were replaced by their corresponding amide to reduce structural perturbations. Alanine, although smaller and neutral, was also used to replace specific acidic residues. In an attempt to identify residues involved in electron transfer, some phenylalanines were replaced either by leucine to maintain hydrophobicity or by alanine. The aromatic ring of phenylalanine contains delocalized $\pi$-electron clouds that enable it to interact with other $\pi$-systems and to transfer electrons. We chose highly conserved phenylalanine that are in the vicinity of the cysteine residues proposed to be coordinated to the iron-sulfur cluster. All the mutant proteins were expressed in *E. coli* using the pKK223-3 expression vector and purified to homogeneity. They were tested for their ability to allow cholesterol side chain cleavage activity and to associate with adrenodoxin reductase and cytochrome $P450_{scc}$.

## Materials and Methods

Site directed mutagenesis was carried out according to the method of Taylor & Eckstein (1985). The cDNA fragment coding for mature adrenodoxin was inserted into pKK223-3 expression vector to transform *E.coli* JM83 cells. The recombinant bovine adrenodoxin was isolated from *E. coli* as previously described (Palin *et al*). The concentration of adrenodoxin was estimated spectrophotometrically using a molar extinction coefficient of $\epsilon_{414}$ : 11 $(mM.cm)^{-1}$. Cholesterol side chain cleavage assays were performed as previously described (Palin *et al*), and pregnenolone was measured by radio-immunoassay. Cytochrome *c* reduction was performed as previously described (Palin *et al*). Reduction was monitored at 550 nm, and electron transfer activity was calculated on the basis of a molar extinction coefficient of $\epsilon_{550}$ : 20 $(mM.cm)^{-1}$ for cytochrome *c*. Spectral titrations were assayed as described (Coghlan & Vickery, 1991) and concentration of free adrenodoxin was calculated as follows:

$$Ferredoxin_{FREE} = Ferredoxin_{TOTAL} - (\Delta A/\Delta A_{MAX}) \times P450_{SCC}$$

The far UV circular dichroism (CD) spectra were made using a Jasco J-710 spectropolarimeter. Ten scans were taken for each sample and then smoothed. The protein concentration for CD was 20$\mu$M in 33mM potassium phosphate buffer.

## Results & Discussion

Amino acids substitutions of bovine adrenodoxin were made in an attempt to identify residues potentially implicated either in electron transfer or in the adrenodoxin reductase and $P450_{scc}$ binding domain. Each mutant was expressed as a soluble protein capable of incorporating the iron-sulfur cluster *in vivo* with the exception of mutation Phe59Ala.

Mutants Glu74Ala, Glu74Gln and Phe43Leu also incorporated the iron-sulfur cluster but only when bacteria were grown at 30°C instead of 37°C. With the exception of mutant Phe59Ala, CD in the far UV confirmed that the substitution that were made did not cause any large changes in the polypeptide backbone structure. It also confirmed that there were no major conformational changes in the vicinity of the iron-sulfur cluster.

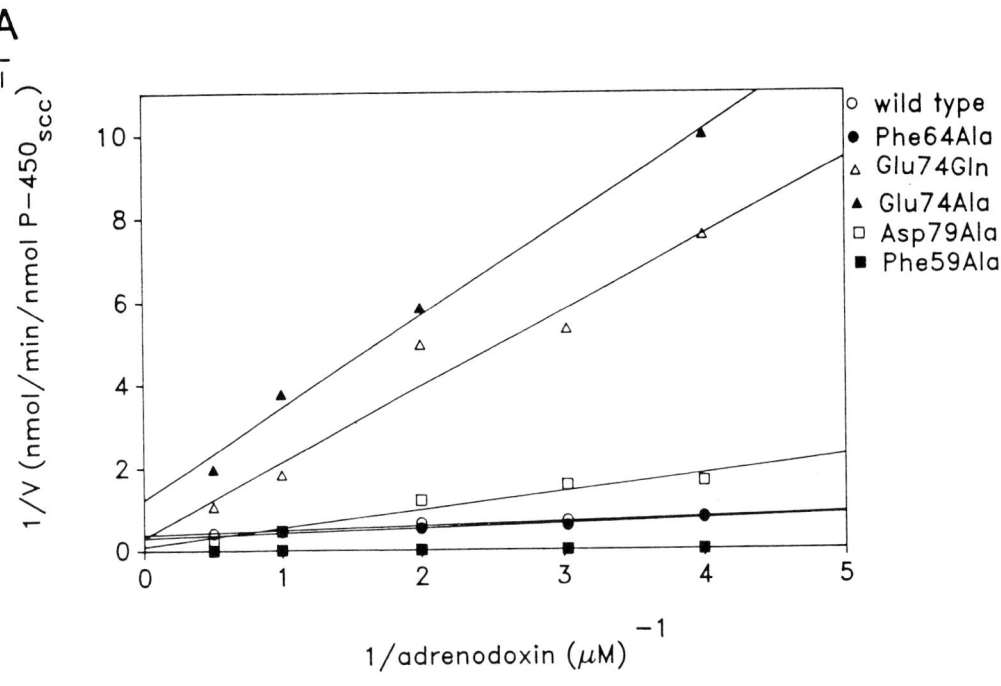

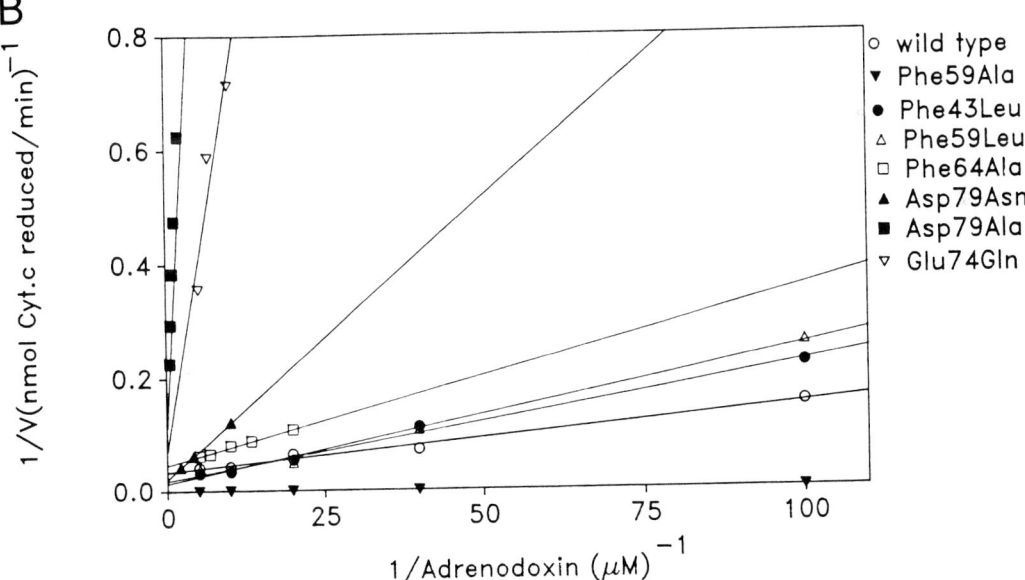

Fig. 1 Kinetics studies of recombinant bovine adrenodoxin **A.** Reciprocal plot showing kinetics of cholesterol transformation to pregnenolone for various concentrations of recombinant bovine adrenodoxin. **B.** Reciprocal plot showing kinetics corresponding to cytochrome *c* reductase activity by recombinant bovine adrenodoxin. Cytochrome *c* reduction was monitored at 550nm.

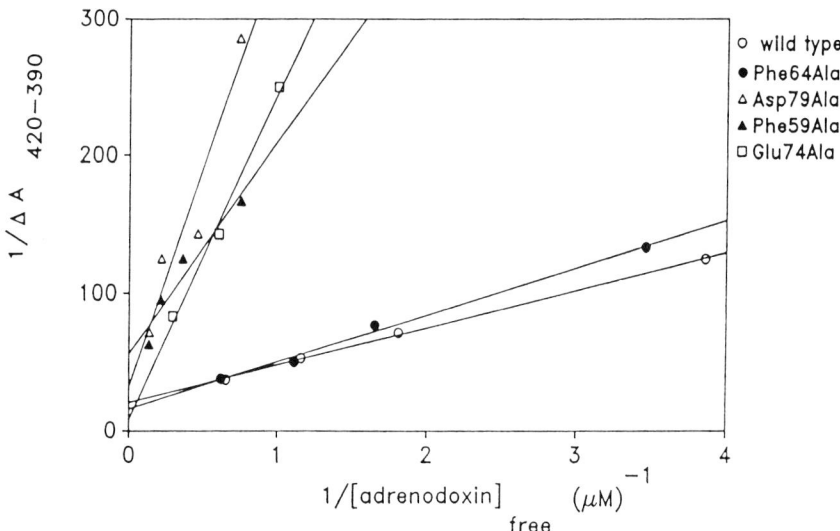

**Fig. 2** Reciprocal plot showing kinetics of P450$_{scc}$ titrations.

Mutant Phe64Ala acted exactly as the wild type adrenodoxin so this residue does not seem to be involved either in electron transfer or in the adrenodoxin reductase and P450$_{scc}$ binding domain. Mutant Phe59Ala yielded an apoprotein that lost all its electron transfer activity despite the fact that it kept an affinity to P450$_{scc}$ which is similar to the wild type adrenodoxin.

Mutants Glu74Ala, Glu74Gln and Asp79Ala seemed to be implicated in the adrenodoxin reductase and P450$_{scc}$ binding domain. These mutants were able to support cholesterol side chain cleavage and cytochrome $c$ reduction, but they all showed lower affinity to their electron transfer partners. Km values were 10 to 70 fold greater than wild type. These results indicate that adrenodoxin acts as a mobile electron shuttle, dissociating from adrenodoxin reductase before reassociating with cytochrome P450$_{scc}$.

Preliminary results on mutants Phe59Leu, Phe43Leu and Asp79Asn showed a lower affinity towards adrenodoxin reductase as confirmed by the cytochrome $c$ reduction assay. Minor differences in V$_{MAX}$ values were observed for these mutants suggesting that the reduced activity are due to a decreased binding affinity towards adrenodoxin reductase and not to a defect in electron transfer.

## References

Coghlan VM and Vickery LE (1991) Site-Specific Mutations in Human Ferredoxin That Affect Binding to Ferredoxin Reductase and Cytochrome P450$_{scc}$. *J. Biol. Chem.* **266**:18606-18612

Geren LM, O'Brien P, Stonehuerner J and Millet F (1984) Identification of Specific Carboxylate Groups on Adrenodoxin That Are Involved in the Interaction with Adrenodoxin Reductase. *J. Biol. Chem.* **259**:2155-2160

Palin MF, Berthiaume L, LeHoux JG, Waterman MR & Sygusch J (1992) Direct Expression of Mature Bovine Adrenodoxin in *Escherichia coli*. *Arch. Biochem. Biophys.* **295**:126-131

Taylor JW, Ott J and Eckstein F (1985) *Nucleic Acids Res.* **13**:8765-8785

# Cytochrome P450 obtusifoliol 14α-methyl demethylase from *Zea mays*: enzymology and inhibition

Maryse Taton, Florence Salmon, Alain Rahier

*Institut de biologie moléculaire des plantes, CNRS UPR 406, Département d'enzymologie moléculaire et cellulaire, 28, rue Goethe, 67083 Strasbourg Cedex, France*

The number of cytochrome P-450-dependent oxygenases which have been identified in higher plants is still limited, and the function and specificity of plant P-450s are much less well known than in animals and fungi, probably because of the low content and supposed lability of P-450 in cell-free plant systems.
Plant sterol biosynthesis involves the removal of the 14α-methyl group of obtusifoliol, catalysed by the cytochrome P-450 dependent obtusifoliol 14α-methyl demethylase ($P450_{OBT.14DM}$)
The demethylation requires NADPH and molecular oxygen and is photoreversibly inhibited by carbon monoxide. Rahier & Taton (1986). Fig :1.

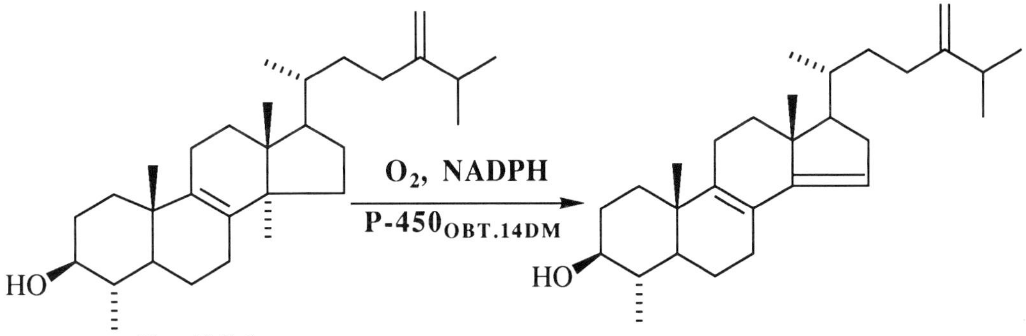

Obtusifoliol
Fig :1 Reaction catalysed by Cyt $P450_{OBT.14DM}$

Sterol 14-demethylase constitutes the only known example of P-450-dependent enzyme carrying out an identical metabolic function in different eukaryotes. Thus studies on sterol 14-demethylases from different organisms should provide interesting informations about evolution of P-450 monooxygenases and differentiation of sterol synthesis pathway.
A variety of important agrochemicals such as fungicides and plant growth regulators have been recently shown to inhibit $P450_{OBT.14DM}$ in correlation with plant growth inhibition effects. The phytotoxic effects of such derivatives has been ascribed to the damaging effects of excessive amounts of $\Delta^8$-14α-methyl-sterols and (or) depletion of typical $\Delta^5$-sterols in the plant cell membranes. This property has recently led to the search and development of potent herbicides having $P450_{OBT.14DM}$ as their primary site of action.

Therefore, P450$_{OBT.14DM}$ is of interest because of its role in plant metabolism and because it is a target for agronomically important compounds.

## Critical molecular features for recognition by P-450$_{OBT.14DM}$

Plant P450$_{OBT.14DM}$ displays a high degree of substrate specificity which is clearly distinct from that of the animal and yeast sterol 14-demethylase. Three distant domains of the sterol substrate are governing demethylation by P450$_{OBT.14DM}$. Rahier & Taton (1990) Salmon et al (1992) ..

P-450$_{OBT.14DM}$ has probably a specific apolar binding site for the side chain
the $\Delta^8$-unsaturation is an absolute requirement for substrate demethylation
the 3-hydroxy group plays a critical role in the enzyme-substrate interaction
the binding site beyond the C3 position contains a cleft which cannot accomodate a 4β methyl substituent present in lanosterol or eburicol, the substrates of 14-demethylase respectively in mammals and yeast.

The strict substrate specificity of P-450$_{OBT.14DM}$ is consistent with the constitutive nature of this new plant P-450 and provides further evidences for the plurality of specific P-450 and associated monooxygenase activities in higher plants. The different substrate specificity of the plant demethylase raises the question of whether it indicates the presence of large differences in amino acid sequence and structure when compared to that of the mammalian or yeast enzymes. Finally, the strict substrate specificity of P-450$_{OBT.14DM}$ constitutes another example of an obligatory sequence of enzymatic steps during plant sterol biosynthesis. These results sustain the hypothesis according to which plant sterol metabolism pathway is linear and does contain only very few parallele routes. Fig. : 2. Taton & Rahier (1991).

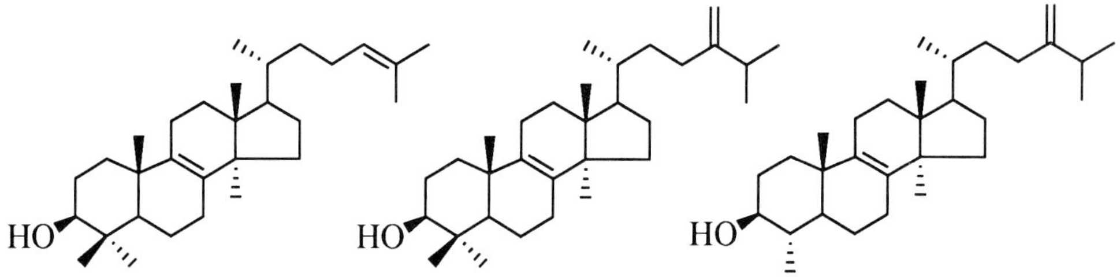

| Lanosterol | Eburicol | Obtusifoliol |
|---|---|---|
| *Mammalians Yeasts* | *Filamentous Fungi* | *Higher Plants* |

Fig :2 Substrate selectivity of P450$_{OBT.14DM}$ from different organisms .

## Inhibition of P450$_{OBT.14DM}$ by type I steroidal competitive inhibitors.

P450$_{OBT.14DM}$ shows high selectivity for obtusifoliol. Analysis of interaction of P450$_{OBT.14DM}$ with substrate analogue inhibitors and elucidation of structural requirements for the inhibitory action is therefore of great relevance. 7-oxo-sterol derivatives have been shown to be potent inhibitors of cholesterol synthesis in various animal cells and yeast both *in vitro* and *in vivo*. Therefore we synthetized and tested the inhibitory potency of a series of 7-oxo-obtusifoliol analogues.
7-oxo-24ζ(24')-dihydroobtusifoliol and 7-oxo-24(25)-dihydro-29-nor-lanosterol are potent competitive inhibitors of P450$_{OBT.14DM}$, binding 125-200 times more tightly than the corresponding substrates.
Inhibition of P450$_{OBT.14DM}$ by these analogues shows strict structural requirements. Particularly 7-oxo-24(25)-dihydro-lanosterol is not effective on the plant 14-demethylase while it inhibits strongly the 14DM from yeast.

The activity of these compounds, while not fully understood at this time, may be a consequence of the ability of these structures to mimic the putative demethylase 14α-oxygenated intermediates involved in the demethylation process and to be more stable towards metabolization by P450$_{OBT.14DM}$
Finally it would be interesting to evaluate the dual action of these derivatives on plant sterol biosynthesis, that is inhibition of P450$_{OBT.14DM}$ and regulatory inhibition of HMG CoA reductase.
However the spectral binding data dit not correlate with the corresponding inhibition data indicating that spectroscopic analysis performed with maize microsomes are not suitable for the determination of the specific inhibition of P450$_{OBT.14DM}$ by azoles.
A very good correlation between the $I_{50}$ values measured for the *in vitro* inhibition of the plant 14DM and the herbicidal activity *in vivo* is observed, suggesting that the plant 14DM is the primary site of action of these imidazoles.
To sustain this hypothesis, we studied the selectivity of CGA 214372 towards obtusifoliol 14-demethylase
among other P450 species present in maize,
among sterol 14-demethylases from various organisms.
The results gave strong evidence that P450$_{OBT.14DM}$ is a herbicidal target.

## Inhibition of P450$_{OBT.14DM}$ by non steroidal type II binding inhibitors

Azoles, pyrimidines and pyridines belong to a large group of fungicides, plant growth regulators and, more recently, herbicides used for plant protection in agriculture.
The heterocyclic $sp^2$ nitrogen atom is able to ligand the ferric iron atom of the heme of P-450.
We have directly demonstrated the potent inhibition of plant P-450$_{OBT.14DM}$ by a series of such derivatives. We have also measured the spectral binding properties of these toxicants. Maize microsomes show characteristic type II binding interactions expected for such azole derivatives.
Recently, in the course of a herbicide screening program, a novel class of herbicidally active imidazole carbonic esters has been discovered by the firms Ciba-Geigy and Janssen. We have demonstrated that this series of compounds are extremely potent inhibitors of the plant P450$_{OBT.14DM}$ both *in vitro* in a maize microsomal preparation and *in vivo*. The complete inhibition of the plant 14DM, achieved with a $1 \times 10^{-7}$M concentration of the R-(-) isomer of methyl 1-(2,2-dimethylindan-1-yl) imidazole-5-carboxylate (CGA 214372) would suggest binding interactions in the range of stoïchiometric concentrations of enzyme and this derivative.
The molecular features of a variety of analogues of CGA 214372 were related to their potency as inhibitors of P450$_{OBT.14DM}$ *in vitro*, allowing delineation of the key structural requirements governing inhibition of the demethylase. The binding affinities of the above-mentionned series of analogues of CGA 214372 to the total P450 fraction were also investigated and the corresponding Kd values determined. Fig. : 3. Rahier & Taton (1992) ; Salmon *et al.* (1992).

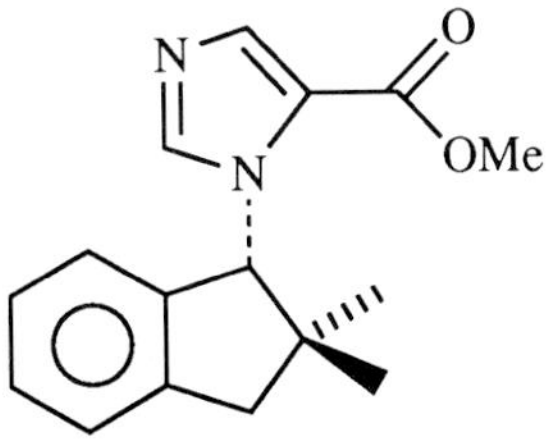

CGA 214372
non-steroidal inhibitor

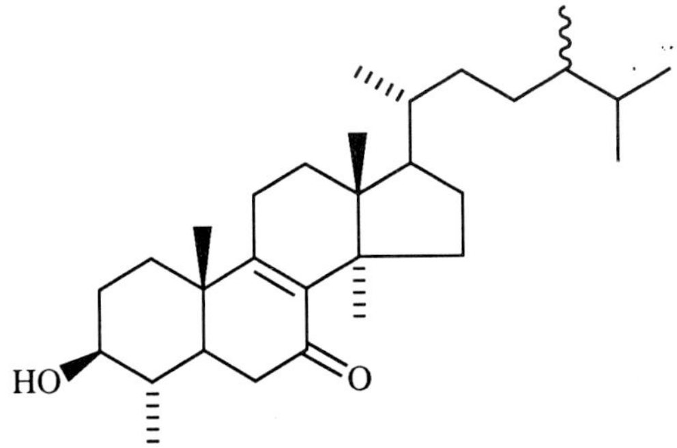

7-oxo-24ξ($24^1$)-dihydroobtusifoliol
steroidal-inhibitor

Fig: 3 Inhibition of P450 $_{OBT.14DM}$ by different type of inhibitors

## Conclusion

The results discussed prove that $P450_{OBT.14DM}$ is a target enzyme for azole herbicide (and antifungal).
They show that the mammalian, fungal and plant 14DM show major differences in
substrate selectivity, sensitivity towards the 7-oxo-steroïd competitive inhibitors sensitivity towards imidazole inhibitors ,
indicating that P450 enzymes having the same catalytic function and metabolic role are not necessarily identical. However strong variation in substrate and inhibitor specificities may result from limited amino-acid replacement insofar as they are located in the vicinity of the binding site.
Finally, the example discussed here demonstrate that both plant sterol biosynthesis and P-450 systems can be exploited in the search for compounds having potential agrochemical applications.

## References

Rahier A. and Taton M. (1986) Biochem. Biophys. Res. Commun. 140 (3), 1064-1072
Taton M. and Rahier A. (1991) Biochem. J. 277, 483-492
Rahier A. and Taton M. (1990) Bioch. Soc. Trans. 18, 52-56
Salmon F., Taton M., Benveniste P. and Rahier A. (1992) Arch. Biochem. Biophys. 297, (1), 123-131
Rahier A. and Taton M. (1992) Biochim. Biophys. Acta 1125, 215-222

# Mechanism of 9-hydroxy ellipticine induced inhibition of hepatic aryl hydrocarbon hydroxylase

Matthew J. Olnes[1,2], Paolo B. DePetrillo[1], Rabinder N. Kurl[1,2]

[1]Program in Clinical Pharmacology and [2]Graduate Program in Pathobiology, Brown University School of Medicine, Providence, RI 02912, USA

The ellipticines constitute a series of aromatic, planar alkaloids that exert their antineoplastic effects through inhibition of topoisomerase II by sequence-specific intercalation with DNA, followed by entrapment of the DNA-topoisomerase II "cleavable complex" [1-3]. This process in turn results in double-stranded DNA breakage. Ellipticine and its metabolite 9-hydroxy ellipticine (9-OHE) have been demonstrated to specifically bind to the Aryl hydrocarbon (Ah) receptor in rat hepatic [4-6] and lung [7] cytosols. In rat hepatic microsomes, both compounds have been reported to inhibit aryl hydrocarbon hydroxylase (AHH) [8], and it has been reported that 7-hydroxy ellipticine inhibits induction of AHH by the Ah receptor agonist benzo(a)pyrene [9].

It has been suggested that protein phosphorylation by protein kinase C (PKC) may play a role in regulating Ah receptor transformation and induction of CYP1A1 mRNA [10-14]. We recently demonstrated that low concentrations of 9-OHE (19nM-1.9uM) exert agonist effects on Ah receptor transformation, while higher concentrations of the drug (19uM-190uM) inhibit Ah receptor transformation *in vitro* [7]. In this communication we examined the involvement of protein kinase C in the inhibitory effect exerted by 9-OHE at high concentrations. We report that low concentrations of 9-OHE stimulated PKC activity in rat hepatic cytosol, while concentrations of 9-OHE which diminish Ah receptor transformation inhibited PKC activity through a mechanism independent of intact cells or nuclei. We further demonstrate that pre-treatment of intact cells with phorbol ester prevented 9-OHE antagonism of Ah receptor transformation. These results suggest that modulation of Ah receptor transformation by 9-OHE may be mediated through alterations in PKC activity.

MATERIALS AND METHODS

*Preparation of hepatic cytosol. and gel shift assays.* Male Sprague-Dawley rats (70-100g) from Taconic Farms (NY) were sacrificed under anesthesia and liver cytosol was prepared as described previously [7] The complementary oligodeoxyribonucleotides corresponding to the XRE-3 region of the CYP1A1 (5'- CGACCTCGGAGTTG CGTGAGAACAGCC-3') gene were synthesized, annealed, and [$^{32}$P]-labeled at the 5' ends using T-4 polynucleotide kinase. Whole cell extract was initially incubated for 2 hr at 22° with TCDD, 9-OHE, or vehicle. Thereafter, 0.05mg aliquots of cellular protein were analyzed for DNA binding by gel retardation assay as described [7].

*Measurement of cytosolic protein kinase C activity and protein dephosphrylation.* 200ul of extraction buffer (20mM Tris pH 7.5, 0.5mM EDTA, 0.5mM EGTA, 0.5% Triton X-100, 25ug/ml aprotinin, 25ug/ml leupeptin) was added to 0.20ml of liver cytosol (protein concentration of 10mg/ml) and vortexed for 15 sec at high speed. The sample was placed on ice for 20 min. Subsequently, the sample was diluted 1:5 with extraction buffer and 0.025ml of the resulting solution was added to each tube for quantitation of PKC activity as described [7]. For dephosphorylation assays, rat hepatic cytosol was phosphorylated in the presence of 0.2uCi [$^{32}$P]-ATP, 2.5mM $Ca^{2+}$, and 5mM $Mg^{2+}$ at 22°C for 30 min. Unincorporated [$^{32}$P]-ATP was removed by gel exclusion chromatography using an Excellulose GF-5 column (Pierce, Rockville, IL). The eluate was incubated with either vehicle, TCDD, 9-OHE, or 25U calf intestinal alkaline phosphatase for 2 hr at 22°C. Aliquots were then spotted onto phosphocellulose discs, washed as described for the PKC assay, and quantitated by scintillation counting with 5ml of Cytoscint ES (ICN Costa Mesa, CA).

RESULTS AND DISCUSSION

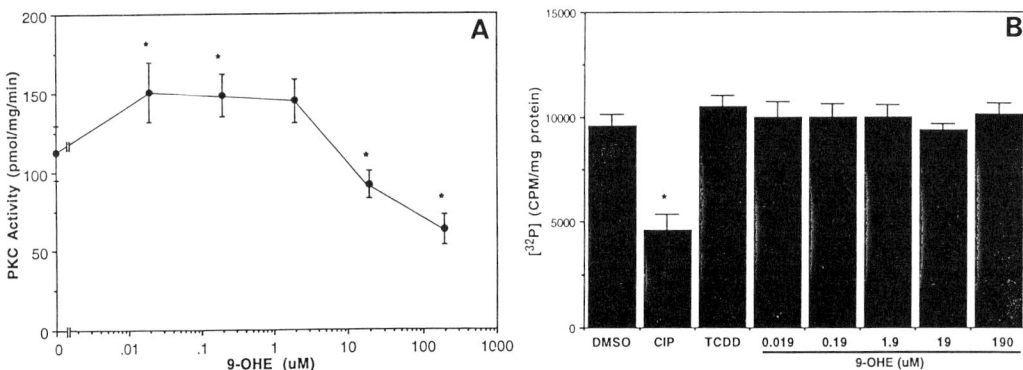

**Figure 1.** Effect of 9-OHE on protein kinase C and protein dephosphorylation activities. Rat hepatic cytosol was incubated in vitro with the indicated concentrations of either DMSO or 9-OHE for 2hr at 22°C. Protein kinase C (A) and protein dephosphorylation (B) assays were performed as described in Materials and Methods. Values represent means ± SEM from triplicate determinations, and asterix denotes statistical significance from control (p<0.05) as determined by a Fisher PLSD test.

9-Hydroxy ellipticine has been demonstrated to bind specifically to the Ah receptor in rat hepatic cytosol and lung [5,6]. We previously performed gel shift assays using [$^{32}$P]-labelled XRE-3 probe and rat hepatic cytosol treated with various concentrations of 9-OHE, and observed that 9-OHE transforms the Ah receptor to the XRE-binding state only at low (19-190nM) concentrations [7]. It has been suggested that ligand-induced transformation of the Ah receptor to the DNA-binding state may be regulated by PKC-dependent phosphorylation [10-14]. To examine the changes in phosphorylation state of proteins within rat hepatic cytosol upon incubation with 9-OHE, an *in vitro* PKC phosphorylation assay was performed. Cytosol was incubated for 2 hr in the presence or absence of increasing concentrations of 9-OHE, and then partially purified by DEAE-celluolose chromatography. As shown in Fig. 1A, extracts incubated with low concentrations of 9-OHE (19nM-190nM) significantly increased PKC activity relative to control, while higher concentrations of 9-OHE (19uM-190uM) significantly inhibited PKC-dependent phosphorylation in a concentration-specific manner.

Recently it was reported that addition of phosphatases to cultured hepatoma extracts inhibited binding of the liganded Ah receptor to its cognate response element [10,11]. To examine the effects of TCDD and 9-OHE on protein dephosphorylation, in vitro dephosphorylation assays were performed on rat hepatic cytosol incubated in the presence or absence of DMSO, 15nM TCDD, or the indicated concentrations of 9-OHE. As shown in Fig. 1B, treatment of cytosol with either TCDD or 9-OHE resulted in no detectable deviations from control dephosphosphorylation, while calf intestinal phosphatase (CIP) dramatically increased dephosphorylation within the cytosol.

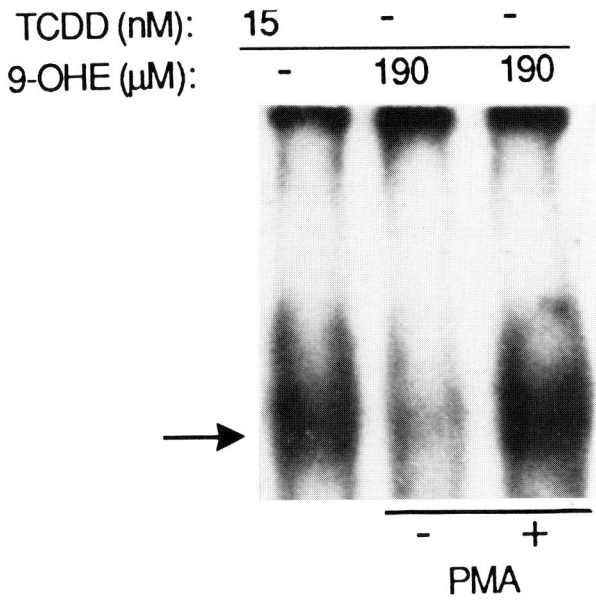

**Figure. 2.** Effect of PMA on 9-hydroxy ellipticine-induced Ah receptor antagonism in human leukemic cell extracts. Extracts were prepared from RW Leu-4-D3 cells grown in the presence (lane 3) or absence (lane 2) of 10nM PMA for 24hr prior to harvest. Thereafter, extracts were treated with 15nM TCDD (Lane 1), or 190uM 9-OHE (lanes 2,3) and analyzed by gel retardation assay as described in Materials and Methods. Ah receptor-XRE-3 complex is indicated by an arrow.

To investigate the involvement of PKC in 9-OHE mediated inhibition of Ah receptor transformation, we tested if phorbol 12-myristate 13-acetate (PMA) would rescue inhibition of Ah receptor-XRE complex formation in response to 9-OHE. Human monocytic cells were incubated with 10nM PMA for 24 hr, and then whole cell extracts were harvested and incubated *in vitro* with either 15nM TCDD, or 190uM 9-OHE. As shown in figure 2, extracts incubated in the presence of TCDD exhibited an Ah receptor-XRE complex (lane 1), and XRE binding was inhibited in the presence of 190uM 9-OHE. However, in cells pre-treated with PMA, 9-OHE mediated antagonism of XRE complex formation was abrogated. Taken together, these results suggest that 9-OHE inhibits Ah receptor transformation to the XRE binding state through a mechanism involving inhibition of PKC. These findings may be relevent to the mechanism by which 9-OHE exerts its inhibitory effect on AHH expression.

REFERENCES

1. Festy B, Poisson J, Paoletti C, A new DNA intercalating drug: methoxy-9-ellipticine. *FEBS Lett.* **17**: 321-325, 1971.
2. LePecq JB, Dat-Xuong N, Grosse C, and Paoletti C, A new antitumoral agent (9-hydroxy ellipticine). Possibility of a rational design of anticancer drugs in the series of DNA intercalating agents. *Proc. Natl. Acad.Sci. USA.* **71**: 5078-5082, 1974.
3. Multon E, Riou JF, LeFeuvre D, Ahomadegbe JC, and Riou G, Topoisomerase II-mediated DNA cleavage activity induced by ellipticines on the human tumor cell line N417. *Biochem. Pharmacol.* **38**:2077-2086, 1989.
4. Bigelow SW, and Nebert DW, The Ah regulatory gene product. Survey of nineteen polycyclic aromatic compounds and fifteen benzo(a)pyrene metabolites' capacity to bind to the cytosolic Ah receptor. *Toxicol. Letts.* **10**: 109-118, 1982.
5. Roy M, Fernandez N, and Lesca P, Binding characteristics of 4-S proteins from rat and mouse liver- high affinity of ellipticines. *Eur. J. Biochem.* **172**: 593-599, 1988.
6. Kurl RN, Chaudhary KC, and Villee CA, Characterization and control of cytosolic binding proteins for 2,3,7,8-tetrachlorodibenzo-p-dioxin (TCDD) in the rat lung. *Pharmacol.* **33**: 181-189, 1986.
7. Kurl RN, DePetrillo PB, and Olnes MJ, Inhibition of Ah (Dioxin) Receptor Transformation by 9-Hydroxy Ellipticine: Involvement of Protein Kinase C. *Biochem. Pharmacol.* **46**: 1745-1753, 1993.
8. Fernandez N, Roy M, and Lesca P, Binding characteristics of Ah receptors from rats and mice before and after separation from hepatic cytosols. *Eur. J. Biochem.* **172**: 585-592, 1988.
9. Lesca P, Lecointe P, Paoletti C, and Mansuy D, Ellipticines as potent inhibitors of aryl hydrocarbon hydroxylase: their binding to microsomal cytochromes P450 and protective effect against benzo(a)pyrene mutagenicity. *Biochem. Pharmacol.* **27**: 1203-1209, 1978.
10. Pongratz I, Stromstedt PE, Mason GGF, and Poellinger L, Inhibition of the specific DNA binding activity of the dioxin receptor by phosphatase treatment. *J. Biol. Chem.* **266**: 16813-16817, 1991.
11. Carrier F, Owens RA, Nebert DW, and Puga A, Dioxin-dependent activation of murine Cyp1A1 gene transcription requires protein kinase C-dependent phosphorylation. *Mol. Cell. Biol.* **12**: 1856-1863, 1992.
12. Okino ST, Pendurthi UR, and Tukey RH, Phobol esters inhibit the dioxin receptor-mediated transcriptional activation of the mouse Cyp1a1 and Cyp1a2 genes by 2,3,7,8-tetrachlorodibenzo-p-dioxin. *J. Biol. Chem.* **267**: 6991-6998, 1992.
13. Reiners JJ Jr, Cantu AR, and Schöller A, Phorbol ester-mediated suppression of cytochrome P450 CYP1a1 induction in murine skin: involvement of protein kinase C. *Biochem. Biophys. Res. Commun.* **186**: 970-976, 1992.
14. Berghard A, Gradin K, Pongratz I, Whitelaw M, and Poellinger L, Cross-coupling of signal transduction pathways: the dioxin receptor mediates induction of cytochrome P-450IA1 expression via a protein kinase C-dependent mechanism. *Mol. Cell. Biol.* **13**: 677-689, 1993.

# Induction of rat liver cytochrome P450 3A1 by metyrapone

Matthew C. Wright, Alan J. Paine

DH Department of Toxicology, St Bartholomew's Hospital Medical College, West Smithfield, London EC1 7ED, UK

> **Summary**
>
> Metyrapone induces the expression of CYP3A protein and CYP3A1 mRNA in rats and in rat hepatocytes cultured in a serum-free and glucocorticoid-free medium. Metyrapone and dexamethasone bind to a microsomal "receptor". The identity of the microsomal "receptor" is unclear but may be a cytochrome P450.

## INTRODUCTION

The rat cytochrome P450 sub-family 3A is known to contain at least two members (Gonzalez, 1990). Rat liver CYP3A1[*] is expressed at low to undetectable levels in adult rat liver (Gonzalez et al., 1986) but induced by glucocorticoids (most potently by the glucocorticoid dexamethasone), phenobarbital and macrolide antibiotics (Schuetz et al., 1984; Schuetz et al., 1986; Wrighton et al., 1985). CYP3A2 is a male-specific form which is constitutively expressed in adult rat liver (Gonzalez et al., 1986). The sex-specific expression of CYP3A2 has been suggested in part to be negatively-regulated by growth hormone (Waxman et al., 1990).

The mechanism of induction of CYP3A1 by dexamethasone has been suggested to be independent of the glucocorticoid receptor since many inducers of CYP3A1 do not also induce the expression of the glucocorticoid receptor-dependent gene tyrosine aminotransferase (TAT) (Schuetz and Guzelian, 1984). The mechanism of induction of CYP3A1 has been shown to be transcriptional (Gonzalez et al., 1986; Burger et al., 1992) although a recent report indicated that a post-transcriptional mechanism of regulation may also operate in adult rats but not immature rats (Telhada et al., 1992).

The effect of metyrapone on the CYP3A sub-family has been investigated because metyrapone potentiates the induction of TAT by glucocorticoids through a mechanism unrelated to any effects on glucocorticoid metabolism (Wright et al., 1993). The mechanism of glucocorticoid potentiation may therefore be related to the non-classical mechanism of glucocorticoid-dependent of CYP3A1 gene expression.

## MATERIALS AND METHODS

RU38486 (Mefipristone) was a gift from Roussel-Uclaf, Romainville, France. $^3$H-Dexamethasone (89-90Ci/mmole) was purchased from Amersham Int., UK. Polyclonal antisera to CYP3A was supplied by Dr. D. Waxman, Dept. Biological

---

[*]The nomenclature employed is that recommended by Nelson et al. (1993).

Chemistry and Molecular Pharmacology, Harvard Medical School, Boston, Mass. U.S.A. Purified CYP3A1 and CYP3A2 were supplied by Pf. C.R. Wolf, Dundee University, UK and Dr J. Halpert, Dept. Pharmacology and Toxicology, University of Arizona, Arizona, U.S.A respectively. An oligonucleotide specific for CYP3A1 (Gonzalez et al., 1986) was synthesised by Promega, Southampton, UK.

250-300g Male Sprague-Dawley CD rats were employed in all studies. Administration of metyrapone (100mg/kg body weight daily) and microsomal liver fractions were prepared and Western Blotting performed essentially as previously described (Shean and Paine, 1990) except that enhanced chemiluminescence (ECL kit, Amersham International, UK) was used for immunodetection on western blots. Isolation of mRNA and northern blotting were performed as previously described (Padgham and Paine, 1993). Hepatocytes were isolated by collagenase perfusion (Shean and Paine, 1990) and cultured in William's Medium E supplemented with 2nM insulin, 10nM glucagon and 50µg/ml gentamycin in a humidified atmosphere of 5% carbon dioxide in air at 37°C on Lux™ 150mm diameter plates (18 million cells/plate). Where indicated, hydrocortisone 21-hemisuccinate (100µM) and metyrapone (0.5mM) were dissolved directly in the medium.

Receptor-ligand binding studies were performed in cell fractions prepared from liver perfused with cold HBSS buffer. The liver was disrupted with a potter homogeniser in TS buffer (10mM Tris/HCl and 250mM sucrose pH 7.4) and cell fractions prepared by differential centrifugation. Cell fractions were diluted with TS buffer containing 5mM dithiothrietol, 10mM molybdate and 10mM trisodium citrate. After 6 hours incubation at 0°C (time for maximal specific binding), free ligand was removed by the dextran charcoal method as previously described (Poland and Glover, 1976). Non-specific binding of dexamethasone was determined by co-incubation of a 200 fold excess concentration of unlabelled dexamethasone. Radioactivity was determined by scintillation counting.

**RESULTS**

Gene Expression
Metyrapone administration to rats increases the levels of microsomal immunodetectable CYP3A. CYP3A1 and CYP3A2 are electrophoretically and immunochemically indistinguishable and so the relative contributions of each form cannot be determined by Western Blotting. However, using purified CYP3A2 as standard, the levels of increase in immunoreactive protein are increased approximately 4-fold in metyrapone treated animals compared to controls at both 24 and 72 hours. Using a CYP3A1 form-specific oligonucleotide, levels of CYP3A1 mRNA are induced from almost undetectable levels in rats treated with metyrapone suggesting that the mechanism of induction of CYP3A1 is operating at least in part at the pre-translational level.

In order to determine whether metyrapone induces CYP3A1 by blocking the metabolism of glucocorticoids in vivo, rat hepatocytes were cultured in a serum-free medium supplemented with a glucocorticoid (100µM hydrocortisone 21-hemisuccinate) and metyrapone. After 72 hours of culture the constitutive levels of immunodetectable CYP3A (presumably CYP3A2) have been lost and hydrocortisone 21-hemisuccinate does effect the rate of this loss. However, metyrapone alone increases the levels of immunodetectable CYP3A at 72 hours by 4.7-fold (using CYP3A2 as standard) compared to isolated cells. Co-addition of hydrocortisone 21-hemisuccinate with metyrapone enhances the levels of immunodetectable CYP3A apparently stimulated by metyrapone. Examination of mRNA expression in these cultures indicates however that both hydrocortisone 21-hemisuccinate and metyrapone induce CYP3A1 mRNA. Medium containing both these components further increases the levels of expression of CYP3A1 mRNA. It is clear however, that metyrapone alone may induce the expression CYP3A1 by a direct effect on the liver.

## Receptor-Ligand Studies

Since dexamethasone is a potent inducer of CYP3A1, radiolabelled dexamethasone was employed to analyze specific binding to rat liver fractions and by the co-incubation of unlabelled metyrapone, a common binding site examined. Table 1 indicates that specific binding of dexamethasone is observed in 2000g liver supernatants, cytosolic and microsomal fractions. The glucocorticoid receptor antagonist RU38486 and metyrapone reduce the binding of dexamethasone in 2000g supernatants. RU38486 potently blocks the specific binding of dexamethasone in cytosolic fractions whereas metyrapone is ineffective. In contrast, metyrapone blocks the specific binding of dexamethasone to microsomal

Table 1. Effect of Metyrapone and RU38486 on the Specific Binding of Dexamethasone in Rat Liver Cell Fractions.

| [dex] nM | additions to incubate | % specific binding compared to dex alone | | |
|---|---|---|---|---|
| | | 2000g sup'tant | cytosol | microsomes |
| 1 | - | 100 ± 3.2 | 100 ± 4.7 | 100 ± 5.8 |
| 1 | 100μM Metyrapone | 100 ± 4.7 | 100 ± 1.5 | 45 ± 16.6* |
| 1 | 10μM RU38486 | 46 ± 6.5* | n/detect* | 104 ± 5.7 |
| 10 | - | 100 ± 2.6$^a$ | 100 ± 7.7$^b$ | 100 ± 10.5$^c$ |
| 10 | 100μM Metyrapone | 72 ± 2.9* | 105 ± 4.0 | 57 ± 13.9* |
| 10 | 10μM RU38486 | 51 ± 6.3* | n/detect* | 123 ± 9.2 |

Cell fractions were incubated with $^3$H-dexamethasone for 6 hours as outlined in methods section except that on addition of dextran/charcoal to 2000g supernatant incubations, samples were centrifuged at 2000g for 3 mins. Data are the mean and standard deviation of 3 incubations from the same experiment, which is typical of 3 separate experiments. *Significantly different from control incubation (no additions to incubate) P>5% using Student's unpaired t-test (two tailed). $^a$100% specific binding, 372 x 10$^3$ dpm/ml 2000g supernatant (6mg protein/ml); $^b$100% specific binding, 58.8 x 10$^3$ dpm/ml cytosol (1.2 mg protein/ml); and $^c$100% specific binding, 139 x 10$^3$ dpm/ml microsomes (1.2 mg protein/ml). Dex = dexamethasone.

Figure 1. Dexamethasone Binding to Rat Liver Microsomes.

A. Scatchard Plot

B: Antagonism by Metyrapone

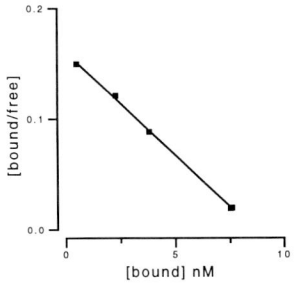
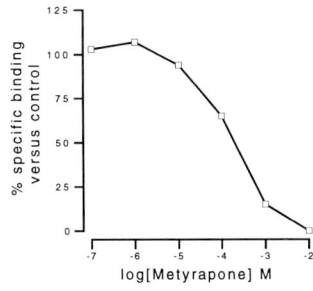

A: Microsomes (1mg protein/ml) were incubated for 6 hours and total and non-specifically bound dexamethasone determined as outlined in methods section. Total ligand concentration in each incubation by removing 10μl and determining radioactivity. Reference to specific activity enables a calculation of labelled dexamethasone concentration. B: Rat liver microsomes were incubated with 10nM $^3$H-dexamethasone for 6 hours with non-specific bound dexamethasone determined by co-incubation with 200 fold excess unlabelled dexamethasone. Metyrapone was added to assay from ethanol stocks (final concentration of ethanol was 1%)

fractions whereas RU38486 has little effect. Figure 1A confirms that dexamethasone binds to rat liver microsomes in a saturable manner and assuming that dexamethasone combines in a 1:1 ratio with the binding site and does not breakdown to different products an affinity constant of approximately 50nM and a binding site concentration of approximately 8 pmoles/mg microsomal protein can be calculated. Figure 1B indicates that metyrapone antagonizes the binding of dexamethasone to rat liver microsomes in a concentration-dependent manner. Solubilised microsomes pre-labelled with dexamethasone co-elutes with haemoproteins on a sephadex G75 (gel-filtration) column suggesting that the "receptor" may be a cytochrome P450.

## DISCUSSION

Metyrapone induces the expression of the glucocorticoid-inducible CYP3A1 by a direct effect of the liver. Binding studies indicate that dexamethasone and metyrapone share a common interaction with a microsomal "receptor". The lack of any binding of metyrapone to cytosolic fraction (glucocorticoid receptor) supports the work of Schuetz and Guzelian (1984) that the mechanism of induction of CYP3A1 is independent of the glucocorticoid receptor. Furthermore, metyrapone does not induce TAT activity in rats or in cultured hepatocytes (Wright and Paine, 1993) although it does potentiate the induction of TAT by glucocorticoids (Wright et al., 1993).

Metyrapone is an inhibitor of several cytochrome P450-mediated reactions by presumably liganding the haem prosthetic group of the cytochrome although it is not a universal inhibitor of cytochrome P450 (Waxman and Walsh, 1985). A block in a cytochrome P450 which metabolises a pro-repressor of CYP3A1 as suggested for CYP1A1 (Nebert et al., 1990) could explain the mechanism of induction of CYP3A1. However, whether this binding site, be it a cytochrome P450 or not, plays a role in the regulation of CYP3A1 at the transcriptional or post-transcriptional level require further examination.

## REFERENCES

Burger H., Schuetz J.D., Schuetz E.G. and Guzelian P.S., (1992). *Proceedings of the National Academy of Sciences USA* **89**, 2145-2149.
Gonzalez F.J., Song B-J. and Hardwick J.P., (1986). *Molecular and Cellular Biology* **6**, 2969-2976.
Gonzalez F.J., (1990). *Pharmacolgy and Therapeutics* **45**, 1-38.
Nebert D.W., Petersen D.D. and Fornace A.J., (1990). *Environmental Health Perspectives* **88**, 13-25.
Nelson D.R., Kamataki T., Waxman D.R., Guengerich F.P., Estabrook R.W., Feyereisen R., Gonzalez F.J., Coon M.J., Gunsalus I.C., Gotoh O., Okuda K. and Nebert D.W., (1993). *DNA Cell Biology* **12**, 1-51.
Padgham C.R.W. and Paine A.J., (1993). *Biochemical Journal* **289**, 621-624.
Poland A. and Glover E., (1976). *The Journal of Biological Chemistry* **251**, 4936-4946.
Schuetz E.G., Wrighton S.A., Barwick J.L. and Guzelian P.S., (1984). *The Journal of Biological Chemistry* **259**, 1999-2006.
Schuetz E.G. and Guzelian P.S., (1984). *The Journal of Biological Chemistry* **259**, 2007-2012.
Schuetz E.G., Wrighton S.A., Safe S.H. and Guzelian P.S., (1986). *Biochemistry* **25**, 1124-1133.
Shean K. and Paine A.J., (1990). *Biochemical Journal* **267**, 715-719.
Telhada M.B., Pereira T.M. and Lechner M., (1992). *Archives of Biochemistry and Biophysics* **298**, 715-725.
Waxman D.J. and Walsh C., (1985). *Biochemistry* **22** (1985) 4846-4855.
Waxman D.J., Ram P.A., Notani G., LeBlanc G.A., Alberta J.A., Morrissey J.J. and Sundseth S.S., (1990). *Molecular Endocrinology* **4**, 447-454.
Wright M.C., Paine A.J., Skett P. and Auld R., (1993). *Journal of Steroid Biochemistry and Molecular Biology*, in press.
Wright M.C. and Paine A.J. (1993). *Human and Experimental Toxicology*, in press.
Wrighton S.A., Maurel P., Schuetz E.G., Watkins P.B., Young B. and Guzelian P.S., (1985). *Biochemistry* **22**, 2171-2178.

# Long-term preservation and induction of cytochromes P450 in co-cultured rat hepatocytes

May Akrawi[1], Vera Rogiers[2], Antoine Vercruysse[2], Ian R. Phillips[3], Elizabeth A. Shephard[1]

[1]Department of Biochemistry and Molecular Biology, University College London, Gower Street, London WC1E 6BT, UK. [2]Department of Toxicology, Vrije Universiteit Brussel, Laarbeeklaan 103, B-1090 Brussels, Belgium. [3]Department of Biochemistry, Queen Mary and Westfield College, University of London, Mile End Road, London E1 4NS, UK

## INTRODUCTION

To maintain their differentiated state *in vitro* hepatocytes require a complex and well-defined environment. If cultured under conventional conditions rat hepatocytes fail to express many of the cytochromes P-450 (CYPs) responsible for metabolizing foreign compounds and fail to induce the expression of members of the CYP2B subfamily in response to the antiepileptic drug phenobarbital (Guillouzo, 1986). However, when hepatocytes are cultured on the basement membrane, matrigel (Schuetz et al., 1988), or on vitrogen-coated plates (Waxman et al., 1990) they are able to sustain the expression and inducibility of several CYPs. Using an alternative approach, in which cell-cell interactions are maintained by culturing rat hepatocytes with primary rat liver epithelial cells, we have shown that hepatocytes retain the ability to express members of the CYP2B subfamily and are able to respond well to phenobarbital (Akrawi et al., 1993a).

Phenobarbital is often co-administered to patients together with another anti-epileptic agent sodium valproate. Although the latter compound is widely used in the treatment of epilepsy, its use has been associated with hepatotoxicity (Zimmerman & Ishak, 1982). Baillie (1988) has shown that sodium valproate is metabolized by CYP2B1. This CYP isoform is known to be substantially increased on exposure to phenobarbital (Omiecinski, 1986). Thus the co-administration of these two drugs may result in an increased metabolism of sodium valproate. In addition, Fisher et al. (1991) have shown that prolonged exposure of rats to sodium valproate results in this drug inducing its own metabolism.

In this paper we describe the use of the co-culture system to examine the effects of sodium valproate on the expression of mRNAs encoding members of the CYP2B subfamily. Because sodium valproate is known to induce peroxisomal proliferation (Dirven et al., 1992) we have also investigated the effect of this compound on the expression of members of the CYP4A subfamily.

## MATERIALS AND METHODS

Hepatocytes were isolated from outbred male Sprague-Dawley albino rats and co-cultured as described previously (Akrawi et al., 1993a). Cells were either untreated, or treated with sodium phenobarbital (2 mM) or sodium valproate (200 µg/ml). In the case of treated cells, sodium valproate and phenobarbital were present in the medium continuously for 7 days after cell attachment. Sodium valproate was kindly donated by Sanofi-Labaz (Brussels, Belgium). Total RNA was isolated from cell pellets as described previously (Akrawi et al., 1993a). CYP2B1 and CYP2B2 mRNAs were detected using an RNase protection assay and antisense RNA probes as previously described (Akrawi et al., 1993a). cDNA clones encoding CYP4A1, 4A2 and 4A3 were synthesized by a reverse transcription/PCR method. The cDNAs were cloned into the plasmid pBluescript. Antisense RNA probes were generated using T3 RNA polymerase and an *in vitro* transcription assay kit (Stratagene). Protected RNA species were detected and quantified as described previously (Akrawi *et al.*, 1993a).

RESULTS AND DISCUSSION

We have found that, in rat hepatocytes co-cultured with rat liver epithelial cells, members of the CYP2B subfamily are induced by phenobarbital (Akrawi *et al.*, 1993a) and by another anti-epileptic drug, valproate (Akrawi *et al.*, 1993b).The induction of the proteins is mediated by increases in the amounts of the corresponding mRNAs. In this system, maximal induction of the mRNAs is obtained after 7 days of co-culture. In these experiments, CYP2B mRNAs were quantified by RNase protection using an antisense RNase protection probe that recognises both CYP2B1 and CYP2B2 mRNAs. To determine the effect of the two inducers on mRNAs encoding CYP2B1 and 2B2, RNA samples were analysed with antisense RNA probes capable of distinguishing between these mRNAs (Fig. 1). The cDNAs used as templates for the synthesis of the probes contained sequences derived from exons 6 and 7 of the CYP2B1 and 2B2 genes, a region in which the genes have only 90% nucleotide sequence similarity.

In co-cultured hepatocytes, the mRNA encoding CYP2B1 was induced 19-fold by phenobarbital and 27-fold by valproate (Fig. 1a). CYP2B2 mRNA was also induced by both compounds, but in each case the extent of induction (3-fold by phenobarbital and 3-fold by valproate) was less than that of the CYP2B1 mRNA (Fig. 1b). The results obtained for the phenobarbital-induction of CYP2B1 and 2B2 mRNAs in co-cultured rat hepatocytes reflect the situation found *in vivo* in that CYP2B1 mRNAs are more inducible than CYP2B2 mRNAs. Neither of the mRNAs was expressed or induced in rat liver epithelial cells.

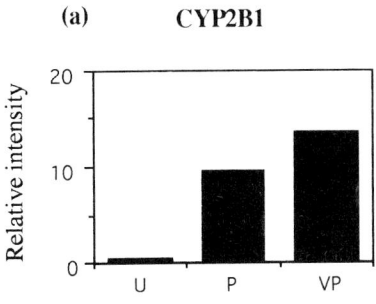
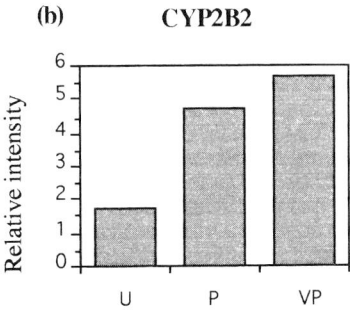
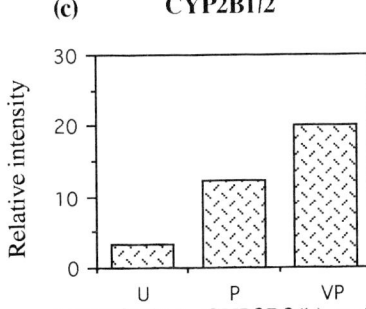

Fig.1. RNase protection assays of CYP2B1(a), CYP2B2(b) and CYP2B1/2(c) mRNAs. Total RNA was extracted from hepatocytes co-cultured for 7 days either in the absence (U) or presence of 2 mM phenobarbital ( P) or 200 µg/ml sodium valproate (VP ). Each hybridization reaction contained 10 µg of total RNA.

The same RNA samples were analysed with an antisense probe that recognized both CYP2B1 and 2B2 mRNAs (Fig. 1c). The combined abundance of CYP2B1/2 mRNAs as determined with this probe, was found to be 4-fold by phenobarbital and 6-fold by valproate. The sum of the abundances of the individual CYP2B1 and 2B2 mRNAs (determined as described above through the use of specific probes) was very similar to that obtained from the use of the non-discriminatory probe.

We have previously found, through the use of a polyclonal antibody raised against CYP4A1 that CYP4A was induced in co-cultured hepatocytes by valproate but not by phenobarbital (Akrawi et al., 1993b). Because the antibody recognizes CYP4A1, 4A2 and 4A3 the effect of valproate on individual members of the CYP4A subfamily could not be determined. To address this problem, we designed and synthesized antisense RNA probes that would discriminate between mRNAs encoding different members of the CYP4A subfamily. Results obtained with these probes demonstrate that, in co-cultured rat hepatocytes, mRNAs for CYP4A1, 4A2 and 4A3 are induced 20-fold, 17-fold and 10-fold respectively, by valproate.

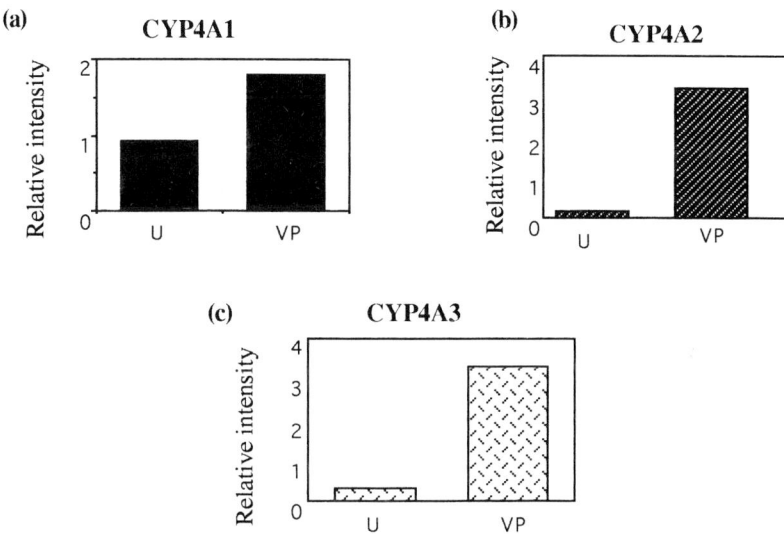

Fig. 2 RNase protection assays using antisense probes specific for CYP4A1(a), CYP4A2(b) and CYP4A3(c). RNA was isolated from hepatocytes co-cultured for 7 days either in the absence (U) or presence of 200μg/ml sodium valproate (VP). Each hybridization reaction contained 20 μg of total RNA.

In conclusion, we have shown that rat hepatocytes co-cultured with rat liver epithelial cells, are capable of inducing CYP2B1 and 2B2 proteins and mRNAs in response to the anti-epileptic drugs, phenobarbital and valproate. The relative inducibility of the two mRNAs by phenobarbital is similar to that found *in vivo*. Valproate, but not phenobarbital, induces also mRNAs encoding three members of the CYP4A subfamily, namely CYP4A1, 4A2 and 4A3.

ACKNOWLEDGEMENTS

Supported by grants from the Commission of the European Community DGXI, the Cancer Research Campaign (U.K.) and the National Fund for Scientific Research (Belgium).

REFERENCES

Akrawi, M., Rogiers V., Vandenberghe, Y., Palmer, C.N.A., Vercruysse, A., Shephard, E.A. & Phillips, I.R. (1993a): Maintenance and induction in co-cultured rat hepatocytes of components of the cytochrome P-450-mediated mono-oxygenase. *Biochem. Pharmacol.* 45, 1583-1591.

Akrawi, M., Shephard, E.A., Phillips, I.R., Vercruysse, A .& Rogiers, V. (1993b): Effects of phenobarbital and valproate on the expression of cytochromes P-450 in co-cultured rat hepatocytes. *Toxic. in Vitro.* 7, 477-480.

Baillie, T.A. (1988): Metabolic activation of valproic acid and drug-mediated hepatotoxicity. Role of the terminal olefin, 2-*n*-propyl -4-pentanoic acid. *Chem. Res. Toxicol..* 1, 195-199.

Dirven, H.A.A.M., Van Den Broek, P.H.H., Peters, J.G.P., Noordhoek, J. & Jongeneelen, F.J. (1992): Microsomal lauric acid hydroxylase activities after treatment of rats with three classical cytochrome P450 inducers and peroxisome proliferating compounds. *Biochem. Pharmacol..* 43, 2621-2629.

Fisher, J.E., Nan, H.Y. & Löscher, W. (1991): Alterations in the renal excretion of valproate and its metabolites after chronic treatment. *Epilepsia .* 32, 146-150.

Guillouzo, A. (1986): Use of cultured hepatocytes for xenobiotic metabolism and cytotoxicity studies. In *Isolated and Cultured Hepatocytes.* Edited by Guillouzo A. and Guguen-Guillouzo C. pp. 313-332. Les Editions INSERM, John Libbey-Eurotext, Paris, London.

Schuetz, E.G., Li, D., Omiecinski, C.J., Muller-Eberhard, U., Kleinman, H.K., Elswick, B.& Guzelian P.S. (1988): Regulation of gene expression in adult rat hepatocyte cultured on a basement membrane matrix. *J. Cell. Physiol.* 134,309-323.

Omiecinski, C. J. (1986): Tissue specific expression of rat mRNAs homologous to cytochromes P-450b and P-450e. *Nucleic Acids Res.* 14, 1525-1539.

Waxman, D.J., Morrissey, J.J., Naik, S. & Jauregui, H.O. (1990): Phenobarbital induction of cytochrome P-450. *Biochem. J.* 271:113-119.

Zimmerman H.J. & Ishak, K.G. (1982): Valproate-induced hepatic injury: analyses of 23 fatal cases. *Hepatology* 2, 591-597.

# Development and prevalidation of an *Escherichia coli* tester strain for genotoxins

Michel Kranendonk[1], Margarida Ruas[1], António Laires[1,2], José Rueff[1]

[1]*Department of Genetics, Faculty of Medical Sciences, New University of Lisbon, R. Junqueira 96, P-1300, Lisbon.* [2]*Faculty of Sciences and Technology, New University of Lisbon P-2825 Monte de Caparica, Portugal*

## INTRODUCTION

In the last few years, a lot of effort was put in the unraveling of mammalian metabolic pathways, involved in the metabolism of genotoxins. We have been involved in studies of metabolic systems for the use in short-term genotoxicity assays (Rueff *et al.*, 1992[a], Rueff *et al.*, 1992[b]). One of the more promising approaches is the direct introduction of genes/cDNA's of mammalian drug metabolizing enzymes in the assay's target cell (Gonzalez *et al.*, 1991, Rodrigues *et al.*, 1993). A number of different expression vectors has been developed for the expression of mammalian genes in bacteria. They contain inducible *E.coli* promoters, the majority based on the lactose operon repressor LacI. The activity of these promoters can be chemically regulated, namely by IPTG. This chemical regulation is lost upon introduction of these plasmids in *S.typhymurium,* a bacterium frequently used in genotoxicity testing (e.g. the Ames Assay (Maron and Ames, 1983)), due to the lack of the *lac* operon in this bacterium (Lampel and Riley, 1982). We have constructed an *Escherichia coli* tester strain to be used in mechanistic and metabolic studies of genotoxins. We started with one of the more used and better characterized *E.coli* K-12 laboratory strains, AB1157. We isolated a lipopolysaccharide defective mutant of strain AB1886, an excision repair deficient mutant of AB1157 and introduced a new constructed mucAB encoding plasmid, the pKR11, resulting in strain MR2101/pKR11. A genotoxicity assay was designed, monitoring the reversion to arginine prototrophy and a preliminary validation against Ames' tester strain TA100 was carried out with a set of diagnostic carcinogens. The results seem to indicate that strain MR2101/pKR11 is an adequate tester strain which can be a useful tool in mutagenic and metabolic studies of genotoxins. Moreover, this tester strain can serve as mother strain for the isolation of specialized tester strains, especially for the heterologous expression of mammalian cytochrome P450 isoenzymes.

## DEVELOPMENT OF TESTER STRAIN MR2101/pKR11

For the development of an *E.coli* tester strain, we started with one of the more used and better characterized *E.coli* K-12 laboratory strain, AB1157, which has been shown to be one of the more mutable *E.coli* strains (Sedgwick *et al.*, 1991). AB1886, a strain isogenic to AB1157 except for *uvrA*, was used in the construction since a large number of genotoxins is more readily detected in strains with an excision repair deficient background (Quillardet and Hofnung, 1988). The different used and constructed strains are described in Table 1.

**Table 1** Bacterial strains and plasmids.

| STRAINS | GENOTYPE | ORIGIN |
|---|---|---|
| *E. coli* | | |
| AB1157 | *thr-1, ara-14, leuB6,* Δ*(gpt-proA)62, lacY1, tsx-33,* qsr⁻, *supE44, galK2,* λ⁻, rac⁻ *hisG4, rfbD1, mgl-51, rpsL31, kdgK51, xyl-5, mtl-1, argE3, thi-1* | B. Bachmann |
| MR1001 | AB1157, LPS defective | This work (U3$^r$ selection on AB1157) |
| AB1886 | AB1157, *uvrA6* | B. Bachmann |
| NK6033 | Δ*(gpt-lac)5, nadA50*::Tn*10, relA1, spoT1, thi-1,* λ⁻ | B. Bachmann |
| MR2000 | AB1886, *nadA50*::Tn*10* | This work (P1[NK6033]×AB1886 to Tet$^r$) |
| PQ33 | *sfi*::Mud(Ap *lac*)cts, *lac*ΔU169, *uvrA, galE, galY,* PhoC, *thr, leu, his, pyrD, thi, trp*::Muc$^+$, *srl300*::Tn*10, rpoB* | M. Hofnung and P. Quillardet |
| MR2100 | AB1886, *galE* | This work (P1[PQ33]×MR2000 to NAD$^+$) |
| MR2101 | MR2100, LPS defective | This paper (C21$^r$ selection on MR2100) |
| DH1 | *supE44, hsdR17, recA1, endA1, gyrA96, thi-1, relA1* | B. Bachmann |
| *S. typhimurium* | | |
| TA100 | *hisG46,* Δ*(gal-uvrB), rfa-1001*/pKM101 | B.N. Ames |
| PLASMIDS | RELEVANT GENETIC MARKERS | ORIGIN |
| pKM101 | *mucAB*$^+$, Amp$^r$, | B.N. Ames |
| pGW270 | *mucAB*$^+$, Amp$^r$, Kan$^r$ | G.C. Walker |
| pKR11 | *mucAB*$^+$, Kan$^r$ | This work (Δ[*Sal*I-*Bam*HI]×pGW270) |

We constructed lipopolysaccharide defective (LPS$^d$) variants of strains AB1157 and AB1886, resulting in MR1001 and MR2101 respectively. Bacteria defective in LPS, the major component of the outermembrane, have an increased permeability for many compounds and have shown to be useful for testing carcinogenic substances (Ames *et al.*, 1973, Quillardet and Hofnung, 1985).

The plasmid pKM101, a mucAB (Amp$^r$) encoding plasmid, has been introduced in a number of different bacterial tester strains for its ability to increase the susceptibility of cells to chemical mutagens (Langer *et al.*, 1981). For this reason we introduced a *mucAB* carrying plasmid (pKR11) in our tester strains. It was necessary to introduce a *mucAB* carrying plasmid, without conveying ampicillin resistance to the cells, because the majority of the prokaryotic expression vectors for heterologous expression convey Amp$^r$. We constructed pKR11 by deletion of the Amp$^r$ encoding region of plasmid pGW270, a (Kan$^r$) deletion derivative of pKM101, which has been shown to have equal effects on SOS mutagenesis as pKM101 (Langer *et al.*, 1981). pKR11 showed equal effects on UV and chemical mutagenesis as pGW270 and thus equals pKM101 (data not shown).

We developed a mutagenicity assay for the constructed strains, monitoring the reversion to arginine prototrophy as test-target. The [*ArgE3*]-strains can revert via all possible transition and transversion substitution mutations to arginine prototrophy (Todd *et al.*, 1979).

We determined the genotoxicity of the carcinogen aflatoxin B1 (AFB1) in the different constructed strains to demonstrate the importance of the genetic make-up of strain MR2101/pKR11 for genotoxicity testing, namely LPS$^d$, excision repair deficiency and the presence of the *mucAB* operon, as shown in Fig. 1.

## PREVALIDATION OF TESTER STRAIN MR2101/pKR11

A prevalidation of the tester strain MR2101/pKR11 was carried out against Ames' tester strain TA100 with a number of diagnostic compounds, namely sodium azide, 2-nitrofluorene (2NF), quercetin (Q), 4-nitroquinoline-1-oxide (4NQO) and pyrene, 7,12-dimethyl benzoanthracene (DMBA), benzo(a)pyrene (B(a)P), aflatoxin B1 (AFB1) and quercetin (Q) with metabolic activation (standard 10% S9 mix (Maron and Ames, 1983)). The mutagenic activities of the different compounds in strain MR2101/pKR11 were plotted against the ones found in TA100 (see Fig. 2), with a good correlation for the carcinogens DMBA, B(a)P, 4NQO and AFB1 (r =0.998). The results obtained with Q (equal response), 2NF (low response) and sodium azide (no response) in comparison to TA100 seem to point out the difference between the genetic targets of the two tester strains. Pyrene, a non-carcinogen, was found negative in both strains.

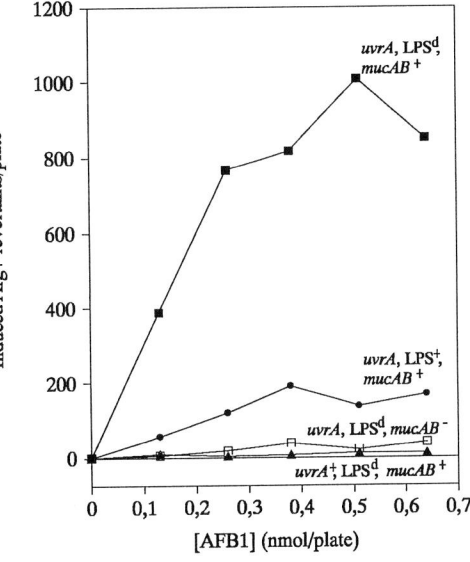

**Fig. 1.** Dose-response curves for AFB1 in MR1001/pKR11, MR2100/pKR11, MR2101 and MR2101/pKR11. Results are shown as means, representing three independent determinations for each dose-level

## PERSPECTIVES

The obtained results seem to characterize strain MR2101/pKR11 as an adequate tester strain which can be a useful tool in mechanistic and metabolic studies of genotoxins. Moreover, MR2101/pKR11 with minor modifications becomes a suitable host for the LacI expression vectors, resulting in a tester bacterium in which controllable expression of mammalian cytochrome P450 isoenzymes can be obtained. This is of special interest because E.coli seem to have a reductase system which is able to support, at least for a part, the enzymatic activity of mammalian cytochrome P450 (Barnes et al., 1991).

**Fig. 2.** Correlation of mutagenic activities of strain MR2101/pKR11 against TA100 for seven compounds. Mutagenic activities were determined as the least square-line of the linear portion of the dose-response curve.

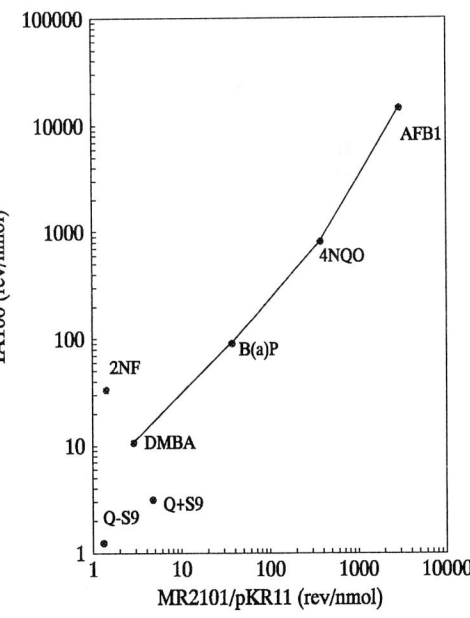

ACKNOWLEDGMENTS

We are grateful to Dr. B.J. Bachmann, Prof. Dr. M. Hofnung and Dr. P. Quillardet for supplying the *E.coli* strains, to Dr. G.C. Walker for giving the plasmid pGW270, to Dr. R.P.P. Fuchs for the phages $U_3$ and $P_1^{vir}$, Dr. M. Llagosterra and Dr. B. Clerch for supplying phage $C_{21}$. This work was supported by the CEC (STEP Programme) and the CIENCIA Programme.

REFERENCES

Ames B.N., Lee F.D. and Durton W.E. (1973) *An improved bacterial test system for the detection and classification of mutagens and carcinogens*. Proc. Natl. Acad. Sci. USA. **70**: 782-786

Barnes H.J., Arlotto M.P and Waterman M.R. (1991) *Expression and enzymatic activity of recombinant cytochrome P450 17α -hydroxylase in Escherichia coli*. Proc. Natl. Acad. Sci. USA. **88**: 5597-5601.

Gonzalez F.J., Crespi C.L. and Gelboin H.V. (1991) *cDNA-expressed human cytochrome P450s: a new age of molecular toxicology and human risk assessment*. Mutat.Res. **247**: 113-127

Lampel K.L. and Riley M. (1982) *Discontinuity of homology of Escherichia coli and Salmonella typhimurium DNA in the lac region*. Mol. Gen. Genet. **186**: 82-86

Langer P.J., Shanabruch W.G. and Walker G.C. (1981) *Functional organization of plasmid pKM101*. J. Bacteriol. **145**: 1310-1316

Maron D.M. and Ames, B.N. (1983) *Revised methods for Salmonella mutagenicity test*. Mutat. Res. **113**: 173-215

Quillardet P. and Hofnung M. (1988) *The screening, diagnosis and evaluation of genotoxic agents with batteries of bacterial tests*. Mutat. Res. **205**: 107-118

Quillardet P., Huisman O., D'Ari R. and Hofnung M. (1985) *The SOS Chromotest, a colorimetric bacterial assay for genotoxins: validation study with 83 compounds*. Mutat. Res., **147**: 79-95

Rodrigues A.S., Duarte-Silva I., Laires A., Cheveca T.., Glatt H.R. and Rueff J. (1993) *The use of CYP1A2 and Acetyltransferase expressing V79 cells in the Metabolic Activation of Aromatic Amines*. Abstract 23rd Annual Meeting of the European Environmental Mutagen Society, Barcelona

Rueff J., Laires A., Gaspar J., Borba H. and Rodrigues A. (1992[a]) *Oxygen species and the genotoxicity of quercetin*. Mutat. Res. **265**: 75-81

Rueff J., Rodrigues a., Laires A. and Gaspar J. (1992[b]) *Activation of promutagens by porphyrinic biomemetic systems*. Mutat. Res. **269**: 243-250

Sedgwick, S.G., Ho, C., and Woodgate, R. (1991[a]) *Mutagenic DNA repair in enterobacteria*. J. Bacteriol. **173**: 5604-5611

Todd P.A., Monti-Bragadin aqnd Glickman B.W. (1979) *MMS mutagenesis in strains of Escherichia coli carrying the R46 mutagenic enhancing plasmid: phenotypic analysis of arg+ revertants*. Mutat. Res. **62**: 227-237

# Coumarin 7-hydroxylation (CYP2A6) polymorphism in a Turkish population

Mümtaz Iscan[1], Hamed Rostami[1], Mesude Iscan[2], Tülin Güray[2], Arja Rautio[3], Olavi Pelkonen[3]

[1]Department of Toxicology, Faculty of Pharmacy, Ankara University. [2]Department of Biology, METU, Ankara, Turkey. [3]Department of Pharmacology and Toxicology, University of Oulu, Oulu, Finland

Coumarin is a plant alkoloid, which has been widely used as a sweetener, fixative, stabilizer and food additive (Egan et al., 1990). Recently, coumarin has been tested as a drug in cancer and several other diseases (Egan et al., 1990). Coumarin is metabolized to 7-hydroxycoumarin (7OHC) by P450 isozyme called P450coh (Raunio et al.,1988) or CYP2A6 (Nebert et al., 1991) in man. The main metabolite 7OHC is rapidly converted by conjugation to the glucuronide and excreted into urine (Moran et al., 1987; Egan et al., 1990). Moreover, CYP2A6 has also been shown to be able to activate the procarcinogens N-nitrozodimethylamine, N-nitrozodiethylamine and aflatoxin $B_1$ (Gonzales et al., 1991). Considerable evidence in mice (Wood and Conney, 1974; Lang et al., 1989), and in humans (Pelkonen et al., 1985) has implicated the polymorphic regulation of this isozyme. Recently, Rautio et al. (1992) developed an in vivo method to measure the activity of CYP2A6 in humans, by measuring the urinary excretion of the metabolite in humans after the administration of a small dose of coumarin (5mg). In 110 healthy Finnish volunteers, there was a great interindividual variability in the formation of 7OHC, but the distribution did not display any polymorphism. Since ethnic variations in other P450 isozymes, e.g. CYP2D6, have been observed in numerous ethnogeographic populations (Kalow, 1982), it is obviously of great importance and interest to study ethnical variability of CYP2A6 in other populations. Therefore, in the present study we aimed to evaluate the pharmacogenetic coumarin hydroxylation profile in a group of 100 healthy Turkish volunteers.

METHODS

100 healthy volunteers (70 males, 30 females), aged between 19 to 56 years, were given 5 mg coumarin p.o. (Venalot$^R$, manufactured by Schaper & Brummer, Ringelheim, FRG), after an overnight fast. Urine samples were collected before and 2, 4 and 8 hours after the drug administration. The in vivo determination of 7OHC was performed by the spectrofluorometric method, essentially as described previously (Rautio et al.,1992), after β-glucuronidase hydrolysis and chloroform extraction. Urine samples were mixed with β-glucuronidase solution and incubated in a metabolic shaker for two hours at 37°C. Thereafter the incubates were extracted with chloroform, and chloroform phase was measured in 1 M glycine buffer (pH 10.4) immediately by a spectrofluorometer at 365 and 454 nm excitation and emission wavelengths, respectively.

RESULTS AND DISCUSSION

In the preliminary studies no urinary excretion of 7OHC was noted after the first 8 h (8-24 h). The 7OHC excreted in 2 and 4 hours was 80.9 ± 8.7 % (mean ± SD) and 96.3 ± 6.0 % (mean ± SD) of the total excretion. Since the greatest variation between individuals occured during the first 2 h and since the amount excreted in 2 hours was relatively constant in the same individual in repeated tests, the

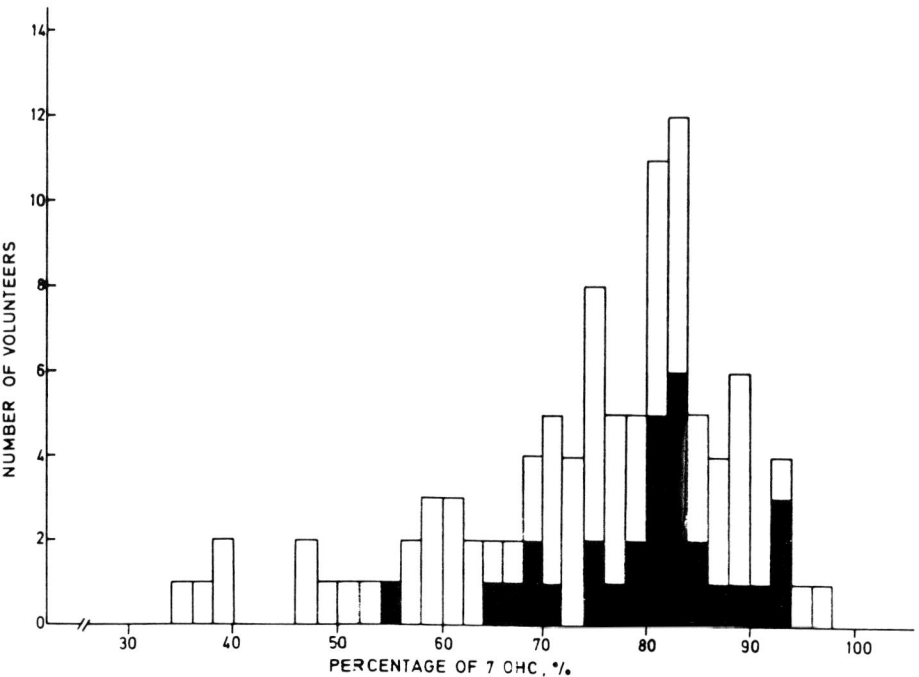

Fig. 1. The frequency distribution of 100 individuals with respect to percentage of urinary excretion of 7-hydroxycoumarin in 2 h out of total excretion ('2 h coumarin test'). Shaded area indicates females.

percentage of excreted 7OHC during the first 2 h out of total excretion in 8 h, was chosen as the index of 7OHC which is named as '2 h coumarin test'. Urine collecting time was 8 h, because no excretion was noted after the first 8 h. In 100 volunteers, the rate of excretion of 7OHC, '2 h coumarin test' values, varied from 34.9 to 97.8 % (74.7 ± 13.5 %; mean ± SD). Four individuals were observed to have relatively 'slow' coumarin test values (34-40 %) and appeared to constitute a group (Fig. 1). When the coumarin test values were analyzed statistically, it was observed that the difference between the subgroups of the population was significant (Mann-Whitney U test, $p<0.001$). Hence, a polymorphism has been shown in the rate of formation of 7OHC in a Turkish population. 4 % of the Turkish subjects were slow hydroxylators of coumarin. Since CYP2A6 is also able to activate the procarcinogens N-nitrosodimethylamine, N-nitrosodiethylamine and aflatoxin $B_1$ (Gonzales et al., 1991), slow hydroxylators of coumarin are likely to be more resistant whereas the fast hydroxylators are more prone to the carcinogenic effects of these chemicals. Thus, this test may be useful in evaluating the susceptibility of the individuals and populations to these cancer causing chemicals which in turn contribute to the understanding of the individual and population differences in environmental and chemical effects on the incidence of carcinogen and nutritional related human cancer.

REFERENCES

Egan, D., O'Kennedy, R., Moran, E., Cox, D., Prosser, E. & Thornes, R.D. (1990): The pharmacology, metabolism, analysis, and applications of coumarin and

coumarin-related compounds. Drug Metab. Rev. 22: 503-529.
Gonzales, F.J., Crespi, C.L. & Gelboin, H.V. (1991): cDNA-expressed human cytochrome P450s: a new age of molecular toxicology and human risk assessment. Mutation Res. 247: 113-127.
Kalow, W. (1982): Ethnic differences in drug metabolism. Pharmacokinet. 7:373-391.
Lang, M.A., Juvonen, R., Jarvinen, P., Honkakoski, P. & Raunio, H. (1989): Mouse liver P450coh: genetic regulation of the pyrazole-inducible enzyme and comparison with other P450 isozymes. Arch. Biochem. Biophys. 271: 139-148.
Moran, E., O'Kennedy, R. & Thornes, R.D. (1987): Analysis of coumarin and its urinary metabolites by high-performance liquid chromatography. J. Chromatogr. 416: 165-169.
Nebert, D.W., Nelson, D.R., Coon, M.R., Estabrook, R.W., Feyereisen, R., Fujii-Kuriyama, Y., Gonzales, F.J., Guengerich, F.P., Gunsalus, I.C., Johnson, E.F., Loper, J.C., Sato, R., Waterman, M.R. & Waxman, D.J. (1991): The P450 superfamily: Update on new seqeunces, gene mapping and recommended nomenclature. DNA Cell Biol. 10:1-14.
Pelkonen, O., Sotaniemi, E.A. & Ahokas, J.T. (1985): Coumarin 7-hydroxylase activity in human liver microsomes. Properties of the enzyme and interspecies comparisons. Br. J. Clin. Pharmacol. 19: 59-66.
Raunio, H., Kojo, A., Juvonen, R., Honkakoski, P., Jarvinen, P., Lang, M.A., Vahakangas, K., Gelboin, H.V., Park, S.S. & Pelkonen, O. (1988): Mouse hepatic cytochrome P-450 isozyme induction by 1,4-bis 2-(3,5-dichloropyridyloxy) benzene, pyrazole, and phenobarbital. Biochem. Pharmacol. 37:4141-4147.
Rautio, A., Holger, K., Kojo, A., Salmela, E. & Pelkonen, O. (1992): Interindividual variability of coumarin 7-hydroxylation in healthy volunteers. Pharmacogenetics 2: 227-233.
Wood, A.W. & Conney, A.H. (1974): Genetic variation in coumarin hydroxylase activity in the mouse (Mus musculus). Science 185: 612-614.

# Exon-7 point mutation (m2; 4889A→G) in human CYP1A1 gene as susceptibility factor for lung cancer

N. Drakoulis, I. Cascorbi, J. Brockmöller, C.R. Gross, Ivar Roots

*Institute of Clinical Pharmacology, Charité, Humboldt-University, 10098 Berlin, Germany*

Genetic differences in the metabolism of chemical carcinogens have been suggested in the past to be associated with individual predisposition to cancer. CYP1A1 metabolically activates precarcinogens in cigarette smoke, such as benzo(a)pyrene, which is also an inducer of CYP1A1. Two point mutations have been reported, *m1* in the 3'-flanking region, and *m2* within exon 7, the latter leading to isoleucine to valine exchange at codon 462. In the Japanese a close correlation was found between both *m1* and *m2* with bronchogenic carcinoma. We have studied 142 lung cancer and 171 reference patients in an ethnically homogeneous German group for *m1*- and *m2*-mutations by RLFP and allele-specific PCR, respectively. No difference was found in the distribution of *m1*-alleles between lung cancer and controls, the allele-frequency was 8.5 % and 7.3 % (odds ratio =1.17). In contrast, the frequency of *m2* in lung cancer patients was twofold higher (6.7 %) than in the reference group (3.2 %) (odds ratio=2.16; $P$=0.03); the odds ratio in squamous cell =2.51; $P$=0.05). A close, but not obligatory genetic linkage of *m1* and *m2* has been observed. The results concerning *m1* conform with previous studies in Caucasians, describing no association between *m1* and lung cancer. The analysis of *m2* (odds ratio for heterozygotes = 1.72, $P$ = n.s.), estimated odds ratio for homozygotes = 11.2, $P$ = n.s.), however, showed a gene-dose dependent association of this mutation and lung cancer risk especially of *m2*-alleles not linked to *m1*.

## INTRODUCTION

Several carcinomas are supposed to be associated with the exposure to metabolically activated precarcinogens [Roots *et al.*, 1992]. These substances are activated by metabolic enzymes such as cytochromes P450. Cytochrome P450 1A1 (CYP1A1) is expressed in lung tissue and hydroxylates polycyclic aromatic hydrocarbons. The various isoenzymes of CYP differ in their substrate specificity and are induced by such substrates by by a complex reaction with the aryl hydrocarbon (*Ah*) receptor. Since cytochromes P-450 may either detoxify carcinogens, or transform precarcinogens to ultimate carcinogens, the genetic variability of *CYP* may lead to interindividual difference in susceptibility to carcinogens.

A polymorphism of inducibility has been reported previously: approximately 10% of the Caucasian population exhibits high CYP1A1 inducibility [Petersen *et al.*, 1991] leading to an increased activation of polycyclic hydrocarbon precarcinogens to reactive ultimate carcinogens; thus this trait may result greater risk of cigarette smoke induced bronchogenic carcinoma [Kouri *et al.*, 1982].

Two point mutations associated to *CYP1A1* have been recently demonstrated [Kawajiri *et al.*, 1990; Hayashi *et al.*, 1991]. The mutation discovered first, *m1*, is situated in the 3'-flanking region (i.e. 1,194 bp downstream from exon 7) and represents a thymine to cytosine

($T$ to $C$) transition, providing a new restriction endonuclease cleavage site for *Msp*I [Nakachi et al., 1991]. The second mutation, *m2*, is a replacement of adenine by guanine ($A$ to $G$) and is localised in the coding region at position 4,889 within exon 7 [Hayashi et al., 1991]. This mutation results an amino acid substitution of isoleucine to valine (Ile to Val) at residue 462 in the heme-binding region of *CYP1A1*.

The prevalence of both mutations of *CYP1A1* in the Japanese population is significantly higher among lung cancer patients [Kawajiri et al., 1993], indicating that the presence of these traits represents a risk factor towards lung cancer. Mutations *m1* and *m2* mostly occur jointly so that the impact of each of them can not easily be discerned. In the European (Caucasian) population, the allelic frequency of *m1* is low (around 11 %) and an association with lung cancer was not detected [Tefre et al., 1991; Hirvonen et al., 1992].

We have detected both mutations of *CYP1A1* by polymerase chain reaction and restriction fragment length polymorphism to determine their frequency in an ethnically uniform German study group. We have investigated the frequency of both these mutations in bronchial carcinoma patients in order to elucidate whether these *CYP1A1* polymorphisms can be established as a susceptibility factor for bronchial carcinoma in the Caucasian population.

## MATERIALS and METHODS

The study includes 142 patients with histologically verified bronchogenic carcinoma and as a control group 171 healthy individuals or hospital patients with various other pulmonary diseases but no known malignancy. Both lung cancer and control patients were selected within the same time period between 1991 and 1992 from the Department of Pulmonology of the Zehlendorf Hospital and from the Klinikum Steglitz in Berlin, Germany.

Genomic DNA was isolated from 10 ml blood samples by standard phenol/chloroform extraction. The identification of *CYP1A1* genotypes was performed as described previously [Drakoulis et al., 1993]. Briefly, templates of the possible mutation sites were amplified. Fragments, containing the possible mutation *m1* in the 3´-flanking region was carried out by restriction fragment length polymorphism analysis (RFLP) after amplification of a 335-bp fragment by polymerase chain reaction. The wild type, *wt1*, was defined as the absence, where the *m1* allele provided the *Msp*I specific GGCC motif. The *m2* mutation within exon 7 was determined by the use of allele specific PCR.

The frequencies of alleles or genes were compared by calculating odds ratios and their 95%-confidence limits. The exact Fisher test was used to evaluate statistical significance and the statistical level of significance was fixed at $P = 0.05$; statistical power and sample size estimations were calculated.

## RESULTS

The distribution of *CYP1A1* allele conformations in 142 lung cancer and 171 reference patients is demonstrated in Tab. 1. The combination *wt1/wt1* with *wt2/m2* was significantly overrepresentated as compared to all remaining genotypes (odds ratio = 10.15 (95%-confidence limits = 1.33 - 453; $P = 0.009$). The allelic frequencies of *wt1* and *m1* among lung cancer patients and controls showed no statistically significant difference (Tab. 2). The division into histologically defined subclasses of bronchogenic carcinoma showed a particular, statistically nonsignificant increase of the *m1*-allele in squamous-cell cancer patients. The observed allele frequencies for *wt2* and *m2* were 0.968 and 0.032, respectively, for the controls and 0.933 and 0.067 for lung cancer cases. Among lung cancer patients, *m2* was more frequent (odds ratio = 2.16); especially the subgroup of squamous-cell cancer showed a significant overrepresentation of the *m2*-allele (odds ratio = 2.51, $P = 0.05$).

The reference patients showed a tight (about 90 %) but not obligatory linkage of *m2* to *m1*. In contrast, among lung cancer patients half of the carriers of *wt2/m2* provided no *m1*-mutation. The odds ratio of the genetic constellation (*wt2/m2, wt1/wt1*) in lung cancer and reference patients compared with all remaining genotypes was 10.15 (95%-confidence limits 1.33 - 453, *P* = 0.009).

Tab. 1. Frequencies of polymorphic *CYP1A1* allelic conformations in 142 lung cancer and 171 control patients. *wt1* = wild type (6235T), *m1* = 6235C; *wt2* = wild type (4889A), *m2* = 4889G

|  | wt1/wt1 (%) | wt1/m1 (%) | m1/m1 (%) | Sum (%) |
|---|---|---|---|---|
| Lung cancer |  |  |  |  |
| wt2/wt2 | 78.2 | 9.9 | 0.0 | 88.0 |
| wt2/m2 | 5.6* | 4.2 | 0.7 | 10.6 |
| m2/m2 | 0.0 | 1.4 | 0.0 | 1.4 |
| Sum | 83.8 | 15.5 | 0.7 | 100.0 |
| References |  |  |  |  |
| wt2/wt2 | 84.8 | 8.8 | 0.0 | 93.6 |
| wt2/m2 | 0.6 | 5.8 | 0.0 | 6.4 |
| m2/m2 | 0.0 | 0.0 | 0.0 | 0.0 |
| Sum | 85.4 | 14.6 | 0.0 | 100.0 |

Tab. 2. Allele frequency of *CYP1A1* mutation *m1* and *m2* in histologically classified lung cancer patients and reference group.

|  | n | mutant (%) | wild-type (%) | Odds ratio | 95%-Conf.-limits | P |
|---|---|---|---|---|---|---|
| m1 |  |  |  |  |  |  |
| Lung cancer | 142 | 8.5 | 91.5 | 1.17 | 0.62-2.19 | n.s. |
| Squamous ca. | 52 | 11.5 | 88.5 | 1.65 | 0.73-3.57 | n.s. |
| Large ca. | 23 | 8.7 | 91.3 | 1.21 | 0.29-3.75 | n.s. |
| Small ca. | 26 | 5.8 | 94.2 | 0.78 | 0.14-2.69 | n.s. |
| Adenoca. | 34 | 5.9 | 94.1 | 0.79 | 0.19-2.41 | n.s. |
| Mixed | 7 | 7.1 | 92.9 | 0.98 | 0.02-7.02 | n.s. |
| Reference | 171 | 7.3 | 92.7 | 1.00 | - | - |
| m2 |  |  |  |  |  |  |
| Lung cancer | 142 | 6.7 | 93.3 | 2.16 | 0.96-5.11 | 0.033 |
| Squamous ca. | 52 | 7.7 | 92.3 | 2.51 | 0.85-7.05 | 0.050 |
| Large ca. | 23 | 6.5 | 93.5 | 2.10 | 0.36-8.36 | n.s. |
| Small ca. | 26 | 5.8 | 94.2 | 1.84 | 0.32-7.29 | n.s. |
| Adenoca. | 34 | 7.4 | 92.6 | 2.39 | 0.63-7.75 | n.s. |
| Mixed | 7 | 0.0 | 100.0 | 0.00 | 0.00-10.5 | n.s. |
| Reference | 171 | 3.2 | 96.8 | 1.00 | - | - |

## DISCUSSION

Similar to other studies among Caucasians [Tefre et al., 1992; Hirvonen et al., 1993; Shields et al., 1993], no association was found between m1 and lung cancer risk. These findings were in contrast to an about fourfold higher frequency of the m1-allele in a Japanese population [Nakachi et al., 1991]. However, due to the rarity of the m1-trait among Caucasians, for a statistical proof of a lack of association, high numbers are required to keep the statistical power at levels below $\beta = 0.10$.

The point mutation m2 was more frequent in lung cancer patients than in controls, especially among patients with squamous-cell carcinoma, as also described by Kawajiri et al. [1993] for the Japanese population. The estimated frequency of m2/m2 by the Hardy-Weinberg-equation in the reference group was 0.0011. This proposes that homozygous m2-carriers may possess a 13-fold risk for lung cancer. These data suggest that mutation m2 in exon 7 is indeed a susceptibility factor for lung cancer also in the Caucasians. Interestingly, in lung cancer patients singular A to G transitions (m2) were found in 8 of 17 cases without linkage to the T to C transition (m1) ($P = 0.04$). This overrepresentation of single m2 without linkage to m1 may be a special risk constellation. The genetic combination (wt2/m2, wt1/wt1) was highly overrepresented in lung cancer ($P = 0.009$). Thus no asssociation was found between presence of m1-alleles and lung cancer, but in contrast, m2-alleles proved as a hereditary risk factor, espcially if not linked with m1-alleles.

## REFERENCES

Drakoulis, N., Cascorbi, I., Brockmöller, J., Gross, C.R., Roots, I. (1993): Polymorphisms in the human CYP1A1 gene as susceptibility factors for lung cancer: Exon-7 mutation (4889 A to G), and a T to C mutation in the 3'-flanking region. Clin. Investig. in press.

Hayashi, S., Watanabe, J., Nakachi, K., Kawajiri, K. (1991): Genetic linkage of lung cancer-associated MspI polymorphism with amino acid replacement in the heme binding region of the human cytochrome P450IA1 gene. J. Biochemistry 110:407-411

Hirvonen, A., Husgafvel-Pursiainen, K., Anttila, S., Karjalainen, A., Sorsa, M., Vainio, H. (1992): Metabolic cytochrome P450 genotypes and assessment of individual susceptibility to lung cancer. Pharmacogenetics 2:259-263

Kawajiri, K., Nakachi, K., Imai, K., Watanabe, J., Hayashi, S. (1993): The CYP1A1 gene and cancer susceptibility. Crit. Rev. Oncol. Hematol. 14:77-87

Kouri, R.E., McKinney, C.E., Slomianry, D.J., Snodgrass, D.R., Wray, N.P., McLemore, T.L. (1982): Positive correlation between high aryl hydrocarbon hydroxylase activity and primary lung cancer as analyzed in cryopreserved lymphocytes. Cancer Res. 42:5030-5037

Nakachi, K., Imai, K., Hayashi, S., Watanabe, J., Kawajiri, K. (1991): Genetic susceptibility to squamous-cell carcinoma of the lung in relation to cigarette smoking dose. Cancer Res. 51: 5177-5180

Petersen, D.D., McKinney, C.E., Ikeya, K., Smith, H.H., Bale, A.E., McBride, O.W., Nebert, D.W. (1991): Human CYP1A1 gene: cosegregation of the enzyme inducibility phenotype and an RFLP. Am. J. Hum. Genet. 48:720-725

Roots, I., Brockmöller, J., Drakoulis, N., Loddenkemper, R. (1992): Mutant genes of cytochrome P-450IID6, glutathione S-transferase class Mu, and arylamine N-acetyltransferase in lung cancer patients. Clin. Investig. 70:307-319

Shields, P.G., Sugimura, H., Caporaso, N.E., Petruzelli, S.F., Bowman, E.D., Trump, B.F., Weston, A., Harris, C.C. (1992): Polycyclic aromatic hydrocarbon-DNA adducts and the CYP1A1 restriction fragment length polymorphism. Environ. Health Perspect. 98:191-194

Tefre, T., Ryberg, D., Haugen, A., Nebert, D. W., Skaug, V., Brøgger, A.. Børresen, A. L. (1991): Human CYP1A1 (cytochrome $P_1450$) gene: lack of association between the MspI restriction fragment length polymorphism and incidence of lung cancer in a Norwegian population. Pharmacogenetics 1: 20-25

# A role for Asp251 in the activation of oxygen by cytochrome P450$_{cam}$

Nancy Counts Gerber, Stephen G. Sligar

*Department of Biochemistry, University of Illinois, Urbana, Illinois, USA 61801*

The region corresponding to Thr252 is in P-450cam is completely conserved as either a Thr or Ser across all P-450 sequences (Nelson and Strobel, 1987). It has been found that only when a hydroxyl is at position 252 (Ser, Thr) is P-450cam capable of efficiently hydroxylating substrate (Imai, et al., 1989, Martinis, et al., 1989). The remaining reducing equivalents in the mutant enzymes are found in $H_2O_2$ as a result of a side reaction that uncouples the enzyme. A water molecule occupying the position of the Thr side-chain in the T252A mutant has been proposed to be the source of the uncoupling by de-stabilizing the oxy-complex (Raag, et al., 1991). The other residue in this region that is highly conserved in the P-450 enzymes corresponds to Asp or Glu at position 251 in the P450cam protein. Mutation of this residue in P450cam, P450d or aromatase leads to proteins whose rate of catalysis is dramatically decreased, while the ability to make product was retained (Gerber and Sligar, 1992, Ishigooka, et al., 1992, Shimada, et al., 1990, Zhou, et al., 1992). A comparison of the wild-type and D251N mutant rate constants for all isolatable steps in the P-450cam reaction cycle (Gerber and Sligar, 1992) demonstrated that the major kinetic effect was on the combined second electron transfer, O-O bond scission, and substrate hydroxylation step. This communication focuses on the further characterization of the P-450cam mutant D251N to determine more precisely the mechanism of Asp251 in P-450cam catalysis.

Effects of pH on P-450 catalysis

The effect of pH on wild-type and mutant P-450cam is shown in Figure 1. The effect of pH on the steady-state reaction rate of T252A reflects the effect of pH on the steps prior to, but not including, oxygen bond cleavage, since the T252A mutant uncouples the reaction before the O-O bond scission step. The stronger dependence of the rate of wild-type catalysis on pH compared to T252A implies that the effect of pH on the wild-type enzyme occurs primarily after the input of the second electron. The K178Q protein displays a somewhat stronger dependence of rate on pH, and one that is altered from the wild-type enzyme. There are several possibilities for the origin of this effect one of which is that Lys178 is a catalytically significant ionizable group that is active in the wild-type system. It seems unlikely that Lys178 is a proton donor in the wild-type system due to the fact that the K178Q mutant still retains a significant amount of activity. It is more likely that Lys178 is not a proton donor but rather acts to increase the nucleophilicity or basicity of another residue in P-450cam. Upon removal of the basic Lys178 residue the pKa of an acidic residue near Lys178 would be expected to shift to higher values. This residue may be either Arg186 or Asp251, although the pKa would appear to be too low for an arginine side-chain. Both Lys178 and Arg186 form salt linkages with Asp251 in the P-450cam crystal structure and mutation of these residues in P-450cam causes large changes in activity (Gerber and Sligar, 1992, Shimada, et al., 1990). The most dramatic pH effect is seen in the D251N mutation. At high pH the activity of the mutant enzyme is indistinguishable from the background and as the pH decreases from 8 to 5 the activity increases by over an order of magnitude. Although this is still an order of magnitude lower the wild-type rate at pH 5, it represents a large recovery of activity lost due to the mutation. One explanation for this effect is to view Asp251 as a catalytically important proton donor where upon mutation to Asn the protein must have an input of solvent protons to function. The proximal electron "push" from the axial cysteine would be expected to be sufficient to allow some degree of O-O bond cleavage in the absence of the distal "pull" from Asp251.

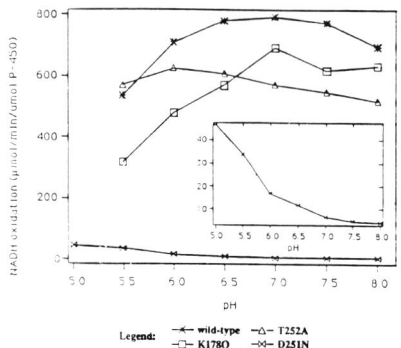

Figure 1: Effect of pH on the steady-state rates for NADH oxidation in the reconstituted P-450cam system. The insert shows more detail of the effect of pH on the D251N mutant. Rates were measured in 50 mM potassium phosphate, 0.5 mM camphor, 2 μM Fp, 250 μM NADH, 10 μM Pd and 1 μM P-450cam at 22°C.

### Effects of the D251N Mutation on First Electron Transfer

All protein samples and buffers were thoroughly degassed under vacuum to remove oxygen and prevent the reaction from progressing past the first electron transfer step. The concentration of P-450 was held constant while the concentration of Pdx was varied and the increase in absorbance at 450 nm was monitored due to the reduction of P-450cam and the oxidation of Pdx (Davies and Sligar, 1992). The rate for the reaction of Pdx with wild-type P-450cam showed saturation behavior with increased Pdx concentration indicating that a Pdx-P-450 complex is involved in the rate limiting step of the reaction. The fit of the data to a first-order kinetic equation indicates that the protein-protein association is not the rate-limiting step of the reaction. A Hanes plot of the first electron transfer data gives an apparent Km of 3.0 μM for the mutant D251N compared to a Km of 1.6 μM for wild-type enzyme. The calculated Vmax of 172 $s^{-1}$ of D251N is almost 3-times that of wild-type. One explanation for the change in Vmax is a change in the redox potential for the mutant protein relative to wild-type. If a negatively charged residue near the heme-center is changed to a neutral one, this might increase the driving force for electron transfer and lead to a faster electron transfer rate. The measured Km of 1.6 μM is nearly the same as the Kd measured by Davies and Sligar for the association of reduced Pdx with oxidized P-450cam (Davies and Sligar, 1992) and would indicate that the measured Km for the reaction reflects the true Kd for the complex. One explanation for the change in Km between wild-type and D251N is that a change in the electrostatic potential energy on the surface of the protein has affected the interaction between the two proteins. Since the interaction of P-450cam and Pdx is known to have an electrostatic component (Hintz and Peterson, 1981, Stayton and Sligar, 1990) a change in the surface potential energy might lead to a weaker interaction between Pdx and P-450cam.

### Single Turnover of D251N mutant

The rate constant for the single-turnover of the D251N was measured by mixing equimolar amounts of reduced Pdx and reduced P-450 with an oxygenated buffer and monitoring the appearance of oxidized P-450 at 395 nm. A rate constant of 0.15 $s^{-1}$ was measured for the single turnover of mutant D251N, consistent with the results of the steady-state experiments. When the wild-type protein was studied without a trap present, an increase in absorbance was seen below 405 nm and a decrease above 405 nm. This corresponds to the formation of reduced substrate-bound P-450cam as the P-450 returns to the resting state of the enzyme after catalysis. The D251N mutant, however, displays different behavior. Below 410 nm the absorbance decreased at a rate comparable to the rate of wild-type turnover while above 410 nm the absorbance increased at the same rate. There were no further change observable in the mutant sample over a period of several milliseconds. The absorbance changes observed are consistent with a coupled oxidation of Pdx and reduction of oxy-P-450cam and correlate with the net extinction changes expected from the conversion of oxy-D251N to the steady-state intermediate.

The reaction was further conducted as a function of Pdx concentration and the kinetic traces were fit to a single exponential. Saturating behavior with increased Pdx concentration was seen as in the first electron transfer reaction, which indicates that the Pdx-P-450 complex is important in the rate-limiting step of the reaction. When the data is plotted as the concentration of Pdx divided by the initial velocity of the reaction verses the concentration of [Pdx] an apparent $K_m$ and $V_{max}$ can be calculated. Since little is known of the kinetics of the forward and reverse reactions, the $K_m$ may not be a true dissociation constant. The calculated $V_{max}$ is 68 s$^{-1}$ and $K_m$=1.5 µM. If the measured $K_m$ is reflective of a P-450:Pdx complex then the D251N complex for the conversion of the oxy-bound D251N to the intermediate is tighter than for the first electron transfer. This is in contrast with results seen in for the wild-type system in which the complex for the second electron transfer is actually weaker than for the first (Brewer and Peterson, 1988). After flow is stopped, the absorbance spectra of the P-450 resembles the spectral intermediate seen in steady-state experiments and has an absorbance maximum at 419 nm. If this material is taken from the stopped flow, extracted with methylene chloride and analyzed for product, close to 100% coupling of reducing equivalents to product is seen. This indicates that the system is supplied with sufficient reducing equivalents and little autooxidation or other processes are occurring.

## DISCUSSION

A model has been proposed to explain the effects of the D251N mutant on P-450 catalysis and is shown in Figure 2. Although the Asp251 side-chain is not close enough to a modeled heme-bound oxygen to interact with it directly, its side-chain can adopt a rotamer configuration that would allow it to interact via a bridging water molecule. A configuration similar to this can be seen in the crystal structure of the 2-phenylimidazole bound P-450cam (Poulos and Howard, 1987) in which the side-chain of Asp251 is rotated so that the salt linkage to Lys178 is broken and a new interaction formed with a water molecule. The Arg186 linkage is maintained and the water molecule is now close enough to interact with the Thr252 side-chain oxygen. This structural model is consistent with the data on the mutation of both Lys178 and Arg186. In the model shown in Figure 2, Asp251 is acting as a proton donor to either Thr252 which is H-bonded to the distal oxygen, or to the distal oxygen directly. The bracket around the bridging water molecule indicates that a water molecule may or may not be necessary for Asp251 to interact with the bound oxygen. This interaction of Asp251 might allow Thr252 to donate a proton to the oxygen, preventing formation of a transient alkoxide. The electron attracting ability of Asp251 along with the electron donating property of the proximal cysteine ligand allows for efficient and rapid oxygen bond cleavage, while Thr252 stabilizes the bound oxygen against autooxidation. The Arg186 salt bridge provides Asp251 a link to solvent; when Asp251 is mutated proton donation from Thr252 would be extremely unfavorable and would only happen very slowly and at a pH where a sufficient supply of solvent protons is available to stabilize the side-chain against alkoxide formation.

Another effect of the mutation of Asp251 may be to alter the charge distribution of the protein in such a way as to disrupt either the interaction between the P-450 and Pdx and slow the second electron transfer or to inhibit a conformational change occurring when the Pdx binds to P-450. EPR experiments designed to monitor the oxidation of Pdx during the reaction forming the mutant D251N spectral intermediate have indicated that an electron is transferred from Pdx to the mutant D251N at a rate faster than that of the overall turnover of oxygenated-P-450. The results seen in this study argues for a role for Asp251, Lys178 and Arg186 in a proton relay to protonate the bound dioxygen prior to O-O scission and water release. Such proton shuttles that allow delivery of protons to buried active sites without water access are becoming recognized as a common theme in protein structure-function (Meyer, 1992). The shuttle would allow the buried Asp251 to link to solvent via Lys178 and/or Arg186 and the Asp residue would then interact either directly with the bound oxygen or through a hydrogen bond with Thr252. Computer modeling shows that the Asp251 residue can adopt a conformation that would allow it to form a hydrogen bond with Thr252, which has been proposed to interact with the bound dioxygen. This altered conformation would still allow the Asp251 side-chain to hydrogen bond to the Arg186 side-chain. The interaction of Asp251 with the distal oxygen might aid in pulling electron density towards the distal oxygen atom, thereby facilitating heterolytic oxygen-oxygen bond cleavage.

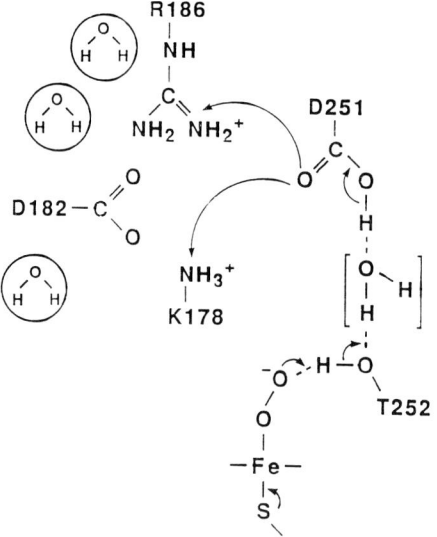

Figure 2: Schematic of the proposed mechanism for oxygen bond cleavage in P-450cam. Arrows indicate the direction of electron flow. Brackets around the water bridging Asp251 and Thr252 indicate the lack of evidence as to the existence of this molecule.

REFERENCES

Brewer, C. B. and Peterson, J. A. (1988): Single turnover kinetics of the reaction between oxycytochrome P-450cam and reduced putidaredoxin. *J. Biol. Chem.* 263, 791-798.

Davies, M. D. and Sligar, S. G. (1992): Genetic variants in the putidaredoxin-cytochrome P-450cam electron transfer complex: Indentification of the residue responsible for redox-state-dependent conformers. *Biochemistry* 31, 11383-11389.

Gerber, N. C. and Sligar, S. G. (1992): Catalytic mechanism of cytochrome P-450: Evidence for a distal charge relay. *J. Am. Chem. Soc.* 114, 8742-8743.

Hintz, M. J. and Peterson, J. A. (1981): The kinetics of reduction of cytochrome P-450cam by reduced putidaredoxin. *J. Biol. Chem.* 256, 6721-6728.

Imai, M., Shimada, H., Watanabe, Y., Matsushima-Hibiya, Y., Makino, R., Koga, H., Horiuchi, T. and Ishimura, Y. (1989): Uncoupling of the cytochrome P-450cam monooxygenase reaction by a single mutation, Threonine-252 to Alanine or Valine: A possible role of the hydroxy amino acid in oxygen activation. *Proc. Natl. Acad. Sci. U.S.A.* 86, 7823-7827.

Ishigooka, M., Shimizu, T., Hiroya, K. and Hatano, M. (1992): Role of Glu318 at the putative distal site in the catalytic function of cytochrome P450d. *Biochemistry* 31, 1528-1531.

Martinis, S. A., Atkins, W. M., Stayton, P. S. and Sligar, S. G. (1989): A conserved residue of cytochrome P-450 is involved in heme-oxygen stability and activation. *J. Am. Chem. Soc.* 111, 9252-9253.

Meyer, E. (1992): Internal water molecules and H-bonding in biological macromolecules: A review of structural features with functional implications. *Protein Science* 1, 1543-1562.

Nelson, D. R. and Strobel, H. W. (1987): Evolution of cytochrome P-450 proteins. *Mol. Biol. Evol.* 4, 572-593.

Poulos, T. L. and Howard, A. J. (1987): Crystal structures of metyrapone and phenylimidizole-inhibited complexes of cytochrome P-450cam. *Biochemistry* 26, 8165-8174.

Raag, R., Martinis, S. A., Sligar, S. G. and Poulos, T. L. (1991): Crystal structure of the cytochrome P-450 active site mutant Thr252Ala. *Biochemistry* 30, 11420-11429.

Shimada, H., Makino, R., Imai, M., Horiuchi, T. and Ishimura, Y. (1990): Mechanism of oxygen activation by cytochrome P-450cam. In International Symposium on Oxygenases and Oxygen Activation, ed. S. Yamamoto and Y. Ishimura, pp. 133-136. Yamada Science Foundation.

Stayton, P. S. and Sligar, S. G. (1990): The cytochrome P-450cam binding surface defined by site-directed mutagenesis and electrostatic modeling. *Biochemistry* 29, 7381-7386.

Zhou, D., Korzekwa, K. R., Poulos, T. and Chen, S. (1992): A site-directed mutagenesis study of human placental aromatase. *J. Biol Chem.* 267, 762-768.

# Activation and regulation of aromatic hydrocarbon-inducible *Cyp*1a1 and *Cyp*1a2 gene expression in mouse hepatocytes in primary culture

Nobuo Nemoto, Junko Sakurai

*Department of Experimental Pathology, Cancer Institute, 1-37-1, Kami-Ikebukuro, Toshima-ku, Tokyo 170, Japan*

Regulation of expression of Cyp1a1 and Cyp1a2 genes was investigated in mouse hepatocytes for up to 5 days after transferring to either monolayer or spheroid(multicellular aggregate) primary culture. Expression of 3−methylcholanthrene(MCA)−induced CYP1A1 mRNA remained high during the observation period under both culture conditions. In contrast, while levels of CYP1A2 mRNA in spheroid culture were also appreciable throughout, at the same levels as in vivo, they rapidly decreased in monolayer culture to become negligible. Increase in intracellular cyclic nucleotide content induced CYP1A1 mRNA in either spheroid or monolayer cultures, but not CYP1A2 mRNA. Amounts of CYP1A1 mRNA after MCA were elevated in the presence of cycloheximide(CHI). In contrast, its presence did not increase either constitutive or MCA−induced CYP1A2 mRNA. The present observations indicate that the spheroid culture is suitable for studying the mechanism of Cyp1a2 gene expression and that, while the Cyp1a1 gene is regulated by a cyclic nucleotide−dependent system and a short−lived suppressor protein, but the same mechanisms do not operate to control Cyp1a2 gene expression.

Since the liver is fortified with several kinds of metabolizing enzymes, primary culture systems have been developed for investigation of drug metabolism in adult hepatocytes(Sirica & Pitot, 1980). However, rapid decline of total P450 contents in culture systems has been reported(Paine, 1990). A typical example is CYP1A2, which is constitutively expressed and is inducible by treatment with aromatic hydrocarbons in vivo. When hepatocytes are transferred to monolayer culture, its content decreases rapidly with scarce potential for induction. A culture system, in which it could be induced for at least a few days, would therefore be of advantage. The present paper deals with expression of Cyp1a2 and Cyp1a1 genes, in mouse hepatocytes during formation of spheroids.

EXPERIMENTAL PROCEDURES

The livers of female C57BL/6NCrj mice were perfused with collagenase and then Percoll isoden-

---

This work was partly supported by a Grant−in−Aid for Cancer Research from the Ministry of Education, Culture and Science, Japan, and by the Smoking Research Foundation.

sity centrifugation was done for isolation of viable hepatocytes(Nemoto & Sakurai, 1991). The cells were dispersed in Waymouth MB 752/1 medium without serum, and allowed to seed in dishes with or without collagen−coating. Total RNA prepared from the hepatocytes was applied for Northern−blot hybridization, in which cDNA probes of Cyp1a1 and Cyp1a2(gifts from Dr. D. W. Nebert, Univ. of Cincinnati) were employed for detection of the generated mRNAs. Enzyme activities of AHH(aryl hydrocarbon hydroxylase) and AAH(acetanilide 4−hydroxylase) were measured as described(Nemoto & Sakurai, 1993).

RESULTS

Differences in Cyp1a1 and Cyp1a2 expression between monolayer and spheroid cultures(Fig. 1)
When mouse hepatocytes were seeded on collagen−coated dishes, they formed monolayers within 24 h. In contrast, if dishes were not coated with collagen, hepatocytes did not spread in spite of anchoring to the dish and subsequently aggregated. The aggregates gradually grew and detached to form floating spherical aggregates(spheroid) after 72 h. However, at day 5 of cultivation half of the aggregates were still anchored to the dish. Treatment with MCA prominently induced both

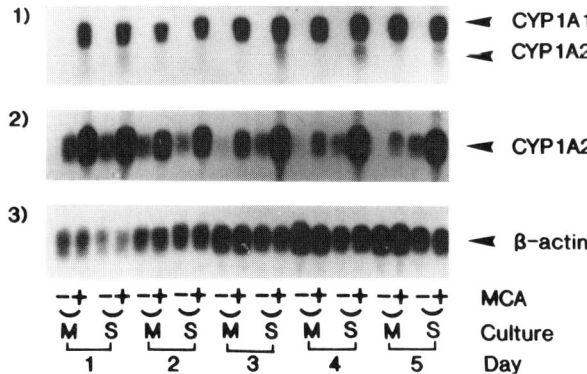

Fig. 1. Mouse hepatocytes in either monolayer(M) or spheroid(S) culture were treated with MCA at the the indicated culture days. Total RNA was prepared 10 h after the start of the treatment. cDNA probes for hybridization were mouse Cyp1a1(1) and Cyp1a2(2), and chicken beta−actin(3).

CYP1A1 and CYP1A2 mRNAs in either type culture at day 1. However, thereafter, in monolayer −cultured hepatocytes induction of CYP1A2 mRNA rapidly decreased. In contrast, induction levels of CYP1A1 mRNA were high even at day 5. In spheroid culture, although the induction pattern of CYP1A1 mRNA was the same as that of monolayer−cultured cells, CYP1A2 mRNA expression levels were still maintained at day 5. Induced amounts of CYP1A2 mRNA were compatible with those after in vivo treatment with MCA.

Induction of enzyme activity(Table 1)
Treatment with MCA or TCDD was started at the 3rd day of cultivation and enzyme activities were determined 48 h later. The ratios of AAH against AHH were significantly elevated in spheroid culture.

Effects of cyclic nucleotide on Cyp1a gene expression(Fig. 2)
Induction of CYP1A1 mRNA was observed with both dibutyryl cAMP(dBcAMP) and dibutyryl cGMP(dBcGMP) under either of the culture conditions and was further enhanced by simultaneous addition of the two nucleotides. Forskolin also additionally increased either of the cyclic nucleotide

|  | AHH | AAH | AAH/AHH |
|---|---|---|---|
| Monolayer: | | | |
| Control | 1.6 ± 0.3 | n.d. | – |
| MCA | 230.9 ± 10.0 | 172.9 ± 27.1 | 0.75 ± 0.15 |
| TCDD | 336.5 ± 31.2 | 304.6 ± 64.2 | 0.90 ± 0.12 |
| Spheroid: | | | |
| Control | 3.1 ± 0.3 | n.d. | – |
| MCA | 243.9 ± 19.2 | 325.2 ± 21.7 | 1.34 ± 0.10 ($p<0.02$) |
| TCDD | 282.7 ± 27.0 | 368.2 ± 30.8 | 1.31 ± 0.10 ($p<0.02$) |

Table 1. Enzyme activities are expressed as pmol/mg protein/min. n.d.: not detected.

induced CYP1A1 mRNA expressions. However, CYP1A2 mRNA expression was not influenced by the nucleotide alone or in combination. H-8, or N-[2-(methylamino)ethyl]-5-isoquinolinesulfonamide, a specific inhibitor for a cyclic nucleotide dependent protein kinase, did not affect the induction of CYP1A1 or CYP1A2 mRNAs.

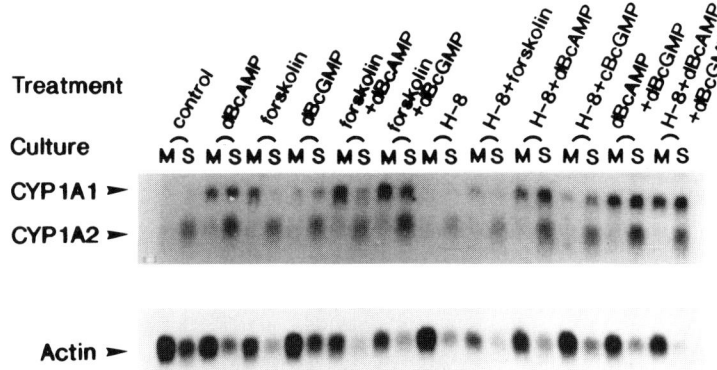

Fig. 2. Mouse hepatocytes in either monolayer(M) or spheroid(S) culture at day 4 were treated with the indicated compounds and total RNAs prepared 10 h later.

Effects of CHI on Cyp1a gene expression(Fig. 3)

Elevation of MCA-induced CYP1A1 mRNA was observed after addition of CHI. The effects were prominent with wide range of CHI concentrations. In contrast, no effects of CHI on CYP1A2 mRNA amounts were observed in either spheroid or monolayer cultured hepatocytes.

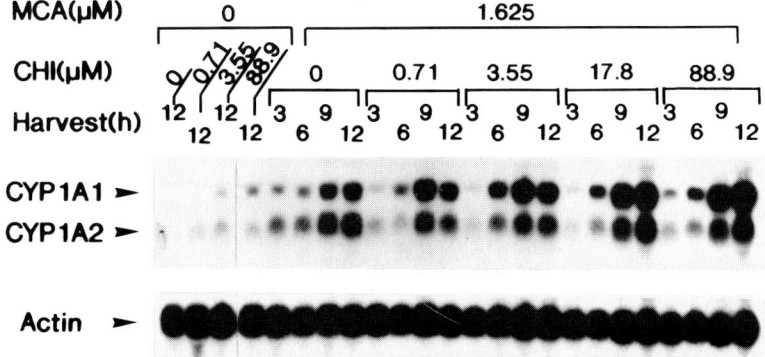

Fig. 3. Hepatocytes at day 4 of spheroid culture were treated with the indicated concentrations of MCA and/or CHI. Total RNA was prepared at the indicated times.

DISCUSSION

Up till now, hepatocytes in conventional primary monolayer culture have not been considered as a suitable tool for investigation of Cyp1a2 gene expression, because of the observed rapid decline in both constitutive expression and responsiveness to inducers(Paine et al, 1990). Multicellular spheroid culture of rat hepatocytes has been found to maintain several differentiation functions (Koide et al, 1989). However, little information is available on conditions for formation of spheroids by mouse hepatocytes in primary culture, but we found their cultivation on dishes without any cell attachment factors and the time-course of their generation was morphologically identical to that of rat cells. Within 72 h the aggregates anchored to the dish surface, but subsequently half of them detached to become free, floating, by day 5. Induction of CYP1A1 and CYP1A2 mRNA in floating spheroid after this day was very low. Our observations do not allow any firm conclusions presently as to whether tight cell-cell interaction or communication in spheroids may be necessary for maintenance of in vivo differentiated functions.

A prominent elevation of AHH activity was observed by increasing intracellular cAMP content in hamster kidney cells(Yamasaki et al, 1975). We also found elevation of enzyme activities, including AHH, by dBcAMP or forskolin, an activator of adenyl cyclase, in either type culture(Nemoto & Sakurai, 1992). Moreover, both dBcAMP and dBcGMP induced CYP1A1 mRNA expression in the present experiments. Cyclic nucleotides are involved in intracellular signal transduction, cAMP stimulating protein kinase A and cGMP protein kinase G, both of which subsequently activate a regulator protein(s) by phosphorylation. However, H-8 did not suppress expression of CYP1A1 mRNA due to addition of dBcAMP or forskolin. Therefore, induction of CYP1A1 mRNA by cyclic nucleotides might be independent of the known signal transduction pathway. In clear contrast, expression of CYP1A2 mRNA was not increased by dBcAMP, but was rather reduced by forskolin. Elevated expression of CYP1A1 mRNA by CHI, a potent inhibitor of protein synthesis, has been observed in several cultured cells(Whitlock & Gelboin, 1973;Teifeld et al., 1989). We also observed an increase in aromatic hydrocarbon-induced CYP1A1 mRNA amounts in monolayer-cultured mouse hepatocytes(Nemoto & Sakurai, 1991;idem, 1992). The present investigations revealed, however, that addition of CHI did not elevate induction of Cyp1a2 gene expression by aromatic hydrocarbon.

In conclusion, although earlier attempts to maintain CYP1A2 expression or induction beyond the first one or two days after transferring hepatocytes to primary culture were not successful, the present observations demonstrate that use of spheroid culture prolongs these functions, facilitating investigations of the regulatory mechanisms, especially at the transcriptional level.

REFERENCES

Koide, N. et al(1989): Biochem. Biophys. Res. Commun., 161, 385-391.
Nemoto, N. & Sakurai, J.(1991): Carcinogenesis, 12, 2115-2121.
Nemoto, N. & Sakurai, J.(1992): Carcinogenesis, 13, 2249-2254.
Nemoto, N. & Sakurai, J.(1993): Jpn. J. Cancer Res., 84, 272-278.
Paine, A.J.(1990): Chem.-Biol. Interactions, 74, 1-31.
Sirica, A.E. & Pitot, H.C.(1980): Pharmacol. Rev., 31, 205-228.
Teifeld, R.M. et al(1989): DNA, 8, 329-338.
Whitlock, Jr., J.P. & Gelboin, H.V.(1973): J. Biol. Chem., 248, 6114-6121.
Yamasaki, H. et al(1975): J. Biol. Chem., 250, 7766-7770.

# Species differences in the *in vitro* metabolism of probe substrates for CYP2 enzymes

O.M. Corcoran, R. Hyland, B.C. Jones, D.A. Smith

*Department of Drug Metabolism Pfizer Central Research, Sandwich, Kent, CT13 9NJ, UK*

INTRODUCTION

The CYP2 family has been identified as one of the major families involved in the metabolism of many clinically important compounds such as β-blockers, antiarrhythmics and anticoagulants. This family currently contains eleven members (A-K) of which only the first five have been shown to play a role in xenobiotic metabolism.

There is a wide range in activities of P450 isozymes across the species most commonly used for toxicological studies. This study employed substrates which have had one of their routes of metabolism characterised as being specifically mediated by one of the CYP2 family in man and compares the enzyme activity in the major species used in the pre-clinical evaluation of pharmaceuticals (mouse, rat, rabbit and dog) with those observed in man. The probe substrates used were coumarin (7-hydroxylation mediated by CYP2A6 in man), phenytoin (4-hydroxylation mediated by CYP2C9 in man), bufuralol (1'-hydroxylation mediated by CYP2D6 in man) and chlorzoxazone (6-hydroxylation mediated by CYP2E1 in man).

METHODS

Hepatic microsomal fractions were prepared from mouse, rat, rabbit, dog and human samples. The incubation conditions for all the assays were similar, differences between the methods are detailed in table 1. In gerneral the incubations contained 50mM Tris-HCl buffer (pH 7.4), 5mM $MgCl_2$, 5μM $MnCl_2$, 1mM β-NADPH, 5mM isocitrate, 1 unit isocitric dehydrogenase and microsomal protein. Samples were preincubated for 5 minutes at 37°C and the reaction initiated by addition of substrate.

Table 1.

| Activity | Substrate conc. (mM) | Protein conc. (mg/ml) | Incubation time (min) | Method of analysis |
|---|---|---|---|---|
| Coumarin 7-hydroxylase | 5000 | 0.25 | 10 | HPLC-Fluorimetry |
| Phenytoin 4-hydroxylase | 200 | 2 | 60 | HPLC-uv |
| Bufuralol 1'-hydroxylase | 10 | 0.25 | 30 | HPLC-Fluorimetry |
| Chlorzoxazone 6-hydroxylase | 100 | 0.5 | 15 | HPLC-uv |

Results

The results of these studies, expressed as pmol of metabolite formed/ min/ mg microsomal protein are illustrated in Figs 1, 2, 3 and 4 for coumarin 7-hydroxylase, phenytoin 4-hydroxylase, bufuralol 1'-hydroxylase and chlorzoxazone 6-hydroxylase respectively.

Fig 1.

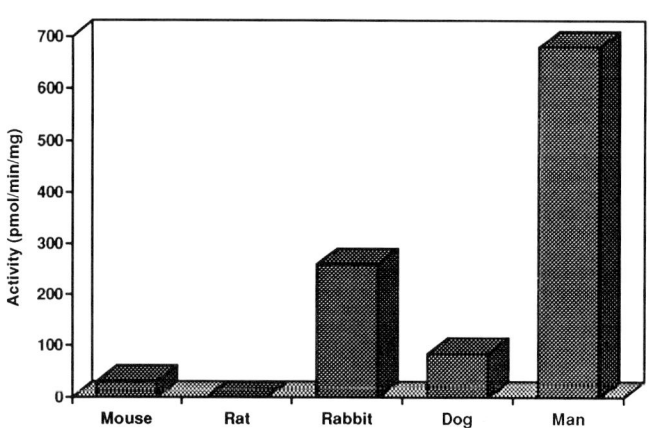

Fig 2.

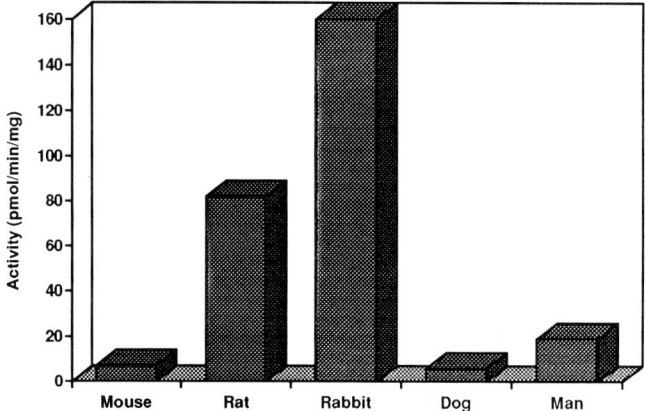

Fig 3.

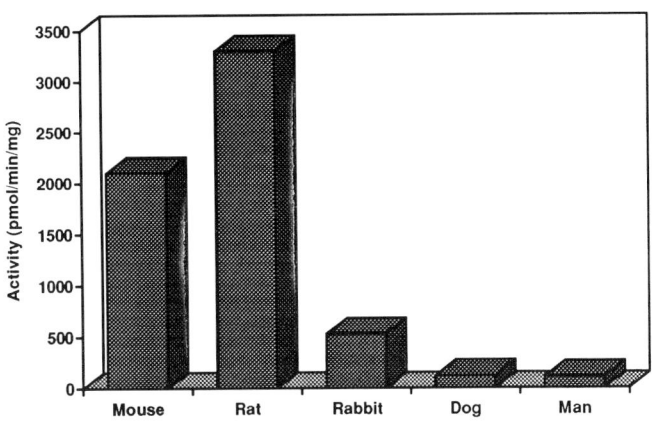

Fig 4.

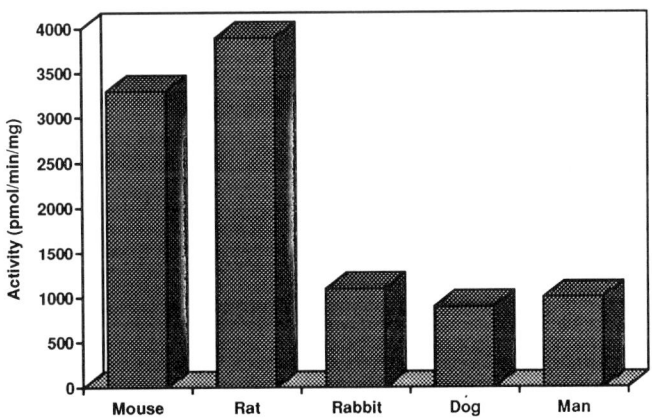

CONCLUSIONS

These results show that there is variation in the rate of metabolism of probe substrates for human CYP2A, CYP2C, CYP2D and CYP2E across the species investigated. Of the probes used the least variation was seen with chlorzoxazone 6-hydroxylation and the greatest with coumarin 7-hydroxylation. However, all the species catalysed the probe reactions with the exception of coumarin 7-hydroxylation in the rat.

The difference in coumarin 7-hydroxylation rates between rat and man is in accordance with the results of other workers (Fentem and Fry, 1992). This is supported by *in vivo* data which shows no detectable 7-hydroxy coumarin in the urine of rats after coumarin administration, whilst man excretes 70-90% of administered coumarin as the 7-hydroxy metabolite (Shilling et al., 1969). The low level of phenytoin 4-hydroxylase activity in the dog confirms earlier observations (Smith, 1991) that the low level or absence of a CYP2C9-like enzyme in the dog explains the low hydroxylation rates of many acidic compounds in this species. Similarly the high rate of bufuralol 1'-hydroxylase (CYP2D6-like) activity in the rat explains the very high rates of aromatic hydroxylation of many arylalkyl amines in this species.

REFERENCES

Fentem, J. H. & Fry, J. R. (1992): Metabolism of coumarin by rat, gerbil and human liver microsomes. *Xenobiotica* 22, 357-367.

Shilling, W. H. , Crampton, R. F. & Longland, R. C. (1969): Metabolism of coumarin in man. *Nature* 221 664-665.

Smith, D. A. (1991): Species differences in metabolism and pharmacokinetics: Are we close to an understanding? *Drug Metab. Rev.* 23 355-373

# Resonance raman spectra of NADPH cytochrome P450 reductase and its complex with cytochrome P450 2B4

Olga Alexandrova, Yury Ivanov, Valentin Uvarov

*Laboratory of physico-chemical methods, Institute of biomedical chemistry RAMS, Pogodinskaya 10, Moscow, Russia, 119832*

INTRODUCTION

The structure and function of the enzymes cytochrome P-450 and cytochrome P-450 reductase are being intensively studied at present. The interest in these proteins is determined by the fact that they are the major components of the hydroxylation system which participates in detoxication of various compounds, including pharmaceutical drugs. The understanding of the structure of these enzymes will allow a better insight into the mechanisms whereby this system operates. One of the basic methods in molecular structure studies is Resonance Raman spectroscopy. The structural study of cytochrome P-450 by use of this approach is described by many authors, in particular by Hildebrandt *et. al.* (1989). At the same time, cytochrome P-450 reductase has received little attention of investigators. Our work deals with the structure of the oxidized form of cytochrome P-450 reductase and the structural changes occuring in the cytochrome P-450 reductase upon complex formation with the oxidized form of cytochrome P-450.

MATERIALS AND METHODS

Cytochrome P-450 and cytochrome P-450 reductase isolated from rabbit liver microsomes, were kindly provided by Dr. Cohen Ya.M. The monomeric forms of the enzymes (33 nm/ml in 100 mM phosphate buffer) were used throughout. The complex of cytochrome P-450 with cytochrome P-450 reductase was obtained by mixing the both enzymes. The sampling cuvette was cooled to $4°C$ to diminish thermal destruction from laser emission. RR spectra were recorded on a DILOR Model Z-24 Raman spectrometer. For exitation was used the emission of $Ar^+$ laser (Spectra Physics Model 2020) at the 457.9 nm. Spectral slit width was 3.7 $cm^{-1}$. Laser power at the sample was 25 mW. The calibration of the spectrometer was carried out at 457.9 and 488.0 nm by use of $Ar^+$ laser. 27 scans with the scanning parameter $1c/1cm^{-1}$ was accumulated. Smoothing of spectra was carried out according to the Fourier transforms described in Ivanov et al.,(1993).

## RESULTS AND DISCUSSION

RR spectra of cytochrome P-450, cytochrome P-450 reductase and their complex are presented in Fig.1. Of interest is a small shift of characteristic lines of cytochrome P-450 reductase upon complex formation. Notice, for instance, lines X and VI (the numbering system is according to Sugiyama et al.,(1985)), which are assigned exclusively to vibrational modes of ring III of the isoalloxasine cycle in the flavin groups (Abe and Kyogoku (1987)). Line X with a frequency of 1249.2 $cm^{-1}$ and line VI with that of 1403.3 $cm^{-1}$ are shifted towards higher frequences upon complex formation. Line VII, with a frequency of 1348 $cm^{-1}$, as characteristic of the vibrational modes of ring II, was also found to shift towards higher frequences. The tendency for the above lines to shift towards higher frequences is consistent with the fact that during complex formation there occurs the strengthening of hydrogen bonds between rings II and III of flavins groups and protein globule. It would be interesting to compare the position of lines X and VI in FAD and FMN at the 1255-1256 $cm^{-1}$ and 1407-1408 $cm^{-1}$ (Schmidt et al., (1983)) respectively, with those for cytochrome P-450 reductase. It will be seen that the frequencies of the appropriate lines of the flavin groups in the protein globule are lower than in water. That is in agreement with the fact that hydrogen bonds of ring III of the above groups in the protein globule are much weaker than in water solution. By way of discussion, it is necessery to note the distribution in the contours of cytochrome P-450 reductase and its complex in the 1200-1300 $cm^{-1}$ region. Ivanova (1987) believes that in the case of glutathion reductase the different contours for this molecule in this region as compared with the FAD spectra in water are due to the presence of 2 FAD domains in the protein and also to the differences in orientation and tightness of hydrogen bonds of isoalloxasine rings. In our case, it is possible , since cytochrome P-450 reductase has two prosthetic groups, FAD and FMN, the differences in this spectral region for cytochrome P-450 reductase and its complex are associated as well with changes in the mutual orientation of FAD and FMN in the protein globule. The position of Line XI (with the frequency 1224 $cm^{-1}$), which characterizes the vibrations of bonds in ring I of the isoalloxasine cycle, remains unchanged, indicating the virtual invariability of force constant of this bonds upon complex formation. As regards cytochrome P-450, the only line which has no significant overlap with cytochrome P-450 reductase is band $\nu_4$, with the frequency 1372 $cm^{-1}$, which upon complex formation shifts towards lower frequencies in accordance with the increase in electron density on the orbital of porphyrin ring of cytochrome P-450. Because of the significant overlap of spectral bands of cytochrome P-450 and cytochrome P-450 reductase it was impossible, by use of the RR method, to obtain information regarding the changes in the P-450 heme iron's spin states; however, the analysis of absorption spectra of the enzyme complex binding reveals the changes in the cytochrome P-450's spin states.

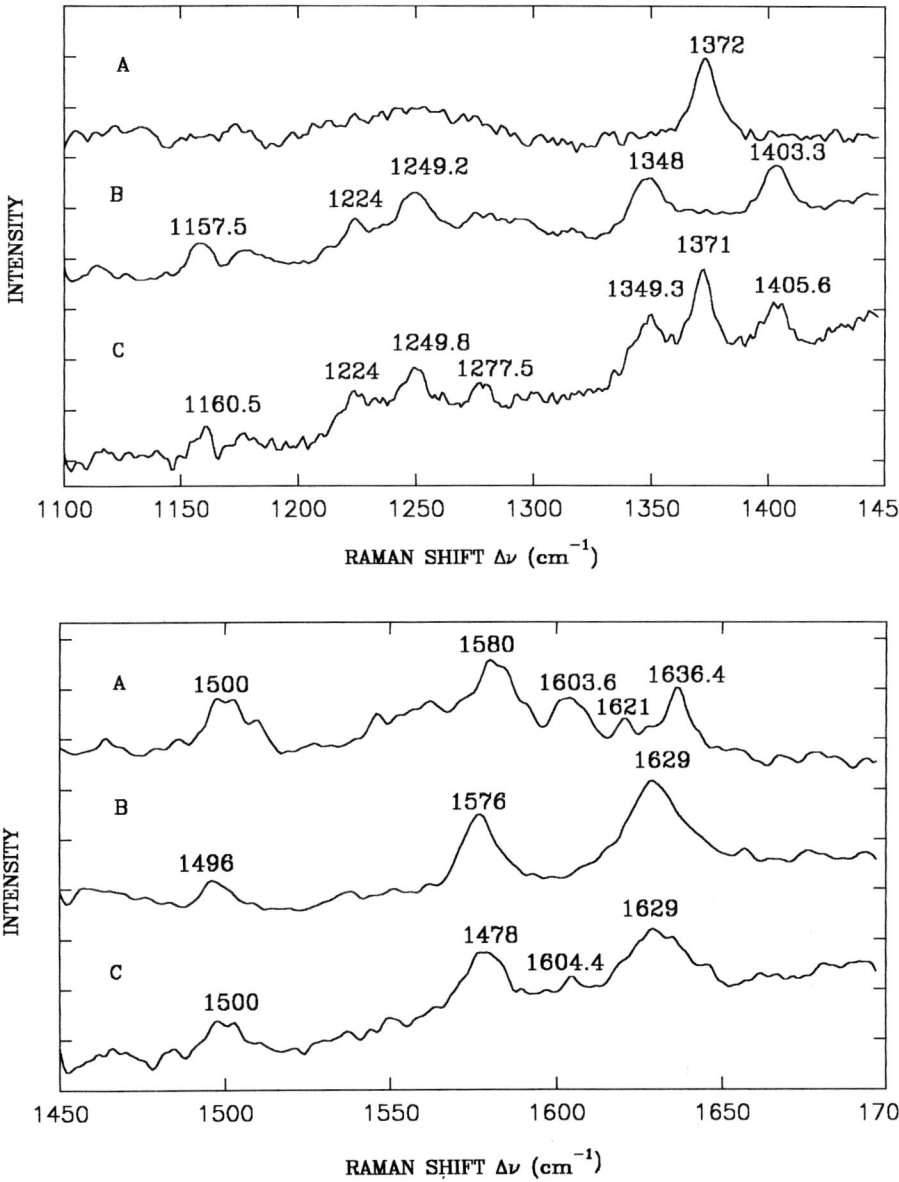

Fig. 1. Resonance Raman spectra of cytochrome P-450 (A), oxidized NADPH–cytochrome P-450 reductase (B) and their complex (C) in the 1100–1450 cm$^{-1}$ region (top), in the 1450–1700 cm$^{-1}$ region (bottom).

Thus our study has shown that:
- in the native form of cytochrome P-450 reductase the hydrogen bonds between the protein globule and ring III of the isoalloxasine cycle are weak;
-with complex formation, there occur changes in the positions of lines characterizing the tightness of hydrogen bonds between the isoalloxazine ring of prosthetic groups and the protein globule in the cytochrome P-450 reductase and lines, characterizing the electron density on the orbital of the porphyrine ring of the cytochrome P-450 heme; at the same time, formation of the complex induces changes in the spin states of the cytochrome P-450 heme iron. This allows the conclusion as to the mutual induction of the changes in the microenvironment of prosthetic groups and consequently the conformational restructuring of the region of the active site;
- complex formation of cytochrome P-450 reductase with cytochrome P-450 is likely to be accompanied by changes in the mutual orientation of FAD and FMN in the protein globule.

REFERENCES

Abe, M. and Kyogoku Y. (1987): Vibrational analysis of flavin derivatives: normal coordinate treatments of lumiflavin. *Spectrochemica Acta* 43A: 1027-1037.

Ivanov, Yu.D., Tsotov, A.A., Rukavishnikov, I.G. and Uvarov, V.Yu. (1993): Determination of secondary structure of epoxide hydrolase by Raman spectroscopy. *BBA* 1162: 217-220.

Ivanova, T.M. (1987): Resonance combination scattering of light in study of structure and function of flavins and flavoproteins. *Uspehi Khimii* LVI, issue 2: 322-352.

Hildebrandt, P., Greinert, R., Stier, A., Taniguchi, H. (1989): Resonance Raman study on the structure of the active sites of microsomal cytochrome P-450 isozymes LM2 and LM4. *Eur. J. Biochem.* 186: 291-302.

Schmidt, J., Coudron, P., Thompson, A.W., Watters, K.L., and McFarland T.M. (1983): Hydrogen bounding between flavin and protein: a Resonance Raman study. *Biochem.* 22: 76-84.

Sugiyama, T., Nisimoto, Y., Mason, H.S., Loehr, T.M. (1985): Flavins of MADPH-cytochrome P-450 reductase: evidence for structural alteration of flavins in their one-electron-reduced semiquinone states from Resonanse Raman spectroscopy. *Biochem.* 24: 3012-3019.

# Effects of acute and chronic acetone administration on cytochrome P450 2E1 mRNA in murine liver using *in situ* hybridization

P.G. Forkert, A.C. Jackson, S. Chen, S.P. Tam

*Departments of Anatomy, Medicine and Biochemistry, Queen's University, Kingston, Ontario, Canada K7L 3N6*

Cytochrome P450 2E1 is highly induced by a variety of chemicals including solvents such as ethanol or acetone. The degree of induction appears to be related to conditions under which the reaction occurs. The inductive response is exacerbated in some cases and in particular after chronic exposure to ethanol (Mostafa et al., 1981; Forkert et al., 1991), whereas other treatments such as repeated exposure to acetone have not produced cumulative increases in 2E1-dependent enzyme activity (Sipes et al., 1978). Regardless of the magnitude of 2E1 induction, the induced enzyme is localized primarily in the centrilobular region of the hepatic lobule as is the constitutive form, and has been described in human and rat (Tsutsumi et al., 1989) as well as murine liver (Forkert et al., 1991). However, recent studies have shown that high level pyridine exposure produces an enhanced level of 2E1 protein that is manifested in a larger hepatocyte population expressing 2E1 than is found in untreated rats (Hotchkiss et al., 1993). In human liver, the distribution of the 2E1 protein coincides with that of the mRNA (Takahashi et al., 1993).

In the present studies, we have investigated the effects of acute and chronic acetone exposure in mice, and have used *in situ* hybridization and quantitative image analysis to examine the distribution of 2E1 mRNA within the hepatic lobule. In conjunction with these studies, we have determined total 2E1 mRNA content and magnitudes of enzyme induction by acetone exposure.

## MATERIALS AND METHODS

<u>Animal treatment</u>. Male CD-1 mice were administered acetone using an acute or chronic regimen. Acute exposure consisted of acetone (5 ml/kg) given intragastrically and mice were killed 24 h later. Chronic treatment consisted of exposure to 1% acetone in drinking water for 8 days. Control mice were untreated or treated with tap water. All mice were freely provided with food for the duration of the experiment.

<u>In situ hybridization</u>. A $^3$H-labeled RNA probe was prepared from a cDNA clone (Dr. F. J. Gonzalez) containing the coding sequence for 2E1 mRNA (Song et al., 1986). *In situ* hybridization was performed using procedures described previously (Jackson and Wunner, 1991). Computerized image analysis was used to quantitate distribution of 2E1 mRNA transcripts in liver sections from control and acetone-treated mice. Images were viewed under

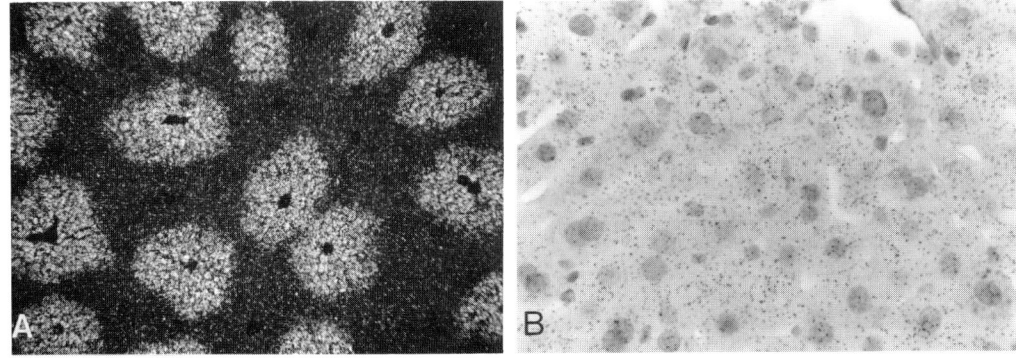

Fig. 1. Intralobular distribution of 2E1 mRNA in murine liver after *in situ* hybridization using a ³H-labeled RNA probe. Dark-field illumination of a representative liver section exposed to acetone chronically shows that 2E1 mRNA transcripts are primarily localized in the centrilobular zone (A). Bright-field illumination of the same liver section shows 2E1 mRNA transcripts in centrilobular hepatocytes (B). Original magnification: x 65 (A); x 450 (B).

dark-field illumination and processed by thresholding to produce optimal signal-to-noise ratios; a threshold was set above which specific mRNA transcripts were quantitated. After thresholding, total areas adjacent to individual central veins which contained 2E1 mRNA transcripts were measured. In addition, fractions of areas in relation to total areas containing transcripts were measured. Eighty centrilobular regions from each of 4 mice from control and acetone-treated groups were scored. Measurements were accomplished using a 486 computer-based videodensitometry system and commercial software (JAVA, Jandel, Inc.).

Northern blotting. A cDNA insert (1.6 kb) containing a portion of the 2E1 gene (Song *et al.*, 1986) was labeled by nick translation using [$\alpha^{32}$P]dATP (3000 Ci/mMol). Northern blots were prepared using techniques described in previous studies (Tam, 1991)

Protein immunoblotting. Microsomes were prepared and protein immunoblotting performed according to procedures detailed in previous studies (Forkert *et al.*, 1991). Densitometric scanning was used to quantitate the intensities of the bands, and the extent of 2E1 protein induction determined.

Hydroxylation of *p*-nitrophenol (PNP). PNP hydroxylation was determined by measuring the amount of 4-nitrocatechol formed, using the method previously described (Kim et al. (1988). Cytochrome P450 and protein levels were determined using the methods of Omura and Sato (1964) and Lowry *et al.* (1951), respectively.

Statistical analysis. Data were analyzed using a one-way ANOVA and the Student-Newman Keuls test.

RESULTS

Distribution of 2E1 mRNA . *In situ* hybridization revealed 2E1 mRNA transcripts in liver sections from control and acetone-treated mice. Signals for 2E1 mRNA transcripts were found primarily in the centrilobular zone of the hepatic lobule, whereas only sparse transcripts were visible in the periportal zone (Fig. 2). This pattern of distribution was exhibited in liver from control as well as acetone-treated mice. Pretreatment of liver sections with RNase prior to hybridization eliminated the signals, and hybridization with control probes revealed low backgrounds.

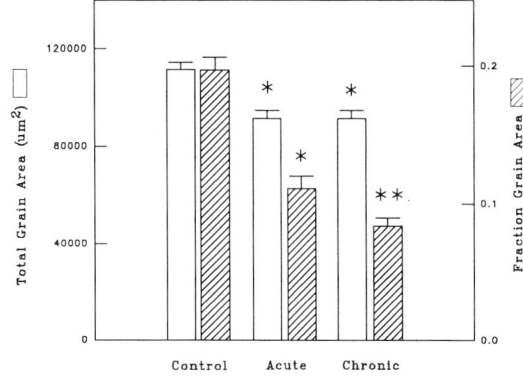

Fig. 2. Effect of acute and chronic acetone exposure on total and fraction of acinar areas occupied by grains representing 2E1 mRNA transcripts. Values are expressed as mean ± S.E.M. *Significantly different from control; **significantly different from control and acute acetone exposure ($p < 0.05$).

Quantitative image analysis. Total areas within individual hepatic lobules that contained 2E1 mRNA transcripts were, in comparison to the control, significantly decreased in acetone-treated mice (Fig. 2). These areas were approx. 20% less in mice exposed to acetone acutely as well as chronically. Fractions of total areas which were positively labeled for 2E1 mRNA were also significantly decreased in mice treated with acetone. However, significantly greater reduction of 2E1 mRNA transcripts was evoked by chronic (60% of control) than by acute (40% of control) acetone exposure (Fig. 2). Thus, at the level of the hepatic lobule, acetone exposure under both acute and chronic conditions evoked significantly lower 2E1 mRNA expression than in untreated mice.

Northern blot analysis. The cDNA probe for 2E1 hybridized to a single mRNA species of approximately 2 kb on northern blots prepared from total RNA samples from livers of control and acetone-treated mice, and confirmed the specificity of the probe in murine liver.

Induction of 2E1 protein. A single band of $M_r$ 51 000 was detected by protein immunoblotting. Densitometric quantitation revealed induced increases of 4.4- and 5.3-fold by acute and chronic acetone exposure, respectively.

PNP hydroxylation. PNP hydroxylase activities in acetone-treated mice were significantly different from those in control mice. Moreover, chronic acetone exposure evoked significantly higher catalytic activities than acute exposure; rates of PNP hydroxylation were elevated by 1.7- and 2.8-fold after acute (0.62 nmol/nmol P450/min) and chronic (0.82 nmol/nmol/P450/min) treatment, respectively.

DISCUSSION

The *in situ* hybridization experiments reported herein found 2E1 mRNA transcripts to be localized most abundantly in centrilobular hepatocytes and to considerably lesser extents in midzonal and periportal hepatocytes (Fig. 1). This pattern of distribution coincided with the intralobular sites observed for the protein (Forkert *et al.*, 1991). Quantitation of 2E1 mRNA transcripts by computer-assisted imaging revealed that expression in acetone-treated mice differed from that of control mice in two respects: one, the hepatocyte population expressing 2E1 mRNA was reduced, as assessed by significantly diminished areas occupied by transcripts; two, expression of the mRNA was significantly decreased in hepatocytes from acetone-treated mice, in comparison with constitutive expression (Fig. 2). Furthermore, chronic acetone exposure evoked significantly greater diminution of 2E1 mRNA expression than acute exposure. This differential effect was accomplished by decreased 2E1 mRNA expression within hepatocytes rather than in further decrease in cell population expressing the mRNA. This phenomenon is the converse of that observed in human liver; in recently-drinking alcoholics, the number as well as acinar areas occupied by 2E1 mRNA transcripts were significantly increased in liver biopsy specimens compared to levels in control subjects (Takahashi et al., 1993).

The decreased expression of 2E1 mRNA found in the liver after acetone exposure occurred in conjunction with significant increases of 2E1 protein and PNP hydroxylation. It is of interest in this regard that the significantly greater diminution of 2E1 mRNA expression observed after chronic *versus* acute acetone exposure coincided with higher 2E1 enzyme induction. This finding of higher 2E1 induction after more prolonged and persistent acetone exposure supports the proposed mechanism of protein stabilization (Song et al. 1989). In summary, our results in murine liver indicate that 2E1 induction by acetone occurs in conjunction with a more restricted distribution and reduction of 2E1 mRNA content, suggesting the importance of post-transcriptional events in the inductive response.

REFERENCES

Forkert, P. G., Massey, T. E., Jones, A. B., Park, S. S., Gelboin, H. V., and Anderson, L. M. (1991): Distribution of cytochrome CYP2E1 in murine liver after ethanol and acetone administration. *Carcinogenesis* 12: 2259-2268.

Hotchkiss, J. A., Kim, S. G., Novak, R. F., and Dahl, A. R. (1993): Enhanced hepatic expression of P450IIE1 following inhalation exposure to pyridine. *Toxicol. Appl. Pharmacol.* 118: 98-104.

Jackson, A. C., and Wunner, W. H. (1991): Detection of rabies virus genomic RNA and mRNA in mouse and human brains by using *in situ* hybridization. *J. Virol.* 65: 2839-2844

Kim, S. G., Williams, D. E., Schuetz, E. G., Guzelian, P. S., and Novak, R. F. (1988): Pyridine induction of cytochrome P-450 in the rat: role of P-450j (alcohol-inducible form) in pyridine N-oxidation. *J. Pharmacol. Exp. Ther.* 246: 1175-1182.

Lowry, O. H., and Rosebrough, N. J., Farr, A. L., and Randall, R. T. (1951): Protein measurement with the Folin phenol reagent. *J. Biol. Chem.* 193: 265-275.

Mostafa, M. H., Ruchirawat, M., and Weisburger, E. K. (1981): Effect of indole on N-nitrosodimethylamine demethylase in rats treated with carbon tetrachloride. *Fd. Cosmet. Toxicol.* 19: 717-721.

Omura, T. and Sato, R. (1964): The carbon monoxide-binding pigment of liver microsomes. I. Evidence for its hemoprotein nature. *J. Biol. Chem.* 239: 2370-2378.

Sipes, I. G., Slocumb, M. L., and Holtzman, G. (1978): Stimulation of microsomal dimethylnitrosamine-N-demethylase by pretreatment of mice with acetone. *Chem.-Biol Interactions* 21: 155-166.

Song, B.-J., Gelboin, H. V., Park, S.-S., Yang, C. S., and Gonzalez, F. J. (1986): Complementary DNA and protein sequences of ethanol-inducible rat and human cytochrome P-450s. *J. Biol. Chem.* 261: 16689-16697.

Song, B.-J., Veech, R. L., Park, S. S., Gelboin, H. V., and Gonzalez, F. G. (1989): Induction of rat hepatic N-nitrosodimethylamine demethylase by acetone is due to protein stabilization. *J. Biol. Chem.* 264: 3568-3572.

Takahashi, T., Lasker, J. M., Rosman, A. S., and Lieber, C. S. (1993): Induction of cytochrome P-4502E1 in the human liver by ethanol is caused by a corresponding increase in encoding messenger mRNA. *Hepatology* 17: 236-245.

Tam, S.-P. (1991): Effects of gemfibrozil and ketoconazole on human apolipoprotein A1, B and E levels in 2 hepatoma cell lines, HepG2 and Hep3B. *Atherosclerosis* 91: 51-61.

Tsutsumi, M., Lasker, J. M., Shimizu, M., Rosman, A. S., and Lieber, C. S. (1989): The intralobular distribution of ethanol-inducible P450IIE1 in rat and human liver. *Hepatology* 10: 437-446.

# Downregulation of the Ah receptor following 2, 3, 7, 8-tetrachlorodibenzo-*p*-dioxin (TCDD) treatment in Hepa-1 cells in culture

Patricia A. Harper[1], John V. Giannone[2], Markus R. Probst[3], Allan B. Okey[2]

[1]Division of Clinical Pharmacology and Toxicology, The Hospital for Sick Children, Toronto, Ontario, Canada, M5G 1X8. [2]Department of Pharmacology, University of Toronto, Toronto, Ontario, Canada, M5S 1A8. [3]Laboratory of Biomedical and Environmental Sciences UCLA, Los Angeles, California, USA 90024-1786

## INTRODUCTION.

The AhR (aromatic hydrocarbon receptor) regulates expression of several drug-metabolizing enzymes including CYP1A1 (Whitlock, 1990). In the unliganded state the AhR is present in the cytosol of cells. Ligands such as 2,3,7,8-tetrachlorodibenzo-*p*-dioxin (TCDD) bind to cytosolic AhR, transforming the receptor to a DNA-binding protein which translocates to the nucleus where it binds to specific DNA enhancer sequences (Whitlock, 1990; Landers and Bunce 1991; Okey *et al.*, 1993). Thus, the AhR is a ligand-dependent activator of gene transcription.

The DNA-binding form of the receptor has been shown to consist of at least two proteins, the AhR (Ema *et al.*, 1992; Burbach *et al.* 1992) and a protein known as the AhR nuclear translocator (ARNT; Reyes *et al.*, 1992). The AhR is the ligand-binding subunit, however ARNT protein is necessary for function and both proteins interact with the regulatory enhancer sequences (Probst *et al.*, 1993).

The concentration of AhR in the nucleus increases in a time-dependent manner when cells are exposed to [$^3$H]TCDD in culture (Prokipcak and Okey, 1991). After this initial rise in the first hour, AhR in the nucleus decreases. We have shown that this decrease is not the result of recycling of AhR back to the cytosol but rather is a decrease in the total cellular AhR pool. We postulated that the loss might have been the result of a loss in the ligand-binding ability of the receptor.

Recently, we examined this apparent ligand-dependent decrease of [$^3$H]TCDD•AhR complex by western blot using a polyclonal antibody that recognizes the AhR. This antibody was developed by immunizing rabbits with a synthetic peptide-KLH conjugate corresponding to 20 amino acids from the N-terminal of the AhR (Bradfield *et al.*, 1991). Western blot analysis showed that TCDD exposure resulted in a decrease in total cellular AhR protein not just of ligand-binding ability, and that this loss of AhR may be the result of proteolytic degradation.

In the present report we have extended our observations to examine the fate of the second member of the nuclear AhR complex, i.e. ARNT protein, after ligand exposure. Since ARNT does not bind ligand, we used an anti-ARNT antibody to detect ARNT protein by immunoblot.

## MATERIALS and METHODS

*Cell culture.* Hepa-1c1c9 (Hepa-1) cells were grown in alpha-MEM (antibiotic-free) supplemented with 10% fetal bovine serum.

*Nuclear Translocation.* Hepa-1 cells were incubated in monolayer with 2 nM [$^3$H]TCDD for 0 h to 20 h at 37°C. The cells were collected and nuclear extracts and cytosols were prepared as previously described (Prokipcak and Okey, 1991), and analyzed by sucrose gradient centrifugation for [$^3$H]TCDD binding or western blot.

*Antibody.* A fusion protein consisting of the C-terminal fragment coding for amino acids 399-777 of human ARNT fused to glutathione-S-transferase was used to immunize New Zealand White rabbits. Nonspecific antibodies were removed by incubating the crude antiserum with immobilized bacterial extract expressing GST only. Specific anti-ARNT antibodies were affinity purified using immobilized GST-ARNT (Probst *et al.*, 1993).

*Western Blot.* Proteins were resolved by discontinuous polyacrylamide gel electrophoresis (6% polyacrylamide), and transferred to nitrocellulose. After blocking, the membranes were incubated with antibody for 1 h at room temperature, washed and antigen-antibody complexes were detected with goat-anti-rabbit antibody coupled to horseradish peroxidase (1:20,000,) using the Amersham ECL immunochemiluminescent system.

## RESULTS and DISCUSSION

Previously, we measured AhR content in Hepa-1 cells exposed to TCDD in culture, by sucrose gradient centrifugation to determine ligand-binding ability, and western blot using anti-AhR antibodies to detect AhR protein. There was a rapid drop in cytosolic AhR in Hepa-1 cells after exposure to TCDD. Conversely, nuclear AhR in the same cells increased for a period of 1 to 2 h and then steadily decreased, such that total AhR was decreased to about 20% that of pre-TCDD levels after 6 hours.

We repeated this experiment to determine the fate of ARNT protein after TCDD exposure, Fig 1.

In the absence of ligand (time=0), ARNT protein is found in the cytosol and not the nuclear extract. Levels of ARNT in cytosol drop during the first hour after exposure to TCDD, accompanied by an increase in nuclear ARNT levels. The high levels of nuclear ARNT were transient and after about 4 h the amount decreased and approached that seen in unexposed cells. This decrease in nuclear ARNT was paralleled by an increase in cytosolic ARNT, which returned to levels detected in cytosol from cells unexposed to TCDD. These results are in distinct contrast to the results obtained for the AhR. AhR underwent a translocation from cytosol to the nucleus, but after about 2 h post-TCDD exposure, both cytosolic and nuclear AhR were dramatically decreased and remained so for at least 20 hours.

The subcellular distribution of ARNT has not been clearly elucidated. In the experiments presented here ARNT is detected in the cytosolic fraction from cells not exposed to TCDD whereas, in cells exposed to TCDD, ARNT is transiently localized in the nuclear compartment. Two major mechanisms can be hypothesized to account for these observations: (*i*) ARNT is weakly associated with the nucleus in the absence of TCDD•AhR complex and is released into the cytosolic compartment upon cellular fractionation or (*ii*) ARNT is a cytosolic protein and upon TCDD•AhR transformation it "translocates" to the nucleus as part of the transformed AhR complex. The experiments described here can not distinguish between these two posibilities. Recently Probst *et al.* (1993) have shown that ARNT does not co-immunoprecipitate with cytosolic AhR. The subcellular localization of ARNT and AhR *in vivo* remain to be resolved.

## *ARNT Expression in Cells Exposed to TCDD for 1 to 20 h*

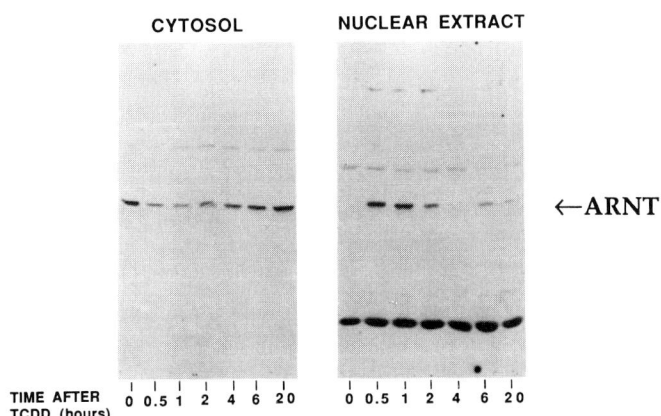

Fig.1: Levels of ARNT Detected by Immunochemical Staining in Cytosolic and Nuclear Extracts from Cells Treated for Increasing Times with [$^3$H]TCDD at 37°C. *Hepa-1 cells were incubated in culture with 2 nM [$^3$H]TCDD at 37°C. After different periods of incubation, cytosolic and nuclear extracts were prepared, resolved by denaturing gel electrophoresis (6% polyacrylamide), and the proteins visualized by western blot using anti-ARNT antibodies*

The downregulation of AhR was sensitive to the action of actinomycin D. Incubating cell with TCDD and actinomycin D preserved the levels of nuclear AhR. We repeated this experiment, to determine the effect of actinomycin D on ARNT protein, Fig 2.

In the absence of ligand (time=0), ARNT protein is found in the cytosol and not the nuclear extract. Levels of ARNT in cytosol drop during the first hour after exposure to TCDD, accompanied by an increase in nuclear ARNT levels (time=1 h). Treatment with actinomycin D and TCDD resulted in increased nuclear-associated ARNT. After 6 h exposure to TCDD, cytosolic levels of ARNT have returned to pre-TCDD levels accompanied by a decrease in ARNT levels in the nucleus. However, for cells treated with actinomycin D and TCDD, nuclear levels of ARNT remained elevated and the cytosolic levels did not return to that detected in unexposed cells. These data indicate that exposure to TCDD in the presence of actinomycin D resulted in the nuclear retention of ARNT protein. These results are similar to those obtained for the AhR. Whereas total AhR was lost in cells incubated for 6 h with TCDD alone, nuclear AhR levels were maintained for upto 6 hours in cells incubated with TCDD plus actinomycin D.

**SUMMARY and CONCLUSION.**
In the absence of ligand, ARNT and AhR are found in cell cytosol extracts. One to 2 h following exposure of Hepa-1 cells to TCDD, both AhR and ARNT are extractable from cell nuclei, however, at this point their fates diverge. AhR is rapidly lost from the nucleus and does not reappear in the cytosol, whereas ARNT is lost from the nucleus, but does reappear in the cytosol. If AhR degradation is inhibited by the action of actinomycin D, such that

levels of AhR are maintained in the nucleus, then levels of ARNT also are maintained in the nucleus.

## Effect of Actinomycin D on ARNT Expression.

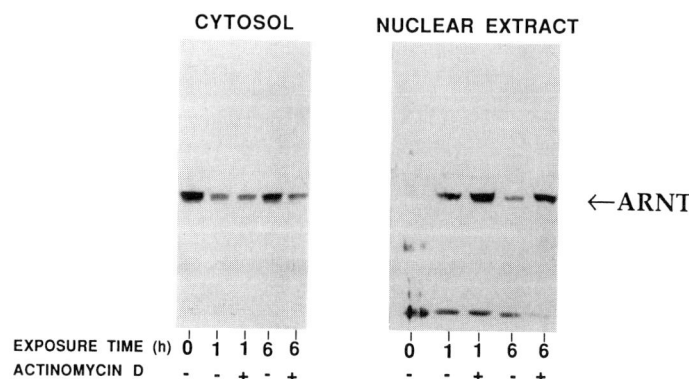

Fig.2: **Effect of Actinomycin D on Levels of ARNT Detected by Immunochemical Staining in Cytosolic and Nuclear Extracts from Cells Treated with [³H]TCDD at 37°C.** *Hepa-1 cells were incubated in culture with 2 nM [³H]TCDD in the presence or absence of 10 µM actinomycin D at 37°C. After 0 h, 1 h or 6 h of incubation, cytosolic and nuclear extracts were prepared, resolved by denaturing gel electrophoresis (6% polyacrylamide) and the proteins visualize by western blot using anti-ARNT antibodies.*

## REFERENCES.

Bradfield, C. A., Glover, E., and Poland, A. (1991) Purification and N-terminal amino acid sequence of the Ah receptor from C57BL/6J mouse. *Mol. Pharmacol.* 39: 13-19

Burbach, K. M., Poland, A., and Bradfield, C. C. (1992) Cloning of the Ah-receptor cDNA reveals a distinctive ligand-activated transcription factor. *Proc. Natl. Acad. Sci. USA* 89: 8185-8189.

Ema, M., Watanabe, N., Chujoh, Y., Matsushitu, N., Gotoh, O., Funae, Y., and Fujii-Kuriyama, Y. (1992) cDNA cloning and structure of a mouse putative Ah receptor. *Biochem. Biophys. Res. Commun.* 184: 246-253.

Landers, J. P., and Bunce, N. J. (1991) The Ah receptor and the mechanism of dioxin toxicity. *Biochem. J.* 276: 273-287.

Okey, A. B, Riddick, D. S., Harper, P. A. (1993) The Ah receptor: mediator of the toxicity of 2,3,7,8-tetrachlorodibenzo-*p*-dioxin (TCDD) and related compounds. *Toxicol. Lett.* (in press)

Probst, M. R., Reisz-Porszasz, S., Agbunag, R. V., Ong, M. S., and Hankinson, O. (1993) Role of the aryl hydrocarbon receptor nuclear translocator protein in aryl hydrocarbon (dioxin) receptor action. *Mol. Pharmacol.* 44: 511-518.

Prokipcak, R. D., and Okey, A. B. (1991) Downregulation of the Ah receptor in mouse hepatoma cells treated in culture with 2,3,7,8-tetrachlorodibenzo-*p*-dioxin. *Can. J. Physiol. Pharmacol.* 69: 1204-1210.

Reyes, H., Reisz-Porszasz, S., and Hankinson, O. (1992) Indentification of the Ah receptor nuclear translocator protein (ARNT) as a component of the DNA binding form of the Ah receptor. *Science (Washington D.C.)I* 256: 1193-1195.

Whitlock, J. P. (1990) Genetic and molecular aspects of 2,3,7,8-tetrachlorodibenzo-*p*-dioxin action. *Annu. Rev. Pharmacol. Toxicol.* 30: 251-277.

# Lidocaine metabolism as a liver function test

Pentti Arvela[1], Arja Rautio[1], Eero A. Sotaniemi[2]

[1]Departments of Pharmacology and Toxicology and [2]Internal Medicine, University of Oulu, 90220 Oulu, Finland and Deaconess Institute of Oulu, Finland

## Summary

The usefulness of lidocaine metabolite monoethylglycinexylide (MEGX) blood levels to reflect liver function was investigated in patients with liver disease (n=36) and controls (n=12). The severity of the liver disease was assessed by determining serum aminoterminalpropeptide (PIIINP). The MEGX values in the patients with severe (4.6 + 3.0 ng/ml), moderate (19.1 + 11.6 ng/ml) and mild (32.8 + 14.2 ng/ml) liver disease were reduced as compared to the controls (69.6 + 13.2 ng/ml) and related to the PIIINP values ($r2=0.595$, $p<0.001$). Use of drugs known to be metabolized by CYP4503A4 reduced MEGX values whereas the metabolite formation was independent on kidney function. Liver condition and drug therapy must be recognized when considering the MEGX values.

## INTRODUCTION

Assessment of liver function is essential in many clinical conditions. Recent developments allow a degree of quantification of liver metabolism through determination of P450 isozyme activity by using probe drugs. Formation of MEGX after iv infusion of lidocaine reflects CYP3A4 activity (Oellerlich et al. 1987) and the test have used to separate liver patients from controls (Bargetzi et al. 1989). We evaluated the usefulness of hepatic CYP3A4 activity for the prediction of liver function by comparing MEGX values in patients with histologically verified liver disease and matched controls. The activity of liver disease was assessed by determining the serum content of aminoterminalpropeptide (PIIINP) (Sotaniemi et al. 1986). We also developed a new assay method to measure MEGX content in serum.

## METHODS

We developed a specific GC-MS method to determine MEGX levels in serum. Shortly 1 ml of sample or standard was alkalized with 1 N NaOH and extracted with ethyl acetate. Tocainide (500ng) was used as an internal standard. After evaporating the organic phase to dryness, the residue was derivatized with acetic acid anhydride in pyridine (3:2). After evaporating again to dryness the derivative was dissolved in methanol. An aliquot (2ul) was injected splitness into a HP 5971 capillary GC/MS. The column used was a 25 m long HP-1 capillary column. The oven temperature was programmed from 80° C to 280° C (20° C/min) and the ionization voltage was 70 eV. We used a selective ion monitoring (SIM) mode for MEGX ions 100, 128, 248 and those for tocainide 121, 148 and 234. The concentration of PIIINP was measured radioimmunologically (Risteli & Risteli 1990).

SUBJECTS

Thirty-six patients with alcoholic liver disease (Table 1) were investigated. They had been drinking for 10 years and the alcohol abuse had disturbed their social life. All had been hospitalized several timed due to acute intoxication. There were 26 patients with cirrhosis and 10 subjects with fibrosis with or without fatty liver. None was receiving drugs known to alter CYP3A4 activity. Eight subjects were smokers.

As controls, twelve age and sex matched subjects who had normal liver and kidney function were investigated.

RESULTS

GC/MS -method. Figure 1 gives the calibration curve of MEGX as measured by GC/MS methos. Comparison of GC/MS method with a commercial fluorescent polarization immunoassay (FPIA) demonstrated a excellent correlation between the variables (Fig. 2). FPIA was carried out using ABBOT TDX assay kit

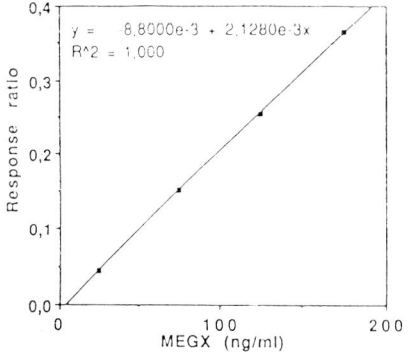

FIG.1: CALIBRATION CURVE OF MEGX

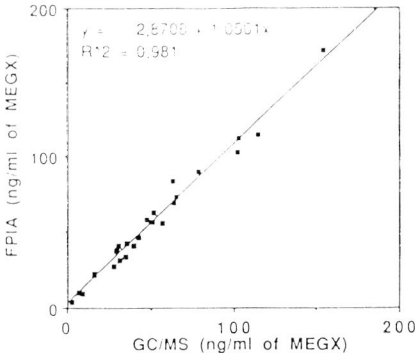

FIG.2: COMPARISON OF GC/MS-METHOD AND A COMMERCIAL FLUORESCENT POLARIZATION IMMUNOASSAY (FPIA)

CYP3A4 activities and serum PIIINP in liver disease. The fibrotic process was related to MGX values (Table 1). Comparison of serum PIIINP values with MEGX revealed an inverse relationship between the variables ($r = -0.772$, $p < 0.001$).

DISCUSSION

Liver structure and function change with time, and no single test (biochemical, histological, clinical) provides information about the nature and severity of the liver disease. Serum PIIINP, related to fibrogenesis (Risteli & Risteli, 1990), reflects type III collagen turnover in liver. In normal liver collagen synthesis and degradation are in balance. In diseased liver the synthesis of collagen is stimulated by the inflammatory and necrotic process (Clement et al. 1993). Collagen production then exceeds the liver degradation ability and the fibres together with other extracellular matrix components accumulated in the extracellular spaces. This leads to disturbances in hepatic microcirculation and to arteriovenous shunts (Schaffner & Popper, 1963), which reduce the substrate availability to already lowered enzyme system. Our study demonstrated an inverse correlation between CYP3A4 activities and serum PIIINP, as seen between total cytochrome P450 activity and collagen antigens (Sotaniemi et al. 1986). The serum collagen antigens can be utilized when evaluating the severity of the liver disease on drug metabolizing ability, as shown by correlation between MEGX and PIIINP values.

The GC/MS-method used proved to be suitable for routine monitoring of MEGX in human serum. Comparison with the immunological method indicated a good correlation and better selectivity. The GC/MS-equipment is expensive but if one is available the price of MEGX-measurements is much cheaper than with the commercial immunological methos.

Table 1. Clinical data, serum MEGX and PIIINP values in liver patients and control subjects

| Subjects | Age years | Sex | BMI kg/m2 | MEGX ng/ml | PIIINP µg/L |
|---|---|---|---|---|---|
| 1. Liver disease | | | | | |
| severe (n=12) | 52 ± 9 | 8 M | 27.5 | 4.6$^d$ | 14.8$^d$ |
| | | 4 F | 3.9 | 3.0 | 3.4 |
| moderate (n=12) | 50 ± 9 | 8 M | 27.4 | 19.1$^d$ | 6.6$^d$ |
| | | 4 F | 2.9 | 11.6 | 1.7 |
| mild (n=12) | 56 ± 8 | 8 M | 27.2 | 32.8$^d$ | 3.4 |
| | | 4 F | 4.0 | 14.2 | 0.5 |
| 2. Controls (n=12) | 53 ± 10 | 7 M | 26.3 | 69.6 | 3.4 |
| | | 5 F | 3.8 | 13.2 | 0.5 |

p-values compared with control subjects a = 0.001

## REFERENCES

Bargetzi, M.J., Aoyama, T., Conzalez, F.J., and Meyer, U.A. (1989): Lidocaine metabolism in human liver microsomes by cytochrome P450IIIA4. *Clin. Pharmacol. Ther.* 46, 521-527.

Clement, B., Loreal, O., Levavasseur, F., and Guillongo, A. (1993): A new challenger in hepatic fibrosis. *J. Hepatol.* 18, 1-4.

Oellerlich, M., Raude, E., Brudelski, M., Schulz, M., Schmidt, P.W., Ringe, B., Lamesch, P., Pichlmayr, R., Raith, H., Scheruhn, M., Wrenger, M., and Wittekind, C.H. (1987): Monoethylglycinexylidide formation kinetics: A novel approach to assessment of liver function. *J. Clin. Chem. Clin. Biochem.* 25, 845-853.

Risteli, L., and Risteli, J. (1990): Non-invasive methods for detection of organ fibrosis. In *Connective tissue in health and disease*, ed. M. Rojkind, pp. 61-98. Boca Raton, FL: CRC press.

Schaffner, F., and Popper, H. (1963): Capillarization of hepatic sinusoids in man. *Gastroenterology* 44, 239-242.

Sotaniemi, E.A., Niemelä, O., Risteli, L., Stenbäck, F., Pelkonen, R.O., Lahtela, J.T., and Risteli, J. (1986): Fibrotic process and drug metabolism in alcoholic liver disease. *Clin. Pharmacol. Ther.* 40, 46-55.

# Formation of cytochrome P450 metabolite complexes: isozyme specificity and use for isozyme quantitation

Peter H. Roos, Georgia Günther, Walter G. Hanstein

Institute for Physiological Chemisry, Ruhr-University Bochum, D-44780 Bochum, FRG

Formation of P450 metabolite complexes by drugs is of pharmacological relevance because it results in enzymatic inactivation of the P450 species involved. For assessment of drug/drug interferences, the isozyme specificity of complex formation needs to be studied. Knowledge of the isozyme specificity is also required when P450 complex formation is used for selective spectroscopic quantitation of P450 isozymes in microsomes. In this respect we investigated two drugs, the anti-Parkinson drug orphenadrine and the macrolide antibiotic triacetyloleandomycin (TAO). P450-dependent metabolism of orphenadrine and TAO leads to formation of reactive C-nitroso derivatives after primary N-demethylations. Complex formation with orphenadrine has been reported to occur with CYP2B1 and CYP2C6 (Reidy et al., 1989; Reidy & Murray, 1990). In order to assess the suitability of orphenadrine for spectroscopic isozyme quantitation, we studied kinetics and extent of complex formation in liver microsomes of untreated and inducer treated rats. Our findings that high levels of complex forming isozymes are present in microsomes of untreated male rats and that their expression can be only slightly modulated by inducers are not in accordance with the conclusions of others (Reidy et al., 1989; Reidy & Murray, 1990). Our results suggest the involvement of a male-dominant P450 isozyme. Metabolite complex formation of TAO with CYP3A isozymes is long known (Mansuy et al., 1981). We established an assay for quantitative *in vitro* complex formation (Roos et al., 1993) and quantified CYP3A isozymes by this method in untreated and inducer treated rats.

MATERIALS & METHODS

*Animals and animal treatment.* Male and female Sprague Dawley rats (200-250g) were used. Induction regimes for phenobarbital (PB), hexachlorobenzene (HCB) and dexamethasone (DEX) were as described (Roos et al., 1993).
*'In vitro' formation of the orphenadrine metabolite complex.* 3 ml samples containing 1 nmol microsomal P450, Tris-HCl (50 mM, pH 7.5), KCl (50 mM), $MgCl_2$ (10 mM), isocitrate (5 mM), isocitrate dehydrogenase (0.5 U/ml), catalase (230 U/ml) and NADP (350 µM) were preincubated for 1 min at 37°. A baseline was recorded and the reaction started by addition of orphenadrine (0.1 - 100 µM final concentra-

tion). Complex formation was recorded by repetitive scanning in the wavelength range of 510 nm to 400 nm up to 30 min with cycles of 3 min. A difference spectrum was calculated by subtraction of the baseline and the absorbance difference at 453 nm and 500 nm was determined.

*Preparation of microsomes, testosterone hydroxylation, quantitation of P450 and protein and 'in vitro' formation of the TAO-metabolite complex*: Roos et al. (1993).

RESULTS

We find extensive complex formation with orphenadrine in liver microsomes of untreated male rats, i.e. about 34% based on total cytochrome P450 (Fig. 1). The absorption maximum of the complex at 453 nm is identical to that of the complex obtained with microsomes of PB-treated rats (Fig. 3). The CO-binding capacity of P450 in microsomes of untreated rats decreases with increasing amounts of orphenadrine metabolite complex formed. Due to its stability towards CO, the extinction coefficient of the complex could be determined by complex formation in the presence of different orphenadrine concentrations (1-10 μM) and subsequent CO-binding. Based on the coefficient for the P450-CO-complex (91 $mM^{-1}cm^{-1}$), a value of 52 $mM^{-1}cm^{-1}$ has been calculated. Kinetics of complex formation appears very similar in microsomes of untreated or inducer treated rats (Fig. 2). In addition, however, a small amount of a complex with an absorption maximum at 436nm is formed with microsomes of PB-treated, but not of untreated or DEX-treated rats. This complex appears within 3 min after addition of orphenadrine. It is not due to orphenadrine binding which elicits a type-II spectrum with a maximum at 426 nm (Fig. 3B). Because the absorption of the additional complex overlaps with that of the metabolite complex proper, the initial lag phase in complex formation observed with microsomes of untreated rats seems to be absent (Fig. 2). The expression of complex forming isozymes remains unaffected by rat treatment with phenobarbital or hexachlorobenzene whereas a slight induction (1.5-fold) is observed with dexamethasone. In microsomes of untreated female rats, only low levels of complex forming isozymes are present (3%, Fig. 1).

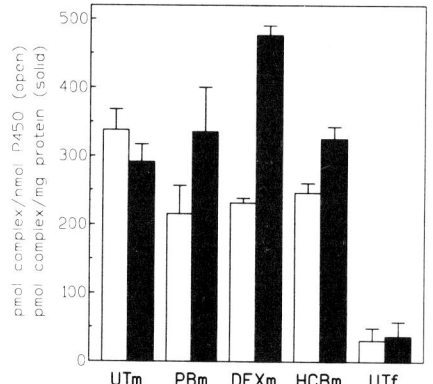

Fig. 1. Quantitation of *in vitro* formed orphenadrine metabolite complex in liver microsomes of untreated and inducer treated rats.

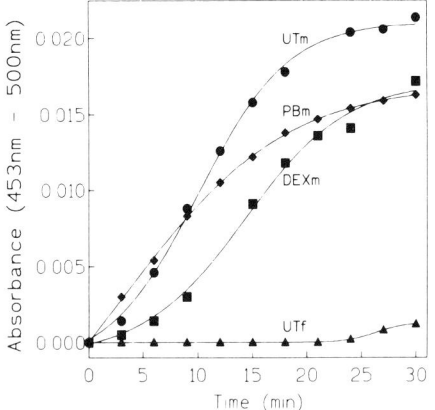

Fig. 2. Kinetics of complex formation with orphenadrine in liver microsomes of untreated and inducer treated rats. [P450] = 1 μM.

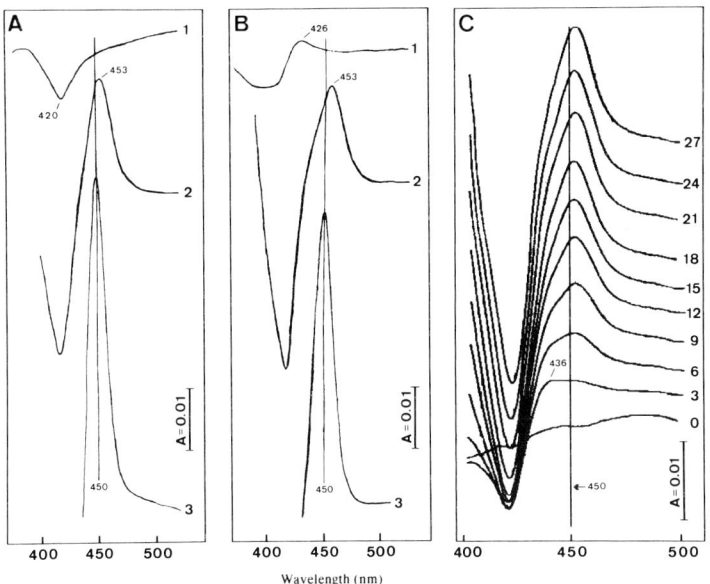

Fig. 3. Spectra obtained with microsomal P450 of untreated (A) and phenobarbital-treated rats (B). (A,B) 1: Binding of orphenadrine to oxidized P450. [P450] = 0.5 µM, [orphenadrine] = 50 µM. 2: Orphenadrine metabolite formed in the presence of NADPH. [P450] = 1 µM, [orphenadrine] = 50 µM. 3: Carbon monoxide complex with dithionite reduced P450. [P450] = 1 µM. (C) Formation of an orphenadrine metabolite complex in microsomes of phenobarbital treated rats: Repetitive scans from 0 to 27 min. [P450] = 1µM, [orphenadrine] = 50 µM.

Isozymes of the P450 subfamily 3A can be quantitated by complex formation with triacetyloleandomycin (TAO). The specificity of this reaction for CYP3A-isozymes has been shown by correlation of CYP3A-levels with 3A-specific enzymatic activities, i.e. 2ß and 6ß-hydroxylation of testosterone and TAO N-demethylation in microsomal preparations (Roos et a., 1993). Complete *in vitro* complexation of the isozymes is only achieved in the presence of NADH in addition to NADPH. Microsomal CYP3A-levels of untreated and inducer treated rats are presented in Table 1.

Table 1. Total P450 and CYP3A content of liver microsomes of untreated and inducer treated rats.

| Inducer, sex | UTm | PBm | DEXm | HCBm | TAOm | UTf |
|---|---|---|---|---|---|---|
| Total P450 | 0.90 | 1.34 | 1.90 | 1.80 | 4.20 | 0.82 |
| CYP3A | 0.11 | 0.37 | 1.25 | 0.20 | 3.31 | 0.01 |

Total P450 was determined as carbon monoxide complex, CYP3A as TAO metabolite complex. Values give nmol/mg protein. UT, untreated; m, male; f, female.

The quantitations are based on an extinction coefficient of 69.0 ± 1.0 $mM^{-1}cm^{-1}$ for the metabolite complex which has been determined by ferricyanide dissociation of *in vitro* formed complex and subsequent monitoring of CO-binding to the dithionite reduced sample (Roos et al., 1993).

DISCUSSION

While the isozyme specificity of complex formation with TAO is sufficiently proven, the isozymes responsible for orphenadrine metabolite complexation remain to be identified. Our data show that CYP2B-isozymes are not primarily involved in orphenadrine metabolite complex formation, as suggested by Reidy and coworkers, because (1) high levels of complex forming isozymes are found in microsomes of untreated male rats which do not contain considerable amounts of CYP2B1 (analyzed by immunoblotting, not shown) and (2) phenobarbital treatment dramatically increases CYP2B content but fails to induce orphenadrine complex forming isozymes. Together with the fact that only low amounts of orphenadrine complex are formed in microsomes of untreated female rats, our data suggest that a male-specific or male-dominant P450 is primarily involved in complex formation with orphenadrine. Reidy et al. (1989) observed inhibition of CYP2B-specific activities, i.e. testosterone 16β-hydroxylation and pentoxyresorufin O-dealkylation, in the presence of orphenadrine and NADPH. It is not clear, however, whether the orphenadrine metabolite complex is responsable for this effect or whether the inhibition is due to other orphenadrine metabolites with low $K_I$-values for CYP2B isozymes. The appearance of a complex with an absorption maximum at 436 nm shows that, besides metabolite complex formation, additional interactions occur during orphenadrine metabolism especially in microsomes of phenobarbital treated rats. Absence of the 436 nm-complex in microsomes of untreated and dexamethasone-treated rats suggests that CYP2B isozymes may be involved. Inhibition of testosterone 16α-hydroxylase (Reidy et al., 1989, and own data, not shown) after orphenadrine complex formation, high levels of complex forming enzymes in male and low levels in female microsomes support the idea that CYP2C11 is involved in metabolite complex formation with orphenadrine.

REFERENCES

Mansuy, D., Delaforge, M., LeProvost, E., Flinois, J.P., Columelli, S., and Beaune, P. (1981): Induction of cytochrome P-450 in rat liver by the antibiotic troleandomycin: partial purification and properties of cytochrome P-450-troleandomycin metabolite-complexes. *Biochem. Biophys. Res. Commun.* 103, 1201-1208.

Reidy, G.F., Mehta, I., and Murray, M. (1989): Inhibition of oxidative drug metabolism by orphenadrine: *in vitro* and *in vivo* evidence for isozyme specific complexation of cytochrome P-450 and inhibition kinetics. *Mol. Pharmacol.* 35, 736-743.

Reidy, G.F., and Murray, M. (1990): Evidence for complexation of P-450 IIC6 by an orphenadrine metabolite. *Biochem. Biophys. Res. Commun.* 166, 772-779.

Roos, P.H., Golub-Ciosk, B., Kallweit, P., Kauczinski, D. & Hanstein, W. (1993): Formation of ligand and metabolite complexes as a means for selective quantitation of cytochrome P450 isozymes. *Biochem. Pharmacol.* 45, 2239-2250.

# Risk assessment and bioavailability for mammals of soil-bound polycyclic aromatic hydrocarbons

Peter H. Roos[1], Walter G. Hanstein[1], Dagmar Strotkamp[1],
Manfred van Afferden[2], Dietmar Tappe[3]

[1]Institute for Physiological Chemistry, Ruhr-University Bochum, D-44780 Bochum, FRG.
[2]DMT-Gesellschaft für Forschung und Prüfung mbH, D-45307 Essen, FRG. [3]Umweltagentur Ruhrgebiet, Ruhr-University Bochum, D-44780 Bochum, FRG

Soils contaminated with environmental toxicants, such as polycyclic aromatic hydrocarbons (PAH), constitute a potential risk for human health. Risk assessment is usually based on chemical analyses for hazardous compounds and the effects of water-extractable components on biological activities of microorganisms and plants. However, neither method allows for evaluation of the bioavailability of the toxicants for mammalian species. It is an open question to what extent soil-bound PAHs are extracted by the gastro-intestinal fluids and whether the soil matrix affects the extractability of the PAHs. Therefore, we studied the effect of ingestion of chemically characterized, PAH contaminated soils on the induction of microsomal CYP1A-isozymes in rats. These isozymes are expressed constitutively at low levels, are induced by PAHs (Thomas et al., 1983) and are responsible for the activation of procarcinogens (Guengerich, 1992).

MATERIALS & METHODS

*Animals and animal treatment.* Male Sprague Dawley rats were fed with a commercial diet supplemented with 5% of ground soil.
*Preparation of microsomes.* Liver microsomes were prepared as described by Guengerich (1977). For preparation of duodenum microsomes, cells were scraped off, homogenized with a potter and sonicated for 6 seconds. After centrifugation for 20 min at 9.000xg, the supernatant was taken as crude microsomal preparation.
*Immunoblotting* was done as described by Towbin et al. (1979).
*Ethoxyresorufin O-deethylase (EROD).* The activity was determined fluorometrically by continuous monitoring. Excitation: 530 nm, emission: 585 nm. For further details see Roos et al. (1993). NADPH was generated from NADP by isocitrate/isocitrate dehydrogenase in the presence of $Mg^{2+}$.
*PAH analyses.* Qualitative and quantitative analyses of soil bound PAHs were performed by HPLC of toluene extracts (Table 1) and include the following compounds: 2-ring: naphthaline; 3-ring: acenaphthene, acenaphthylene, anthracene, fluorene, phenanthrene; 4-ring: benz[a]anthracene, chrysene, fluoranthene, pyrene; 5-ring: benzo[b]fluoranthene, benzo[k]fluoranthene, benzo[a]pyrene, dibenzo[ah]anthracene; 6-ring: benzo[ghi]perylene, indeno[1,2,3,cd]pyrene.

Table 1. PAH content of the soil samples (mg/kg soil)

| Soil | 2-ring | 3-ring | 4-ring | 5-ring | 6-ring |
|------|--------|--------|--------|--------|--------|
| I    | 4      | 65     | 36     | 7      | 3      |
| II   | 39     | 52     | 42     | 17     | 9      |
| VIa  | 192    | 1560   | 484    | 114    | 47     |
| VIb  | 10     | 18     | 78     | 37     | 14     |
| VII  | 3      | 1      | 3      | 1      | 1      |
| VIII | 12     | 1162   | 637    | 2      | <0.1   |
| IX   | 130    | 334    | 523    | 189    | 84     |

RESULTS

Extractability of soil-bound PAHs by the intestinal fluid, their uptake and disposition in the liver is shown by the selective induction of CYP1A-isozymes in this organ as measured by ethoxyresorufin O-deethylase activity (Table 2, Fig. 1) and detection by immunoblotting with CYP1A-antibodies (Fig. 2). The increase in CYP1A-expression is mainly dependent on the amount of 5 and 6-ring PAHs and is not related to the total PAH content in the soil samples (Table 2, Fig. 1). Sample VIII with high concentrations of 3-ring PAHs but only negligible amounts of 5- and 6-ring PAHs does not lead to significant 1A-induction in the liver after 1 and 4 (Table 2) or even 10 and 14 weeks of ingestion (not shown). Soil contamination with high amounts of 5- and 6-ring PAHs leads up to a 180-fold increased expression of EROD-activity. A linear correlation is obtained between log(EROD) and the content of 5,6-ring PAHs (Fig. 1). In contrast to CYP1A-induction in the liver, there is no simple relation between PAH-content and composition in the samples and expression in duodenum microsomes (Table 2).

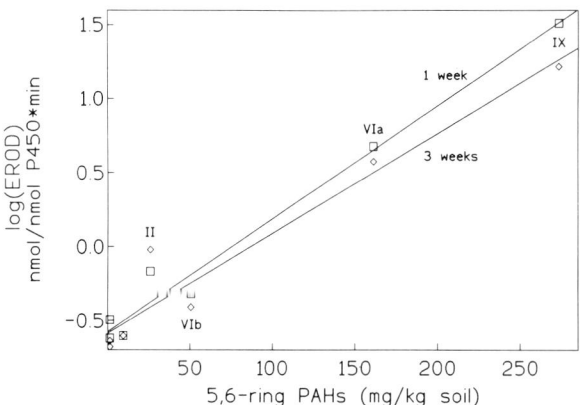

Fig. 1. EROD activity in liver microsomes of soil-fed rats as a function of soil contamination with 5- and 6-ring PAHs (semi-logarithmic presentation). Rats were fed for 1 or 3 weeks with the soil-supplemented diet. Roman numerals correspond to the soil samples (see Table 1).

TABLE 2. EROD-activities in liver and duodenum microsomes: Effect of ingested PAH contaminated soils.

| Soil | Week | EROD-activity nmol/min*mg protein | | PAH-contamination mg PAH/kg soil | |
|---|---|---|---|---|---|
| | | Liver | Duodenum | 5,6-ring | 2,3,4-ring |
| I | 1 | 0.32 ± 0.06 | 0.07 ± 0.05 | 9.6 | 104.8 |
| | 3 | 0.23 ± 0.04 | 0.03 ± 0.03 | | |
| II | 1 | 0.87 ± 0.17 | 0.11 ± 0.01 | 26.3 | 133.1 |
| | 3 | 0.95 ± 0.18 | 0.30 ± 0.01 | | |
| VIa | 1 | 6.20 ± 1.38 | 0.02 | 161.1 | 2236.6 |
| | 3 | 5.32 ± 1.45 | 0.08 ± 0.03 | | |
| VIb | 1 | 0.55 ± 0.15 | 0.06 ± 0.03 | 50.7 | 105.7 |
| | 3 | 0.50 ± 0.12 | 0.01 | | |
| VII | 1 | 0.35 ± 0.01 | 0.003 | 1.5 | 6.8 |
| | 3 | 0.27 ± 0.02 | 0.008 | | |
| VIII | 1 | 0.27 ± 0.04 | 0.01 | 1.5 | 1811.3 |
| | 4 | 0.20 ± 0.05 | 0.001 | | |
| IX | 1 | 57.56 ± 4.50 | 0.02 | 273.6 | 987.1 |
| | 3 | 21.08 ± 5.93 | 0.04 ± 0.02 | | |
| ---- | | 0.22 | 0.001 | | |
| ßNF* | | 20.30 | 0.581 | | |

Rat diet: + 5% of PAH contaminated soils. Treatment time: 1 or 3 weeks. ßNF = ß-naphthoflavone (CYP1A inducer). *Rat treatment: i.p. injections with 40 mg ßNF/kg body weight for 3 consecutive days, ßNF in corn oil.

DISCUSSION

The selective induction of liver microsomal CYP1A isozymes shows that ingested soil-bound PAHs are extracted by the intestinal fluid, then pass the intestinal permeability barrier and thus exhibit their effects on biological functions. Under the conditions described above, effective induction is observed with higher condensated PAHs only and a good correlation is found between CYP1A expression and the content of 5 and 6-ring PAHs. However, the 2,3 and 4-ring components have also to be considered for risk assessment, because they can be transformed to toxic or mutagenic compounds by CYP1A isozymes. Residual amounts of PAHs in decontaminated soil, i.e. sample VIb compared to VIa, show a lower induction effect than expected. This is probably due to reduced bioavailability of the residual PAHs caused by tight binding to the soil matrix. Although expression of CYP1A can be increased in duodenal mucosa cells (Peters & Kremers, 1989), the response to extracted PAHs is low compared to the induction in hepatocytes. Furthermore, there is no good correlation between PAH-components and EROD-activity in duodenum microsomes. The causes for these effects are not known. Because structure, function and induction mechanisms of CYP1A-isozymes are very similar in different organisms, indicating clear orthology for mammalian species (Nelson et al., 1993), an extrapolation of results from rats to humans may be justified. In conclusion, induction of liver microsomal CYP1A is a relevant parameter for mammalian risk assessment concerning PAH-contaminated soils. Our future studies will focus on clinical application and design of an extraction medium for contaminated soils with an efficacy and selectivity equivalent to *in vivo* obtained bioavailability. The latter should allow for a realistic estimation of risk for humans by 'bio-equivalent' extraction of soils.

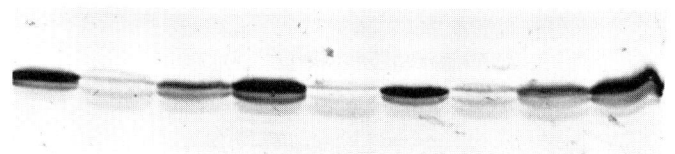

IX   VIa   VIb   ßNF   UT   IX   VIa   VIb   ßNF

Fig. 2. Detection of CYP1A1 by immunoblotting in liver microsomes of rats fed with a soil-supplemented diet. Roman numerals correspond to the soil samples (see Tables 1 and 2). Results from 2 individuals each are shown. UT: untreated rats. ßNF: ß-naphthoflavone-treated rats. Amount of applied microsomal protein: 30 µg each.

REFERENCES

Guengerich, F.P. (1977): Separation and purification of multiple forms of microsomal cytochrome P450. Partial characterization of three apparently homogeneous cytochromes P450 prepared from livers of phenobarbital and 3-methylcholanthrene treated rats. *J. Biol. Chem.* 252, 3970-3979.

Guengerich, F.P. (1992): Metabolic activation of carcinogens. *Pharmac. Ther.* 54, 17-61.

Nelson, D.R., Kamataki, T., Waxman, D.J., Guengerich, F.P., Estabrook, R.W., Feyereisen, R., Gonzalez, F.J., Coon, M.J., Gunsalus, I.C., Gotoh, O., Okuda, K. & Nebert, D.W. (1993): The P450 superfamily: update on new sequences, gene mapping, accession numbers, early trivial names of enzymes, and nomenclature. *DNA Cell Biol.* 12, 1-51.

Peters, W.H.M. & Kremers, P.G. (1989): Cytochromes P-450 in the intestinal mucosa of man. *Biochem. Pharmacol.* 38, 1535-1538.

Roos, P.H., Golub-Ciosk, B., Kallweit, P., Kauczinski, D. & Hanstein, W.G. (1993): Formation of ligand and metabolite complexes as a means for selective quantitation of cytochrome P450 isozymes. *Biochem. Pharmacol.* 45, 2239-2250.

Thomas, P.E., Reik, L.M., Ryan, D.E. & Levin, W. (1983): Induction of two immunochemically related rat liver cytochrome P-450 isozymes, cytochrome P-450c and P-450d, by structurally diverse xenobiotics. *J. Biol. Chem.* 258, 4590-4598.

Towbin, H.T., Staehelin, T. & Gordon J. (1979) Electrophoretic transfer of proteins from polyacrylamide gels to nitrocellulose sheets: procedure and some applications. *Proc. Natl. Acad. Sci. USA* 76, 4350-4354.

ACKNOWLEDGEMENTS

Financial support of the project by the Umweltagentur Ruhrgebiet is gratefully acknowledged.

# Expression of components of the cytochrome P450 mediated mono-oxygenase system using recombinant baculoviruses

Philippe Clair[1], Elizabeth A. Shephard[2], Ian R. Phillips[1]

[1]Department of Biochemistry, Queen Mary and Westfield College, University of London, Mile End Road, London E1 4NS, UK. [2]Department of Biochemistry and Molecular Biology, University College London, Gower Street, London WC1E 6BT, UK

## INTRODUCTION

The identification of the cytochrome P-450 isoform(s) responsible for the metabolism of a particular medicament is a prerequisite to predict and avoid drug interactions *in vivo*. Owing to species differences in foreign compound metabolism, animal models are often unsatisfactory for direct *in vivo* testing. The use of human microsomes or primary cultures of hepatocytes is complicated by inter-individual variations in the expression of many P450s, requires correlations of enzyme activities and concentrations (Beaume *et al.*, 1988) and frequently relies on the use of specific inducers or inhibitors of P450s. The presence in human microsomes of closely related P450s, together with the limited availability of suitable human material poses challenging problems for the purification of these proteins. The cDNA-mediated expression of individual components of the mono-oxygenase system in heterologous system offers an attractive alternative capable of providing an abundant source of highly enriched, well-defined enzyme suitable for determining the enzymatic activity spectrum, identifying metabolites, establishing structure-activity relationships and preparing form-specific antibodies. In the work reported here, insect Sf9 cells were used as recipients for recombinant baculovirus-mediated expression of eight enzymes of the drug-metabolising mono-oxygenase system.

## MATERIALS AND METHODS

Methods for the construction, expression and maintenance of recombinant baculoviruses were essentially as described by Summers *et al.* (1987) and O'Reilly *et al.* (1992).

<u>Generation of recombinant baculoviruses:</u>
Human cytochrome P-450 reductase: The coding sequence of a full-length cDNA isolated in our laboratory (Shephard *et al.*, 1992) was modified by PCR amplification using the self-correcting, thermostable DNA polymerase from *P. furiosis* (Stratagene), to obtain convenient restriction sites for subcloning into the baculovirus transfer vector pAcC12. A truncated sequence coding for the residues 47-679 of the reductase also was inserted into pAcC12.

Human CYP3A4: The coding sequence was obtained by a reverse transcription PCR amplification from Poly(A)$^+$ RNA isolated from a primary culture of human hepatocytes treated with the inducer rifampicin (Daujat *et al.*, 1991). The amplified cDNA was subcloned into the baculovirus transfer vector pAcC4. Partial sequencing and restriction mapping of the insert confirmed identity of the CYP3A4 coding sequence.

Rabbit CYP3A6: A PstI fragment containing the coding sequence of CYP3A6 (Dalet *et al.*, 1988) was subcloned into pAcC12.

Human CYP27: An EcoRI fragment containing the coding sequence was isolated from a full length cDNA (Andersson *et al.*, 1989) obtained from H. Dahlbäck and subcloned into pAcC12.

Human CYP2A6: Work on CYP2A6 is described elsewhere in these proceedings (Nanji *et al.*, 1993)

Human flavin-containing mono-oxygenases: Coding sequences for FMO 1, 3 and 4, obtained from cDNA clones isolated in our laboratory (Dolphin *et al.*, 1991-1992), were inserted into pAcC4.

Each recombinant baculovirus was generated by colipofection of Sf9 cells with 5 µg of the transfer vector and 0.5 µg of Bsu36I-linearised AcMNPV DNA (Baculogold, AMS Biotechnologies). Recombinant viral plaques were identified by a visual screening procedure. The size of the cDNA sequence contained within each recombinant viral genome was verified by PCR amplification from total DNA extracted from infected Sf9 cells.

Characterization of expressed proteins

Recombinant baculoviruses were obtained in high titer by two rounds of amplification in Sf9 cells, then used to infect 100 ml suspension cultures of fresh cells. The following sub-cellular fractions were prepared 48 to 60 hours post-infection (h.p.i.): nuclear-mitochondrial pellet, post-mitochondrial supernatant, microsomal fraction, and a cytosolic fraction. The fractions were analysed by SDS polyacrylamide gel electrophoresis, western blotting, spectroscopy and enzyme activity assays.

Enzymatic assays

Cytochrome P450 reductase activity was assayed by measuring cytochrome c reductase activity, according to Vermilion *et al.* (1974). The ability of cytochrome P450 reductases to reduce cytochrome P450 was assessed by their capacity to support the formation of 16α, 16β hydroxytestosterone (OHT), and androstenedionne (Ad) in reconstituted systems containing purified CYP2B1 (Phillips *et al.*, 1983), [$^{14}$C]-testosterone and a NADPH-regenerating system. The ability of CYP3A4 and CYP3A6 to produce 6β-OHT was evaluated in a system containing Sf9 microsomes, purified rat P450 reductase (Shephard *et al.*, 1983b), [$^{14}$C]-testosterone and a NADPH-regenerating system. Separation and identification of the hydroxytestosterones were carried out according to Waxman (1991). The function of FMOs was assessed by measuring thiobenzamide-S-oxidation (TBZ-SO), (Cashman *et al.*, 1981) in reactions containing Sf9 microsomes and a NADPH-regenerating system.

RESULTS AND DISCUSSION

Isolation of recombinant viruses and expression of the apoproteins

From each recombination experiment, at least twelve independent viral plaques were isolated and tested for the presence of the appropriate full-length cDNA insert by PCR. In each case, all isolates were positive. Lysates of cells infected during the second passaging of each recombinant virus contained an additional polypeptide of the expected size, when compared with a lysate prepared from wild-type AcMNPV-infected cells. The expression of CYP3As, and of the full-length and truncated cytochrome P450 reductase was confirmed by the use of antibodies directed respectively against CYP3A6 and rat P-450 reductase (Shephard *et al.*, 1983a). No signal was obtained with the lysates of cells infected with the wild-type virus.

Cellular localisation of expressed polypeptides

Analysis of the sub-cellular fractions by SDS-polyacrylamide gel electrophoresis revealed that CYP3A4, CYP3A6, CYP2A6, full-length P450 reductase, FMO1 and FMO3 were located mainly in the microsomal fraction. Truncated P450 reductase was clearly a cytosolic protein, and CYP27 was found mainly in the mitochondrial fraction. We were not able to detect FMO4. When infected with the FMO4 recombinant virus, Sf9 cells lysed 24 to 36 h.p.i., whereas the other viruses provoked the cell lysis between 96-120 h.p.i. We suspect that the expression of FMO4 is deleterious and may kill cells before they are able to accumulate detectable amounts of the protein.

P450 holoenzyme content of the cells

In initial experiments, despite the abundance of the expressed CYP polypeptides, we were unable to detect CO-reduced difference spectra with the cell lysates. Addition of haemin (6μM) to the medium during the course of infection allowed the formation of cytochrome P450 holoenzyme. The spectra obtained showed a strong 'P420' component which, as it increased with haemin concentration in a non-saturable fashion, was the result of non-specific adsorption of haemin to cellular components. The presence of this spectral component may interfere with spectrophotometric P450/drug interaction studies. Moreover, haemin was toxic to the cells, even when provided as methaemalbumin. To optimise haem incorporation into P450, 24 hours before infection the culture medium was supplemented with 5-ALA and various chelators loaded with $Fe^{3+}$ (Iron-citrate, ferric-ammonium-citrate, iron-choline-citrate, iron-dextran, iron EDTA and iron pyridoxal isonicotinoïl hydrazone). Concentrations used ranged between 1 to 1000 μM for both 5-ALA and $Fe^{3+}$. This study showed that i) the lack of spectrally detectable P450 was due to a deficiency of the medium, rather than to aggregation or missfolding of the expressed polypeptides, ii) the limiting factor is protoporphyrin synthesis rather then iron, iii) 5-ALA is not toxic to Sf9 cells up to 1 mM, whereas iron toxicity strongly depends on the chelator used. iv) a concentration of 100 μM for both 5-ALA and iron supports the same level of P450 production as obtained with haemin but without the appearance of a 'P420 shoulder'. The Sf9 cells are thus able to synthesise sufficient haem for the expressed apoproteins, provided that the appropriate precursors are present in the medium. Using these conditions, we were able to obtain expression of CYP3A6, CYP3A4 and CYP27 holoenzymes at concentrations of 0.44μM, 0.14 μM, 0.25 μM (μmol P450 per liter of culture) respectively.

Catalytic activities of expressed proteins

Both the full-length and truncated forms of cytochrome P450 exhibited a strong cytochrome c reductase activity. Active full-length reductase was located mainly in the microsomal fraction (65%, spec. act. 2.16 μmol cyt.c reduced/mg microsomal proteins/min). The truncated form was present mainly in the cytosolic fraction (91%, spec. act. 1.94 μmol cyt.c reduced/mg microsomal proteins/min). From the total activities present in the cell lysates and the specific activities of the purified proteins, the level of expression of both proteins is estimated to be 57-60 mg of active enzyme per liter of culture. Both enzymes were able to drive the CYP2B1-catalysed formation of 16α-OHT, 16βOHT and Androstenedionne, but with markedly different efficiencies. The microsomal fractions of Sf9 cells containing spectrally active CYP3As catalysed the formation of 6βOHT. The expressed CYP2A6 is an active coumarin hydroxylase. Experiments with microsomal fractions of Sf9 cells expressing CYPs showed a requirement for the addition of an exogenous P450 reductase (either purified or. as microsomal fraction of insect cells expressing the reductase) in order to obtain maximal reaction rates. An endogenous cytochrome P450 reductase activity was detected in the microsomal fraction of Sf9 insect cells, but this activity was rate-limiting and unstable over a period of 20-30 min in the conditions of incubation (37°C). The catalytic abilities of CYP27 have not yet been tested. The expressed FMO1 and 3 were very active (spec. acts. respectively of 24 and 10 nmol TBZ-SO/mg microsomal proteins/min). Stereoselective N-oxidation activities of these enzymes are currently under investigation. Taken together, these results show that the Sf9-baculovirus expression system is capable of supporting high-level expression of active NADPH-cytochrome P450 reductase, microsomal and probably mitochondrial cytochromes P-450, and flavin-containing mono-oxygenases. However, for the formation of active P450s the medium must be supplemented with haem or its precursors. P450-mediated catalytic activities can be measured directly on microsomes prepared from infected Sf9 cells, but there is insufficient endogenous reductase to saturate the expressed cytochromes P-450s. This deficiency can be overcome by adding purified reductase or microsomal fractions of Sf9 cells infected with a reductase-expressing virus. It should be noted that Sf9 microsomes contain low endogenous NADPH-dependant steroidogenic activities, and thus care should be exercised in choosing the appropriate controls.

ACKNOWLEDGEMENTS

This work was supported by the Glaxo Group Research, the Cancer Reseaech Campaign and the Wellcome Trust.

REFERENCES

Andersson, S., Davis, D. L., Dahlback, H., Jornvall. H. & Russell, D. W. (1989) Cloning, structure, and expression of the mitochondrial cytochrome P-450 sterol 26-hydroxylase, a bile acid biosynthetic enzyme. *J.Biol.Chem.* **264**, 822-829.

Beaune, P. H. & Guengerich, F. P. (1988): Human drug metabolism *in vitro*. *Pharmac. Ther.* **37**, 193-211.

Cashman, J. R. & Hanzlik, R. P. (1981): Microsomal oxidation of thiobenzamide. A photometric assay for the flavin-containing-monooxygenase. *Biochem. Biophys. Res. Commun.* **98**, 147-153.

Dalet, C., Clair, P., Daujat, M., Fort, P., Blanchard, J. M. & Maurel, P. (1988): Complete sequence of cytochrome P450 3c cDNA and presence of two mRNA species with 3' untranslated regions of different lengths. *DNA* **7**, 36-46.

Daujat, M., Pichard, L., Fabre, I., Pineau, T., Fabre, G., Bonfils, C. & Maurel-P (1991):Induction protocols for cytochromes P450IIIA in vivo and in primary cultures of animal and human hepatocytes. *Methods.Enzymol.* **206**,345-353.

Dolphin, C. T., Shephard, E. A., Povey, S., Palmer, C. N. A., Ziegler, D. M., Ayesh, R., Smith , R. L., & Phillips, I. R. (1991): Cloning, primary sequence and chromosomal mapping of a human flavin-containing monooxygenase (FMO1). *J. Biol. Chem.* **266**, 12379-12385.

Dolphin, C. T., Shephard, E. A., Povey, S., Smith, R. L. & Phillips, I. R. (1992): Cloning, primary sequence and chromosomal localization of human FMO2, a new member of the flavin-containing monooxygenase familly. *Biochem. J.* **287**, 261-267.

Nanji, M., Clair, P., Phillips, I. R. & Shephard, E. A. (1993): Expression of cDNAs encoding human and marmoset CYP2As. in *Proc. VIII th International Conference on Cytochrome P450, Lisbon, October 1993.*

O'Reilly, D., Miller, L. K., Luckow, V. A., Ed. (1992) *Baculovirus expression vectors: a laboratory manual. ISBN 0-7167-7017-2.* W. H. Freeman & Co. New york.

Phillips, I. R., Shephard, E. A., Bayney, R. M., Pike, S. F., Rabin, B. R., Heath, R. and Carter, N. (1983): Induction and repression of the major phenobarbital-induced cytochroma P-450 measured by radioimmunoassay. *Biochem J.* **212**, 55-64.

Shephard, E. A., Palmer, C. N., Segall, H. J. & Phillips, I. R. (1992) Quantification of cytochrome P450 reductase gene expression in human tissues. *Arch. Biochem. Biophys.* **294,** 168-172.

Shephard, E. A., Phillips, I. R., Bayney, R. M., Pike, S. F. & Rabin, B. R. (1983a): Quantification of NADPH: cytochrome P-450 reductase in liver microsomes by a specific radioimmunoassay technique. *Biochem. J.* **211**, 333-340.

Shephard, E. A., Pike, S. F., Rabin, B. R. & Phillips, I. R. (1983b):A rapid one-step purification of NADPH-Cytochrome c (P-450) reductase from rat liver microsomes. *Anal. Bioch.* **129**, 430-433

Summers, M. D. & Smith, G. E. (1987): A manual of methods for baculovirus vectors and insect cell culture procedures. *Texas Agricultural Experiment Station Bulletin.* **1555**.

Vermillon, J. L. & Coon, M. J. (1974) *Biochem. Biophys. Res. Commun.* **60,** 1315-1322.

Waxman, D. J.(1991): P450-catalysed steroid hydroxylation: assay and product identification by thin-layer chromatography. *Methods in Enzymol.* **206**, 462-476.

# Advantages of the human over yeast NADPH-P450 reductase on supporting human P450 activities in yeast

Philippe Urban, Alain Perret, Gilles Truan, Denis Pompon

Centre de génétique moléculaire, Laboratoire 2420 du CNRS, associé à l'Université de Paris-VI, 91190 Gif-sur-Yvette, France

Due to the multiplicity and inducibility of mammalian P450s, the necessity to understand the precise activities of a particular P450 makes the availability of tools allowing easy access to individual P450 isoforms crucial. This goal was fullfiled by heterologous expression in various systems. The yeast expression system presents the advantages of producing sufficient amounts of the functional heterologous P450 (Urban et al., 1990; Guengerich et al., 1991). Recognition between P450 reductase and P450s implies a dual mechanism involving both hydrophobic interactions probably mediated by the membrane and direct charge-pairing interactions between positive charges on P450 and negatively charged P450 reductase. The yeast P450 reductase presents a limited identity of its amino acid sequence with those of mammalian enzymes (33% with the human enzyme). Therefore, the potential coupling of the yeast P450 reductase with heterologous P450s, belonging to a distantly related species, was not straightforward. Despite that, the yeast enzyme is able to couple with heterologous P450s. However, several previous studies have demonstrated the limiting effects on P450-catalysed heterologous activities both of low amounts of the yeast P450 reductase and of its inefficient coupling (Urban et al., 1990; Murakami et al., 1990). Therefore, ways for enhancing the monooxygenase activities of P450s expressed in yeast have been actively sought. Three schools developed complementary approaches based i) on the construction of gene expressing P450 - P450 reductase fusion protein as in H. Ohkawa's laboratory (Shibata et al., 1990), ii) on the incorporation of two expression cassettes, one for the P450 of interest and the other for human P450 reductase, on the same plasmid as in H. P. Eugster's laboratory (Eugster et al., 1992), and iii) on the 'humanized' strain concept developed by ourselves (Truan et al., 1993). This paper presents the preliminary comparison of the specific activities, both in living cells and in microsomes, of two mammalian P450s of the 1A family produced in three isogenic strains different in the type and in the amount of their P450 reductase.

## CONSTRUCTION OF THE W(hR) STRAIN

Formally, production of human P450 reductase in yeast requires to substitute the sequence coding for the yeast P450 reductase by the human one. Practically, the coding sequence for human P450 reductase was PCR-amplified from total human liver cDNAs, then cloned in pYeDP60 expression vector which places it under the transcriptional control of *GAL10-CYC1* promoter, a galactose-induced artificial promoter, and *PGK* termination yeast sequences. The human P450 reductase when produced in yeast was found microsomal and catalytically functional in supporting cytochrome *c* reduction. In a following step, the human *CPR* expression cassette was integrated by homologous recombination in the genomic DNA of W(N) cells at the *CPR* locus, leading to the W(hR) strain (See Fig. 1).

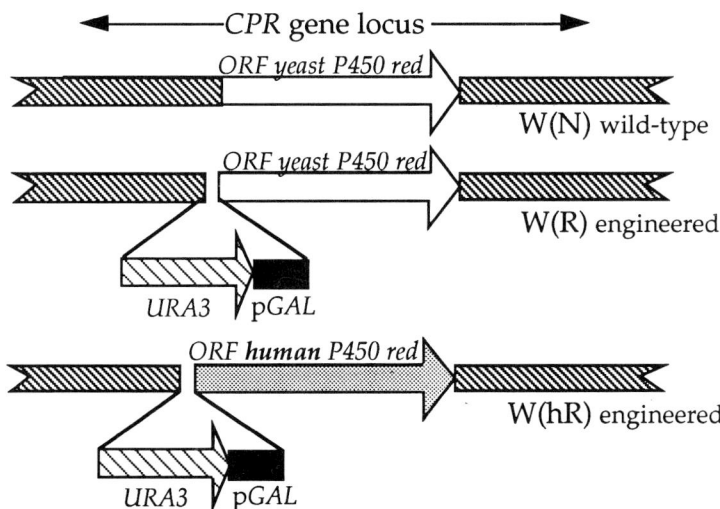

Fig. 1. Structures of the *CPR* locus in the haploid strains W(N), W(R) and W(hR).

W(hR) cells do not produce any yeast P450 reductase, their original *CPR* gene locus being disrupted, but they produce solely the human enzyme when grown on galactose. The human P450 reductase was found rather inefficient in coupling with endogenous yeast P450s resulting in a growth rate slower for W(hR) cells than that of W(N) wild-type cells.

## BIOCONVERSION IN TRANSFORMED W(hR) CELLS

When measured using cytochrome *c* as an acceptor, the specific activity of human P450 reductase in W(hR) microsomes was found much lower than that measured for yeast P450 reductase in wild-type yeast (See Table 1). This low specific activity suggests that the human enzyme is produced at rather low amounts in yeast W(hR) cells. This low expression is not due to an intrinsic factor that would make expression of human P450 reductase activity in yeast low. Indeed the over-expression of human P450 reductase in yeast from a multicopy plasmid-based expression cassette was found rather high (specific activity : 370 ± 40 nmol cyt. $c$ reduced per min per mg protein). Therefore, the low expression level observed is due to the fact that a single copy of the human P450 reductase expression cassette is present in each W(hR) cell. The three isogenic strains W(N), W(R) and W(hR) were transformed by expression vectors for human and mouse P450 1A1. Several clones of each transformation assay were tested for bioconversion of ethoxyresorufin in resorufin, an activity typical of P450s 1A1 absent in control yeast. The bioconversion rate measured on living W(hR) cells was found to be significantly higher than that measured with W(N) cells which produce yeast P450 reductase, and slightly lower than that measured with W(R) cells which overproduce yeast P450 reductase (See Table 1).

**Table 1.** Comparison of the relative efficiency of human and yeast P450 reductases to support mammalian 1A1-catalysed bioconversion.

| Strain | P450 reductase type | Cytochrome $c$ reductase activity (nmol/mg per min) | Relative P450 1A1 EROD Bioconversion rate | |
|---|---|---|---|---|
| | | | Human 1A1 | Mouse 1A1 |
| W(N) | Yeast | 100 ± 20 | 1 | 1 |
| W(R) | Yeast (overproduced) | 2200 ± 200 | 5 ± 1 | 10 ± 1 |
| W(hR) | Human | 35 ± 5 | 6 ± 2 | 7 ± 2 |

These results indicate that low levels of expression of the human P450 reductase in W(hR), as judged by NADPH-cytochrome $c$ reductase activity, support bioconversions catalysed by mammalian P450s as efficiently as the overproduced yeast P450 reductase in W(R) cells. The engineered yeast W(hR) strain thus appears the best choice for bioconversion purposes with mammalian P450s, even though the human P450 reductase is produced in W(hR) at rather low amounts.

## P450 ACTIVITIES IN W(hR) MICROSOMAL FRACTIONS

Spectral and activity determinations carried out on yeast microsomal fractions permit to reach two independent parameters that simultaneously affect the rate of bioconversion. One parameter is the spectral content of yeast microsomes in heterologous P450 that is determined by differential spectroscopy on reduced iron - carbon monoxide complex. The fact that some yeast strains, when respiring, are inherently devoid of any spectrally detectable endogenous P450s makes possible the use of yeast microsomal fractions as such for spectral screenings specific of the heterologous P450. The other parameter is the turnover number for each of the activities catalyzed by the heterologous P450, it indicates the efficiency with which the microsomal P450 reductase couples with the heterologous P450. The three isogenic strains W(N), W(R) and W(hR) were transformed by plasmid for expression of human and murine P450s 1A1. The transformed cells were grown in minimal synthetic medium and the cells were dirupted and microsomal fractions prepared. Preliminary spectral results indicate that the level of P450 production in W(hR) cells is higher than that observed in the two other strains. One possibility could be that over-production of the yeast P450 reductase woul be accompanied by toxic effects that would lower microsomal content in heterologous P450. Preliminary activity ethoxy-resorufin O-deethylase (EROD) assays indicate that the low amounts of human P450 reductase in W(hR) microsomes seems to support turnover numbers for both P450s 1A1 with a comparable or slightly higher efficiency than that in W(N) strain, for a P450 reductase activity 3-fold lower. Therefore, the human P450 reductase is more efficient than the yeast enzyme in reducing mammalian P450s. However, the highest turnover numbers were obtained with microsomes from W(R) cells. Therefore, the production of the P450 of interest in W(R) cells appears to be the best choice for activity screening purposes.

## REFERENCES

Eugster, H.P., Bärtsch, S., Würgler, F.E., & Sengstag, C. (1992): Functional co-expression of human oxidoreductase and cytochrome P450 1A1 in Saccharomyces cerevisiæ results in increased EROD activity. *Biochem. Biophys. Res. Commun.* 185, 641-647.

Guengerich, F.P., Brian, W.R., Sari, M.A., & Ross, J.T. (1991): Expression of mammalian cytochrome P450 enzymes using yeast-based vectors. *Methods Enzymol.* 206, 130-145.

Murakami, H., Yabusaki, Y., Sakaki, T., Shibata, M., & Ohkawa, H. (1990): Expression of cloned yeast NADPH-cytochrome P450 reductase gene in Saccharomyces cerevisiæ. *J. Biochem. (Tokyo)* 108, 859-865.

Shibata, M., Sakaki, T., Yabusaki, Y., Murakami, H., & Ohkawa, H. (1990): Genetically engineered P450 monooxygenases: construction of bovine P450 c17 / yeast reductase fused enzyme.*DNA Cell Biol.* 9, 27-36.

Truan, G., Cullin, C., Reisdorf, P., Urban, P., & Pompon, D. (1993): Enhanced in vivo monooxygenase activities of mammalian P450s in engineered yeast cells producing high levels of NADPH-P450 reductase and human cytochrome b5. *Gene* 125, 49-55.

Urban, P., Cullin, C., & Pompon, D. (1990): Maximizing the expression of mammalian cytochrome P450 monooxygenase activities in yeast cells. *Biochimie* 72, 463-472.

# Inducing effect of porphyrinogenic agents on mouse liver Cyp2a-5 expression

Pirkko Salonpää, Kai Krause, Olavi Pelkonen, Hannu Raunio

*Department of Pharmacology and Toxicology, University of Oulu, SF-90220 Oulu, Finland*

## Summary

1. Mouse hepatic coumarin 7-hydroxylase (COH) activity and the levels of CYP2A5 steady-state mRNA are substantially increased by the porphyrinogenic agents aminotriazole, thioacetamide, and griseofulvin.

2. mRNA contents of aminolevulinate synthase, ferrochelatase and heme oxygenase are also increased to a variable extent.

3. The induction is blocked by heme, if heme arginate is administered simultaneously with the inducing agents.

4. These results suggest that there could be a heme-sensitive repressor regulating the transcription of the *Cyp2a-5* gene.

## INTRODUCTION

Coumarin 7-hydroxylase (COH) activity is catalyzed by the *Cyp2a-5* gene product (CYP2A5) in mice. Mouse hepatic *Cyp2a-5* expression is often increased in conditions in which other P450 isoforms are repressed, e.g. after the administration of liver damaging agents. Cobalt, a heme oxygenase inducer, increases Cyp2a-5 expression (Hahnemann et al., 1992), an effect that is reversed compared with CYP2B1/2. To study the involvement of agents affecting heme metabolism in *Cyp2a-5* expression we measured the effects of several porphyrinogenic agents on Cyp2a-5 mRNA levels and on some key enzymes in heme metabolism. To assess the effects of heme, the prosthetic group of cytochrome P450, on heme metabolism intermediates and CYP2A5 stimulation, mice were treated with hemearginate to intervene with the action of heme-depleting compounds.

## METHODS

Inbred DBA/2 and C57BL/6 mice given i.p. injections of the various agents. Heme arginate (Normosang®) was a gift from Dr. M. Kangasaho.

A part of the liver from each mouse was used to prepare 10000 g supernatants for enzyme activity assays. The Bradford method was used for protein content determinations. COH activity was measured by the method of Aitio (1978) using 100 µM coumarin as substrate.

cDNA probes for aminolevulinate synthase, ferrochelatase and heme oxygenase were prepared by reverse transcriptase-PCR and TA cloning. The sequences of these cDNAs were used to design gene-specific 20-mer primers. PCR template was prepared from RNA utilizing reverse transcriptase enzyme. The full-length cDNAs were amplified and cloned into pCRII (Invitrogen) vector. The livers within each group were pooled and homogenized in guanidine thiocyanate and total cellular RNA was prepared by cesium chloride centrifugation method. 25 µg RNA was separated on a 1 % agarose gel and transferred and fixed to Qiabrane nylon membrane (Qiagen), which was hybridized with $^{32}$P-labelled cDNAs. An end-labelled 24-mer oligonucleotide probe hybridizing to 18S ribosomal RNA was used as a loading control and the values were corrected against the corresponding 18S band intensities.

## RESULTS

In D2 mice aminotriazole and thioacetamide increased hepatic COH activity 5.3- and 5.5-fold and in B6 mice 9.8- and 6.7-fold, respectively. Griseofulvin increased COH activity about 9.8-fold in D2 mice and 11.7-fold in B6 mice. The COH response was most pronounced 48 hr after thioacetamide (10 mg/kg i.p.) or griseofulvin (1 g/kg i.p.) treatment and 24 hr after aminotriazole(1 g/kg) treatment in both strains (Figure 1). Pentoxyresorufin O-depentylase (PROD) and ethoxyresorufin O-deethylase (EROD) activities were unchanged or decreased after aminotriazole, thioacetamide and griseofulvin injection, indicating isoform selectivity of induction.

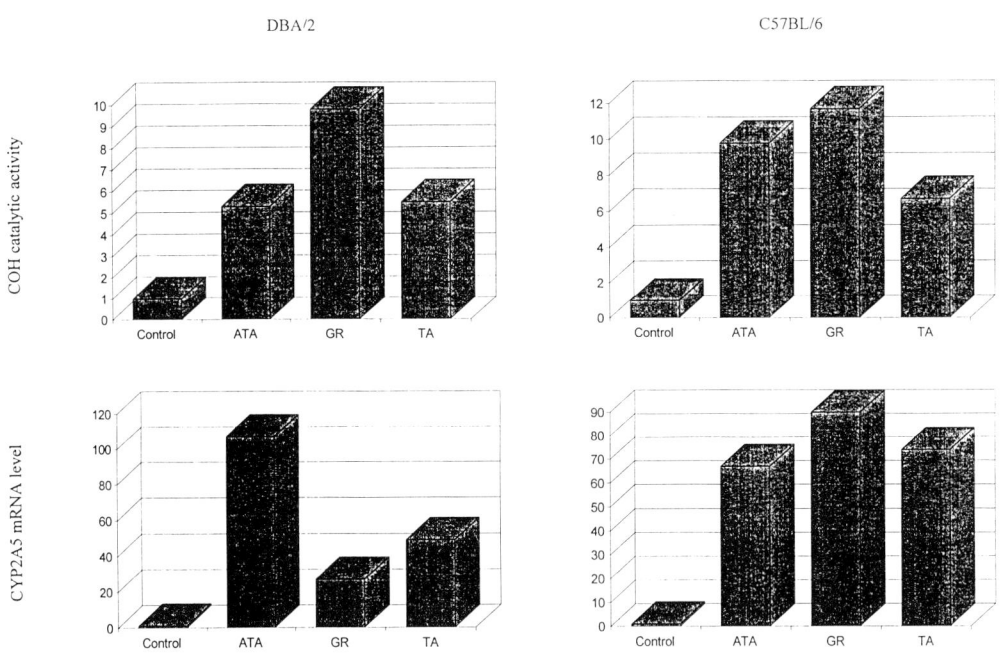

Figure 1. COH catalytic activity and CYP2A5 mRNA levels after aminotriazole, griseofulvin and thioacetamide treatment in mouse liver.

Aminotriazole, thioacetamide and griseofulvin increased CYP2A5 mRNA in D2 mice 107-, 49-, and 27-fold 24 hr after injection, respectively. In B6 mice CYP2A5 mRNA was increased 67-, 74- and 90-fold after aminotriazole (6 hr), thioacetamide (12 hr) and griseofulvin (72 hr) (Figure 1). All these time-points represent the maximal level of CYP2A5 mRNA induction.

The mRNA contents of aminolevulinate synthase, ferrochelatase and heme oxygenase were also increased to a variable extent by these agents. Aminotriazole elevateted aminolevulinate synthase mRNA 2.1-fold and 2.4-fold in D2 and B6 mice, respectively. The most pronounced effect of aminotriazole and thioacetamide was on ferrochelatase mRNA in B6 mice (10.0-fold and 4.9-fold increases). Griseofulvin increased the mRNA contents of aminolevulinate synthase and ferrochelatase about 1.6-fold and 3.3-fold in D2 mice, and 3.0-fold and 3.2-fold in B6 mice, respectively.

That the porphyrinogenic agents used in this study actually increased the contents of hepatic heme metabolism intermediates was verified by directly measuring them fluorimetrically.

Administration of heme arginate led to a decrease in COH activity in the D2 mouse liver (Table 1). Combined treatment with aminotriazole and thioacetamidethe and heme arginate caused a reduction of the COH activity to below the control level.

Table 1. Effects of heme arginate on COH induction in D2 mouse liver

| Treatment | COH activity | Fold induction |
|---|---|---|
| Control | 4.75* | 1.0 |
| Heme arginate | 3.99 | 0.8 |
| ATA | 24.98 | 5.3 |
| ATA + Heme arginate | 2.56 | 0.5 |
| TA | 16.25 | 3.4 |
| TA + Heme arginate | 1.51 | 0.3 |

The dose of heme arginate was 30 mg/kg. Aminotriazole (ATA) and thioacatamide (TA) were administered i.p. at the dose of 1 g/kg and 10 mg/kg, respectively.
*Mean of 4 mice

## DISCUSSION

The present data show that the porphyrinogenic agents aminotriazole, thioacetamide and griseofulvin cause a dramatic elevation in COH catalytic activity and the amount of hepatic CYP2A5 mRNA in D2 and B6 mice. Heme is known to regulate the levels of several hemeproteins and enzymes in the heme biosynthetic pathway. It is suspected to play a regulatory role in the expression of one or more cytochrome P450 genes (Waxman and Azaroff 1992).

The present results strongly suggest that heme plays a major role in the regulation of Cyp2a-5, since 1) COH catalytic activity and CYP2A5 mRNA levels are markedly elevated by agents disrupting the heme biosyhthetic pathway and 2) administration of heme in the form of heme arginate reduces both constitutive and especially induced COH activity.

Several cis elements and trans factors affecting them have been found in the 5' upstream regions of the chicken *CYP2H1* (Hahn et al 1991), rat *CYP2B1/2* (Upadhya et al. 1992) and Bacillus megaterium P450BM-3 (Shaw and Fulco 1992) genes. It is likely that similar mechanisms are operable also in the mouse *Cyp2a-5* gene.

# REFERENCES

Aitio, A. (1978): A simple and sensitive assay of 7-hydroxycoumarin deethylation. Anal. Biochem. 85, 488-491.

Bradford, M.M. (1976): A rapid and sensitive method for the quantitation of microgram quantities of protein utilizing the principle of protein-dye binding. Anal. Biochem. 72, 248-254.

Hahn, C.N., Hansen A.J. & May B.K. (1991): Transciptional regulation of the chicken CYP2H1 gene. J. Biol. Chem. 266, 17031-17039.

Hahnemann, B., Salonpää, P., Pasanen, M., Mäenpää, J., Honkakoski, P., Juvonen, R., Lang, M.A., Pelkonen, O. & Raunio, H. (1992): Effect of pyrazole, cobalt and phenobarbital on mouse liver cytochrome P-450 2a-4/5 (Cyp2a-4/5) expression.Biochem. J. 286, 289-294.

Shaw, G.-C. & Fulco, A.J. (1992) Barbiturate-mediated regulation of expression of the cytochrome $P450_{BM-3}$ gene of *Bacillus megaterium* by Bm3R1 protein. J. Biol. Chem. 267, 5515-5526.

Upadhya, P.,Venkateswara Rao, M., Venkateswar, V., Rangarajan, P.N. & Padmanaban G. (1992): Identification and functional characterization of cis-acting positive DNA element regulating CYP 2B1/B2 gene transcription in rat liver. Proc. Natl. Sci. USA 20, 557-562.

Waxman, D.J. & Azaroff, L. (1992): Phenobarbital induction of cytochrome P-450 gene expression. Biochem. J. 281, 577-592.

# Function of residue-365 in mouse P4502A5 (P450coh): site directed mutagenesis, spectral and enzymatic properties

R.O. Juvonen[1,2], M. Iwasaki[2], M. Negishi[2]

[1]Department of Pharmacology and Toxicology, University of Kuopio, PO Box 627, Kuopio, Finland. [2]Laboratory of Reproductive and Developmental Toxicology, National Institute of Environmental Health Sciences, PO Box 12233, Research Triangle Park, NC 27709, USA

## INTRODUCTION

Mouse P4502A5 (P450coh) specifically catalyzes 7-hydroxylation of coumarin. Mouse P4502A4 (P45015α) differing only in 11 amino acid residues from P4502A5 is a steroid 15α-hydroxylase exhibiting no coumarin 7-hydroxylase acitivity (Juvonen et al., 1988, Lindberg et al., 1989, Negishi et al., 1989). Our site-directed mutagenesis study have shown that three residues 117, 209 and 365, are the critical amino acids in these P450s determining the activity and substrate specificity of these P450s (Lindberg and Negishi, 1989).

Here we have concentrated on to study the role of residue 365 which has lozalized between the predicted helix K and ß-sheet 3 in alingment studies (Nelson and Strobel 1988, Gotoh 1992). Our own alignment matches residue-365 with leusine-294 in P450101A (Iwasaki et al. 1993). In Gotoh's alignment this region belongs to putative substrate recognizing region 5 corresponding area interacting with camphor in P450101A. By site-directed mutagenesis of P45017 amino acid residues in this region are identified critical for the catalysis (Kitamura et al. 1991). Also the decrease in bufuralol hydroxylation of the allelic variant 2D1v have been shown to due to the mutation in this region (Matsunaga et al. 1991). We mutated methionine 365 to aspartic acid, asparagine and lysine residue in P4502A5 to elucidate if the properties of absorption spectra and monooxygenase reaction are dependent on this residue.

## MATERIALS AND METHODS

We mutated P450coh cDNA using a pSELECTtm-1 vector (Promega, Madison, Wi, U.S.A.) and primers 5'- AGATTTGCAGACAAGATCCCCATGGGC, AGATTTGCAGACGA-CATCCCCATGGGC and AGATTTGCAGACAATATCCCCATGGGC for M365K, M365D and M365N, respectively, as previously described (Juvonen et al., 1991). The mutated cDNAs were ligated to pAAH5 vector and transfected to Saccharomyces cerevisiae AH22. The wild type and mutant P450cohs were purified from yeast microsomes, their spectral and enzymatic activities measured as described previously (Iwasaki et al.,1991, Juvonen et al. 1992).

## RESULTS

Spin equilibrium of all P450s was dependent on the salt concentration. At 10 mM potassium phosphate buffer pH 7.4 in room temperature wild type P4502A5 was 91 %, M365D 80 %

and M365N 69 % in high spin state but M365K was mostly in low spin state 80 % (Table 1.).

| TABLE 1. SPECTRAL NATURE OF MUTATED P4502A5. Absolute spectra of mutated and purified P4502A5 was recorded and the absorbance ratio of 390 nm and 417 nm calculated which indicates the spin equilibrium of the P450s. | | |
|---|---|---|
| P4502A5 | RESIDUE 365 | RATIO A390/A417 |
| wild type | Met | 1.48 |
| M365D | Asp | 1.05 |
| M365N | Asn | 0.99 |
| M365K | Lys | 0.84 |

The mutations in residue 365 changed monooxygenase activity of P4502A5 (Table 2.). M365N had two times higher coumarin 7-hydroxylase and three times higher testosterone 15$\alpha$-hydroxylase activitity than wild type P4502A5 while M365D had only one tenth and one fifth of these activities. In addition to reaction rate M365N also increased the specificity of testosterone hydroxylation because the ratio between metabolites of 15$\alpha$- and 7$\alpha$- hydroxy testosterone went up to 19 from the eight of the wild type.

| TABLE 2. MONOOXYGENASE ACITIVITIES OF MUTATED P4502A5S. Testosterone 15$\alpha$- and 7$\alpha$-hydroxylation and coumarin 7-hydroxylation (coh) activities were reconstituted using wild type and mutated P4502A5. Concentration of both substrates was 100 $\mu$M. | | | | | |
|---|---|---|---|---|---|
| P4502A5 | Residue 365 | COH | Testosterone 15$\alpha$ | 7$\alpha$ | RATIO 15$\alpha$/7$\alpha$ |
| | | nmol /min/nmol P450 | | | |
| wild | Met | 5.3 | 1.1 | 0.14 | 7.9 |
| M365D | Asp | 0.4 | 0.16 | 0.02 | 8 |
| M365N | Asn | 12.3 | 3.3 | 0.17 | 19.4 |
| M365K | Lys | 2.5 | 1.1 | 0.15 | 6.3 |

DISCUSSION

Our data indicate that the residue-365 of P4502A5 resides close to the putative 6th axial position of heme because the spin equilibrium of P4502A5 is dependent on the nature of this residue. This residue also constitutes a part of substrate binding pocket because the monooxygenase rate and specificity of testosterone hydroxylation are altered by changing the amino acid of the position. Based on the alignment of subfamily 2 with bacterial P450101A Gotoh (1992) has localized six putative substrate recognizing regions. Our data support the idea that residue 365 belongs to substrate recognizing sequence 5.

# REFERENCES

Gotoh, O. (1992): Substrate recognition sites in cytochrome P450 family 2 (Cyp2) proteins inferred from comparative analyses of amino acid and coding nucleotide sequences. J. Biol. Chem. 267, 83-89.

Iwasaki, M., Darden, T.A., Pedersen, L-G., Davis, D.G., Juvonen R.O., Sueyoshi T. and Negishi, M. (1993): Engineering mouse P450coh to a novel corticosterone 15α-hydroxylase and modeling steroid-binding orientation in the substrate pocket. J. Biol. Chem. 268,759-762.

Juvonen, R.O., Shkumatov, V.M. and Lang, M.A. (1988): Purification and characterization of a liver microsomal cytochrome P-450 isoenzyme with a high affinity and metabolic capacity for coumarin from pyrazole treated D2 mice. Eur. J. Biochem 171,205-211.

Kitamura, M., Buczko, E. and Dufau, M.L. (1991): Dissociation of hydroxylase and lyase activities by site-directed mutagenesis of the rat P45017α. Mol. Endocrinol. 5,1373-80.

Lindberg, R.L.P. and Negishi, M. (1989): Alteration of mouse cytochrome P450coh substrate specificity by mutation of a single amino acid residue. Nature 339,632-634.

Matsunaga, E., Zeugin, T., Zanger, U.M., Aoyama, T., Meyer, U.A. and Gonzalez, F.J. (1990): Sequence requirements for cytochrome P-450IID1 catalytic activity: a single amino acid change (Ile380Phe) specifically decreases Vmax of the enzyme for bufuralol but not debrisoquine hydroxylation. J. Biol. Chem. 265,17197-17201.

Negishi, M., Lindberg, R., Burkhart, B., Ichikawa, P., Honkakoski, P. and Lang, M. (1989): Mouse steroid 15α-hydroxylase gene family: Identification of type II P-45015α as coumarin 7-hydroxylase. Biochem. 28,4169-4172.

Nelson, D.R. and Strobel, H.W. (1988): Secondary structure prediction of 52 membrane-bound cytohromes P450 shows a strong structural similarity to P450cam. Biochem. 28:656-660.

# Looking for the gene encoding the fungal cytochrome P450 involved in 11β-hydroxylation of steroids: application of PCR techniques

Radovan Komel[1], Damjana Rozman[1], Marko Vitas[1], Katja Drobič, Steven L. Kelly[2]

[1]Institute of Biochemistry, Medical Faculty, Vrazov trg 2, 61000 Ljubljana, Slovenia. [2]Kebs Institute, The University of Sheffield, Western Bank, Sheffield S10 2TN, GB

## Summary

Genomic DNA from the filamentous fungus *Cochliobolus lunatus* m118 which is an important 11β-hydroxylator of steroids, was amplified using degenerate PCR primers which were designed according to sequence homology of cytochrome P450 amino acid sequences as found in different eucaryotic organisms. The PCR products generated were subjected to Southern blot analysis with those obtained after RT-PCR from mRNAs isolated from *C.lunatus* cells pooled after cultivation with/without steroid inducer. However there was some doubt about the inducibility of steroidal P450 hydroxylase in this case from CO difference spectra for microsomes from the fungus grown with or without inducer. On the other side there was a difference in the fragment pattern as obtained after RT-PCR of both mRNAs, and the DNA fragments produced are the subject of a DNA sequence determination.

## INTRODUCTION

*Cochliobolus lunatus* m118 is an important hydroxylator of steroids at the 11β position which is required for corticosteroid activity (Hoerhold et al., 1986). The microsomal cytochrome P450 monooxygenase system that catalyzes this reaction has recently been described by Jaenig et al. (1992), suggesting that microbial steroid hydroxylations might be catalyzed by a cytochrome P450 based reaction mechanism as occurs in higher eucaryotes. The system seems to be inducible. An identical conclusion was reached from the work of Hudnik Plevnik and Breskvar (1991) who succeeded in purifying and characterizing the cytochrome P450 involved in 11α-hydroxylation of steroids in Rhizopus nigricans. Nevertheless, the only fungal 11β-hydroxylase of steroids purified to homogeneity up to now was that from *Curvularia lunata* recently reported by Suzuki et al. (1993). However there are still no reports about identification of the relevant gene structure(s) in any of the fungal hydroxylators of steroids. The aim of the present work was, starting from the known data about amino acid sequence homology of cytochrome P450 as found in different higher eucaryotes, to use PCR methods in order to obtain some indications about the gene structure of the cytochrome P450 involved in 11β-hydroxylation of steroids in *C. lunatus*.

## MATERIALS AND METHODS

*Strains and growth conditions.* *C.lunatus* m118 was obtained from the strain collection of the Friedrich Schiller University, Jena FRG. Media, growth conditions and storage of strains have been described previously (Rozman and Komel, 1992).

*Isolation of DNA and preparation of cDNA.* DNA isolation followed the methods described previously (Rozman and Komel, 1991). RNA was prepared from 1 g of frozen powdered mycelium by the guanidinium thiocyanate method according to Promega

Protocols and Application Guide. Half of *C. lunatus* cells were incubated with progesterone (Dermastia et al., 1991) prior to RNA isolation procedure. cDNA ynthesis was performed from the total RNA by the *RiboClone* System (Promega) using 0.5 μg of degenerate downstream primer per μg of RNA.

PCR and RT-PCR. The 50μl reaction micture consisted of 500 ng of genomic DNA or cDNA, 1000 pmol of each degenerate primers, 0.2 mM dNTP solution, and 2.5 U of Taq polymerase (Promega) in 1 X reaction buffer with $MgCl_2$. Initial conditions for PCR amplification were as follows: denaturation 4 min at 94°C, 35 cycles of denaturation at 94°C 1 min, annealing 1 min at 46°C, extension 2 min at 72°C, final extension at 72°C 7 min.

Hybridization analysis. Fragments obtained after first PCR amplification were separated on 2 % low gelling temperature agarose, excised, purified with Magic PCR Prep DNA Purification System (Promega) and reamplified. A contact blot of purified DNA fragments in 2 % agarose gel onto Immobilon S (Millipore) membranes was performed. Hybridization analysis was performed under standard conditions at 42°C, using hybridization buffer with formamide and dextran sulfate. As a probe the 267 bp fragment, obtained after RT PCR, at concentration $1.9 \times 10^5$ cpm/ml was used.

Microsomal preparation. The methods used for microsomal preparations are essentially those of Ballard et al. (1990). The main difference is that the fungus was grown in 3 % glucose, 1 % CSL (Sigma) medium and that mycellial innoculation was used. 60 ml precultures were used which had been grown for 24h. Induction was by 0.1 mg/ml progesterone.

**RESULTS AND DISCUSSION**

In principle, structures of primers for amplification of DNA with polymerase chain reaction are determined by nucleotide sequences harboured within both ends of the DNA fragment to be amplified, which should represent a serious limit of the method being as such restricted only to well-known sequences (Ochman et al., 1988). This barrier can be overcome by using a pair of "universal" primers which should allow amplifications of a chosen DNA fragment starting from DNAs of organisms which are not closely related. In fact, universal primers proved to be very successful in investigations of DNA sequences of various organisms belonging to different classes and even to different stems (Kocher et al., 1989; Kocher and White, 1989; Erlich, 1989). As there is still no information about amino acid or DNA sequence for a related fungal cytochrome P450, we performed amplification of genomic DNA from the well-known 11β-hydroxylator of steroids, *C. lunatus* m118. Degenerate PCR primers were used which were designed according to sequence homology of cytochrome P450 amino acid sequences as found in different eucaryotic organisms. We have considered criteria for homology matching (Lewis and Moereels, 1992), especialy preservation of heme-binding and substrate binding regions. The amino acid sequences of rat, bovine and human enzymes were aligned as found in the work of Nonaka et al. (1989) who indicated the putative steroid binding site starting at the position 361 and the heme binding site at the position 441. Within these sequences LPLLRAA and RQCLGRR were found to be highly conserved and were chosen for construction of a related pair of "universal" PCR primers limiting a 267 bp cDNA to be amplified. In the genomic DNAs presented the encoding sequence is interrupted by two relatively small introns, but nothing was known about their possible existence within the related fungal DNA sequence. In addition, nothing is known about codon usage in *C.lunatus*. The only studied fungal codon usage is that of *Aspergillus nidulans* (Lloyd and Sharp, 1991) which presented a highly degenerate pattern of codons. Therefore for PCR amplifications we gave preference to utilize a mixture of 18-mer highly degenerate oligonucleotides for each primer. Non-stringent conditions, as successfully used in PCR amplification of AChE genomic sequence from rat (Komel and Grubič, 1993 a, b) and as described in Materials and Methods, were selected for amplification of *C.lunatus* genomic DNA in order to check the accuracy of the primers and to determine starting conditions for further RT-PCR from RNAs. As shown in Fig.1, PCR amplification of the genomic DNA, in several independent experiments, generated five main products which were submitted to further analysis.

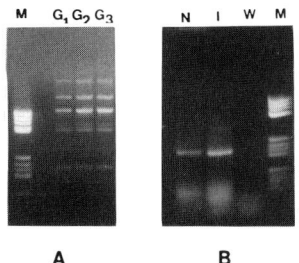

**Fig.1. A**: PCR amplification of *C.lunatus* genomic (G1-G3) DNA. **B**: RT-PCR of its RNA from induced (I) and non-induced (N) cells. M- pBR322/*Hae III* digest. W- water.

Although several authors (Žakelj-Mavrič et al. 1990; 1991; Jaenig et al., 1992) reported the inducibility of the steroid 11β-hydroxylase in *C.lunatus*, the comparison of CO difference spectra of microsomal preparations from cells grown in the presence of inducer or without inducer did not give a significant difference in P450 content in growth medium or in buffer. An example of a reduced CO-difference spectrum obtained with microsomes from untreated mycelium was 17.2+2 pmol/mg protein in comparison to 7+2 pmol/mg protein for microsomes from treated mycelia. For mycelia resuspended in buffer for 3 h the level of microsomal P450 was 19.6+1pmol/mg protein and on incubation with progesterone was 18.8+8 pmol/mg.

A typical Type I substrate binding spectrum could be obtained with all the microsomal fractions on addition of progesterone irrespective of pretreatment with progesterone. Figure 2B shows a typical spectrum, in this case obtained from microsomal samples from untreated mycelium, with maxima at 387 and 367 nm and a minimum at 412 nm. These results imply the presence of a constitutive cytochrome P450 capable of progesterone metabolism and calls into question the inducibility of steroid metabolising cytochrome P450, at least with these conditions and strain. Cycloheximide inhibition as reported in previous papers (Žakelj-Mavrič et al., 1990; 1991) might result from rapid protein turnover rather than induction.

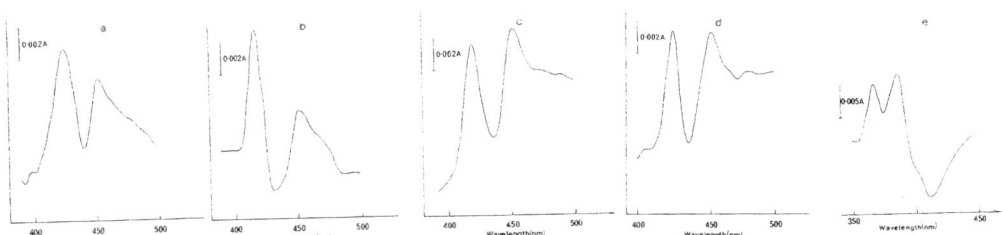

**Fig.2.** Reduced CO difference spectra of microsomal preparations from *C.lunatus*: uninduced microsomes after growth (17+2 pmol P450/mg microsomal protein) (**a**); induced in growth medium (7.3+2 pmol P450/ mg microsomal protein) (**b**); uninduced when incubated in buffer (0.75 mM Na3PO4, 0.2 mM EDTA, 0.04 mM reduced glutathione, pH 5.5) for 3h (19.6+1 pmol P450/mg microsomal protein) (**c**), induced in buffer (18.8+8 pmol P450/mg microsomal protein) (**d**); untreated microsomes type I substrate binding spectrum (**e**).

Nevertheless in vitro activity (Km, Vmax, etc.) should be examined before excluding induction of the activity from a constitutive background. As in the former reports the 11β-hydroxylases as well as the 11α-hydroxylases from various filamentous fungi have been characterized as steroid inducible enzymes (cf. Undisz et al., 1992), we have still performed RT-PCR amplifications of mRNAs isolated from both steroid treated and steroid untreated mycelium. The conditions were settled where the same two pronounced DNA fragments (240 and 267 bp) were generated in both cases and the third (380 bp) appeared only after amplification of mRNA from induced cells (Fig.1) The autoradiogram resulting from hybridization analysis where the 267 bp DNA fragment was used as hybridization probe is shown on Fig. 3. After stringent washing the 267 bp fragment hybridized strongly with all three DNA fragments generated by RT-PCR but not with those obtained after PCR of the genomic DNA. Therefore all three DNAs from the RT-PCR are the subject of DNA sequence determination which is in progress.

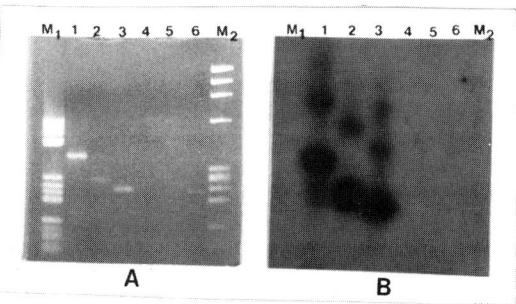

**Fig.3. A**: Isolated DNA fragments from RT-PCR of *C.lunatus* RNA (1- 380bp; 2- 267bp; 3- 240bp) and from PCR of its genomic DNA (4- 550bp; 5- 460bp; 6- 240bp). **B**: The corresponding Southern blot hybridized with the 267 bp fragment as the DNA probe. $M_1$- pBR322/*HaeIII*; $M_2$- ΦX174/*Hae III*.

## ACKNOWLEDGEMENT

This work was founded by the Ministry of Science and Technology of the Republic of Slovenia.

## REFERENCES

Ballard, S.A., Kelly, S.L., Ellis, S.W., and Troke, P.F. (1990): Interaction of microsomal cytochrome P450 from *Aspergillus fumigatus* with fluconazole and itraconazole. *J.Med.Vet.Mycol.* 28, 327-334.

Dermastia, M., Rozman, D. and Komel, R. (1991) Heterologous transformation of *Cochliobolus lunatus*. *FEMS Microbiol. Lett.*77, 145-150.

Erlich, H.A., ed. (1989): *PCR Technology. Principles and Application for DNA Amplification*. New York: Stockton Press.

Hoerhold, C., Undisz, K., Groh, H., Sahm, R., Schade, W., and Komel, R. (1986): Bioconversion of steroids by *Cochliobolus lunatus*. - I. Transformation of Reichstein's compound S with cell-free preparations of *C.lunatus*. *J.Basic Microbiol.* 26, 335-339.

Hudnik-Plevnik, T., and Breskvar, K. (1991): In *Frontiers in Biotransformation*, ed. K.Ruckpaul & H.Rein, pp.150-168: Academie Verlag Berlin.

Jaenig, G.R., Pfeil, D., Müller - Frohne, M., Reimer, M., Hennig M., Schwarze, W., Ruckpaul, K. (1992): Steroid 11β-hydroxylation by a fungal microsomal cytochrome P-450. *J.Steroid Biochem. Molec. Biol.* 43, 1117-1123.

Kocher, T.D., Thomas, W.K., Meyer, A., Edwards, S.V., Paabo, S., Villabanca, F.X., and Wilson, A.C. (1989): Dynamics of mitochondrial evolution in animals: amplification and sequencing with conserved primers. *Proc.Natl.Acad.Sci.USA* 86, 6196-6200.

Kocher, T.D., and White, T.J. (1989): Evolutionary analysis via PCR. In *PCR Technology: Principles and Application for DNA Amplification*, ed.H.A. Erlich, pp.317-326. New York: Stockton Press.

Komel, R., and Grubič, Z. (1993a): Synthesis of the DNA probe for the determination of rat AChE mRNA. *Chem. Biol. Interactions*, in press.

Komel, R., and Grubič, Z. (1993b): R.norvegicus gene for acetylcholinesterase. EMBL, Accession No.: X71089 RNACHE.

Lewis, D.F.V., and Moereels, H. (1992): The sequence homologies of cytochromes P-450 and active-site geometries. *J.Computer-Aided Mol.Design* 6, 235-252.

Lloyd, A.T., and Sharp, P.M. (1991): Codon usage in *Aspergillus nidulans*. *Mol.Gen.Genet.* 230, 288-294.

Nonaka, Y., Matsukawa, N., Morohashi, K., Omura, T., Ogihara, T., Teraoka, H., and Okamoto, M. (1989): Molecular cloning and sequence analysis of cDNA encoding rat adrenal cytochrome P-45011β. *FEBS Lett.* 255, 21-26.

Ochman, H., Gerber, A.S., and Hartl, D.L. (1988): Genetic application of an inverse polymerase chain reaction. *Genetics* 120, 621-623.

Rozman, D., and Komel, R. (1991): Altered expression of the steroid bioconverting pathway in pAN 7-1 transformants of *Cochliobolus lunatus*. *Curr. Genet.* 20, 385-389.

Rozman, D., and Komel, R. (1992): Transformation of *Cochliobolus lunatus* with pUT 720 changes the steroid hydroxylating ability of the fungus. *Curr. Genet.* 22, 123-127.

Suzuki, K., Sanga, K., Chikaoka, Y., and Itagaki, E. (1993): Cytochrome P-450lun of *Curvularia lunata* catalyzing steroid 11β-hydroxylation: purification and characterization. At 2nd International Symposium on Cytochrome P450 of Microorganisms and Plants, Tokyo.

Undisz, K., Groh, H., Stopsack, H., and Hoerhold-Schubert, C. (1992): Bioconversion of steroids by *Cochliobolus lunatus*. - II. 11β-hydroxylation of 17a,21-dihydroxypregna-1,4-diene-3,20-dione 17-acetate in dependence of the inducer structure. *J.Steroid Biochem. Molec.Biol.* 43, 543-547.

Žakelj-Mavrič, M., and Belič, I. (1991): Induction of steroidal 11β-hydroxylase of *Cochliobolus lunatus*. *J.Steroid Biochem.Molec.Biol.* 38, 117-118.

Žakelj-Mavrič, M., Plemenitaš, A., Komel, R., and Belič, I. (1990): 11b-hydroxylation of steroids by Cochliobolus lunatus. *J.Steroid Biochem.* 35, 627-629.

# Interaction of verapamil and cimetidine in rat liver microsomes

Rainer Wacke, Ulrike Fischer, Jens Ahrend, Bernd Drewelow

*Institute of Pharmacology and Toxicology, University of Rostock, Schillingallee 70, D-18057 Rostock, Federal Republic of Germany*

## INTRODUCTION

Previous *in vivo* investigations of the interaction between the calcium channel blocker verapamil - a high clearance drug undergoing extensive metabolism by cytochrome P450 - and the $H_2$-antagonist cimetidine and its clinical relevance yielded contradictory results. Several authors reported an increase in the bioavailability of verapamil (Smith et al. 1984, Mikus et al 1991) or a decrease in the clearance (Loi et al. 1985) when cimetidine was given simultaneously in therapeutic doses. Other authors did not find such interactions (Abernethy et al. 1985, Wing et al. 1985).
Attempts to explain the extent of inhibition of drug metabolism by cimetidine during therapeutic use in humans using $K_i$- and $IC_{50}$-values from *in vitro* experiments were unsatisfactory. These values are at least tentimes higher ($K_i$: 166 - 312µM, $IC_{50}$: 78 - 435µM, unpublished data from our lab with verapamil and cimetidine) than therapeutic cimetidine concentrations in human serum and liver (10µM and 20µM, respectively).
The aim of our study was to find out whether the intrinsic clearance is more suitable for describing the *in vivo* drug interaction by means of *in vitro* data.
Verapamil and cimetidine seem to be suitable as model substances because the metabolism of verapamil is well known (Eichelbaum et al. 1978, Kroemer et al. 1991, Kroemer et al. 1993) and data regarding the cimetidine-induced inhibition of drug metabolism are plentiful.

## MATERIALS AND METHODS

Microsomes were prepared from 8 rat livers as described by Meier et al. (1983). The protein content of individual preparations was estimated according to Lowry et al. (1951). Incubation of verapamil in the presence of increasing concentrations of cimetidine (5, 10, 25, 50 and 100µM) in a NADPH-regenerated system (Kronbach et al. 1987) was carried out at 37°C in a final volume of 250µl containing 200µg microsomal protein. The reaction was stopped after 15 min by adding perchloric acid (Kroemer et al. 1992).

The incubation mixture was extracted by diethylether (Haparat & Kates 1979), and HPLC assay was carried out according to a modified method described by Kroemer et al. (1992).
The formation of the verapamil metabolites D617 (N-dealkylation), D702 (O-demethylation), D703 (O-demethylation) and norverapamil (N-demethylation) was quantified with and without cimetidine. $V_{max}$ and $K_m$ was used to determine the intrinsic clearance ($Cl_{int} = V_{max}/K_m$). All data are presented as means $\pm$ S.D. Statistical significant differences were evaluated by paired t-tests.
The Dixon and Hill-plots were used to determine $K_i$- and $IC_{50}$ values of cimetidine-induced inhibition of verapamil metabolism.

## RESULTS

Kinetic data for the formation of verapamil metabolites are listed in Table 1. The order of $Cl_{int}$ was D702 < D703 < D617 < norverapamil.

Table 1: Kinetic data from incubation of verapamil and cimetidine in the presence of the microsomal fraction of 8 rat livers

| | Cimetidine | | | | | |
|---|---|---|---|---|---|---|
| | 0 µM | 5 µM | 10 µM | 25 µM | 50 µM | 100 µM |
| **Norverapamil** | | | | | | |
| $K_m$ (µM) | 58.6 ± 34.5 | 69.7 ± 43.0 | 62.9 ± 42.1 | 112.8 ± 118.3 | 69.0 ± 40.4 | 99.1 ± 60.0 |
| $V_{max}$ (pmol/mg/min) | 1496 ± 1090 | 1513 ± 1203 | 1353 ± 1053 | 2108 ± 1945 | 1347 ± 1346 | 1291 ± 1100 |
| $Cl_{int}$ | 23.7 ± 18.5 | 20.9 ± 13.4 | 18.9 ± 12.6 | 19.3 ± 16.0 | 13.7 ± 10.5 | 14.3 ± 12.9 |
| **D-702** | | | | | | |
| $K_m$ (µM) | 69.4 ± 43.3 | 75.4 ± 37.9 | 66.2 ± 35.6 | 88.1 ± 70.6 | 88.0 ± 94.7 | 100.5 ± 59.8 |
| $V_{max}$ (pmol/mg/min) | 308 ± 187 | 344 ± 252 | 258 ± 169 | 334 ± 289 | 241 ± 153 | 175 ± 86 |
| $Cl_{int}$ | 5.01 ± 2.09 | 4.31 ± 1.48 | 4.14 ± 1.63 | 4.75 ± 2.94 | 3.83 ± 2.63 | 1.99 ± 0.80* |
| **D-703** | | | | | | |
| $K_m$ (µM) | 49.2 ± 51.1 | 85.2 ± 66.0 | 73.1 ± 62.4 | 108.7 ± 116.3 | 90.6 ± 71.0 | 235.6 ± 261.5 |
| $V_{max}$ (pmol/mg/min) | 289 ± 178 | 323 ± 228 | 257 ± 204 | 320 ± 260 | 276 ± 182 | 290 ± 216 |
| $Cl_{int}$ | 5.40 ± 2.43 | 5.04 ± 2.42 | 4.76 ± 2.01 | 3.79 ± 1.47 | 3.85 ± 1.51 | 2.85 ± 2.17* |
| **D-617** | | | | | | |
| $K_m$ (µM) | 73.0 ± 44.9 | 94.9 ± 51.0 | 83.6 ± 59.7 | 128.1 ± 106.0 | 137.8 ± 109.3 | 117.0 ± 86.4 |
| $V_{max}$ (pmol/mg/min) | 846 ± 611 | 996 ± 766 | 796 ± 638 | 929 ± 807 | 871 ± 741 | 784 ± 672 |
| $Cl_{int}$ | 10.58 ± 6.76 | 9.87 ± 5.93 | 9.50 ± 6.38 | 6.94 ± 4.38 | 5.36 ± 3.13* | 6.31 ± 4.66 |

*p < 0,05 Incubation without cimetidine vs incubation with cimetidine

We observed a cimetidine concentration dependent inhibition of the verapamil metabolism. The decrease in $Cl_{int}$ amounted to 20 - 25% for 25µM cimetidine and 45 - 55% for 100µM, respectively (Fig. 1 - 4; the $Cl_{int}$ is expressed as per cent of the corresponding control incubation without cimetidine).

## DISCUSSION

The kinetic *in vitro* data ($V_{max}$, $K_m$ and $Cl_{int}$ /Tab. 1) from rat liver microsomes are in close agreement with *in vitro* data from human liver microsomes reported by Kroemer et al. (1992). The kinetic data they analyzed closely resemble those in Table 1 as far as the magnitudes of the relative contributions of the various metabolic pathways are concerned, but the order of D617 and norverapamil is reversed (D617 > norverapamil). This example demonstrates that it is possible to obtain comparable *in vitro* data from various species, including humans. The predictability of *in vivo* drug metabolism from *in vitro* data has been shown recently for verapamil (Kroemer et al. 1992). Similar studies on the interaction potential of drugs that interfere with cytochrome P450-mediated metabolism have recently been published (Chiu, 1993, Birkett et al. 1993).

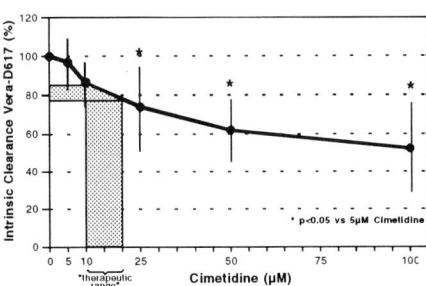

Fig. 1
Inhibition of N-Dealkylation of Verapamil (D617 formation) by Cimetidine

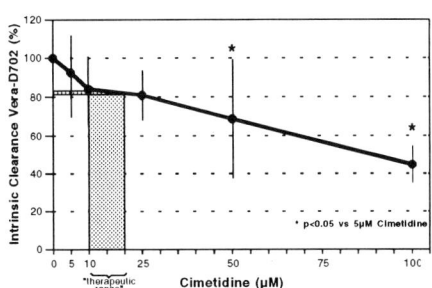

Fig.2
Inhibition of O-Demethylation of Verapamil (D702 formation) by Cimetidine

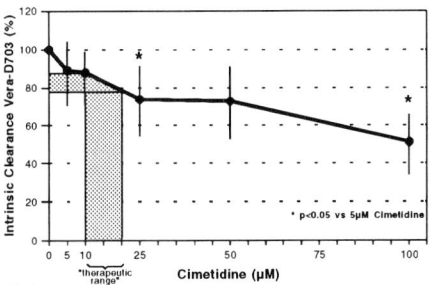

Fig.3
Inhibition of O-Demethylation of Verapamil (D703 formation) by Cimetidine

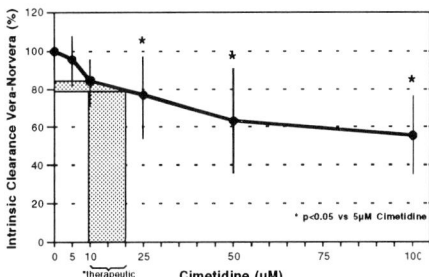

Fig.4
Inhibition of N-Demethylation of Verapamil (Norverapamil formation) by Cimetidine

At 25$\mu$M cimetidine, which corresponds to therapeutic concentrations in human liver (20$\mu$M), we registered an inhibition of the verapamil metabolism expressed as a reduction in $Cl_{int}$ by 10 - 35% (Fig. 1-4). This is in accordance with *in vivo* observations of pharmacokinetics in humans (Smith et al. 1984, Loi et al. 1985, Mikus et al 1991). Similar data have been published for cimetidine interactions with other drugs metabolized by cytochrome P450. For instance, cimetidine lowers the diazepam clearance by 45% (Pasanen et al. 1986). Schellens et al. (1989) reported a decreased clearance for nifedipine (32%) and sparteine (55%) when cimetidine was coadministered in therapeutic doses.

This approach which takes into account **the intrinsic clearance model** seems to be suitable for in vitro investigations of drug interactions at the level of the cytochrome P450 mediated metabolism.

# REFERENCES

Abernethy, D. R., Schwartz, J. B., Todd, E. L. (1985): Lack of interation between verapamil and cimetidine. *Clin. Pharmacol. Ther.* 38, 342-349

Birkett, D. J., Mackenzie, P. I., Veronese, M. E., Miners, J. O. (1993): In vitro approaches can predict human drug metabolism. *TiPS* 14, 292-294

Chiu, S. H. L. (1993): The use of in vitro metabolism studies in the understanding of new drugs. *J. Pharmacol. Toxicol. Meth.* 29, 77-83

Eichelbaum, M., Ende, M., Remberg, G., Schomerus, M., Dengler, H. J. (1978): The metabolism of DL-(C14)-verapamil in man. *Drug Metab. Dispos.* 7, 145-148

Haparat, S. A., Kates, R. E. (1979): Rapid high-pressure liquid chromatographic analysis of verapamil in blood and plasma. *J. Chromatograph.* 170, 385-390

Kroemer, H. K., Beaune, P., Henderson, C. J., Wolf, C. R., Heidemann, H. (1991): Identification of cytochrome P-450 isoenzymes involved in the metabolism of verapamil. *Naunyn-Schmiedebergs Arch. Pharmacol.* 343 (Suppl), R124

Kroemer, H. K., Echizen, H., Heidemann, H., Eichelbaum, M. (1992): Predictability of the in vivo metabolism of verapamil from in vitro data: Contribution of individual pathways and stereoselectiv aspects. *J. Pharmacol. Exp. Ther.* 260, 1052-1057

Kroemer, H. K., Gautier, J. C., Beaune, P., Henderson, C. J., Wolf, C. R., Eichelbaum, M. (1993): Identification of P450 enzymes involved in metabolism of verapamil. *Naunyn-Schmiedebergs Arch. Pharmacol. in press*

Kronbach, T., Mathys, D., Gut, J., Catin, T., Meyer, U. A. (1987): High performance liquid chromatography assays for bufuralol 1'hydroxylase, debrisoquine 4-hydroxylase, and dextromethorphane O-demethylase in microsomes and purified cytochrome P-450 isozymes of human liver. *Anal. Biochem.* 162, 24-32

Loi, C. M., Rollins, D. E., Dukes, G. E., Peat, M. A. (1985): Effect of cimetidine on verapamil disposition. *Clin. Pharmacol. Ther.* 37, 654-657

Lowry, O. H., Rosebrough, N. J., Farr, A. L., Randall, R. J. (1951): Protein measurement with the folin phenol reagent. *J. Biol. Chem.* 193, 265-275

Meier, P. J., Müller, H. K., Dick, B., Meyer, U. A. (1983): Hepatic monooxygenases activities in subjects with genetic defect in drug oxidation. *Gastroenterology* 85, 682-692

Mikus, G., Eichelbaum, M., Fischer, C., Gumulka, S., Klotz, U., Kroemer, H. K. (1990): Interaction of verapamil and cimetidine: Stereochemical aspects of drug metabolism, drug disposition and drug action. *J. Pharmacol. Exp. Ther.* 253, 1042-1048

Pasanen, M., Arvela, P., Pelkonen, O., Sotaniemi, E., Klotz, U. (1986): Effect of five structurally diverse H2-receptor antagonists on drug metabolis. *Biochem. Pharmacol.* 35, 4457-4461

Schellens, H. J. M., van der Wart, J. H F., Brugman, M., Breimer, D. D. (1989): Influence of enzyme induction and inhibition on the oxidation of nifedipine, sparteine, mephenytoin and antipyrine in humans as assessed by a "cocktail" study design. *J. Pharmacol. Exp. Ther.* 249, 638-645

Smith, M. S., Benyunes, M. C., Bjornsson, T. D., Shand, D. G., Pritchett, E. L. C. (1984): Influence of cimetidine on verapamil kinetics and dynamics. *Clin. Pharmacol. Ther.* 36, 551-554

Wing, L. M., Miners, J. O., Lillywhite, K. J. (1985): Verapamil disposition - effects of sulphinpyrazone and cimetidine. *Br. J. Clin. Pharmacol.* 19, 385-391

# Stopped flow kinetics under high pressure: an approach to explore the transition state of haemoproteins in elementary reactions

Reinhard Lange, Irmgard Heiber-Langer, Nicole Bec, Claude Balny

INSERM U.128, BP 5051, route de Mende, 34033 Montpellier, France

## Summary

We have characterized the transition state occurring in the course of the binding of CO to ferrous P450 and chloroperoxidase in comparison to other haemoproteins. Our approach was a stopped flow kinetic analysis under high pressure (up to 2 kbar). The activation volume, $\Delta V^{\ddagger}$, appeared to depend on the nature of the fifth axial haem ligand: when this ligand was a thiolate, as in the case of cytochromes P450 scc (CYP11A1), LM2 (CYP2B6), LM3c (CYP3A6), Cyp2a (testosterone 7α-hydroxylase) or of chloroperoxidase, the pressure dependence of the bimolecular binding rate was very small and positive: $\Delta V^{\ddagger}$ = +1 to +6 ml/mole. When, on the other hand, the sulphur ligand bond was disturbed or broken, as in P420, or when it was replaced by a nitrogen from histidine, as in the case of lactoperoxidase, horse radish peroxidase or cytochrome P460, then $\Delta V^{\ddagger}$ became strongly negative (-10 to -40 ml/mole). Our results suggest that the transition state of ferrous thiolate ligand haemoproteins has a similar molecular conformation as the enzyme at rest.

The complexity of P450 catalysed reactions, such as oxygen activation and peculiar interaction with substrates constitutes a challenge for both structural and dynamic investigations. Now, whereas we have some structural insight from methods such as crystallography and molecular cloning, the question of the reaction mechanism, that is of how the protein structure behaves in the course of the reaction, still remains. However, most studies are involved only in either structural features - as the determination of amino acid sequence and of three-dimensional protein structure - or in the temporal resolution of the reaction. Traditionally, the transition state of a reaction is assessed by a temperature dependent kinetic analysis, which yields the activation enthalpy, $\Delta H^{\ddagger}$, and entropy, $\Delta S^{\ddagger}$, through the Arrhenius relation.

It is interesting, however, to connect this energetic information to structural features. Our approach is high pressure stopped-flow operating at different temperatures, a technique which we developed some years ago, and have applied since to the study of several enzyme systems (Balny et al., 1984, Lange et al., 1993). This method allows one to determine the activation volume of a reaction (that is, the volume change of the transition state with respect to the ground state) from the dependence of the rate constant, k, on pressure (up to 200 MPa) according to

$$\delta \ln k / \delta p = \Delta V^{\ddagger} / RT \qquad (1).$$

The magnitude of the activation volume can result from conformational changes in major or restricted regions of the protein. It can also reflect a peculiar interaction of the transition state with solvent molecules. Here we applied this method to the binding reaction of a small, neutral ligand, carbon monoxide, and compared the results with those obtained with other haemoproteins.

## Materials and Methods

The various P450 samples: CYP11A1(scc), CYP2B4(LM2) and CYP3A6(LM3c) were isolated according to published procedures; Cyp2a(testosterone 7α-hydroxylase) was a kind gift from Dr. M. Negishi, NIEHS. P420 was obtained from P450scc after treatment with butanol. The other haemoproteins were commercial. Kinetic measurements were performed in a home built thermostated stopped-flow device which operated in the 0.1 to 300 MPa pressure range. For the CO-binding reaction, the haemoproteins were reduced by 2-5 mM dithionite under argon and mixed with an equal volume of a CO containing solution in the stopped-flow apparatus which was interfaced to an Aminco DW2 spectrophotometer.

## Results and Discussion

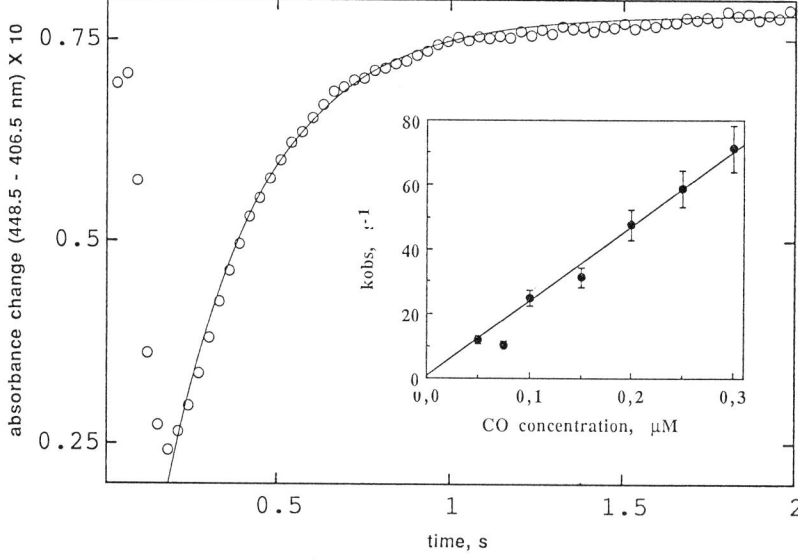

Fig. 1 Time course of the CO - binding to P450scc after stopped-flow mixing at high pressure; [CO] = 5 μM; t = 5 °C, p = 130 MPa; the monoexponential fit is shown by the solid line through the data points. Insert: CO dependence of the rate constant.

An example of CO binding under high pressure (130 MPa) is given in Fig. 1. The kinetics were well fitted by a monoexponential time course. For all haemoproteins we observed a linear relationship between $k_{obs}$ and the CO concentration. The kinetics were therefore interpreted in terms of a one-step binding model, where $k_{obs} = k_1[CO] + k_{-1}$. Since the back rate $k_{-1}$ was always very small, $k_{obs}$ reflected essentially the binding rate $k_1$. As shown in Fig. 2, the pressure effect on the binding

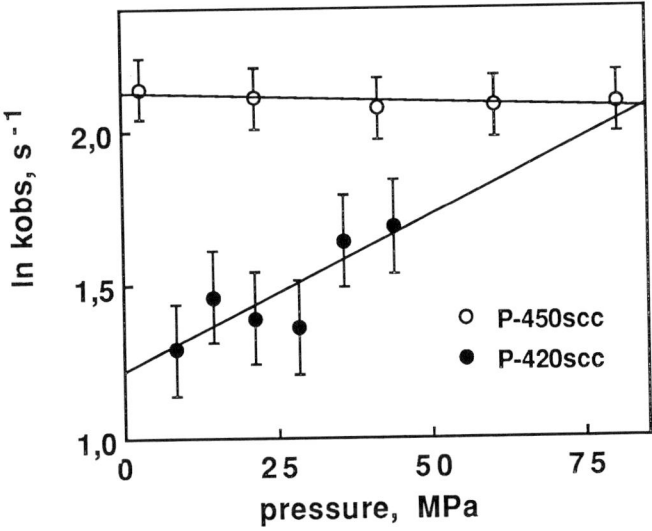

Fig. 2 Pressure dependence of the CO binding rate

was very different for the various enzymes. Interestingly, for cytochrome P-450scc the slope of $\ln k_{obs} = f(p)$ was opposite to that obtained with P-420scc. More generally,

| Enzyme | fifth axial haem ligand | | $\Delta V^{\ddagger}$ | Reference |
|---|---|---|---|---|
| | | | ml mol$^{-1}$ | |
| P-450scc | S$^-$ (cys) | | 2 ± 2 | Lange et al. (1993) |
| P-450 LM2 | S$^-$ (cys) | | 3 ± 2 | id. |
| P-450 LM3 | S$^-$ (cys) | | 6 ± 5 | id. |
| P-450 7α | S$^-$ (cys) | | 2 ± 2 | id. |
| chloroperoxidase | S$^-$ (cys) | | 1 ± 2 | id. |
| lactoperoxidase | N (his) | | - 10 ± 3 | id. |
| horse-radish-peroxidase | N (his) | | - 24 ± 4 | Balny and Travers (1989) |
| P-460 | N (his) | | - 36 ± 6 | Balny and Hooper (1988) |
| P-420 | unknown | (a) | - 25 ± 5 | Lange et al. (1993) |
| | | (b) | - 11 ± 6 | id. |
| | | (c) | - 42 ± 8 | id. |

TABLE 1 : Activation volume of CO binding to haemoproteins; (a) to (c): phase 1 to 3; P460, hydroxylamine oxidoreductase from *Nitrosomonas Europaea*.

haemoproteins such as P450 or chloroperoxidase which have a thiolate sulphur as the fifth iron ligand, were characterized by very small, positive activation volumes. On the other hand, the volume of the transition state of haemoproteins with an axial histidine nitrogen ligand, was strongly negative (see Table 1). Furthermore, the activation volume of the N-class enzymes was strongly dependent on the composition of the solvent (Balny and Hooper, 1988), whereas the activation volume of the S-class enzymes was independent on the solvent.

These results suggest that for enzymes with an axial $S^-$ ligand, the transition state of the CO binding reaction has a similar structure as the ferrous ground state. For this reactional state, the activation may be due to a very particular electronic distribution at the iron center, rather than to protein conformational changes, or to interaction with the solvent. Apparently, the volume of this transition state is not correlated to the conformational fluctuations which were found for the ferric state (Lange et al., 1992, 1992a). Support for this hypothesis comes from the P420 data: when the iron-sulphur bond was broken or disturbed, the enzyme activation was characteristic of a haemoprotein of the N-class (the activation volume was strongly negative).

Interestingly, these features of the transition state were revealed only by an analysis of the pressure dependence of the CO binding rate constant. The determination of the activation enthalpy and activation entropy alone was not informative for the transition state of this reaction. If we want to begin to understand what happens in the crucial moment of a reaction, the approach by high pressure stopped-flow appears therefore promising. The necessary technology being now available, one may speculate to see in the near future further investigations of P450 elementary reactions under high pressure.


## References

Balny, C., Saldana, J.L., and Dahan, N. (1984): High pressure stopped flow spectrometry at low temperature. *Anal. Biochem.* 139: 178 - 189.

Balny, C., and Hooper, A.B. (1988): Effect of solvent, pressure and temperature on the reaction rates of the multiheme hydroxylamine oxidoreductase: evidence for conformational change. *Eur. J. Biochem.* 176: 273-279.

Balny, C. and Travers, F. (1989): Activation thermodynamics of the binding of carbon monoxide to horseradish peroxidase. Role of pressure, temperature and solvent. *Biophys. Chem.* 33: 237 - 244.

Lange, R., Larroque, C., and Anzenbacher, P. (1992): The cholesterol-side-chain-cleaving cytochrome P450 spin-state equilibrium. 1. Thermodynamic analysis. *Eur. J. Biochem.* 207: 69-73.

Lange, R., Pantaloni, A., and Saldana, J.L. (1992a): The cholesterol-side-chain-cleaving cytochrome P450 spin-state equilibrium. 2. Conformational analysis. *Eur. J. Biochem.* 207: 75-79.

Lange, R., Heiber-Langer, I., Bonfils, C., Fabre, I., Negishi, M., and Balny, C. (1993): Activation volume and energetic properties of the binding of CO to hemoproteins. *Biophys. J.*, in press.

# A three-dimensional model of P450arom using P450BM-P as a template

S. Graham-Lorence[1], J.A. Peterson[1], E.R. Simpson[2]

The [1]Department of Biochemistry and The [2]Cecil and Ida Green Center for Reproductive Biological Sciences, University of Texas, Southwestern Medical Center, Dallas, Texas, USA

## INTRODUCTION

The enzyme system which converts androgens to estrogens is know as aromatase which comprises the ubiquitous NADPH reductase and cytochrome P450AROM.[1] It is found predominantly in the granulosa cells of the ovary, in adipose tissue, and in placenta, and is believed to play a role in development of breast cancer.[2,3] Because of the need for enzyme-specific inhibitors[4,5], we are interested in the structure/function relationship of P450AROM in order to have a better understanding of the substrate recognition regions, the access channel, active site/heme pocket of P450AROM, and the NADPH reductase docking region. However, no eukaryotic P450 structure has been determined to date because of the difficulty of crystallizing membrane-bound proteins. Therefore, we have modeled P450AROM using P450BM-P as the class II P450 prototype[6].

In the past, several groups have modeled microsomal P450s using P450cam[7] as a template since it is a soluble P450 and therefore able to be crystalized; however, P450cam is a class I P450 requiring both an iron-sulfur protein and an NADH reductase. On the other hand, the soluble protein P450BM-P, the P450 domain of the fusion enzyme P450BM3, is a class II P450 requiring an FAD/FMN-containing NADPH reductase similar to microsomal P450s, e.g. P450AROM, and thus, *a priori*, P450BM-P would be a more suitable prototype for class II P450s.

## METHODS

### Sequence Alignments

Initially, to do this modeling of P450AROM a sequence alignment was generated. This was accomplished by first obtaining a computer generated alignment of over 100 P450s using the University of Wisconsin GCG "Pileup" program which first clusters proteins using a pairwise alignment to determine the proteins with highest sequence similarity, and then using the order generated in the sequence alignment, aligns the proteins using the Needlman-Wunch algorithm. Following the computer-generated sequence alignment, P450cam and P450BM-P were aligned according to their three-dimensional structure and P450AROM was hand aligned with them using the computer generated alignment as a starting point as show in Figure 1. From the alignment of P450arom with P450BM-P we were able to obtain 18% identity and 30% similarity. The alignment with P450cam was not as fruitful. We obtained 15% identity and 22% similarity.

### Computer Modeling

Modeling of P450AROM was conducted using InsightII from Biosym on a Silicon Graphics workstation. The model of P450arom was generally not built by molecular replacement of P450BM-P; rather it was constructed by building a structure. The strategy was to build the most structurally conserved regions first, and then, "fit in" the less structurally conserved regions in a complementary fashion. The amino-terminal 68 residues and the carboxy-terminal 10 residues were not included in this model of P450AROM.

```
arom  MVLEMLNPIHYNITSIVPEAMPAATMPVLLLTGLFLLVWNYEGTSSIPGPGYCMGIG...PLISHGRFLWM... GIGSACNYYNRVYG..... EFMRVWIS..GEETLIISK  99
BM3   ...........................................TIKEMPQP.KTFGELKNLPLLTDK..........  PVQALMKIADELG.. EIFKFEA...PGRVTRYLS.  53
cam   ..................................TTETIQSNANLAPLPPHV.PEHLVFDFDMYNPSNLSA...... GVQEAWAVLQ ESNVPDLVWTRCN...GGH.WIAT.  66
terp  ..................................DARATIPEHIARTVILPQGYAD..... DEVIPYPAFKWLRD.... EQPLAMAHIEGYDPMWIATK  55
                                                                 A' HELIX            A Helix          β1-1        β1-2

arom  SSSMFHIMKH NHYSSRFGSKL....... GLQCIGMH .......... EKGIIFNNN.... PELWKTTR ..... PFFMKALSG PGLVRMVTVCAESLKTHLDR... 174
BM3   SQRLIKEAC....DESRFDKNL....... SQALKFVRDFAG........DGLFTSWTH... EKNWKKAHNILL..PSFSQQAMK. GYHAMMVDIAVQLVQKWER... 132
cam   RGQLIREAYED....YRHFSS.ECPFI..... PREAGEAY ........DFIPTSMDH. PEQRQFRALANQVVGMPVVDK...... LENRIQELACSL.IESLRPQ. 145
terp  HADVMQIG.....KQPGLFSNAEGSEILYD. QNNEAFMRSIS.GGCPHVIDSLTSMDPP. THTAYRGLTLNWFQ.PASIRKL..... EENIRRIAQASVQRLL.... 144
      B Helix           β1-5            B' Helix              C Helix                  D Helix

arom  .....LEEV. TNESGYVDVLTLLRRVML ...DTSNTLFLRIPL DESAIVVKIQ.GYFDAWQAL..LIKPDIFFKISW...... LYKKYEKSVKDLKDAIEVLIAEKRC. 264
BM3   LNADEHIEV. PEDMTR.LTLDTIGLCGF ..NYRFNSFYRDQPH PFITSMVRALDEAMNKLQ....RANPDDP....... AYDENKRQFQEDIKVMNDLVDKIIADRKAS 226
cam   ....GQCNF.. TEDYAEPFPIRIFMLLAGL. PEE.......... DIPHLKYLTDQMT........RPDGSM.......... TFAEA.KEALYDYLIPIIEQRRQK 214
terp  .DFDGECD.. FMTDCALYYPLHVVMTAL. ..GVPED........ DEPLMLKLT.......QDFFGVHEPDEQAVAAPRQSADE AARRHE.TIATFYDYFNGFTVDRRS. 229
        β3-1      E Helix                          F Helix                                   G Helix
                                                                *                                              *   * *

arom  RISTEEKLEECMD ...FATELILA .EKRGDL..... TRENVNQCILEMLIAAPDTMSVSLFFMLFLIAK HPNVEEAIIKEIQTVI GERD. IKIDDIQK LKVMENFIYESMRYQ 367
BM3   GEQSDD....... ...LLTHML .NGKDPETGEPL. DDENIRYQIITFLIAGHETTSGLLSFALYFLVK NPHVLQKAAEEAARVL VDPVP. SYKQVKQ LKYVGMVLNEALRLW 325
cam   .PGT........ ...DAISIVAN .GQVNGRPI.... TSDEAKRMCGLLLVGGLDTVVNFLSFSMEFLAKS PEHRQELIE ........RPE............ RIPAACEELLRRF 292
terp  .CPKDD...... ...VMSLLA .NSKLDGNYID... DKYINAYYVAIATAGHDTTSSSSGGAIIGLSR NPEQLALAKS .......DPAL........... IPRLVDEAVRWT 307
         H Helix β5-1   β5-2              I Helix                       J Helix          J' Helix           K Helix
                                                                                           *

arom  .PVVDLVMRKALEDDVIDG. YPVKKGTNIILN IGRMH..RLEFFPKP.NEFTL.ENFAKN.......VPYRYFQPFGFGPRGCA GKYIAMVMMKAILVTLLR... RFHVKTLQGQCV 469
BM3   .PTAPAFSLYAKEDTVLGGEYPLEKGDELMVL IPQLH.RDKTIWGDDVEEFR.PERFEN.PSAI....PQHAFKPFGNGGRACI GQQFALHEATLVLGMMLK... HFDFEDHTNYEL 431
cam   .SLVADGRILTSDYEFHG. .VQLKKGDQILLPQ...MLSGL............DERENACPMHVDFSRQKVSHTTFGHGSHLCL GQHLARREIIVTLKEWLTRI. PDFSIA.PGAQI 389
terp  .APVKSFMRTALADTEVRG.QNIKRGDRIMLS YPSAN............RDEEVFSNPDEFDITRFPNRHLGFGWGAHMCL GQHLAKLEMKIFFEEL..... LPKLKSVELSGP 405
       β1-4       β2-1   β2-2   β1-3 K' Helix                          | Heme-Binding|    L Helix              β3-3

arom  ESIQKIH...DLSLHPDETK NMLEMIF. TPRNSDRCLEH         503
BM3   DIKETLTL..KPEG...... FVVKAKSKKIPLGGIPSPSTEQSAKKVR 471
cam   .QHKSGIVS.GVQA...... LPLVW... DPATTKAV            414
terp  PRLVATNFVGGPKN...... VPIRFTKA                     427
       β4-1    β4-2       β3-2
```

**Figure 1.** Alignment of P450$_{AROM}$ with P450cam, P450terp, and P450BM-P. The helical regions of the three structures, and the corresponding regions of P450$_{AROM}$, are labeled and indicated by underlining. β-strands of P450BM-P are also underlined and labeled, and the corresponding regions in P450$_{AROM}$ are indicated where possible. Double underlining of residues in P450cam, P450terp, and P450BM-P indicate which are in the heme pocket, and astericks above the residues show the highly conserved residues.

## The Conserved Regions

First, the structurally conserved regions of P450AROM were built based on the conserved four-helix bundle found in P450cam, P45terp, and P450BM-P; then, the two conserved β-sheets, and the heme-binding region were added. Figure 2 shows an overlay of the conserved regions of P450cam and P450BM-P

In P450cam, P450terp, and P450BM-P, the four-helix bundle contains three parallel helices and one antiparallel helix; they are helices D, I, and L, and the antiparallel helix E. From the alignments, four helices were built and then individually overlaid on the corresponding peptide backbone of P450BM-P. These were minimized to remove any sidechain bumping. The heme and the heme-binding region were then added. On subsequent minimizations the conserved cysteine and the heme were held fixed. The conserved sheets in the β-sheet domain were then built by molecular replacement of P450BM-P, except that in P450AROM the turn in sheet β2 has three residues rather than four residues. This was then minimized.

The less conserved regions were added by constructing the helices and then placing these helices in a location similar to that of P450BM-P. These helices were then moved and rotated to align hydrophobic patches and maximize charge-paired residues. These regions consisted of helices F, G, J', and K. Finally, the non-conserved regions were constructed and added.

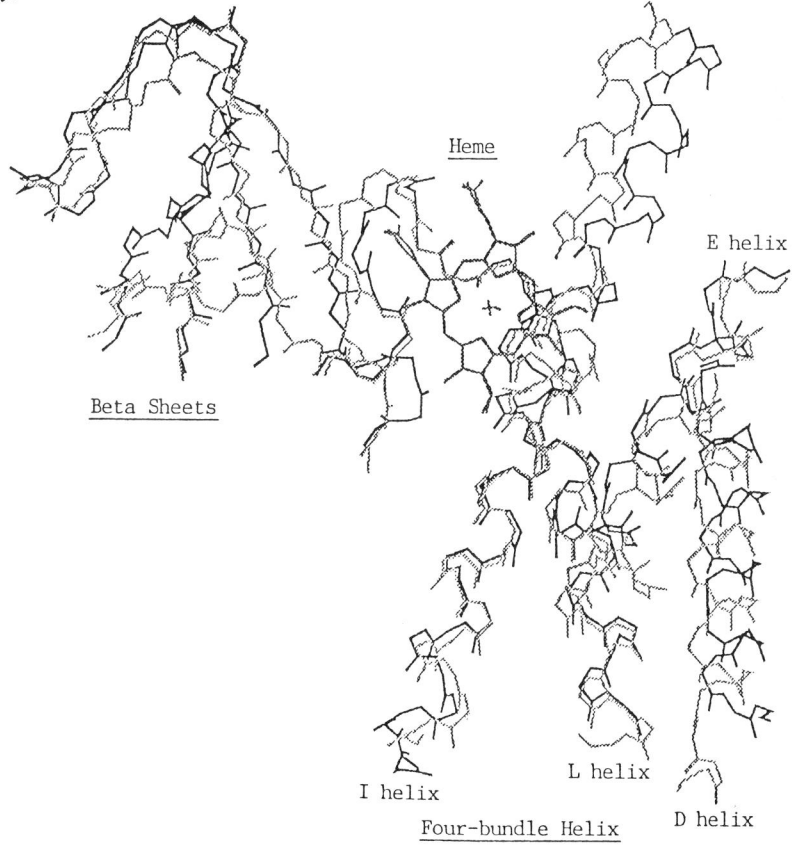

**Figure 2** shows an overlay of the highly conserved regions of P450cam and P450BM-P.

# THE MODEL

## The Redox-partner Binding Region

The redox-partner binding region in P450BM-P takes on a different shape compared to that of P450cam. In P450cam, the binding pocket is very shallow with four basic residues in the center. A similar configuration is found in P450terp. In contrast, the redox-partner binding region in P450BM-P is shaped more like a bowl. This is not surprising since the redox-binding partner for P450cam is an 11 kDa iron-sulfur protein, but a 66 kDa protein in P450BM-P. The deeper shaped pocket is also found in the P450AROM model. This is a result of two major inserts in P450AROM similar to the ones found in P450BM-P. These inserts form the J' helix and a $3_{10}$ helix which corresponds to helix $f$ in P450BM-P and form part of the rim of the bowl/binding region. On sequence alignments, these two regions form gaps in P450cam.

Additionally, in P450AROM and to a lesser extent in P450BM-P, there is a cluster of 10 aromatic residues in the redox binding region. These residues come from several different regions: F407 and F413 from the $f$ helix insert; F418, Y424, F426, and F427 from the $f$ helix to the heme binding region; F430 and F432 in the heme binding region; Y361 at the C-terminal end of helix K; and F104 from the B-B' loop.

## The Substrate Access Channel

The mouth of the access channel takes on a different shape than that of P450BM-P. In P450BM-P, the mouth of the pocket is essentially round; however, it was found after construction of the P450AROM model that the mouth was flatter and wider like a slit. This in fact would more readily accommodate a steroid molecule. In contrast to P450BM-P, the residues around the entrance of P450AROM are hydrophobic and aromatic except for E92 which is part of the $\beta 1$ sheet. (Unfortunately, figures of the model demonstrating this and other points could not be included since they are color and do not show well in black and white.)

# CONCLUSIONS

While P450cam may be a good model for class I P450s, P450BM-P is a better model for class II P450s which require an NADPH reductase rather than a redoxin as the docked partner. In this model, it is apparent in the comparison between P450cam, and P450BM-P that the shape and charge distribution is different. Additionally from this model, it appears that there is a largely hydrophobic region around the access channel in P450AROM which may serve as a good membrane docking region in contrast to the corresponding region in P450BM-P which is more charged. Finally, it appears from this exercise, that one can build a model of a microsomal P450 based on P450BM-P, and from this model learn a great deal about charge distribution and tertiary interaction not apparent from the sequence alignments and secondary structure predictions.

References

1. Thompson, E.A., Jr., and Sitteri, P.K. J. Biol. Chem. 249: 5373-5378, 1974.
2. McNatty, K.P., Baird, D.T., Bolton, A., Chambers, P., Corker, C.S.,and MacLean, H. J. Endocrinol 71:77-85, 1976.
3. Ackerman, G.E., Smith M.E., Mendelson, C.R., MacDonald, P.C., and Simpson, E.R.. Clin Endocrinol Metab 53:412-417, 1981.
4. Santen, R. J Clin Endocrinol Metab 38:476-479, 1974.
5. Trunet, P.F., Mueller, P., Hatnager, A.S., Dickes, I., Monnet, G., White, G.. J Clin Endocrinol Metab 77:319-323, 1993.
6. Ravichandivan,K., Boddupalli,S., Haseman, C., Peterson, J., Deisenhofer. Science 261:731-736,1993.
7. Poulos, T.L., Finzel, B.C., Howard, A.J. J Mol Biol 195:687-700, 1987.

# Ortho- and para-hydroxylation of pentachlorophenol in aquatic plants: possible involvement of cytochrome P450

Sashwati Roy, Osmo Hänninen

Department of Physiology, University of Kuopio, PO Box 1627, SF-70211, Kuopio, Finland

Pentachlorophenol (PCP) is a well known industrial byproduct which was widely used until recently as a wood preservative. Significant amounts of the compound have often been found in the sediments and water of the aquatic reservoirs contaminated with pulp and paper mill effluents[1].

High toxicity and mutagenicity of PCP in living organisms has attracted considerable interest to study the metabolism and consequent biological effect of the compounds in environment. The uptake of environmental chemicals into plants is the first step for accumulation and metabolism in the food chain[1]. Metabolism of foreign compounds in plants may take place through transformation, conjugation and subsequently compartmentation.

We have previously reported that hydroxylation and metylations are one of the major mechanisms of PCP metabolism in aquatic plants[2]. Responses of some of the major enzymes of oxidative (*e.g.* peroxidase) and conjugative (*e.g.* glutathione S-transferase) systems of aquatic plants to PCP have also been studied[2]. Involvement of cytochrome P-450 (cyt P-450) in oxidative hydroxylation reaction and conversion of carbon-halogen bond by rat liver microsomes and bacterial cyt P-450 have been intensively studied[8,9]. The aim of the present study was to investigate a possible involvement of cytochrome P-450 system in the hydroxylation of PCP in aquatic plants.

## MATERIAL AND METHODS

*Eichhornia crassipes* [(Mart) Solms], an aquatic plant, was selected for the present study because this plant is widely used in the treatment of waste water in tropical and subtropical climates. The plants were exposed to 2 μM PCP for 48 hours (hrs). Control set of plants were maintained in distilled water. Leaf samples were collected after 48 hours of exposure to 2 μM PCP.

*Preparation of microsomes*

Microsomes were prepared by the procedure of Russell[3] modified to include protease inhibitors[4]. Plant tissues were homogenized in three volumes of buffer (0.1 M sodium phosphate, pH 7.5; 0.25 M sucrose, 1 mM EDTA, 10 mM ß-mercaptoethanol, 2% Polyclar AT (w/v), 5 μg/ml leupeptin and 5 μg/ml pepstatin). The homogenate was filtered through cheese cloth and centrifuged at 12,000 *g* for 10 min. The resultant supernatant was recentrifuged at 100,000 *g* for 60 min. The microsomal pellet was resuspended in 50 mM sodium phosphate (pH 7.5), 1 mM ß-mercaptoethanol, 5 μg/ml leupeptin and 5 μg/ml pepstatin.

*Cytochrome P-450 content*

The total cytochrome P-450 (cyt P-450) content from the leaf microsomal pellet was determined spectrophotometrically by the method of Omura and Sato[5]. The concentrations of cyt P-450 were calculated from reduced carbon monoxide difference spectrum, using an extinction coefficient 91 mM$^{-1}$ cm$^{-1}$ for $A_{450-490}$. The protein content of the microsomal fraction was determined by the method of Lowry et al.[6]

## RESULTS AND DISCUSSION

Significant levels (41.4 ± 8.6 pmoles/mg protein) of cytochrome P-450 were detected in plant leaves exposed to PCP for 48 hours (Fig 1B). However, in control leaves the levels of cytochrome P-450 were either very low (9.2 ± 2.8) or undetectable (Fig 1A). The quantitation of cytochrome P-450 in plants is difficult due to its low levels and instability. In the present study, protease inhibitors (leupeptin and pepstatin) were used to stabilize the cytochrome P-450. Leupeptin, which inhibits a wide range of proteases, has been found to stabilize a number of P-450 monoxygenases[4].

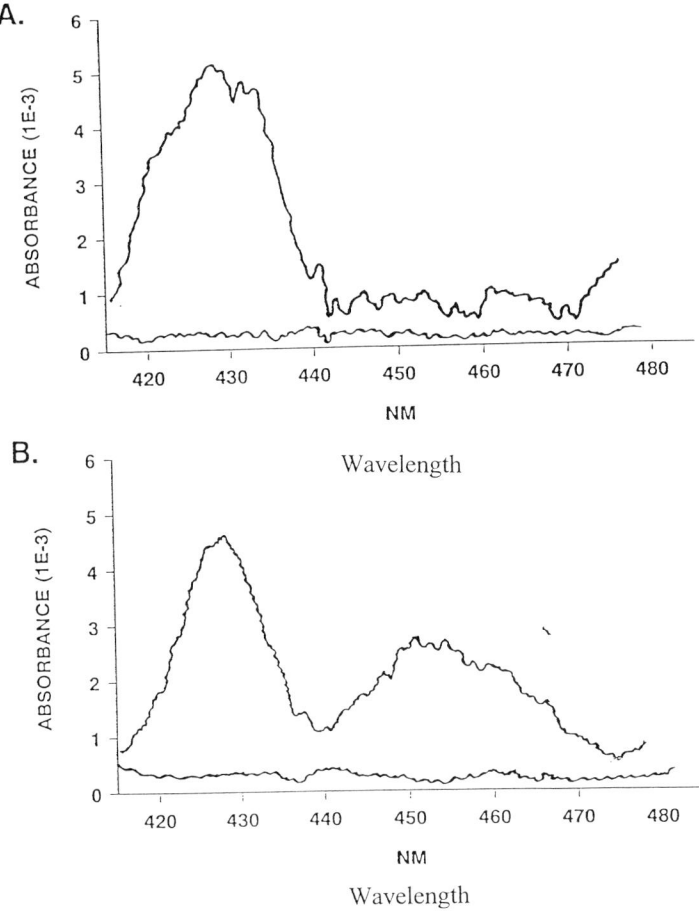

Fig. 1. Carbon monoxide difference spectra of leaf microsomal fractions prepared from A. control (unexposed) and B. exposed (2 µM pentachlorophenol for 48 hours) plants.

In higher plants, the involvement of cyt P-450 in the metabolism of several endogenous (e.g., fatty acids) and exogenous (e.g., herbicide Mouron) have been reported[7]. Induction of cyt P-450 have been reported in response to wounding in Jerusalem artichoke tissue[7]. Furthermore, a time course dependent increase of cyt P-450 content have also been reported in Jarusalem artichoke tissue exposed to phenobarbital, herbicides and $Mn^{++}$ ions[7].

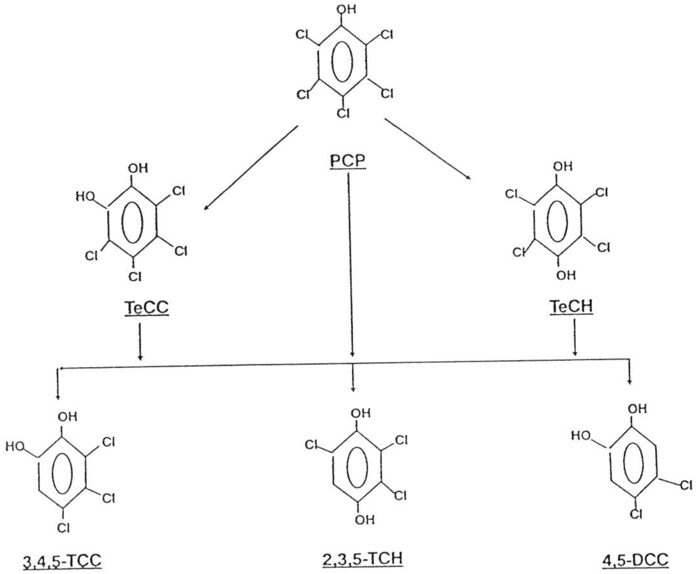

Fig. 2. Major *ortho-* and *para-*hydroxylated (tetrachlorocatechol, TeCC; tetrachlorohydroquinone, TeCH; trichlorocatechol, TCC; trichlorohydroquinone, TCH and dichlorocatechol, DCC) by-products of pentachlorophenol (PCP) metabolism in *E. crassipes*[2].

Direct evidences for the involvement of cyt P-450 in dechlorination of PCP have been reported in the cells of *Rhodococcus chlorophenolicus*. An induction in cyt P-450 content was observed in the cells of *R. chlorophenolicus* treated with PCP[8]. Hydroxylations are one of the major reactions in xenobiotic metabolism which are facilitated by cyt P-450. The major dechlorinated byproducts of PCP metabolism in *R. chloriphenolicus* observed were *para-* hydroxylated PCP[8]. Similarly, ortho- and *para-* substituted chloro-hydroxyphenols (chloro-catechols and -hydroquinones) have also been detected as major transformed products of PCP metabolism in *E. crassipes*[2] (Fig. 2). Detection of such hydroxylated products together with significant increase in the cyt P-450 content in the microsomes of the treated leaf tissue (Fig 1B) suggest a possible involvement of cytochrome P-450 in the metabolism of PCP in *E. crassipes*.

**REFERENCES**

1. Roy S., and Hänninen O. (1993) Biochemical monitoring of the aquatic environment: possibilities and limitations, In: *Ecotoxicology monitoring*, Mervyn Richardson (ed.) VCH publishers Ltd. London, U.K., invited review, pp. 119-135.
2. Roy S., and Hänninen O. (1993) Pentachlorophenol: Uptake/elimination kinetics and metabolism in an aquatic plant. *Environ. Toxicol. Chemistry* 13(4) (in press).
3. Russell D.W. (1971) The metabolism of aromatic compounds in higher plants. Properties of the cinnamic acid 4-hydroxylase of pea seedlings and some aspects of its metabolic and developmental control. *J. Biol. Chem.* 246: 3870-3878.
4. Stewart C.B., and Schuler M.A. (1989) Antigenic crossreactivity between bacterial and plant cytochrome P-450 monooxygenases. *Plant Physiol.* 90: 534-541.
5. Omura T. and Sato R.(1964) The carbon monoxide-binding pigment of liver microsomes. I. Evidence for its hemoprotein nature. *J. Biol. Chem.* 239: 2370-2378.
6. Lowry O.H., Rosbrough J., Farr A.L., Randall R.J. (1951) Protein measurement with folin phenol reagent. *J. Biol. Chem.* 193: 265-275.
7. Reichhart D., Salaun J-P, Benveniste I., and Durst F. (1980) Time course of induction of cytochrome P-450, NADPH-cytochrome *c* reductase, and cinnamic acid hydroxylase by phenobarbital, ethanol, herbicides, and manganese in higher plant microsomes. *Plant Physiol.* 66: 600-604.
8. Uotila J.S., Salkinoja-Salonen M.S., and Apajalahti J.H.A. (1991) Dechlorination of pentachlorophenol by membrane bound enzymes of *Rhodococcus chlorophenolicus* PCP-1. *Biodegradation* 2: 25-31.
9. Tanaka K., Kurihara N, and Nakajima M. (1979) Oxidative metabolism of tetrachlorocyclohexene, pentachlorocyclohexenes, and hexachlorocyclohexenes with microsomes from rat liver and house fly abdomen. *Pestic. Biochem. Physiol.* 10:79-95.

# Cytochrome P450 1A1 expression during the differentiation process in the Caco-2 cell line

Sophie Charrasse, Isabelle Fabre, Martine Daujat, Patrick Maurel, Christian Larroque

*INSERM U.128, CNRS, route de Mende, BP 5051, 34033 Montpellier, France*

## Summary

Cytochrome P450 1A1 was demonstrated to be inducible by various polycyclic hydrocarbons in the Caco-2 cell line in its steady state growth. However this human colon adenocarcinoma cell line is known to evolve from an undifferentiated to a well-differentiated "enterocyte like" organisation. The expression of P4501A1 during this differentiation process is studied in the presence of its typical inducers (TCDD, ßNF). Three steps of the P450 1A1 expression are analyzed: the nuclear transcription, the messenger and the protein accumulation. Involvement of the Ah receptor in the regulation of CYP1A1 gene expression in the Caco-2 cell line is evidenced.

## INTRODUCTION:

The small intestine provides an essential route for exposure to xenobiotics by the ingested nutriments. Enzymatic biotransformation in intestinal mucosa cells has the potential to facilitate the excretion of foreign compounds through the intestinal lumen. This first pass metabolism prevents many xenobiotics from being distributed into the blood flow. Thus, some enzymatic systems essential for detoxication processes were described in both intestinal tractus and Caco-2 cell line derived from a human colon adenocarcinoma (Peters and Roelofs 1989). The expression of cytochromes P450 in such tissue exposed to many xenobiotics is well documented (Kaminsky and Fasco. 1992) Among them, the 1A1 isoform has received a particular attention because of its ability to activate (pre)carcinogenics compounds (Newaz et al 1983, Boulenc et al 1992). In this paper we present data on the CYP1A1 gene regulation at various stages of the Caco-2 cell line transformation. Briefly, the cells remains undifferentiated up to seven days after seeding (J 7), differentiate during five days (J 12) then present the pattern of well differentiated enterocytes at J 15. By both classical microsomal activity or Western blots assays and specific mRNA analyses we study the cytochrome P4501A1 level in ßNF and TCDD Caco-2 treated cells. These results permit some conclusions to be made on the influence of differentiation events on the cytochrome P4501A1 expression.

## EXPERIMENTAL:

*Cells and microsomes preparation*:
Human adenocarcinoma cells were seeded at $3.10^5$ cells/ $cm^2$ and grown in 25 $cm^2$ plastic flasks in Dulbecco's Modified Eagle Medium (GIBCO) at 37°C under a 5 % $CO_2$ atmosphere. The cells were treated at various time after seeding (J 0) with $5\ 10^{-6}$ to

$50.10^{-6}$ M ß-naphthoflavone (ßNF) or $10^{-11}$ to $10^{-9}$ M dioxine (TCDD). For microsome preparation, Caco-2 cells were washed twice with ice-cold ph 7.4 phosphate buffer then harvested in 0.53 M EDTA. Cell suspension was spun twice and the pellet finally suspended in a phosphate buffer (0.1M, pH 7.4, EDTA 1mM) supplemented with proteases inhibitors for subsequent homogeneisation and centrifugations.

*Proteins assays :* 7-ethoxyresorufine-O-deethylase activity was measured by a fluorimetric method as decribed by Boulenc et al (1992). For Western-blots analysis, 50µg of microsomal proteins were submitted to electrophoresis, transfert onto nitro-cellulose sheet then developed in the presence of 0.1 mg/ ml polyclonal anti-rabbit cytochrome P4501A1 antibodies.

*RNA preparation, cDNA probes and Northern blots:* Caco-2 cells were scraped and lysed in a denaturing solutions and total RNA prepared according to Chomczynski (1987). For the Northern blot analysis, 25 µg of total RNA were run on a 1.2% agarose gel then blotted onto Zeta probe membrane and hybridized with pPCR1A1, pRED (cytochrome P450reductase specific) and pGAPDH cDNA probes as detailed by Daujat et al. (1992).

## RESULTS AND DISCUSSION:

*ßNF, TCDD concentration and induction time dependence on the 1A1 specific mRNA accumulation:.*

The maximum induction was obtained for $50\ 10^{-6}$M ßNF, higher concentration exerts an evident toxicity on the Caco-2 cells. With TCDD the maximum is reached for $10^{-9}$M ; although non toxic, higher concentrations do not increase the level of 1A1 messenger.

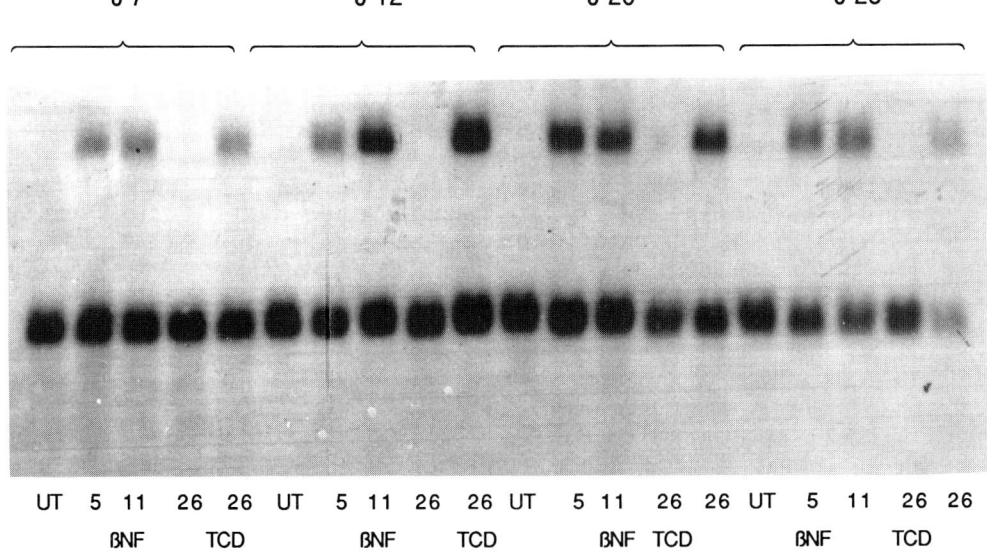

Figure 1: Northern blot analysis of total RNA from Caco-2 cells treated with various compounds (ßNF, $20\ 10^{-6}$ M ß-naphthoflavone; TCD, $10^{-9}$M TCDD; UT, untreated) during the differentiation process (seeding = J 0). Upper signal : 1A1 specific messenger (2.6kb) , lower signal GAPDH messenger.

The time dependence for the messenger accumulation under the influence of the inducers is presented in the the figure 1. Clearly, the induction becomes maximum between 5 and 11 hours after the addition of ßNF with a greater sensibility for the cells 12 to 20 days after plating. The very low level of the messenger after 26 h induction remains unexplained. This observation is not valid with TCDD for which the induction reaches the maximum after 26 hours treatment.

As previously described for human or rabbit hepatocytes (Daujat et al. 1992) the concomitant addition of cycloheximide and ßNF produces a superinduction. Whereas mRNA remains undetectable in untreated cells at any stage of their differentiation, the addition of cycloheximide alone produces a significant transcription of the CYP1A1 gene (data not shown)

To confirm the transcriptional activation of the gene, run on experiments were conducted on isolated nuclei from cells untreated or treated with 50 µM ßNF and/or 10µg:ml cycloheximide for various time. Results are presented in figure 2.

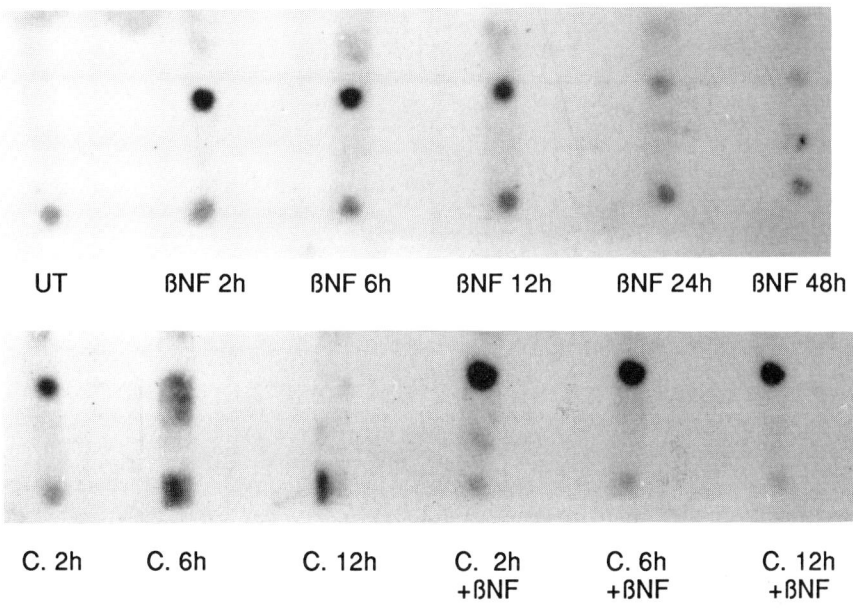

Figure 2: Analysis o the RNA synthetized in isolated Caco-2 cells nuclei in the presence of various compounds Spots are probed with pPCR1A1 probe (upper arrow) or pRED (lower arrow) (C = cycloheximide)

The transcription signal is maximum 2 to 6 hours after induction. Here again the effect of cycloheximide alone or in conjonction with ßNF has a maximun effect in a similar time period. These results are in agreement with the corresponding messenger accumulation as described above.

In experiments using microsomes from either treated or untreated Caco-2 cells, the accumulation of 1A protein is analysed (data not shown). Using both western blots and

7-ethoxy-resorufine o-deethylase activity, the maximum protein induction is attained 96 hours after ßNF administration, whatever the inducer concentration. Interestingly, maximum induction occurs faster (48 h) when $10^{-9}$M TCDD is used as inducer.

The aryl -hydrocarbon receptor (Ah receptor) is generally accepted as the essential element in the regulation of the CYP1A1 gene by typical inducers with the exception of omeprazole (Daujat et al. 1992). When a 9S enriched fraction of cytosol from Caco-2 cell is incubated with [$^3$H] TCDD and analysed on a sucrose density gradient, a peak of radioactivity is clearly obtained in the 9S region (figure 3) reflecting the TCDD-Ah receptor complex. The co-incubation of [$^3$H] TCDD and ßNF results in a lowering of the bound radioactivity due to a competition between these two products for the same receptor. Therefore, in the presence of TCDD or ßNF as inducers, the regulation mechanism of the 1A1 gene in the Caco-2 cell line acts via the Ah receptor as previously demonstrated in the liver.

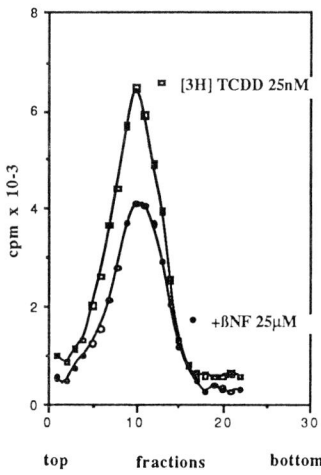

Figure 3 :Sucrose density gradient of the 9S enriched fraction of cytosol from Caco-2 cells incubated with 25 nM [$^3$H] TCDD or 25 nM [$^3$H] TCDD + 25µM ßNF

## REFERENCES

Boulenc X., Bourrie M., Fabre I., Roque C., Joyeux H., Berger Y. and Fabre G. (1992): Regulation of cytochrome P4501A1 gene expression in a human intestinal cell line, Caco-2. *J. Pharmacol. Exp. Ther.* 263,1471-178.

Chomczynski P. and Sacchi N. (1987) Single step method of RNA isolation by acid guanidinium thiocyanate-phenol-chloroform extraction *Anal. Biochem.*, 162, 156-159.

Daujat M., Peryt B., Lesca P., Fourtanier G., Domergue J. and Maurel P. (1992) *Biochem. Biophys. Res. Commun* 188, 820-825

Kaminsky L. S. and Fasco M.J. (1992) Small intestinal cytochrome P450 *Critical Rev. Toxicol.* 21, 407-421

Newaz S.N., Fang W.F. and Strobel H.W. (1983) Metabolism of carcinogen 1,2-dimethyl-hydrazine by isolated human colon microsomes and human colon tumor cells in culture *Cancer Res.* 52,794-798

Peters W. H. M., and Roelfs H. M. J. (1989) Time-dependent activity and expression of glutathion S-transferase in the human adenocarcinoma cell line Caco-2. *Biochem. J.* 264,613-616

# Substrate-regulated, cAMP-dependent phosphorylation of cytochrome P450 3A1

Souren Mkrtchian[1], Erik Eliasson[1], James R. Halpert[2], Magnus Ingelman-Sundberg[1]

[1]Department of Medical Biochemistry and Biophysics, Karolinska Institute, 171 77 Stockholm, Sweden. [2]Department of Pharmacology and Toxicology, College of Pharmacy, University of Arizona, Tucson, AZ, USA

## Introduction

Protein phosphorylation has been recognized to be very important for the regulation of enzymatic activity, conformational stability and in some cases for susceptibility to proteolysis. Posttranslational modification by phosphorylation has been proposed to play a key role in short-term regulation of several P450 enzymes in gene family 2 [1]. Our studies on the ethanol-inducible P4502E1 have shown cAMP-dependent CYP2E1 phosphorylation in hepatocytes with concomitant denaturation, loss of enzymatic activity and rapid degradation, possibly via specific microsomal proteolytic system [2,3]. CYP2E1-specific substrates can abolish these effects, protecting P450 molecules from phosphate incorporation and subsequently from targeting them to the proteolytic pathway [3].
A member of an other P450 gene family, glucocorticoid-inducible cytochrome P4503A1 has been found to be substrate/ligand-stabilised [4]. This implies that CYP3A1 could also be modified by apoprotein posttranslational modification what prompted us to investigate the regulation of CYP3A1 by substrates/ligands and phosphorylation [5].

## Results and discussion

Hepatocytes, isolated from rats treated with dexamethasone to induce CYP3A1 were exposed to glucagon and membrane-permeable cAMP-derivatives. The results of experiments revealed increased phosphorylation of CYP3A1 under conditions increasing cellular level of cAMP [5]. Similar increase was obtained in 8-Br-cAMP[1]-stimulated cells. In both cases phosphorylation of CYP3A1 was accompanied with loss of immunodetectable enzyme. Incubation of hepatocytes with CTZ and other CYP3A1 substrates/ligands caused protection of enzyme from phosphorylation and inhibition of CYP3A1 degradation [5].

The data obtained from experiments on liver microsomes isolated from DEX-treated rats are summarised in Table I. Addition of cAMP to the microsomes enhanced phosphate incorporation by 85% indicating

---

[1]8-Br-cAMP, 8-bromo cyclic adenosin monophosphate; CTZ, clotrimazole; DEX, dexamethasone; ER, endoplasmic reticulum; PEG, polyethylenglycol; PKA, catalytic subunit of cAMP-dependent protein kinase.

the presence of a microsomal cAMP-dependent kinase active on CYP3A1. NADPH further enhanced the phosphorylation, as well as the catalytic subunit of the cAMP-dependent protein kinase. In all these experiments CTZ efficiently inhibited the phosphorylation.

The spectrophotometric monitoring of CO-reduced difference spectrum under the same conditions revealed a rapid phosphorylation-dependent decrease in P450 absorbance, whereas CTZ had pronounced protective effect. This could be interpreted as a phosphorylation-mediated denaturation of the CYP3A1 molecule, and was furthermore supported by the observation that the CYP3A1-specific 6b-hydroxylation of testosterone was inhibited under the conditions stimulating phosphorylation (Table 1).

**Table 1. Phosphorylation, enzymatic activity and denaturation of microsomal CYP3A1.**
Liver microsomes from of DEX-treated rats were incubated for 30 min at $37^o$. The extent of CYP3A1 phosphorylation was analysed by immunoprecipitation, SDS-PAGE and scintillation counting of gel slices including the precipitated P450. CYP3A1 catalytic activity was assayed by incubation of microsomes with $^{14}C$-testosterone and further fractionation and determination of metabolites by TLC-chromatography. Reduced CO spectrum was recorded spectrophotometrically between 400 and 500 nm. All data are expressed in relation to results obtained in MgATP (or MgATP+NADPH)[a] incubations.[2]

|  | Phosphorylation | | Testosterone 6β-hydroxylase activity | | Reduced CO spectrum | |
|---|---|---|---|---|---|---|
|  | - CTZ | + CTZ | - CTZ | + CTZ | - CTZ | + CTZ |
| MgATP (10/5 mM) | 100±1 | 81±10 | 100 |  |  |  |
| +cAMP (5µM) | 179±45 | 133±17 |  |  |  |  |
| +NADPH (0.5 mM) | 185±28 | 133±17 | 81±9 |  | 100±25[a] |  |
| +Emulgen (0.1%) | 129±13 |  |  |  |  |  |
| +PKA (50 U) | 784±103 | 545±40 | 66±12 | 102±0.5 |  |  |
| +NADPH+cAMP | 278±47 | 172±17 |  |  |  |  |
| +NADPH+PKA | 1822±250 | 487±110 |  |  | 42±4.7 | 107.1±4.7 |

To determine the phosphorylation site of CYP3A1 the latter was purified to homogeneity employing several chromatographic steps, phosphorylated and digested in three different incubations by chymotrypsin, trypsin and V8 proteases and resulting phosphopeptides were purified by HPLC and subjected to radiosequencing. The radioactivity in all three cases was eluting in cycles, corresponding to serine in position 393 [5]. These data taken together with the results of TLC- separation of the chymotrypsin-generated peptides hydrolysate indicate the existence of a single phosphorylation site on CYP3A1, namely $Ser^{393}$. A homologous residue in P450cam has been described to be included in a β-structure interacting with the heme [6].

Experiments on isolated DEX-microsomes showed that the denaturation of CYP3A1 holoenzyme caused by phosphorylation is apparently essential for accelerated degradation in endoplasmic reticulum. Kinetic studies on the apoprotein level (Western blot) revealed the disappearance of CYP3A1, with simultaneous formation of high molecular weight aggregates which later (21 hours) are degraded by ER-resident protease(s). This process also appears to be NADPH and phosphorylation-dependent [5].

Although many experiments performed in microsomes were carried out in the presence of exogenous protein kinase, it is obvious that CYP3A1 could be phosphorylated also by an endogenous kinase present in the ER (Table 1). Albeit the classification of such kinase remains uncertain we have shown its

---

[2]All data included in Table 1 are taken from [5].

sensitivity toward the specific inhibitor of cAMP-dependent protein kinase (PKI). In cAMP-stimulated microsomes phosphorylation of CYP3A1 was 30% lower in presence of inhibitor than in control (Fig.1). This provides additional evidence that the endogenous kinase could be analogous to the cytosolic cAMP-dependent enzyme.

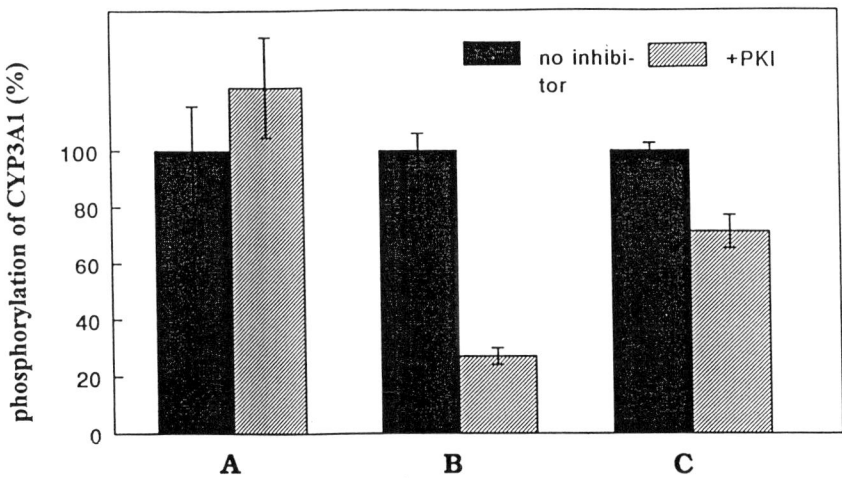

Fig.1. Inhibition of phosphorylation of CYP3A1. Incubation conditions and evaluation of CYP3À1 phosphorylation were described in legend to Table 1. **A**, no additions; **B**, +PKA; **C**, +cAMP. PKI, peptide that specifically binds to catalytic subunit of cAMP-dependent protein kinase thereby displacing regulatory subunit.

We are currently purifying this ER-kinase. Briefly the procedure includes the sodium cholate solubilisation of microsomes, their PEG-fractionation and ion-exchange chromatography on DEAE-Sepharose. Preliminary results show the existence of several protein kinase activity peaks, but only one was active on CYP3A1 This indicates the existence of relatively specific P450-oriented protein kinase in the endoplasmic reticulum (Fig.3).

In conclusion, it appears that glucocorticoid-inducible CYP3A1 is phosphorylated by exogenous cAMP-dependent protein kinase as well as by endogenous ER-bound enzyme with subsequent denaturation, heme loss and drop in catalytic activity. Misfolded apoprotein molecules then aggregate into high molecular weight complexes, which are recognized and digested by ER-resident proteases. Recently two such P450-active enzymes were purified and characterised in our laboratory [7]. Substrates/ligands protect CYP3A1 from both phosphorylation and denaturation pointing out the involvement of substrates in regulation of P450 turnover. It seems like the substrate- and phosphorylation-dependent regulatory mechanism for control of cytochrome P450 turnover could be a general mechanism for posttranslational regulation of various isoforms of P450.

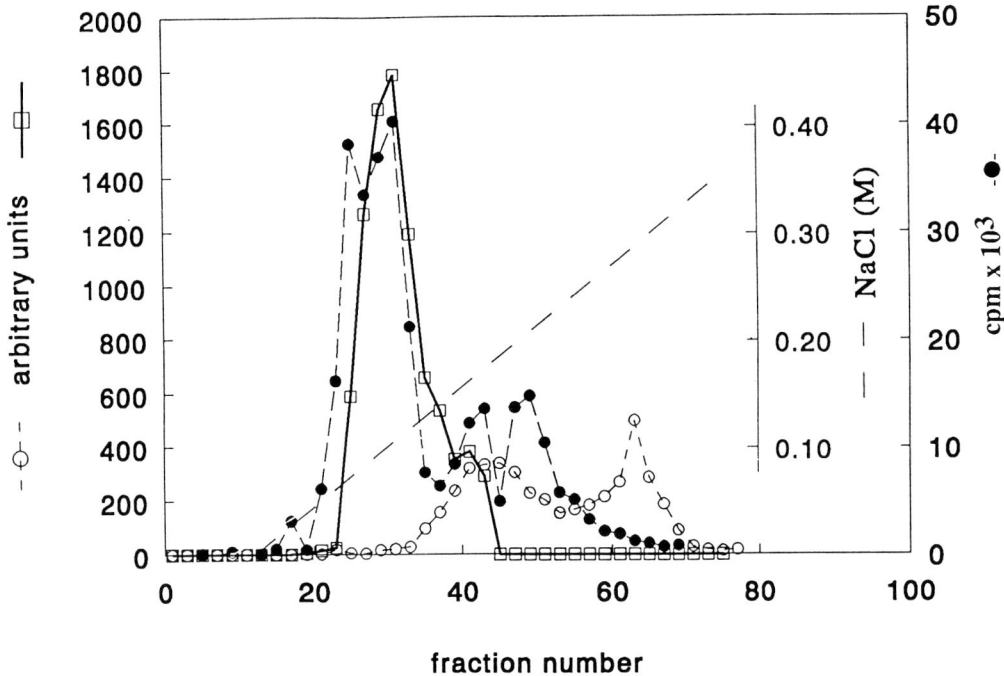

Fig 2. **DEAE-Sepharose chromatography of sodium cholate solubilized and PEG-fractionated microsomes.** Protein kinase activity was assayed by incubating 25 µl of fractions with 10 mM $MgCl_2$, 10 µM [$^{32}P$]ATP (0.5 µCi) and three different substrates : (1) -●-, synthetic 13-aminoacids long peptide containing RRAS sequence (specific for serine/threonine kinases). Activity was determined by phosphocellulose paper method and expressed in cpm; (2), - O - protamine sulphate. Activity was measured by dot blot method with subsequent autoradiography and densitometry. ; (3) -□ -, purified CYP3A1. Incorporation of [$^{32}P$] into CYP3A1 was evaluated by immunoprecipitation, SDS-PAGE, autoradiography and densitometry.

**References**

1. Pyerin, W., and Taniguchi, H. (1989) Phosphorylation of hepatic phenobarbital-inducible cytochrome P-450. EMBO J. 8: 3003-3010
2. Eliasson, E., Johansson,I. and Ingelman-Sundberg,M. (1990) Substrate-, hormone and cAMP-regulated cytochrome P450 degradation. Proc.Natl.Acad.Sci.USA. 87,3225-3229
3. Eliasson, E., Mkrtchian,S, and Ingelman-Sundberg,M. (1992) Isozyme-specific and substrate-regulated intracellular degradation of cytochrome P450 (2E1) involving MgATP-activated, rapid proteolysis in the endoplasmic reticulum mebranes. J.Biol.Chem. 257, 15765-15789
4. Watkins, P.B., Wrighton, S.A., Scheutz, E.G., Maurel, P., and Guzelian, P.S. (1986) Macrolide antibiotics inhibit the degradation of the glucocorticoid-responsive cytochrome P450p in rat hepatocytes *in vivo* and in primary monolayer culture. J. Biol Chem. 261: 6264-6271
5. Eliasson, E.,Mkrtchian,S, Halpert,J.R. and Ingelman-Sundberg,M., submitted for publication.
6. Poulos, T.L., Finzel, B.C. Gunsalus, I.C., Wagner, G.C., and Kraut, J. (1985) The 2.6-Å crystal structure of *Pseudomonas putida* cytochrome P450. J. Biol. Chem. 260: 16122- 16130.
7. Zhukov, A., Werlinder, V. and Ingelman-Sundberg, M., submitted for publication.

# Growth hormone regiospecifically represses the expression of CYP2B1/2 and CYP3A in liver

Teija Oinonen, Kai O. Lindros

*Biomedical Research Center, Alko Ltd, POB 350, 00101 Helsinki, Finland*

INTRODUCTION

In the mature mammalian liver many genes are expressed in a zonated fashion within the acinus, the microcirculatory unit of the liver (see Gebhardt, 1992). The mechanisms regulating this spatial gene expression are poorly known. This zonated expression is particularly prominent with respect to the cytochrome P450 (CYP) genes, most of which are constitutively expressed and induced in the perivenous region (Bühler et al., 1992). The total hepatic expression of a number of CYP genes is affected by pituitary dependent hormones, like growth hormone (GH) and thyroid hormones. These hormones repress, for example, the expression of CYP2B1, 2B2 and 3A2 (Murayama et al. 1991, Waxman et al. 1990). To study whether hormonal signals are regulating the spatial expression of CYP genes, we have investigated the effect of hypophysectomy and GH and triiodothyronine ($T_3$) administration. Zonation was studied by classical immunohistochemical analysis and by analysis of cell lysates obtained from either the periportal (pp) and perivenous (pv) acinar region. Lysates were collected after zone-restricted digitonin-induced cell lysis during *in situ* liver perfusion. Lysates were analysed for their content of CYP2B1/2 apoprotein and mRNA and 3A1/2 apoprotein by immunoblotting and cDNA PCR. We found that hypophysectomy permitted and GH prevented the expression of these CYPs in the periportal region, especially in female rats.

MATERIALS AND METHODS

Hypophysectomized and sham operated male and female Sprague Dawley rats, obtained from Møllegaard, Ejby, Denmark, were treated with human recombinant GH (Norditropin, Nordisk Gentofte A/S, Denmark), infused subcutaneously by osmotic minipump for 7 days (0.01 IU/h) (Oinonen et al. 1993). $T_3$ was injected daily for 7 days (50 ug/kg i.p.). Periportal (pp) and perivenous (pv) cell lysates were obtained, total RNA isolated, reverse transcribed to cDNA and PCR amplified and immunoblots were performed and analysed as in Saarinen et al. (1993) and Oinonen et al. (1993). Antiserums recognized both CYP2B1 and 2B2 or CYP3A1 and 3A2 and PCR primers were specific to 2B1 and 2B2. Immunohistochemistry was done as described in Bühler et al. (1992).

RESULTS

After hypophysectomy there was a marked increase of CYP2B1/2 protein and mRNA in cell lysates from both the periportal and the perivenous region. This increase was striking in the pp eluates and resulted in similar levels of

CYP2B1/2 apoprotein and mRNA in pp and pv eluates (Oinonen et al., 1993). The effect was similar in males and females. Treatment of female rats with GH almost completely counteracted the effect of hypophysectomy, both with respect to total expression and with respect to zonation. In males continuous infusion of GH (mimicking the female *in vivo* GH secretion pattern) caused a qualitatively similar, but less prominent effect. The amount of CYP2B1/2 apoprotein in pp eluates decreased significantly more (to about one-seventh) than in pv eluates (to about one-half), thus partially restoring the normally prevailing perivenous zonation pattern. On the other hand, CYP2B1/2 mRNA decreased to approximatly the same extent in pp and pv eluates.

The expression of CYP3A, that also normally exhibits a perivenous zonation pattern, was affected by hypophysectomy and subsequent GH administration much in the same way as that of CYP2B1/2. While pp lysates from control animals were almost devoid of CYP3A protein, those obtained from hypophysectomized rats exhibited dramatically increased amounts of CYP3A protein, equalling those obtained from pv lysates. In females GH infusion repressed CYP3A and normalized the distribution to that seen in controls. In males the amount of both pp and pv CYP3A protein was decreased.

The effect of hypophysectomy and GH treatment on the zonation of CYP 2B1/2 and CYP 3A was also observed by immunohistochemical analysis. While in female controls only a 2-4 cell layer thick ring surrounding the terminal hepatic venules were strongly stained, in hypophysectomized animals staining approached, but did not fully reach, the terminal portal veins. Growth hormone treatment seemed to partly counteract this effect. In male controls the perivenous region exhibiting staining was larger, but the effect of hypophysectomy was nevertheless obvious.

Since the counteracting effect of the female-type GH administration on CYP2B1/2 expression in the periportal region was less efficient in males than in females, we investigated the possible effect of $T_3$, another pituitary regulated repressor of CYP2B1/2. However, although daily injections of $T_3$ indeed repressed both CYP2B1/2 protein and mRNA levels, the effect was approximately the same in both pp and pv eluates and there was no gender difference (table 1).

**Table 1.** The effect of hypophysectomy, GH and $T_3$ on the distribution of CYP2B1/2 betveen pp and pv cell lysates determined by videodensitometric quantitation of CYP2B1/2 immunoblots and CYP2B1/2 cDNA PCR. The values are means $\pm$ SD from 4 or 5 pairs of pp and pv cell lysates from control, hypophysectomized (Hx) and hypophysectomized GH-treated (Hx+GH), $T_3$ treated (Hx+$T_3$) and Hx+GH+$T_3$ treated male rats. 100%=pv control value. *,** and ***=p<0.05, 0.01 and 0.001, respectively, for statistical significance (Student's t-test) of pp-pv difference within each group. b.d. below detection limit.

| treatment | 2B1/2 apoprotein | | 2B1/2 mRNA | |
|---|---|---|---|---|
| control pp | 14±13 | * | 21±27 | ** |
| control pv | 100±77 | | 100±36 | |
| Hx pp | 208±70 | | 222±17 | * |
| Hx pv | 238±38 | | 159±31 | |
| Hx+GH pp | 37±43 | * | 117±48 | |
| HX+GH pv | 85±10 | | 149±22 | |
| Hx+$T_3$ pp | 13±8 | | 51±54 | |
| Hx+$T_3$ pv | 19±16 | | 48±24 | |
| Hx+GH+$T_3$ pp | 3±5 | * | b.d. | * |
| Hx+GH+$T_3$ pv | 11±1 | | 15±11 | |

CONCLUSIONS

Our results provide strong evidence that GH represses the expression of CYP2B1/2 and CYP3A in the periportal region in the liver. After hypophysectomy high expression of these CYPs was observed in the normally almost silent periportal region. This was counteracted by the continuous infusion of GH mimicking the female type of GH secretion. The mode of GH dosing may explain why in the males the suppressing effect of GH was less prominent in the periportal region than in females. In males the GH secretion is pulsatile. Continuous infusion of GH has on the other hand proved to more efficiently repress the expression of CYP2B1/2 than intermittent injections (Murayama et al. 1989).

GH could repress the expression of genes specifically in the periportal region by several mechanisms, all of which would be based upon gradients along the portocentral axis in the sinusoid. Information on the possible sinusoidal GH gradient or on zonation of GH receptor in the liver is, to our knowledge, not yet available. Such gradients have been reported for EGF and its receptor as well as for several hormones including insulin, glucagon and thyroxine (see Gebhardt 1992).

In addition to the CYP genes many other liver genes also exhibit a zonated expression pattern. Our present data suggest that GH and other hormones have an important role in establishing these zonated patterns, either by repressing or stimulating expression in specific regions.

REFERENCES

Bühler R., Lindros, K.O., Nordling, Å., Johansson I. and Ingelman-Sundberg, M. (1992) Zonation of cytochrome P450 isozyme expression and induction in rat liver. *Eur. J. Biochem*. 204, 407-412

Gebhardt, R. (1992): Metabolic zonation of the liver: Regulation and implications for liver function. *Pharmac. Ther*. 53, 275-352

Murayama, N., Shimada, M., Yamazoe, Y. and Kato, R. (1991): Difference in the susceptibility of two phenobarbital-inducible forms, P450IIB1 and P450IIB2, to thyroid hormone- and growth hormone-induced suppression in rat liver: Phenobarbital-inducible P450IIB2 suppression by thyroid hormone acting directly, but not through the pituitary system. *Mol. Pharm*. 39, 811-817

Oinonen T., Nikkola E. and Lidros K.O. (1993): Growth hormone mediates zone-specific gene expression in liver. *FEBS Lett*. 327, 237-240

Saarinen J., Saarelainen R. and Lindros K.O. (1993): A rapid method to study heterogenous gene expression in liver by direct assay of messenger RNA from periportal and perivenous cell lysates. *Hepatology* 17, 466-469

Waxman, D.J., Ram, P.A., Notani, G., LeBlanc, G.A., Alberta, J.A., Morrissey, J.J. and Sundseth, S.S. (1990): Pituitary regulation of the male-specific steroid 6ß-hydroxylase P-450 2a (gene product IIIA) in adult rat liver. influence of growth hormone and thyroxine acting at the pretranslational level. Mol. Endocrinol. 4, 447-454.

# Developmentally programmed changes in CYP3A1 induction by dexamethasone in male rat liver

Teresa Pereira[1], Maria Celeste Lechner[2]

*Laboratory of Biochemistry, Gulbenkian Institute of Science, Oeiras, Portugal*

## INTRODUCTION

CYP3A genes constitute particularly interesting models for studying mammalian gene regulation since they code for highly homologous proteins but showing markedly different patterns of expression. This family of enzymes are catalytically active steroid hydroxylases expressed in the liver which are up-regulated by both hormonal and synthetic steroids (Lu and *at al.*, 1972; Gonzalez and *at al.*, 1986) such as Dexamethasone (Dex) and the anti-glucocorticoid Pregnenolone-16α-carbonitrile (PCN).

CYP3A1 first described as a non constitutive hepatic enzyme, has been shown to be expressed at residual levels in the liver in the absence of stimulation by specific inducers (Shimada and *at al.*, 1989; Ribeiro and Lechner, 1992).

After cloning the 5' terminal and upstream contiguous flanking region of the CYP3A1 gene (26 Kb cosCYP/3A1 clone -EMBL-X62086) we demonstrated by run-on transcription assay a strong and early activation of CYP3A1 to be brought about in the liver following the *in vivo* administration of Dex to immature 21 day-old male rats but not in the adults although an accumulation of the mRNA is observed in both cases (Telhada and *at al.*, 1992).

This paradoxical response illustrates the recognized complexity of the regulatory mechanisms triggered by glucocorticoids as well as by anti-glucocorticoids that up-regulate CYP3A1 gene expression in the liver cell (Schuetz and Guzelian, 1984).

In order to identify the possible importance of trancriptional and post-transcriptional events in the adaptive response of the CYP3A1 gene to Dex we characterized the time course activation of CYP3A1 gene by run-on transcription assay in function of rat liver development. The primary or secondary nature of the *in vivo* inductive response to Dex was further investigated by simultaneous blockade of the protein synthesis by Cycloheximide (Cyc) administration to both immature and adult animals.

## METHODS

Male Wistar rats, immature (21 day-old) and adult (90 day old), were treated with Dexamethasone 21-phosphate (Dex) (80 mg/Kg b.w.) i.g., and sacrificed at different time points (0.5 to 24 hours) after drug administration. Animals receiving Cycloheximide (Cyc) were given 5 mg/Kg b.w., 0.5 h prior to Dex.
Run-on transcription assay was performed in the isolated nuclei and the incorporated [a-32P] UTP probed with a CYP3A1 subgenomic fragment corresponding to part of the unique intron 1 as previously described (Telhada and *at al.*, 1992).
CYP3A1 mRNA relative concentration was measured in the Li/Urea extracted total liver RNA by Northern blot analysis using a specific oligonucleotide probe as described.

## RESULTS

### Effect of Dexamethasone on CYP3A1 gene expression

Our results have showing that CYP3A1 mRNA concentration markedly increases, in both immature and adult rat liver attainig comparable induced steady-state levels at 18-24h after Dex administration (results not shown).
Besides we demonstrate that the significant increases of the mRNA are systematically produced following transcription activation of the CYP3A1 gene which starts to be detected 0.5 hour after Dex, attaining a maximum at 2.0 hours in the immature rat liver but not in the adult were a delayed response is observed characterized by a maximum transcription activation at 12h, as shown in Fig.1.

### Effect of Cycloheximide on CYP3A1 induction by Dexamethasone

The effect of Cycloheximide on CYP3A1 gene expression was analysed at the time-points corresponding to maximum inductive response previously determined in the immature (2h) and in the adult (12h) rat liver. Cyc was given 0.5 h prior to Dex treatment.
Inhibition of protein synthesis by Cyc did not prevent an early inductive response of CYP3A1 gene to Dex in the immature rat liver although a relative decrease (40%) in the intensity of the response was observed.
Conversely, in the adult rat liver the induction of CYP3A1 gene transcription by Dex previously observed was almost abbolished (90%) by Cyc.

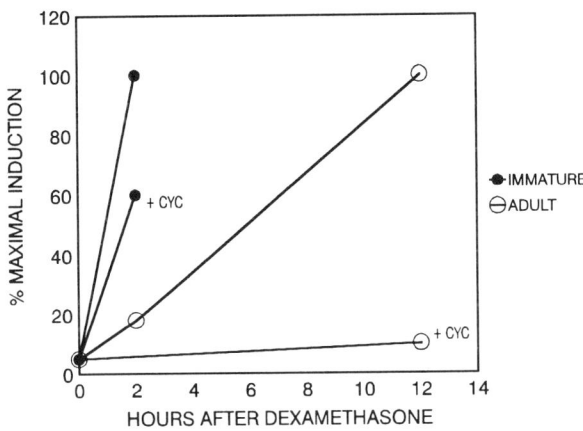

Fig 1 - Relative transcription activation of the CYP3A1 gene determined by run-on assay in the immature and in the adult rat liver following *in vivo* administration of Dexamethasone. Effect of Cyc treatment (+Cyc).

## DISCUSSION

Our results demonstrate that CYP3A1 gene is transcriptionally activated in rat liver by *in vivo* administration of Dex regardless to the developmental stage of the animals.

The markedly different time-lag of the adaptive response set-up observed as a function of development early (0.5-2.0 h) in the immature, and late (12.0 h) in the adults, strongly suggests that a complex array of regulatory mechanisms is triggered by Dex in the liver cell which results in a marked up-regulation of CYP3A1 gene expression.

CYP3A1 gene transcription activation by Dex is known to differ from the "classical" glucocorticoid receptor mediated mechanisms mainly driven by canonical GRE observed for genes such as the TAT (Shuetz and Guzelian, 1984; Burger and *at al.*, 1992). CYP3A1 is however a particularly sensitive gene to synthetic glucocorticoids induction responding in an almost all-or-none way to either glucocorticoids or anti-glucocorticoids. Conversely, the prototype TAT gene is actively transcribed *in vivo* under the physiological influence of endogenous glucocorticoid hormones and inhibited by anti-glucocorticoids.

We demonstrate that *in vivo* accumulation of CYP3A1 mRNA in the liver is a direct consequence of transcription activation mechanisms. However the pattern of response to the synthetic glucocorticoid is deeply affected by development.

Both primary and secondary responses to Dexamethasone clearly occur, which are dominant in the immature and in the adult rat liver respectively.

These may involve glucocorticoid receptor or new regulatory proteins recognizing the synthetic glucocorticoids and anti-glucocorticoids as activactor ligands.

The fact that *de novo* protein synthesis is necessary to the inductive response in the adults suggests that some labile protein participates in the control of this adaptive response which may depend on more complex signalling pathways ledding to gene activation through secondary messenger molecules.

Work is in progress to identify structural regulatory elements involved in glucocorticoid control at the CYP3A1 locus.

**REFERENCES**

Burger, H.-J, Schuetz, J.D., Schutz, E., and Guzelian, P.S. (1992): Paradoxical transcription activation of rat liver cytochrome-P450 3A1 by dexamethasone and the antiglucocorticoid pregnenolone 16α-carbonitrile: Analysis by transiente transfection into primary monolayer cultures of rat hepatocytes. *Proc. Natl. Acad. Sci. USA* 89, 2145-2149.

Gonzalez, F.J., Song, B.J., and Hradwick, J.P. (1986): Pregnenolone 16α-carbonitrile- Inducible P-450 Gene Family: Gene Conversion and Differential Regulation. *Mol. Cell. Biol.* 6, 2969-2976.

Lu, A.Y.H., Somogyi, A., West, S., Kuntzman, R., and Conney, A.H. (1972): Pregnenolone 16α-carbonitrile: A New Type of Inducer of Drug-Metabolising Enzymes. *Arch. Biochem. Biophys.* 152, 457-462.

Ribeiro, V.L., and Lechner, M.C. (1992): Cloning and Characterization of a Novel CYP3A1 Allelic Variant: Analysis of CYP3A1 and CYP3A2 Sex-Hormone-Dependent Expression Reveals that the CYP3A2 is Regulated by Testosterone. *Arch. Biochem. Biophys.* 293, 147-152.

Schuetz, E.G., and Guzelian, P.S. (1984): Induction of Cytochrome P450 by Glucocorticoids in Rat Liver. II Evidence that Glucocorticoids regulate induction of Cytochrome P450 by a nonclassical Receptor Mechanism. *J. Biol. Chem.* 259, 2007-2012.

Shimada, M., Nagata, K., Murayama, N., Yamazoe, Y., and Kato, R. (1989): Role of growth hormone in modulating the constitutive and phenobarbital-induced levels of two P-450 6β (testosterone 6β-hydroxilase) mRNAs in rat liver. *J. Biochem.* 106, 1003-1034.

Telhada, M.B., Pereira, T.M., and Lechner, M.C. (1992): Effect of Dexamethasone and Phenobarbital on Run-on Transcription Rate and CYP3A mRNA Concentration in Rat Liver: Changes during Development. *Arch. Biochem. Biophys.* 298, 715-725.

*Teresa Pereira is recipient of a PhD Fellowship from JNICT, Portugal.*

# Different mechanisms of regioselection of fatty acid hydroxylation by two laurate ($\omega$-1)-hydroxylating P450s

Tomohiko Fukuda[1], Yoshio Imai[1], Masayuki Komori[1], Masahiko Nakamura[2], Emi Kusunose[3], Kiyoshi Satouchi[4], Masamichi Kusunose[4]

[1]Department of Veterinary Science, University of Osaka Prefecture, Sakai, Osaka 593. [2]Institute for Protein Research, Osaka University, Suita, Osaka 565. [3]Toneyama Institute for Tuberculosis Research, Osaka City University Medical School, Toyonaka, Osaka 560. [4]Department of Food Science and Technology, Faculty of Engineering, Fukuyama University, Fukuyama, Hiroshima 729-02, Japan

INTRODUCTION  Both rabbit P450 2E1 and 2C2 catalyze the hydroxylation of laurate predominantly at the $\omega$-1 position at almost the same rate, though the two P450s are 54% identical in the sequence and they have different substrate specificities. The present study was undertaken to clarify how the two P450s recognized the fatty acid as a substrate and selected the hydroxylation position and to examine whether a subtle difference in the environment of the essentially conserved heme domain could be detected between the two P450s.

METHODS  The conserved Thr at the putative distal helix (Thr303 of P450 2E1 and Thr301 of P450 2C2) was replaced with Ser, Lys, and His *via* site-directed mutagenesis. P450 2E1, P450 2C2, and their Ser-, Lys-, and His-mutants were purified from the transformed yeast cells (Imai and Nakamura, 1988, 1989, 1991; Fukuda *et al.*, 1993). The hydroxylation reactions of $C_9$-$C_{18}$ fatty acids were performed in the reconstituted systems containing the purified preparations of the wild-type and the Ser-mutated P450s and the hydroxylation products were analyzed by HPLC and GC-EIMS (Fukuda *et al.*, 1993).

RESULTS AND DISCUSSION
*Substrate selectivity (chain lengths of fatty acids)*—Both P450 2E1 and P450 2C2 hydroxylated not only laurate but also the other fatty acids, but the two P450s showed different selectivities of the chain lengths (Fig. 1). P450 2E1 hydroxylated $C_9$-$C_{18}$ fatty acids at different rates and the highest activity was toward laurate. On the other hand, P450 2C2 hydroxylated $C_9$-$C_{13}$ fatty acids at nearly similar rates, when compared in the total amounts of the products, but the activity toward myristate was only a trace. When the conserved Thr at the putative distal helix (Thr303 of P450 2E1, Thr301 of P450

2C2) was replaced with Ser, the mutated P450s exhibited the same selectivities of the chain lengths of the fatty acids as the wild-type counterparts, though the mutation significantly decreased the hydroxylase activity of P450 2E1 but somewhat increased that of P450 2C2 (data not shown).

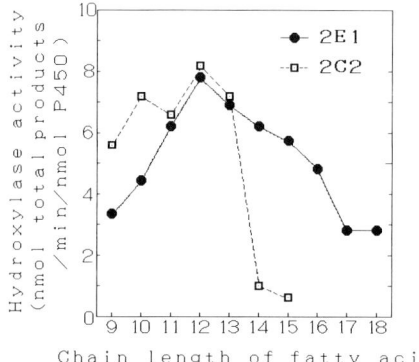

Fig. 1. Hydroxylation of fatty acids by P450 2E1 and P450 2C2. Hydroxylation products were separated by HPLC (Fukuda et al., 1993) and the amount of the total products was obtained by summing those of all the peaks detected.

*Regioselectivity of fatty acid hydroxylation*—When $C_9$-$C_{18}$ fatty acids were examined as the substrate in the reconstituted hydroxylase system containing P450 2E1, the ($\omega$-1)-hydroxylated products were mainly formed together with the $\omega$- and ($\omega$-2)-hydroxy isomers as minor products, independent of the chain lengths of the fatty acids, and the relative distributions for the ($\omega$-1)-isomers were more than 87% (Fig. 2A). On the other hand, P450 2C2 hydroxylated $C_9$-$C_{13}$ fatty acids at different positions dependent on the chain length of fatty acids (Fig. 2C). When undecanoate was used as the substrate, the ($\omega$-1)-hydroxy isomer was formed almost exclusively. Laurate was hydroxylated at the $\omega$-1 position and to a much lesser extent at the $\omega$ and $\omega$-2 positions. The hydroxylation positions were enlarged to the $\omega$-3 position in case with tridecanoate. The proportion of the $\omega$-hydroxylated products was markedly increased with decreasing the chain lengths of the fatty acids. Thus, when pelargonate was used as the substrate, the ratio of the $\omega$- and ($\omega$-1)-hydroxy isomers became about 2:1 and no ($\omega$-2)-hydroxy isomer was detected.

*Effects of the Ser-mutation on Regioselectivity*—When the conserved Thr was replaced with Ser, the regioselectivities of P450 2E1 and P450 2C2 were affected in different manners. The mutated P450 2E1 gave additional products to those with the wild-type P450 and the number of different products was increased with increasing the chain length of the fatty acids (Fig. 2B). The seven products were formed from palmitate and were identified as $\omega$-1, $\omega$-2, $\omega$-3, $\omega$-4, $\omega$-5, $\omega$-6, and $\omega$-7 monohydroxypalmitate by GC-EIMS analysis. The relative distribution of the regioisomers from $\omega$-1 to $\omega$-7 was calculated to be 14:15:27:32:10:2:1. Although the mutation enlarged the hydroxylation positions, the position of $\leq$C6 of the fatty acids was not hydroxylated by the mutated P450 2E1 at all and the $\omega$ hydroxy analog was detected only in trace amounts with $C_{12}$-$C_{18}$ fatty acids. When caprate was used as the substrate, only one additional metabolite ($\omega$-3 hydroxycaprate) to those with the wild-type P450 was seen and no additional product in the case of pelargonate.

The mutation of P450 2C2 did not change the hydroxylation positions of $C_9$-$C_{12}$ fatty acids but caused a significant decrease in the proporyion of the $\omega$-1 hydroxy analog in the total products (Fig. 2D). When pelargonate was used as the substrate of the mutated P450 2C2, the $\omega$-hydroxy isomer was produced 5.7 times as much as the ($\omega$-1)-hydroxy isomer. In the case of tridecanoate, the mutation increased the relative amount of the ($\omega$-3)-hydroxy isomer and, in addition, trace amounts of the ($\omega$-4)-hydroxy isomer was also detected. On the other hand, the relative distribution for the ($\omega$-1)-hydroxy isomer in the total products of undecanoate was still about 95%.

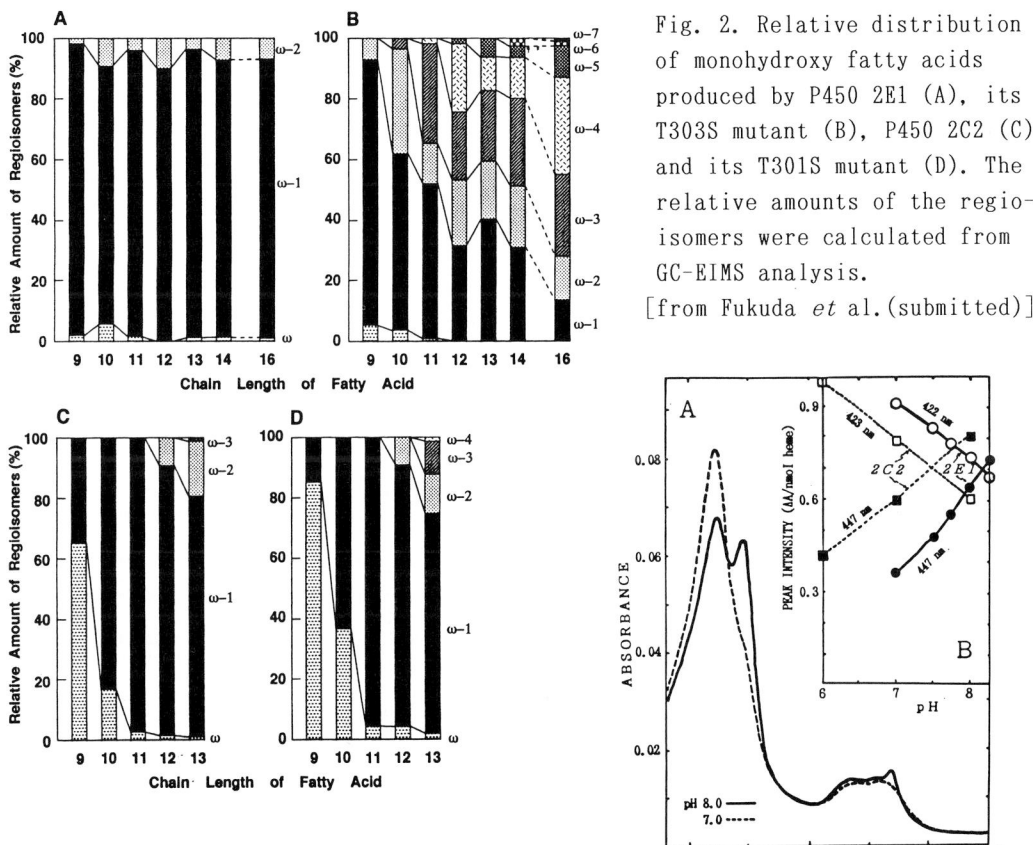

Fig. 2. Relative distribution of monohydroxy fatty acids produced by P450 2E1 (A), its T303S mutant (B), P450 2C2 (C) and its T301S mutant (D). The relative amounts of the regioisomers were calculated from GC-EIMS analysis.
[from Fukuda et al. (submitted)]

Fig. 3. Effect of pH on absorption spectrum of the ferrous oxygenated P450s. $\Delta A$($\lambda$ max-500nm) for the mutants of P450 2E1 and P450 2C2 (in 200 mM potassium phosphate buffer-20% glycerol-0.01% Emulgen 913) was calculated from the spectra as shown in A and in the reference (Imai and Nakamura, 1991), respectively.

*Mechanism of regioselection by P450 2E1 and P450 2C2*—These results indicate that P450 2E1 and P450 2C2 recognize fatty acids as the substrate and determine the hydroxylation position in different manners. P450 2E1 may be taken as $\omega$-1 hydroxylase of fatty acids, though this P450 has the broad substrate specificity. Various compounds having alkyl chain or aromatic ring but no carboxy group are hydroxylated at a position in the hydrocarbons by P450 2E1, suggesting that alkyl or aryl group is important to be recognized as the substrate by this P450. On the other hand, P450 2C2 may be regarded as 10-hydroxylase of $C_{11}$ fatty acid, undecanoate. 10-undecynoate was a very efficient suicide substrarte of P450 2C2 but P450 2E1 was less sensitive to it than P450 2C2, while *n*-alkyl alcohols inhibited the laurate hydroxylase activity of P450 2E1 but did not that of P450 2C2 (data not shown). As the selectivities of the fatty acids were loosened by the Thr to Ser-mutation, the $\gamma$-methyl group of the Thr may contribute to the limitation of the hydroxylation position by P450 2E1 and P450 2C2 but in clearly different ways between the two P450s.

*Spectral properties of the Lys-mutated P450s*—The conserved Thr to Lys-mutated P450 2E2 and P450 2C2 had absorption spectra characteristic of the nitrogenous-ligand bound form of P450, while the Thr to His-mutated P450s showed those typical of the low-spin type of P450 in the ferric state, suggesting that the $\varepsilon$-amino nitrogen of the Lys is located at the right position for coordination as the 6th ligand but the imidazole nitrogen of the His cannot. This difference is probably ascribed to the difference in the topology of the nitrogen atom and the heme iron between the Lys- and His-mutated P450s. Absorption spectra of the ferrous Lys-mutated P450s had two Soret peaks at around 447 and 422 nm and their relative heights were dependent on pH, indicating the existence of two interconvertible states which are in equilibrium (Fig. 3A). The equilibrium between the two states of the mutated P450 2E1 appears to be shifted toward the 422 nm state, compared with the mutated P450 2C2 (Fig. 3). The pyridine complex of P450 2E1 showed similar spectral properties to those of the Lys-mutated P450, while ferrous P450 2C2 was very unstable on binding to pyridine (data not shown). The CO complex of the Lys-mutated P450 2E1 was considerably stable, while that of the Lys-mutated P450 2C2 was very unstable and rapidly converted to its denatured form (data not shown). These findings suggest a certain spatial difference in the essentially conserved structure of the distal heme domain between P450 2E1 and P450 2C2. P450 2E1 seems to have a larger substrate pocket than P450 2C2.

REFERENCES

Imai,Y. & Nakamura,M. (1988) *FEBS Lett.* 234, 313-315
Imai,Y. & Nakamura,M. (1989) *Biochem. Biophys. Res. Commun.* 158, 717-722
Imai,Y. & Nakamura,M. (1991) *J. Biochem.* 110, 884-888
Fukuda,T. *et al.* (1993) *J. Biochem.* 113, 7-12

# Use of the paraxathine/caffeine ratio in plasma and in saliva for CYP1A2 phenotyping

Uwe Fuhr[1], Karl-Ludwig Rost[2], Reinald Engelhardt[1], A. Horst Staib[1]

[1]Department of Clinical Pharmacology, University Hospital Frankfurt am Main, FRG. [2]Institute of Clinical Pharmacology, Klinikum Steglitz, Free University of Berlin, FRG

## INTRODUCTION

Procedures to determine CYP1A2 activity *in vivo* by use of a caffeine (137X) probe comprise 137X clearance (Cl), based on both plasma (P) and saliva (S) concentrations, ratios of urinary metabolites, the $^{13}C$-137X breath test, and the paraxanthine/caffeine ratio (17X/137X) (R) in P (Kalow & Tang, 1993; Tanaka et al., 1992). However, in intraindividual comparison in patients undergoing partial hepatectomy, only clearance and the 17X/137X ratio in plasma and in saliva were shown to be highly correlated to CYP1A2 activity *in vitro*.

Besides potential shortcomings of any of the methods (for a recent review, see Kalow & Tang, 1993), the use of the 17X/137X ratio in plasma or in saliva may offer considerable practical advantages. These are simple analytical methods, no need sample extraction procedures, of synthesis of reference substances which are not commercially available, of a highly sophisticated technical equipment, or of strict rules of conduct for subjects. Sample withdrawal is restricted to one point, and, as an additional benefit, saliva sampling is non-invasive.

To date, there are only sparse data available about the relationship between CYP1A2 activity and the 17X/137X plasma or saliva ratio. Caffeine clearance is established as an appropriate reference parameter for determining CYP1A2 activity *in vivo* (Kalow & Tang, 1993). Therefore, we included measurement of these ratios into four clinical studies where caffeine clearance and derived parameters were used for CYP1A2 phenotyping, in order to assess them for CYP1A2 phenotyping in man.

## MATERIAL & METHODS

The design of the clinical studies was as follows:
STUDY I: 24 healthy male volunteers, all nonsmokers; one period without comedication; *parameters*: Clp, Rp 3 and 6 hours postdose.

STUDY II: 12 healthy volunteers, 7f/5m, 6 smokers/6 nonsmokers; period A without comedication, period B with coadministration of grapefruit juice, *parameters:* Clp, Cls; Rs 3 hours postdose, Rp and Rs 5 hours postdose, ratios of caffeine and its metabolites in spontaneous urine 5 hours postdose.

STUDY III: 25 liver tumor patients, 14f/11m, 5 smokers/20 nonsmokers; period A within two days prior to surgery, period B appr. one week after surgery, with multiple comedications; *parameters:* Clp, Cls; Rp and Rs 3 and 6 hours postdose, ratios in spontaneous urine 6 hours postdose (n = 16 - 19 data pairs).

STUDY IV: 18 healthy volunteers, 7f/11m, 2 smokers/16nonsmokers; period A without comedication (n = 10 data pairs available), period B coadministration of 40 or 120 mg omeprazole per day (n = 17 data pairs available), *parameters:* Clp, $^{13}$C-caffeine breath test (BT), ratios in urine sampled 6-8 hours postdose.

The correlations between plasma clearance and other CYP1A2 parameters were estimated by linear regression analysis with variables weighted by their inverse values. Explorative estimation of correlation in pooled data was done by nonlinear regression analysis.

RESULTS

Table 1. Correlation between caffeine plasma clearance and pharmacokinetic parameters of caffeine used to estimate CYP1A2

| Correlation to Clp (ml/min/kg) of: | | Study I | Study IIA | Study IIB | Study IIIA | Study IIIB | Study IVA | Study IVB |
|---|---|---|---|---|---|---|---|---|
| Cls (ml/min/kg) | n | | 12 | 12 | 16 | 16 | | |
| | r | not done | 0.947 | 0.975 | 0.989 | 0.985 | not done | |
| | p | | 0.000 | 0.000 | 0.000 | 0.000 | | |
| $R_{P3h}$ | n | 24 | | | 19 | 16 | 10 | 17 |
| | r | 0.653 | not done | | 0.843 | 0.949 | 0.628 | 0.629 |
| | p | 0.000 | | | 0.000 | 0.000 | 0.051 | 0.007 |
| $R_{P5h}$ | n | | 12 | 12 | | | | |
| | r | not done | 0.911 | 0.854 | not done | | not done | |
| | p | | 0.000 | 0.000 | | | | |
| $R_{P6h}$ | n | 24 | | | 19 | 16 | 10 | 17 |
| | r | 0.903 | not done | | 0.886 | 0.955 | 0.869 | 0.719 |
| | p | 0.000 | | | 0.000 | 0.000 | 0.001 | 0.002 |
| $R_{S3h}$ | n | | 12 | 12 | 19 | 16 | | |
| | r | not done | 0.846 | 0.747 | 0.852 | 0.915 | not done | |
| | p | | 0.000 | 0.006 | 0.000 | 0.000 | | |
| $R_{S5h}$ | n | | 12 | 12 | | | | |
| | r | not done | 0.875 | 0.901 | not done | | not done | |
| | p | | 0.000 | 0.000 | | | | |
| $R_{S6h}$ | n | | | | 18 | 16 | | |
| | r | not done | not done | | 0.874 | 0.950 | not done | |
| | p | | | | 0.000 | 0.000 | | |
| $R_{S7h}$ | n | | 12 | 12 | | | | |
| | r | not done | 0.943 | 0.910 | not done | | not done | |
| | p | | 0.000 | 0.000 | | | | |
| Breath Test (% of dose) | n | | | | | | 10 | 17 |
| | r | not done | not done | | not done | | 0.623 | 0.679 |
| | p | | | | | | 0.054 | 0.003 |

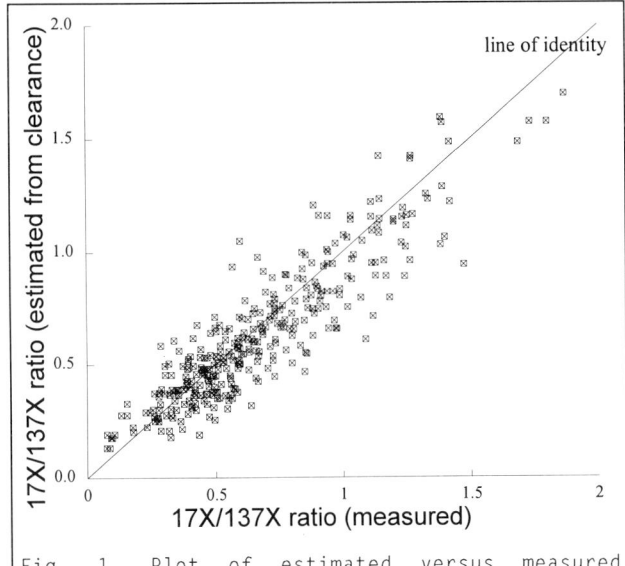

Fig. 1. Plot of estimated versus measured 17X/137X ratios both in plasma and in saliva (all studies), based on results of eq. 1

Correlation coefficients r for urine ratios ranged from 0.236 to 0.744 for 17X/137X, from 0.015 to 0.810 for (17X+17U)/137X, from 0.271 to 0.703 for (1X+1U+AFMU)/17U, and from 0.232 to 0.662 for (1X+1U+AFMU+AAMU)/17U. Non-linear regression calculated exploratively between pooled 17X/137X ratios and 137X clearance for all studies and for all samples, using eq. 1: **R = a + b • Clp (ml/min/kg) • postdose sampling time (h)**, gave parameters of a=0.0736 and b=0.0769, with an correlation coefficient of r = 0.975. Inclusion of parameters for the difference plasma - saliva was not favored by the Akaike criterion.

## DISCUSSION

The close correlations observed between plasma or saliva 17X/137X ratios and caffeine plasma clearance strongly support the validity of the 17X/137X ratios for CYP1A2 phenotyping. These ratios reflected caffeine clearance better than the urinary ratios tested (Kalow & Tang, 1993). By using equation (1), it seems possible to estimate caffeine clearance from any single plasma or saliva sample 3-7 hours postdose more precisely than by urinary ratios or by the breath test.

The increase of correlation coefficients for the 17X/137X ratios with postdose time, presumably caused by the decreasing influence of dietary intake of caffeine prior to the test, suggests that sampling should be done 5 or more hours postdose. For practial reasons, and to avoid problems of caffeine assay sensitivity when using simple HPLC devices, sampling should not be later than 7 hours postdose. Saliva samples should be preferred to plasma samples due to its non-invasive character.

The expenditure of time, money, and equipment for CYP1A2 phenotyping is reduced considerably by the use of paraxanthine/caffeine ratios in saliva, without loss of precision.

## REFERENCES

Kalow, W. & Tang, B.-K. (1993): The use of caffeine for enzyme assays: A critical appraisal. *Clin. Pharmacol. Ther. 53*, 503-514.

Tanaka, E., Ishikawa, A., Yamamoto, Y., Osada, A., Tsuji, K., Fukao, K., Misawa, S., and Iwasaki, Y. (1992) A simple useful method for the determination of hepatic function in patients with liver cirrhosis using caffeine and its three major dimethylmetabolites. *Int. J. Clin. Pharmacol. Ther. Toxicol. 30*, 336-341.

# Human CYP1A2 activity: relationship between caffeine clearance *in vivo* and its 3-demethylation *in vitro*

Uwe Fuhr[1], Reinald Engelhardt[1], Michael Sachs[2], Karl-Ludwig Rost[5], Paul Schymanski[1], Christoph Huschka[1], Dieter Liermann[3], Heimo Wissing[4], A. Horst Staib[1]

[1]Department of Clinical Pharmacology. [2]Department of General Surgery. [3]Center of Radiology. [4]Center of Anaesthesia, University Hospital, Frankfurt am Main, FRG. [5]Institute of Clinical Pharmacology, Klinikum Steglitz, Free University of Berlin, FRG

## INTRODUCTION

The 3-demethylation of caffeine (137X) *in vitro* at concentrations below 100 µM is mediated in human liver microsomes almost exclusively by CYP1A2 (Tassaneeyakul et al., 1993). This reaction is responsible for 80 % of primary caffeine metabolism in man *in vivo*. The major fraction of 1-demethylation and a part of 7-demethylation are also mediated by CYP1A2 (Tassaneeyakul et al., 1993; Fuhr et al., 1992). Based on this rationale, caffeine is applied as a probe to determine CYP1A2 activity *in vivo*. Pharmacokinetic parameters comprise caffeine clearance, calculated either from plasma or from saliva concentrations, ratios of urinary metabolites, the $^{13}C$-caffeine breath test, and the paraxanthine/caffeine ratio in plasma (for a recent review, see Kalow & Tang, 1993).

However, all these parameters have potential shortcomings, of which the most important are: (a) CYP1A1 is also capable to 3-demethylate caffeine (Tassaneeyakul et al., 1993). This inducible enzyme is usually not expressed to a relevant extent in human liver, but was detected in other tissues. Information on the whole body content of CYP1A2 and CYP1A1 are not available. (b) Urinary flow and interethnic differences in renal function may influence the excretion of caffeine and its metabolites (Kalow & Tang, 1993). (c) Mutual competitive inhibition of caffeine, paraxanthine and other metabolites at the CYP1A2 binding site may occur. Thus, although all parameters at least in part reflect CYP1A2 activity and are highly intercorrelated, definition of the error components is urgently needed (Kalow & Tang, 1993). The present investigation was designed to directly validate caffeine parameters used for human CYP1A2 phenotyping by intraindividual comparison to CYP1A2 activity *in vitro*.

## MATERIAL & METHODS

25 patients determined for partial hepatectomy (11 m, 14 f, age 56 ± 11 years, 5 smokers/20 nonsmokers, weight 72 ± 14 kg [mean ± SD]), in most cases due to primary liver neoplasm or liver metastasis of other carcinoma, gave their informed

consent to participation in the study. Liver size (without tumor) was measured using computer assisted tomograms, recorded within one month prior to surgery.

A 165 mg 137X test dose was administered within two days before surgery. Blood samples (before administration of 137X and 3, 6, and 9 hours thereafter), saliva samples (before administration and 2, 3, 6, 9, 12 and 15 thereafter), and a urine sample 6 hours after administration were drawn. The following pharmacokinetic parameters were calculated: 137X clearance ($Cl_{PS}$) based on plasma concentrations, and on saliva concentrations, transformed to plasma concentrations using the intraindividual saliva/plasma ratio; caffeine clearance ($Cl_S$), based on saliva concentrations, transformed to plasma values using a saliva/plasma ratio of 0.79 (Fuhr et al., 1993); paraxanthine/caffeine ratio in plasma ($R_P$) and in saliva ($R_S$) 3 and 6 hours postdose; urinary ratios of 137X and its metabolites [17X/137X; (17X+17U)/137X; (1X+1U+AFMU)/17U; (1X+1U+AFMU+AAMU)/17U] (Kalow & Tang, 1993).

If available, a macroscopically unaffected liver sample was derived from the excised lobe during surgery and frozen immediately. Microsomes were prepared by standard methods. The yield of microsome suspension was documented by several weighing steps during preparation. Caffeine (concentrations: 62.5 - 2000 µM) was incubated for 15 min at 37 °C in triplicate. Enzyme constants were calculated by nonlinear regression analysis assuming a low and a high affinity binding site, with the $K_M$ for the low affinity site fixed at 30 mM (Tassaneeyakul et al., 1993). Intrinsic clearance was calculated as $K_M/V_{max}$ and corrected for liver size.

Concentrations of 137X and its metabolites were determined by HPLC. The relationship between intrinsic clearance for caffeine 3-demethylation in vitro and pharmacokinetic parameters of caffeine in vivo was established using linear regression analysis with variables weighted by their inverse values.

## RESULTS

$K_M$ for 137X 3-demethylation varied from 202 to 655 µM, intrinsic clearance from 5.5 to 46.3 ml/min. 137X clearance in vivo ranged from 0.33 to 3.02 ml/min/kg body weight. The correlation between intrinsic clearance and caffeine metabolism is characterised in table 1. Relationship between intrinsic clearance and clearance in vivo is presented in fig. 1.

Table 1. Correlation between intrinsic clearance (ml/min for whole liver) for caffeine 3-demethylation and pharmacokinetic parameters of caffeine

| parameter | n | r | p |
|---|---|---|---|
| clearance (plasma/saliva; ml/min/kg) | 18 | 0.801 | 0.000 |
| clearance (saliva; ml/min/kg) | 16 | 0.821 | 0.000 |
| 17X/137X ratio in plasma 3 hours postdose | 18 | 0.702 | 0.001 |
| 17X/137X ratio in plasma 6 hours postdose | 18 | 0.777 | 0.000 |
| 17X/137X ratio in saliva 3 hours postdose | 17 | 0.700 | 0.002 |
| 17X/137X ratio in saliva 6 hours postdose | 17 | 0.734 | 0.001 |
| 17X/137X in urine | 17 | 0.440 | 0.077 |
| (17X+17U)/137X in urine | 17 | 0.392 | 0.119 |
| (1X+1U+AFMU)/17U in urine | 17 | 0.562 | 0.019 |
| (1X+1U+AFMU+AAMU)/17U in urine | 17 | 0.532 | 0.028 |

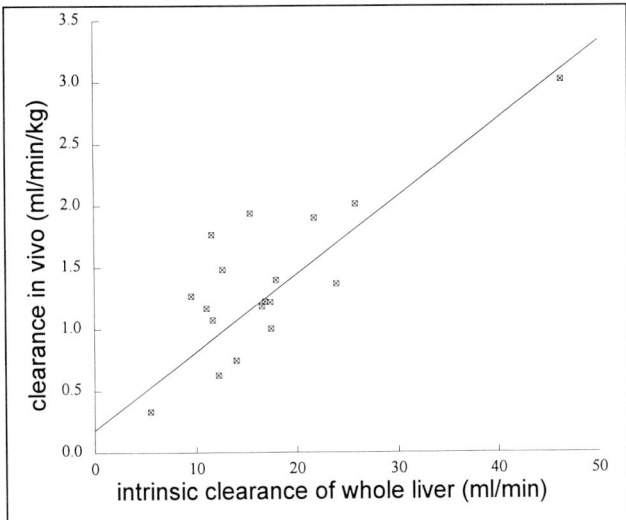

Fig. 1. Correlation between intrinsic clearance for caffeine 3-demethylation *in vitro* and caffeine clearance (CLPS[plasma/saliva]) *in vivo*

## DISCUSSION

Our results show that CYP1A2 activity is the major determinant of 137X clearance *in vivo*. A fraction of $r^2=0.642$ of the variability of this *in vivo* parameter is explained by interindividual diversity of intrinsic clearance. Similar coefficients for the ratios 17X/137X in plasma and in saliva were found. The residual variablility may mainly be due to the wide range of co-medications, and to the inaccuracy of the methods used for the numerous measurements necessary to establish the comparison, which approximate a coefficient of variation of 5% each, rather than by additional physiological factors contributing to caffeine metabolism. Therefore, CYP1A1 and other enzymes might not be relevant for primary caffeine metabolism *in vivo* in the patients tested, or their expression might be coregulated with CYP1A2. Lower correlation coefficients for the ratios investigated in spontaneous urine 6 hours postdose may be overcome by strictly defined sampling periods (Kalow & Tang, 1993).

In conclusion, caffeine clearance *in vivo* and the ratios 17X/137X six hours postdose, derived either from plasma or from saliva concentrations, provide valid estimates of individual CYP1A2 activity in subjects without induction of CYP1A1.

## REFERENCES

Fuhr, U., Klittich, K., and Staib, A.H. (1993): Inhibitory effect of grapefruit juice and the active component naringenin on CYP1A2 dependent metabolism of caffeine in man. Br. J. Clin. Pharmacol. 35, 431-436.

Fuhr, U., Doehmer, J., Battula, N., Wölfel, C., Kudla, C., Keita, Y., and Staib, A.H. (1992): Biotransformation of caffeine and theophylline in mammalian cell lines genetically engineered for expression of single cytochrome P450 isoforms. Biochem. Pharmacol. 43, 225-235.

Kalow, W. & Tang, B.-K. (1993): The use of caffeine for enzyme assays: A critical appraisal. Clin. Pharmacol. Ther. 53: 503-514.

Tassaneeyakul, W., Mohamed, Z., Birkett, D.J., McManus, M.E., Veronese, M.E., Tukey, R.H., Quattrochi, L.C., Gonzalez, F.J., and Miners, J.O. (1992): Caffeine as a probe for human cytochromes P450: validation using cDNA-expression, immunoinhibition and microsomal kinetic and inhibitor techniques. Pharmacogenetics 2, 173-183.

# Prediction and experimental confirmation of the cytochromes b5 and P450 peptide mapping

V. Zgoda, D. Proschlyakov, O. Ipatova, V. Prosorovsky, G. Bachmanova

Institute of Biomedical Chemistry, Pogodinskaya str., 10, Moscow, Russia

Peptide mapping is an universal method for study of proteins` structure and function. The method of multidimentional peptide mapping using high performance liquid chromatography (HPLC) is proposed in this work as combining a good resolution with rapidity and economisity of analysis.

For prediction of multidimentional peptide maps an original program for IBM-compatible computers was created. This program allows to predict sequences of peptides, appeared during enzymatic hydrolysis of protein with known sequence, and to calculate their molecular weights, spectral properties and retention times during separation of those peptides using reversed-phase (RP) HPLC. For this program a data bank of amino acids` extinction coefficients in conditions of RP HPLC was created and opportunity to predict spectral properties of peptides with known sequence was confirmed. An opportunity to predict a retention time of peptides in RP HPLC was confirmed also.

The multidimentional peptide maps of P450 2B4, 1A2 and b5 using Lysylendopeptidase were created on the basis of molecular weight (gel filtration), hydrophobicity and absorbance ratio 280/210 nm (RP HPLC) of peptides.

To confirm of correctness this prediction the analysis of the primary structure of some peptides has been carried out using a "MilliGen/Biosearch ProSequencer 6600" sequenator.

The experimental and predicted peptide maps of P450 2B4, 1A2 and b5 were compared. It was found, that correlation between experimental and predicted peptides was good. The results lead to the conclusion that multidimentional mapping using HPLC can be used for study of proteins with known sequence.

MATERIALS AND METHODS.
Cytochrome b5, P450 2B4, 1A2 preparations were isolated as described earlier [1, 2, 3]. The proteins showed a single band on SDS-PAGE.

---

This work was supported in part by Grants from Russian Foundation for Fundamental Investigations.

b5, 2B4, 1A2 cleavage by Lysylendopeptidase (LEP). For cleavage 0.5 mg of carboxymethylated protein was dissolved in 0.2 ml 50 mM borate buffer, pH 9.2, contained 6 M Gu-HCl. The solution was diluted with 50 mM borate buffer, pH 9.2 up to 2.5 M Gu-HCl. Solution of LEP in the same buffer was added to give a 1:25 molar ratio and 2.0-2.5 M Gu-HCl. The cleavage proceeded at 25°C for 16-20 hours.

Range of hydrolysis was appraised using the method of NH2-groups determination [4].

Gel-Filtration chromatography was carried out on G-50 (fine) (600*9 mm) column and 6 M urea, 0.1 formic acid, pH 2.9 as a mobile phase. Flow rate 0.2 ml/min [5].

Reverse-Phase HPLC was performed using Hi-Pore RP-318 C-18 column (250 * 4.6 mm, Bio-Rad).

Predicted retention time of peptides by Sasagawa et al. [6] were calculated in the following form:

$$t_r = A*\ln(1 + \sum_j D_j n_{ij}) + C$$

where A, C are slope and intercept of line, ones were defined by using experimental values of retention times of markers, $D_j$ is the retention constant of amino acid j and $n_{ij}$ is number of residues of amino acid j in peptide i.

Peptide spectral characteristic calculation. Molar extinction coefficient at 210 and absorbance ratio at 280/210 nm of peptides were obtained by summarizing both acid molar extinction coefficients of amino acids and peptide bonds.

RESULTS AND DISCUSSION

Creation of three-dimentional peptide map.

The hydrolysis of protein is a first stage of peptide maps creating. Selection of protease and conditions of hydrolysis, are found of importance for peptide map be repeatable. In present work we used bacterial protease from Achromobacter (Lysylendopeptidase) possessing both high specificity in the Lys-x peptide bond and stability in detergent conditions. An optimized method of NH2-group registration in reaction with fluorescamine was applied for titration of b5 hydrolysis by LEP, and conditions for stable hydrolysis were selected. Using the conditions have been described in "Material and Methods" the yield of hydrolysis was about 90 %.

Then the digested protein was separated sequentially using GF and RP-HPLC. During the GF chromatography fractions (by 1 ml each) were collected. Molecular weight of fractions were defined from calibration curve of GF column. In detergent conditions that we used the peak positions are correlated with molecular weight of peptides [5]. The next step was separation of some collected fractions in RP-HPLC (see Fig. 1). Molecular weight of peptide was defined using value of molecular weight of the same fraction in which the square of peak was the biggest.

By the same procedure we have defined molecular weight, and retention time of peptides in RP-HPLC for each fragment of b5, P450 2B4, 1A2.

The ratio of peptide coefficient extinction at two wavelengths was chosen as a third independent parameter. We used the absorbance ratio A280/A210 that is a measure of aromatic acids content in protein. This parameter was measured during RP-HPLC.

Data of three-dimensional peptide map of b5 is shown in Table 1.

Prediction of peptide map's parameters.

To predict sequences of peptides and to calculate their molecular weights appeared during enzymatic hydrolysis of protein with known sequence, isn't a challenging task.

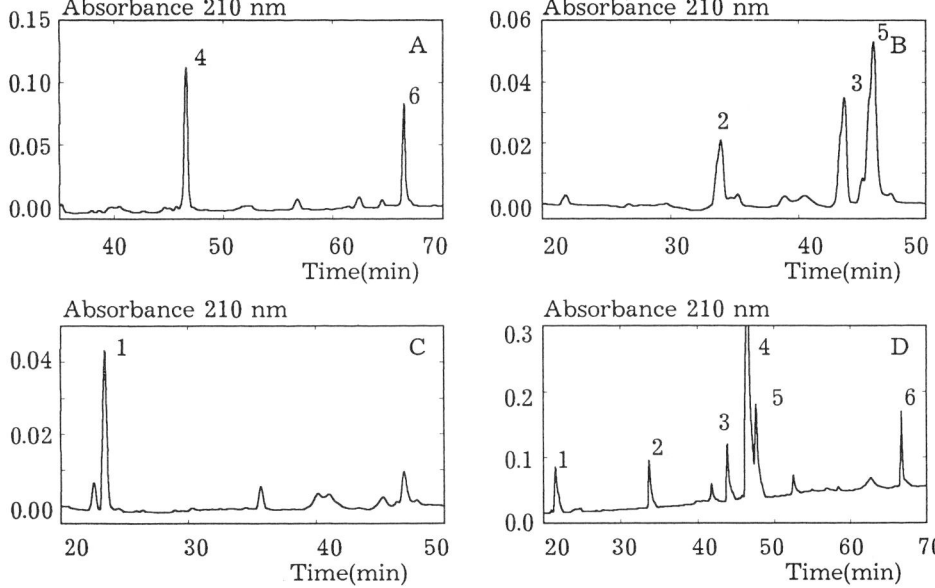

Fig. 1. RP-HPL chromatograms of fractions that were obtained during separations on G-50 column (a-c) and RP-HPL chromatogram of unseparated digested b5 (d). 1 ml of fraction after GF or 0.1 ml (5 nmole) of digested b5 were separated using RP-HPLC C18 column. Chromatograms of 4-5 kD, 1-2kD, and 0.1-1 kD fractions are shown in Figures a, b, c, respectively.

The methods of predicting the peptides' retention of known composition in RP-HPLC have been reported previously [6]. Sasagawa et al. noted that under different conditions of separation (e. g., alternative gradient slope) only the intercept and the slope of straight lines of relationship between calculated and observed retention times change, but not coefficients. Consequently, fitting the intercept and the slope of straight lines with the peptides of known sequences, we can use the Sasagawa constant for predicting the peptide's retention time. Fig. 2 shows the plot of the observed retention time of peptides of known sequences as a function of $\ln(1+H)$, where H is a summing of Sasagawa's retention constants.

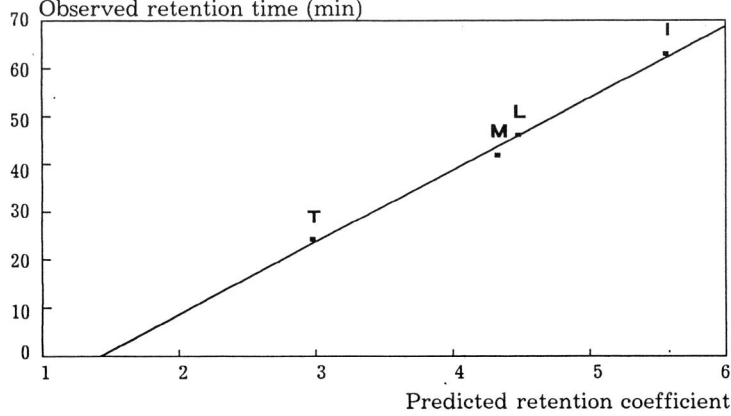

Fig. 2. Relationship between observed retention time of marker peptides and their retention constant calculated according to Sasagawa et al. [6].

855

According to [7], the peptides spectral properties are defined by such chromophores as peptide bonds, amino acid residues, and factors of protein's secondary structure. Considering conditions, which are applied in RP-HPLC separation and the small length of peptides, the probability of existing of helix and sheet is a little. From the above reasoning, we calculated molar extinction coefficient at 210 nm and absorbance ratio at 280/210 nm of peptides by summarizing molar extinction coefficients both of amino acid and peptide bonds. The results of calculations for all parameters of b5 are present in Table 1.

TABLE I
COMPARISON OF PREDICTED AND EXPERIMENTAL DATA.

| No. | Observed[*] / Predicted[**] | Molecular weight (D) | Retention Time (min.) | A280/A210 | Result of sequencing |
|---|---|---|---|---|---|
| 1. | Peak-1 | 5000 | 66.8 | 0.2 | |
| 2. | 94-133 | 4620±7% | 67.6±1 % | 0.47 | |
| 3. | Peak-2 | 5000 | 46.5 | 0.06 | 39-76 |
| 4. | 39-76 | 4370±12% | 50±7.4 % | 0.03 | |
| 5. | Peak-3 | 1500 | 47.5 | 0 | 77-90 |
| 6. | 77-90 | 1660±11 % | 49±3.2 % | 0.02 | |
| 7. | Peak-4 | 1500 | 43.9 | 0.14 | 10-17 |
| 8. | 10-17 | 1110±25 % | 42.2±4 % | 0.4 | |
| 11. | Pcak-5 | 1000 | 33.7 | 0.12 | 33-38 |
| 12. | 33-38 | 738±26 % | 37.7±11% | 0.38 | |
| 13. | Pcak-6 | 500 | 21.6 | 0 | |
| 14. | 91-93 | 346±30 % | 25.1±16% | 0.04 | |

* Peak numbered according to fig. 1d
** Number of amino acid residue of b5 sequence.

The theoretical and observed data was compared and the more suitable to each other experimental and predicted fragments were defined. Results of analysis are presented in Table 1.

To confirm correctness of this identification the analysis of the primary structure of 4 peptides of b5 and 2 peptides of P450 2B4 has been carried out using a sequenator.

For 5 sequenced peptides was found a one-to-one correspondence between theoretical and experimental results.

REFERENCES
1. Sparts,L., and Strittmater,P., Proc.Natl.Acad.Sci. USA 68, (1971), 1042-1046.
2. Imai Y., Hashimoto-Yutsudo C., et. al., J. Biochem., 88, (1980), 489-503.
3. Alterman M.A., Biomed. Chromatogr., 4, 1990, 221-222.
4. Bohlen P.,Stain S., et. al., Arh. Bioch. Bioph. 155,213-220, 1973.
5. Waterfield M.D.,Scrace G.T. In Biological/Biomedical Applications of Liquid Chromatography, Hawk G.L., (ed.), Chromatographic Scince Series, v. 18, Marcel Dekker, New York, Basel, (1981), 135-158.
6. Sasagawa T., Okuyama T., Teller D.C., J.Chromatogr., 240, (1982), 329.
7. Wetlaufer D.B., Adv. Protein Chem., 17, (1962), 303.

# Modulation of CYP3A1 expression by heme in transfected COS-1 cells

Vera Ribeiro, Maria Celeste Lechner

*Laboratório de Bioquímica, Instituto Gulbenkian de Ciência, Apartado 14, 278 Oeiras, Portugal*

## INTRODUCTION

Heme has been demonstrated to play a general role in protein synthesis, acting by inhibiting phosphorylation of the initiation factor eIF2. Besides, the biogenesis and correct folding of Cytochrome P450 depend on the availability of its prosthetic group, the heme. Actually, heme has long been reported to play a positive regulatory role in the *in vivo* transcription elongation of CYP2B and CYP1A (Ravishankar and Padmanaban, 1986; Dwarki *et al.*, 1987). However, the inhibition of heme biogenesis does not prevent induction of CYP mRNA by PB (Srivastava *et al.*, 1989; Hamilton *et al.*, 1988; Sinclair *et al.*, 1990), although an impairment of CYP2B as well as CYP3A protein has been observed (Sinclair *et al.*, 1990).

Therefore, the effect of heme on the expression of liver microsomal CYPs remains a controversial question.

In order to identify the relative importance of transcriptional and post-transcriptional events in the modulation of CYP3A1 biogenesis by heme, we analyzed the effects on the expression of CYP3A1 in transfected COS-1 cells. The different levels of the CYP3A1 biogenesis were investigated by determination of the time-course accumulation of the mRNA and of the protein, for a period of 5 hours, starting 48 h and 72 h post-transfection, respectively (Ribeiro and Lechner, 1994).

## MATERIAL AND METHODS

CYP3A1 full-length cDNA (EMBL Data Library Accession Number X64401) originally cloned in pBluescript IISK+/- (Ribeiro and Lechner, 1992) was subcloned into *Cla*I / *Xba*I-digested pCMV4 (Andersson *et al.*, 1989).
COS-1 cells grown in Dulbecco's Minimal Essential Medium supplemented with 10% fetal calf serum were transfected by the DEAE-dextran method (400 µg/ml), in the presence of chloroquine (100 µM), followed by a shock with 10% DMSO.
Total RNA was prepared by the LiCl/urea method (Auffray and Rougeon, 1980) and the relative concentration of CYP3A1 mRNA determined by Northern Blot analysis, and hybridization to the CYP3A1 cDNA, $^{32}$P-labeled by random priming.
Protein concentration was estimated by Western Blot analysis of total cell lysates prepared using triple-detergent lysis buffer (Sambrook *et al.*, 1989). Polyclonal specific antibody was kindly supplied by Dr. M.A.Correia, from UCSF. Detection of antigen-antibody complexes was performed by incubation with $^{125}$I-labeled protein A.

The time-course variation (5 hours) of the relative concentration of mRNA and protein was evaluated in the COS-1 cells 47 and 71 h after transfection, as a function of heme addition (100 µM) to the culture medium (Aziz and Munro, 1987). The stability of both CYP3A1 mRNA and protein was estimated, over the same 5 hours period, by including in the culture medium the transcription and translation inhibitors, actinomycin D (0,2 µM) and cycloheximide (1,8 µM).

## RESULTS

The effect of heme on the protein accumulation was analyzed in steady-state conditions and after inhibition of protein synthesis by cycloheximide. Transfection assays of COS-1 cells with 10 µg of recombinant pCMV4-CYP3A1 full-length cDNA resulted in the expression of CYP3A1 protein, which attained steady-state levels at 72-76 h. Supplementation of the culture medium with 100 µM heme brought about a marked increase in the concentration of CYP3A1 protein, which was observed over the whole studied period (5 hours), as represented in Fig.1. Inhibition of protein synthesis by cycloheximide resulted in a progressive decrease in the protein concentration, observed over the same interval (72-78 h), regardless of heme addition.

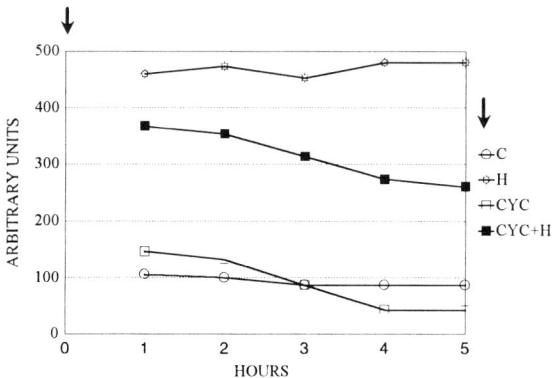

Fig. 1. Plot of the CYP3A1 protein relative concentration, evaluated from Western Blots, 72 to 76 h after transfection. The figures are represented as a percentage of control values (C), at different time points, 1 to 5 hours after supplementation of the culture medium with heme (H), cycloheximide (CYC), or cycloheximide plus heme (CYC+H).

The effect of heme on the CYP3A1 mRNA concentration was investigated by Northern Blot analysis over a 5 hours period. The results summarized in Fig. 2 demonstrate that a net decrease in the steady-state level of CYP3A1 mRNA is carried on by heme. However, a less marked relative decay in CYP3A1 mRNA concentrations is observed in the heme-treated cells, along the period 48-52 h post-transfection.

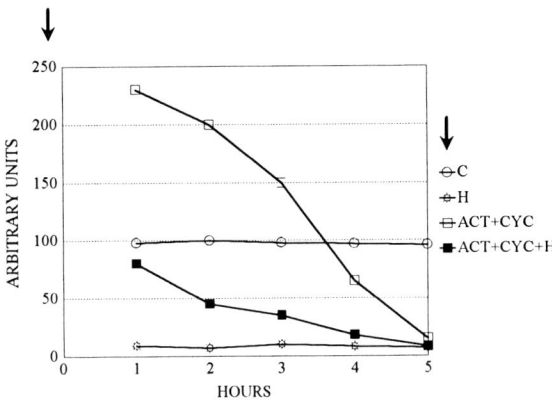

Fig. 2. Plot of the CYP3A1 mRNA relative concentration, determined from Northern Blots, 48 to 52 h after transfection. The figures are represented as a percentage of control values (C), at different time points, 1 to 5 hours after supplementation of the culture medium with heme (H), actinomycin D and cycloheximide (ACT+CYC), or actinomycin D plus cycloheximide in the presence of heme (ACT+CYC+H).

## CONCLUSIONS

The results here summarized demonstrate that heme exerts a net positive control on the biosynthesis of CYP3A1 protein. In COS-1 transfected cells, this protein accumulates at significantly higher steady-state levels after heme addition to the culture medium. Conversely, a net decrease in the steady-state level of CYP3A1 mRNA is observed. This decrease in the abundance of the mRNA over the period 48-52h after transfection in the presence of heme does not prevent the over accumulation of the translated protein, observed 72-76h post-transfection.

Taken together, these observations suggest that heme may lead to a marked increase in the translation efficiency carrying on a negative feedback control on the synthesis of the CYP3A1 mRNA.

## REFERENCES

Andersson, S., Davis, D.L., Dahlbäck, H., Jörnvall, H., Russell, D.W. (1989): Cloning, Structure and Expression of the Mitochondrial Cytochrome P450 Sterol 26-Hydroxylase, a Bile Acid Biosynthetic Enzyme. *J. Biol. Chem.* 264, 8222-8229.

Auffray, C., Rougeon, F.(1980): Purification of mouse immunoglobulin heavy-chain messenger RNAs from total myeloma tumor RNA. *Eur. J. Biochem.* 107, 303-314.

Aziz, N. and Munro, H.N. (1987): Iron regulates ferritin mRNA translation through a segment of its 5' unstranslated region. *Proc. Natl. Acad. Sci. USA* 84, 8478-8482.

Dwarki, V.J., Francis, V.N.K., Bhat, G.J. and Padmanaban, G. (1987): Regulation of Cytochrome P450 messenger RNA and apoprotein levels by heme. *J. Biol. Chem.* 262, 16958 - 16962.

Hamilton, J.W., Bement, W.J., Sinclair, P.R., Sinclair, W.J.Wetterhahn, K.W. (1988): Expression of 5-aminolevulinate synthase and cytochrome P450 mRNAs in chicken embryo hepatocytes in vivo and in culture. Effect of porphyrinogenic drugs and heme. *Biochem. J.* 255, 267-275.

Ravishankar, H. and Padmanaban, G. (1986): Regulation of Cytochrome P450 gene expression. Studies with a cloned probe. *J. Biol. Chem.* 260, 1588-1592.

Ribeiro, V.L., Lechner, M.C. (1992): Cloning and characterization of a novel CYP3A1 allelic variant: analysis of CYP3A1 and CYP3A2 sex-hormone-dependent expression reveals that the CYP3A2 gene is regulated by testosterone. *Arch. Biochem. Biophys.* 293, 147-152.

Ribeiro, V.L., Lechner, M.C.: Paradoxical effects of heme on CYP3A1 expression in transfected COS-1 cells", submitted.

Sambrook, J., Fritsch, E.F., Maniatis, T. (1989): Molecular Cloning: a Laboratory Manual. 2nd edition, Cold Spring Harbor Laboratory Press, Cold Spring Harbor.

Sinclair, P.R., Schuetz, E.G., Bement, W.J., Haugen, S.A., Sinclair, J.F., May, B.K., Li, D., Guzelian, P.S. (1990): Role of heme in phenobarbital induction of cytochromes P450 and 5-aminolevulinate synthase in cultured rat hepatocytes maintained on an extracellular matrix. *Arch. Biochem. Biophys.* 282, 386 - 392.

Srivastava, G., Bawden, M.J., Hansen, A.J., May, B.K. (1989): Heme may not be a positive regulator of Cytochrome P450 gene expression. *Eur. J. Biochem.* 178, 689-692.

**ACKNOWLEDGMENTS**

The authors are grateful to Dr. M. A. Correia, University of California, S.Francisco, USA, for providing the polyclonal CYP3A1 antibody.

Vera Ribeiro is recipient of a fellowship from JNICT.

# Limited proteolysis of cytochrome P450c21 from bovine adrenal cortex microsomes

Victor M. Guzov, Igor N. Zelko, Svetlana A. Bylinskaya, Sergey A. Usanov

*Institute of Bioorganic Chemistry, Academy of Sciences of Belarus, Zhodinskaya Street 5/2, Minsk, 220141, Republic of Belarus*

INTRODUCTION

Limited proteolysis was shown to be a useful tool to study membrane topology and domain structure of cytochromes P450 (P450s) (Chashchin et al., 1985). Therefore, we have chosen this approach to perform a comparative investigation on molecular organization and the membrane topology of mitochondrial and microsomal steroidogenic P450s. Although they catalyze similar types of reactions these two types of P450s have different mechanisms of incorporation into the membrane and different membrane topology. The aim of the present study is to use proteolytic digestion in conjunction with immunochemical detection to elucidate structural similarity between mitochondrial (P450scc) and microsomal (P450c21) steroidogenic P450s.

METHODS

Microsomes from bovine adrenal cortex were isolated using differential centrifugation. To purify P450c21, microsomes were solubilized with 1% Na-cholate and 0.01% Emulgen 913 followed by two-step hydrophobic chromatography using Cholate-Sepharose and Aminohexyl-Agarose. The protein was electrophoretically homogeneous and characterized by the specific content of 14-15 nmol P450/mg protein. Antibodies against P450c21 were raised in rabbits. Limited trypsinolysis of P450c21 in microsomes was performed in 50 mM Na-phosphate buffer, pH 7.4, and that of purified P450c21 - in 10 mM Na-phosphate buffer, pH 7.4, containing 20% glycerol, 0.01% Emulgen 913, 0.1 mM EDTA, 0.1 mM DTT - at trypsin:protein ratio of 1:25 (w/w) and 30°C. Trypsinolysis was stopped by the addition of 5-fold excess of soybean trypsin inhibitor (STI). The products formed were separated by SDS-electrophoresis and subjected to immunoblotting using anti-P450c21 antibodies. To analyze 21-hydroxylase activity we used either purified P450c21 supplemented with rat liver reductase or adrenocortical microsomes using progesterone or 17-hydroxyprogesterone as substrate. The reaction was initiated by the addition of NADPH. Steroids were extracted with ethyl acetate and separated by the reversed-phase HPLC.

RESULTS AND DISCUSSION

It is generally accepted that microsomal P450s from liver have single N-terminal membrane-binding region with the rest of the molecule facing the cytosol.

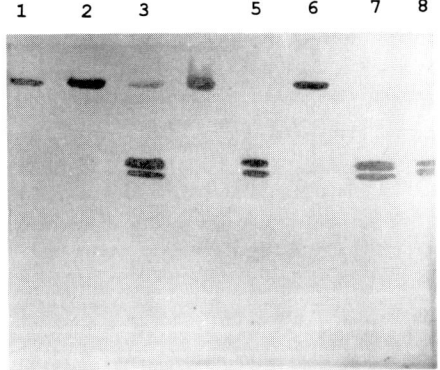

Fig. 1. Immunoblotting with anti-P450c21 antibodies following SDS-electrophoresis in 15% PAAG of the microsomes after trypsinolysis. Lane 1 - without treatment; lanes 2 and 4 - microsomes+trypsin+STI; lanes 3 and 5 - microsomes+trypsin; lane 6 - microsomes+trypsin +STI+washing with 0.5 M $Na_2CO_3$ and 0.5 M NaCl; lane 7 - microsomes+trypsin+salt washing; lane 8 - microsomes+1% Na-cholate+trypsin.

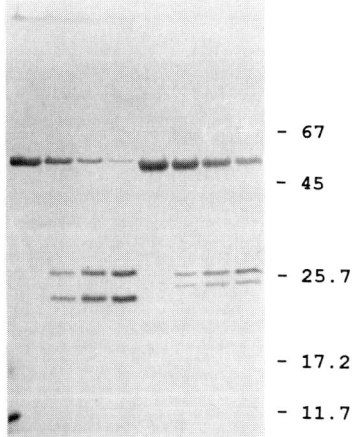

Fig. 2. SDS-electrophoresis in 12% PAAG of the fragments formed upon limited proteolysis of the purified P450scc (lanes 1-4) and P450c21 (lanes 5-8). Lanes 1 and 5 contain proteins treated with trypsin in the presence of STI. Limited proteolysis has been done during 10 min (lanes 2 and 6), 30 min (lanes 3 and 7), 60 min (lanes 4 and 8).

Mitochondrial P450scc, however, has several membrane-spanning domains (Usanov et al., 1990). Moreover, mitochondrial P450s are synthesized as precursors having N-terminal signal sequence which is cleaved upon translocation into mitochondria, while microsomal P450s have non-cleavable N-terminal membrane-targeting signal. Thus, the question arises: is there any structural similarity between mitochondrial and microsomal P450s which makes them able to catalyze similar (steroid-hydroxylating) reactions?

Previous limited proteolysis studies revealed the unique domain structure of mitochondrial P450scc which is cleaved by trypsin to form N- and C-terminal fragments without the loss of enzymatic activity (Chashchin et al., 1984). To understand the similarities and differences in molecular organization between mitochondrial and microsomal P450-dependent monooxygenases, limited trypsinolysis of adrenocortical P450c21 has been performed.

The treatment of adrenocortical microsomes with the trypsin:protein ratio of 1:25 (w/w) for 30 min at 30°C leads to the formation of two main fragments with Mr of 25 and 29 kDa which are recognized by anti-P450c21 antibodies (fig. 1).

These data are in agreement with the results of Ohta et al. (1992) which revealed formation of similar fragments upon trypsinolysis of P450c21 in liposomes. However, we have also observed formation of the same fragments upon trypsinolysis of either solubilized microsomes or purified P450c21 (fig. 2), while Ohta et al. (1992) found complete digestion of P450c21 with trypsin upon solubilization of liposomal membrane. This discrepancy may be due to lower amount of trypsin used in our case. Anyway, the data of the present study indicate that two major fragments formed under the conditions of limited trypsinolysis are protected from the protease not only due to membrane surrounding but also due to specific three-dimensional structure of the protein which masks potential sites of the trypsin attack. This situation is similar to that one of P450scc (Chashchin et al., 1985).

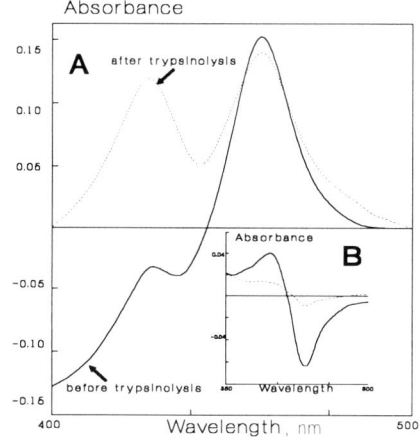

Fig. 3. The effect of limited proteolysis on spectral properties of P450c21 in microsomes. A - CO-binding, B - substrate (17-hydroxyprogesterone) binding.

To estimate the effect of trypsinolysis on spectral and catalytic properties of P450c21, adrenocortical microsomes were subjected to limited trypsinolysis. Spectral properties were assessed by CO-binding spectra of the reduced hemeprotein and substrate-induced difference spectra.

Trypsinolysis of P450c21 leads to conversion of the active P450 to its inactive form - cytochrome P420 and disappearance of substrate-induced spectral changes (fig. 3). This is also accompanied by corresponding decrease in enzymatic activity as determined upon reconstitution of the electron transport chain using purified P450c21.

Enzymatic activity measured at the level of microsomes was also decreased (Fig. 4) what is correlated well with the decrease in substrate-induced spectral changes. The decrease of 21-hydroxylase activity in microsomes may be due to the disturbance of P450c21 structure itself or inactivation of cytochrome P450-reductase which is known to loose the ability to reduce P450 after cleavage of the N-terminus by trypsin (Black & Coon, 1982). We are currently performing the experiments to discriminate these two possibilities.

There are two possible mechanisms for the formation of two fragments during proteolysis: 1) cleavage in the middle part of the molecule yielding N- and C-terminal fragments, 2) sequential digestion of either C- or N-terminal portion of the molecule yielding two N- or C-terminal fragments of the different length. The first probability seems to be much more likely what is supported by the kinetics of trypsinolysis (both fragments appear simultaneously during trypsinolysis, fig. 2) and good correlation between the sum of their Mr values and Mr of the whole molecule. The fragments formed can not be solubilized by the treatment with $Na_2CO_3$ with or without reducing agent (2-mercaptoethanol) (fig. 1) suggesting that they are deeply embedded into the membrane.

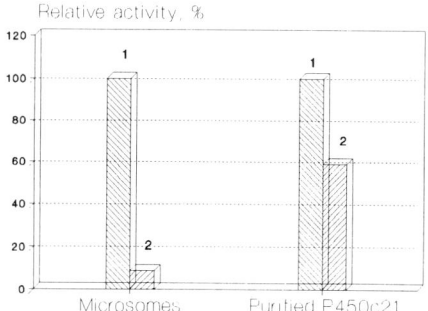

Fig. 4. The effect of trypsinolysis on 21-hydroxylase activity. 1 - before trypsinolysis, 2 - after trypsinolysis.

Thus, the data obtained assume the presence of membrane-bound regions of P450c21 not only in N-terminus, but also in C-terminal region of the molecule.

To theoretically localize possible fragments formed upon trypsinolysis, we have developed a computer program which analyzes the sites of trypsinolysis and estimates the fragments formed with respect to hydrophilicity and mobility according to approaches of Hopp (1986) and Van Regenmortel & Daney de Marcillac (1988). Computer analysis has revealed that P450c21 is most likely cleaved at position 266 yielding two fragments with Mr values corresponding to experimental data. However, to prove that the direct sequencing of the fragments formed is necessary. Such sequencing is currently in progress.

Similar results of trypsinolysis were obtained with liver microsomal P450 2B1 which is cleaved by trypsin at position 277 yielding fragments of 27.3 and 23.9 kDa what is accompanied by disappearance of substrate-induced spectral changes (Tsokos et al., 1992) yet Tsokos et al. (1992) didn't estimate enzymatic activity of the cleaved protein. However, single-point cleavage of another liver microsomal P450 - P450 1A1 - didn't reduce substrate-induced spectral changes. The position of the cleavage point was predicted to be localized in the region corresponding to interhelical turn between helices I and H of P450cam (Tsokos et al., 1992) which is situated on the surface of P450. Position 266 of P450c21 according to the sequence alignment of Nelson & Strobel (1988) also corresponds to this interhelical turn.

Thus, the microsomal P450c21 from adrenal cortex seems to have intermediate structure which has similarity with both mitochondrial P450scc (more than one membrane-binding region) and hepatic microsomal P450 2B1 (similar domain structure as revealed by limited trypsinolysis accompanied by structural disturbances in heme-binding region).

REFERENCES

Black, S.D., & Coon, M.J. (1982): Structural features of liver microsomal NADPH-cytochrome P-450 reductase. Hydrophobic domain, hydrophilic domain, and connecting region. *J. Biol. Chem.* 257, 5929-5938.

Chashchin, V.L., Shkumatov, V.M., Usanov, S.A., Vasilevsky, V.I., Turko, I.V. & Akhrem, A.A. (1985): Domain structure of adrenocortical cytochrome $P-450_{scc}$: Localization of functionally important sites in the polypeptide chain. *Biomed. Biochim. Acta* 44, 665-677.

Nelson, D.R., & Strobel, H.W. (1988): On the membrane topology of vertebrate cytochrome P-450 proteins. *J. Biol. Chem.* 263, 6038-6050.

Hopp, T.P. (1986): Protein surface analysis. Methods for identifying antigenic determinants and other interacting sites. *J. Immunol. Meth.* 88, 1-18.

Ohta, Y., Kawato, S., Tagashira, H., Takemori, S., & Kominami, S. (1992): Dynamic structures of adrenocortical cytochrome P-450 in proteoliposomes and microsomes: Protein rotation study. *Biochemistry* 31, 12680-12687.

Tsokos, D.C., Omata, Y., Robinson, R.C., Krutzsch, H.C., Gelboin, H.V., & Friedman, F.K. (1992): A proteolytically sensitive region common to several rat liver cytochromes P450: Effect of cleavage on substrate binding. *Biochemistry* 31, 7155-7159.

Usanov, S.A., Chernogolov, A.A., & Chashchin, V.L. (1990): Is cytochrome P-450scc a transmembrane protein? *FEBS Lett.* 272, 33-35.

Van Regenmortel, M.H.V., & Daney de Marcillac, G. (1988): An assessment of prediction methods for locating continuos epitopes in proteins. *Immunology Lett.* 17, 95-108.

# Interaction of FITC with P450s' N-terminus

Victoria V. Shumyantseva, Galina P. Kuznetsova, Valentine Yu. Uvarov, Alexander I. Archakov

*Institute of Biomedical Chemistry, Pogodinskaya 10, Moscow 119832, Russia*

## Summary

The paper describes the reactivity of fluorescein isothiocyanate (FITC) towards the N-terminus of cytochromes P450 2B4 and 1A2 in solution, in the natural membrane of microsomes and in proteoliposomes (cholate and ultrasonic). The N-termini of isolated cytochromes P450 2B4 and 1A2 were stoichiometrically modified in solution with FITC. The N-termini of the hemoproteins in phenobarbital and methylcholantrene treated microsomes, as well as those in cytochromes P450 2B4 and 1A2 containing cholate and ultrasonic proteoliposomes could not be modified with FITC but became accessible to this reagent when the microsomal and liposomal membranes were dissoled with Triton X-100. It appears therefore that the N-terminus of microsomal or proteoliposomal cytochromes P450 2B4 and 1A2 spans the membrane only once and faces the vesicles interior. Based on considerable differences in the amino acid sequences of these hemoproteins' N-terminal fragments, it was suggested that of major importance in orientation of N-terminal residues in the membrane is not the hydrophobic segment itself but rather the positively charged fragment, following it. Apparently the positively charged cluster not only interrupts the transfer of the polypeptide chain throught the membrane but, by interacting with the negatively charged phosphate groups of phosphlipids, pushes through the membrane those polar fragment of cytochrome P450 1A2 that precedes the hydrophobic membrane segment.

## INTRODUCTION

The microsomal monooxygenase system comprises several membrane-bound proteins. The membrane topology of cytochrome P450, the key enzyme of the microsomal monooxygenase system, is yet unclear. Analysis of amino acids sequences of P450 2B4 and 1A2 revealed that N-terminal fragments of 1A and 2B subfamilies have similar motifs with "minus" (Asp, Glu) before the hydrophobic fragment and "plus"

(Lys, Arg, His) after it (Archakov et al., 1992). Our results indicate that N-terminal amino groups of purified P450 2B4 and 1A2 in solution are modified with FITC. In intact microsomes the N-terminal amino group does not interact with FITC but becomes accessible to the action of this reagent after the treatment of the microsomes with Triton X-100, when the membrane has been made leaky with detergent. Incubation of P450 2B4- and 1A2-containing cholate and ultrasonic proteoliposomes with FITC did not lead to the modification of the N-terminal amino group. Like with intact microsomes, treatment of proteoliposomes with Triton X-100 renders the N-terminal amino group accessible to the fluorescent reagent. Based on the results obtained, it was concluded that the N-terminal transmembraneous fragments of P450 2B4 and 1A2 face the luminal side of microsomal membrane.

## EXPERIMENTAL

Cytochrome P450 2B4 was purified from the microsomes of phenobarbital-treated rabbits by the modified method of Imai et al. (1980). P450 1A2 was isolated from microsomes of methylcholanthrene-induced rabbits as described by Hashimoto (1976). SDS-electrophoresis was carried out using a 10% polyacrylamide gel. Proteoliposomes were obtained by the method of cholate dialysis or by the incubation of P450 2B4 or 1A2 with sonificated liposomes. Modification of the N-terminal amino group by FITC was carried out according to Vergeres et al. (1991).

## RESULTS AND DISCUSSION

Since the experimental data on the membrane topology of N-terminal region of P450 2B4 are controversail (Vergeres et al., 1991, Bernhardt et al., 1988) and the membrane topology of P450 1A2 is yet unknown, we studied the localization of the N-terminus of these enzymes with the site-specific reagent FITC. The advantage of FITC are that it reacts with deprotonated amino residues. Therefore at pH 7.4 only $\alpha$-amino group of the N-terminus is modified because the pK values of the $\epsilon$-amino group of lysine residues are between 9.5-10.5. FITC interacts with $\alpha$-amino group only in water phase, but not in the membrane itself owing to its inability to penetrate the phospholipid bilayer with unimpaired permeability. To modify the N-terminal amino groups localized on the inner side of membrane, they must be made penetratable for FITC by adding low concentration of detergent. Purified P450 2B4 and 1A2 are modified with FITC stoichiometrically in solution. Hemoprotein:FITC ratious were 1:1.2 and 1:1.1 respectively, as follows from the absorbtion spectra of the modified enzymes. It was identified that FITC reacts only with N-terminal amino groups of P450 2B4 and 1A2 (methionine or alanine, respectevly, (Heinemann et al., 1983; Ozols, 1986) as was confirmed by direct Edman-like degradation of FITC-modified P450 2B4 and 1A2.

Phenobarbital and methylcholanthrene-induced microsomes were not modified by FITC in 0.05 M $KP_i$ buffer, containing 20% glycerol, pH 7.4 at 30-fold molar excess FITC over P450 in the microsomes Electrophoregrammes show that the bands, corresponding to solubilized forms of P450 2B4 and 1A2, intensively fluoresced while proteins of phenobarbital and methylcholanthrene microsomes did not interact with FITC. However, when the microsomal membrane was treated with Triton X-100 (1 mg/ml) FITC could modify the N-terminus of P450 in both phenobarbital and methylcholanthrene microsomes.

Proteoliposomes are often used for clarification of the membraneous environment of P450. In view of this we have undertaken to study the orientation of the N-terminal fragment of P450 2B4 and 1A2 in the membranes of proteoliposomes in order to compare these data with data on microsomal membranes. Proteoliposomes, obtained by cholate dialysis and by the method of ultrasonic treatment and containing 2B4 and 1A2, respectively, were treated with FITC and analyzed by SDS-PAGE. Treatment of the membranes with Triton X-100 led to the interaction of FITC with P450 2B4 and 1A2. Thus, the results obtained provide evidence in favour of a model in which the N-terminal fragment of P450 2B4 and 1A2 spans the membrane only once and where the N-terminal amino groups of P450 1A2 and 2B4 are localized on the inner side of the microsomal or proteoliposomal membrane.

The question arises why N-terminal fragments of P450 1A2 and 2B4 differing in their amino acids sequences have the same membrane localization of the N-terminus. The presumable membrane fragment of P450 2B4 is localized in the stretch 1-21 and its hydrophobicity as calculated by the method of Eisenberg et al. is 0.83. The membrane fragment of P450 1A2 with hydrophobicity 0.87 is localized in the stretch 11-27. In this hemoprotein hydrophobicity of the first ten amino acids is only 0.3 and it is not clear how their transfer to the inner side of membrane is occurs. It was suggested that in the case of P450 2B4 we have a single transmembrane anchor where the N-terminus is localized on the inner membrane while with P450 1A2 the structure is looplike and the N-terminus is on the outer membrane (Ivanov et al., 1992). However, this suggestion was not confirmed by experimental data. It seems that the localization of the N-terminal fragment is basically determined not by the membrane spanning sequence but rather by the positively charged sequence following it. The N-terminal sequence of P450 2B4 is

```
1        10        20        30        40
MEFSLLLLLAFLAGLLLLLLFRGHPKAHGRLPPGPSPLPVL,            1A2

1        10        20        30
AMSPAAPLSVTELLLVSAVFSLPWAVRASRPKVPKELKR.
```

Membrane-flanking sequence is localized in the stretch 22-30 for P450 2B4 and in the stretch 27-32 for 1A2. In both cases it contains a cluster comprising five positively charged amino acids RGHPPKAHGR for P450 2B4 and three positively charged amino acids RASRPK for 1A2. It is highly likely that such positively charged fragment, interacting with the negative charges of phospholipids' phosphate groups on the membrane surface, not only interrupts the process of incorporation of the polypeptide chain into the membrane but is also able to push the polar fragment of P450 1A2-AMSPAAPLSVTE, followed by the membrane spanning sequence through the bilayer.

# REFERENCES

Archakov, A.I., Bachmanova, G.I., Sandler, M.K., Tutochkin, I.Yu. and Lisitsa, A.V. (1992): Cytochrome P-450 database and its scientific application. In Cytochrome P-450, Biochemistry and Biophysics, eds. Archakov, A.I. and Bachmanova, G.I., pp.673-679. Moscow, INCO-TNC.

Bernhardt, R., Kraft, R. and Ruckpaul, K. (1988): A simple determination of the sideness of the $NH_2$-terminus in the membrane bound cytochrome P-450 $LM_2$. Biochem. Int. 17,1143-1150.

Eisenberg, D., Schwarz, E., Komarony, M. and Wall, R. (1984): Analysis of membrane and surface protein sequences with the hydrophobic moment plot. J. Mol. Biol. 179, 125-142.

Hashimoto, C. and Imai, Y. (1976):Purification of a substrate complex of cytochrome P-450 from liver microsomes of 3-methylcholanthrene treated rabbits. Biochem. Biophys. Res. Commun. 68, 821-827.

Heinemann, F.S. and Ozols, J. (1983): The complete amino acid sequence of rabbit phenobarbital-induced liver microsomal cytochrome P-450. J. Biol. Chem. 261, 3965-3979.

Imai, J., Hashimoto,Y.C., Sakake, H., Girardin, A. and Sato, R. (1980): Multiple forms of cytochrome P-450 purified from liver microsomes of 3-methylcholantrene and phenobarbital-treated rabbits. J.Biochem. 88, 489-503.

Ivanov, A.S. and Archakov, A.I.(1992): Membrane anchor domain of cytochrome P450 2B4 and 1A2 : computer molecular design. J. Basic Clin. Physiol. Pharmacol. 3, 203.

Ozols, J. (1986): Complete amino acid sequence of a cytochrome P-450 Isolated from naphtoflavone-induced rabbit liver microsomes. J. Biol. Chem. 261, 3965-3979.

Vergeres, G., Witerhalter, K.N. and Richter, Ch. (1991): Localization of the N-terminal methionine of rat liver cytochrome P-450 in the lumen of the endoplasmic reticulum. Bichem. Biophys. Acta, 1063, 236-241.

# Structural requirements for ferredoxin associated electron transfer evaluated by analysis of Tyr-82 and His-56 bovine adrenodoxin mutants

Vita Beckert, Rudolf Dettmer, Rita Bernhardt

*Max-Delbrück-Center for Molecular Medicine, Robert-Rössle-Straße 10, D-13122 Berlin, FGR*

## INTRODUCTION

Adrenodoxin (Adx) is a low molecular weight (~14 kDa) iron-sulfur protein of the [2Fe-2S] ferredoxin type. Adx is involved in two electron transfer systems in the inner mitochondrial membrane of adrenal cortex containing NADPH-dependent adrenodoxin reductase (AdR), adrenodoxin and cytochromes P450 - $P450_{scc}$ (CYP11A1) and P45011ß (CYP11B1).

The mechanism of protein-protein interaction among Adx and its redox partners and the structural basis of the recognition of the respective electron donor and acceptor are not completely understood.

Two different models have been proposed for the electron transport chain from NADPH-dependent adrenodoxin reductase (AdR) to mitochondrial cytochromes P450. The formation of a stable ternary complex among Adx, AdR and P450 was proposed to be an essential step for Adx-mediated electron transfer (Kido & Kimura, 1979; Usanov et al., 1986; Hara & Kimura, 1989). In the other model Adx functions as a mobile electron shuttle (Lambeth et al., 1979, 1984; Hanukoglu & Jefcoat, 1980) initially forming a binary complex with AdR where it accepts an electron, and then a tight complex with P450 transferring thereby the reduction equivalent. There is little known about the involvement of amino acid residues into the recognition and interaction site of Adx with AdR and P450. On the basis of chemical modification studies and site-directed mutagenesis techniques it was shown that the interaction site with the redox partners in Adx appears to be around negatively charged amino acids - Glu-74, Asp-76, Asp-79 and Asp-86 (Geren et al., 1984; Coghlan & Vickery, 1991, 1992). The unique tyrosine in position 82 of Adx had been proposed to be involved in electron transfer and/or AdR binding by chemical modification studies (Taniguchi & Kimura, 1975, 1976).

We used site-directed mutagenesis of Tyr-82 and His-56 to characterize the role of these residues in electron transfer as well as in electron donor and acceptor interaction.

## MATERIAL AND METHODS

Oligonucleotide-directed mutagenesis was carried out using the vector pKKAdx (Uhlmann et al., 1992) to replace Tyr-82 by phenylalanine, leucine or serine. His-56 was replaced by arginine, glutamine or threonine. Expression in *Echerichia coli* and purification of recombinant Adx proteins were carried out as previously described (Uhlmann et al., 1992). AdR , CYP11A1 and CYP11B1 were isolated from bovine adrenocortical mitochondria according to Akhrem et al.,1979.

*EPR spectroscopy* was carried out on a Varian E3 spectrometer using whole E. coli cells with expressed proteins which were reduced by dithionite.

*CD spectra* were recorded on a Jasco J720 spectropolarimeter at room temperature in the ultraviolet and visible region .

*Redox potentials* of Tyr-82 mutants were measured by the dye photoreduction method with Safranin T as indicator and mediator (Sligar et al.,1979). All data were analyzed according to the Nernst equation.

*Fluorescence spectra* were taken with an RF-5001 PC spectrofluorometer by exciting at 270 nm. Measurements were carried out in 10 mM phosphate buffer, pH 7.4, at room temperature.

*Cytochrome c reduction mixture* (1 ml) contained 0.2µM AdR, 100 µM horse heart cytochrome c, and variable amounts of recombinant Adx. The reduction of cytochrome c after addition of 100 µM NADPH was monitored at 550 nm, and the activity was calculated on the basis of a molar extinction coefficient of 20 $(mM \cdot cm)^{-1}$ for cytochrome c.

The reaction mixture for *11ß-hydroxylation assays* contained 0.4 µM CYP11B1, 0.4 µM AdR, 100 µM deoxycorticosterone, the NADPH-regenerating system (64 µM glucose-6-phosphoric acid disodium salt, and 2 U of glucose oxidase), and variable amounts of Adx in 0.5 ml 50 mM potassium phosphate buffer, pH 7.4. The reaction was started with 60 µM NADPH and after incubation at 37 °C for 10 min the reaction was stopped with dichloromethane, which simultaneously extracts the steroids. The extracted steroids were analyzed by reverse-phase HPLC on Nucleosil-gel 5C18 column with a solvent system of acetonitrile: phosphoric acid (10 mM) (6:4, v/v) at a flow rate of 1ml/min.

*The cholesterol side-chain cleavage activity* was measured in the reconstituted assay system according to Sugano et al. (1989). The incubation mixture contained 0.5 µM CYP11A1, 0.2 µM AdR, 100 µM cholesterol, and NADPH-regenerating system. After 10 min incubation at 37 °C, the reaction was stopped by heating up to 98 °C. The steroids were conversed into 3-one-4-en form by incubation for 10 min with cholesterol oxidase (0.4 U), and then extracted with dichloromethane. The analyses of steroids was carried out under the same conditions as for CYP11B1.

*Differential spectral titration* was performed according to Kido & Kimura (1979). Binding of Adx was followed by a high spin shift of the P450 heme iron from 417 to 392 nm, caused by Adx-induced cholesterol binding.

Values of kinetic parameters were determined by least squares linear regression analysis of the data from three or four separate experiments.

## RESULTS AND DISCUSSION

The importance of the unique Tyr-82 residue in bovine Adx for its functioning as electron mediator as well as the location of this residue with respect to the [2Fe-2S] cluster and to the binding domain of Adx with AdR and P450s were analyzed by site-directed mutagenesis studies.

Adx mutants Y82F, Y82L, Y82S, H56R, H56Q and H56T were expressed each in yields similar to the wild type protein (50 mg/l cell culture). All mutants were expressed as holoprotein and exhibited absorption spectra characteristic for correctly assembled [2Fe-2S] cluster (data not shown). The mutations were confirmed by nucleotide sequencing of the whole cDNA as well as by mass spectrometry and amino acid analysis.

To check whether Tyr-82 mutations affect the structure of the [2Fe-2S] cluster the circular dichroism (CD) and EPR signals were analyzed. The CD and EPR spectra did not show any significant changes of the mutant proteins as compared to the wild type (Beckert et al., J. Biol. Chem., in press). The redox potentials of wild type Adx and Tyr-82 mutants were estimated to be -274 mV ($\pm$ 5 mV), supporting the above mentioned data that there is no influence of the mutation on the [2Fe-2S] cluster. Contrary to the bacterial electron transmitting systems (Adman et al.,1973) Tyr-82 in Adx is thus obviously not located in the immediate iron-sulfur cluster environment. This result is consistent with data from fluorescence measurements of Adx and $^1$H NMR studies (Lim & Kimura, 1981; Greenfield et al., 1989).

However, an involvement of tyrosine 82 in electron transfer cannot be excluded on the basis of these data only. The participation of Tyr-82 of Adx in intra- or intermolecular electron transfer from AdR to P450 was tested by analyzing the influence of replacement of this amino acid on the efficiencies of cytochrome c reduction and P450-dependent substrate conversion.

Assuming the cytochrome c reduction as to reflect the binding affinity of Adx to AdR neither a change of the $V_{max}$ values nor a significant decrease in the $K_m$ values could be observed to occur in the Y82 mutants in our studies (data not shown). Thus, Y82 seems neither to be directly involved in AdR binding nor do mutations of this residue cause conformational changes affecting the affinity to AdR.

Determination of the hydroxylating activities of P45011A1 and P45011B1 reconstituted with Adx mutants, however, indicated marked changes

The $K_m$ values in the P45011B1-dependent conversion of deoxycorticosterone to corticosterone vary by factor of up to 4. Although possessing unchanged $V_{max}$ the mutants Y82F and Y82L due to changes in the $K_m$ values reveal higher activities then the corresponding wild type Adx, being 180% and 85% higher, respectively, under our standard reconstitution conditions (0.2 nmol P45011B1: 0.5 nmol Adx: 0.2 nmol AdR) (Fig. 1. A).

The effects of the replacement are less pronounced when the reconstitution is performed with P45011A1 (Fig. 1. B). The Vmax is also unchanged when using Y82 mutants instead of wild type Adx as electron mediator. The Km values in this reaction differ only by factor of up to 1.5. In addition, besides Y82S also Y82L exhibits a decreased reaction velocity with cholesterol as substrate when compared to wild type Adx. This behaviour corresponds well with the affinities of Adx mutants for P45011A1 estimated by spectral titration method (Beckert et al., J. Biol. Chem., in press).

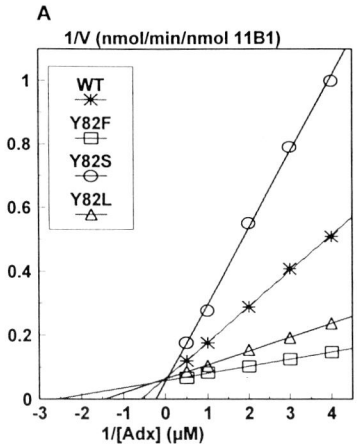 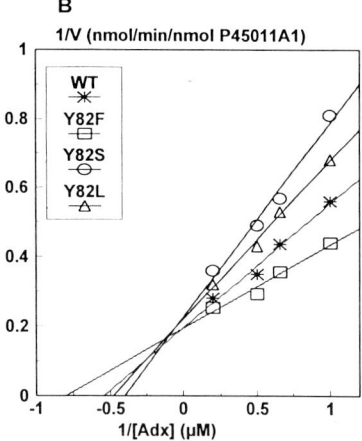

**Fig. 1.** P45011B1-dependent conversion of deoxycorticosterone to corticosterone (A) and P45011A1-dependent cholesterol conversion to pregnenolone (B) by Tyr-82 mutants of Adx. Data are present from experiments using wild type Adx (∗), and mutants Y82F (□), Y82S (O), and Y82L (Δ).

These results indicate that Tyr-82 does not play an essential role in intra- or intermolecular electron transfer. But the fact that various mutations in position 82 cause different effects on the interaction of mutant adrenodoxins with P45011A1 and 11B1 (Fig. 1. A, B) seems to support a direct involvement of Tyr-82 in the interaction with P450s. The possibility of an indirect effect of the mutations on P450 binding, however, cannot be ruled out. On the other hand, a change in the conformation of the recognition region caused by Tyr-82 replacement would be expected to affect not only P450, but also AdR binding. When comparing the secondary structure of the bacterial ferredoxin, putidaredoxin, which was obtained on the basis of 2D-NMR measurements (Ye et al., 1992), with predictions for Adx, it turned out that on the basis of a combination of different prediction methods (Köpke, K., unpublished results) the similarity between Adx and putidaredoxin is especially high around the recognition region (Fig.2).

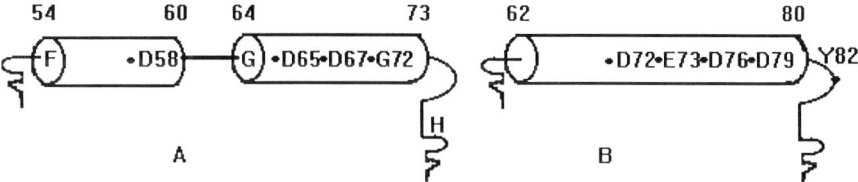

Fig 2. Schematic representation of secondary structure around the recognition region in oxidized putidaredoxin according to Ye et al., 1992, (A), and oxidized adrenodoxin (B). Letters correspond to boldface lettering scheme for the primary structure in Ye et al., 1992. Numbers correspond to the position of amino acid residues involved in the structure element.

The recognition region of putidaredoxin is located on helix G, spanning residues 64-73. This region is followed by a turn, spanning residues 79-82 (Ye et al., 1992). For Adx the corresponding recognition region spans residues 62 till 80, followed by a turn from position 86 till 89. Thus, Y82 is located just between the negatively charged recognition site and the ß-turn in a conserved region.

Protein studies using $^1$H NMR suggested that His-56 of Adx is located at a position close to Ser-88 and thus proximal to the binding domain of Adx for the redox partners between Glu-74 and Asp-86 (Miura & Ichikawa, 1992). To check whether His-56 is also close to Tyr-82 we recorded the fluorescence of tyrosine 82 in dependence on different replacements of the histidine residue.
It can been seen from Fig. 3 that the intensity of Adx tyrosine fluorescence strongly depends on the nature of the residue in position 56 being more then doubled in case of H56R. This indicates that replacement of His-56 leads to changes in the microenvironment of Tyr-82 and that His-56 appears to be in the immediate vicinity of Tyr-82 and thus to the intermolecular interface of Adx with redox partners. Whether the influence of His-56 on Tyr-82 fluorescence properties is due to direct interactions between His-56 and Tyr-82 or to small conformational changes remains to be elucidated.

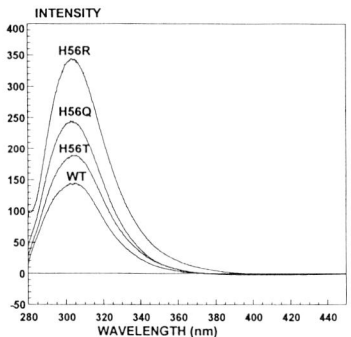

**Fig. 3.** Fluorescence emission spectra of 80 μM wild type Adx and His-56 mutants in 10 mM potassium phosphate buffer, pH7.4 at room temperature.

Further studies, especially the resolution of the three-dimensional structure of Adx and its complexes with P450 and AdR, are necessary to define the detailed role of tyrosine 82 and histidine 56 in protein-protein interaction.

### ACKNOWLEDGEMENTS
This work was supported by a grant from the Deutsche Forschungsgemeinschaft.
We thank Dr. R. Misselwitz for help in fluorescence measurements and Mrs. Ch. Jaeger for expert technical assistance. The help of Dr. H. Schrauber in secondary structure predictions is greatly acknowledged.

### REFERENCES
Adman, E.T., Sieker, L.C., and Jensen, L.H. (1973) *J. Biol.Chem.* **248**, 3987-3996
Beckert, V., Dettmer, R., and Bernhardt, R., *J. Biol. Chem.* in press
Carter, C.W., Kraut, J.J., Freer, S.T., Xuong, N., Alden, R.A., and Bartsch, R.G. (1974) *J. Biol. Chem.* **13**, 4212-4225
Coghlan, V.M., and Vickery, L.E. (1991) *J. Biol. Chem.* **266**, 18606-18612
Coghlan, V.M., and Vickery, L.E. (1992) *J. Biol. Chem.* **267**, 8932-8935
Geren, L.M., O'Brien, P., Stonehuerner, J., and Millett, F. (1984) *J. Biol. Chem.* **259**, 2155-2160
Greenfield, N.J., Wu, X., and Jordan, F. (1989) *Biochim. Biophys. Acta* **995**, 246-254
Hanukoglu, I., and Jefcoate, C.R. (1980) *J. Biol. Chem.* **255**, 3057-3061
Hara, T., and Kimura, T. (1989) *J. Biochem.* **105**, 601-605
Lambeth, J.D., Seybert, D.W., and Kamin, H. (1979) *J. Biol. Chem.* **254**, 7255-7264
Lambeth, J.D., Geren, L.M., and Millett, F. (1984) *J. Biol. Chem.* **259**, 10025-10029
Landt, O., Grunert, H.-P., and Hahn, U. (1990) *Gene* **96**, 125-128
Lim, B.T., and Kimura, T. (1981) *J. Biol. Chem.* **256**, 4400-4406
Sligar, S.G., Cinti, D.L., Gibson, G.G., and Schenkman, J.B. (1979) *Biochem. Biophys. Res. Commun.* **90**, 925-932
Sugano, S., Morishima, N., Ikeda, N., and Horie, S. (1989) *Analytical Biochem.* **182**, 327-333
Taniguchi, T., and Kimura, T. (1975) *Biochemistry* **14**, 5573-5578
Taniguchi, T., and Kimura, T. (1976) *Biochemistry* **15**, 2849-2853
Uhlmann, H., Beckert, V., Schwarz, D., and Bernhardt, R. (1992) *Biochem. Biophys. Res. Commun.* **188**, 1131-1138
Usanov, S.A., Turko, I.V., Chashchin, V.L., and Akhrem, A.A. (1986) *Bioorg. Khim. (USSR)* **12**, 185-194
Ye, X.M., Pochapsky, T., and Pochapsky, S.S. (1992) *Biochemistry* **31**, 1961-196

# Nonchemical induction of rat liver CYP1A1

Vyacheslav V. Lyakhovich, Alevtina Yu. Grishanova, Olga A. Gromova, Lyudmila F. Gulyaeva

*Institute of Molecular Pathology and Ecological Biochemistry, Russian Academy of Medical Sciences, Novosibirsk 630117, Russia*

We showed the CYP1A1 induction in rat liver under contact skin application of ultrasound (US) or the long-term exposure to cold. The rate of the 7-ethoxyresorufin-O-deethylase (EROD) activity increased from 100 in control to 300 pmol/min/mg in US-treated rats. Western blot analysis also revealed CYP1A1 induction in US-liver absent in untreated animals. The mRNA level for CYP1A1 in the liver of US-treated rats increased from 0.5 to 6.0 fmol mRNA/mcg total RNA, while it did not change for CYP1A2. 6-fold increase of EROD was observed in the liver microsomes of rats exposed to cold for 10 days. Significant increase was also observed for 7-methoxyresorufin-O-demethylase (MROD) activity. The CYP1A1 mRNA level was higher in cold-exposure rats than in control. The CYP1A2 mRNA level did not increase.
The obtained results give evidence for nonchemical induction of CYP1A1 which is known to be induced by polycyclic aromatic hydrocarbons (PAH). The increase of the mRNA level for CYP1A1 suggests transcriptional regulation of this process.

## INTRODUCTION

It is widely accepted that CYP1A1 is not constitutively expressed in rat liver, but is inducible by PAH like benzo[a]pyrene. CYP1A genes are considered to be transcriptionally activated and regulated by a mechanism involving the ligand-dependent Ah-receptor. We have previously assessed the presence of CYP1A1 in rat liver under contact skin application of the low intensity ultrasound or the long-term exposure of rats to cold using data of EROD activity which is specific for CYP1A1 as well as data of Western blot analysis with monoclonal antibodies reactive with CYP1A1/A2 (Gromova et al., 1992; Grishanova et al., 1992). The aim of this study was to find out whether the induction of rat liver CYP1A1 is transcriptionally controlled in ultrasound-treated or cold exposed rats.

## MATERIALS AND METHODS

The right side under the ribs of Wistar male rats has been exposed to the contact skin application of ultrasound (0.2 WT/sm, 5 min, 5 times daily). Intact rats were used as control. Another group of male Wistar rats was exposed to cold (+5 C) for 1, 5, 10 or 16 days. Rats kept at 26 C were used as a control. The catalytic marker activity measurements and immunochemical studies were performed as described by Gulyaeva et al. (1989). We used two types of monoclonal antibodies obtained by the hybridoma technology after immunization of BALB/c mice with P-450IA1 and IA2 antigens (clones 14A10 and 14H5). Clone 14H5 specifically recognized two proteins of IA family and clone 14A10 recognized P450IA1 only. The solution hybridization analysis of total hepatic mRNA was carried out using specific [32P]-oligodeoxyribonucleotides for CYP1A1 and CYP1A2 mRNA. We used solution hybridization technique modified for small quantities of mRNA (Raval et al., 1991).

## RESULTS AND DISCUSSION

Data of catalytic studies of rat liver microsomes under ultrasonic skin application on liver region are summarized in Table 1.

Table 1. Contents of cytochromes P-450 and monooxygenase activities in rat liver US-microsomes (mean ± SE, n=5)

|  | Control | Ultrasound |
|---|---|---|
| P-450 total (a) | $0.882 \pm 0.038$ | $0.780 \pm 0.047^{**}$ |
| P-450IA1/A2 (a) | $0.044 \pm 0.002$ | $0.103 \pm 0.014^{*}$ |
| 7-ethoxyresorufin O-deethylase (b) | $0.111 \pm 0.027$ | $0.322 \pm 0.060^{*}$ |
| Androstenedione hydroxylase 7$\alpha$-OH | $0.584 \pm 0.045$ | $0.716 \pm 0.133$ |
| (b) 16$\beta$-OH | $0.328 \pm 0.032$ | $0.354 \pm 0.034$ |
| 16$\alpha$-OH | $0.682 \pm 0.121$ | $0.494 \pm 0.110^{***}$ |
| 6$\beta$-OH | $0.924 \pm 0.066$ | $0.662 \pm 0.008^{*}$ |
| Chlorozoxasone hydroxylase (b) | $0.559 \pm 0.098$ | $0.617 \pm 0.102$ |

(a)-nmol/mg; (b)-nmol/min/mg; * $p¡0.001$; ** $p¡0.01$; *** $p¡0.05$

One can see that the most notable changes are observed in the EROD activity specific for P-450IA subfamily. The rate of the EROD increased from 111 pmol/min/mg protein in control up to 322 pmol/min /mg in ultrasound treated rats. There is slight decrease of 16$\alpha$- and 6$\beta$-androstenedione hydroxylase activities which is specific for P-450IIC11 and IIIA, and absence of any variations in the marker activities for P-450IIA1, IIB1, IIE1 (7$\alpha$ and 16$\beta$ hydroxylation of androstenedione and hydroxylation of clorozoxasone respectively). Data of catalytical quantitation of P450IA1 (EROD), P450IA2 (MROD) in liver microsomes of rats exposed to cold suggest the expression of P450IA1 during the cold exposure (Table 2).

Table 2. The effects of exposure of rats to cold on liver metabolism of P450 isoenzyme-specific substrates

| Substrates | Cold exposure (days) | | | | |
|---|---|---|---|---|---|
|  | 0 | 1 | 5 | 10 | 16 |
| Ethoxy-R (a) | $21.6 \pm 7.0$ | $^{*}73.9 \pm 11.6$ | $^{*}65.9 \pm 5.7$ | $^{*}124.5 \pm 22.2$ | $^{*}102.2 \pm 27.2$ |
| Pentoxy-R (a) | $2.7 \pm 0.6$ | $4.5 \pm 1.3$ | $^{*}7.6 \pm 0.3$ | $^{*}9.7 \pm 0.8$ | $6.6 \pm 1.4$ |
| Methoxy-R (a) | $5.6 \pm 1.5$ | $10.6 \pm 2.0$ | $^{*}25.5 \pm 1.7$ | $^{*}16.6 \pm 0.6$ | $10.2 \pm 2.2$ |
| Bph (b) | $2.1 \pm 0.2$ | $2.8 \pm 0.5$ | $3.8 \pm 0.6$ | $^{*}5.8 \pm 0.8$ | $4.7 \pm 0.1$ |
| Ery (b) | $2.8 \pm 0.5$ | $4.5 \pm 0.5$ | $^{*}6.2 \pm 0.4$ | $^{*}7.8 \pm 0.6$ | $^{*}6.7 \pm 0.6$ |
| p-NP (b) | $0.6 \pm 0.2$ | $0.9 \pm 0.2$ | $1.3 \pm 0.2$ | $^{*}1.5 \pm 0.1$ | $1.1 \pm 0.2$ |

7-alkoxyresorufins (R), benzphetamine (Bph), erythromycin (Ery), p-nitrophenol (pNP).
Values given as a) pmol and b) nmol product/min/mg protein (mean ± SE, n=5),
* $p¡0.05$ compared to the control group.

It is shown than the catalytic activity of all P-450 isoforms examined in liver microsomes increased gradually, reached the maximum level on the 10th day followed by a decrease. 6-fold increase of the EROD activity was observed in liver microsomes of rats exposed to cold for 10 days. Significant increase was also observed for MROD (P450IA2), 7-pentoxyresorufin O-dealkylase (P450IIB), benzphetamine N-demethylase (P450IIB, P450IIIA) and erythromycine N-demethylase (P450IIIA1/2) activities. It suggests that the expression of these isoenzymes was markedly activated during the cold exposure.
Western blot analysis with Mab 14H5 and 14A10 demonstrated the appearance of P-450IA1 in liver microsomes under ultrasonic application as well as during the exposure of rats to cold. P450IA2 alone was revealed in control microsomes.
P450IA1 induction was unexpected in liver microsomes of cold exposured and US-treated rats in the absence of environmental inducers. To confirm the expression of CYP1A1 and CYP1A2 the levels of their specific mRNAs

were measured. The results are shown below (Tables 3 and 4).

Table 3. The mRNA level of CYP1A1 and CYP1A2 in the liver of US-treated rats (fmol mRNA/mcg of total RNA)

| Hours after US treatment | CYP1A1 | CYP1A2 |
|---|---|---|
| 8 | 5.0 | 1.6 |
| 16 | 6.5 | 6.8 |
| 23 | 4.0 | 1.9 |
| 48 | 2.6 | 5.2 |
| control | 0.5 | 4.0 |
| Aroclor 1254-treated | 45.0 | 110.0 |

The mRNA level for CYP1A1 in US-liver increased from 0.5 to 6.5 fmol of mRNA/total RNA, while it did not change for CYP1A2.

Table 4. Microsomal EROD and MROD activities, levels of P450IA1- and P450IA2 -specific mRNAs in individual livers from rats after 16 days of cold exposure.

|  | EROD | MROD | P450IA1 mRNA | P450IA1 | P450IA2 mRNA |
|---|---|---|---|---|---|
| Cold-0 | 31.4 | 1.6 | 0.20 | n.d. | 4.7 |
| Cold-16 |  |  |  |  |  |
| 1 | 114.2 | 14.1 | 0.40 | 13 | 6.4 |
| 2 | 142.3 | 9.6 | 0.77 | 35 | 1.5 |
| 3 | 50.2 | 6.7 | 0.20 | n.d. | 0.9 |

Values given as pmoles resorufin/min/mg protein, fmol P450IA1 RNA/mcg total liver mRNA, arbitrary optical density units/40 mcg of microsomal protein (P450IA1), n.d. - not detected.

Our data suggest that increase the CYP1A1 induction is apparently due to transcriptional activation leading to increased levels of specific mRNAs and P450IA proteins.

Taking together, our results give evidence for the non-chemical induction of P450IA1. It is long known that the cytosolic Ah-receptor mediates induction of cytochrome P450IA1 by PAH-type compounds. Inducers of P450IA1 are initially bound to the Ah-receptor, after that the inducer- receptor complex translocates into the nuclear compartment. The underlying molecular mechanism associated with the potentiation of P-450IA1 induction in rat liver microsomes by the physical factors remains to be elucidated. It seems likely that endogenous substrates (metabolites or hormones) are initially bound to the Ah-receptor and then trigger the activation of the P450IA1 gene under the abnormal physiological conditions in absence of environmental chemical inducers.

# References

[1] Grishanova A.Yu., Gulyaeva L.F., Zolotaryova T.A., Lyakhovich V.V. (1992) : Nonchemical induction of rat liver cytochrome P-450IA1.*J. Basic and Clin. Physiol. and Pharmacol.* 3S:188..

[2] Gromova O.A., Polyakova N.E., Kozlovskaya N.E., Vavilin V.A., Grishanova A.Yu., Kolpakov A.R., Lyakhovich V.V. (1992) :*Effects of exposure to cold on cytochromes P450 and monooxygenase* activities in rat liver microsomes. J. Basic and Clin. Physiol. and Pharmacol. 3S:189.

[3] Gulyaeva L.F., Khatsenko O.G., Gerasimov K.E., Mitrofanov D.V., Mishin V.M. (1989) : Identification of the molecular forms of cytochrome P-450 isolated from liver microsomes of rats induced by phenobarbital and 3-methylcholanthrene.*Biokhimiya.* 54: 487-494.

[4] Raval P., Patrick L.I., Bresnic E. (1991) : Induction of cytochromes P450IA1 and P450IA2 as determined by solution hybridization.*Biochem. Pharmacol.* 41:1719.

# Inhibition of cyclosporine metabolism in mouse and human microsomes after cytochrome P450 inactivation by Cannabidiol

Walter Jaeger[1], Lester M. Bornheim[2], Maria A. Correia[2], Leslie Z. Benet[1]

*Departments of [1]Pharmacy and [2]Pharmacology, University of California, San Francisco, 513 Parnassus Avenue, San Francisco, California 94143, USA*

Cyclosporine (CyA), a cyclic endecapeptide, is a powerful immunosuppressant drug (Fig. 1). In human medicine, it is widely used in transplantation to prevent rejection of the transplanted organ and for the treatment of autoimmune disorders. CyA is predominantly metabolized by the cytochrome P450-dependent monooxygenase (P450 3A) system in the liver and intestinal ER membranes. The primary metabolites are the monohydroxylated derivatives (AM1, AM9 and the intramolecular cyclized AM1c) as well as AM4N, its N-demethylated derivative.

Cannabidiol (CBD), a nonpsychoactive major constituent of marijuana, has been shown to effect the in vivo and in vitro metabolism of different drugs in several animal species, including humans. CBD-mediated inhibition of drug metabolism results from the selective inactivation of hepatic P450s belonging to the 2C and 3A subfamilies (Bornheim *et al.*, 1991). Human liver contains at the least 4 related P450 3A isozymes. At least 2 members of this subfamily have been detected in rat liver. P450 3A2 is constitutively expressed in male rat liver, whereas P450 3A1 is only expressed after induction by steroids and macrolide antibiotics. Although rat P450s 3A1 and 3A2 are functionally and immunochemically very similar, CBD selectively inactivates constitutively expressed P450 $3A_C$ in the mouse, but has no effect on the steroid-inducible P450$_S$ (Bornheim *et al.*, 1990). The following studies describe the effect of CBD on mouse liver microsomal CyA metabolism. In additional experiments, mouse and human liver microsomes were preincubated with CBD to determine its effects on the relative formation of CyA metabolites. Clinical interactions of CyA with CBD may result in higher blood CyA levels and an increase of its toxic side effects.

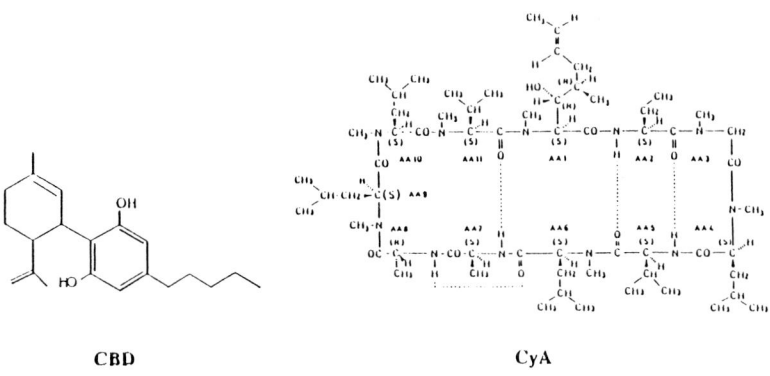

Fig. 1. Chemical structures of CBD and CyA

## METHODS

### Animals and Microsomal Preparation:
Liver microsomes from male mice, untreated or pretreated with CBD (120 mg/kg for 2 hours), were prepared using standard procedures.

### Inhibition of CyA Metabolism:
*In Vivo.* Liver microsomes (0.2 mg protein/ml) from CBD pretreated mice were incubated with CyA (30 µM), NADPH (1mM), isocitrate (5mM) and 0.5U/ml isocitrate dehydrogenase (for 30 min at 37°C) and the formation of AM1, AM9, AM1c and AM4N monitored by HPLC.

*In Vitro.* Microsomes from untreated male mouse or human liver (2 mg protein/ml) were preincubated at 37°C for 0, 10, 20 and 30 min, in a buffer containing NADPH (1mM) and CBD (64 µM) in a total volume of 1 ml. An aliquot (0.2 mg protein) was taken out and incubated at 37°C for 30 min with CyA (30 µM), NADPH (1mM) and NADPH-regenerating system. Incubation, metabolite extraction, HPLC-system and quantitation were performed as described previously (Prueksaritanont *et al.*, 1993).

## RESULTS AND DISCUSSION

CBD pretreatment (120 mg/kg for 2 h) of male mice inhibited the liver microsomal formation of mono-hydroxylated CyA metabolites AM1, AM9, AM1c and the N-demethylated derivative, AM4N by 60 %, 82 %, 86% and 66 %, respectively. The metabolite formation was liniar in respect to microsomal protein concentration (0.1-0.3 mg/ml). CyA is mainly metabolized by cytochromes P450 3A, and according to recent results possibly by P450 2C in ethinyl estradiol pretreated rats. CBD selectively inhibits hepatic P450s belonging to the 2C and 3A subfamilies. The lesser CBD-elicited decrease of AM1 formation could be due to a catalytic involvement of a different isozyme resistance to CBD inactivation (Fig. 2).

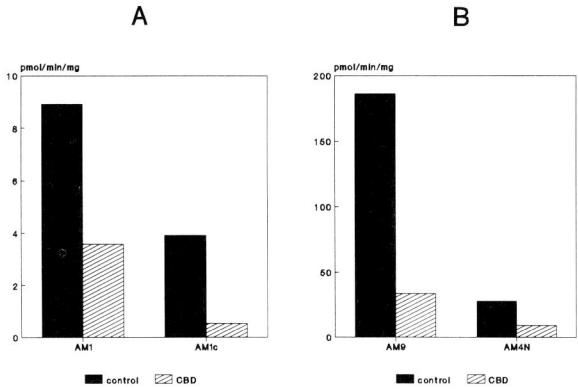

Fig. 2. Inhibition of CyA-metabolism by CBD treatment of male mice: A (AM1, AM1c), B (AM9, AM4N)

Incubation of liver microsomes from uninduced mice with CBD without preincubation (time 0), moderately inhibited CyA metabolism by ~ 29 % due to competitive inhibition. Preincubation with CBD showed a time-dependent decrease (non-competitive inhibition) of the CyA biotransformation of 45 % for AM1, 61 % for AM9 and 48 % for AM4N (Fig. 3). Due to interference with free CBD, no value for AM1c could be calculated.

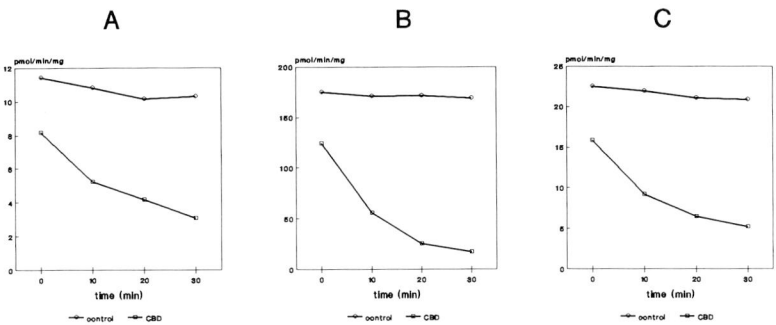

**Fig. 3.** CBD-mediated inhibition of mouse liver microsomal CyA metabolism; A (AM1), B (AM9), C (AM4N)

Human liver microsomes, preincubated with CBD (64 µM) exhibited comparable values of ~ 25 % of competitive inhibition (time 0) for all metabolites. AM1 was decreased by 51%, AM9 by 58 % and AM4N by 48 %, respectively (Fig. 4).

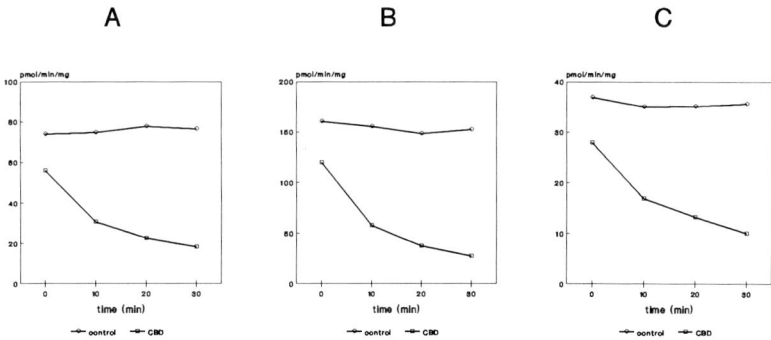

**Fig. 4.** CBD-mediated inhibition of human liver microsomal CyA metabolism; A (AM1), B (AM9), C (AM4N)

CBD-pretreatment of male mice inhibits hepatic P450 2C and 3A and causes a decrease of CyA metabolism in pretreated male mice. In addition, in vitro experiments using mouse and human liver microsomes reveal a similar profile of CyA metabolism. Higher blood levels of CyA may result in individuals using CBD either therapeutically as an antiepileptic drug or illicitly in marijuana.

## REFERENCES

Bornheim, L.M., and Correia, M.A. (1990): Selective Inactivation of Mouse Liver Cytochrome P-450IIIA by Cannabidiol. *Mol Pharmacol.* 38, 319-326.

Bornheim, L.M., and Correia, M.A. (1991): *Mol Pharmacol.* 40, 228-234.

Prueksaritanont, T., Correia, M.A., Rettie, A.E., Swinney, D.C., Thomas, P.E., and Benet, L.Z. (1993). Cyclosporine Metabolism by Rat Liver Microsomes - Evidence for Involvement of Enzyme(s) Other than Cytochromes P-450 3A. *Drug Metab. Dispos.* 21, 730-737.

# Ethinyl estradiol-mediated induction of a cyclosporin A metabolizing 2C6 isozyme in rats

Walter Jaeger[1], Maria A. Correia[2], Lester M. Bornheim[2], Leslie Z. Benet[1]

Departments of [1]Pharmacy and [2]Pharmacology, University of California, San Francisco, 513 Parnassus Avenue, San Francisco, California 94143, USA

## INTRODUCTION

Cyclosporine (CyA) is a cyclic endecapeptide (Fig. 1) used in transplantation medicine and to treat autoimmune diseases. It undergoes extensive hepatic intestinal metabolism, resulting in three primary metabolites, two hydroxylated products, AM1 (AM1c, a cyclic derivative) and AM9; and one N-demethylated metabolite, AM4N. To date, cytochromes P450 3A have been identified as the major CyA metabolizing enzymes in human and rabbit livers ( Kronbach, *et al.* 1988; Bertault-Peres, *et al.* 1987). Pretreatment of rats with ethinyl estradiol (EE) increased CyA metabolism, but decreased P450s 3A. Immunoblot analysis using monoclonal antibodies raised against purified rat P450s 3A1 and 3A2, and polyclonal antibodies raised against rat P450s 2C6, 2C7, 2C11 and 2E1 revealed an increase of 2C6/2C7 and non detectable levels of P450s 3A1 and 3A2. EE-treatment also enhanced 21-progesterone and (S)-7-warfarin hydroxylases (both selective functional markers for 2C6), 2-3 fold, whereas 6β-testosterone hydroxylase was markedly decreased, ruling out P450s 3A1 and 3A2 for the increased CyA metabolism (Prueksaritanont, *et al.* 1993). The aim of the following work was to purify and characterize the unknown CyA metabolizing enzyme(s) in liver microsomes from EE pretreated adult female rats (containing nondetectable levels of P450 3A). A better knowledge of CyA metabolism is highly desirable in order to predict/elucidate clinical drug interactions that may influence its immunosuppressive activity as well as its toxic side-effects.

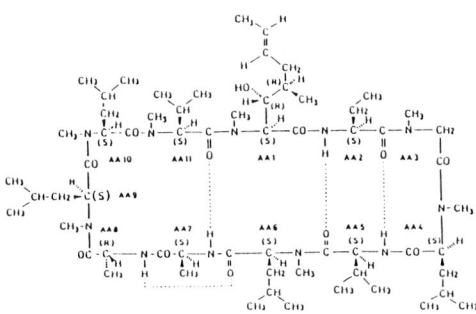

Fig. 1. Chemical Structure of CyA

## METHODS

### Animals and Microsomal Preparation:
Female rats were pretreated with EE (5 mg/kg, sc, for 5 days), the livers removed and microsomes subsequently prepared using standard procedures.

### P450 Purification:
Solubilized microsomes were applied at room temperature to a Whatman DEAE-cellulose (DE-52) column (10 nmol of P450/ml of gel) and eluted with 10 mM potassium phosphate, pH 7.4, 20 % (v/v) glycerol, 0.5 % (w/v) sodium cholate, 0.2 % (v/v) Lubrol PX, and 0.1 mM EDTA (buffer A). Analysis of the eluate by SDS-PAGE (7.5 % acrylamide) and UV-VIS absorbance (280 nm and 417 nm) revealed three major fractions. To check the fractions for their CyA metabolizing activity, nonionic detergent was removed as previously described (Bornheim & Correia, 1989) and CyA metabolism assayed by the formation of 4 metabolites. The fraction of interest was concentrated and dialyzed against buffer containing 3 mM sodium phosphate pH 6.8, 20 % (v/v) glycerol, 0.2 % (v/v) Lubrol PX, and 0.1 mM EDTA (buffer B), before application to a Whatman CM-cellulose (CM-52) column (10 nM P450/ml of gel). P450 was eluted with a linear gradient of 0-100 mM potassium phosphate, pH 6.8, in buffer B. The fraction exhibiting the highest CyA metabolizing activity was concentrated and dialyzed against 10 mM potassium phosphate, pH 7.4, containing 20 % (v/v) glycerol, 0.2 % (v/v) Lubrol PX, and 0.1 mM EDTA (buffer C), before application to a hydroxylapatite column (10 nmol of P450/ml of gel). A step gradient (45 mM, 67.5 mM, 90 mM and 180 mM potassium phosphate in buffer C) was used to separate 2 proteins of approximately $M_r$ 50,000. The purified protein eluting with 67.5 mM buffer C, exhibited the highest CyA metabolizing activity.

### Immunochemical procedures:
Polyclonal antibody was raised against the purified isozyme in rabbits, and used for Western blotting.

### $NH_2$-terminal sequence analyses:
The amino acid sequence of the first 11 residues was determined by Edman degradation.

### Determination of Enzyme Activities. *CyA Metabolism.*
Purified enzyme (25 pmol) or crude P450-fraction (250 pmol) was functionally reconstituted in the presence of dilaurylphosphatidylcholine (50 µg/ml), NADPH-P-450 reductase, $b_5$ (equimolar ratios), NADPH (1mM), isocitric acid (5mM) and isocitric dehydrogenase (0.5 U/ml). The reaction was started by the addition of CyA (30 µM) and its metabolism assessed by the formation of AM1, AM9, AM1c and AM4N after 30 min of incubation at 37°C. Extraction and HPLC-analysis were performed as previously described (Prueksaritanont *et al.*, 1993).

*Progesterone Metabolism.* The incubation mixture contained 1 mg of microsomal protein or 50 pmol purified enzyme (functionally reconstituted as described above), NADPH (1mM), NADPH regenerating system and 250 nmol of progesterone containing 1 µCi of [$^{14}C$]progesterone in a final volume of 1 ml. The reaction was stopped after 10 min incubation at 37°C by the addition of 6 ml of ethyl acetate, extracted and analyzed by HPLC.

## RESULTS AND CONCLUSIONS

EE treatment of female rats enhanced CyA metabolism despite the reduction of 6β-OH testosterone hydroxylase, a functional marker for P450s 3A1/3A2, and the absence of immunodetectable P450s 3A, ruling out these isoenzymes in the increased biotransformation. To identify the isozyme involved, liver microsomes from EE-pretreated female rats were solubilized and purified via anion (DE 52)/cation (CM 52)-exchange/hydroxylapatite chromatography. An enzyme with high CyA-metabolizing activity, homogeneous, >95% pure with an apparent molecular weight of 50,000 by SDS-PAGE (Fig. 2), the reduced-CO absorbance maximum of 452 nm, and specific content of 11.5 nmol/mg protein was obtained. N-terminal amino acid sequence analysis of the first 11 residues (Table 1) revealed no differences to those of the 2C6 isozymes previously reported (Wolf, *et al.* 1986). The purified enzyme reacted positively with polyclonal 2C6-antibody (Fig. 3) but negatively with polyclonal rat 3A antibody after immunoblotting.

**Table 1.** NH2-terminal amino acid sequence comparison of purified enzyme with rat P450s 2C and 3A isozymes

| P450 isozyme | Residue |
|---|---|
| Purified enzyme | M D L V M L L V L T L |
| 2C6 | M D L V M L L V L T L |
| 2C7 | M D L V T F L V L T L |
| 3A1 | M D L L S A L T L E T |
| 3A2 | M D L L S A L T L E T |

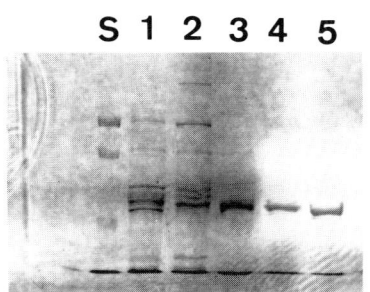

**Fig. 2.** SDS-PAGE of hepatic microsomes, chromatographic column fractions and purified enzyme: STD (protein standard: 45.0, 66.2 and 97.4 kDa), 1 (control microsomes), 2 (EE-microsomes), 3 (DE-52 eluate), 4 (CM-52 eluate), 5 (hydroxylapatite)

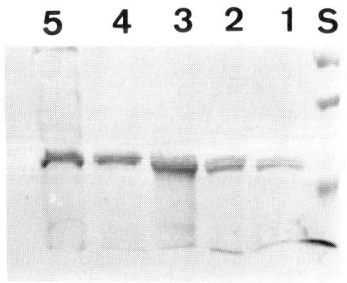

**Fig. 3.** Immunoblot of hepatic microsomes, chromatographic column fractions and purified enzyme: STD (protein standard: 45.0, 66.2 and 97.4 kDa), 1 (control microsomes), 2 (EE-microsomes), 3 (DE-52 eluate), 4 (CM-52 eluate), 5 (hydroxylapatite)

When functionally reconstituted with NADPH-P450 reductase, cytochrome b5, lipid and NADPH, the isolated P450 2C6 isozyme exhibited ~ a 10-fold higher specific CyA-metabolizing activity (AM9-formation) and a 20-fold increase in the formation in 21-OH progesterone formation (Table 2). Since female rats metabolize a host of "P450 3A-specific" substrates in the absence of detectable 3A levels, further studies are in progress to examine the role of this purified enzyme in their metabolism.

**Table 2.** Progesterone hydroxylase activities in microsomes and purified enzyme (nmol hydroxyprogesterione /nmol P450/min)

|                    | 16α   | 6β   | 21    | 2α   |
|--------------------|-------|------|-------|------|
| Control Microsomes | 1.14  | 1.32 | 0.10  | 0.11 |
| EE-microsomes      | 2.31  | 0.06 | 0.56  | 1.76 |
| Purified enzyme    | 19.61 | 3.24 | 11.18 | -    |

## REFERENCES

Bertault-Peres, P., Bonfils, C., Fabre, G., Just, S., Cano, J.-P., and Maurel, P. (1987): Metabolism of cyclosporin A. II. Implication of the macrolide antibiotic inucible cytochrome P-450 3c from rabbit liver microsomes. *Drug Metab. Dispos.* 17, 197-207.

Bornheim, L.M., and Correia, M.A. (1989): Purification and characterization of a mouse liver cytochrome P-450 induced by cannabidiol. *Mol. Pharmacol.* 36, 377-383.

Kronbach, T., Fischer, V., and Meyer U.A. (1988): Cyclosporine metabolism in human liver: identification of a cytochrome P450 II gene family as the major cyclosporine-metabolizing enzyme explains interactions of cyclosporine with other drugs. *Clin. Pharmacol. Ther.* 43, 630-635.

Prueksaritanont, T., Correia, M.A., Rettie, A.E., Swinney, D.C., Thomas, P.E., and Benet, L.Z. (1993): Cyclosporine Metabolism by Rat Liver Microsomes - Evidence for Involvement of Enzyme(s) Other than Cytochromes P-450 3A. *Drug Metab. Dispos.* 21, 730-737.

Wolf, C.R., Seilman, S., Oesch, F., Mayer, R.T., and Burke, M.D. (1986): Multiple forms of cytochrome P-450 related to forms induced marginally by phenobarbital. *Biochem. J.* 240, 27-33.

# Heterologous expression and structure-function analysis of CYP2A10 and CYP2A11, which differ in only eight amino acids but have strikingly different activities toward testosterone and coumarin

Xinxin Ding, Hwei-Ming Peng, Minor J. Coon

*Department of Biological Chemistry, University of Michigan Medical School, Ann Arbor, Michigan, USA 48109-0606*

## INTRODUCTION

Recently, cDNA clones encoding two highly similar P450s, designated 2A10 and 2A11, were obtained from a rabbit nasal cDNA library (Peng *et al.*, 1993). Both 2A10 and 2A11 contain 494 amino acids and correspond to purified P450 NMa (Ding & Coon, 1988) in the first 20 amino acids. The predicted amino acid sequences of the two proteins differ from each other in only eight positions and are over 80% identical to the sequences of 2A3, 2A4, 2A5, 2A6, and 2A7 (see Nelson *et al.*, 1993 for sequence information). After heterologous expression of 2A10 and 2A11 cDNAs in *E. coli*, both enzymes are active toward several xenobiotic substrates, although 2A10 is generally more active than 2A11. In the present study, structure-function analysis was carried out to determine which of the eight different amino acid residues are important for the 10-fold greater activity of 2A10 toward coumarin (C) and more than 20-fold greater activity in the conversion of testosterone (T) to androstenedione. Available data suggest that Leu[104] and Val[117] may be important for the androstenedione formation and coumarin 7-hydroxylation activities of 2A10 and that mutation of Arg[372]→His results in a significant increase in the turnover number of T hydroxylation by 2A11.

## EXPERIMENTAL PROCEDURES

Chimeras of 2A10 and 2A11 were produced using restriction endonuclease sites common to both cDNAs. Chimera A was constructed using *Apa* I, which cleaves at nucleotides 110 and 223, and chimera B with *Bgl* I, which cleaves at nucleotides 105 and 381. Site-directed mutagenesis of 2A11 was carried out by polymerase chain reaction (PCR) with primers containing the desired mutation and 2A11 cDNA as the template. The PCR fragments were generated in a Perkin-Elmer Cetus DNA thermal cycler using *Taq* DNA polymerase. Each of the 30 thermal cycles included a denaturation step at 94 °C for 1 min, an annealing step at 68°C for 1 min, and an extension step at 72 °C for 3 min. For Thr[120]→Ser mutant, the primers used were 5'-AAGCTCTGGGGGAAGCTGCCC-3' (upstream) and 5'-AAGCGCCGCAGC-GGCCTGGCGCGCTCCCAGCTG-3' (downstream), with the underlined residue indicating the nucleotide mutated. The resulting fragment was then digested with *Bgl* I and ligated with similarly digested 2A11 cDNA. The same strategy was used to construct Gln[104]→Leu and Ala[117]→Val mutants with downstream primers 5'-AAGCGCCGCAGCGGCCTGGCGCGCTC-CCAGGTGCTGAATGCCACGCCGTAGCCTTTGAAGAGCCAGTCGAAGGTGGCCAGC-3' and 5'-AAGCGCCGCAGCGGCCTGGCGCGCTCCCAGGTGCTGAACACC-3', respectively. Mutant Arg[372]→His was constructed by replacing an *Nco* I/*Bst* E2 fragment in 2A11 cDNA with an *Nco* I/*Bst*

E2 linker (5'-CATGGGCCTGGCCC<u>A</u>CAGG-3'/5'-CCGGACCGGG<u>T</u>GTCCCAGTG-3') that contained the mutation. The DNA sequence of the PCR-generated segments was determined to ensure the fidelity of *Taq* DNA polymerase extension of the amplified primers, and all mutations were confirmed by sequence analysis of the resulting DNA constructs. Oligonucleotides were synthesized on automated instruments from Applied Biosystems by the DNA facility at the University of Michigan. *E. coli* strain XL1-blue was transformed with the expression plasmids as described (Larson *et al.*, 1991; Peng *et al.*, 1993), and was used for heterologous expression of the recombinant P450s. Partial purification of recombinant P450 from solubilized bacterial membranes and determination of catalytic activity in a reconstituted system toward C and T were as described (Peng *et al.*, 1993).

RESULTS AND DISCUSSION

The distinct substrate specificity of 2A10 and 2A11 toward testosterone and coumarin makes these two enzymes of interest for examination of the amino acid residues involved in substrate binding and oxygenation. One of the regions of greatest dissimilarity between these cytochromes is at amino acid residues 117-120; marked dissimilarity in this region is also found between P450 2A1 and 2A2 (Matsunaga *et al.*, 1988), 2A4 and 2A5 (Burkhart *et al.*, 1985; Squires & Negishi, 1988), 2C4 and 2C5 (Kronbach & Johnson, 1991), and 2B1 and 2B2$^V$ (Aoyama *et al.*, 1989). In the case of P450 2A4 and 2A5, Val$^{117}$ in the former cytochrome appears to be critical for coumarin 7-hydroxylation (Lindberg & Negishi, 1989). In addition, the corresponding region in 2C4 and 2C5 (amino acid residues 113-115) was found to be important in progesterone 21-hydroxylation (Kronbach & Johnson, 1991). Gotoh (1992) has recently proposed six substrate-recognition regions in P450 family 2 proteins, based on group-to-group alignment of CYP 2 sequences and those of bacterial P450s, including P450 101A, for which substrate-binding residues have been identified by x-ray crystallography (Poulos *et al.*, 1985). Interestingly, as shown in Fig. 1, five of the eight amino acid differences between 2A10 and 2A11 fall into those proposed substrate recognition sites (SRSs).

```
CYP2A10    MLASGLLLAALLACLTVMILLSVWRQRKLWGKLPPGPTPLPFIGNYLQLN    50
CYP2A11    MLASGLLLAALLACLTVMILLSVWRQRKLWGKLPPGPTPLPFIGNYLQLN    50

CYP2A10    TEQMYDSLMKISERYGPVFTIHLGPRRIVVLCGQEAVKEALVDQAEDFSG   100
CYP2A11    TEQMYDSLMKIRDRYGPVFTIHLGPRRIVVLCGQEAVKEALVDQAEDFSG   100
                      **

CYP2A10    RGELATFDWLFKGYGVVFSSWERARPLRRFAISTLRDFGVKRGIEERIQ    150
CYP2A11    RGEQATFDWLFKGYGVAFSTWERARPLRRFAISTLRDFGVKRGIEERIQ    150
              *            *  *

CYP2A10    EEAGFLIEAFRDTRGAFIDPTFFLSRTVSNVISSIVFGDRFDYEDKEFLS   200
CYP2A11    EEAGFLIEAFRDTRGAFIDPTFFLSRTVSNVISSIVFGDRFDYEDKEFLS   200

CYP2A10    LLRMMLGSFQFTATPTGQLYEMFYSVMKHLPGPQQQAFKELEGLRDFIAK   250
CYP2A11    LLRMMLGSFQFTATPTGQLYEMFYSVMKHLPGPQQQAFKELEGLRDFIAK   250

CYP2A10    KVERNQRTLDPNSPRDFIDSFLIRMQEEKKDPKSEFHMKNLVMTTLNLFF   300
CYP2A11    KVERNQRTLDPNSPRDFIDSFLIRMQEEKKDPKSEFHMKNLVLTTLNLFF   300
                                                      *

CYP2A10    AGTETVSTTMRYGFLLLMKHPDVEAKVHEEIDRVIGRNRQPKFEDRAKMP   350
CYP2A11    AGTETVSTTMRYGFLLLMKHPDVEAKVHEEIDRVIGRNRQPKFEDRAKMP   350

CYP2A10    YTEAVIHEIQRFTDMIPMGLAHRVTRDTKFRDFLLPKGAEVFPMLGSVLK   400
CYP2A11    YTEAVIHEIQRFTDMIPMGLARRVTRDTKFRDFLLPKGTEVFPMLGSVLK   400
                                *                *

CYP2A10    DPKFFSKPREFYPQHFLDEKGQFKKSDAFMPFSVGKRYCLGEGLARMELF   450
CYP2A11    DPKFFSKPREFYPQHFLDEKGQFKKSDAFMPFSVGKRYCLGEGLARMELF   450

CYP2A10    LFFTTIMQNFRFRSQQAPQDIDVSPKHVGFATIPRTYTMSFVPR    494
CYP2A11    LFFTTIMQNFRFRSQQAPQDIDVSPKHVGFATIPRTYTMSFVPR    494
```

Fig. 1. Sequence alignment of P450 2A10 and 2A11. The positions of eight amino acid differences are indicated by an *asterisk*, and the double-underlined areas are the SRSs proposed by Gotoh (1992).

Structure-function analysis indicated that not all of the eight different amino acid residues are important for the 10-fold greater activity of 2A10 toward coumarin and more than 20-fold greater activity in the conversion of T to androstenedione in a reconstituted system, as shown in Table 1. Mutation of $Arg^{62}$ and $Asp^{63}$ of 2A11 to corresponding residues in 2A10 (chimera A), or mutation of $Thr^{120}$ to Ser (as found in 2A10) did not change the activities of 2A11 toward C or T, but mutation of $Arg^{62}$, $Asp^{63}$, $Gln^{104}$, $Ala^{117}$, and $Thr^{120}$ of 2A11 to corresponding residues in 2A10 (chimera B) resulted in a protein which is equally as active as 2A10 in C hydroxylation and approximately half as active as 2A10 in androstenedione formation. Thus, at least two of the five residues ($Leu^{104}$ and $Val^{117}$) located in the proposed SRSs may be involved in substrate binding with C and T. Additional studies are in progress with $Gln^{104} \rightarrow Leu$ and $Ala^{117} \rightarrow Val$ mutants of 2A11 to determine whether one or both of the amino acids at these positions are responsible for the higher rates of 2A10 in C hydroxylation and androstenedione formation.

Table 1. Catalytic activities of wild-type and mutant P450 2A10 and 2A11 cytochromes. For coumarin 7-hydroxylation, the reaction mixtures contained 50 μmol of potassium phosphate buffer, pH 7.4, 0.05 nmol of P450, 0.2 nmol of reductase, 30 μg of dilauroylglyceryl-3-phosphorylcholine, 0.5 μmol of coumarin added in 12.5 μl of methanol, and 1 μmol of NADPH as the final addition in a total volume of 1.0 ml. The incubation was carried out at 37 °C for lengths of time that represented the initial linear rate of product formation. For testosterone hydroxylation, the reaction mixtures contained 5 nmol of $[1,2,6,7-^3H(N)]$testosterone (4.7 Ci/mmol, added in 10 μl of methanol), 1 μmol of ascorbic acid, and other components as described above. The values reported are the average of duplicate determinations, with a standard error less than 10% of the mean. The specific content of P450 for the partially purified recombinant proteins ranged from 1.5 to 3.0 nmol/mg of protein. The underlined residues correspond to those of 2A10. X1 designates an unidentified metabolite of T, and 15α, 11β, and 17 designate 15α- and 11β-hydroxytestosterone and androstenedione, respectively.

| Enzyme | Amino Acid No. | | | | | | | Coumarin hydroxylase | Testosterone hydroxylase | | |
|---|---|---|---|---|---|---|---|---|---|---|---|
| | 62 | 63 | 104 | 117 | 120 | 293 | 372 | 389 | | X1+15α | 11ß | 17 |
| | | | | | | | | | nmol/min/nmol | pmol/min/nmol | | |
| 2A10 | S | E | L | V | S | M | H | A | 1.78 | 34 | 65 | 217 |
| 2A11 | R | D | Q | A | T | L | R | T | 0.17 | 69 | 22 | <10 |
| A | S | E | Q | A | T | L | R | T | 0.18 | 61 | 17 | <10 |
| B | S | E | L | V | S | L | R | T | 1.80 | 17 | 13 | 95 |
| 120 | R | D | Q | A | S | L | R | T | 0.18 | 58 | 32 | <10 |

Interestingly, mutation of $Arg^{372}$ in 2A11 to His (as found in 2A10) resulted in an increase in the rates of formation of all testosterone metabolites, but gave the same metabolite profile as that of 2A11 (Table 2). In experiments not presented, both 2A11 and the $Arg^{372} \rightarrow His$ mutant showed higher activity toward T at pH 6.7 than at pH 7.4, with the mutant cytochrome being more active than the wild-type at either pH. Furthermore, the difference in turnover number toward T appears to be due to intrinsic properties of the P450 proteins, but not to other components in the partially purified enzyme preparations. This was demonstrated by the finding that, when 2A11 and the $Arg^{372} \rightarrow His$ mutant were mixed before reconstitution with P450 reductase and phospholipid, the combined activity was equivalent to the sum of activities of the two cytochromes determined individually (not shown) Thus, it appears that, although the residue at position 372 is not involved in determining the regiospecificity of testosterone hydroxylation by 2A10 and 2A11, it is important for enzyme turnover. It remains to be determined whether the increase in activity is a result of altered substrate binding, electron transfer, or oxygen insertion. According to sequence alignments made by Gotoh (1992), members of CYP family 2 have either Arg or His at this position, suggesting their possible importance in function.

Table 2. Testosterone hydroxylation by 2A10, 2A11 and Arg$_{372}$→His mutant of 2A11. The assay conditions are the same as described in Table 1 except that a different batch of P450 reductase, which had lower activity, was used. The values reported represent the mean ± standard error, with the number of experiments performed (n) indicated in parentheses.

| Substrate Concentration | Enzyme | X1+15α | 11ß | 17 |
|---|---|---|---|---|
| | | *pmol/min/nmol* | | |
| 5 µM | 2A10, wild-type (n=8) | 19 ± 7 | 35 ± 16 | 166 ± 38 |
| | 2A11, wild-type (n=12) | 60 ± 16 | 27 ± 9 | < 5 |
| | 2A11, Arg$^{372}$→His (n=12) | 111 ± 22 | 61 ± 16 | 10 ± 4 |
| 25 µM | 2A11, wild-type (n=2) | 166 ± 43 | 74 ± 2 | <30 |
| | 2A11, Arg$^{372}$→His (n=2) | 280 ± 4 | 155 ± 11 | <30 |

*Acknowledgment*--We thank Drs. Steven J. Pernecky and Alfin D. N. Vaz for helpful discussions and Ms. Valerie Fujita and Dr. Jiang-Jun Sheng for technical assistance. This research was supported by National Institutes of Health Grant DK-10339.

REFERENCES

Aoyama, T., Korzekwa, K., Nagata, K., Adesnik, M., Reiss, A., Lapenson, D.P., Gillette, J., Gelboin, H.V., Waxman, D.J., & Gonzalez, F.J. (1989): Sequence requirements for cytochrome P-450IIB1 catalytic activity; alteration of the stereospecificity and regioselectivity of steroid hydroxylation by a simultaneous change of two hydrophobic amino acid residues to phenylalanine. *J. Biol. Chem.* 264, 21327-21333

Burkhart, B. A., Harada, N., & Negishi, M. (1985): Sexual dimorphism of testosterone 15α-hydroxylase mRNA levels in mouse liver. *J. Biol. Chem.* 260, 15357-15361

Ding, X. & Coon, M. J. (1988): Purification and characterization of two unique forms of cytochrome P-450 from rabbit nasal microsomes. *Biochemistry* 27, 8330-8337

Gotoh, O. (1992): Substrate recognition sites in cytochrome P450 family 2 (CYP2) proteins inferred from comparative analyses of amino acid and coding nucleotide sequences. *J. Biol. Chem.* 267, 83-90

Kronbach, T. & Johnson, E. F. (1991): An inhibitory monoclonal antibody binds in close proximity to a determinant for substrate binding in cytochrome P450IIC5. *J. Biol. Chem.* 266, 6215-6220

Larson, J. R., Coon, M. J., & Porter, T. D. (1991): Alcohol-inducible cytochrome P-450IIE1 lacking NH$_2$-terminal segment retains catalytic activity and is membrane-bound when expressed in *Escherichia coli*. *J. Biol. Chem.* 266, 7321-7324

Lindberg, R. L. P. & Negishi, M. (1989): Alteration of mouse cytochrome P450$_{coh}$ substrate specificity by mutation of a single amino-acid residue. *Nature (Lond.)* 339, 632-634

Matsunaga, T., Nagata, K., Holsztynska, E. J., Lapenson, D. P., Smith, A., Kato, R., Gelboin, H. V., Waxman, D.J., & Gonzalez, F.J. (1988): Gene conversion and differential regulation in the rat P450IIA gene subfamily: purification, catalytic activity, cDNA and deduced amino acid sequence, and regulation of an adult male-specific testosterone 15α-hydroxylase. *J. Biol. Chem.* 263, 17995-18002

Nelson, D. R., Kamataki, T, Waxman, D. J., Guengerich, F. P., Estabrook, R. W., Feyereisen, R., Gonzalez, F. J., Coon, M. J., Gunsalus, I. C., Gotoh, O., Okuda, K., & Nebert, D. (1993): The P450 superfamily: update on new sequences, gene mapping, accession numbers, early trivial names of enzymes, and nomenclature. *DNA Cell Biol.*, 12, 1-51

Peng, H.-M., Ding, X., & Coon, M. J. (1993): Isolation and heterologous expression of cloned cDNAs for two rabbit nasal microsomal proteins, CYP2A10 and CYP2A11, that are related to nasal microsomal cytochrome P450 form a. *J. Biol. Chem.* 268, 17253-17260

Poulos, T. L., Finzel, B. C., Gunsalus, I. C., Wagner, G. C., & Kraut, J. (1985): The 2.6-Å crystal structure of *Pseudomonas putida* cytochrome P450 P-450. *J. Biol. Chem.* 260, 16122-16130

Squires, E. J. & Negishi, M. (1988): Reciprocal regulation of sex-dependent expression of testosterone 15α-hydroxylase (P-450$_{15α}$) in liver and kidney of male mice by androgen: evidence for a single gene. *J. Biol. Chem.* 263, 4166-4171

# Dynamic topology and electron transfer interactions of microsomal cytochrome P450 in liver and adrenal cortex: protein rotation study

Yoshihiro Ohta, Makoto Yamada, Suguru Kawato

*Institute of Physics, College of Arts and Sciences, University of Tokyo at Komaba, Meguro, Tokyo 153, Japan*

To visualize dynamic electron transfer interactions of proteins in the monooxygenase system in the membrane, we have applied a time-resolved absorption anisotropy measurements. The rotational diffusion of cytochrome P-450 is particularly sensitive to formation and dissociation of transient electron transfer complex in membranes. Rotation of cytochrome P-450 is measured by observing the absorption anisotropy due to photolysis of the P-450 · CO complex by a vertically polarized laser flash at 532 nm from a Nd/YAG laser. Decay of absorption anisotropy due to the rotation of the cytochromes was analyzed with the equation:

$$r(t) = r_1 \exp(-t/\phi) + r_2 \exp(-4t/\phi) + r_3 \quad (Eq.\ 1)$$

where rotation occurs about an axis perpendicular to the membrane plane with the average rotational relaxation time $\phi$ (Inversely proportional to the speed of rotation). The residual time-independent anisotropy $r_3/r(0)$ is related to the immobile cytochromes. The tilt angle of the heme plane $\theta_N$ from the membrane plane was calculated by

$$r_3/r(0) = 1/4(3\cos^2\theta_N - 1)^2 \quad (Eq.\ 2)$$

when all cytochrome P-450 molecules are rotating, even in the presence of multiple rotating species.

Proteoliposomes were prepared by cholate dialysis procedures. Successful incorporation of P-450 and other proteins was demonstrated by ultracentrifugation of vesicles in sucrose density gradient.

## RESULTS & DISCUSSION

### Rabbit Liver Microsomal P-450(IIB4 and IA2) in Proteoliposomes (Collaboration with Drs. G.I. Bachmanova, and A. I. Archakov at Inst. Biol. Med. Chem., Moscow)

Cytochrome P-450IA2 was reconstituted in PC/PE/PS=10:5:1 vesicles with/without NADPH-cyt. P-450 reductase or cyt.$b_5$. 80% of cytochrome P-450 IA2 rotates with $\phi = 237\ \mu$sec and 20% of P-450IA2 was immobilized. By the co-reconstitution with reductase, the mobile P-450IA2 was increased by 9% from 80 to 89%. However $\phi = 300$ $\mu$ sec was not changed considerably. By the presence of cytochrome $b_5$, the rotating population was increased by 16% form 80% to 96% with $\phi = 250$ $\mu$ sec. These results indicate that P-450IA2 forms a transient association not only with the reductase but also with cyt.$b_5$, resulting in dissociation of P-450IA2 oligomers. On the other hand, neither the reductase nor cyt. $b_5$ significantly affected the mobility of P-450IIB4 ($\phi = 190$ $\mu$ sec). It should be noted that all P-450IIB4 are rotating in all cases. Taken together these results with chemical cross-linking experiments the P-450IIB4 is probably diffusing independently of the reductase and $b_5$.

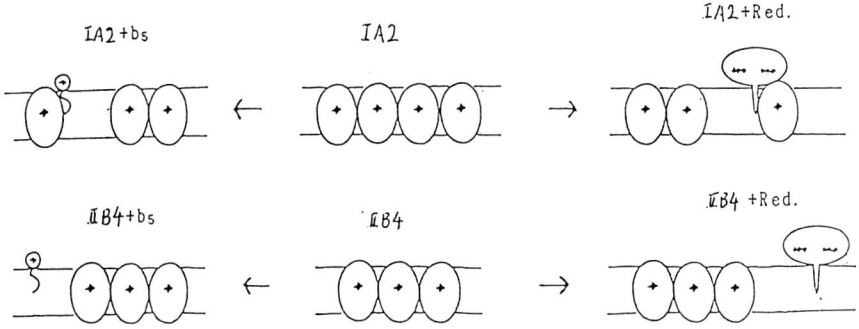

Fig.1: Schematic Model Illustrating the Distribution and Protein-protein Interactions for Cytochrome P-450IA2(upper) and P-450IIB4(lower)

Cytochrome P-450IA2 forms a transient complex with the reductase or $b_5$, resulting in mobilization of P-450IA2 due to dissociation of P-450 oligomers. Cytochrome P-450IIB4 is rotating without complex formation with redox partners.

Genetically Engineered Rat Liver Microsomal P-450 (Collaboration with Drs. T. Sakaki, Y. Yabusaki, H. Ohkawa at Sumitomo Chem. Co. Ltd.)

We have successfully examined the rotational mobility of rat liver P-450IA1 genetically expressed in yeast microsomes in the absence and presence of genetically co-expressed yeast reductase. When molar ratio of reductase to P-450IA1 was increased from 1:17 to 2:1, a significant mobilization of P-450IA1 was observed from 28% to 43% in mobile population with $\phi \sim 1100$ $\mu$ sec, implying that P-450IA1 forms a transient complex with the reductase. Electrons may be transported within this complex.

Truncated P-450IA1 lacking the N-terminal hydrophobic segment(2-30) has genetically expressed in yeast microsomal membranes. This truncated P-450IA1 rotates slowly with $\phi = 1020$ $\mu$ sec, indicating that the shortened P-450IA1 is incorporated properly into the membrane and that N-terminal segment is not solely responsible for attachment to the membrane. This cytochrome was not removed from the membrane by the high salt treatment.

These results provide the evidence that the additional segments of P-450IA1 are involved in the membrane binding.

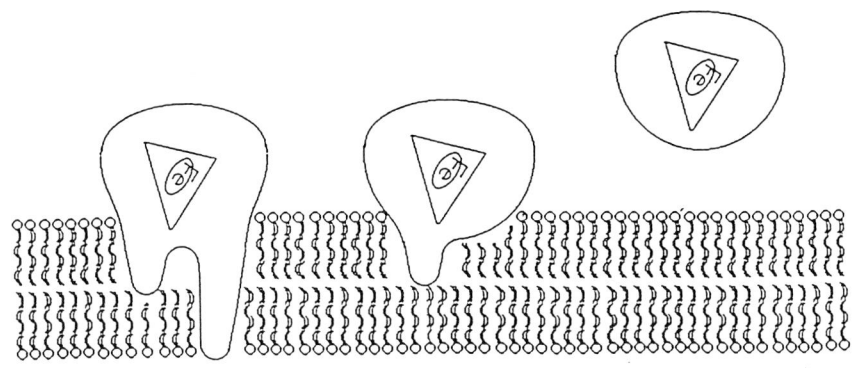

Fig.2: Membrane Topology of Cytochrome P-450IA1
A Schematic model illustrating the topology of cytochromes P-450IA1 in the full-length form (left), P-450IA1 lacking N-terminal hydrophobic segment (center), and P-450cam (right). N-terminally truncated P-450IA1 tightly binds to the membrane, showing the slow rotation characteristics ($\phi \sim 1000\ \mu$ sec) similar to that of the full-length P-450IA1. Therefore, P-450IA1 probably has another membrane anchor segments in addition to the N-terminal segment. Water soluble P-450cam rotates much faster ($\phi < 1\ \mu$ sec) than other P-450IA1s.

Membrane Topology of Adrenocortical Microsomal P-450 (Collaboration with Drs. S. Kominami and S. Takemori at Hiroshima Univ.)
Purified P-45017$\alpha$,lyase and P-450C21 were reconstituted in PC/PE/PS=5:3:1 vesicles. Coexistence of mobile(80%) and immobile(20%) populations of P-450 has been observed ($\phi$=140-170 $\mu$ sec). The co-reconstitution with the reductase completely mobilized the cytochrome P-450C21, while the mobility of P-45017$\alpha$,lyase was not affected by the presence of reductase. These results indicate that P-450C21 forms a transient association with the reductase, whereas P-45017$\alpha$,lyase diffuses independently of the reductase. Electrons may be transferred from the reductase to P-45017$\alpha$,lyase by collisions between these proteins. The exhaustive trypsinolysis of these two P-450s has shown that P-45017$\alpha$, lyase is not trypsinized at all. Though P-450C21 was cleaved at both ends of I-helix, this trypsinized P-450C21 was observed to be fully active and rotational mobility was almost the same as the intact P-450C21. These results demonstrate that these two adrenocortical P-450s are deeply embedded in the lipid bilayer and probably have a different topology from that of liver microsomal P-450IA/IIB.

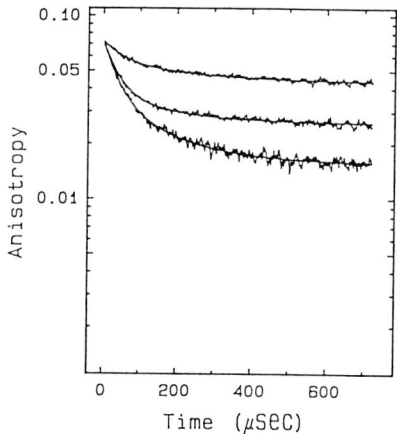

Fig.3: Time-Dependent Absorption Anisotropy of Cytochrome P-450C21 in PC/PE/PS Vesicles

(Middle) only P-450C21, (Lower) Addtion of reductase mobilized P-450C21, (Upper) antireductase IgG added to the sample shown in the lower curve immobilized P-450C21.

After mobilizing all P-450 molecules by the presence of 730 mM NaCl, the heme angles were determined to be either 47° or 63° for P-45017α,lyase and either 38° or 78° for P-450C21.

|   | P-45017α,lyase | P-450C21, | P-450IA2, | P-450IIB4 |
|---|---|---|---|---|
| $\theta_N$ | 47° or 63° | 38° or 78° | 47° or 63° | 55° |

## REFERENCES

Iwase, T., Sakaki, T., Yabusaki, Y., Ohkawa, H., Ohta, Y., and Kawato, S. Biochemistry (1991) 30 8347-8351

Ohta, Y., Kawato S., Tagashira, H., Takemori, S., and Kominami, S. Biochemistry (1992) 31 12680-12687

Ohta, Y., Sakaki, T., Yabusaki, Y., Ohkawa, H., and Kawato, S. J. Biol. Chem. (1993) in press

# Diverse functions of aromatase cytochrome P450: catecholestrogen synthesis, cocaine N-demethylation, and other selective drug metabolisms

Yoshio Osawa, Tadayoshi Higashiyama, Carol Yarborough

Endocrine Biochemistry Department, Medical Foundation of Buffalo Research Institute, 73 High St., Buffalo, NY 14203, USA

## Summary

Aromatase cytochrome P450 has long been thought to catalyze only one of the many human placental P450 enzyme functions, namely that of aromatization of androgens to produce estrogens. However, we found that aromatase is capable of catalyzing several other specific reactions. When we incubated [2-$^3$H, 4-$^{14}$C]estradiol with reconstituted aromatase and NADPH, we found that 2-hydroxy[4-$^{14}$C]estradiol was produced (Osawa et al., 1993). In a similar manner, we incubated cocaine, aminopyrine, dopamine, morphine, diazepam, epinephrine, and others with aromatase and assessed N-demethylase activity by the Nash method, and incubated [G-$^3$H, 3,6-$^{14}$C]benzo[a]pyrene and assessed arylhydrocarbon hydroxylase (AHH) activity by the radioisotope partition method. Further kinetic analyses on compounds which showed positive reactions gave Km and Vmax values of 1.58 µM and 0.64 min$^{-1}$ for estradiol 2-hydroxylation, 240 µM and 8.0 min$^{-1}$ for cocaine N-demethylation, and 280 µM and 12.4 min$^{-1}$ for aminopyrine N-demethylation. Diazepam, dopamine, tyramine and methadone inhibited androstenedione aromatization (Ki = 21, 160, 150, and 560 µM, respectively). Diazepam and morphine did not undergo N-demethylation. Aromatase showed no metabolic activity on benzo[a]pyrene. The results indicate that aromatase is a diverse function enzyme which may be involved in the metabolism of various hormones and drugs.

## INTRODUCTION

The human placenta is known to perform many cytochrome P450 catalyzed enzyme functions and it has been thought that aromatase cytochrome P450 is specific for aromatization of androgens to estrogens, and that it has no other function. Indeed, one of the criteria for purity of aromatase has been the absence of benzphetamine N-demethylase activity which is present in abundance in crude preparations. However, we recently found that purified aromatase is capable of catalyzing estrogen 2-hydroxylation (Osawa et al., 1993). This led us to question whether aromatase could catalyze additional reactions. In the present study, we show that in addition to aromatization and catechol formation, purified aromatase also catalyzes N-demethylation of various drugs.

EXPERIMENTAL

Purification of aromatase P450. Microsomal aromatase P450 was purified according to previously described procedures (Osawa et al., 1987, Yoshida & Osawa, 1991) using washed placental microsomes. We extended the procedure to purify aromatase cytochrome P450 from the washed precipitates of the whole tissue homogenate, using the same monoclonal antibody immunoaffinity chromatography procedure.

Assay procedures. For the assay of xenobiotic N-demethylase activity, purified aromatase P450 (18.7 nM) was reconstituted with bovine liver NADPH-cyt P450 reductase (40.9 nM) and dilauroyl L-α-phosphatidylcholine (7.8 μM) in 100 mM potassium phosphate buffer containing 1% glycerol and 0.0038% Emulgen 913. The reconstituted system was prewarmed in the presence of various concentrations of putative xenobiotic substrate and the reaction was initiated by the addition of NADPH (0.5 mM) in a total volume of 1.0 ml and incubated at 37°C for 10 and 30 min. The formation of formaldehyde was determined by the Nash method (Werringloer, 1978).

To determine aromatase activity, various concentrations of [1-$^3$H, 4-$^{14}$C]androstenedione were incubated with the reconstituted aromatase with or without putative xenobiotic inhibitor. The concentrations of P450, reductase, phospholipid, and glycerol were one-fifth of those used in the N-demethylase assay. The $^3$H-water method was used to determine the quantity of 1β-$^3$H elimination upon aromatization (Osawa et al., 1987).

Aryl hydrocarbon hydroxylase (AHH)assays were performed following the procedure of Van Cantfort et al. (1977), using [G-$^3$H, 3,6-$^{14}$C]benzo[a]pyrene as substrate.

RESULTS AND DISCUSSION

Our progress in developing a monoclonal antibody-based immunoaffinity chromatography method for the purification of aromatase P450 (Osawa et al., 1987, Yoshida & Osawa, 1991) led to interest in other cytochrome P450 enzyme functions in the human placenta. We found that microsomal aromatase P450 also functions as estrogen 2-hydroxylase (Osawa et al., 1993), leading us to question whether aromatase is capable of catalyzing other reactions. We purified aromatase P450 from the total tissue homogenate of human placenta and tested several putative substrates throughout the purification procedure. As indicated in Table 1, the presence of cocaine N-demethylase activity coincided with that of aromatase activity, while aminopyrine N-demethylase activity was present in both the aromatase and non-aromatase fractions, and aryl hydrocarbon hydroxylase activity was totally separate from aromatase activity.

Kinetic analysis of cocaine N-demethylase activity of purified and reconstituted aromatase gave Km of 240 μM, Vmax of 95 nmol/min•mg and a turnover rate of 8.0 min$^{-1}$ (Fig. 1). The Km value for aminopyrine was 280 μM and its Vmax and turnover rate were 146 nmol/min•mg and 12.4 min$^{-1}$, respectively. Diazepan did not undergo N-demethylation, and morphine and benzo[a]pyrene as well were unaffected by aromatase.

Cocaine N-demethylase activity was suppressed by our aromatase specific monoclonal antibody in the same manner as aromatase activity, suggesting that the two activities belong to the same P450. We found that the compounds which served as substrates for aromatase also functioned as inhibitors of androstenedione aromatization. Although diazepam did not undergo demethylation, it was capable of aromatase inhibition, with a Ki = 21 μM (Fig. 2). Dopamine, tyramine, and methadone inhibited aromatase activity with Ki values of 160, 150 and 560 μM, respectively. Our results indicate that aromatase is a diverse function enzyme which may be involved in the metabolism of various drugs, and whose function may in turn be affected by the presence of these drugs. Since aromatase is present in greater quantities in females, especially during pregnancy, the effectiveness of these drugs may differ significantly between women and men.

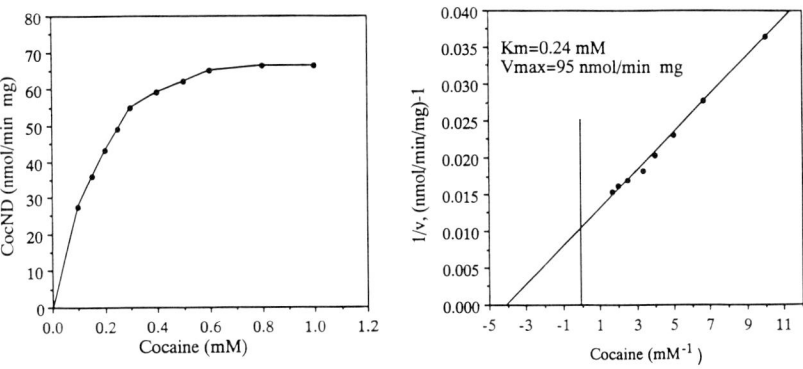

Fig. 1. Kinetic analysis of cocaine N-demethylase activity exhibited by purified aromatase.

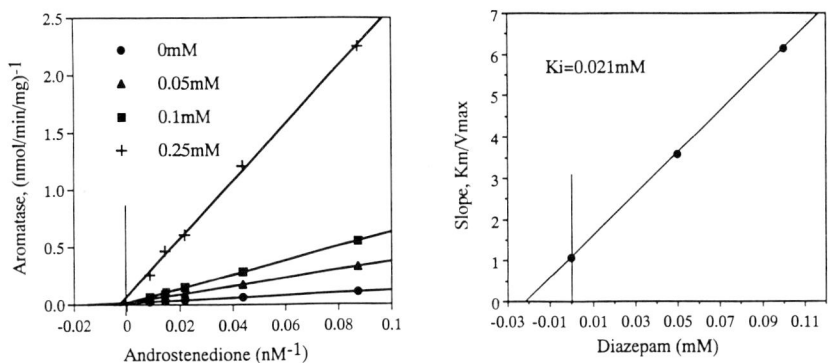

Fig. 2. Kinetic analysis of the inhibition by diazepam of androstenedione aromatization.

TABLE 1. PURIFICATION OF AROMATASE P450 FROM HUMAN PLACENTAL HOMOGENATE

| Preparation | Protein | | P450 | | | Aromatase | | |
|---|---|---|---|---|---|---|---|---|
| | mg | % | nmol/mg | nmol | % | nmol/min•mg | nmol/min | % |
| Tissue Homogenate | 21662 | 100 | - | - | - | 0.04 | 868 | 100 |
| Washed ppt | 14543 | 67 | - | - | - | 0.045 | 658 | 76 |
| Extract | 2837 | 13 | 0.067 | 190 | 100 | 0.209 | 593 | 68 |
| MAb-pass | 2747 | 13 | 0.045 | 124 | 65 | 0.047 | 128 | 15 |
| MAb-G25 | 3.88 | 0.02 | 8.15 | 31.8 | 17 | 53.5 | 207 | 24 |
| HA | 2.33 | 0.01 | 9.57 | 22.3 | 12 | 88.6 | 206 | 24 |
| Purified P450arom | 2.0 | 0.01 | 10.77 | 19.3 | 10 | 102 | 203 | 22 |

| Preparation | Cocaine-ND | | | APND | | | AHH | | |
|---|---|---|---|---|---|---|---|---|---|
| | nmol/min•mg | nmol/min | % | nmol/min•mg | nmol/min | % | pmol/min•mg | nmol/min | % |
| Tissue Homogenate | 0.04 | 869 | 100 | 0.022 | 477 | | 0.21 | 3.25 | |
| Washed ppt | 0.055 | 806 | 93 | 0.031 | 448 | 94 | 0.21 | 3.05 | 94 |
| Extract | 0.327 | 928 | 107 | 0.262 | 720 | 151 | 2.31 | 6.55 | 202 |
| MAb-pass | 0 | 0 | 0 | 0.128 | 352 | 74 | 3.23 | 8.87 | 273 |
| MAb-G25 | 103 | 400 | 46 | 56.8 | 220 | 43 | 0 | 0 | 0 |
| HA | 105 | 244 | 28 | 71.7 | 167 | 35 | - | - | - |
| Purified P450arom | 129 | 256 | 29 | 100 | 199 | 42 | 0 | 0 | 0 |

ACKNOWLEDGEMENTS

This research was supported in part by USPHS NIH Research Grant HD04945 and a grant from the Helen Woodward Rivas Memorial Fund. The authors thank Mrs. Margaret Cegielski for manuscript preparation.

REFERENCES

Osawa, Y., Higashiyama, T., Shimizu, Y, and Yarborough, C. (1993): Multiple functions of aromatase and the active site structure; aromatase is the placental estrogen 2-hydroxylase. *J. Steroid Biochem. Molec. Biol.* 44, 469-480.

Osawa, Y., Yoshida, N., Fronckowiak, M. and Kitawaki, J. (1987): Immunoaffinity purification of aromatase cytochrome P450 from human placental microsomes, metabolic switching from aromatization to 1β- and 2β monohydroxylation, and recognition of aromatase isozymes. *Steroids* 50, 11-28.

Van Cantfort, J., DeGraeve, J. and Gielen, J. (1977): Radioactive assay for aryl hydrocarbon hydroxylase. Improved methods and biological importance. *Biochem. Biophys. Res. Commun.* 79, 505-512.

Werringloer, J. (1978): Assay of formaldehyde generated during microsomal oxidation reactions. *Methods Enzym.* 52c, 297-302.

Yoshida, N. and Osawa, Y. (1991): Purification of human placental aromatase cytochrome P450 with monoclonal antibody and its characterization. *Biochemistry* 30, 3003-3010.

# Computer modeling of cytochromes P450cam interaction with camphor and its derivatives by Monte Carlo method

Yury A. Lyulkin, Alexis S. Ivanov, Vladlen S. Skvortsov, Andrey B. Rumyantsev

*Laboratory of Computers in Biochemistry, Institute of Biomedical Chemistry, Pogodinskaya str. 10, Moscow, 119832, Russia*

## Summary

The interaction of cytochrome P450cam with its substrate (camphor and camphor derivatives) was investigated. The 3-D coordinates of the cytochrome P450cam and its complexes with camphor and camphor derivatives were taken from PDB. The Monte Carlo method was used for analysis of the possible enzyme active site conformations with different substrates. The original computer program DOCKMASTER with accelerated Monte Carlo algorithm was designed for solving such problems. The samples of enzyme-substrate conformations were obtained during stochastic molecular modelling. The distance between hem's geometrical center and substrate's carbon atoms, as well as substrate orientations were found. The distribution functions for conformation numbers both via this distance and via substrate orientation were obtained. The distance distribution functions have the sharp picks. There is the pronounced anisotropy of the orientation substrate distribution as well. These results allow to predict the product formation specificity for camphor derivatives.

INTRODUCTION

The investigation of the substrate specificity mechanism of cytochrome P450's family is the important problem of biochemistry. Molecular modeling of the conformational and orientational mobility of substrates allows to understand the enzyme specificity. It was suggested, that various parts of substrates are subjected to oxidation in dependence on substrate orientation in active site. If substrate mobility is very low, then enzyme specificity must be pronounced. In other case enzyme was characterized by very poor specificity. The main aim of given study is the investigation of the cytochrome P450cam substrate specificity by molecular modeling.

## MATERIALS AND METHODS

Accelerated Monte Carlo method:
In our investigation stochastic modeling method was used to study the orientational and conformational mobility of macromolecules. This method is based on collective degree of freedom. Information about collective motion was obtained by statistical analysis of local conformational atomic mobility in macromolecules. Accelerated Monte Carlo method allows to carry out the molecular modeling in time intervals which are larger then one for molecular dynamics. The geometrical criteria which were suggested in [Poulsen, 1992], were used to study enzyme specificity. Probability density of the orientational mobility was obtained in dependence on distances between substrate hydrogen atoms and ferril oxygen and angles which are defined by three atoms (substrate carbon C - substrate hydrogen H - ferril oxygen O).

## RESULTS

In molecular modeling statistical information about substrate mobility was obtained. It was shown (fig.1, 2) that probability density of the substrate displacement and rotations are unimodal. Thus camphor molecule has only one predominant orientation in active site. Therefore cytochrome P450cam must be monospecific. For other complexes (cytochrome P450cam + norcamphor/thiocamphor) orientational distributions are more flat and multimodal. Therefore molecules of norcamphor and thiocamphor have some predominant orientations. These results allow to make conclusion, that cytochrome P450cam do not show high specificity for these substrates.

## LITERATURE

Paulsen M.D., Orstein R.L. (1992) Predicting the product specificity coupling of cytochrome P-450cam. *J. Computer-Aided Molecular Design.* 21, 449-460.

# Inhibition of cytochrome P450 enzymes by cytokines in human hepatocyte cultures

Ziad Abdel Razzak, Laurent Corcos, André Guillouzo

INSERM U.49, Hôpital Pontchaillou, 35033 Rennes Cedex, France

Hepatic drug biotransformation is impaired during inflammation and infection. Total cytochrome P-450 (CYP) content and CYP-associated enzyme activities are depressed in parallel (Renton et al. 1990). Cytokines can decrease CYP levels, both *in vivo* and *in vitro* (Pous et al., 1990 ; Sujita et al. 1990). Recently, we provided the first demonstration that various cytokines can affect major CYP genes in adult human hepatocytes maintained in primary culture (Abdel-Razzak et al., 1993). The most efficient cytokines were interleukin 1β (IL1β), interleukin 6 (IL6) and tumor necrosis factor α (TNFα). Interferon γ (IFNγ) was also effective on some enzymes. A wide range of responses was observed for a given cytokine. This prompted us to ask whether cytokines could suppress induction by classical CYP inducers, such as polycyclic aromatic hydrocarbons (PAHs). Moreover, since transforming growth factor β (TGFβ) has recently been shown to affect the expression profile of some liver plasma proteins synthesized during the acute phase response, its effects on some CYP enzymes were also analyzed.

## MATERIALS AND METHODS

**Recombinant cytokines and chemicals.** All cytokines were human recombinant molecules. IL1β ($5 \times 10^8$ U/mg protein), IL6 ($10^7$ U/mg protein) and TNFα ($2 \times 10^7$ U/mg protein) were purchased from Genzyme (Cambridge, UK). TGFβ was purchased from British Bio-Technology (Oxon, UK). 3-methylcholantrene (3-MC), benzo(a)pyrene (BaP), 7-ethoxyresorufin, nifedipine and chlorzoxazone were from Sigma (St Louis, MO, USA).

**Hepatocyte isolation and culture.** Human liver fragments were resected from tissue surrounding primary or secondary tumors. All experimental procedures were done in compliance with French laws and regulations and were approved by the National Ethics Committee. Hepatocytes were isolated and cultured as previously described (Guguen-Guillouzo et al., 1982). Cytokines and PAHs were added 15 to 24 hr after cell seeding, at the first medium renewal. This medium, deprived of serum and supplemented with 1 µM hydrocortisone hemisuccinate, was renewed every day thereafter. IL1β, IL6 and TNFα were added to achieve a final concentration of 100, 50 and 50 U/ml, respectively. TGFβ was added at the concentration of 2 ng/ml. BaP and 3-MC were added at the concentrations of 1 µM and 5 µM, respectively. At such concentrations, neither the cytokines nor the inducers induced morphological alterations throughout the incubation period.

**Isolation of RNA and blot analysis.** Total RNA was prepared and analyzed as described (Corcos and Weiss, 1988). cDNA probes were as described previously (Abdel-Razzak et al., 1993). Hybridization signals were quantified by densitometry.

**Monooxygenase assays.** CYP enzyme activities were measured in living cells immediately after treatment. The CYP1A-associated 7-ethoxyresorufin-O-deethylase (EROD) activity was measured essentially according to Lubet et al. (1985). Oxidation of nifedipine, an activity which is mainly supported by CYP3A4, was measured as described by Guengerich et al. (1986). Hydroxylation of chlorzoxazone, an activity supported by CYP2E1 and CYP1A1 (Carriere et al., 1993), was measured according to Peter et al. (1990).

# RESULTS AND DISCUSSION

## EFFECT OF CYTOKINES ON BASAL CYP GENE EXPRESSION

*1-Effects of cytokines on CYP mRNA levels* : Cells from four donors were analyzed following exposure to IL1, IL6 and TNFα. Data for three of them have already been reported (Abdel-Razzak et al., 1993). Conspicuous variations were observed in CYP mRNA levels for both untreated and cytokine-treated hepatocytes. Therefore, results are expressed as the average of ratios of cytokine-treated versus control individual values. The three cytokines had a suppressive effect on all mRNA species studied, i.e. CYP1A2, CYP2C, CYP2E1 and CYP3A (Fig. 1). IL1β was the most potent inhibitor, inducing a 30 to 70 per cent decrease. TNFα and IL6 produced 25 to 52 per cent and 44 to 73 per cent decreases, respectively. These results agree with those obtained for rat hepatocyte primary cultures, as well as for human hepatoma cells (Fukuda et al., 1992 ; Sujita et al., 1990). The mechanism(s) by which cytokines down regulate CYP expression is (are) not known. It has been shown that the promoter regions of human CYP1A1/1A2 genes contain consensus IL6 response elements (Kawajiri et al., 1986 ; Ikeya et al., 1989). We do not know, however, whether IL6 acts via these sequences to elicit the effects reported here. No specific trend was observed regarding possible differences in gene sensitivity to cytokines.

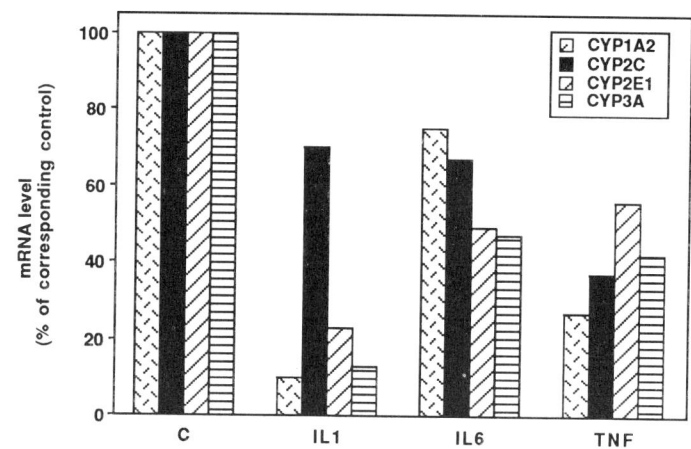

**Figure 1 : EFFECT OF CYTOKINES ON CYP mRNAs.** Hepatocyte cultures from four donors were treated with cytokines for 72 hr. For each CYP, results from different individuals are expressed as per cent of their corresponding control (untreated hepatocytes). The average values are presented without the SEM because of large interindividual variations (see text). IL1 refers to IL1β, TNF to TNFα and TGF to TGFβ. Results were obtained by densitometry of autoradiographs.

*2-Effects of cytokines on CYP enzyme activities* : Five different cell populations were studied. Again, despite large inter-individual variations, all three enzyme activities analyzed, namely EROD, nifedipine oxidase and chlorzoxazone 6-hydoxylase activities were suppressed by the three inflammatory mediators, IL1β, IL6 and TNFα (Fig. 2). EROD activity was decreased by 13 to 70 per cent by IL1β, 17 to 54 per cent in 3 out of 5 samples by IL6, and 32 to 83 per cent by TNFα. Nifedipine oxidase was decreased by 8 to 90 per cent by IL1, 11 to 52 per cent by IL6, and 24 to 80 per cent by TNFα. Chlorzoxazone 6-hydroxylase activity was decreased 18 to 55 per cent by IL1β, 20 to 60 per cent by IL6, and 20 to 87 per cent by TNFα. TGFβ was also analyzed for its ability to influence one CYP enzyme, i.e. the CYP1A-associated EROD activity. Similarly to other cytokines, TGFβ induced an average 60 per cent decrease in EROD activity level. These data show that cytokines were effective on both mRNA and enzyme activities. This favors the conclusion that the mediators suppressed CYP enzymes by acting at a pretranslational step.

## EFFECT OF CYTOKINES ON CYP1A INDUCTION BY PAHs

When human hepatocytes were exposed to 3-MC for three days, a strong increase in EROD activity was observed (Fig. 3). Exposure of the cells to 3-MC and IL1β, IL6 or TNFα, which suppressed basal enzyme activities (Fig. 2), produced only a limited inhibitory effect on the strong (18 fold) induction of EROD activity by 3-MC. By contrast, when added together with 3-MC, TGFβ exerted a strong inhibition of the 3-MC response. This effect was investigated further by measuring CYP1A mRNA levels in two cell populations treated simultaneously with 3-MC and TGFβ.

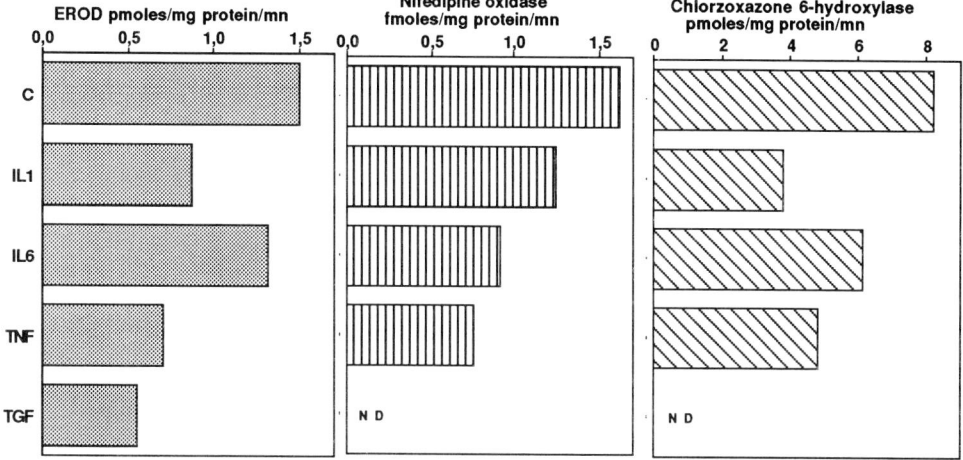

**Figure 2 : EFFECT OF CYTOKINES ON CYP-ASSOCIATED ENZYME ACTIVITIES** : Enzyme activities were assayed on living hepatocytes from five donors. For each cell population, measurement was done on 3 to 8 dishes. Cultures were treated for 72 hr with different cytokines. The average of five experiments is presented without SEM because of large inter-individual variations. Enzyme activities in control untreated cells were between 0.92 and 1.95 pmoles/mg protein/min, 0.62 and 4.43 fmoles/mg protein/min and 6 and 11.9 pmoles/mg protein/min, for EROD, nifedipine oxidase and chlorzoxazone 6-hydroxylase, respectively. Differences in treated cultures were statistically significant (Student's t-test).

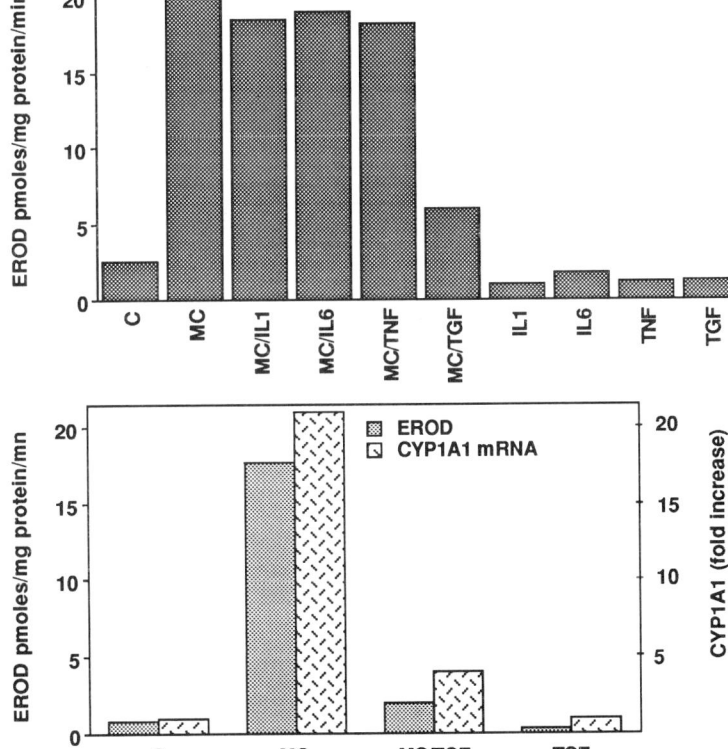

**Figure 3 : EFFECT OF CYTOKINES ON CYP1A INDUCTION** : Cultures were treated with cytokines and 3-MC for 72 hr before analysis of EROD activity.

**Figure 4 : EFFECT OF TGFβ ON CYP1A mRNA INDUCTION BY 3-MC** : mRNA levels are presented as per cent of control untreated hepatocytes. The average presented is from 2 cell populations.

As observed for EROD activity, 3-MC-mediated induction of CYP1A1 mRNA was strongly antagonized by TGFβ. Since the basal level of CYP1A1 mRNA was barely detectable, we could not accurately determine if TGFβ also affected the basal mRNA accumulation.

In the case of rat hepatocytes, it has been shown that, contrary to what we found for human cells, IL1β, but not TGFβ, was able to antagonize the dioxin-mediated induction of the CYP1A gene (Barker et al., 1992). Although much lower, BaP induction was also blocked by TGFβ (data not shown). The fact that, in our cell system, BaP, unlike 3-MC, is metabolized, might account for the relatively poor inductive effect of BaP. However, since both PAHs are known to bind the Ah receptor during the inductive process, it appears likely that TGFβ is able to antagonize the "classical" PAH receptor-mediated pathway of CYP1A induction. Analysis of transcription rates, as well as of binding of the liganded receptor to its target DNA response elements in the promoter region of CYP1A genes should help clarifying this point. The ability of TGFβ to antagonize PAH induction in human hepatocytes might lead to uncover some aspects of the negative regulation of CYP1A genes (Lusska et al., 1992).

## REFERENCES

Abdel-Razzak, Z., Loyer, P., Fautrel, A., Gautier, J.C., Corcos, L., Turlin, B., Beaune, P.H. & Guillouzo, A. (1993) : Cytokines down-regulate expression of major cytochrome P-450 enzymes in adult human hepatocytes in primary culture. *Mol. Pharmacol.* (in press).

Barker, C.W., Fagan, J.B. & Pasco,. D.S. (1992) : Interleukin-1-β suppresses the induction of P-450 1A1 and P-450 1A2 in isolated hepatocytes. *J. Biol. Chem.* 267:8050-8055.

Carriere, V., Goasduff, T., Ratanasavanh, D., Morel, F., Gautier, J.C., Guillouzo, A. Beaune, P.H. & Berthou, F. (1993) : Both cytochromes P-450 2E1 and 1A1, are involved in the metabolism of chlorzoxazone. *Chem. Res. Toxicol.* (in press).

Corcos, L. & Weiss, M.C. (1988) : Phenobarbital, dexamethasone and benzanthracene induce several cytochrome P-450 mRNAs in rat hepatoma cells. *FEBS Lett.* 233, 37-40.

Fukuda, Y., Ishida, N., Noguchi, T., Kappas, A. & Sassa, S. (1992) : Interleukin-6 down regulates the expression of transcripts encoding cytochrome P-450IA1, IA2 and IIIA3 in human hepatoma cells. *Biochem. Biophys. Res. Commun.* 184:960-965.

Guguen-Guillouzo, C., Campion, J.P., Brissot, P., Glaise, D. Launois, B., Bourel, M. & Guillouzo, A. (1982) : High yield preparation of isolated human adult hepatocytes by enzymatic perfusion of the liver. *Cell. Biol. Int. Rep.* 6:625-628.

Guengerich, F.P., Martin, M.V., Beaune, P.H., Kremers, P., Wolff, T. & Waxman, D.J. (1986) : Characterization of rat and human liver microsomal cytochrome P-450 forms involved in nifedipine oxidation, a prototype for genetic polymorphism in oxidative drug metabolism. *J. Biol. Chem.* 261:5051-5060.

Ikeya, K., Jaiswal, A.K., Owens, R., Jones, J.E., Nebert, D.W. & Kimura, S. (1989) : Human cytochrome CYP1A2 : Sequence, gene structure, comparison with the mouse and rat orthologous genes, and differences in liver 1A2 mRNA expression. *Mol. Endocrinol.* 3:1399-1408.

Kawajiri, K., Watanabe, J., Gotoh, O., Tagashira, Y., Sogawa, K. & Fujii-Kuryama, Y. (1986) : Structure and drug inducibility of the human cytochrome P-450c gene. *Eur. J. Biochem.* 159:219-225.

Lubet, R.A., Mayer, R.T., Cameron, J.W., Nims, R.W., Burke, M.D., Wolff, T. & Guengerich, F.P. (1985) : Dealkylation of pentoxyresorufin : A rapid and sensitive assay for measuring induction of cytochrome(s) P-450 by phenobarbital and other xenobiotics in the rat. *Arch. Biochem. Biophys.* 238:43-48.

Lusska, A., Wu, L. & Whitlock, J. P. (1992) : Superinduction of CYP1A1 transcription by cycloheximide. *J. Biol. Chem.* 267:15146-15151.

Peter, R., Bocker, R., Beaune, P.H., Iwasaki, M., Guengerich, F.P. & Yang, C.S. (1990) : Hydroxylation of chlorzoxazone as a specific probe for human liver cytochrome P-450IIE1. *Chem. Res. Toxicol.* 3:566-573.

Pous, C., Giroud, J.P., Damais, C., Raichvarg, D. & Chauvelot-Moachon, L. (1990) : Effect of recombinant human interleukin-1-β and tumor necrosis factor-α on liver cytochrome P-450 and serum α-1-acid glycoprotein concentrations in the rat. *Drug Metab. Dispos.* 18:467-470.

Renton, K.W. & Knickle, L.C. (1990) : Regulation of hepatic cytochrome P-450 during infectious disease. *Can. J. Physiol. Pharmacol.* 68:777-781.

Sujita, K., Okuno, F., Tanaka, Y., Hirano, Y., Inamoto, Y., Eto, S. & Arai, M. (1990) : Effect of interleukin-1 on the levels of cytochrome P-450 involving interleukin-1 receptor on the isolated hepatocytes of rat. *Biochem. Biophys. Res. Commun.* 168:1217-1222.

# VIII CYTOCHROME P450'S: AGENDA FOR ACTION, ADAPTATION AND APPLICATION

# The P450s: agenda for action, adaptation, and application

Ronald W. Estabrook

Department of Biochemistry, University of Texas, Southwestern Medical Center at Dallas, Dallas, Texas 75235-9038, USA

Research on the physical, chemical and biological properties of the cytochromes P450 is at a crossroads. At this meeting we have witnessed the remarkable advances that have occurred in recent years. New knowledge of the crystal structure of P450s has taken a major leap forward; the technology for the heterologous expression of P450s permits now the synthesis of large quantities of pure P450 proteins, from both common as well as exotic sources, and opens new opportunities for the study of chemical mechanisms; and the expansion of interest into other arenas of biology, such as plant, insect, and fish tissues, offers the promise of a wealth of new information expanding our understanding of the functions of P450s and the regulation of their expression. These new directions will offer a renewed burst of activity which will expand the magnitude of contributions to our lexicon of P450s. The study of P450s can be divided into four segments (Figure 1). The initial period of discovery and identification of function; the mastering of techniques for purification of P450s and their use as specific antigens; the introduction of methods of molecular biology permitting characterization of structure/function properties; and, lastly, the proposed expansion of interest into phylogeny and the associated discovery of new roles for P450s in biology.

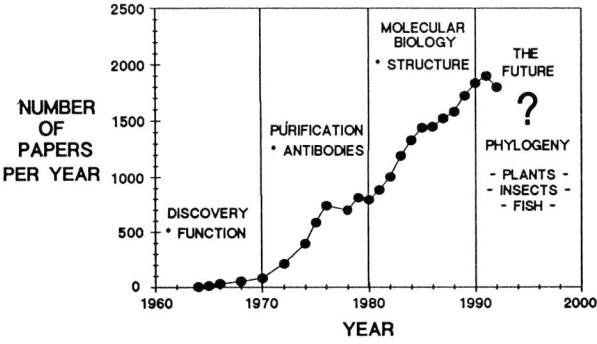

Figure 1. The growth in number of publications on P450.

Any attempt to encapsulate all the excellent science presented during the intense four days of this meeting would be a futile exercise. Therefore I have selected six topics to emphasize in this brief summary.

THE DISCOVERY OF NEW P450s. It is apparent that we have not exhausted cataloging the inventory of P450s functional in biological processes. The current listing identifies nearly 300 different P450s that have been cloned and sequenced. The vast majority of these are associated with mammalian tissues, in particular from rodents. This is appropriate because of the interest in the role of P450s in pharmacology, carcinogenesis, and endocrinology. However, today we see a shift in the direction of discovery into other areas of biology. Accompanying this shift of direction is a change in the approach taken to identify new P450s. In the past it has been easiest to purify a P450 (from mammalian tissues), use the protein as an antigen for the preparation of antibodies, which are then used as probes to screen for the expression of immunoreactive proteins from cDNA transformed cells. In this way one can identify clones for a specific P450 and determine the unique nucleotide sequence characteristic of a P450. But many sources of current interest have P450s in very low concentration, frequently associated with tissues not available in unlimited quantities. The new strategy for identifying P450s is to apply the polymerase chain reaction (PCR) method using degenerate oligonucleotide primers spanning the heme-binding region of P450s. This type of "fishing" has yielded remarkable results in the study of plant and insect P450s. A large number of clones can be identified whose sequences for the heme-binding segment fits the description of a P450. Clearly, this is a very powerful approach which reveals the extensive distribution of P450-type proteins in biology. One wonders what will happen when this approach is applied to a study of mammalian P450s. Will the present efforts to sequence the human genome also reveal new P450s? Are there a large number of unknown P450s, present in mammalian tissues at low concentrations, whose presence has been overlooked and whose functions remain to be defined?

As new methods of discovering new P450s develop, new problems are generated. It is estimated that we know the enzymatic function for less than <u>fifty</u> percent of the P450s that have already been cloned and sequenced. Adding an even greater number of members with unknown function to this inventory of P450s shifts the burden of discovery to the arena of the enzymologist. The question of how to characterize the enzymatic function of a new P450 becomes primary. Fortunately, the techniques of heterologous expression of P450s in yeast, bacteria, or tissue culture cells have developed in parallel with the discovery of cDNA sequences for new P450s. These techniques provide the ability to transform cells with the cDNA for a specific P450 and to overexpress enzymatically active recombinant P450s that can be isolated and purified for study. So, one adds a second pond for experimental "fishing"; one that requires the enzymologist to screen a battery of substrates in the search for a function for a new P450.

The search for a function of a newly cloned P450 sequence introduces additional questions that require consideration. At what level is an enzymatic activity for a P450 considered meaningful? Currently we are indiscriminate in assigning an activity to a P450 - regardless of the magnitude of such an activity. Is it reasonable to assign a reaction as one catalyzed by a P450 when the activity is expressed by a turnover number for the P450 of once every minute? or hour? or even once every day? We will have to decide on limits before assigning an activity as an identifying property for a specific P450.

The application of the methods of heterologous expression, in particular when using bacteria, provides a means to generate large amounts (grams) of purified P450s. This offers the chemist the availability of reagent quantities of these proteins which can then be used for the study of mechanism(s) of catalysis. Assigning a function to a new P450 will undoubtably require an understanding of the mechanism(s) of catalysis. Studies of this type, combined with increased knowledge gained from the crystallographic studies of P450s (see below), brings us one step closer to the advent of molecular engineering of "designer P450s", i.e. artificial P450s constructed to catalyze specific reactions with defined substrates.

THE PHYSICAL CHARACTERISTICS OF P450s. Knowledge of the crystal structures of two more P450s adds critical new information to the arguments about the extrapolation of structural similarities between different P450s. It is apparent that there are many common features shared by the three known P450 structures - but there are also many differences which reflect the variation of amino acid sequences and differences in forces dictating the folding of the various proteins. Great interest surrounds the desire to understand the geometry of the "active site" of P450. This is appropriate since the unique chemistry of oxygen activation and substrate metabolism, which is the hallmark of a P450 catalyzed reaction, occurs in this confined pocket of the molecule. We are getting closer to understanding the role of specific amino acids in this "substrate pocket" during the process of catalysis - but the establishment of generalities is still encumbered by exceptions. The goal is to understand the rules governing structural determination in order to establish the pattern of the archetype P450, i.e. the original pattern from which all other things of the same type are made.

The function of a P450 is intimately linked to the electron transport process in which the flavoprotein, NADPH-P450 reductase, or an iron-sulfur protein, serve as necessary reactants. The factors regulating protein-protein interactions required for these reactions has served as a fertile field of research (and controversy). It is accepted that electron donation for the reduction of the substrate-bound ferric form of P450 frequently serves as the rate limiting step of catalysis. As our knowledge of the crystal structures of P450s increases, a growing awareness of differences in the surface profiles of each protein becomes more apparent. As yet common features of surface structure have not emerged that permit a clearer definition of "docking sites" for specific proteins, beyond the identification of clusters of charges that may serve as facilitating factors for protein - protein interactions. In a similar vein, we now know almost nothing about the rules that dictate the proper conformational folding of a P450. What amino acids are critical for the transitions involved in the folding of the protein? What is the extent of molecular "breathing" that provides a dynamic flexibility to the surface features of the protein? One approach to this problem is the successful engineering of cDNAs permitting the high level expression in E. coli of artificial fusion proteins containing the heme-domain of different P450s linked to the flavin-domains of different NADPH-P450 reductases. Evaluation of the enzymatic properties of these self-sufficient catalysts may better delineate the limits of surface characteristics required for effective interactions.

What is the influence of those segments of a P450 protein which participate in the "binding" of the protein to a membrane? Would the presence of these "sticky" hydrophobic domains modify the crystal structure of a protein?. The three P450s crystallized and analyzed to date do not have a hydrophobic N-terminal segment. Will we learn that other conformational states prevail for membrane-bound P450s which influence the physical and catalytic properties of the P450s? The more we learn the more questions remain unanswered.

THE POWER OF BIOLOGY. The expansion of interests to P450s in different phyla brings into view many questions about changes of the progenitor P450 in evolutionary development. What was the role of the primordial P450? Can we learn of invariant segments of P450 structure by comparison of amino acid sequences for orthologous proteins across species. Can we use this information to compliment structural studies for defining essential elements and motifs of the P450 molecule? Present efforts to define essential structural elements influencing catalysis suggest a key role played by one or a very few amino acids. The design of mutagenesis studies to test this "single hit" theory for the role of a limited number of unique amino acids should greatly profit from a comparison of proteins with common function that are broadly distributed phylogenetically.

The P450s were originally recognized by their characteristic absorbance band at 450nm when the reduced hemoprotein complexes with carbon monoxide. This spectrophotometric property served for many years as a special fingerprint for defining a P450. In the last ten years a new definition of P450 has been introduced - the presence of a unique amino acid sequence associated with the heme-binding site on the protein. In the last year great interest has developed in the P450-like properties of the natural mammalian fusion protein that catalyzes the conversion of arginine to citrulline with the formation of nitric oxide - NOS, nitric oxide synthase. This fusion protein contains a flavin-domain analogous to NADPH-P450 reductase and a heme-domain which behaves in many ways like a P450. NOS fulfills the earlier definition of a P450, *i.e.* the ferrous hemoprotein binds CO to form an absorbance band about 450nm; But, the sequence similarity of the heme-binding segment of the protein does not meet the present definition required by the newer definition. This is not an unusual property specific only to NOS. It has been recognized for many years that reduced chloroperoxidase forms an absorbance band at about 450 nm in the presence of CO. It is established that thromboxane synthase differs in its enzymatic properties from a "typical" P450. The pigment H450 has been described with a cyano-thiolate ligand. We propose here the need to consider an over-arching hierarchy in which the common property is the presence of a heme-iron thiolate ligand for the protein (Figure 2). In this way the P450s share some but not all characteristics with other types of proteins.

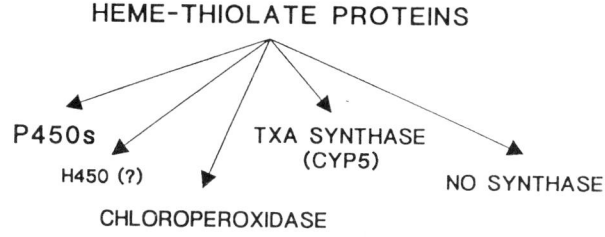

Figure 2. A proposed hierarchy for pigments forming an absorbance band at 450nm.

HOW MANY ANGELS CAN DANCE ON THE HEAD OF A PIN? Highly reactive products formed as the result of reactions catalyzed by many different P450s has placed special interest on the role of P450s in toxicology and chemical carcinogenesis. As a part of the evaluation of newly developed drugs, regulatory agencies now require a study of the effect of such chemicals on P450s. Huge amounts of money are committed each year to the Superfund cleanup of toxic waste dumps; the chemical and pharmacuetical industry invests large amounts of money and personnel in evaluating the safety of new products. The P450s play a central role in the decision making for establishing risk assessment regulations. One criteria for assessing the influence of a chemical on the P450 system is to measure the extent of induction of P450s following exposure of an animal to the chemical.

A major challenge is to understand the mechanism(s) by which a chemical can induce the synthesis of one (or a very few) P450s in a tissue specific manner. Elegant studies have defined the role of receptors in transmitting the signal from an initiating chemical to a regulatory element of a gene. There are a plethora of transcription factors that are known. Deletion analysis of the 5'-upstream regulatory region of genes have pin-pointed sequences that might serve as cis-acting elements for the binding of transcription factors. Even so, the control of synthesis of a P450 is complex, and those seeking generalizations which can be applied as fundamental principals may frequently have an empty feeling associated with a lack of fulfillment. Attention is now turning toward a better understanding of the pleiotropic effect of chemical inducers. The application of differential/subtractive hybridization techniques will increase our understanding of the coordinate increased expression of a number of proteins, including the P450s. These studies will permit a comparison of the concomitant response to regulatory elements of a number of different genes. Evaluation of changes associated with changing patterns of expression during development combined with an understanding of down-regulation and the underlying control of sexual dimorphism will all identify the most favorable systems to be studied. Particular interest should focus on the role of cellular proliferation on the expression of selected P450s. Clearly, understanding the regulation of expression of P450s is of major importance. New approaches will offer new insights beyond the currently limited empirical approach which has dominated these studies to date.

POSSIBLE COMMERCIAL APPLICATIONS OF THE P450s. The oxygen chemistry catalyzed by the P450s, combined with the great diversity of organic structures that can serve as substrates, makes the application of the P450s to processes of speciality chemical synthesis a logical extension of current research. Further, the P450s combine the advantages associated with biocatalysts for organic syntheses, *i.e.* chemoselectivity including stereoselectivity and enantioselectivity. The ability to engineer cDNAs for the large scale production of recombinant self-sufficient catalytic units, modified to contain binding elements such as multiple histidine domains or biotin-binding domains, which can be used for immobilization of the P450s, places the practical consideration of commercial application of the P450s in the realm of reality. In Dallas we have initiated studies to construct such a regenerating reactor using the photochemistry of ruthenium salts or deazaflavin as the means of introducing energy for the function of a P450. A schematic representation of this reactor is shown in Figure 3.

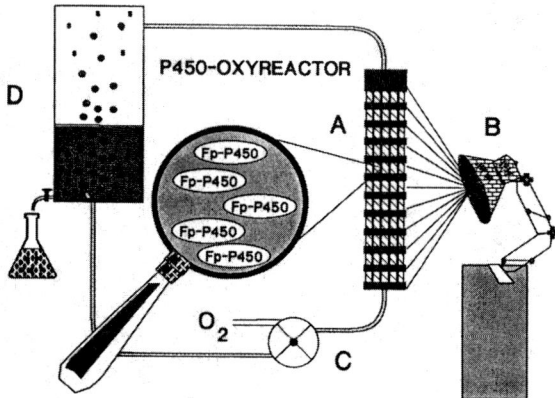

Figure 3. The P450 OXYreactor. Fusion proteins engineered to contain a P450-domain and a flavin-domain (A) are linked to an immobilizing matrix suitable for superfusion of a substrate (D). Light (B) activates anaerobic electron transport with P450 catalyzed reactions initiated by the introduction of oxygen (C).

THE FUTURE. The quality of a meeting is judged by the number of new ideas one hears and the number of unanswered questions that stimulate thoughts of new experiments upon returning to the laboratory. This has been a very successful meeting. My own list of unanswered questions include:

1. What are the factors influencing protein folding to form a functional P450?
2. What are the mechanisms of electron tunnelling for reduction of P450?
3. What is the influence of oxygen affinity on pathways of metabolism?
4. Will P450s from new biological sources reveal new chemistry?
5. What more will we learn from new crystal structures of P450s?
6. How extensive are polymorphisms and the microheterogeneity of structure?
7. What is the role of P450s in establishing patterns of behavioral biology?
8. What will be the application of P450s to gene therapy?

Each of us has our own list. Our next meeting will reveal our success as we progress toward fulfilling these areas of opportunity.

---

This research was supported in part by grants from the National Institutes of Health (GM 16488-25) and the Robert A. Welch Foundation (I-0959).

---

# Index des auteurs

## A
Achstetter T., 527, 583
Ahrend J., 809
Aikens J., 373
Akiyoshi-Shibata M., 429
Akrawi M., 737
Albin N., 217
Alexandrova O., 765
Amar C., 193
Amichot M., 689
Andersen J.F., 31
Apletalina E.V., 571
Arand M., 147
Archakov A.I., 15, 183, 319, 395, 445, 603, 865
Arvela P., 519, 777
Asperger O., 643

## B
Bachmanova G.I., 395, 477, 853
Bairos V., 717
Baldwin B.C., 209
Ballet F., 189
Balny C., 813
Barnett C.R., 547
Beaune P., 189, 193, 201
Bec N., 813
Beckert V., 387, 869
Bellamine A., 233, 515
Benet L.Z., 877, 881
Benson D., 373
Benveniste I., 23, 481
Berenbaum M.R., 249
Bernard H., 615
Bernhardt R., 387, 869
Bestervelt L.L., 285
Bhatia A., 497
Boddupalli S.S., 271
Bone A.J., 547
Bonierbale E., 193
Bornheim L.M. 877, 881
Brake P., 121
Brockmöller J., 749
Brown R., 705
Brun A., 689
Buège-Kirn, 481

Bykhovskaya E.A., 445
Bylinskaya S.A., 861

## C
Campleman S.L., 213
Caria H., 205
Cariño F.A., 31
Cascorbi I., 749
Cauet G., 527, 583
Chabot G.G., 217
Challine D., 189
Chang T.K.H., 103
Charrasse S., 825
Chaveca T., 205
Chen G., 103
Chen S., 769
Chernogolov A., 311, 327, 555
Chow T.Y.K., 667
Chung B.C., 323
Ciaramella G., 587, 655
Clair P., 713, 789
Cohen M.B., 31
Constantino L., 677
Conte F.A., 535
Cook V.A., 197
Coon M.J., 285, 885
Corcoran O.M., 761
Corcos L., 899
Correia M.A., 877, 881
Cuany A., 689
Cullin C., 433

## D
Dansette P.M., 193, 697
Darden T.A., 225
Daujat M., 825
Davila J.C., 459
Davydov D.R., 477, 559
Dawson J.H., 441
De Voss J.J., 409
de-la-Torre R., 575
Dedinsky I.R., 603
Defaye G., 455
Degryse E., 527, 583
Degtyarenko K.N., 15, 319
Deisenhofer J., 271

Delorme C., 455
DePetrillo P.B., 729
Deprez E., 403, 531
Descatoire V., 201
Dettmer R., 869
Di Primo C., 531
Ding X., 885
Douzou P., 531
Drakoulis N., 749
Drewelow B., 809
Drobnič K., 805
Duarte-Silva I., 205
Ducharme L., 615
Dumas B., 527, 583
Durham B., 379
Durst F., 23, 481, 689

**E**
Edwards M., 587, 655
Egger H., 611
Elia A.P., 507
Eliasson E., 829
Elovaara E., 567
Ema M., 75
Engelhardt R., 845, 849
Epinat J.C., 433
Estabrook R.W., 241, 905

**F**
Fabre I, 825
Feldmann G., 201
Fernandez L.M., 663
Feyereisen R., 3, 31
Figueiredo M.H., 717
Fischer U., 809
Fisher C.R., 535
Fisher C.W., 241
Flatt P.R., 547
Forkert P.G., 769
Forrest L.A., 663
Fruetel J., 409
Fujita V.S., 285
Fujii-Kuriyama Y., 75, 89
Fukuda T., 841
Fulco A.J., 37
Furh U., 845, 849

**G**
Gaedigk A., 503
Gama M.J., 125
Gast K., 555
Gautier J.C., 189
Gelboin H.V., 567
Gerber N.C., 373, 753
Geren L., 379
Giannone J.V., 773
Gieschen H., 467, 693

Gillam E.M.J., 97
Giovanni-Gervasi P., 623
Glatt H., 147
Gonzalez F.J., 59
Gordon Gibson G., 115
Gotoh O., 279
Gouyette A., 217
Graham-Lorence S., 271, 471, 817
Grant D.M., 503
Grishanova A.Y., 485, 685, 873
Gromova O.A., 873
Gross C.R., 749
Grumbach M.M., 535
Guengerich F.P., 97, 201
Guillouzo A., 899
Gulyaeva L.F., 485, 685, 873
Günther G., 781
Güray T., 745
Gustafsson J.Å., 51, 81
Guzelian P.S., 109
Guzov V.M., 861

**H**
Hahm S., 379
Hakkola J., 129
Halpert J.R., 829
Hänninen O., 821
Hanstein W.G., 781, 785
Hanukoglu I., 357
Hara M., 67
Hara T., 417
Harada D., 365
Harada N., 141, 315, 651
Harper P.A., 551, 773
Hasemann C., 271
Hasenfratz M.P., 481
Hashimoto H., 89
Hatcher M., 705
Hawksworth G., 523
Hayashi S.I., 155
He J.S., 37
Hedman B., 441
Heiber-Langer I., 813
Helvig C., 689
Henrique D., 717
Herzig G., 611
Hevel J.M., 43
Higashiyama T., 893
Hildebrand M., 467, 693
Hodgson K.O., 441
Hoener B.A., 213
Holmans P.L., 241
Homberg J.C., 193
Honda S.I., 67, 141
Horiuchi T., 299
Horne V., 523
Horst Staib A., 849

Hsu L.C., 323
Hui Bon Hoa G., 403, 531, 559
Hung C.F., 249
Huschka C., 849
Husemann M., 709
Hyland R., 761

## I
Ichikawa Y., 701
Ikawa S., 599
Ikegwuonu F., 121
Ikushima H., 647
Ikushiro S.I., 365
Iley J., 677
Imai Y., 841
Ingelman-Sundberg M., 623, 829
Ioannides C., 547
Ipatova O.M., 571, 853
Iscan Mesude, 745
Iscan Mümtaz, 745
Ishikawa H., 89
Ishimura Y., 299, 659
Ito Y., 535
Ivanov A.S., 493, 897
Ivanov Y., 765
Iwasaki M., 225, 437, 801

## J
Jackson A.C., 769
Jaeger W., 877, 881
Janot F., 217
Janssen P.A.J., 671
Jefcoate C., 121
Jenkins C.M., 293
Jones B.C., 519, 761
Jung C., 543, 595
Juvonen R.O., 511, 631, 801

## K
Kadkhodayan S., 441
Kanaeva I.P., 395, 477
Kärgel E., 307
Kato R., 89
Katoh S., 651
Kawajiri K., 155
Kawato S., 889
Kayser H., 619
Kazmaier M., 233
Kelly D.E., 209
Kelly S.L., 209, 805
Kikuchi H., 599
Kim B.R., 97
Knushko T.V., 395
Koen Y.M., 395
Koener J.F., 31
Kojo A., 511
Kolesanova E.F., 183

Kolyada L.N., 445
Komel R., 805
Kominami S., 365, 429
Komori M., 841
Köpke K., 543
Koymans L.M.H., 671
Kozin S.A., 183, 603
Kraft R., 327
Kramer K., 655,
Kranendonk M., 741
Krause K., 797
Kraut J., 379
Kritsky A.M., 477
Krueger V., 311
Kuban R.J., 421
Kühn-Velten W.N., 425
Kurl R.N., 729
Kusunose E., 841
Kusunose M., 841
Kuznetsova G., 865

## L
Lainé R., 451
Laires A., 205, 741
Laitinen S., 511
Lamb D.C., 209
Lange R., 813
Larroque C., 825
Larsen M., 121
Lechner M.C., 125, 717, 837, 857
Lecoeur S., 189
Lee Y.H., 59
Leeder J.S., 197
Lefebvre A., 615
LeHoux J.G., 615, 721
LeHoux J., 615
Lesot A., 481
Lewi P.J., 671
Liang Q., 37
Liermann D., 849
Lindros K.O., 833
Liu H.S., 441
Liu R., 379
Loeper J., 201
Loida P., 463
Longo V., 623
López-de-Briñas E., 575, 579
Lopez-Garcia P.M., 193, 697
Lottermoser K., 643
Lozano J.J., 575, 579
Lu A.Y.H., 257
Lyakhovich V.V., 485, 685, 873
Lyulkin Y.A., 493, 897

## M
Mäenpää J., 129, 631
Manaut F., 579

Mansuy D., 193, 697
Marg A., 421
Marletta M.A., 43
Marquet A., 455
Massaad L., 217
Matsushita N., 75
Maurel P., 825
Maurice M., 201
Mazzuco K., 497
McLean M., 373, 379
Menzel R., 307
Meyer C.A., 459
Meyer U.A., 163
Mihara K., 315
Mihimara K., 266
Miller M., 379
Millett F., 379
Mitjavila M.T., 623
Mizuno S., 647
Mkrtchian S., 829
Mode A., 51
Moereels H., 671
Molitor E., 563
Monostory K., 639
Monteiro M.J., 205
Morohashi K.I., 67
Mukai K., 659
Muntane J., 623
Murayama N., 89

# N
Nagata K., 89
Nakajima T., 567
Nakamura H., 503
Nakamura M., 841
Nakatsuka M., 459
Nakayama K., 89
Nanji M., 713
Nebert D.W., 3
Negishi M., 225, 437, 801
Nemoto N., 757
Newlands A., 523
Nikityuk O.V., 395
Nirodi C., 133
Nishimori K., 647
Nomura M., 67

# O
O'Sullivan M.J., 705
Oesch F., 147
Oesch-Bartlomovicz B., 147
Ogawa H., 315
Ohkawa H., 429
Ohta Y., 889
Oinonen T., 833
Okey A.B., 551, 773
Olken N.M., 43

Olnes M.J., 729
Omura T., 67, 266, 315
Ortiz de Montellano P.R., 409
Osawa Yoichi, 459
Osawa Yoshio, 893
Osterman-Golkar S., 567

# P
Padgham C.R.W., 539
Padmanaban G., 133
Paine A.J., 539, 733
Palin M.F., 721
Parise O., 217
Pasanen M., 129, 511
Pedersen L.G., 225
Peilak G., 379
Pelkonen O., 129, 511, 519, 631, 745, 797
Pellinen P., 511
Peng H.M., 885
Pereira T., 837
Pereira M.P., 125
Pernecky S.J., 285
Pessayre D., 201
Perret A., 793
Petchkovski E.V., 485, 685
Peterson J.A., 271, 409, 471, 817
Petri T., 709
Pfeil W., 543
Phillips I.R., 137, 507, 587, 655, 663, 713, 737, 789
Piffeteau A., 455
Pikuleva I.A., 293
Platt K.L., 147
Pompon D., 233, 433, 451, 515, 793
Pons C., 193
Postlind H., 591
Prabhu L., 133
Prapaipong H., 249
Probst M.R., 773
Proschlyakov D., 853
Prosorovsky V., 853
Pufahl R.A., 43
Puga A., 173
Purkunen R., 129

# R
Rahier A., 725
Raikhinstein M., 357
Rakoto J.S., 627
Ram N., 133
Rane A., 129
Rapoport R., 357
Raunio H., 129, 511, 631
Rautio A., 519, 631, 745, 777, 797
Razzak Z.A., 899
Repecko S., 31
Ribeiro V., 125, 717, 857

Ribeiro-Pinto L.F., 681
Ribrag V., 217
Richards M.K., 43
Richter W., 555
Riddick D.S., 541
Riihimäki V., 567
Riley R.J., 197
Ristau O., 543, 595
Rodrigues A.S., 205
Rogiers V., 737
Roos P.H., 781, 785
Roots I., 635, 749
Rost K.L., 635, 845, 849
Rostami H., 745
Rougeulle C., 433
Roy S., 821
Rozman D., 805
Ruas M., 741
Ruckpaul K., 421
Rueff J., 205, 741
Rumyantsev A.B., 493, 897

## S
Sachs M., 849
Sadekova S.I., 667
Sagami I., 599
Sakaguchi M., 265, 315
Sakaki T., 429
Sakurai J., 757
Salaun J.P., 689
Salaün J.P., 23
Salmela E., 519
Salmon F., 725
Salomon B., 467, 693
Salonpää P., 797
Sanders D., 471
Sanz F., 575, 579
Satouchi K., 841
Schenkman J.B., 349
Schuler M.A., 249
Schulze H., 543, 595
Schunk W.H., 307, 643
Schuster I., 611
Schwarz D., 311, 327, 555
Schymanski P., 849
Scott J.A., 31
Sevrukova I.F., 395
Shen S., 341
Shephard E.A., 137, 507, 587, 655, 663, 713, 737, 789
Shervington A., 137, 663
Shet M.S., 241
Shimada H., 299, 659
Shimada T., 97
Shimozawa O., 315
Shirane N., 409
Shumyantseva V.V., 445, 865

Simpson E.R., 535, 817
Sklan D., 357
Skvortsov V.S., 493, 897
Sligar S.G., 373, 379, 387, 463, 531, 753
Smith D.A., 523, 761
Smith M.T., 213
Snyder M.J., 31
Sogawa K., 89, 75
Sono M., 441
Sorsa M., 567
Sotaniemi E.A., 777
Spagnoli R., 527, 583
Staib A.H., 845
Stayton P., 379
Stepanova N.V., 395, 477
Stier A., 311
Strobel H.W., 341
Strömstedt M., 81
Strotkamp D., 785
Sueyoshi T., 225
Sui Z., 409
Sultana S., 133
Suslick K., 373
Swann P.F., 681
Swinburne S., 705
Sygusch J., 721
Szutowski M.M., 627

## T
Takagi Y., 141, 651
Takayama K., 67
Takemori S., 365
Takeshima M., 417
Tam S.P., 769
Tamaki S.J., 213
Tan L., 667
Tappe D., 785
Taton M., 725
Telhada M.M., 125
Thomas H., 563
Tollet P., 51
Toussaint C., 217
Truan G., 233, 433, 793
Tsujita M., 701
Tute M., 523

## U
Uhlmann H., 387
Unno M., 299
Urban P., 233, 451, 793
Usanov S.A., 311, 327, 437, 861
Usuda M., 599
Uvarov V.Y., 445, 571, 765, 865

## V
Valadon P., 193
van Afferden M., 785

Vercruysse A., 737
Vereczkey L., 639
Véronneau S., 615
Vitas M., 805
Vorisek G., 611
Voznesensky A.I., 349

## W
Wacke R., 809
Waechter F., 563
Wang X.J., 539
Warner M., 81
Watanabe J., 155
Watanabe M., 599
Waterman M.R., 293
Waxman D.J., 103
Werck D., 23
Weymann J., 563
White K.A., 43
White R.E., 333

Willie A., 379
Winter J., 619
Wissing H., 849
Wolff C., 307
Wright M.C., 733
Wyss A., 81

## Y
Yabusaki Y., 429
Yamada K., 651
Yamada M., 889
Yamazaki T., 365
Yamazoe Y., 89
Yao Y., 173
Yarborough C., 893
Yoshida S., 81

## Z
Zelko I.N., 861
Zgoda V., 853

LOUIS-JEAN
avenue d'Embrun, 05003 GAP cedex
Tél. : 92.53.17.00
Dépôt légal : 365 — Avril 1994
Imprimé en France